Evaluation Guide for M

Merrill Algebra One provides a comprehens
Merrill Algebra One was high school teacher
tested. This contemporary text provides every opportunity for student success.

To further strengthen the presentation of the text material, many special features have been included in the text. To examine selected examples of the features of *Merrill Algebra One,* please turn to the following pages.

(pp. 82, 178) **Student Annotations** explain steps in solving problems and help students identify important concepts.

(pp. 153, 305) **Color Tintblocks** highlight important concepts in the lessons.

(pp. 197, 208) **Mini-Reviews** and **Cumulative Reviews** help students maintain skills introduced in previous chapters.

(pp. 92–96, 319) **Problem Solving** skills are developed throughout the textbook using a basic four-step plan. Additional strategies for solving problems are presented on special gold-tinted pages.

(pp. 331–333) **Review and Test** material is provided in the Vocabulary, Chapter Summary, Chapter Review, and Chapter Test at the end of each chapter.

(pp. 528–542) The **BASIC Appendix** provides a short course in BASIC.

(pp. 546–562) The **Algebraic Skills Review** includes numerous exercises which can be used for remediation and review throughout the year.

(pp. 216, 339) **Applications** pages provide insights into the use of mathematics in everyday life.

(pp. 184, 412) **Excursions in Algebra** provide short topics in the history, development, and uses of algebra.

(pp. 21, 115) **Reading Algebra** focuses on the interpretation of mathematical symbolism and terminology.

(pp. 163, 192) **Using Calculators** features instruct students in the use of the calculator as related to concepts taught in the lesson.

(pp. 204, 330) **Using Computers** topics help students understand the uses and limitations of computers. Programs are shown that relate to the content of the chapter.

(pp. 209, 403) **Standardized Test Practice Questions** help students prepare for college-entrance tests.

(pp. 210-215) Annotations in the **Teacher's Annotated Edition** show objectives for each lesson and answers to exercises. Margin notes beside the annotated pages provide readiness activities, discussion questions, chalkboard exercises, answers, quizzes, and suggested daily assignments.

Teacher's Guide and Tests for *Merrill Algebra One*

Authors
Foster • Rath • Winters

Merrill Publishing Co.
Columbus, Ohio

Contents

Permission is granted to teachers of *Merrill Algebra One* to reproduce and use tests in this Teacher's Guide entirely or in part.

ISBN 0-675-05477-X

Published by
Merrill Publishing Company
Columbus, Ohio 43216

Printed in the United States of America

5 6 7 8 9 10 11 12 13 14 15 RING 00 99 98 97 96 95 94 93 92 91

Preface

This third edition of *Merrill Algebra One* is designed for use by students in first-year high school algebra courses. The text, which was developed in the classroom by experienced high school teachers, is based on the successful prior editions of *Merrill Algebra One*. The goals of the text are to develop proficiency with mathematical skills, to expand understanding of mathematical concepts, to improve logical thinking, and to promote success. To achieve these goals, the following strategies are used.

Build upon a Solid Foundation. This program develops and utilizes the learning spiral. Students first review those concepts basic to the understanding of algebra. Students' understanding is thus solidified before the introduction of more difficult concepts.

Utilize Sound Pedagogy. *Merrill Algebra One* covers in logical sequence all topics generally presented at this level. Concepts are introduced when they are needed. Each concept presented is then used within that lesson and in later lessons.

Gear Presentation for Learning. An appropriate reading level has been maintained throughout the text. Furthermore, many photographs, illustrations, charts, graphs, and tables provide visual aids for the concepts and skills presented. Hence, students are able to read and learn with increased understanding.

Use Relevant Real-Life Applications. Applications are provided not only for practice but also to aid understanding of how concepts are used.

The text offers a variety of aids for the students.

Student Annotations Help students identify important concepts as they study.

Selected Answers Allow students to check their progress as they work. These answers are provided at the back of the text.

Mini-Review Provides students with a quick review of skills and concepts taught previously.

Vocabulary Enables students to focus on increasing their mathematical vocabulary.

Chapter Summary Provides students with a compact listing of major concepts presented within each chapter.

Chapter Review Permits students to review each chapter by working sample problems from each lesson.

Chapter Test Enables students to check their own progress.

Cumulative Review Helps students to maintain their skills in applying algebraic concepts.

Standardized Test Practice Questions Help students to become familiar with types of questions that appear on standardized tests.

The following special features, which appear periodically throughout the text, provide interesting and useful extra topics.

Using Calculators Instructs students in using a calculator. The use of the calculator is related to concepts taught within the chapter.

Reading Algebra Provides students with the instruction needed to read and interpret mathematical symbolism.

Problem Solving Illustrates several helpful strategies for solving mathematical problems.

Applications Provide insights into the uses of mathematics in everyday life, from a variety of disciplines.

Using Computers Provides a computer program and exercises related to the objectives of the chapter.

Excursions in Algebra Enliven and help maintain student interest by providing interesting side trips. Topics are varied and include history, glimpses into development of algebra, enrichment concepts, and puzzles.

The textbook contains an appendix on **BASIC**, which provides instruction in writing programs using the BASIC computer language.

The **Algebraic Skills Review** at the back of the text provides review exercises to help students maintain algebraic skills. These exercises can be used for remediation, to supplement exercises in the lessons, or as review material at the end of the school year.

Teachers and students familiar with the prior editions of this text will be pleased to see that the clarity of explanations and careful sequencing of topics have been retained in this new edition of *Merrill Algebra One*.

Components of Merrill Algebra

Student Texts

Merrill Algebra One and *Merrill Algebra Two with Trigonometry* were written by experienced classroom teachers. The clear explanations and examples allow students to grasp new concepts easily. Each chapter includes a balance of lesson material, review and test exercises, applications and extension features, and instruction in technology. Problem-solving skills are developed throughout both textbooks.

Teacher's Annotated Editions

Each *Teacher's Annotated Edition* includes a 96-page Teacher's Guide, and annotated student pages. Beside each annotated page is a margin containing Chalkboard Exercises, answers, and teaching suggestions.

Teacher's Resource Books

Each *Teacher's Resource Book* provides teachers with an abundance of materials including masters for quizzes, tests, cumulative review, extension, extra practice, problem solving, and applications. Pages are keyed to the appropriate chapter or lesson of the student textbook.

Solutions Manuals

Each *Solutions Manual* contains a complete solution to each problem in the student textbook. Solutions are displayed in a format similar to the examples in the student textbooks.

Student Features

Lessons

The textbook is divided into 16 chapters. Each chapter contains from 7-10 easily managed lessons. Lessons are logically organized. Color and art are used to enhance the readability of the material.

Tintblocks highlight important concepts and ideas.

Completely worked *examples* illustrate the lesson's objectives and provide students with models for solving problems. *Student annotations* provide explanations for steps in problem solving.

Exploratory Exercises can be used in class to review or clarify concepts, or as part of the assignment. *Written Exercises* provide ample practice to facilitate mastery of the lesson's objectives. Exercises progress in difficulty from the beginning to the end of the exercise set. *Challenge Exercises* appear in some lessons and are extensions of the topics presented in the lesson.

12-9 The Distance Formula

The Pythagorean Theorem can be used to find the distance between any two points in the coordinate plane. For example, find the distance between (−5, 2) and (4, 5).

In the diagram at the right, the distance to be found is the length of segment d. Notice that a right triangle can be formed by drawing lines parallel to the axes from (−5, 2) and (4, 5). The coordinates of the point where these lines meet are (4, 2).

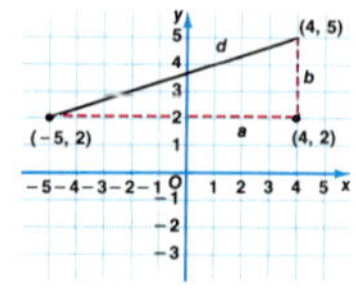

The length of side a of the triangle is the difference of the x-coordinates, $4 - (-5)$, or 9 units. The length of side b of the triangle is the difference of the y-coordinates, $5 - 2$, or 3 units.

Now the Pythagorean Theorem can be used to find the length of d, that is, the distance between (−5, 2) and (4, 5).

$$d^2 = 9^2 + 3^2$$
$$d^2 = 81 + 9$$
$$d^2 = 90$$
$$d = \sqrt{90}$$
$$d = 3\sqrt{10}$$

The distance between (−5, 2) and (4, 5) is $3\sqrt{10}$ units. The method for finding the distance between any two points can be generalized as the following formula.

The distance between any two points with coordinates (x_1, y_1) and (x_2, y_2) is given by the following formula:
$$d = \sqrt{(x_2 - x_1)^2 + (y_2 - y_1)^2}$$

The Distance Formula

Example

1 **Find the distance between the points with coordinates (8, −2) and (5, 3).**

$d = \sqrt{(x_2 - x_1)^2 + (y_2 - y_1)^2}$ *Let $x_2 = 5$, $x_1 = 8$, $y_2 = 3$, and $y_1 = -2$.*

$d = \sqrt{(5 - 8)^2 + [3 - (-2)]^2}$

$d = \sqrt{(-3)^2 + 5^2}$

$d = \sqrt{9 + 25}$

$d = \sqrt{34}$ *This is approximately 5.831 units.*

Exploratory Exercises

Simplify.

1. $3x + 4x - 5 - 6$
2. $12 - 3y + 12y - 23$
3. $3y + 5y - 8 - 34$
4. $17 - (3a + 4) - 6a$
5. $6y - 8y + 34 - 34y$
6. $8r + 34 + 39r - 8r$
7. $12(q + 4) - 2q$
8. $5(x - 3) + 22 - 8x$
9. $-3(x - 12) + 5(x - 23)$
10. $-9(9 - y) - 8(12 - y) + 23$

Written Exercises

Solve each equation.

1. $3x + 8x - 7 = 21x - 5$
2. $9x - 8 + 4x = 7x + 16$
3. $4y + 8y - 7 = 30y + 19$
4. $8y + 16 + 8y = 21 - 9y$
5. $-8 + 12a = 14a - 23 + 5a$
6. $7a + 23 - 12a = 34 + 23a$
7. $-9a + 34 - 2a = 28 - 5a + 27$
8. $9a - 11 = 23 + 24a - 27a$
9. $11(a - 3) + 5 = 2a + 44$
10. $-3(2a - 12) + 48 = 3a - 3$
11. $26 + 4(3a - 5) = 14a - 15$
12. $-13 + 3(3a + 11) = -33 - 3a$
13. $57 + 2a = 3a - 5(a - 9)$
14. $29 - 3a = 2(3a - 4) + 3$
15. $3a - 35 = 4(a + 12)$
16. $13 - 3a = 23(2a - 3)$
17. $2(5w - 12) = 6(-2w + 3) + 2$
18. $-11(3a - 4) = -3(-4a - 14)$
19. $-6(12 - 2w) = 7(-2 - 3w)$
20. $15(3a - 4) = -4(3a - 3)$
21. $-5(2x - 13) + 12 = 3(2x - 5)$
22. $7(x - 12) = 13 + 5(3x - 4)$
23. $\frac{1}{2}(2x - 34) = \frac{2}{3}(6x - 27)$
24. $\frac{3}{4}(8z - 12) = \frac{5}{6}(12z - 18)$
25. $19 - (2y + 3) = 2(y + 3) + y$
26. $2(a + 2) + 3a = 13 - (2a + 2)$
27. $x(x + 2) + 3x = x(x - 3)$
28. $w(w + 12) = w(w + 14) + 12$
29. $a(a - 6) + 2a = 3 + a(a - 2)$
30. $q(2q + 3) + 20 = 2q(q - 3)$
31. $x(x + 8) - x(x + 3) - 23 = 3x + 11$
32. $y(y - 12) + y(y + 2) + 25 = 2y(y + 5) - 15$

Challenge

Solve each equation for x or y.

33. $2x = 5a$
34. $4aby = 8a^2bc$
35. $30b^2c^2d = 10bc^2y$
36. $y - 2a = 0$
37. $4n + 3y = 2m$
38. $ax + 3 = b$
39. $2x - 7 = 8 + a$
40. $2y + b = 3 - 5y$
41. $7ax - 2d = 3ax + 6d$
42. $4x + 5mn = 7mn + 2x$
43. $by - c = a - d$
44. $3(y - 2a) = 24a$
45. $a(x + b) = 3ab + 5$
46. $2(3a - 2x) = 5a + x$
47. $4(y + 3a) = 4a - y$

Review and Testing

Each chapter includes *Vocabulary*, *Chapter Summary*, *Chapter Review*, and *Chapter Test*.

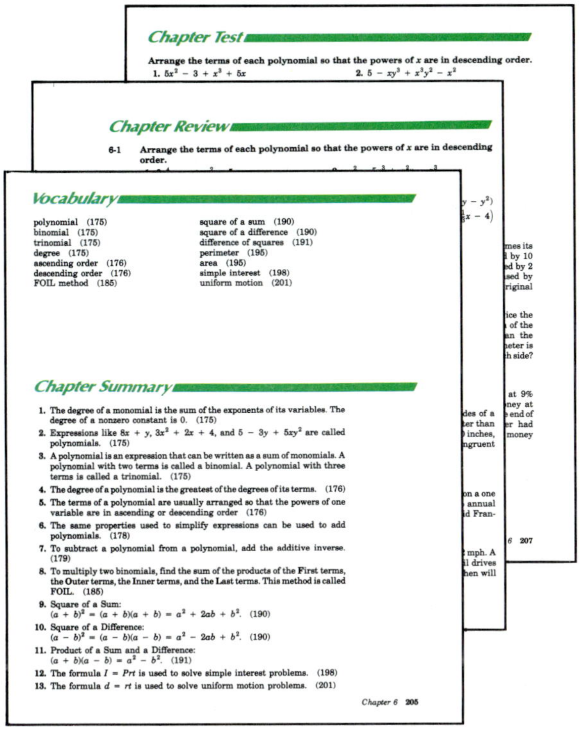

Mini-Reviews appear about 2 times in each chapter. *Cumulative Reviews* appear after even-numbered chapters. These features contain exercises from previous chapters to help students maintain their skills.

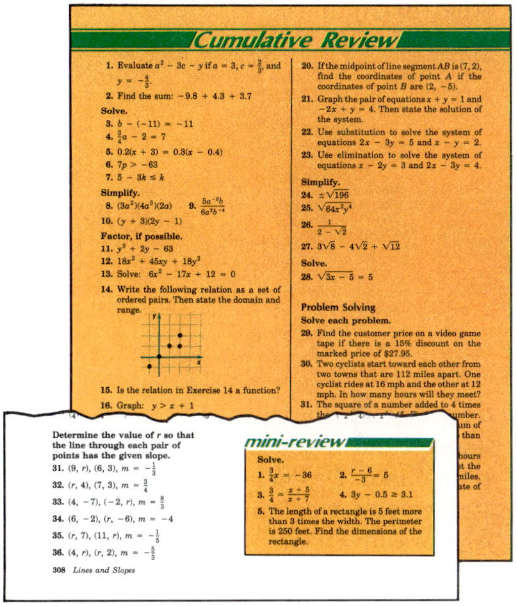

The *Algebraic Skills Review* (pages 546-562) has exercises for review and remediation of algebraic topics.

Standardized Test Practice Questions help students prepare for pre-college tests. These pages appear after each even-numbered chapter.

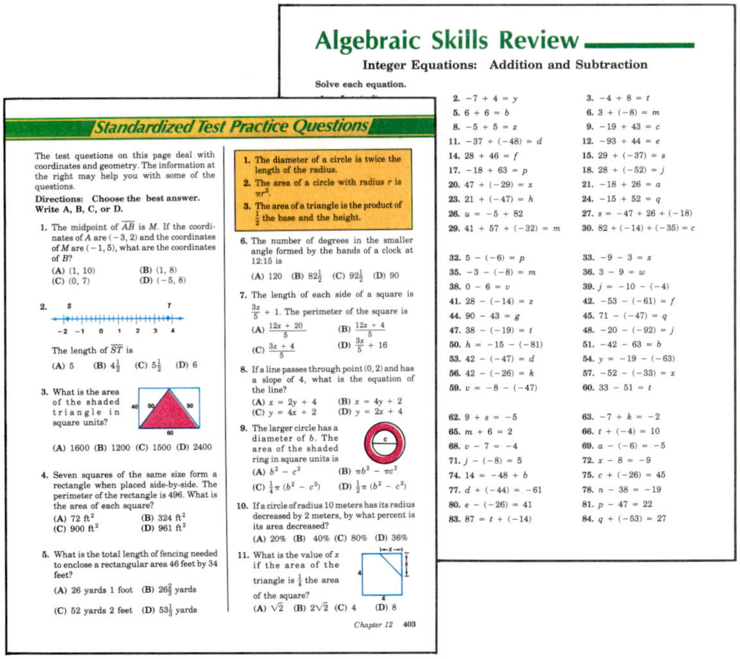

Student Features (Continued)

Applications and Extensions

Each chapter includes an *Applications* page, which shows how mathematics is used in other disciplines.

Reading Algebra pages help students understand algebra terminology and symbolism. *Problem Solving* pages present strategies for solving problems. A Reading Algebra or Problem Solving page appears in each chapter.

Excursions in Algebra provide short topics in the development and use of algebra.

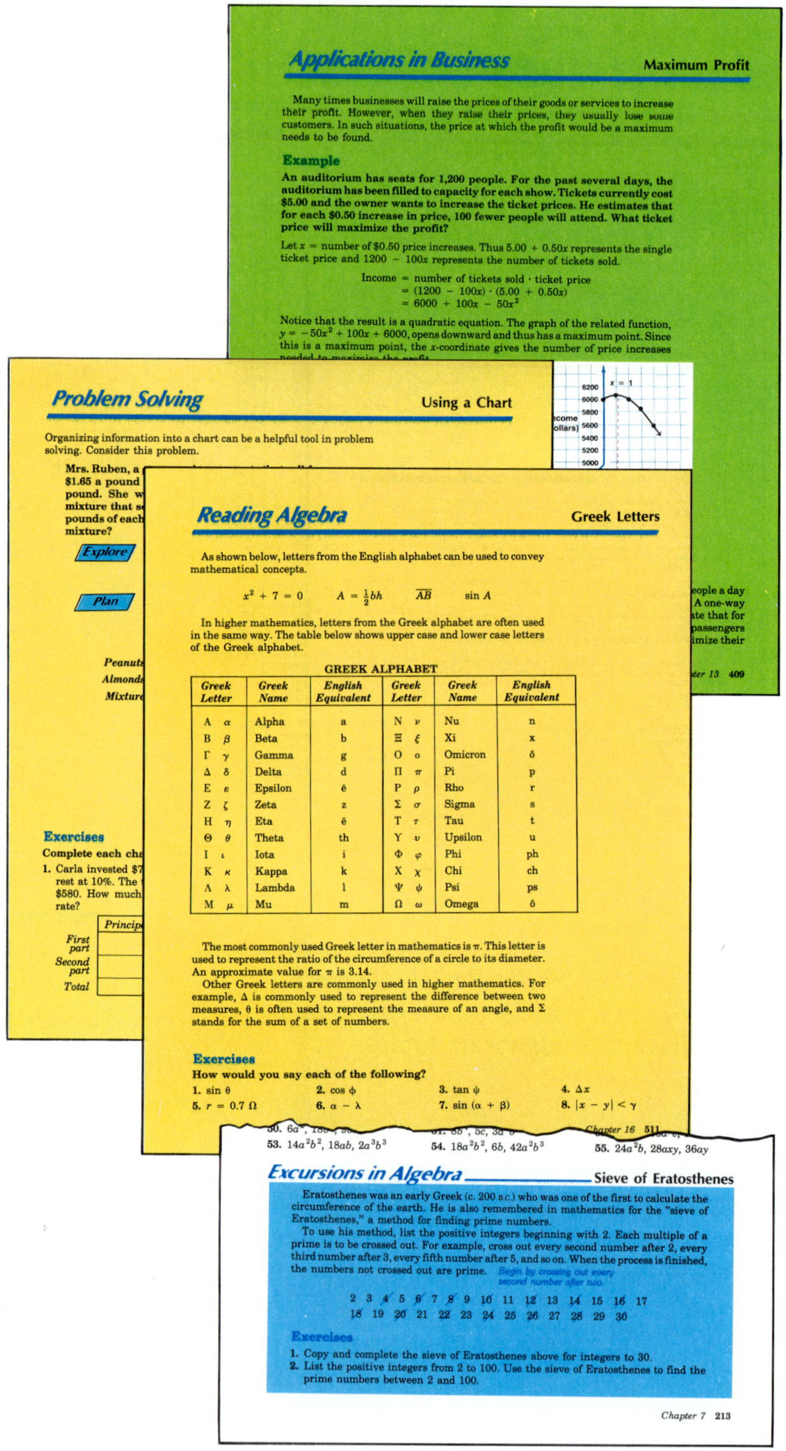

Applications in Business — Maximum Profit

Many times businesses will raise the prices of their goods or services to increase their profit. However, when they raise their prices, they usually lose some customers. In such situations, the price at which the profit would be a maximum needs to be found.

Example

An auditorium has seats for 1,200 people. For the past several days, the auditorium has been filled to capacity for each show. Tickets currently cost \$5.00 and the owner wants to increase the ticket prices. He estimates that for each \$0.50 increase in price, 100 fewer people will attend. What ticket price will maximize the profit?

Let x = number of \$0.50 price increases. Thus $5.00 + 0.50x$ represents the single ticket price and $1200 - 100x$ represents the number of tickets sold.

$$\begin{aligned}\text{Income} &= \text{number of tickets sold} \cdot \text{ticket price}\\ &= (1200 - 100x) \cdot (5.00 + 0.50x)\\ &= 6000 + 100x - 50x^2\end{aligned}$$

Notice that the result is a quadratic equation. The graph of the related function, $y = -50x^2 + 100x + 6000$, opens downward and thus has a maximum point. Since this is a maximum point, the x-coordinate gives the number of price increases ...

Problem Solving — Using a Chart

Organizing information into a chart can be a helpful tool in problem solving. Consider this problem.

Reading Algebra — Greek Letters

As shown below, letters from the English alphabet can be used to convey mathematical concepts.

$x^2 + 7 = 0$ $\quad A = \frac{1}{2}bh$ $\quad \overline{AB}$ $\quad \sin A$

In higher mathematics, letters from the Greek alphabet are often used in the same way. The table below shows upper case and lower case letters of the Greek alphabet.

GREEK ALPHABET

Greek Letter	Greek Name	English Equivalent	Greek Letter	Greek Name	English Equivalent
Α α	Alpha	a	Ν ν	Nu	n
Β β	Beta	b	Ξ ξ	Xi	x
Γ γ	Gamma	g	Ο ο	Omicron	ŏ
Δ δ	Delta	d	Π π	Pi	p
Ε ε	Epsilon	ĕ	Ρ ρ	Rho	r
Ζ ζ	Zeta	z	Σ σ	Sigma	s
Η η	Eta	ē	Τ τ	Tau	t
Θ θ	Theta	th	Υ υ	Upsilon	u
Ι ι	Iota	i	Φ φ	Phi	ph
Κ κ	Kappa	k	Χ χ	Chi	ch
Λ λ	Lambda	l	Ψ ψ	Psi	ps
Μ μ	Mu	m	Ω ω	Omega	ō

The most commonly used Greek letter in mathematics is π. This letter is used to represent the ratio of the circumference of a circle to its diameter. An approximate value for π is 3.14.

Other Greek letters are commonly used in higher mathematics. For example, Δ is commonly used to represent the difference between two measures, θ is often used to represent the measure of an angle, and Σ stands for the sum of a set of numbers.

Exercises

How would you say each of the following?

1. $\sin \theta$ 2. $\cos \phi$ 3. $\tan \psi$ 4. Δx
5. $r = 0.7\ \Omega$ 6. $\alpha - \lambda$ 7. $\sin(\alpha + \beta)$ 8. $|x - y| < \gamma$

Chapter 16 511

Excursions in Algebra — Sieve of Eratosthenes

Eratosthenes was an early Greek (c. 200 B.C.) who was one of the first to calculate the circumference of the earth. He is also remembered in mathematics for the "sieve of Eratosthenes," a method for finding prime numbers.

To use his method, list the positive integers beginning with 2. Each multiple of a prime is to be crossed out. For example, cross out every second number after 2, every third number after 3, every fifth number after 5, and so on. When the process is finished, the numbers not crossed out are prime.

2 3 4 5 6 7 8 9 10 11 12 13 14 15 16 17 18 19 20 21 22 23 24 25 26 27 28 29 30

Exercises

1. Copy and complete the sieve of Eratosthenes above for integers to 30.
2. List the positive integers from 2 to 100. Use the sieve of Eratosthenes to find the prime numbers between 2 and 100.

Chapter 7 213

Technology

Using Calculators features appear at the end of one or more lessons in each chapter. These features explain how calculators may be used with the lesson, and provide practice exercises.

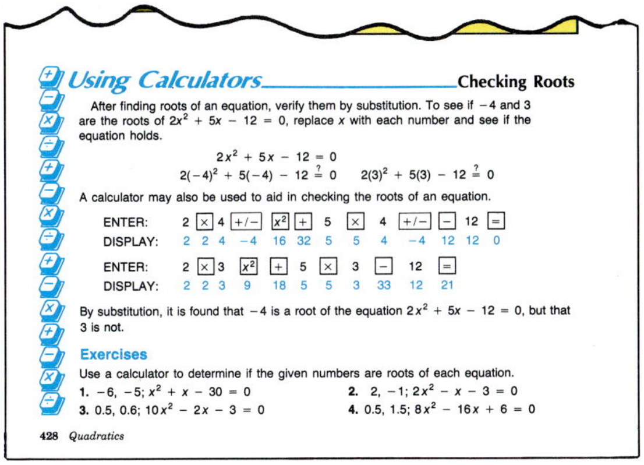

Using Calculators — Checking Roots

After finding roots of an equation, verify them by substitution. To see if −4 and 3 are the roots of $2x^2 + 5x - 12 = 0$, replace x with each number and see if the equation holds.

$$2x^2 + 5x - 12 = 0$$
$$2(-4)^2 + 5(-4) - 12 \stackrel{?}{=} 0 \qquad 2(3)^2 + 5(3) - 12 \stackrel{?}{=} 0$$

A calculator may also be used to aid in checking the roots of an equation.

ENTER:	2	×	4	+/−	x^2	+	5	×	4	+/−	−	12	=
DISPLAY:	2	2	4	−4	16	32	5	5	4	−4	12	12	0

ENTER:	2	×	3	x^2	+	5	×	3	−	12	=
DISPLAY:	2	2	3	9	18	5	5	3	33	12	21

By substitution, it is found that −4 is a root of the equation $2x^2 + 5x - 12 = 0$, but that 3 is not.

Exercises

Use a calculator to determine if the given numbers are roots of each equation.

1. −6, −5; $x^2 + x - 30 = 0$
2. 2, −1; $2x^2 - x - 3 = 0$
3. 0.5, 0.6; $10x^2 - 2x - 3 = 0$
4. 0.5, 1.5; $8x^2 - 16x + 6 = 0$

428 *Quadratics*

Using Computers features appear after the last lesson of each chapter, and include a computer program and exercises related to the chapter. These programs are available on a diskette. No previous programming experience is necessary.

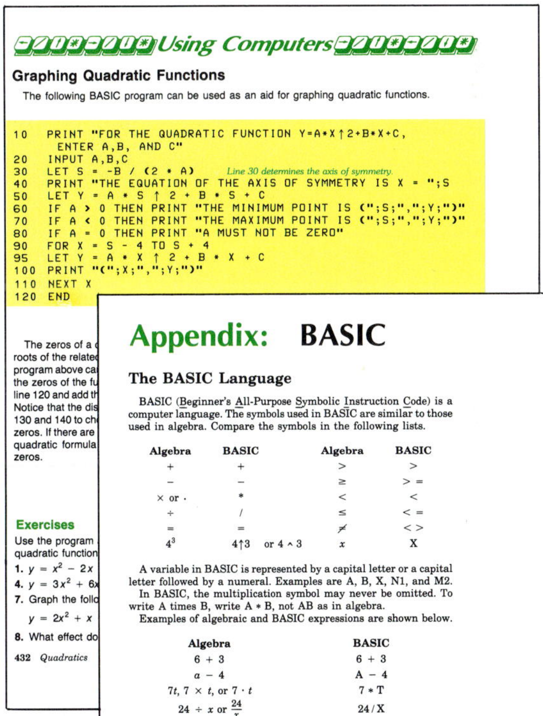

Using Computers

Graphing Quadratic Functions

The following BASIC program can be used as an aid for graphing quadratic functions.

```
10   PRINT "FOR THE QUADRATIC FUNCTION Y=A*X↑2+B*X+C,
      ENTER A,B, AND C"
20   INPUT A,B,C
30   LET S = -B / (2 * A)
40   PRINT "THE EQUATION OF THE AXIS OF SYMMETRY IS X = ";S
50   LET Y = A * S ↑ 2 + B * S + C
60   IF A > 0 THEN PRINT "THE MINIMUM POINT IS (";S;",";Y;")"
70   IF A < 0 THEN PRINT "THE MAXIMUM POINT IS (";S;",";Y;")"
80   IF A = 0 THEN PRINT "A MUST NOT BE ZERO"
90   FOR X = S - 4 TO S + 4
95   LET Y = A * X ↑ 2 + B * X + C
100  PRINT "(";X;",";Y;")"
110  NEXT X
120  END
```

Line 30 determines the axis of symmetry.

The zeros of a … roots of the relate… program above ca… the zeros of the fu… line 120 and add th… Notice that the dis… 130 and 140 to ch… zeros. If there are … quadratic formula… zeros.

Exercises

Use the program … quadratic function…

1. $y = x^2 - 2x$ …
4. $y = 3x^2 + 6x$ …
7. Graph the follo… $y = 2x^2 + x$ …
8. What effect do…

432 *Quadratics*

The *BASIC Appendix* (pages 528-542) provides a short course in the BASIC programming language.

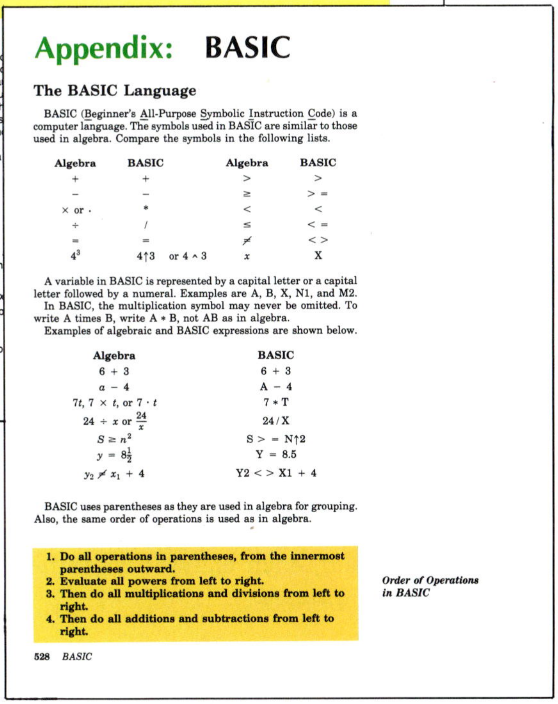

Appendix: BASIC

The BASIC Language

BASIC (Beginner's All-Purpose Symbolic Instruction Code) is a computer language. The symbols used in BASIC are similar to those used in algebra. Compare the symbols in the following lists.

Algebra	BASIC	Algebra	BASIC
+	+	>	>
−	−	≥	> =
× or ·	*	<	<
÷	/	≤	< =
=	=	≠	< >
4^3	4↑3 or 4 ^ 3	x	X

A variable in BASIC is represented by a capital letter or a capital letter followed by a numeral. Examples are A, B, X, N1, and M2.

In BASIC, the multiplication symbol may never be omitted. To write A times B, write A * B, not AB as in algebra.

Examples of algebraic and BASIC expressions are shown below.

Algebra	BASIC
$6 + 3$	6 + 3
$a - 4$	A − 4
$7t$, $7 \times t$, or $7 \cdot t$	7 * T
$24 \div x$ or $\frac{24}{x}$	24 / X
$S \geq n^2$	S > = N↑2
$y = 8\frac{1}{2}$	Y = 8.5
$y_2 \neq x_1 + 4$	Y2 < > X1 + 4

BASIC uses parentheses as they are used in algebra for grouping. Also, the same order of operations is used as in algebra.

Order of Operations in BASIC

1. **Do all operations in parentheses, from the innermost parentheses outward.**
2. **Evaluate all powers from left to right.**
3. **Then do all multiplications and divisions from left to right.**
4. **Then do all additions and subtractions from left to right.**

528 *BASIC*

Teacher Features

- *Objectives* are stated at the beginning of each lesson.
- *Readiness Activities* and/or *Discussion Questions* are suggested for most lessons to check prerequisite skills and deepen students' understanding of concepts.
- *Chalkboard Exercises* keyed to the examples, can be used during class to develop concepts and practice skills.
- *Teaching suggestions* and hints are keyed to specific content on the student pages.

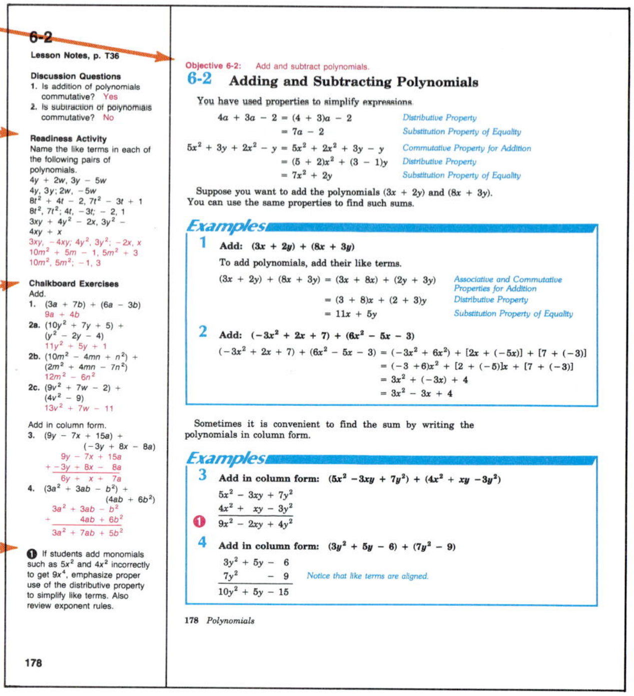

6-2

Lesson Notes, p. T36

Discussion Questions

1. Is addition of polynomials commutative? Yes
2. Is subtraction of polynomials commutative? No

Readiness Activity

Name the like terms in each of the following pairs of polynomials.

$4y + 2w$, $3y - 5w$
$4y, 3y; 2w, -5w$
$8t^2 + 4t - 2$, $7t^2 - 3t + 1$
$8t^2, 7t^2; 4t, -3t; -2, 1$
$3xy + 4y^2 - 2x$, $3y^2 - 4xy + x$
$3xy, -4xy; 4y^2, 3y^2; -2x, x$
$10m^2 + 5m - 1$, $5m^2 + 3$
$10m^2, 5m^2; -1, 3$

Chalkboard Exercises

Add.

1. $(3a + 7b) + (6a - 3b)$ $9a + 4b$

2a. $(10y^2 + 7y + 5) + (y^2 - 2y - 4)$ $11y^2 + 5y + 1$

2b. $(10m^2 - 4mn + n^2) + (2m^2 + 4mn - 7n^2)$ $12m^2 - 6n^2$

2c. $(9v^2 + 7w - 2) + (4v^2 - 9)$ $13v^2 + 7w - 11$

Add in column form.

3. $(9y - 7x + 15a) + (-3y + 8x - 8a)$

$$\begin{array}{r} 9y - 7x + 15a \\ + \underline{-3y + 8x - 8a} \\ 6y + x + 7a \end{array}$$

4. $(3a^2 + 3ab - b^2) + (4ab + 6b^2)$

$$\begin{array}{r} 3a^2 + 3ab - b^2 \\ + \underline{4ab + 6b^2} \\ 3a^2 + 7ab + 5b^2 \end{array}$$

❶ If students add monomials such as $5x^2$ and $4x^2$ incorrectly to get $9x^4$, emphasize proper use of the distributive property to simplify like terms. Also review exponent rules.

178

Objective 6-2: Add and subtract polynomials.

6-2 Adding and Subtracting Polynomials

You have used properties to simplify expressions.

$$\begin{aligned} 4a + 3a - 2 &= (4 + 3)a - 2 && \text{Distributive Property} \\ &= 7a - 2 && \text{Substitution Property of Equality} \\ 5x^2 + 3y + 2x^2 - y &= 5x^2 + 2x^2 + 3y - y && \text{Commutative Property for Addition} \\ &= (5 + 2)x^2 + (3 - 1)y && \text{Distributive Property} \\ &= 7x^2 + 2y && \text{Substitution Property of Equality} \end{aligned}$$

Suppose you want to add the polynomials $(3x + 2y)$ and $(8x + 3y)$. You can use the same properties to find such sums.

Examples

1 **Add: $(3x + 2y) + (8x + 3y)$**

To add polynomials, add their like terms.

$$\begin{aligned} (3x + 2y) + (8x + 3y) &= (3x + 8x) + (2y + 3y) && \text{Associative and Commutative Properties for Addition} \\ &= (3 + 8)x + (2 + 3)y && \text{Distributive Property} \\ &= 11x + 5y && \text{Substitution Property of Equality} \end{aligned}$$

2 **Add: $(-3x^2 + 2x + 7) + (6x^2 - 5x - 3)$**

$$\begin{aligned} (-3x^2 + 2x + 7) + (6x^2 - 5x - 3) &= (-3x^2 + 6x^2) + [2x + (-5x)] + [7 + (-3)] \\ &= (-3 + 6)x^2 + [2 + (-5)]x + [7 + (-3)] \\ &= 3x^2 + (-3x) + 4 \\ &= 3x^2 - 3x + 4 \end{aligned}$$

Sometimes it is convenient to find the sum by writing the polynomials in column form.

Examples

3 **Add in column form: $(5x^2 - 3xy + 7y^2) + (4x^2 + xy - 3y^2)$**

$$\begin{array}{r} 5x^2 - 3xy + 7y^2 \\ \underline{4x^2 + xy - 3y^2} \\ 9x^2 - 2xy + 4y^2 \end{array}$$

❶

4 **Add in column form: $(3y^2 + 5y - 6) + (7y^2 - 9)$**

$$\begin{array}{r} 3y^2 + 5y - 6 \\ \underline{7y^2 \qquad - 9} \\ 10y^2 + 5y - 15 \end{array}$$

Notice that like terms are aligned.

178 *Polynomials*

- *Suggested Daily Assignments* are provided for fundamental, average, and enriched classes.
- *Answers* are provided for all exercises.
- In addition, each chapter contains two or three *quizzes*.

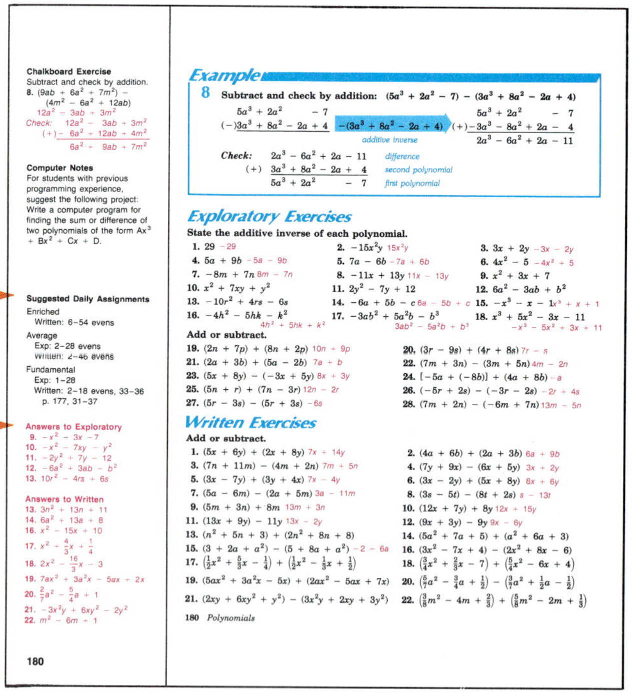

Chalkboard Exercise

Subtract and check by addition.

8. $(9ab + 6a^2 + 7m^2) - (4m^2 - 6a^2 + 12ab)$ $12a^2 - 3ab + 3m^2$

Check:
$$\begin{array}{r} 12a^2 - 3ab + 3m^2 \\ (+) \underline{-6a^2 + 12ab + 4m^2} \\ 6a^2 + 9ab + 7m^2 \end{array}$$

Computer Notes

For students with previous programming experience, suggest the following project: Write a computer program for finding the sum or difference of two polynomials of the form $Ax^3 + Bx^2 + Cx + D$.

Suggested Daily Assignments

Enriched
Written: 6–54 evens

Average
Exp: 2–28 evens
Written: 2–46 evens

Fundamental
Exp: 1–28
Written: 2–18 evens, 33–36
p. 177, 31–37

Answers to Exploratory

9. $-x^2 - 3x - 7$
10. $-x^2 - 7xy - y^2$
11. $-2y^2 + 7y - 12$
12. $-6a^2 + 3ab - b^2$
13. $10r^2 - 4rs + 6s$

Answers to Written

13. $3n^2 + 13n + 11$
14. $6a^2 + 13a + 8$
16. $x^2 - 15x + 10$
17. $x^2 + \frac{4}{3}x + \frac{1}{4}$
18. $2x^2 - \frac{16}{3}x - 3$
19. $7ax^2 + 3a^2x - 5ax + 2x$
20. $\frac{2}{7}a^2 - \frac{5}{4}a + 1$
21. $-3x^2y + 6xy^2 - 2y^2$
22. $m^2 - 6m + 1$

180

Example

8 **Subtract and check by addition: $(5a^3 + 2a^2 - 7) - (3a^3 + 8a^2 - 2a + 4)$**

$$\begin{array}{r} 5a^3 + 2a^2 \qquad - 7 \\ (-)\underline{3a^3 + 8a^2 - 2a + 4} \end{array} \quad -(3a^3 + 8a^2 - 2a + 4) \quad \begin{array}{r} 5a^3 + 2a^2 \qquad - 7 \\ (+)\underline{-3a^3 - 8a^2 + 2a - 4} \\ 2a^3 - 6a^2 + 2a - 11 \end{array}$$

additive inverse

Check:
$$\begin{array}{rl} 2a^3 - 6a^2 + 2a - 11 & \text{difference} \\ (+) \underline{3a^3 + 8a^2 - 2a + 4} & \text{second polynomial} \\ 5a^3 + 2a^2 \qquad - 7 & \text{first polynomial} \end{array}$$

Exploratory Exercises

State the additive inverse of each polynomial.

1. 29 -29
2. $-15x^2y$ $15x^2y$
3. $3x + 2y$ $-3x - 2y$
4. $5a + 9b$ $-5a - 9b$
5. $7a - 6b$ $-7a + 6b$
6. $4x^2 - 5$ $-4x^2 + 5$
7. $-8m + 7n$ $8m - 7n$
8. $-11x + 13y$ $11x - 13y$
9. $x^2 + 3x + 7$
10. $x^2 + 7xy + y^2$
11. $2y^2 - 7y + 12$
12. $6a^2 - 3ab + b^2$
13. $-10r^2 + 4rs - 6s$
14. $-6a + 5b - c$ $6a - 5b + c$
15. $-x^3 - x - 1$ $x^3 + x + 1$
16. $-4h^2 - 5hk - k^2$ $4h^2 + 5hk + k^2$
17. $-3ab^2 + 5a^2b - b^3$ $3ab^2 - 5a^2b + b^3$
18. $x^3 + 5x^2 - 3x - 11$ $-x^3 - 5x^2 + 3x + 11$

Add or subtract.

19. $(2n + 7p) + (8n + 2p)$ $10n + 9p$
20. $(3r - 9s) + (4r + 8s)$ $7r - s$
21. $(2a + 3b) + (5a - 2b)$ $7a + b$
22. $(7m + 3n) - (3m + 5n)$ $4m - 2n$
23. $(5x + 8y) - (-3x + 5y)$ $8x + 3y$
24. $[-5a + (-8b)] + (4a + 8b)$ $-a$
25. $(5n + r) + (7n - 3r)$ $12n - 2r$
26. $(-5r + 2s) - (-3r - 2s)$ $-2r + 4s$
27. $(5r - 3s) - (5r + 3s)$ $-6s$
28. $(7m + 2n) - (-6m + 7n)$ $13m - 5n$

Written Exercises

Add or subtract.

1. $(5x + 6y) + (2x + 8y)$ $7x + 14y$
2. $(4a + 6b) + (2a + 3b)$ $6a + 9b$
3. $(7n + 11m) - (4m + 2n)$ $7m + 5n$
4. $(7y + 9x) - (6x + 5y)$ $3x + 2y$
5. $(3x - 7y) + (3y + 4x)$ $7x - 4y$
6. $(3x - 2y) + (5x + 8y)$ $8x + 6y$
7. $(5a - 6m) - (2a + 5m)$ $3a - 11m$
8. $(3s - 5t) - (8t + 2s)$ $s - 13t$
9. $(5m + 3n) + 8m$ $13m + 3n$
10. $(12x + 7y) + 8y$ $12x + 15y$
11. $(13x + 9y) - 11y$ $13x - 2y$
12. $(9x + 3y) - 9y$ $9x - 6y$
13. $(n^2 + 5n + 3) + (2n^2 + 8n + 8)$
14. $(5a^2 + 7a + 5) + (a^2 + 6a + 3)$
15. $(3 + 2a + a^2) - (5 + 8a + a^2)$ $-2 - 6a$
16. $(3x^2 - 7x + 4) - (2x^2 + 8x - 6)$
17. $\left(\frac{1}{2}x^2 + \frac{5}{3}x - \frac{1}{4}\right) + \left(\frac{1}{2}x^2 - \frac{1}{3}x + \frac{1}{2}\right)$
18. $\left(\frac{3}{4}x^2 + \frac{2}{3}x - 7\right) + \left(\frac{5}{4}x^2 - 6x + 4\right)$
19. $(5ax^2 + 3a^2x - 5x) + (2ax^2 - 5ax + 7x)$
20. $\left(\frac{5}{7}a^2 - \frac{3}{4}a + \frac{1}{2}\right) - \left(\frac{3}{7}a^2 + \frac{1}{2}a - \frac{1}{2}\right)$
21. $(2xy + 6xy^2 + y^2) - (3x^2y + 2xy + 3y^2)$
22. $\left(\frac{3}{8}m^2 - 4m + \frac{2}{3}\right) + \left(\frac{5}{8}m^2 - 2m + \frac{1}{3}\right)$

180 *Polynomials*

Teacher's Guide

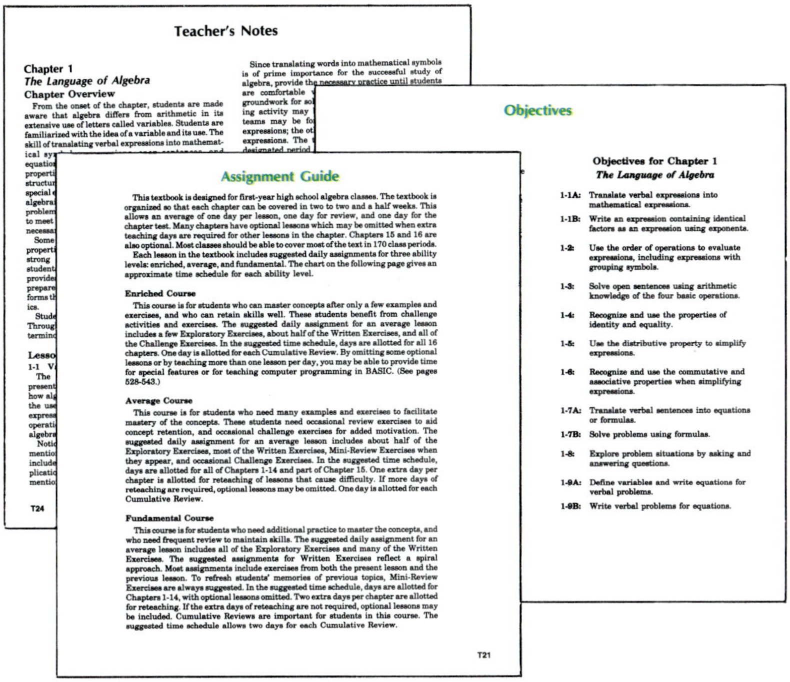

Teacher's Notes

The Teacher's Guide has Teacher's Notes that include a chapter overview with notes and suggestions for each lesson. Enrichment ideas are included for many lessons. **(T24)**

Objectives and Assignment Guide

Objectives are stated for each lesson. An assignment guide provides information about gearing the course for students of different abilities. **(T12, T21)**

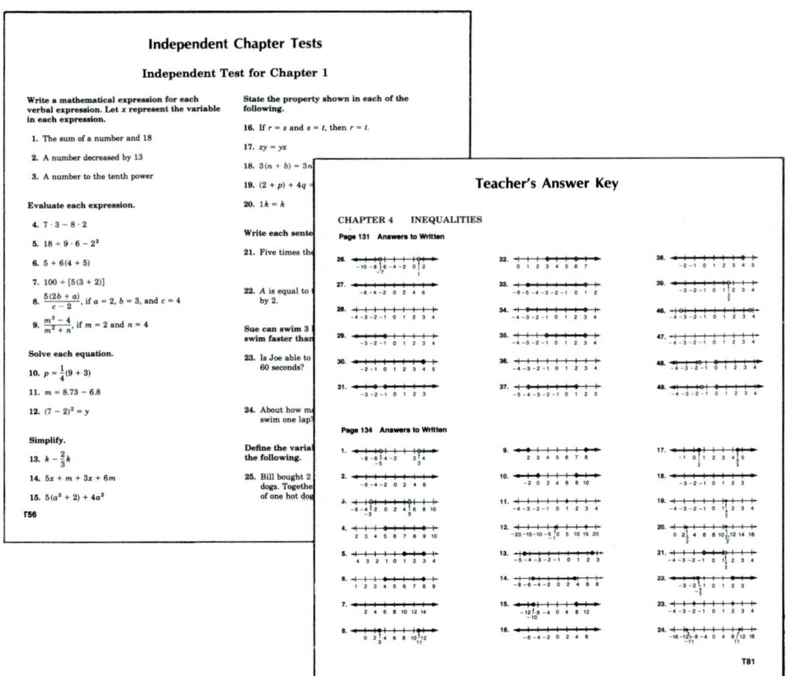

Tests

An independent test for each chapter as well as two semester tests are included. Answers are printed on pages T76-T80. **(T56)**

Teacher's Answer Key

The Teacher's Answer Key provides answers that do not appear on the annotated page or in the margin. **(T81)**

Objectives

About the Objectives
Each lesson of Merrill Algebra One has one or more learning objectives. The objective numbers correspond to the lesson numbers. If two objectives are given for a lesson, they are designated with the letters A and B.

Objectives for Chapter 1
The Language of Algebra

1-1A: Translate verbal expressions into mathematical expressions.

1-1B: Write an expression containing identical factors as an expression using exponents.

1-2: Use the order of operations to evaluate expressions, including expressions with grouping symbols.

1-3: Solve open sentences using arithmetic knowledge of the four basic operations.

1-4: Recognize and use the properties of identity and equality.

1-5: Use the distributive property to simplify expressions.

1-6: Recognize and use the commutative and associative properties when simplifying expressions.

1-7A: Translate verbal sentences into equations or formulas.

1-7B: Solve problems using formulas.

1-8: Explore problem situations by asking and answering questions.

1-9A: Define variables and write equations for verbal problems.

1-9B: Write verbal problems for equations.

Objectives for Chapter 2
Adding and Subtracting Rational Numbers

2-1A: State the coordinate of a point on the number line.

2-1B: Graph integers on a number line.

2-2: Add integers by using the number line.

2-3A: Find absolute value of a number.

2-3B: Add integers without using the number line.

2-4: Add rational numbers by using skills in adding integers and positive rational numbers.

2-5A: Add three or more rational numbers.

2-5B: Simplify expressions that contain rational numbers.

2-6: Subtract rational numbers by using additive inverses.

2-7: Solve equations involving addition by using the addition and subtraction properties of equality.

2-8: Solve equations involving subtraction.

2-9: Solve verbal problems by writing and solving equations.

Objectives for Chapter 3
Multiplying and Dividing Rational Numbers

3-1: Use the rules for multiplying positive and negative numbers.

3-2: Use the rules for dividing positive and negative numbers to simplify expressions.

3-3: Solve equations by using the multiplication and division properties of equality.

3-4: Solve equations involving more than one operation.

3-5A: Solve equations containing variables on both sides.

3-5B: Solve equations containing grouping symbols.

3-6: Solve equations containing fractions and decimals.

3-7: Use equations to solve problems.

3-8A: Solve equations using the means-extremes property of proportions.

3-8B: Use ratios and proportions to solve problems.

3-9: Solve percent problems.

Objectives for Chapter 4
Inequalities

4-1A: Learn the meaning of inequality symbols.

4-1B: Graph inequalities on the number line.

4-2: Solve inequalities involving addition or subtraction.

4-3: Solve inequalities involving multiplication or division.

4-4: Solve inequalities involving more than one operation.

4-5A: Use the comparison property to compare rational numbers.

4-5B: Determine which of two items is a better buy.

4-6: Solve compound sentences and graph the solution set.

4-7: Solve open sentences containing absolute value and graph the solution sets.

4-8: Use inequalities to solve verbal problems.

Objectives for Chapter 5
Powers

5-1: Multiply monomials.

5-2: Simplify expressions involving powers of monomials.

5-3A: Simplify expressions involving quotients of monomials.

5-3B: Simplify expressions containing negative exponents.

5-4A: Express numbers in scientific notation and decimal notation.

5-4B: Find products and quotients of numbers expressed in scientific notation.

5-5A: Solve verbal problems (age problems) that contain extra information.

5-5B: Recognize problems which do not contain enough information.

5-6A: Solve problems involving percent of increase or percent of decrease.

5-6B: Solve problems involving discount or sales tax.

5-7: Solve mixture problems by organizing the information in a chart.

Objectives for Chapter 6
Polynomials

6-1A: Find the degree of a polynomial.

6-1B: Arrange the terms of a polynomial so that the powers of a certain variable are in ascending or descending order.

6-2: Add and subtract polynomials.

6-3: Multiply a polynomial by a monomial by using the distributive property.

6-4A: Use the FOIL method to multiply two binomials.

6-4B: Multiply any two polynomials by using the distributive property.

6-5: Develop and use the patterns for $(a + b)^2$, $(a - b)^2$, and $(a + b)(a - b)$.

6-6: Solve equations that require addition, subtraction or multiplication on one or both sides of the equal sign before the equation can be solved.

6-7: Solve perimeter and area problems using equations.

6-8: Solve simple interest problems using the formula $I = Prt$.

6-9: Solve uniform motion problems using the formula $d = rt$.

Objectives for Chapter 7
Factoring

7-1A: Find the prime factorization of an integer.

7-1B: Find the greatest common factor (GCF) for a set of numbers or monomials.

7-2: Use the GCF and the distributive property to factor polynomials.

7-3: Recognize and factor differences of squares.

7-4: Factor perfect square trinomials.

7-5: Factor trinomials of in form $x^2 + bx + c$.

7-6: Factor polynomials containing four terms by grouping.

7-7: Factor trinomials of the form $ax^2 + bx + c$.

7-8: Review and summarize the various methods of factoring.

Objectives for Chapter 8
Applications of Factoring

8-1: Use the zero product property to solve equations in factored form.

8-2: Review the various methods of factoring.

8-3: Solve simple equations by factoring and using the zero-product property. (These equations contain two or three nonzero terms, and can be easily factored.)

8-4: Solve equations by factoring and using the zero-product property. (These equations contain three or four nonzero terms.)

8-5: Solve integer problems by factoring.

8-6: Solve problems that involve velocity and area by factoring.

8-7: Simplify rational expressions.

Objectives for Chapter 9
Functions and Graphs

9-1: Graph ordered pairs.

9-2A: Identify the domain and range of a relation.

9-2B: Show relations as sets of ordered pairs and mappings.

9-3A: Solve linear equations for a specific variable.

9-3B: Solve an equation for a given domain.

9-4: Graph linear equations.

9-5A: Determine whether a given relation is a function.

9-5B: Calculate functional values for a given function.

9-6: Graph inequalities in the coordinate plane.

9-7: Write an equation for a relation, given a chart of values.

Objectives for Chapter 10
Lines and Slopes

10-1: Find the slope of a line, given the coordinates of two points on the line.

10-2A: Write a linear equation in standard form given the coordinates of a point on the line and the slope of the line.

10-2B: Write a linear equation in standard form given the coordinates of two points on the line.

10-3A: Write an equation in slope-intercept form given the slope and y-intercept of the line.

10-3B: Determine the slope and y-intercept of the graph of a given equation.

10-3C: Determine the x- and y-intercepts of the graph of a given equation.

10-4: Graph linear equations using the x- and y-intercepts or the slope and y-intercept.

10-5A: Write a linear equation in slope-intercept form given the slope and the coordinates of a point on the line.

10-5B: Write a linear equation in slope-intercept form given the coordinates of two points of the line.

10-6: Write an equation of a line that is parallel or perpendicular to the graph of a given equation and that passes through a given point.

10-7: Find the coordinates of the midpoint of a line segment in the coordinate plane given the coordinates of the endpoints.

Objectives for Chapter 11
Systems of Open Sentences

11-1: Solve systems of equations by graphing.

11-2: Determine whether a system of equations has one solution, no solutions, or infinitely many solutions by graphing.

11-3: Solve systems of equations by the substitution method.

11-4: Solve systems of equations by the elimination method using addition.

11-5: Solve systems of equations by the elimination method, using multiplication along with addition.

11-6: Solve systems of inequalities by graphing.

11-7: Use systems of equations to solve digit problems.

11-8: Use systems of equations to solve uniform-motion problems and mixture problems.

Objectives for Chapter 12
Radical Expressions

12-1: Simplify radicals by using the product and quotient properties of square roots.

12-2: Identify irrational numbers.

12-3: Use the divide-and-average method to compute square roots.

12-4A: Simplify irrational square roots.

12-4B: Simplify radical expressions that contain variables.

12-5: Simplify expressions involving division of radical expressions.

12-6: Simplify expressions involving addition and subtraction of radical expressions.

12-7: Solve simple radical equations in which only one radical contains a variable.

12-8A: Use the Pythagorean Theorem to find the length of the third side of a right triangle given the lengths of the other two sides.

12-8B: Determine whether a triangle is a right triangle given the length of its sides.

12-9: Find the distance between two points in the coordinate plane.

Objectives for Chapter 13
Quadratics

13-1A: Find the equation of the axis of symmetry and the coordinates of the maximum or minimum point of a quadratic function.

13-1B: Graph a quadratic function.

13-2: Locate the roots of a quadratic equation by graphing the related function.

13-3: Solve quadratic equations by completing the square.

13-4: Solve quadratic equations by using the quadratic formula.

13-5: Find the value of the discriminant and use it to determine the nature of the roots of a quadratic equation.

13-6: Use a variety of methods to solve quadratic equations.

13-7: Solve verbal problems using quadratic equations.

13-8A: State the sum and product of the roots of a quadratic equation.

13-8B: Write a quadratic equation given its roots.

Objectives for Chapter 14
Rational Expressions

14-1: Simplify algebraic fractions.

14-2: Multiply algebraic fractions.

14-3: Divide algebraic fractions.

14-4: Divide a polynomial by a binomial.

14-5: Add and subtract algebraic fractions with like denominators.

14-6: Add and subtract algebraic fractions with unlike denominators.

14-7: Simplify mixed expressions and complex fractions.

Objectives for Chapter 15
Applications of Rational Expressions

15-1: Solve equations containing rational expressions.

15-2A: Solve work problems using the formula $rt = w$.

15-2B: Solve uniform-motion problems involving current or wind speed using the formula $rt = d$.

15-3: Solve formulas for a specified variable.

15-4: Solve problems involving direct variations.

15-5: Solve problems involving inverse variations.

15-6: Solve lever problems by applying the product rule for inverse variations.

15-7: Calculate resistances of circuits by applying knowledge of rational expressions.

15-8: Solve problems involving joint variations, quadratic direct variations, or combined variations.

Objectives for Chapter 16
Trigonometry

16-1A: Find the complement and supplement of an angle.

16-1B: Find the measure of the third angle of a triangle given the measures of the other two angles.

16-2: Find the measures of the sides of a 30°-60° right triangle given the measure of one side.

16-3: Use proportions to find the missing measures for similar triangles.

16-4: Compute the sine, cosine, and tangent of an angle of a right triangle given the measures of the sides.

16-5: Read trigonometric tables.

16-6: Use trigonometric ratios to solve right triangles.

16-7: Use trigonometric ratios to solve verbal problems.

Assignment Guide

This textbook is designed for first-year high school algebra classes. The textbook is organized so that each chapter can be covered in two to two and a half weeks. This allows an average of one day per lesson, one day for review, and one day for the chapter test. Many chapters have optional lessons which may be omitted when extra teaching days are required for other lessons in the chapter. Chapters 15 and 16 are also optional. Most classes should be able to cover most of the text in 170 class periods.

Each lesson in the textbook includes suggested daily assignments for three ability levels: enriched, average, and fundamental. The chart on the following page gives an approximate time schedule for each ability level.

Enriched Course

This course is for students who can master concepts after only a few examples and exercises, and who can retain skills well. These students benefit from challenge activities and exercises. The suggested daily assignment for an average lesson includes a few Exploratory Exercises, about half of the Written Exercises, and all of the Challenge Exercises. In the suggested time schedule, days are allotted for all 16 chapters. One day is allotted for each Cumulative Review. By omitting some optional lessons or by teaching more than one lesson per day, you may be able to provide time for special features or for teaching computer programming in BASIC. (See pages 528-543.)

Average Course

This course is for students who need many examples and exercises to facilitate mastery of the concepts. These students need occasional review exercises to aid concept retention, and occasional challenge exercises for added motivation. The suggested daily assignment for an average lesson includes about half of the Exploratory Exercises, most of the Written Exercises, Mini-Review Exercises when they appear, and occasional Challenge Exercises. In the suggested time schedule, days are allotted for all of Chapters 1-14 and part of Chapter 15. One extra day per chapter is allotted for reteaching of lessons that cause difficulty. If more days of reteaching are required, optional lessons may be omitted. One day is allotted for each Cumulative Review.

Fundamental Course

This course is for students who need additional practice to master the concepts, and who need frequent review to maintain skills. The suggested daily assignment for an average lesson includes all of the Exploratory Exercises and many of the Written Exercises. The suggested assignments for Written Exercises reflect a spiral approach. Most assignments include exercises from both the present lesson and the previous lesson. To refresh students' memories of previous topics, Mini-Review Exercises are always suggested. In the suggested time schedule, days are allotted for Chapters 1-14, with optional lessons omitted. Two extra days per chapter are allotted for reteaching. If the extra days of reteaching are not required, optional lessons may be included. Cumulative Reviews are important for students in this course. The suggested time schedule allows two days for each Cumulative Review.

Suggested Time Schedule

Chapter	Number of Lessons	Optional Lessons	Suggested Number of Teaching Days		
			Enriched	Average	Fundamental
1	9	none	11	12	13
2	9	none	11	12	13
3	9	3-6	11	12	12
4	8	4-5, 4-7	10	11	10
5	7	5-7	9	10	10
6	9	6-6	11	12	12
7	8	none	10	11	12
8	7	8-2, 8-6	8	10	9
9	7	9-7	9	10	10
10	7	10-7	9	10	10
11	8	11-6, 11-7	10	11	10
12	9	12-3	11	12	12
13	8	13-3, 13-5, 13-8	10	11	9
14	7	14-7	9	10	10
15	8	all	10	5	—
16	7	all	9	—	—
	Cumulative Reviews		8	7	14
	Semester Reviews and Tests		4	4	4
	Total Number of Days		170	170	170

Teaching Problem Solving

One of the challenges of teaching mathematics is helping students learn to solve problems. Problem solving involves more than solving verbal problems—it also involves the ability to analyze a new problem and discover a way to solve it.

Algebra is a very useful tool for solving problems. Many verbal problems can be easily solved by translating the problem into an equation. This involves many skills which must be developed gradually—reading and exploring the problem, defining a variable, writing an equation, solving the equation, interpreting the result, and checking the solution. In this text, these skills are introduced in lessons 1-8, 1-9, 2-9, and 3-7. Other lessons throughout the text reinforce and build upon these skills. The following problem-solving plan is used throughout *Merrill Algebra.*

1. *EXPLORE the problem.* Read the problem and identify what is given and what is asked. Jot down important facts from the problem. Sometimes it is helpful to draw a chart or diagram. Think about how the facts are related.

   If an equation will be used to solve the problem, choose a variable to represent one of the unspecified numbers in the problem. Read the problem again and use the variable in writing expressions for other unspecified numbers in the problem.
2. *PLAN the solution.* Many different strategies may be used. If an equation will be used to solve the problem, read the problem again. Decide how the unspecified numbers relate to other given information. Write an equation to represent the relationship.
3. *SOLVE the problem.* This involves doing the mathematics and interpreting the answer. If an equation was written, solve the equation and interpret the solution. State the answer to the verbal problem.
4. *EXAMINE the solution.* Check whether the answer makes sense with the conditions of the problem. If not, check your mathematics again. If the mathematics was correct, a mistake was made in "setting up" the problem. In that case, explore the problem again and try a different approach.

Not all problems in real-life can be solved by using an equation. Sometimes there is no "clear-cut" method of solution. Thus, it is important for students to learn alternative strategies for solving problems. This text includes numerous *Problem Solving* features that illustrate various strategies, such as guess-and-check (p. 50). Other strategies are introduced in *Problem Solving* features on pages 96, 147, 247, 319, 355, 419, 440, and 475. The four-step problem-solving plan can be used to solve *any* type of problem, using *any* strategy.

It is important to create a classroom environment which encourages problem-solving abilities. The following suggestions should be helpful in building students' confidence in solving problems:

- Encourage students to try a variety of strategies. Praise them for suggesting ways to approach a problem, even if the suggestions do not lead to a correct result. Also, spend time "looking back" at problems which were previously solved and discuss alternative ways to solve the problems.
- Assume that students will make mistakes when solving verbal problems. An evaluation system which gives partial credit for thoughtful attempts is recommended. For example, you could award 4 points for a correct solution, 3 points for a solution which has the right process but had a small mathematical error, 2 points for a thoughtful attempt with a mistake in "setting up" the solution, and 1 point for other meaningful attempts.
- Allow sufficient time for the class to explore problems. In some cases it is preferable to spend twice the time on half the problems. Although discussions may seem to move slowly, the extra time will give students more confidence.
- Throughout the year, emphasize the steps for solving problems. A chart listing the steps could be displayed on the bulletin board, along with a problem for students to solve in their spare time. Problems could be chosen from this text, or from other sources such as *The Mathematics Teacher.*

If the suggestions on this page are followed, problem solving will become more rewarding and enjoyable.

Teacher's Notes

Chapter 1
The Language of Algebra
Chapter Overview

From the onset of the chapter, students are made aware that algebra differs from arithmetic in its extensive use of letters called variables. Students are familiarized with the idea of a variable and its use. The skill of translating verbal expressions into mathematical symbols, expressions, open sentences, and equations is developed. Relative number rules and properties help students realize the organization and structure of algebra. The distributive property is given special emphasis because of its vital role in performing algebraic functions. The positive approach to verbal problem solving should instill confidence in students to meet the challenge of solving problems so vital and necessary in today's present computer age.

Some lessons, like the one on identity and equality properties, may be covered rapidly if students have a strong mathematical background. Even if some students have weaker backgrounds, the chapter provides the necessary introduction and practice to prepare them for the successful study of algebra which forms the basis of study for almost all other mathematics.

Students will learn many new terms in this chapter. Throughout the course, emphasize the need for correct terminology.

Lesson Notes

1-1 Variables and Expressions

The ideas of a variable and an expression are presented so students may be helped to understand how algebra makes arithmetic more general through the use of letters. Students discover that language expressions characteristic of each of the mathematical operations are the only tools required to write algebraic expressions.

Notice that the definition of an expression does not mention exponents. Point out that exponents may be included in expressions because they indicate multiplication. (Only natural number exponents need to be mentioned at this point.)

Since translating words into mathematical symbols is of prime importance for the successful study of algebra, provide the necessary practice until students are comfortable with this skill, which lays the groundwork for solving verbal problems. The following activity may be used to provide practice. Two teams may be formed—one team supplies verbal expressions; the other team writes the mathematical expressions. The teams exchange positions after a designated period of time.

Discuss the examples in the lesson. For Example 4, some students may think $(y - 9)^3$ is a reasonable alternative. Tell your students not to insert parentheses into an expression unless they are indicated. None of the expressions in the exercises need parentheses. Students will learn about parentheses later in this chapter.

1-2 Evaluating Expressions

Students become familiar with the order of operations upon which mathematicians have agreed so each mathematical expression has a unique value. Emphasize the importance of having exactly one correct way to evaluate an expression. Point out that computers and most calculators follow the same order of operations as in algebra. You may want to demonstrate a few examples on the chalkboard to show how confusion results from having no rules of order before formally introducing the order of operations.

Students' awareness that basic grouping symbols indicate a change in the order of operations is important. Emphasize that expressions within grouping symbols must be treated as a single value, or as a quantity.

Point out that there are three ways to indicate multiplication in algebra: ab, $a(b)$, and $a \cdot b$. Emphasize that only the parentheses are correct for certain expressions containing numbers. For example, $3(4)$ cannot be written as 34, and $3(4 + 7)$ cannot be written as $3 \cdot 4 + 7$. Make sure your students can indicate multiplication correctly.

1-3 Open Sentences and Equations

Familiarize students with the idea that open sentences cannot be determined to be true or false until the variable is replaced. In addition to the examples provided in the lesson, you may wish to present these open sentences as additional examples: $12 - x = 2$, $n(12) = 144$, $5 \div c < 2$, and $4y + 8 > 27$.

Emphasize that solving an open sentence means finding replacements for the variable so true sentences result. Be sure students understand these replacements are called solutions and an open sentence may have zero, one, or many solutions. Zero solutions occasionally baffle students. You may ask "How can $1.00 be evenly divided among 3 people?" It cannot be; thus, there is no solution or zero solutions.

Guide students in discovering that algebra is the basic tool for solving an equation, a mathematical sentence containing the equals sign. In this lesson students rely on arithmetic knowledge to solve equations. A parallel reference can be made to elementary school when students were asked to "find the missing addend for $\square + 3 = 8$." Formal methods for solving equations are presented later.

The idea of a replacement set may be new to some students. The replacement set may also be called the domain of a variable. Students who do the Challenge Exercises come to realize that replacing each variable in an open sentence by one of the numbers in its replacement set may or may not result in a true statement. Only those numbers from the replacement set making the open sentence true are called the solutions.

You may want to explain that open sentences that contain the symbols $<$, $>$, $\leq$, $\geq$, or $\neq$ are called inequalities. They will be discussed in Chapter 4.

1-4 Identity and Equality Properties

Properties of identity are introduced by arithmetic examples and then formalized in algebraic terms. Emphasize the difference between the Additive Identity Property and the Multiplicative Property of Zero, as many students make the mistake of thinking that expressions such as $7 \cdot 0$ equal 0.

Point out to students the differences between the Transitive and Substitution Properties of Equality. Stress the important use of the Substitution Property to simplify expressions.

Students may ask "Why learn the names of properties?" One response is to tell them that knowing the vocabulary will help them understand explanations and discussions of mathematics, just as knowing the vocabulary of football (forward pass, split-formation, and so on) helps them understand football.

The following activity may be used to help students learn the properties. Ask students to write five equations for each of the properties presented in the lesson without writing the properties shown. Then have students trade papers and write the names of the properties that have been shown by their classmate's equations.

Emphasize that this lesson focuses on properties of equality. The next lessons will focus on properties of operations (addition and multiplication).

1-5 Distributive Property

Two distributive properties are introduced, the distributive property of multiplication over addition, and the distributive property of multiplication over subtraction. The distributive property for subtraction will not be needed after integers are introduced. In the next chapter students will learn to write subtraction expressions as addition expressions. Then only one distributive property will be needed. For example:

$$\begin{aligned} 4(9 - 6) &= 4(9 + (-6)) \\ &= 4(9) + 4(-6) \\ &= 4(9) - 4(6) \end{aligned}$$

The distributive property is one of the most useful properties in algebra. Students can demonstrate the distributive property using ten pennies, chips, or paper squares in a multiplication problem like $2 \times 5 = 2(2 + 3) = (2 \times 2) + (2 \times 3) = 4 + 6 = 10$. Five objects are placed in two rows, one above the other. Students may lay a pencil between the second and third objects in the rows to see how the distributive property works. If students appear to be having difficulty in fully understanding the distributive property, allow students to take turns demonstrating the property with small objects on an overhead projector. You may want to limit the total number of objects a student selects to no more than 25.

The emphasis of this lesson should be combining terms rather than naming the properties. Many students will be able to simplify expressions without showing as many steps as are shown in the example.

Since students have not yet studied the associative property, the expressions to be simplified have like terms in adjacent positions only.

You may want to give examples of problems similar to Written Exercises 19-24, 27, 32, and 33 before assigning these exercises.

1-6 Commutative and Associative Properties

You may want to review briefly the distributive property before discussing the commutative and associative properties for addition and multiplication. Most students have had sufficient background with these two properties. The significance in demonstrating these properties is to show how important they are in simplifying algebraic expressions.

The method of justifying each step as shown in the examples provides good preparation for doing two-column proofs in geometry.

Encourage students to carefully study the summary chart of properties used when simplifying expressions.

1-7 Using Formulas

Emphasize that a formula is an equation that can be used to solve many similar problems because it states a rule for the relationship between certain quantities. Ask students to name some common formulas. Encourage students to share their home or school experiences of using formulas for area and perimeter. Point out that when studying science, students will need to use formulas frequently.

Reinforce the importance of being able to translate language expressions into mathematical symbols. Transfer this knowledge to stress the importance of being able to translate sentences into equations or formulas. It is generally agreed that mathematical development of students correlates to a great extent with their ability to read mathematics. Influence students to read problems with great care as careless reading usually results in incorrect solutions.

When discussing Example 2 point out that the height of a triangle is often called the *altitude* of the triangle. Any side of the triangle can be used for the base; the altitude is the shortest distance from the base to the opposite vertex of the triangle.

Enrichment

Provide students with the opportunity to build a formula from known relationships. For example, have them invent a formula to determine the final score for a team in a football game. Suggest the following variables: S for final score, t for the number of touchdowns, p for the number of points after touchdowns, f for the number of field goals, and y for the number of safeties. The formula would then be $S = 6t + p + 3f + 2y$.

1-8 Exploring Verbal Problems

So students may develop a positive attitude toward solving verbal problems, this lesson focuses on providing students with practice in exploring facts about verbal problems. The emphasis at this stage of algebra is *not* the solution of the problem. It is acquiring a skill for reading problems carefully and understanding the information that they contain. Students learn questioning techniques through the several models provided. This section may prove to be the most important of all in establishing a "questioning attitude."

In the exercises, the purpose of the questions is to help students learn to study problems carefully. It may be helpful for students to use a chart to organize information given for each exercise.

In the next lesson students discover that skillful analysis of the facts leads to recognition of which quantities may be represented by variables so equations may be written to solve the problems. Challenge students to write and analyze their own verbal problems. Seek to establish confidence through positive reinforcement when exploring verbal problems with students.

1-9 Problem Solving: Writing Equations

Reemphasize the importance of first reading and exploring a problem before deciding how to solve it. Again, the emphasis is *not* on solving the verbal problems. It is acquiring the skills needed to translate verbal problems into equations.

Discuss thoroughly the four steps in the problem-solving plan. Stress that only the first two steps will be discussed in detail in this lesson. Through these two steps, students become aware that when one of the unspecified numbers is defined as a variable, other unspecified numbers may be expressed in terms of the variable so an equation can be written to represent the relationship according to the information provided in the problem.

Students will find it interesting that verbal problems can be developed from equations. You may want a student to write a defined variable and equation on a chalkboard. Then ask another student to reconstruct a word problem from it. This is also an excellent way to quiz. Both approaches are necessary.

To conclude the chapter, encourage students to come up with reasons why it is important to be able to solve problems.

Chapter 2
Adding and Subtracting Rational Numbers

Chapter Overview

The fundamental knowledge and skills required for successful experiences in problem solving are expanded through use of the number line. Positive and negative numbers, addition of integers, absolute value, rational numbers, and additive inverses are all introduced using the familiar model of a number line. The positive approach used to reach the ultimate goal of being able to solve problems successfully is accomplished through building the students' knowledge and skills in small steps.

Lesson Notes

2-1 Integers on the Number Line

Traditionally at the elementary level, raised signs are used to denote signed numbers. Thus, *negative two* and *positive three* would be denoted as $^-2$ and $^+3$ respectively. In this way, students are guided not to confuse the *sign of a number* with the operations *addition* and *subtraction*.

By the time students get to algebra this distinction is assumed to be clear. The signs are then lowered and *negative two* and *positive three* are denoted -2 and $+3$. This may present an opportunity for you to point out that there are different conventions or styles of notation in mathematics. For although we all have our personal tastes, as long as properly defined, no one style of notation is any more correct than another.

Enrichment

Define and discuss *closure*. A set is *closed* with respect to a particular operation if, when the operation is applied to elements of the set, the result is also an element of the set. Discuss the operations on whole numbers and decide whether the set is closed for each operation.

Is the set of whole numbers closed under:

a. Addition? *Yes, since the sum of any two whole numbers is also a whole number.*

b. Subtraction? *No, since the difference is not always a whole number. For example, 2 − 3 is not a whole number.*

c. Multiplication? *Yes, since the product of any two whole numbers is also a whole number.*

d. Division? *No, since the quotient is not always a whole number. For example,* $1 \div 2 = \frac{1}{2}$*, which is not a whole number.*

2-2 Addition on the Number Line

Use the number line to introduce the addition of integers. Permit students who know the rules of football to explain the diagram to those who may not know. The examples provided should help students formalize and better understand rules for adding integers.

When showing addition on a number line, emphasize that negative numbers are indicated by arrows pointing to the left, and positive numbers are indicated by arrows pointing to the right. The first arrow must start at the origin. The second arrow starts at the arrow-head of the first arrow. The sum is at the head of the second arrow.

2-3 Adding Integers

Students learn the concept of absolute value through the use of the number line. Since absolute value is used repeatedly in algebra, be certain students understand that absolute value is a distance from 0 on the number line so it can never be negative. They need to feel at ease with the idea that the distance may be to the left or to the right from 0 on the number line.

Provide enough practice so students develop the ability to apply the rules for adding integers with the same or different signs. Challenge students to develop and play a game involving the adding of integers.

Enrichment

Another definition of absolute value is

$$|x| = x \text{ if } x \geq 0$$
$$|x| = -x \text{ if } x < 0$$

Some students are confused by the second part of the definition. Provide numeric examples.

2-4 Adding Rational Numbers

Students extend their knowledge of points other than integers on a number line. In exploring the definition of a rational number, be sure students understand division by 0 is not defined in mathematics. If you wish to elaborate more fully on the meaning of a rational number, relate a to the numerator which signifies the total number of parts taken as a unit and b to the denominator which signifies the total number of parts into which a unit has been cut. Students must further understand that the size of each of the parts of the numerator and denominator are equal. We cannot talk of a fraction whose numerator is the number of nickels and denominator the number of dimes. Some students may be helped to think of the numerator as the part of a whole and the denominator as the whole.

Provide sufficient practice in working with rational numbers in the form of fractions or decimals. Students will have no difficulty in realizing how impractical a number line is in adding rational numbers. Help students transfer their skills used in adding integers to adding rational numbers.

2-5 More About Addition

You may want to review briefly the associative, commutative, and distributive properties. Extend their use to working with integers.

To be certain students understand that only like terms can be added using the distributive property, have them cite examples of terms that may be added and examples that may not. A good example that leads to discussion and understanding is "Add 1 nickel and 1 dime." Students usually say "15 cents." Now emphasize that each was converted to cents then added, after all 1 + 1 is still 2. You may say "1 coin plus 1 coin" to emphasize the need for like terms. To explore further, ask why $\frac{1}{2}$ and $\frac{1}{3}$ cannot be added. They, too, are unlike terms. That is why we need a common denominator to make them like terms.

It may be interesting to note that by mathematical definition, addition is a *binary* operation. To add more than two numbers requires, technically, the repeated application of the binary operation. This need not be stressed for students. Make the transition to adding 3 or more terms as informal as possible.

2-6 Subtraction

The introduction of the additive inverse property is used to direct students to the rule for subtracting rational numbers. In this way, students' prior experience in adding integers is utilized. The concept of inverse operations is also strengthened.

To help students better understand subtracting a negative integer, introduce an everyday example such as the following:

If I owe someone \$16 and that person "takes away" \$7 of my debt, I would still owe \$9.

$$\begin{aligned} -16 - (-7) &= -16 + 7 \\ &= -9 \end{aligned}$$

Encourage students to explain the three different meanings of the minus sign in algebra so they fully understand. One use indicates the operation of subtraction, a second use indicates direction on a number line for negative numbers and a third use indicates the additive inverse or opposite of a number.

Review, if necessary, the idea of absolute value when students evaluate the algebraic expressions.

2-7 Solving Equations Involving Addition

The scale serves as a model for the introduction of the addition and subtraction properties of equality. Students should have no difficulty in understanding that the same number must be added to or subtracted from both sides of an equation for an equivalent equation to result. You may want to use more chalkboard examples such as the ones below to illustrate the properties of equality.

Addition:

$$\begin{aligned} 15 &= 8 + 7 \\ 15 + 3 &\stackrel{?}{=} (8 + 7) + 3 \\ 18 &= 18 \end{aligned}$$

Subtraction:

$$\begin{aligned} 16 &= 4 + 12 \\ 16 - 5 &\stackrel{?}{=} (4 + 12) - 5 \\ 11 &= 11 \end{aligned}$$

Stress the most efficient means of solving problems by encouraging students to select and use the properties that will isolate the variable on the left or right side of the equals sign. Challenge students to demonstrate why this is the most efficient approach to solving algebraic equations. Point out that because every number has an additive inverse, this is also a useful tool in solving equations.

Build confidence in solving problems by demonstrating how easy it is to check solutions by simple substitution. Remind students to check their solutions to the Written Exercises.

2-8 Solving Equations Involving Subtraction

Before discussing the solving of equations involving subtraction, answer any questions the students may have thus far. You may also want to take time to allow students to share observations or discoveries which they may have made in their study of algebra. Use positive reinforcement when students transfer prior knowledge and skills to the lesson. You might want to have students demonstrate again the three ways in which the minus sign is used in algebra as pointed out in the notes for Lesson 2-6.

2-9 Problem Solving: Rational Numbers

Problem solving provides students with the opportunity to experience success with a purpose. While guiding students through the elevation problem, help them recall the exploring skills learned in Chapter 1. Review the four steps of the problem-solving plan:

1. Explore the problem.
2. Plan the solution.
3. Solve the problem.
4. Examine the solution.

You may want to point out that while a problem may have only one solution, there may be more than one way to solve it. Continue to build students' confidence in solving problems by indicating that even if they discover the incorrect solution was arrived at after checking it, they have learned something. They have learned one answer that is not correct and they may need to think of a different approach. They may learn that a small error has been made. It is important students realize that finding incorrect solutions is inevitable when solving problems and does not indicate failure. Have students discuss the various ways of solving problems learned thus far as alternate approaches that may be used.

Chapter 3
Multiplying and Dividing Rational Numbers

Chapter Overview

Multiplying and dividing signed numbers logically follows the adding and subtracting of signed numbers. The idea of inverse operation is expanded upon to help develop a strong foundation of skills in problem solving. Various types of equations are presented including the false equation. The study of ratio, proportion, and percent round out the chapter's presentation. So students do not get the false impression that all quantities and situations in the world are equal, inequalities will be presented in the next chapter.

Lesson Notes

3-1 Multiplying Rational Numbers

Guide students to discover the rules used when multiplying two numbers with different or same signs. Review, if necessary, the additive inverse property and extend their understanding to the multiplicative property of 1. Allow students to explain the sign rules concerning the multiplication of an even and odd number of negative factors.

Enrichment

Have students show multiplication graphically, using parallel number lines as described below. (1) Draw two parallel number lines. Draw a perpendicular line that intersects each number line at 0. (2) For the first factor, draw a line through that point on the upper number line and 1 on the lower number line. Label the point of intersection on the perpendicular line P. (3) For the second factor, draw a line through P and that point on the lower number line. The point at which this line intersects the upper number line represents the product. The example below shows a graph of $3 \cdot 2 = 6$.

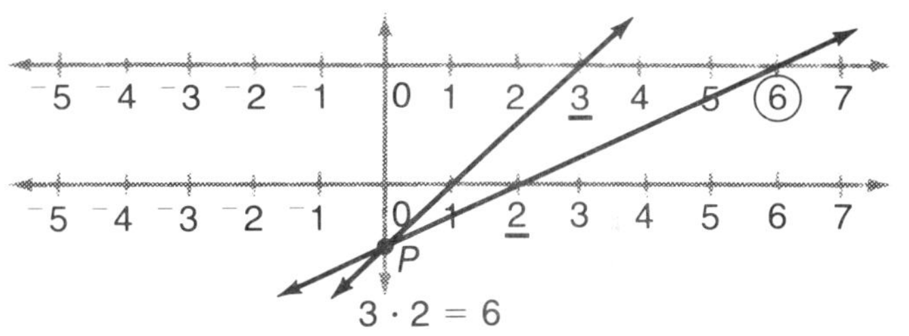

3-2 Dividing Rational Numbers

Strengthen students' understanding of what an inverse operation is. The text states that multiplication and division are inverse operations. More specifically, the inverse of multiplying by a certain number is dividing by the same number.

Rules for dividing integers are developed by relating to students' prior experience with multiplication of integers. Emphasize that the rules for multiplication and division of positive and negative numbers are identical.

After discussing the idea of multiplicative inverses or reciprocals, stress their important use in dividing fractions.

3-3 Solving Equations Using Multiplication and Division

Encourage students to explore and consider the facts presented in the example problem. You may want to write the problem on the chalkboard and let students arrive at an equation before asking them to look at the lesson.

Ask students to recall how an equation relates to a scale in balance. Allow students to tell how this relates to using the multiplication property of equality to solve equations. Be sure students develop the habit of checking the solution. Reemphasize that the most practical approach to solving each equation is to isolate the variable on the left or right side of the equals sign.

The division rule (page 79) and the multiplication property of equality (page 81) can be used to derive the division property of equality. The division property of equality is not necessary for solving equations, since you can always multiply by the inverse instead. However, it does seem simpler to divide in many cases.

Enrichment

Have students consider what is done to solve an equation that involves more than one operation. Have them contemplate equations like: (1) $2x - 5 = 23$, (2) $3x + 8 = 14$, (3) $-2x - 5 = 25$, (4) $\frac{x}{4} + 7 = 6$, (5) $\frac{a + 10}{3} = 4$, and (6) $\frac{a - 7}{4} = 6$.

3-4 Solving Equations Using More Than One Operation

Remind students that the goal of solving an equation is to make the variable stand alone. They must "undo" the operations on the variable. A generally safe procedure is to reverse the order of operations as described below.

1. Undo addition and subtraction.
2. Undo multiplication and division.
3. Undo parentheses.

For example, to solve the equation $\frac{a + 10}{3} = 6$, students should undo the division first, since the fraction bar implies parentheses, that is $\frac{a + 10}{3} = (a + 10) \div 3$.

Most importantly, students should always apply the same operations to both sides of the equation. If they have the steps in a different order, they may still get the correct answer, but the solution will be more difficult. For example, consider the equation $3x + 7 = 13$. Suppose both sides are first divided by 3.

$$\begin{aligned} 3x + 7 &= 13 \\ \frac{3x + 7}{3} &= \frac{13}{3} \\ x + \frac{7}{3} &= \frac{13}{3} \\ x &= \frac{13}{3} - \frac{7}{3} \\ x &= \frac{6}{3} \\ &= 2 \end{aligned}$$

Until students can solve equations with ease, require that the steps be written out in detail.

3-5 More Equations

The positive approach taken thus far in solving equations has developed confidence in students so that information presented in the lesson should not discourage them. Before discussing the lesson's examples, you may want to challenge students to propose their own suggestions to solve equations containing grouping symbols, equations with no solutions, and identity equations. Help students become comfortable with the idea that not all equations are solvable.

3-6 Still More Equations

Point out to students that they can solve equations containing fractions and decimals without first eliminating the fractions and decimals. Also point out that, in general, not eliminating them makes it more difficult to solve the equation. You may want to emphasize this by having students attempt to solve the equations in the examples without first eliminating the fractions and decimals. Have them discuss the difficulties they encounter.

Solving equations containing more than one variable for a specific variable will help students in future work. For example, when writing linear equations in slope-intercept form, the equation must be solved by y.

3-7 Problem Solving: Using Equations

Review the four steps in the problem-solving plan: explore, plan, solve, and examine. Emphasize the importance of checking the answers to all problems.

Students may think it confusing to represent consecutive odd integers as x, $x + 2$, $x + 4$, and so on. Demonstrate with numerical examples.

$$\begin{aligned} 7 \quad &= 7 \\ 7 + 2 &= 9 \quad \text{consecutive odd integers} \\ 7 + 4 &= 11 \end{aligned}$$

As students do the Exploratory and Written Exercises, observe which concepts they are experiencing difficulty in applying. Plan to review or reteach, if necessary, those concepts.

3-8 Ratio and Proportion

Students should find the idea of proportions interesting and easy to work with because the product of the extremes equals the product of the means. This property of proportions can be proved using the multiplication property of equality. You may want to assign this proof to your students.

Ask students to recall properties that can be used to solve equations and encourage them to add the property of proportions to the list. You may want to point out that proportions may be written in more than one way. For example, the proportions $\frac{x}{3} = \frac{5}{8}$, $\frac{3}{x} = \frac{8}{5}$, $\frac{x}{5} = \frac{3}{8}$, and $\frac{5}{x} = \frac{8}{3}$ are all equivalent. One way to demonstrate this is to use two dominoes in vertical and horizontal positions.

Emphasize the importance of consistency in writing proportions to avoid confusion. That is, the two ratios in a proportion should compare the same types of quantities, as in Example 5, where both ratios compare miles to gallons.

Point out to students that ratio and proportion are essential to the study of similar triangles and proportional segments of triangles in geometry and trigonometry. Challenge some students as a group to research ways in which ratio and proportion are used in the business world and in sports. Allow them time to make a presentation to the class.

3-9 Percent

Students are very familiar with the term percent, but may not have in-depth understanding of its meaning. You may want to determine the extent of their understanding before discussing the lesson.

Students should be familiar with the percent equivalents of halves, thirds, fourths, fifths, and eighths. They should be capable of recognizing the number in either decimal, fraction, or percent form.

You may wish to show students how to change fractions to percents by using division. For example, $\frac{3}{5} = 3 \div 5 = 0.6$, and $0.6 = \frac{60}{100} = 60\%$.

Help students learn strategies for solving percent problems. Practice translating language expressions into algebraic percent equations. If you prefer, you could ask students to restate all percent problems in this lesson in the following form:

$$a \text{ is } b\% \text{ of } c$$

Then the equation will always have the form $a = \frac{b}{100} \cdot c$, with a variable appearing in the unknown position.

Time permitting, lead a class discussion on the applications of percent students may have experienced in real life.

Chapter 4
Inequalities

Chapter Overview

Many relations existing between numbers cannot be expressed as equalities. Therefore, students need to learn to work with inequalities. Simple inequalities are introduced and their solution sets are shown graphically on a number line. Inequalities are solved using the four basic operations. Practice is also provided in solving inequalities containing more than one operation. A short cut method using cross products is presented for comparing rational numbers. Compound sentences are solved and their solution sets shown graphically. Open sentences containing absolute value are interpreted as compound sentences. To conclude the chapter, students apply knowledge and skills developed in using inequalities to solve real-life problems.

Lesson Notes

4-1 Inequalities and their Graphs

Students should be very familiar with the $<$ and $>$ inequality symbols. However, the inequality symbols $\not>$, $\leq$, and $\geq$ may be less familiar to them. Students may experience difficulty in determining if an expression such as $3.2 \leq 3.2$ is true or false. Thus, be sure students understand the meaning of and how to read each inequality symbol.

You may want to point out the distinction between the verbal phrase *is less than*, $<$, and the phrase *less than* which indicates subtraction. Similarly, point out the distinction between *is greater than*, $>$, and *greater than*, which indicates addition.

Numbers are compared on a number line in such a way that students should have no difficulty graphing solution sets of true inequalities. The solution sets are written in words to help students understand the symbolism and graphs better. Emphasize what closed and open circles show on a graph as well as the heavy arrow.

4-2 Solving Inequalities Using Addition and Subtraction

Students solve inequalities through applying the addition and subtraction properties for inequalities. They should recognize the similarity to methods used to solve equations.

In the examples, the answers are stated as solution sets. You may prefer to have students omit this step of the solution. That is, the answer to Example 1 would be written as "$m > 5$" instead of {all numbers greater than 5}.

Ask students why checking an inequality could be very time-consuming. Be sure they realize that only two or three solutions need to be checked. Encourage students to check an inequality as if it were an equation, then check at least one other solution. That is, if the answer to be checked is $x \leq 4$, they should check $x = 4$ and one or two numbers less than 4. They could also check a number greater than 4 to make sure it is not a solution.

You may want to point out that working with inequalities is much like working with equations. The most efficient method involves isolating the variable on the left or right side of the inequality symbol. Some inequalities are easiest to solve by isolating the variable on the right side of the inequality. For example, solve $3y - 4 < 4y$.

$$-4 < y$$

Discuss why this inequality is equivalent to $y > -4$.

4-3 Solving Inequalities Using Multiplication and Division

Allow students to state the similarities and differences of inequalities and equations before introducing the solving of inequalities using multiplication and division. Ask students to explain what happens to the direction of an inequality when both sides are multiplied by the same negative number. Challenge some students to illustrate the point on the chalkboard using their own examples.

When introducing the division property for inequalities, have students compare and contrast its use in solving inequalities with the multiplication property.

Enrichment

Have students answer the following question: If $a < b$, what should $\underline{?}$ be replaced with in expression $\frac{c}{a} \underline{?} \frac{c}{b}$?

Answer: $<$ if $c < 0$, $\geq$ if $c \geq 0$

4-4 Inequalities with More Than One Operation

Inequalities containing more than one operation are solved by applying methods students have learned previously. The order of operations performed in solving inequalities is the same used in solving equations. You may want to precede this lesson with a review of solving equations containing more than one operation.

Use Exploratory Exercises 1-16 to have students explain step-by-step how to solve an inequality.

4-5 Comparing Rational Numbers

Comparing rational numbers by changing both fractions to a common denominator is well-known by students. Familiarize them with the cross products method or comparison property which is a shortcut method. Be sure students realize the one limitation—each rational number must be expressed so that the denominator is positive. Stress that the comparison property works for *all* true equations and inequalities.

Enrichment

Discuss why the values of b and d in the comparison property are restricted to positive numbers.

4-6 Compound Sentences

Compound sentences consisting of two inequalities are solved and their solution sets graphed. Be sure it is clear to students that a compound sentence containing *and* is true only if both inequalities are true.

Provide sufficient practice in graphing compound sentences so students understand fully that it is the intersection of the graphs of the two inequalities that is the solution to the graph of a compound sentence containing *and.*

The concept that a compound sentence containing *or* is true if at least one of the inequalities is true may be difficult for some students. Before introducing it, consider some compound sentences containing *or*, and discuss possible solutions. For example, consider the sentence which will be graphed in example 6: $x \geq 3$ or $x < -2$. Is 0 a solution? Is 5 a solution? Is -2 a solution? Be sure it is clear to students that it is the *union* of the graphs of two inequalities that is the solution to the graph of a compound sentence containing *or.*

It is important that students understand the concepts presented in this lesson before attempting to understand the concepts in lesson 4-7.

4-7 Open Sentences Containing Absolute Value

You may want to review the concept of absolute value. The open sentences solved in this lesson contain absolute value. Help students discover that open sentences containing absolute value can be interpreted as compound sentences by referring them to the graphs on page 132.

Be sure students can interpret each of the three open sentences on page 132 before doing the example. Have them interpret several other examples like $|x| = 5$, $|x| < 5$, and $|x| > 5$. Stress the connection between *less than* and *and.* Also stress the connection between *greater than* and *or.* Notice that the *less than* condition usually confines the solutions between two values; the *greater than* condition usually results in a graph which includes both ends of the number line.

Exploratory Exercises 1-16 provide practice in translating open sentences into compound sentences.

4-8 Problem Solving: Inequalities

Emphasize the usefulness of inequalities in addition to equations in solving verbal and real-life problems. Suggest that students look through newspapers and/or catalogs to find situations which suggest inequalities. For example, mail order catalogs often have a minimum purchase, such as \$10. This means that the total of the order must be *at least* \$10 or $x \geq \$10$. Also, many coupons specify a maximum amount, such as *limit 4.* This means *at most* 4 or $y \leq 4$.

Provide several inequality statements using the language expressions at least, at most, and between until you are certain students know their meanings.

Determine whether students recall why inequality problems have more than one solution. Have students check the two examples presented in the lesson by checking two or three possible solutions.

Chapter 5
Powers

Chapter Overview

The first part of the chapter uses a step-by-step approach to develop the power rules for multiplying and dividing monomials. These rules are then applied to expressions written in scientific notation. Problem-solving skills and techniques are expanded upon in the last part of the chapter. Students have the opportunity to review proportion and percent as related to problem solving. Students learn to apply chart-making skills to their problem-solving skills.

Lesson Notes

5-1 Multiplying Monomials

Remind students that they are continuing to expand their knowledge of the language of algebra. Stress the importance of understanding the vocabulary of algebra.

Be sure students understand what constitutes a monomial. Explain that the difference between a term and a monomial is that terms may contain variables in the denominator. That is, a term may represent a *quotient* of variables, but a monomial may not. Thus, $\frac{x}{y}$ is a term but it is not a monomial.

Later in this chapter, students will learn that $\frac{x}{y}$ is equivalent to $x(y^{-1})$. The expression $x(y^{-1})$ is *not* a monomial since its simplest form contains a variable in the denominator. You may want to discuss this with your students after they have been introduced to negative exponents in Lesson 5-3.

Guide students to simplify the lesson examples by applying the product of powers rule and the commutative and associative properties when necessary. When finding the product of monomials, remind students that a variable without an expressed power is considered to be the first power. Emphasize that in multiplication the bases must be the same to add exponents. Have students cite examples of like bases and unlike bases to be certain they understand.

Some common errors made when multiplying monomials are shown below.

$$3^2 \cdot 3^3 \neq 9^5$$
$$3^2 \cdot 4^3 \neq 12^5$$

To help correct such errors, stress that expressions to be multiplied must have the *same* base, and it is left untouched. Only the exponents are added. Even though students appear to have the idea clearly in mind when working with variables, many students will make the mistake for expressions with constants.

5-2 Powers of Monomials

Students learn the rule for a power of a power. Challenge students to tell how and why this rule differs from the product of powers rule. You may want to encourage any students who are experiencing difficulty to write monomials with powers in the expanded form.

Students use the associative and commutative properties of multiplication to discover the power of a product rule. This rule is then combined with the power of a power rule into the power of a monomial rule. Challenge students to tell how the combining of the two rules is similar to applying the distributive property for numbers.

5-3 Dividing Monomials

The rule for dividing powers is developed through understanding of the expanded method in simplifying a monomial divided by a monomial. Encourage students to compare and contrast both methods of dividing monomials with several chalkboard examples.

Stress that the rule for dividing powers is true for all integers m and n. However, when m is less than n, the result has a negative exponent. Negative exponents are introduced on page 151. If you prefer, you may introduce the following rule, which causes a positive exponent when m is less than n:

$$\frac{a^m}{a^n} = \frac{1}{a^{n-m}}$$

If necessary, illustrate more examples in addition to the lesson example to help students understand that any nonzero number to the zero power equals 1. After studying the example presented in the lesson, students should have little difficulty in understanding the definition of a^{-n}. Students should discover that the skills used to simplify, multiply, and divide algebraic fractions are the basis for dividing monomials.

5-4 Scientific Notation

Science and technology have ushered in the computer age. Scientific notation has become knowledge that all should understand. The students' strong background in working with exponents should assure success in reading and writing numbers in scientific notation. Be sure students understand the meaning of a negative exponent. Have students develop two charts—one showing positive powers of 10 and the other showing negative powers of 10.

Positive Powers of 10

$10^1 = 10$
$10^2 = 100$
$10^3 = 1000$
$10^4 = 10{,}000$
$10^5 = 100{,}000$
$10^6 = 1{,}000{,}000$

Compare the exponent and the number of places between the decimal point and the one.

Negative Powers of 10

$10^{-1} = 0.1$
$10^{-2} = 0.01$
$10^{-3} = 0.001$
$10^{-4} = 0.0001$
$10^{-5} = 0.00001$
$10^{-6} = 0.000001$

Compare the absolute value of the exponent and the number of places to the right of the decimal point.

To clarify the relationship between the exponent of 10 and the placement of the decimal point, you may wish to move the decimal point one place at a time, as shown below.

$$\begin{aligned} 38{,}245 &= 3824.5 \times 10 \\ &= 382.45 \times 10 \times 10 \\ &= 38.245 \times 10 \times 10 \times 10 \\ &= 3.8245 \times 10 \times 10 \times 10 \times 10 \end{aligned}$$

There are 4 tens.

$$= 3.8245 \times 10^4$$

5-5 Problem Solving: Age Problems

Review problem-solving skills. Help students recall that the first step is to read and explore the problem carefully while analyzing each sentence. Then decisions are made concerning the facts and how they relate. Stress once again flexibility in problem solving. Students develop more confidence when they realize there are more ways than one to solve a problem. You may want students to put on the chalkboard as many different correct ways to solve a given problem as they can. This should help them appreciate the methods.

Extend students perceptions to the realization that many problems have extra information while others have insufficient information. Help them to be comfortable with the idea that they may have to conclude there is not enough information presented in a problem to solve it. Point out that in real-life situations, people often have to do research to find the information needed to solve a problem.

Point out that the age problems introduced in this lesson are not intended to be *practical*; their purpose is to provide practice in writing and solving equations.

5-6 Problem Solving: Percents

If necessary, have students practice changing fractions and decimals into percents. It is equally important they practice changing percents into decimals and fractions. You may also want to spend some time reviewing proportion.

Remind students to use problem-solving skills when using percents in the real-life problems and situations presented in the lesson. The Exploratory Exercises and Written Exercises 1-6 provide continued practice in translating verbal expressions into mathematical symbols and expressions. Note any verbal expressions with which students are experiencing difficulty and provide extra practice.

Changes in the national economy are often reported using percent of increase or decrease. Ask students to bring in newspapers and/or magazines which give such reports, and discuss the reports in class.

Enrichment

Have students study the brackets on a sales tax chart. On a tax chart for $5\frac{1}{2}\%$, the sales tax on 95¢ is 6¢. However, $5\frac{1}{2}\%$ of 95¢ is 5.225¢. Why do states usually round to the higher number of cents?

5-7 Problem Solving: Mixtures

This lesson presents mixture problems. Many students have difficulty with this topic. Although students have studied the necessary prerequisites, you may want to save this lesson for later in the year after they have had more practice with writing and solving equations.

For mixture problems, it is often helpful to organize the information in chart form. Help students with chart-making skills if they lack experience in this area. Explain that the charts will be comparison charts in which the elements will be numeric or statistical in nature. They need to understand charts are visual summaries of the important relationships under study.

Enrichment

Solve each problem.

1. Jean has half as many dimes as nickels, and two and one-half times as much money in quarters as in dimes. She has 12 coins in all. How many of each coin does she have? *3 quarters, 3 dimes, 6 nickels*
2. Lester has 10 times as much money in quarters as in dimes, and 2 more nickels than dimes. He has $1.25 in all. How many of each coin does he have? *4 quarters, 1 dime, 3 nickels*

Chapter 6
Polynomials

Chapter Overview

Polynomials are introduced in a step-by-step approach so students fully understand this essential topic. The chapter systematically proceeds through the four basic operations with polynomials. The FOIL method is appropriately introduced when the multiplication of two binomials is presented. Two special products are presented. The last three sections of the chapter emphasize applying previously learned problem-solving skills. Uniform motion, perimeter and area, and simple interest problems are explored.

Lesson Notes

6-1 Polynomials

Direct students to turn to their glossaries and read the definition of monomial. Discuss the difference between term and monomial. For example, the term $\frac{3}{b}$ is not a monomial since a monomial cannot contain a variable in the denominator of a fraction.

Students should realize that a monomial is a specific type of term that can be expressed as a product of numbers and variables with nonnegative integral exponents.

Point out the meanings of the prefixes mono-, bi-, tri-, and poly-. After students examine the types of polynomials in the chart, they should understand that a polynomial can be made up of any number of monomials.

Reemphasize that the key to success in algebra is learning its language. Be sure students understand what is meant by the degree of a monomial and the degree of a polynomial. Encourage students to study the arrangement chart so they fully understand what is meant by ascending and descending order of polynomial terms. Point out that descending order is used more frequently.

6-2 Adding and Subtracting Polynomials

Students use the same properties to add polynomials as they used to simplify expressions. After they become good at combining like terms, you may not want students to justify each step.

Some students may find it easier to use the column form when finding sums of polynomials. You may wish to introduce the idea of placeholders in Example 4. For example, the placeholder $+0y$ can be written under $+5y$ in Example 4.

Students learn to subtract polynomials by adding the additive inverse. Review, if necessary, the additive inverse concept. The Exploratory Exercises provide the necessary practice on this topic. Have students show at the chalkboard how the additive inverses in the chart are found.

Encourage students to check subtraction by addition. In this way adding and subtracting skills are strengthened.

6-3 Multiplying a Polynomial by a Monomial

Students use the distributive property to multiply a polynomial by a monomial. Remind students that the bases must be alike to add exponents. For example, $3^2 \cdot 3^3 = 3^5$, but $2^2 \cdot 3^3 \neq 6^5$.

If you discuss the Excursion on page 184, select volunteers to research what is meant by non-commutative algebra and the axiomatic method.

Enrichment

Simplify.

1. $-x[3 - x(2 + x[x + 2])] - 2x[5 - 3x(2 - x[2x - 2])]$

Answer: $-11x^4 + 14x^3 + 18x^2 - 13x$

2. $x^a(2x^{2a} - 3x^a) - x^{2a}(x^{3a} + x^a)$

Answer: $-x^{5a} + x^{3a} - 3x^{2a}$

6-4 Multiplying Polynomials

In this lesson, students use another application of the distributive property to multiply two binomials. The FOIL method is used so students remember to multiply each term of the first binomial by each term of the second binomial. The FOIL method also helps students to multiply binomials mentally and write just the result. This is the goal of the method. Through positive reinforcement, help students to use the FOIL method automatically when multiplying two binomials.

Point out that when multiplying polynomials in column form, the polynomial with the least number of terms is arranged to be the multiplier. You may prefer to have your students use the following form of column multiplication with ones, tens, and hundreds related to constants, first degree terms and second degree terms.

$$\begin{array}{rrrr}
 & x^2 + & 7x - & 6 \\
 & & x + & 5 \\
\hline
 & 5x^2 + & 35x - & 30 \\
x^3 + & 7x^2 - & 6x & \\
\hline
x^3 + & 12x^2 + & 29x - & 30
\end{array}$$

6-5 Some Special Products

Squaring a binomial is introduced by an application. After students check the result by using the FOIL method, ask them to show which step was eliminated when using this shortcut method. Students should have little difficulty comparing the square of a sum with the square of a difference.

The Written Exercises should provide the practice students need to feel confident. Encourage the use of the FOIL method to verify the products. However, since the forms $(a + b)^2$ and $(a - b)^2$ are used frequently, students should memorize their patterns.

Students should find the product of a sum and a difference easy to apply. Make sure students realize the product of a sum and a difference cannot be used to find the product of certain sums and differences. For instance, $(x^2 - 7)(x + 7)$ cannot be found this way because x^2 and x are not identical terms.

Enrichment

The special products for the square of a sum and for the product of a sum and a difference can be used to show students how to multiply numbers mentally. Use the following examples.

Square of a Sum:

1. $(101)^2 = (100 + 1)^2$
$= (100)^2 + 2(100) + 1$
$= 10{,}000 + 200 + 1$
$(101)^2 = 10{,}201$

2. $(31)^2 = (30 + 1)^2$
$= 30^2 + 2(30) + 1$
$= 900 + 60 + 1$
$(31)^2 = 961$

Product of a Sum and a Difference:

3. $36 \cdot 44$
$= (40 - 4)(40 + 4)$
$= 40^2 - 4^2$
$= 1600 - 16$
$= 1584$

4. $53 \cdot 67$
$= (60 - 7)(60 + 7)$
$= 60^2 - 7^2$
$= 3600 - 49$
$= 3551$

6-6 Solving Equations

Equations that require addition, subtraction, or multiplication of polynomials to be solved are presented in this lesson. You may want to present the following steps for solving these equations.

1. Simplify both sides of the equation.
2. Get all variables on the same side of the equation.
3. "Undo" all additions or subtractions that are on the same side of the equation as the variable.
4. "Undo" all multiplications or divisions that are on the same side of the equation as the variable.

6-7 Problem Solving: Perimeter and Area

Determine the students' understanding of perimeter and area before beginning the lesson. Have students name common units used to measure perimeter and area. Elicit perimeter and area formulas from the students. For example, students may respond with $A = lw$ for the area of a rectangle or $P = 2l + 2w$ for the perimeter of a rectangle.

Enrichment

Present the following problem:

A circular pool has a sidewalk 5 ft wide around it. The area of the walkway is 1,100 ft^2. Use $\pi = \frac{22}{7}$ to find the radius of the pool. *$32\frac{1}{2}$ ft*

6-8 Problem Solving: Simple Interest

Students learn the simple interest formula, $I = Prt$, and how to use it so solve simple interest problems. Be sure students know what each letter in the formula represents.

Encourage students with regular savings accounts to check the interest amounts recorded in their accounts for the past year. Ask what information they will need and how they will go about checking the amounts.

Make sure students understand that the rate and time must correspond in units of time. For example, if rate is annual, time must be expressed in years.

6-9 Problem Solving: Uniform Motion

Problem-solving skills are extended to solving problems of uniform motion. Be sure students understand that in this type of problem an object is moving at a constant speed (rate).

Note that rates are seldom as constant as implied in these problems. Faster and slower speeds occur. However, the formula works if the *average* rate is given.

Students learn the formula $d = rt$. Be sure they can identify what each letter represents.

After exploring each problem carefully, students often find it helpful to draw a diagram or organize relevant information in a chart. If students feel drawing a diagram is elementary, point out that even scientists do not hesitate to use these tools.

Help students recall that the actual solution of a uniform motion problem is a number with an appropriate unit of measure. The solution is not the equation containing the isolated variable on the left or right side of the equals sign.

Enrichment

Have students write and solve their own uniform-motion problems.

Chapter 7 *Factoring*

Chapter Overview

Prime factorization is the first topic of factoring introduced. The concept of greatest common factor (GCF) is also discussed. By using the GCF and the distributive property, students learn how to factor out a common monomial from a polynomial expression. Patterns for factoring the difference of squares and perfect square trinomials are introduced. Trinomials of the form $x^2 + bx + c$ where c is either positive or negative are presented next.

In the following lesson, polynomials are factored by

grouping in pairs and then factoring a monomial from each group. This lesson is important since students will use these skills to factor trinomials of the form $ax^2 + bx + c$. The gradual development of skills should help students approach factoring with less apprehension. The summary chart for factoring methods will help students organize the factoring process.

Lesson Notes

7-1 Factors, Greatest Common Factors

Demonstrate prime factorization with several different integers. Have students practice so they can become comfortable with organizing factors in factor trees.

You may want to discuss the difference between factoring a number and finding the prime factorization of a number.

For example, 20 can be factored as:

$1 \cdot 20$	$-1 \cdot -20$	$2 \cdot 10$
$-2 \cdot -10$	$4 \cdot 5$	$-4 \cdot -5$

But the prime factorization of 20 is $2^2 \cdot 5$. The prime factorization is unique.

Discuss the difference between prime and composite numbers. Be sure students understand that 1 is not considered prime. Encourage students to show how the set of prime numbers less than 50 is determined. Select volunteers to determine prime and composite numbers between 50 and 100. As students practice finding the prime factorization of composite numbers, encourage them to develop a chart listing characteristics of integers and their factors. For example, a number is divisible by 2 if the last digit is even or 0; a number is divisible by 5 if the last digit is 5 or 0; a number is divisible by 3 if the sum of the digits is divisible by 3.

Students will use their knowledge of prime factorization to find the greatest common factor (GCF). The GCF is used in factoring and simplifying fractions.

Another method for finding the greatest common factor other than the method shown in the text is to follow the steps below.

1. List the factors.
2. Find the common factors.
3. Determine the greatest of these common factors.

An example is given below.

factors of 20: 1, 2, 4, 5, 10, 20
factors of 24: 1, 2, 3, 4, 6, 8, 12, 24
common factors: 2, 4
greatest common factor: 4

Students may be interested to know that numbers, such as 30 and 77, that have no common factor other than 1 are called relatively prime.

7-2 Factoring Using the Distributive Property

The factoring method in this lesson can be summarized as follows:

1. Find the GCF of all terms.
2. Write each term as the product of the GCF and its other factors.
3. Apply the distributive property.

Encourage students to check for a GCF before attempting to factor all polynomials. Point out what it means when the GCF is 1. They need to realize such an expression is not factorable. Exploratory Exercise 8 is an example of such an expression.

Encourage students to check the factoring process by multiplying the factors.

7-3 Factoring Differences of Squares

A geometric development is used to introduce factoring the difference of squares. Students will understand the development better if they follow it by cutting the squares out of paper. This approach may be used by having students use a 6-inch square. Have them draw the smaller square 2 inches on a side. Be sure they label sides and parts.

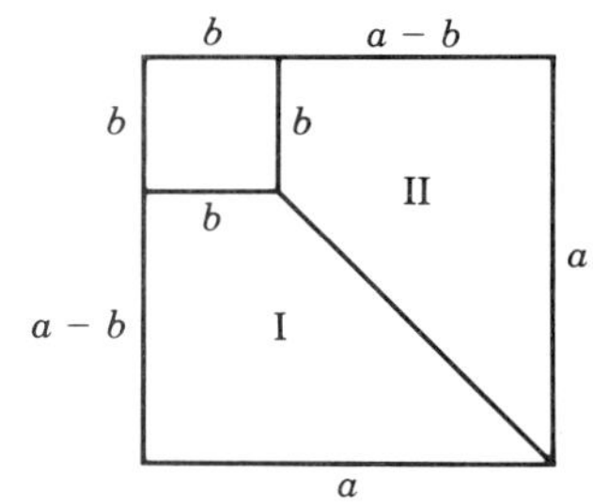

First, have students cut out the smaller square. By doing this they are subtracting the area b^2. Then have them label the back side of part II. When parts I and II are fit together, students should see that $(a + b)(a - b) = a^2 - b^2$.

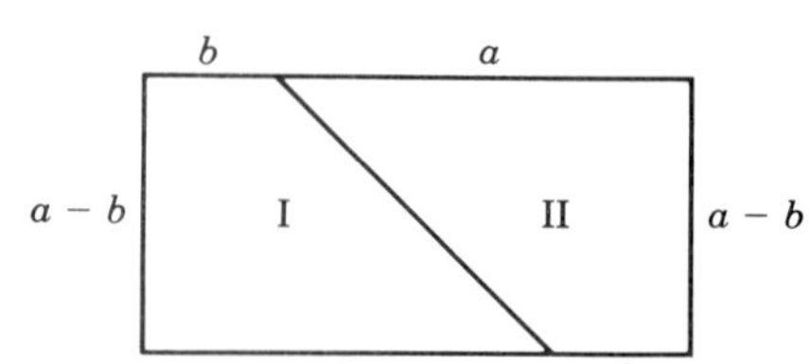

Emphasize to students that in order to factor a binomial as a difference of squares *both* terms must be squares connected by a subtraction symbol. The expression $a^2 + b^2$ *cannot* be factored with real numbers.

Enrichment

The following can be used to show factoring the sum or difference of two cubes.

$a^3 + b^3 = (a + b)(a^2 - ab + b^2)$
$a^3 - b^3 = (a - b)(a^2 + ab + b^2)$

Have students factor each of the following.

1. $m^3 + 27$
Ans: $= m^3 + 3^3$
$= (m + 3)(m^2 - 3m + 3^2)$
$= (m + 3)(m^2 - 3m + 9)$

2. $x^3 - 8$
Ans: $= x^3 - 2^3$
$= (x - 2)(x^2 + 2x + 2^2)$
$= (x - 2)(x^2 + 2x + 4)$

3. $64y^3 - 1$
Ans: $= (4y)^3 - 1^3$
$= (4y - 1)[(4y)^2 + 4y + 1^2]$
$= (4y - 1)(16y^2 + 4y + 1)$

Page 227 has an Excursion on factoring differences of cubes, with another example and exercises.

7-4 Perfect Squares and Factoring

Make sure students are familiar with the meaning of perfect squares. Explain that a perfect square trinomial is a trinomial that is equal to the square of a binomial. Suggest that students memorize the patterns of perfect square trinomials. Recognizing these patterns is an aid to factoring. Students must decide whether a trinomial is a perfect square before attempting to use one of the patterns. Therefore, emphasize the form of a perfect square trinomial and encourage students to ask the three questions used in example 1. Reinforce this skill by using the Exploratory Exercises.

7-5 Factoring $x^2 + bx + c$

Illustrate how factoring a trinomial of form $x^2 + bx + c$ depends upon finding factors of c whose sum is b.

When c is positive in the quadratic $x^2 + bx + c$, the factors of c are either both positive or both negative, depending on the sign of b. If b is positive, the factors are positive. If b is negative, the factors are negative. In $x^2 + 7x + 12$ both factors are positive since both b and c are positive. In $x^2 - 9x + 18$ both factors are negative since b is negative and c is positive. When c is negative in the quadratic $x^2 + bx + c$, one factor of c must be negative and the other positive.

The table below can be used for reference when factoring quadratics of the form $x^2 + bx + c$.

$c > 0$	b is pos.	both factors positive
$c > 0$	b is neg.	both factors negative
$c < 0$	b is pos.	factor with larger absolute value is positive
$c < 0$	b is neg.	factor with larger absolute value is negative.

Encourage students to use tables like those given in the examples to write and test the factors. If necessary, review the FOIL method briefly. Have students tell why the FOIL method should be used to check factoring. Students should add to their algebra vocabulary the term *prime polynomial.*

7-6 Factoring by Grouping

In this lesson, polynomials with four terms are factored by grouping the terms in pairs and then factoring a monomial from each group. Students will need practice in order to recognize which terms have a GCF and can be grouped together.

Be sure students realize that not every polynomial with four terms can be factored this way. For example, $x^2 + 5a - ab + b^2$ cannot be factored by grouping. Assure them that the polynomials in this lesson can be factored by grouping.

Remind students to check factoring by the FOIL method.

7-7 Factoring $ax^2 + bx + c$

Have students compare and contrast the method used to factor trinomials of the form $ax^2 + bx + c$ with the one used to factor trinomials of the form $x^2 + bx + c$. Since in $x^2 + bx + c$, $a = 1$, the method used to factor $x^2 + bx + c$ is a special case of the more general method outlined in this section for $ax^2 + bx + c$.

In factoring trinomials of the form $ax^2 + bx + c$, students are actually finding two terms that satisfy *both* of the following conditions.

1. Their product is the same as the product of the first and last terms.
2. Their sum is the same as the middle term.

You may want to show students how to factor trinomials using the "trial-and-error" method. To factor $8y^2 - 6y - 9$, students should write it as a product of two binomials. Each binomial should have a y-term and a constant. First, consider the possible ways to factor the first term.

$$8y^2 = 8y \cdot y \qquad 8y^2 = 2y \cdot 4y$$

Using these possibilities, begin writing the factored form.

$(8y \qquad)(y \qquad) \qquad (2y \qquad)(4y \qquad)$

Next, consider the possible ways to factor -9.

$$-9 = -1 \cdot 9 \qquad -9 = 1 \cdot -9 \qquad -9 = -3 \cdot 3$$

Using trial and error, position the factors of -9 such that $-6y$ is the sum of the product of the inner terms and the product of the outer terms. The possible arrangements are as follows.

$(8y - 1)(y + 9)$	$(2y - 1)(4y + 9)$
$(8y + 9)(y - 1)$	$(2y + 9)(4y - 1)$
$(8y + 1)(y - 9)$	$(2y + 1)(4y - 9)$
$(8y - 9)(y + 1)$	$(2y - 9)(4y + 1)$
$(8y - 3)(y + 3)$	$(2y - 3)(4y + 3)$
$(8y + 3)(y - 3)$	$(2y + 3)(4y - 3)$

Use FOIL to determine the correct factorization, $(2y - 3)(4y + 3)$. After much practice, students will be able to make better choices and will not have to list all possibilities in order to find the correct factorization.

7-8 Summary of Factoring

Allow sufficient time for students to study the factoring summary chart thoroughly. Stress the importance of the information in providing a systematic approach to the factoring process. Be sure students understand that they should check each factoring method according to the order shown in the chart. Select one or two volunteers to make a large chart like the one in the text to hang in the classroom.

Enrichment

Factor completely.

1. $y^2 + \frac{7}{6}y - \frac{1}{2}$

Ans: $= y^2 + \frac{7}{6}y - \frac{3}{6}$

$= \frac{1}{6}(6y^2 + 7y - 3)$

$= \frac{1}{6}(2y + 3)(3y - 1)$

2. $x^2 - \frac{7}{4}x - \frac{1}{2}$

Ans: $= x^2 - \frac{7}{4}x - \frac{2}{4}$

$= \frac{1}{4}(4x^2 - 7x - 2)$

$= \frac{1}{4}(4x + 1)(x - 2)$

Chapter 8
Applications of Factoring

Chapter Overview

The factoring skills presented in Chapter 7 form the foundation of this chapter. These skills are reviewed in Lesson 8-2. Equations in factored form are solved prior to equations that must be factored before they can be solved. Students apply their factoring skills to solve integer, velocity, and area problems. In the last lesson, students are introduced to simplifying rational expressions.

Lesson Notes

8-1 Solving Equations Already in Factored Form

In this lesson the zero product property is applied to equations in factored form. In the next lesson students will have to factor expressions before applying this property. In this way, students can develop an understanding of the zero product property.

Emphasize to students that they should always check their solutions. Some solutions are not reasonable and should be disregarded.

8-2 Factoring Review

This lesson is basically a repeat of Lesson 7-8. Review with students the importance of checking each factoring method according to the order shown in the chart.

8-3 Solving Some Equations by Factoring

In this lesson, factoring and the zero product property are used to solve polynomial equations. Explain to students that all of the equations in this lesson are called quadratic equations since they can be written in the form $ax^2 + bx + c = 0$.

Students should follow the steps listed below when solving a polynomial equation.

1. Write the equation in standard form ($ax^2 + bx + c = 0$).
2. Factor $ax^2 + bx + c$ completely.
3. Set each factor equal to zero (zero product property).
4. Solve each resulting equation.
5. Check the solutions.

You may want to show at least one example with each step labeled.

Review problem solving skills. Remind students

that one solution of a quadratic equation is often unreasonable and does not answer the question in the problem. Students should reread to make sure their answer satisfies the question in the problem.

Enrichment

Two trapezoids have equal altitudes, h. The bases of one are m and n. The bases of the second trapezoid are s and t. Find the sum of their areas.

Answer:
$$A_1 = \frac{1}{2}h(m + n)$$
$$A_2 = \frac{1}{2}h(s + t)$$
$$A_1 + A_2 = \frac{1}{2}h(m + n) + \frac{1}{2}h(s + t)$$
$$A_1 + A_2 = \frac{1}{2}h(m + n + s + t)$$

8-4 Solving More Equations by Factoring

Remind students to always look for the GCF first when factoring. Point out that the equation in example 3 would be nearly impossible to solve if the GCF had not been factored out first.

Illustrate what happens if the zero-product property is not used correctly. For example, some students may try to solve $x(x - 2) = 8$ by using $x = 8$ or $x - 2 = 8$. Show students that this is incorrect by asking "If $ab = 8$, does a or b have to be 8?"

8-5 Problem Solving: Integer Problems

Consecutive integer problems that involve products are solved. Students solved consecutive integer problems involving sums in Chapter 3. Review problem-solving skills utilized in setting up equations for consecutive integers.

Students may find that trial and error can be used to solve these problems. However, the objective is for students to learn the method presented.

8-6 Problem Solving: Velocity and Area

This section consists of verbal problems involving velocity and area. These problems provide practical applications for factoring.

Students use the formula $h = 144t - 16t^2$ to solve accelerated motion problems and the formula $A = lw$ to solve area problems that involve rectangles. Be sure they do not confuse the motion problems with the uniform-motion problems presented in Chapter 6. Students should be encouraged to label all answers to these problems using the appropriate units of measure.

Enrichment

Introduce students to the quadratic formula. Then have them use it to solve the following problem:
A photograph is 10 cm long and 8 cm wide. Suppose you want to reduce each dimenson by the same amount so that the area is half the area of the original photo. By what amount should you reduce each dimension? $(9 - \sqrt{41})$ cm

8-7 Simplifying Rational Expressions

Algebraic fractions are defined. Students are then taught how to simplify such fractions by factoring and eliminating common factors. Emphasize to students that only common factors can be eliminated. Use the following examples to illustrate this concept.

$$\frac{6}{9} = \frac{2 \cdot 3}{3 \cdot 3} = \frac{2}{3} \qquad \frac{5}{6} \neq \frac{2 + 3}{3 + 3} \neq \frac{2}{3}$$

$$\text{Thus, } \frac{y + 5}{y + 3} \neq \frac{5}{3}$$

You may want to remind students that division by zero is undefined. Thus, it is essential to state that the denominator is assumed not to be equal to zero. Otherwise, the fraction would be undefined.

The concepts presented in this lesson will be reviewed and expanded upon in Chapter 14.

Chapter 9
Functions and Graphs
Chapter Overview

A clear step-by-step presentation helps students to understand the relationship between the algebraic use of numbers and symbols and the geometric concepts of points and lines on a plane. Information presented builds on students' prior knowledge and experience of graphs on a number line and extends the concepts to graphs in the plane. The concept of function is introduced both algebraically as a special kind of relation and geometrically through the vertical line test. Problem-solving skills are enhanced in the last section when students learn techniques in finding equations from relations.

Lesson Notes

9-1 Ordered Pairs

Emphasize the one-to-one correspondence between ordered pairs of numbers and points in the plane.

Make students aware that the arrowhead on each axis indicates positive direction but that the lines extend infinitely in both directions. Be sure they understand that the two axes do not lie in any quadrant. Encourage students to propose reasons.

Emphasize the importance of order. Be sure students understand that the first component of an ordered pair is the x-coordinate and that the second component is the y-coordinate.

As an activity, students could draw straight line figures such as triangles and quadrilaterals on graph paper, then develop a set of ordered pairs which, when graphed in a specific order, produce the figure. The sets of ordered pairs could then be given out to other members of the class.

9-2 Relations

Students should become proficient in showing a relation as a set of ordered pairs, a table, a mapping, or a graph. Refer to the domain and range of relations frequently to ensure student familiarity with these important terms.

9-3 Equations as Relations

Students discover that equations may also be used to describe certain relations. The special type of relation called a function will be introduced in Lesson 9-5. By transforming the equation so that the range variable is on one side by itself and then using the domain elements to determine the range elements, students are prepared for work with functions.

9-4 Graphing Relations

Before beginning this lesson, a quick review of the terminology and definitions presented in the first three lessons of this chapter may be helpful to students.

In this lesson, the transition from finite domains to the infinite domain of the set of real numbers is made. Students should realize that there are an infinite number of solutions to equations whose domain is the set of real numbers.

Point out that there is a one-to-one correspondence between the solutions to an equation and the points of the line that are the graph of that equation. If you wish, extend skill in graphing by introducing intercept graphing. Finding points where lines cross the axes is a more selective process. The concept of intercepts is formally introduced in Lesson 10-3. However, "point-by-point" graphing should be mastered first. Better students will enjoy graphing the parabolas in the Challenge Exercises.

9-5 Functions

Introduce the special type of relation called a function. Compare and contrast several examples of relations that are and are not functions.

At first encourage all students to use their pencils to represent a vertical line in testing for a function. Then encourage graphic thinking in performing the test. Reemphasize the importance of learning the language of algebra when you introduce the symbol $f(x)$. Point out the importance of reading the symbol correctly and of understanding what it represents. Up to this lesson students have used symbols to represent variables, numbers, or operations. If necessary, have students practice writing functional values for various values of x.

Enrichment

Give $f(x) = 3x - 5$ and $g(x) = x^2 - x$, determine the value for each of the following.

1. $f(3) + g(2)$ *6*
2. $f(4) - g(2)$ *5*
3. $3[f(1)] - g(6)$ *-36*
4. $[f(3)]^2$ *16*
5. $f(a) - g(a)$ *$-a^2 + 4a - 5$*
6. $g(a) + 2[f(a)]$ *$a^2 + 5a - 10$*
7. $f(2a) + g(3a)$ *$9a^2 + 3a - 5$*
8. $[f(a)]^2$ *$9a^2 - 30a + 25$*
9. $f(a^2)$ *$3a^2 - 5$*
10. $f(a^2) - [g(a)]^2$ *$a^4 - 2a^2 + 4a^2 - 5$*

9-6 Graphing Inequalities in Two Variables

Review graphing inequalities in one variable as was presented in Chapter 4. Help students transfer this knowledge to graphing inequalities in two variables. Encourage students to tell what added graphing skills are required to graph two-variable inequalities. They should see that in this lesson, the boundaries of all inequalities are lines whereas in Chapter 4, the boundary for an inequality on the number line was shown as an open circle or a dot. (Recall that the open circle indicates that the point was not included.)

Be sure students establish the correct boundary line as well as decide whether to use a broken or a solid line. Students must also decide which half-plane is the graph. Encourage them to develop the habit of testing the coordinates of the origin to determine whether the inequality is satisfied in cases where the origin is not a point on the boundary. If the coordinates of the origin satisfy the inequality, the origin must be contained in the graph. Be sure students study Example 3 on page 292 carefully so they can determine the graph when the origin is on the boundary.

Another test can be used to determine the correct half-plane to include as part of the graph. When the variable y is isolated on the left side of an inequality, the *greater than* symbol indicates that the half-plane *above* the boundary should be included. The *less than* symbol indicates that the half-plane *below* the boundary should be included. For example, in the graph of $y > x + 3$ the half-plane above the boundary should be included. In the graph of $y < 5x$ the half-plane below the boundary should be included.

9-7 Finding Equations from Relations

Students will find this section interesting and challenging. This lesson provides the groundwork for finite differences.

Encourage students to treat the patterns found in this section as puzzles rather than problems. By doing this, this lesson should encourage creativity in problem solving.

In the next chapter, students will learn to write the equation of a line when given two points on the line. Then students could use a more systematic approach to do the exercises in this lesson. If you prefer, you could teach this lesson after Chapter 10.

Have students who can write equations for relations easily share their reasonings on how they derived the equations. Have all students write exercises similar to those in the text to exchange with other students.

Make sure students realize that not all relations can be represented as equations. The relations shown in the charts in this lesson are all functions, and they can be represented by *linear* equations. Point out that the equation is *not* equivalent to the relation *unless* the domain of the equation is restricted to the values in the domain of the relation.

Enrichment

Most of the relations in this lesson can be described by linear equations (degree one). Notice that the differences between successive values of y are constant for such relations when the x values are consecutive. If a relation can be described by a higher-degree equation, the differences will *not* be constant. By studying *finite differences*, you can determine the degree of an equation which describes a relation. The x-values must be consecutive numbers. Study the differences shown below the following chart.

x	1	2	3	4	5
y	1	8	27	64	125

1st +7 +19 +37 +61

2nd +12 +18 +24

3rd +6 +6

Since the third differences are constant, a third-degree equation can represent the relation. The equation is $y = x^3$.

Chapter 10
Lines and Slopes
Chapter Overview

The slope of a line is defined as the ratio of the change in y to the corresponding change in x. Several methods are presented to determine slope. The equation of a line is transformed to both standard and slope-intercept forms. Linear equations are graphed by using intercepts and the slope-intercept method. Various methods for writing linear equations are shown. Definitions of parallel and perpendicular lines are presented in terms of slope. A formula for the midpoint of a segment in the plane is introduced.

Lesson Notes

10-1 Slope

A concrete example is presented to explain the meaning of slope. Students learn to determine slope of a line given any two points on the line. They discover that slope is a ratio calculated by dividing the change in y-coordinates (rise) by the corresponding change in x-coordinates (run). Emphasize the significance of the word *corresponding* in the definition of slope so that students do not find the negative of the slope.

Provide practice in estimating slope by tilting a meter stick at different angles and having students estimate the slope shown. Present them with both positive and negative slopes. Check by placing the stick on a coordinate system and finding the quotient of the change in y and the change in x. Be sure students also recognize the no-slope position.

Discuss the slopes of vertical and horizontal lines. The slope of a horizontal line is 0, since there is no change in the y-values. For a vertical line, there is no change in the x-values. Since division by 0 is undefined, a vertical line has no slope. If you prefer, you could say that a vertical line has *undefined* slope.

10-2 Equations of Lines in Point-Slope and Standard Form

In this lesson, students learn the point-slope form of a linear equation. Emphasize to students that the name "point-slope form" is used because you need to know the coordinates of one point on the line and the slope of the line to use this form to write the equation of the line.

Review the standard form of a linear equation with students. Remind students that the standard form of a linear equation is not unique. For example, the equivalent equations $2x - 3y = 6$, $-2x + 3y = -6$, and $4x - 6y = 12$ are all in standard form.

10-3 Slope-Intercept Form

The slope-intercept form of a linear equation is developed using the point-slope form of a linear equation. Provide students with practice in finding the slope and y-intercept from both the equation and its graph.

Students are then shown how to find the slope and y-intercept of a line whose equation is in standard form. Emphasize that an alternate way of finding the slope and y-intercept in such cases is to write the equation in slope-intercept form.

Enrichment

The following exercises are similar to the Challenge Exercises.

1. Find the coordinates of a point on the graph of $3x - 4y = -20$, if the y-coordinate is 4 more than the x-coordinate. (4, 8)
2. Find the coordinates of a point on the graph of $4x - y = -2$, if the y-coordinate is 11 more than the x-coordinate. (3, 14)
3. Find the coordinates of a point on the graph of $x + 2y = 11$, if the y-coordinate is 5 times the x-coordinate. (1, 5)
4. Find the coordinates of a point on the graph of $7x + 3y = 2$, if the x-coordinate is 16 more than the y-coordinate. (5, −11)

10-4 Graphing Linear Equations

Students learn to graph a linear equation by using x- and y-intercepts. Ask students to tell why this method does not work when both intercepts are 0. Then have a student demonstrate on the chalkboard what to do in such cases.

When graphing equations using the slope-intercept method, students may have difficulty when the slope is negative. Discuss examples of negative slopes such as -3 and $-\frac{1}{4}$. A slope of -3 means $\frac{-3}{1}$. Starting at the intercept point, another point can be found by moving 3 units *down* and 1 unit to the right. A slope of $-\frac{1}{4}$ means $\frac{-1}{4}$. Starting at the intercept point, another point can be found by moving 1 unit *down* and 4 units to the right.

10-5 Writing Slope-Intercept Equations of Lines

Discuss what correct information is needed about a line to write an equation for the line. You may want to ask students such questions as "Is only knowing the slope of a line enough to write the equation of the line?" and "Is only knowing the coordinates of one point on a line enough to write the equation of a line?"

Enrichment

Whenever variables form a linear relationship, an equation can be written from two ordered pairs. Introduce the following problem:

> A certain mail-order business sells posters. All posters are priced the same, and a flat fee is charged for shipping and handling. Someone ordered 3 posters for a total charge of \$12.50. Someone else ordered 7 posters for a total charge of \$26.50. Find the price for each poster and the charge for shipping and handling.

To solve this problem, let x = the number of posters and y = the total charge. Use the ordered pairs (3, 12.50) and (7, 26.50) to write an equation. The equation is $y = 3.5x + 2$. Thus, the posters cost \$3.50 each and the shipping and handling charge is \$2.

Ask students to write similar problems for their classmates to solve.

10-6 Parallel and Perpendicular Lines

Two equations in the same variables are compared. Definitions of parallel and perpendicular lines are presented. Information presented in this section will help assure success when students learn to graph systems of equations in Chapter 11.

10-7 Midpoint of a Line Segment

Knowledge of line segments is expanded upon through work with endpoints and midpoint. Before introducing the formula for the midpoint of a line segment, show an example on the chalkboard. Draw a line segment from (2, 1) to (4, 7). Have students guess the coordinates of the midpoint. The correct coordinates are (3, 4). Have them compare the x-coordinates of the endpoints and the midpoint. Students should notice that 3 is halfway between 2 and 4, or the *average* of 2 and 4. Then have them compare the y-coordinates of the endpoints and the midpoint. Show additional examples, if necessary, before formalizing the concept.

If necessary, provide additional examples like Example 2 on page 328 for practice. Students who later take a course in analytic geometry will benefit.

Point out that exercises 37-44 present a practical application of finding midpoints.

Enrichment

Find the coordinates of P on line segment AB if P is three-fourths of the distance from A to B.

1. $A(8, 4), B(12, 12)$ (11, 10)
2. $A(-3, 9), B(5, 1)$ (3, 3)
3. $A(-3, 2), B(5, 4)$ $\left(3, 3\frac{1}{2}\right)$
4. $A(2, -6), B(9, 5)$ $\left(7\frac{1}{4}, 2\frac{1}{4}\right)$

Chapter 11
Systems of Open Sentences
Chapter Overview

Various methods for solving systems of equations are presented in this chapter. Graphing is the first method presented. Algebraic methods of substitution, elimination using addition, and elimination using multiplication are presented as alternative methods of solving systems of equations. Systems of inequalities are solved by graphing. Digit problems, uniform-motion problems, and mixture problems are solved by systems of equations.

Lesson Notes

11-1 Graphing Systems of Equations

Review graphing a linear equation. Discuss the meaning of a system of equations. Emphasize that a system of equations is graphed in the same coordinate plane. Students discover that the graphs may intersect. They learn that the solution of this system is the coordinates of the point of intersection.

Students should see that the graphing method demands accuracy and is practical only for integer solutions. To help ensure accuracy, students should use grid paper for all graphs in this chapter. Ask students to explain why accuracy is essential when solving systems by graphing.

11-2 Systems of Equations

You may want to place more emphasis on the terms consistent, inconsistent, dependent, and independent. Review a system of equations that has one ordered pair as its solution. Describe this system as consistent and independent. Introduce a system that has no solution (parallel lines) as inconsistent and independent. Remind students that parallel lines will never intersect no matter how far they are extended. Emphasize that independent means the lines (graphs) are distinct. Lastly, introduce a system that is consistent and dependent as one in which the graphs coincide. Ask students to explain why such a system has infinitely many solutions.

After discussing Example 2, present the following summary:

1. If the *slopes are different* and the *y-intercepts are the same or different*, the lines intersect. Therefore, the system has *one solution*. (The system is independent and consistent.)
2. If the *slopes are the same* and the *y-intercepts are different*, the lines are parallel. The system has *no solutions*. (The system is independent and inconsistent.)
3. If the *slopes are the same* and the *y-intercepts are the same, the lines coincide. Therefore, the system has infinitely many solutions*. (The system is dependent and consistent.)

11-3 Substitution

Explain that other methods can be used to solve most systems of linear equations in two variables. Introduce the algebraic method called substitution. Most students find this method the easiest to understand. Some students may benefit from making a chart showing the steps used in the substitution method.

11-4 Elimination Using Addition

Introduce the elimination method of solving a system of two linear equations in two variables. The steps for using the elimination method are summarized below.

1. Add the equations to eliminate one of the variables. This will produce a value for one variable.
2. Substitute the known value into one of the original equations to determine the value of the second variable. (It does not matter which of the original equations is used—choose the one with the easiest arithmetic.)

Point out that all systems of equations can be solved by adding the equations. However, it is often necessary to multiply one or both of the equations by a constant before adding. In this lesson, the systems can be solved *without* using multiplication. Elimination using multiplication will be introduced in the next lesson.

After students have carefully studied the examples on page 345, ask students to explain why elimination method is such an appropriate name.

Enrichment

Use elimination to solve the system of equations $x + y = 5$ and $x = 3 - y$. Graph the equations to explain the result. The system has no solution. The graphs are parallel lines.

11-5 Elimination Using Multiplication

Present the example of a system of equations that can be solved by the elimination method if one of the equations is first multiplied by a specific number. Emphasize that one or both equations may be multiplied by a number.

To determine the simplest way to solve a system of equations by elimination, students should first look at the coefficient of x in each equation, to see whether one is a multiple of the other. Then students should look at the coeffients of y to see whether one is a multiple of the other. Emphasize that the object is to write an equivalent system of equations that can be solved directly by addition or subtraction. Sometimes it is necessary to multiply *both* equations by constants before adding or subtracting.

11-6 Graphing Systems of Inequalities

Systems of inequalities are solved by graphing. Emphasize that the solution of a system of inequalities is a region that contains an infinite number of ordered pairs. Some students may need to review methods for graphing inequalities in two variables presented in Lesson 9-6. Be sure students realize a system of inequalities may have no solution. In such a system the graphs do not intersect so no ordered pair will satisfy the system of inequalities.

Emphasize that an inequality containing absolute value is graphed by graphing the corresponding system of inequalities. Remind students that the inequality $|x| \leq y$ corresponds to the system $x \leq y$ *and* $x \geq -y$. The solution set is represented by the *intersection* of the graphs of each sentence. The inequality $|x| \geq y$ corresponds to the system $x \geq y$ *or* $x \leq -y$. The solution set is represented by the *union* of the graphs of each sentence.

Enrichment

Linear programming is a procedure for finding the maximum or minimum value of a function, given certain restrictions on the variables. For example, suppose you want to find the maximum value of $3x + 4y$ for each graph given in the Challenge Exercises. For exercise 27, the vertices of the shaded region are (0, 0), (6, 0), and (0, 3). Find the value of $3x + 4y$ at each vertex.

Vertex	Value
(0, 0)	$3(0) + 4(0) = 0$
(6, 0)	$3(6) + 4(0) = 18$
(0, 3)	$3(0) + 4(3) = 12$

The function could also be evaluated for other ordered pairs in the region. However, you will discover that the maximum or minimum value occurs at a vertex. The maximum value of $3x + 4y$ for the shaded region is 18. Encourage students to find the maximum value of $3x + 4y$ for the shaded region in exercise 28. The maximum value is 12.

11-7 Problem Solving: Digit Problems

Digit problems are solved by using systems of equations. Provide practice in expressing a two-digit number in the form $10t + u$. This skill will assure students' success in solving digit problems.

Notice that Exercises 16-20 involve three-digit numbers. Have students develop an expression that represents a three-digit number.

11-8 Problem Solving: Using Systems of Equations

Uniform motion and mixture problems are solved by using systems of equations. Students solved uniform motion problems earlier in Lesson 6-9. They solved mixture problems earlier in Lesson 5-7.

Review problem-solving skills. Be sure that students clearly define two variables. Encourage them to use a chart and the steps given in Examples 1 and 2.

Chapter 12
Radical Expressions

Chapter Overview

Radicals are presented to prepare students for future work with quadratic equations. Irrational numbers including irrational square roots are presented as completing the number line. Square roots are approximated, simplified, added, and subtracted. In addition, students learn to use the table of squares, evaluate radical expressions, rationalize denominators, and solve radical equations. Then important applications of square root are introduced, the Pythagorean Theorem and the distance formula.

Lesson Notes

12-1 Square Roots

Explain that the inverse operation of squaring a positive number is finding the principal square root of the number. Ask students to recall other inverse operations. Encourage students to expand their language of algebra by learning the definition of square root and all the related terms presented in the lesson.

Emphasize the differences in the following symbols:

$$\sqrt{x} \qquad -\sqrt{x} \qquad \pm\sqrt{x}$$

Stress that the principal square root is nonnegative, which means positive or zero. Make sure that students understand that $\sqrt{64} = 8$, not ± 8.

If necessary, review finding the prime factorization of integers. Provide sufficient practice so students become proficient with both product and quotient properties of square roots in simplifying numerical square roots. Limit examples to perfect squares so students can concentrate on understanding the concept of square root.

In Lesson 12-2, students will be taught to use the table of squares and square roots on page 544. If you wish, suggest that students write their own short table of squares to help in simplifying square roots in this lesson.

12-2 Irrational Numbers and Real Numbers

Irrational numbers are introduced through presentation of square roots. You may want to have students reflect upon the kinds of numbers they have encountered in algebra up to this point.

Students learn that any and all points on a number line correspond to the set of real numbers. Illustrate how to graph irrational numbers by using decimal approximations. Then introduce the table of square roots on page 544. Take time to help students learn how to read the table. Point out that values in the square root tables had to be computed by someone, using a procedure such as the one in Lesson 12-3. A table is merely a convenient list of previously computed answers.

After consideration of the lesson example, have students practice graphing irrational numbers by finding their decimal rational approximations in the square root table. Then, give students practice in using the table to determine whether the square root of a number is rational or irrational. If a number is a perfect square, it will be listed in the column of squares and therefore has a rational square root. Help students know how to determine the closest value of an irrational square root.

Discuss repeating and non-repeating decimals. Repeating decimals represent rational numbers. A bar can be used to indicate digits that repeat in sequence, as shown below.

$$0.46\overline{8} = 0.4688888888\ldots$$
$$8.\overline{175} = 8.1751751751\ldots$$

Three dots (ellipses) are used after a decimal to indicate that the digits continue. When three dots are used, there is no way to be absolutely certain whether the decimal represents a rational number or an irrational number. The digits may begin to repeat after 20 or 30 digits. Or, the repeating pattern may change to random digits. Suggest the following guidelines for deciding whether the number is rational or irrational.

1. If a decimal has an obvious repeating pattern, assume that the pattern continues, and that the

decimal represents a rational number.
Examples: 0.555555. . .($0.\overline{5}$)
6.353535. . .($6.\overline{35}$)

2. If a decimal has a non-repeating pattern or no pattern, assume that the decimal represents an irrational number.
Examples: 4.7447444744447. . .
3.952764013785. . .
Point out that nonrepeating patterns are not typical of irrational numbers. Most decimals for irrational numbers have *no* pattern.

12-3 Approximating Square Roots

The divide-and-average method for approximating square roots is presented in a step-by-step approach. When discussing step 3, write the symbol for *is approximately equal to* on the chalkboard. You might want to remind students of the symbols chart on page 543. Be sure students equate accuracy with the number of times the method is applied to get an approximation. You may want to point out that even calculators and computers must calculate the value of a square root and that many do it by a divide-and-average method.

12-4 Simplifying Square Roots

To simplify square roots containing irrational numbers and variables, students apply prior skills. Using prime factorization uncomplicates the process. Emphasize that with the modern use of calculators and computers, radicals can be approximated quickly and easily. However, it is more mathematically accurate to leave radicals in simplified radical form. Hence, it is still worthwhile for students to learn to manipulate radicals with the product and quotient properties.

Be sure students understand that to ensure positive results absolute value is used when needed. Emphasize that absolute value is often necessary, since $\sqrt{x^2}$ is not always equal to x. Study the following examples:

If $x = 3$ then $\sqrt{3^2} = \sqrt{9} = 3$.

If $x = -5$ then $\sqrt{(-5)^2} = \sqrt{25} = 5$. However, absolute value is not necessary for $\sqrt{x^3}$ since the radicand must be nonnegative to have a real square root. That is, x and x^3 have the same sign. If x^3 is positive, then x is also positive.

$\sqrt{x^3} = x\sqrt{x}$

12-5 Dividing Radical Expressions and Rationalizing Denominators

Students divide radicals by using the quotient property of square roots. Emphasize that rationalizing the denominator means simplifying radicals so that the denominator becomes a rational number. You may want to have students find the quotients of Examples 1 and 2 in the lesson without first simplifying to demonstrate why simplifying is a more practical approach.

Point out to students that calculators and computers may be used to approximate many radical expressions. However, many radical expressions are not numerical. Students serious about math should learn to manipulate such expressions.

12-6 Adding and Subtracting Radical Expressions

Encourage students to compare the adding and subtracting of radicals with the adding and subtracting of monomials. They should have no difficulty understanding that the distributive property is helpful in adding and subtracting radicals when the radicands are alike. So that students become comfortable with the idea that some radical expressions can be simplified no further, have students provide additional examples to those supplied in the text.

Enrichment

Test the hypothesis that $\sqrt{a + b}$ is never equal to $\sqrt{a} + \sqrt{b}$.

Ans: $(\sqrt{a + b})^2 = (\sqrt{a} + \sqrt{b})^2$, $a > 0$ and $b > 0$

$$\begin{aligned} a + b &= a + 2\sqrt{ab} + b \\ 0 &= 2\sqrt{ab} \\ 0 &= 4ab \\ 0 &= ab \end{aligned}$$

Therefore if $a = 0$ or $b = 0$, $\sqrt{a + b} = \sqrt{a} + \sqrt{b}$.

12-7 Radical Equations

To solve radical equations students should first isolate the radical expression on the left or the right of the equals sign. You may want to elicit ideas from the students on how they might eliminate the radical to solve the equation before presenting the lesson. Encourage students to explain why squaring both sides of an equation does not necessarily produce results that satisfy the original equation. Emphasize the importance of checking all solutions when solving radical equations.

12-8 The Pythagorean Theorem

To introduce the lesson, write the pronunciation of Pythagorean on the chalkboard. Also, select a volunteer to research and report to the class some interesting information about Pythagoras.

To be certain students understand fully what $c^2 = a^2 + b^2$ means, have students practice reading the theorem in sentence form as follows: The square of the length of the hypotenuse of a right triangle equals the sum of the squares of the lengths of the other two sides. Have students illustrate geometrically the Pythagorean Theorem by cutting the appropriate areas of square units from graph paper for a right triangle whose hypotenuse length is 10 and whose legs are 6 and 8, respectively, in length. Ask students to suggest practical applications of the Pythagorean Theorem.

Enrichment

An application of the Pythagorean Theorem is the construction of line segments for irrational numbers. To construct $\sqrt{2}$, using a number line, first construct a perpendicular line at 1. On the perpendicular line, mark the point that is 1 unit above the vertical number line. Draw the line segment from this point to the zero point. This segment is $\sqrt{2}$ units long.

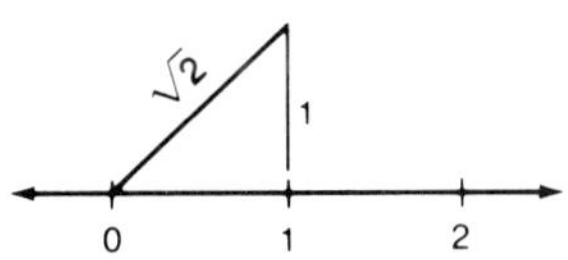

Use a compass to place $\sqrt{2}$ on the number line.

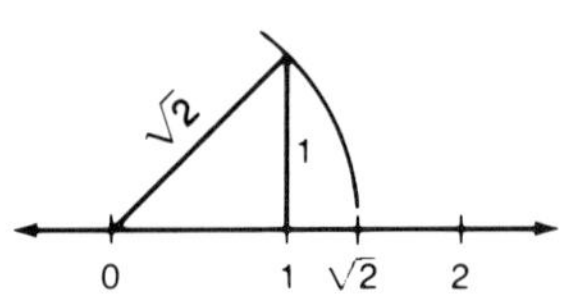

To construct $\sqrt{3}$, use a similar procedure to locate a point one unit above $\sqrt{2}$ on the number line. The segment from this point to 0 is $\sqrt{3}$ units long.

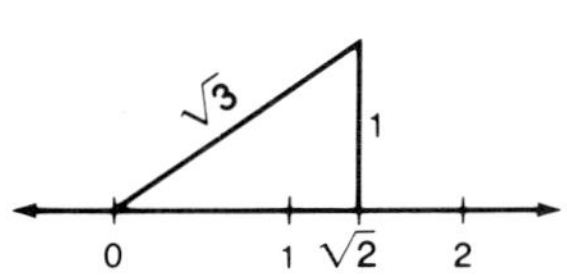

Use a compass to place $\sqrt{3}$ on the number line and repeat the procedure to construct $\sqrt{4}$. If constructions were done accurately, $\sqrt{4}$ will correspond to 2 on the number line.

12-9 The Distance Formula

The distance formula is directly developed from the Pythagorean Theorem. Students use the distance formula to determine the distance between two points in a coordinate plane. This important application of the Pythagorean Theorem is used to solve problems that otherwise would be most difficult.

Enrichment

Have students derive the distance formula using the Pythagorean Theorem.

Chapter 13
Quadratics

Chapter Overview

The quadratic function and related terminology are introduced. Students have solved quadratic equations by factoring in Chapter 7. In this chapter, they graph a quadratic function and determine the solutions of the related quadratic equation from the graph. The term roots is formally presented. Then quadratic equations are solved by completing the square or the quadratic formula. The discriminant is used to determine the type of roots of a quadratic equation. Problem-solving skills are enhanced by means of a summary table. The rule concerning the sum and product of roots is discovered after specific examples are studied.

Lesson Notes

13-1 Graphing Quadratic Functions

The definition of a quadratic function is presented. Students are familiarized with the general shape of the graphs of quadratic functions. Provide the necessary practice until students can draw the graphs easily, since in the next lesson they will be required to locate roots of a quadratic equation by graphing.

After Examples 1 and 2 have been discussed, have students summarize the common characteristics of quadratic functions. Be sure students use terms such as axis of symmetry and minimum or maximum point and are incorporating them into their algebra vocabularies.

The equation of the axis of symmetry can be developed from the quadratic formula. This exercise is provided as an enrichment in the lesson notes for Lesson 13-4.

13-2 Solving Quadratic Equations by Graphing

Have students recall factoring methods for solving quadratic equations. Familiarize them with the meaning of the term roots.

In this lesson, students are asked to *locate* the roots of equations. Since graphing does not yield exact results, explain that they should find the *approximate* value of the roots. If the root does not appear to be an integer, they should state the two integers which bound (lie on either side of) the root. If a root appears to be an integer, it can be checked by substitution.

Provide the practice the students need to locate the roots of a quadratic equation on a graph as well as state a quadratic function shown by a graph. Be sure students understand that the roots are the x-intercepts of the graph.

13-3 Completing the Square

Quadratic equations that are not perfect squares are solved by a method called completing the square. Encourage students to study carefully the three-step method presented.

Point out that the method of completing the square is useful not only for solving equations. It is also important in the study of conic sections in *Merrill Algebra Two*, and in other topics of advanced mathematics.

Encourage them to memorize the procedure as it will simplify the use of this method. This method will be used to derive the quadratic formula in Lesson 13-4.

13-4 The Quadratic Formula

Guide students through the step-by-step development of the quadratic formula by completing the square. If you prefer, you could show the quadratic formula without derving it, then demonstrate its use on equations with known roots.

Only real solutions are considered in this text. The imaginary unit is introduced in the special feature on page 423.

Encourage student to write the quadratic formula each time they use it. Recall of the formula should become a habit. Emphasize that the formula can be used to solve any quadratic equation.

Enrichment

You may wish to show better students the following development for the equation of the axis of symmetry. The axis of symmetry lies halfway between any two points with the same y-coordinate. When $y = 0$, $x_1 = \frac{-b+\sqrt{b^2-4ac}}{2a}$ and $x_2 = \frac{-b-\sqrt{b^2-4ac}}{2a}$.

The average of x_1 and x_2 is $\frac{x_1 + x_2}{2}$ or $\frac{-2b}{2a} \div 2 = -\frac{b}{2a}$.

13-5 The Discriminant

The discriminant in the quadratic formula is identified. It provides information about the nature of the roots of a quadratic equation. Point out to students that this information can be useful when graphing a quadratic function.

After students study the three situations in the presentation, help them summarize the nature of roots of a quadratic equation by referring to the summary chart on page 395. The one real root of the discriminant $b^2 - 4ac = 0$ is often referred to as a double solution or root. For example, the equation $x^2 - 6x + 9 = 0$ is equivalent to $(x - 3)(x - 3) = 0$, which has a double root, 3. This is shown graphically as a point of tangency for the graph of $y = x^2 - 6x + 9$.

Enrichment

In advanced courses, students will learn that an nth degree equation has n roots (not necessarily distinct). Thus, all quadratic equations have two roots. A quadratic equation which appears to have one root actually has a double root (two identical roots). A quadratic equation which appears to have no roots actually has two *imaginary* roots (they are not real numbers). Imaginary roots are introduced in the Excursion on page 423.

13-6 Solving Quadratic Equations

Students have studied a variety of methods for solving quadratic equations. So students may organize these methods in a systematic manner, have them study carefully the summary table.

Remind students that the graphing method yields approximate solutions. Roots which appear to be integral should be checked by substitution. Point out that factoring and completing the square are practical only for the cases listed under "comments" in the chart. In other cases it is usually quicker to use the quadratic formula.

Ask students to tell what technique to use when coefficients are in fractional or decimal form.

13-7 Problem Solving: Quadratic Equations

Verbal problems are solved by using previously learned techniques. The equations solved will be quadratic.

Be sure students are using the steps in the problem-solving plan.

13-8 The Sum and Product of Roots

Students discover interesting relationships between the sum and products of roots and the quadratic equation. After studying Example 2, students should have no difficulty seeing how the sum-and-product-of-roots rule is derived. Students will find it interesting to write quadratic equations when the roots are known.

Chapter 14
Rational Expressions

Chapter Overview

Algebraic fractions are simplified using the same procedure used to simplify numerical fractions. Numerical and algebraic fractions are then multiplied, divided, added, and subtracted. The four operations are presented in this order of difficulty to build confidence. Algebraic fractions with like denominators are dealt with before those with unlike denominators. Lastly, mixed expressions and complex fractions are introduced.

Lesson Notes

14-1 Simplifying Algebraic Fractions

Before beginning this lesson, it may be beneficial to students to review some of the methods of factoring they have learned. Some examples are given below. Page references are also given.

Factoring using the distributive property (page 214):

$$2a + 4c = 2(a + 2c)$$

$$\begin{aligned} 9x^3 + 12x^2y - 6xy &= 3x(3x^2) + 3x(4xy) - 3x(2y) \\ &= 3x(3x^2 + 4xy - 2y) \end{aligned}$$

Factoring differences of squares (page 217):

$$\begin{aligned} 4x^2 - 25 &= (2x)^2 - (5)^2 \\ &= (2x - 5)(2x + 5) \end{aligned}$$

Factoring perfect square trinomials (page 221):

$$\begin{aligned} 9m^2 + 30m + 25 &= (3m)^2 + 2(3m)(5) + (5)^2 \\ &= (3m + 5)^2 \\ 16y^2 - 16y + 4 &= (4y)^2 - 2(4y)(2) + (2)^2 \\ &= (4y - 2)^2 \end{aligned}$$

Factoring trinomials (pages 224 and 232):

$$\begin{aligned} r^2 - r - 12 &= (r - 4)(r + 3) \\ 4y^2 + 16y + 7 &= 4y^2 + (14 + 2)y + 7 \\ &= 4y^2 + 14y + 2y + 7 \\ &= (4y^2 + 14y) + (2y + 7) \\ &= 2y(2y + 7) + 1(2y + 7) \\ &= (2y + 1)(2y + 7) \end{aligned}$$

Have students recall that a fraction indicates division and division by zero is undefined. Students should have no difficulty determining values of the variable that must be excluded from the domain of the variable.

Review simplifying common numerical fractions. Emphasize using the same procedure to simplify algebraic fractions. When a fraction is in simplest form, the numerator and denominator have no common factors other than 1 or -1. Encourage students to study Example 4 carefully so they become aware of factors that are opposites and how to deal with them.

14-2 Multiplying Fractions

Review multiplying numerical fractions. Encourage students to generalize the method. Have them apply the same method to multiply algebraic fractions. If necessary, review the GCF (Lesson 7-2) when studying Example 2.

Have students recall the shortcut method used for simplifying and finding products at the same time. The method is sometimes called cancelling. Use Exploratory Exercises 9-15 in class to have students explain the steps they would use to simplify and find the products at the same time.

Review the different ways to express a negative fraction.

$$\frac{-a}{b} = \frac{a}{-b} = -\frac{a}{b}$$

14-3 Dividing Fractions

Help students recall what is meant by multiplicative inverses or reciprocals. Guide them in generalizing a method for dividing all rational numbers. Students should be able to apply the same method when dividing algebraic fractions.

Enrichment

Find the appropriate value of each expression if very large numbers are substituted for x.

1. $\frac{1}{x}$ *0* **2.** $\frac{1}{x+1}$ *0* **3.** $\frac{x}{x+1}$ *1* **4.** $\frac{x^2+x}{x^2-1}$ *1*

14-4 Dividing Polynomials

Parallel the long division process used for dividing numbers with the long division process used for dividing polynomials. Encourage students to discuss the similarities and differences.

Be sure students realize that the divisor is a factor of the dividend when the remainder is zero. Discuss how to express the quotient when there is a nonzero remainder.

Students should study Example 2 carefully so they know how to rename and expression in order of decreasing degree of the variable, using 0 coefficients when needed.

Enrichment

Introduce students to the synthetic division method of dividing polynomials. An example is given below.

$(x^2 + 7x + 10) \div (x + 5)$

$$\begin{array}{r|rrr} -5 & 1 & 7 & 10 \\ & & -5 & -10 \\ \hline & 1 & 2 & 0 \end{array}$$

The quotient is $x + 2$.

14-5 Adding and Subtracting Fractions with Like Denominators

Algebraic fractions with like denominators are added and subtracted. Have students generalize the methods used with numerical fractions. Help them apply the methods to working with algebraic fractions. They should have no difficulty adding or subtracting the numerators and placing the result over the same denominator. Be sure they do not add or subtract the denominators. Point out that Example 4 shows an expression in which the denominators are additive inverses.

14-6 Adding and Subtracting Fractions with Unlike Denominators

Algebraic fractions with unlike denominators are added and subtracted. Discuss renaming fractions so the denominators are alike. Be sure students understand the meaning of LCD. Ask them to tell why the computation is easier if the LCD is used instead of the product of the denominators.

14-7 Mixed Expressions and Complex Fractions

Mixed expressions and complex fractions are introduced. Have students compare changing mixed numerals to improper fractions and changing mixed expressions to algebraic fractions. Be sure students understand that any mixed expression may be expressed as an improper fraction. Encourage them to write the improper fraction in simplest form.

Guide students through the development of the generalization concerning the simplifying of complex fractions. Using the LCD approach to simplifying complex fractions reinforces properties of fractions.

Chapter 15
Applications of Rational Expressions

Chapter Overview

Equations containing rational expressions are solved by the same method used to solve equations containing fractions. Such equations are used to solve work problems. Formulas that contain rational expressions are presented. Formal definitions of direct variation and inverse variation are given. Lever problems are studied as an example of inverse variation. The scientific formulas for resistance in series, parallel, and series-parallel electrical circuits are used as examples of rational equations. The chapter closes with a discussion of joint variations, quadratic direct variations, and combined variations. Throughout the entire chapter students are made aware of the many applications of rational expressions.

Lesson Notes

15-1 Solving Fractional Equations

Fractional equations are solved using the LCD. Remind them that to eliminate the fractions in an equation, the terms can be multiplied by a common denominator. Point out that this is an application of the multiplication property of equality.

Review Lesson 14-1 in which students must state the values of the variable that must be excluded from the domain of the variable. This may help students understand Examples 3 and 4 more easily.

15-2 Problem Solving: Work and Uniform Motion

Verbal problems involving work and distance are solved. Emphasize that when solving work problems students should think in terms of the portion of a job completed in one unit of time.

Use Exploratory Exercises 1-4 to help students analyze work problems. Reinforce skills in solving fractional equations. Use Exploratory Exercise 5 to help them analyze distance problems involving the speed of the wind.

15-3 Using Formulas

Often in applied mathematics it is useful, convenient, or even necessary to solve a formula for a specific variable. This lesson is intended to prepare students for such demands.

Ask students to explain what a formula is and why it is used. Emphasize that each formula states a rule for a particular relationship. Have students recall the most recent formula learned from Lesson 13-4, the quadratic formula. Encourage them to name other formulas they have learned.

15-4 Direct Variation

Direct variation is introduced and defined. Emphasize that a direct variation is described by an equation of the form $y = kx$, where k is the constant of variation and is not zero. The equivalent form $k = \frac{y}{x}$ is often helpful in solving problems when the constant of variation must first be determined. Have students tell why the equation for direct variation is also a linear equation.

Allow sufficient time for students to study the text which relates direct variations to proportions. Encourage consistency when using proportions to solve problems involving direct variation.

Enrichment

Have students draw a graph using the table on page 478. Then have them bring in graphs from newspapers and magazines that represent direct variations.

15-5 Inverse Variation

To introduce inverse variations, you may want to discuss the following problem.

The area of a rectangle is 36 square inches. What are the possible integer values of its length and width?

Answer:

The formula for the area of a rectangle is $A = lw$. Therefore, $36 = lw$. The following table shows the possible values of the length and width.

l	36	18	12	9	6
w	1	2	3	4	6

Notice that l and w are inversely related. That is, $l = \frac{36}{w}$ or $w = \frac{36}{l}$

Use the following chart to clarify inverse variation. Show how the time required to travel 100 miles decreases as the speed of the vehicle increases.

Speed m/hr	*Hours required to travel 100 miles*
25	4.0
30	3.3
35	2.9
40	2.5
45	2.2
50	2.0
55	1.8

Emphasize that an inverse variation is described by an equation of the form $xy = k$, where k is the constant of variation and is not zero. The equivalent form $y = \frac{k}{x}$ shows that y varies directly as the reciprocal of x.

Many problems require two steps—first finding the value of k and then using this value to find the value of x or y. In Example 2 have students compare the two methods shown—the product rule and the proportion.

Have students develop a chart to illustrate the similarities and differences of direct variation and inverse variation. This will help students avoid confusing the two formulas.

Point out to students that many quantities in mathematics and the physical sciences are related in inverse variation. Skill in interpreting and using inverse variation is highly useful.

Enrichment

Have students draw a graph using the table on page 481. Before completing the graph, students should find at least six more ordered pairs that are solutions to $rt = 600$.

15-6 Lever Problems

The property of levers is one example of inverse variation. A lever is a bar with a pivot point called the fulcrum. For simplicity, the lever is assumed to be weightless in the problems presented. Be sure students understand that of two weights, the lesser weight must be positioned farther from the fulcrum for a lever system to balance.

15-7 Rational Expressions in Science

Series and parallel circuits are explained to introduce a scientific formula that contains rational expressions. The total resistance in a series circuit is the sum of the individual resistances. Students need to realize that R_3, R_4, R_5, and so on could be added to the formula presented in the text. The total resistance of a parallel circuit decreases as each new resistor is added. The reciprocal of the total resistance is equal to the sum of the reciprocals of the individual resistances. Example 3 provides an example of a series-parallel circuit.

Enrichment

Have students solve the formula

$$\frac{1}{R_T} = \frac{1}{R_1} + \frac{1}{R_2 + R_3} \text{ for } R_3.$$

Answer:

$$\begin{aligned} R_1(R_2 + R_3) &= R_T(R_2 + R_3) + R_TR_1 \\ R_1R_2 + R_1R_3 &= R_TR_2 + R_TR_3 + R_TR_1 \\ R_1R_3 - R_TR_3 &= R_TR_2 + R_TR_1 - R_1R_2 \\ R_3(R_1 - R_T) &= R_TR_2 + R_TR_1 - R_1R_2 \\ R_3 &= \frac{R_TR_2 + R_TR_1 - R_1R_2}{R_1 - R_T} \end{aligned}$$

15-8 Other Variations

Joint variation, quadratic direct variation, and combined variation are introduced. In all three cases, formulas or situations that most students are familiar with are used to illustrate the variation. Emphasize the difference between the proportion for a joint variation, $\frac{z_1}{x_1y_1} = \frac{z_2}{x_2y_2}$, and the proportion for a combined variation, $\frac{z_1y_1}{x_1} = \frac{z_2y_2}{x_2}$.
Use the Exploratory Exercises to help emphasize the differences among all three variations.

Chapter 16
Trigonometry
Chapter Overview

The chapter begins with an introduction to angles. The relationship between the angles and sides of a 30°–60° right triangle is presented. Similar triangles are defined in terms of corresponding angles and corresponding sides. The trigonometric ratios of sine, cosine, and tangent are studied in relation to right triangles only. Instruction is given on how to use the table of trigonometric ratios. Trigonometric ratios are used to solve triangles and verbal problems with practical application.

Lesson Notes

16-1 Angles

This lesson introduces the concept of angle and the use of a protractor in measuring angles. Discuss the term *angle*. An angle is formed by two rays (paths extending infinitely in one direction) that have a common endpoint. You may want to mention that the common endpoint is called the *vertex* of the angle, and the rays are called the *sides* of the angle.

Other concepts from geometry including supplementary and complementary angles are introduced. Students discover the sum of the angles of different triangles. Examples 3 and 4 identify the equilateral and isosceles triangles.

In the Written Exercises, students are asked to find the complement and supplement of given angles. Tell students to assume that an angle greater than 90° has no complement.

Enrichment

In geometry, students will be introduced to a proof for the sum of the angles of a triangle. You may want to tell students about the proof. The diagram below shows a triangle with a line drawn parallel to the base, through the opposite vertex.

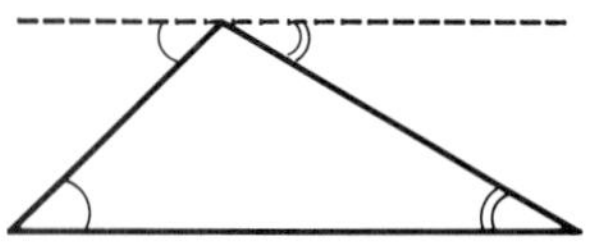

The proof follows from the fact that alternate interior angles are congruent, and the sum of the measures of the angles that form a straight angle is 180°.

16-2 30°–60° Right Triangles

From examination of an equilateral triangle, students learn what is meant by a 30°–60° right triangle. The measures of the sides opposite the 30° and 60° angles in 30°–60° right triangles are discussed. You may want to have students use a protractor and a ruler to draw 30°–60° right triangles and measure the sides. Encourage students to memorize the relationship among the sides of a 30°–60° right triangle. The following diagram may be helpful.

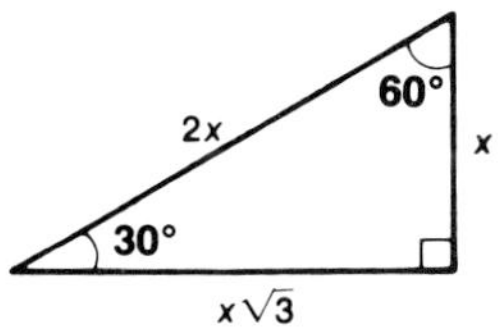

The relationship may also be shown as follows.

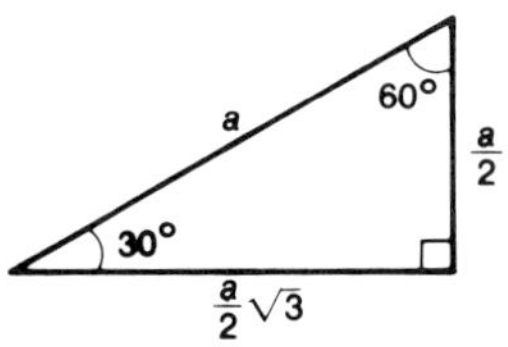

The Pythagorean Theorem is used to find the length of the side opposite the 60° angle. This theorem was discussed in Lesson 12-8. Some students may need to review the lesson.

16-3 Similar Triangles

Corresponding angles and corresponding sides are explained and used in the definition of similar triangles. Again, students could draw triangles with the same angle measures. Then they can measure the sides and compare corresponding sides to see that the measures are proportional. Be sure that students compare corresponding sides when writing proportions. If this is done, students should have little difficulty using proportions to find the missing measures of similar triangles.

16-4 Trigonometric Ratios

The trigonometric ratios of sine, cosine, and tangent are discussed in relation to right triangles. Students should realize that the definitions given here for these ratios apply only to right triangles. Point out the convenience of memorizing the definitions of trigonometric ratios. Have students explain in their own words why a trigonometric ratio like sine 30° will always be the same regardless of the triangle's size.

After discussing Example 1, you may also want to discuss the sine and cosine of a 45°–45° right triangle.

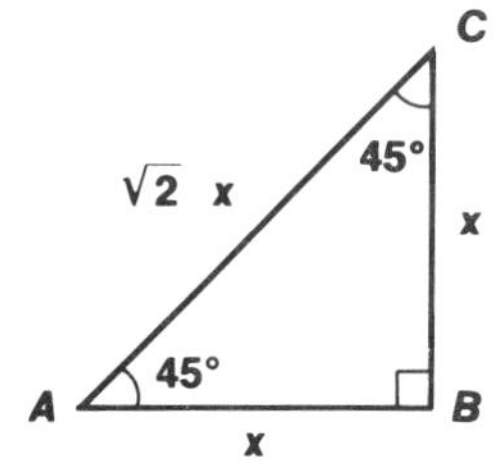

$$\sin A = \frac{x}{\sqrt{2}x} = \frac{1}{\sqrt{2}} = \frac{\sqrt{2}}{2}$$

In this case, the cosine of A is equal to the sine of A.

16-5 Using Tables

Students learn to use the table of trigonometric ratios given on page 545. Be sure they can locate and read this table. Some students may be helped by using a ruler beneath the angle for easier reading of the trigonometric ratio. All students need to realize the ratios in decimal form are approximations. Review, if necessary, how to express ratios as decimals.

Example 3 provides an example in which the measure of an angle can be only an approximation. Ask students to suggest ways to obtain more accurate approximations.

Enrichment

Suggest that students graph the sine and cosine functions from 0° to 90°. The horizontal axis could be labeled 0°, 10°, 20°, . . . , 90°. The vertical axis could be labeled 0, 0.1, 0.2, 0.3, . . . , 1. Values are given in the chart on page 545.

16-6 Solving Right Triangles

The trigonometric ratios are used to find missing measures of a right triangle. Emphasize that you can sometimes use more than one ratio to find a missing measure. To help emphasize this, have students explain why the same result is found in answer to the last question on page 514 as was found in Example 2.

16-7 Problem Solving: Using Trigonometry

Verbal problems are solved by using trigonometric ratios. The diagram of a realistic situation shows what the angles of elevation and depression are. Challenge students to determine the approximate height of the school's flagpole or the school building by using trigonometry.

Independent Chapter Tests

Independent Test for Chapter 1

Write a mathematical expression for each verbal expression. Let x represent the variable in each expression.

1. The sum of a number and 18
2. A number decreased by 13
3. A number to the tenth power

Evaluate each expression.

4. $7 \cdot 3 - 8 \cdot 2$
5. $18 \div 9 \cdot 6 - 2^2$
6. $5 + 6(4 + 5)$
7. $100 \div [5(3 + 2)]$
8. $\frac{5(2b + a)}{c - 2}$, if $a = 2$, $b = 3$, and $c = 4$
9. $\frac{m^3 - 4}{m^2 + n}$, if $m = 2$ and $n = 4$

Solve each equation.

10. $p = \frac{1}{4}(9 + 3)$
11. $m = 8.73 - 6.8$
12. $(7 - 2)^2 = y$

Simplify.

13. $k - \frac{2}{3}k$
14. $5x + m + 3x + 6m$
15. $5(a^2 + 2) + 4a^2$

State the property shown in each of the following.

16. If $r = s$ and $s = t$, then $r = t$.
17. $xy = yx$
18. $3(n + b) = 3n + 3b$
19. $(2 + p) + 4q = 2 + (p + 4q)$
20. $1k = k$

Write each sentence as an equation.

21. Five times the sum of k and 4 equals thirty.
22. A is equal to the product of b and h divided by 2.

Sue can swim 3 laps in 60 seconds. She can swim faster than Joe.

23. Is Joe able to swim 3 laps in more or less than 60 seconds?
24. About how many seconds does it take for Sue to swim one lap?

Define the variable, then write an equation for the following.

25. Bill bought 2 hot dogs and Jean bought 5 hot dogs. Together they spent \$3.50. Find the price of one hot dog.

Independent Test for Chapter 2

Name the set of numbers graphed.

1. 

Graph each set of numbers on a number line.

2. $\{-3, -1, 2\}$

3. {all integers greater than -3}

Find each sum.

4. $-6 + (-8)$

5. $64 + (-47)$

6. $\frac{7}{8} + \left(-\frac{3}{40}\right)$

7. $2\frac{4}{9} + \left(-\frac{7}{36}\right)$

8. $-9.818 + (-7.623)$

9. $3x + (-5x) + 7x$

10. $15 + (-4) + 9 + (-11)$

11. $3x + (-9x + 4x)$

Find each difference.

12. $-3 - 6$

13. $5.92 - (-3.99)$

14. $\left(-\frac{2}{3}\right) - \left(-\frac{4}{9}\right)$

15. $10b - 19b$

Evaluate each of the following.

16. $b + (-3)$, if $b = 9$

17. $m + \frac{1}{2}$, if $m = \frac{3}{4}$

Solve each equation.

18. $b + 18 = 43$

19. $s + (-9) = -15$

20. $y - 5 = -9$

21. $n - (-3) = 6$

22. $b - (-8) = -9$

Solve each problem.

23. The difference of two integers is 5. The greater integer is $+3$. What is the other integer?

24. The sum of the two integers is -16. One of the integers is 9. Find the other integer.

25. On Tuesday, the Sussex River was 7 feet below normal. During the next three days, heavy rains raised the water level 12 feet. How many feet above normal was the river after the heavy rains?

Independent Test for Chapter 3

Find each product.

1. $3(-6)$
2. $-4c(6d)$
3. $(-1.5)(-3.1)$
4. $\left(\frac{3}{7}\right)\left(-\frac{4}{3}\right)$

Simplify each expression.

5. $(-2y)(-7z) + (7y)(-7z)$
6. $(2)(-6)(1)(-5)(-3)$
7. $4(3b + 9b) - 5(5b - 7b)$

Simplify.

8. $38 \div (-2)$
9. $\frac{-81}{-9}$
10. $\frac{9x + 18}{9}$

Solve each equation.

11. $-\frac{1}{3}t = 17$
12. $\frac{n}{4} = 15$
13. $-4n = -72$
14. $-2x + 6 = 26$
15. $\frac{m}{-3} + 8 = 89$
16. $\frac{n - 6}{4} = 12$

Solve each equation.

17. $\frac{y}{4} - 6 = 15$
18. $5.4y + 8.2 = 9.8y - 2.8$
19. $6(m + 5) = -9(m - 2)$
20. $\frac{m}{7} = \frac{m - 1}{16}$

Solve each problem.

21. Find two consecutive even integers whose sum is 58.

22. If a gardener gets 3 bushels of beans from one row, how many rows must be grown to get 12 bushels of beans?

23. Change $\frac{2}{9}$ to a percent.

24. What number is 72% of 300?

25. A salesman earns \$85 per week plus 6% commission on his sales. If his salary for last week was \$448, what were his sales?

Independent Test for Chapter 4

State whether each sentence is *true* or *false*.

1. $15 < -8\frac{2}{9}$

2. $-5.3 > -12.7$

3. $-6 \leq -6$

Graph the solution set of each inequality on a number line.

4. $x \geq 4$

5. $-3 < x$

Solve each inequality.

6. $n - 8 > 5$

7. $5x < 4x - 9$

8. $10x - 18 < 7x + 6$

9. $\frac{n}{4} < -2$

10. $\frac{5}{7}y \leq -15$

11. $-6x \leq -36$

12. $(-0.9)y < 0.18$

13. $5c + 6 > 36$

14. $6z + 9 > 3z - 6$

15. $(1.2)(x + 5) \geq (1.1)(x + 6)$

State which ratio is greater.

16. $-\frac{2}{7}, -\frac{7}{8}$

Write a ratio for the unit cost.

17. 10 pounds of potatoes for \$1.69

Replace each $\underline{?}$ with <, >, or = to make each sentence true.

18. $-\frac{14}{17} \underline{?} -\frac{2}{5}$

19. $\frac{5}{17} \underline{?} \frac{25}{85}$

Graph the solution set of each inequality.

20. $-4 < w < 3$

21. $-2 \leq x \leq 3$

22. $|3x + 4| \leq 16$

23. $|6x - 5| \geq 31$

Solve each problem.

24. The sum of two positive consecutive integers is at most 9. What are the integers?

25. Jim and Bill plan to spend at most \$50 on their trip. They spent \$38 for gas. How much can they spend for food?

Independent Test for Chapter 5

Simplify.

1. $(4y^4)(5y^2)$
2. $(m^2n^3)(m^5n^4)$
3. $(4b^3)(-2b^4c^3)(3b^2c^2)$
4. $(8c^2d^3)(-4c^5d^6)$
5. $(6w^3z^4)(-3w^2z^5)(8w^9z^8)$
6. $(-5w^2v^4)(-4w^3v^8)$
7. $(x^5)^4$
8. $(x^3y^5)^3$
9. $(-4x^6y^4z^8)^3$
10. $\dfrac{a^5b^4}{a^3b^3}$
11. $\dfrac{7s^4t^5}{14s^8t}$
12. $\dfrac{20m^5n^2}{10m^9n^2}$

Express each number in scientific notation.

13. 59,462
14. 617.92
15. 0.000000000016
16. 0.0000572

Evaluate. Express each result in scientific notation.

17. $(3 \times 10^7)(5 \times 10^{-3})$
18. $\dfrac{4.37 \times 10^{-6}}{1.9 \times 10^{-2}}$
19. $(0.00161)(470{,}000)$

Solve each problem.

20. Kelley is 22 years older than Darrell. The sum of their ages is 72. What are their ages?
21. Yvonne is 2 years younger than Donna. Donna is 5 years older than Angie. How old is Angie?
22. The number of deer increased from 67 to 99. What was the percent of increase?
23. Sue's stamp collection had a value of \$3500 last year. The value has increased by 10% since then. Find the amount of increase and the value now.
24. Karen paid \$45.90 for a new pair of shoes. She got them at a 15% discount. What was the price of the shoes without the discount?
25. How many mL of a 75% solution of sodium chloride should be added to 50 mL of a 20% solution to make a 50% solution?

Independent Test for Chapter 6

State the degree of the polynomial.

1. $6x^3y^5 - 8x^2y^2 + 7x^5y^2$

Arrange the terms of the polynomial so that the powers of x are in ascending order.

2. $3x^2 + 6x + 13x^3 - 5$

Arrange the terms of the polynomial so that the powers of x are in descending order.

3. $15x^2 - 6x^3 + 3x - 6$

Find each sum.

4. $(7w^2 + 3x - 5) + (8w^2 - 5x + 2)$
5. $(6x^2 - 5x + 4) + (-2x^2 - 4x - 8)$
6. $$\begin{array}{rr} & 10b^2 - 7bc + 3c^2 \\ (+) & \underline{5b^2 + 4bc - 9c^2} \end{array}$$

Find each difference.

7. $(9w^2 + 5wx - 8x^2) - (6w^2 + 8wx - 9x^2)$
8. $$\begin{array}{rr} & 15c^2 + 6cd - 8d^2 \\ (-) & \underline{8c^2 - 5cd + 3d^2} \end{array}$$

Find each product.

9. $5n^2(3n^2 - 6n + 4)$
10. $(3x^2y)(6x^3y - 5xy^3 + 4x^2y^2)$
11. $(x + 8)(x + 9)$
12. $(3x + 5)(2x - 6)$

Find each product.

13. $\left(\frac{1}{4}b - \frac{2}{3}\right)\left(\frac{1}{2}b + \frac{1}{3}\right)$
14. $(2y + 3)(4y^2 - 5y + 8)$
15. $(b^2 - 2b + 6)(3b^2 + b - 3)$
16. $$\begin{array}{r} m^2 + 3m + 6 \\ \underline{2m - 7} \end{array}$$
17. $(c + 5)^2$
18. $(5m + 6n)(5m - 6n)$
19. $(4x - 3)^2$
20. $(m + 4)(m - 2)$
21. $(4m + 3n)^2$

Solve.

22. $5(4 - 2y) = 3y + 7$

Solve each problem.

23. Bob leaves school at the same time as Jane. Bob drives 4 mph faster than Jane. After 2 hours of driving in opposite directions, they are 116 miles apart. How fast is Bob driving?
24. Charlene is making a picture frame. The frame will have a uniform width of 2 cm and its length will be 3 cm greater than its width. If the area of the frame will be 124 square centimeters, what will be the inside dimensions of the picture frame?
25. Eric invested \$10,000. If the annual interest rate is 15%, how much interest will he earn in 6 months?

Independent Test for Chapter 7

Find the prime factorization of each of the following.

1. 36

2. -54

Find the GCF of each of the following.

3. 8, 24

4. 72, 48, 96

5. $27a^2b^3$, $75a^4b$

Factor each expression.

6. $28m^2 + 18m$

7. $25m^6n^4 + 20m^4 - 30m^3p$

8. $7b^3c^3 + 28b^2c + 14b^2$

9. $9x^2 - 25y^2$

10. $192a^2 - 75b^2$

11. $\frac{1}{9}x^2 - \frac{9}{16}y^2$

12. $w^2 - 16w + 64$

Factor each expression.

13. $36m^2 + 60mn + 25n^2$

14. $49x^2 - 42xy + 9y^2$

Factor each expression.

15. $n^2 + 19n + 84$

16. $x^2 - 10xy + 21y^2$

17. $k^2 + 3k - 108$

18. $a^2 - ab - 42b^2$

19. $12mw + 9mz + 8nw + 6nz$

20. $14sx - 12xt + 21sy - 18ty$

21. $12ac + 14ad - 30bc - 35bd$

22. $5x^2 + 9x - 18$

23. $3m^2 + 7m - 20$

24. $15k^2 - 17k - 42$

25. $4xy^2 + 8xy - 60x$

Independent Test for Chapter 8

Factor each of the following.

1. $m^2 + 2m - 35$
2. $b^2 - 20b + 96$
3. $nx^2 + 4nx + 4n$
4. $3mx^2 - 75m$

Solve each equation.

5. $6w(4w - 28) = 0$
6. $(3s + 6)(2s - 10) = 0$
7. $t^2 - 121 = 0$
8. $18s^2 + 72s = 0$
9. $3x^2 = 15x$
10. $4x^2 = 36$
11. $m^3 - 144m = 0$
12. $x^2 - 2x - 35 = 0$
13. $n^3 - n^2 - 72n = 0$

Solve the problem.

14. Jim's father gave him the following problem. "The product of 5 times my salary and 61,000 less than 4 times my salary is 0. What is my salary?" How much is Jim's father's salary?
15. Find two consecutive integers whose product is 132.
16. Find two consecutive odd integers whose product is 195.

Solve each problem.

17. Find two consecutive even integers whose product is 360.
18. While playing on a trampoline, Jim leaves the surface at a velocity of 16 feet per second. How many seconds will it take for him to return to the surface of the trampoline? Use $h = vt - 16t^2$.
19. The length of a rectangle is 7 cm more than its width. The area of the rectangle is 260 cm^2. Find the length and width.
20. A print is 11 inches by 14 inches. It is to be placed in a frame which will cover an equal strip on all four edges of the print. The frame will cover 46 square inches. What will be the visible dimensions of the print?

Simplify each expression.

21. $\dfrac{2x^5y}{18x^4y^2z}$
22. $\dfrac{3b^2}{3b(b + 9)}$
23. $\dfrac{x - y}{x^2 - y^2}$
24. $\dfrac{5x^2 - 25x}{x^2 - 10x + 25}$
25. $\dfrac{3y^2 - 16y + 5}{3y^2 - 18y + 15}$

Independent Semester Test for Chapters 1-8 (Page 1)

Solve each equation.

1. $t = 13 - 5 + 9^2$
2. $m = 16 \div 4 + 9 - 2 \cdot 5$
3. $4.9 - b = 2.3$
4. $32 = w - (-61)$
5. $4n = -62$
6. $\frac{-5c + 8}{-2} = 21$
7. $3y^2 + 5y - 12 = 0$
8. $x^2 - 16 = 0$
9. $2x^2 + 4 = 9x$
10. $27x + 10 = -5x^2$
11. $x^2 + x = 6$

Simplify.

12. $\frac{1}{3}\left(\frac{1}{2}k^2 + 4k\right) + \frac{5}{6}k^2 - \frac{1}{3}k$
13. $0.4(0.2x + 0.3y) + 1.1x$
14. $4(5x + (-2x)) - 3(7x - 4x)$
15. $\frac{48c - 18d}{-6}$

Simplify.

16. $(7c^5d^9)(-13c^4d^3)$
17. $\left(-\frac{1}{4}x^2y^3z^3\right)(-20z^3y^2x)$
18. $(-6x^5y^3z^9)^3$
19. $(4mn)^2 + (3mn)(2mn)$
20. $\frac{35a^7b^4c}{-5a^5b^7c^6}$

Solve each inequality.

21. $x - 17.43 < -4.31$
22. $-\frac{2}{5}t \geq \frac{3}{7}$
23. $3p + 7 < -8p + 19$
24. $-10(y - 2) \geq 10(y + 6)$

Evaluate. Express each result in scientific notation.

25. $\frac{3.6 \times 10^4}{1.8 \times 10^{-7}}$
26. $(4 \times 10^3)(2.5 \times 10^7)$
27. $(0.00056)(4000)$
28. $\frac{0.00000294}{42{,}000}$

Independent Semester Test for Chapters 1-8 (Page 2)

Find each sum, difference, or product.

29. $(4x^2 + 9x + 12) + (5x^2 + x - 10)$

30. $\left(\frac{4}{3}c^2 - \frac{2}{7}c + \frac{1}{6}\right) - \left(\frac{1}{3}c^2 + \frac{5}{7}c + \frac{1}{6}\right)$

31. $-5mn(3m^2 - 4mn + n^2)$

32. $(4a + 7)(2a - 3)$

33. $(3y + 4z)^2$

34. $(2x - 6y)^2$

35. $(2x - 8y)(2x + 8y)$

Factor.

36. $28x^5y - 21x^2y^6$

37. $\frac{1}{9}b^2 - \frac{4}{25}c^2$

38. $x^2 + 18x + 81$

39. $k^2 - 10k + 25$

40. $r^2 + 4rs - 12s^2$

41. $t^2 + 10t + 16$

42. $m^2 - 5mp + 6p^2$

43. $2sw + 6sz - 5tw - 15tz$

44. $5x^2 + 30nx + 45n^2$

Solve each problem.

45. Stan charges \$5.00 for a private swimming lesson. He has 7 private lessons per week. How much does Stan earn each week for giving private swimming lessons?

46. June walked 25 km in a walk-a-thon. This was 6 km more than she walked in another walk-a-thon. How far did she walk in the other walk-a-thon?

47. Find three consecutive odd integers whose sum is 93.

48. Alex saves \$25 per month. How long will it take him to save \$400?

49. 133 is what percent of 190?

50. Erika is 4 years younger than Lena. The sum of their ages is 16. What are their ages?

Independent Test for Chapter 9

State the quadrant in which the point for each ordered pair is located.

1. $(-3, -5)$
2. $(6, -6)$
3. $(-4, 7)$
4. $(1, 9)$

Graph each of the following.

5. $A(-2, 3)$
6. $D(-3, -1)$
7. $F(4, 2)$
8. $K(3, -2)$

State each relation as a set of ordered pairs. Then state the domain and range of each.

9.

x	y
5	7
7	9
12	15
13	16
18	27

10.

x	y
1	4
2	5
3	7
4	8
5	9

State which of the ordered pairs given are solutions of the equation.

11. $y = 2x - 3$
 a. $(3, 3)$ **b.** $(1, 2)$ **c.** $(5, 7)$ **d.** $(4, 4)$

12. $t = s - 3$
 a. $(-4, 1)$ **b.** $(3, -6)$ **c.** $(-2, -5)$ **d.** $(2, -1)$

Solve each of the following equations if the domain is {1, 4, 6, 9}.

13. $y = 3x + 2$
14. $x = y + 2$

Graph each equation.

15. $3x - 2y = 9$
16. $5x + 4y = 15$

State whether or not each relation is a function.

17. $\{(2, 6), (3, 9), (4, 7), (5, 3), (6, 15)\}$
18. $\{(4, -1), (3, -3), (-2, 4), (3, 3), (0, 1), (2, 4)\}$

Given $f(x) = 4x + 3$, determine the value of each of the following.

19. $f(5)$
20. $f\left(\frac{1}{2}\right)$
21. $f(-1)$

Graph each inequality.

22. $4x + 3y \leq 9$
23. $2x - 5y > 16$

Write an equation for the relationship between the variables in each chart. Then copy and complete each chart.

24.

x	1	3	5	7	9	11
y	8	10	12		16	

25.

x	2	3	4	6	9	11	12
y	3	5	7	11	17		

Independent Test for Chapter 10

Determine the slope of the line passing through each pair of points named.

1. $(2, 9), (-3, 5)$
2. $(10, 12), (3, 6)$

Determine the value of r so that the line through each pair of points has the given slope.

3. $(24, 5), (r, 2), m = \frac{3}{16}$
4. $(5, r), (4, 15), m = 3$

State the slope and y-intercept of the graph of the equation.

5. $y = 7x - 4$

Find the x- and y-intercepts for each of the following.

6. $y = \frac{2}{3}x - 2$
7. $y = \frac{5}{9}x - 5$

Write each of the following in standard form and slope-intercept form.

8. $2x - 5 = y$
9. $7y + 8x = 12$

Write an equation in slope-intercept form for the line passing through the given point and having the given slope.

10. $(2, 5), \frac{4}{5}$
11. $(-3, 7), -\frac{2}{7}$

Write an equation in standard form for the line through each of the following pairs of points.

12. $(2, 7), (-5, 4)$
13. $(1, -2), (3, 6)$

Write a linear equation in slope-intercept form for the line with the given slope and y-intercept.

14. $m = 2, b = -7$
15. $m = -5, b = \frac{2}{3}$

Graph each of the following.

16. $y = 4x - 9$
17. $3x + 2y = 15$

Write an equation in slope-intercept form for the line that has each pair of intercepts below.

18. x-intercept, 4; y-intercept, -3
19. x-intercept, -2; y-intercept, 6

Write an equation for the line that is parallel to the graph of each equation below and that passes through the indicated point. Use slope-intercept form.

20. $4x + 3y = 6; (-3, 8)$
21. $x = \frac{1}{2}y; (2, 10)$

Write an equation for the line that is perpendicular to the graph of each equation below and that passes through the indicated point. Use slope-intercept form.

22. $7x - 2y = 9; (2, 1)$
23. $y = 5x - 3; (-5, 4)$

Find the coordinates of the midpoint of $\overline{AB}$ given the following.

24. $A(2, 6), B(9, 8)$
25. $A(10, -2), B(2, 7)$

Independent Test for Chapter 11

Solve each system of equations by graphing.

1. $y = 2x$
   $x - y = 7$

2. $2x + 3y = 12$
   $x + y = 4$

3. $3x + 3y = 24$
   $x - y = 6$

4. $x + y = 9$
   $4x - 2y = 18$

Graph each pair of equations. Then state whether the system has one solution, no solutions, or infinitely many solutions.

5. $3x - y = -2$
   $3x - y = 5$

6. $x - 3y = 14$
   $3x - 9y = 42$

Solve each system of equations by substitution.

7. $3x + 2y = 6$
   $y - x = 5$

8. $2y - x = 9$
   $4x + 5y = 4$

9. $2x - 6y = 10$
   $5x + 6y = 8$

Solve each system of equations by elimination.

10. $2y + 3x = 8$
    $5y - 3x = 6$

11. $x + y = 3$
    $x - y = 13$

12. $7x + 4y = 26$
    $-4x + 4y = -7$

13. $3x + 2y = 14$
    $x - y = 6$

14. $2x + 5y = 8$
    $x + 3y = 3$

15. $6x + 7y = 12$
    $3x + 2y = 3$

Solve each system of inequalities by graphing.

16. $y > 2x + 5$
    $y > -4x - 3$

17. $y \geq 3x + 5$
    $y > x - 1$

18. $y < 3x + 6$
    $y > x + 2$

Use a system of equations to solve each problem.

19. The farmers cooperative mixed two kinds of grain. One kind was worth \$3.00 per bushel and the other was worth \$1.50 per bushel. If a mixture of 200 bushels was worth \$2.625 per bushel, how much of each kind did they use?

20. The Waco Oil Company mixes two grades of gasoline. One kind costs \$1.00 per gallon and the other costs \$1.50 per gallon. They want 1000 gallons worth \$1.25 per gallon. How much of each kind should they use?

21. Candy worth \$2.50 per pound is mixed with candy that is worth \$2.20 per pound. One hundred pounds of the mixture is worth \$2.29 per pound. How much of each kind is included in the mixture?

22. A two-digit number is 6 times its units digit. The sum of the digits is 9. Find the number.

23. The units digit of a two digit number is 5 more than its tens digit. The number is 36 more than the sum of its digits. Find the number.

24. A plane took 1 hour 20 minutes to go 600 miles when flying with the jetstream and 2 hours 40 minutes to fly against it the same distance. Find the plane's rate in calm air and the rate of the jetstream.

25. Two people start driving toward each other on the same highway at the same time from towns 380 miles apart. One car travels 15 miles per hour faster than the other. What is the rate of each car if they meet in 4 hours?

Independent Test for Chapter 12

Simplify each radical.

1. $-\sqrt{81}$
2. $\sqrt{0.09}$
3. $\pm\sqrt{\frac{36}{169}}$
4. $\sqrt{324}$

Use the table on page 530 to find a value for each expression.

5. $\sqrt{37}$
6. $\sqrt{79}$
7. $\sqrt{800}$
8. $\sqrt{\frac{38}{36}}$

Approximate each square root. Use the divide-and-average method twice for each exercise.

9. $\sqrt{15}$
10. $\sqrt{162}$

Simplify.

11. $\sqrt{112}$
12. $\sqrt{50x^7y^5}$
13. $\sqrt{6}\cdot\sqrt{24}$
14. $\frac{\sqrt{50}}{\sqrt{2}}$

Simplify.

15. $(\sqrt{8}+\sqrt{3})(\sqrt{8}-\sqrt{3})$
16. $\frac{1}{\sqrt{15}+2}$
17. $5\sqrt{7}+18\sqrt{7}$
18. $2\sqrt{50}-3\sqrt{72}$

Solve and check.

19. $\sqrt{b}=13$
20. $\sqrt{2x-9}-7=3$
21. $\sqrt{7x+9}-5=0$

Determine whether the triangle is a right triangle.

22. 4, 6, 8

Solve the problem.

23. Find the length of the diagonal of a square if each side is 6 feet long.

Find the distance between the pair of points.

24. $(5, -2), (3, 6)$

Find two possible values for *y* for the pair of points that are the given distance apart.

25. $(6, 2), (5, y);\ d=\sqrt{5}$

Independent Test for Chapter 13

Find the equation of the axis of symmetry and the maximum or minimum point for the graph of each quadratic function.

1. $y = 3x^2 + 6x - 7$
2. $y = 5x^2 - 15x + 2$
3. $y = -6x^2 - 2x + 9$
4. $y = -4x^2 + 8x - 6$

Locate the roots of each equation by graphing the related function.

5. $x^2 - x - 12 = 0$
6. $x^2 + 11x + 18 = 0$
7. $x^2 - 12x + 36 = 0$
8. $x^2 + 25 = 0$

Solve each equation by completing the square.

9. $x^2 - 24x - 6 = 0$
10. $x^2 + 30x - 8 = 0$
11. $x^2 + 10x + 4 = 0$

Use the quadratic formula to solve each equation.

12. $3x^2 - 5x - 1 = 0$
13. $x^2 - 2x - 35 = 0$
14. $2x^2 + 9x + 3 = 0$
15. $x^2 - 2x - 15 = 0$

Use the discriminant to determine the number of real roots for each equation.

16. $x^2 + 3x - 54 = 0$
17. $x^2 + 16x + 64 = 0$
18. $x^2 + x + 21 = 0$
19. $x^2 - 14x + 45 = 0$

Solve each problem.

20. Find the length and width of the rectangle whose length is 7 more than the width. The area is 18 cm^2.
21. Find the length and width of the rectangle whose width is 1 less than the length. The area is 20 cm^2.

State the sum and product of the roots of each equation.

22. $x^2 + 2x - 63 = 0$
23. $4x^2 - 5x - 7 = 0$

Find a quadratic equation having the given roots.

24. $-6, 12$
25. $\sqrt{5}, -\sqrt{5}$

Independent Test for Chapter 14

Simplify each algebraic fraction.

1. $\dfrac{x^3}{x^8}$

2. $\dfrac{108c^4d^9}{9c^{12}d^3}$

3. $\dfrac{12x^{16}y^9}{8x^5y^{11}}$

4. $\dfrac{m^2 + 9m + 18}{m^2 - 6m - 27}$

5. $\dfrac{3b^2 + 23b + 14}{b^2 + 2b - 35}$

6. $\dfrac{x^2 - 25}{x^2 - 13x + 40}$

Find each product in simplest form.

7. $\dfrac{ac}{ad} \cdot \dfrac{e}{c}$

8. $\dfrac{3x^2 + 14x + 15}{x^2 + 2x - 3} \cdot \dfrac{x^2 + x - 2}{3x^2 - 10x - 25}$

9. $\dfrac{x^2 + 8x + 16}{x^2 - 4x - 12} \cdot \dfrac{x^2 - x - 30}{2x^2 + 11x + 12}$

Find each quotient in simplest form.

10. $\dfrac{2y^3}{x + 1} \div \dfrac{8y}{x + 1}$

11. $\dfrac{n + 3}{n - 7} \div \dfrac{n^2 + 8n + 12}{n^2 - 2n - 35}$

12. $\dfrac{p^2 - p - 6}{p^2 + p - 2} \div \dfrac{p^2 + p - 12}{p^2 + 3p - 4}$

Find each quotient.

13. $x + 4 \overline{\smash{)}\, 6x^2 + 31x + 28}$

14. $2m + 3 \overline{\smash{)}\, 2m^2 + 13m + 18}$

Find each sum or difference. Express each result in simplest form.

15. $\dfrac{x}{x^2 - y^2} + \dfrac{y}{x^2 - y^2}$

16. $\dfrac{2x}{3x + 5} + \dfrac{4x + 6}{3x + 5}$

17. $\dfrac{3n + 5}{n - 7} - \dfrac{2n - 6}{n - 7}$

18. $\dfrac{3}{7} + \dfrac{2}{9}$

19. $\dfrac{n + 5}{2n + 4} + \dfrac{n - 6}{3n + 6}$

20. $\dfrac{2x + 3y}{x^2 - 2xy + y^2} - \dfrac{3}{x - y}$

Simplify.

21. $\dfrac{\frac{4}{7}}{\frac{7}{8}}$

22. $\dfrac{2\frac{1}{4}}{4\frac{2}{5}}$

23. $5 + \dfrac{2x - y}{2x + y}$

24. $\dfrac{x + 2 - \dfrac{2x}{x + 2}}{x + \dfrac{4}{x + 2}}$

25. $\dfrac{x - 2 - \dfrac{1}{x - 2}}{x - 3}$

Independent Test for Chapter 15

Solve each equation.

1. $\frac{x-7}{4} + \frac{x}{2} = 14$

2. $\frac{6x}{x+5} + \frac{2x+1}{x+3} = 8$

3. $x - \frac{9}{x-2} = \frac{x-3}{2-x}$

4. $\frac{x}{x-1} + \frac{4}{x^2-1} = 2$

5. $\frac{x+5}{x} + \frac{x-2}{x} = 7$

6. $\frac{y}{y+4} - \frac{y}{y-4} = \frac{-48}{y^2-16}$

7. $\frac{3p}{p^2-5p+6} = \frac{1}{p-3} + \frac{10}{p-2}$

8. $\frac{y^2-5y+4}{2y^2+4y-6} + \frac{1}{y+3} = 0$

Solve each formula for the variable indicated.

9. $KE = \frac{1}{2}mv^2$, for m

10. $V_2 = \frac{P_1V_1}{P_2}$, for V_1

11. $V_2 = \frac{V_1T_2}{T_1}$, for T_2

12. $V_2 = \frac{P_1V_1T_2}{P_2T_1}$, for P_1

Solve each problem.

13. Jim can stuff 1000 envelopes in 5 hours. Sue can stuff 1000 envelopes in 3 hours. How long will it take to stuff 1000 envelopes if they work together?

14. A gas storage tank can be filled from one pipeline in 7 hours and by another pipeline in 5 hours. How many hours will it take to fill the tank if both pipelines are used?

15. If y varies directly as x and $y = 7$ when $x = 5$, find y when $x = 35$.

Solve each problem.

16. If y varies inversely as x and $y = 15$ when $x = 2$, find y when $x = 6$.

17. Four gallons of gas cost $4.84. How much will nine gallons of gas cost?

18. Bob drove 4 hours at a rate of 50 mph. How long would the same distance take Bob if he drove at 40 mph?

19. Weights of 42 pounds and 58 pounds are placed on a lever. The two weights are 20 feet apart, and the lever is balanced. How far from the fulcrum is the 58-pound weight?

20. If y varies directly as x^2 and $y = 6$ when $x = 5$, find x when $y = 24$.

21. A fulcrum is placed 2 feet from the end of 7-foot lever. Will a 150-pound man on the other end of the lever be able to lift a 750-pound load of bricks?

22. Two coils with resistances of 7 ohms and 9 ohms are connected in parallel. What is the total resistance?

23. Four coils with resistances of 2 ohms, 6 ohms, 10 ohms, and 14 ohms are connected in series. What is the total resistance?

24. Given the formula $\frac{1}{R_T} = \frac{1}{R_1} + \frac{1}{R_2 + R_3}$ for the resistance of a circuit, determine the resistance R_1 if $R_T = 2.5$ ohms, $R_2 = 3$ ohms, and $R_3 = 7$ ohms.

25. If z varies jointly as x and y, and $z = 15$ when $x = 3$ and $y = 7$, find y when $x = 189$ and $z = 45$.

Independent Test for Chapter 16

Find the complement and supplement of each angle whose measure is listed below.

1. 88° **2.** 121° **3.** $(x - 3)°$

4. $(y + 47)°$ **5.** $4x°$

In each of the following, the measures of two angles of a triangle are given. Find the measure of the third angle.

6. 13°, 59° **7.** 94°, 60° **8.** 42°, 132°

Find the missing measures of each of the 30°–60° right triangles described below. (Use $\sqrt{3} \approx 1.732$.)

	Hypotenuse	Side Opposite 30° Angle	Side Opposite 60° Angle
9.	______	5 cm	______
10.	20 ft	______	______
11.	______	______	18 cm

Triangles *ABC* and *DEF* are similar. For each set of measures, find the missing side measures.

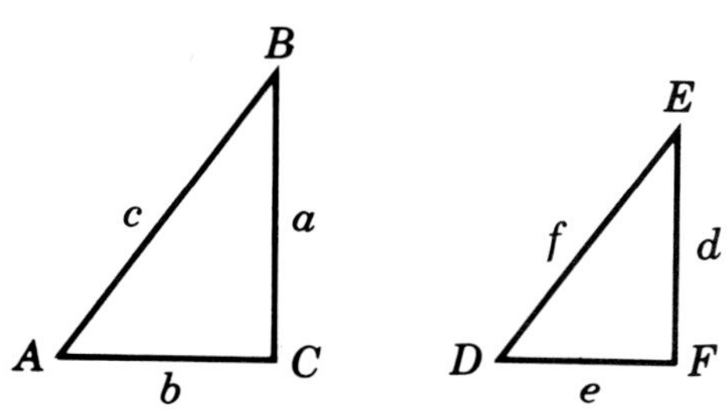

12. $a = 4, b = 3, c = 5, d = 2$

13. $a = 12, b = 9, c = 15, f = 10$

14. $b = 5, c = 7, d = 3, e = 6$

Use the triangle below. Express each of the following ratios as a fraction.

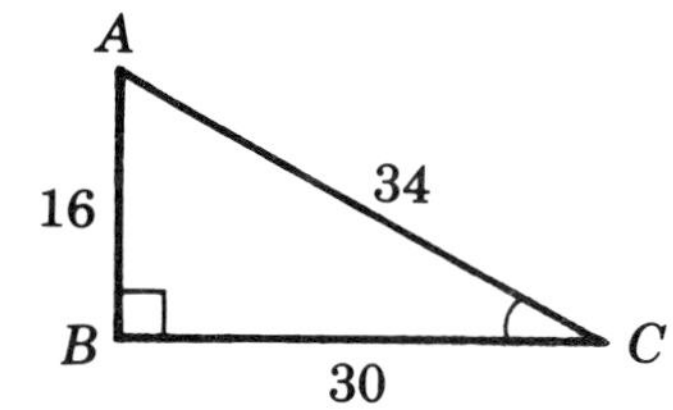

15. $\sin C$

16. $\tan A$

17. $\cos A$

18. $\cos C$

19. $\sin A$

20. $\tan C$

Solve each of the following right triangles. (Angle *C* is a right angle.)

21. angle $A = 20°$, $a = 5.13$

22. $a = 6$, $b = 9$

23. angle $B = 22°$, $c = 14$

Solve each problem.

24. A road rises 250 feet vertically over a horizontal distance of 700 feet. What is the angle of elevation of the road?

25. The diagonal of a rectangle is 30 inches long and makes an angle of 15° with a side of the rectangle. Find the length and width of the rectangle.

Independent Semester Test for Chapters 9-16 (Page 1)

State the domain and range for each of the following relations. State whether or not each relation is a function.

1. $\{(1, 9), (2, 12), (3, 15), (4, 22)\}$

2. $\{(5, -4), (4, 4), (3, -4), (2, 5), (4, 6)\}$

Solve each equation if the domain is {1, 2, 3}.

3. $y = 3x - 7$

4. $5x + 2y = 13$

Determine the slope of the line passing through the following points.

5. $(-8, 1), (2, -5)$

6. $(15, 7), (4, 5)$

Write an equation for the line satisfying the following conditions. Use slope-intercept form.

7. $m = 5$ and passes through $(3, 7)$

8. Passes through $(2, -7)$ and $(9, 15)$

9. Slope $= 6$ and y-intercept $= -2$

Solve each system of equations.

10. $3x - 4y = 15$
$5x + 4y = 17$

11. $3x - 2y = 4$
$x + 5y = 7$

12. $4x - y = 10$
$6x + 2y = 8$

Simplify.

13. $\sqrt{27a^5b^4}$

14. $\sqrt{3} \cdot \sqrt{12}$

15. $(\sqrt{9} + \sqrt{5})(\sqrt{9} - \sqrt{5})$

16. $\dfrac{6}{6 + \sqrt{5}}$

Solve each equation.

17. $\sqrt{5x} = 11$

18. $\sqrt{b} = 4\sqrt{7}$

Use the Pythagorean Theorem to find each missing measure.

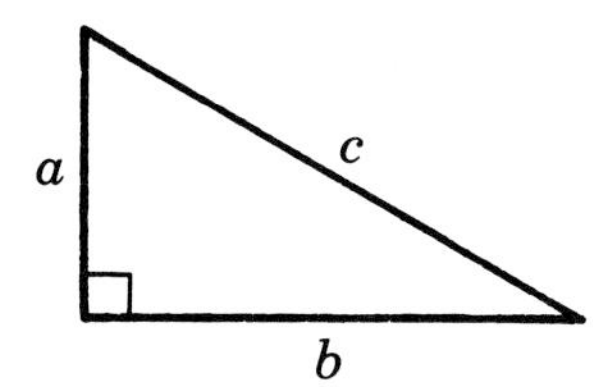

19. Find b if $a = 2$ and $c = 9$.

20. Find c if $a = 4$ and $b = 9$.

Find the distance between each pair of points.

21. $(7, 6), (3, 3)$

22. $(-4, -6), (5, -7)$

Find the equation of the axis of symmetry and the maximum or minimum point for the graph of each quadratic function.

23. $y = 5x^2 - 4x + 8$

24. $y = -2x^2 + 6x + 21$

Independent Semester Test for Chapters 9-16 (Page 2)

Solve each equation.

25. $x^2 + 5x - 6 = 0$

26. $y^2 + 6y + 9 = 144$

27. $x^2 + 3x + 7 = 0$

Simplify.

28. $\dfrac{72b^3c^4}{63b^5c^2}$

29. $\dfrac{x^2 + 5x + 6}{x^2 - 5x - 14}$

Perform the indicated operations and simplify.

30. $\dfrac{x^2 + 12x + 27}{x^2 + 2x - 48} \cdot \dfrac{x^2 + x - 56}{x^2 - x - 12}$

31. $\dfrac{m + 4}{m - 3} \div \dfrac{m^2 + 9m + 20}{m^2 - m - 6}$

32. $\dfrac{5m + 4}{2m - 6} + \dfrac{3m}{4m - 12}$

33. $\dfrac{6x}{2x - 3} - \dfrac{9}{2x - 3}$

Solve each equation.

34. $\dfrac{x}{6} + \dfrac{x}{6} = 4$

35. $\dfrac{-2}{y - 4}$

Find the complement and supplement of each angle whose measure is listed below.

36. $(x + 30)°$

37. $45°$

Find the missing measures of each of the 30°–60° right triangles described below. (Use $\sqrt{3} \approx 1.732$.)

	Hypotenuse	Side Opposite 30° Angle	Side Opposite 60° Angle
38.	6 cm	______	______
39.	______	6 in.	______

Triangles *ABC* and *DEF* are similar. For each set of measures, find the missing side measures.

40. $a = 8$, $b = 15$, $c = 9$, $d = 4$

41. $b = 2$, $c = 7$, $e = 1$, $d = 5$

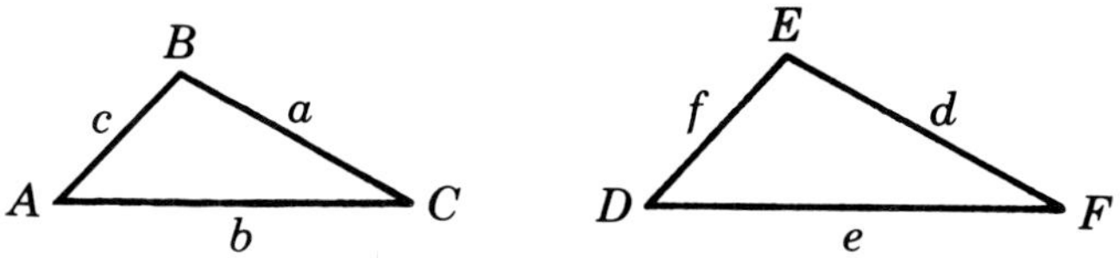

Solve each of the following right triangles. (Angle *C* is a right angle.)

42. $a = 12$, $b = 16$

43. angle $A = 23°$, $a = 15$

Graph each of the following.

44. $2x + 3y = 7$

45. $x + y = 5$
$3x + 2y = 6$

Solve each problem.

46. A two digit number is 12 times its tens digit. The sum of the digits is 6. Find the number.

47. Find the dimensions of a rectangle having a perimeter of 26 cm, if the width is 3 cm less than the length.

48. Jan can do a job in 2 hours. Marla can do the same job in 3 hours. If they work together, how long will it take to complete the job?

49. If y varies inversely as x and $y = 5$ when $x = 3$, find y when $x = 7$.

50. A road rises 120 feet vertically over a horizontal distance of 480 feet. What is the angle of elevation of the road?

Answers to Independent Tests

Chapter 1, page T56 **1.** $x + 18$ **2.** $x - 13$ **3.** x^{10} **4.** 5 **5.** 8 **6.** 59 **7.** 4 **8.** 20 **9.** $\frac{1}{2}$ **10.** 3 **11.** 1.93 **12.** 25 **13.** $\frac{1}{3}k$ or $\frac{k}{3}$ **14.** $8x + 7m$ **15.** $9a^2 + 10$ **16.** transitive property of equality
17. commutative property for multiplication
18. distributive property
19. associative property for addition
20. multiplicative identity property
21. $5(k + 4) = 30$ **22.** $A = \frac{bh}{2}$ or $A = bh \div 2$ **23.** more
24. 20 **25.** Let x = price of one hot dog. The equation is $2x + 5x = 350$ or $2x + 5x = 3.50$.

Chapter 2, page T57 **1.** $\{-2, 0\}$

2. 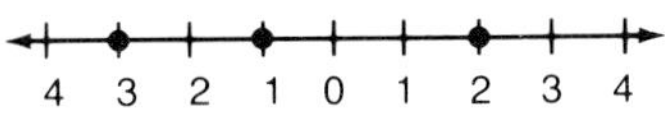

3. 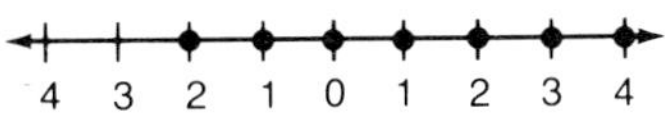

4. -14 **5.** 17 **6.** $\frac{4}{5}$ **7.** $2\frac{1}{4}$ **8.** -17.441 **9.** $5x$ **10.** 9 **11.** $-2x$ **12.** -9 **13.** 9.91 **14.** $-\frac{2}{9}$ **15.** $-9b$ **16.** 6 **17.** $\frac{5}{4}$ **18.** 25 **19.** -6 **20.** -4 **21.** 3 **22.** -17 **23.** -2 **24.** -25
25. 5 feet

Chapter 3, page T58 **1.** -18 **2.** $-24cd$ **3.** 4.65 **4.** $-\frac{4}{7}$ **5.** $-35yz$ **6.** -180 **7.** $58b$ **8.** -19 **9.** 9 **10.** $x + 2$ **11.** -51 **12.** 60 **13.** 18 **14.** -10 **15.** -243 **16.** 54 **17.** 84 **18.** 2.5 **19.** $-\frac{4}{5}$ **20.** $-\frac{7}{9}$ **21.** 28, 30 **22.** 4 rows **23.** $22\frac{2}{9}\%$ **24.** 216 **25.** \$6050

Chapter 4, page T59 **1.** false **2.** true **3.** true

4. 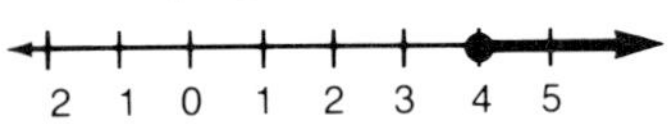

5. 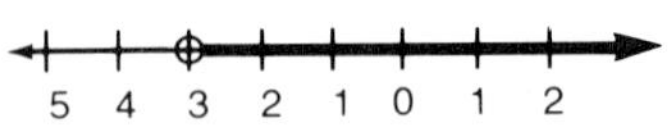

6. $n > 13$ **7.** $x < -9$ **8.** $x < 8$ **9.** $n < -8$ **10.** $y \leq -21$ **11.** $x \leq 6$ **12.** $y > -0.2$ **13.** $c > 6$ **14.** $z > -5$ **15.** $x \geq 6$ **16.** $-\frac{2}{7}$ **17.** $\frac{169}{10}$ **18.** $<$ **19.** $=$

20. 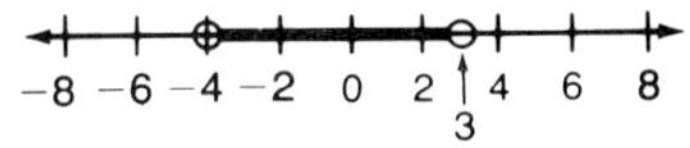

21. 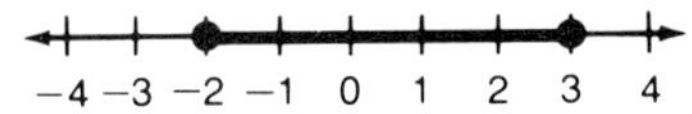

22. 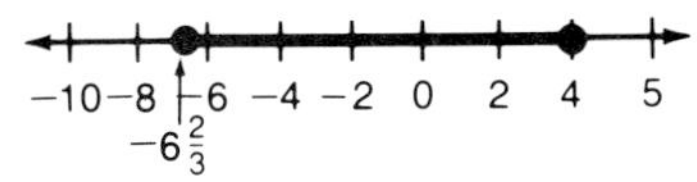

23. 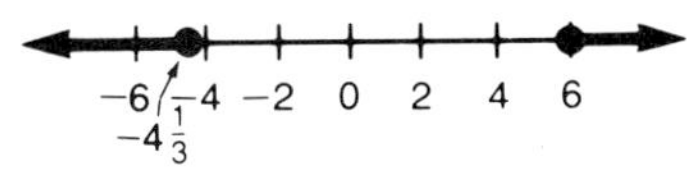

24. 1, 2; 2, 3; 3, 4; 4, 5 **25.** \$12 or less

Chapter 5, page T60 **1.** $20y^6$ **2.** m^7n^7 **3.** $-24b^9c^5$ **4.** $-32c^7d^9$ **5.** $-144w^{14}z^{17}$ **6.** $20w^5v^{12}$ **7.** x^{20} **8.** x^9y^{15} **9.** $-64x^{18}y^{12}z^{24}$ **10.** a^2b **11.** $\frac{t^4}{2s^4}$ **12.** $\frac{2}{m^4}$ **13.** 5.9462×10^4 **14.** 6.1792×10^2 **15.** 1.6×10^{-11} **16.** 5.72×10^{-5} **17.** 1.5×10^5 **18.** 2.3×10^{-4} **19.** 7.567×10^2 **20.** Kelley, 47; Darrell, 25 **21.** NEI **22.** 47.8% **23.** increase, \$350; value now, \$3850 **24.** \$54 **25.** 60 mL

Chapter 6, page T61 **1.** 8 **2.** $-5 + 6x + 3x^2 + 13x^3$ **3.** $-6x^3 + 15x^2 + 3x - 6$ **4.** $15w^2 - 2x - 3$ **5.** $4x^2 - 9x - 4$ **6.** $15b^2 - 3bc - 6c^2$ **7.** $3w^2 - 3wx + x^2$ **8.** $7c^2 + 11cd - 11d^2$ **9.** $15n^4 - 30n^3 + 20n^2$ **10.** $18x^5y^2 - 15x^3y^4 + 12x^4y^3$ **11.** $x^2 + 17x + 72$ **12.** $6x^2 - 8x - 30$ **13.** $\frac{1}{8}b^2 - \frac{1}{4}b - \frac{2}{9}$ **14.** $8y^3 + 2y^2 + y + 24$ **15.** $3b^4 - 5b^3 + 13b^2 + 12b - 18$ **16.** $2m^3 - m^2 - 9m - 42$ **17.** $c^2 + 10c + 25$ **18.** $25m^2 - 36n^2$ **19.** $16x^2 - 24x + 9$ **20.** $m^2 + 2m - 8$ **21.** $16m^2 + 24mn + 9n^2$ **22.** $y = 1$ **23.** 31 mph **24.** width = 12 cm, length = 15 cm **25.** \$750

Chapter 7, page T62 **1.** $2^2 \cdot 3^2$ **2.** $-1 \cdot 2 \cdot 3^3$ **3.** 8 **4.** 24 **5.** $3a^2b$ **6.** $2m(14m + 9)$ **7.** $5m^3(5m^3n^4 + 4m - 6p)$ **8.** $7b^2(bc^3 + 4c + 2)$ **9.** $(3x + 5y)(3x - 5y)$ **10.** $3(8a + 5b)(8a - 5b)$ **11.** $\left(\frac{1}{3}x + \frac{3}{4}y\right)\left(\frac{1}{3}x - \frac{3}{4}y\right)$ **12.** $(w - 8)^2$ **13.** $(6m + 5n)^2$ **14.** $(7x - 3y)^2$ **15.** $(n + 7)(n + 12)$ **16.** $(x - 3y)(x - 7y)$ **17.** $(k - 9)(k + 12)$ **18.** $(a - 7b)(a + 6b)$ **19.** $(3m + 2n)(4w + 3z)$ **20.** $(2x + 3y)(7s - 6t)$ **21.** $(2a - 5b)(6c + 7d)$ **22.** $(5x - 6)(x + 3)$ **23.** $(3m - 5)(m + 4)$ **24.** $(5k + 6)(3k - 7)$ **25.** $4x(y + 5)(y - 3)$

Chapter 8, page T63 **1.** $(m + 7)(m - 5)$ **2.** $(b - 8)(b - 12)$ **3.** $n(x + 2)^2$ **4.** $3m(x + 5)(x - 5)$ **5.** 0, 7 **6.** -2, 5 **7.** -11, 11 **8.** 0, -4 **9.** 0, 5 **10.** -3, 3 **11.** -12, 0, 12 **12.** -5, 7 **13.** 0, -8, 9 **14.** $15,250 **15.** -12, -11 or 11, 12 **16.** -15, -13 or 13, 15 **17.** -20, -18 or 18, 20 **18.** 1 second **19.** length = 20 cm, width = 13 cm **20.** length = 12 cm, width = 9 cm **21.** $\frac{x}{9yz}$ **22.** $\frac{b}{b + 9}$ **23.** $\frac{1}{x + y}$ **24.** $\frac{5x}{x - 5}$ **25.** $\frac{3y - 1}{3(y - 1)}$

Semester Test 1-8, page T64 **1.** 89 **2.** 3 **3.** 2.6 **4.** -29 **5.** $-\frac{31}{2}$ **6.** 10 **7.** $\frac{4}{3}$, -3 **8.** 4, -4 **9.** $\frac{1}{2}$, 4 **10.** $-\frac{2}{5}$, -5 **11.** 2, -3 **12.** $k^2 + k$ **13.** $1.18x + 0.12y$ **14.** $3x$ **15.** $-8c + 3d$ **16.** $-91c^9d^{12}$ **17.** $5x^3y^5z^6$ **18.** $-216x^{15}y^9z^{27}$ **19.** $22m^2n^2$ **20.** $\frac{-7a^2}{b^3c^5}$ **21.** $x < 13.12$ **22.** $t \le -\frac{15}{14}$ **23.** $p < \frac{12}{11}$ **24.** $y \le -2$ **25.** 2×10^{11} **26.** 1×10^{11} **27.** 2.24×10^0 **28.** 7×10^{-11} **29.** $9x^2 + 10x + 2$ **30.** $c^2 - c$ **31.** $-15m^3n + 20m^2n^2 - 5mn^3$ **32.** $8a^2 + 2a - 21$ **33.** $9y^2 + 24yz + 16z^2$ **34.** $4x^2 - 24xy + 36y^2$ **35.** $4x^2 - 64y^2$ **36.** $7x^2y(4x^3 - 3y^5)$ **37.** $\left(\frac{1}{3}b - \frac{2}{5}c\right)\left(\frac{1}{3}b + \frac{2}{5}c\right)$ **38.** $(x + 9)^2$ **39.** $(k - 5)^2$ **40.** $(r - 2s)(r + 6s)$ **41.** $(t + 2)(t + 8)$ **42.** $(m - 2p)(m - 3p)$ **43.** $(2s - 5t)(w + 3z)$ **44.** $5(x + 3n)^2$ **45.** $35 **46.** 19 km **47.** 29, 31, 33 **48.** 16 months **49.** 70% **50.** Lena = 10 years old, Erika = 6 years old

Chapter 9, page T66 **1.** III **2.** IV **3.** II **4.** I

5-8.

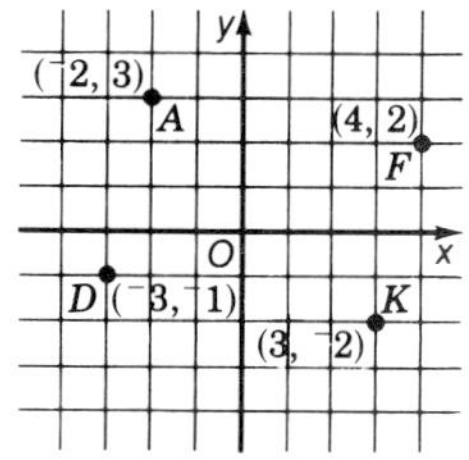

9. {(5, 7), (7, 9), (12, 15), (13, 16), (18, 27)}; {5, 7, 12, 13, 18}; {7, 9, 15, 16, 27} **10.** {(1, 4), (2, 5), (3, 7), (4, 8), (5, 9)}; {1, 2, 3, 4, 5}; {4, 5, 7, 8, 9} **11.** a., c. **12.** c., d. **13.** {(1, 5), (4, 14), (6, 20), (9, 29)} **14.** {(1, -1), (4, 2), (6, 4), (9, 7)}

15–16.

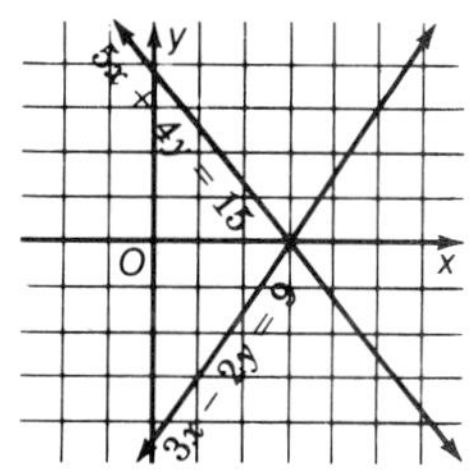

17. yes
18. no
19. 23
20. 5
21. -1

22.

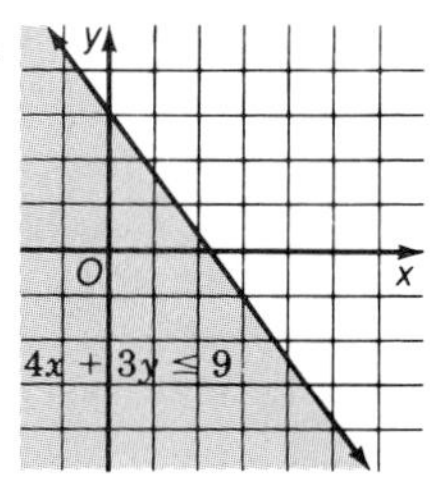

23.

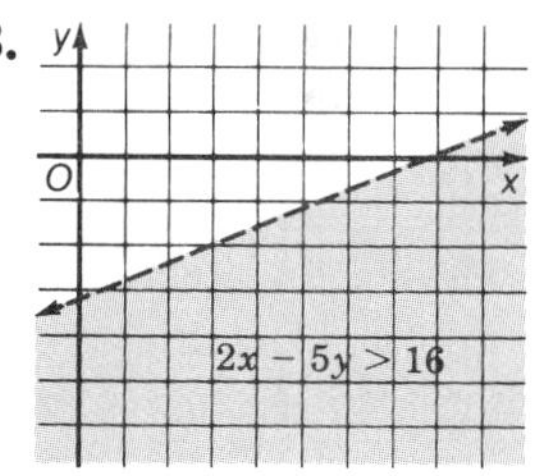

24. $y = x + 7$; 14, 18

25. $y = 2x - 1$; 21, 23

Chapter 10, page T67 **1.** $\frac{4}{5}$ **2.** $\frac{6}{7}$ **3.** 8 **4.** 18 **5.** 7, -4, **6.** 3, -2 **7.** 9, -5 **8.** $2x - y = 5$, $y = 2x - 5$ **9.** $8x + 7y = 12$, $y = -\frac{8}{7}x + \frac{12}{7}$

10. $y = \frac{4}{5}x + \frac{17}{5}$ **11.** $y = -\frac{2}{7}x + \frac{43}{7}$

12. $3x - 7y = -43$ **13.** $4x - y = 6$

14. $y = 2x - 7$ **15.** $y = -5x + \frac{2}{3}$

16–17. 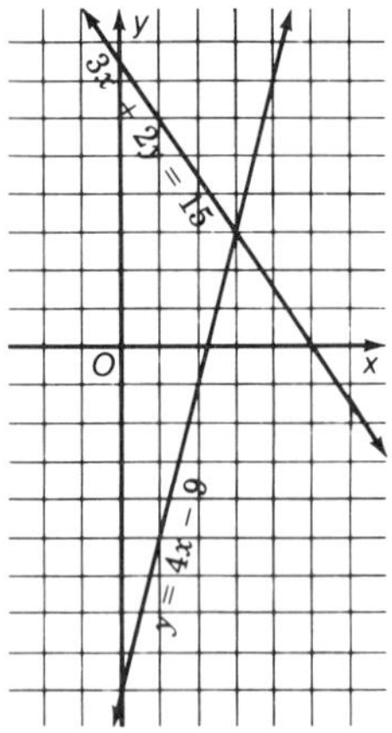

18. $y = \frac{3}{4}x - 3$

19. $y = 3x + 6$

20. $y = -\frac{4}{3}x + 4$

21. $y = 2x + 6$

22. $y = -\frac{2}{7}x + \frac{11}{7}$

23. $y = -\frac{1}{5}x + 3$

24. $\left(5\frac{1}{2}, 7\right)$ 25. $\left(6, 2\frac{1}{2}\right)$

Chapter 11, page T68

1. 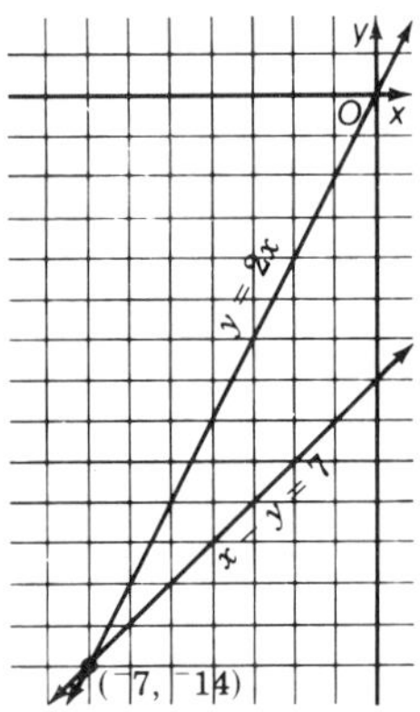

2. 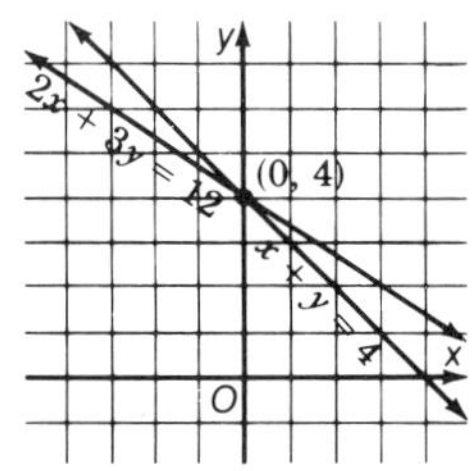

3. 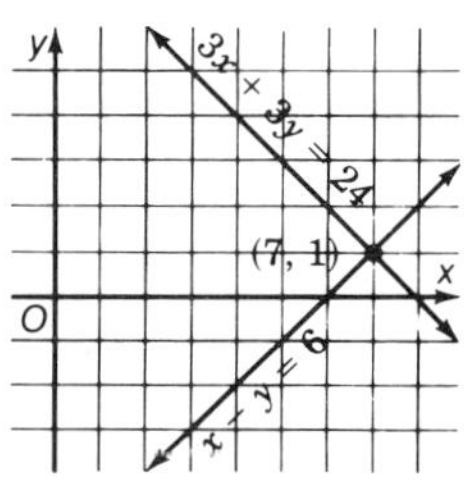

4. 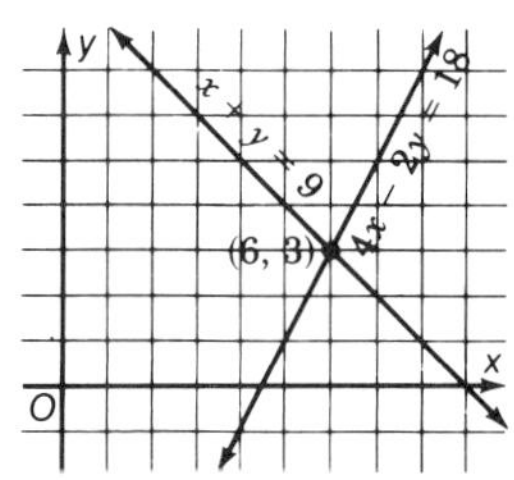

5. 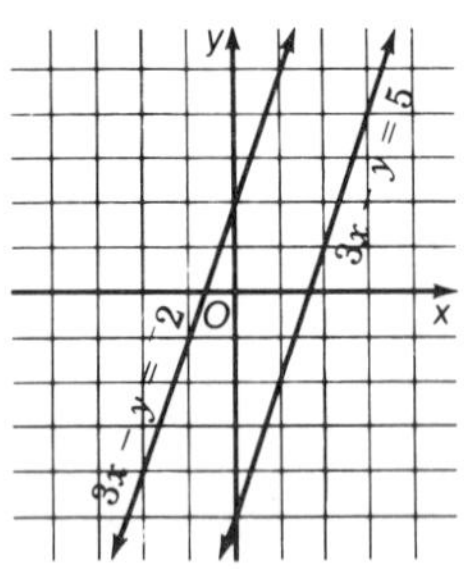

no solutions

6. 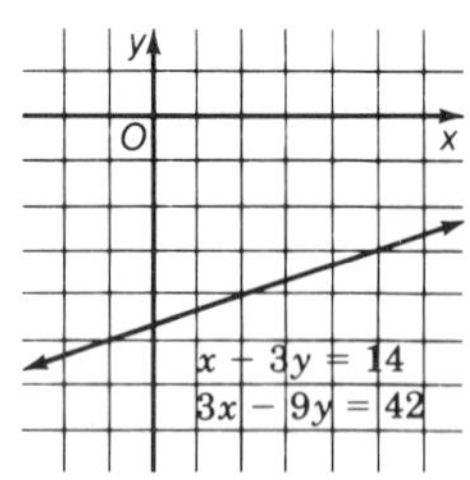

infinitely many solutions

7. $\left(-\frac{4}{5}, \frac{21}{5}\right)$ 8. $\left(-\frac{37}{13}, \frac{40}{13}\right)$ 9. $\left(\frac{18}{7}, -\frac{17}{21}\right)$
10. $\left(\frac{4}{3}, 2\right)$ 11. $(8, -5)$ 12. $\left(3, \frac{5}{4}\right)$
13. $\left(\frac{26}{5}, -\frac{4}{5}\right)$ 14. $(9, -2)$ 15. $\left(-\frac{1}{3}, 2\right)$

16. 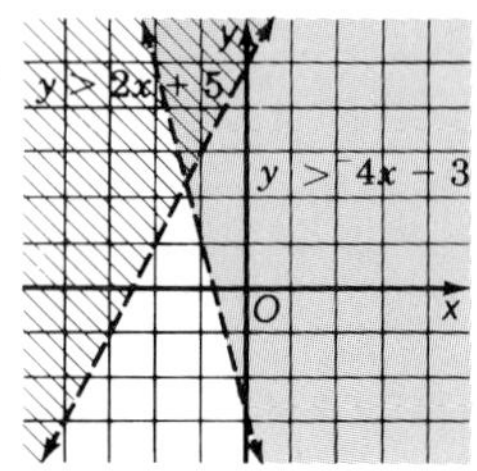

17. 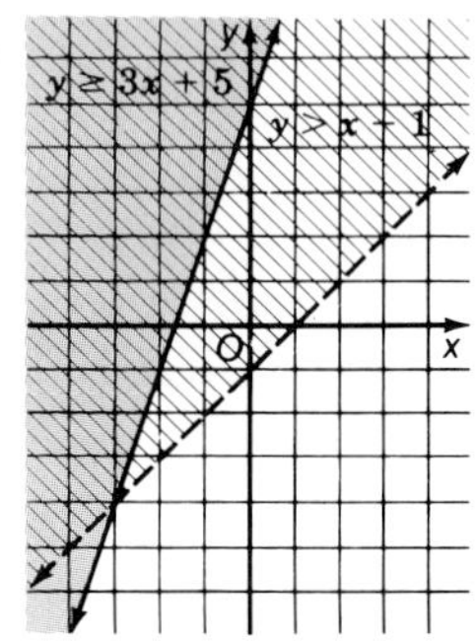

18. 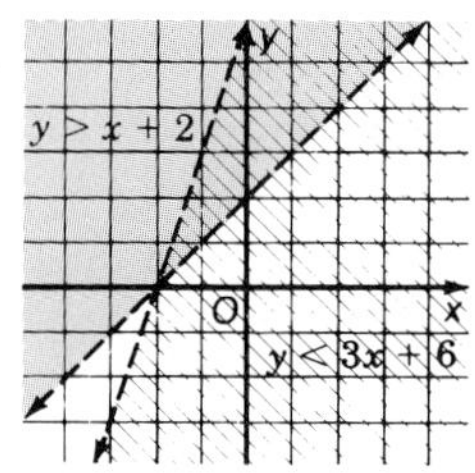

19. 150 bu, 50 bu 20. 500 gal, 500 gal
21. 30 lb, 70 lb 22. 36 23. 49
24. 337.5 mph, 112.5 mph 25. 40 mph, 55 mph

Chapter 12, page T69 1. -9 2. 0.3 3. $\pm\frac{6}{13}$
4. 18 5. 6.083 6. 8.888 7. 28.28 8. 1.027
9. 3.87 10. 12.73 11. $4\sqrt{7}$ 12. $5x^3y^2\sqrt{2xy}$
13. 12 14. 5 15. 5 16. $\frac{\sqrt{15} - 2}{11}$ 17. $23\sqrt{7}$
18. $-8\sqrt{2}$ 19. 169 20. $\frac{109}{2}$ 21. $\frac{16}{7}$ 22. no
23. $6\sqrt{2}$ ft 24. $2\sqrt{17}$ 25. 0, 4

Chapter 13, page T70 1. $x = -1$; $(-1, -10)$
2. $x = \frac{3}{2}$; $\left(\frac{3}{2}, -\frac{37}{4}\right)$ 3. $x = -\frac{1}{6}$; $\left(-\frac{1}{6}, \frac{55}{6}\right)$
4. $x = 1$; $(1, -2)$ 5. $-3, 4$ 6. $-2, -9$ 7. 6
8. no real solution 9. $12 \pm 5\sqrt{6}$ 10. $-15 \pm \sqrt{233}$ 11. $-5 \pm \sqrt{21}$ 12. $\frac{5 \pm \sqrt{37}}{6}$ 13. 7,

-5 **14.** $\frac{-9 \pm \sqrt{57}}{4}$ **15.** 5, -3 **16.** 2 **17.** 1 **18.** 0 **19.** 2 **20.** 9 cm, 2 cm **21.** 5 cm, 4 cm **22.** -2, -63 **23.** $\frac{5}{4}$, $-\frac{7}{4}$ **24.** $x^2 - 6x - 72 = 0$ **25.** $x^2 - 5 = 0$

Chapter 14, page T71 **1.** $\frac{1}{x^5}$ **2.** $\frac{12d^6}{c^8}$ **3.** $\frac{3x^{11}}{2y^2}$ **4.** $\frac{m + 6}{m - 9}$ **5.** $\frac{3b + 2}{b - 5}$ **6.** $\frac{x + 5}{x - 8}$ **7.** $\frac{e}{d}$ **8.** $\frac{x + 2}{x - 5}$ **9.** $\frac{x^2 + 9x + 20}{2x^2 + 7x + 6}$ **10.** $\frac{y^2}{4}$ **11.** $\frac{n^2 + 8n + 15}{n^2 + 8n + 12}$ **12.** 1 **13.** $6x + 7$ **14.** $m + 5 + \frac{3}{2m + 3}$ **15.** $\frac{1}{x - y}$ **16.** $\frac{6x + 6}{3x + 5}$ **17.** $\frac{n + 11}{n - 7}$ **18.** $\frac{41}{63}$ **19.** $\frac{5n + 3}{6n + 12}$ **20.** $\frac{-x + 6y}{x^2 - 2xy + y^2}$ **21.** $\frac{32}{49}$ **22.** $\frac{45}{88}$ **23.** $\frac{12x + 4y}{2x + y}$ **24.** 1 **25.** $\frac{x - 1}{x - 2}$

Chapter 15, page T72 **1.** 21 **2.** $-\frac{23}{7}$ **3.** 4, -3 **4.** 3, -2 **5.** $\frac{3}{5}$ **6.** 6 **7.** 4 **8.** 2 **9.** $m = \frac{2KE}{v^2}$ **10.** $V_1 = \frac{P_2V_2}{P_1}$ **11.** $T_2 = \frac{T_1V_2}{V_1}$ **12.** $P_1 = \frac{P_2T_1V_2}{V_1T_2}$ **13.** $\frac{15}{8}$ hr **14.** $\frac{35}{12}$ hr **15.** 49 **16.** 5 **17.** \$10.89 **18.** 5 hr **19.** $8\frac{2}{5}$ ft **20.** 10 **21.** no **22.** $3\frac{15}{16}$ ohms **23.** 32 ohms **24.** $3\frac{1}{3}$ ohms **25.** $\frac{1}{3}$

Chapter 16, page T73 **1.** 2°, 92° **2.** none, 59° **3.** $(93 - x)°$, $(183 - x)°$ **4.** $(43 - y)°$, $(133 - y)°$ **5.** $(90 - 4x)°$, $(180 - 4x)°$ **6.** 108° **7.** 26° **8.** 6° **9.** 10 cm, 8.66 cm **10.** 10 ft, 17.32 ft **11.** 20.78 cm, 10.39 cm **12.** $e = \frac{3}{2}$, $f = \frac{5}{2}$ **13.** $d = 8$, $e = 6$ **14.** $a = \frac{5}{2}$, $f = \frac{42}{5}$ **15.** $\frac{8}{17}$ **16.** $\frac{15}{8}$ **17.** $\frac{8}{17}$ **18.** $\frac{15}{17}$ **19.** $\frac{15}{17}$ **20.** $\frac{8}{15}$ **21.** angle $B = 70°$, $c = 15$, $b \approx 14.1$ **22.** angle $A \approx 34°$, angle $B \approx 56°$, $AB \approx 10.8$ **23.** angle $A = 68°$, $b \approx 5.24$, $a \approx 12.98$ **24.** 20° **25.** 29 in., 7.8 in.

Semester Test 9-16, page T74 **1.** {1, 2, 3, 4}, {9, 12, 15, 22}, yes **2.** {5, 4, 3, 2}, {-4, 4, 5, 6}, no **3.** {(1, -4), (2, -1), (3, 2)} **4.** $\left\{(1, 4), \left(2, \frac{3}{2}\right), (3, -1)\right\}$ **5.** $-\frac{3}{5}$ **6.** $\frac{2}{11}$ **7.** $y = 5x - 8$ **8.** $y = \frac{22}{7}x - \frac{93}{7}$ **9.** $y = 6x - 2$ **10.** $\left(4, -\frac{3}{4}\right)$ **11.** (2, 1) **12.** (2, -2) **13.** $3a^2b^2\sqrt{3a}$ **14.** 6 **15.** 4 **16.** $\frac{36 - 6\sqrt{5}}{31}$ **17.** $\frac{121}{5}$ **18.** 112 **19.** 8.775 **20.** 9.849 **21.** 5 **22.** 9.055 **23.** $x = \frac{2}{5}$; $\left(\frac{2}{5}, \frac{36}{5}\right)$ **24.** $x = \frac{3}{2}$; $\left(\frac{3}{2}, \frac{51}{2}\right)$ **25.** -6, 1 **26.** -15, 9 **27.** no real solution **28.** $\frac{8c^2}{7b^2}$ **29.** $\frac{x + 3}{x - 7}$ **30.** $\frac{x^2 + 2x - 63}{x^2 - 10x + 24}$ **31.** $\frac{m + 2}{m + 5}$ **32.** $\frac{13m + 8}{4m - 12}$ **33.** 3 **34.** 12 **35.** -8 **36.** $(60 - x)°$, $(150 - x)°$ **37.** 45°, 135° **38.** 3 cm, 5.196 cm **39.** 12 in., 10.392 in. **40.** $e = 7\frac{1}{2}$, $f = 4\frac{1}{2}$ **41.** $a = 10$, $f = 3\frac{1}{2}$ **42.** $c = 20$, angle $A \approx 37°$, angle $B \approx 53°$ **43.** $b \approx 35.34$, $c \approx 38.39$, angle $B = 67°$

44.

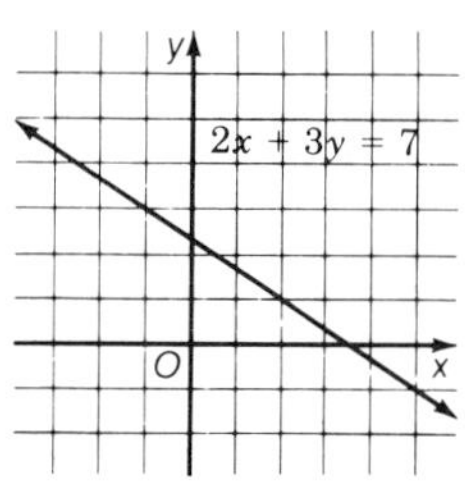

45.

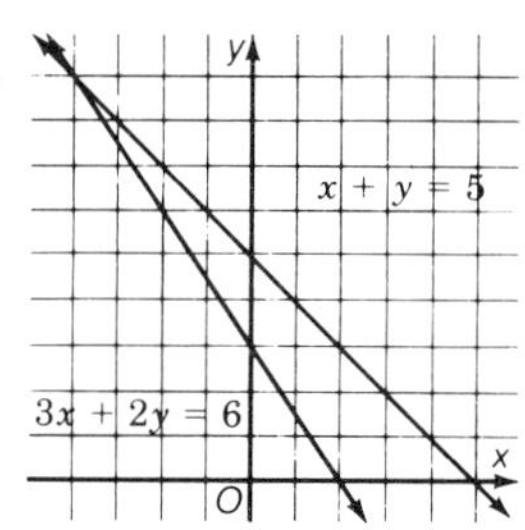

46. 24 **47.** 8 cm, 5 cm **48.** $1\frac{1}{5}$ hours **49.** $y = \frac{15}{7}$ **50.** 14°

Teacher's Answer Key

CHAPTER 4 INEQUALITIES

Page 131 Answers to Written

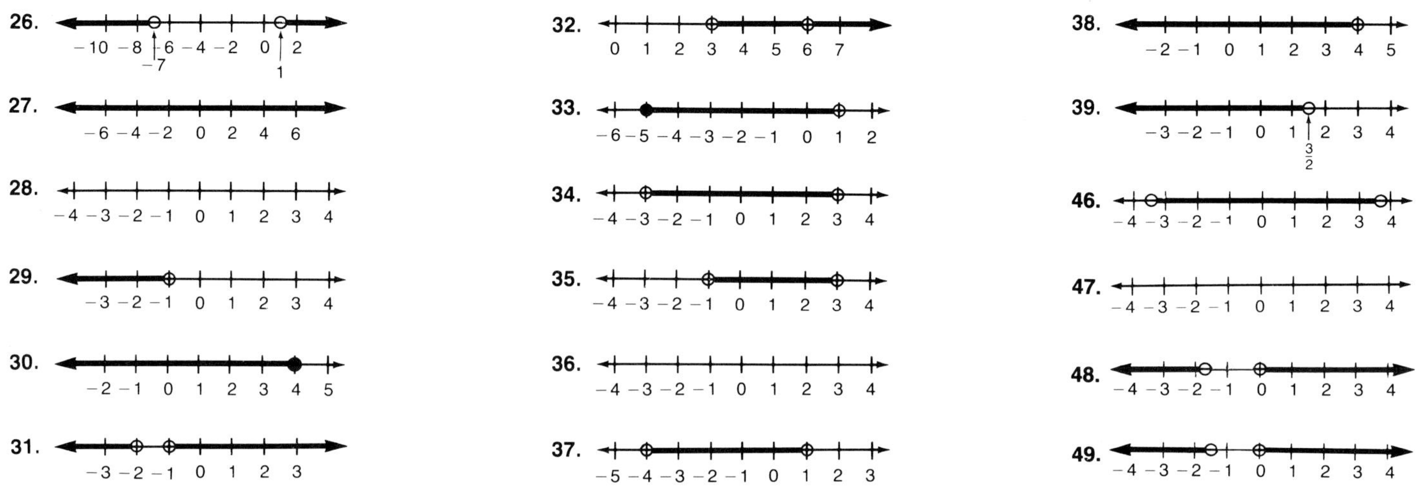

Page 134 Answers to Written

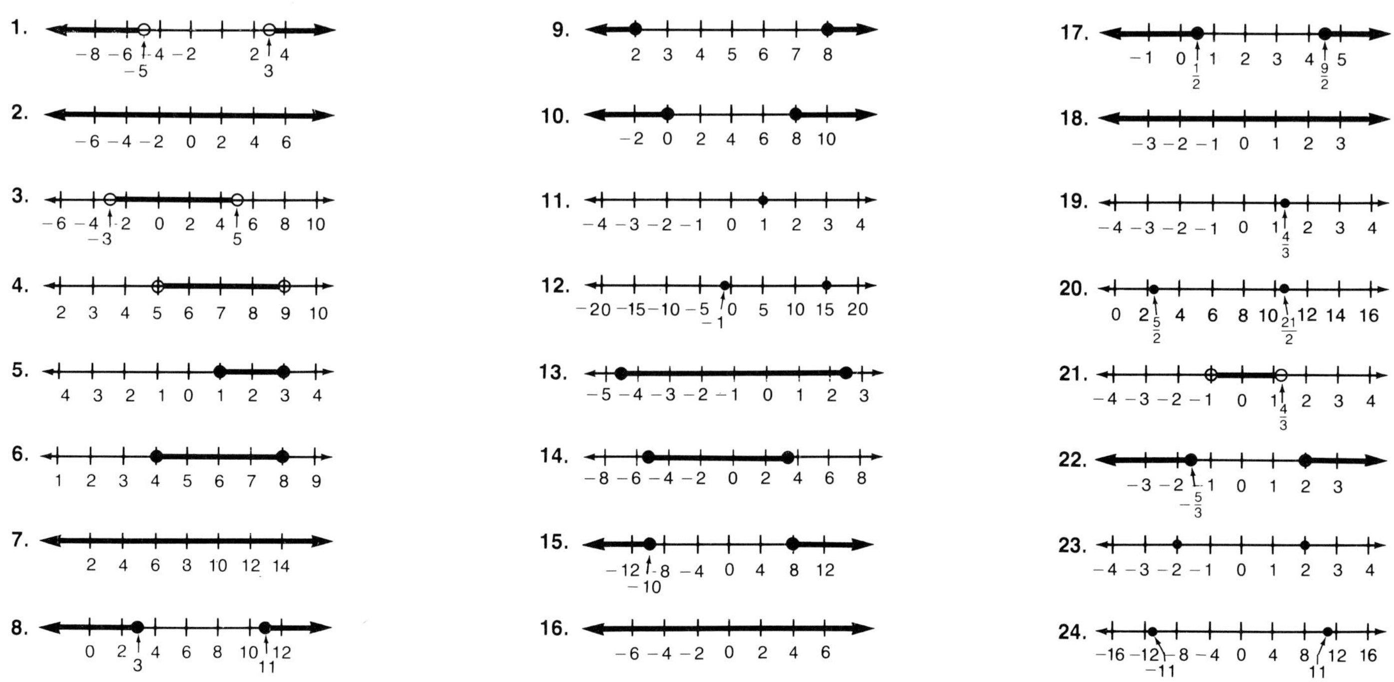

Page 285 Answers to Written

13. 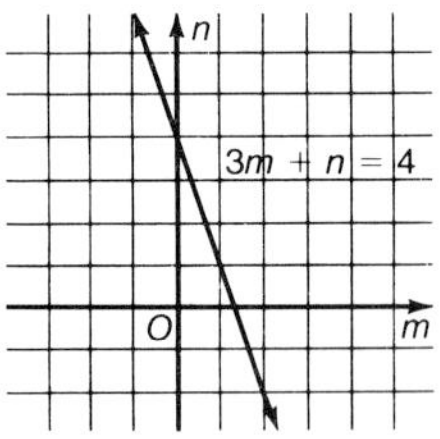

14. 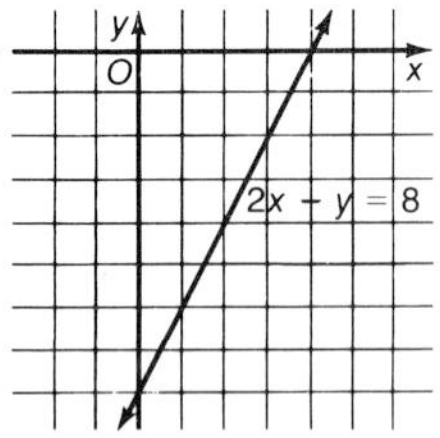

15. 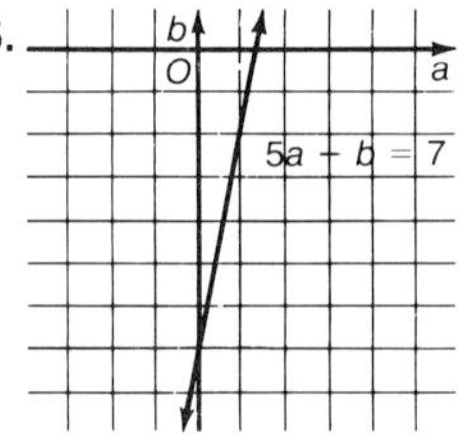

16. 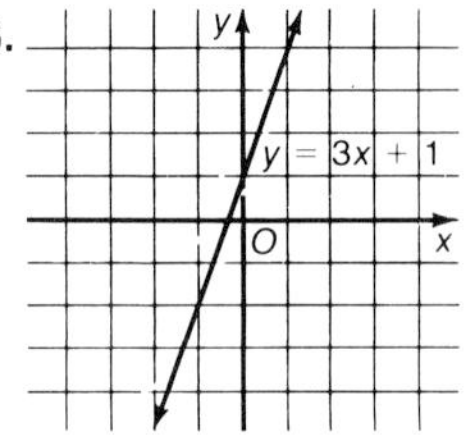

17. 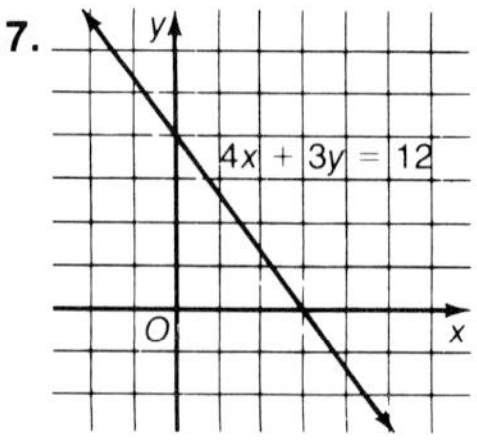

18. 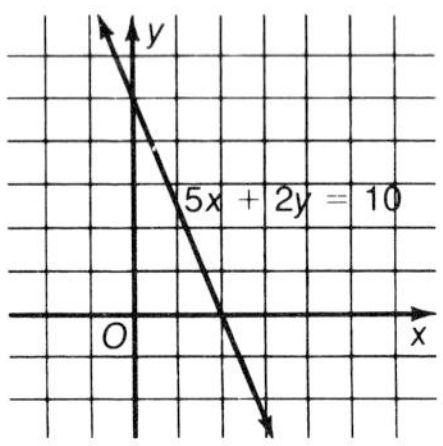

19. 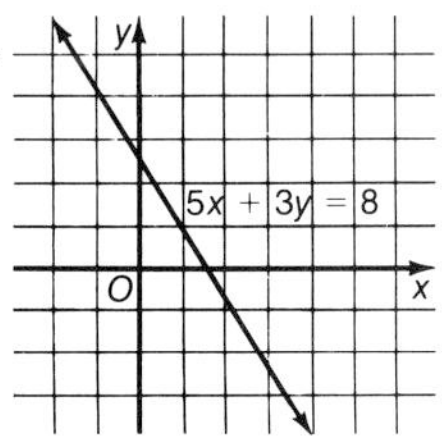

20. 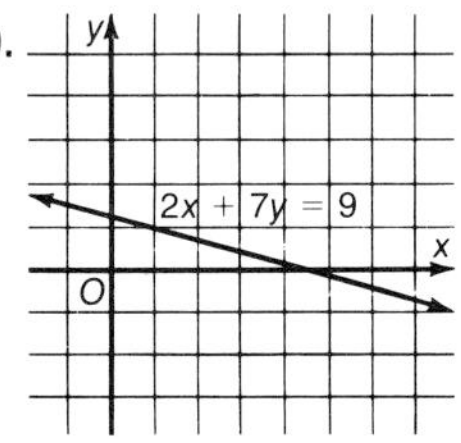

21. 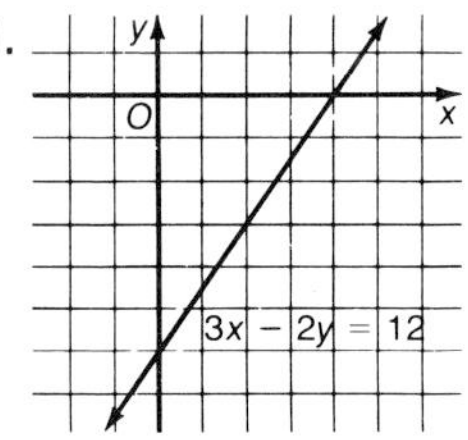

22. 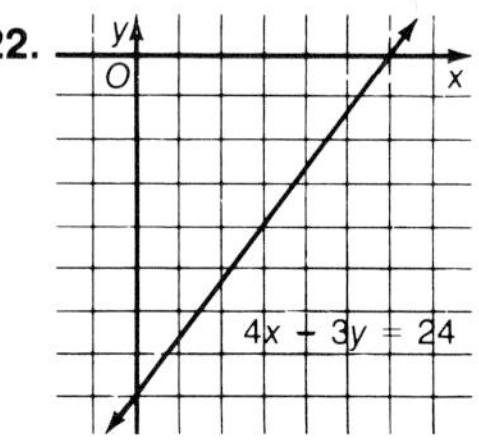

23. 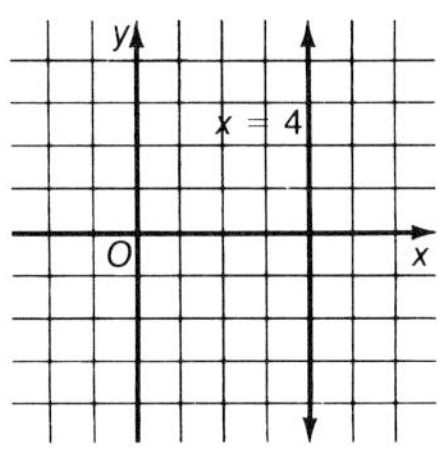

24. 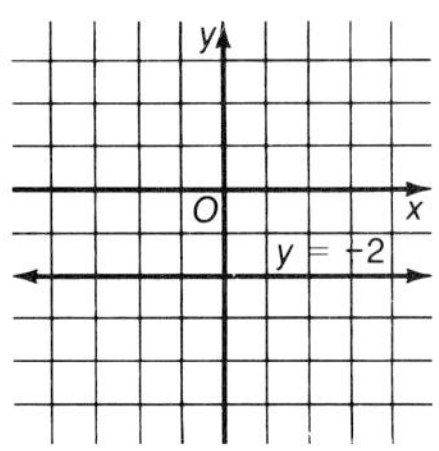

25. 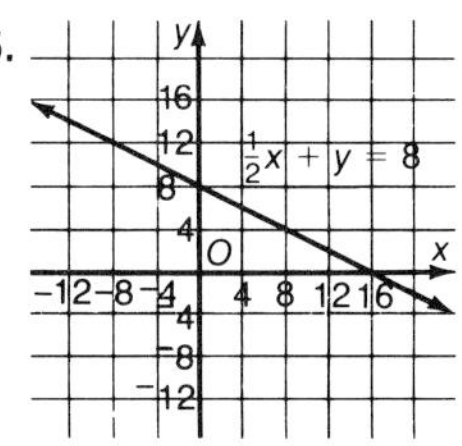

26. 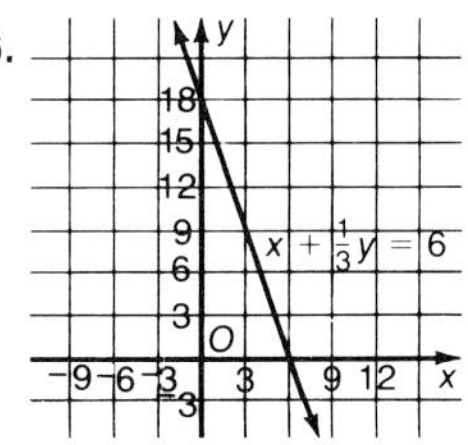

27. 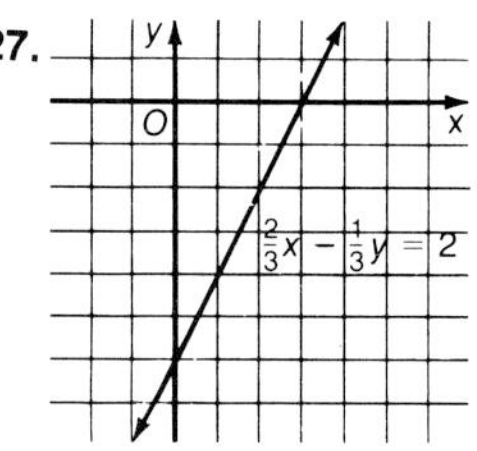

28. 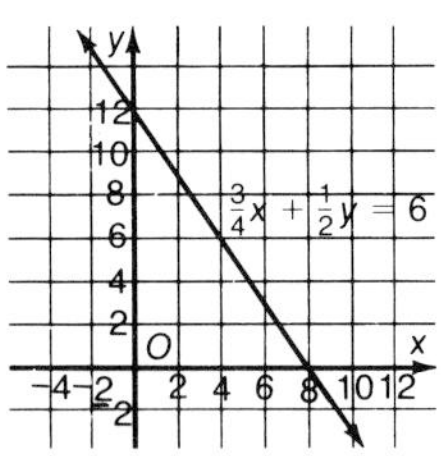

29. 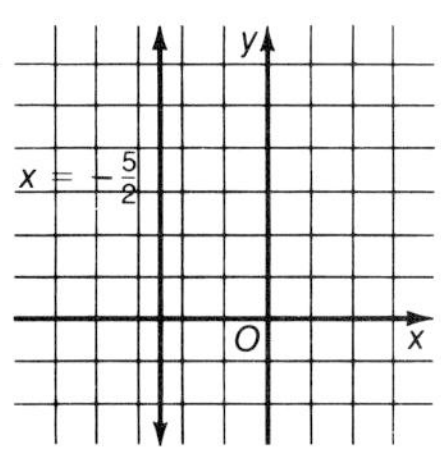

30. 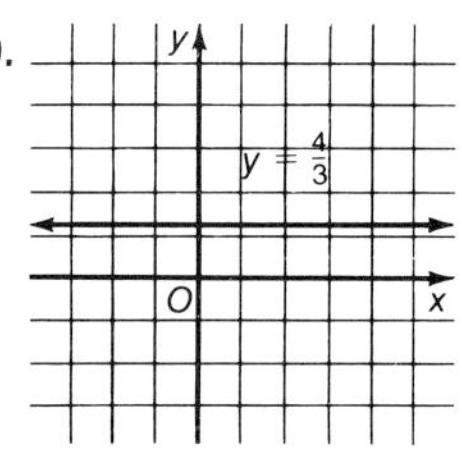

31. 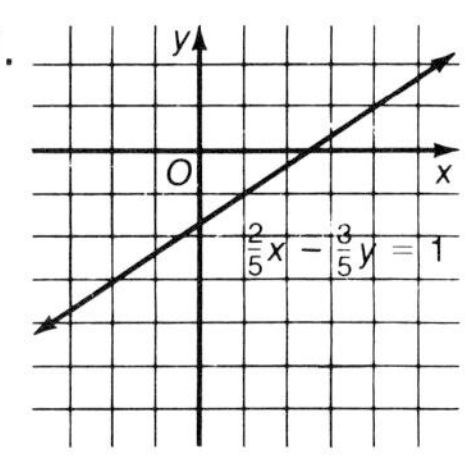

32. 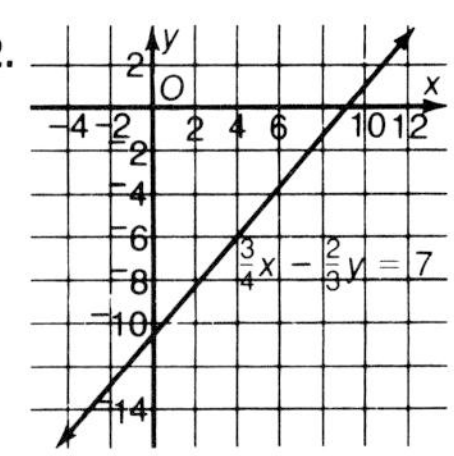

33. 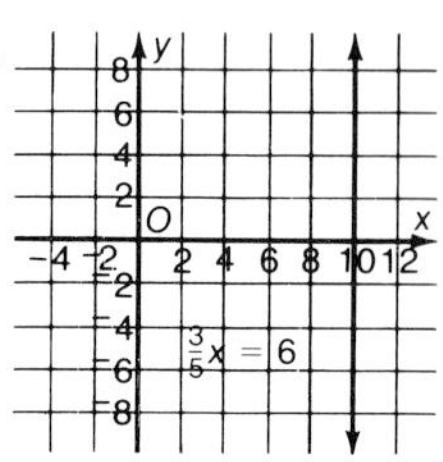

34. 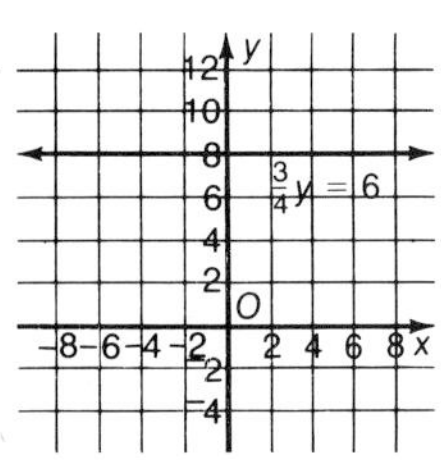

35. 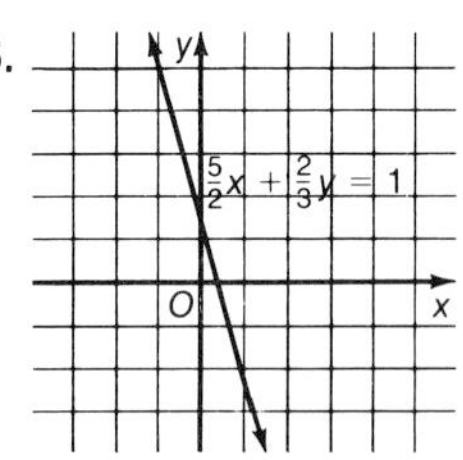

36. 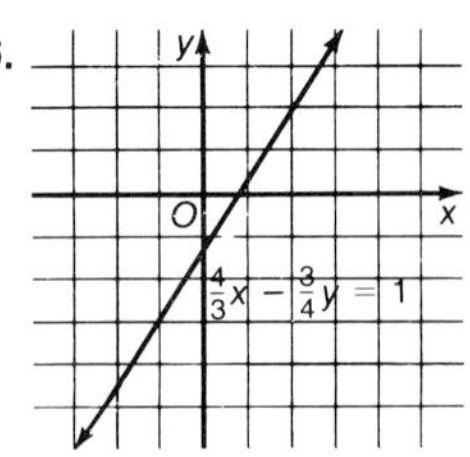

Page 285 Answers to Challenge

37.
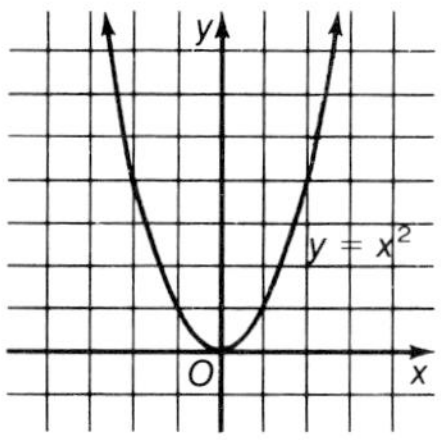

38.
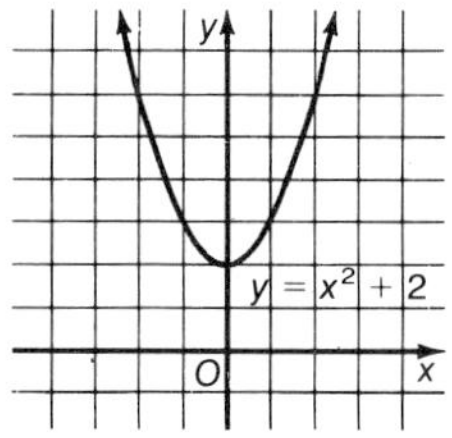

39.
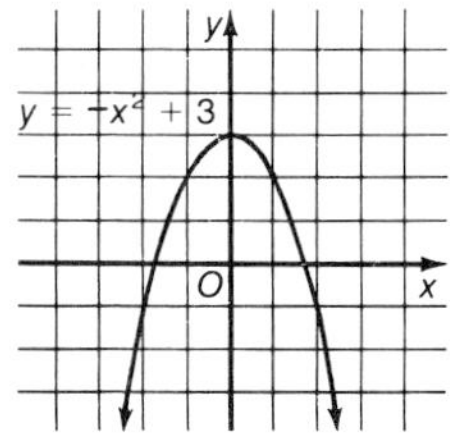

40.
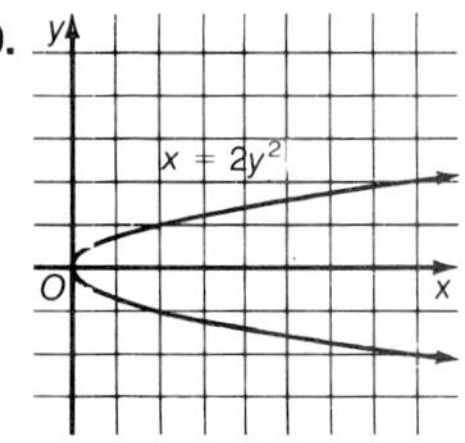

41.
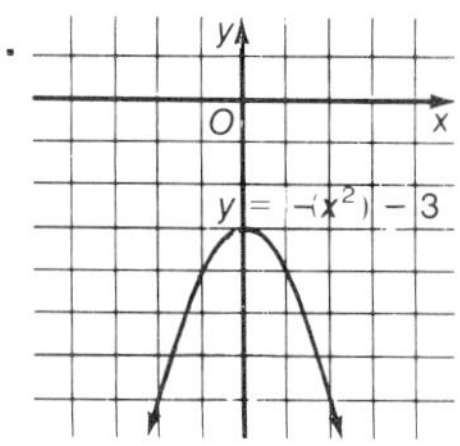

42.
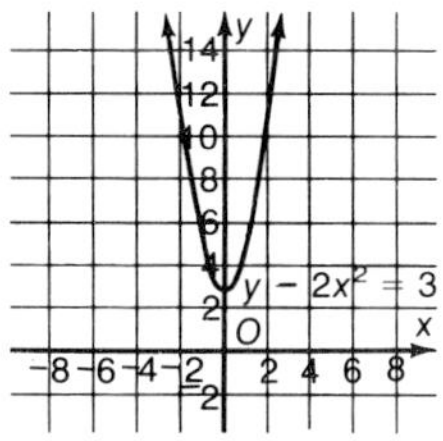

43.
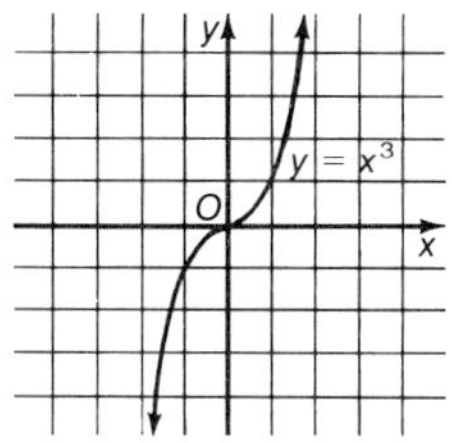

44.
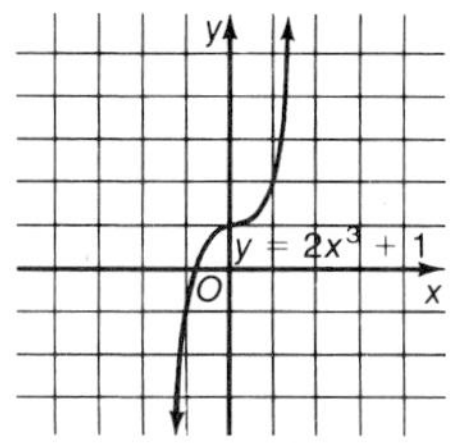

45.
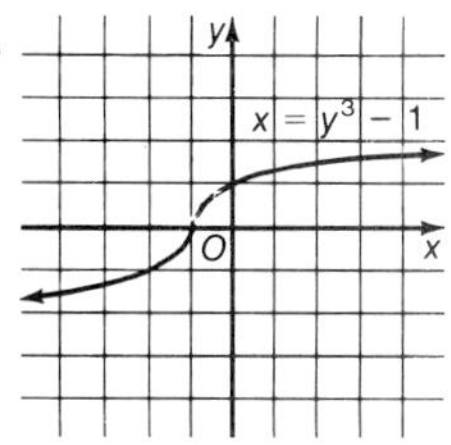

Page 293 Answers to Written

1.
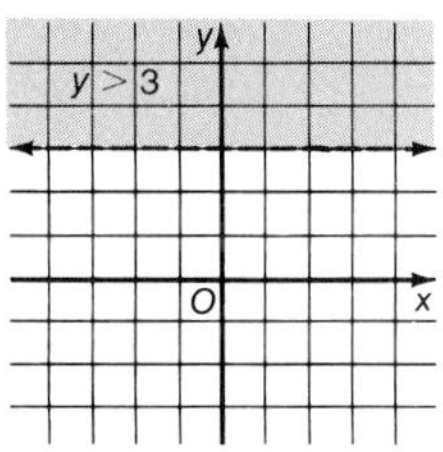

2.
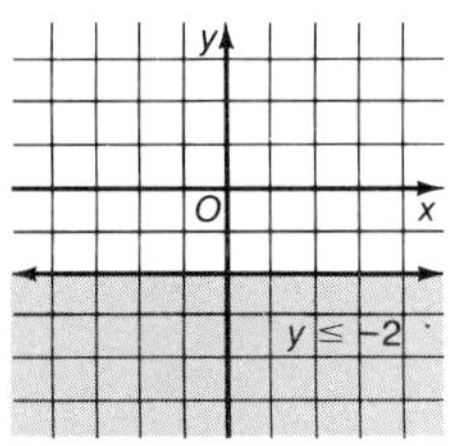

3.
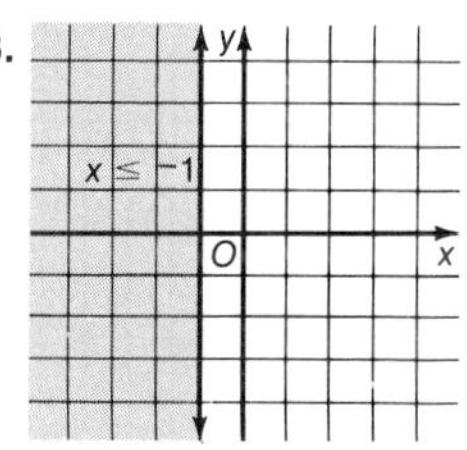

4.
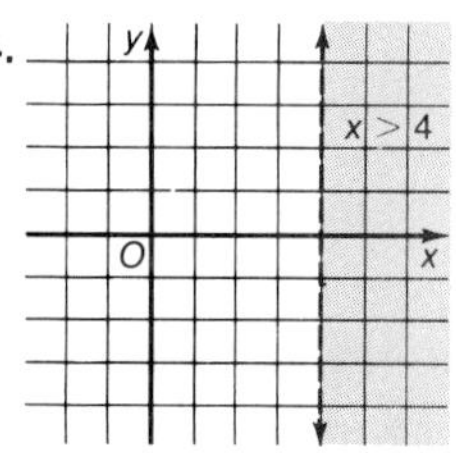

5.
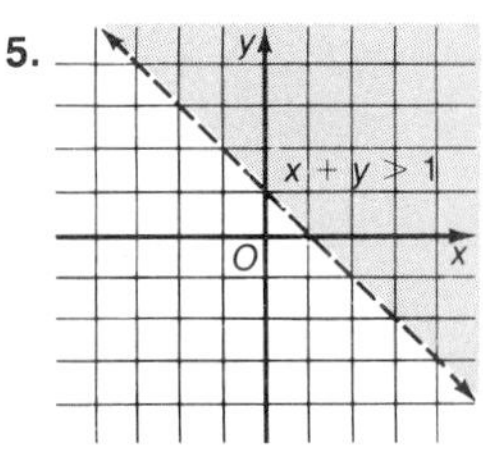

6.
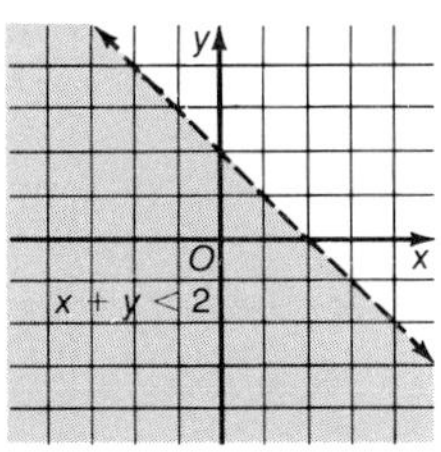

7.
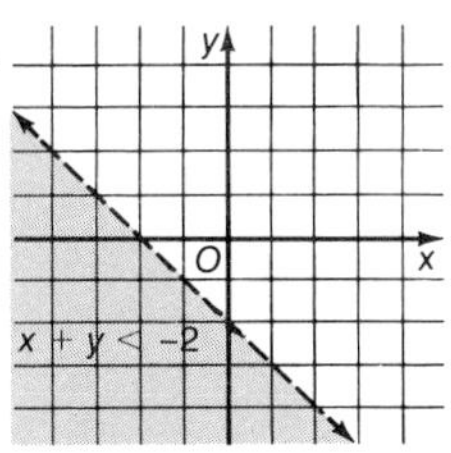

8.
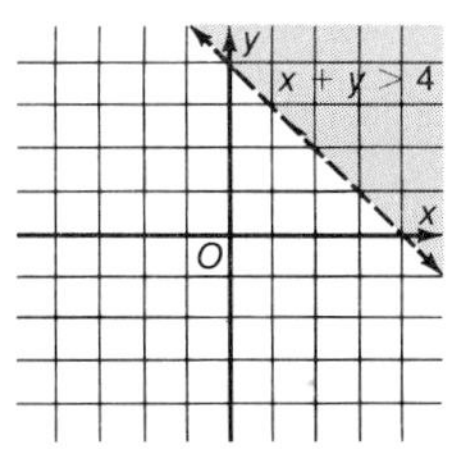

9.
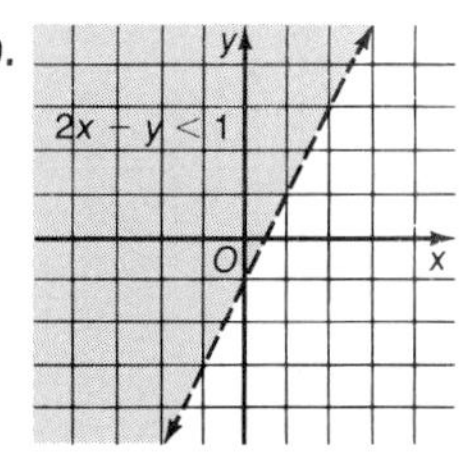

10.
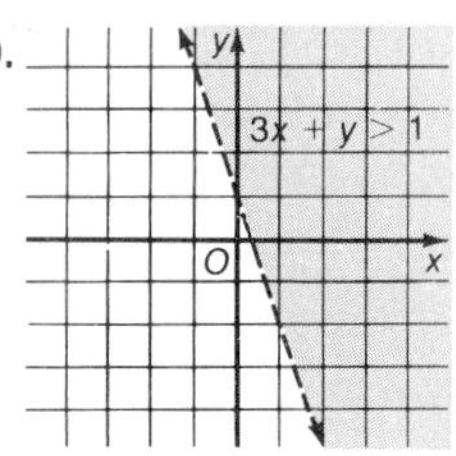

11.
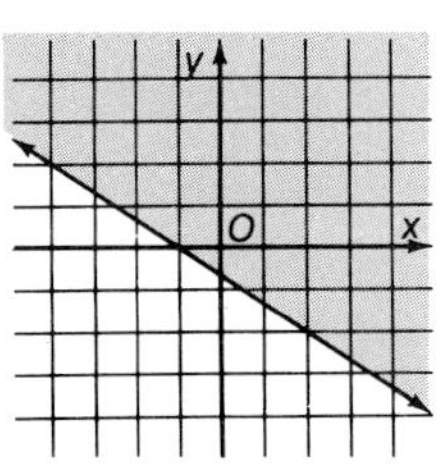

12.
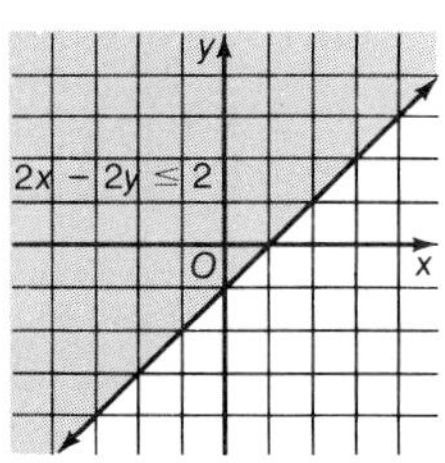

13.
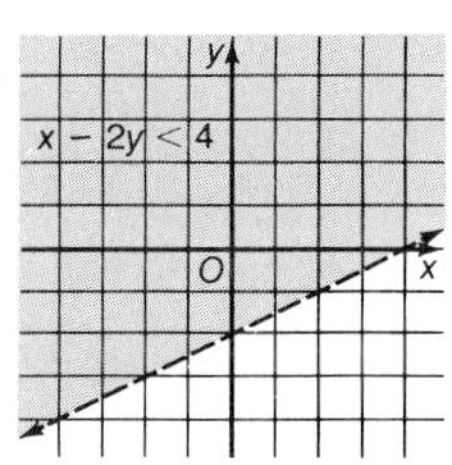

14.
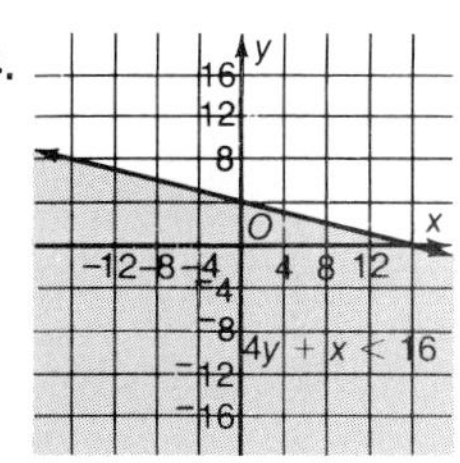

15.
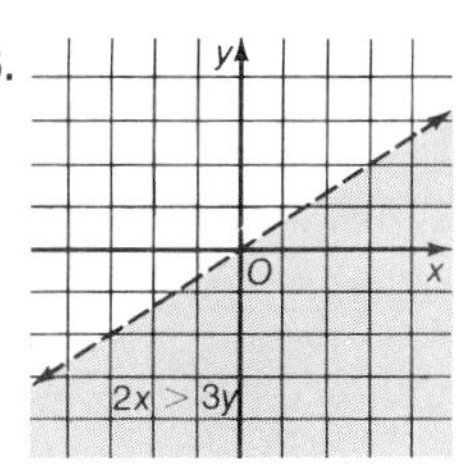

16.
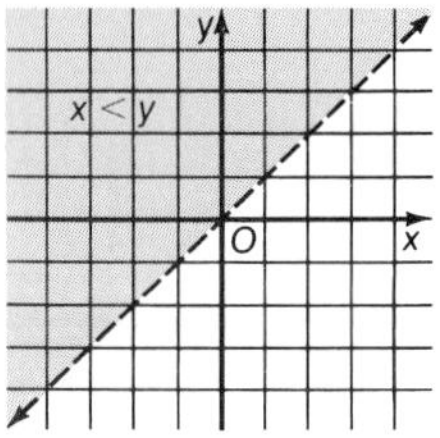

17.
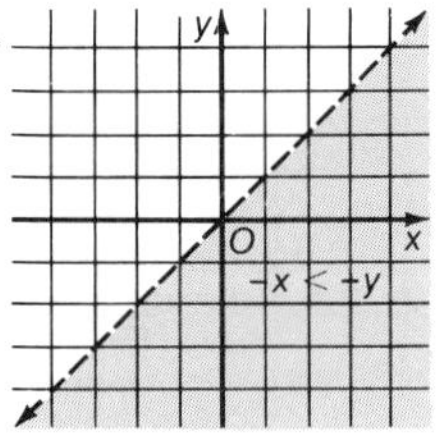

18.
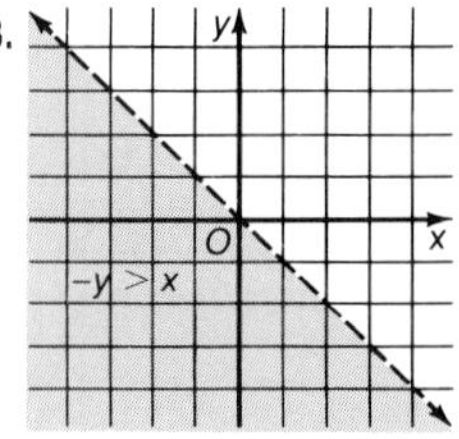

19.
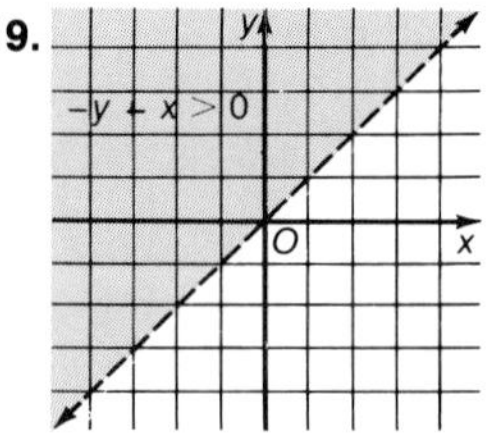

20.
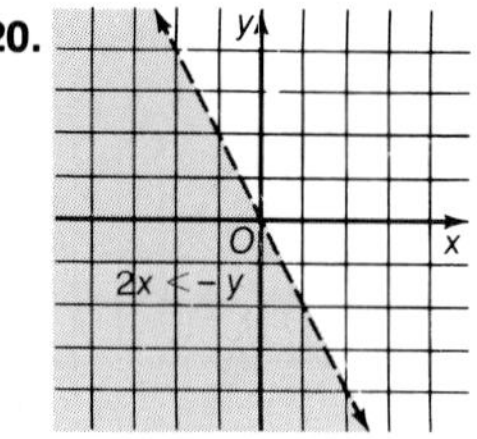

21.
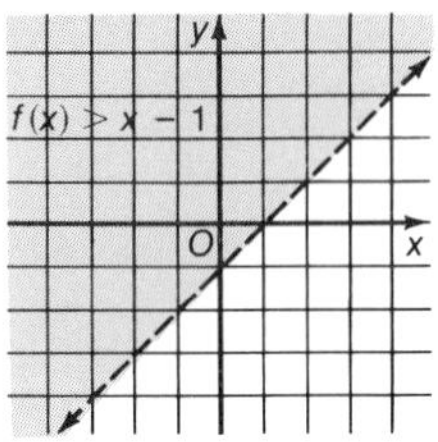

22.
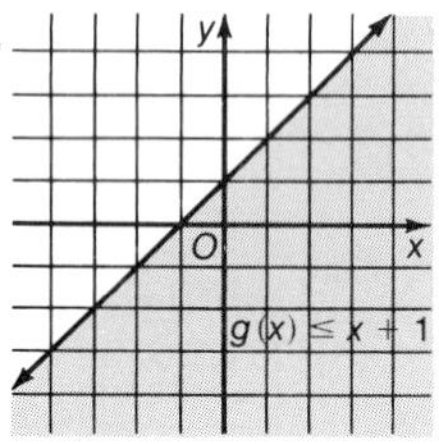

23.
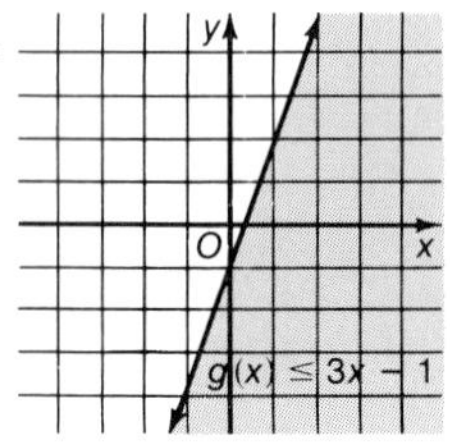

24.
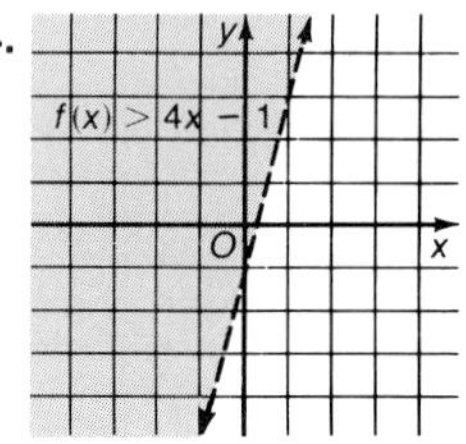

Page 293 Answers to Challenge

25.
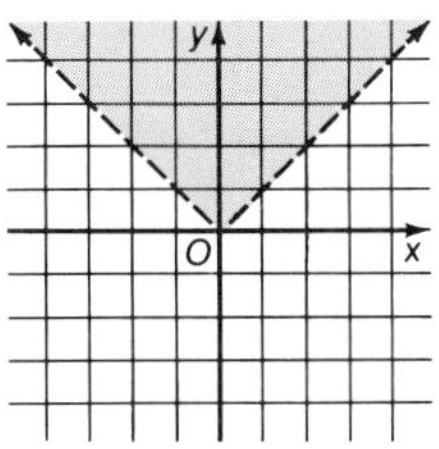

26.
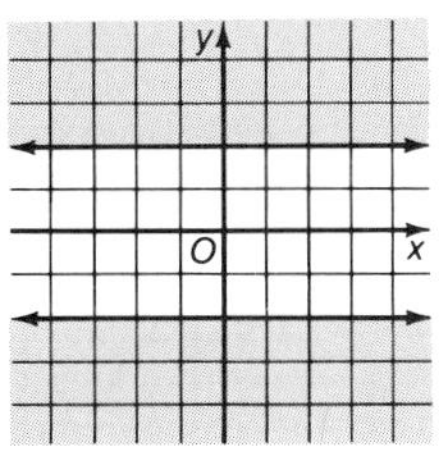

27.
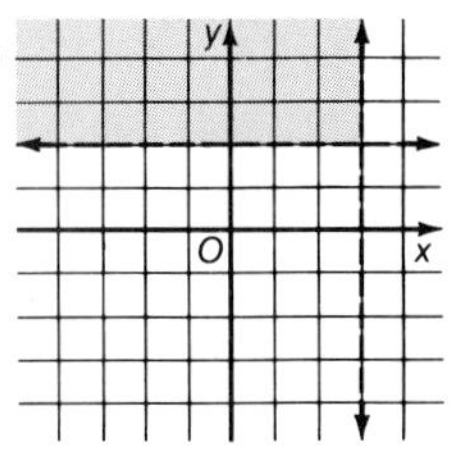

28.
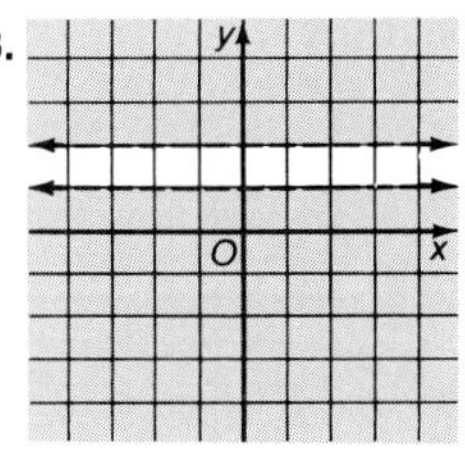

Page 301 Answers to Chapter Test

13.
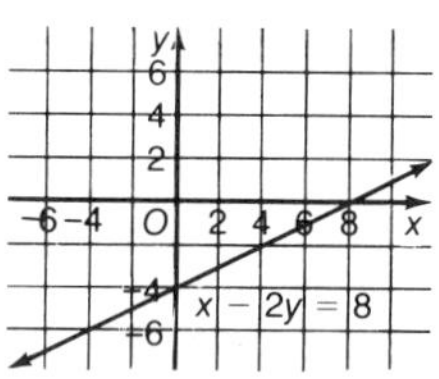

14.
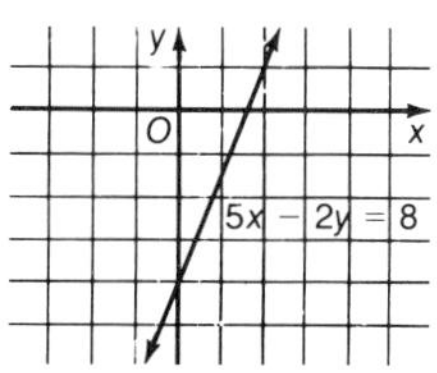

15.
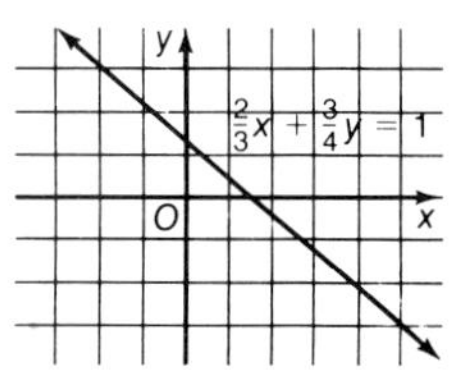

23.
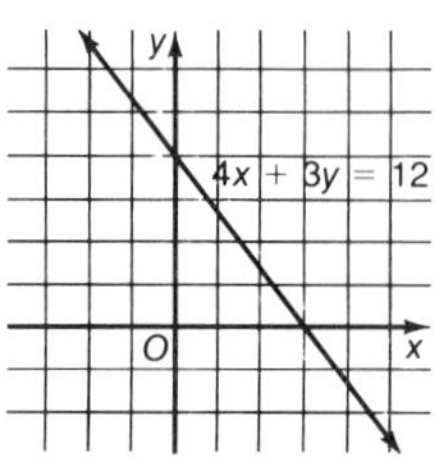

24.
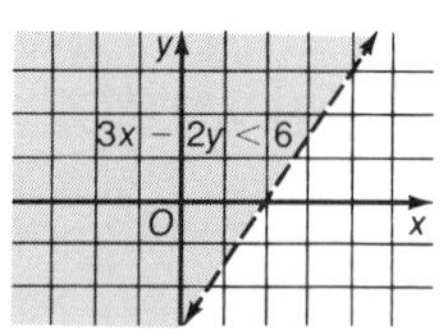

25.
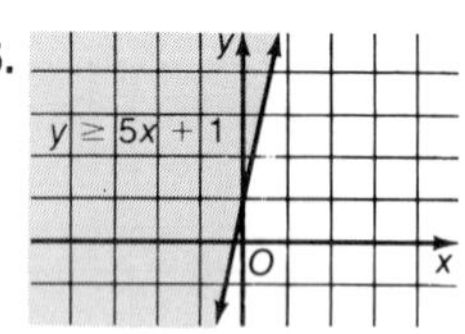

26.
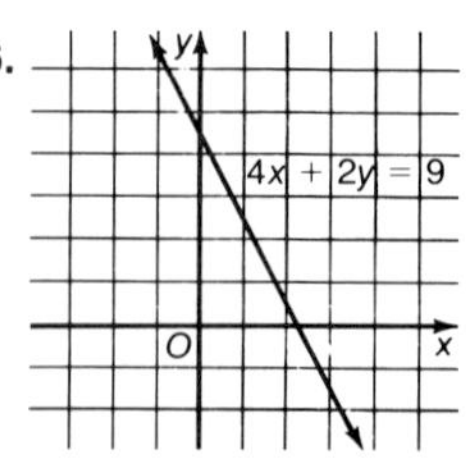

Page 303 Answers

2.

Number of Students at Broadview High School

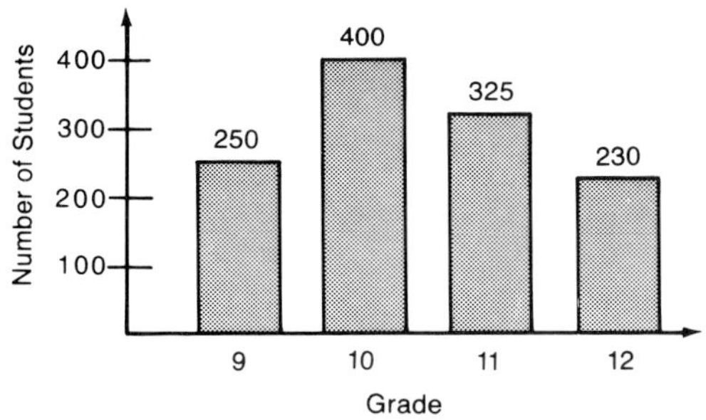

Number of Students at Broadview High School

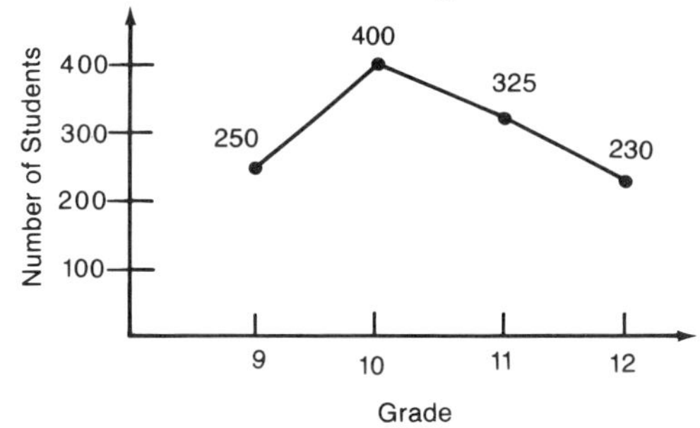

3.

Percent of 18-Year-Olds with High School Diplomas

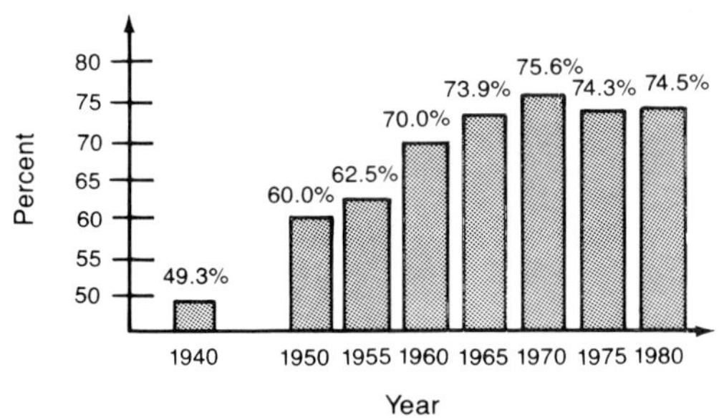

Percent of 18-Year-Olds with High School Diplomas

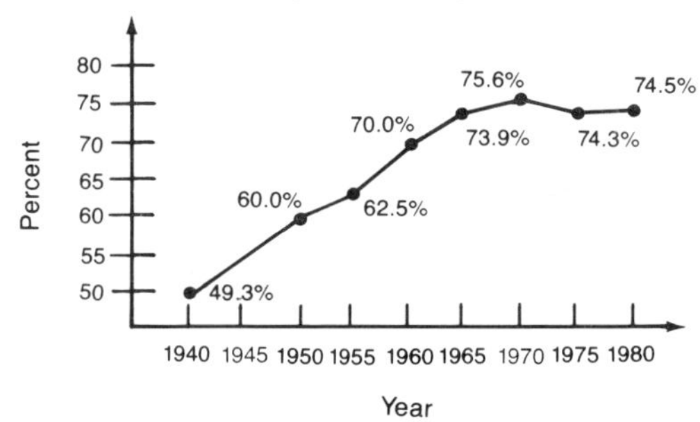

4.

Postal Rates for First Class Mail

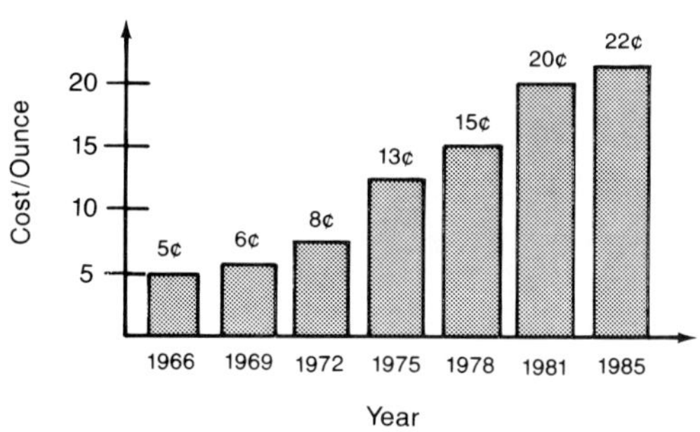

Postal Rates for First Class Mail

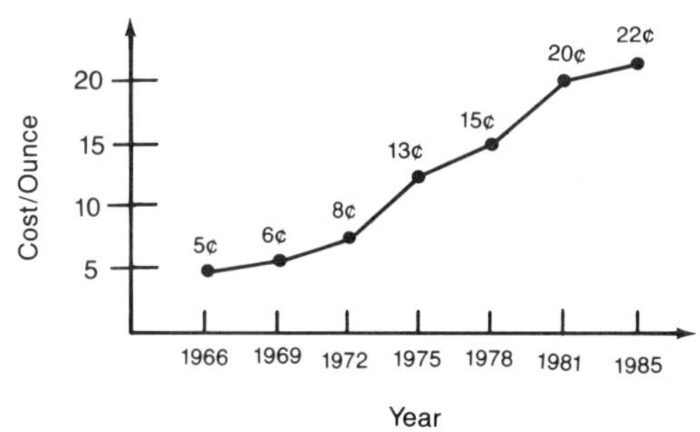

Page 315 Answers to Written

34. $m = \frac{3}{2}, b = \frac{7}{2};$
$y = \frac{3}{2}x + \frac{7}{2}$

35. $m = -\frac{5}{2}, b = 4;$
$y = -\frac{5}{2}x + 4$

36. $m = \frac{2}{3}, b = -\frac{7}{3};$
$y = \frac{2}{3}x - \frac{7}{3}$

37. $m = \frac{1}{2}, b = \frac{3}{2};$
$y = \frac{1}{2}x + \frac{3}{2}$

38. $m = \frac{11}{2}, b = -16;$
$y = \frac{11}{2}x - 16$

39. $m = \frac{3}{5}, b = -\frac{18}{5};$
$y = \frac{3}{5}x - \frac{18}{5}$

CHAPTER 10 LINES AND SLOPES

Page 318 Answers to Written

1.
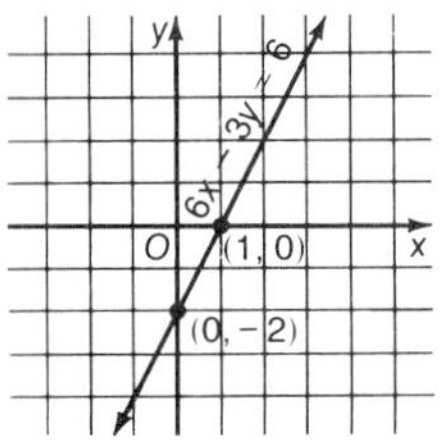

2.
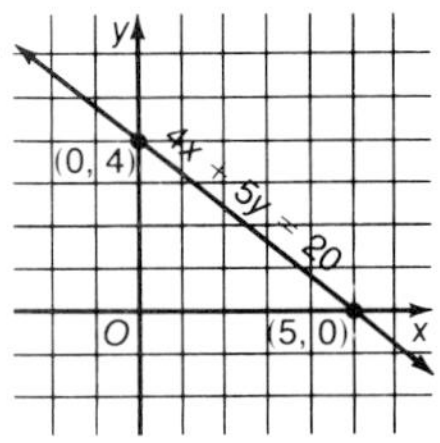

3.
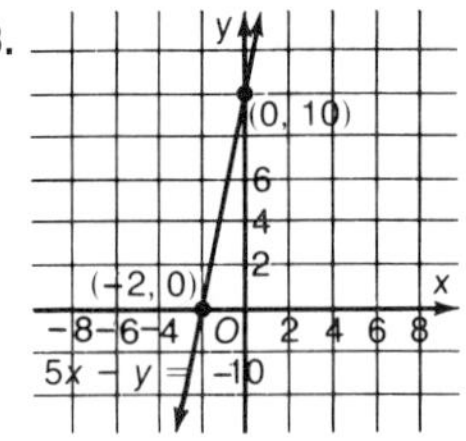

4.
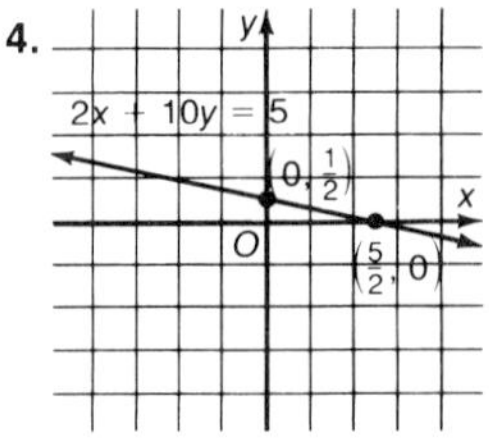

5.
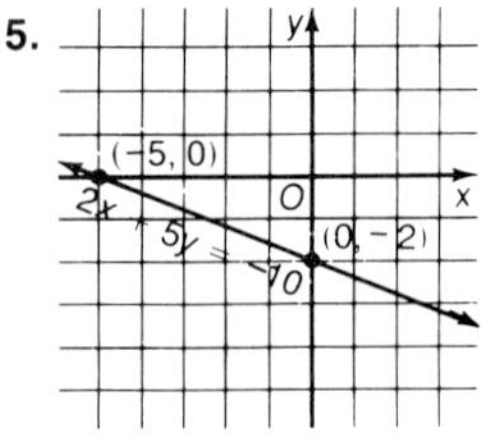

6.
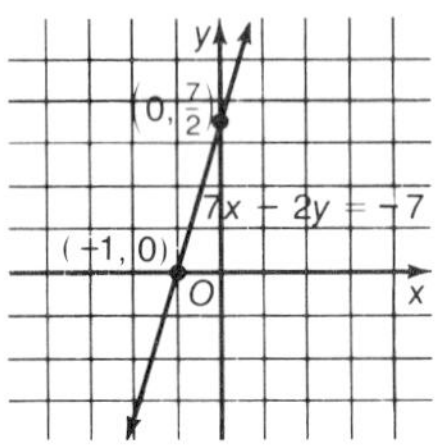

7.
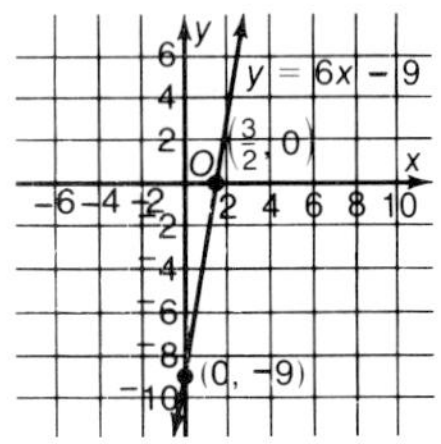

8.
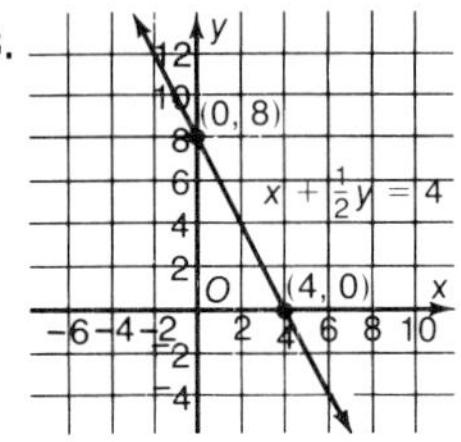

9.
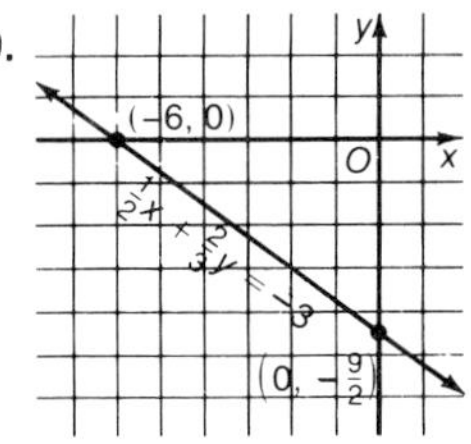

10.
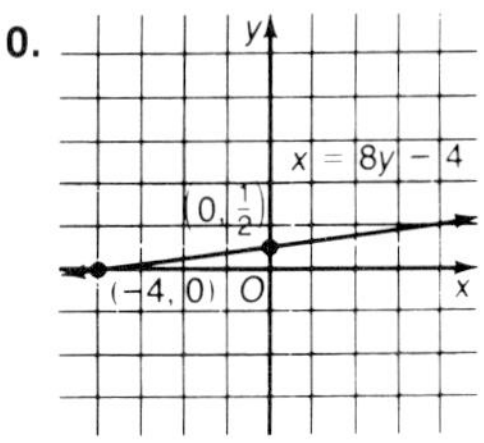

11.
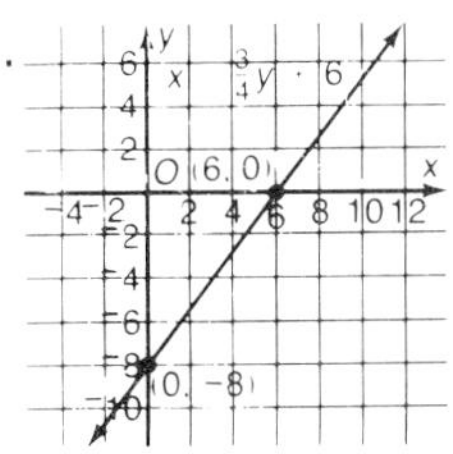

12.
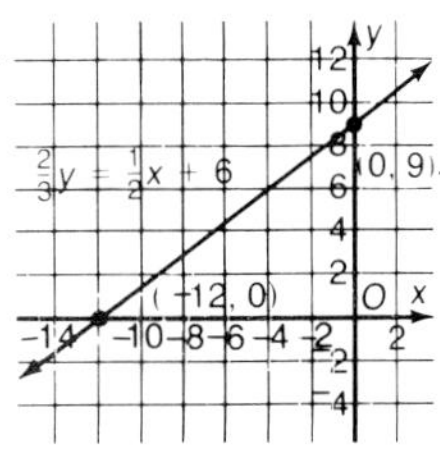

13.
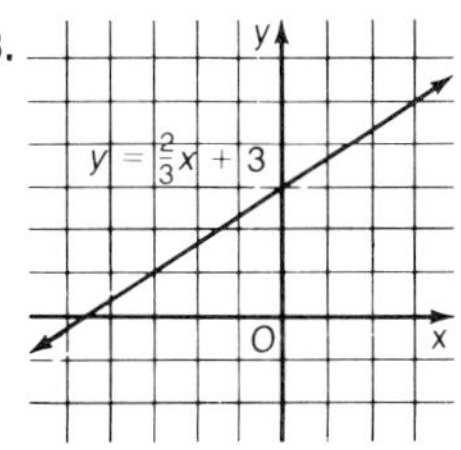

14.
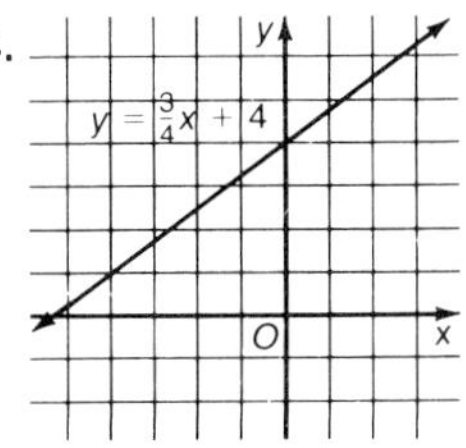

15.
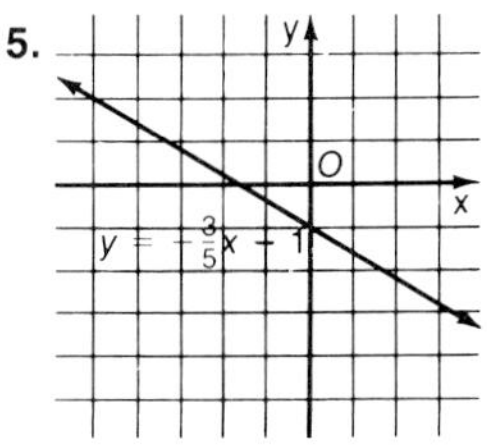

16.
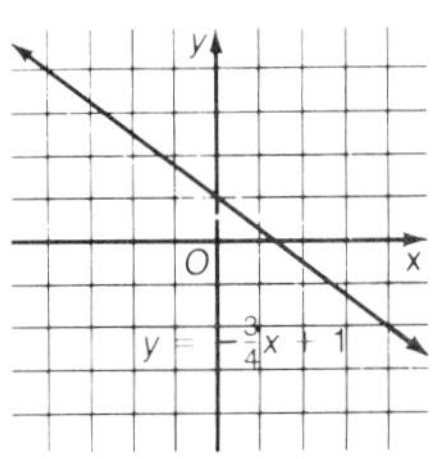

17.
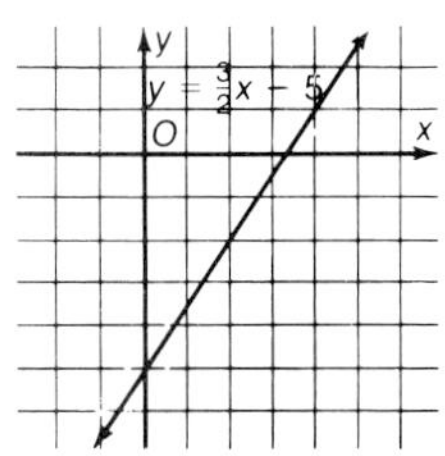

18.
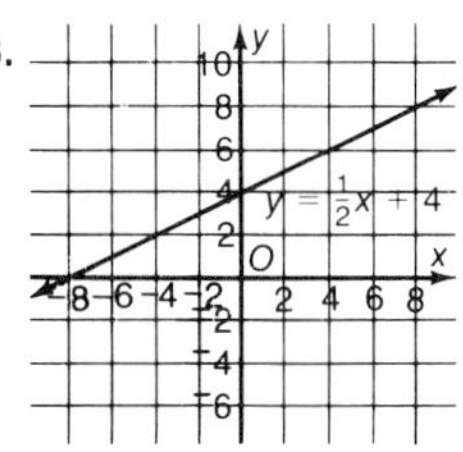

19.
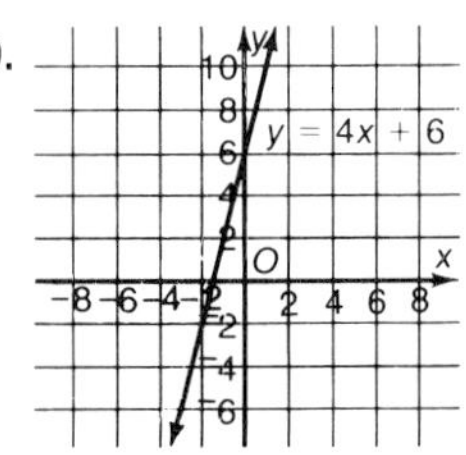

20.
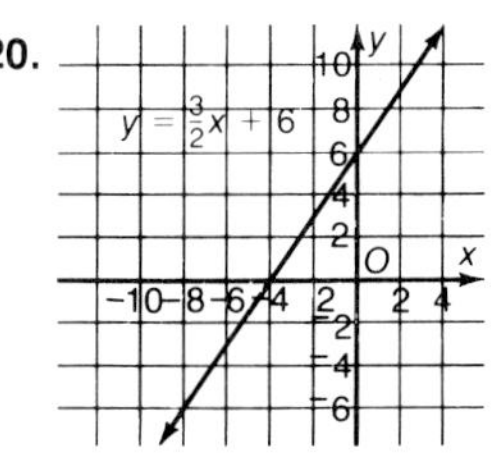

21.
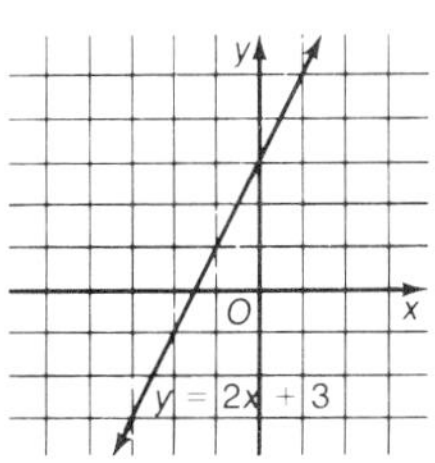

22.
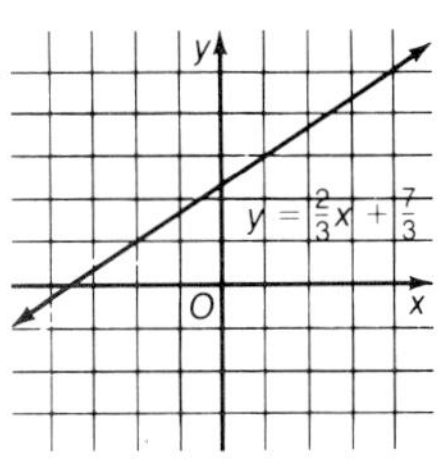

23.
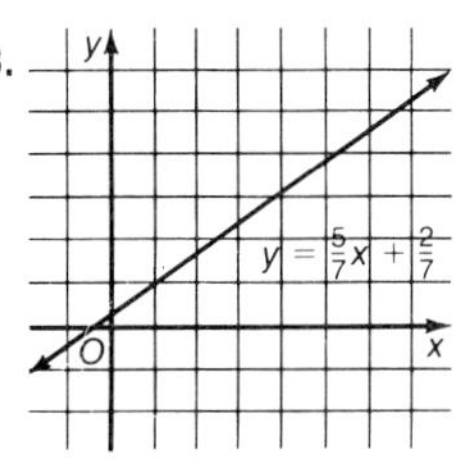

24.
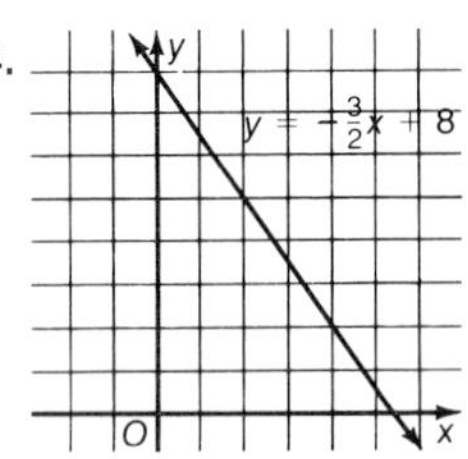

Page 338 Answers to Written

1. 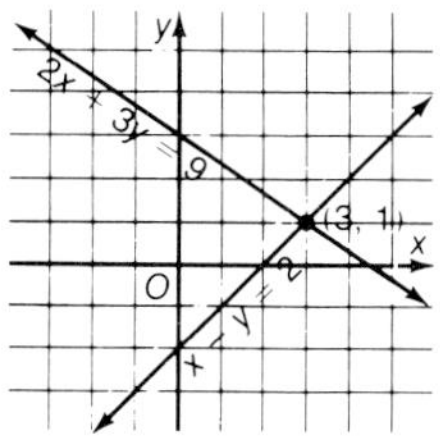

2. 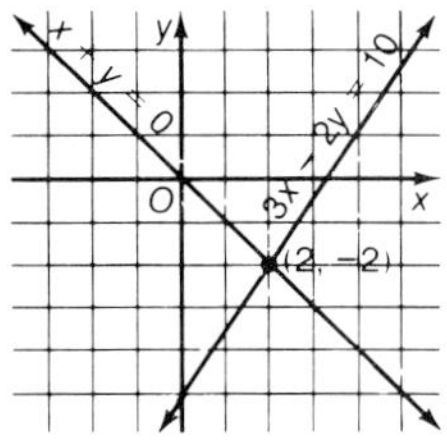

3. 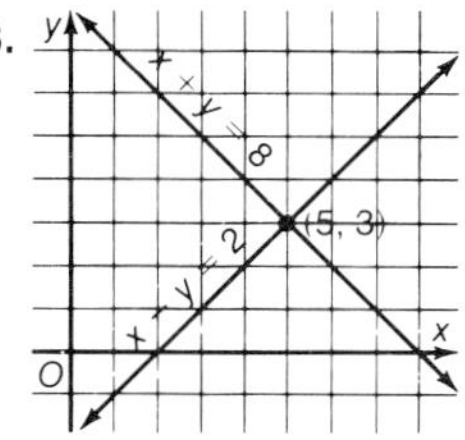

4. 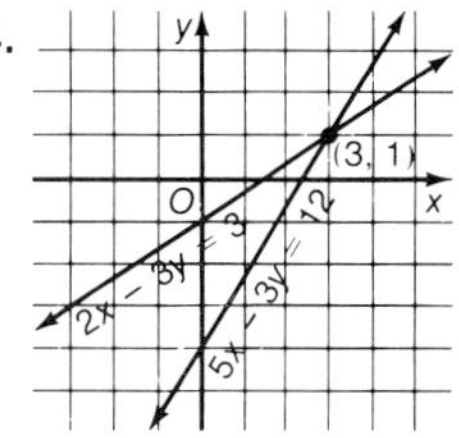

5. 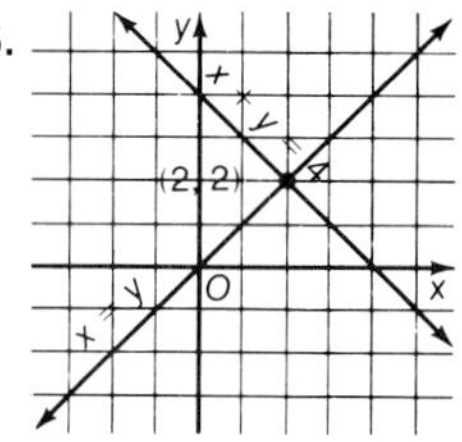

6. 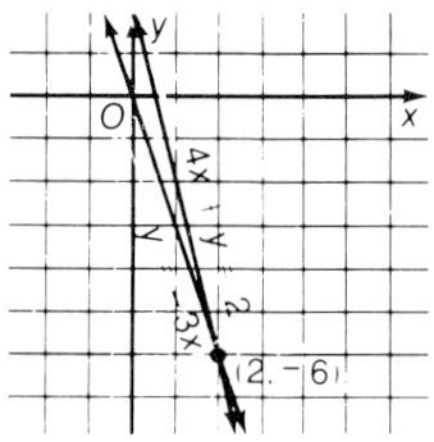

7. 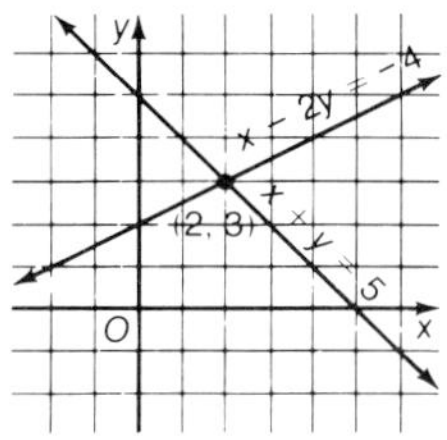

8. 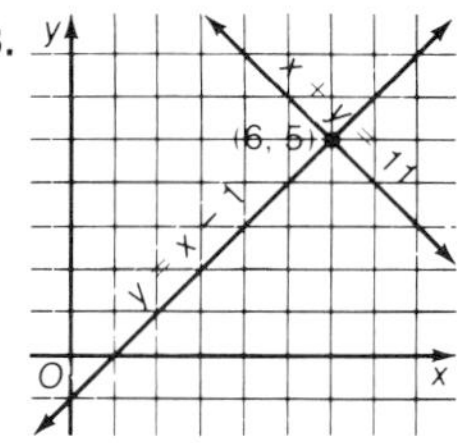

9. 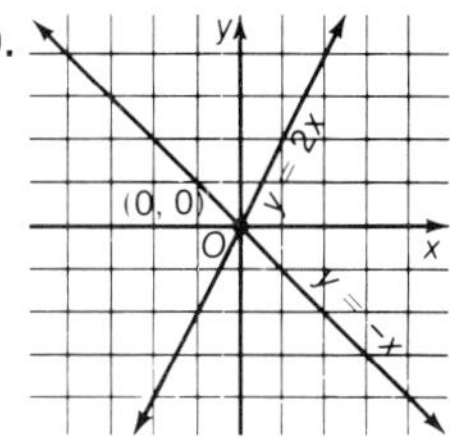

10. 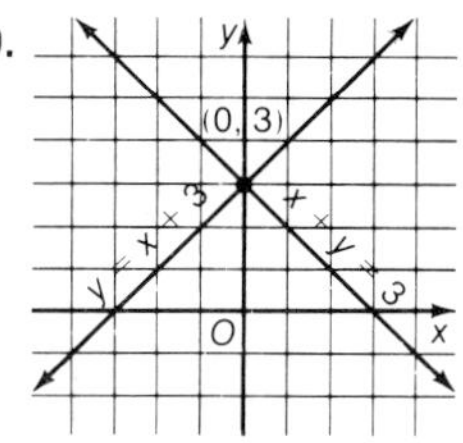

11. 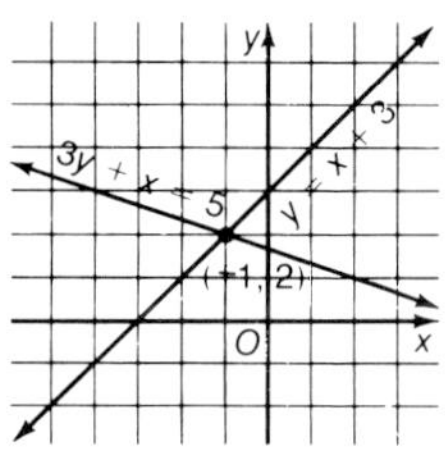

12. 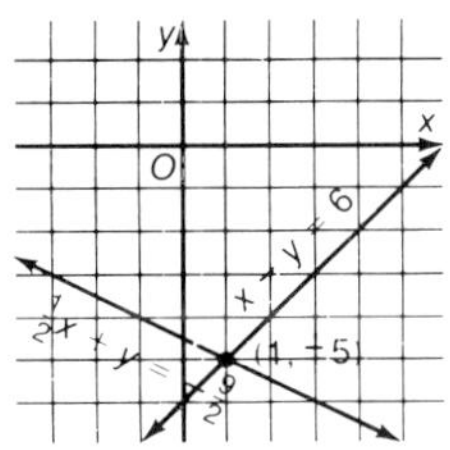

13. 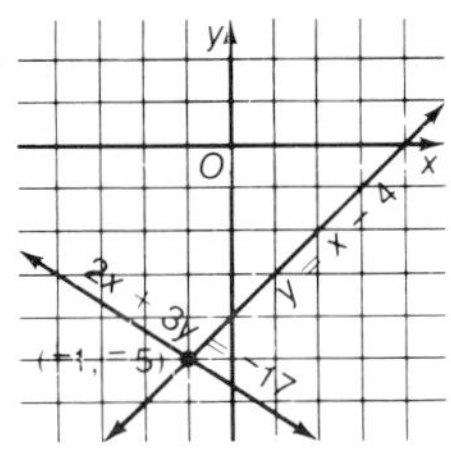

14. 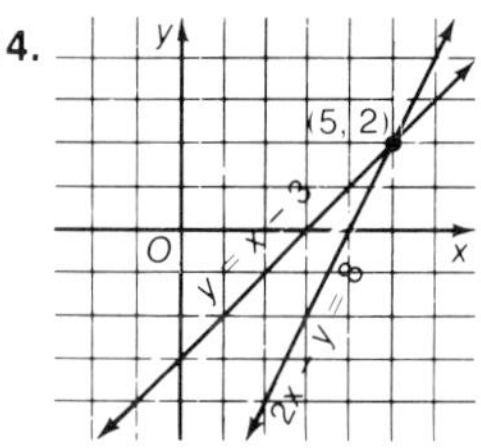

15. 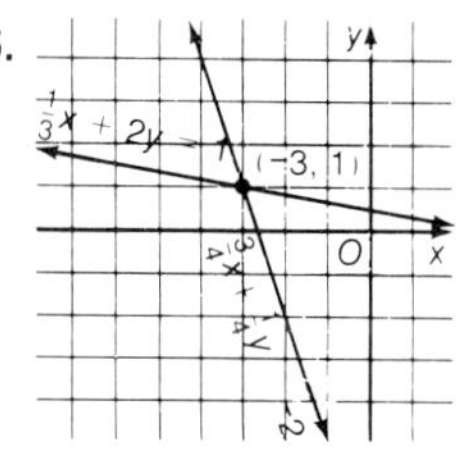

16. 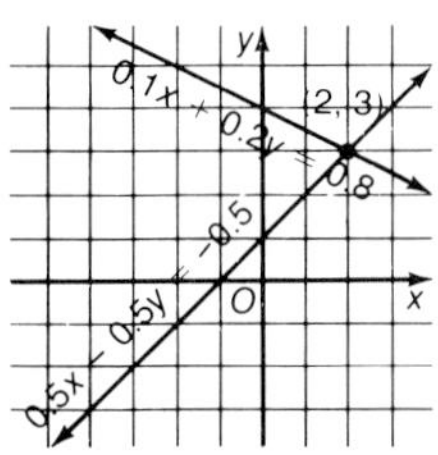

17. 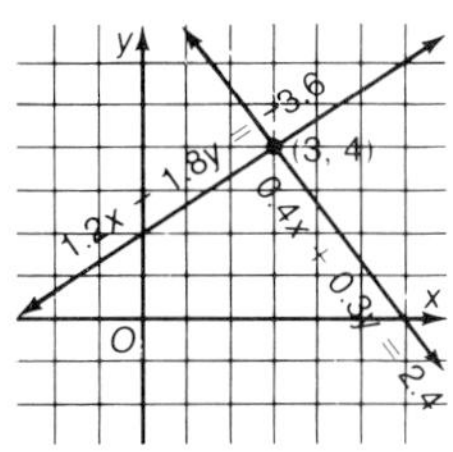

18. 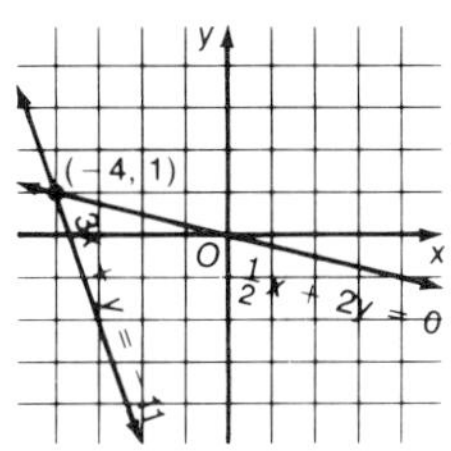

19. 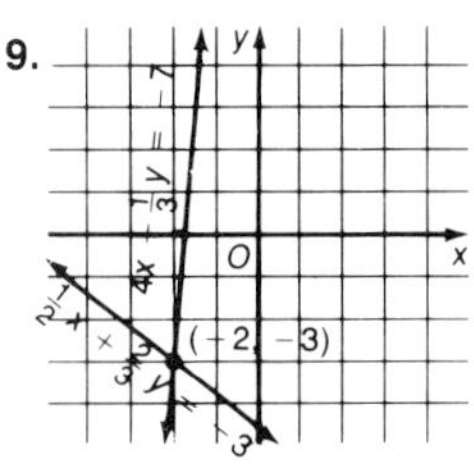

20. 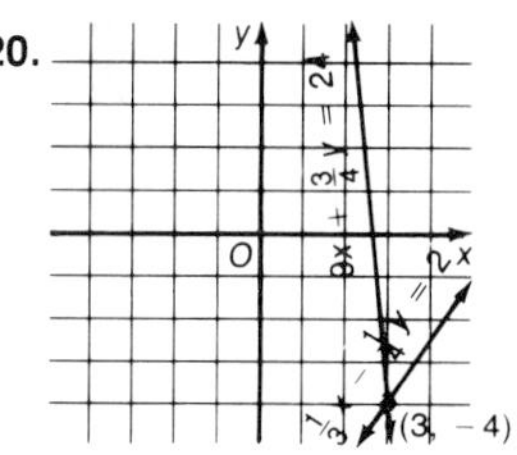

21. 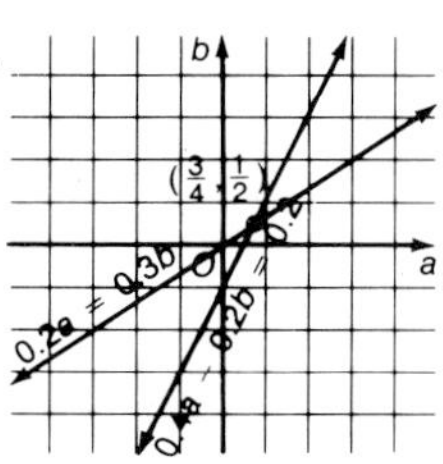

22. 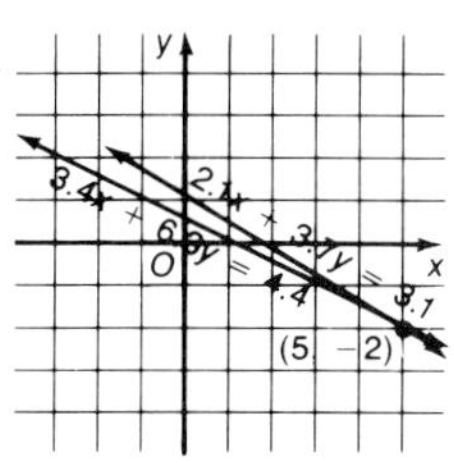

23. 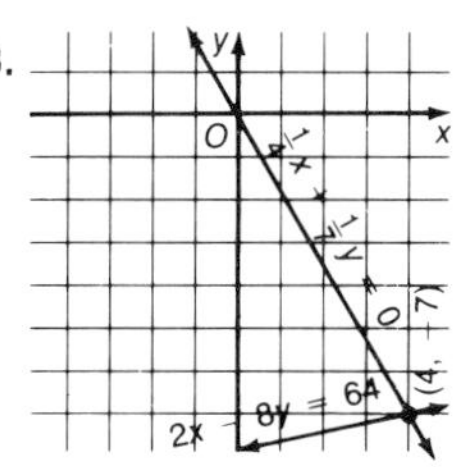

24. 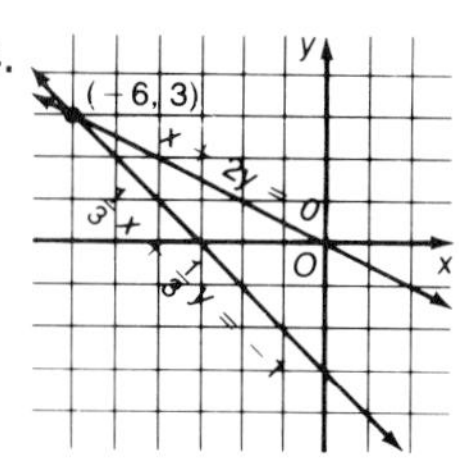

Page 341 Answers to Written

1. 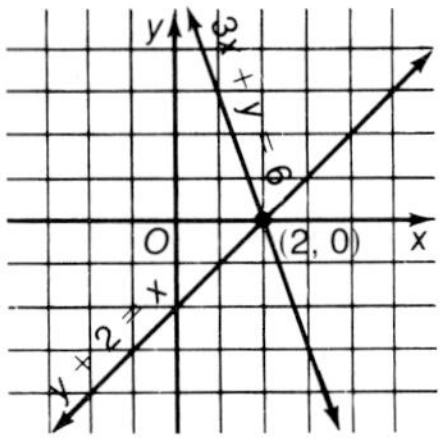

2. 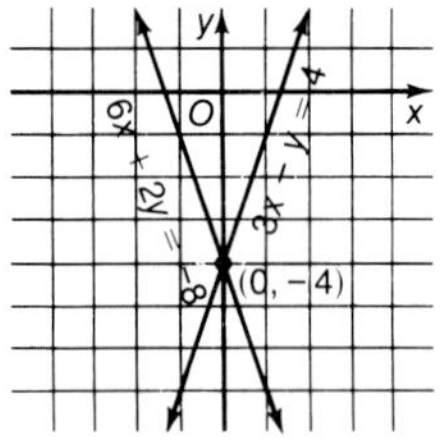

3. 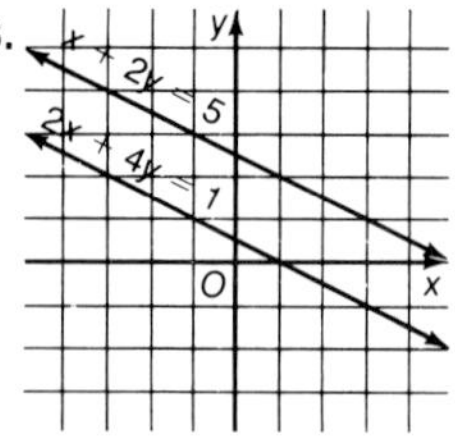

4. 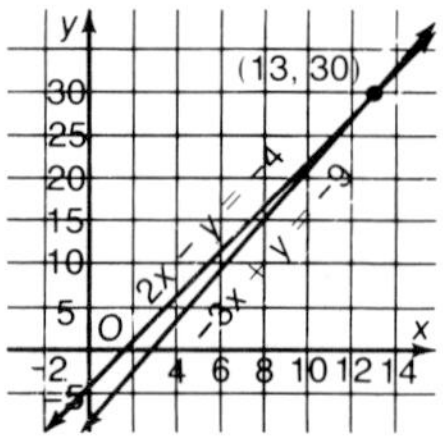

5. 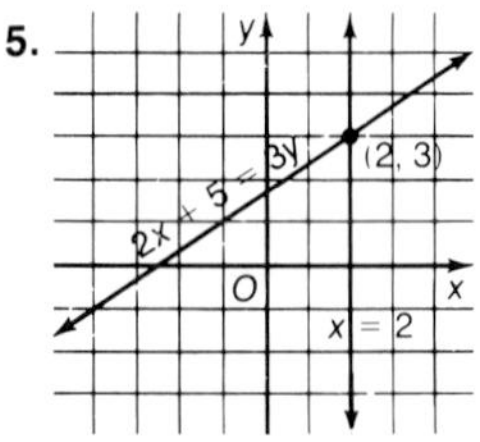

6. 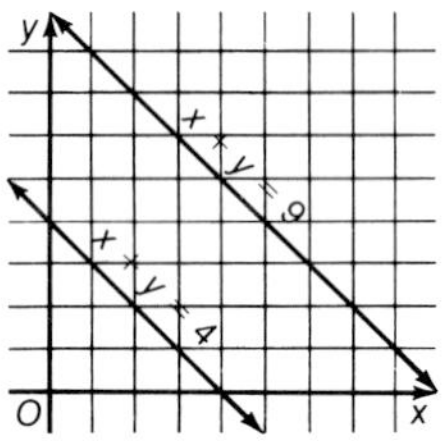

7. 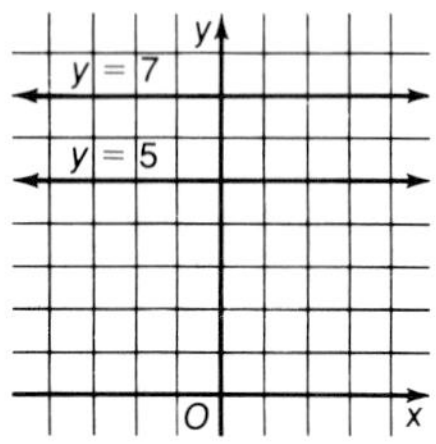

8. 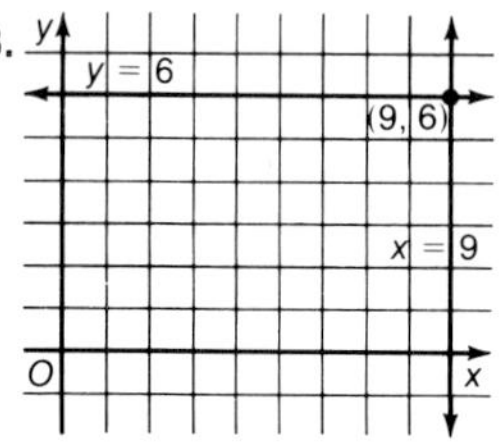

9. 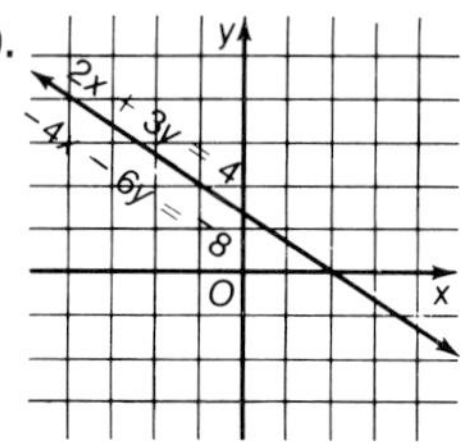

10. 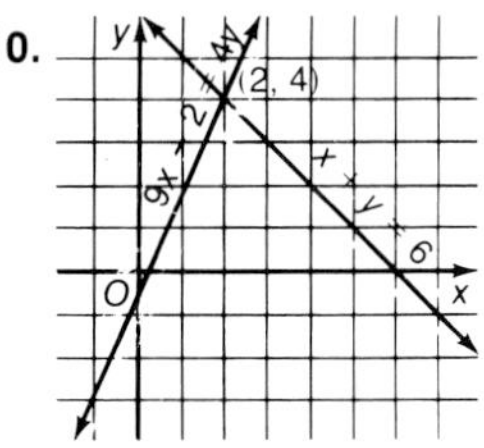

11. 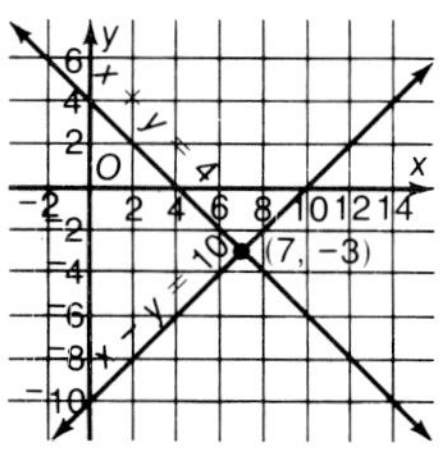

12. 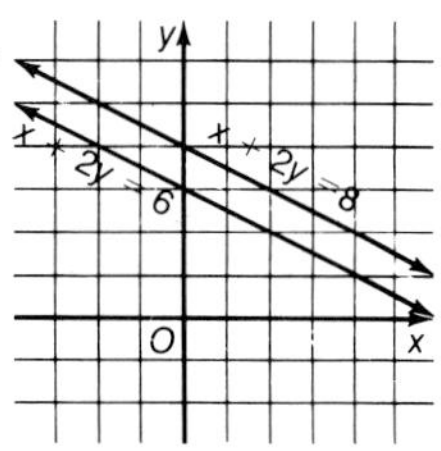

13. 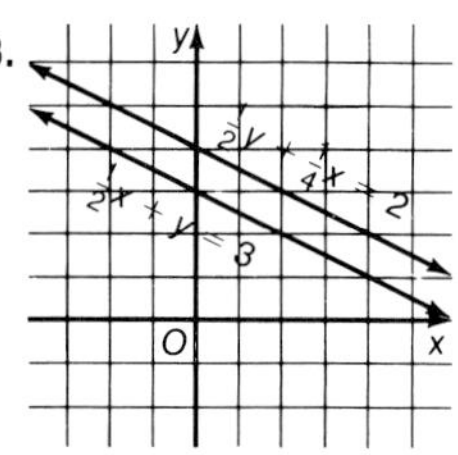

14. 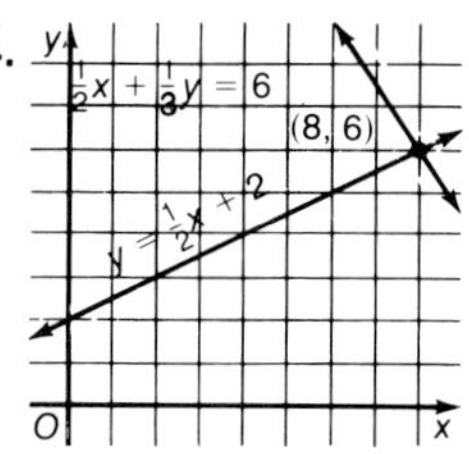

15. 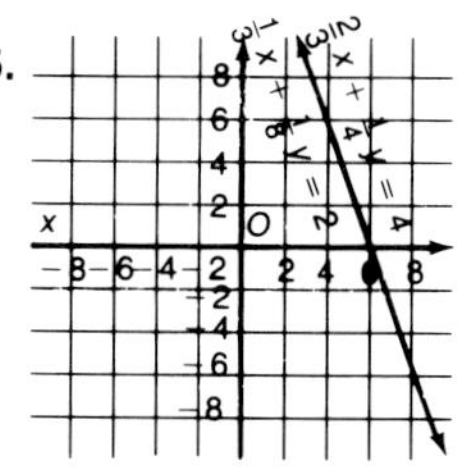

16. 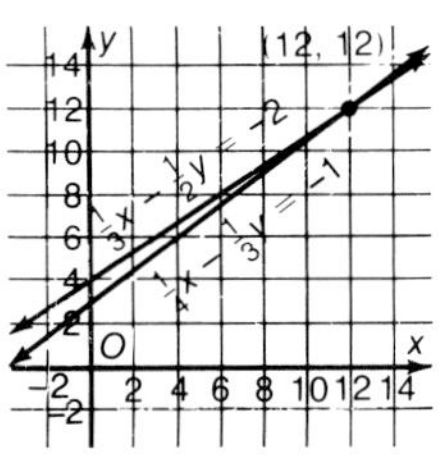

17. 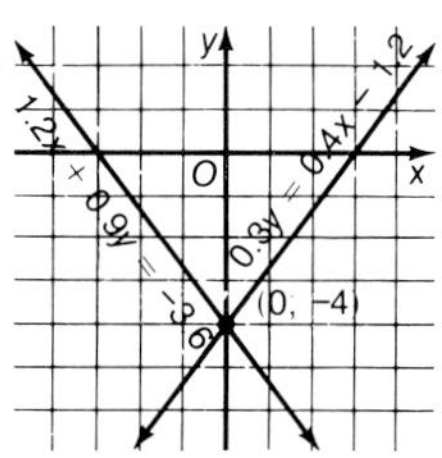

18. 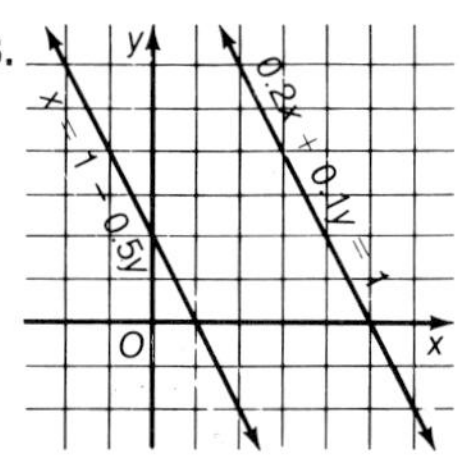

Page 343 Answers to Exploratory

6. $x = 5 - y$
   $y = 5 - x$

7. $x = -5 - y$
   $y = -5 - x$

8. $x = \frac{1}{3}(8 - 5y)$
   $y = \frac{1}{5}(8 - 3x)$

9. $x = \frac{3}{2}\left(10 + \frac{1}{2}y\right)$
   $y = 2\left(\frac{2}{3}x - 10\right)$

10. $x = \frac{1}{1.6}(3.2 - 2.3y)$
    $y = \frac{1}{2.3}(3.2 - 1.6x)$

11. $x = -\frac{1}{0.75}(0.8y + 6)$
    $y = -\frac{1}{0.8}(0.75x + 6)$

12. $x = \frac{3}{2}\left(3 + \frac{4}{5}y\right)$
    $y = \frac{5}{4}\left(\frac{2}{3}x - 3\right)$

Page 354 Answers to Written

1. 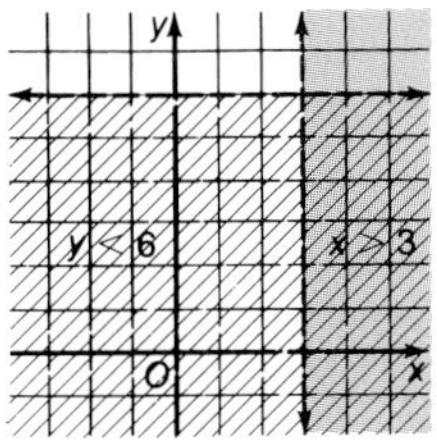

2. 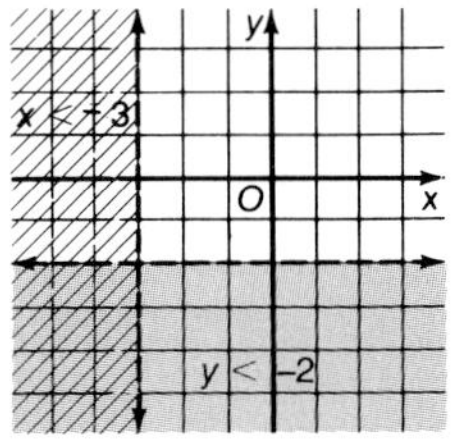

3. 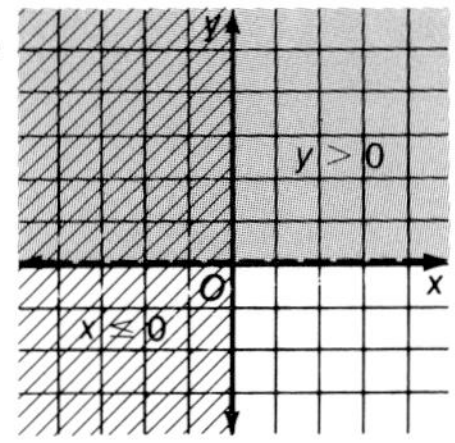

4. 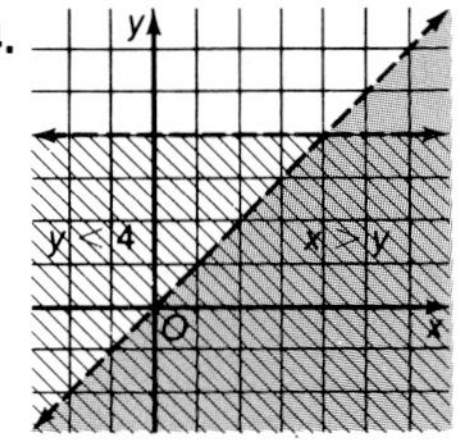

5. 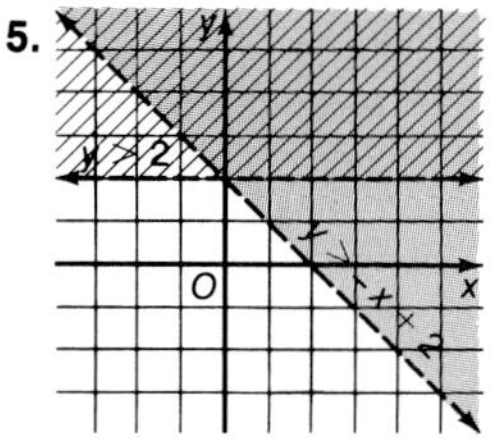

6. 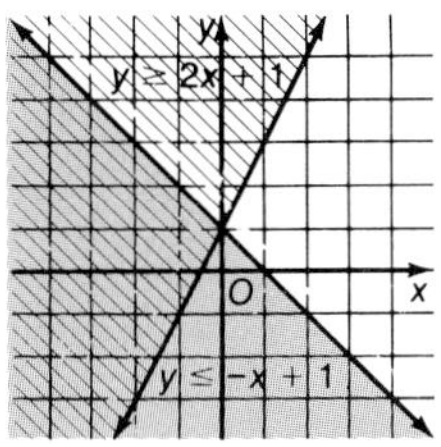

7. 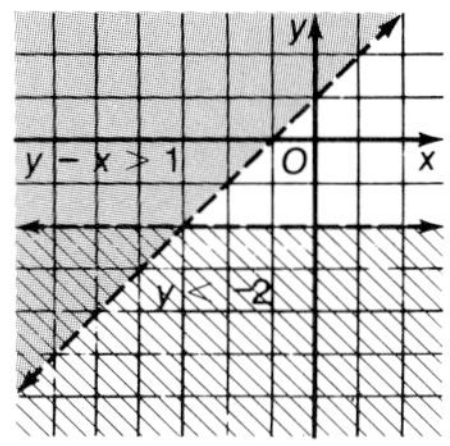

8. 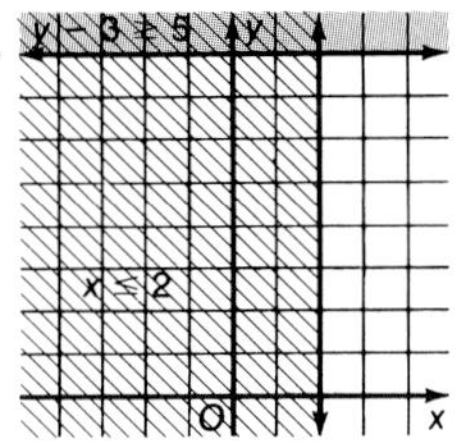

9. 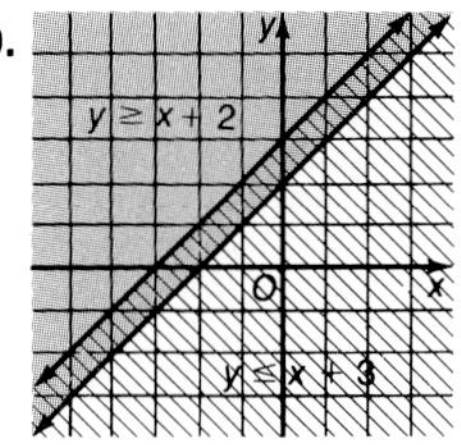

10. 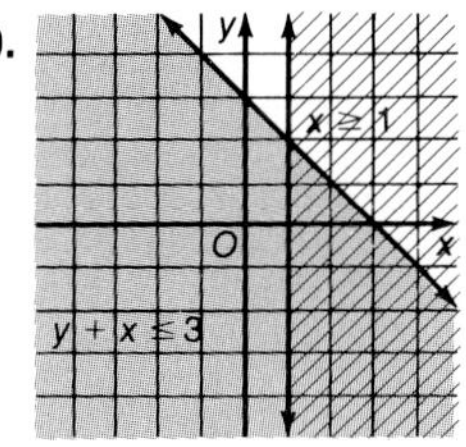

11. 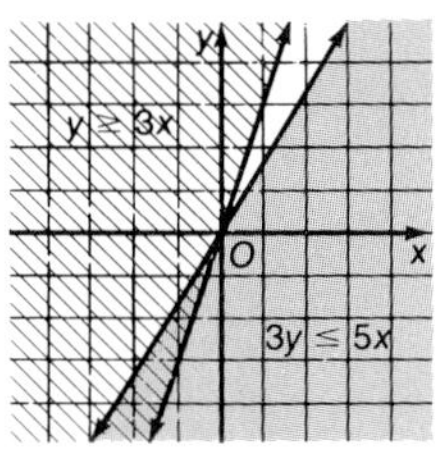

12. 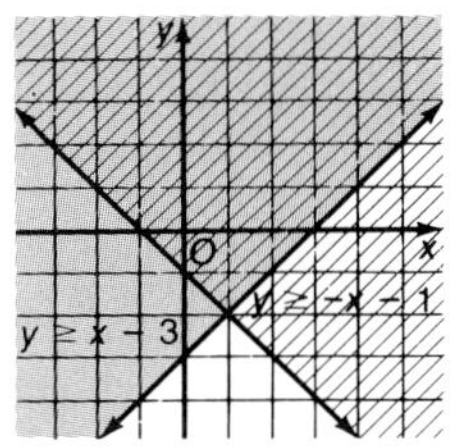

13. 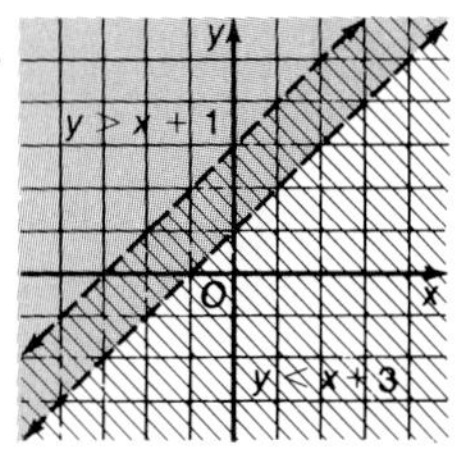

14. 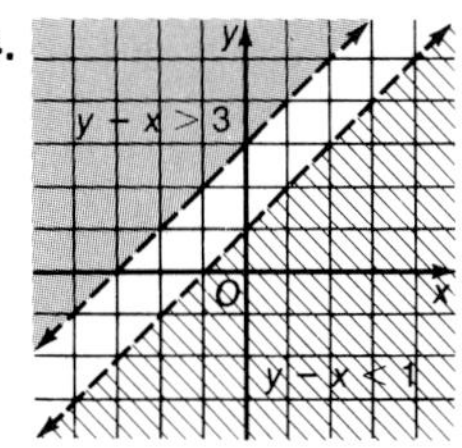

15. 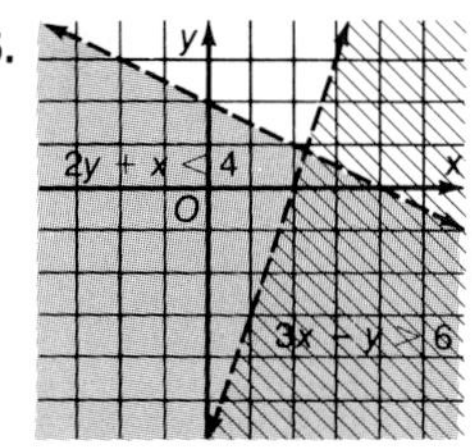

16. 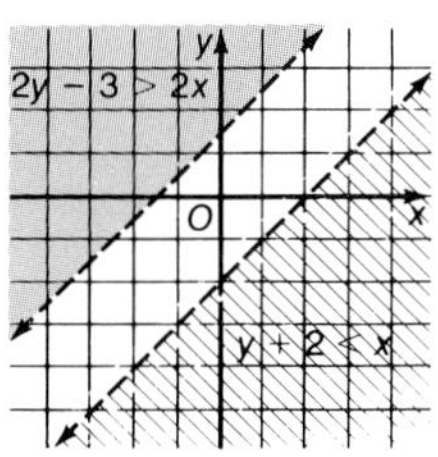

17. 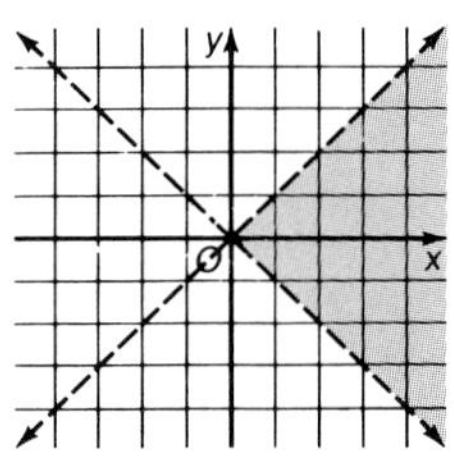

18. 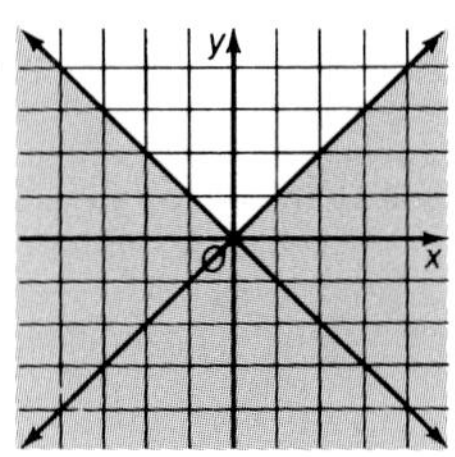

19. 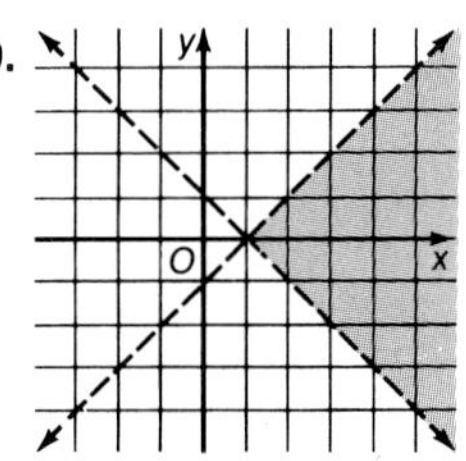

20. 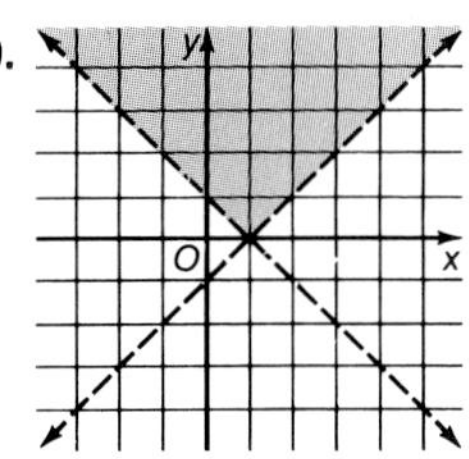

CHAPTER 13 QUADRATICS

Page 408 Answers to Written

19. 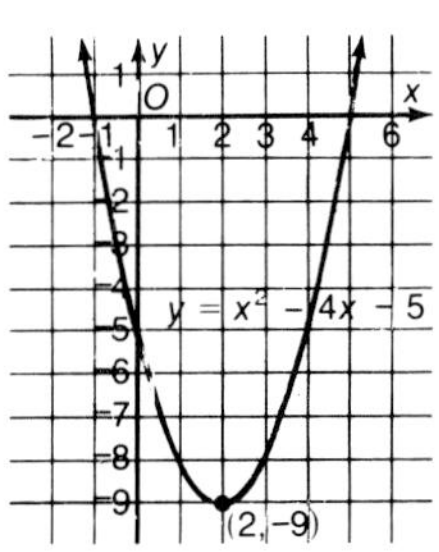

20. 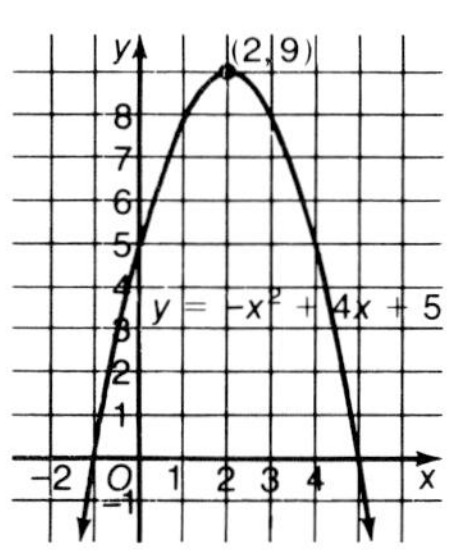

21. 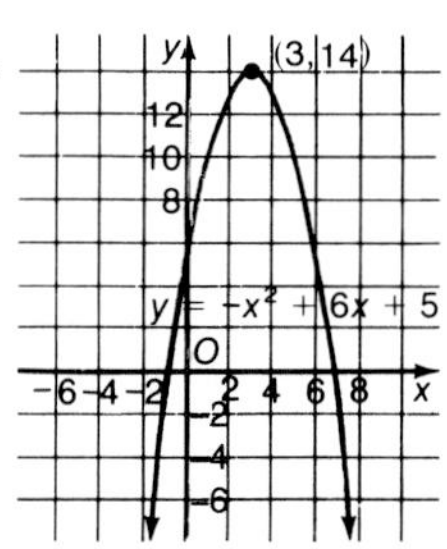

22. 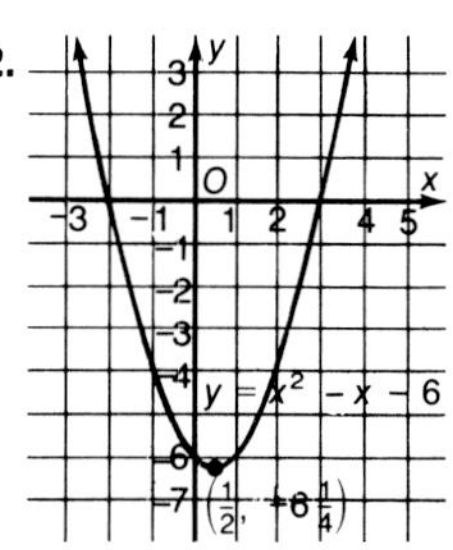

23. 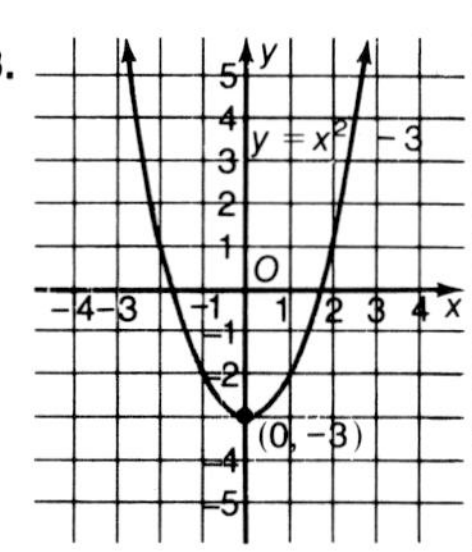

24.

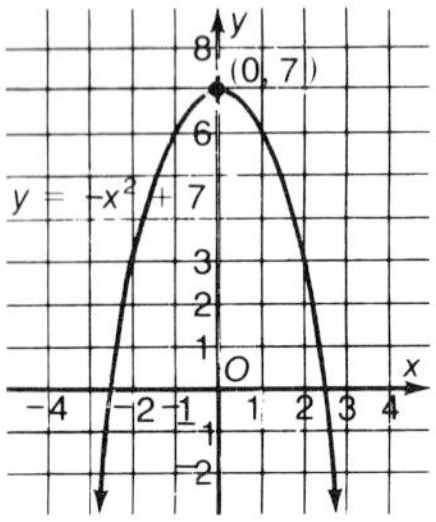

25.

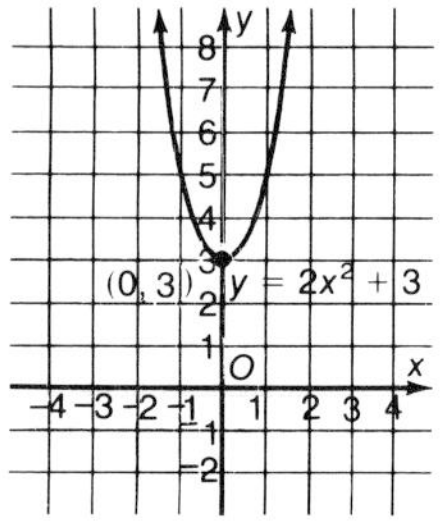

26.

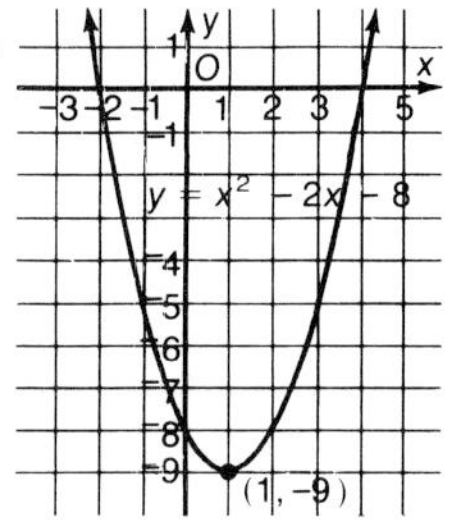

27.

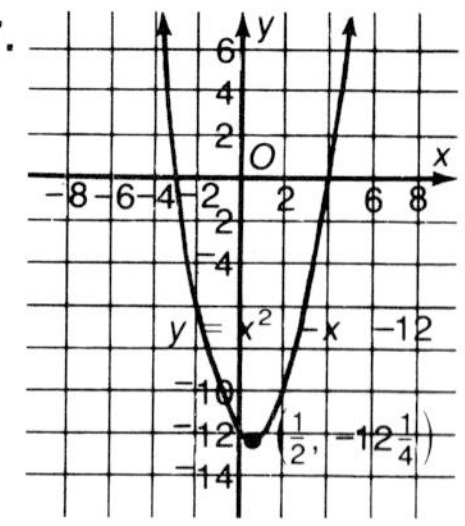

28.

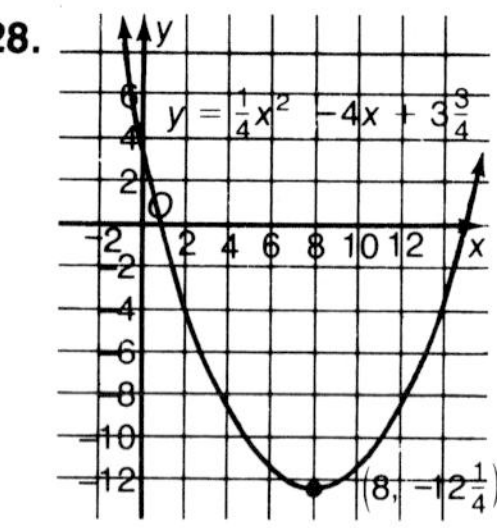

29.

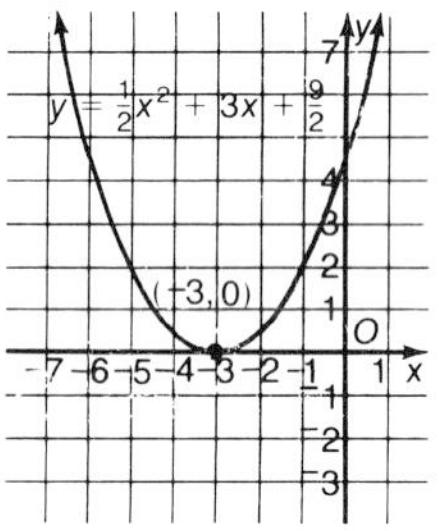

30.

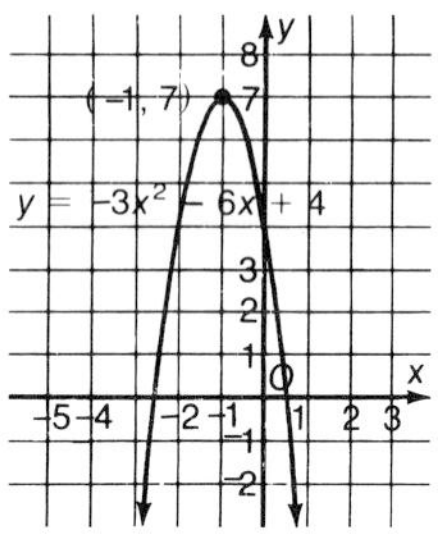

Page 412 Answers to Written

1.

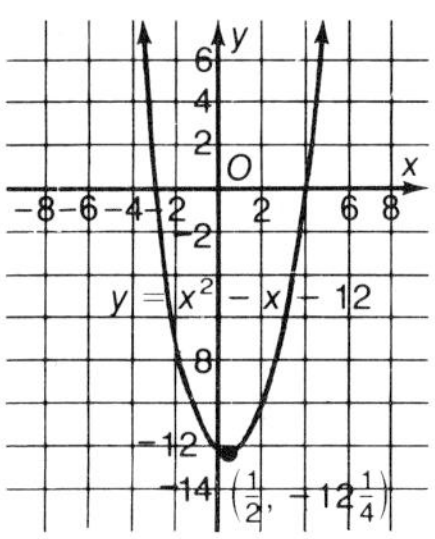

2.

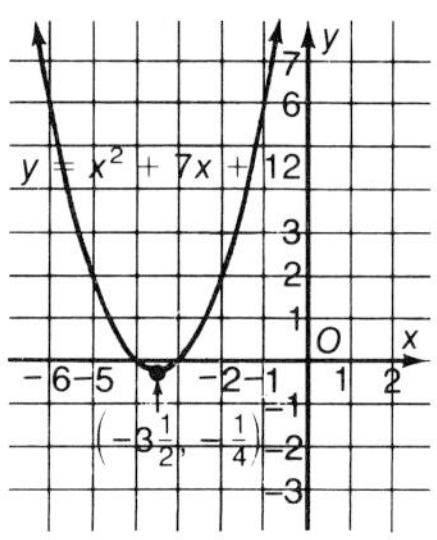

3.

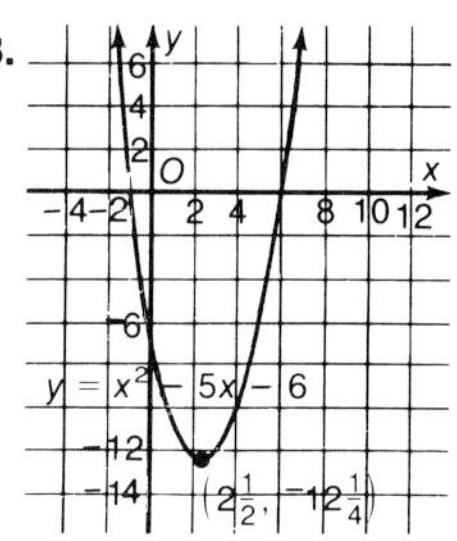

4.

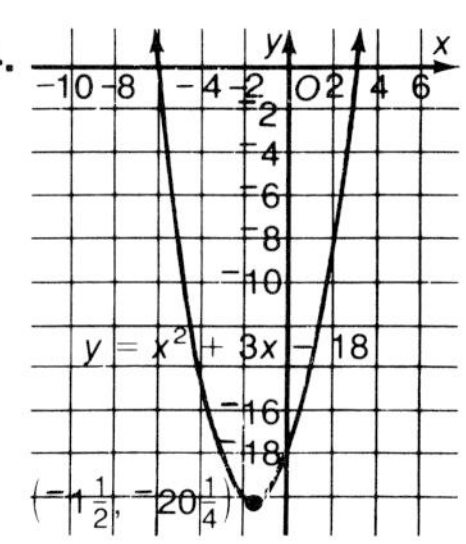

5.

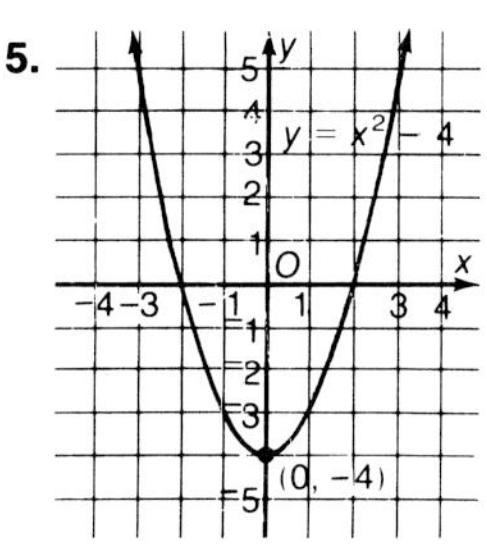

6.

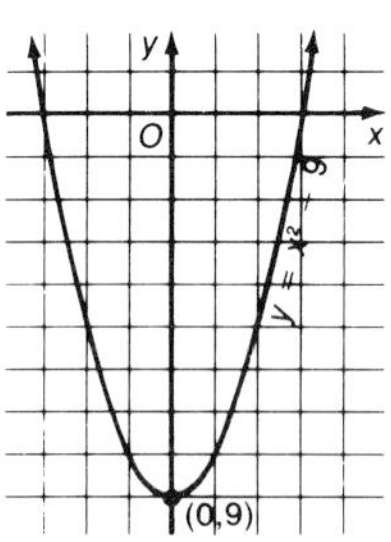

7.

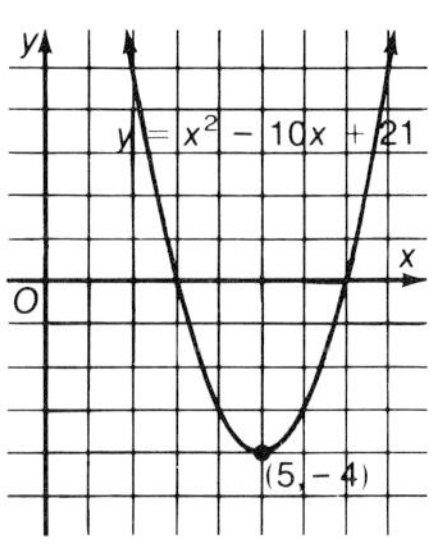

8.

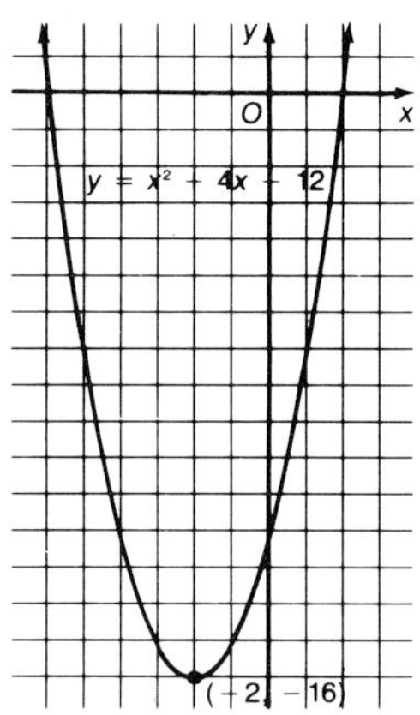

9.

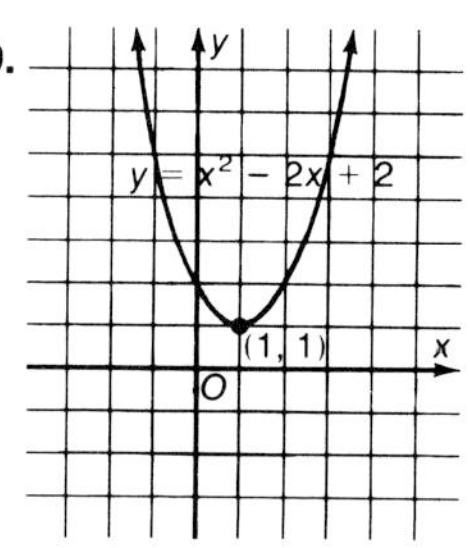

10.

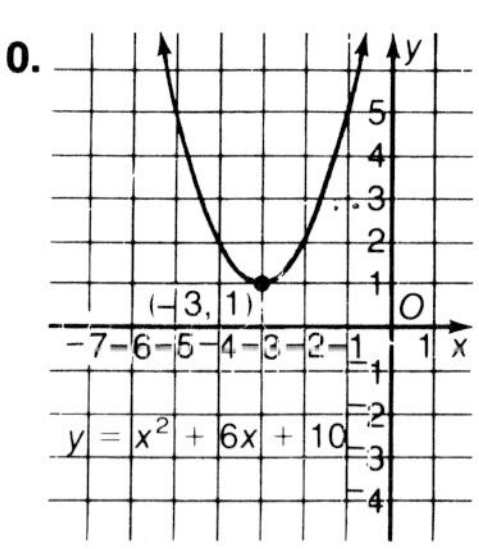

11.

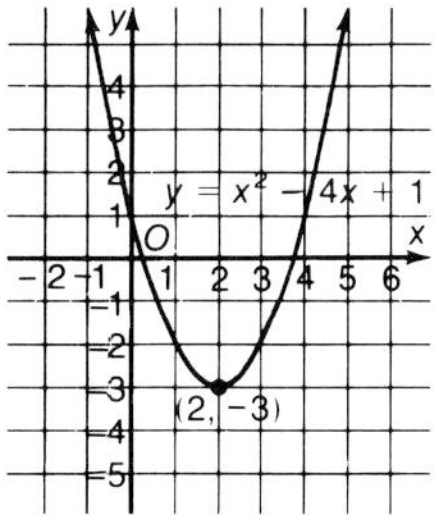

12.

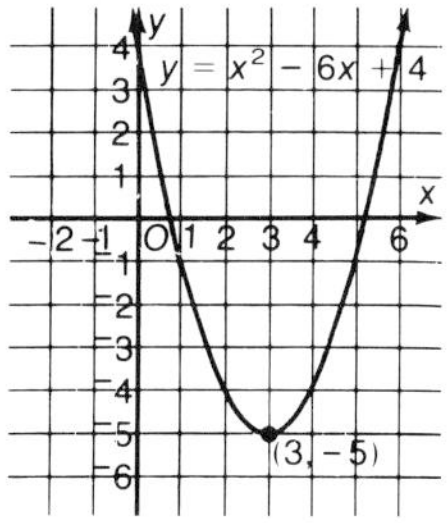

13.

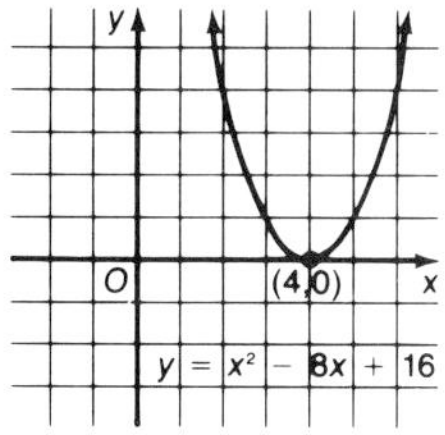

14.

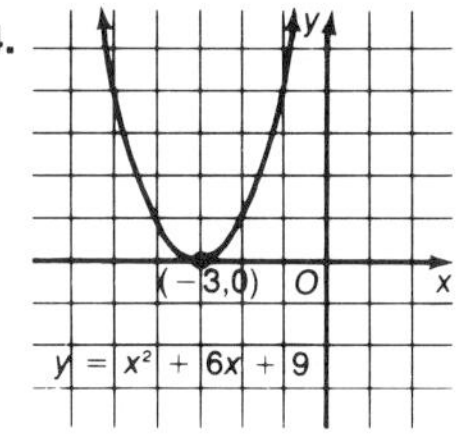

15.

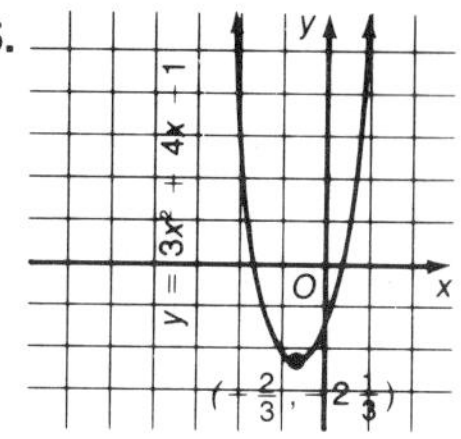

16.

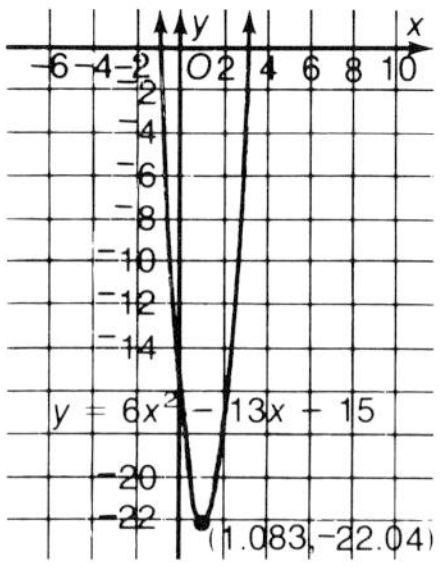

17.

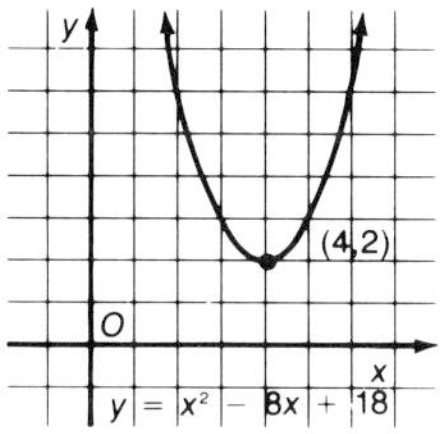

18.

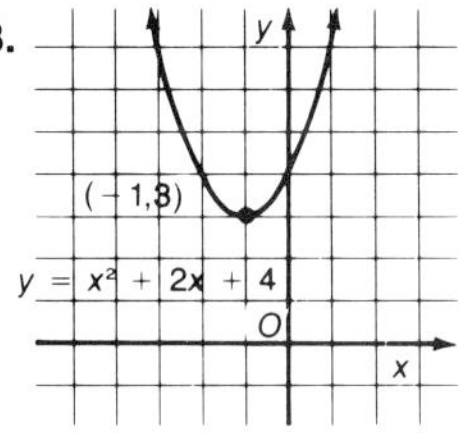

19.

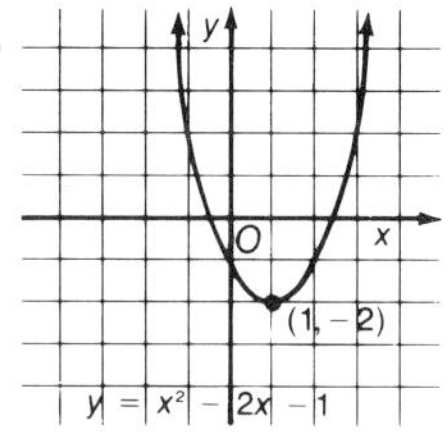

20.

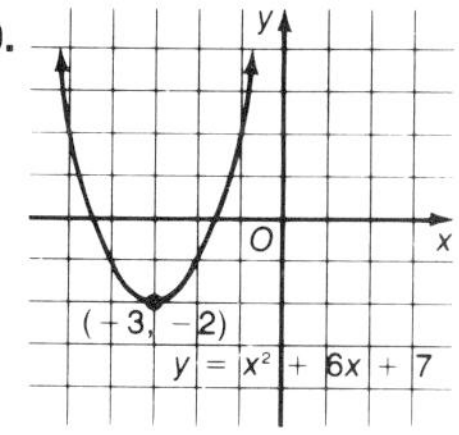

21.

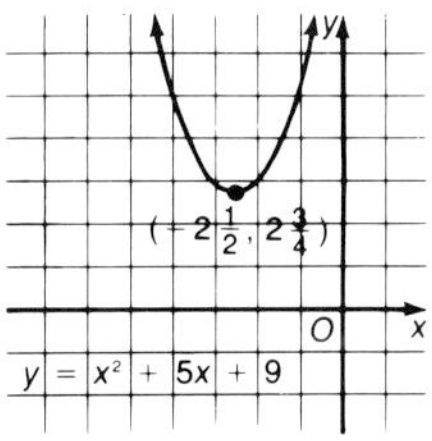

22.

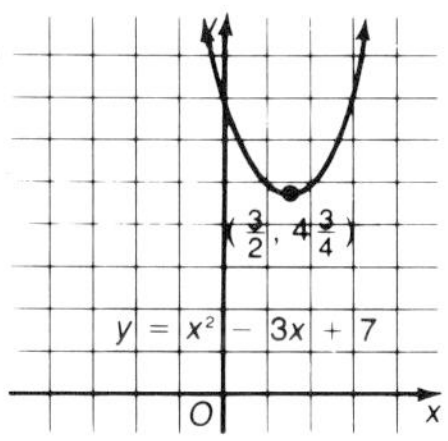

23.

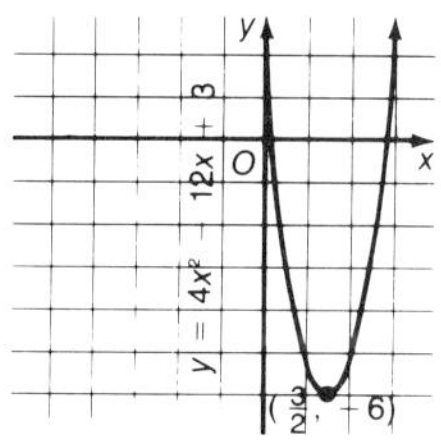

24.

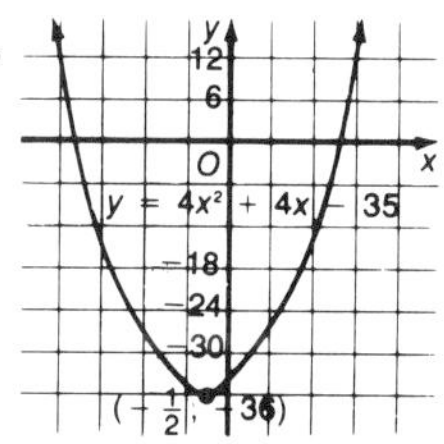

Notes

Notes

Notes

Notes

Notes

Merrill Algebra One

Foster • Rath • Winters

Merrill Publishing Co.

Columbus, Ohio

Authors

Alan G. Foster is chairman of the Mathematics Department at Addison Trail High School, Addison, Illinois. He has taught mathematics courses at every level of the high school curriculum. Mr. Foster obtained his B.S. from Illinois State University and his M.A. in mathematics from the University of Illinois, with additional work at Northern Illinois University, Purdue, Northwestern, and Princeton. Mr. Foster is active in professional organizations at local, state, and national levels, frequently speaking and conducting workshops. He is a past president of the Illinois Council of Teachers of Mathematics. Mr. Foster is a co-author of *Merrill Geometry.*

James N. Rath has 25 years of classroom experience in teaching mathematics at every level of the high school curriculum. He has also taught calculus at the college level. Mr. Rath is a former head of the Mathematics Department at Darien High School, Darien, Connecticut. He earned his B.A. in Philosophy from the Catholic University of America. He obtained both his M.Ed. and his M.A. in mathematics from Boston College. He developed a minicourse in BASIC for the Darien Public Schools. Mr. Rath is a member of various professional organizations in which he is active at local, state, and national levels. He is a co-author of *Merrill Pre-Algebra.*

Leslie J. Winters is chairperson of the Mathematics Department at John F. Kennedy High School, Granada Hills, California. He has taught mathematics at every level from junior high to college. He received his B.A. in mathematics from Pepperdine University, his B.S. in secondary education from the University of Dayton, his M.S. in secondary education from the University of Southern California, and his M.A. in mathematics from Boston College. Mr. Winters is a frequent speaker at local, state, and national conferences. He is a past president of the California Mathematics Council–Southern Section. Mr. Winters was one of two recipients of the 1983 Presidential Award for Excellence in Mathematics Teaching in the state of California.

ISBN 0-675-05476-1
Published by
Merrill Publishing Company
Columbus, Ohio 43216

Printed in the United States of America

Consultants

Jack Price
Superintendent of Schools
Palos Verdes, California

Margaret J. Kenney
Assistant to the Director
Mathematics Institute
Boston College
Chestnut Hill, Massachusetts

Reviewers

Ronald L. Ashby
Secondary Math Chairperson
Terre Haute North Vigo High School
Terre Haute, Indiana

Carl Blair
Mathematics and
Computer Science Teacher
Mustang Public Schools
Mustang, Oklahoma

Dorothy Coleman
Mathematics Teacher
John C. Fremont High School
Los Angeles, California

Diane M. Fite
Mathematics Teacher
Temple High School
Temple, Texas

Mary Hines
Mathematics Department Chairperson
Manzano High School
Albuquerque, New Mexico

Charles E. Loyet
Coordinator of Mathematics/Testing
Carman/Ainsworth Community Schools
Flint, Michigan

Eileen McGinty
Mathematics Teacher
Robert E. Lee High School
Baton Rouge, Louisiana

Harry J. Needham
Mathematics Teacher
Yuma High School
Yuma, Arizona

Winston J. Rose
Mathematics Department Chairperson
Masconomet Regional School
Topsfield, Massachusetts

Sylvia Shugart
Mathematics Teacher and
Department Chairperson
Clovis High School
Clovis, New Mexico

Staff

Editorial

Series Editor: Evangeline Seltzer; *Project Editor:* Susan Danko; *Editor:* Jane Seliga; *Photo Editor:* Lindsay Gerard; *Production Editor:* Joy Dickerson

Art:

Book Designer and Project Artist: Larry W. Collins; *Artist:* Jeffrey A. Clark

Photo Credits

Cover Photo: Steve Traudt/Tom Stack & Associates

2, Cobalt Productions; **9,** Doug Martin; **15,** Cobalt Productions; **22,** Dave Seltzer; **28,** Cobalt Productions; **38,** Geri Murphy; **62,** Image Workshop; **63,** Mark Newman/Tom Stack & Associates; **74,** Tim Courlas; **81,** Ruth Dixon; **86,** David Frazier; **101,** Doug Martin; **110,** Leo de Wys; **116,** Cobalt Productions; **135,** Image Workshop; **143,** File Photo; **153,** Kitt Peak; **164,** R. Laird/FPG; **174,** McGee/FPG; **210,** Dave Seltzer; **244,** NASA; **278,** Cobalt Productions; **304,** David Frazier; **323,** Tim Courlas; **336,** Kenfre, Inc./International Stock Photo; **370,** Guy Sauvage/Photo Researchers; **388,** Jerry Cooke/Photo Researchers; **404,** Joseph R. Pearce/DRK Photo; **436,** Joseph A. DiChello; **458,** Tim Courlas; **468,** Ric Helstrom/Tom Stack & Associates; **487,** Tersch; **498,** Tim Courlas; **520,** Tom McGuire.

Preface

This third edition of *Merrill Algebra One* is designed for use by students in first-year high school algebra courses. The text, which was developed in the classroom by experienced high school teachers, is based on the successful prior editions of *Merrill Algebra One*. The goals of the text are to develop proficiency with mathematical skills, to expand understanding of mathematical concepts, to improve logical thinking, and to promote success. To achieve these goals, the following strategies are used.

Build upon a Solid Foundation. This program develops and utilizes the learning spiral. Students first review those concepts basic to the understanding of algebra. Students' understanding is thus solidified before the introduction of more difficult concepts.

Utilize Sound Pedagogy. *Merrill Algebra One* covers in logical sequence all topics generally presented at this level. Concepts are introduced when they are needed. Each concept presented is then used within that lesson and in later lessons.

Gear Presentation for Learning. An appropriate reading level has been maintained throughout the text. Furthermore, many photographs, illustrations, charts, graphs, and tables provide visual aids for the concepts and skills presented. Hence, students are able to read and learn with increased understanding.

Use Relevant Real-Life Applications. Applications are provided not only for practice but also to aid understanding of how concepts are used.

The text offers a variety of aids for the students.

Student Annotations	Help students identify important concepts as they study.
Selected Answers	Allow students to check their progress as they work. These answers are provided at the back of the text.
Mini-Review	Provides students with a quick review of skills and concepts taught previously.
Vocabulary	Enables students to focus on increasing their mathematical vocabulary.
Chapter Summary	Provides students with a compact listing of major concepts presented within each chapter.
Chapter Review	Permits students to review each chapter by working sample problems from each lesson.
Chapter Test	Enables students to check their own progress.
Cumulative Review	Helps students to maintain their skills in applying algebraic concepts.
Standardized Test Practice Questions	Help students to become familiar with types of questions that appear on standardized tests.

The following special features, which appear periodically throughout the text, provide interesting and useful extra topics.

Using Calculators	Instructs students in using a calculator. The use of the calculator is related to concepts taught within the chapter.
Reading Algebra	Provides students with the instruction needed to read and interpret mathematical symbolism.
Problem Solving	Illustrates several helpful strategies for solving mathematical problems.
Applications	Provide insights into the uses of mathematics in everyday life, from a variety of disciplines.
Using Computers	Provides a computer program and exercises related to the objectives of the chapter.
Excursions in Algebra	Enliven and help maintain student interest by providing interesting side trips. Topics are varied and include history, glimpses into development of algebra, enrichment concepts, and puzzles.

The textbook contains an appendix on **BASIC** which provides instruction in writing programs using the BASIC computer language.

The **Algebraic Skills Review** at the back of the text provides review exercises to help students maintain algebraic skills. These exercises can be used for remediation, to supplement exercises in the lessons, and as review material at the end of the school year.

Teacher and students familiar with the prior editions of this text will be pleased to see that the clarity of explanations and careful sequencing of topics have been retained in this new edition of *Merrill Algebra One*.

Table of Contents

1 The Language of Algebra — 2

2 Adding and Subtracting Rational Numbers — 38

3 Multiplying and Dividing Rational Numbers — 74

4 Inequalities — 110

5 Powers 144

6 Polynomials 174

7 Factoring ___ 210

Lessons

Review and Testing

Applications and Extensions

Technology

8 Applications of Factoring ___ 244

Lessons

Review and Testing

Applications and Extensions

Technology

11 Systems of Open Sentences 336

12 Radical Expressions 370

13 Quadratics 404

14 Rational Expressions 436

To the Student

When you read this book, you should look for special features. These features will help you in studying algebra. Consider the following features when studying section.

section title →

1-2 Evaluating Expressions

The numerical expression $5 \cdot 6 + 3$ might be computed in two ways. Only one is correct.

Multiply 5 and 6. Add 3.	Add 6 and 3. Multiply by 5.
$5 \cdot 6 + 3 = 30 + 3$	$5 \cdot 6 + 3 = 5(9)$
$= 33$	$= 45$

Each numerical expression should have a *unique* value. Therefore, the following order of operations has been agreed upon.

important concept →

Order of Operations

1. Evaluate all powers.
2. Do all multiplications and divisions from left to right.
3. Do all additions and subtractions from left to right.

Thus, using the order of operations, the value of $5 \cdot 6 + 3$ is 33.

Follow the steps in the example.

Example 1

Evaluate $12 \div 3 \cdot 5 - 4^2$.

$12 \div 3 \cdot 5 - 4^2 = 12 \div 3 \cdot 5 - 16$ — *Evaluate 4^2.*
$= 4 \cdot 5 - 16$ — *Divide 12 by 3.*
$= 20 - 16$ — *Multiply 4 by 5.*
$= 4$ — *Subtract 16 from 20.*

An algebraic expression can be evaluated when the value of each variable is known. The variables are simply replaced by their values. Then, the value of the numerical expression is calculated.

Example 2

Evaluate $2x^2 + 5xy$ if $x = 3$ and $y = \frac{2}{3}$.

$2x^2 + 5xy = 2 \cdot 3^2 + 5(3)\left(\frac{2}{3}\right)$ — *Substitute 3 for x and $\frac{2}{3}$ for y.*
$= 2 \cdot 9 + 5(3)\left(\frac{2}{3}\right)$ — *Evaluate 3^2.*
$= 18 + 10$ — *Multiply.*
$= 28$ — *Add.*

student annotations →

In mathematics, *grouping symbols* such as parentheses are used to clarify or change the order of operations.

$5 \cdot 6 + 3$ **means** $30 + 3$ or 33

$5(6 + 3)$ **means** $5 \cdot 9$ or 45

At the end of each chapter, use the **Vocabulary** and **Chapter Summary** to review the key ideas. Then use the problems in the **Chapter Review** to review methods of working problems. If you follow these steps, your study of algebra should be interesting and enjoyable.

About the Photo

Although deafness has no legal definition, it usually refers to the inability to hear and understand speech. The range of human speech is from 400 to 3,000 vibrations per second. For this reason, many deaf people never learn to speak well enough to be understood. Instead, they learn a sign language or other techniques to communicate.

Many deaf people learn the finger alphabet. It is one of the special systems deaf people use to communicate. Most use a combination of finger spelling and mannual communication called sign language.

Many deaf people earn college degrees and enter a wide variety of occupations. Some of the finest classical music was written by the German composer Ludwig van Beethoven after he became deaf.

Stress that learning to communicate well, no matter what language it happens to be, usually assures success. Likewise, learning the language of algebra well is fundamental to achievement and success in a math specialty that forms the basis of study for almost all other mathematics.

Teacher Resource Book Materials for Chapter 1 See pages 65-86.

Chapter Test, Form I (Multiple Choice)	65-67
Chapter Test, Form II (Free Response)	68-69
Quizzes	70-71
Extra Practice	72-74
Application: Formulas (Use after 1-7.)	75-78
Extension: Solution Sets (Use after 1-3.)	79
Problem Solving	80
Cumulative Review	81
Answers	82-86

CHAPTER 1

The Language of Algebra

Many deaf people use sign language to express thoughts and ideas. Each hand signal that they use has a meaning. In mathematics, numbers and math symbols have meanings. In this chapter, you will learn about symbols used in algebra.

Objective 1-1A: Translate verbal expressions into mathematical expressions.
Objective 1-1B: Write an expression containing identical factors as an expression using exponents.

1-1 Variables and Expressions

In algebra, verbal expressions are often translated into mathematical expressions. Consider the following sentence.

Kathy has three more dollars than Nick.

From the information given, you do not know exactly how much money Nick has or how much Kathy has. However, you can find the amount Kathy has if you know the amount Nick has.

If Nick has 6 dollars, then Kathy has $6 + 3$ dollars.
If Nick has 10 dollars, then Kathy has $10 + 3$ dollars.
If Nick has n dollars, then Kathy has $n + 3$ dollars.

The letter n is a **variable**, and $n + 3$ is an algebraic **expression**. In algebra, variables are symbols that are used to represent unspecified numbers. Any letter may be used as a variable.

6 + 3 and 10 + 3 are numerical expressions.

An algebraic expression consists of one or more numbers and variables with the operations of addition, subtraction, multiplication, and/or division. Some algebraic expressions are shown below.

$ab + 4$ $\quad \frac{s}{t} - 1$ $\quad 6a(4n)$ $\quad 8rs \cdot 3k$

In a multiplication expression, the quantities being multiplied are called **factors** and the result is called the **product**. In a division expression, the **dividend** is divided by the **divisor**. The result is called the **quotient**.

$4 \times 5 \times 8 = 160$ (4, 5, 8: *factors*; 160: *product*)

$12 \div 3 = 4$ (12: *dividend*; 3: *divisor*; 4: *quotient*)

In algebraic expressions, a dot or parentheses are often used instead of × to indicate multiplication. When variables are used to represent factors, the multiplication sign is usually omitted. Each of the following expressions represents the product of a and b.

ab $\quad a \cdot b$ $\quad a(b)$

A fraction bar is often used to indicate division.

$\frac{x}{3}$ **means** $x \div 3$

1-1

Lesson Notes, p. T24

Readiness Activity
What operation is indicated by each of the following words?
sum addition
quotient division
product multiplication
difference subtraction

Discussion Questions
State two different verbal expressions for each mathematical expression. Answers may vary.
1. $8 + a$
2. $9x$
3. $7 - 3y$
4. $\frac{b^3}{4}$
5. Notice that $4^2 = 16$ and $2^4 = 16$. Does $3^2 = 2^3$? Does $5^2 = 2^5$? no; no

Chalkboard Exercises
Write an algebraic expression for each verbal expression.

1a. *m* increased by 5
$m + 5$

1b. *z* less than 6
$6 - z$

1c. 7 times the product of *x* and *t*
$7xt$

1d. the ratio of *a* to *t*
$\frac{a}{t}$

Computer Notes
Discuss how products, quotients, and powers are shown in BASIC.

Algebra	BASIC
$10 \div 5$	10/5
$3 \cdot 4$	3*4
2^6	2↑6

PRINT statements can be used to evaluate expressions. For example, the following command can be used to evaluate 2^6.
PRINT 2↑6
For more information about using the BASIC language to evaluate expressions, see pages 528-529 of the appendix.

To solve verbal problems in mathematics, words must be translated into mathematical symbols. Thus, "the product of 3 and 4" becomes "$3 \cdot 4$." The following chart shows some of the words that are used to indicate mathematical operations.

ADDITION	SUBTRACTION	MULTIPLICATION	DIVISION
the sum of	the difference of	the product of	the quotient of
increased by	decreased by	multiplied by	divided by
plus	minus	times	the ratio of
more than	less than		
the total of	subtracted from		
added to			

Example 1

Write an algebraic expression for each verbal expression.

a. *x* decreased by 9
The algebraic expression is $x - 9$.

b. the sum of *a* and *b*
The algebraic expression is $a + b$.

c. a number *k* divided by a number *n*
The algebraic expression is $\frac{k}{n}$.

In an expression like 10^3, the 10 names the **base** and the 3 names the **exponent**. An exponent indicates the number of times the base is used as a factor.

3^4 means $3 \cdot 3 \cdot 3 \cdot 3$ $\qquad$ a^2 means $a \cdot a$

An expression in the form x^n is called a power.

Study the following chart.

Symbols	Words	Meaning
8^1	8 to the first power	8
8^2	8 to the second power or 8 squared	$8 \cdot 8$
8^3	8 to the third power or 8 cubed	$8 \cdot 8 \cdot 8$
8^4	8 to the fourth power	$8 \cdot 8 \cdot 8 \cdot 8$
$6n^5$	6 times *n* to the fifth power	$6 \cdot n \cdot n \cdot n \cdot n \cdot n$

Examples

2 **Evaluate 2^5.**

$2^5 = 2 \cdot 2 \cdot 2 \cdot 2 \cdot 2$
$= 32$

3 **Evaluate 29^2.**

$29^2 = 29 \cdot 29$
$= 841$

4 **Write an algebraic expression for the following verbal expression: the cube of a number y decreased by nine.**

The algebraic expression is $y^3 - 9$.

Exploratory Exercises

Write an algebraic expression for each verbal expression.

1. the product of x and 7 $7x$
2. the quotient of r and s $\frac{r}{s}$
3. the total of a and 19 $a + 19$
4. a number c increased by 10 $c + 10$
5. a number b to the third power b^3
6. 7 less than w $w - 7$
7. 2 divided by a number k $2 \div k$ or $\frac{2}{k}$
8. a number h cubed h^3
9. 25 squared 25^2
10. 16 multiplied by m $16m$
11. a number n decreased by 4 $n - 4$
12. 9 to the fifth power 9^5

State a verbal expression for each of the following.

13. x^2
14. n^4
15. 5^3
16. y^5
17. n^1
18. z^7

Written Exercises

Write each of the following as an expression using exponents.

1. $5 \cdot 5 \cdot 5$ 5^3
2. $4 \cdot 4 \cdot 4 \cdot 4$ 4^4
3. $7 \cdot a \cdot a \cdot a \cdot a$ $7a^4$
4. $2(m)(m)(m)$ $2m^3$
5. $\frac{1}{2} \cdot a \cdot a \cdot b \cdot b \cdot b$
6. $\frac{3}{4} \cdot x \cdot y \cdot y \cdot y \cdot y \cdot y$
7. $5 \cdot 5 \cdot 5 \cdot x \cdot x \cdot y$
8. $3 \cdot 3 \cdot a \cdot a \cdot a$

Write an algebraic expression for each verbal expression. Use x as the variable.

9. a number increased by 17 $x + 17$
10. seven times a number $7x$
11. the cube of a number x^3
12. a number to the sixth power x^6
13. twice the square of a number $2x^2$
14. one-half the square of a number $\frac{1}{2}x^2$
15. 8 times a number decreased by 17 $8x - 17$
16. twice a number decreased by 25 $2x - 25$
17. 94 increased by twice a number $94 + 2x$
18. twice the cube of a number $2x^3$
19. three-fourths of the square of a number $\frac{3}{4}x^2$
20. the product of 7 and the cube of a number $7x^3$
21. the cube of a number increased by seven $x^3 + 7$
22. 29 increased by the square of a number $29 + x^2$

Evaluate each expression.

23. 2^4 16
24. 6^2 36
25. 5^3 125
26. 10^4 10,000
27. 4^5 1024
28. 3^6 729

Chalkboard Exercises

2. Evaluate 3^5. 243
3. Evaluate 15^2. 225

Write an algebraic expression for each verbal expression.

4a. a number n to the eighth power n^8
4b. the square of a number b increased by 27 $b^2 + 27$
4c. the sum of two numbers x and y $x + y$

Suggested Daily Assignments

Enriched
Written: 2–28 evens

Average
Exp: 2–18 evens
Written: 2–28 evens

Fundamental
Exp: 1–18
Written: 2–28 evens

Answers to Exploratory

13. x squared or x to the second power
14. n for the fourth power
15. 5 cubed or 5 to the third power
16. y to the fifth power
17. n to the first power
18. z to the seventh power

Answers to Written

5. $\frac{1}{2}a^2b^3$
6. $\frac{3}{4}xy^5$
7. 5^3x^2y
8. 3^2a^3

1-2

Lesson Notes, p. T24

Discussion Question
Insert parentheses in different places in the following expression and evaluate. What values can you obtain?
$5 + 2 \times 3 + 1 + 6 \times 0$
Answers may vary.

Chalkboard Exercises
Evaluate each expression.
1a. $14 \div 7 \cdot 2 - 3$ 1
1b. $64 \div 8 - 3 \cdot 2$ 2
1c. $13 - 2^4 \div 4 + 2$ 11

Evaluate $2x^2 + 5xy$ if x and y have the following values.
2a. $x = 4$ and $y = \frac{3}{4}$ 47
2b. $x = \frac{1}{2}$ and $y = 3$ 8
2c. $x = 7$ and $y = \frac{3}{35}$ 101

❶ Point out to students that *unique* means "one and only one."

❷ The formal definition of the Substitution Property of Equality is given on page 13. Most students should be able to grasp this concept easily.

Objective 1-2: Use the order of operations to evaluate expressions, including expressions with grouping symbols.

1-2 Evaluating Expressions

The numerical expression $5 \cdot 6 + 3$ might be computed in two ways. Only one is correct.

Multiply 5 and 6. Add 3.	Add 6 and 3. Multiply by 5.
$5 \cdot 6 + 3 = 30 + 3$	$5 \cdot 6 + 3 = 5(9)$
$= 33$	$= 45$ ❶

Each numerical expression should have a *unique* value. Therefore, the following order of operations has been agreed upon.

Order of Operations

1. **Evaluate all powers.**
2. **Do all multiplications and divisions from left to right.**
3. **Do all additions and subtractions from left to right.**

Thus, using the order of operations, the value of $5 \cdot 6 + 3$ is 33.

Example 1

Evaluate $12 \div 3 \cdot 5 - 4^2$.

$12 \div 3 \cdot 5 - 4^2 = 12 \div 3 \cdot 5 - 16$ *Evaluate 4^2.*
$= 4 \cdot 5 - 16$ *Divide 12 by 3.*
$= 20 - 16$ *Multiply 4 by 5.*
$= 4$ *Subtract 16 from 20.*

An algebraic expression can be evaluated when the value of each variable is known. The variables are simply replaced by their values. Then, the value of the numerical expression is calculated.

Example 2

Evaluate $2x^2 + 5xy$ if $x = 3$ and $y = \frac{2}{3}$.

$2x^2 + 5xy = 2 \cdot 3^2 + 5(3)\left(\frac{2}{3}\right)$ *Substitute 3 for x and $\frac{2}{3}$ for y.* ❷
$= 2 \cdot 9 + 5(3)\left(\frac{2}{3}\right)$ *Evaluate 3^2.*
$= 18 + 10$ *Multiply.*
$= 28$ *Add.*

In mathematics, *grouping symbols* such as parentheses are used to clarify or change the order of operations.

$5 \cdot 6 + 3$ means $30 + 3$ or 33

$5(6 + 3)$ means $5 \cdot 9$ or 45

The basic grouping symbols used are parentheses, ( ), brackets, [ ], and the fraction bar. Parentheses and brackets indicate that the expression within is to be treated as a single value and should be evaluated first. A fraction bar indicates that the numerator and denominator should each be treated as a single value. ❸

When more than one grouping symbol is used, start evaluating expressions within the innermost grouping symbols.

Examples

3 **Evaluate $3(2 + 5)^2 \div 7$.**

$3(2 + 5)^2 \div 7 = 3(7)^2 \div 7$ — *Add 2 and 5.*
$= 3(49) \div 7$ — *Evaluate 7^2.*
$= 147 \div 7$ — *Multiply 3 by 49.*
$= 21$ — *Divide 147 by 7.*

4 **Evaluate $\frac{2}{3}[8(a - b)^2 + 3b]$ if $a = 5$ and $b = 2$.**

$\frac{2}{3}[8(a - b)^2 + 3b] = \frac{2}{3}[8(5 - 2)^2 + 3 \cdot 2]$ — *Substitute 5 for a and 2 for b.*
$= \frac{2}{3}[8(3)^2 + 3 \cdot 2]$ — *Subtract 2 from 5.*
$= \frac{2}{3}[8 \cdot 9 + 3 \cdot 2]$ — *Evaluate 3^2.*
$= \frac{2}{3}[72 + 6]$ — *Multiply 8 by 9 and 3 by 2.*
$= \frac{2}{3}[78]$ — *Add 72 and 6.*
$= 52$ — *Multiply $\frac{2}{3}$ by 78.*

5 **Evaluate $\frac{b^2 - 2}{b + x^3}$ if $b = 5$ and $x = 3$.**

$\frac{b^2 - 2}{b + x^3} = \frac{5^2 - 2}{5 + 3^3}$ — *Substitute 5 for b and 3 for x.*
$= \frac{25 - 2}{5 + 27}$ — *Simplify the numerator and the denominator.*
$= \frac{23}{32}$

Exploratory Exercises

State how to evaluate each expression. Do *not* evaluate.

1. $3 + 2 \cdot 4$	**2.** $2 \cdot 4 + 3$	**3.** $8 \div 4 \cdot 2$	**4.** $8 \cdot 2 \div 4$
5. $12 - 6 \cdot 2$	**6.** $(12 - 6) \cdot 2$	**7.** $(9 - 3)^2$	**8.** $9 - 3^2$
9. $9^2 - 3^2$	**10.** $3(3^2 - 3)$	**11.** $7 \cdot 3^2$	**12.** $7^2 \cdot 2$
13. $4(5 - 3)^2$	**14.** $8 \div 2 + 6 \cdot 2$	**15.** $(8 + 6) \div 2 + 2$	**16.** $8 + 6 \div (2 + 1)$

❸ Braces are sometimes used as grouping symbols. For example, $2\{3[5 + 4(7 - 5)]\}$.

Chalkboard Exercises

Evaluate eac.. expression.

3a. $3(3 + 7)^2 \div 5$ 60
3b. $27 \div 3(5 - 3)^2$ 36
4a. Evaluate $\frac{2}{3}[8(a - b)^2 + 3b]$ if $a = 6$ and $b = 3$. 54

Evaluate $\frac{a^2 - 5}{a + x^2}$ if a and x have the following values.

5a. $a = 6$ and $x = 2$ $\frac{31}{10}$ or $3\frac{1}{10}$
5b. $a = 5$ and $x = 4$ $\frac{20}{21}$

Suggested Daily Assignments

Enriched
Written: 2–46 evens

Average
Exp: 2–16 evens
Written: 1–24, 26–46 evens

Fundamental
Exp: 1–16
Written: 2–36 evens
p. 5, 1–27 odds

Answers to Exploratory

1. Multiply 2 by 4. Then add 3.
2. Multiply 2 by 4. Then add 3.
3. Divide 8 by 4. Then multiply by 2.
4. Multiply 8 by 2. Then divide by 4.
5. Multiply 6 by 2. Then subtract from 12.
6. Subtract 6 from 12. Then multiply by 2.
7. Subtract 3 from 9. Then square the result.
8. Square 3. Then subtract from 9.
9. Square 9. Square 3. Then subtract.
10. Square 3. Subtract 3. Then multiply by 3.
11. Square 3. Then multiply by 7.
12. Square 7. Then multiply by 2.
13. Subtract 3 from 5. Square the result. Then multiply by 4.

Answers to Exploratory (continued)

14. Divide 8 by 2. Multiply 6 by 2. Then add.
15. Add 8 and 6. Divide by 2. Then add 2.
16. Add 2 and 1. Divide 6 by 3. Then add 8.

Written Exercises

Evaluate each expression.

1. $3 + 8 \div 2 - 5$ 2
2. $4 + 7 \cdot 2 + 8$ 26
3. $12 \div 4 + 15 \cdot 3$ 48
4. $3 \cdot 6 - 12 \div 4$ 15
5. $5(9 + 3) - 3 \cdot 4$ 48
6. $7(8 - 4) + 11$ 39
7. $29 - 3(9 - 4)$ 14
8. $4(11 + 7) - 9 \cdot 8$ 0
9. $12 \cdot 6 \div 3 \cdot 2 \div 8$ 6
10. $16 \div 2 \cdot 5 \cdot 3 \div 6$ 20
11. $288 \div [3(9 + 3)]$ 8
12. $196 \div [4(11 - 4)]$ 7
13. $5^3 + 6^3 - 5^2$ 316
14. $6(4^3 + 2^2)$ 408
15. $\frac{38 - 12}{2 \cdot 13}$ 1
16. $\frac{9 \cdot 4 + 2 \cdot 6}{7 \cdot 7}$ $\frac{48}{49}$
17. $\frac{9 \cdot 3 - 4^2}{3^2 + 2^2}$ $\frac{11}{13}$
18. $\frac{2 \cdot 8^2 - 2^2 \cdot 8}{2 \cdot 8}$ 6
19. $\frac{2}{3}(16) - \frac{1}{3}(6)$ $\frac{26}{3}$ or $8\frac{2}{3}$
20. $\frac{3}{4}(6) + \frac{1}{3}(12)$ $\frac{17}{2}$ or $8\frac{1}{2}$
21. $\frac{1}{2}(8 + 30) - 4$ 15
22. $25 - \frac{1}{3}(18 + 9)$ 16
23. $0.2(0.6) + 3(0.4)$ 1.32
24. $7(0.2 + 0.5) - 0.6$ 4.3

Evaluate if $a = 6$, $b = 4$, $c = 3$, $d = \frac{1}{2}$, $e = \frac{2}{3}$, $x = 0.2$, and $y = 1.3$.

25. $a + b^2 + c^2$ 31
26. $3ab - c^2$ 63
27. $8(a - c)^2 + 3$ 75
28. $\frac{a^2 - b^2}{2}$ 10
29. $\frac{2ab - c^3}{7}$ 3
30. $12d + bc$ 18
31. $d(a + b) - c$ 2
32. $a(8 - 3e) + 4d$ 38
33. $ax + bc$ 13.2
34. $5y + 3$ 9.5
35. $x^2 + 6d$ 3.04
36. $c^2 + y^2$ 10.69
37. $(100x)^2 + 10y$ 413
38. $(15x)^3 - y$ 25.7
39. $\frac{a + b^2}{3bc}$ $\frac{11}{18}$
40. $\frac{4d^2}{4ce}$ $\frac{1}{8}$
41. $\frac{8b^2c}{x}$ 1920
42. $\frac{6ab^2}{x + y}$ 384

Write an algebraic expression for each verbal expression. Then, evaluate the expression if $a = 3$, $b = \frac{1}{2}$, and $c = 0$.

43. twice the sum of a and b $2(a + b)$; 7
44. twice the product of a and b $2(ab)$; 3
45. the square of b increased by c $b^2 + c$; $\frac{1}{4}$
46. the cube of a decreased by b $a^3 - b$; $26\frac{1}{2}$

Using Calculators — Order of Operations

What is the result if you enter the following on your calculator?

ENTER: 25 [−] 6 [×] 3 [=]

If the display shows 7, the calculator followed the correct order of operations. It multiplied 6 and 3 first, then subtracted the result from 25.

If the display shows 57, the calculator performed the operations in the order that the keys were pressed. This result is incorrect. How can you get the correct value?

ENTER: 6 [×] 3 [=] *The display should show 18.*

Then ENTER: 25 [−] 18 [=]

The display should show 7, the correct value of the expression $25 - 6 \times 3$.

Exercises Evaluate each expression.

1. $70 - 6 \times 8 + 10$ 32
2. $300 - 4 \times 32$ 172
3. $61 \times 4 + 3 \times 47$ 385

About Using Calculators
Almost all calculators made today follow the correct order of operations. Make sure your students know whether or not their calculators follow the correct order of operations.

For more exercises on evaluating expressions, see exercises 1-36 on page 554.
For review exercises on multiplying and dividing fractions, see exercises 1-24 on page 549.

Objective 1-3: Solve open sentences using arithmetic knowledge of the four basic operations.

1-3 Open Sentences and Equations

Which of the following sentences are true? Which are false?

A dog has three ears.

California is larger than Rhode Island.

$3 \times 7 = 21$

27 divided by 3 is less than 7.

$\frac{3}{4} + \frac{1}{2} = 1\frac{1}{4}$

Note that each of the previous sentences are either true or false. However, it is not possible to say whether the following sentences are true or false.

Words	Open Sentences
A number x plus six is equal to 8.	$x + 6 = 8$
A number n is less than 10.	$n < 10$
Seven is greater than twice a number c.	$7 > 2c$

Before you can determine whether these sentences are true or false, you must know what numbers will replace the variables x, n, and c. Mathematical sentences such as $x + 6 = 8$ are called **open sentences**. In an open sentence such as $x + 6 = 8$, the variable must be replaced in order to determine if the sentence is true or false. ❶

Open sentences are neither true nor false.

Finding the replacements for the variable that make a sentence true is called solving the open sentence. Each replacement is called a solution of the open sentence.

Solving an Open Sentence

Is 11 a solution of the open sentence $n < 10$?
Is 3 a solution of the open sentence $x + 6 = 8$?
Try naming some solutions of the three open sentences listed above. ❷

A mathematical sentence that contains an **equals sign**, =, is called an **equation**. Sometimes you can solve an equation by simply performing the indicated operations.

1-3

Lesson Notes, p. T24

Readiness Activity
Fill in the blank to make each sentence true.

$4 \cdot \underline{\ 2\ } = 8.$
$\underline{\ 17\ } + 5 = 22.$
$15 = 30 \div \underline{\ 2\ }.$
$8 = \underline{\ 12\ } - 4.$

Discussion Question
What are some open sentences that have 2 as a solution?
Answers may vary. Some possible answers are $3x = 6$, $b + 5 = 7$, $y > 1$.

❶ In mathematics, sentences that can be classified as true or false are called *statements*. In general, *open sentences* are true for some replacements and false for others. The original sentence is neither true nor false.

❷ Remind students that "fill in the blank" questions are really open sentences for which they must find the solution set. Some examples of such questions are given below.
__________ was the first president of the United States.
A pentagon has __________ sides.

Chalkboard Exercises

1. Solve: $m = \frac{7 \cdot 7 + 2}{10 + 7}$ 3
2. Solve: $\frac{4.2 + 2(3.9)}{3.1 + 2.9} = k$ 2

Suggested Daily Assignments

Enriched
Exp: 2–22 evens
Written: 2–24 evens, 25–33

Average
Exp: 2–22 evens
Written: 2–26 evens

Fundamental
Exp: 1–22
Written: 2–24 evens
p. 8, 1–35 odds

Quiz
Lessons 1-1 to 1-3

Write a mathematical expression for each verbal expression.

1. Six less than twice a number $2n - 6$
2. One-third the cube of a number $\frac{1}{3}n^3$

Write an expression using exponents for the following.

3. $2 \cdot 2 \cdot 2 \cdot 2 \cdot x \cdot x \cdot x \cdot x \cdot x$ 2^4x^5

Evaluate the following.

4. $12 - 2 \cdot 5 + 18 \div 3$ 8
5. $34 - [6(5 - 3)]$ 22
6. $\frac{2x^2 - y^3}{xy + 3}$ if $x = 3$ and $y = 2$. $\frac{10}{9}$

Solve each equation.

7. $\frac{1}{8} + \frac{2}{3} = y$ $\frac{19}{24}$
8. $k = \frac{8.6 - 1.16}{(1.2)(2)}$ 3.1

Examples

1 **Solve: $a = 9.4 - 3.06$**

$a = 9.4 - 3.06$
$= 9.40 - 3.06$
$= 6.34$

The solution is 6.34.

2 **Solve: $\frac{5 \cdot 3 + 3}{4 \cdot 2 - 2} = t$**

$\frac{5 \cdot 3 + 3}{4 \cdot 2 - 2} = t$

$\frac{15 + 3}{8 - 2} = t$

$\frac{18}{6} = t$

$3 = t$

The solution is 3.

Exploratory Exercises

State whether each sentence is true or false.

1. The capital of the U.S. is Houston. false
2. Birds have wings. true
3. $3(11 - 5) > 18$ false
4. $15 \div 3 + 7 < 13$ true
5. $0.01 + 0.01 = 0.0002$ false
6. $\frac{2}{3}(6) = 9 - 5$ true
7. $\frac{3 + 15}{6} = \frac{1}{2}(6)$ true
8. $\frac{1}{2} + \frac{3}{4} = \frac{3}{2} + \frac{1}{4}$ false
9. $3 - 1\frac{1}{2} = \frac{1}{2}$ false
10. $21 - 12 > 9$ false
11. $17 + 3 < 100 \div 4$ true
12. $0.11 + 1.1 > 1.2$ true
13. $10\left(\frac{2}{5}\right) = \frac{1}{5} \cdot 2 \cdot 10$ true
14. $0.101 > 0.110$ false
15. $7 = 7(0)$ false
16. $0 + 7 = 7$ true

Replace the variable to make each open sentence true.

17. x is President of the U.S. (name of current president)
18. There are y states in the U.S. 50
19. $18 + y = 20$ 2
20. $3 \cdot a = 24$ 8
21. t is an author of this book. Foster, Rath, or Winters
22. $\frac{1}{2}(m) = 10$ 20

Written Exercises

Solve each equation.

1. $x = 8 + 3$ 11
2. $y = 7 - 4$ 3
3. $x = 12 - 0.03$ 11.97
4. $x = 6 + 0.28$ 6.28
5. $a = \frac{3}{4} \cdot 12$ 9
6. $\frac{5}{6} \cdot 18 = m$ 15
7. $8.2 - 6.75 = a$ 1.45
8. $9.6 + 4.53 = b$ 14.13
9. $a = \frac{12 + 8}{4}$ 5
10. $m = \frac{64 + 4}{17}$ 4
11. $\frac{21 - 3}{12 - 3} = x$ 2
12. $\frac{14 + 28}{4 + 3} = y$ 6
13. $14.8 - 3.75 = a$ 11.05
14. $29.7 - 5.86 = t$ 23.84
15. $x = \frac{84 \div 7}{18 \div 9}$ 6
16. $x = \frac{96 \div 6}{8 \div 2}$ 4
17. $\frac{2}{13} + \frac{5}{13} = x$ $\frac{7}{13}$
18. $\frac{3}{7} + \frac{2}{7} = y$ $\frac{5}{7}$
19. $\frac{5}{8} + \frac{1}{4} = x$ $\frac{7}{8}$
20. $\frac{7}{12} + \frac{1}{3} = x$ $\frac{11}{12}$
21. $a = 3\frac{1}{2} \div 2$ $\frac{7}{4}$
22. $m = 3\frac{1}{3} \div 3$ $\frac{10}{9}$
23. $r = 5\frac{1}{2} + \frac{1}{3}$ $5\frac{5}{6}$
24. $s = 6\frac{3}{4} + \frac{1}{6}$ $6\frac{11}{12}$

Challenge

The set of numbers that a variable may represent is called a *replacement set*. Find all numbers from the replacement set $\{4, 5, 6, 7, 8\}$ that are solutions to each open sentence.

25. $x + 2 > 7$ {6, 7, 8}

26. $10 - x < 7$ {4, 5, 6, 7, 8}

27. $\frac{x + 3}{2} < 5$ {4, 5, 6}

28. $x - 3 > \frac{x + 1}{2}$ {8}

29. $\frac{2(x - 2)}{3} = \frac{4}{7 - 5}$ {5}

30. $9x - 20 = x^2$ {4, 5}

31. $0.3(x + 4) \leq 0.4(2x + 3)$ {4, 5, 6, 7, 8}

32. $1.3x - 12 < 0.9x + 4$ {4, 5, 6, 7, 8}

33. $\frac{2x + 1}{7} \geq \frac{x + 4}{5}$ {8}

About the Excursion
Be sure to emphasize the importance of braces. Many students will have difficulty with the idea that the symbol 1 represents an element, while {1} represents a set.

Excursions in Algebra — Sets

A collection of objects or numbers is often called a **set**. A set can be shown by using braces.

$\{1, 2, 3\}$ is *the set of numbers 1, 2, 3.*

Each object or number in a set is called an **element**.

1 *is an element of* $\{1, 2, 3\}$.
$1 \in \{1, 2, 3\}$

Sets are often named by capital letters.

Set A is $\{1, 2, 3\}$.
$A = \{1, 2, 3\}$

If every element of set A is also an element of set B, then set A is a **subset** of set B.

$\{1, 2\}$ *is a subset of* $\{1, 2, 3\}$. $\{1\}$ *is a subset of* $\{1, 2, 3\}$.
$\{1, 2\} \subset \{1, 2, 3\}$ $\{1\} \subset \{1, 2, 3\}$

The set with no elements is the **null** or **empty set**. It is shown by { } or $\emptyset$. The empty set is a subset of every set.

Exercises

Write in symbols.

1. the set of numbers 3, 4, 5, 6

2. the set of numbers 6, 7, 8, 9, 10, 11

3. 3 is an element of the set of numbers 3, 4, 5, 6.

4. 6 is an element of the set of integers between and including 5 and 10.

Find all subsets of each set to show that the following are true.

5. $\{1, 2, 4\}$ has 8 subsets.

6. $\{5, 6, 7, 8\}$ has 16 subsets.

Answers to Excursion

1. {3, 4, 5, 6}
2. {6, 7, 8, 9, 10, 11}
3. $3 \in \{3, 4, 5, 6\}$
4. $6 \in \{5, 6, 7, 8, 9, 10\}$
5. {1}, {2}, {4}, {1, 2}, {1, 4}, {2, 4}, {1, 2, 4}, $\emptyset$
6. {5}, {6}, {7}, {8}, {5, 6}, {5, 7}, {5, 8}, {6, 7}, {6, 8}, {7, 8}, {5, 6, 7}, {5, 6, 8}, {5, 7, 8}, {6, 7, 8}, {5, 6, 7, 8}, $\emptyset$

1-4

Lesson Notes, p. T25

Readiness Activity
True or false?
3 + 0 = 3 true
1 = 4 · 1 false
0 = 6 · 0 true
If 3 + 2 = 5 and 5 = 7 − 2, then 3 + 2 = 7 − 2. true

Discussion Question
(Use after the Identity Properties are introduced.) Multiplying by 0 gives a product of 0 (Mult. Prop. of Zero). Does dividing by 0 give a quotient of 0?
No, division by 0 is undefined.

Objective 1-4: Recognize and use the properties of identity and equality.

1-4 Identity and Equality Properties

Solve each equation.

$6 + a = 6$ $\qquad b + 15 = 15$ $\qquad 4780 + c = 4780$

The solution of each equation is 0. The sum of any given number and 0 is the given number. The **additive identity** of any number is 0.

For any number *a*, $a + 0 = 0 + a = a$.

Additive Identity Property

Solve each equation.

$5x = 5$ $\qquad 18y = 18$ $\qquad z \cdot 655 = 655$

The solution of each equation is 1. The product of any given number and 1 is the given number. The **multiplicative identity** of any number is 1.

For any number *a*, $a \cdot 1 = 1 \cdot a = a$.

Multiplicative Identity Property

What role does 0 play in multiplication?

$0(11) = 0$ $\qquad 5 \cdot 0 = 0$ $\qquad 8 \cdot 3 \cdot 0 = 0$ $\qquad 4 \cdot 0 \cdot 27 \cdot 8 = 0$

When a factor is 0, the product is 0.

For any number *a*, $a \cdot 0 = 0 \cdot a = 0$.

Multiplicative Property of Zero

Think about the following statements.

1. $5 = 5$
2. If $7 + 3 = 10$, then $10 = 7 + 3$.
3. If $10 - 2 = 8$ and $8 = 5 + 3$, then $10 - 2 = 5 + 3$.

You have assumed statements like these to be true from your experiences in arithmetic. Actually, these are examples of the properties of equality.

The following properties are true for any numbers *a*, *b*, and *c*.
Reflexive Property: $a = a$
Symmetric Property: If $a = b$, then $b = a$,
Transitive Property: If $a = b$ and $b = c$, then $a = c$.

Properties of Equality

Which property corresponds to each of the three numbered statements given above?

Another property of equality is the substitution property. This property permits substitution of a quantity for its equal. You use this property regularly. For example, you would simplify the expression $8 + (3 + 9)$ as follows.

$8 + (3 + 9) = 8 + 12$ *Substitute 12 for (3 + 9).*
$= 20$ *Substitute 20 for 8 + 12.*

For any numbers a and b, if $a = b$ then a may be replaced by b. ***Substitution Property of Equality***

Exploratory Exercises

State the property shown in each of the following.

1. $3 = 3$ reflexive
2. $7 + 6 = 7 + 6$ reflexive
3. If $7 = 2 + 5$, then $2 + 5 = 7$. symmetric
4. If $6 = 3 + 3$, then $3 + 3 = 6$. symmetric
5. If $8 = 6 + 2$ and $6 + 2 = 5 + 3$, then $8 = 5 + 3$. transitive
6. If $8 + 1 = 9$ and $9 = 3 + 6$, then $8 + 1 = 3 + 6$. transitive
7. $3 + (2 + 1) = 3 + 3$ substitution
8. $9 + 5 = (6 + 3) + 5$ substitution

Solve each equation.

9. $0 + x = 7$ 7
10. $a \cdot 1 = 5$ 5
11. $7b = 7$ 1
12. $0(18) = n$ 0

Written Exercises

State the property shown in each of the following.

1. If $8 + 1 = 9$, then $9 = 8 + 1$. symmetric
2. $0 \cdot 36 = 0$ mult. property of zero
3. If $5 + 1 = 6$ and $6 = 4 + 2$, then $5 + 1 = 4 + 2$. transitive
4. $9 + (2 + 10) = 9 + 12$ substitution
5. $1(87) = 87$ multiplicative identity
6. $0 + 17 = 17$ additive identity
7. $14 + 16 = 14 + 16$ reflexive
8. $6 \cdot 1 = 6$ multiplicative identity
9. $14 \cdot 0 = 0$ mult. property of zero
10. If $3 = 4 - 1$, then $4 - 1 = 3$. symmetric
11. $6 + 8 = 6 + 8$ reflexive
12. If $9 + 1 = 10$ and $10 = 5(2)$, then $9 + 1 = 5(2)$. transitive
13. $(9 - 7)(5) = 2(5)$ substitution
14. $62 = 62$ reflexive
15. If $11 - 5 = 4 + 2$, then $4 + 2 = 11 - 5$.
16. If $a = 6 + 1$ and $6 + 1 = b$, then $a = b$.
17. $7(0) = 0$ mult. property of zero
18. $3 + 5 + 7 = 3 + 5 + 7$ reflexive
19. $abc = 1abc$ multiplicative identity
20. If $a + b = c + d$, then $c + d = a + b$. symmetric

mini-review

Evaluate each expression.

1. 4^3 64
2. $5(7 + 4) - 3$ 52
3. $\frac{16 + 8}{4 - 2}$ 12

Solve.

4. $\frac{1}{2} + \frac{1}{6} = m$ $\frac{2}{3}$
5. $r = 2.3 + 12$ 14.3

Suggested Daily Assignments

Enriched
Exp: 9–12
Written: 1–20
Mini–Review: 1–5

Average
Exp: 1–12
Written: 1–20
Mini–Review: 1–5

Fundamental
Exp: 1–12
Written: 2–20 evens
p. 10, 1–23 odds
Mini-Review: 1–5

Answers to Written
15. symmetric
16. transitive

About the Excursion

Students who enjoy science benefit greatly from an understanding of the metric system. This section tries to develop an intuitive feeling for metric length, rather than emphasizing conversion factors. You may want to have students do research in order to answer the following questions.

1. How was the length of a meter selected?
2. In the early 1800's the United States considered adopting the metric system. Why was the conventional system retained?
3. What are the advantages and disadvantages of the metric system?

Excursions in Algebra — Length in the Metric System

During the 1790's, a group of French scientists developed a new system of measurement called the **metric system**. The basic unit of length is the **meter**.

The metric system is a *decimal* system. That is, units increase or decrease in size by multiples of ten. A prefix is attached to the word *meter* to define larger and smaller units. The table shows some metric prefixes and their meanings.

Prefix	Symbol	Increase or decrease in unit
mega	M	1,000,000 (one million)
kilo	k	1,000 (one thousand)
hecto	h	100 (one hundred)
deka	da	10 (ten)
deci	d	0.1 (one-tenth)
centi	c	0.01 (one-hundredth)
milli	m	0.001 (one-thousandth)
micro	μ	0.000001 (one-millionth)

When the prefix *centi* is added to *meter*, the result is *centimeter*. The symbol for centimeter is cm. A centimeter is *one-hundredth* of a meter.

1 cm = 0.01 m 1 m = 100 cm *m is the symbol for meter.*

25 cm = 0.25 m 3.5 m = 350 cm

A door is about 1 m wide.

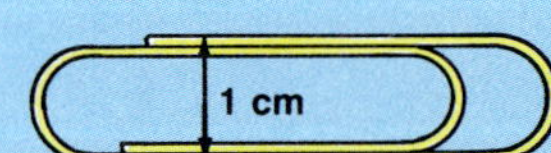

A paper clip is about 1 cm wide.

Exercises

1. Name 3 things that are best measured in meters. Answers will vary.
2. Name 3 things that are best measured in centimeters. Answers will vary.

Complete. *km stands for kilometer and mm stands for millimeter.*

3. 35 cm = <u>0.35</u> m	**4.** 57 cm = <u>0.57</u> m	**5.** 136 cm = <u>1.36</u> m
6. 395 cm = <u>3.95</u> m	**7.** 3 m = <u>300</u> cm	**8.** 4.7 m = <u>470</u> cm
9. 1 km = <u>1000</u> m	**10.** 4 km = <u>4000</u> m	**11.** 5000 m = <u>5</u> km
12. 1 cm = <u>10</u> mm	**13.** 14 cm = <u>140</u> mm	**14.** 55 mm = <u>5.5</u> cm

Objective 1-5: Use the distributive property to simplify expressions.

1-5 The Distributive Property

Angie and Maria are clerks in a local department store. Each one earns \$4.65 per hour. Angie works 24 hours per week while Maria works 32 hours per week. What are their total earnings?

This problem can be solved in two ways. You could find Angie's earnings and Maria's earnings, then add to find the total earnings. Or, you could find the total number of hours worked, then multiply by the hourly wage.

Angie's Earnings	+	Maria's Earnings	=	Total Earnings
(\$4.65)(24)	+	(\$4.65)(32)	=	(\$4.65)(24 + 32)
\$111.60	+	\$148.80	=	(\$4.65)(56)
		\$260.40	=	\$260.40

The total earnings of \$260.40 are distributed between Angie and Maria, each being paid \$4.65 per hour for each hour worked. This is an example of the **distributive property**. ❶

For any numbers a, b, and c
1. $a(b + c) = ab + ac$ and $(b + c)a = ba + ca$.
2. $a(b - c) = ab - ac$ and $(b - c)a = ba - ca$.

Distributive Properties

Example 1

Use two ways to evaluate 8(5 + 4).

$$8(5 + 4) = 8(9) = 72$$

$$8(5 + 4) = 8(5) + 8(4) = 40 + 32 = 72$$

The distributive property is needed to simplify algebraic expressions containing **like terms**. A **term** of an expression is a number, a variable, or a product or quotient of numbers and variables. Some examples of terms are $5x^2$, $\frac{ab}{4}$, $\frac{x}{y}$, and $7k$. Some pairs of like terms are $5x$ and $3x$, $2xy$ and $7xy$, and $7ax^2$ and $11ax^2$.

Like terms are terms that contain the same variables, with corresponding variables raised to the same power.

Definition of Like Terms

1-5

Lesson Notes, p. T25

Readiness Activity
Which of the following expressions have the same value?

(7 + 6)5	3 · 4 + 3 · 2
8 · 6 + 9 · 6	4(2 + 5)
3(4 + 2)	7 · 5 + 6 · 5
4 · 2 + 4 · 5	(8 + 9)6

Discussion Questions
Show that each of the following sentences is false for some real numbers. (Hint: Replace each variable with a number.)

1. $7m - m = 7$
2. $3(n + 4) = 3n + 4$
3. $5 + x = 5x$
4. $6y + 2 = 8y$

❶ Graph paper can be used to illustrate the distributive property.
4(2 + 3) = 4(2) + 4(3)

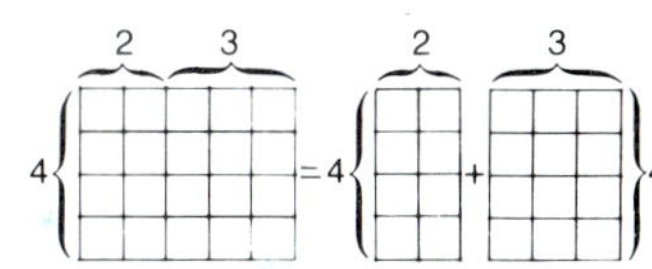

Chalkboard Exercises
Use two methods to evaluate each of the following expressions.

1a. 11(8 + 2)
11(8 + 2) = 11(10) = 110
11(8 + 2) = 11(8) + 11(2) = 88 + 22 = 110

1b. 6(7 − 3)
6(7 − 3) = 6(4) = 24
6(7 − 3) = 6(7) − 6(3) = 42 − 18 = 24

Chalkboard Exercises

Simplify.

2a. $16a + 7a$ $23a$

2b. $6.1y - 3.2y$ $2.9y$

3. $6xy^2 + 7xy^2$ $13xy^2$

Name the coefficient of each term.

4a. y^2 1

4b. $\frac{r}{4}$ $\frac{1}{4}$

4c. $\frac{5x^2}{7}$ $\frac{5}{7}$

Simplify.

5a. $13a^2 + 8a^2 + 6b$ $21a^2 + 6b$

5b. $6a^2 + 12a^2 + 4d - d$ $18a^2 + 3d$

Simplify.

6a. $\frac{3}{4}y + \frac{y}{4}$ y

6b. $\frac{5}{6}m - \frac{m}{6} + 4m^2$ $4m^2 + \frac{2}{3}m$

❷ Emphasize that the coefficient of $\frac{b}{2}$ is $\frac{1}{2}$. This is often difficult for students.

An expression in **simplest form** has no like terms and no parentheses.

Examples

2 **Simplify: $5a + 7a$**

$5a + 7a = (5 + 7)a$ *Use the Distributive Property: $ba + ca = (b + c)a$.*

$= 12a$ *Substitution Property*

3 **Simplify: $6x^2y - 2x^2y$**

$6x^2y - 2x^2y = (6 - 2)x^2y$ *Distributive Property*

$= 4x^2y$ *Substitution Property*

Like terms may also be defined as terms which are the same, or which differ only in their coefficients. The **coefficient**, or **numerical coefficient**, is the numerical part of a term. For example, the coefficient of $6ab$ is 6. The coefficient of rs is 1 since by the multiplicative identity property $1 \cdot rs = rs$.

Examples

4 **Name the coefficient of each term.**

a. $19g^2h$ The coefficient is 19.

b. xy^2z^2 The coefficient is 1 since $1xy^2z^2 = xy^2z^2$.

c. $\frac{mn}{2}$ The coefficient is $\frac{1}{2}$ since $\frac{mn}{2}$ can be written as $\frac{1}{2}(mn)$.

d. $\frac{2x^3}{5}$ The coefficient is $\frac{2}{5}$ since $\frac{2x^3}{5}$ can be written as $\frac{2}{5}(x^3)$.

5 **Simplify: $8n^2 + n^2 + 7n + 3n$**

$8n^2 + n^2 + 7n + 3n = 8n^2 + 1n^2 + 7n + 3n$ *Multiplicative Identity Property*

$= (8 + 1)n^2 + (7 + 3)n$ *Distributive Property*

$= 9n^2 + 10n$

6 **Simplify: $\frac{b}{2} + b$** ❷

$\frac{b}{2} + b = \frac{1}{2}b + 1b$ *Multiplicative Identity Property*

$= \left(\frac{1}{2} + 1\right)b$ *Distributive Property*

$= \left(\frac{1}{2} + \frac{2}{2}\right)b$ *Substitution Property*

$= \frac{3}{2}b$ *Substitution Property*

Exploratory Exercises

State the coefficient in each term.

1. $5m$ 5 **2.** $3am$ 3 **3.** $0.2mp$ 0.2 **4.** a^2q 1

5. $\frac{1}{5}mn^2$ $\frac{1}{5}$ **6.** $\frac{am}{3}$ $\frac{1}{3}$ **7.** $0.5abm$ 0.5 **8.** $\frac{4am^2}{5}$ $\frac{4}{5}$

Name the like terms in each list of terms.

9. $6b, 6bc, bc$ $6bc, bc$ **10.** $7a, 7a^2, 29a, 7a^3$ $7a, 29a$ **11.** $4xy, 5xy, 6xy^2, 6x^2y$ $4xy, 5xy$

12. $2rs^2, r^2s, rs^2$ $2rs^2, rs^2$ **13.** $\frac{m^2n}{2}, 3mn, 5m^2n, \frac{mn^2}{4}$ $\frac{m^2n}{2}, 5m^2n$ **14.** $a^3, b^3, 3a^2, 4b^3$ $b^3, 4b^3$

15. $v^3, 3v^4, v^2, 4v^3$ $v^3, 4v^3$ **16.** $\frac{3d^2}{4}, \frac{d^2}{2}, d^3, \frac{3}{4}d$ $\frac{3d^2}{4}, \frac{d^2}{2}$ **17.** $2cd^2, 3d^2c, 5c^2d$ $2cd^2, 3d^2c$

Use the distributive property to write each expression in a different form.

18. $6(a + b)$ $6a + 6b$ **19.** $3(a + c)$ $3a + 3c$ **20.** $8(3x - 7)$ $24x - 56$ **21.** $4(7 - 2a)$ $28 - 8a$

22. $ax + ay$ $a(x + y)$ **23.** $cx + dx$ $x(c + d)$ **24.** $4m - 4n$ $4(m - n)$ **25.** $3r - kr$ $r(3 - k)$

Written Exercises

Simplify.

1. $3a + 7a$ $10a$ **2.** $9c + 7c$ $16c$ **3.** $13a + 5a$ $18a$

4. $21x - 10x$ $11x$ **5.** $5am - 4am$ am **6.** $4np + 10np$ $14np$

7. $18mn + 20mn$ $38mn$ **8.** $6xy + 28xy$ $34xy$ **9.** $15x^2 + 7x^2$ $22x^2$

10. $23y^2 + 32y^2$ $55y^2$ **11.** $10a + 30a + a$ $41a$ **12.** $16b + 17b - b$ $32b$

13. $9y^2 + 13y^2 + 3$ $22y^2 + 3$ **14.** $11a^2 - 11a^2 + 12a^2$ $12a^2$ **15.** $16a + 17b + 6$

16. $14a^2 + 13b^2 + 27$ **17.** $5a + 7a + 10b - 5b$ $12a + 5b$ **18.** $16a + 21a + 30b - 7b$

19. $2(a + b) + b$ $2a + 3b$ **20.** $3(x + 2y) - 2y$ $3x + 4y$ **21.** $5x + 3(x - y)$ $8x - 3y$

22. $9a + 14(a + 3)$ $23a + 42$ **23.** $6(5a + 3b - 2b)$ $30a + 6b$ **24.** $2(3a + 2a + b)$ $10a + 2b$

25. $5ab^2 + 2a^2b + ab^2$ **26.** $3xy^3 - 2y^3 + 5xy^3$ **27.** $4(3x + 2) + 2(x + 3)$

28. $\frac{3}{4}y + \frac{x}{4} + 3x$ $\frac{3}{4}y + \frac{13}{4}x$ **29.** $x^2 + \frac{7}{8}x - \frac{x}{8}$ $x^2 + \frac{3}{4}x$ **30.** $1.4a^2 + 3.8a + 5a$

31. $0.38k^2 + 0.6k^2 + 8.5k$ $0.98k^2 + 8.5k$ **32.** $m + 3(0.2m + 2.8m^3)$ $1.6m + 8.4m^3$ **33.** $\frac{n}{2} + 3\left(\frac{n}{4} + n^2\right)$ $\frac{5}{4}n + 3n^2$

Using Calculators — Simplifying Expressions

Use a calculator to do the arithmetic when simplifying expressions.

Example: Simplify $374n + 582n + 52.3y + 36y$.

$374n + 582n + 52.3y + 36y$

$= (374 + 582)n + (52.3 + 36)y$ *Distributive Property*

$= 956n + 88.3y$

Exercises Simplify.

1. $1436x^2 - 789x^2 + 5689x^2$ $6336x^2$ **2.** $8946y^3 + 9744y - 438y$ $8946y^3 + 9306y$

3. $485m^2n + 203m^2n + 168m^2 - 87m^2$ $688m^2n + 81m^2$ **4.** $5.86rs^3 - 4.06rs^3 + 0.92r^3s + 8.69r^3s$ $1.8rs^3 + 9.61r^3s$

Suggested Daily Assignments

Enriched
Exp: 2–24 evens
Written: 1–33 odds

Average
Exp: 2–24 evens
Written: 1–28

Fundamental
Exp: 1–25
Written: 2–26 evens
p. 13, 1–19 odds

Answers to Written

15. $16a + 17b + 6$
16. $14a^2 + 13b^2 + 27$
18. $37a + 23b$
25. $2a^2b + 6ab^2$
26. $8xy^3 - 2y^3$
27. $14x + 14$
30. $1.4a^2 + 8.8a$

About Using Calculators
Point out to students that by using a calculator to simplify expressions such as these, they can concentrate more on the algebra and less on the routine arithmetic.

1-6

Lesson Notes, P. T26

Readiness Activity

True or false?

$17 + 1 = 1 + 17$ true

$4 \cdot (6 \cdot 7) = (4 \cdot 6) \cdot 7$ true

$2 \cdot 5 = 5 \cdot 2$ true

$(1 + 3) + 10 = 1 + (3 + 10)$ true

Discussion Questions

1. Does $a - b = b - a$ for all numbers a and b? no
2. Does $\frac{a}{b} = \frac{b}{a}$ for all numbers a and b? no
3. Does $(a - b) - c = a - (b - c)$ for all numbers a, b, and c? no
4. Does $(a \div b) \div c = a \div (b \div c)$ for all numbers a, b, and c? no

Chalkboard Exercise

1. Simplify $5a + 2(3 + a)$ indicating all properties used.

$5a + 2(3 + a)$
$= 5a + (6 + 2a)$ distributive
$= 5a + (2a + 6)$ commutative +
$= (5a + 2a) + 6$ associative +
$= (5 + 2)a + 6$ distributive
$= 7a + 6$ substitution

Objective 1-6: Recognize and use the commutative and associative properties when simplifying expressions.

1-6 Commutative and Associative Properties

The order in which two numbers are added or multiplied does *not* change their sum or product.

$6 + 5 = 5 + 6$ $\quad 12 + 9 = 9 + 12$ $\quad 37 + 82 = 82 + 37$

$4 \cdot 5 = 5 \cdot 4$ $\quad 7 \cdot 2 = 2 \cdot 7$ $\quad 83 \cdot 37 = 37 \cdot 83$

For any numbers a and b,
$a + b = b + a$ and $a \cdot b = b \cdot a$.

Commutative Properties

How do you evaluate $5 + 7 + 4$? You could add the 5 and 7 and then add the 4 to the sum, or you could add the 7 and 4 and then add the 5.

$$5 + 7 + 4 = (5 + 7) + 4 = 12 + 4 = 16$$

$$5 + 7 + 4 = 5 + (7 + 4) = 5 + 11 = 16$$

The sum is 16 in each case.

How do you evaluate $3 \cdot 8 \cdot 7$?

$$3 \cdot 8 \cdot 7 = (3 \cdot 8) \cdot 7 = 24 \cdot 7 = 168$$

$$3 \cdot 8 \cdot 7 = 3 \cdot (8 \cdot 7) = 3 \cdot 56 = 168$$

Notice that the product is 168 in each case.

Thus, the way you group, or associate, three numbers does not change the sum or product.

For any numbers a, b, and c,
$(a + b) + c = a + (b + c)$ and $(ab)c = a(bc)$.

Associative Properties

The commutative, associative, and distributive properties are used when simplifying expressions.

Example 1

Simplify $3x + (5 + 6x)$ indicating all properties used.

$3x + (5 + 6x) = 3x + (6x + 5)$ *Commutative Property for Addition*
$= (3x + 6x) + 5$ *Associative Property for Addition*
$= (3 + 6)x + 5$ *Distributive Property*
$= 9x + 5$ *Substitution Property*

Notice that in Example 1 the commutative and associative properties were used to group like terms.

Example

2 **Simplify $5n + 2(n^2 + 4n) + n^2$ indicating all properties used.**

$5n + 2(n^2 + 4n) + n^2 = 5n + (2n^2 + 8n) + n^2$ *Distributive Property*
$= 5n + (8n + 2n^2) + n^2$ *Commutative Property for Addition*
$= (5n + 8n) + (2n^2 + n^2)$ *Associative Property for Addition*
$= (5n + 8n) + (2n^2 + 1n^2)$ *Multiplicative Identity*
$= (5 + 8)n + (2 + 1)n^2$ *Distributive Property*
$= 13n + 3n^2$ *Substitution Property*

The chart below summarizes the properties that are used when simplifying expressions.

The following properties are true for any numbers a, b, and c.		
	Addition	**Multiplication**
Commutative:	$a + b = b + a$	$ab = ba$
Associative:	$(a + b) + c = a + (b + c)$	$(ab)c = a(bc)$
Identity:	0 is the identity. $a + 0 = 0 + a = a$	1 is the identity. $a \cdot 1 = 1 \cdot a = a$
Zero:		$a \cdot 0 = 0 \cdot a = 0$
Distributive:	$a(b + c) = ab + ac$ and $(b + c)a = ba + ca$; $a(b - c) = ab - ac$ and $(b - c)a = ba - ca$	
Substitution:	If $a = b$, then a may be substituted for b.	

Exploratory Exercises

State the property shown in each of the following.

1. $(8 + 4) + 2 = 8 + (4 + 2)$
2. $5 + 3 = 3 + 5$
3. $8(a + b) = 8a + 8b$
4. $(3 + 8) + x = 11 + x$
5. $3x = x \cdot 3$
6. $(a + b) + 3 = a + (b + 3)$
7. $5(ab) = (5a)b$
8. $cb + ab = (c + a)b$
9. $10(x + y) = 10(y + x)$
10. $3(a + 2b) = (a + 2b) \cdot 3$
11. Is $12 - 8 = 8 - 12$? Is $27 - 10 = 10 - 27$? Is subtraction a commutative operation?
12. Is $24 \div 6 = 6 \div 24$? Is $36 \div 9 = 9 \div 36$? Is division a commutative operation?

State the property that justifies each step.

13. Simplify: $6a + (8b + 2a)$
 a. $6a + (8b + 2a) = 6a + (2a + 8b)$ commutative property for addition
 b. $= (6a + 2a) + 8b$ associative property for addition
 c. $= (6 + 2)a + 8b$ distributive property
 d. $= 8a + 8b$ substitution property of equality

Chalkboard Exercise

2. Simplify $4k^2 + 3(4k + k^2) + 2k$ indicating all properties used.
$4k^2 + 3(4k + k^2) + 2k$
$= 4k^2 + (12k + 3k^2) + 2k$ distributive
$= 4k^2 + (3k^2 + 12k) + 2k$ commutative +
$= (4k^2 + 3k^2) + (12k + 2k)$ associative +
$= (4 + 3)k^2 + (12 + 2)k$ distributive
$= 7k^2 + 14k$ substitution

Suggested Daily Assignments

Enriched
 Exp: 2–14 evens
 Written: 2–32 evens
 Mini-Review: 1–5

Average
 Exp: 2–14 evens
 Written: 1–26
 Mini-Review: 1–5

Fundamental
 Exp: 1–14
 Written: 2–26 evens
  p. 17, 1–23 odds
 Mini-Review: 1–5

Answers to Exploratory

1. associative property for addition
2. commutative property for addition
3. distributive property
4. substitution property
5. commutative property for multiplication
6. associative property for addition
7. associative property for multiplication
8. distributive property
9. commutative property for addition
10. commutative property for multiplication
11. no, no, no
12. no, no, no

Answers to Written
1. commutative property for addition
2. multiplicative identity
3. associative property for addition
4. commutative property for addition
5. commutative property for multiplication
6. associative property for multiplication
7. additive identity property
8. distributive property
9. commutative property for addition
10. associative property for addition
11. distributive property
12. additive identity
13. multiplicative identity property
14. commutative property for multiplication
21. $3a + 13b + 2c$

14. Simplify: $8a^2 + (8a + a^2) + 7a$
   a. $8a^2 + (8a + a^2) + 7a = 8a^2 + (a^2 + 8a) + 7a$ commutative property for addition
   b. $= (8a^2 + a^2) + (8a + 7a)$ associative property for addition
   c. $= (8a^2 + 1a^2) + (8a + 7a)$ multiplicative identity
   d. $= (8 + 1)a^2 + (8 + 7)a$ distributive property
   e. $= 9a^2 + 15a$ substitution property of equality

Written Exercises

State the property shown in each of the following.

1. $5a + 2b = 2b + 5a$
2. $1 \cdot a^2 = a^2$
3. $(a + 3b) + 2c = a + (3b + 2c)$
4. $x^2 + (y + z) = x^2 + (z + y)$
5. $ax + 2b = xa + 2b$
6. $(3 \cdot x) \cdot y = 3 \cdot (x \cdot y)$
7. $29 + 0 = 29$
8. $5(a + 3b) = 5a + 15b$
9. $5a + 3b = 3b + 5a$
10. $5a + \left(\frac{1}{2}b + c\right) = \left(5a + \frac{1}{2}b\right) + c$
11. $(m + n)a = ma + na$
12. $0 + 7 = 7$
13. $1(a + b) = a + b$
14. $3m + nq = 3m + qn$

Simplify.

15. $5a + 6b + 7a$ $12a + 6b$
16. $8x + 2y + x$ $9x + 2y$
17. $3x + 2y + 2x + 8y$ $5x + 10y$
18. $x^2 + 3x + 2x + 5x^2$ $6x^2 + 5x$
19. $\frac{2}{3}x^2 + 5x + x^2$ $\frac{5}{3}x^2 + 5x$
20. $\frac{3}{4}a^2 + 5ab + \frac{1}{2}a^2$ $\frac{5}{4}a^2 + 5ab$
21. $3a + 5b + 2c + 8b$
22. $2x^2 + 3y + z^2 + 8x^2$ $10x^2 + 3y + z^2$
23. $5 + 7(ac + 2b) + 2ac$ $5 + 9ac + 14b$
24. $3(4x + y) + 2x$ $14x + 3y$
25. $3(x + 2y) + 4(3x + y)$ $15x + 10y$
26. $6(2x + y) + 2(x + 4y)$ $14x + 14y$
27. $\frac{3}{4} + \frac{2}{3}(x + 2y) + x$ $\frac{3}{4} + \frac{5}{3}x + \frac{4}{3}y$
28. $\frac{3}{5}\left(\frac{1}{2}x + 2y\right) + 2x$ $\frac{23}{10}x + \frac{6}{5}y$
29. $0.2(3x + 0.2) + 0.5(5x + 3)$ $3.1x + 1.54$
30. $0.3(0.2 + 3y) + 0.21y$ $1.11y + 0.06$
31. $3[4 + 5(2x + 3y)]$ $12 + 30x + 45y$
32. $4[1 + 4(5x + 2y)]$ $4 + 80x + 32y$

mini-review

Simplify.

1. $6 \cdot x \cdot x \cdot y \cdot y \cdot y \cdot y$ $6x^2y^4$
2. $\frac{3}{4}[17 - (11 - 6)]$ 9
3. $x^2 - \frac{1}{4}x^2 + 3y^2$ $\frac{3}{4}x^2 + 3y^2$

Solve.

4. $f = (0.85)(100)$ 85
5. $9c = 9$ 1

Excursions in Algebra — History

The use of the term *associative* is due to Sir William Rowan Hamilton (1805-1865). Hamilton was a mathematician who studied and taught at Trinity College in Dublin, Ireland. He worked in astronomy and physics as well as higher forms of algebra. Some of his algebraic work is used in exactly the same form today. At the age of 30 Hamilton was knighted. He holds the honor of being the first foreign associate named to the United States Academy of Sciences.

About the Excursion
History excursions such as this one introduce students to people who were involved in the development of mathematics. You may want to encourage some students to do further research on Hamilton.

Reading Algebra

Parentheses

Suppose you are to write each of the following in symbols.

Words	Symbols
Three times x plus y	$3x + y$
Three times the sum of x and y	$3(x + y)$

In the second expression, parentheses are used to show that the *sum*, x plus y, is multiplied by three. In algebraic expressions, terms enclosed by parentheses are treated as one quantity. The expression $3(x + y)$ can be read *three times the quantity x plus y*. Suppose you are to read each expression below.

Symbols	Words
$a + 5^2$	a plus five squared
$(a + 5)^2$	the quantity a plus 5 squared

The phrase *the quantity* tells you that the sum, $a + 5$, is squared. To avoid confusion, read $(a - b)^2$ as *a minus b the quantity squared*. Read $a - b^2$ as *a minus b squared.*

Try an experiment. Have a classmate close his or her book. Read the following expressions to your classmate. Then have him or her write the symbols for each expression.

a. Seven times the quantity b minus 4
b. The quantity 6 plus x cubed
c. Two times the quantity a plus 3 times the quantity b minus 9

The correct answers are $7(b - 4)$, $(6 + x)^3$, and $2(a + 3)(b - 9)$, respectively. Compare your classmate's answers with the correct answers. How do they compare?

In verbal problems, look for key words that indicate that parentheses are to be used. Sometimes the words *sum, difference, quantity,* and *total* signal the use of parentheses. Study the following examples.

Words	Symbols
Four divided by the difference of a number and 6	$4 \div (n - 6)$
The quantity a plus b divided by x	$(a + b) \div x$

Exercises

Write each expression in symbols.

1. Eight times four plus x squared $8 \cdot 4 + x^2$
2. Eight times the quantity four plus x squared $8(4 + x^2)$
3. Add n to your age and double it. $2(a + n)$

Write each algebraic expression in words.

4. $3a + 2$ **5.** $3(a + 2)$ **6.** $\frac{5}{9}(F - 32)$
7. $(r + s) - (r - s)$ **8.** $8 \div (4 - c)^2$ **9.** $8 \div 4 - c^2$

About Reading Algebra
This section is intended to prepare students for translating verbal problems into algebraic equations. Be sure to emphasize the importance of looking for key words in the problem.

Answers to Exercises
4. Three times a plus two.
5. Three times the quantity a plus two.
6. Five ninths times the quantity F minus 32.
7. The sum of r and s minus the difference of r and s.
8. Eight divided by the square of the quantity four minus c.
9. Eight divided by four minus the square of c.

Quiz
Lessons 1-4 to 1-6
Name the property shown in each of the following.
1. $256 \cdot 1 = 256$ multiplicative identity
2. $(11 + 4)3 = 15 \cdot 3$ substitution

Simplify.
3. $4(3c + 8) - 7c + 5$ $5c + 37$
4. $11xy + 4x^2 + 5xy$ $16xy + 4x^2$
5. $9a^2 + 3b + 8(a^2 + 2b)$ $17a^2 + 19b$

State the property shown in each of the following.
6. $(765 + 832) \cdot 0 = 0$ multiplicative property of zero
7. $(8c + d)6 = 48c + 6d$ distributive property
8. $a^2 + (c + d) = (c + d) + a^2$ commutative property for addition

1-7

Lesson Notes, p. T26
Enrichment, p. T26

Readiness Activity
Write a verbal expression for each mathematical expression. Answers may vary.

$7 + x$	r^3
$2x - 3$	$3 - y^2$

Discussion Questions
Write an algebraic expression for each verbal expression.

1. the square of the sum of x and a $(x + a)^2$
2. the sum of the square of x and the square of a $x^2 + a^2$
3. the product of m, n, and the square of p mnp^2
4. the square of the product of m, n, and p $(mnp)^2$

Compare your answers to questions 1 and 2, and to questions 3 and 4.

Chalkboard Exercises

1. Find the area of a rectangle of length 3.5 mm and width 2.4 mm. 8.4 mm^2
2. Find the area of a triangle with a base of 5 m and a height of 7.5 m. 18.75 m^2

Objective 1-7A: Translate verbal sentences into equations or formulas.
Objective 1-7B: Solve problems using formulas.

1-7 Using Formulas

Marie wants to find the area of the picture shown at the right. The length is 51 cm and the width is 41 cm. She knows that the area of a rectangle is equal to the product of the length and width. She uses the formula $A = lw$ to express this relationship. The variables A, l, and w represent the measures of the area, length, and width.

$$A = lw$$
$$= 51 \cdot 41$$
$$= 2091$$

Marie finds that the area is 2091 square centimeters (cm^2).

A **formula** is an equation that states a rule for the relationship between certain quantities.

Examples

1 **Find the area of a rectangle of length 17 cm and width 13 cm.**

$$A = lw$$
$$A = 17 \cdot 13$$
$$= 221$$

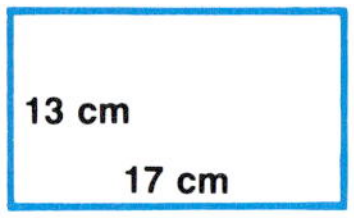

The area is 221 cm^2.

2 **The formula for the area of a triangle is $A = \frac{1}{2}bh$. Find the area of a triangle with a base (b) of 6 units and a height (h) of 9 units.**

$$A = \frac{1}{2}bh$$
$$A = \frac{1}{2}(6)(9)$$
$$= 3 \cdot 9$$
$$= 27$$

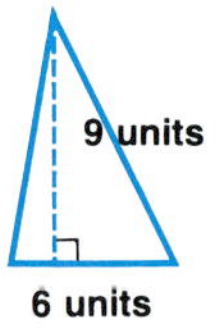

The area of the triangle is 27 square units.

Many sentences can be written as equations or formulas. Use variables to represent the unspecified numbers or measures referred to in the sentence. Then write the verbal expressions as algebraic expressions. Some verbal expressions that suggest the equals sign are listed below.

is	is equal to	is as much as
equals	is the same as	is identical to

Example

3 **Translate each sentence into an equation or formula.**

a. The number z equals twice the sum of x and y.
The equation is $z = 2(x + y)$.

b. The area of a circle is equal to the product of π and the square of the radius (r).
The formula is $A = \pi r^2$.

Exploratory Exercises

Write a formula for each of the following. Use the variables indicated.

1. The area (A) of a square is the square of the length of one of its sides (s). $A = s^2$
2. The perimeter (P) of a parallelogram is twice the sum of the lengths of two adjacent sides (a and b). $P = 2(a + b)$
3. The perimeter (P) of a square is the product of 4 and the length of a side (s). $P = 4s$
4. The circumference (C) of a circle is the product of 2, π, and the radius (r). $C = 2\pi r$

Written Exercises

The formula for the surface area (S) of a rectangular solid as shown at the right is $S = 2(lh + wh + lw)$. Copy and complete the charts.

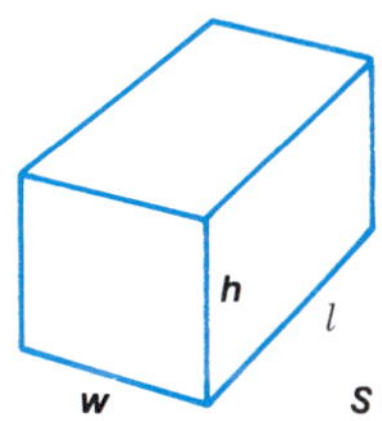

$S = 2(lh + wh + lw)$

	l	w	h	S
1.	5	8	6	236
2.	18	10	4	584
3.	$20\frac{1}{2}$	3	8	499
4.	$5\frac{1}{2}$	12	$3\frac{1}{2}$	$254\frac{1}{2}$

	l	w	h	S
5.	12.9	11	4.6	503.68
6.	21.8	6.5	9.7	832.42

The formula for the area (A) of a trapezoid as shown at the right is $A = \frac{1}{2}h(a + b)$. Copy and complete the charts.

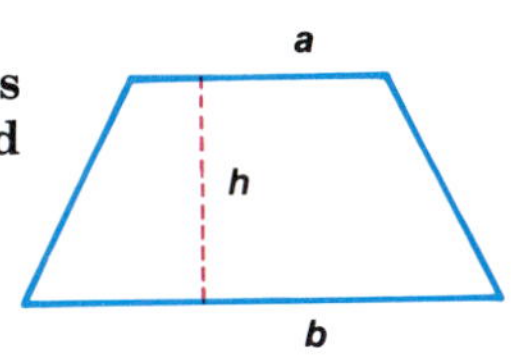

$A = \frac{1}{2}h(a + b)$

	h	a	b	A
7.	6	24	19	129
8.	11	37	23	330
9.	12	24	40	384
10.	10	19	54	365
11.	$\frac{5}{8}$	$\frac{3}{4}$	$\frac{1}{2}$	$\frac{25}{64}$

	h	a	b	A
12.	$3\frac{1}{3}$	12	$8\frac{1}{4}$	$33\frac{3}{4}$
13.	4	18.9	12.7	63.2
14.	2.4	8.25	3.15	13.68

Chalkboard Exercises

Translate each sentence into an equation or formula.

3a. The distance (d) is equal to the product of the rate (r) and the time (t). $d = rt$

3b. The number n divided by the number y equals the number z. $\frac{n}{y} = z$

3c. The volume of a cone (V) is equal to the area of its base (a) times its height (h) divided by 3. $V = \frac{ah}{3}$

Suggested Daily Assignments

Enriched
Exp: 1–4
Written: 2–38 evens, 39–41

Average
Exp: 1–4
Written: 2–22 evens, 23–36

Fundamental
Exp: 1–4
Written: 1–16, 23–28
p. 20, 1–25 odds

The formula for the area (A) of the shaded region of the figure below is $A = \frac{1}{2}\pi a^2 - a^2$. Find the area of the region for the following values of a. Use 3.14 for π. Round each answer to the nearest whole number.

15. 6 21 **16.** 4 9 **17.** 81 3740 **18.** 64 2335

19. 3.8 8 **20.** 5.6 18 **21.** 18.3 191 **22.** 27.4 428

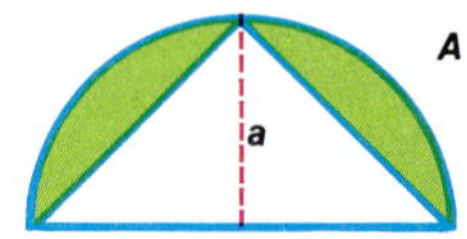

Translate each sentence into an equation.

23. Twice x increased by the square of y is equal to z. $2x + y^2 = z$

24. The square of a decreased by the cube of b is equal to c. $a^2 - b^3 = c$

25. The sum of x and the square of a is equal to n. $x + a^2 = n$

26. The square of the sum of x and a is equal to m. $(x + a)^2 = m$

27. The number r equals the cube of the difference of a and b. $r = (a - b)^3$

28. The number b equals x decreased by the cube of m. $b = x - m^3$

29. The square of the product of a, b, and c is equal to k. $(abc)^2 = k$

30. The product of a, b, and the square of c is f. $abc^2 = f$

31. Y is the sum of twice m and the square of n. $y = 2m + n^2$

32. A is equal to the sum of m and the square of n. $A = m + n^2$

33. Z is equal to 29 decreased by the product of x and y. $z = 29 - xy$

34. R is the product of a and m decreased by z. $R = am - z$

The distance (d) traveled is equal to the rate (r) times the time (t). The formula for this is $d = rt$. Use this formula to solve each problem.

35. Find the distance from Danville to the beach if it takes 3 hours to drive there at an average rate of 50 miles per hour. 150 mi

36. George runs for 30 minutes every day. Find the distance he runs if he averages 660 feet per minute. 19,800 ft

37. The speed of sound through air is about 330 meters per second. Find the distance between Carl and an explosion if it takes 10 seconds for the sound to reach him. 3300 m

38. The speed of light is about 300,000 kilometers per second. Find the distance between Karen and a flash of light if it takes 6 seconds for Karen to see the light. 1,800,000 km

Challenge

Write a formula for the area of each figure.

39.

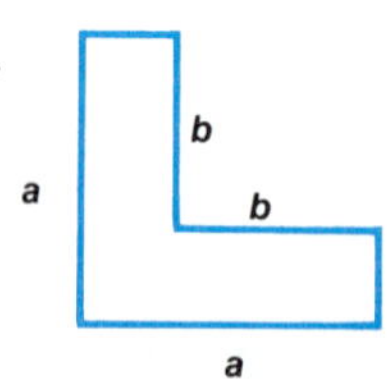

$A = a^2 - b^2$

40.

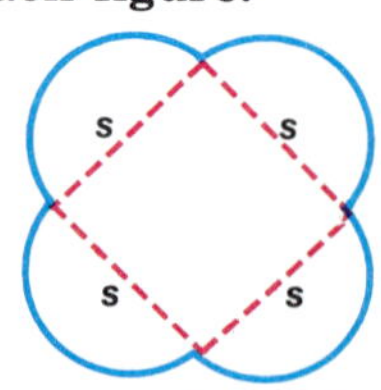

$A = s^2 + \frac{\pi}{2}s^2$

41.

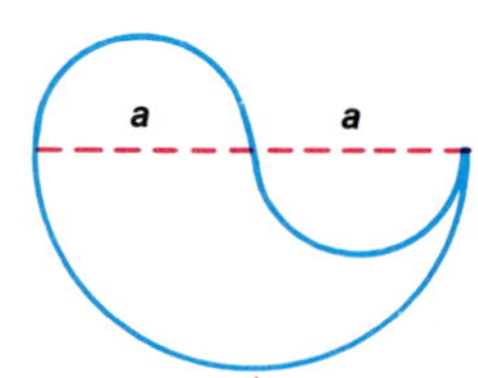

$A = \frac{1}{2}\pi a^2$

Objective 1-8: Explore problem situations by asking and answering questions.

1-8 Exploring Verbal Problems

Many problems in algebra are solved by translating the verbal problem into symbols. To do this accurately, you must explore the problem until you completely understand the relationships among the given information. You can explore a problem situation by asking and answering questions. Study the following examples.

Examples

1 **Peggy Richards has \$3.25 in nickels and dimes. She has 7 more dimes than nickels.**

Questions	*Answers*
a. Does she have more nickels or dimes?	**a.** more dimes
b. How much money does she have in all?	**b.** \$3.25
c. Do you know how many nickels and dimes she has?	**c.** no
d. If she has n nickels, how many dimes does she have?	**d.** $7 + n$ or $n + 7$
e. If she has x dimes, how much money does she have in dimes?	**e.** $10x$ cents

2 **One day John and Inger picked peaches for 4 hours. In all they picked 30 baskets of peaches. Inger picked 6 baskets less than John.**

Questions	*Answers*
a. How many baskets were picked in all?	**a.** 30
b. How long did John and Inger work?	**b.** 4 hours
c. Who worked longer?	**c.** They worked the same.
d. Who picked more?	**d.** John
e. If Inger picked n baskets, how many did John pick?	**e.** $n + 6$
f. If John picked r baskets, how many did Inger pick?	**f.** $r - 6$

Written Exercises

For each problem, answer the related questions.

1. Mr. Limotta checked his cash register at the end of the day. He found that he had 7 fewer \$5 bills than \$1 bills. He had eleven \$10 bills and no larger bills. In all he had \$267.
   a. Did he have more \$5 bills or \$1 bills? \$1 bills
   b. How many more? 7
   c. How much money did he have in all? \$267
   d. How many \$20 bills did he have? none

1-8

Lesson Notes, p. T26

Discussion Question

John has 6 fewer pencils than Mary. If Mary has x pencils, how many does John have? $x - 6$ Is the order of the terms in this answer important? yes

Chalkboard Exercises

1. Julie purchased a package of nails for \$1.18, a package of sandpaper for \$2.09, and two sheets of plywood. She spent \$19.25 altogether, not counting tax.
   a. Which costs more, the nails or the sandpaper? sandpaper
   b. How much did she spend in all? \$19.25
   c. If one sheet of plywood costs x dollars, how much did she spend on plywood? $2x$ dollars
   d. Would two pieces of plywood cost more or less than \$19.25? less

2. Mike is 6 years older than Ray. In 10 years the sum of their ages will be 72.
   a. Who is younger? Ray
   b. If Ray is y years old now, how old is Mike? $y + 6$
   c. If Mike is m years old now, how old will he be in 10 years? $m + 10$
   d. What is the sum of their ages now? 52

Suggested Daily Assignments

Enriched
  Written: 2–12 evens

Average
  Written: 1–12

Fundamental
  Written: 1–8
    p. 24, 29–36

e. How much money did he have in \$1 and \$5 bills together? \$157
f. When did he check his cash register? end of the day
g. If he had n \$5 bills, how much money did he have in \$5 bills? $5n$ dollars

2. Michelle has 20 books on crafts and cooking. She also has 21 novels. She has 6 more cookbooks than craft books.
   a. Does she have more cookbooks or craft books? cookbooks
   b. How many more? 6
   c. Does she have more novels than craft books? yes
   d. How many books does she have in all? 41
   e. If she has n cookbooks, how many craft books has she? $20 - n$
   f. Of what kind of book does she have the least? crafts

3. Two breakfast cereals, Kornies and Krispies, together cost \$3.59. One of them costs 7¢ more than the other.
   a. Which one costs more? does not say
   b. What is the difference in their prices? 7¢
   c. How much would two boxes of each cost? \$7.18
   d. Would two boxes of the more expensive cereal cost more or less than \$3.59? more
   e. If the more expensive cereal costs n cents, what is the cost of the other cereal? $(n - 7)$¢ or $(359 - n)$¢

4. Marvin is 6 inches taller than Ira and Ira is 4 inches shorter than Bruce.
   a. Who is the tallest? Marvin
   b. Who is the shortest? Ira
   c. Is Marvin taller or shorter than Bruce? taller
   d. How much taller than Ira is Marvin? 6 inches
   e. How much taller than Bruce is Marvin? 2 inches

5. Brenda Summer has 3 more nickels than dimes and 5 fewer pennies than nickels. She has 70 coins in all.
   a. Of which coin does she have the most? nickels
   b. Of which coin does she have the fewest? pennies
   c. How many fewer dimes than nickels does she have? 3
   d. How many fewer pennies than dimes does she have? 2
   e. How much money does she have in all? does not say
   f. Does she have more money in nickels or dimes? dimes

6. The Vegetable Mart offers corn at 18¢ per ear, cucumbers at 12¢ each, and tomatoes at 59¢ per basket. Norma has only \$3 to spend and wants one basket of tomatoes and 5 fewer ears of corn than cucumbers.
   a. How many baskets of tomatoes does she want? 1
   b. How much more is a basket of tomatoes than an ear of corn? 41¢
   c. After buying the tomatoes, how much money does she have remaining? \$2.41
   d. Will she buy more cucumbers or ears of corn? cucumbers
   e. If she buys n cucumbers, how many ears of corn will she buy? $n - 5$

7. Chris can mow the lawn in 4 hours and Mark can do it in 3 hours.
   a. Alone, how much of the lawn will Chris mow in 3 hours? $\frac{3}{4}$
   b. Alone, how much of the lawn will Mark mow in an hour? $\frac{1}{3}$
   c. In n hours, how much of the lawn will Chris mow? $\frac{n}{4}$
   d. If they both start mowing at the same time, using two mowers, can they finish the lawn in 2 hours? yes

8. Phoebe goes to Pluto's Platters to buy her stereo records. She bought 3 more rock records than classical and 2 less western records than rock. Including 5 jazz records, she bought 18.
   a. Did she buy more classical than rock? no
   b. Did she buy more rock than western? yes
   c. Which did she buy the most of? rock
   d. How many rock, classical, and western records did she buy? 13
   e. If she bought n classical records, how many rock records did she buy? $n + 3$

9. Ninety-six students signed up for football at East High School. Eight are not eligible because of grades. There are eight fewer sophomores than juniors. There are 26 seniors and no freshmen.
   a. Are there more sophomores than juniors? no
   b. If there are x juniors, how many sophomores are there? $x - 8$
   c. Are there more juniors or seniors? juniors
   d. Which class has the most players? juniors

10. Craig said, "I am 24 years younger than my mom and the sum of our ages is 68 years."
    a. How old was Craig's mother when Craig was born? 24
    b. How much older than Craig is his mother now? 24
    c. What will be the sum of their ages in 5 years? 78
    d. How old was Craig's mother when Craig was 10 years old? 34
    e. In ten years, how much younger than his mother will Craig be? 24

11. Doug was working some math problems. He thought that he could finish 3 pages of 24 problems per page in 2 hours. After $1\frac{1}{2}$ hours, he had finished 2 pages.
    a. How many problems had Doug completed in $1\frac{1}{2}$ hours? 48
    b. How many problems are there in all? 72
    c. At the rate he is working, will he finish the three pages in 2 hours? no

12. Darryl is 7 years younger than Ernie. In 5 years the sum of their ages will be 75 years.
    a. Who is older? Ernie
    b. In 5 years, who will be younger? Darryl
    c. If Darryl is n years old now, how old will he be in 5 years? $n + 5$
    d. Ernie is x years old now. How old is Darryl? $x - 7$
    e. Who was younger seven years ago? Darryl
    f. What is the sum of their ages now? 65 years

About Applications
Base two and hexadecimal arithmetic provide a huge supply of problems that promote abstract thinking on the part of the student. The application to computer science helps make it more meaningful to students.

Applications in Computer Science — Hexadecimals

Laura Sheppard is a computer programmer for a large bank. She uses data from the computer to find any errors in her programs. The data is in the hexadecimal numeration system.

The hexadecimal numeration system is a base 16 system. It uses the sixteen symbols 0, 1, 2, 3, 4, 5, 6, 7, 8, 9, A, B, C, D, E, and F. Notice that $B_{sixteen} = 11_{ten}$.

The position of each symbol gives its place value as a power of sixteen. Study how the hexadecimal numeral 9F3 is changed to a decimal numeral.

$$\begin{aligned} 9F3_{sixteen} &= 9 \cdot 16^2 + F \cdot 16^1 + 3 \\ &= 9 \cdot 256 + 15 \cdot 16 + 3 \\ &= 2304 + 240 + 3 \\ &= 2547_{ten} \end{aligned}$$

The hexadecimal numeration system provides a shorter way of representing numbers than the binary system. Suppose you want to change a binary numeral to a hexadecimal. Separate the numeral into groups of 4 digits starting at the right. Each group can be changed to one hexadecimal digit.

$$11\wedge0100\wedge1101\wedge0010 = \underbrace{11}_{3}\ \underbrace{0100}_{8}\ \underbrace{1101}_{D}\ \underbrace{0010}_{2}$$

$1101_{two} = 13_{ten} = D_{sixteen}$

Exercises

Change each hexadecimal numeral to a decimal numeral.

1. $53_{sixteen}$ 83 **2.** $C6_{sixteen}$ 198 **3.** $2E5_{sixteen}$ 741 **4.** $10A3_{sixteen}$ 4259

Change each binary numeral to a hexadecimal numeral.

5. 1001_{two} $9_{sixteen}$ **6.** 10101100_{two} $AC_{sixteen}$ **7.** 100101010_{two} $12A_{sixteen}$ **8.** 10111010111_{two} $5D7_{sixteen}$

Challenge

9. Subtract the hexadecimals: 6E7 − 542 $1A5_{sixteen}$

Objective 1-9A: Define variables and write equations for verbal problems.
Objective 1-9B: Write verbal problems for equations.

1-9 Problem Solving: Writing Equations

Four steps that can be used to solve verbal problems are listed below.

Problem-Solving Plan

1. **Explore the problem.**
2. **Plan the solution.**
3. **Solve the problem.**
4. **Examine the solution.**

Explore To solve a verbal problem, first read the problem carefully and *explore* what the problem is about.

- Identify what information is given.
- Identify what you are asked to find.
- Choose a variable to represent one of the unspecified numbers in the problem. This is called defining the variable.
- Use the variable in writing expressions for other unspecified numbers in the problem.

Plan Next, *plan* how to solve the problem. One way to solve a problem is to use an equation.

- Read the problem again. Decide how the unspecified numbers relate to other given information.
- Write an equation to represent the relationship.

Example 1 Write an equation for the following problem.

One number is 12 greater than a second number. The sum of the two numbers is 86. Find the numbers.

Explore Read the problem carefully.
Let n = the lesser number. ❶ ❷
Then $n + 12$ = the greater number.

Plan The problem states the sum of the two numbers is 86.
So, the equation is $n + (n + 12) = 86$.

1-9

Lesson Notes, p. T26

Discussion Questions

Alan is twice as old as Barbara. The sum of their ages is 63.

1. Write an equation if b represents Barbara's age. $b + 2b = 63$
2. Write an equation if a represents Alan's age. $a + \frac{a}{2} = 63$

Chalkboard Exercise

Write an equation for the following problem.

1. One number is 7 less than a second number. The product of the two numbers is 60. Find the numbers. Let n = the larger number. $n(n - 7) = 60$

❶ Point out that variables can often be defined in more than one way. In example 1, you could let n equal the greater number. Then $n - 12$ equals the lesser number. The equation would be $n + (n - 12) = 86$.

❷ Point out that the symbol = is usually read "is equal to." However, in sentences beginning with "let" it should be read "equal." For example, the sentence "Let n = the lesser number," is read "Let n equal the lesser number." The sentence "Then $n + 12$ = the greater number." is read "Then n plus twelve is equal to the greater number."

Chalkboard Exercises

Write an equation for the following problem.

2. How old is Mark if 3 times his age decreased by 4 is 83?
Let m = Mark's age.
$3m - 4 = 83$

Write a problem based on the given information.

3. Let x = Andrea's age now.
$x + 8 = 26$
One possible solution is: "In eight years Andrea will be 26. How old is she now?"

4. Let x = distance Lynn drove south.
$x + 60$ = distance Sarah drove north.
$x + (x + 60) = 248$
One possible solution is: "Lynn drove a certain distance south. Sarah drove 60 miles farther north. At the end of the drive they were 248 miles part. How far did Lynn drive?"

Suggested Daily Assignments

Enriched
Exp: 1–10
Written: 2–34 evens
Mini-Review: 1–5

Average
Exp: 1–10
Written: 1–12, 17–35 odds
Mini-Review: 1–5

Fundamental
Exp: 1–10
Written: 1–12, 18–34 evens
p. 27, 9, 10
Mini-Review: 1–5

Example

2 **Write an equation for the following problem.**

Jennifer Winter is in the ninth grade. Six years ago, twice her age was 16 years. How old is she now?

Explore Let j = Jennifer's age now.
Then $j - 6$ = Jennifer's age 6 years ago.

Plan The problem states that twice her age six years ago was 16 years. So, the equation is $2(j - 6) = 16$.

Verbal problems can be developed from equations. First, you must know what the variable in the equation represents. Then you can use the equation to establish the conditions of the problem. Study the following examples.

Examples

3 **Write a problem based on the given information.**

Let x = Mary's age now.

$x - 5 = 17$

Since Mary's age is represented by x, then $x - 5$ must be her age 5 years ago, which was 17.

Thus, the following problem can be written.

If Mary's age 5 years ago was 17, how old is she now?

4 **Write a problem based on the given information.**

Let w = Steve's weight in kilograms.

$w - 12$ = Perry's weight in kilograms.

$w + (w - 12) = 118$

The following problem can be written.

Perry weighs 12 kg less than Steve. The sum of their weights is 118 kg. How much does Steve weigh? How much does Perry weigh?

Exploratory Exercises

Write an expression for each of the following.

1. Mr. Jackson is now 49 years old. How old was he n years ago? $49 - n$
2. The sum of two numbers is 18. The lesser number is t. What is the greater number? $18 - t$
3. The length of a rectangle is 4 more than the width. The width is w. Find the length. $w + 4$
4. Sheldon is 8 years older than Jane. If Sheldon is n years old, how old is Jane? $n - 8$
5. Harold has 8 more than twice as many red marbles as blue marbles. He has t blue marbles. How many red marbles does he have? $2t + 8$

6. This year's senior class has 117 fewer students than last year's class. This year's class has 947 students. How many were in last year's class? 947 + 117
7. The cost of gasoline has tripled in the past 8 years. Gasoline now costs \$1.59 per gallon. What was the cost 8 years ago? \$1.59 ÷ 3 or $\frac{\$1.59}{3}$
8. Tess types 42 words per minute. How many minutes would it take for her to type a 3000-word paper? 3000 ÷ 42 or $\frac{3000}{42}$
9. Anthony is paid \$5.65 per hour. How much is he paid for working n hours? $\$5.65n$
10. Buzz graduated from high school y years ago at the age of 17. How old is he now? $17 + y$

Written Exercises

Define the variable, then write an equation for each problem.

1. A number increased by 24 is 89. Find the number.
2. Twice a number decreased by 84 is 40. Find the number.
3. A number decreased by 19 is 83. Find the number.
4. 67 decreased by twice a number is 39. Find the number.
5. How old is Tyrone if twice his age increased by 17 is 53?
6. Twenty-seven years ago Clarice was 21. How old is she now?
7. Peggy is 3 years older than Melissa. The sum of their ages in 4 years will be 59 years. How old is Peggy now?
8. Three times Cecile's age 4 years ago is 42. How old is Cecile now?
9. Bill is 5 inches taller than Bob and the sum of their heights is 137 inches. How tall is Bob?
10. Twice Mary Lou's height increased by 17 inches is 141 inches. How tall is Mary Lou?
11. Bob's dad is 27 years older than Bob. The sum of their ages 5 years ago was 45 years. How old is Bob now?
12. The sum of Mrs. Black's age and her daughter's age is 56 years. In 8 years, Mrs. Black will be twice as old as her daughter. How old is her daughter now?
13. In a football game, Michael gained 134 yards running. This was 17 yards more than the previous game. How many yards did he gain in both games?
14. Ponderosa pines grow about $1\frac{1}{2}$ feet each year. If a pine is now 17 feet tall, about how long will it take the tree to become $33\frac{1}{2}$ feet tall?
15. For 6 consecutive weeks, Connie lost the same amount of weight. Six weeks ago she weighed 145 pounds. She now weighs 125 pounds. How many pounds did Connie lose each week?
16. Each week for 7 weeks Sav-A-Buk stores reduced the price of a sofa by \$18.25. The final reduced price was \$252.50. What was the original price?

Write a problem based on the given information. Answers will vary. Sample answers are given.

17. Let x = Olivia's age now.
$x + 7 = 29$
18. Let a = Quincy's age now.
$2(a - 7) = 58$
19. Let x = greater of two numbers.
$x - 7$ = lesser of two numbers.
$x + (x - 7) = 33$
20. Let x = lesser of two numbers.
$x + 29$ = greater of two numbers.
$x + (x + 29) = 135$

Answers to Written

1. Let n = the number; $n + 24 = 89$
2. Let n = the number; $2n - 84 = 40$
3. Let n = the number; $n - 19 = 83$
4. Let n = the number; $67 - 2n = 39$
5. Let a = Tyrone's age; $2a + 17 = 53$
6. Let c = Clarice's age now; $c - 27 = 21$
7. Let p = Peggy's age now; $(p + 4) + (p - 3 + 4) = 59$
8. Let c = Cecile's age now; $3(c - 4) = 42$
9. Let x = Bob's height in inches; $x + (x + 5) = 137$
10. Let m = Mary Lou's height in inches; $2m + 17 = 141$
11. Let x = Bob's age now; $(x - 5) + (x + 27 - 5) = 45$
12. Let x = daughter's age now; $(56 - x) + 8 = 2(x + 8)$
13. Let y = number of yards gained in both games; $134 + (134 - 17) = y$
14. Let x = number of years for a tree to become $33\frac{1}{2}$ feet tall; $17 + 1\frac{1}{2}x = 33\frac{1}{2}$
15. Let p = number of pounds lost each week; $145 - 6p = 125$
16. Let p = the original price; $p = 252.50 + 7\,(18.25)$
17. In 7 years, Olivia will be 29 years old. How old is she now?
18. Twice Quincy's age 7 years ago is 58. How old is Quincy now?
19. The sum of two numbers is 33. One number is 7 less than the other. Find the numbers.
20. The sum of two numbers is 135. One number is 29 more than the other. Find the numbers.

Answers to Written

21. The sum of Anne's age and her mother's age is 58. Anne's mother is 26 years older than Anne. How old is Anne?
22. Ellen is 4 years older than Willie. In 10 years, the sum of their ages will be 54. How old is Willie now?
23. Terry's car weighs 250 pounds more than Seth's car. The sum of the weights of both cars is 7140 pounds. How much does each car weigh?
24. Sven is 7 inches taller than Roxanne. The sum of Sven's height and twice Roxanne's height is 193 inches. How tall is Roxanne?
25. There are 33 students enrolled in Mr. Wyatt's class. If 5 are absent, how many are present?
26. There are 7 fewer students in Jim's class than in Alan's class. There are 189 students in both classes. How many students are in each class?
27. Reggie is 31 cm shorter than Elroy. The sum of Elroy's height and twice Reggie's height is 502 cm. How tall is Elroy?
28. Yvette has 64 coins. She has 17 fewer pennies than nickels. How many nickels does she have?
29. Bonnie drove 80 miles further west than Cheryl drove east. Together they drove 520 miles. How far did each drive?
30. Danny's book has 27 fewer pages than Ira's book. Together their books have 873 pages. How many pages are in each book?
31. Alex won 12 fewer games than Greg. Greg won twice as many games as Alex. How many games did Greg win?

Write a problem based on the given information. Answers will vary. Sample answers are given.

21. Let n = Anne's age now.
$n + 26$ = Anne's mother's age now.
$n + (n + 26) = 58$

22. Let w = Willie's age now.
$w + 4$ = Ellen's age now.
$(w + 10) + (w + 4 + 10) = 54$

23. Let x = weight of Seth's car in pounds.
$x + 250$ = weight of Terry's car in pounds.
$x + (x + 250) = 7140$

24. Let h = Roxanne's height in inches.
$h + 7$ = Sven's height in inches.
$2h + (h + 7) = 193$

25. Let x = the number of students present in Mr. Wyatt's class.
$x + 5 = 33$

26. Let n = the number of students in Alan's class.
$n - 7$ = the number of students in Jim's class.
$n + (n - 7) = 189$

27. Let m = Elroy's height in centimeters.
$m - 31$ = Reggie's height in centimeters.
$m + 2(m - 31) = 502$

28. Let n = number of nickels that Yvette has.
$n - 17$ = number of pennies that Yvette has.
$n + (n - 17) = 64$

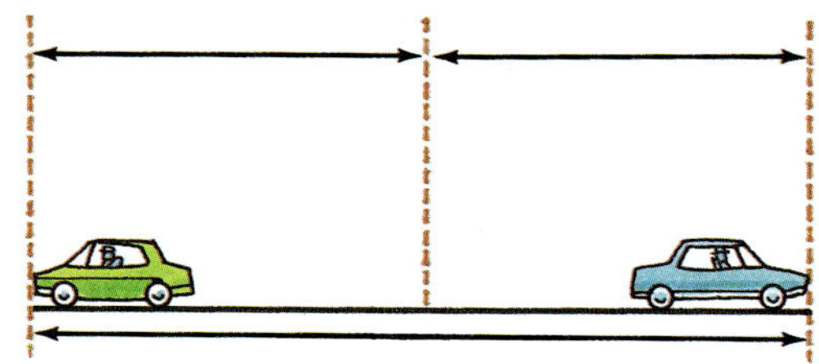

29. Let x = distance that Cheryl drove east.
$x + 80$ = distance that Bonnie drove west.
$x + (x + 80) = 520$

30. Let p = number of pages in Ira's book.
$p - 27$ = number of pages in Danny's book.
$p + (p - 27) = 873$

31. Let n = number of games that Greg won.
$n - 12$ = number of games that Alex won.
$n = 2(n - 12)$

32. Let x = Carol's height in inches 5 years ago.
$x + 7$ = Carol's height in inches now.
$x + 3$ = Martha's height in inches now.
$(x + 7) + (x + 3) = 124$

33. Let n = number of \$5 bills that Lloyd has.
$n - 7$ = number of \$10 bills that Lloyd has.
$n + 3$ = number of \$1 bills that Lloyd has.
$n + (n + 3) + (n - 7) = 29$

34. Let d = number of dimes that Teresa has.
$d + 4$ = number of nickels that Teresa has.
$d - 9$ = number of quarters that Teresa has.
$d + (d + 4) + (d - 9) = 37$

35. Let x = number of albums that Anna has.
$x - 3$ = number of albums that Barbara has.
$x + 4$ = number of albums that Chris has.
$x + (x - 3) + (x + 4) = 43$

mini-review

Write *true* or *false*.

1. $5^2 = 5 + 5$ false
2. $7(0) = 7$ false
3. $1(4.2) = 4.2$ true
4. $3(4 \times 1) = 3(4) + 3(1)$ false
5. $7.2 + (3 + 2.8) = (7.2 + 2.8) + 3$ true

Using Computers

Formulas

The formula for the area of a rectangle is $A = \ell w$. To find the area of a rectangle with length 51 cm and width 34 cm, you assign the variable ℓ the value 51 and the variable w the value 34. The area of the rectangle is (51)(34) or 1734 cm^2.

In the BASIC computer language, READ-DATA statements are used to assign values to variables. Consider the following program.

```
10  READ L,W
20  DATA 51,34
30  LET A = L * W
40  PRINT A
```

The computer locates the data and reads them in the order in which they are listed.

In line 30 the variable A is assigned the value of (51)(34).

When this program is executed, the value of A, 1734, is printed.

READ-DATA statements are very useful when working with many calculations. For example, the following program can be used to help complete the chart for exercises 1-6 on page 23.

```
 10  PRINT "L","W","H","S"
 20  PRINT
 30  READ L,W,H
 40  IF L = 0 THEN 110
 50  DATA 5,8,6,18,10,4,20.5,3,8
 60  DATA 5.5,12,3.5,12.9,11,4.6
 70  DATA 21.8,6.5,9.7,0,0,0
 80  LET S = 2 * (L * H + W * H + L * W)
 90  PRINT L,W,H,S
100  GOTO 30
110  END
```

When the value of L is zero, the program will end.

Notice that more than one set of data can be included in each DATA statement.

A LET statement is used in line 80 to write the formula as a BASIC statement.

Enter and run this program on your computer. Compare the output with the chart that you have already completed.

Exercises

Write each formula as a BASIC statement. 2. 10 LET P = 2*(L + W) 3. 10 LET I = P*R*T

1. $A = s^2$ 10 LET A = S↑2 **2.** $P = 2(\ell + w)$ **3.** $I = P \cdot r \cdot t$

4. $A = \frac{1}{2}bh$ 10 LET A = 1/2*B*H **5.** $D = rt$ 10 LET D = R*T **6.** $A = \frac{1}{2}h(a + b)$

7. Write a program similar to the one above to print a chart for exercises 7-14 on page 23. Enter and run the program.

8. Write a program to find the area for each of the figures described in exercises 15-22 on page 24. Enter and run the program.

9. Write a program to find the distance in problems 35-38 on page 24. Enter and run the program.

32. Carol is 7 inches taller now than she was 5 years ago. Martha is 3 inches taller now than Carol was 5 years ago. The sum of their heights now is 124 inches. How tall was Carol 5 years ago?

33. Lloyd has 29 bills. He has 7 fewer \$10 bills than \$5 bills, and 3 more \$1 bills than \$5 bills. How many \$5 bills does he have?

34. Teresa has 37 coins. She has 4 more nickels than dimes and 9 fewer quarters than dimes. How many dimes does she have?

35. Barbara has 3 fewer albums than Anna, and Chris has 4 more albums than Anna. Together Anna, Barbara, and Chris have 43 albums. How many albums does Anna have?

About Using Computers

These programs are adaptable to many different computers. Consult your computer's manual for its abilities and limitations.

Answers to Exercises

6. 10 LET A = $\frac{1}{2}$ * H * (A + B)

7.
```
 10  PRINT "H", "A", "B",
     "AREA"
 20  PRINT
 30  READ H, A, B
 40  IF H = 0 THEN 110
 50  DATA 6, 24, 19, 11,
     37, 23, 12, 24, 40
 60  DATA 10, 19, 54,
     .625, .75, .5, 3.33333,
     12, 8.25
 70  DATA 4, 18.9, 12.7,
     2.4, 8.25, 3.15, 0, 0, 0
 80  LET AREA = 1/2 * H *
     (A + B)
 90  PRINT H, A, B, AREA
100  GOTO 30
110  END
```

8.
```
10 READ A
20 IF A = 0 THEN 80
30 DATA 6, 4, 81, 64,
   3.8, 5.6
40 DATA 18.3, 27.4, 0
50 LET AREA = 1/2 * 3.14
   * A↑2 - A↑2
60 PRINT "AREA = ";
   AREA
70 GOTO 10
80 END
```

9.
```
10 READ R, T
20 IF R = 0 THEN 80
30 DATA 50, 3, 660, 30,
   330, 10
40 DATA 300000, 6, 0, 0
50 LET D = R * T
60 PRINT "DISTANCE
   = "; D
70 GOTO 10
80 END
```

Quiz
Lessons 1–7 to 1–9

Write each of the following as an equation.

1. Twice the sum of three and x is equal to z.
   $2(3 + x) = z$
2. The sum of nine and twice x is equal to y.
   $9 + 2x = y$
3. Three times a number decreased by 6 is 39.
   $3n - 6 = 39$

Define a variable and write an equation for the following problem.

4. Ann is five years older than Larry. Two years ago the sum of their ages was 67. How old is Larry now?
   Let L = Larry's age now
   $L + 5$ = Ann's age now
   $(L - 2) + [(L + 5) - 2] = 67$

Vocabulary

variable (3)
expression (3)
factor (3)
product (3)
dividend (3)
divisor (3)
quotient (3)
base (4)
exponent (4)
open sentence (9)
solution (9)
equals sign (9)
equation (9)
additive identity (12)
multiplicative identity (12)
multiplicative property of zero (12)
distributive property (15)
properties of equality:
  reflexive (12)
  symmetric (12)
  transitive (12)
  substitution (13)
term (15)
like terms (15)
simplest form (16)
coefficient (16)
commutative property for addition (18)
commutative property for multiplication (18)
associative property for addition (18)
associative property for multiplication (18)
formula (22)

Chapter Summary

1. Variables are symbols that are used to represent unspecified numbers. (3)
2. An algebraic expression consists of one or more numbers and variables with the operations of addition, subtraction, multiplication, and/or division. (3)
3. In a multiplication expression, the quantities being multiplied are called factors and the result is called the product. In a division expression, the dividend is divided by the divisor. The result is called the quotient. (3)
4. Parentheses or a dot can be used instead of × to indicate multiplication. When variables are used to represent factors, the multiplication sign is usually omitted. (3)
5. An exponent indicates the number of times the base is used as a factor. (4)
6. Order of Operations:
   1. Evaluate all powers.
   2. Do all multiplications and divisions from left to right.
   3. Do all additions and subtractions from left to right. (6)

7. Grouping symbols, such as parentheses and brackets, are used to clarify or change the order of operations. Evaluate expressions within grouping symbols first. (7)
8. In an open sentence, the variable must be replaced in order to determine if the sentence is true or false. (9)
9. Finding the replacements for the variable that make a sentence true is called solving the open sentence. Each replacement is called a solution of the open sentence. (9)
10. A mathematical sentence that contains an equals sign, =, is called an equation. (9)
11. The following properties of equality are true for any numbers a, b, and c.
    Reflexive: $a = a$ (12)
    Symmetric: If $a = b$, then $b = a$. (12)
    Transitive: If $a = b$ and $b = c$, then $a = c$. (12)
    Substitution: If $a = b$, then a may be replaced by b. (13)
12. Like terms are terms that contain the same variables, with corresponding variables raised to the same power. (15)
13. The coefficient, or numerical coefficient, is the numerical part of a term. (16)
14. The following properties are true for any numbers a, b, and c. (19)

	Addition	Multiplication
Commutative:	$a + b = b + a$	$ab = ba$
Associative:	$(a + b) + c = a + (b + c)$	$(ab)c = a(bc)$
Identity:	0 is the identity. $a + 0 = 0 + a = a$	1 is the identity. $a \cdot 1 = 1 \cdot a = a$
Zero:		$a \cdot 0 = 0 \cdot a = 0$
Distributive:	$a(b + c) = ab + ac$ and $(b + c)a = ba + ca$ $a(b - c) = ab - ac$ and $(b - c)a = ba - ca$	

15. A formula is an equation that states a rule for the relationship between certain quantities. (22)
16. To understand a verbal problem, you should explore the problem situation by asking and answering questions. (25)
17. A four-step problem-solving plan is given below. (29)
    1. Explore the problem.
    2. Plan the solution.
    3. Solve the problem.
    4. Examine the solution.

Answers to Chapter Review
19. additive identity property
20. multiplicative identity property
21. multiplicative property of zero
22. symmetric property of equality
23. reflexive property of equality
24. transitive property of equality
29. commutative property for addition
30. associative property for multiplication
31. associative property for addition
32. commutative property for multiplication

Chapter Review

1-1 **Write each of the following as an expression using exponents.**

1. $a \cdot a \cdot a \cdot a$ a^4 **2.** $15 \cdot x \cdot x \cdot x \cdot y \cdot y$ $15x^3y^2$

Write an algebraic expression for each verbal expression. Use *x* as the variable.

3. twice a number decreased by 17 $2x - 17$ **4.** a number increased by 14 $x + 14$
5. the product of 8 and a number $8x$ **6.** twice the cube of a number $2x^3$

1-2 **Evaluate if $a = 5$, $b = 8$, $c = \frac{2}{3}$, $d = \frac{1}{2}$, and $e = 0.3$.**

7. ab^2 320 **8.** $3ac - bd$ 6 **9.** $(2a - b)^2$ 4
10. bcd $\frac{8}{3}$ **11.** $5e^2$ 0.45 **12.** $b + c + 2d$ $9\frac{2}{3}$

1-3 **Solve each equation.**

13. $a = 29 - 5^2$ 4 **14.** $5(6) - 3(5) = y$ 15 **15.** $8 - 2 \cdot 3 = b$ 2
16. $r = (29 - 5)^2$ 576 **17.** $w = (0.2)(8 + 3)$ 2.2 **18.** $m = \frac{2}{3}\left(3 - \frac{1}{2}\right)$ $\frac{5}{3}$

1-4 **State the property shown in each of the following.**

19. $7 + 0 = 7$ **20.** $2(1) = 2$ **21.** $11 \cdot 0 = 0$
22. If $a + b = 5$ then $5 = a + b$. **23.** $r = r$
24. If $a = 7$ and $7 = 9 - 2$, then $a = 9 - 2$.

1-5 **Simplify.**

25. $10x + x$ $11x$ **26.** $9a - 7a$ $2a$ **27.** $5b + 3(b + 2)$ $8b + 6$ **28.** $9(r + s) - 2s$ $9r + 7s$

1-6 **State the property shown in each of the following.**

29. $5(a + c) = 5(c + a)$ **30.** $10(ab) = (10a)b$
31. $4 + (x + y) = (4 + x) + y$ **32.** $a(b + c) = (b + c)a$

Simplify.

33. $6a + 7b + 8a + 2b$ $14a + 9b$ **34.** $\frac{3}{4}a^2 + \frac{2}{3}ab + ab$ $\frac{3}{4}a^2 + \frac{5}{3}ab$

1-7 **Write each sentence as an equation.**

35. Eighteen decreased by the square of d is equal to f. $18 - d^2 = f$
36. The number c equals the cube of the product of 2 and x. $c = (2x)^3$

1-8 **Two cans of vegetables together cost $1.08. One of them costs 10¢ more than the other.**

37. Would 2 cans of the less expensive vegetable cost more or less than $1.08? less
38. How much would 3 cans of each cost? $3.24

1-9 **Define the variable, then write an equation for each problem.**

39. Carol weighs 8 pounds less than Claudia. Together they weigh 182 pounds. How much does Carol weigh? Let x = Carol's weight; $x + (x + 8) = 182$
40. Three times a number decreased by 21 is 57. Find the number. Let n = the number; $3n - 21 = 57$

Chapter Test

Write an algebraic expression for each verbal expression. Use x as the variable.

1. a number increased by 17 $x + 17$
2. twice the square of a number $2x^2$
3. the sum of a number and its cube $x + x^3$
4. twice Nica's age decreased by 23 years $2x - 23$

Evaluate.

5. $(12 - 10)^4$ 16
6. $(17 - 4)^2$ 169
7. $0.7(1.4 + 0.6)$ 1.4
8. $\frac{3}{4}(8 + 28)$ 27
9. $23 - 12(1.5)$ 5
10. $6 + 3(3.4)$ 16.2

Evaluate if $m = 8$, $n = 3$, $p = \frac{3}{4}$, $q = \frac{2}{3}$, and $r = 0.5$.

11. $(mn)^2$ 576
12. pq^2 $\frac{1}{3}$
13. $n + r^2$ 3.25

Solve each equation.

14. $v = \frac{6^2 - 2^3}{7}$ 4
15. $8(0.03) - 0.05 = y$ 0.19
16. $\frac{3}{4} - \left(\frac{1}{2}\right)^2 = k$ $\frac{1}{2}$

State the property shown in each of the following. 18. commutative property for addition

17. $7a = 7a$ reflexive
18. $7(m + 2n) = 7(2n + m)$
19. $34 \cdot 1 = 34$ multiplicative identity
20. $3(2a + b) = 6a + 3b$ distributive
21. If $m = a + b$ then $a + b = m$. symmetric
22. $3(2) = 2(3)$ commutative for mult.
23. $a + (2b + 5c) = (a + 2b) + 5c$ associative property for addition
24. If $6 = 2 \cdot 3$ and $2 \cdot 3 = 8 - 2$, then $6 = 8 - 2$. transitive

Simplify.

25. $n + 5n$ $6n$
26. $2.5x - x + y + 3.5y$ $1.5x + 4.5y$
27. $5a^2 + 7a + 3a + 11a^2$ $16a^2 + 10a$
28. $4an + \frac{2}{3}am + 8an + \frac{1}{3}am$ $12an + am$

Write each sentence as an equation.

29. The product of π and the square of r is A. $\pi r^2 = A$
30. The sum of a, b, and c is equal to P. $a + b + c = P$

Define the variable, then write an equation for each problem.

31. Peggy is two years younger than Seymour. The sum of their ages is 68. How old is Peggy?
32. 79 decreased by 5 times a number is 49. Find the number.

31. Let x = Peggy's age; $x + (x + 2) = 68$

32. Let n = the number; $79 - 5n = 49$

About the Photo

Positive and negative numbers are used in many everyday measurements which involve not only size but direction. Examples are temperature, altitude, and stock.

Positive and negative numbers are also used to represent feet above or below sea level. Sea level is position of air and sea interface. It is constantly changing at every location due to tides, atmospheric pressure and wind conditions. Therefore, sea level is defined in terms of mean sea level which is the height of the surface of the sea averaged over all stages of the tide over a long period of time.

Teacher Resource Book Materials for Chapter 2
See pages 87-104.

Chapter Test, Form I (Multiple Choice)	87-88
Chapter Test, Form II (Free Response)	89-90
Quizzes	91-92
Extra Practice	93-95
Application: Using Time Zones (Use after 2-6.)	96-97
Extension: Counterexamples (Use after 2-8.)	98
Problem Solving	99
Cumulative Review	100
Answers	101-104

CHAPTER 2

Adding and Subtracting Rational Numbers

Fresh water freezes at 0° Celsius. However, ocean water freezes at two degrees *less than zero*, or $-2°C$. The symbol "$-$" is used in expressing numbers less than 0. In this chapter you will learn more about such numbers.

Objective 2-1A: State the coordinate of a point on the number line.
Objective 2-1B: Graph integers on a number line.

2-1 Integers on the Number Line

Consider these subtraction sentences.

$6 - 2 = 4$	$5 - 1 = 4$
$6 - 4 = 2$	$5 - 3 = 2$
$6 - 6 = 0$	$5 - 5 = 0$
$6 - 8 = \underline{\ ?\ }$	$5 - 7 = \underline{\ ?\ }$

There are no answers to $6 - 8$ and $5 - 7$ among the set of whole numbers. Thus, the whole numbers lack *closure* for subtraction. That is, the answer to a subtraction problem involving whole numbers is not always a whole number. ❶

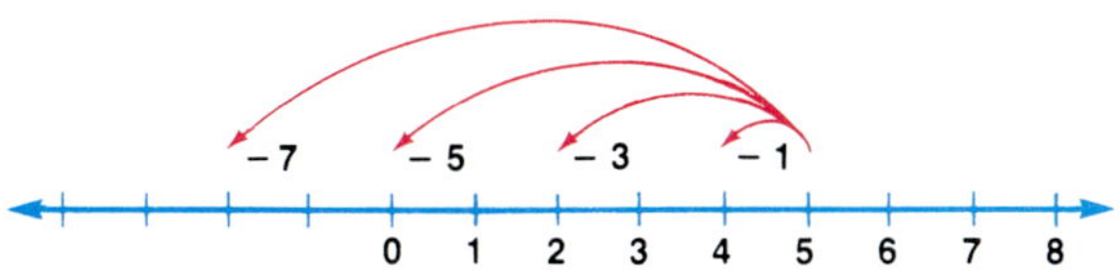

The number line shows that the answer for $5 - 7$ should be *two less than* zero. "Less than" is like taking away, or subtraction. You can write the number *two less than zero* as -2. This is an example of a **negative** number.

You can make a number line which includes negative numbers. Draw a line. Choose a starting point on the line and mark it 0. Then mark off equal distances from this point. To avoid confusion, name points to the right of 0 using the *positive sign* (+) and to the left of zero using the *negative sign* (−), as shown below.

0 is neither positive nor negative.

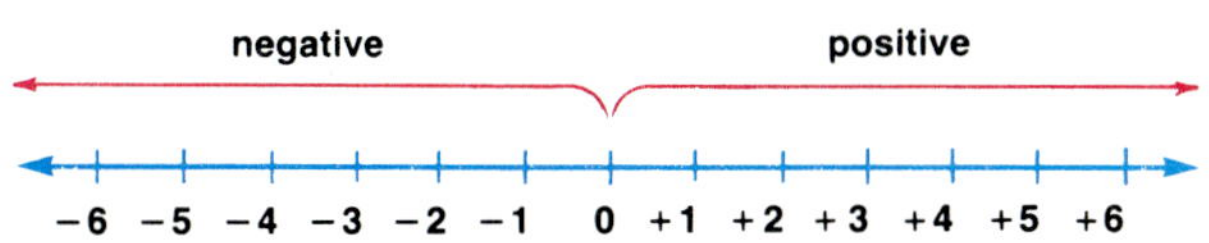

-3 is read *negative* 3 $\qquad$ $+4$ is read *positive* 4

Usually +4 is written as 4, without the positive sign.

The set of numbers used to name the points marked on the number line above is called the set of **integers**. This set can be written $\{\ldots, -5, -4, -3, -2, -1, 0, 1, 2, 3, 4, 5, \ldots\}$ where . . . means continued indefinitely.

To graph a set of numbers means to locate the points named by those numbers on the number line. The number that corresponds to a point on the number line is called the **coordinate** of the point.

2-1

Lesson Notes, p. T27
Enrichment, p. T27

Discussion Questions
(Use after Example 4.)
Graph the set $\left\{-4\frac{1}{4}, -1.5, 0, 1.5, 4\frac{1}{4}\right\}$ on a number line. What do you notice about the distance between the origin and -1.5, and between the origin and 1.5? Can you name another pair of numbers for which this is true?

It is the same. Answers may vary.

❶ Remind students that the set of whole numbers is the set $\{0, 1, 2, \ldots\}$. Note that braces are used to indicate a set of numbers.

Chalkboard Exercises

1. Name the coordinates of points, *A*, *B*, *C*, and *D* on the graph in Example 1.
   A, −5; *B*, 1; *C*, 4; *D*, −3

Graph each set of numbers on a number line.

2a. {−3, −1, 1, 3}

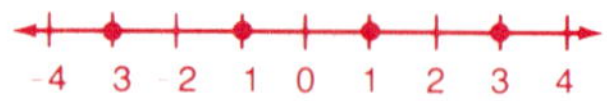

2b. {−2, 1, 4}

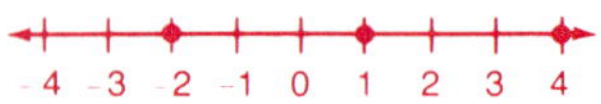

2c. {−5, −4, −3, −2}

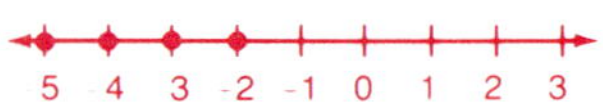

3. Name the set of numbers graphed.

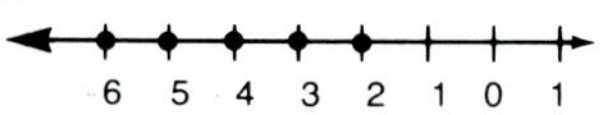

{. . . , −6, −5, −4, −3, −2}

4. Graph the set of all integers greater than −2.

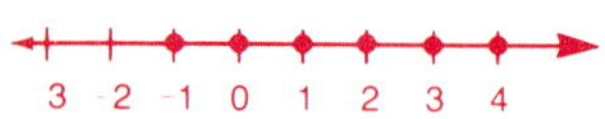

Examples

1 Name the coordinate of point *E*.

The coordinate of *E* is −1.

2 Graph {−2, +3, +4} on a number line.

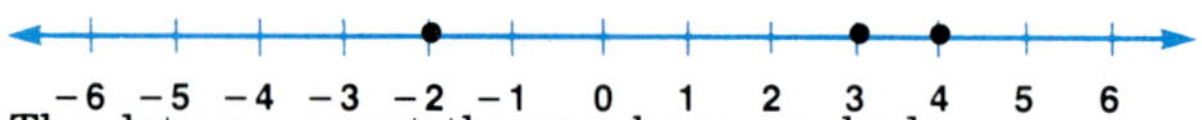

The dots represent the numbers graphed.

3 Name the set of numbers graphed.

The bold arrow means that the graph continues indefinitely.

The set of numbers graphed is {−3, −2, −1, 0, 1, 2, 3, 4, . . .}.

4 Graph the set of all integers less than 3.

Less than 3 means that 3 is not included.

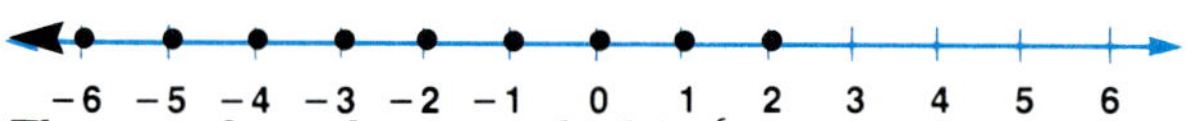

The set of numbers graphed is {. . . , −3, −2, −1, 0, 1, 2}.

Integers are used to describe many real-life situations.

Football: A loss of 5 yards is −5.
A gain of 3 yards is +3.

Business: A profit of \$540 is +540.
A loss of \$130 is −130.

Weather: 20° above zero is +20.
8° below zero is −8.

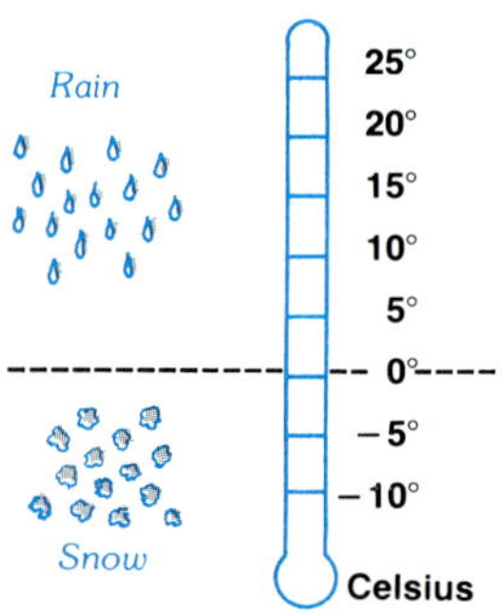

Exploratory Exercises

State the coordinate of each point.

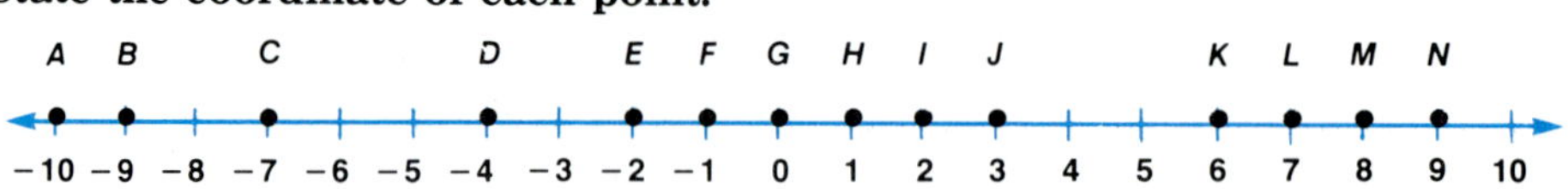

1. *J* 3 **2.** *H* 1 **3.** *A* −10 **4.** *D* −4 **5.** *K* 6 **6.** *L* 7
7. *C* −7 **8.** *F* −1 **9.** *B* −9 **10.** *M* 8 **11.** *E* −2 **12.** *I* 2

Suggested Daily Assignments

Enriched
Exp: 20–28 evens
Written: 2–28 evens, 29–32

Average
Exp: 1–27 odds
Written: 1–28

Fundamental
Exp: 1–28
Written: 2–28 evens

Name the set of numbers graphed.

13. 

14. 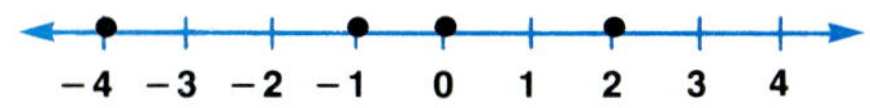

15. 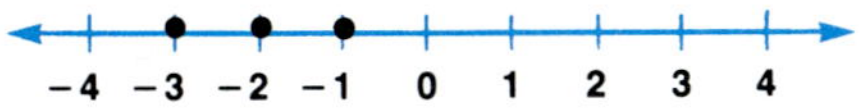

16. 

17. 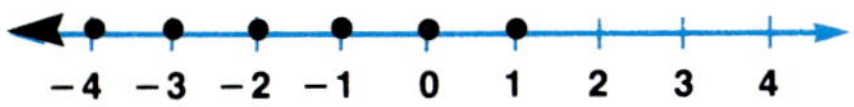

18. 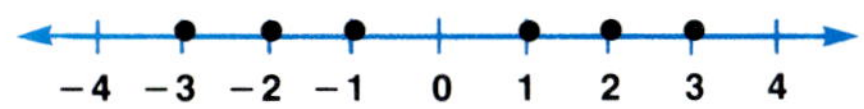

Name an integer to describe each situation.

19. 3 yard loss −3
20. 12 yard gain +12
21. 650 meters above sea level +650
22. 189 meters below sea level −189
23. up 12 floors +12
24. down 3 floors −3
25. \$450 loss −450
26. \$325 profit +325
27. 37° above zero +37
28. 25° below zero −25

Written Exercises

Name the set of numbers graphed.

1. 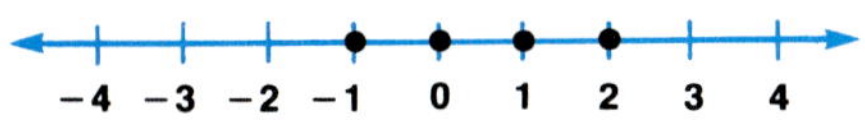

2. 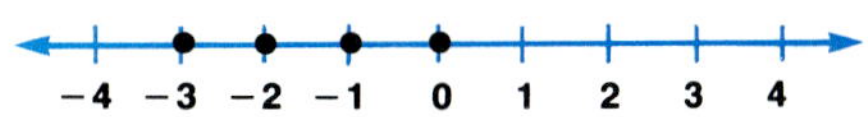

3. 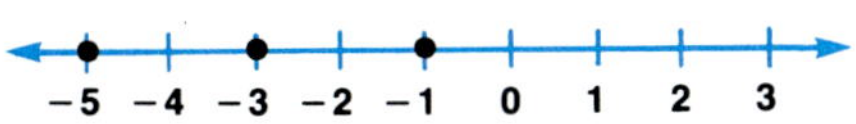

4. 

5. 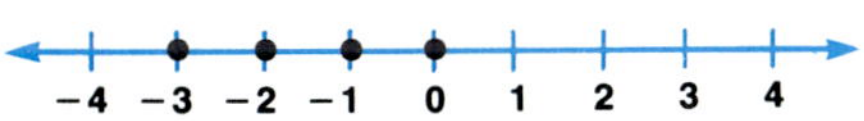

6. 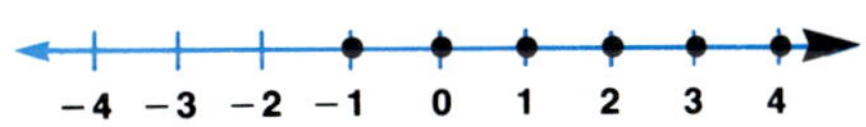

7. 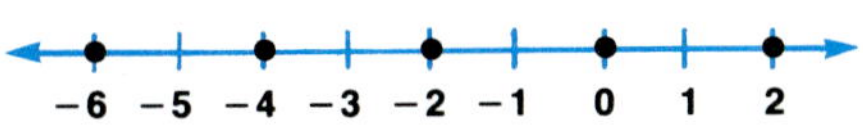

8. 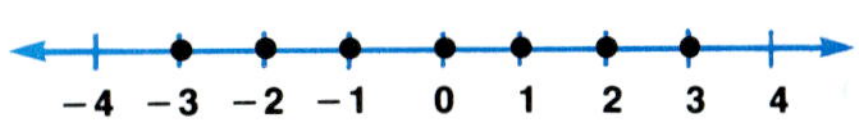

Graph each set of numbers on a number line.

9. {−1, 1, 3, 5}
10. {0, 2, 6}
11. {−2, 2, 4}
12. {−2, −4, −6}
13. {−2, 1, 4}
14. {−3, 0, 2}
15. {−4, −3, −2, 1}
16. {−3, −2, 2, 3}
17. {. . . , −3, −2, −1, 0}
18. {−3, −2, −1, . . .}
19. {4, 5, 6 . . .}
20. {. . . , −1, 0, 1}
21. {all integers less than −2}
22. {all integers greater than 3}

Find the next three numbers in each pattern.

23. 33, 25, 17, 9, 1, −7.
24. −6, −3, 0, 3, 6, 9.
25. 19, 11, 3, −5, −13, −21.
26. 37, 26, 15, 4, −7, −18.
27. 22, 14, 6, −2, −10, −18.
28. 12, 3, −6, −15, −24, −33.

Challenge

Find the next three numbers in each pattern.

29. 1, 3, 9, 27, 81, 243, 729.
30. 1, $\frac{1}{2}$, $\frac{1}{4}$, $\frac{1}{8}$, $\frac{1}{16}$, $\frac{1}{32}$.
31. 0, 1, −2, 3, −4, 5, −6, 7, −8.
32. 1, 1, 2, 3, 5, 8, 13, 21, 34.

Answers to Exploratory

13. {−3, 1, 2}
14. {−4, −1, 0, 2}
15. {−3, −2, −1}
16. {−2, −1, 0, 1, 2, 3, . . .}
17. {. . . , −4, −3, −2, −1, 0, 1}
18. {−3, −2, −1, 1, 2, 3}

Answers to Written

1. {−1, 0, 1, 2}
2. {−3, −2, −1, 0}
3. {−5, −3, −1}
4. {1, 3, 5}
5. {−3, −2, −1, 0}
6. {−1, 0, 1, 2, 3, 4, . . .}
7. {−6, −4, −2, 0, 2}
8. {−3, −2, −1, 0, 1, 2, 3}

9. 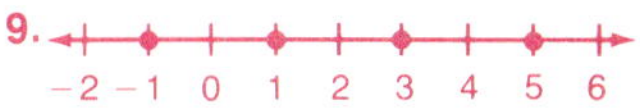

10. 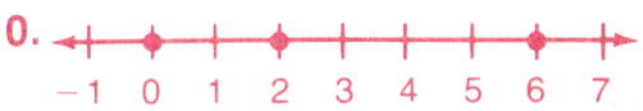

11. 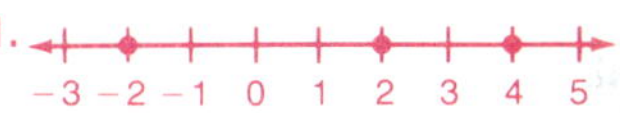

12. 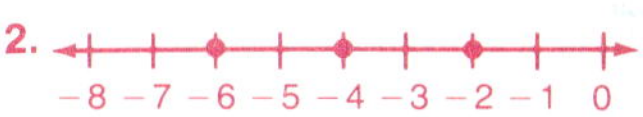

13. 

14. 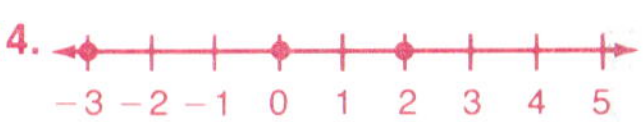

15. 

16. 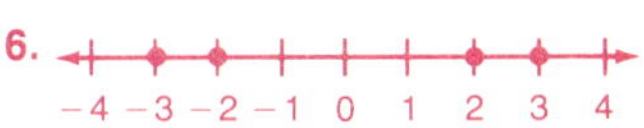

17. 

18. 

19. 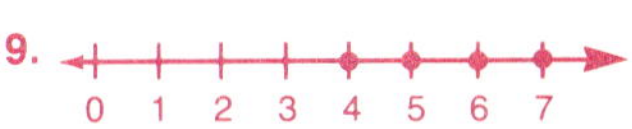

20. 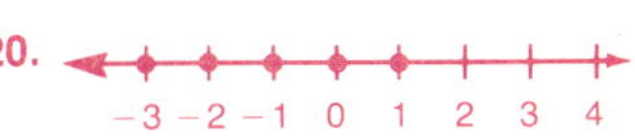

21. 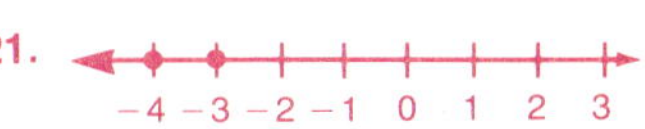

22. 

2-2

Lesson Notes, p. T27

Discussion Questions
Study the following sets of addition exercises. Describe the pattern in each set. Then complete the last two exercises in each set.

1. $3 + 2 = 5$
$3 + 1 = 4$
$3 + 0 = 3$
$3 + (-1) = 2$
$3 + (-2) = 1$
$3 + (-3) = 0$
$3 + (-4) = -1$
$3 + (-5) = \underline{-2}$
$3 + (-6) = \underline{-3}$

2. $2 + 1 = 3$
$2 + 0 = 2$
$2 + (-1) = 1$
$2 + (-2) = 0$
$2 + (-3) = -1$
$2 + (-4) = -2$
$2 + (-5) = -3$
$2 + (-6) = \underline{-4}$
$2 + (-7) = \underline{-5}$

The first addend is the same in each case, while the second addend and sum decrease by 1 each time.

Chalkboard Exercises
Find each sum on a number line.

1a. $-9 + 6$ -3
1b. $-21 + 11$ 10
2a. $-21 + (-18)$ -39
2b. $-10 + (-35)$ -45
3a. $16 + (-14)$ 2
3b. $17 + (-6)$ 11

Objective 2-2: Add integers by using the number line.

2-2 Addition on the Number Line

In Saturday's football game the Brookfield Warriors lost 3 yards on one play. On the next play they gained 7 yards. The diagram at the right shows that the net gain was 4 yards.

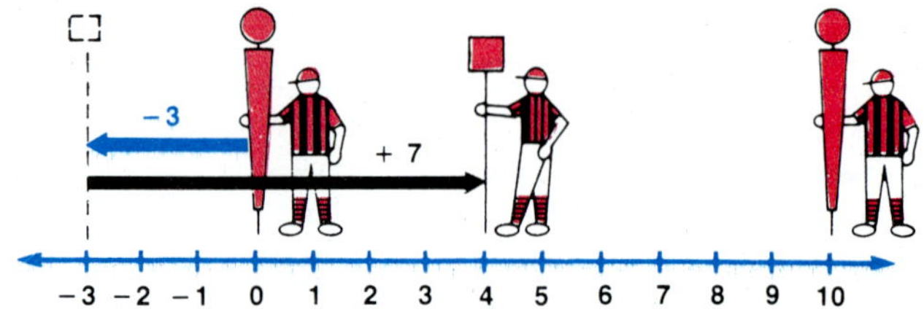

A number line is often used to show addition of integers. For example, to find the sum of 4 and −6, follow these steps.

Step 1 Draw an arrow, starting at 0 and going to 4.

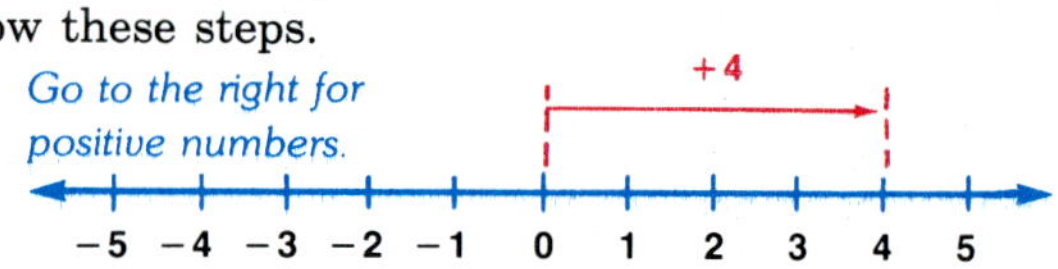

Step 2 Starting at 4, draw an arrow 6 units to the left.

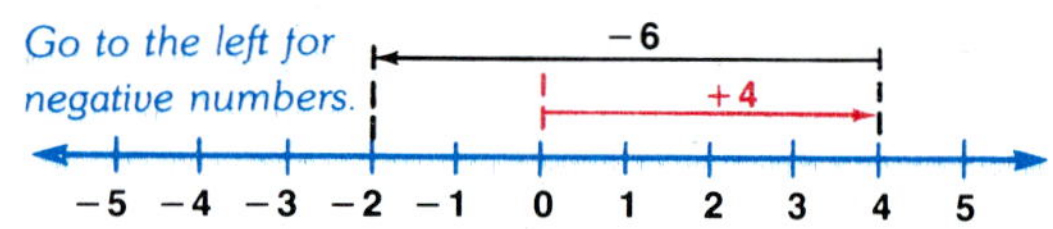

Step 3 The sum is shown at the head of the second arrow.

$$4 + (-6) = -2$$

Note that parentheses are used in the expression $4 + (-6)$ so that the sign of the number is not confused with the operation symbol.

Examples

1 **Find $-4 + 6$ on a number line.**

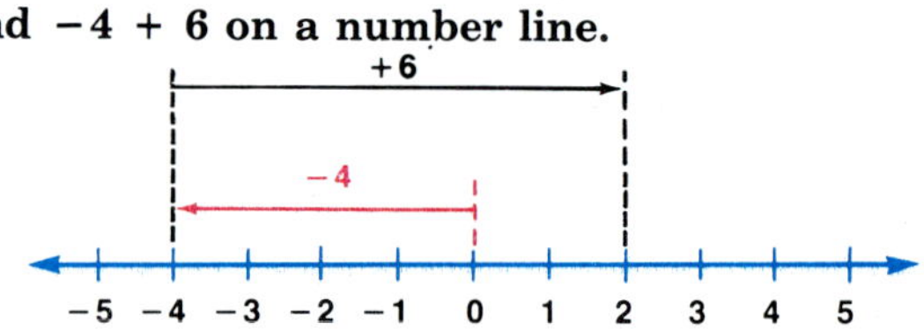

$-4 + 6 = 2$

2 **Find $-2 + (-5)$ on a number line.**

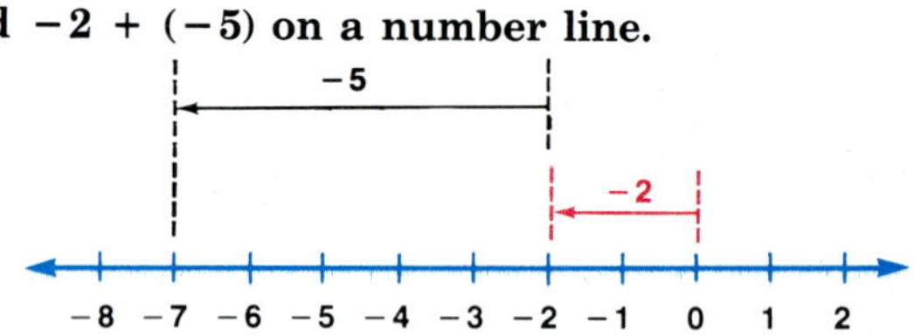

$-2 + (-5) = -7$

3 **Find $4 + (-3)$ on a number line.**

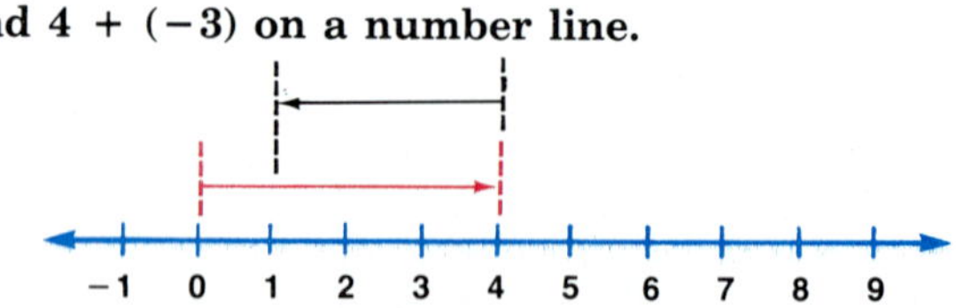

$4 + (-3) = 1$

Exploratory Exercises

State the corresponding addition sentence for each diagram.

1. 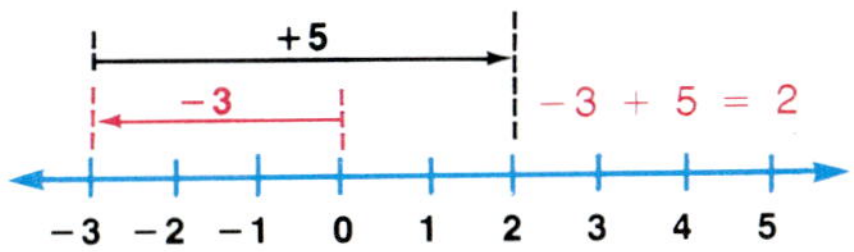

−3 + 5 = 2

2. 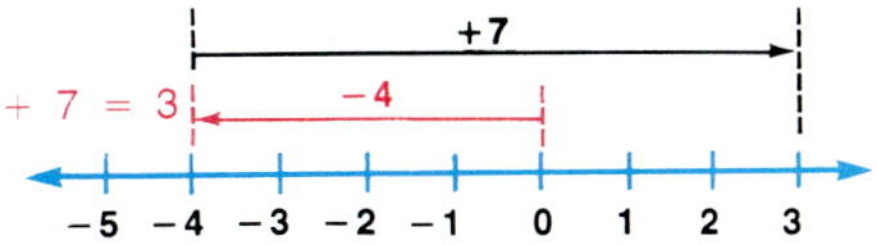

−4 + 7 = 3

3. 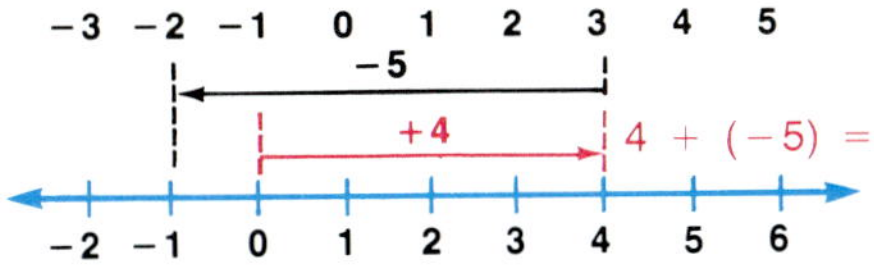

4 + (−5) = −1

4. 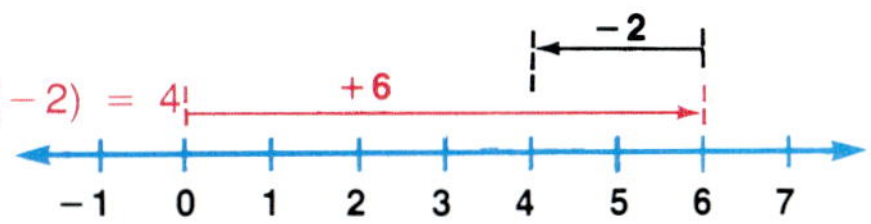

6 + (−2) = 4

5. 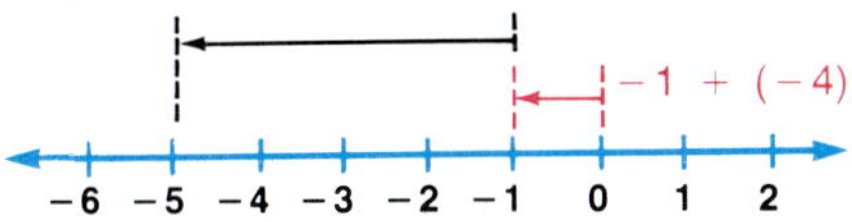

−1 + (−4) = −5

6. 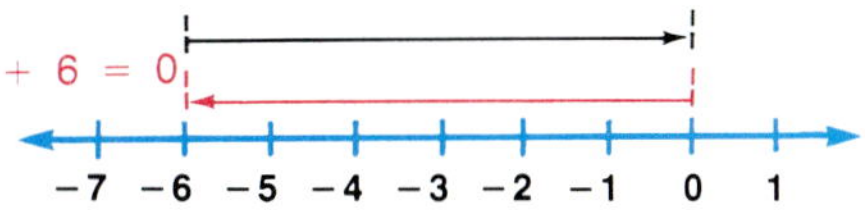

−6 + 6 = 0

7. 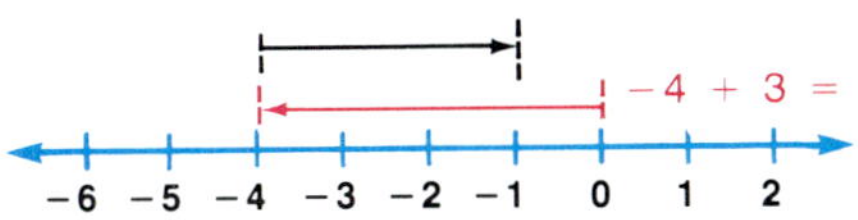

−4 + 3 = −1

8. 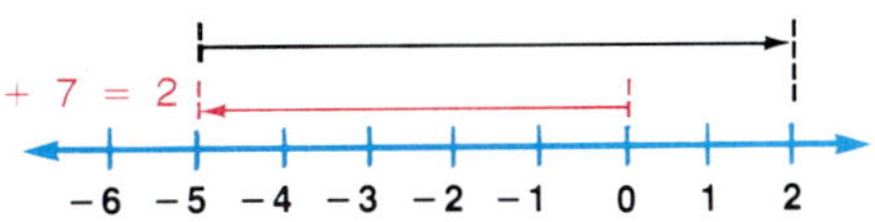

−5 + 7 = 2

9. 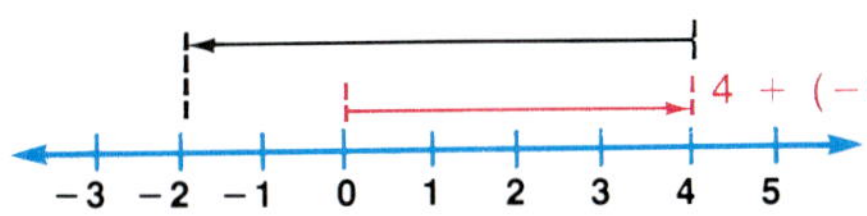

4 + (−6) = −2

10. 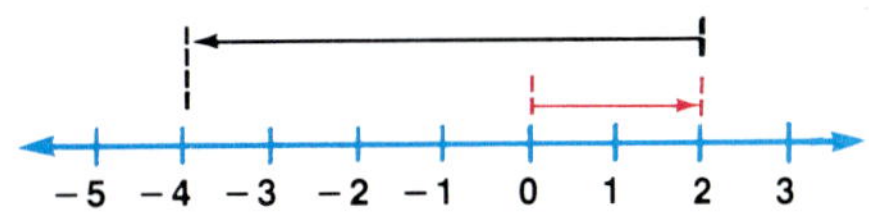

2 + (−6) = −4

11. How is the meaning of the parentheses in $4 + (-6)$ different from those used in $4 + (5 - 3)$?
The parentheses in 4 + (−6) show the sign of the number. The parentheses in 4 + (5 − 3) indicate that 5 − 3 should be evaluated first.

Written Exercises

Find each sum. If necessary, use a number line.

1. $7 + 6$ 13	**2.** $9 + 7$ 16	**3.** $-9 + (-7)$ −16	**4.** $-8 + (-11)$ −19
5. $-5 + 0$ −5	**6.** $0 + (-9)$ −9	**7.** $-9 + 4$ −5	**8.** $6 + (-11)$ −5
9. $4 + (-4)$ 0	**10.** $-11 + 11$ 0	**11.** $9 + (-5)$ 4	**12.** $-3 + 9$ 6
13. $2 + (-7)$ −5	**14.** $-11 + 10$ −1	**15.** $9 + (-4)$ 5	**16.** $-14 + 6$ −8
17. $-3 + 12$ 9	**18.** $8 + 3$ 11	**19.** $4 + (-7)$ −3	**20.** $-11 + 5$ −6
21. $7 + (-7)$ 0	**22.** $-491 + 491$ 0	**23.** $0 + 12$ 12	**24.** $-13 + 0$ −13
25. $-6 + (-11)$ −17	**26.** $-13 + (-9)$ −22	**27.** $-15 + (-11)$ −26	**28.** $-18 + (-24)$ −42

For each problem, state an addition sentence. Then solve the problem.

29. One night in Burlington, Vermont, the temperature was −24°C. During the day the temperature rose 12°C. What was the temperature then? −24 + 12 = −12; −12°C

30. A traffic helicopter ascended to a height of 400 meters to monitor traffic. Then it descended 300 meters. How high was it then? 400 + (−300) = 100; 100 m

31. An elevator went up to the 36th floor. Then it came down 11 floors. At what floor was it then? 36 + (−11) = 25; 25th floor

32. A scuba diver was exploring at a depth of 75 meters. She then went up 30 meters. At what depth was she then? −75 + 30 = −45; −45m

Suggested Daily Assignments

Enriched
- Written: 2–32 evens

Average
- Exp: 1–11 odds
- Written: 1–32

Fundamental
- Exp: 1–11
- Written: 1–28
  - p. 41, 1–27 odds

Quiz
Lessons 2–1 to 2–2

1. Graph {−4, −2, 0, 2} on a number line.

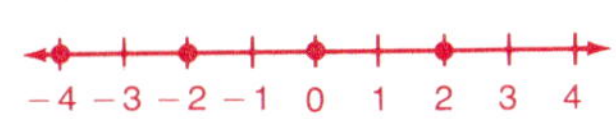

2. Name an integer to describe a loss of $650. −650
3. State an addition sentence for the following problem. Jan's investment lost $85 the first quarter and earned a profit of $97 the next quarter. Find the total profit or loss for two quarters.
−85 + 97 = 12
4. Solve the problem in Exercise 3.
The profit was $12.

2-3

Lesson Notes, p. T27

Discussion Questions

Find each sum.

1. $-4 + 4$ 0
2. $11 + (-11)$ 0
3. $23 + (-23)$ 0
4. $-75 + 75$ 0
5. $-300 + 300$ 0
6. $456 + (-456)$ 0
7. What do you notice about the addends in each exercise above?
   They are both the same distance from the origin.

Chalkboard Exercises

Find each sum.

1a. $-182 + (-183)$ -365

1b. $-235 + (-362)$ -597

❶ Another way to state the definition for absolute value is as follows: "The distance a number is from 0 on the number line is called its absolute value."

❷ Emphasize that the absolute value of a number is always positive. Absolute value will be used in the rules for adding signed numbers.

❸ Students may not remember the meaning of the term *addend*. Remind them that an addend is one of the numbers being added.

Using Computers

A computer program for finding absolute value is given on page 541 of the appendix. The program uses the internal function ABS(X).

Objective 2-3A: Find absolute value of a number.
Objective 2-3B: Add integers without using the number line.

2-3 Adding Integers

It would be difficult to use a number line to add -713 and 425. You need a mathematical concept called absolute value to make the job easier.

Looking at -4 and 4 on the number line, you can see that they are different numbers. However, they are the same number of units from 0. The numbers -4 and 4 have the same **absolute value**.

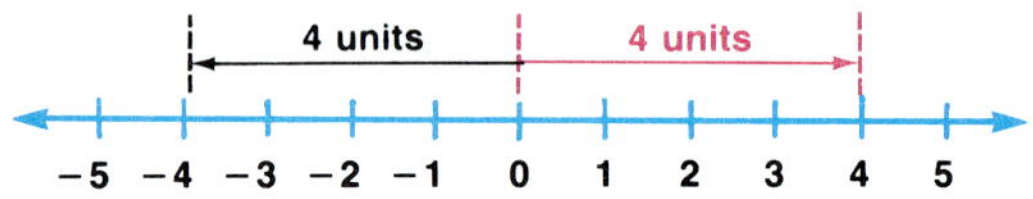

❶

Definition of Absolute Value

The absolute value of a number is the number of units it is from 0 on the number line.

❷

$|-4|$ represents the absolute value of -4.

$|-4| = 4$

$|4|$ represents the absolute value of 4.

$|4| = 4$

$|0| = 0$

Absolute value can be used to find the sum of integers. Consider the following facts learned from the number line.

$+6 + (+7) = +13$

Notice that the sign of each addend is positive. The sum is positive.

$-6 + (-7) = -13$

Notice that the sign of each addend is negative. The sum is negative.

❸

These and other similar examples suggest the following rule.

Adding Integers with the Same Sign

To add integers with the same sign, add their absolute values. Give the sum the same sign as the addends.

Example 1

Add: $-5 + (-7)$

$-5 + (-7) = -(|-5| + |-7|)$ *Both numbers are negative.*

$= -(5 + 7)$ *Add absolute values.*

$= -12$ *The sum is negative.*

Study these facts learned from the number line.

$4 + (-9) = -5$

$|4| = 4$ and $|-9| = 9$.

Notice that $9 - 4$ is 5. Which integer, 4 or -9, has the greater absolute value? The sum has the same sign as this addend.

$-5 + 8 = 3$

$|-5| = 5$ and $|8| = 8$.

Notice that $8 - 5$ is 3. Which integer, -5 or 8 has the greater absolute value? The sum has the same sign as this addend.

These and other similar examples suggest the following rule.

To add integers with different signs, subtract the lesser from the greater absolute value. Give the result the same sign as the addend with the greater absolute value.

Adding Integers with Different Signs

Examples

2 **Add: $3 + (-7)$**

$3 + (-7) = -(|-7| - |3|)$ *-7 has the greater absolute value. The sum is negative.*

$= -(7 - 3)$ *Subtract absolute values.*

$= -4$

3 **Add: $9 + (-2)$** ❹

$9 + (-2) = +(|9| - |-2|)$ *$+9$ has the greater absolute value. The sum is positive.*

$= +(9 - 2)$ *Subtract absolute values.*

$= +7$

Exploratory Exercises

State the absolute value of each integer.

1. $+8$ 8 **2.** $+14$ 14 **3.** -6 6 **4.** -24 24 **5.** 17 17

6. 321 321 **7.** -21 21 **8.** -271 271 **9.** 0 0 **10.** 59 59

State the sign of each sum.

11. $-6 + (-11)$ − **12.** $-14 + (-15)$ − **13.** $+8 + (+9)$ + **14.** $13 + 21$ +

15. $-8 + (+16)$ + **16.** $-11 + (+7)$ − **17.** $+8 + (-35)$ − **18.** $+7 + (-27)$ −

19. $18 + (-3)$ + **20.** $17 + (-38)$ − **21.** $-27 + 31$ + **22.** $-17 + 12$ −

Complete each statement.

23. If $n > 0$, then $|n| =$ _?_. n **24.** If $n < 0$, then $|n| =$ _?_. $-n$

25. If $n > 0$, then $|-n| =$ _?_. n **26.** If $n < 0$, then $|-n| =$ _?_. $-n$

❹ After students have done a few exercises like Examples 2 and 3, they should be able to skip the first step of the solution. That is, they will not need to write down the absolute value symbols.

Chalkboard Exercises

Find each sum.

2a. $49 + (-55)$ -6

2b. $-99 + 67$ -32

3a. $1982 + (-1304)$ 678

3b. $-182 + 183$ 1

Suggested Daily Assignments

Enriched
Exp: 23–26
Written: 1–65 odds

Average
Exp: 2–22 evens, 23–26
Written: 1–24, 26–64 evens

Fundamental
Exp: 1–24
Written: 1–24
p. 43, 29–32

Written Exercises

Find each sum.

1. $+7 + (+9)$ 16
2. $18 + 22$ 40
3. $-6 + (-13)$ -19
4. $-13 + (-8)$ -21
5. $-3 + (+16)$ 13
6. $14 + (-9)$ 5
7. $-5 + 31$ 26
8. $-10 + 4$ -6
9. $24 + (-3)$ 21
10. $-15 + 12$ -3
11. $13 + (-21)$ -8
12. $-8 + 11$ 3
13. $-18 + (-11)$ -29
14. $-23 + (-47)$ -70
15. $82 + (-78)$ 4
16. $43 + (-67)$ -24
17. $38 + (-47)$ -9
18. $-25 + 47$ 22
19. $63 + (-47)$ 16
20. $-21 + 52$ 31
21. $102 + (-12)$ 90
22. $-104 + 16$ -88
23. $-93 + 39$ -54
24. $97 + (-79)$ 18

Evaluate each expression if $a = -5$, $k = 3$, and $m = -6$.

25. $a + 13$ 8
26. $-5 + k$ -2
27. $15 + m$ 9
28. $8 + a$ 3
29. $k + (-18)$ -15
30. $m + (-31)$ -37
31. $-17 + a$ -22
32. $-21 + k$ -18
33. $m + 6$ 0
34. $k + (-7)$ -4
35. $a + (-13)$ -18
36. $m + 17$ 11
37. $|m|$ 6
38. $|a|$ 5
39. $|m + 4|$ 2
40. $|7 + a|$ 2
41. $|-8 + k|$ 5
42. $|a + k|$ 2
43. $|k + m|$ 3
44. $|k| + |m|$ 9
45. $-|3 + a|$ -2
46. $-|k + 8|$ -11
47. $-|-24 + m|$ -30
48. $-|a + (-11)|$ -16

Find each sum.

49. $+374 + 239$ 613
50. $521 + 124$ 645
51. $-374 + (-165)$ -539
52. $-179 + (-826)$ -1005
53. $-582 + 379$ -203
54. $573 + (-336)$ 237
55. $1982 + (-1482)$ 500
56. $-1492 + 876$ -616
57. $294 + |-49|$ 343
58. $|-931| + (-643)$ 288
59. $|871 + (-284)|$ 587
60. $|-285 + (-641)|$ 926
61. $|-857 + 931|$ 74
62. $-|-423 + (-148)|$ -571
63. $|-429| + 243$ 672
64. $-197 + |-483|$ 286
65. $-\big||-843| + |-231|\big|$ -1074

About Using Calculators
Emphasize that to enter a negative number on a calculator, you must first press the absolute value of the number and then press the change-sign key. To enter a positive number, simply press the number.

Using Calculators — Change-Sign Key

Most calculators have a **change-sign key**, labeled [+/−]. When this key is pressed, the calculator changes the sign of the number in the display.

You can use the change-sign key to enter negative numbers. For example, you can enter −7 by pressing 7 and then pressing the change-sign key.

Example: Use the calculator to add −9 and 5.

Press the following sequence of keys.

9 [+/−] [+] 5 [=]

The sum is −4.

Exercises

Use a calculator to find each sum.

1. $-25 + (-18)$ -43
2. $173 + (-200)$ -27
3. $-618 + 52$ -566

For more exercises on adding integers, see exercises 1-30 on page 546.

Objective 2-4: Add rational numbers by using skills in adding integers and positive rational numbers.

2-4 Adding Rational Numbers

Some points on a number line cannot be named by integers. For example, the number line below is separated into fourths to show a sample of numbers that appear in the form of common fractions and decimals. The numbers shown on this number line are all **rational numbers**.

A rational number is a number that can be expressed in the form $\frac{a}{b}$, where a and b are integers and b is not equal to 0.

Definition of Rational Numbers

Examples of rational numbers expressed in the form $\frac{a}{b}$ are shown in this chart.

Rational Number	Fraction
3	$\frac{3}{1}$
$-2\frac{3}{4}$	$-\frac{11}{4}$
0.125	$\frac{1}{8}$
0	$\frac{0}{1}$
0.333 . . .	$\frac{1}{3}$

Terminating decimals and repeating decimals can be expressed as fractions.

Addition of rational numbers can be represented on a number line by following the same steps as addition of integers.

Consider the addition $\frac{1}{4} + \left(-\frac{3}{8}\right)$. First replace $\frac{1}{4}$ with $\frac{2}{8}$. Then add $\frac{2}{8}$ and $-\frac{3}{8}$.

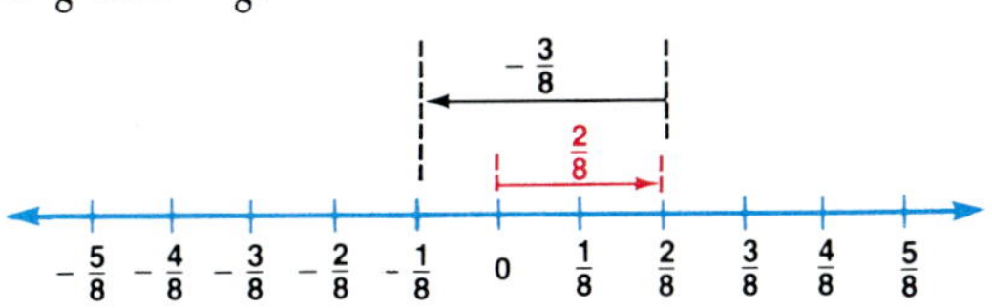

The sum of $\frac{1}{4}$ and $-\frac{3}{8}$ is $-\frac{1}{8}$.

Using the number line to add rational numbers is difficult. However, the rules used to add integers and to add positive rational numbers can be used to add all rational numbers.

2-4

Lesson Notes, p. T27

Readiness Activity

Find each sum.

$\frac{1}{2} + \frac{2}{3}$ $1\frac{1}{6}$

$\frac{6}{7} + \frac{2}{21}$ $\frac{20}{21}$

1.46 + 2.7 4.16

0.006 + 0.839 0.845

Discussion Question

Give several examples of rational numbers that are not integers.

Some possible answers are $\frac{1}{3}$, $-2\frac{1}{2}$, 6.7.

Chalkboard Exercises

Find each sum.

1a. $-\frac{3}{8} + \left(-\frac{4}{5}\right)$ $-1\frac{7}{40}$

1b. $\frac{2}{15} + \left(-\frac{4}{10}\right)$ $-\frac{4}{15}$

2a. $-11.9 + 11.95$ 0.05

2b. $44.3 + (-50)$ -5.7

3a. $-1\frac{1}{3} + 3\frac{2}{5}$ $2\frac{1}{15}$

3b. $5\frac{1}{3} + \left(-8\frac{3}{7}\right)$ $-3\frac{2}{21}$

Suggested Daily Assignments

Enriched
- Written: 2–30 evens, 31–38
- Mini-Review: 1–5

Average
- Exp: 2–22 evens
- Written: 1–30
- Mini-Review: 1–5

Fundamental
- Exp: 1–22
- Written: 1–18
  - p. 46, 25–40, 53–59
- Mini-Review: 1–5

Examples

1 **Add:** $-\frac{3}{8} + \left(-\frac{5}{16}\right)$

$-\frac{3}{8} + \left(-\frac{5}{16}\right) = -\frac{6}{16} + \left(-\frac{5}{16}\right)$ *Replace $-\frac{3}{8}$ with $-\frac{6}{16}$.*

$= -\left(\left|-\frac{6}{16}\right| + \left|-\frac{5}{16}\right|\right)$ *Since the numbers have the same sign, add their absolute values.*

$= -\left(\frac{6}{16} + \frac{5}{16}\right)$

$= -\frac{11}{16}$ *The sum is negative.*

2 **Add:** $1.354 + (-0.765)$

$1.354 + (-0.765) = +(|1.354| - |-0.765|)$ *1.354 has the greater absolute value.*

$= +(1.354 - 0.765)$ *Since the addends have different signs, find the difference of their absolute values.*

$= 0.589$ *The sum is positive.*

3 **Add:** $-1\frac{2}{3} + 4\frac{3}{4}$

$-1\frac{2}{3} + 4\frac{3}{4} = -1\frac{8}{12} + 4\frac{9}{12}$ *Replace $-1\frac{2}{3}$ with $-1\frac{8}{12}$ and $4\frac{3}{4}$ with $4\frac{9}{12}$.*

$= +\left(\left|4\frac{9}{12}\right| - \left|-1\frac{8}{12}\right|\right)$ *$4\frac{9}{12}$ has the greater absolute value.*

$= +\left(4\frac{9}{12} - 1\frac{8}{12}\right)$ *Since the addends have different signs, find the difference of their absolute values.*

$= +3\frac{1}{12}$ or $3\frac{1}{12}$ *The sum is positive.*

Summary of Rules for Addition of Rational Numbers

1. **If the numbers have the same sign, add their absolute values. Give the sum the same sign as the numbers.**
2. **If the numbers have different signs, subtract the lesser from the greater absolute value. Give the result the same sign as the number with the greater absolute value.**

Exploratory Exercises

State the absolute value of each number.

1. $+\frac{3}{4}$ $\frac{3}{4}$ **2.** $+\frac{9}{8}$ $\frac{9}{8}$ **3.** -1.76 1.76 **4.** -2.53 2.53 **5.** $-\frac{3}{11}$ $\frac{3}{11}$

6. -16.7 16.7 **7.** 1.82 1.82 **8.** $-\frac{7}{16}$ $\frac{7}{16}$ **9.** $+\frac{3}{82}$ $\frac{3}{82}$ **10.** 1.48 1.48

State the sign of each sum.

11. $-\frac{3}{4} + \frac{7}{8}$ +
12. $\frac{15}{16} + \left(-\frac{13}{16}\right)$ +
13. $-1.354 + 1.265$ −
14. $37.42 + (-45.36)$ −
15. $-\frac{12}{31} + \frac{13}{62}$ −
16. $-\frac{9}{13} + \left(-\frac{3}{13}\right)$ −
17. $-394.1 + 427.6$ +
18. $-85.32 + 76.3$ −
19. $-\frac{5}{12} + \left(-\frac{5}{24}\right)$ −
20. $\frac{2}{5} + \left(-\frac{2}{7}\right)$ +
21. $-0.0034 + 0.034$ +
22. $-0.0759 + 0.4$ +

Written Exercises

Find each sum.

1. $-\frac{11}{9} + \left(-\frac{7}{9}\right)$ -2
2. $\frac{17}{21} + \left(-\frac{13}{21}\right)$ $\frac{4}{21}$
3. $\frac{5}{11} + \left(-\frac{6}{11}\right)$ $-\frac{1}{11}$
4. $-\frac{8}{13} + \left(-\frac{11}{13}\right)$ $-\frac{19}{13}$
5. $-\frac{7}{12} + \frac{5}{12}$ $-\frac{1}{6}$
6. $-\frac{9}{28} + \frac{13}{28}$ $\frac{1}{7}$
7. $4.57 + (-3.69)$ 0.88
8. $-4.8 + 3.2$ -1.6
9. $-1.7 + (-3.9)$ -5.6
10. $-2.31 + 7.62$ 5.31
11. $-38.9 + 24.2$ -14.7
12. $-0.007 + 0.06$ 0.053
13. $\frac{3}{4} + \left(-\frac{7}{12}\right)$ $\frac{1}{6}$
14. $-\frac{2}{7} + \frac{3}{14}$ $-\frac{1}{14}$
15. $-\frac{3}{8} + \frac{5}{24}$ $-\frac{1}{6}$
16. $-\frac{5}{6} + \left(-\frac{7}{12}\right)$ $-\frac{17}{12}$
17. $-\frac{1}{8} + \left(-\frac{5}{2}\right)$ $-\frac{21}{8}$
18. $\frac{2}{3} + \left(-\frac{2}{9}\right)$ $\frac{4}{9}$
19. $-1.543 + 2.165$ 0.622
20. $-3.948 + 4.826$ 0.878
21. $-0.376 + (-0.289)$ -0.665
22. $-0.006 + 0.0052$ -0.0008
23. $-0.0005 + (-0.3)$ -0.3005
24. $0.0007 + (-0.001)$ -0.0003
25. $-\frac{3}{5} + \frac{5}{6}$ $\frac{7}{30}$
26. $\frac{3}{8} + \left(-\frac{7}{12}\right)$ $-\frac{5}{24}$
27. $-\frac{4}{15} + \frac{3}{4}$ $\frac{29}{60}$
28. $-\frac{9}{4} + \left(-\frac{4}{3}\right)$ $-\frac{43}{12}$
29. $\frac{17}{5} + \left(-\frac{13}{4}\right)$ $\frac{3}{20}$
30. $-\frac{7}{15} + \left(-\frac{5}{12}\right)$ $-\frac{53}{60}$

Challenge

Find each sum. Express each result in decimal form.

31. $-\frac{4}{5} + (-3.8)$ -4.6
32. $-0.37 + \left(-\frac{21}{8}\right)$ -2.995
33. $-5\frac{3}{4} + 6.25$ 0.5
34. $-8.66 + 6\frac{7}{8}$ -1.785
35. $7.43 + \left(-\frac{9}{4}\right)$ 5.18
36. $4\frac{7}{8} + -3.754$ 1.121
37. $-11\frac{7}{16} + 7.225$ -4.2125
38. $-3.276 + \left(-\frac{15}{8}\right)$ -5.151

mini-review

Simplify.

1. $7(x + y) + 2y$ $7x + 9y$
2. $7a^2 + 3(a^2 + b) + 6a$ $10a^2 + 3b + 6a$

Evaluate.

3. $2^4 - 2^2 \div 2^2$ 15

Write an equation for each sentence.

4. The square of the difference of a and b is equal to twice c. $(a - b)^2 = 2c$
5. The circumference (C) of a circle is equal to the product of π and the diameter (d). $C = \pi d$

For review exercises on adding and subtracting fractions, see exercises 1-24 on page 548.
For review exercises on adding and subtracting decimals, see exercises 1-30 on page 550.

About Problem Solving
Developing good guess-and-check procedures can be very beneficial. Emphasize the importance of recording guesses. Sometimes a pattern that develops may be found studying the recorded guesses.

Problem Solving

Guess and Check

Many different strategies can be used to solve problems. One strategy is to write an equation. Another important strategy is called *guess and check.* To use this strategy, guess the answer to the problem, then check whether the guess is correct. If the first guess is incorrect, guess again until you find the right answer. Often, the results of one guess can help you make a better guess. Always keep a record of your guesses so you do not make the same guess twice.

Sometimes there will not be one right answer. You may have to find the best answer.

Example: **Insert parentheses so that a true equation results.**

$$10 - 4 \cdot 2 - 1 = 3$$

Copy the expression to the left of the equals sign.
Then insert parentheses and evaluate the expression.

$10 - 4 \cdot (2 - 1) = 6$ *Try again since the value does not equal 3.*
$(10 - 4) \cdot 2 - 1 = 11$ *Try again.*
$10 - (4 \cdot 2 - 1) = 3$ *Correct!*

Exercises

Use the guess-and-check strategy to solve each problem.

1. When a certain number is multiplied by itself the product is 2916. Find the number. 54

2. Find an odd number between 10 and 25 which is divisible by 3, but is not divisible by 5. 21

3. Using each of the digits 1, 2, . . ., 6 only once, write two whole numbers whose product is as large as possible. 631 × 542

4. If it costs a nickel each time you cut and weld a link, what is the minimum cost to make a chain out of 5 links? 10¢

5. Paper plates can be purchased in packages of 15 or 25. Joe bought 7 packages and got 125 plates. How many packages of 25 did he buy? 2

6. The cube of a certain whole number is close to 4000. Find the number. 16

7. Copy the figure at the right. Then fill in the digits 1, 2, . . ., 8 in such a way that no two consecutive numbers are in boxes which touch at a point or side.

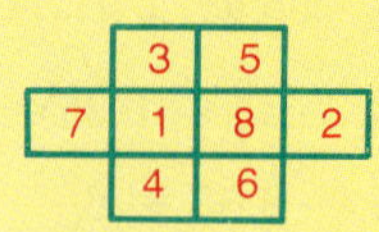

8. Gina and Jim raise cats and birds. They counted all the heads and got 10. They counted all the feet and got 34. How many birds and cats do they have? 3 birds, 7 cats

For each of the following, insert parentheses so that a true equation results.

9. $4 \cdot 5 - 2 + 7 = 19$ (5 − 2)

10. $25 - 4 \cdot 2 + 3 = 5$ (2 + 3)

11. $3 + 5 \cdot 8 - 2 = 48$ (3 + 5)(8 − 2)

12. $3 + 6 \cdot 4 \cdot 2 = 54$ (3 + 6 · 4)

Objective 2-5A: Add three or more rational numbers.
Objective 2-5B: Simplify expressions that contain rational numbers.

2-5 More About Addition

Previously you have added pairs of numbers. To add three or more rational numbers, first group the numbers in pairs. Use the commutative and associative properties to rearrange the addends if necessary.

Examples

1 Add: $(-4) + 5 + (-6)$

$-4 + 5 + (-6) = (-4 + 5) + (-6)$ *Group two of the addends.*
$= 1 + (-6)$
$= -5$ *Do you get the same result if you find the sum $-4 + [5 + (-6)]$?*

2 Add: $-\frac{4}{3} + \frac{5}{8} + \left(-\frac{7}{3}\right)$

$-\frac{4}{3} + \frac{5}{8} + \left(-\frac{7}{3}\right) = \left[-\frac{4}{3} + \left(-\frac{7}{3}\right)\right] + \frac{5}{8}$ *Commutative and Associative Properties.*
$= -\frac{11}{3} + \frac{5}{8}$
$= -\frac{88}{24} + \frac{15}{24}$
$= -\frac{73}{24}$
$= -3\frac{1}{24}$

3 Add: $28.32 + (-56.17) + 32.41 + (-75.13)$

Group the positive numbers and group the negative numbers.

$28.32 + (-56.17) + 32.41 + (-75.13) = 28.32 + 32.41 + (-56.17) + (-75.13)$
$= (28.32 + 32.41) + [(-56.17) + (-75.13)]$
$= 60.73 + (-131.30)$
$= -70.57$

You can use the distributive property and what you know about rational numbers to simplify expressions that have like terms. ❶

Example

4 Simplify: $-5x + a + 7x + (-4a)$

$-5x + a + 7x + (-4a) = (-5x + 7x) + [1a + (-4a)]$ *Group the like terms.*
$= (-5 + 7)x + [1 + (-4)]a$ *Use the distributive property.*
$= 2x + (-3a)$

2-5

Lesson Notes, p. T28

Readiness Activity

Find each sum.
$(-8 + -2) + 4$ -6
$-6 + (-4 + 5)$ -5
$(3 + 6) + -2$ 7
$12 + (17 + -28)$ 1

Discussion Question

Suggest two different ways to add $-5 + 51 + 26 + -13$. (Hint: use the commutative and associative properties, and group the numbers in pairs.) Answers will vary.

Chalkboard Exercises

Find each sum.
1. $6 + (-2) + (-14)$ -10
2a. $\frac{5}{8} + \left(-\frac{3}{4}\right) + \frac{15}{16}$ $\frac{13}{16}$
2b. $-\frac{3}{5} + \left(-\frac{2}{3}\right) + \frac{3}{8}$ $-\frac{107}{120}$
3a. $-4.13 + (-6.18) + (-9.31)$ -19.62
4a. $-3a + (-2a) + 9a$ $4a$
4b. $5y + (-12y) + (-18y)$ $-25y$
4c. $6k + (-5a) + (-17k) + 8a$ $3a + (-11k)$

❶ Some students may need to review the meaning of like terms.

Suggested Daily Assignments

Enriched
Written: 2–40 evens, 41–44

Average
Exp: 1–12
Written: 2–32 evens, 33–44

Fundamental
Exp: 1–12
Written: 2–40 evens
p. 49, 19–30

Exploratory Exercises

Find each sum.

1. $2 + (-6) + 4$ 0
2. $5 + 2 + (-5)$ 2
3. $0.6 + (-0.3) + (-0.4)$ −0.
4. $-0.2 + 0.4 + (-0.7)$ −0.5
5. $\frac{1}{2} + \left(-\frac{1}{4}\right) + \frac{3}{4}$ 1
6. $-\frac{1}{5} + \frac{3}{5} + \left(-\frac{4}{5}\right)$ $-\frac{2}{5}$
7. $-4m + 8m$ $4m$
8. $-8x + (-10x)$ $-18x$
9. $r + (-2r) + 5r$ $4r$
10. $-k + 6k + (-3k)$ $2k$
11. $-b + y + 4b + (-5y)$ $3b - 4y$
12. $s + (-5t) + 9t + 8s$ $9s + 4t$

Written Exercises

Find each sum.

1. $4 + (-12) + (-18)$ −26
2. $7 + (-11) + 32$ 28
3. $8 + (-15) + 13$ 6
4. $-41 + (-78) + 51$ −68
5. $83 + (-19) + 16$ 80
6. $16 + (-9) + (-94)$ −87
7. $-3a + 12a + (-14a)$ $-5a$
8. $5y + (-12y) + (-21y)$ $-28y$
9. $14b + (-21b) + 37b$ $30b$
10. $5x + (-21x) + 29x$ $13x$
11. $-3z + (-17z) + (-18z)$ $-38z$
12. $9m + 43m + (-16m)$ $36m$
13. $-\frac{3}{4} + \frac{5}{12} + \left(-\frac{5}{4}\right)$ $-\frac{19}{12}$
14. $\frac{7}{3} + \left(-\frac{5}{6}\right) + \left(-\frac{2}{3}\right)$ $\frac{5}{6}$
15. $-\frac{2}{7} + \frac{3}{14} + \frac{3}{7}$ $\frac{5}{14}$
16. $-\frac{3}{5} + \frac{6}{7} + \left(-\frac{2}{35}\right)$ $\frac{1}{5}$
17. $\frac{5}{8} + \left(-\frac{3}{16}\right) + \left(-\frac{7}{24}\right)$ $\frac{7}{48}$
18. $\frac{3}{4} + \left(-\frac{5}{8}\right) + \frac{3}{32}$ $\frac{7}{32}$
19. $12 + (-17) + 36 + (-45)$ −14
20. $81 + (-31) + (-9) + 62$ 103
21. $-75 + 47 + 32 + (-16)$ −12
22. $28 + (-56) + 32 + (-75)$ −71
23. $6.7 + (-8.1) + (-7.3)$ −8.7
24. $-7.6 + 1.8 + (-3.5)$ −9.3
25. $-37.12 + 42.18 + (-12.6)$ −7.54
26. $-4.13 + (-5.18) + 9.63$ 0.32
27. $-7.9 + (-5.3) + (-4.2)$ −17.4
28. $-9.7 + 5.3 + 4.2$ −0.2
29. $-14a + 36k + 12k + (-83a)$ $-97a + 48k$
30. $16b + (-22xy) + 31b + (-36xy)$ $47b - 58xy$
31. $13mp + (-3ps) + 76mp + (-21ps)$ $89mp - 24ps$
32. $-17px + 22bg + 35px + (-37bg)$ $18px - 15bg$

33. $12a + (-9a)$ $3a$
34. $-15m + 6m$ $-9m$
35. $-23w + 47w$ $24w$
36. $-13c + (-28c)$ $-41c$
37. $5.8k + (-3.6k)$ $2.2k$
38. $-7.9s + (-3.8s)$ $-11.7s$
39. $-0.23x + (-0.5x)$ $-0.73x$
40. $0.81h + (-0.93h)$ $-0.12h$

Solve each problem.

41. Julie Thomas shot rounds in a recent golf tournament in which her scores relative to par were −3, +2, −4, and −1. What was her score for the tournament relative to par? −6

42. Last week, the following day-to-day changes in the noon temperatures were recorded: Sun., +3; Mon., −7; Tues., −5; Wed., +4; Thurs., +6; Fri., −3; Sat., −8. What was the net change for the week? 42. −10

43. On Tuesday, Johnny Lomax wrote checks for \$35.76 and \$41.32. On Wednesday, he deposited \$135.59. Friday he wrote a check for \$63.17. What was the net increase or decrease in his account for the week? −4.66 (decrease)

44. Last week, the following day-to-day changes for the Dow-Jones stock averages were recorded: Mon., $+5\frac{3}{8}$; Tues., $-6\frac{1}{4}$; Wed., $+11\frac{1}{8}$; Thurs., $+3\frac{5}{8}$; Fri., $-7\frac{1}{2}$. What was the net change for the week? $+6\frac{3}{8}$

Objective 2-6: Subtract rational numbers by using additive inverses.

2-6 Subtraction

There are many patterns in nature. On an oak leaf, points match each other *opposite* a center line. Similar patterns are helpful in understanding subtraction.

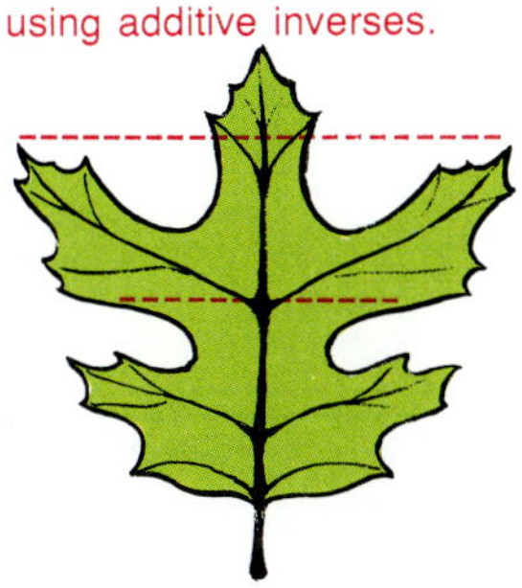

Notice that -3 and 3 are the same distance but in *opposite* directions from 0 on the number line.

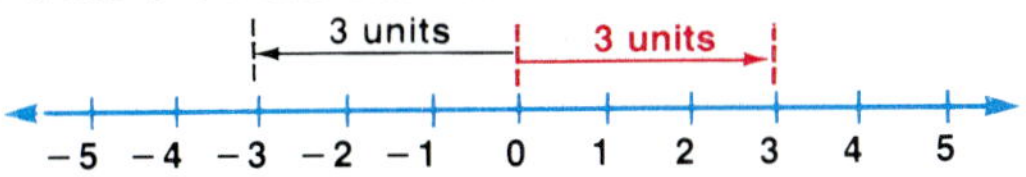

What is the result when you add two numbers such as 3 and -3?

$3 + (-3) = 0 \quad -7 + 7 = 0 \quad -\frac{4}{7} + \frac{4}{7} = 0 \quad 19.3 + (-19.3) = 0$

If the sum of two numbers is 0, the numbers are called **additive inverses** or **opposites**.

-3 is the additive inverse, or opposite, of 3.

7 is the additive inverse, or opposite, of -7.

$\frac{4}{7}$ is the additive inverse, or opposite, of $-\frac{4}{7}$.

-19.3 is the additive inverse, or opposite, of 19.3

Is 3 the additive inverse of −3?

Zero is its own opposite.

These and many other similar examples suggest the following property.

❶

For every number a, $a + (-a) = 0$

Additive Inverse Property

If a is negative, then (−a) is positive.

Additive inverses are used in the subtraction of rational numbers. Observe the following patterns.

Subtraction	***Addition***	***Subtraction***	***Addition***
$7 - 3 = 4$	$7 + (-3) = 4$	$8 - 2 = 6$	$8 + (-2) = 6$

additive inverses (3 and −3; 2 and −2)

same result (4 and 4; 6 and 6)

2-6

Lesson Notes, p. T28

Readiness Activity

True or false?

$4 - 1 = 4 + (-1)$ true

$8 - 6 = 8 + (-6)$ true

$2.2 + (-1.9) = 2.2 - 1.9$ true

$7x + (-5x) = 7x - 5x$ true

Discussion Questions

Find the value of $a + (-a)$ for each of the following values of a.

1. 3 0 **2.** 0 0

3. -5 0

4. When a is a negative number, what is the sign of $-a$? positive

❶ Emphasize to students that $(-a)$ is read "the opposite of a" and not "negative."

Chalkboard Exercises

Find each difference.

1. $19 - (-9)$
   28
2. $34 - 59$
   -25
3. $-12 - (-4)$
   -8

4a. $-31.5 - (-13.2)$
   -18.3

4b. $13.81 - -13.4$
   27.21

5. $\frac{1}{5} - \left(-\frac{1}{10}\right)$
   $\frac{3}{10}$

It appears that subtracting a number is equivalent to adding its additive inverse. These and other similar examples suggest the rule for subtracting rational numbers.

Subtraction Rule

To subtract a rational number, add its additive inverse. For rational numbers a and b, $a - b = a + (-b)$.

Examples

1 **Subtract: $5 - (-8)$**

$5 - (-8) = 5 + (+8)$ *To subtract -8, add $+8$.*

$= 13$

2 **Subtract: $6 - 11$**

$6 - 11 = 6 + (-11)$ *To subtract 11, add -11.*

$= -5$

3 **Subtract: $-4 - (-7)$**

$-4 - (-7) = -4 + (+7)$ *To subtract -7, add $+7$.*

$= 3$

4 **Subtract: $7.32 - (-6.85)$**

$7.32 - (-6.85) = 7.32 + (+6.85)$ *To subtract -6.85, add $+6.85$.*

$= 14.17$

5 **Subtract: $-\frac{7}{8} - \left(-\frac{3}{16}\right)$**

$-\frac{7}{8} - \left(-\frac{3}{16}\right) = -\frac{7}{8} + \frac{3}{16}$ *To subtract $-\frac{3}{16}$ add $+\frac{3}{16}$.*

$= -\frac{14}{16} + \frac{3}{16}$

$= -\frac{11}{16}$

The distributive property is used to simplify expressions that contain like terms as shown in the next example.

Example

6 Simplify: $7x - 12x$

$7x - 12x = 7x + (-12x)$ *To subtract 12x, add $-12x$.*

$= [7 + (-12)]x$ *Distributive Property*

$= -5x$

Exploratory Exercises

State the additive inverse of each of the following.

1. $+6$ -6
2. -6 6
3. 5 -5
4. -14 14
5. -13 13
6. a $-a$
7. $-a$ a
8. $-2y$ $2y$
9. 0 0
10. $4x$ $-4x$
11. 3.7 -3.7
12. -7.4 7.4
13. $\frac{3}{7}$ $-\frac{3}{7}$
14. $-\frac{4}{9}$ $\frac{4}{9}$
15. $-\frac{8}{17}$ $\frac{8}{17}$

State the number named.

16. $-(-5)$ 5
17. $-(+7)$ -7
18. $-(4)$ -4
19. $-(-13)$ 13
20. $-(-36)$ 36
21. $-(-7)$ 7
22. $-(18)$ -18
23. $-(-16)$ 16
24. $-(-43)$ 43
25. $-(+56)$ -56
26. $-(-1.8)$ 1.8
27. $-(+7.1)$ -7.1
28. $-\left(\frac{8}{11}\right)$ $-\frac{8}{11}$
29. $-\left(-\frac{3}{4}\right)$ $\frac{3}{4}$
30. $-(6.5)$ -6.5

State an addition expression for each of the following.

31. $8 - 13$ $8 + (-13)$
32. $-18 - 7$ $-18 + (-7)$
33. $-17 - 8$ $-17 + (-8)$
34. $11 - (-2)$ $11 + 2$
35. $-1.7 - (-1.5)$ $-1.7 + 1.5$
36. $\frac{3}{8} - \left(-\frac{2}{3}\right)$ $\frac{3}{8} + \frac{2}{3}$
37. $9y - 3y$ $9y + (-3y)$
38. $-8a - (-7a)$ $-8a + 7a$

Written Exercises

Find each difference.

1. $27 - 19$ 8
2. $47 - 32$ 15
3. $52 - 37$ 15
4. $8 - 13$ -5
5. $13 - 31$ -18
6. $27 - 43$ -16
7. $17 - (-23)$ 40
8. $29 - (-25)$ 54
9. $18 - (-34)$ 52
10. $-21 - (-14)$ -7
11. $-23 - (-12)$ -11
12. $-47 - 35$ -82
13. $\frac{3}{5} - \left(-\frac{1}{5}\right)$ $\frac{4}{5}$
14. $\frac{1}{6} - \frac{2}{3}$ $-\frac{1}{2}$
15. $-\frac{1}{2} - \left(-\frac{3}{4}\right)$ $\frac{1}{4}$
16. $-5\frac{7}{8} - 2\frac{3}{4}$ $-8\frac{5}{8}$
17. $0 - 21$ -21
18. $69 - (-95)$ 164
19. $7.3 - (-4.2)$ 11.5
20. $-4.5 - 8.6$ -13.1
21. $-67.1 - (-38.2)$ -28.9
22. $89.3 - (-14.2)$ 103.5
23. $-72.5 - 81.3$ -153.8
24. $3.81 - (-4.65)$ 8.46
25. $\begin{array}{r} 19 \\ -36 \\ \hline \end{array}$ -17
26. $\begin{array}{r} 61 \\ -(-43) \\ \hline \end{array}$ 104
27. $\begin{array}{r} -5.2 \\ -(-3.8) \\ \hline \end{array}$ -1.4
28. $\begin{array}{r} -1.7 \\ -(+3.9) \\ \hline \end{array}$ -5.6
29. $19m - 12m$ $7m$
30. $8h - 23h$ $-15h$
31. $-18p - 4p$ $-22p$
32. $-21a - 3a$ $-24a$
33. $24b - (-9b)$ $33b$
34. $-31x - (-33x)$ $2x$
35. $-52z - (-17z)$ $-35z$
36. $41y - (-41y)$ $82y$

For more exercises on subtracting integers, see exercises 31-60 on page 546.
For more exercises on adding and subtracting rational numbers, see exercises 25-48 on page 548 and exercises 31-60 on page 550.

Chalkboard Exercise

6. $-19p - (-3p)$ $-16p$

Suggested Daily Assignments

Enriched
- Written: 2–54 evens
- Mini-Review: 1–5

Average
- Exp: 2–38 evens
- Written: 2–54 evens
- Mini-Review: 1–5

Fundamental
- Exp: 1–38
- Written: 1–36
  - p. 52, 1–17 odds, 41–44
- Mini-Review: 1–5

Quiz
Lessons 2-3 to 2-6

Find each sum.

1. $-26 + (-32)$ -58
2. $17 + (-8)$ 9
3. $\frac{-11}{25} + \frac{19}{25}$ $\frac{8}{25}$
4. $0.567 + 0.2068$ 0.7738
5. $-3.6 + 1.9$ -1.7
6. $-1\frac{1}{4} + \left(-3\frac{3}{5}\right)$ $-4\frac{17}{20}$
7. Evaluate $k + (-36)$ if $k = 51$. 15

Find each sum or difference.

8. $\frac{1}{4} + \left(-\frac{2}{3}\right) + \left(-\frac{3}{8}\right)$ $\frac{-19}{24}$
9. $8y - (-3y)$ $11y$
10. Evaluate $-16 - k$, if $k = -1.5$. -14.5
11. Evaluate $x - 85$, if $x = 56$. -29

About the Excursion

The excursion provides extra practice in adding signed numbers. Ask students to copy the magic squares at the bottom of the page. Have students look at the upper left square of the first magic square at the bottom. The 3 was obtained by adding some number to the 1 in the magic square above. What number was added? 2

Explain that a new magic square can be obtained by adding 2 to each number in the magic square above. Discuss how the entries in the other two magic squares can be obtained. In the second magic square, the missing entries can be obtained by adding -3. In the third magic square, the missing entries can be obtained by adding -7. After the students have filled in all the missing entries, have them check whether they have actually obtained magic squares.

Evaluate each of the following.

37. $m - 7$, if $m = 5$ -2

38. $a - 12$, if $a = -8$ -20

39. $p - 14$, if $p = 72$ 58

40. $y - 0.5$, if $y = -0.8$ -1.3

41. $a - (-7)$, if $a = 1.9$ 8.9

42. $b - (-0.5)$, if $b = -13$ -12.5

43. $h - (-1.3)$, if $h = -18$ -16.7

44. $k - (-12)$, if $k = 2.7$ 14.7

45. $w - 3.7$, if $w = -1.8$ -5.5

46. $\frac{3}{5} - n$, if $n = -\frac{7}{5}$ 2

47. $\frac{11}{2} - m$, if $m = -\frac{5}{2}$ 8

48. $\frac{8}{13} - a$, if $a = -\frac{9}{13}$ $\frac{17}{13}$

49. $\frac{11}{4} - x$, if $x = \frac{27}{8}$ $-\frac{5}{8}$

50. $-\frac{8}{5} - y$, if $y = \frac{12}{5}$ -4

51. $-\frac{12}{7} - z$, if $z = \frac{16}{21}$ $-\frac{52}{21}$

52. $-\frac{13}{6} - k$, if $k = -\frac{11}{12}$ $-\frac{5}{4}$

53. $-\frac{18}{17} - q$, if $q = \frac{31}{34}$ $-\frac{67}{34}$

54. $-\frac{33}{38} - r$, if $r = -\frac{16}{19}$ $-\frac{1}{38}$

mini-review

Find each sum.

1. $-83 + 24$ -59

2. $-0.25 + (-1.873)$ -2.123

3. $-3x + 5x + (-9x)$ $-7x$

State the property that is shown.

4. $6(x + 3y) = 6(3y + x)$

commutative property for addition

Define a variable, then write an equation for the following.

5. 55 decreased by three times a number is 31. Find the number.

Let n = the number; $55 - 3n = 31$

Excursions in Algebra — Magic Squares

Magic squares are square arrays of numbers. The sum of the numbers in each row, column, and diagonal is the same.

1	2	−3
−4	0	4
3	−2	−1

The sum of the numbers in each row, column, and diagonal of this magic square is 0.

New magic squares can be formed by adding the same number to each entry in the magic square above. In the squares below, how was each entry obtained? Find each new magic square.

Add 2.

3	4	−1
−2	2	6
5	0	1

Add −3.

−2	−1	−6
−7	−3	1
0	−5	−4

Add −7.

−6	−5	−10
−11	−7	−3
−4	−9	−8

Objective 2-7: Solve equations involving addition by using the addition and subtraction properties of equality.

2-7 Solving Equations Involving Addition

A scale, as shown below on the left is in balance when both sides hold equal weights. If you add weight to only one side, as shown on the right, then the scale is no longer in balance.

However, if you add the same weight to both sides, the scale will balance.

Think of an equation as a scale in balance. If the same number is added to both sides of the equation, then the result is an equivalent equation. **Equivalent equations** are equations that have the same solution. ❶

$$11 = x + 3$$ *The solution to this equation is 8.*

$$11 + 5 = (x + 3) + 5$$ *Add 5 to both sides.*

$$16 = x + 8$$ *The solution to this equation is also 8.*

The property that is used to add the same number to both sides of an equation is called the **addition property of equality**.

For any numbers a, b, and c,
if $a = b$, then $a + c = b + c$.

Addition Property of Equality

2-7

Lesson Notes, p. T28

Readiness Activity

Here are two true equations.

$$2 = 2$$
$$-19 = -19$$

Is each equation still true if you:

(a) add 3 to both sides? yes

(b) add -3 to both sides? yes

(c) subtract 3 from both sides? yes

(d) subtract -3 from both sides? yes

Discussion Questions

State whether each variable represents a positive or negative number.

1. $x + 2 = -5$ negative
2. $-4 + a = -10$ negative
3. $2\frac{1}{2} = y + 1\frac{1}{3}$ positive
4. $-32 = -40 + b$ positive
5. $n + -16.2 = 0$ positive
6. $-5\frac{3}{4} + p = 2\frac{3}{8}$ positive

❶ Emphasize that isolating the variable is the basic method of solving equations. Be sure students thoroughly understand the concept of equivalent equations. It is the clue to solutions of all equations!

Chalkboard Exercises

Solve each equation.

1. $p + 156 = -92$ -248
2. $s + 81 = -7$ -88

❷ The subtraction property of equality is not necessary for solving equations since you can always add the opposite instead. However, it does seem simpler to subtract in many cases.

❸ Explain the symbol $\stackrel{?}{=}$ to your students. For example $a \stackrel{?}{=} b$ means "Is a equal to b?"

Also, if the same number is *subtracted* from both sides of an equation, the result is an equivalent equation.

$$11 = x + 3$$
$$11 - 3 = (x + 3) - 3$$ *Subtract 3 from both sides.*
$$8 = x$$ *Why is this equation equivalent to $11 = x + 3$?*

The property that is used to subtract the same number from both sides of an equation is called the **subtraction property of equality**. ❷

For any numbers a, b, and c, if $a = b$, then $a - c = b - c$.

Subtraction Property of Equality

Both the addition and subtraction properties of equality can be used to solve equations involving addition.

Examples

1 **Solve: $k + 15 = -6$**

$$k + 15 = -6$$ *Use the subtraction property of equality.*
$$k + 15 - 15 = -6 - 15$$ *Subtract 15 from both sides.*
$$k + 0 = -21$$
$$k = -21$$

To check that -21 is the solution, substitute -21 for k in $k + 15 = -6$.

Check: $k + 15 = -6$

$-21 + 15 \stackrel{?}{=} -6$ ❸

$-6 = -6$ *The check verifies that the solution to the equation is -21.*

2 **Solve: $r + 16 = -7$**

$$r + 16 = -7$$ *Use the addition property of equality and the additive inverse.*
$$r + 16 + (-16) = -7 + (-16)$$ *Add -16 to both sides.*
$$r + 0 = -23$$ *The sum of 16 and -16 is 0.*
$$r = -23$$

Check: $r + 16 = -7$

$-23 + 16 \stackrel{?}{=} -7$

$-7 = -7$ *The check verifies that the solution to the equation is -23.*

Example

3 Solve: $m + \left(-\frac{3}{4}\right) = -\frac{1}{2}$

$$m + \left(-\frac{3}{4}\right) + \frac{3}{4} = -\frac{1}{2} + \frac{3}{4}$$

$$m + 0 = -\frac{2}{4} + \frac{3}{4}$$

$$m = \frac{1}{4}$$ The solution is $\frac{1}{4}$.

Check: $m + \left(-\frac{3}{4}\right) = -\frac{1}{2}$

$$\frac{1}{4} + \left(-\frac{3}{4}\right) \stackrel{?}{=} -\frac{1}{2}$$

$$-\frac{1}{2} = -\frac{1}{2}$$

Exploratory Exercises

State the number you add to both sides of each equation to solve it.

1. $y + 21 = -7$ −21
2. $13 + x = -16$ −13
3. $y + (-5) = 11$ 5
4. $z + (-9) = 34$ 9
5. $-10 + k = 34$ 10
6. $y + 13 = 45$ −13

State the number you subtract from both sides of each equation to solve it.

7. $m + 16 = 14$ 16
8. $k + 9 = -16$ 9
9. $t + 5 = 8$ 5
10. $y + 9 = -53$ 9
11. $z + (-3) = -8$ −3
12. $x + (-4) = -37$ −4

Written Exercises

Solve each equation.

1. $m + 10 = 7$ −3
2. $y + 16 = 7$ −9
3. $5 + a = -14$ −19
4. $9 = x + 13$ −4
5. $k + 11 = -21$ −32
6. $b + 15 = -32$ −47
7. $-11 = a + 8$ −19
8. $18 + m = -57$ −75
9. $14 + c = -5$ −19
10. $y + 3 = -15$ −18
11. $p + 12 = -4$ −16
12. $w + 42 = -51$ −93
13. $r + (-8) = 7$ 15
14. $z + (-17) = 0$ 17
15. $z + (-18) = 34$ 52
16. $-12 + b = 12$ 24
17. $-15 + d = 13$ 28
18. $x + (-7) = 36$ 43
19. $r + (-11) = -21$ −10
20. $y + (-13) = -27$ −14
21. $h + (-13) = -5$ 8
22. $d + (-6) = -9$ −3
23. $-11 = k + (-5)$ −6
24. $-23 = -19 + n$ −4
25. $-4.1 = m + (-0.5)$ −3.6
26. $0 = t + (-1.4)$ 1.4
27. $y + 2.3 = 1.5$ −0.8
28. $x + 4.2 = 1.5$ −2.7
29. $2.4 = m + 3.7$ −1.3
30. $4.4 = b + 6.3$ −1.9
31. $y + \frac{7}{16} = -\frac{5}{8}$ $-\frac{17}{16}$
32. $x + \frac{4}{9} = -\frac{2}{27}$ $-\frac{14}{27}$
33. $-\frac{7}{6} + k = \frac{5}{6}$ 2
34. $-\frac{5}{7} + w = \frac{5}{7}$ $\frac{10}{7}$
35. $m + \frac{5}{9} = -\frac{3}{5}$ $-\frac{52}{45}$
36. $m + \left(-\frac{7}{8}\right) = \frac{5}{12}$ $\frac{31}{24}$

Using Calculators — Checking Solutions

A calculator can help you check the solution to an equation. For example, enter the following sequence to check if −2.4 is the solution to $x + (-7.3) = -9.7$.

ENTER: 2.4 [+/−] [+] 7.3 [+/−] [=]

Because −9.7 shows in the display screen, −2.4 is the solution.

Exercises Use a calculator to check the solutions for Written Exercises 19-30.
See student's work.

Chalkboard Exercises

3a. $-13 + b = 10$ 23

3b. $-\frac{3}{4} + w = \frac{4}{5}$ $1\frac{11}{20}$

3c. $m + \frac{-3}{8} = \frac{3}{5}$ $\frac{39}{40}$

3d. $a + -5 = 6$ 11

Suggested Daily Assignments

Enriched
Written: 2–36 evens

Average
Exp: 1–11 odds
Written: 1–36

Fundamental
Exp: 1–12
Written: 2–36 evens
p. 56, 37–54

2-8

Lesson Notes, p. T28

Readiness Activity

Rewrite using subtraction

$x + (-10)$ $q + (-12)$

$x - 10$ $q - 12$

Rewrite using addition

$m - (-14)$ $a - (-1)$

$m + 14$ $a + 1$

Discussion Questions

Solve each equation.

1. $-x = -4$ $x = 4$

2. $5 = -y$ $y = -5$

3. $-a = 17$ $a = -17$

4. $-2\frac{1}{2} = -b$ $b = 2\frac{1}{2}$

Chalkboard Exercises

Solve each equation.

1a.	$v - 21 = -78$	-57
1b.	$y - 12 = 20$	32
2a.	$y - (-13) = 14$	1
2b.	$s - (-13.1) = 14.2$	1.1
3.	$y + \left(-\frac{2}{3}\right) = \frac{4}{5}$	$1\frac{7}{15}$

❶ Remind students that $a - b = a + (-b)$ and $a + (-b) = a - b$.

❷ The equation in Example 2 can be solved by adding -8 to both sides, that is, by performing the inverse operation. However, most students can solve it more easily by first writing the equation in a simpler form, as shown.

❸ The equation in Example 3 can be solved directly by adding 7.5 to both sides, as was done in Lesson 2-7. However, students should feel at ease with writing $y + (-7.5)$ as $y - 7.5$.

Objective 2-8: Solve equations involving subtraction.

2-8 Solving Equations Involving Subtraction

Many equations involving subtraction may be solved by using the addition property of equality. ❶

Example

1 **Solve: $m - 9 = -13$.**

$$m - 9 = -13$$
$$m - 9 + 9 = -13 + 9 \quad \textit{Add 9 to both sides.}$$
$$m = -4$$

Check:
$$m - 9 = -13$$
$$-4 - 9 \stackrel{?}{=} -13$$
$$-13 = -13$$

The solution is -4.

Sometimes an equation can be solved more easily if it is first rewritten in a different form. Recall that subtracting a number is the same as adding its inverse. For example, the equation $b - (-8) = 23$ may be rewitten as $b + 8 = 23$. Also, $y + (-7.5) = -12.2$ may be rewritten as $y - 7.5 = -12.2$.

Examples

2 **Solve: $b - (-8) = 23$** ❷

This equation is equivalent to $b + 8 = 23$.

$$b + 8 = 23$$
$$b + 8 - 8 = 23 - 8 \quad \textit{Subtract 8 from both sides.}$$
$$b = 15$$

Check:
$$b - (-8) = 23$$
$$15 - (-8) \stackrel{?}{=} 23$$
$$23 = 23$$

The solution is 15.

3 **Solve: $y + (-7.5) = -12.2$**

This equation is equivalent to $y - 7.5 = -12.2$ ❸

$$y - 7.5 = -12.2$$
$$y - 7.5 + 7.5 = -12.2 + 7.5 \quad \textit{Add 7.5 to both sides.}$$
$$y = -4.7 \quad \textit{Check this result.}$$

The solution is -4.7.

Solving some equations requires two steps. Study the following example.

Example

4 Solve: $-8 - k = 13$ ❹

$$\begin{aligned} -8 - k &= 13 \\ -8 - k + k &= 13 + k && \text{Add } k \text{ to both sides.} \\ -8 &= 13 + k \\ -13 + (-8) &= -13 + 13 + k && \text{Add } -13 \text{ to both sides (or subtract 13).} \\ -21 &= k \end{aligned}$$

Check:

$$\begin{aligned} -8 - k &\stackrel{?}{=} 13 \\ -8 - (-21) &\stackrel{?}{=} 13 \\ 13 &= 13 \end{aligned}$$

The solution is -21.

Exploratory Exercises

Rename the following expressions by using their inverse operations.

1. $m + (-8)$ $m - 8$ **2.** $r + (-12)$ $r - 12$ **3.** $y - (-11)$ $y + 11$ **4.** $k - (-12)$ $k + 12$
5. $b + (-17)$ $b - 17$ **6.** $a - (-21)$ $a + 21$ **7.** $n - (-23)$ $n + 23$ **8.** $x + (-36)$ $x - 36$
9. $z + (-31)$ $z - 31$ **10.** $s - (-18)$ $s + 18$ **11.** $p - (-47)$ $p + 47$ **12.** $q + (-62)$ $q - 62$

Written Exercises

Solve each equation.

1. $a - 15 = -32$ -17 **2.** $h - 26 = -29$ -3 **3.** $y - 7 = -32$ -25
4. $r - 21 = -37$ -16 **5.** $y + (-7) = -19$ -12 **6.** $x + (-8) = -31$ -23
7. $b + (-14) = 6$ 20 **8.** $k + (-13) = 21$ 34 **9.** $d - (-27) = 13$ -14
10. $m - (-13) = 37$ 24 **11.** $r - (-31) = 16$ -15 **12.** $t - (-16) = 9$ -7
13. $-8 + m = -11$ -3 **14.** $-7 + y = -12$ -5 **15.** $g + 27 = 8$ -19
16. $j + 13 = -27$ -40 **17.** $k - 12 = -14$ -2 **18.** $n + (-5) = 27$ 32
19. $m - (-21) = 9$ -12 **20.** $-9 + s = 17$ 26 **21.** $p + (-23) = -14$ 9
22. $x - (-33) = 14$ -19 **23.** $-12 + z = -36$ -24 **24.** $x - 13 = 45$ 58
25. $d - 27 = -63$ -36 **26.** $y + (-18) = 7$ 25 **27.** $w - (-37) = 28$ -9
28. $13 - m = 41$ -28 **29.** $16 - y = 37$ -21 **30.** $65 = 12 - x$ -53
31. $41 = 32 - r$ -9 **32.** $-7 = -16 - k$ -9 **33.** $-27 = -6 - p$ 21
34. $-14 - a = -21$ 7 **35.** $-27 - b = -7$ -20 **36.** $-19 - s = 41$ -60
37. $r - 6.5 = -9.3$ -2.8 **38.** $y - 7.3 = 5.1$ 12.4 **39.** $p - (-1.3) = -7.1$
40. $s - (-7.1) = 2.4$ -4.7 **41.** $-1.43 + w = 0.89$ 2.32 **42.** $-0.0056 + z = 0.065$
43. $m - \frac{6}{7} = \frac{3}{14}$ $\frac{15}{14}$ **44.** $a - \frac{2}{3} = -\frac{4}{9}$ $\frac{2}{9}$ **45.** $b + \left(-\frac{3}{4}\right) = \frac{4}{5}$ $\frac{31}{20}$
46. $j + \left(-\frac{3}{5}\right) = \frac{5}{3}$ $\frac{34}{15}$ **47.** $x - \left(-\frac{5}{6}\right) = \frac{2}{7}$ $-\frac{23}{42}$ **48.** $h - \left(-\frac{7}{8}\right) = \frac{5}{9}$ $-\frac{23}{72}$

For more exercises on solving equations, see exercises 61-84 on page 546, exercises 49-72 on page 548, and exercises 61-90 on page 550.

Chalkboard Exercises
Solve each equation.
4a. $-31 = -7 - p$ 24
4b. $-25 = -11 - t$ 14
4c. $-56 - r = 71$ -127

❹ When solving equations like the one in Example 4, some students may end up with equivalent equations like the ones in the Discussion Questions. Point out that the minus sign preceding the variable can be read as "the opposite of." For example, $-k = 21$ means the opposite of k is positive 21. Therefore, k is negative 21, that is $k = -21$.

Suggested Daily Assignments

Enriched
Written: 2–48 evens

Average
Exp: 1–12
Written: 2–48 evens

Fundamental
Exp: 1–12
Written: 1–24
p. 59, 1–36 odds

Answers to Written
39. -8.4
42. 0.0706

About Applications
Point out to students the practical use of signed numbers in discussing temperature. You may want to have students come up with other practical uses of signed numbers.

Applications in Meteorology

Wind Chill Factor

Anna Lopez is a meteorologist. When forecasting the weather, Anna predicts the high and low temperatures for each day. In addition, she predicts the equivalent temperatures due to the wind-chill factor, which depends on the actual temperature and the speed of the wind. Anna uses a chart similar to this to predict the wind-chill factor.

Wind Chill Chart

Wind speed in mph	*Actual temperature (°Fahrenheit)*								
	50	**40**	**30**	**20**	**10**	**0**	**−10**	**−20**	**−30**
	Equivalent temperature (°Fahrenheit)								
0	50	40	30	20	10	0	−10	−20	−30
5	48	37	27	16	6	−5	−15	−26	−36
10	40	28	16	4	−9	−21	−33	−46	−58
15	36	22	9	−5	−18	−36	−45	−58	−72
20	32	18	4	−10	−25	−39	−53	−67	−82
25	30	16	0	−15	−29	−44	−59	−74	−88
30	28	13	−2	−18	−33	−48	−63	−79	−94

In Chicago on January 11, the low temperature was −10°F. If the wind speed was 15 mph, the wind-chill factor made the temperature equivalent to −45°F.

Exercises

Use the wind chill chart to find each equivalent temperature.

1. 30°F, 10 mph 16°F
2. 20°F, 20 mph −10°F
3. −10°F, 15 mph −45°F
4. −20°F, 5 mph −26°F
5. 0°F, 15 mph −36°F
6. 40°F, 30 mph 13°F
7. Find the difference between the equivalent temperatures in problems **1** and **2**. 26°F
8. Find the difference between the equivalent temperatures in problems **3** and **4**.
9. Find the difference between the equivalent temperatures in problems **5** and **6**. −49°F
10. Find the difference between the equivalent temperatures in problems **2** and **6**.

8. −19°F 10. −23°F

Objective 2-9: Solve verbal problems by writing and solving equations.

2-9 Problem Solving: Rational Numbers

Before you solve a problem, you must study the problem carefully. Consider the following problem and its solution.

Death Valley is about 86 meters below sea level. The top of Mt. McKinley is 6194 meters above sea level. What is the difference in these elevations?

Explore *Read the problem to find out what is asked. Identify important facts from the problem.*

The problem asks for the difference in elevations.

Death Valley	86 m below sea level
Mt. McKinley	6194 m above sea level

Then think about how the facts are related. Sometimes it is helpful to draw a chart or diagram.

Elevation above sea level can be represented with a positive number. Elevation below sea level can be represented with a negative number.

Mt. McKinley	6194 m
Sea Level	0 m
Death Valley	−86 m

Plan *Decide how to solve the problem.*

To solve the problem, find the difference between 6194 and −86. Recall that difference is the result of subtraction. Find $6194 - (-86)$.

Solve *Do the computation and answer the problem.*

$6194 - (-86) = 6280$

The difference in elevations is 6280 m.

Examine *See if the solution makes sense for the given problem. If not, try another way to solve the problem.*

Since Mt. McKinley is about 6200 m above sea level and Death Valley is about 90 m below sea level, the distance between elevations should be approximately 6290 m. The solution 6280 makes sense.

2-9

Lesson Notes, p. T29

Readiness Activity

Define the variable, then write an equation for the following problem. Rhonda has 7 more than twice as many books as her sister. Together they have 16 books. How many books does Rhonda have?

Let x = number of books Rhonda's sister has

$2x + 7 + x = 16$

Discussion Questions

1. The elevation of a mountain is 8000 ft *above* sea level, and a valley is 200 ft *below* sea level. Will the difference in elevation be greater or less than 8000? greater
2. The temperature Monday morning was −10°F. The next day it was 8°F lower. Must the temperature rise or fall to get to +15°F? rise

Chalkboard Exercises

Solve each problem by following the four steps below.

a. Explore the problem.
b. Plan the solution.
c. Solve the problem.
d. Examine the solution.

1. Nancy swam 4 laps. She is working toward a goal of 10 laps. How many more laps must she swim to reach her goal?
   a. Let x = laps she must swim to reach goal.
   b. $4 + x = 10$
   c. $4 + x = 10$
   $4 - 4 + x = 10 - 4$
   $x = 6$
   Nancy must swim 6 laps more.
   d. If Nancy swims 6 more laps, she will have swum 4 + 6 or 10 laps.

2. The sum of two integers is 199. One integer is -62. What is the other integer?
   a. Let x = the other integer.
   b. $-62 + x = 199$
   c. $-62 + x = 199$
   $-62 + 62 + x = 199 + 62$
   $x = 261$
   The other integer is 261.
   d. If 261 is added to -62, the sum is 199.

Suggested Daily Assignments

Enriched
Exp: 2–10 evens
Written: 2–30 evens

Average
Exp: 1–10
Written: 1–10, 12–30 evens

Fundamental
Exp: 1–10
Written: 1–20
p. 61, 25–36

Sometimes a problem can be solved by translating it into a mathematical equation. Study this example.

Example 1

Write an equation and solve the problem.

Jerry Flaherty has bench pressed 90 kilograms. He is working toward a goal of pressing 115 kilograms. How many more kilograms does he need to reach his goal?

Explore *Read the problem to find out what is asked, and define a variable.*
The problem asks for how many more kilograms are needed to reach the goal.
Let x = the number of kilograms.

Plan *Write an equation.*
An amount must be added to 90 kg to reach the goal of 115 kg.
$90 + x = 115$

Solve *Solve the equation and answer the problem.*
$90 + x = 115$
$90 - 90 + x = 115 - 90$
$x = 25$

Jerry needs to press 25 kg more.

Examine *Examine the solution.*
If Jerry presses 25 more kilograms, he will have pressed 90 + 25, or 115 kilograms.

Exploratory Exercises

State the answer to each question.

1. Frieda traveled 110 kilometers by boat up the Rhine River last summer. She then traveled 320 kilometers down along the river by train. How far downstream was she from her starting point?
   a. What is asked? How far downstream was she from her starting point?
   b. How far up the river did Frieda travel? 110 km
   c. How far down the river did she travel? 320 km
   d. How can these distances be represented using positive and negative numbers?
   e. Define a variable. Then state an equation for the problem.
   Let d = distance from starting point; $110 + (-320) = d$.

d. Let upstream be the positive direction, downstream negative.

2. The sum of two integers is -23. One of the integers is $+9$. What is the other integer?
   a. What is asked? What is the second integer in the sum of two integers?
   b. What is one of the integers? 9
   c. What is the sum of the integers? -23
   d. What mathematical operation is indicated in the problem? addition
   e. Define a variable. Then state an equation for the problem. Let x = the missing integer; $9 + x = -23$.

3. The perimeter of a triangle is $35\frac{3}{8}$ inches. One side measures $10\frac{1}{8}$ inches. A second side measures $12\frac{3}{16}$ inches. What does the third side measure? measure of third side
   a. What is the perimeter of the triangle? $35\frac{3}{8}$
   b. What are the measures of the known sides? $10\frac{1}{8}$, $12\frac{3}{16}$
   c. Draw a diagram to represent the problem.
   d. What is unknown? third side
   e. What do you need to know about the perimeter to solve this problem?
   f. Define a variable. Then state an equation for the problem.

4. An English assignment covers 182 pages. Kristina has 79 pages yet to read. How many pages has she read?
   a. What is asked? How many pages has Kristina read?
   b. How many pages does the assignment cover? 182
   c. How many pages has Kristina yet to read? 79
   d. Define a variable. Then state an equation for the problem. Let p = pages already read; $p + 79 = 182$.

For each problem, define a variable. Then state an equation.

5. An elevator started at the first floor and went up 14 floors. Then it went down 9 floors. At what floor was it then located?

6. Fairmeadow High School has 1283 students. On Friday 1116 were present. How many were absent?

7. Thirteen subtracted from a number is -5. Find the number.

8. The sum of a number and 12 is equal to 47. Find the number.

9. In a mid-season slump the Yankees scored only 17 runs in 9 games. Their opponents scored 41 runs. By how many runs did their opponents outscore them?

10. A traffic helicopter descended 160 meters to observe road conditions. It leveled off at 225 meters. What was its original altitude?

Written Exercises

1-10. Solve the problems in Exploratory Exercises 1-10.

Read and explore each problem. Make a diagram if necessary. Then define a variable, write an equation, and solve the problem.

11. Jeff Simons sold 27 cars last month. This amount is 36 less than the same time period one year ago. What were his sales one year ago? 63

12. A rancher lost 47 cattle because of the summer heat. His herd now numbers 396. How large was the herd before the summer heat? 443

Answers to Exploratory

3c.

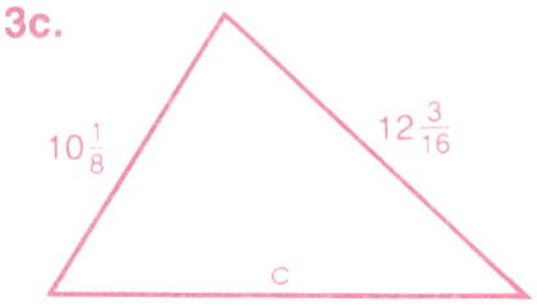

3e. The perimeter is equal to the sum of the lengths of the three sides.

3f. Let x = the measure of the third side; $10\frac{1}{8} + 12\frac{3}{16} + x = 35\frac{3}{8}$

5. Let f = floor; $1 + 14 + (-9) = f$

6. Let a = number absent; $1116 + a = 1283$

7. Let n = the number; $n - 13 = -5$

8. Let n = the number; $n + 12 = 47$

9. Let r = runs; $41 - 17 = r$

10. Let a = original altitude; $a - 160 = 225$

Answers to Written

1. -210 (210 km downstream)

2. -32 3. $13\frac{1}{16}$ 4. 103
5. 6 6. 167 7. 8
8. 35 9. 24 10. 385

13. The temperature at mid-afternoon was 12°C. By early evening the temperature was −7°C. What was the temperature change? −19°C

14. On February 2 the highest temperature in the United States was 87°F in Miami, Florida. The lowest temperature was −19°F in Rice Lake, Wisconsin. What was the difference between the high and low temperatures? 106°

15. Gary Carson skied down the slalom run in 131.3 seconds. This was 21.7 seconds faster than his sister. What was her time? 153 seconds

16. Lisa Thorson skied down the slalom run in 139.8 seconds. This was 13.7 seconds slower than her best time. What is her best time? 126.1 seconds

17. Agnes Graham bought a jacket on sale and saved \$27.35. The regular price was \$81.79. What was the sale price? \$54.44

18. The total of Barbara Orfanello's gas bill and electric bill for January was \$210.87. Her electric bill was \$95.25. How much was her gas bill? \$115.62

19. Thirteen less than some number is 64. What is the number? 77

20. Some number subtracted from 161 yields 87. What is the number? 74

21. The sum of two integers is −32. One of the integers is +6. What is the other integer? −38

22. The difference of two integers is 26. The lesser integer is −11. What is the greater integer? 15

23. A scuba diver swam to a depth of 53 meters below sea level. Then he saw a shark 10 meters above him. At what depth was the shark? −43m (43m below sea level)

24. The entrance to a South Dakota gold mine is 3273 feet above sea level. The mine is 4386 feet deep, measured from the entrance. What is the elevation of the bottom of the mine? −1113 feet

25. Eagle's Bluff is about 143 meters above sea level. From the peak of the bluff to the floor of Sulphur Springs Canyon is 217 meters. How far below sea level is Sulphur Springs Canyon? −74m (74m below sea level)

26. A 4-sided city lot has a perimeter of 340.8 feet. Two sides have lengths of 60.23 feet. A third side has a length of 109.13 feet. What is the length of the fourth side? 111.21 feet

27. Four cave explorers descended to a depth of 112 meters below the cave entrance. They discovered a large cavern whose ceiling was 27 meters above them. At what depth below the cave entrance was the cavern ceiling? −85m (85m below cave entrance)

28. The area of a triangular courtyard is 1520.2 square feet. The area occupied by a circular fountain in the middle is 132.7 square feet. A walkway covers 253.6 square feet. If the remaining area is used for gardens, how much area is set aside for gardens? 1133.9 ft^2

29. Shares of stock in Olympia Motors were listed at $37\frac{3}{4}$ per share. The shares then dropped $2\frac{1}{8}$ points. What was the new listing? $35\frac{5}{8}$

30. A triangle has sides of $16\frac{3}{8}$, $14\frac{1}{16}$, and $17\frac{3}{4}$ inches. What is its perimeter? $48\frac{3}{16}$

Using Computers

Temperature

Temperature is commonly measured in Celsius degrees or Fahrenheit degrees. Celsius and Fahrenheit thermometers are shown at the right.

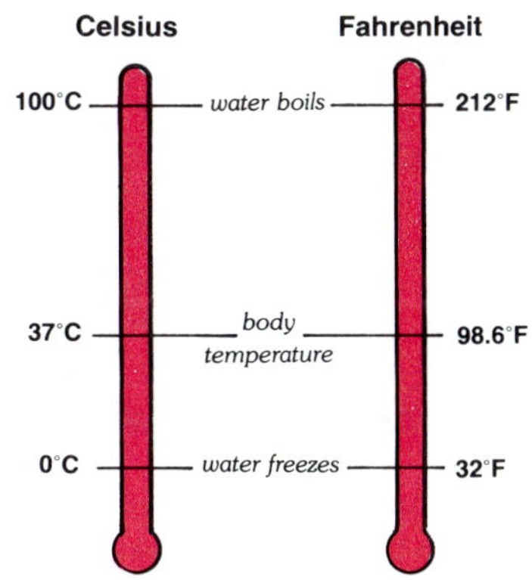

The following program converts a Fahrenheit reading to the equivalent Celsius reading.

```
10  INPUT F
20  LET C = 5/9*(F-32)
30  PRINT "F = ";F,"C = ";C
40  END
```

Note how the formula $C = \frac{5}{9}(F - 32)$ *is written in line 20.*

When you input the Fahrenheit degrees shown on the thermometer, the corresponding Celsius degrees will be printed.

There is one temperature at which the value of F is the same as C. The program can be modified so that the computer will evaluate all integer values of F from −50 to 50 and print only the value of F that is the same as C.

```
10  FOR F = -50 TO 50
20  LET C = 5/9*(F-32)
25  IF C< >F THEN 35
30  PRINT "F = ";F,"C = ";C
35  NEXT F
40  END
```

In line 25, the symbol < > represents not equal.

Exercises

1. At what temperature is the value of F the same as C? −40
2. If line 25 was removed from the program, how would the output change?
3. Change lines 10 and 25 of the program to determine if there is a Fahrenheit temperature that is 100 more than the corresponding Celsius temperature.
4. Modify the program to determine if there is a Fahrenheit temperature that is 50 more than the corresponding Celsius temperature.

About Using Computers
As an extension of this feature, have students write a program that converts a Celsius reading to the equivalent Fahrenheit reading.

Answers to Exercises

2. A value for F and C would be printed every time the loop is executed.
3. 10 FOR F = −200 TO 200
   25 IF C + 100 <> F THEN 35
4. 10 FOR F = −100 TO 100
   25 IF C + 50 <> F THEN 35

Quiz
Lessons 2–7 to 2–9

Solve each equation.

1. $-28 = m + 16$ -44
2. $y - (-22) = 65$ 43
3. $6x + 12 = 4x$ -6
4. $-5 - t = -7$ 2

Define a variable for each of the following. Then use an equation to solve the problem.

5. The morning temperature was 25°F and in the evening it was −6°F. By how many degrees did the temperature decrease?
   Let x = temp. change.
   $25 - x = -6$
   $x = 31$
   The temperature decreased 31 degrees.
6. The sum of two integers is −74. One integer is −16. Find the other integer.
   Let x = other integer.
   $-16 + x = -74.$
   $x = -58$
   The other integer is −58.

Vocabulary

negative (39)
integers (39)
coordinate (39)
absolute value (44)
rational number (47)
additive inverse (53)
opposite (53)
equivalent equations (57)

Chapter Summary

1. On a number line, points to the right of zero are named with a positive sign (+) and points to the left of zero are named with a negative sign (−). (39)
2. The set of integers can be written $\{\ldots, -5, -4, -3, -2, -1, 0, 1, 2, 3, 4, 5, \ldots\}$ where . . . means continued indefinitely. (39)
3. To graph a set of numbers means to locate the points named by those numbers on the number line. (39)
4. A number line can be used to add integers. Move to the right for positive integers and to the left for negative integers. (42)
5. Definition of Absolute Value: The absolute value of a number is the number of units it is from 0 on the number line. (44)
6. Adding Integers with the Same Sign: To add integers with the same sign, add their absolute values. Give the sum the same sign as the addends. (44)
7. Adding Integers with Different Signs: To add integers with different signs, subtract the lesser from the greater absolute values. Give the result the same sign as the addend with the greater absolute value. (45)
8. A rational number is a number that can be expressed in the form $\frac{a}{b}$, where a and b are integers and b is not equal to zero. (47)
9. The rules used to add integers can also be used to add rational numbers. (47)
10. If the sum of two numbers is 0, the numbers are called additive inverses or opposites. (53)
11. Additive Inverse Property: For every number a, $a + (-a) = 0$. (53)
12. To subtract a rational number, add its additive inverse. For rational numbers a and b, $a - b = a + (-b)$. (54)
13. Addition Property of Equality: For any numbers a, b, and c, if $a = b$ then $a + c = b + c$. (57)

14. Subtraction Property of Equality: For any numbers a, b, and c, if $a = b$ then $a - c = b - c$. (58)
15. Equations involving addition may be solved by using the subtraction property of equality or by using the additive inverse. (58)
16. Equations involving subtraction may be solved by using the addition property of equality. Sometimes it is easier to write and solve an equivalent equation involving addition. (60)
17. Before you solve a problem, you must explore the problem and plan carefully. It is often helpful to draw a chart or diagram. Sometimes a problem can be solved by translating it into a mathematical equation. (64)

Answers to Chaper Review

5. 

6. 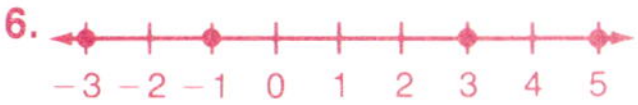

7. 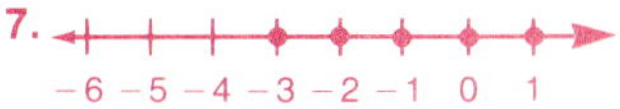

8. 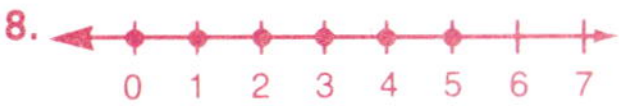

Chapter Review

2-1 **Name the set of numbers graphed.**

1. 

$\{-3, 1, 2, 5\}$

2. 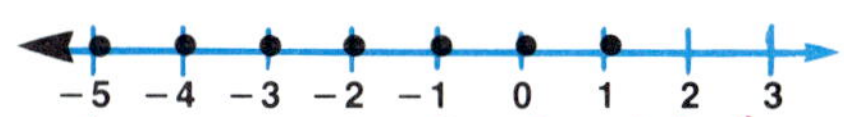

$\{\ldots, -5, -4, -3, -2, -1, 0, 1\}$

3. 

$\{3, 4, 5, \ldots\}$

4. 

$\{-2, 0, 3\}$

Graph each set of numbers on a number line.

5. $\{-2, 1, 4\}$ **6.** $\{5, 3, -1, -3\}$
7. $\{-3, -2, -1, 0, \ldots\}$ **8.** $\{\ldots, 3, 4, 5\}$

2-2 **Use a number line to find each sum.**

9. $4 + (-5)$ -1 **10.** $-3 + 6$ 3 **11.** $-2 + (-4)$ -6 **12.** $-4 + 1$ -3

2-3 **Find each sum.**

13. $17 + (-9)$ 8 **14.** $-9 + (-12)$ -21 **15.** $-12 + 8$ -4
16. $13 + (-51)$ -38 **17.** $-74 + 21$ -53 **18.** $-17 + (-31)$ -48

2-4 **Find each sum.**

19. $\frac{6}{7} + \left(-\frac{13}{7}\right)$ -1 **20.** $-\frac{5}{8} + \frac{3}{16}$ $-\frac{7}{16}$ **21.** $-7.5 + 10.2$ 2.7
22. $-0.37 + (-0.812)$ -1.182 **23.** $3.707 + (-0.058)$ 3.649 **24.** $-4.375 + 5.432$ 1.057

2-5 **Find each sum.**

25. $84 + (-31) + 13$ 66
26. $65 + (-13 + 28)$ 80
27. $-4b + 17b + (-36b)$ $-23b$
28. $a + 9a + (-12a)$ $-2a$
29. $\frac{7}{3} + \frac{5}{12} + \left(-\frac{5}{6}\right)$ $\frac{23}{12}$
30. $-\frac{4}{3} + \frac{5}{6} + \left(-\frac{7}{3}\right)$ $-\frac{17}{6}$
31. $-6.7 + 8.1 + (-5.3)$ -3.9
32. $-5 + (-3.7) + 9.4$ 0.7
33. $-13pq + 41k + 12k + (-38pq)$ $-51pq + 53k$
34. $61mb + (-51pg) + (-48mb) + 13pg$ $13mb - 38pg$

2-6 **Find each difference.**

35. $14 - 36$ -22
36. $8 - (-5)$ 13
37. $-7 - (-11)$ 4
38. $-13 - 16$ -29
39. $3.72 - (-8.65)$ 12.37
40. $0.702 - 0.98$ -0.278
41. $-13x - (-7x)$ $-6x$
42. $-4.5y - 8.1y$ $-12.6y$
43. $\begin{array}{r} 36 \\ -19 \\ \hline \end{array}$ 17
44. $\begin{array}{r} -61 \\ -(-43) \\ \hline \end{array}$ -18
45. $\begin{array}{r} -2.8 \\ -(+3.9) \\ \hline \end{array}$ -6.7
46. $\frac{4}{5} - \left(-\frac{9}{5}\right)$ $\frac{13}{5}$
47. $-\frac{4}{7} - \frac{3}{14}$ $-\frac{11}{14}$
48. $-\frac{3}{2} - \frac{5}{4}$ $-\frac{11}{4}$

2-7 **Solve each equation.**

49. $k + 13 = 5$ -8
50. $19 = y + 7$ 12
51. $z + 15 = -9$ -24
52. $m + (-5) = -17$ -12
53. $p + (-7) = 31$ 38
54. $19 = -8 + d$ 27

2-8 **Solve each equation.**

55. $x - 16 = 37$ 53
56. $15 - y = 9$ 6
57. $r - (-5) = -8$ -13
58. $m - (-4) = 21$ 17
59. $-13 = 6 - k$ 19
60. $y + (-9) = -35$ -26

2-9 **For each problem, define a variable. Then use an equation to solve the problem.**

61. Some number added to -16 is equal to 39. What is the number? 55
62. When 9 is subtracted from another integer, the difference is -13. What is the other integer? -4
63. The temperature at sunrise was $-2°$ F. At noon the temperature was 15°F. What was the temperature change? 17°F
64. Darlene's score at the end of the first half of a game was -7.8. At the end of the game her score was 19.2. How many points did she score during the second half of the game? 27

Chapter Test

Graph each set on a number line.

1. $\{1, 2, -5\}$

2. $\{\ldots, -2, -1, 0, 1\}$

Find each sum.

3. $-11 + (-13)$ -24

4. $-15 + 23$ 8

5. $36 + (-42)$ -6

6. $1.654 + (-2.367)$ -0.713

7. $\frac{3}{7} + \left(-\frac{9}{7}\right)$ $-\frac{6}{7}$

8. $-\frac{1}{8} + \frac{3}{4}$ $\frac{5}{8}$

9. $12x + (-21x)$ $-9x$

10. $-7 + (-6) + 37$ 24

11. $[4 + (-13)] + (-12)$ -21

12. $\frac{5}{8} + \left(-\frac{3}{16}\right) + \left(-\frac{3}{4}\right)$ $-\frac{5}{16}$

13. $-7.9 + 3.5 + 2.4$ -2

14. $18b + 13xy + (-46b)$ $13xy - 28b$

Find each difference.

15. $12 - 19$ -7

16. $-14 - 3$ -17

17. $21 - (-7)$ 28

18. $-41 - (-52)$ 11

19. $-\frac{7}{16} - \frac{3}{8}$ $-\frac{13}{16}$

20. $6.32 - (-7.41)$ 13.73

Solve each equation.

21. $m + 13 = -9$ -22

22. $k + 16 = -4$ -20

23. $y + (-3) = 14$ 17

24. $x + (-6) = 13$ 19

25. $k - (-3) = 28$ 25

26. $-5 - k = 14$ -19

27. $r - (-1.2) = -7.3$ -8.5

28. $b - \frac{2}{3} = -\frac{5}{6}$ $-\frac{1}{6}$

Solve each problem.

29. The sum of two integers is -11. One integer is 8. Find the other integer. -19

30. Joe's golf score was 68. This was 4 less than Marie's golf score. What was Marie's score? 72

Answers to Chapter Test

1. 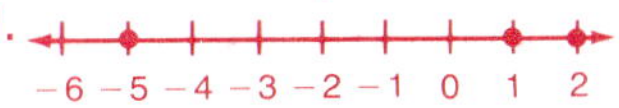

2. 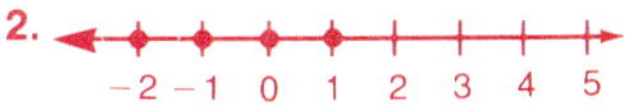

Using the Cumulative Review
If students have difficulty with a particular topic, have them review the appropriate lesson. You may want to provide copies of the solutions from the *Algebra One Solutions Manual*. This Cumulative Review covers Chapters 1-2.

Other Cumulative Reviews appear on pages 142, 208, 268, 334, 402, 466, and 526.

Answers

20.

21.

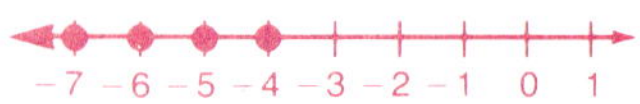

Cumulative Review

1. Write a mathematical expression for twice a number decreased by 17. $2n - 17$
2. Write the expression $3 \cdot 3 \cdot 3 \cdot a \cdot a \cdot b \cdot b \cdot b$ using exponents. $3^3 \cdot a^2 \cdot b^3$

Evaluate.

3. $3 + 2(4) - 10 \div 2$ 6
4. $[(4 + 8)^2 \div 4 \times 2] + 3 \times 2$ 78
5. $4(5b - 3a)^2$ if $a = 6$ and $b = 4$ 16
6. $3ab - 4c$ if $a = 2$, $b = 3$, and $c = \frac{1}{2}$ 16
7. Solve: $x = \frac{3 + 7}{10}$ 1
8. What number is called the additive identity? 0

State the property shown.

9. $(3 + 8)4 = (11)4$ substitution
10. $(7 \cdot x) \cdot a = (x \cdot 7) \cdot a$ commutative of mult.
11. $y + 8 = y + 8$ reflexive
12. $6(7a + 3) = 42a + 18$ distributive

Simplify.

13. $4mn + 12mn$ $16mn$
14. $16x^2 + 12xy - 13x^2$ $3x^2 + 12xy$
15. $3(a + 0.2b) - 0.4b$ $3a + 0.2b$
16. $4(x + 3y) + 2(2x + y)$ $8x + 14y$

17. $y + n^3 = x$

Write each sentence as an equation.

17. The sum of y and the cube of n equals x.
18. The square of a side (s) is equal to the area (A). $s^2 = A$
19. Name the set of numbers graphed.

{0, 1, 2, 3}

Graph each set of numbers on a number line.

20. $\{-3, -1, 2, 5\}$
21. $\{\ldots, -6, -5, -4\}$
22. Find $4 + (-6)$ on a number line. −2

Find each sum or difference.

23. $36 + (-73)$ −37
24. $-8 + (-21)$ −29
25. $-9.3 + 4.7$ −4.6
26. $-\frac{1}{2} + \left(-\frac{1}{6}\right) + \left(-\frac{2}{3}\right)$ $-\frac{4}{3}$
27. $-21 - (-28)$ 7
28. $3.6 - 7.9$ −4.3
29. $-3a + 5m + 16m + (-25a)$ $-28a + 21m$
30. $-7t - 16t$ $-23t$

Solve.

31. $r + 17 = 23$ 6
32. $-13 + a = 14$ 27
33. $y - 8 = -12$ −4
34. $-20 = m - (-14)$ −34
35. $7 - x = -10$ 17
36. $m - \frac{1}{2} = \frac{2}{5}$ $\frac{9}{10}$

Problem Solving

For the following problem, answer the related question.

37. Heather is 4 years older than Brian. How old was Brian when Heather was 11 years old? 7 years old

Define the variable, then write an equation for the following problem.

38. A number decreased by 11 is 40. Find the number. Let n = the number
$n - 11 = 40$

Read and explore the following problems. Then define a variable, write an equation, and solve.

39. Ryan scored 32 points in a basketball game. This was 8 points more than the last game. How many points did he score in the last game? Let p = points scored last game
$p + 8 = 32$, $p = 24$ points
40. The sum of two integers is −38. One of the integers is −56. What is the other integer? Let x = the other integer
$x + (-56) = -38$
$x = 18$

Standardized Test Practice Questions

The test questions on this page deal with number concepts and basic operations. The information at the right may help you with some of the questions.

Directions: Choose the best answer. Write A, B, C, or D.

1. Which of the following numbers is *not* a prime number?

(A) 17 **(B)** 23 **(C)** 37 **(D)** 87

2. How many integers between 325 and 400, inclusive, are divisible by 4?

(A) 18 **(B)** 19 **(C)** 20 **(D)** 24

3. How many integers between 99 and 201 are divisible by 2 or 5?

(A) 60 **(B)** 61 **(C)** 70 **(D)** 71

4. How many integers are between, but not including, 5 and 1995?

(A) 1988 **(B)** 1989
(C) 1990 **(D)** 2000

5. A person is standing in line, thirteenth from the front and eleventh from the back. How many people are in the line?

(A) 22 **(B)** 23 **(C)** 24 **(D)** 25

6. What number is missing from the sequence 2, 5, 15, 18, 54, 171, 174, 522?

(A) 36 **(B)** 56 **(C)** 57 **(D)** 398

7. What digit is represented by $\triangle$ in this subtraction problem?

$$\begin{array}{r} 80\square \\ -602 \\ \hline \triangle 98 \end{array}$$

(A) 1 **(B)** 2 **(C)** 3 **(D)** 4

8. How many fourths of a mile will a car travel on a 400-mile trip.

(A) 100 **(B)** 400 **(C)** 1000 **(D)** 1600

---

1. A number is *divisible* by another number if it can be divided exactly with no remainder. A number is divisible by each of its factors.

2. A number is *prime* if it has no factors except itself and 1.

7 is a prime number.
8 is not *prime* since it has factors other than itself and 1.
2 and 4 are factors of 8.

---

9. 8(916) + 916 =

(A) 4(916) + 3(916)
(B) 5(916) + 4(916)
(C) 6(916) + 4(916)
(D) 3(916) + 4(916)

10. Nancy has 9 more marbles than James. If she gives him 13 of her marbles, how many more will he have than she?

(A) 4 **(B)** 11 **(C)** 17 **(D)** 21

11. A person has 100 green, 100 orange, and 100 yellow jelly beans. How many jars can be filled if each jar must contain 8 green, 5 orange, and 6 yellow jelly beans?

(A) 12 **(B)** 15 **(C)** 16 **(D)** 25

12. A hospital patient must be given medication every 5 hours, starting at 10 A.M. Thursday. On what day will the patient first receive medication at noon?

(A) Thursday **(B)** Friday
(C) Saturday **(D)** Sunday

13. Which of the following is the difference of two consecutive prime numbers less than 30?

(A) 5 **(B)** 6 **(C)** 7 **(D)** 8

14. The sum of five consecutive integers is always divisible by

(A) 2 **(B)** 4 **(C)** 5 **(D)** 10

Using The Standardized Test Practice Questions

The questions on this page may be used to help students prepare for pre-college tests. These questions usually require careful analysis and a thorough understanding of the concepts.

This page could be used as an overnight assignment for better students. Other students may benefit most if only a few questions are assigned at a time.

After students have chosen an answer for each problem, you should discuss how each problem can be solved, or provide copies of the solutions from the *Algebra One Solutions Manual*.

Other practice questions appear in this text on pages 143, 209, 269, 335, 403, 467, and 527. Also see pages 50-58 in the *Teacher Resource Book*.

About the Photo

Commission earned on new car sales is one application of percent familiar to students. In this chapter, students will learn to use equations to solve a variety of problems. Emphasize that many practical problems can be solved by using algebra.

Multiplication and division of signed numbers are also introduced in this chapter. As a result, students can apply algebra to applications with negative numbers. An application which relates to automobiles is acceleration. If the acceleration is negative, it is called deceleration. This topic is introduced in the special feature on page 86.

Teacher Resource Book Materials for Chapter 3 See pages 105-122.

3-1

Lesson Notes, p. T29
Enrichment, p. T29

Readiness Activity

Find each sum.

$-4 + (-4)$ -8; $2(-4)$

$-1 + (-1) + (-1) + (-1) + (-1)$

-5; $5(-1)$

$-3 + (-3) + (-3)$ -9; $3(-3)$

Choose a product from the list below that describes each sum above.

$3(-3)$ $2(-4)$ $5(-1)$

CHAPTER 3

Multiplying and Dividing Rational Numbers

Steve Bernard earns 3% commission on sales of all new cars. To earn $850.00 per week in commission, what must his weekly sales be? In this chapter, you will learn to use equations to solve problems such as this.

Objective 3-1: Use the rules for multiplying positive and negative numbers.

3-1 Multiplying Rational Numbers

How can you multiply positive and negative numbers? One way is to consider multiplication as repeated addition. For example, $4(-2)$ may be expressed as $(-2) + (-2) + (-2) + (-2)$. ❶

$$(-2) + (-2) + (-2) + (-2) = -8$$
$$4(-2) = -8$$

The product of *positive* 4 and *negative* 2 is *negative* 8.

Another way to find the product of a positive number and a negative number is to consider the following example.

$0 = 7(0)$	*Multiplication Property of Zero*
$0 = 7\,[5 + (-5)]$	*Substitution and Additive Inverse Property*
$0 = 7(5) + 7(-5)$	*Distributive Property*
$0 = 35 + 7(-5)$	*Substitution*

Using the additive inverse property, 0 is equal to $35 + (-35)$. Therefore $7(-5)$ must equal -35. Using the commutative property, $(-5)(7)$ must also equal -35. These and many other similar examples suggest the following rule.

The product of two numbers that have different signs is negative.

Multiplying Two Numbers with Different Signs

Examples

1 Simplify: $9(-4)$

$9(-4) = -36$

2 Simplify: $(-3a)(4b)$

$(-3a)(4b) = (-3)(4)ab$ — *Commutative and Associative Properties*
$= -12ab$

3 Simplify: $(0.5)(-6)$

$(0.5)(-6) = -3$

You know that the product of two positive numbers is positive. What is the sign of the product of two negative numbers?

$0 = -6(0)$	*Multiplicative Property of Zero*
$0 = -6[8 + (-8)]$	*Substitution*
$0 = -6(8) + (-6)(-8)$	*Distributive Property*
$0 = -48 + (-6)(-8)$	*Substitution*

The additive inverse of -48 is 48. Thus, $(-6)(-8)$ must equal 48. The product of *negative* 6 and *negative* 8 is *positive* 48.

Discussion Questions

Study the following sets of multiplication exercises. Describe the pattern in each set. Then complete the last two exercises in each set.

1. $3 \times 3 = 9$
   $3 \times 2 = 6$
   $3 \times 1 = 3$
   $3 \times 0 = 0$
   $3 \times -1 = \underline{-3}$
   $3 \times -2 = \underline{-6}$

   The products are decreasing by 3.

2. $-3 \times 3 = -9$
   $-3 \times 2 = -6$
   $-3 \times 1 = -3$
   $-3 \times 0 = 0$
   $-3 \times -1 = \underline{3}$
   $-3 \times -2 = \underline{6}$

   The products are increasing by 3.

Chalkboard Exercises

Simplify.

1. $16(-3)$ -48

2a. $(-4b)(5c)$ $-20bc$

2b. $(8k)(-3m)$ $-24\,km$

3. $(-0.6)(0.35)$ -0.21

❶ Many multiplication expressions, such as $\frac{1}{2}(-6)$ *cannot* be shown as repeated addition. The number properties can be used to discover the product. Discuss the properties for the steps below.

$0 = \frac{1}{2}(0)$

$0 = \frac{1}{2}(6 + -6)$

$0 = \frac{1}{2}(6) + \frac{1}{2}(-6)$

$0 = 3 + \frac{1}{2}(-6)$

Thus, $\frac{1}{2}(-6) = -3.$

Chalkboard Exercises

Simplify.

4. $(-10)(-15)$
150

5a. $(-0.3)(-1.7)$
0.51

5b. $\left(-\frac{1}{2}\right)\left(-\frac{1}{4}\right)$
$\frac{1}{8}$

6. $(-4x)(-7y) + (-3x)(12y)$
$-8xy$

7a. $(-6)(5)(-3)$
90

7b. $(-3)(-2)(-4)(2)$
-48

7c. $(-6)(-1)(-2)(-3)$
36

❷ Before studying this example you may wish to review the distributive property.

These and many other similar examples lead to the following rule.

The product of two numbers that have the same sign is positive.

Multiplying Two Numbers with the Same Sign

Examples

4 **Multiply: $(-4)(-13)$**

$(-4)(-13) = 52$ *Both factors are negative, so the product is positive.*

5 **Multiply: $\left(-\frac{2}{3}\right)\left(-\frac{1}{5}\right)$**

$\left(-\frac{2}{3}\right)\left(-\frac{1}{5}\right) = \frac{2}{15}$ *Why is the product positive?*

6 **Simplify: $6x(-7y) + (-3x)(-5y)$**

$$\begin{aligned} 6x(-7y) + (-3x)(-5y) &= -42xy + 15xy \\ &= (-42 + 15)xy \quad \textit{Distributive Property} \\ &= -27xy \end{aligned}$$

❷

What is the result if a number is multiplied by -1?

$-1(3) = -3$ $\frac{4}{5}(-1) = -\frac{4}{5}$ $-1(-4) = 4$ $(-0.3)(-1) = 0.3$

These examples are summarized in the multiplicative property of -1.

The product of any number and -1 is its additive inverse.
$-1(a) = -a$ and $a(-1) = -a$

Multiplicative Property of −1

To find the product of 3 or more numbers, first group the numbers in pairs.

Example

7 **Simplify: $(-2)(-3)(-1)(4)$**

$$\begin{aligned} (-2)(-3)(-1)(4) &= 6(-1)(4) \\ &= -6(4) \\ &= -24 \end{aligned}$$

First multiply -2 by -3.
Then multiply 6 by -1.

Recall that the product of *two* negative factors is positive. Why are the following statements true?

The product of an *even* number of negative factors is positive.
The product of an *odd* number of negative factors is negative.

Exploratory Exercises

Determine if each product is positive or negative.

1. $3(-2)$ −
2. $(-2)(-8)$ +
3. $7(8)$ +
4. $3(-5)$ −
5. $(-6)(-5)$ +
6. $(-9)(7)$ −
7. $(-11)(10)$ −
8. $(-9)(-8)$ +
9. $\left(\frac{2}{3}\right)\left(-\frac{1}{4}\right)$ −
10. $\left(\frac{7}{3}\right)\left(\frac{7}{3}\right)$ +
11. $\left(-\frac{7}{9}\right)\left(\frac{9}{14}\right)$ −
12. $\left(-\frac{3}{4}\right)\left(\frac{4}{3}\right)$ −
13. $(-3)(4)(-2)$ +
14. $(-6)(4)(-3)$ +
15. $\left(\frac{3}{5}\right)\left(\frac{2}{3}\right)(-3)$ −
16. $\left(-\frac{4}{5}\right)\left(-\frac{1}{5}\right)(-5)$ −
17. Under what conditions is ab positive?
18. Under what conditions is ab negative?
19. If a^2 is positive, what can you conclude about a?
20. If a^3 is positive, what can you conclude about a? a is positive
21. If ab is negative, what is the sign of the inverse of ab? positive
22. Under what conditions is ab equal to zero? when $a = 0$ or $b = 0$ or both a and b are zero

Written Exercises

1-16. Find each product for Exploratory Exercises 1-16.

Find each product.

17. $5(12)$ 60
18. $(-6)(11)$ -66
19. $\left(\frac{3}{5}\right)\left(-\frac{4}{7}\right)$ $-\frac{12}{35}$
20. $\left(-\frac{7}{8}\right)\left(-\frac{1}{3}\right)$ $\frac{7}{24}$
21. $4\left(-\frac{7}{8}\right)$ $-\frac{7}{2}$
22. $(-5)\left(-\frac{2}{5}\right)$ 2
23. $\frac{3}{5}(5)(-2)\left(-\frac{1}{2}\right)$ 3
24. $\frac{2}{11}(-11)(-4)\left(-\frac{3}{4}\right)$ -6
25. $\left(-\frac{1}{3}\right)\left(-\frac{3}{4}\right)\left(-\frac{4}{5}\right)$ $-\frac{1}{5}$
26. $\left(-\frac{7}{12}\right)\left(\frac{6}{7}\right)\left(-\frac{3}{4}\right)$ $\frac{3}{8}$
27. $(-4)(0)(-2)(-3)$ 0
28. $(4)(-2)(-1)(-3)$ -24

Simplify each expression.

29. $3(-4) + 2(-7)$ -26
30. $-3(2) + (-2)(-3)$ 0
31. $-4(4) + (-2)(-3)$ -10
32. $(-7)(8) + 8(-7)$ -112
33. $5(-2) - 3(-8)$ 14
34. $4(-1) - 2(-6)$ 8
35. $4(7) - 3(11)$ -5
36. $8(9) - 6(3)$ 54
37. $\frac{2}{3}\left(\frac{1}{2}\right) - \left(-\frac{3}{2}\right)\left(\frac{2}{3}\right)$ $\frac{4}{3}$
38. $\left(-\frac{1}{2}\right)\left(-\frac{1}{2}\right) - \frac{2}{3}\left(\frac{3}{2}\right)$ $-\frac{3}{4}$
39. $\left(\frac{5}{8}\right)\left(-\frac{1}{2}\right) - \left(\frac{5}{8}\right)\left(\frac{1}{2}\right)$ $-\frac{5}{8}$
40. $\frac{5}{6}\left(\frac{6}{7}\right) - \left(\frac{5}{6}\right)\left(-\frac{6}{7}\right)$ $\frac{10}{7}$
41. $5(3t - 2t) + 2(4t - 3t)$ $7t$
42. $3(5x + 3x) - 4(2x + 6x)$ $-8x$
43. $4[3x + (-2x)] - 5(3x + 2x)$ $-21x$
44. $6(3x + 7x) + 7[8x + (-4x)]$ $88x$
45. $\frac{1}{2}(6x + 8x) - \frac{1}{3}(6x + 9x)$ $2x$
46. $-\frac{2}{7}(21x + 35a) + \frac{4}{7}(35x - 21a)$ $14x - 22a$
47. $\frac{3}{4}(4a - 12b) + \frac{1}{8}(16a + 48b)$ $5a - 3b$
48. $\frac{5}{6}(-24a + 36b) + \left(-\frac{1}{3}\right)(60a - 42b)$
49. $\frac{1}{2}\left(-\frac{1}{3}a + \frac{2}{3}b\right) + \frac{2}{3}\left(\frac{1}{2}a - \frac{3}{4}b\right)$ $\frac{1}{6}a - \frac{1}{6}b$
50. $\frac{1}{2}(-3a - 4b) + \left(-\frac{2}{9}\right)(-4a - 7b)$ $-\frac{11}{18}a - \frac{4}{9}b$

48. $-40a + 44b$

mini-review

Add or subtract.

1. $-3 - (-4)$ 1
2. $51 + (-93)$ -42
3. $-0.187 + (-8.15)$ -8.337
4. $-3.2x - 8.4x$ $-11.6x$
5. $-5 + (-4.8) + 10.4$ 0.6

For more exercises on multiplying integers, see exercises 1-30 on page 547.
For more exercises on dividing integers, see exercises 31-60 on page 547.
For more exercises on multiplying and dividing rational numbers, see exercises 25-48 on page 549, and exercises 31-60 on page 551.

Suggested Daily Assignments

Enriched
- Exp: 17–22
- Written: 2–50 evens
- Mini-Review: 1–5

Average
- Exp: 2–16 evens, 17–22
- Written: 2–50 evens
- Mini-Review: 1–5

Fundamental
- Exp: 1–22
- Written: 2–50 evens
- Mini-Review: 1–5

Answers to Exploratory

17. both a and b are positive or both a and b are negative
18. if a is negative and b is positive *or* b is negative and a is positive
19. a can be either positive or negative

Answers to Written

1. -6	2. 16	3. 56
4. -15	5. 30	6. -63
7. -110	8. 72	9. $-\frac{1}{6}$
10. $\frac{49}{9}$	11. $-\frac{1}{2}$	12. -1
13. 24	14. 72	15. $-\frac{6}{5}$
16. $-\frac{4}{5}$		

3-2

Lesson Notes, p. T29

Discussion Question
Fill in each blank with the appropriate number, if it exists.

1. $\frac{1}{11} \cdot \underline{11} = 1$
2. $\frac{7}{5} \cdot \underline{\frac{5}{7}} = 1$
3. $-\frac{2}{3} \cdot \underline{-\frac{3}{2}} = 1$
4. $2 \cdot \underline{\frac{1}{2}} = 1$
5. $-6 \cdot \underline{-\frac{1}{6}} = 1$
6. $0 \cdot \underline{\quad} = 1$

no solution

Chalkboard Exercises
Simplify.
1. $81 \div -9$ $\quad -9$
2. $\frac{-56}{-7}$ $\quad 8$

❶ You may want to use the format shown below.

$5 \cdot (-6) = -30$

$5 = \frac{-30}{-6}$

$-6 = \frac{-30}{5}$

If you use this format, remind students that a fraction bar indicates division.

Objective 3-2: Use the rules for dividing positive and negative numbers to simplify expressions.

3-2 Dividing Rational Numbers

Study the following examples.

$\frac{60}{6} \cdot 6 = 10 \cdot 6 = 60$ *Multiplying by 6 undoes dividing by 6.*

$\frac{38 \cdot 2}{2} = \frac{76}{2} = 38$ *Dividing by 2 undoes multiplying by 2.*

These and other examples show that multiplication and division are inverse operations. You can use this fact to derive rules for dividing positive and negative numbers.

$3 \cdot 4 = 12$ so $12 \div 4 = 3$

same signs — *positive quotient*

$5 \cdot (-6) = -30$ so $-30 \div (-6) = 5$ ❶

The quotient of two numbers with the same sign is *positive*.

What happens if the numbers have different signs?

$-7 \cdot 9 = -63$ so $-63 \div 9 = -7$

different signs — *negative quotient*

$-8 \cdot (-6) = 48$ so $48 \div (-6) = -8$

The quotient of two numbers with different signs is *negative*.

The quotient of two numbers is positive if the numbers have the same sign. The quotient of two numbers is negative if the numbers have different signs.

Division of Rational Numbers

Notice that the rule for finding the sign of the quotient of two numbers is the same as the rule for finding the sign of their product.

Examples

1 Simplify: $\mathbf{-36 \div 3}$

Since the numbers have different signs, the quotient is negative.
$-36 \div 3 = -12$

2 Simplify: $\frac{-45}{-9}$

$\frac{-45}{-9} = 5$ *The fraction bar indicates division. The quotient is positive.*

Two numbers whose product is 1 are **multiplicative inverses** or **reciprocals**.

The reciprocal of $\frac{4}{9}$ is $\frac{9}{4}$ because $\frac{4}{9} \cdot \frac{9}{4} = 1$.

The reciprocal of -5 is $-\frac{1}{5}$ because $-5\left(-\frac{1}{5}\right) = 1$

The reciprocal of 1 is 1 because $1 \cdot 1 = 1$

Zero has no reciprocal because the product of zero and any rational number is zero, not 1.

These examples can be summarized in the multiplicative inverse property.

For every nonzero number a, there is exactly one number $\frac{1}{a}$, such that $a\left(\frac{1}{a}\right) = \frac{1}{a}(a) = 1$.

Multiplicative Inverse Property

Recall that it is possible to subtract a number by adding its opposite. In a similar manner, it is possible to divide a number by multiplying by its reciprocal.

Consider this example.

$$\frac{2}{3} \div \frac{3}{4} = \frac{2}{3} \cdot \frac{4}{3} \qquad \textit{\(\frac{3}{4}\) and \(\frac{4}{3}\) are reciprocals.}$$
$$= \frac{8}{9}$$

❷

In general, *any* division expression can be changed to an equivalent multiplication expression.

For all numbers a and b, with $b \neq 0$,

$$a \div b = \frac{a}{b} = a\left(\frac{1}{b}\right) = \frac{1}{b}(a).$$

Division Rule

Examples

3 Divide: $-\frac{3}{4} \div 8$

$$-\frac{3}{4} \div 8 = -\frac{3}{4} \cdot \frac{1}{8} \qquad \textit{Multiply by the reciprocal of 8.}$$
$$= -\frac{3}{32}$$

4 Simplify: $\frac{4a + 32}{4}$ ❸

$$\frac{4a + 32}{4} = (4a + 32) \div 4$$
$$= (4a + 32)\left(\frac{1}{4}\right)$$
$$= 4a\left(\frac{1}{4}\right) + 32\left(\frac{1}{4}\right)$$
$$= a + 8$$

Chalkboard Exercises

Simplify.

3a. $7 \div \frac{2}{3}$ $\quad \frac{21}{2}$ or $10\frac{1}{2}$

3b. $-\frac{5}{6} \div 4$ $\quad -\frac{5}{24}$

3c. $-\frac{7}{9} \div -6$ $\quad \frac{7}{54}$

4a. $\frac{7a + 49}{-7}$ $\quad -a + (-7)$

4b. $\frac{-14 - 21b}{7}$ $\quad -2 - 3b$

4c. $\frac{-20 - 28x}{-4}$ $\quad 5 + 7x$

❷ You may also wish to show the quotient $\frac{2}{3} \div \frac{3}{4}$ as follows.

$$\frac{2}{3} \div \frac{3}{4} = \frac{\frac{2}{3}}{\frac{3}{4}}$$
$$= \frac{\frac{2}{3} \cdot \frac{4}{3}}{\frac{3}{4} \cdot \frac{4}{3}}$$
$$= \frac{\frac{2}{3} \cdot \frac{4}{3}}{1}$$
$$= \frac{2}{3} \cdot \frac{4}{3}$$
$$= \frac{8}{9}$$

❸ You could show the following alternative:

$$\frac{4a + 32}{4} = \frac{4(a + 8)}{4}$$
$$= 4\left(\frac{1}{4}\right)(a + 8)$$
$$= a + 8$$

Suggested Daily Assignments

Enriched
  Written: 2–44 evens

Average
  Exp: 2–18 evens
  Written: 1–44

Fundamental
  Exp: 1–18
  Written: 1–24
    p. 77, 29–49 odds

Quiz
Lessons 3-1 to 3-2
Simplify each expression.

1. $-8 \cdot \left(\frac{-5}{2}\right)\left(\frac{1}{6}\right)(-10)$ $-33\frac{1}{3}$
2. $2(14a - 6c) - 3(7a + 9c)$ $7a - 39c$
3. $-126 \div 9$ -14
4. $\frac{42y - 63z}{-7}$ $-6y + 9z$
5. $\frac{\frac{-4}{5}}{-6}$ $\frac{2}{15}$

About Using Calculators
Be sure you emphasize that when the reciprocal key is pressed, the only number for which the reciprocal will be found is the number in the display.

Answers

7. The original number is displayed. When n is even, the original number is displayed. When n is odd, the reciprocal is displayed.
8. Most calculators display an error message. This happens because 0 has no reciprocal. The expression $\frac{1}{0}$ is undefined.

Exploratory Exercises

State the reciprocal of each number.

1. 3 $\frac{1}{3}$
2. -5 $-\frac{1}{5}$
3. 0 none
4. -1 -1
5. -14 $-\frac{1}{14}$
6. 7 $\frac{1}{7}$
7. $\frac{2}{3}$ $\frac{3}{2}$
8. $\frac{1}{15}$ 15
9. $-\frac{3}{11}$ $-\frac{11}{3}$
10. $-\frac{1}{11}$ -11
11. $\frac{10}{7}$ $\frac{7}{10}$
12. $\frac{21}{5}$ $\frac{5}{21}$
13. $-\frac{3}{5}$ $-\frac{5}{3}$
14. $-\frac{8}{15}$ $-\frac{15}{8}$
15. $3\frac{1}{4}$ $\frac{4}{13}$
16. $2\frac{1}{8}$ $\frac{8}{17}$
17. $-2\frac{3}{7}$ $-\frac{7}{17}$
18. $-6\frac{5}{11}$ $-\frac{11}{71}$

Written Exercises

Simplify.

1. $\frac{-30}{-5}$ 6
2. $\frac{-48}{8}$ -6
3. $\frac{30}{-6}$ -5
4. $\frac{-55}{11}$ -5
5. $\frac{70}{5}$ 14
6. $\frac{80}{20}$ 4
7. $\frac{-40}{8}$ -5
8. $\frac{-48}{6}$ -8
9. $\frac{-96}{-16}$ 6
10. $\frac{-84}{-7}$ 12
11. $\frac{-36}{-9}$ 4
12. $\frac{42}{-6}$ -7
13. $\frac{-38}{2}$ -19
14. $\frac{54}{-9}$ -6
15. $\frac{-200}{50}$ -4
16. $\frac{-450}{10}$ -45
17. $\frac{-36a}{-6}$ $6a$
18. $\frac{45b}{9}$ $5b$
19. $\frac{63a}{-9}$ $-7a$
20. $\frac{77b}{-11}$ $-7b$
21. $49 \div 7$ 7
22. $-16 \div 8$ -2
23. $65 \div (-13)$ -5
24. $75 \div (-15)$ -5
25. $\frac{-\frac{5}{6}}{8}$ $-\frac{5}{48}$
26. $\frac{-\frac{3}{4}}{9}$ $-\frac{1}{12}$
27. $\frac{\frac{7}{8}}{-10}$ $-\frac{7}{80}$
28. $\frac{\frac{3}{11}}{-6}$ $-\frac{1}{22}$
29. $\frac{\frac{1}{3}}{4}$ $\frac{1}{12}$
30. $\frac{\frac{3}{8}}{6}$ $\frac{1}{16}$
31. $\frac{7}{-\frac{2}{5}}$ $-\frac{35}{2}$
32. $\frac{11}{-\frac{5}{6}}$ $-\frac{66}{5}$
33. $\frac{-6}{-\frac{4}{7}}$ $\frac{21}{2}$
34. $\frac{-5}{\frac{2}{7}}$ $-\frac{35}{2}$
35. $\frac{-6}{-\frac{4}{9}}$ $\frac{27}{2}$
36. $\frac{-9}{-\frac{10}{17}}$ $\frac{153}{10}$
37. $\frac{3a + 9}{3}$ $a + 3$
38. $\frac{6a + 24}{6}$ $a + 4$
39. $\frac{7a + 35}{-7}$ $-a - 5$
40. $\frac{14a + 56}{-7}$ $-2a - 8$
41. $\frac{20a + 30b}{-2}$ $-10a - 15b$
42. $\frac{-5x + (-10y)}{-5}$ $x + 2y$
43. $\frac{60a - 30b}{-6}$ $-10a + 5b$
44. $\frac{70x - 30y}{-5}$ $-14x + 6y$

Using Calculators — The Reciprocal Key

The key labeled $\frac{1}{x}$ on your calculator is the reciprocal key. When this key is pressed, the calculator replaces the number in the display with its reciprocal.

Exercises

Use a calculator to evaluate each expression. Round each result to the nearest thousandth.

1. $\frac{1}{2.5}$ 0.4
2. $\frac{1}{7}$ 0.143
3. $\frac{1}{3(5)}$ 0.067
4. $\frac{1}{3} \cdot \frac{1}{5}$ 0.067
5. $\frac{1}{0.618}$ 1.618
6. $\frac{1}{5 - 2(0.8)}$ 0.294
7. Enter a number. Then press the reciprocal key twice. What happens? Predict what will happen if you press the key n times.
8. Enter 0 and then press the reciprocal key. What happens? Why?

Objective 3-3: Solve equations by using the multiplication and division properties of equality.

3-3 Solving Equations Using Multiplication and Division

A certain basket holds $\frac{1}{4}$ case of apples. There are 30 apples in the basket. About how many apples would a case hold?

An equation can be used to solve this problem. If x is the number of apples in a case, then the equation is $\frac{1}{4}x = 30$. To solve this equation, you would use the **multiplication property of equality**. It states that if both sides of an equation are multiplied by the same nonzero number, the result is an equivalent equation.

Recall that equivalent equations have the same solution.

For any numbers a, b, and c, if $a = b$, then $ac = bc$.

Multiplication Property of Equality

Example 1

Solve: $\frac{1}{4}x = 30$

$\frac{1}{4}x = 30$ — *The coefficient of x is $\frac{1}{4}$.*

$4\left(\frac{1}{4}x\right) = 4(30)$ — *Multiply both sides by 4, the reciprocal of $\frac{1}{4}$.* ❶

$1x = 120$

$x = 120$

Check: $\frac{1}{4}x = 30$

$\frac{1}{4}(120) \stackrel{?}{=} 30$ — *Replace x with 120.*

$30 = 30$

The solution is 120.

In the equation $\frac{1}{4}x = 30$, the variable x represented the number of apples in a case. Thus, there are about 120 apples in a case.

3-3

Lesson Notes, p. T30
Enrichment, p. T30

Readiness Activity

Here are two true equations.

$-6 = -6$

$21 = 21$

Is each equation still true if you:

(a) multiply both sides by 4? yes

(b) multiply both sides by -4? yes

(c) divide both sides by 3? yes

(d) divide both sides by -3? yes

❶ Emphasize to students that both sides are multiplied by 4 since $4 \cdot \frac{1}{4} = 1$.

Discussion Questions

1. If both sides of an equation are multiplied by zero, is the result always an equivalent equation? Justify your answer. no; answers may vary.
2. In the division property of equality, why must the divisor be a non-zero number? Division by zero is not defined in the set of real numbers.

Chalkboard Exercises

Solve each equation.

1a. $\frac{1}{5}x = 30$ 150

1b. $\frac{1}{7}x = 15$ 105

Chalkboard Exercises

Solve each equation.

2a. $\frac{r}{6} = -12$ -72

2b. $\frac{k}{3} = 25$ 75

3a. $\frac{-2}{3}y = 40$ -60

3b. $\left(4\frac{1}{4}\right)d = 7\frac{1}{2}$ $1\frac{13}{17}$

4a. $37 = -3a$ $-12\frac{1}{3}$

4b. $-52 = 7x$ $-7\frac{3}{7}$

Examples

2 **Solve:** $\frac{k}{7} = -45$

$$\frac{k}{7} = -45$$

$$\frac{1}{7}k = -45 \quad \textit{Substitute } \tfrac{1}{7}k \textit{ for } \tfrac{k}{7}.$$

$$7 \cdot \frac{1}{7}k = 7(-45) \quad \textit{Multiply both sides by 7.}$$

$$k = -315 \quad \textit{Check this result.}$$

The solution is -315.

3 **Solve:** $\left(2\frac{1}{3}\right)m = 3\frac{1}{9}$

$$\left(2\tfrac{1}{3}\right)m = 3\tfrac{1}{9}$$

$$\frac{7}{3}m = \frac{28}{9} \quad \textit{Rewrite the mixed numbers as improper fractions.}$$

$$\frac{3}{7}\left(\frac{7}{3}m\right) = \frac{3}{7}\left(\frac{28}{9}\right) \quad \textit{Multiply both sides by } \tfrac{3}{7}.$$

$$m = \frac{4}{3} \quad \textit{Check this result.}$$

The solution is $\frac{4}{3}$ or $1\frac{1}{3}$.

4 **Solve:** $24 = -2a$

$$24 = -2a$$

$$-\frac{1}{2}(24) = -\frac{1}{2}(-2a) \quad \textit{Multiply both sides by } -\tfrac{1}{2}.$$

$$-12 = a \quad \textit{Check this result.}$$

The solution is -12.

The equation $24 = -2a$ was solved by multiplying both sides by $-\frac{1}{2}$. The same result could have been obtained by dividing both sides by -2. In general, dividing both sides of an equation by the same nonzero number results in an equivalent equation. This is called the **division property of equality**.

The division property of equality is a special case of the multiplication property of equality.

For any numbers a, b, and c, with $c \neq 0$, if $a = b$, then $\frac{a}{c} = \frac{b}{c}$.

Division Property of Equality

Example

5 **Solve:** $-6x = 11$

$-6x = 11$

$\frac{-6x}{-6} = \frac{11}{-6}$ *Divide both sides by −6.*

$x = -\frac{11}{6}$ *Check this result.*

The solution is $-\frac{11}{6}$.

Exploratory Exercises

State the number that you would multiply both sides of each equation by to solve it.

1. $\frac{b}{3} = -6$ 3
2. $\frac{b}{5} = 10$ 5
3. $\frac{3}{4}n = 30$ $\frac{4}{3}$
4. $\frac{5}{7}t = -10$ $\frac{7}{5}$
5. $\frac{4}{9}n = -24$ $\frac{9}{4}$
6. $-\frac{5}{9}x = 15$ $-\frac{9}{5}$
7. $-8n = 24$ $-\frac{1}{8}$
8. $3n = -18$ $\frac{1}{3}$
9. $1 = \frac{k}{9}$ 9
10. $-14 = \frac{k}{2}$ 2

State the number that you would divide both sides of each equation by to solve it.

11. $4x = 24$ 4
12. $28 = 7x$ 7
13. $35 = 4y$ 4
14. $-36 = 4z$ 4
15. $-7x = 21$ −7
16. $-5x = 14$ −5
17. $-8x = -9$ −8
18. $-14x = -30$ −14
19. $-6x = -36$ −6
20. $-10x = 40$ −10

Written Exercises

Solve each equation.

1. $-4r = -28$ 7
2. $-7w = -49$ 7
3. $-8t = 56$ −7
4. $5t = -45$ −9
5. $6x = -42$ −7
6. $11y = -77$ −7
7. $-5s = -85$ 17
8. $-7h = -91$ 13
9. $9x = 40$ $\frac{40}{9}$
10. $-3y = 52$ $-\frac{52}{3}$
11. $3w = -11$ $-\frac{11}{3}$
12. $5c = 8$ $\frac{8}{5}$
13. $6x = 28$ $\frac{14}{3}$
14. $4x = 62$ $\frac{31}{2}$
15. $-11q = -81$ $\frac{81}{11}$
16. $-13s = -64$ $\frac{64}{13}$
17. $434 = -31y$ −14
18. $17b = -391$ −23
19. $\frac{k}{8} = 6$ 48
20. $11 = \frac{k}{5}$ 55
21. $-10 = \frac{b}{-7}$ 70
22. $-13 = \frac{b}{-8}$ 104
23. $\frac{h}{11} = -25$ −275
24. $\frac{h}{12} = -24$ −288
25. $\frac{d}{3} = -30$ −90
26. $\frac{d}{4} = -5$ −20
27. $-65 = \frac{f}{29}$ −1885
28. $-70 = \frac{f}{25}$ −1750
29. $\frac{2}{3}x = -6$ −9
30. $\frac{3}{5}x = 9$ 15
31. $\frac{2}{5}t = -10$ −25
32. $\frac{4}{9}t = 72$ 162
33. $\frac{4}{7}r = 20$ 35
34. $\frac{3}{4}x = -12$ −16
35. $\frac{4}{9}x = -9$ $-\frac{81}{4}$
36. $\frac{9}{11}y = 20$ $\frac{220}{9}$
37. $-\frac{3}{5}y = -50$
38. $-\frac{2}{3}x = -41$ $\frac{123}{2}$
39. $-\frac{11}{8}x = 42$
40. $-\frac{13}{5}y = -22$ $\frac{110}{13}$
41. $3x = 4\frac{2}{3}$ $\frac{14}{9}$
42. $\frac{5}{2}x = -25$ −10
43. $\left(5\frac{1}{2}\right)x = 33$ 6
44. $-5x = -3\frac{2}{3}$ $\frac{11}{15}$
45. $\left(-4\frac{1}{2}\right)x = 36$ −8
46. $\left(3\frac{1}{2}\right)x = -1\frac{1}{6}$ $-\frac{1}{3}$

mini-review

Solve each equation.

1. $y + 14 = -3$ −17
2. $m - (-43) = 17$ −26

Simplify.

3. $\frac{4}{5}\left(-\frac{5}{9}\right)$ $-\frac{4}{9}$
4. $\frac{8a + 32}{-4}$ $-2a - 8$
5. $14a + (-18b) + 6a^2 + 10b$

5. $14a - 8b + 6a^2$

For more exercises on solving equations, see exercises 61-90 on page 547, exercises 49-72 on page 549, and exercises 61-90 on page 551.

Chalkboard Exercises

Solve each equation.

5a. $-9x = -42$ $4\frac{2}{3}$

5b. $4x = -17$ $-4\frac{1}{4}$

Suggested Daily Assignments

Enriched
- Written: 2–46 evens
- Mini-Review: 1–5

Average
- Exp: 2–20 evens
- Written: 1–46
- Mini-Review: 1–5

Fundamental
- Exp: 1–20
- Written: 1–25
  - p. 80, 25–44
- Mini-Review: 1–5

Answers to Written

37. $\frac{250}{3}$ **39.** $-\frac{336}{11}$

3-4

Lesson Notes, p. T30

Readiness Activity
Show the steps you would use to evaluate the following expressions. Describe each step.

1. $4 + 6 \cdot 8$
   $4 + 48$; multiply
   52; add
2. $14 \div 7 - 12$
   $2 - 12$; divide
   -10; subtract
3. $\frac{18 - 2}{4}$
   $\frac{16}{4}$; subtract
   4; divide
4. $-6 \cdot 5 - 20$
   $-30 - 20$; multiply
   -50; subtract

Discussion Questions

1. Solve $13x - 13x = 5$.
   No solution
2. Show two ways to solve $3(n - 1) = 12$.
   a. $3(n - 1) = 12$
      $n - 1 = 4$
      $n = 5$
   b. $3n - 3 = 12$
      $3n = 15$
      $n = 5$

Chalkboard Exercises
Solve each equation.

1a. $7y + 16 = -5$ -3
1b. $6a - 13 - 17$ 5
2a. $\frac{r}{7} + 7 = -2$ -63
2b. $8 + \frac{3}{4}n = -10$ -24

Objective 3-4: Solve equations involving more than one operation.

3-4 Solving Equations Using More than One Operation

Each of these equations can be solved by performing the inverse of the operation indicated in the equation.

$$x + 2 = 5 \qquad 3x = 4.8 \qquad \frac{x}{2} = -11$$

But each of the following equations contains more than one operation.

$$\frac{x}{4} + 7 = 10 \qquad \frac{a + 10}{3} = 6 \qquad -5x - 7 = 28$$

These equations can be solved using the same methods that were used to solve the first set of equations. However, more than one step is involved.

Recall the order of operations for evaluating an expression. To solve an equation with more than one operation, undo the operations in reverse order.

Examples

1 **Solve: $3x + 7 = 13$**

$3x + 7 = 13$ — *Addition of 7 is indicated.*

$3x + 7 - 7 = 13 - 7$ — *Therefore, subtract 7 from both sides.*

$3x = 6$ — *Multiplication by 3 is also indicated.*

$\frac{3x}{3} = \frac{6}{3}$ — *Therefore, divide both sides by 3.*

$x = 2$

Check:

$$3x + 7 \stackrel{?}{=} 13$$
$$3(2) + 7 \stackrel{?}{=} 13$$
$$6 + 7 \stackrel{?}{=} 13$$
$$13 = 13$$

The solution is 2.

2 **Solve: $\frac{x}{4} + 9 = 6$**

$\frac{x}{4} + 9 = 6$

$\frac{x}{4} + 9 - 9 = 6 - 9$ — *First subtract 9 from both sides. Why?*

$\frac{x}{4} = -3$

$4\left(\frac{x}{4}\right) = 4(-3)$ — *Then multiply both sides by 4. Why?*

$x = -12$ — *Check this result.*

The solution is -12.

Examples

3 **Solve: $7 = 14 - 3x$**

$$
\begin{aligned}
7 &= 14 - 3x \\
7 - 14 &= 14 - 3x - 14 \\
7 - 14 &= 14 - 14 - 3x \quad ❶ \\
-7 &= -3x \\
\frac{-7}{-3} &= \frac{-3x}{-3} \\
\frac{7}{3} &= x
\end{aligned}
$$

Check this result.

The solution is $\frac{7}{3}$ or $2\frac{1}{3}$.

4 **Solve: $\frac{d-4}{3} = 5$**

$$
\begin{aligned}
\frac{d-4}{3} &= 5 \\
3\left(\frac{d-4}{3}\right) &= 3(5) \\
d - 4 &= 15 \\
d - 4 + 4 &= 15 + 4 \\
d &= 19
\end{aligned}
$$

Check this result.

The solution is 19.

Exploratory Exercises

State the steps that you would use to solve each equation.

1. $3x - 7 = 2$
2. $5n - 4 = -7$
3. $2x + 5 = 13$
4. $8 + 3x = 5$
5. $\frac{a+2}{5} = 10$
6. $\frac{a-7}{15} = -30$
7. $\frac{3}{11}x + 2 = 5$
8. $-\frac{4}{13}y - 7 = 6$

Written Exercises

Solve and check each equation.

1. $4t - 7 = 5$ 3
2. $6 = 4n + 2$ 1
3. $4 + 7x = 39$ 5
4. $7n - 4 = 17$ 3
5. $5n + 3 = 9$ $\frac{6}{5}$
6. $34 = 8 - 2t$ -13
7. $5 - 3x = 32$ -9
8. $2 - 7s = -19$ 3
9. $-3x - 7 = 18$ $-\frac{25}{3}$
10. $-4y + 2 = 29$ $-\frac{27}{4}$
11. $\frac{x}{2} + 5 = 7$ 4
12. $\frac{m}{3} - 6 = 14$ 60
13. $\frac{y}{3} + 6 = -45$ -153
14. $\frac{a}{6} - 7 = 13$ 120
15. $\frac{b}{3} + 2 = -21$ -69
16. $\frac{c}{-4} - 8 = -42$ 136
17. $\frac{c}{-3} - 1 = 26$ -81
18. $\frac{c}{-9} + 18 = 40$ -198
19. $\frac{d+5}{3} = -9$ -32
20. $\frac{3+n}{7} = -5$ -38
21. $\frac{m-5}{4} = 5$ 25
22. $\frac{t-6}{7} = 8$ 62
23. $\frac{s-8}{-7} = 16$ -104
24. $7 = \frac{r-8}{6}$ 50
25. $\frac{4d+5}{7} = 7$ 11
26. $\frac{4r+8}{16} = 7$ 26
27. $\frac{7n+(-1)}{8} = 8$ $\frac{65}{7}$
28. $\frac{-3n-(-4)}{-6} = -9$ $-\frac{50}{3}$
29. $0.2n + 3 = 8.6$ 28
30. $0.5x + 1.5 = 12$ 21
31. $8 - 1.2s = -1.6$ 8
32. $4 + 2.3y = -5.2$ -4
33. $8 = 4.1n - 24.8$ 8
34. $-7 = 2.6t + 8.6$ -6
35. $\frac{3}{4}n - 3 = 9$ 16
36. $\frac{1}{3}x + 4 = 6$ 6
37. $3 - \frac{n}{3} = 7$ -12
38. $5 + \frac{t}{4} = 8$ 12
39. $5 - \frac{2}{3}n = 7$ -3
40. $8 + \frac{3}{4}n = 26$ 24

Challenge

41. $2[x + 3(x - 1)] = 18$ 3
42. $4(2x - 7) + 3(x - 1) = 46$ 7
43. $\frac{1}{4}\left[\frac{3}{2}(x - 2) + 2x\right] = -6$ -6
44. $-\frac{1}{2}\left[4 - 3\left(x + \frac{1}{5}\right)\right] = -\frac{4}{5}$ $\frac{3}{5}$

For more exercises on solving equations with more than one operation see exercises 1-21 on page 560.
For review exercises on multiplying and dividing decimals, see exercises 1-30 on page 551.

Chalkboard Exercises

3a. $3 = 11r - 30$ 3
3b. $15 = 29 - 2x$ 7
4a. $\frac{y-6}{7} = 2$ 20
4b. $\frac{a+14}{4} = 3$ -2

❶ Point out to students that the associative property was used in this step.

Suggested Daily Assignments

Enriched
Written: 2–28 evens, 29–44

Average
Exp: 2–8 evens
Written: 1–28, 29–39 odds

Fundamental
Exp: 1–8
Written: 1–24
p. 83, 26-46 evens

Answers to Exploratory

1. Add 7 to both sides. Then divide both sides by 3.
2. Add 4. Then divide by 5.
3. Subtract 5. Then divide by 2.
4. Subtract 8. Then divide by 3.
5. Multiply by 5. Then subtract 2.
6. Multiply by 15. Then add 7.
7. Subtract 2. Then multiply by $\frac{11}{3}$.
8. Add 7. Then multiply by $-\frac{13}{4}$.

About Applications
This feature provides reinforcement not only in computations, but in translating verbal problems to equations. Acceleration is often given in meters per second per second, abbreviated m/s^2. You may wish to show units with the formula.

$$a(\text{m/s}^2) = \frac{(f - s)\text{m/s}}{(t)\text{s}}$$

You may want to have students write a computer program to solve the problems on this page. A typical program is given below.

```
10 PRINT "ENTER FINAL
   SPEED, STARTING
   SPEED, AND TIME"
20 INPUT F,S,T
30 LET A = (F-S)/T
40 IF A < 0 THEN 60
50 PRINT "THE
   ACCELERATION IS";A
55 GO TO 70
60 PRINT "THE
   DECELERATION IS";A
70 END
```

Applications in Physics

Acceleration

Suppose an automobile is moving at a speed of 50 mph. Wishing to pass another car, the driver increases the speed to 55 mph in 10 seconds. The driver has accelerated the car.

Acceleration is the rate at which speed is changing with respect to time. To compute acceleration (a) you divide the change in speed by the time (t) needed to make the change. The change in speed is the difference between the final speed (f) and the starting speed (s). This can be summarized in the following formula.

$$a = \frac{f - s}{t}$$

Example: **A race car goes from 44 m/s to 77 m/s in 11 seconds. Find the change in speed per second.**

$$a = \frac{f - s}{t}$$

$$a = \frac{77 - 44}{11} \text{ or } 3$$ *Substitute 77 for f, 44 for s, and 11 for t.*

The car accelerates 3 meters per second each second or 3 m/s^2.

Consider what happens when the racing car slows down from 77 m/s to 44 m/s during the 11 seconds. The same equation can be used. Subtracting the starting speed from the final speed gives a negative value. This means that the acceleration is negative. *Negative acceleration is deceleration.*

$$a = \frac{44 - 77}{11} = -3$$

The acceleration is -3 m/s^2.

Exercises

For each problem, find the acceleration.

1. A motorcycle goes from 2 m/s to 14 m/s in 6 seconds. 2 m/s^2
2. A skateboard goes from 5 m/s to 0 m/s in 1 second. -5 m/s^2
3. A train is traveling for 1 hour at a constant speed of 50 mph. 0 m/s^2
4. A car starts from a standstill. It accelerates to 40 mph in 10 seconds. 4 mph per seco
5. A jet plane decreases its speed from 500 km/h to 350 km/h in 30 seconds. -5 km/h per second
6. A car traveling at 35 mph is stopped in 5 seconds. -7 mph per second

Objective 3-5A: Solve equations containing variables on both sides.
Objective 3-5B: Solve equations containing grouping symbols.

3-5 More Equations

Many equations contain variables on both sides. To solve such equations, first use the addition or subtraction property of equality to write an equivalent equation that has all the variables on one side. Then finish solving the equation by using the methods shown earlier in this chapter.

Examples

1 **Solve:** $5n - 7 = 3n + 2$

$5n - 7 = 3n + 2$ ❶

$5n - 3n - 7 = 3n - 3n + 2$ *Subtract 3n from both sides*

$2n - 7 = 2$

$2n - 7 + 7 = 2 + 7$ *Add 7 to both sides.*

$2n = 9$

$\frac{2n}{2} = \frac{9}{2}$ *Divide both sides by 2.*

$n = \frac{9}{2}$ or $4\frac{1}{2}$

Check: $5n - 7 = 3n + 2$

$5\left(\frac{9}{2}\right) - 7 \stackrel{?}{=} 3\left(\frac{9}{2}\right) + 2$

$\frac{45}{2} - \frac{14}{2} \stackrel{?}{=} \frac{27}{2} + \frac{4}{2}$

$\frac{31}{2} = \frac{31}{2}$

The solution is $\frac{9}{2}$ or $4\frac{1}{2}$.

2 **Solve:** $\frac{2}{5}x + 3 = \frac{1}{5}x - 7$

$\frac{2}{5}x + 3 = \frac{1}{5}x - 7$

$\frac{2}{5}x - \frac{1}{5}x + 3 = \frac{1}{5}x - \frac{1}{5}x - 7$ *Subtract $\frac{1}{5}x$ from both sides.*

$\frac{1}{5}x + 3 = -7$

$\frac{1}{5}x + 3 - 3 = -7 - 3$ *Subtract 3 from both sides.*

$\frac{1}{5}x = -10$

$5\left(\frac{1}{5}x\right) = 5(-10)$ *Multiply both sides by 5.*

$x = -50$ *Check this result.*

The solution is -50.

3-5

Lesson Notes, p. T30

Readiness Activity

Simplify.

$5x + 7x$ $12x$

$-12n - 14n$ $-26n$

$4(m - 3) + 2m$ $6m - 12$

$\frac{1}{3}(15t + 12)$ $5t + 4$

Discussion Questions

(Use after example 5)

1. Write two equations that have no solutions. Answers may vary.
2. Write two equations that are identities. Answers may vary.

Chalkboard Exercises

Solve and check each equation.

1a. $7a - 3 = 5a + 5$ 4

1b. $6m + 4 = 9m + 15$ $-3\frac{2}{3}$

2a. $\frac{3}{7}x - 3 = \frac{1}{7}x + 3$ 21

2b. $\frac{6}{11}y - 2 = \frac{9}{11}y - 14$ 44

❶ This detailed solution may no longer be necessary for better students. Discuss the following abbreviated solution.

$5n - 7 = 3n + 2$

$5n = 3n + 9$

$2n = 9$

$n = \frac{9}{2}$ or $4\frac{1}{2}$

Chalkboard Exercises

Solve and check each equation.

3a. $4(3x + 5) + 1 = 3 + 5(2x + 2)$ -4

3b. $7 - 2(4 - 3y) = 4(y + 6)$ $\frac{25}{2}$ or $12\frac{1}{2}$

4. $4x + 7 = 2(2x - 8)$ no solutions

5. $7(2x + 1) = 2(7x + 6) - 5$ identity

❷ Students usually have the least difficulty with equations similar to this one if they apply the distributive property first. Then they can finish solving the equation by applying the properties of equality.

❸ More specifically, an identity is an equation that is true for all values of the variable *except* the values for which the equation is not defined. However, at this point in the text, the equations are defined for *all* values of the variable. Later in the text, students will learn to solve equations such as $\frac{5}{x - 1} = \frac{3}{x - 1} + \frac{2}{x - 1}$. This equation is an identity, even though it is not defined when x is 1.

Many equations also contain grouping symbols. When solving equations of this type, first use the distributive property to remove the grouping symbols.

Example

3 **Solve: $2(2r + 5) + 1 = 5 - 2(3 - r)$** ❷

$$
\begin{aligned}
2(2r + 5) + 1 &= 5 - 2(3 - r) \\
4r + 10 + 1 &= 5 - 6 + 2r && \textit{Use the distributive property.} \\
4r + 11 &= -1 + 2r && \textit{Simplify.} \\
4r - 2r + 11 &= -1 + 2r - 2r \\
2r + 11 &= -1 \\
2r + 11 - 11 &= -1 - 11 \\
2r &= -12 \\
\frac{2r}{2} &= \frac{-12}{2} \\
r &= -6 && \textit{Check this result.}
\end{aligned}
$$

The solution is -6.

Some equations may have *no* solutions. Other equations may have *all numbers* in their solution sets. An equation that is true for every value of the variable is called an **identity**. ❸

Examples

4 **Solve: $2x + 5 = 2x - 3$**

$$
\begin{aligned}
2x + 5 &= 2x - 3 \\
2x + 5 - 5 &= 2x - 3 - 5 \\
2x &= 2x - 8 \\
2x - 2x &= 2x - 8 - 2x \\
0 &= -8 && \textit{This is a false statement.}
\end{aligned}
$$

This equation has no solutions.

5 **Solve: $3(x + 1) - 5 = 3x - 2$**

$$
\begin{aligned}
3(x + 1) - 5 &= 3x - 2 \\
3x + 3 - 5 &= 3x - 2 \\
3x - 2 &= 3x - 2 \\
3x - 2 + 2 &= 3x - 2 + 2 && \textit{Add 2 to both sides.} \\
3x &= 3x \\
3x - 3x &= 3x - 3x && \textit{Subtract } 3x \textit{ from both sides.} \\
0 &= 0 && \textit{This statement is true for all values of } x.
\end{aligned}
$$

This equation is an identity.

Exploratory Exercises

State the steps that you would use to solve each equation.

1. $3x + 2 = 4x - 1$ **2.** $6x - 8 = 9 + 2x$ **3.** $4y - 3 = -5y + 7$ **4.** $2y + 7 = -3y - 7$

5. $3(x + 1) = 7$ **6.** $4(x - 1) = 5$ **7.** $6(2x - 3) = -9$ **8.** $-7(x - 3) = -4$

Written Exercises

Solve and check each equation.

1. $6x + 7 = 8x - 13$ 10

3. $3 - 4x = 10x + 10$ $-\frac{1}{2}$

5. $17 + 2x = 21 + 2x$ no solutions

7. $6x + 3 = 6x + 3$ identity

9. $3(x + 2) = 12$ 2

11. $7(x - 3) - 2 = 5$ 4

13. $7 + 2(x + 1) = 2x + 9$ identity

15. $6 = 3 + 5(x - 2)$ $\frac{13}{5}$

17. $4(x - 2) = 4x$ no solutions

19. $6(x + 3) = 4(x - 1)$ -11

21. $-3(x + 5) = 3(x - 1)$ -2

23. $7 - 3x = x - 4(2 + x)$ no solutions

25. $2(x - 3) + 5 = 3(x - 1)$ 2

27. $2x - 4(x - 5) = 2(10 - x)$ identity

29. $\frac{1}{3}(x + 6) = 7$ 15

31. $3(x - 5) = \frac{1}{5}(10x - 25)$ 10

33. $\frac{2}{3}n + 8 = \frac{1}{3}n - 2$ -30

35. $3 + \frac{2}{5}x = 11 - \frac{2}{5}x$ 10

37. $2.1k - 50 = 7.3k - 3.2$ -9

39. $8n - 11.35 = 1.2n + 17.21$ 4.2

41. $3(4x - 8) = -7(2x - 3)$ $\frac{45}{26}$

42. $16(2x - 5) = -2(5x - 18)$ $\frac{58}{21}$

43. $3(5 - t) = 14$ $\frac{1}{3}$

44. $10\left(t - \frac{3}{5}\right) = 8$ $\frac{7}{5}$

45. $8 + 10x = 5x + 11$ $\frac{3}{5}$

46. $5\left(2w - \frac{3}{5}\right) = w - 3$ 0

47. $3(2x + 1) = 6(x - 1)$ no solutions

48. $12(4 - y) = 57$ $-\frac{3}{4}$

49. $9(2 + w) = 33$ $\frac{5}{3}$

50. $2(3v - 1) = 6v - 2$ identity

2. $3x - 5 = 7x + 7$ -3

4. $13 - 8x = 5x + 2$ $\frac{11}{13}$

6. $-5x - 1 = -5x - 1$ identity

8. $18 - 4x = 42 - 4x$ no solutions

10. $3(x - 5) = -6$ 3

12. $3x - 2(x + 3) = x$ no solutions

14. $6(x + 2) - 4 = -10$ -3

16. $-2(2x - 3) = 6 - 4x$ identity

18. $-5 = 4 - 2(x - 5)$ $\frac{19}{2}$

20. $8(x - 9) = -8(x + 5)$ 2

22. $5x - 7 = 5(x - 2) + 3$ identity

24. $4(2x - 1) = -10(x - 5)$ 3

26. $5x + 4 = 7(x + 1) - 2x$ no solutions

28. $6(x - 2) = 5(x - 11) - 21$ -64

30. $5 - \frac{1}{2}(x - 6) = 4$ 8

32. $4(2x - 8) = \frac{1}{7}(49x + 70)$ 42

34. $\frac{3}{4}n + 16 = 2 - \frac{1}{8}n$ -16

36. $\frac{1}{2}t + 6 = \frac{1}{3}t$ -36

38. $18 - 3.8x = 7.36 - 1.9x$ 5.6

40. $3.1x - 2.78 = 7.22 - 1.9x$ 2

mini-review

Write an equation.

1. The product of π and the square of r is A. $\pi r^2 = A$

State the property.

2. $a + 0 = a$ **3.** $x(yz) = (yz)x$

Simplify.

4. $4.75 - (-7.63)$ 12.38 **5.** $a + (-12a)$ $-11a$

Suggested Daily Assignments

Enriched
- Written: 2–50 evens
- Mini-Review: 1–5

Average
- Exp: 2–8 evens
- Written: 2–50 evens
  - p. 85, 41
- Mini-Review: 1–5

Fundamental
- Exp: 1–8
- Written: 1–41 odds
  - p. 85, 29–34
- Mini-Review: 1–5

Answers to Exploratory

Answers will vary. Sample answers are given.

1. Add 1 to both sides. Subtract $3x$ from both sides.
2. Add 8 to both sides. Subtract $2x$ from both. Divide both by 4.
3. Add 3 to both sides. Add $5y$ to both sides. Divide both sides by 9.
4. Subtract 7 from both sides. Add $3y$ to both. Divide both sides by 5.
5. Multiply $x + 1$ by 3. Subtract 3 from both sides. Divide both sides by 3.
6. Multiply $x - 1$ by 4. Add 4 to both sides. Divide both sides by 4.
7. Multiply $2x - 3$ by 6. Add 18 to both sides. Divide both sides by 12.
8. Multiply $x - 3$ by -7. Subtract 21 from both sides. Divide both sides by -7.

Answers to Mini-Review

2. additive identity property
3. commutative property for multiplication

3-6

Lesson Notes, p. T30

Discussion Questions

1. What happens when you multiply both sides of the equation $\frac{a}{16} + \frac{1}{2} = \frac{1}{8}$ by 16? Can the same result be achieved by multiplying both sides by 32? Why is multiplying by 16 better than multiplying by 32?
The fractions are eliminated; yes; the numbers are smaller.
2. What happens when you multiply both sides of the equation $3.2x - 1.5 = 6.8$ by 10? Can the same result be achieved by multiplying by 10? Why is multiplying by 10 better than multiplying by 100?
The decimals are eliminated; yes; the numbers are smaller.

Chalkboard Exercises

Solve and check each equation.

1a. $\frac{a}{2} + \frac{a}{5} = \frac{21}{5}$ 6

1b. $\frac{3r}{4} - \frac{r}{3} = \frac{15}{4}$ 9

2a. $3.1d - 68 = 55 - 5.1d$ 15

2b. $28 - 2.2y = 11.6y + 262.6$ -17

Objective 3-6: Solve equations containing fractions and decimals.

3-6 Still More Equations

You can solve equations containing fractions and decimals by first using the multiplication property of equality to eliminate the fractions and decimals.

Examples

1 **Solve:** $\frac{2x}{5} + \frac{x}{4} = \frac{26}{5}$

$$\frac{2x}{5} + \frac{x}{4} = \frac{26}{5}$$ *The least common denominator is 20.*

$$20\left(\frac{2x}{5} + \frac{x}{4}\right) = \left(\frac{26}{5}\right)20$$ *Multiply both sides of the equations by 20.*

$$20\left(\frac{2x}{5}\right) + 20\left(\frac{x}{4}\right) = \left(\frac{26}{5}\right)20$$ *Use the distributive property.*

$$8x + 5x = 104$$ *The fractions are eliminated.*

$$13x = 104$$

$$x = 8$$

Check: $\frac{2x}{5} + \frac{x}{4} = \frac{26}{5}$

$$\frac{2(8)}{5} + \frac{8}{4} = \frac{26}{5}$$

$$\frac{16}{5} + \frac{8}{4} \stackrel{?}{=} \frac{26}{5}$$

$$\frac{64}{20} + \frac{40}{20} \stackrel{?}{=} \frac{104}{20}$$

$$\frac{104}{20} = \frac{104}{20}$$

The solution is 8.

2 **Solve:** $2.1x + 45.2 = -7.3 - 8.4x$

$$2.1x + 45.2 = -7.3 - 8.4x$$

$$10(2.1x + 45.2) = 10(-7.3 - 8.4x)$$ *Multiply both sides of the equation by 10.*

$$10(2.1x) + 10(45.2) = 10(-7.3) - 10(8.4x)$$

$$21x + 452 = -73 - 84x$$ *The decimals are eliminated.*

$$21x + 84x + 452 = -73 - 84x + 84x$$

$$105x + 452 - 452 = -73 - 452$$

$$105x = -525$$

$$x = -5$$ *Check this result.*

The solution is -5.

Some equations contain more than one variable. It is often necessary to solve such equations for a specific variable.

Example

3 **Solve for x: $ax + b = dx + c$**

$$ax + b = dx + c$$
$$ax + b - b = dx + c - b$$
$$ax = dx + c - b$$
$$ax - dx = dx + c - b - dx$$
$$(a - d)x = c - b$$
$$\frac{(a-d)x}{a-d} = \frac{c-b}{a-d}$$

Division by zero is undefined. Therefore $a - d \neq 0$.

$$x = \frac{c-b}{a-d}$$

Exploratory Exercises

State the number that each side of the equation must be multiplied by to remove the fractions or decimals. Then rewrite the equation.

1. $\frac{3}{4}x - 7 = 8 + \frac{2}{3}x$
2. $\frac{2}{5}x = 7 - \frac{3}{4}x$
3. $4t - 7 = \frac{5t}{2} - 3$
4. $1.2s + 8.1 = 3.5 - 2s$
5. $5.2z = 3 + 1.7z$
6. $8.17y = 4.2 - 3.7y$

Written Exercises

Solve and check each equation.

1. $\frac{y+5}{3} = 7$ 16
2. $\frac{x-7}{5} = 12$ 67
3. $\frac{3n-2}{5} = \frac{7}{10}$ $\frac{11}{6}$
4. $0.2x + 1.7 = 3.9$ 11
5. $5.3 - 0.3x = -9.4$ 49
6. $1.9s + 6 = 3.1 - s$ -1
7. $\frac{3}{4}x - 4 = 7 + \frac{1}{2}x$ 44
8. $\frac{3}{8} - \frac{1}{4}x = \frac{1}{2}x - \frac{3}{4}$ $\frac{3}{2}$
9. $\frac{5}{8}x + \frac{3}{5} = x$ $\frac{8}{5}$
10. $3y - \frac{4}{5} = \frac{1}{3}y$ $\frac{3}{10}$
11. $\frac{1}{3}y + 3 = \frac{1}{2}y$ 18
12. $\frac{2}{5}x - 1 = \frac{1}{4}x$ $\frac{20}{3}$
13. $\frac{4y+3}{7} = \frac{9}{14}$ $\frac{3}{8}$
14. $\frac{4-x}{5} = \frac{1}{5}x$ 2
15. $\frac{7+3t}{4} = -\frac{1}{8}t$ -2
16. $1.3x - 4 = 2.1x + 8$ -15
17. $0.8 - 0.2z = 0.3z - 4$ 9.6
18. $2.1y + 7 = 1.2y - 2$ -10
19. $\frac{2}{3}x + \frac{1}{2} = \frac{1}{3}x + 4$ $\frac{21}{2}$
20. $\frac{3y}{2} - y = 4 + \frac{1}{2}y$ no solutions
21. $\frac{x}{2} - \frac{1}{3} = \frac{x}{3} - \frac{1}{2}$ -1

Solve for x.

22. $5x = y$ $\frac{y}{5}$
23. $x + r = 2d$ $2d - r$
24. $\frac{x+a}{3} = c$ $3c - a$
25. $\frac{d+x}{e} = f$ $ef - d$
26. $ax + b = c$ $\frac{c-b}{a}$
27. $ex - 2y = 3z$ $\frac{3z+2y}{e}$

Solve for y.

28. $ay - b = c$ $\frac{c+b}{a}$
29. $c - 2y = d$ $\frac{c-d}{2}$
30. $ay + z = am - n$ $\frac{am-n-z}{a}$
31. $a(y + 1) = b$ $\frac{b}{a} - 1$
32. $\frac{x}{y} = 4$ $\frac{x}{4}$
33. $\frac{3}{5}y + a = b$ $\frac{5}{3}(b - a)$

Chalkboard Exercises

3a. Solve for x: $\frac{x-5}{a} = 3$
$3a + 5$

3b. Solve for y: $\frac{x+y}{-4k} = 7a$
$-28ak - x$

Suggested Daily Assignments

Enriched
Written: 1–33 odds

Average
Exp: 1–5 odds
Written: 2–32 evens

Fundamental
Exp: 1–6
Written: 1–24
p. 89, 2–20 evens

Answers to Exploratory

1. 12; $9x - 84 = 96 + 8x$
2. 20; $8x = 140 - 15x$
3. 2; $8t - 14 = 5t - 6$
4. 10; $12s + 81 = 35 - 20s$
5. 10; $52z = 30 + 17z$
6. 100; $817y = 420 - 370y$

Quiz
Lessons 3-3 to 3-6
Solve each equation.

1. $28 = -7r$ -4
2. $-\frac{4}{9}x = -36$ 81
3. $6a + 12 = -42$ -9
4. $\frac{y}{9} - 6 = 11$ 153
5. $-2 = \frac{-4t - (-6)}{-5}$ -1
6. $32 - 4x = -32 + 8x$ $5\frac{1}{3}$
7. $6(3 + 2t) - (3t - 4) = 7t - 2$ -12

3-7

Lesson Notes, p. T30

Readiness Activity

Consider the following list of integers.

$-3, -2, -1, 0, 1, 2, 3$

What can you add to -3 to get -2? 1

What can you add to 1 to get 2? 1

What can you add to each integer to get the next integer in the list? 1

Discussion Questions

1. What numbers can be expressed as the sum of two consecutive even integers?
   2, 6, 10, 14, . . .
   $4k + 2$ for $k = 0, 1, 2, \ldots$
2. What numbers can be expressed as the sum of two consecutive odd integers?
   4, 8, 12, . . .
   $4k$ for $k = 1, 2, \ldots$

Objective 3-7: Use equations to solve problems.

3-7 Problem Solving: Using Equations

In cross-country, a team's score is found by adding the place numbers of the first five finishers. For example, five runners for a team finished 2nd, 6th, 8th, 9th, and 13th. Their team's score is found as follows.

$$2 + 6 + 8 + 9 + 13 = 38$$

Rick and Tad ran in an invitational meet. Rick finished 3 places ahead of Tad. Their combined score was 27. In what place did each runner finish?

Use the problem solving plan to find the solution.

Explore *Define a variable.*

Let x = Rick's place number.
$x + 3$ = Tad's place number.

Plan *Write an equation.*

The combined score was 27.

$$x + (x + 3) = 27$$

Solve *Solve the equation and answer the problem.*

$$x + (x + 3) = 27$$
$$2x + 3 = 27$$
$$2x + 3 - 3 = 27 - 3$$
$$2x = 24$$
$$\frac{2x}{2} = \frac{24}{2}$$
$$x = 12$$

Since x represents Rick's place number, Rick finished in 12th place. Since $x + 3$ represents Tad's place number, Tad finished in 15th place.

Examine *Check whether the answer makes sense.*

Read the problem again and compare your answer to the statements given in the problem. For the problem above you could ask the following questions.

1. If Rick finished 12th and Tad finished 15th, did Rick finish 3 places ahead of Tad?
2. Was their combined score 27?

Example 1

Use an equation to solve the problem. Then examine the solution.

Rylynn wants to buy a 10-speed bicycle which costs $117. This is $12 more than three times the amount that she saved last month. How much did she save last month?

Explore Let x = amount of money saved last month.

Plan Cost is $12 more than 3 times amount saved last month.

$$\underbrace{117}_{\text{Cost}} \; \underbrace{=}_{\text{is}} \; \underbrace{12}_{\$12} \; \underbrace{+}_{\text{more than}} \; \underbrace{3x}_{\text{3 times amount saved last month}}$$

Solve

$$117 = 12 + 3x$$
$$117 - 12 = 12 + 3x - 12$$
$$105 = 3x$$
$$\frac{105}{3} = \frac{3x}{3}$$
$$35 = x$$

Rylynn saved $35 last month.

Examine Is $117 equal to $12 more than three times $35?

$$117 \stackrel{?}{=} 3(35) + 12$$
$$117 \stackrel{?}{=} 105 + 12$$
$$117 = 117$$

Since this is a true statement, $35 is the correct answer.

Consecutive numbers are numbers in counting order such as 3, 4, 5. Beginning with an even integer and counting by two gives **consecutive even integers**. For example, $-6, -4, -2, 0, 2, 4$ are consecutive even integers. Beginning with an odd integer and counting by two gives **consecutive odd integers**. For example, $-3, -1, 1, 3, 5$ are consecutive odd integers.

Example 2

Find three consecutive even integers whose sum is -12.

Let x = the least even integer.
$x + 2$ = the next greater even integer.
$x + 4$ = the greatest of the three even integers.

$$x + (x + 2) + (x + 4) = -12 \quad \textit{The sum of the integers is } -12.$$
$$3x + 6 = -12$$
$$3x = -18$$
$$x = -6$$

Therefore, $x + 2 = -4$ and $x + 4 = -2$. The integers are -6, -4, and -2.

Chalkboard Exercises

Use an equation to solve each problem.

1a. Jan wants to swim one lap in 20 seconds. This is eight more seconds than half the number of seconds she swam her last lap. How many seconds did it take for Jan to swim her last lap?
Jan swam her last lap in 24 seconds.

1b. Eric weighs twelve pounds less than three times what Herb weighs. Together they weigh 180 pounds. How much does each person weight?
Herb, 48 pounds
Eric, 132 pounds

2a. Find three consecutive integers whose sum is 54.
17, 18, 19

2b. Find two consecutive odd integers whose sum is 100.
49, 51

2c. Find three consecutive odd integers such that twice the first decreased by the third is 21.
25, 27, 29

2d. Find two consecutive odd integers such that four times the second decreased by the first is 35.
9, 11

Computer Notes

Pages 537-539 of the BASIC appendix contain instructional material relating to consecutive numbers. Example 2 on page 538 shows a program for printing consecutive even integers from 2 to 100.

Suggested Daily Assignments

Enriched
- Exp: 1–6
- Written: 2–32 evens, 33–36
- Mini-Review: 1–5

Average
- Exp: 2–12 evens
- Written: 1–33 odds
- Mini-Review: 1–5

Fundamental
- Exp: 1–12
- Written: 2–32 evens
  p. 91, 25–33
- Mini-Review: 1–5

Exploratory Exercises

Solve each problem.

1. State three consecutive integers if the greatest one is 4. 2, 3, 4
2. State three consecutive integers if the least one is -2. -2, -1, 0
3. State three consecutive even integers if the least one is -4. -4, -2, 0
4. State three consecutive even integers if the greatest one is 10. 6, 8, 10
5. State four consecutive odd integers if the least one is 13. 13, 15, 17, 19
6. State four consecutive odd integers if the least one is -7. -7, -5, -3, -1

For each sentence, define a variable. Then state an equation.

7. The sum of two consecutive integers is 17. $x + (x + 1) = 17$
8. The sum of three consecutive even integers is 48. $x + (x + 2) + (x + 4) = 48$
9. The sum of two consecutive odd integers is -36. $x + (x + 2) = -36$
10. The sum of four consecutive integers is -46. $x + (x + 1) + (x + 2) + (x + 3) = -$
11. Three times a number increased by 4 is -11. $3x + 4 = -11$
12. Seventeen decreased by twice a number is 5. $17 - 2x = 5$

Written Exercises

For each problem, define a variable. Then use an equation to solve the problem. Some problems have no solutions.

1. Find three consecutive integers whose sum is 87. 28, 29, 30
2. Find three consecutive integers whose sum is 114. 37, 38, 39
3. Find four consecutive integers whose sum is 130. 31, 32, 33, 34
4. Find four consecutive integers whose sum is 278. 68, 69, 70, 71
5. Find two consecutive even integers whose sum is 115. no solutions
6. Find two consecutive even integers whose sum is 138. 68, 70
7. Find two consecutive odd integers whose sum is 64. 31, 33
8. Find two consecutive odd integers whose sum is 35. no solutions
9. Find three consecutive odd integers whose sum is 99. 31, 33, 35
10. Find three consecutive even integers whose sum is 72. 22, 24, 26
11. Twice a number increased by 20 is 62. Find the number. 21
12. Three times a number decreased by 11 is 58. Find the number. 23
13. Karen has 6 more than twice as many newspaper customers as when she started. She now has 98 customers. How many did she have when she started? 46
14. Bonnie sold some stock for \$42 a share. This was \$10 a share more than twice what she paid for it. At what price did she buy the stock? \$16
15. Last year Marc Ames sold 7 sedans more than twice the number of vans Lola Kaplan sold. Marc sold 83 sedans. How many vans did Lola sell? 38
16. One season Reggie Walker scored 9 more runs than twice the number of runs he batted in. He scored 117 runs that season. How many runs did he bat in? 54

17. A rectangular playground is 60 meters longer than it is wide. It can be enclosed by 920 meters of fencing. Find its length.

18. A soccer field is 75 yards shorter than 3 times its width. Its perimeter is 370 yards. Find its dimensions. 65 yd × 120 yd

19. The captain of the cross-country team placed second in a meet. The other four members placed in consecutive order, but farther behind. The team score was 40. In what places did the other members finish?

20. One member of the cross-country team placed fourth in a meet. The other four members placed in consecutive order, but farther behind. The team score was 70. In what places did the other members finish?

21. Brad bought a used bike for $8 more than half its original price. Brad paid $40 for the bike. What was the original price? $64

22. The lengths of the sides of a triangle are consecutive odd integers. The perimeter is 27. What are the lengths of the sides? 7, 9, 11

23. Twice a number increased by 4 times the number is 96. Find the number. 16

24. Twice a number increased by 12 is 31 less than three times the number. Find the number. 43

25. A number increased by 5 less than twice itself is 415. Find the number. 140

26. Five times a number decreased by 4 more than twice the number is 119. Find the number. 41

27. Find four consecutive even integers such that twice the least increased by the greatest is 96. 30, 32, 34, 36

28. Find 4 consecutive odd integers such that the sum of the first and twice the second is 175. 57, 59, 61, 63

29. Twice the greater of two consecutive odd integers is 13 less than three times the lesser. Find the integers. 17, 19

30. Three times the greatest of 3 consecutive even integers exceeds twice the least by 38. Find the integers. 26, 28, 30

31. Find two consecutive integers such that 3 times the first integer plus 2 times the second integer is equal to 107. 21, 22

32. Find two consecutive integers such that 2 times the first integer plus 4 times the second integer is equal to 256. 42, 43

Challenge

33. The length of a rectangle is 40 m less than 2 times its width. Its perimeter is 220 m. Find its dimensions. 50 m × 60 m

34. The length of a rectangle is 35 m more than 3 times its width. Its perimeter is 390 m. Find its dimensions. 40 m × 155 m

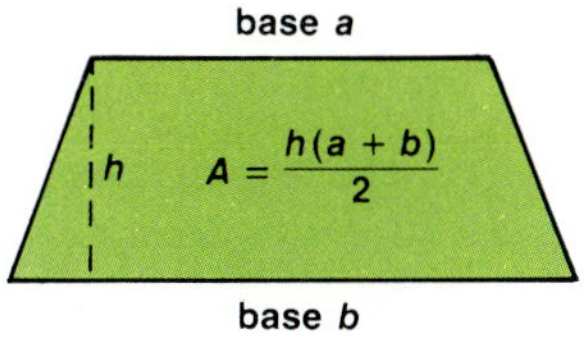

35. A trapezoid has an area of 96 m^2 and a height of 12 m. One base is 2 m longer than the other base. How long is each base? 9 m, 7 m

36. A trapezoid has an area of 117 sq ft and a height of 9 ft. One base is 4 ft shorter than the other base. How long is each base? 11 ft, 15 ft

mini-review

Solve.

1. $x - (-7.4) = 5.92$ -1.48
2. $\frac{7}{8} + y = \frac{11}{20}$ $-\frac{13}{40}$
3. $4x = 62$ $\frac{31}{2}$
4. $0.5x + 1.5 = 12$ 21
5. $6(x + 3) = 4(x - 1)$ -11

Answers to Written

17. 260 meters
19. 8th, 9th, 10th, 11th
20. 15th, 16th, 17th, 18th

About Problem Solving
You may have been encouraging students to draw diagrams and pictures already. This feature emphasizes that some problems can be solved by looking at a diagram or a picture and using very little mathematics.

Problem Solving — Diagrams

Many problems can be solved more easily if you draw a picture or diagram to represent the situation. Sometimes a picture will help you decide how to work a problem. Other times the picture will show the answer to the problem.

The length of a rectangle is twice its width. The perimeter is 39 m. Find the length and width.

Draw one side and mark it x. Then draw an adjacent side twice as long as the first side.

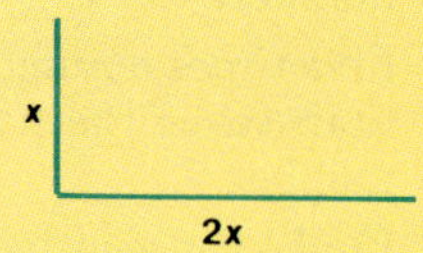

Finish drawing the rectangle. Label all four sides.

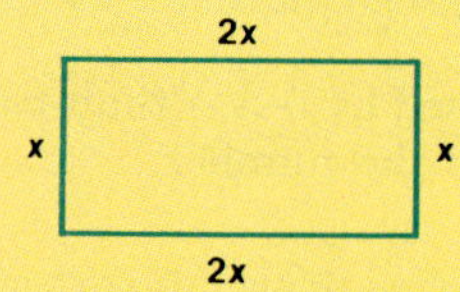

You can use the diagram to write an equation for the problem.

$$x + 2x + x + 2x = 39$$
$$6x = 39$$
$$x = 6.5$$

The width is 6.5 m and the length is twice the width, or 13 m.

Solve each problem. Use pictures and diagrams. See student's diagrams.

1. Two sides of a triangle have the same length. The third side is 2 m longer. If the perimeter of the triangle is 20 m, find the lengths of the sides. 6 m, 6 m, 8 m
2. You can cut a pizza into 7 pieces with only 3 straight cuts. What is the greatest number of pieces you can make with 5 straight cuts? 16 pieces
3. A row of 20 seats has some red seats and some blue seats. The first two are red, the next two are blue, the next two are red, and so on. What color is the 14th seat? red
4. Three spiders are on a 9-ft wall. Susie spider is 4 feet from the top. Sam spider is 7 feet from the bottom. Shirley spider is 3 feet below Sam spider. Which spider is nearest the top of the wall? Sam
5. A row of coins is arranged in the following pattern: 1 dime, followed by 3 nickels, then 5 dimes, 7 nickels, and so on. What is the 40th coin in the row? dime
6. An ant is climbing a 30-ft flagpole. Each day he climbs up 7 feet. Each night he slips back 4 feet. How many days will it take to reach the top of the flagpole? 9 days
7. A staircase with 3 steps was built with 6 blocks. Can a staircase be built with 28 blocks? How many steps would it have? yes, 7 steps

8. There were 9 people at a party. Everyone shook hands with everyone else exactly once. How many handshakes occurred? 36 handshakes

Objective 3-8A: Solve equations using the means-extremes property of proportions.
Objective 3-8B: Use ratios and proportions to solve problems.

3-8 Ratio and Proportion

As gear A revolves 4 times, it will cause gear B to revolve 3 times. Hence, we say that the gear ratio is 4 to 3. ❶

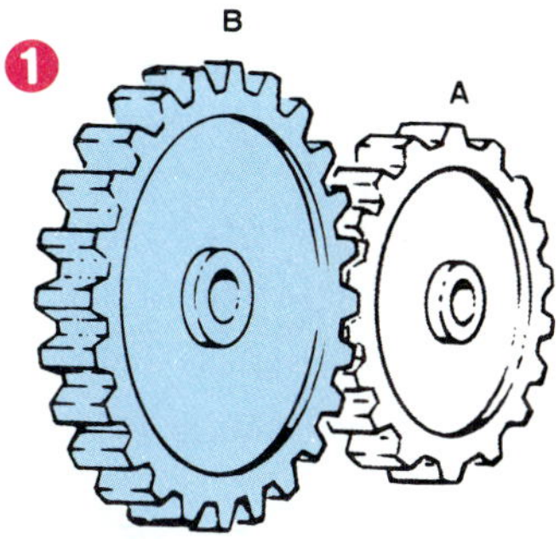

In mathematics, a **ratio** is a comparison of two numbers by division. The gear ratio above can be expressed in the following ways.

$$4 \text{ to } 3 \qquad 4:3 \qquad \frac{4}{3}$$

A ratio is most commonly expressed as a fraction in simplest form.

Examples

1 What is the ratio of 10 m to 6 m?

The ratio is $\frac{10}{6}$ or $\frac{5}{3}$.

2 What is the ratio of 20 inches to 4 feet?

The units should be the same.
4 feet = 48 inches

The ratio is $\frac{20}{48}$ or $\frac{5}{12}$.

An equation of the form $\frac{a}{b} = \frac{c}{d}$ which states that two ratios are equal is called a proportion.

Definition of Proportion ❷

Every proportion consists of four terms.

first ↓ $\frac{a}{b}$ ↑ *second* = *third* ↓ $\frac{c}{d}$ ↑ *fourth*	The first and fourth terms are called the **extremes**. The second and third terms are called the **means**. ❸

In a proportion, the product of the extremes is equal to the product of the means.

If $\frac{a}{b} = \frac{c}{d}$, then $ad = bc$.

Means-Extremes Property of Proportions

You can use this property to solve some equations. The equation must be in the form of a proportion.

3-8

Lesson Notes, p. T31

Readiness Activity

Simplify.

$\frac{6}{8}$ $\frac{3}{4}$ $\frac{12}{15}$ $\frac{4}{5}$ $\frac{20}{16}$ $\frac{5}{4}$ or $1\frac{1}{4}$

$\frac{63}{45}$ $\frac{7}{5}$ or $1\frac{2}{5}$ $\frac{36}{48}$ $\frac{3}{4}$ $\frac{27}{21}$ $\frac{9}{7}$ or $1\frac{2}{7}$

Discussion Questions

1. In the proportion $\frac{a}{b} = \frac{c}{d}$, a is greater than b and a, b, c, and d are nonzero. What can be said about the relationship between c and d? $c > d$
2. In the proportion $\frac{e}{f} = \frac{g}{h}$, e is less than g and e, f, g, and h are nonzero. What can be said about the relationship of f and h? $h > f$

Chalkboard Exercises

Write each ratio as a fraction in simplest form.

1a. 150 km to 25 km 6

2a. 16 inches to 3 feet $\frac{4}{9}$

2b. 11 m to 1600 cm $\frac{11}{16}$

❶ Gear A has 18 teeth and gear B has 24 teeth. The LCM of 18 and 24 is 72. Thus, it takes 4 revolutions of A to equal 3 of B.

❷ The proportion $\frac{a}{b} = \frac{c}{d}$ is generally read "a is to b as c is to d."

❸ Ratios and proportions may also be written in this way:
$a:b = c:d$
In this form it is easy to see that a and d are the *extremes*, since they are at the extreme ends. The *means* are the middle terms.

Chalkboard Exercises

Solve each proportion.

3a. $\frac{x}{5} = \frac{10}{12}$ $4\frac{1}{6}$

3b. $\frac{x}{6} = \frac{-8}{18}$ $-2\frac{2}{3}$

4a. $\frac{3}{7} = \frac{x + 2}{x}$ $-3\frac{1}{2}$

4b. $\frac{3}{5} = \frac{x}{x + 6}$ 9

Use a proportion to solve each problem.

5a. Sue earns \$75 in 8 hours. How much will she earn in 21 hours? \$196.88

5b. The scale on a map is 6 inches to 300 miles. New York and Chicago are 750 miles apart. How far apart are they on the map? 15 inches

❹ You may want to read this proportion as follows: If 96 miles takes 6 gallons, then 152 miles takes how many gallons? Another proportion that can be written is

$$\frac{96 \text{ miles}}{152 \text{ miles}} = \frac{6 \text{ gallons}}{x \text{ gallons}}.$$

Examples

3 **Solve:** $\frac{21}{27} = \frac{x}{18}$

$\frac{21}{27} = \frac{x}{18}$

$21(18) = 27x$ *Means-Extremes Property*

$378 = 27x$

$14 = x$ *Check this result.*

The solution is 14.

4 **Solve:** $\frac{x}{5} = \frac{x + 3}{10}$

$\frac{x}{5} = \frac{x + 3}{10}$

$10x = 5(x + 3)$ *Means-Extremes Property*

$10x = 5x + 15$ *Distributive Property*

$5x = 15$

$x = 3$ *Check this result.*

The solution is 3.

Proportions are often used to solve problems.

Example

5 **A trip of 96 miles required 6 gallons of gasoline. At that rate, how many gallons would be required for a 152-mile trip?**

Explore Let x = gallons required for 152-mile trip.

Plan Write a proportion for the problem.

$\frac{96 \text{ miles}}{6 \text{ gallons}} = \frac{152 \text{ miles}}{x \text{ gallons}}$ *Note that both ratios compare miles to gallons.* ❹

Solve $\frac{96}{6} = \frac{152}{x}$

$96x = 912$

$x = 9\frac{1}{2}$ *Check this result.*

A trip of 152 miles would require $9\frac{1}{2}$ gallons of gasoline.

Examine A trip of approximately 100 miles required 6 gallons of gasoline. A trip of 150 miles would require about $1\frac{1}{2}$ times the amount of gasoline or 9 gallons. Therefore, a solution of $9\frac{1}{2}$ gallons is reasonable.

Exploratory Exercises

Write each ratio as a fraction in simplest form.

1. 3 grams to 11 grams $\frac{3}{11}$
2. 7 feet to 3 feet $\frac{7}{3}$
3. 21 meters to 16 meters $\frac{21}{16}$
4. 16 cm to 5 cm $\frac{16}{5}$
5. 12 ounces to 6 ounces $\frac{2}{1}$
6. 15 km to 5 km $\frac{3}{1}$
7. 8 feet to 28 inches $\frac{24}{7}$
8. 4 pounds to 100 ounces $\frac{16}{25}$
9. 16 cm to 40 mm $\frac{4}{1}$
10. 72 mm to 90 cm $\frac{2}{25}$

Written Exercises

Solve each proportion.

1. $\frac{3}{4} = \frac{x}{8}$ 6
2. $\frac{3}{15} = \frac{b}{45}$ 9
3. $\frac{2}{10} = \frac{1}{y}$ 5
4. $\frac{10}{a} = \frac{20}{28}$ 14
5. $\frac{6}{8} = \frac{7}{x}$ $\frac{28}{3}$
6. $\frac{x}{9} = \frac{7}{16}$ $\frac{63}{16}$
7. $\frac{9}{m} = \frac{15}{10}$ 6
8. $\frac{6}{5} = \frac{18}{t}$ 15
9. $\frac{x+2}{5} = \frac{7}{5}$ 5
10. $\frac{6}{14} = \frac{7}{x-3}$ $\frac{58}{3}$
11. $\frac{3}{5} = \frac{x+2}{6}$ $\frac{8}{5}$
12. $\frac{14}{10} = \frac{5+x}{x-3}$ 23

Use a proportion to solve each problem.

13. Stewart earns \$97 in 4 days. At that rate, how many days will it take him to earn \$485? 20
14. Peggy saves \$18 in 4 weeks. How long will it take her to save \$81 at the same rate? 18 weeks
15. Shan Wong used 25 gallons of gasoline in traveling 350 miles. How much gasoline will he use in traveling 462 miles? 33 gallons
16. Andrea Jones drove 244 kilometers in 4 hours. At that rate, how long will it take her to drive 366 kilometers? 6 hours
17. The scale on a map is 1 centimeter to 57 kilometers. Fargo and Bismarck are 4.7 centimeters apart on the map. What is the actual distance between these cities? 267.9 km
18. The scale on a map is 2 centimeters to 5 kilometers. Dove Creek and Kent are 15.75 kilometers apart. How far apart are they on the map? 6.3 cm
19. One gear has 36 teeth. The ratio of this gear to a second gear is 4 to 3. How many teeth does the second gear have? 27
20. One gear has 45 teeth. The ratio of this gear to a second gear is 3 to 2. How many teeth does the second gear have? 30

Using Calculators — Solving Proportions

It is sometimes helpful to use a calculator when solving a proportion. Follow these steps to solve the proportion $\frac{6}{2.56} = \frac{9}{m}$.

ENTER: 9 [×] 2.56 [÷] 6 [=]

Multiply the values shown diagonally. Then divide by the third value.

The solution is 3.84.

Exercises Solve each proportion by using a calculator.

1. $\frac{x}{4.8} = \frac{5}{16}$ 1.5
2. $\frac{2.4}{3.6} = \frac{q}{1.8}$ 1.2
3. $\frac{s}{9.6} = \frac{7}{16}$ 4.2
4. $\frac{19.2}{a} = \frac{7}{29.4}$ 80.64

For more exercises on solving equations in proportion form, see exercises 53-58 on page 560.

Suggested Daily Assignments

Enriched
- Exp: 1–10
- Written: 1–20

Average
- Exp: 1–10
- Written: 1–20
  - p. 95, 18–26 evens, 34

Fundamental
- Exp: 1–10
- Written: 1–12
  - pp. 94–5, 1–19 odds

About Using Calculators
By now some of your students may have realized that solving a proportion involves one multiplication and one division. This feature shows a systematic way of performing that procedure using a calculator.

About the Excursion
Point out to students that 1 cm^3 = 1 milliliter = 1 gram. Also, 1000 cm^3 = 1 liter = 1 kilogram.
As an extension of this feature, remind students that the formula for the volume of a rectangular prism is $V = l \cdot w \cdot h$ (Volume = length · width · height). Give students the dimensions of several rectangular prisms and have them state the volume of the prism, the capacity of the prism and mass of the water that would fill the prism.

Excursions in Algebra — Capacity in the Metric System

The basic unit of length in the metric system is the meter. A unit of capacity, the **liter**, is defined in terms of the meter.

The large cube shown on this page is 10 centimeters along each edge. A cube this size has a volume of 1000 cubic centimeters and a capacity of one **liter** (L). The usual metric prefixes are used with liter.

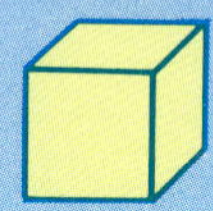
1 cm^3

A **milliliter** is $\frac{1}{1000}$ of a liter. One milliliter of water will fill a cube that is 1 cm along each edge. A thimble holds about 1 milliliter (1 mL).

There is also a relationship between capacity and mass in the metric system. A milliliter of water has a mass of one gram. A liter of water has a mass of one kilogram.

Exercises

Would you use milliliter or liter to measure each of the following?

1. gasoline for a car liter
2. eye drops milliliter
3. a large bottle of soda pop milliliter
4. vanilla for a cake milliliter
5. liquid detergent for laundry liter
6. water for the garden liter

Complete each of the following.

7. 3000 mL = 3 L
8. 2 L = 2000 mL
9. 4.6 L = 4600 mL
10. 2100 mL = 2.1 L
11. 4200 mL = 4.2 L
12. 5.06 L = 5060 mL
13. 0.54 L = 540 mL
14. 236 mL = 0.236 L
15. 20 mL = 0.020 L
16. 3 mL = 0.003 L
17. 3.05 L = 3050 mL
18. 17.98 L = ___ mL 17,980

Challenge

19. How many liters of water will fill a cube that has a volume of 1 cubic meter? 1000 liters

Objective 3-9: Solve percent problems.

3-9 Percent

Les is the quarterback for the JFK High School football team. In last week's game he completed 17 out of 25 passes. What is his rate per 100?

You can solve this problem by using ratios. Write a ratio which is equivalent to $\frac{17}{25}$ and has a denominator of 100.

$$\frac{17}{25} = \frac{17 \cdot 4}{25 \cdot 4} = \frac{68}{100}$$

Les completed passes at a rate of 68 per 100 or 68 percent. The word **percent** means per hundred, or hundredths. The symbol for percent is %.

$$68\% = \frac{68}{100} = 0.68$$

You can use a proportion to change a fraction to a percent, as shown in the following examples.

Examples

1 Change $\frac{3}{5}$ to a percent.

$\frac{3}{5} = \frac{n}{100}$

$5n = 300$ *Means-Extremes Property*

$n = 60$

Thus, $\frac{3}{5}$ is equal to $\frac{60}{100}$ or 60%.

2 Change $\frac{7}{8}$ to a percent.

$\frac{7}{8} = \frac{n}{100}$

$8n = 700$ *Means-Extremes Property*

$n = 87\frac{1}{2}$

Thus, $\frac{7}{8}$ is equal to $87\frac{1}{2}\%$.

Proportions can also be used to solve percent problems. One of the ratios in such proportions is always a comparison of two numbers called the **percentage** and the **base**. If the number 17 is being compared to the number 25, then 17 is the percentage and 25 is the base. The other ratio is formed by expressing the percent as a fraction, called the **rate**. ❶

3-9

Lesson Notes, p. T31

Discussion Questions

State one quick way to find the following percents of any number. Typical answers are given.

1. 50% Take $\frac{1}{2}$ of the number.
2. 25% Take $\frac{1}{2}$ of the number twice.
3. 1% Move the decimal point two places to the left.
4. 10% Move the decimal point one place to the left.

Chalkboard Exercises

Change each ratio or fraction to a percent.

1. $\frac{9}{10}$ 90%

2a. $\frac{5}{6}$ $83\frac{1}{3}\%$

2b. $\frac{11}{8}$ $137\frac{1}{2}\%$

❶ Some students confuse the terms *percentage* and *percent*. If you prefer, you could use the word *part* instead of *percentage*. Then the percent proportion will be:

$$\frac{\text{part}}{\text{base}} = \frac{r}{100}.$$

Chalkboard Exercises

Solve each problem.

3a. 30 is what percent of 50? 60%

3b. 46 is what percent of 80? 57.5%

4a. What number is 42% of 180? 75.6

4b. What number is 26% of 240? 62.4

5a. 70% of what number is 56? 80

5b. 45% of what number is 270? 600

❷ The formula $\frac{P}{B} = R$ can be easily rewritten as $P = R \cdot B$. This form is used when the rate and base are known, and the percentage is to be found. When the rate is given as a percent, you can write it as a decimal before using the formula.
Example: Find 20% of 18.

$$\begin{aligned} P &= R \cdot B \\ &= 0.20(18) \\ &= 3.6 \end{aligned}$$

❸ Emphasize that when a problem asks for the percent, the students must find r. The variable r is *not* used in the equations for examples 4 and 5, to avoid confusion with the variable r in the percent proportion.

❹ The equation can also be written $x = 0.36(150)$.

$$\frac{\textbf{Percentage}}{\textbf{Base}} = \textbf{Rate} \quad \textbf{or} \quad \frac{\textbf{Percentage}}{\textbf{Base}} = \frac{r}{100}$$

Percent Proportion

In the percent proportion, the rate is $\frac{r}{100}$.

❷

Example

3 **50 is what percent of 60?**

$$\frac{\text{Percentage}}{\text{Base}} = \frac{r}{100}$$

$$\frac{50}{60} = \frac{r}{100}$$ *The number 50 is being compared to 60.*

$$5000 = 60r$$

$$83\tfrac{1}{3} = r$$

❸

Thus, 50 is $83\frac{1}{3}$% of 60.

Sometimes it is easier to solve a percent problem by translating the problem into a mathematical equation. Recall that the word "of" suggests multiplication, and the word "is" suggests equality. The problem in Example 3 can be translated into an equation as follows. The percent is represented by $\frac{x}{100}$.

50 is what percent of 60?

$$50 = \frac{x}{100} \cdot 60$$ *Compare this equation to the one in Example 3.*

Examples

4 **What number is 36% of 150?**

$$x = \frac{36}{100} \cdot 150$$

$$x = 54$$

❹

Thus, 54 is 36% of 150.

5 **40% of what number is 30?**

$$\frac{40}{100} \cdot x = 30$$

$$40x = 3000$$

$$x = 75$$

Thus, 40% of 75 is 30.

There are many applications of percent, such as sales tax, discounts, and commission. Example 6 involves commission.

Example

6 **Jim Byars earns 3% commission on sales of all new cars. If he earned $861 in commission last week, what was the dollar amount of his total sales?**

Explore Let x = total sales in dollars.

Plan 3% of total sales equals $861

$$\frac{3}{100} \cdot x = 861$$

Solve

$$\frac{3}{100}x = 861$$
$$3x = 86{,}100$$
$$x = 28{,}700$$ *Check this result.*

The total sales amount was $28,700.

Examine 3% of $28,700 is $861.

Exploratory Exercises

Change each ratio or fraction to a percent.

1. $\frac{31}{100}$ 31%
2. $\frac{9}{100}$ 9%
3. $\frac{3}{10}$ 30%
4. $\frac{1}{25}$ 4%
5. $\frac{4}{5}$ 80%
6. $\frac{7}{20}$ 35%
7. $\frac{3}{8}$ $37\frac{1}{2}\%$
8. $\frac{1}{3}$ $33\frac{1}{3}\%$
9. $\frac{7}{4}$ 175%
10. $\frac{9}{5}$ 180%

Use a proportion to solve each problem.

11. 6 is what percent of 15? 40%
12. 18 is what percent of 60? 30%
13. What percent of 50 is 35? 70%
14. What percent of 17 is 34? 200%
15. 5 is what percent of 40? $12\frac{1}{2}\%$
16. What percent of 60 is 20? $33\frac{1}{3}\%$
17. What percent of 75 is 225? 300%
18. 10 is what percent of 50? 20%

Use an equation to solve each problem.

19. What number is 40% of 80? 32
20. 35 is 50% of what number? 70
21. 17 is 25% of what number? 68
22. What number is 0.3% of 6270? 18.81
23. 14 is what percent of 56? 25%
24. What percent of 72 is 12? $16\frac{2}{3}\%$

Written Exercises

Solve each problem.

1. What is 40% of 60? 24
2. Seventy-five is what percent of 250? 30%
3. Twenty-one is 35% of what number? 60
4. Find 37.5% of 80. 30
5. Fifty-two is what percent of 80? 65%
6. Thirty-six is 45% of what number? 80
7. Find 7.5% of 405. 30.375
8. Find 81% of 32. 25.92
9. Twenty-eight is 20% of what number? 140
10. Sixteen is 40% of what number? 40
11. 19 is what percent of 76? 25%
12. 37 is what percent of 296? $12\frac{1}{2}\%$
13. Find 4% of $6070. $242.80
14. Find 6% of $9.40. $0.564
15. 55 is what percent of 88? $62\frac{1}{2}\%$
16. 88 is what percent of 55? 160%

For more exercises on percents, see page 553.

Chalkboard Exercises

Solve each problem.

6a. Harry Jones earns 3% commission on his sales of algebra books. If he earned $1500 in commission last month, what was the total amount of his sales? $50,000

6b. A store advertised suits at 15% off. What was the original price of a suit that was marked $24 off? $160

Suggested Daily Assignments

Enriched
Written: 2–26 evens, 27–37

Average
Exp: 2–24 evens
Written: 2–36 evens

Fundamental
Exp: 1–24
Written: 2–36 evens
p. 99, 13–20

17. \$7030.50 is 107.5% of how many dollars? \$6,540

18. \$54,000 is 108% of how many dollars? \$50,000

19. Find 112% of \$500. \$560

20. Find 113.4% of \$1000. \$1,134

21. 96 is what percent of 60? 160%

22. 84 is what percent of 96? $87\frac{1}{2}\%$

23. 8 is 20% of what number? 40

24. 90 is 60% of what number? 150

25. Find 0.1% of \$5000. \$5

26. What is 98% of \$140? \$137.20

27. A theater was filled to 75% of capacity. How many of the 720 seats were filled? 540

28. The sales tax on a \$20 purchase was \$0.90. What was the rate of sales tax? $4\frac{1}{2}\%$

29. Janice scored 85% on the last test. She answered 34 questions correctly. How many questions were on the test? 40

30. In a 180-kilogram sample of ore, there was 3.2% metal. How many kilograms of metal were in the sample? 5.76

31. Suppose 6% of 8000 people polled regarding an election expressed no opinion. How many people expressed no opinion? 480

32. Bob Stapleton earns 2% of sales on all truck bodies he sells. Last week he earned \$974. What were his sales? \$48,700

33. Henri Rici paid \$5000 for a car. After one year, its value had decreased by \$1500. By what percent had the car depreciated in value? 30%

34. June Carlos earns \$125 per week in salary and 8% commission on all sales. How much must she sell in order to earn \$200 per week? \$937.50

35. A store advertised tires at 20% off. What was the original price of a tire that was marked \$13 off? \$65

36. The price of a coat was marked down \$6. The original price was \$60. What was the rate of discount? 10%

Challenge

37. Jim Hunts works as a salesman in a men's clothing store. One day, he began with \$75 in the cash register. At the end of the day he had \$1422.79. If he charged 6% sales tax on all items sold, what were his total sales for the day? \$1271.50

Using Computers

Successive Discounts

A store may give a second discount on an item that is already being sold at a discount. Perhaps the item is out of style or slightly damaged. Consider this example.

Example The regular price of a sofa is \$825. What is the final sale price if an additional 20% discount is given on this sofa which is already selling at a discount of 10%?

Regular price	\$ 825	\$742.50	*A 20% discount is the same as 80% of \$742.50.*
A 10% discount is the same as 90% of the regular price.	× 0.90	× 0.80	
	\$742.50	\$594.00	**Sale price**

The two successive discounts can be summarized as follows.

(\$825 × 0.90) × 0.80 = \$594

About Using Computers
Students may be surprised to find that a combined discount gives a lower sale price. Ask them if they can think of any reasons why this is the case. Be sure students understand that successive discounts are being referred to in exercise 6.

The following program will help you determine whether two successive discounts or one combined discount will give a lower sale price.

```
10  PRINT "ENTER ORIGINAL PRICE AND TWO DISCOUNTS
      AS DECIMALS"
20  INPUT P,X1,X2
30  LET S1 = P * (1 - X1) * (1 - X2)      Computes sale price with successive discounts
40  LET S2 = P * (1 - (X1 + X2))          Computes sale price with one combined discount
50  PRINT "TWO SUCCESSIVE DISCOUNTS", "$";S1
60  PRINT "COMBINED DISCOUNT", "$";S2
70  IF S2 < S1 THEN PRINT "ONE COMBINED DISCOUNT HAS A LOWER
      SALE PRICE."
80  IF S1 < S2 THEN PRINT "SUCCESSIVE DISCOUNTS HAVE A LOWER
      SALE PRICE"
90  IF S1 = S2 THEN PRINT "THERE IS NO DIFFERENCE"
100 END
```

Exercises

Find the sale price of each item using successive discounts and one combined discount.

1. Price, \$49.00; Discounts, 20% and 15%
2. Price, \$185; Discounts, 25% and 10%
3. Price, \$12.50; Discounts, 30% and $12\frac{1}{2}$%
4. Price, \$156.95; Discounts, $33\frac{1}{3}$% and 10%
5. What is the relationship between the sale price using successive discounts and one combined discount?
6. Modify the program to determine whether the order in which the discounts are applied affects the sale price.

Vocabulary

multiplicative inverse (79)	ratio (97)	percent (101)
reciprocal (79)	proportion (97)	percentage (101)
identity (88)	extremes (97)	base (101)
consecutive numbers (93)	means (97)	rate (101)

Chapter Summary

1. The product of two numbers that have different signs is negative. (75)
2. The product of two numbers that have the same sign is positive. (76)
3. The product of any number and -1 is its additive inverse.
   $-1(a) = -a$ and $a(-1) = -a$ (76)
4. The quotient of two numbers is positive if the numbers have the same sign, negative if the numbers have different signs. (78)
5. Multiplicative Inverse Property: For every nonzero number a, there is exactly one number $\frac{1}{a}$, such that $\frac{1}{a}(a) = a\left(\frac{1}{a}\right) = 1$. (79)
6. Division Rule: For all numbers a and b, with $b \neq 0$, $a \div b = \frac{a}{b} = a\left(\frac{1}{b}\right) = \frac{1}{b}(a)$. (79)

Answers to Using Computers

1. \$33.32; \$31.85
2. \$124.875; \$120.25
3. \$7.65625; \$7.1875
4. \$94.170047; \$88.938386
5. The sale price is lower if one combined discount is used.
6. 30 S1 = P * (1 − X2) * (1 − X1) The order does not affect the sale price.

Quiz
Lessons 3-7 to 3-9

Define a variable and use an equation to solve each of the following.

1. Find three consecutive integers whose sum is 63.
   Let x = least integer.
   $x + (x + 1) + (x + 2) = 63$
   $x = 20$
   The integers are 20, 21, and 22.
2. Seven less than twice a number is 20 more than 5 times the number. Find the number.
   Let n = the number.
   $2n - 7 = 5n + 20$
   $n = -9$
   The number is -9.

Solve for x.

3. $\frac{x}{6} = \frac{a}{3}$ $2a$
4. $\frac{4 + x}{7} = \frac{5}{9}$ $-\frac{1}{9}$

Use a proportion to solve the following problem.

5. A truck hauls 17 yd^3 of gravel in three trips. How many trips are needed for 27 yd^3?
   $\frac{17}{3} = \frac{27}{t}$; $t = 4.76$
   5 trips needed

Solve each problem.

6. What is 35% of 90? 31.5
7. Thirteen is 20% of what number? 65
8. The price of a necklace was \$65. If \$3.25 tax was added to the purchase price, what was the rate of sales tax? 5%

7. Multiplication Property of Equality: For any numbers a, b, and c, if $a = b$ then $ac = bc$. (81)
8. Division Property of Equality: For any numbers a, b, and c, with $c \neq 0$, if $a = b$ then $\frac{a}{c} = \frac{b}{c}$. (82)
9. An equation that is true for every value of the variable is called an identity. (88)
10. An equation of the form $\frac{a}{b} = \frac{c}{d}$ is called a proportion. (97)
11. In a proportion, the product of the extremes is equal to the product of the means. If $\frac{a}{b} = \frac{c}{d}$, then $ad = bc$. (97)
12. Proportions can be used to solve percent problems. (101)
13. Percent Proportion: $\frac{\text{Percentage}}{\text{Base}} = \frac{r}{100}$ (102)

Chapter Review

3-1 **Find each product.**

1. $(-11)(9)$ -99
2. $(-8)(-12)$ 96
3. $\frac{3}{5}\left(-\frac{5}{7}\right)$ $-\frac{3}{7}$

Simplify each expression.

4. $-3(7) + (-8)(-9)$ 51
5. $\frac{1}{2}(6a + 8b) - \frac{2}{3}(12a + 24b)$ $-5a - 12b$

3-2 **Simplify.**

6. $\frac{-54}{6}$ -9
7. $\frac{63b}{-7}$ $-9b$
8. $\frac{\frac{4}{5}}{-7}$ $-\frac{4}{35}$
9. $\frac{-12}{-\frac{2}{3}}$ 18
10. $\frac{33a + 66}{-11}$ $-3a - 6$

3-3 **Solve each equation.**

11. $-7r = -56$ 8
12. $23y = 1035$ 45
13. $\frac{3}{4}x = -12$ -16
14. $\frac{x}{5} = 7$ 35

3-4

15. $3x - 8 = 22$ 10
16. $-4y + 2 = 32$ $-7\frac{1}{2}$
17. $0.5n + 3 = -6$ -18
18. $-6 = 3.1t + 6.4$ -4
19. $\frac{x}{-3} + 2 = -21$ 69
20. $\frac{r - 8}{-6} = 7$ -34

3-5

21. $5a - 5 = 7a - 19$ 7
22. $-3(x + 2) = -18$ 4
23. $4(2y - 1) = -10(y - 5)$ 3
24. $11.2n + 6 = 5.2n$ -1

3-6

25. $\frac{2}{3}x + 5 = \frac{1}{2}x + 4$ -6
26. $2.9m + 1.7 = 3.5 + 2.3m$ 3
27. $\frac{3t + 1}{4} = \frac{3}{4}t - 5$ no solutions

Solve for x.

28. $x + r = q$ $q - r$
29. $\frac{x + y}{c} = d$ $cd - y$

Solve for y.

30. $5(2a + y) = 3b - y$ $\frac{3b - 10a}{6}$
31. $\frac{2y - a}{3} = \frac{y + 3b}{4}$ $\frac{4a + 9b}{5}$

3-7 32. Find two consecutive even integers whose sum is 94. 46, 48

33. Find three consecutive odd integers whose sum is 81. 25, 27, 29

34. Find four consecutive integers such that the sum of the second and twice the third is 68. 21, 22, 23, 24

35. Four times a number decreased by twice the number is 100. What is the number? 50

3-8 **Solve each proportion.**

36. $\frac{6}{15} = \frac{n}{45}$ 18 37. $\frac{4}{8} = \frac{11}{t}$ 22 38. $\frac{5}{6} = \frac{n-2}{4}$ $\frac{16}{3}$

3-9 **Change each fraction to a percent.**

39. $\frac{6}{10}$ 60% 40. $\frac{7}{8}$ $87\frac{1}{2}\%$ 41. $\frac{17}{100}$ 17%

Solve each problem.

42. 9 is what percent of 15? 60%

43. What number is 60% of 80? 48

44. Twenty-one is 35% of what number? 60

45. 84 is what percent of 96? $87\frac{1}{2}\%$

Chapter Test

Simplify each expression.

1. $\frac{8(-3)}{2}$ -12

2. $(-5)(-2)(-2) - (-6)(-3)$ -38

3. $\frac{2}{3}\left(\frac{1}{2}\right) - \left(-\frac{3}{2}\right)\left(-\frac{2}{3}\right)$ $-\frac{2}{3}$

4. $\frac{3}{4}(8x + 12y) - \frac{5}{7}(21x - 35y)$ $-9x + 34y$

5. $\frac{70a - 42b}{-14}$ $-5a + 3b$

6. $\frac{\frac{11}{5}}{-6}$ $-\frac{11}{30}$

Solve each equation.

7. $-3y = 63$ -21

8. $\frac{3}{4}y = -27$ -36

9. $3x + 1 = 16$ 5

10. $5n + 7 = 28 + 2n$ 7

11. $5(8 - 2n) = 4n - 2$ 3

12. $3(n + 5) - 6 = n + 5$ -2

13. $7x + 9 = 3(x + 3)$ 0

14. $-2(3n-5)+3n=2-n$ 4

15. $\frac{3}{4}n - \frac{2}{3}n = 5$ 60

16. $\frac{t-7}{4} = 11$ 51

17. $\frac{2r-3}{-7} = 5$ -16

18. $8r - \frac{r}{3} = 46$ 6

Solve for x.

19. $\frac{x+y}{b} = c$ $bc - y$

20. $yx + a = c$ $\frac{c-a}{y}$

Solve each proportion.

21. $\frac{7}{8} = \frac{5}{t}$ $\frac{40}{7}$

22. $\frac{9}{11} = \frac{x-3}{x+5}$ 39

23. $\frac{y+2}{8} = \frac{7}{5}$ $\frac{46}{5}$

24. $\frac{2}{5} = \frac{x-3}{-2}$ $\frac{11}{5}$

Solve each problem.

25. Find 6.5% of 80. 5.2

26. 42 is what percent of 126? $33\frac{1}{3}\%$

27. 84 is 60% of what number? 140

28. Find three consecutive odd integers whose sum is 93. 29, 31, 33

29. Twice a number increased by 12 is 31 less than three times the number. Find the number. 43

30. Find two consecutive integers such that twice the lesser integer, increased by the greater integer, is 50. no solutions

About the Application
Other *Applications in Probability* features can be found on pages 172 and 242. *Applications in Statistics* features can be found on pages 302 and 368. Each feature is a two-page lesson that can be used at any time during the year. If you prefer, the five probability and statistics features could be taught as a unit.

It is important to emphasize that each possible outcome from fair or unbiased coins, dice, and spinners are all equally likely to occur.

Applications in Probability

Simple Events

When a fair coin is tossed, only 2 **outcomes** are possible. The coin will land either heads or tails. Each outcome is *equally likely* to happen.

The probability that a tossed coin will land heads is $\frac{1}{2}$. This means that there is one way for a successful outcome (a head) to occur out of two possible outcomes (a head or a tail). The results of similar experiments lead to the following definition.

Probability of a Successful Event

If all of the outcomes of an event are equally likely, the probability of that event occurring is

$$P(\text{event}) = \frac{\textbf{number of favorable outcomes}}{\textbf{total number of possible outcomes}}.$$

When one die is rolled, there are 6 possible outcomes. Thus, the probability of rolling a 3 is $\frac{1}{6}$. The probability of rolling a number between 1 and 6, inclusive, is $\frac{6}{6}$ or 1. The probability of rolling a 7 is $\frac{0}{6}$ or 0.

Probabilities of 1 and 0

The probability of an impossible event is 0.
The probability of an event that is certain to occur is 1.

Example 1: **A five-person committee is comprised of 3 women and 2 men: Annette, Barbara, Claudia, David, and Eddie. A chairperson is selected at random.** *Selected at random means that it is equally likely for each person to be selected.*

a. What is the probability that the chairperson is Annette?

$P(\text{Annette}) = \frac{1}{5}$ $\quad \frac{\text{1 favorable outcome}}{\text{5 possible outcomes}}$

The probability that the chairperson is Annette is $\frac{1}{5}$.

b. What is the probability that the chairperson is a man?

$P(\text{man}) = \frac{2}{5}$ $\quad \frac{\text{2 men}}{\text{5 possible outcomes}}$

The probability that the chairperson is a man is $\frac{2}{5}$.

Example 2: **A card is chosen at random from a standard deck of 52 playing cards. What is the probability of choosing a king or queen?**

There are 4 kings and 4 queens in a standard deck of 52 cards.

$P(\text{king or queen}) = \frac{8}{52} = \frac{2}{13}$

The probability that the card is a king or queen is $\frac{2}{13}$.

Exercises

The spinner shown at the right is equally likely to stop at any of the six numbers. It is spun once. Find the probability for each of the following.

1. 1 $\frac{1}{6}$
2. 3 $\frac{1}{6}$
3. an even number $\frac{1}{2}$
4. an odd number $\frac{1}{2}$
5. a number less than 5 $\frac{2}{3}$
6. a number greater than 5 $\frac{1}{6}$
7. 3 or 4 $\frac{1}{3}$
8. 1, 2, or 3 $\frac{1}{2}$

The spinner shown at the right is divided into four sectors. It is spun once. Find the probability for each of the following.

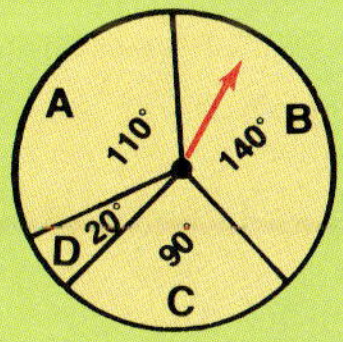

9. $P(A)$ $\frac{11}{36}$
10. $P(B)$ $\frac{7}{18}$
11. $P(C)$ $\frac{1}{4}$
12. $P(D)$ $\frac{1}{18}$
13. $P(A \text{ or } B)$ $\frac{25}{36}$
14. $P(\text{not } C)$ $\frac{3}{4}$

A card is chosen at random from a standard deck of 52 playing cards. Find the probability of choosing each of the following.

15. the king of hearts $\frac{1}{52}$
16. a red ace $\frac{1}{26}$
17. a queen $\frac{1}{13}$
18. the jack of spades $\frac{1}{52}$
19. a club $\frac{1}{4}$
20. not a 7 $\frac{12}{13}$

21. Copy the chart at the right. Flip a coin 50 times, recording your results in groups of 10 as indicated in the chart. Find the ratio of heads to the number of flips for each group of 10 and record this ratio. After 50 flips, find the total number of heads and tails. Determine the ratio of heads to flips for the total and record the ratio. Compare your results to your classmates' results. What ratio of heads to flips is expected? Answers will vary.

Number of Flips	Heads	Tails	$\frac{\text{Heads}}{\text{Flips}}$
1-10			
11-20			
21-30			
31-40			
41-50			
Totals			

About the Photo

Point out to students that Tim must score *more than* a certain number of points in order to win the competition. The words *more than* indicate an inequality, specifically, one involving the greater than symbol (>). You may want to have students answer the question asked after they have completed lesson 4-8.

Teacher Resource Book Materials for Chapter 4 See pages 123-138.

CHAPTER 4

Inequalities

Tim has scores of 9.7, 9.9, 9.4, and 9.2 in the Olympic tryouts. He has one more trial. The leading opponent has completed the competition with an average score of 9.52. What must Tim score on his final trial if he is to win the competition? In this chapter, you will learn how to solve problems such as this using inequalities.

Objective 4-1A: Learn the meaning of inequality symbols.
Objective 4-1B: Graph inequalities on the number line.

4-1 Inequalities and their Graphs

You know that 3 is less than 8. This can be shown in two ways.

$3 < 8$ means 3 *is less than* 8.
$8 > 3$ means 8 *is greater than* 3.

Note that the < and > point to the numeral for the lesser number.

Any mathematical sentence containing < or > is called an **inequality**.

Think about $3 < 8$. This sentence is true. Is $3 > 8$ true? Is $3 = 8$ true? Note that only one of the three sentences is true. This can be summarized by the following property. ❶

For any two numbers *a* and *b*, exactly one of the following sentences is true.

$$a < b \qquad a = b \qquad a > b$$

Comparison Property ❷

Examples

1 Is $4 < 5\frac{1}{2}$ true or false?

$4 < 5\frac{1}{2}$ means 4 *is less than* $5\frac{1}{2}$. This sentence is true.

The inequalities $4 < 5\frac{1}{2}$ and $5\frac{1}{2} > 4$ have the same meaning.

2 Is $9 > 4 + 3 + 2 + 1$ true or false?

$9 \overset{?}{>} 4 + 3 + 2 + 1$ *Is 9 greater than the sum of 4, 3, 2, and 1?*

$9 \overset{?}{>} 10$ *Simplify the right side of the inequality.*

Since 9 is not greater than 10, the sentence is false.

Recall that there are three properties of equality. Equality is reflexive, symmetric, and transitive. Do inequalities have these same properties?

Decide whether the sentences below are true or false.

Reflexive:	$6 > 6$	*False*
Symmetric:	If $4 > 3$, then $3 > 4$.	*False*
Transitive:	If $6 > 2$ and $2 > 1$, then $6 > 1$.	*True*

Check whether the transitive property is true for other examples.

4-1

Lesson Notes, p. T31

Discussion Questions

Suppose $a < b$ and $b < c$. State whether each sentence is true or false.

1. $a < c$ true
2. $a \leq c$ true
3. $a > c$ false
4. $a \geq c$ false
5. $a = c$ false

Chalkboard Exercises

State whether each sentence is true or false

1a. $9 > 8\frac{7}{9}$ true

1b. $-6 > -5\frac{7}{8}$ false

2a. $7 < 2 + (-7) + 9$ false
2b. $-5 < 6 + 3 - 10$ true

❶ Students may benefit from making concrete comparisons involving height, weight, length, and so on to prove that exactly one of the three comparison sentences is true.

❷ The comparison property is sometimes called the trichotomy property.

Computer Notes

The inequality symbols for BASIC are given on page 528. In BASIC, inequalities are often used with IF-THEN statements. For instructional material on IF-THEN statements, see pages 534-536 of the appendix.

Chalkboard Exercises

State whether each sentence is true or false.

3. If $-6 > -7$ and $-7 > -12$ then $-6 > -12$. true

4a. $3.7 \leq 11.5 - 7.9$ false

4b. $5.6 \leq 10.9 - 5.3$ true

The relation $>$ is not reflexive or symmetric. However, it is transitive. Explore some similar examples to verify that $<$ is transitive.

Transitive Property of Order

For all numbers a, b, and c,
1. If $a < b$ and $b < c$, then $a < c$.
2. If $a > b$ and $b > c$, then $a > c$.

Example 3

State whether the following sentence is true or false.

If $-3 < 1$ and $1 < 4$, then $-3 < 4$.

The sentence is true by the transitive property of order.

The symbols $\not>$, $\leq$, and $\geq$ can also be used when comparing numbers. The following chart shows several inequality symbols and their meanings.

Symbol	Meaning
$<$	is less than
$>$	is greater than
$\not>$	is not equal to
$\leq$	is less than or equal to
$\geq$	is greater than or equal to

Example 4

Is $5.1 \leq 7 + 0.9$ true or false?

$5.1 \overset{?}{\leq} 7 + 0.9$ *Is 5.1 less than or equal to 7 + 0.9?*

$5.1 \leq 7.9$

Since 5.1 is less than 7.9, the sentence is true.

Consider the graphs of -3, -1, $2\frac{1}{2}$, and 4.5, shown on the number line below.

The following statements can be made about the numbers and their graphs.

The graph of -3 is to the left of the graph of -1.	$-3 < -1$ ❸
The graph of -1 is to the left of the graph of $2\frac{1}{2}$.	$-1 < 2\frac{1}{2}$
The graph of $2\frac{1}{2}$ is to the right of the graph of -3.	$2\frac{1}{2} > -3$
The graph of 4.5 is to the right of the graph of $2\frac{1}{2}$.	$4.5 > 2\frac{1}{2}$

If a and b represent any numbers and the graph of a is to the left of the graph of b, then $a < b$. If the graph of a is to the right of the graph of b, then $a > b$.

Comparing Numbers on the Number Line

Recall that an equation such as $x + 6 = 7$ is an open sentence. An inequality can also be an open sentence. The set of all replacements for the variable which make the inequality true is called the **solution set** of the inequality.

What replacements for x make $x < 5$ true? All numbers less than 5 make the inequality true. This can be shown by the solution set {all numbers less than 5}, read *the set of all numbers less than 5*. Study the graph of this solution set in Example 5 below.

Examples

5 **Graph the solution set of $x < 5$.**

The graph extends indefinitely in the negative direction.

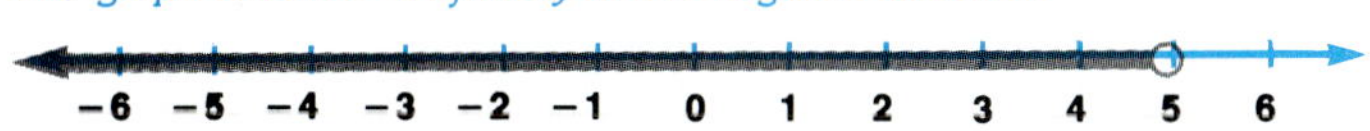

The circle shows that 5 is *not* included in the solution set.
The heavy arrow shows that all numbers to the left of 5 *are* included.

6 **Graph the solution set of $y \neq -1.5$.**

The solution set is {all numbers except -1.5}.

The circle indicates that -1.5 is not included in the graph.

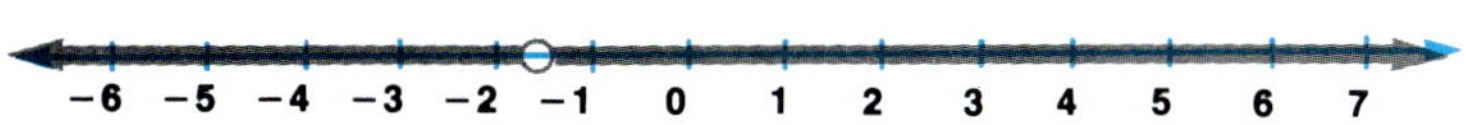

7 **Graph the solution set of $n \geq -\frac{3}{2}$.**

The solution set is $\left\{-\frac{3}{2}\right.$ and all numbers greater than $\left.-\frac{3}{2}\right\}$.

The dot at $-\frac{3}{2}$ indicates that $-\frac{3}{2}$ is included in the graph.

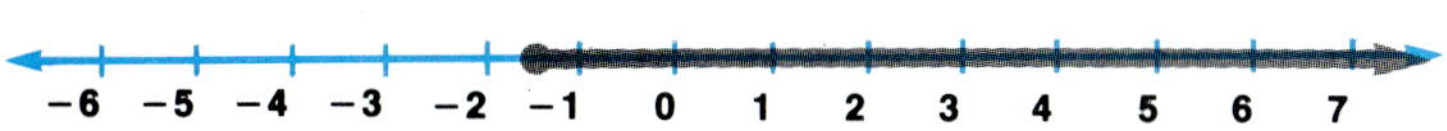

Chalkboard Exercises

Graph the solution set of each inequality on a number line.

5a. $y > -6$

5b. $x < 4$

6. $y \neq -3$

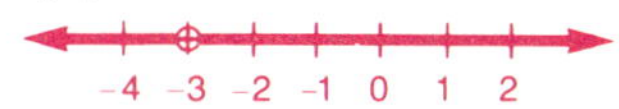

7a. $y \leq -\frac{1}{2}$

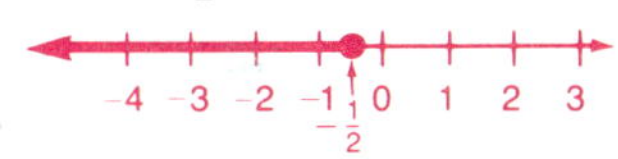

7b. $x \geq 5\frac{2}{3}$

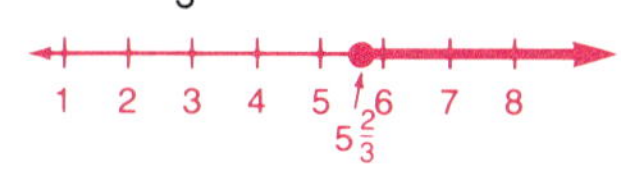

❸ Stress the importance of which direction a number is graphed in relation to another number. For example, the inequality $-1 < -2$ is false, but $-1 > -2$ is true. On the number line, -1 is to the right of -2.

Suggested Daily Assignments

Enriched
Written: 2–44 evens

Average
Exp: 2–20 evens, 21–26
Written: 1–10, 12–44 evens

Fundamental
Exp: 1–26
Written: 2–44 evens

Answers to Exploratory

15. {all numbers greater than 3}
16. {all numbers less than −5}
17. {all numbers less than 6}
18. {all numbers greater than −1}
19. {−4 and all numbers greater than −4}
20. {all numbers except 0}

Answers to Written

33. 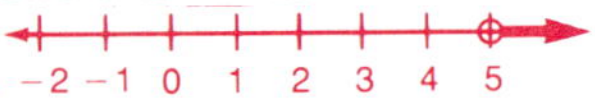

34. 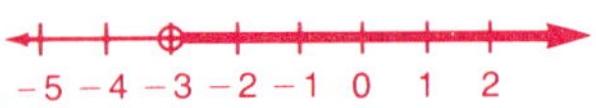

35. 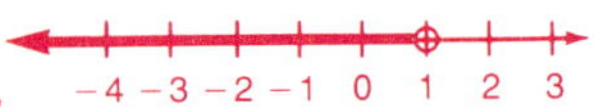

36. 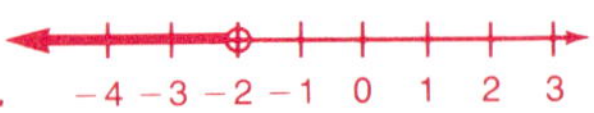

37. 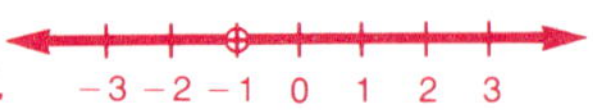

38. 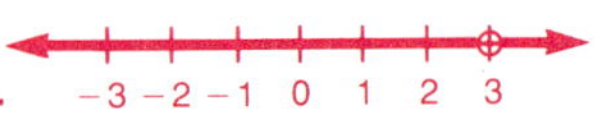

39. 

40. 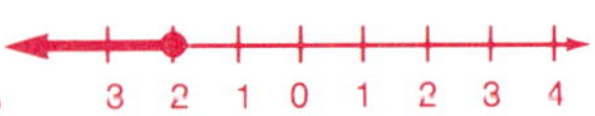

41. 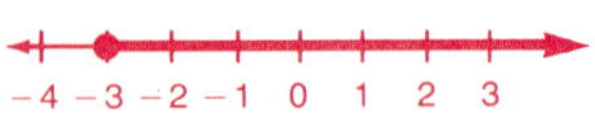

42. 

43. 

44. 

Exploratory Exercises

State whether each sentence is true or false.

1. $7 < 4$ false
2. $-3 < 3$ true
3. $2 > 0$ true
4. $-9 > -4$ false
5. $4\frac{1}{2} \neq 6\frac{1}{2}$ true
6. $-\frac{8}{3} \leq \frac{8}{3}$ true
7. $\frac{1}{2} \geq \frac{1}{2}$ true
8. $-5\frac{2}{3} \neq -6\frac{2}{3}$ true
9. $-4 < 2 - 8$ false
10. $-8 + 4 < -5$ false
11. $-0.5 - 3 \geq -4$ true
12. $-0.6 \leq -1 + 0.4$ true
13. If $4 > 3$ and $3 < 5$, then $4 > 5$. false
14. If $1 < 3$, then $1 > 3$. false

State the solution set of each inequality.

15. $n > 3$
16. $y < -5$
17. $z < 6$
18. $y > -1$
19. $y \geq -4$
20. $m \neq 0$

State whether each number is included on the graph below.

21. -1 yes
22. 4 no
23. 3.9 no
24. -4.5 yes
25. $-1\frac{1}{2}$ yes
26. $2\frac{1}{2}$ no

Written Exercises

State an inequality for each graph.

1. 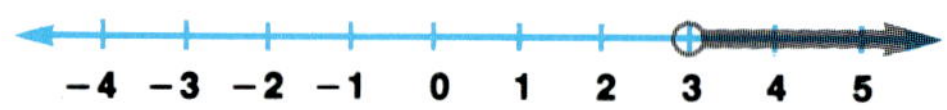 $x > 3$

2. 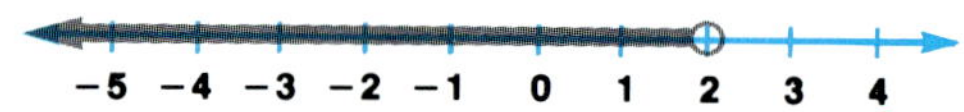 $x < 2$

3. 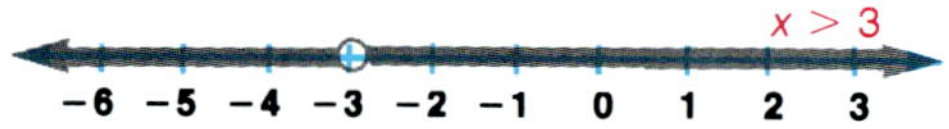 $x \neq -3$

4. 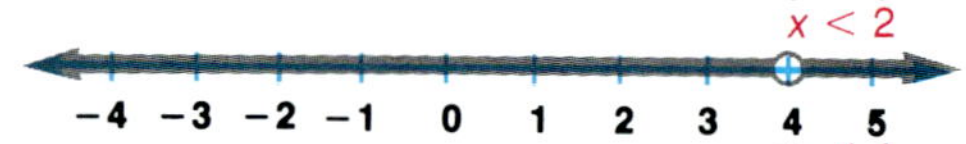 $x \neq 4$

5. 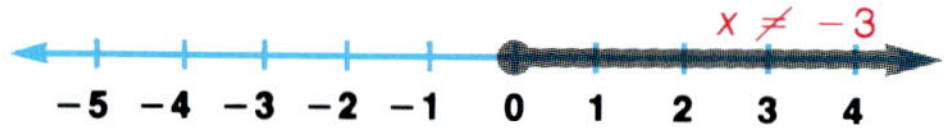 $x \geq 0$

6. 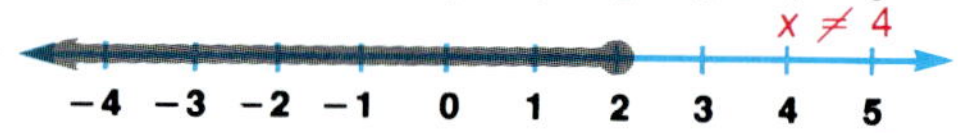 $x \leq 2$

7.  $x < -3$

8. 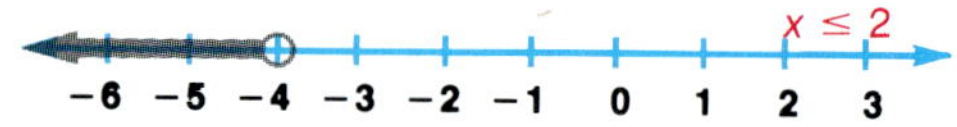 $x < -4$

9. 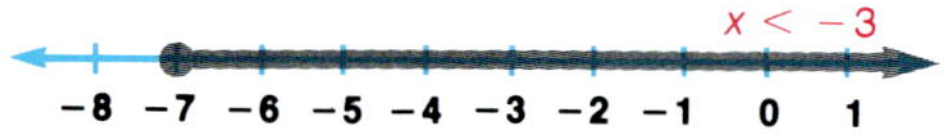 $x \geq -7$

10. 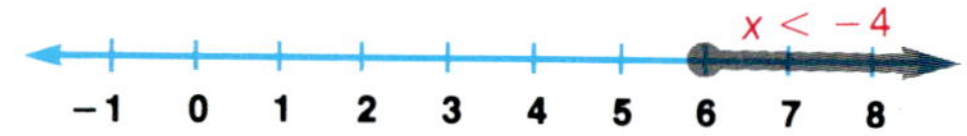 $x \geq 6$

Replace each __?__ with <, >, or = to make each sentence true.

11. -5 __?__ 7 <
12. -2 __?__ -3 >
13. 6 __?__ $4 + 2$ =
14. $-7 - 2$ __?__ -9 =
15. -5 __?__ $0 - 3$ <
16. 3 __?__ $\frac{15}{3}$ <
17. 10 __?__ $\frac{27}{3}$ >
18. 12 __?__ $-15-(-27)$ =
19. 8 __?__ $4.1 + 3.9$ =
20. 5 __?__ $8.4 - 1.5$ <
21. -7 __?__ $-\frac{3.6}{0.6}$ <
22. $\frac{5.4}{18}$ __?__ $-4 + 1$ >
23. $(-7.5)(0.4)$ __?__ 3.1 <
24. $(6.3)(-0.4)$ __?__ 25.2 <
25. $\frac{4}{3}(6)$ __?__ $4\left(\frac{3}{2}\right)$ >
26. $8\left(\frac{3}{4}\right)$ __?__ $6\left(\frac{2}{3}\right)$ >
27. $3 + 1.7$ __?__ $\frac{9.6}{2}$ <
28. $\frac{2.7}{3}$ __?__ $2(4.5)$ <
29. If $-6 < -2$ and $-2 < 5$, then -6 __?__ 5. <
30. If $-6 < -2$ and $-2 < 5$, then 5 __?__ -6. >
31. If $2\frac{2}{3} > -1\frac{1}{2}$ and $-1\frac{1}{2} > -3$, then -3 __?__ $2\frac{2}{3}$. <
32. If $4.5 > 2.1$ and $2.1 > -3.6$, then 4.5 __?__ -3.6. >

Graph the solution set of each inequality on a number line.

33. $n > 5$
34. $x > -3$
35. $m < 1$
36. $y < -2$
37. $x \neq -1$
38. $y \neq 3$
39. $y \leq 6$
40. $n \leq -2$
41. $a \geq -3$
42. $x \geq -2$
43. $x < -10$
44. $m < 3$

For more exercises like 11-28, see exercises 1-16 on page 555.

Reading Algebra — Compound Statements

In algebra a compound statement consists of two simple statements that are connected by the words *and* or *or*. You must read compound statements very carefully. Understanding the meaning of the words *and* and *or* will help you decide if a compound statement is true or false. Study the statements below.

The star is red *and* the square is blue.

For this compound statement to be true, both parts must be true. Since the star is red, the first part is true. However, the square is not blue, so the second part is false. Therefore, the compound statement is false.

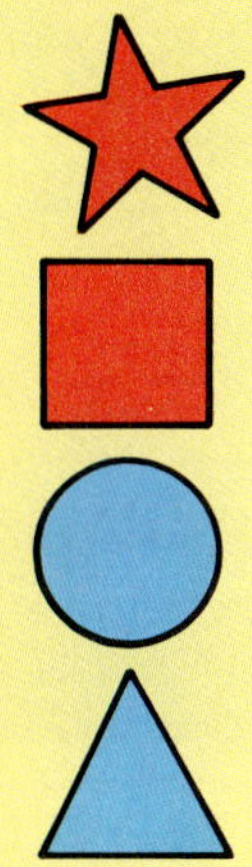

The star is red *or* the square is blue.

For this compound statement to be true, at least one part needs to be true. You know that the star is red, so the compound statement is true.

The compound statements described above can be defined as follows.

A compound statement formed by joining two statements with the word *and* is called a conjunction. For a conjunction to be true, both statements must be true. — ***Conjunction***

A compound statement formed by joining two statements with the word *or* is called a disjunction. For a disjunction to be true, at least one statement must be true. — ***Disjunction***

Exercises

Decide whether each statement is a disjunction or a conjunction. Write true or false.

1. The circle is blue and the triangle is yellow.
2. The triangle is yellow or the circle is blue.
3. The square is red and the triangle is blue.
4. The circle is orange and the star is not red.
5. The star is blue or the triangle is red.
6. The triangle is blue or the circle is blue.
7. $3 < 4$ or $7 < 6$
8. $-4 > 0$ and $2 < 5$
9. $6 > 0$ and $-6 < 0$
10. $2 = 0$ or $-2 > -3$
11. $7 \neq 7$ and $6 > 4$
12. $0 > -4$ or $2 < -2$

About Reading Algebra
You may want to present this feature prior to discussing Lesson 4-6 as it is intended to prepare students for working with conjunctions and disjunctions. Since this presentation gives a more detailed explanation of the use of "and" and "or" than is given in Lesson 4-6, students will find this feature very helpful.

Answers to Exercises
1. conjunction, false
2. disjunction, true
3. conjunction, true
4. conjunction, false
5. disjunction, false
6. disjunction, true
7. disjunction, true
8. conjunction, false
9. conjunction, true
10. disjunction, true
11. conjunction, false
12. disjunction, true

4-2

Lesson Notes, p. T32

Readiness Activity

Here are two true inequalities.

$-1 \leq 5$

$-6 > -10$

Is each inequality still true if you:

(a) add 1 to both sides? yes

(b) add -1 to both sides? yes

(c) subtract 1 from both sides? yes

(d) subtract -1 from both sides? yes

❶ Point out that c can be negative.

Discussion Questions

1. Name some values for a and b that make all three of the following inequalities true. $a < 0$, $b < 0$, $a - b < 0$ Answers will vary.
2. What must the relationship between a and b be for these inequalities to be true? $a < b$

Objective 4-2: Solve inequalities involving addition or subtraction.

4-2 Solving Inequalities Using Addition and Subtraction

Kristina had \$267.23 in her savings account and Mike had \$134.53 in his account. They each received \$10 from their grandmother and deposited it in their accounts. Whose account has more money?

Kristina		Mike
\$267.23	>	\$134.53
\$267.23 + \$10.00		\$134.53 + \$10.00
\$277.23	>	\$144.53

Kristina had more money at the beginning and at the end. Notice that adding the same number to both sides of the inequality did not change the truth of the inequality.

Suppose Kristina and Mike each withdrew, or subtracted, \$25 from their accounts. Kristina would still have more money than Mike.

You know that the sentence $-1 < 3$ is true. Is $-1 + 1 < 3 + 1$ a true sentence? Is $-1 - 2 < 3 - 2$ a true sentence? These examples suggest the following rules that may be used in solving inequalities.

For all numbers a, b, and c,
1. If $a > b$, then $a + c > b + c$ and $a - c > b - c$.
2. If $a < b$, then $a + c < b + c$ and $a - c < b - c$.

Addition and Subtraction Properties for Inequalities ❶

Examples

1 **Solve: $m + 7 > 12$**

$$m + 7 > 12$$
$$m + 7 - 7 > 12 - 7 \quad \text{Subtract 7 from both sides.}$$
$$m > 5$$

The solution set is {all numbers greater than 5}.

Check: First check whether 5 is the correct *boundary*. Is 5 the value of m in $m + 7 = 12$?

$$m + 7 \stackrel{?}{=} 12$$
$$5 + 7 \stackrel{?}{=} 12$$
$$12 = 12$$

Next substitute one or two numbers greater than 5, such as 6 and 10, into the inequality. For numbers greater than 5, the inequality should be true. ❷

$$m + 7 > 12 \qquad m + 7 > 12$$
$$6 + 7 \stackrel{?}{>} 12 \quad \text{Try 6} \qquad 10 + 7 \stackrel{?}{>} 12 \quad \text{Try 10}$$
$$13 > 12 \ ✓ \qquad 17 > 12 \ ✓$$

2 **Solve: $9y - 14 < 8y + 3$**

$$9y - 14 < 8y + 3$$
$$9y - 8y - 14 < 8y - 8y + 3 \quad \text{Subtract } 8y \text{ from both sides.}$$
$$y - 14 < 3$$
$$y - 14 + 14 < 3 + 14 \quad \text{Add 14 to both sides.}$$
$$y < 17$$

The solution set is {all numbers less than 17}. *Try checking 10 and 16.*

Exploratory Exercises

State the number you would add to both sides to solve each inequality.

1. $r + 7 < 21$ -7 **2.** $y + 2 < -16$ -2 **3.** $m - 13 \geq 41$ 13 **4.** $y - 18 \geq -3$ 18
5. $4y < 3y + 12$ $-3y$ **6.** $7x < 6x - 12$ $-6x$ **7.** $6z > 5z - 21$ $-5z$ **8.** $7n > 6n - 4$ $-6n$
9. $12 > m - (-3)$ -3 **10.** $16 \leq k - (-9)$ -9 **11.** $14 < b - 7$ 7 **12.** $9 > p - 21$ 21

Written Exercises

Solve each inequality.

1. $a + 2 < 10$ $a < 8$ **2.** $x + 3 < -17$ $x < -20$ **3.** $r - 9 \geq 23$ $r \geq 32$ **4.** $y - 15 \geq -2$ $y \geq 13$
5. $5b < 4b + 8$ $b < 8$ **6.** $6y < 5y - 11$ $y < -11$ **7.** $3m > 2m - 19$ **8.** $9z > 8z - 8$
9. $y + 2 \leq -6$ $y \leq -8$ **10.** $x + 7 \leq 10$ $x \leq 3$ **11.** $n - 3 > 8$ $n > 11$ **12.** $x - 5 > 3$
13. $6 + n > 40$ $n > 34$ **14.** $3 + y \geq -4$ $y \geq -7$ **15.** $-7 + n \leq -16$ **16.** $-8 + y < 10$
17. $n - (-5) < -6$ $n < -11$ **18.** $y - (-4) > -7$ $y > -11$ **19.** $-12 < s - (-3)$ $-15 < s$ **20.** $-21 > r - (-8)$ $-29 > r$

Chalkboard Exercises

Solve each inequality.

1a. $y + 3 > 11$
$y > 8$

1b. $t - 9 < 6$
$t < 15$

1c. $3t + 11 \leq 4$
$t \leq -2\frac{1}{3}$

2a. $2y - 10 < y + 6$
$y < 16$

2b. $5y - \frac{1}{3} < 6y + 4\frac{2}{3}$
$-5 < y$

❷ When checking the solution, you could also show students that 5 and other numbers less than 5 do *not* check. For example, the inequality $5 + 7 > 12$ is false.

Suggested Daily Assignments

Enriched
Written: 2–44 evens
Mini-Review: 1–5

Average
Exp: 2–12 evens
Written: 1–43 odds
Mini-Review: 1–5

Fundamental
Exp: 1–12
Written: 1–24
p. 114, 1–37 odds
Mini-Review: 1–5

Answers to Written

7. $m > -19$
8. $z > -8$
12. $x > 8$
15. $n \leq -9$
16. $y < 18$

Answers to Written
21. $t \le 34.6$
22. $y < 62.9$
23. $m > 6.38$
24. $p \ge 14.16$

Solve each inequality.

21. $t + 13.1 \le 47.7$
22. $y + 18.7 < 81.6$
23. $m - 1.73 > 4.65$
24. $p - 7.35 \ge 6.81$
25. $\frac{3}{4} + r > \frac{7}{4}$ $r > 1$
26. $r + \frac{5}{3} \le \frac{2}{3}$ $r \le -1$
27. $x - \frac{1}{3} < \frac{1}{2}$ $x < \frac{5}{6}$
28. $x + \left(-\frac{1}{4}\right) \ge -\frac{7}{4}$ $x \ge -\frac{3}{2}$
29. $9f - 6 > 10f$ $-6 > f$
30. $8x - 31 > 7x$ $31 < x$
31. $6n > 5n + 6$ $n > 6$
32. $3m < 2m - 7$ $m < -7$
33. $14 + 7x > 8x$ $14 > x$
34. $-3 + 14z < 15z$ $-3 < z$
35. $6a + 4 \ge 5a$ $-4 \le a$
36. $9y - 6 > 10y$ $-6 > y$
37. $2r - 2.1 < -8.7 + r$ $r < -6.6$
38. $5s - 6.5 < -13.4 + 4s$ $s < -6.9$
39. $12.37 + 4z > 181.3 + 3z$ $z > 168.93$
40. $16.83 + p < 14.65 + 2p$ $p > 2.18$
41. $y + \frac{3}{8} \le \frac{7}{24} + 2y$ $\frac{1}{12} \le y$
42. $7x + \frac{3}{16} < -\frac{11}{16} + 6x$ $x < -\frac{7}{8}$
43. $2r - \frac{3}{4} < r + \frac{5}{3}$ $r < \frac{29}{12}$
44. $y + \frac{16}{9} > 2y + \frac{5}{6}$ $\frac{17}{18} > y$

mini-review

Use the formula $d = rt$ to solve the following problem.

1. Kim can drive from her home to Denver in $6\frac{1}{2}$ hours. If her average speed is 52 miles per hour, how far does she live from Denver? 338 miles

Solve.

2. In Anchorage, Alaska, the average high temperature in January is −7°C. The average high temperature in July is 18°C. Find the difference between the two temperatures. 25°C
3. Find four consecutive integers whose sum is 110. 26, 27, 28, 29
4. 36 is 150% of what number? 24

Solve for *x*.

5. $ax + m = c - 2m$ $\frac{c - 3m}{a}$

Excursions in Algebra — Set-Builder Notation

You have written the solutions of an inequality as a set. For example, the solution set of $16 + x > 19$ is {all numbers greater than 3}. Another way of writing such a solution set is called **set-builder notation.**

Example: **Solve $x + 4 < 16$. Write the solution set in set-builder notation.**

$$x + 4 < 16$$
$$x + 4 - 4 < 16 - 4 \quad \textit{Subtract 4 from both sides.}$$
$$x < 12$$

The solution set in set-builder notation is $\{x \mid x < 12\}$, read *the set of all numbers x such that x is less than 12.*

Exercises

Solve each inequality. Write the solution set in set-builder notation.

1. $x + 5 < 7$ $\{x \mid x < 2\}$
2. $6 + n \ge 9$ $\{n \mid n \ge 3\}$
3. $4x \le 5x - 4$ $\{x \mid x \ge 4\}$
4. $-2 + m > 2m + 5$ $\{m \mid m < -7\}$
5. $9x + 4 \le 13x - 7$ $\left\{x \mid x \ge \frac{11}{4}\right\}$
6. $5(y - 3) \ge 15$ $\{y \mid y \ge 6\}$

About the Excursion
This feature introduces students to a notation that is used later in the textbook. Point out to students that the solution set mentioned in the first paragraph can be written $\{x \mid x > 3\}$. As an extension of this excursion, have students write some of the solution sets for the Written Exercises in set-builder notation.

Objective 4-3: Solve inequalities involving multiplication or division.

4-3 Solving Inequalities Using Multiplication and Division

You know that the sentence $-4 < 6$ is true. Suppose both sides of the inequality are multiplied by the same positive number. Study the following examples. Are the resulting sentences true?

$$-4 < 6 \qquad\qquad -4 < 6$$
$$-4(3) \stackrel{?}{<} 6(3) \qquad\qquad -4\left(\tfrac{1}{2}\right) \stackrel{?}{<} 6\left(\tfrac{1}{2}\right)$$
$$-12 \stackrel{?}{<} 18 \quad \textit{True} \qquad\qquad -2 \stackrel{?}{<} 3 \quad \textit{True}$$

Notice that dividing both sides by 2 would have given the same result as multiplying by $\frac{1}{2}$.

The inequalities $-12 < 18$ and $-2 < 3$ are true. Thus, if both sides of a true inequality are multiplied by the same positive number, the result is also true.

What happens if both sides of an inequality are multiplied by the same negative number?

$$-4 < 6 \qquad\qquad -4 < 6$$
$$-4(-3) \stackrel{?}{<} 6(-3) \qquad\qquad -4\left(-\tfrac{1}{2}\right) \stackrel{?}{<} 6\left(-\tfrac{1}{2}\right)$$
$$12 \stackrel{?}{<} -18 \quad \textit{False} \qquad\qquad 2 \stackrel{?}{<} -3 \quad \textit{False}$$

The inequality $12 < -18$ is false, but $12 > -18$ is true. Also, $2 < -3$ is false, but $2 > -3$ is true. Thus, when both sides of an inequality are multiplied by the same negative number, the direction of the inequality must be reversed.

For all numbers a, b, and c,
1. If c is positive and $a < b$, then $ac < bc$.
If c is positive and $a > b$, then $ac > bc$.
2. If c is negative and $a < b$, then $ac > bc$.
If c is negative and $a > b$, then $ac < bc$.

Multiplication Property for Inequalities ❶

You can use this property to solve inequalities.

Example

1 **Solve:** $\frac{k}{4} > 13$

$$\frac{k}{4} > 13$$
$$4\left(\frac{k}{4}\right) > 4(13) \qquad \textit{Multiply both sides by 4.} ❷$$
$$k > 52$$

The solution set is {all numbers greater than 52}. ❸

4-3

Lesson Notes, p. T32
Enrichment, p. T32

Readiness Activity
Here are two true inequalities.

$16 > -2$
$-8 \leq -4$

Is each inequality still true if you:

(a) multiply both sides by 2? yes
(b) multiply both sides by -2? no
(c) divide both sides by 2? yes
(d) divide both sides by -2? no

Discussion Question
If $-a < b$, can a and b both be negative? no

Chalkboard Exercises
Solve each inequality.

1a. $\frac{x}{3} < 10$ $x < 30$
1b. $\frac{y}{4} > 11$ $y > 44$
1c. $10 < \frac{x}{5}$ $50 < x$ or $x > 50$

❶ Additional numerical examples may be used to reinforce the multiplication property for inequalities.

❷ Remind students that $\frac{k}{4} = \frac{1}{4}k$, and that $\frac{4}{1}$ or 4 is the reciprocal of $\frac{1}{4}$.

❸ Suggest to students that they check several numbers in the solution set to see if true statements result.

Chalkboard Exercises
Solve each inequality.

2a. $\frac{a}{-4} < 6$ $a > -24$

2b. $\frac{m}{-5} > 8$ $m < -40$

3a. $\frac{2}{3}x \geq 6$ $x \geq 9$

3b. $\frac{-3}{4}y \leq 9$ $y \geq -12$

4a. $-8p \geq -96$ $p \leq 12$

4b. $-7v \leq -105$ $v \geq 15$

4 Reinforce the concept of division of inequalities by showing more numerical examples.

$6 < 24$ $\quad 24 > 6$

$\frac{6}{2} < \frac{24}{2}$ $\quad \frac{24}{-2} < \frac{6}{-2}$

$3 < 12$ $\quad -12 < -3$

Examples

2 **Solve:** $\frac{a}{-7} > -2$ *Another way to write* $\frac{a}{-7}$ *is* $-\frac{a}{7}$.

$\frac{a}{-7} > -2$

$-7\left(\frac{a}{-7}\right) < -7(-2)$ *Multiply both sides by −7 and reverse the direction of the inequality.*

$a < 14$

The solution set is {all numbers less than 14}.

3 **Solve:** $\frac{4}{3}x \leq -12$ *Another way to write* $\frac{4}{3}x$ *is* $\frac{4x}{3}$.

$\frac{4}{3}x \leq -12$

$\frac{3}{4}\left(\frac{4}{3}x\right) \leq \frac{3}{4}(-12)$ *The reciprocal of* $\frac{4}{3}$ *is* $\frac{3}{4}$. *Multiply both sides by* $\frac{3}{4}$.

$x \leq -9$

The solution set is {all numbers less than or equal to −9}.

Recall that $\frac{a}{b}$ (or $a \div b$) is equivalent to $a\left(\frac{1}{b}\right)$ for all numbers when b is *not* zero. Thus, the multiplication property for inequalities can also apply to division. When solving inequalities, you can multiply (or divide) both sides by the same positive number. You can also multiply (or divide) both sides by the same negative number if you reverse the direction of the inequality.

Division Property for Inequalities **4**

For all numbers a, b, and c,

1. If c is positive and $a < b$, then $\frac{a}{c} < \frac{b}{c}$.

If c is positive and $a > b$, then $\frac{a}{c} > \frac{b}{c}$.

2. If c is negative and $a < b$, then $\frac{a}{c} > \frac{b}{c}$.

If c is negative and $a > b$, then $\frac{a}{c} < \frac{b}{c}$.

The following example shows two methods to solve the inequality $-6m \geq -72$.

Choose the method which is easiest for you.

Example

4 **Solve:** $-6m \geq -72$

a. $-6m \geq -72$

$\left(-\frac{1}{6}\right)(-6m) \leq \left(-\frac{1}{6}\right)(-72)$ *Multiply both sides by* $-\frac{1}{6}$. *Change ≥ to ≤.*

$m \leq 12$

b. $-6m \geq -72$

$\frac{-6m}{-6} \leq \frac{-72}{-6}$ *Divide both sides by −6. Change ≥ to ≤.*

$m \leq 12$

The solution set is {all numbers less than or equal to 12}.

Example

5 **Solve: $(-0.4)x < 0.6$**

$(-0.4)x < 0.6$

$\frac{(-0.4)x}{-0.4} > \frac{0.6}{-0.4}$ *Divide both sides by -0.4 and reverse the direction of the inequality.*

$x > -1.5$

The solution set is {all numbers greater than -1.5}.

Exploratory Exercises

State the number by which you would multiply both sides of each inequality to solve it. Then state if the direction of the inequality reverses.

1. $\frac{r}{3} > -4$ 3, no **2.** $\frac{s}{-6} < -11$ -6, yes **3.** $\frac{d}{-8} < -9$ -8, yes **4.** $\frac{d}{-11} > 3$ -11, yes

5. $3y < 21$ $\frac{1}{3}$, no **6.** $-6z < 18$ $-\frac{1}{6}$, yes **7.** $-4x > 8$ $-\frac{1}{4}$, yes **8.** $5a < -20$ $\frac{1}{5}$, no

9. $\frac{1}{6}j \geq -9$ 6, no **10.** $-\frac{1}{8}h \leq 10$ -8, yes **11.** $-\frac{s}{11} < 2$ -11, yes **12.** $-\frac{t}{12} < -4$

13. $\frac{4}{3}k > 16$ $\frac{3}{4}$, no **14.** $\frac{2}{7}m < 12$ $\frac{7}{2}$, no **15.** $-\frac{5n}{3} \leq -10$ $-\frac{3}{5}$, yes **16.** $-\frac{3r}{8} \geq 9$ $-\frac{8}{3}$, yes

Written Exercises

Solve each inequality.

1. $14p < 84$ $p < 6$ **2.** $-16q > -128$ $q < 8$ **3.** $-17s > 119$ $s < -7$ **4.** $4z < -6$ $z < -\frac{3}{2}$

5. $-2r \geq 35$ $r \leq -\frac{35}{2}$ **6.** $23b > 276$ $b > 12$ **7.** $-8s < -34$ $s > \frac{17}{4}$ **8.** $5m \leq -17$

9. $\frac{h}{-18} > -25$ $h < 450$ **10.** $\frac{k}{-32} \leq 50$ $k \geq -1600$ **11.** $\frac{b}{8} \leq -16$ $b \leq -128$ **12.** $\frac{a}{-7} < -35$

13. $7r < -4.9$ $r < -0.7$ **14.** $-\frac{1}{4}m \geq 19$ $m \leq -76$ **15.** $-\frac{1}{5}y < -\frac{2}{3}$ $y > \frac{10}{3}$ **16.** $-3n \leq -6.6$

17. $-\frac{5s}{8} \geq \frac{15}{4}$ $s \leq -6$ **18.** $\frac{2}{5}z < \frac{4}{3}$ $z < \frac{10}{3}$ **19.** $-\frac{3}{4}k > \frac{6}{7}$ $k < -\frac{8}{7}$ **20.** $\frac{4z}{7} > -\frac{2}{5}$ $z > -\frac{7}{10}$

21. $-5.1m < -3.57$ $m > 0.7$ **22.** $6.1g < 3.66$ $g < 0.6$ **23.** $0.7t \leq -0.98$ $t \leq -1.4$ **24.** $-1.2x \leq 4.08$ $x \geq -3.4$

25. $-13z > -1.04$ $z < 0.08$ **26.** $\frac{3b}{4} < \frac{2}{5}$ $b < \frac{8}{15}$ **27.** $-\frac{5x}{6} < \frac{2}{3}$ $x > -\frac{4}{5}$ **28.** $1.8z > -54$ $z > -30$

29. $\frac{13}{18} \leq \frac{2}{3}w$ $\frac{13}{12} \leq w$ **30.** $-2.58 > 4.3n$ $-0.6 > n$ **31.** $51.3 < -5.7a$ $-9 > a$ **32.** $-\frac{3}{14} \geq -\frac{5m}{7}$ $\frac{3}{10} \leq m$

Challenge

Solve each inequality. Round solutions to the nearest hundredth.

33. $8.7x < 40$ $x < 4.60$ **34.** $0.07x \geq 0.93$ $x \geq 13.29$ **35.** $-0.41y \leq 2.5$ $y \geq -6.10$ **36.** $-3.1y > -4.2$ $y < 1.35$

37. $843a \geq 25$ $a \geq 0.03$ **38.** $-207a < 30.2$ $a > -0.15$ **39.** $-5.04 \leq -15b$ $0.34 \geq b$ **40.** $3x < 5x$ $x > 0$

State the conditions under which each sentence is true.

41. If $x > y$, then $x^2 > y^2$. $x > 0$, $x > |y|$ **42.** If $x < y$, then $x^2 < y^2$. $y > 0$, $y > |x|$

Chalkboard Exercises

Solve each inequality.

5a. $(-0.7)t < 4.9$
$t > -7$

5b. $(-1.6)n > -11.2$
$n < 7$

Suggested Daily Assignments

Enriched
Written: 2–32 evens, 33–42

Average
Exp: 2–16 evens
Written: 2–36 evens

Fundamental
Exp: 1–16
Written: 1–20
p. 118, 25–36

Answer to Exploratory

12. -12, yes

Answers to Written

8. $m \leq -\frac{17}{5}$

12. $a > 245$

16. $n \geq 2.2$

Calculator Notes

You may wish to allow students to use a calculator for the Challenge Exercises.

4-4

Lesson Notes, p. T32

Readiness Activity

Solve each equation.

$12 + 4c = 20$ 2

$10t - 6 = 4t + 18$ 4

$4(2x + 1) = x - 17$ -3

$2z - 8 + z = 13$ 7

Discussion Questions

1. Name an inequality that has no solution. Answers will vary.
2. Name an inequality that is true for all numbers. Answers will vary.

Chalkboard Exercises

Solve each inequality.

1a. $12 + 3a > 18$ $a > 2$

1b. $15 - 5t < 10$ $t > 1$

2a. $10x + 5 < 6x - 15$ $x < -5$

2b. $6m + 12 > 8m - 15$ $m < 13\frac{1}{2}$

3a. $\frac{1}{3}(3m + 6) < \frac{1}{3}(6m - 9)$ $m > 5$

3b. $9n - 2 > 4n + 0.6$ $n > 0.52$

3c. $3.2n + 7 \geq 4.2n - 16$ $n \leq 23$

❶ Another approach would be to eliminate the decimal by multiplying both sides by 10.

Calculator Notes

Students can use a calculator to check answers to many inequalities.

Chalkboard Exercises

4a. $5(2x + 7) \leq -4(x + 5)$ $x \leq -3\frac{13}{14}$

4b. $-2(3x + 7) \leq 3x - (x + 1)$ $x \geq -1\frac{5}{8}$

Quiz, Lessons 4-1 to 4-4

See margin on page 126.

Objective 4-4: Solve inequalities involving more than one operation.

4-4 Inequalities with More Than One Operation

You can solve inequalities containing more than one operation by applying the methods you have already used.

Examples

1 **Solve: $16 - 5b > 29$**

$$16 - 5b > 29$$
$$16 - 16 - 5b > 29 - 16 \quad \text{Subtract 16 from both sides.}$$
$$-5b > 13$$
$$\frac{-5b}{-5} < \frac{13}{-5} \quad \text{Divide both sides by } -5 \left(\text{or multiply by } -\tfrac{1}{5}\right). \text{ Change } > \text{ to } <.$$
$$b < -\frac{13}{5}$$

The solution set is $\{$all numbers less than $-\frac{13}{5}\}$.

2 **Solve: $9x + 4 < 13x - 7$**

$$9x + 4 < 13x - 7$$
$$9x - 13x + 4 < 13x - 13x - 7 \quad \text{Subtract } 13x \text{ from both sides.}$$
$$-4x + 4 < -7$$
$$-4x + 4 - 4 < -7 - 4 \quad \text{Subtract 4 from both sides.}$$
$$-4x < -11$$
$$\frac{-4x}{-4} > \frac{-11}{-4} \quad \text{Divide both sides by } -4 \left(\text{or multiply by } -\tfrac{1}{4}\right) \text{ Change } < \text{ to } >.$$
$$x > \frac{11}{4}$$

The solution set is $\{$all numbers greater than $\frac{11}{4}\}$.

Use the distributive property to eliminate grouping symbols.

Example

3 **Solve: $(0.7)(m + 3) \leq (0.4)(m + 5)$**

$$(0.7)(m + 3) \leq (0.4)(m + 5)$$
$$0.7m + 2.1 \leq 0.4m + 2.0 \quad \text{Use the distributive property.}$$
$$0.7m + 2.1 - 2.1 \leq 0.4m + 2.0 - 2.1 \quad \text{Subtract 2.1 from both sides.}$$
$$0.7m \leq 0.4m - 0.1$$
$$0.7m - 0.4m \leq 0.4m - 0.4m - 0.1 \quad \text{Subtract } 0.4m \text{ from both sides.}$$
$$0.3m \leq -0.1$$
$$\frac{0.3m}{0.3} \leq \frac{-0.1}{0.3} \quad \text{Divide both sides by 0.3.}$$
$$m \leq -\frac{1}{3}$$

The solution set is $\{$all numbers less than or equal to $-\frac{1}{3}\}$. ❶

Example

4 **Solve:** $-3(2x - 7) \geq 4x - (x - 3)$

$-3(2x - 7) \geq 4x - (x - 3)$	
$-3(2x - 7) \geq 4x + (-1)(x - 3)$	*To subtract, add the inverse.*
$-3(2x) + (-3)(-7) \geq 4x + (-1)x + (-1)(-3)$	*Use the distributive property.*
$-6x + 21 \geq 4x + (-x) + 3$	*Simplify.*
$-6x + 21 \geq 3x + 3$	*Combine 4x and −x.*
$-3x + (-6x) + 21 \geq (-3x) + 3x + 3$	*Add −3x to both sides.*
$-9x + 21 \geq 3$	
$-9x + 21 + (-21) \geq 3 + (-21)$	*Add −21 to both sides.*
$-9x \geq -18$	
$x \leq 2$	*Divide both sides by −9. Change ≥ to ≤.*

The solution set is {all numbers less than or equal to 2}.

Exploratory Exercises

State the steps you would use to solve each inequality.

1. $3x - 1 > 14$ **2.** $9x + 2 > 20$ **3.** $4y - 7 > 21$ **4.** $-7y + 6 < 48$

5. $32 + 14t < 4$ **6.** $5 + 9a > -67$ **7.** $-12 + 11y \leq 54$ **8.** $-9 + 6r \leq -33$

9. $\frac{z}{4} + 7 \geq -5$ **10.** $\frac{b}{-3} + 5 \leq -13$ **11.** $\frac{d}{-4} - 5 < 23$ **12.** $\frac{m}{3} - 7 > 11$

13. $13 - 2a \leq 15$ **14.** $5 - 6g > -19$ **15.** $2k + 7 > k - 10$ **16.** $5y + 4 > y + 13$

Written Exercises

Solve each inequality.

1. $13r - 11 > 7r + 37$ $r > 8$ **2.** $6a + 9 < -4a + 29$ $a < 2$ **3.** $10p - 14 < 8p - 17$

4. $9q + 2 \leq 7q - 25$ $q \leq -\frac{27}{2}$ **5.** $3y + 7 \leq 4y + 8$ $y \geq -1$ **6.** $5 + 10b > 12b + 10$

7. $0.1x < 0.2x - 8$ $x > 80$ **8.** $0.3x + 6.8 \geq 2.0x$ $x \leq 4$ **9.** $7(x + 8) < 3(x + 12)$

10. $-5(y + 5) \geq 5(y - 1)$ $y \leq -2$ **11.** $3n - 8n + 21 > 0$ $n < \frac{21}{5}$ **12.** $9d - 5 + d < -8$

13. $0.6(y + 7) \geq 0.7(2y + 6)$ $y \leq 0$ **14.** $0.3(m + 4) \leq 0.4(2m + 3)$ $m \geq 0$ **15.** $8c - (c - 5) < c + 17$

16. $10x - 2(x - 4) \leq 0$ $x \leq -1$ **17.** $1.3x - 12 < 0.9x + 4$ $x < 40$ **18.** $4x - 13 > 2.7x + 13$

19. $-3(2x - 8) < 2(x + 14)$ $x > -\frac{1}{2}$ **20.** $5x \leq 10 + 3(2x + 4)$ $x \geq -22$

21. $-4(2x - 3) + 5(2x + 10) \geq 0$ $x \geq -31$ **22.** $y - 3(y + 1) \leq 5y - 10$ $y \geq 1$

23. $6 - (3y + 5) > 4 - (2y + 7)$ $y < 4$ **24.** $3y - 2(8y - 11) > 5 - (2y + 6)$ $y < \frac{23}{11}$

Challenge

25. $\frac{3x + 8}{12} < \frac{5}{12}$ $x < -1$ **26.** $\frac{5y - 4}{3} > \frac{y + 5}{3}$ $y > \frac{9}{4}$ **27.** $\frac{2n + 1}{7} \geq \frac{n + 4}{5}$ $n \geq \frac{23}{3}$ **28.** $\frac{c + 8}{4} \leq \frac{-c + 5}{9}$ $c \leq -4$

For more exercises on solving inequalities, see exercises 17-46 on page 555.

Suggested Daily Assignments

Enriched
Written: 2–24 evens, 25–28

Average
Exp: 2–16 evens
Written: 1–12, 14–24 evens

Fundamental
Exp: 1–16
Written: 1–23 odds
p. 121, 22–32 evens

Answers to Exploratory

Answers will vary. Sample answers are given.

1. Add 1 to both sides. Then divide both sides by 3.
2. Subtract 2, then divide by 9.
3. Add 7, then divide by 4.
4. Subtract 6, then divide by −7 and reverse the inequality symbol.
5. Subtract 32, then divide by 14.
6. Subtract 5, then divide by 9.
7. Add 12, then divide by 11.
8. Add 9, then divide by 6.
9. Subtract 7, then multiply by 4.
10. Subtract 5, then multiply by −3 and reverse the inequality symbol.
11. Add 5, then multiply by −4 and reverse the inequality symbol.
12. Add 7, then multiply by 3.
13. Subtract 13, then divide by −2 and reverse the inequality symbol.
14. Subtract 5, then divide by −6 and reverse the inequality symbol.
15. Subtract 7 and subtract k.
16. Subtract 4 and subtract y, then divide by −6 and reverse the inequality symbol.

Answers to Written

3. $p < -\frac{3}{2}$ **6.** $b < -\frac{5}{2}$

9. $x < -5$ **12.** $d < -\frac{3}{10}$

15. $c < 2$ **18.** $x > 20$

4-5

Lesson Notes, p. T32
Enrichment, p. T32

Readiness Activity

State which fraction in each pair is greater.

$\frac{1}{5}, \frac{3}{5}$ $\frac{3}{5}$ $\frac{7}{10}, \frac{6}{10}$ $\frac{7}{10}$

$\frac{2}{8}, \frac{5}{8}$ $\frac{5}{8}$ $\frac{3}{9}, \frac{5}{9}$ $\frac{5}{9}$

$\frac{3}{4}, \frac{1}{4}$ $\frac{3}{4}$ $\frac{7}{15}, \frac{12}{15}$ $\frac{12}{15}$

Discussion Questions

1. Name some inequalities of the form $\frac{a}{b} < \frac{c}{d}$ in which $a = c$. What is the relationship between b and d?
   If $a > 0$ and $c > 0$, then $b > d$.
   If $a < 0$ and $c < 0$, then $b < d$.
2. If $\frac{a}{b} < \frac{c}{d}$ and $b = d$, what can be said of a and c?
   If $b > 0$ and $d > 0$, then $a < c$.
   If $b < 0$ and $d < 0$, then $a > c$.

Chalkboard Exercises

Replace each ? with <, >, or = to make each sentence true.

1a. $\frac{3}{4} \underline{?} \frac{2}{3}$ >

1b. $\frac{6}{7} \underline{?} \frac{9}{10}$ <

1c. $\frac{21}{30} \underline{?} \frac{7}{10}$ =

2a. $\frac{-2}{5} \underline{?} \frac{-4}{7}$ >

2b. $\frac{-7}{5} \underline{?} \frac{-11}{6}$ >

❶ To further emphasize the ease of using the comparison property, change the fractions to equivalent fractions with a common denominator.

$\frac{6}{11} = \frac{72}{132}$ $\frac{7}{12} = \frac{77}{132}$

Notice that $\frac{72}{132} < \frac{77}{132}$.

Thus, $\frac{6}{11} < \frac{7}{12}$.

Objective 4-5A: Use the comparison property to compare rational numbers.
Objective 4-5B: Determine which of two items is a better buy.

4-5 Comparing Rational Numbers

You know that $\frac{6}{9}$ is less than $\frac{7}{9}$ because the denominators are the same and 6 is less than 7. How can you compare $\frac{3}{8}$ and $\frac{4}{11}$? If you change both fractions so that they have a common denominator of 88, you can see that $\frac{33}{88}$ is greater than $\frac{32}{88}$. Thus, $\frac{3}{8} > \frac{4}{11}$.

A shortcut for comparing two rational numbers is to use cross products. Study the property stated below.

Comparison Property for Rational Numbers

> **For any rational numbers $\frac{a}{b}$ and $\frac{c}{d}$, with $b > 0$, $d > 0$,**
>
> **1. If $\frac{a}{b} < \frac{c}{d}$, then $ad < bc$.**
>
> **2. If $ad < bc$, then $\frac{a}{b} < \frac{c}{d}$.**

This property also holds if < is replaced by >, ≤, ≥, or =.

You can use the multiplication property for inequalities to prove the comparison property.

The following example shows how this property is used to compare $\frac{3}{8}$ and $\frac{4}{11}$.

Find the product of the extremes. $\frac{3}{8} \underline{?} \frac{4}{11}$ *Find the product of the means.*

$3(11) = 33$ $4(8) = 32$

Since $33 > 32$, you can conclude that $\frac{3}{8} > \frac{4}{11}$.

Examples

1 **Replace the ? with <, >, or = to make a true sentence.** $\frac{6}{11} \underline{?} \frac{7}{12}$

Find the cross products.

$\frac{6}{11} \underline{?} \frac{7}{12}$

$72 < 77$

The true sentence is $\frac{6}{11} < \frac{7}{12}$. ❶

2 **Replace the ? with <, >, or = to make a true sentence.** $-\frac{1}{4} \underline{?} -\frac{1}{5}$

Rewrite $-\frac{1}{4}$ as $\frac{-1}{4}$ and $-\frac{1}{5}$ as $\frac{-1}{5}$. Then find the cross products.

$\frac{-1}{4} \underline{?} \frac{-1}{5}$

$-5 < -4$

The true sentence is $-\frac{1}{4} < -\frac{1}{5}$.

The comparison property can help you decide which of two items is the better buy. Study the following example. ❷

Example

3 **Super Saver Mart advertised a 9.4-ounce tube of toothpaste for \$1.61. Is this a better buy than another brand of 6 ounces for 95 cents?**

Compare the unit costs. If the quality of the two items is the same, the item with the smaller unit cost is the better buy. Make sure to express costs using the same unit.

$$\text{unit cost} = \frac{\text{total cost}}{\text{number of units}} \text{ or } \frac{\text{total cents}}{\text{number of ounces}}$$ ❸

$$\text{unit cost of first brand} = \frac{161}{9.4}$$

$$\text{unit cost of second brand} = \frac{95}{6}$$

In each case the unit cost is expressed as cents per ounce.

Compare $\frac{161}{9.4}$ and $\frac{95}{6}$ by finding the cross products.

$$\frac{161}{9.4} \stackrel{?}{—} \frac{95}{6}$$

$$966 > 893$$

Thus, $\frac{161}{9.4} > \frac{95}{6}$.

The lesser number is $\frac{95}{6}$, so 6 ounces for 95¢ is the better buy.

Exploratory Exercises

State which ratio in each pair is greater.

1. $\frac{3}{4}, \frac{4}{5}$ $\frac{4}{5}$	**2.** $\frac{11}{12}, \frac{7}{8}$ $\frac{11}{12}$	**3.** $\frac{9}{10}, \frac{10}{11}$ $\frac{10}{11}$	**4.** $\frac{7}{8}, \frac{8}{9}$ $\frac{8}{9}$
5. $\frac{6}{5}, \frac{7}{6}$ $\frac{6}{5}$	**6.** $\frac{11}{9}, \frac{12}{10}$ $\frac{11}{9}$	**7.** $-\frac{1}{3}, -\frac{1}{4}$ $-\frac{1}{4}$	**8.** $-\frac{1}{7}, -\frac{1}{6}$ $-\frac{1}{7}$
9. $-\frac{9}{7}, -\frac{7}{5}$ $-\frac{9}{7}$	**10.** $-\frac{3}{4}, -\frac{2}{3}$ $-\frac{2}{3}$	**11.** $-\frac{13}{11}, -\frac{15}{13}$ $-\frac{15}{13}$	**12.** $-\frac{6}{5}, -\frac{7}{6}$ $-\frac{7}{6}$
13. $\frac{0.3}{4}, \frac{0.2}{2}$ $\frac{0.2}{2}$	**14.** $\frac{3}{0.4}, \frac{5}{0.6}$ $\frac{5}{0.6}$	**15.** $\frac{0.8}{3}, \frac{0.2}{4}$ $\frac{0.8}{3}$	**16.** $\frac{0.06}{0.4}, \frac{0.9}{5}$ $\frac{0.9}{5}$

Write a ratio for the unit cost of each of the following.

17. 10 ounces of coffee for \$4.59 $\frac{459}{10}$

18. \$3.79 for 8 ounces of coffee $\frac{379}{8}$

Chalkboard Exercises

3a. A 12-ounce package of brand-name nuts costs \$1.89. Is this a better buy than a 16-ounce package of generic nuts for \$2.00? No, the generic nuts are the better buy.

3b. Which is the better buy: a twelve ounce bottle of laundry soap for \$1.00 or a 16-ounce bottle for \$1.35? The twelve ounce bottle.

❷ Encourage students to find unit costs when shopping and later report their findings to the class. You may want to point out that this is but one practical application of algebra in everyday life.

❸ Notice that the ratios for the consumer applications in this lesson are written in the form cost/unit. When the ratios are expressed in this form, the lesser number represents the better buy. If the ratios are written in the form unit/cost instead, the greater number would represent the better buy.

Suggested Daily Assignments

Enriched
- Exp: 2–18 evens
- Written: 1–29 odds
- Mini-Review: 1–5

Average
- Exp: 2–18 evens
- Written: 1–16, 18–28 evens
- Mini-Review: 1–5
- p. 123, 25

Fundamental
- Exp: 1–18
- Written: 2–28 evens
- p. 123, 2–20 evens
- Mini-Review: 1–5

Quiz
Lessons 4–1 to 4–4

1. State an inequality for the following graph.

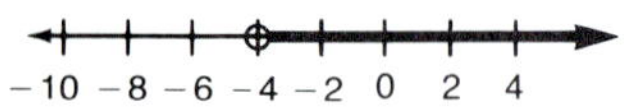

$x > -4$

2. Is $6.2 \geq 3.9 + 2.3$ true or false? true
3. Graph the solution set of $r \leq 2$.

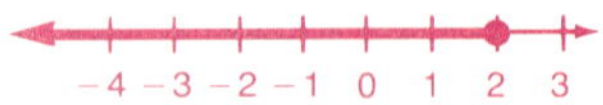

Solve each inequality.

4. $m + 7 < -8$ $m < -15$
5. $y - 3.4 > -4.2$ $y > -0.8$
6. $6w + 5 \leq 2w - 11$ $w \leq -4$
7. $\frac{-2c}{5} \leq \frac{-6}{25}$ $c \geq \frac{3}{5}$
8. $5(8 - 2x) > 24 + 2x$ $x < 1\frac{1}{3}$

Written Exercises

Replace each ? with <, >, or = to make each sentence true.

1. $\frac{6}{7}$? $\frac{7}{8}$ $<$
2. $\frac{8}{7}$? $\frac{9}{8}$ $>$
3. $\frac{7}{19}$? $\frac{6}{17}$ $>$
4. $\frac{8}{15}$? $\frac{9}{16}$ $<$
5. $-\frac{5}{16}$? $-\frac{3}{10}$ $<$
6. $-\frac{4}{11}$? $-\frac{1}{3}$ $<$
7. $-\frac{11}{15}$? $-\frac{9}{11}$ $>$
8. $-\frac{6}{11}$? $-\frac{4}{7}$ $>$
9. $-\frac{2}{3}$? $-\frac{3}{5}$ $<$
10. $-\frac{1}{3}$? $-\frac{2}{7}$ $<$
11. $-\frac{7}{6}$? $-\frac{21}{18}$ $=$
12. $\frac{5}{14}$? $\frac{25}{70}$ $=$
13. $\frac{0.4}{3}$? $\frac{1.2}{2}$ $<$
14. $\frac{1.1}{4}$? $\frac{2.2}{5}$ $<$
15. $-\frac{3.2}{1.06}$? $-\frac{2.8}{2.13}$ $<$
16. $-\frac{5.35}{2.4}$? $-\frac{6.9}{4.01}$ $<$

Write a ratio for the unit cost of each item. Then compare the ratios to determine which of the two items is the better buy. All ratios are expressed in the form cents/units.

17. a 21-ounce can of baked beans for 79¢ or a 28-ounce can for 97¢ $\frac{79}{21}, \frac{97}{28}$
18. a 10-ounce jar of coffee for $4.27 or an 8-ounce jar for $3.64 $\frac{427}{10}, \frac{364}{8}$
19. a 184-gram can of peanuts for 91¢ or a 340-gram can for $1.89 $\frac{91}{184}, \frac{189}{340}$
20. a half-pound bag of cashews for $2.93 or a $\frac{3}{4}$-pound bag for $4.19 $\frac{293}{0.5}, \frac{419}{0.75}$
21. a six-pack of cola containing 2.1 liters for $1.79, or a six-pack containing 1.9 liters for $1.69 $\frac{179}{2.1}, \frac{169}{1.9}$
22. three liters of soda for $2.25 or two liters for $1.69 $\frac{225}{3}, \frac{169}{2}$
23. a 27-ounce loaf of bread for 93¢ or a 20-ounce loaf for 79¢ $\frac{93}{27}, \frac{79}{20}$
24. a dozen extra-large eggs weighing 27 ounces for $1.09, or a dozen large eggs weighing 24 ounces for 99¢ $\frac{109}{27}, \frac{99}{24}$
25. a 100-count package of paper plates for $1.49, or a 75-count package for 98¢ $\frac{149}{100}, \frac{98}{75}$
26. a 25-pound bag of dog food for $9.85 or a 20-pound bag for $6.50 $\frac{985}{25}, \frac{650}{20}$

Solve each problem.

27. At Whittaker's Market a head of lettuce is 79¢. A head of lettuce weighs approximately $\frac{3}{4}$-pound. Dudley's Market sells lettuce at 44¢ for $\frac{1}{2}$-pound. Which store has the better price on lettuce? Dudley's Market
28. Melody Morrison wanted to buy soda for a party. Eight 1-liter bottles of Brand X cost $3.92. Six 1-liter bottles of Brand Y cost $3.08. A 2-liter bottle of Brand Z costs $1.09. Which soda is least expensive per liter? Brand X
29. Mrs. Fischer saw 3 packaged pumpkin pies on sale. The first was 99¢ for a 26-ounce pie, the second was $1.19 for a 28-ounce pie, and the third was $1.25 for a 30-ounce pie. Which pie was least expensive per ounce? 26-ounce pie

mini-review

Define the variable, then write an equation for the following problem. Do not solve.

1. Half of a number increased by 14 is 25. Find the number. Let x = the number; $\frac{1}{2}x + 14 = 25$

Solve.

2. $5t = -3 + 4t$ -3
3. $r - 23 = -10$ 13
4. $-8k = 102$ -12.75

Use a proportion to solve the following problem.

5. In 3 minutes, a printing press printed 1700 pages. How many pages could be printed in an hour? 34,000

Challenge

30. The following is a proof of the first part of the comparison property for rational numbers. State the property that justifies each step of the proof.

Given: $\frac{a}{b} < \frac{c}{d}$, $b > 0$, $d > 0$

Prove: $ad < bc$

Proof: $b\left(\frac{a}{b}\right) < b\left(\frac{c}{d}\right)$ Multiplication Property for Inequalities

$a < \frac{bc}{d}$ Multiplicative Inverse

$a \cdot d < \left(\frac{bc}{d}\right)d$ Multiplication Property for Inequalities

$ad < bc$ Multiplicative Inverse

Write a proof similar to the one from Challenge Exercise 30 for each of the following.

31. If $\frac{a}{b} > \frac{c}{d}$ with $b > 0$ and $d > 0$, then $ad > bc$.

32. If $ad < bc$ with $b > 0$ and $d > 0$, then $\frac{a}{b} < \frac{c}{d}$.

Using Calculators — Comparing Numbers

You can use a calculator to compare rational numbers. First change each fraction to a decimal by dividing the numerator by the denominator. Then compare the decimals.

Example: Change the fractions $\frac{6}{17}$, $\frac{17}{49}$, and $\frac{35}{99}$ to decimals. Then write the fractions in order from least to greatest.

$\frac{6}{17} = 0.3529412$

$\frac{17}{49} = 0.3469388$

$\frac{35}{99} = 0.3535354$

Therefore $\frac{17}{49} < \frac{6}{17} < \frac{35}{99}$.

It is not always necessary to copy all the decimal places shown in the calculator display. Usually, you can round results to the nearest hundredth or to the nearest thousandth.

Exercises

Write the fractions in order from least to greatest.

1. $\frac{17}{21}, \frac{20}{27}, \frac{19}{24}$ $\frac{20}{27} < \frac{19}{24} < \frac{17}{21}$

2. $\frac{5}{13}, \frac{8}{23}, \frac{7}{18}$ $\frac{8}{23} < \frac{5}{13} < \frac{7}{18}$

3. $\frac{11}{15}, \frac{5}{7}, \frac{9}{13}$ $\frac{9}{13} < \frac{5}{7} < \frac{11}{15}$

4. $\frac{8}{19}, \frac{10}{23}, \frac{11}{27}$ $\frac{11}{27} < \frac{8}{19} < \frac{10}{23}$

5. $\frac{17}{19}, \frac{32}{35}, \frac{45}{49}$ $\frac{17}{19} < \frac{32}{35} < \frac{45}{49}$

6. $\frac{3}{14}, \frac{5}{23}, \frac{9}{43}$ $\frac{9}{43} < \frac{3}{14} < \frac{5}{23}$

Use a calculator to determine which of the two items is the better buy.

7. a dozen oranges for \$1.59 or half a dozen for 85¢

8. five pounds of green beans for \$3.50 or 2 pounds for \$1.38

9. a 48-ounce bottle of dish soap for \$2.39 or a 22-ounce bottle for \$1.09

10. a 1-pound package of lunch meat for \$1.98 or a 12-ounce package for \$1.80

Answers to Challenge

The property that justifies each step is given in parenthesis below the step.

31. Given: $\frac{a}{b} > \frac{c}{d}$, $b > 0$, $d > 0$

Prove: $ad > bc$

Proof: $b\left(\frac{a}{b}\right) > b\left(\frac{c}{d}\right)$

(Mult. Property)

$a > \frac{bc}{d}$

(Multiplicative Inverse)

$a \cdot d > \left(\frac{bc}{d}\right)d$

(Mult. Property)

$ad > bc$

(Multiplicative Inverse)

32. Given: $ad < bc$, $b > 0$, $d > 0$

Prove: $\frac{a}{b} < \frac{c}{d}$

Proof: $\frac{ad}{d} < \frac{bc}{d}$

(Division Prop.)

$a < \frac{bc}{d}$

(Multiplicative Inverse)

$a \cdot \frac{1}{b} < \left(\frac{bc}{d}\right)\frac{1}{b}$

(Mult. Property)

$\frac{a}{b} < \frac{c}{d}$

(Multiplicative Inverse)

About Using Calculators

Changing fractions to decimals provides an excellent alternative to the memorization of a rule such as $\frac{a}{b} < \frac{c}{d}$ if $ad < bc$.

In the example, point out that the least decimal approximation indicates the least fraction and the greatest decimal approximation indicates the greatest fraction.

As an extension of this feature, have students compare π with $\frac{355}{113}$.

4-6

Lesson Notes, p. T32

Discussion Question

If the solution of the compound sentence $x < a$ and $x < b$ is the sentence $x < a$, what must be true of a and b?

$a \leq b$

Chalkboard Exercises

Write each compound sentence without *and*.

1a. $x > 5$ and $x < 6$

$5 < x < 6$

1b. $y < 6$ and $2 < y$

$2 < y < 6$

2a. $m \leq -4$ and $m \leq 6$

$-4 \geq m \leq 6$

2b. $n \geq 1$ and $n \leq 5$

$1 \leq n \leq 5$

Objective 4-6: Solve compound sentences and graph the solution set.

4-6 Compound Sentences

Megan Peroni works part-time. Last year she paid \$280 in federal income tax. The tax table that she used is shown at the right. According to the table, her taxable income must have been at least \$4650 but less than \$4700.

Let I represent her income. Then the two inequalities below describe the amount of her income.

$$I \geq 4650 \quad \text{and} \quad I < 4700$$

When considered together, these inequalities form a **compound sentence**. A compound sentence containing *and* is true only if both inequalities are true.

If line 37 (taxable income) is—		And you are—			
At least	But less than	Single	Married filing jointly *	Married filing separately	Head of a household
		Your tax is—			
4,000	4,050	196	69	273	190
4,050	4,100	202	74	280	195
4,100	4,150	208	80	287	201
4,150	4,200	214	85	294	206
4,200	4,250	220	91	301	212
4,250	4,300	226	96	308	217
4,300	4,350	232	102	315	223
4,350	4,400	238	107	322	228
4,400	4,450	245	113	329	234
4,450	4,500	252	118	336	240
4,500	4,550	259	124	343	246
4,550	4,600	266	129	350	252
4,600	4,650	273	135	357	258
4,650	4,700	280	140	364	264
4,700	4,750	287	146	371	270
4,750	4,800	294	151	378	276
4,800	4,850	301	157	385	282
4,850	4,900	308	162	392	288
4,900	4,950	315	168	399	294
4,950	5,000	322	173	406	300

Another way of writing $I \geq 4650$ and $I < 4700$ without *and* is shown below.

$$4650 \leq I < 4700$$

This sentence is read, *I is greater than or equal to 4650 and less than 4700.*

Examples

1 **Write the compound sentence $x > -5$ and $x < 1$ without *and*.**

$x > -5$ and $x < 1$ can be written $-5 < x < 1$ or $1 > x > -5$.

2 **Write the compound sentence $y \geq 0$ and $y \leq 5$ without *and*.**

$y \geq 0$ and $y \leq 5$ can be written $0 \leq y \leq 5$ or $5 \geq y \geq 0$.

The graph of a compound sentence containing *and* is the *intersection* of the graphs of the two inequalities.

❶

Examples

3 Graph the solution set of $x > -5$ and $x < 1$.

Graph $x > -5$.

Graph $x < 1$.

❷ *Find the intersection, that is, where the graphs overlap.*

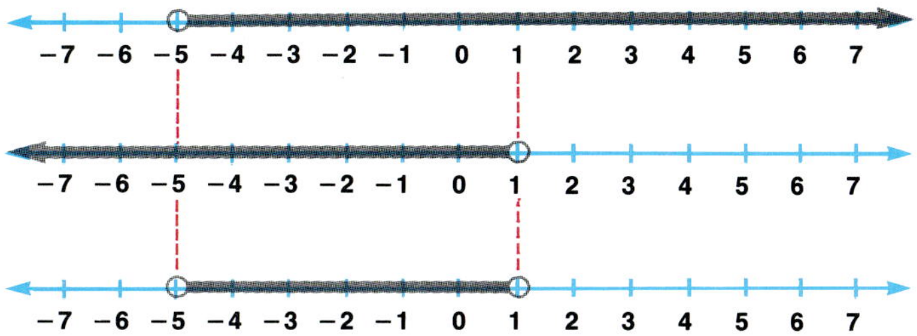

Notice that 1 and −5 are not included in the solution.

4 Graph the solution set of $-3 \le x \le 5$.

Graph $x \ge -3$.

Graph $x \le 5$.

Find the intersection.

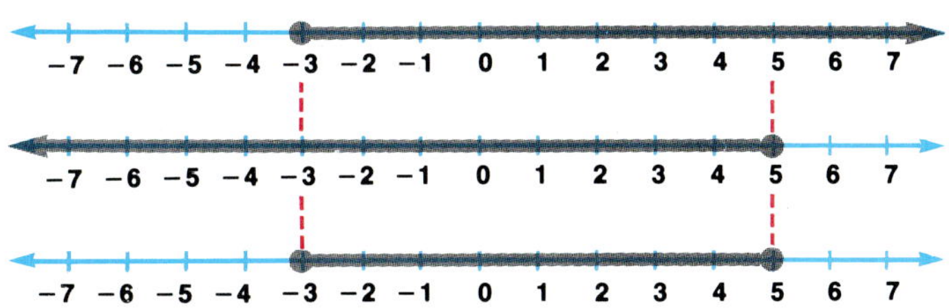

Notice that 3 and 5 are included in the solution.

5 Graph the solution set of $x < 3$ and $x > 5$.

Graph $x < 3$.

Graph $x > 5$.

Find the intersection.

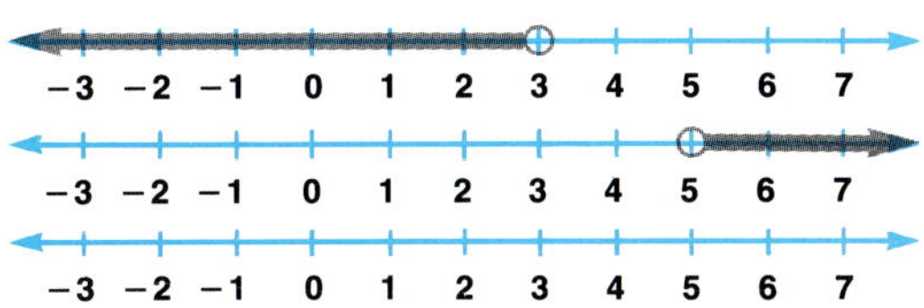

There are *no* points in the intersection of the two graphs. That is, there are no numbers that are *both* less than 3 and greater than 5. Therefore, the intersection is the empty or null set. The symbol $\emptyset$ or { } is used to denote the null set.

A compound sentence may contain *or* instead of *and*. Only one inequality in such a sentence needs to be true for the sentence to be true. The solution of an *or* sentence is the *union* of the solution sets of each inequality. ❸

Chalkboard Exercises

Graph the solution set of each compound sentence.

3. $x > 5$ and $x < 6$

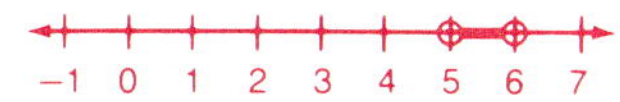

4. $m \ge -4$ and $m \le 6$

5. $y > 6$ and $y < 2$

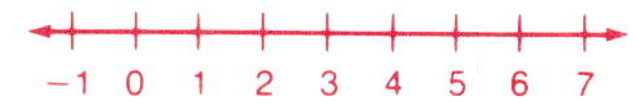

no solutions

❶ *Before* introducing Example 3, discuss possible solutions to the compound sentence in the example: $x > -5$ and $x < 1$. Have students choose a number and substitute it for x. Then have them decide whether the number is a solution to the compound sentence. For example, try 5. Is 5 greater than −5 and less than 1? Note that 5 *is* greater than −5, but it *is not* less than 1. Therefore, the number 5 *is not* a solution to the compound sentence. Other numbers which students could try are −6, 1, −2, and −5.

❷ Make sure students understand how to find the *intersection* of two graphs. The intersection includes all points which belong to *both* graphs.

❸ Point out to students that there is no "shortcut symbolism" for the "or" sentence. For example, the sentence $x \ge 3$ or $x < -2$ *cannot* be written as $-2 > x \ge 3$.

Chalkboard Exercises
Graph the solution set of each compound sentence.

6a. $x > 5$ or $x < -2$

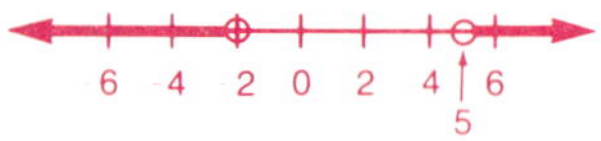

6b. $y < -2$ or $y \geq 4$

Solve each inequality and graph the solution set.

7a. $3y + 4 < 6$ or $2y - 7 < 4y$

7b. $3x + 2 > 8$ or $5x < 3x - 6$

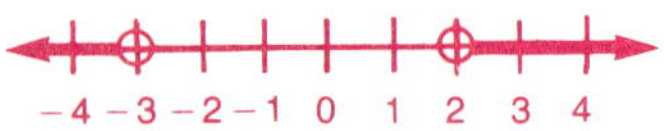

4 Make sure that students understand how to find the *union* of two graphs. The union includes all points which belong to *either* graph. That is, the points on one graph are *combined* with the points on the other graph.

Suggested Daily Assignments

Enriched
  Written: 1–45 odds, 46–49

Average
  Exp: 2–12 evens
  Written: 2–38 evens, 40–45

Fundamental
  Exp: 1–12
  Written: 2–44 evens
    p. 126, 1-27 odds

Examples

6 **Graph the solution set of $x \geq 3$ or $x < -2$.**

Graph $x \geq 3$.

Graph $x < -2$. **4**

Find the union. That is, combine both graphs.

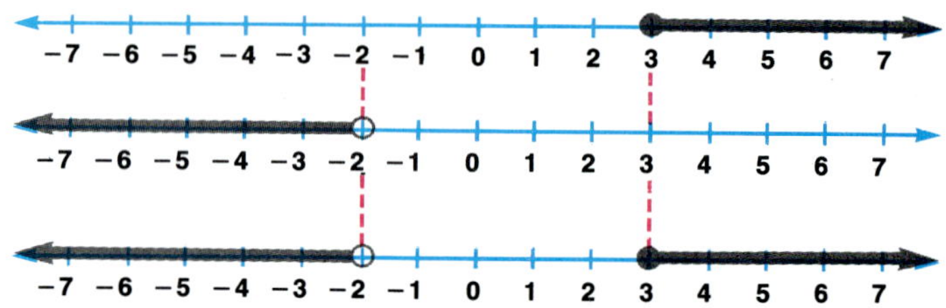

7 **Solve $2y > y - 3$ or $3y < y + 6$. Graph the solution set.**

First solve each inequality for y.

$$\begin{aligned} 2y &> y - 3 \\ 2y - y &> y - y - 3 \\ y &> -3 \end{aligned} \qquad \text{or} \qquad \begin{aligned} 3y &< y + 6 \\ 3y - y &< y - y + 6 \\ 2y &< 6 \\ \frac{2y}{2} &< \frac{6}{2} \\ y &< 3 \end{aligned}$$

Then graph the solution sets.

Graph $y > -3$.

Graph $y < 3$.

The solution set is {all numbers}.

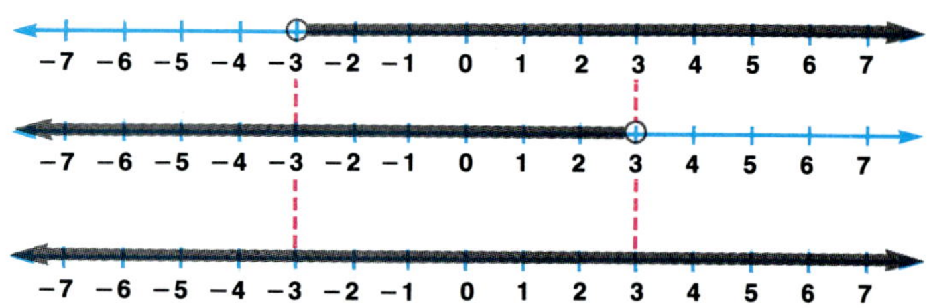

Exploratory Exercises

State whether each compound sentence is true or false.

1. $8 < 3$ and $8 > 3$ false
2. $13 > 9$ and $13 > 12$ true
3. $5 < 7$ and $-8 < -6$ true
4. $9 \neq 0$ and $12 < 17$ true
5. $5 \geq -4$ and $11 \leq 7$ false
6. $5 > -3$ and $-5 > -1$ false
7. $7 > 4$ or $5 < 6$ true
8. $13 > 6$ or $0 < -2$ true
9. $-11 < -19$ or $3 > 0$ true
10. $8 \leq 8$ or $7 > 15$ true
11. $8 \neq 9$ or $16 < -7$ true
12. $-21 > -19$ or $-7 > 0$ false

Written Exercises

Write each compound sentence without *and*.

1. $0 \le m$ and $m < 9$ $0 \le m < 9$

2. $0 < y$ and $y \le 12$ $0 < y \le 12$

3. $p > \frac{3}{4}$ and $p \le \frac{11}{9}$ $\frac{3}{4} < p \le \frac{11}{9}$

4. $r > -\frac{1}{2}$ and $r < \frac{8}{3}$ $-\frac{1}{2} < r < \frac{8}{3}$

5. $z > -\frac{4}{5}$ and $z < \frac{2}{3}$ $-\frac{4}{5} < z < \frac{2}{3}$

6. $y \le \frac{4}{9}$ and $y \ge -\frac{4}{3}$ $-\frac{4}{3} \le y \le \frac{4}{9}$

7. $m < -\frac{6}{5}$ and $m > -\frac{13}{7}$ $-\frac{13}{7} < m < -\frac{6}{5}$

8. $r > -\frac{3}{4}$ and $r \le -\frac{1}{10}$ $-\frac{3}{4} < r \le -\frac{1}{10}$

9. $a \le -2.4$ and $a \ge -4.9$ $-4.9 \le a \le -2.4$

10. $m \ge 0.35$ and $m \le 0.99$ $0.35 \le m \le 0.99$

Graph the solution set of each compound sentence.

11. $m < -7$ or $m \ge 0$
12. $x \ge -2$ and $x \le 5$
13. $n \le -5$ and $n \ge -1$
14. $r > 2$ or $r \le -2$
15. $b > 5$ or $b \le 0$
16. $p < -3$ and $p > 3$
17. $x > -5$ and $x < 0$
18. $d \ge -6$ and $d \le -3$
19. $q \ge -5$ and $q \le 1$
20. $w > -3$ or $w < 1$
21. $d > 0$ or $d < 4$
22. $s \le 8$ or $s \ge 3$
23. $a > 8$ or $a < 5$
24. $r > -4$ or $r \le 0$
25. $p \le 6$ and $p \ge -1$

31. $q < -2$ or $q > -1$
33. $x \ge -5$ and $x < 1$ {all numbers}

Solve each inequality and graph the solution set.

26. $3 + x < -4$ or $3 + x > 4$ $x < -7$ or $x > 1$

27. $-1 + b > -4$ or $-1 + b < 3$

28. $2 > 3t + 2$ and $3t + 2 > 14$ no solutions

29. $9 - 2m > 11$ and $5m < 2m + 9$ $m < -1$

30. $2x + 4 \le 6$ or $x \ge 2x - 4$ $x \le 4$

31. $7 + 3q < 1$ or $-12 < 11q - 1$

32. $x \ne 6$ and $3x + 1 > 10$ $x \ne 6$ and $x > 3$

33. $-2 \le x + 3$ and $x + 3 < 4$

34. $-5 < 4 - 3x < 13$ $x < 3$ and $x > -3$

35. $-3 - x < 2x < 3 + x$ $x > -1$ and $x < 3$

36. $2x - 1 < 2x + 8 < 2x + 4$ no solutions

37. $x - 1 < 2x + 3 < x + 4$ $x > -4$ and $x < 1$

38. $5(x - 3) + 2 < 7$ and $5x > 4(2x - 3)$ $x < 4$

39. $2 - 5(2x - 3) > 2$ or $3x < 2(x - 8)$ $x < \frac{3}{2}$

Write the compound sentence whose solution set is graphed.

40. 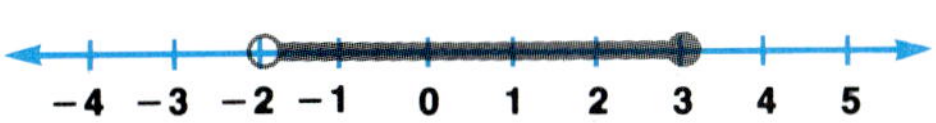

$x > -2$ and $x \le 3$

41. 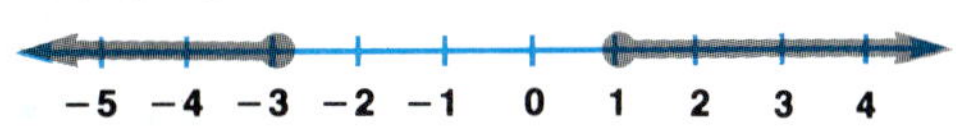

$x \le -3$ or $x \ge 1$

42. 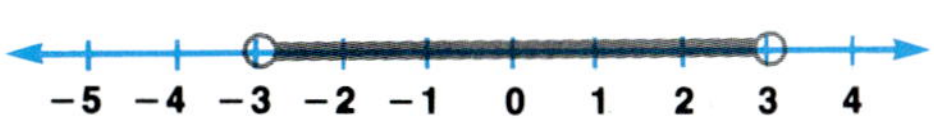

$x > -3$ and $x < 3$

43. 

$x < -3$ or $x > 3$

44. 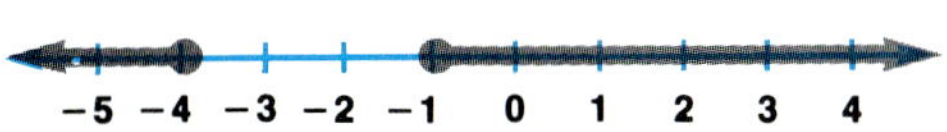

$x \le -4$ or $x \ge -1$

45. 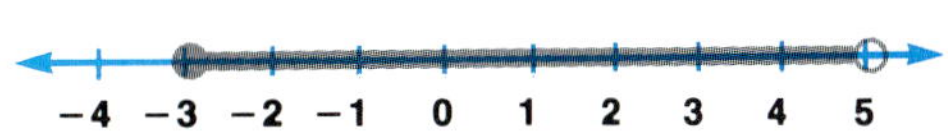

$x \ge -3$ and $x < 5$

Challenge

Solve each inequality and graph the solution set.

46. $2x + \frac{1}{2} < 6$ and $x + 3 > -\frac{1}{2}$ $x < \frac{11}{4}$ and $x > -\frac{7}{2}$

47. $-3x - \frac{2}{3} \ge \frac{7}{3}$ and $-4x - \frac{5}{6} \le \frac{3}{6}$ no solution

48. $\frac{5}{x} + 3 > 0$ $x < -\frac{5}{3}$ or $x > 0$

49. $\frac{3}{x} + 2 > 0$ $x > 0$ or $x < -\frac{3}{2}$

For more exercises on compound inequalities, see exercises 47-66 on page 555.

Answers to Written

11. 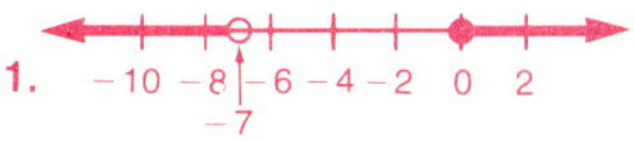

12. 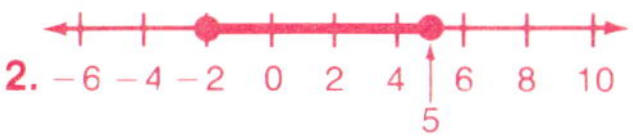

13. 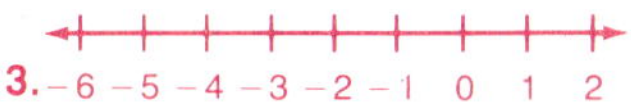

14. 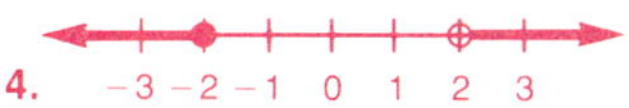

15. 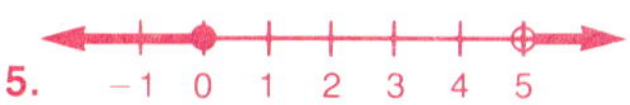

16. 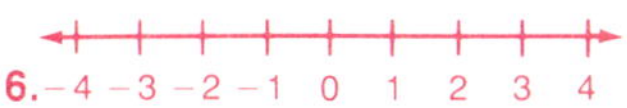

17. 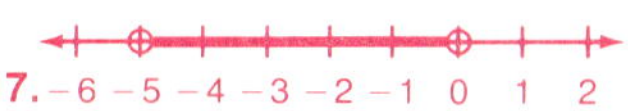

18. 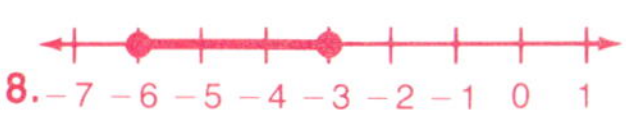

19. 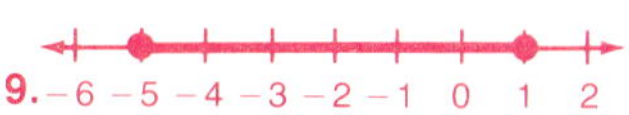

20. 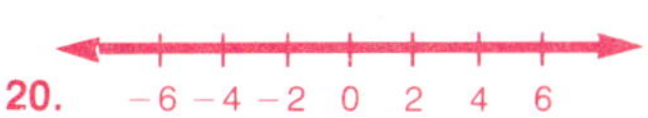

21. 

22. 

23. 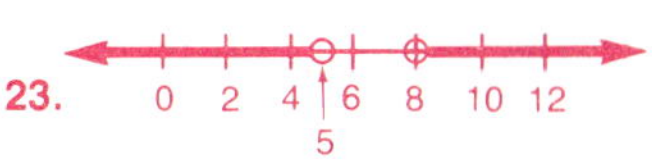

24. 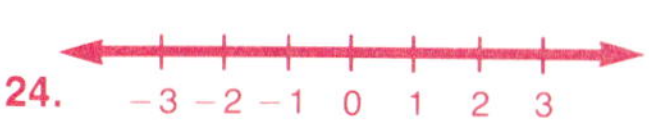

25. 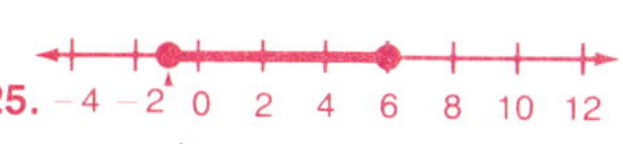

(continued in the Teacher's Answer Key, p. T81)

4-7

Lesson Notes, p. T33

Readiness Activity
Find $|m|$ if:
(a) $m = 5$. 5
(b) $m = -5$. 5
(c) $m > 5$. $|m| > 5$
(d) $m < -5$. $|m| > 5$

Discussion Question
Name some values of b so that the inequality $|x| < b$ has no solutions.
Any b such that $b \leq 0$.

Chalkboard Exercises
Solve each open sentence. Then graph its solution set.

1a. $|4p - 3| = 7$
$\left\{\frac{5}{2}, -1\right\}$

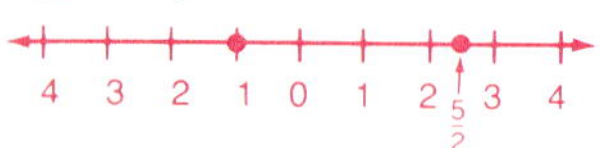

1b. $|4k + 6| = 5$
$\left\{-\frac{1}{4}, -\frac{11}{4}\right\}$

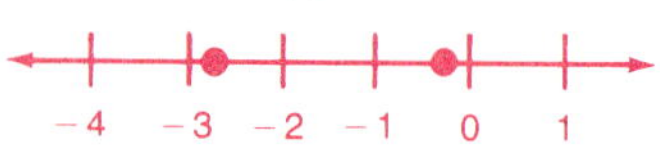

2a. $|3x - 1| \leq 5$
{all numbers greater than or equal to $-\frac{4}{3}$ and less than or equal to 2}

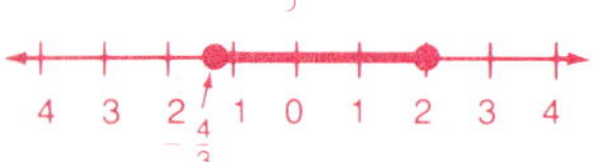

2b. $|2k - 3| \leq 1$
{all numbers greater than or equal to one and less than or equal to two}

3a. $|6y + 6| > 4$
{all numbers less than $-\frac{5}{3}$ or greater than $-\frac{1}{3}$}

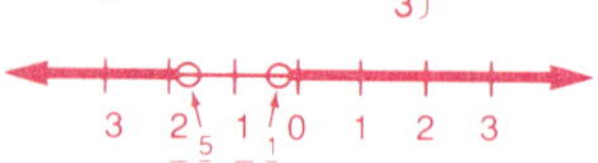

Objective 4-7: Solve open sentences containing absolute value and graph the solution sets.

4-7 Open Sentences Containing Absolute Value

The **absolute value** of a number is the number of units it is from 0 on the number line. An open sentence that contains absolute value must be interpreted carefully. Study the graph and interpretation of each open sentence given below.

$|x| = 2$

The distance from 0 to x is 2 units.
Therefore, $x = -2$ or $x = 2$.

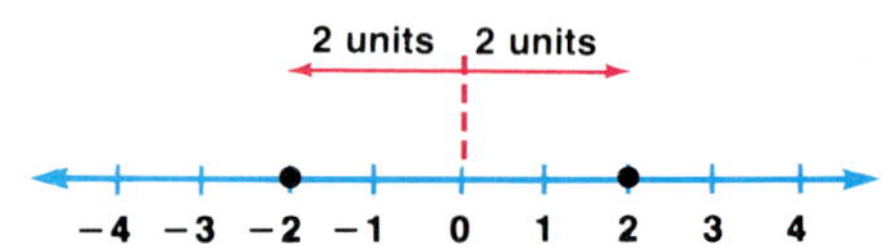

$|x| < 2$

The distance from 0 to x is less than 2 units.
Therefore, $x > -2$ *and* $x < 2$.
This can also be written as $-2 < x < 2$.

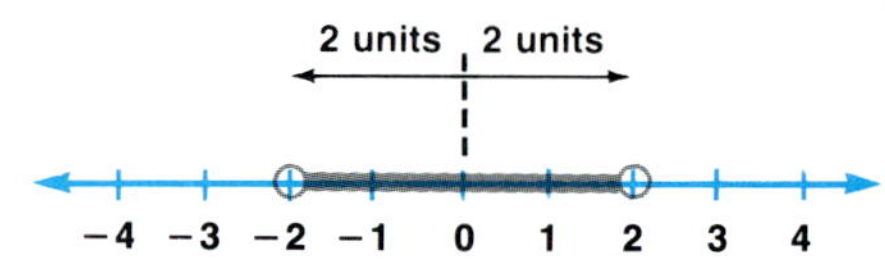

$|x| > 2$

The distance from 0 to x is greater than 2 units.
Therefore, $x < -2$ *or* $x > 2$.

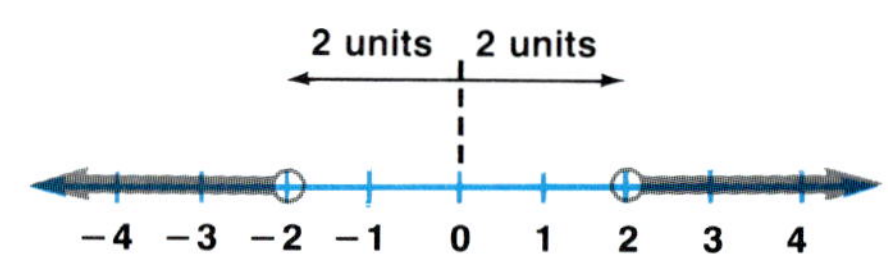

Notice that an open sentence containing absolute value should be interpreted as a compound sentence.

Example 1

Solve: $|3x - 1| = 5$

$|3x - 1| = 5$ means $3x - 1 = 5$ *or* $3x - 1 = -5$.
Solve both equations to find the solution set.

$$3x - 1 = 5 \qquad \text{or} \qquad 3x - 1 = -5$$
$$3x - 1 + 1 = 5 + 1 \qquad 3x - 1 + 1 = -5 + 1$$
$$3x = 6 \qquad 3x = -4$$
$$\frac{3x}{3} = \frac{6}{3} \qquad \frac{3x}{3} = \frac{-4}{3}$$
$$x = 2 \qquad \text{or} \qquad x = -\frac{4}{3}$$

The solution set is $\left\{2, -\frac{4}{3}\right\}$.

Examples

2 **Solve $|2x + 1| < 8$ and graph the solution set.**

$|2x + 1| < 8$ means $2x + 1 > -8$ *and* $2x + 1 < 8$.

$$
\begin{array}{rcl}
2x + 1 > -8 & \text{and} & x + 1 < 8 \\
2x + 1 + (-1) > -8 + (-1) & & 2x + 1 + (-1) < 8 + (-1) \\
2x > -9 & & 2x < 7 \\
\frac{2x}{2} > \frac{-9}{2} & & \frac{2x}{2} < \frac{7}{2} \\
x > -\frac{9}{2} & \text{and} & x < \frac{7}{2}
\end{array}
$$

The solution set is $\{$all numbers greater than $-\frac{9}{2}$ and less than $\frac{7}{2}\}$.

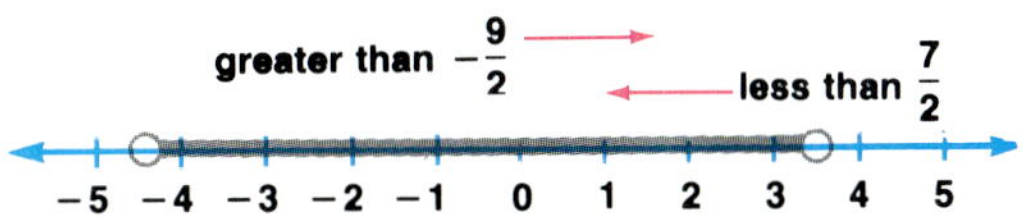

This solution can also be written as $-\frac{9}{2} < x < \frac{7}{2}$.

3 **Solve $|3x + 2| \geq 5$ and graph the solution set.**

$|3x + 2| \geq 5$ means $3x + 2 \leq -5$ *or* $3x + 2 \geq 5$.

$$
\begin{array}{rcl}
3x + 2 \leq -5 & \text{or} & 3x + 2 \geq 5 \\
3x + 2 - 2 \leq -5 - 2 & & 3x + 2 - 2 \geq 5 - 2 \\
3x \leq -7 & & 3x \geq 3 \\
\frac{3x}{3} \leq \frac{-7}{3} & & \frac{3x}{3} \geq \frac{3}{3} \\
x \leq -\frac{7}{3} & \text{or} & x \geq 1
\end{array}
$$

The solution set is $\{$all numbers less than or equal to $-\frac{7}{3}$ or greater than or equal to $1\}$.

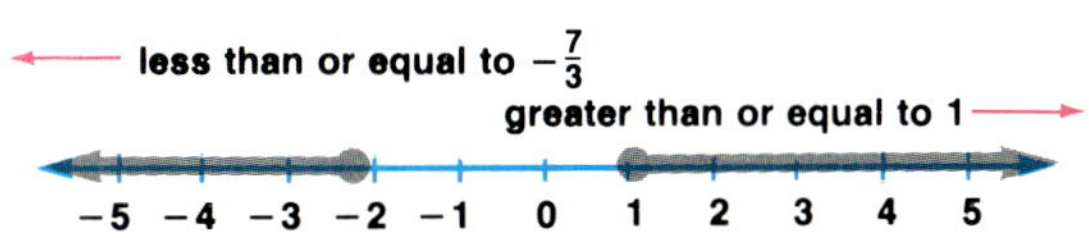

This solution can also be written as $x \leq -\frac{7}{3}$ or $x \geq 1$.

Exploratory Exercises

State each open sentence as a compound sentence.

1. $\lvert x\rvert = 4$	**2.** $\lvert y\rvert = 7$	**3.** $\lvert y\rvert > 3$	**4.** $\lvert x\rvert > 5$
5. $\lvert y\rvert < \frac{5}{2}$	**6.** $\lvert x\rvert < \frac{7}{2}$	**7.** $\lvert x + 2\rvert > 3$	**8.** $\lvert y + 1\rvert > 0$
9. $\lvert x + 2\rvert < 3$	**10.** $\lvert y + 1\rvert < 4$	**11.** $\lvert 2x - 5\rvert \geq 3$	**12.** $\lvert 2x + 3\rvert \geq 5$
13. $\lvert 7 - x\rvert = 4$	**14.** $\lvert 7 + x\rvert = 2$	**15.** $\lvert 3x + 1\rvert > 6$	**16.** $\lvert 3x - 1\rvert < 6$

For more exercises involving absolute value, see exercises 67-82 on page 555.

Chalkboard Exercise

Solve each open sentence. Then graph its solution set.

3b. $|4x - 5| > 3$

$\{$all numbers less than $\frac{1}{2}$ or greater than $2\}$

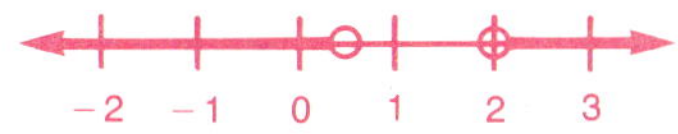

Suggested Daily Assignments

Enriched
- Exp: 2–16 evens
- Written: 2–24 evens, 25-37
- Mini-Review: 1–5

Average
- Exp: 2–16 evens
- Written: 1–24, 31, 32
  - p. 131, 46
- Mini-Review: 1–5

Fundamental
- Exp: 1–16
- Written: 1–16
  - p. 131, 27–45 odds
- Mini-Review: 1–5

Answers to Exploratory

1. $x = 4$ or $x = -4$
2. $y = 7$ or $y = -7$
3. $y > 3$ or $y < -3$
4. $x > 5$ or $x < -5$
5. $y < \frac{5}{2}$ and $y > -\frac{5}{2}$
6. $x < \frac{7}{2}$ and $x > -\frac{7}{2}$
7. $x + 2 > 3$ or $x + 2 < -3$
8. $y + 1 > 0$ or $y + 1 < 0$
9. $x + 2 < 3$ and $x + 2 > -3$
10. $y + 1 < 4$ and $y + 1 > -4$
11. $2x - 5 \geq 3$ or $2x - 5 \leq -3$
12. $2x + 3 \geq 5$ or $2x + 3 \leq -5$
13. $7 - x = 4$ or $7 - x = -4$
14. $7 + x = 2$ or $7 + x = -2$
15. $3x + 1 > 6$ or $3x + 1 < -6$
16. $3x - 1 < 6$ and $3x - 1 > -6$

Answers to Written

See the Teacher's Answer Key, p. T81, for graphs.

1. $y > 3$ or $y < -5$
2. {all numbers}
3. $y > -3$ and $y < 5$
4. $x < 9$ and $x > 5$
5. $y \geq 1$ and $y \leq 3$
6. $x \geq 4$ and $x \leq 8$
7. {all numbers}
8. $x \leq 3$ or $x \geq 11$
9. $y \geq 8$ or $y \leq 2$
10. $x \leq 0$ or $x \geq 8$
11. $\{1\}$
12. $\{-1, 15\}$
13. $x \leq \frac{5}{2}$ and $x \geq -\frac{9}{2}$
14. $x \leq \frac{11}{3}$ and $x \geq -\frac{17}{3}$
15. $x \geq 8$ or $x \leq -10$
16. {all numbers}
17. $y \geq \frac{9}{2}$ or $y \leq \frac{1}{2}$
18. {all numbers}
19. $\left\{\frac{4}{3}\right\}$
20. $\left\{\frac{5}{2}, \frac{21}{2}\right\}$
21. $t < \frac{4}{3}$ and $t > -1$
22. $t \leq -\frac{5}{3}$ or $t \geq 2$
23. $\{2, -2\}$
24. $\{11, -11\}$

About Using Calculators
Ask students how they would solve these inequalities without using a calculator. Have them discuss some difficulties they may encounter.

Answers to Using Calculators

1. $r < 37.76$ and $r > -21.04$
2. $a \geq 15.38$ or $a \leq 10.7$
3. $x \geq -3.97$ and $x \leq 0.38$
4. $t \geq -0.09$ and $t \leq 0.06$
5. $y \leq -2.92$ or $y \geq -2.63$
6. $m < 2.17$ and $m > -1.76$

Written Exercises

Solve each open sentence. Then graph its solution set.

1. $\lvert y + 1\rvert > 4$	**2.** $\lvert x + 1\rvert > -2$	**3.** $\lvert y - 1\rvert < 4$	**4.** $\lvert x - 7\rvert < 2$
5. $\lvert 2 - y\rvert \leq 1$	**6.** $\lvert 6 - x\rvert \leq 2$	**7.** $\lvert 9 - y\rvert \geq -3$	**8.** $\lvert 7 - x\rvert \geq 4$
9. $\lvert 2y - 10\rvert \geq 6$	**10.** $\lvert 12 - 3x\rvert \geq 12$	**11.** $\lvert 3 - 3x\rvert = 0$	**12.** $\lvert 14 - 2y\rvert = 16$
13. $\lvert 4x + 4\rvert \leq 14$	**14.** $\lvert 6x + 6\rvert \leq 28$	**15.** $\lvert 10x + 10\rvert \geq 90$	**16.** $\lvert 9x + 9\rvert \geq -72$
17. $\lvert 2y - 5\rvert \geq 4$	**18.** $\lvert 2y - 7\rvert \geq -6$	**19.** $\lvert 4 - 3x\rvert = 0$	**20.** $\lvert 13 - 2y\rvert = 8$
21. $\left\lvert 3t - \frac{1}{2}\right\rvert < \frac{7}{2}$	**22.** $\left\lvert \frac{1}{2} - 3t\right\rvert \geq \frac{11}{2}$	**23.** $\lvert y\rvert + 6 = 8$	**24.** $\lvert y\rvert - 7 = 4$

Challenge

For each graph write an open sentence containing absolute value.

25.

$\lvert x\rvert = 1$

26.
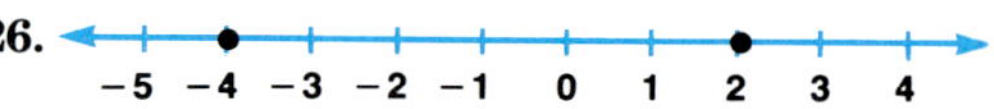

$\lvert x + 1\rvert = 3$

27.
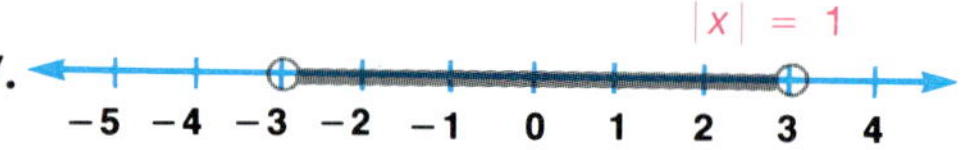

$\lvert x\rvert < 3$

28.
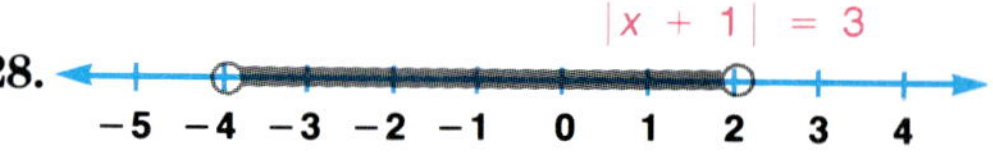

$\lvert x + 1\rvert < 3$

29.

$\lvert x\rvert \geq 2$

30.
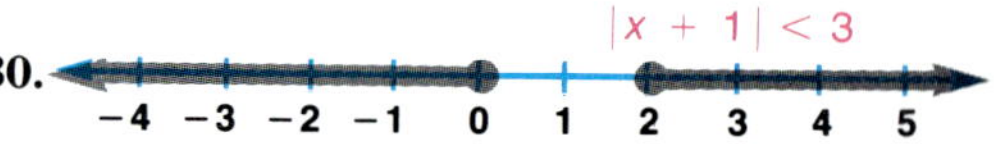

$\lvert x - 1\rvert \geq 1$

Solve each problem. 31. −2, −1, 0, 1, 2 32. −4, −3, −2, −1, 0, 1, 2, 3, 4

31. State all integer solutions of $\lvert x\rvert \leq 2$.

32. State all integer solutions of $\lvert x\rvert \leq 4$.

33. If $\lvert x\rvert \leq a$, how many integer solutions exist? $2a + 1$

34. If $\lvert x\rvert < a$, how many integer solutions exist? $2a - 1$

35. Under what conditions is $-\lvert a\rvert$ negative?

36. Under what conditions is $-\lvert a\rvert$ positive?

37. Suppose $x < 0$, $y > 0$, and $x + y = 0$. Is $\lvert x\rvert > \lvert y\rvert$? no, $\lvert x\rvert = \lvert y\rvert$

35. when a is any real number $\neq 0$

36. It is never positive.

mini-review

Match each expression on the left with one on the right that has the same value.

1. $-6 + (-2)$ c	**a.** $-6 \div 2$
2. $-6\left(\frac{1}{2}\right)$ a	**b.** 3
3. $(-6)(-2)$ e	**c.** $-(6 + 2)$
4. $-6 \div (-2)$ b	**d.** $6 - (-2)$
5. $6 + 2$ d	**e.** $6(2)$

Using Calculators — Inequalities

Use a calculator to help you solve each open sentence. If necessary, round answers to the nearest hundredth.

1. $\lvert 8.36 - r\rvert < 29.4$	**2.** $\lvert 13.04 - a\rvert \geq 2.34$
3. $\lvert 4.96x + 8.9\rvert \leq 10.8$	**4.** $\lvert 194t + 2.36\rvert \leq 14.8$
5. $\lvert 263 + 94.8y\rvert \geq 13.9$	**6.** $\lvert 88.6 - 433m\rvert < 849$

Applications in Industry — Tolerance

Parts used in automobiles must have very precise measurements. Otherwise, they will not work properly. However, it is impossible to produce parts with exact dimensions. Thus, the dimensions of the parts must be between specified limits.

For example, a certain ball bearing that is 1 centimeter in diameter will only work if it is slightly larger or slightly smaller. The diameter may not differ from 1 centimeter by more than 0.001 centimeter. The 0.001 centimeter is called the **tolerance** of the ball bearing. The diameter of the ball bearing must be 1 ± 0.001 centimeter. That is, the diameter can vary between $1 + 0.001$ centimeter and $1 - 0.001$ centimeter. The acceptable diameter can be represented by the following inequality.

$$1 - 0.001 \leq x \leq 1 + 0.001$$
$$0.999 \leq x \leq 1.001$$

The tolerance interval is 0.999 cm to 1.001 cm.

That is, the least possible diameter is 0.999 cm, and the greatest possible diameter is 1.001 cm.

Exercises

Write each expression as an inequality.

1. $x = 3 \pm 0.01$
2. $x = 5 \pm 0.003$
3. $x = 7 \pm 0.0002$
4. $y = 6 \pm 0.0015$
5. $y = 1 \pm 0.15$
6. $y = 2 \pm 0.003$
7. $r = 0.5 \pm 0.0001$
8. $r = 1.5 \pm 0.001$
9. $d = \frac{1}{2} \pm 0.0035$

Solve each problem.

10. A chemical supply company guarantees the precision weighing of its products. They advertise that a certain product weighs 8 oz ± 0.03 oz. What is the tolerance interval?

11. A pane of glass should be 26 inches wide by 32 inches long. The tolerance is $\frac{3}{16}$ inch. Find the tolerance interval for each dimension.

About Applications
This will help prepare students for the use of the ± sign later, as well as provide reinforcement in the use of inequalities and word problems.
You may want to discuss with students the meaning of the expression "precise measurement."

Answers to Exercises

1. $2.99 \leq x \leq 3.01$
2. $4.997 \leq x \leq 5.003$
3. $6.9998 \leq x \leq 7.0002$
4. $5.9985 \leq y \leq 6.0015$
5. $0.85 \leq y \leq 1.15$
6. $1.997 \leq y \leq 2.003$
7. $0.4999 \leq r \leq 0.5001$
8. $1.499 \leq r \leq 1.501$
9. $0.4965 \leq d \leq 0.5035$
10. 7.97 oz to 8.03 oz
11. $25\frac{13}{16}$ in. to $26\frac{3}{16}$ in. wide
$31\frac{13}{16}$ in. to $32\frac{3}{16}$ in. long

4-8

Lesson Notes, p. T33

Discussion Questions

1. Rewrite the first sentence of Written Exercises 1 and 3 using the expressions *greater than or equal to,* or *less than or equal to.*
   Five times a number increased by 12 is greater than or equal to 37.
   The sum of two consecutive positive odd integers is less than or equal to 18.
2. What symbol does the expression *at least* indicate? $\geq$
3. What symbol does the expression *at most* indicate? $\leq$

❶ Point out that figure skating is scored on a scale of 1 to 10. Students may be interested to know that diving and gymnastics are scored similarly.

Chalkboard Exercises

Use an inequality to solve each problem.

1. Cates wants to spend at most \$25 for new stamps for his collection. He has selected a new stamp which costs \$3.50. How much can he spend on other stamps?
   $3.50 + x \leq 25$
   $x \leq 21.50$
   He can spend \$21.50 or less.
2. Jenny has scores of 87, 92, 89, and 97 on algebra quizzes. There will be one more quiz. What must Jenny's fifth score be to have an average score of at least 90?
   $\frac{87 + 92 + 89 + 97 + x}{5} \geq 90$
   $x \geq 85$
   She must score 85 or more.

Objective 4-8: Use inequalities to solve verbal problems.

4-8 Problem Solving: Inequalities

Problems containing the phrases *greater than* or *less than* can often be solved using inequalities. Other phrases that suggest inequalities are *at least, at most,* and *between.* Study the inequalities that correspond to each statement below.

The number x is *at least* 8.	$x \geq 8$	*At least 8 means 8 or greater.*
The number y is *at most* 4.	$y \leq 4$	*At most 4 means 4 or less.*
The number z is *between* 7 and 10.	$7 < z < 10$	*Between 7 and 10 means greater than 7 and less than 10.*
The number w is *between* 7 and 10, *inclusive.*	$7 \leq w \leq 10$	*Between 7 and 10, inclusive, means greater than or equal to 7 and less than or equal to 10.*

Examples

1 **Jerry wishes to spend at most \$12.50 on new equipment for his model railroad. He has chosen a new railroad car that costs \$7.98. How much can he spend on other equipment?**

Explore At most \$12.50 means \$12.50 or less.
Let x = the amount that Jerry can spend for other equipment.

Plan
$$\text{Total spent} \leq 12.50$$
$$7.98 + x \leq 12.50$$

Solve
$$7.98 - 7.98 + x \leq 12.50 - 7.98$$
$$x \leq 4.52$$

Jerry can spend \$4.52 or less for other equipment.

Examine Jerry can spend at most \$4.52 since \$7.98 + \$4.52 = \$12.50. The answer given above is correct. ❶

2 **Roberto has scores of 9.3, 9.2, 9.7, and 8.9 in a figure skating competition. He has one more trial. Roberto will win the competition if his total score is greater than 46.3. What must Roberto's fifth score be?**

Explore You must find Roberto's fifth score so that his total score will be greater than 46.3.
Let x = Roberto's fifth score.

Plan
$$\underbrace{\text{Roberto's total score}} > \underbrace{\text{Opponent's total score}}$$
$$9.3 + 9.2 + 9.7 + 8.9 + x > 46.3$$

Solve
$$37.1 + x > 46.3$$
$$37.1 - 37.1 + x > 46.3 - 37.1$$
$$x > 9.2$$

Roberto's final score must be greater than 9.2. *Examine this solution.*

Written Exercises

Use an inequality to solve each problem.

1. Five times a number increased by 12 is at least 37. What is the number?

2. Seven times a number decreased by 4 times the number is less than 30. What is the number?

3. The sum of two consecutive positive odd integers is at most 18. What are the integers?

4. The sum of two consecutive positive even integers is at most 22. What are the integers?

5. The Two Rivers Tribune pays 5¢ per paper to the carrier. How many papers must the carrier deliver to earn at least $3.50 per day?

6. A bookstore makes a profit of $4.30 on each two-volume set of books sold. How many sets must the store sell to make a profit of at least $175.00?

7. A stove and a freezer weigh at least 260 kg. The stove weighs 115 kg. What is the weight of the freezer?

8. Bill and his father spent at least $110.00 while shopping. Bill spent $47.32. How much did his father spend?

9. Duane earns $12,000 per year in salary and 6% commission on his sales. How much were his sales if his annual income was between $21,000 and $27,000?

10. Sheila bought between 10 and 18 gallons of paint for her house. The paint cost $9.75 per gallon. What was the total cost of the paint?

11. George plans to spend at most $40 for shirts and ties. He bought 2 shirts for $13.95 each. How much can he spend for ties?

12. Jerry plans to spend at most $10.00 on model airplanes and supplies. If he buys two models for $3.49 each, how much will he have left for supplies?

13. If 4 times an integer is increased by 3, the result is between 13 and 25. What is the integer?

14. If 9 less than 6 times a number lies between 31 and 37, what is the number?

15. Glenda has scores of 8.7, 9.3, 8.8, and 9.4 in a figure skating competition. She has one more trial. The leading opponent has a total score of 45.9. What must Glenda's fifth score be if she is to at least tie for first place?

16. Jenny has a total score of 45.9 in five trials of a skating competition. The leading opponent has scores of 9.1, 8.7, 9.5, and 9.3, and has one more trial. What can Jenny's opponent score to assure Jenny of first place?

17. Keith has scores of 9.1, 9.3, 9.6, and 8.7 in a pommel horse competition. He has one more trial. The leading opponent has completed all five trials and has an average score of 9.22. What must Keith score on his final trial if he is to win the competition?

18. Nancy has scores of 9.9, 9.5, 8.9, and 8.7 on the uneven parallel bars. She has one more trial. The leading opponent has completed the competition with an average score of 9.12. What must Nancy score on her final trial if she is to win the competition?

19. The sum of two consecutive even integers is greater than 75. Find the pair with the least sum.

20. Find all sets of three consecutive positive even integers whose sum is less than 30.

Suggested Daily Assignments

Enriched
Written: 2–20 evens

Average
Written: 1–20

Fundamental
Written: 1–19 odds
p. 134, 17–24

Answers to Written

1. The number is greater than or equal to 5.
2. The number is less than 10.
3. 7 and 9; 5 and 7; 3 and 5; or 1 and 3
4. 10 and 12; 8 and 10; 6 and 8; 4 and 6; or 2 and 4
5. He must deliver 70 or more papers per day.
6. The store must sell at least 41 sets.
7. 145 kg or greater
8. $62.68 or more
9. His sales were greater than $150,000 and less than $250,000.
10. between $97.50 and $175.50
11. $12.10 or less
12. $3.02 or less
13. 3, 4, or 5
14. The number is between $6\frac{2}{3}$ and $7\frac{2}{3}$.
15. Her score must be 9.7 or greater.
16. Opponent must score less than 9.3.
17. His score must be greater than 9.4.
18. Her score must be greater than 8.6.
19. 38 and 40
20. 2, 4, and 6; 4, 6, and 8; 6, 8, and 10

About Using Computers
This introduction to the IF-THEN statement gives students more experience with conditional statements, as well as solving comparison problems.
Be sure students understand how the conditional branches in this program work. You may want to have students write the output for some of the exercises before they use the computer. Point out that the computer actually compares the decimal equivalents of the fractions.

Using Computers

Comparisons

In the BASIC language conditional program branches are like railroad branches that are controlled by switches. Whether a train takes a branch, or does not, depends on whether the switch is open or closed. Whether or not a computer takes a conditional branch depends on whether a statement (the condition) is true or false.

A conditional branch uses an IF-THEN statement. In the following program, IF-THEN statements are used to compare two rational numbers.

```
10  INPUT "ENTER THE FIRST NUMERATOR AND DENOMINATOR ";A,B
20  INPUT "ENTER THE SECOND NUMERATOR AND DENOMINATOR ";C,D
30  IF A/B > C/D THEN 60
40  IF A/B < C/D THEN 70
50  PRINT A;"/";B;"=";C;"/";D:GOTO 80
60  PRINT A;"/";B;">";C;"/";D:GOTO 80
70  PRINT A;"/";B;"<";C;"/";D:GOTO 80
80  END
```

A colon is used to connect two statements.

Notice how the comparison property is used in the program. By this property, you know that either $\frac{A}{B} > \frac{C}{D}$, $\frac{A}{B} < \frac{C}{D}$, or $\frac{A}{B} = \frac{C}{D}$. Line 30 checks the first possibility. If the condition is true, the computer branches to line 60. If not, the second possibility is checked in line 40. The branching continues until the appropriate PRINT statement is executed and the program ends.

Exercises

Compare each of the following pairs of rational numbers. If the rational number is negative, enter the negative sign with the numerator.

1. $\frac{9}{11}, \frac{15}{19}$ $\frac{9}{11} > \frac{15}{19}$
2. $\frac{7}{8}, \frac{13}{15}$ $\frac{7}{8} > \frac{13}{15}$
3. $-\frac{7}{8}, -\frac{13}{15}$ $-\frac{7}{8} < -\frac{13}{15}$
4. $\frac{16}{17}, \frac{17}{18}$ $\frac{16}{17} < \frac{17}{18}$
5. $-\frac{7}{6}, -\frac{21}{18}$ $-\frac{7}{6} = -\frac{21}{18}$
6. $\frac{-6}{-5}, \frac{5}{6}$ $\frac{-6}{-5} > \frac{5}{6}$
7. $\frac{-6}{5}, \frac{5}{6}$ $-\frac{6}{5} < \frac{5}{6}$
8. $-\frac{15}{33}, -\frac{22}{45}$ $-\frac{15}{33} > -\frac{22}{45}$
9. Determine which of the following two items is a better buy: a 16-ounce box of crackers for $1.69 or a 9-ounce box for $1.09. a 16-ounce box of crackers for $1.69
10. Determine which of the following two items is a better buy: $1\frac{1}{4}$ pounds of cheese for $3.90 or $1\frac{3}{4}$ pounds for $4.25. $1\frac{3}{4}$ pounds for $4.25

Vocabulary

inequality (111)
comparison property (111)
transitive property of order (112)
solution set (113)
absolute value (132)
properties for inequalities:
- addition (116)
- subtraction (116)
- multiplication (119)
- division (120)

compound sentence (128)

Chapter Summary

1. A mathematical sentence containing $<$ or $>$ is called an inequality. (111)
2. Comparison Property:
   For any two numbers a and b, exactly one of the following sentences is true.
   $$a < b \qquad a = b \qquad a > b \qquad (111)$$
3. Transitive Property of Order:
   For all numbers a, b, and c,
   1. If $a < b$ and $b < c$, then $a < c$.
   2. If $a > b$ and $b > c$, then $a > c$. (112)
4. The symbols $\neq$, $\leq$, and $\geq$ can also be used when comparing numbers. The symbol $\neq$ means *is not equal to*, $\leq$ means *is less than or equal to*, and $\geq$ means *is greater than or equal to*. (112)
5. If a and b represent any numbers and the graph of a is to the left of the graph of b, then $a < b$. If the graph of a is to the right of the graph of b, then $a > b$. (113)
6. Addition and Subtraction Properties for Inequalities:
   For all numbers a, b, and c,
   1. If $a > b$, then $a + c > b + c$ and $a - c > b - c$.
   2. If $a < b$, then $a + c < b + c$ and $a - c < b - c$. (116)
7. Multiplication and Division Properties for Inequalities:
   For all numbers a, b, and c,
   1. If c is positive and $a < b$, then $ac < bc$ and $\frac{a}{c} < \frac{b}{c}$.
      If c is positive and $a > b$, then $ac > bc$ and $\frac{a}{c} > \frac{b}{c}$.
   2. If c is negative and $a < b$, then $ac > bc$ and $\frac{a}{c} > \frac{b}{c}$.
      If c is negative and $a > b$, then $ac < bc$ and $\frac{a}{c} < \frac{b}{c}$.
      (119-120)
8. Comparison Property for Rational Numbers:
   For any rational numbers $\frac{a}{b}$ and $\frac{c}{d}$ with $b > 0$, $d > 0$,
   1. If $\frac{a}{b} < \frac{c}{d}$, then $ad < bc$.
   2. If $ad < bc$, then $\frac{a}{b} < \frac{c}{d}$.
   This property also holds if you replace $<$ by $>$, $\leq$, $\geq$, or $=$. (124)
9. A compound sentence containing *and* is true only if both inequalities are true. (128)
10. A compound sentence containing *or* is true if at least one of the inequalities is true. (129)
11. An open sentence containing absolute value should be interpreted as a compound sentence. (132)

Quiz
Lessons 4-5 to 4-8

1. Replace the ? with $>$, $<$, or $=$ to make a true sentence:
   $-\frac{2}{7}$? $-\frac{4}{9}$
   $-\frac{2}{7} > -\frac{4}{9}$
2. Determine the better buy: three tennis balls for \$4.69 or 7 tennis balls for \$8.50.
   7 for \$8.50

Graph the solution set of each inequality.

3. $6 + m > -2$ and $m - 8 \leq 4$

4. $2y - 4 \leq -10$ or $6 > 8y - 18$

Solve.

5. $|x - 9| \leq 2$
   $7 \leq x \leq 11$
6. $|2c - 1| > 5$
   $c < -2$ or $c > 3$
7. Seven more than the product of four and a number is greater than three. What is the number?
   $7 + 4n > 3$
   $n > -1$
   The number is greater than -1.

Answers to Chapter Review

8. 

9. 

10. 

11. 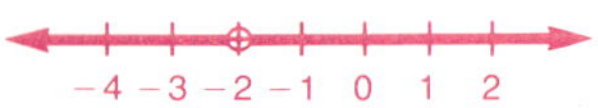

42. 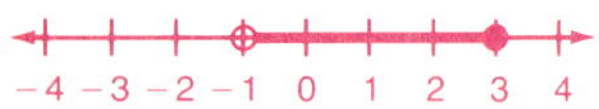

43. 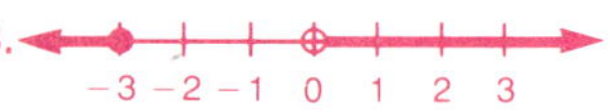

44. 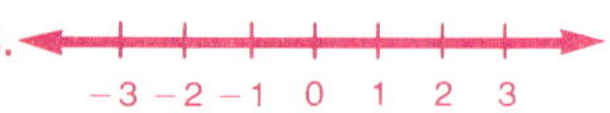

45. 

46. $y > 2$ or $y < -2$

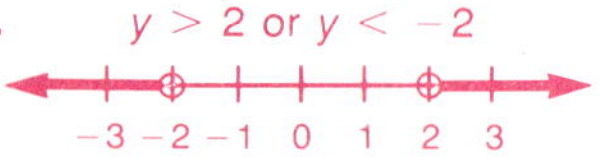

47. $m \le 6$ and $m \ge -4$

48. $t > \frac{5}{2}$ or $t < -2$

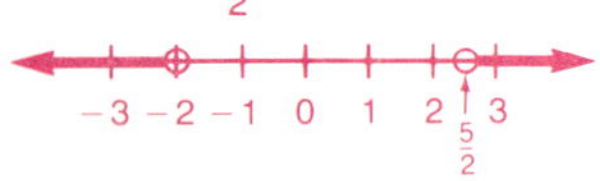

Chapter Review

4-1 **State whether each sentence is true or false.**

1. $4 < -5$ false
2. $13 > 4$ true
3. $7 - 8 \le 0$ true
4. $\frac{6}{2} - 3 \ge 0$ true

Replace each ? with <, >, or = to make each sentence true.

5. $-9 \ \underline{?}\ -11$ $>$
6. $5 \ \underline{?}\ \frac{13}{3}$ $>$
7. $(4.1)(0.2) \ \underline{?}\ 8.2$ $<$

Graph the solution set of each inequality on a number line.

8. $y \ge -4$
9. $x \le 5$
10. $n < 4$
11. $r \ne -2$

4-2 **Solve each inequality.**

12. $m - 4 < 9$ $m < 13$
13. $5y - 6 \ge 4y$ $y \ge 6$
14. $r - 2.3 \ge -7.8$ $r \ge -5$
15. $n + 8 < -3$ $n < -11$
16. $2x + 7 < 3x$ $x > 7$
17. $y + \frac{5}{8} < \frac{11}{24}$ $y < -\frac{1}{6}$

4-3 **Solve each inequality.**

18. $4p > -52$ $p > -13$
19. $\frac{h}{-8} \le 13$ $h \ge -104$
20. $\frac{b}{-7} < -12$ $b > 84$
21. $6a \le -24$ $a \le -$
22. $\frac{4}{3}k > 16$ $k > 12$
23. $-\frac{3}{8}r > 9$ $r < -24$
24. $-7c \le 91$ $c \ge -13$
25. $0.7t < -0.98$ $t <$
26. $-5m > 17$ $m < -\frac{17}{5}$
27. $\frac{2}{3}k > \frac{2}{15}$ $k > \frac{1}{5}$
28. $-0.3x < 4.5$ $x > -15$
29. $\frac{4}{7}z > -\frac{2}{5}$ $z > -\frac{7}{1}$

4-4 **Solve each inequality.**

30. $7m - 12 < 30$ $m < 6$
31. $2r - 0.5 > 3.1$ $r > 1.8$
32. $14a - 11 > 6a + 37$ a
33. $7y + 8 \le 4y - 11$ $y \le -\frac{19}{3}$
34. $4(m - 1) < 7m + 8$ $m > -4$
35. $4z - 11 > 7.3z + 22$ $z < -10$

4-5 **Replace each ? with <, >, or = to make each sentence true.**

36. $\frac{3}{8} \ \underline{?}\ \frac{4}{11}$ $>$
37. $\frac{10}{11} \ \underline{?}\ \frac{11}{12}$ $<$
38. $-\frac{3}{4} \ \underline{?}\ -\frac{7}{9}$ $>$
39. Which is a better buy: 0.75 liter of soda for 89¢ or <u>1.25 liter of soda for $1.31?</u>

4-6 **Write each compound sentence without *and*.**

40. $-4 < y$ and $y \le 7$ $-4 < y \le 7$
41. $-3 \le x$ and $x \le 17$ $-3 \le x \le 17$

Graph the solution set of each compound sentence.

42. $y > -1$ and $y \le 3$
43. $x \le -3$ or $x > 0$
44. $2m + 5 \le 7$ or $2m \ge m - 3$
45. $4r \ge 3r + 7$ and $3r < 33$

4-7 **Solve each inequality. Then graph its solution set.**

46. $|y| > 2$
47. $|m - 1| \le 5$
48. $\left|2t - \frac{1}{2}\right| > \frac{9}{2}$

4-8 **Use an inequality to solve each problem.**

49. If 8 times a number is decreased by 2, the result is between 5 and 15. What is the number? The number is between $\frac{7}{8}$ and $\frac{17}{8}$.
50. Linda plans to spend at most $85 for jeans and shirts. She bought 2 shirts for $15.30 each. How much can she spend on jeans? $54.40 or less

Chapter Test

State whether each sentence is true or false.

1. $4 > -8$ true **2.** $-2 > 0$ false **3.** $0.3 < -0.6$ false

Replace each ? with >, <, or = to make each sentence true.

4. $2 \underline{?} -7$ > **5.** $-4 \underline{?} -3$ < **6.** $\frac{7}{6} \underline{?} \frac{13}{12}$ >

State an inequality for each graph.

7. 

$x \leq 3$

8. 

$x > -2$

Graph the solution set of each inequality on a number line.

9. $y \leq 2$ **10.** $-3 \leq x < 4$ **11.** $y \neq 4$

Solve each inequality.

12. $y - 2 > 11$ $y > 13$

13. $7y > 6y - 11$ $y > -11$

14. $\frac{r}{5} > -4$ $r > -20$

15. $\frac{a}{-3} > -8$ $a < 24$

16. $3z \leq 4.2$ $z \leq 1.4$

17. $5p \geq -19$ $p \geq -\frac{19}{5}$

18. $\frac{3}{4}r + 7 \leq -8$ $r \leq -20$

19. $\frac{2}{3}r > -\frac{7}{12}$ $r > -\frac{7}{8}$

20. $8x + 3 < 13x - 7$ $x > 2$

21. $0.3(m + 4) \leq 0.5(m - 4)$ $m \geq 16$

Graph the solution set of each compound sentence.

22. $x > -3$ and $x < 2$

23. $x \geq 7$ or $x \leq -1$

Solve each inequality. Then graph its solution set.

24. $|n| < 5$

25. $|n| > 3$

26. $|2x - 1| < 5$

27. $|3x - 5| \geq 1$

Use an inequality to solve each problem.

28. Jim plans to spend at most \$35 for record albums and tapes. He bought 2 albums for \$7.95 each. How much can he spend on tapes? \$19.10 or less

29. Seven less than twice a number is less than 83. What is the number? The number is less than 45.

30. The average of four consecutive positive odd integers is less than 20. What are the greatest integers that will satisfy this condition? 15, 17, 19, and 21

Answers to Chapter Test

9. 

10. 

11. 

22. 

23. 

24. $n < 5$ and $n > -5$

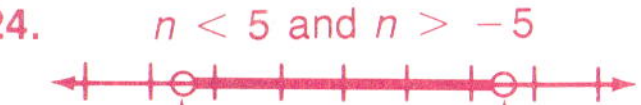

25. $n > 3$ or $n < -3$

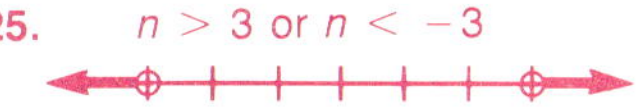

26. $x < 3$ and $x > -2$

27. $x \geq 2$ or $x \leq \frac{4}{3}$

Using the Cumulative Review
If students have difficulty with a particular topic, have them review the appropriate lesson. You may want to provide copies of the solutions from the *Algebra One Solutions Manual*. This Cumulative Review covers Chapters 1-4.

Other Cumulative Reviews appear on pages 72, 208, 268, 334, 402, 466, and 526.

Answers

9.

34.

35.

Cumulative Review

1. Write a mathematical expression for the square of a number increased by twice the number. $x^2 + 2x$
2. Evaluate $\frac{c^2 - a^2}{(c - a)^2}$ if $a = 5$ and $c = 7$. 6
3. Solve: $x = \frac{2}{3} \cdot \frac{3}{4} \div \frac{3}{5}$. $\frac{5}{6}$
4. State the property shown by $0 + 96 = 96$. additive identity

Simplify.

5. $13a + 21a + 14b - 11b$ $34a + 3b$
6. $5x^2 + 2 + 3x - 4x^2$ $x^2 + 3x + 2$

Write each sentence as an equation.

7. Twice x decreased by y is z. $2x - y = z$
8. The product of a, b, and the square of c is f. $abc^2 = f$
9. Graph $\{-1, 0, 1, \ldots\}$ on a number line.

Find each sum or difference.

10. $12 + (-6)$ 6
11. $-11 + (-5)$ -16
12. $-38.6 + 42.73$ 4.13
13. $11a + (-3a)$ $8a$
14. $13 - (-18)$ 31

Solve.

15. $y + (-9) = 19$ 28
16. $p - 13 = 27$ 40
17. $-18.3 = x - 11.6$ -6.7

Simplify.

18. $(-9)(7)(2)$ -126
19. $\frac{3}{4}(12a - 72b) + \frac{2}{5}(10a + 15b)$ $13a - 48b$
20. $\frac{-36}{-12}$ 3
21. $\frac{20x + 30y}{-10}$ $-2x - 3y$

Solve.

22. $\frac{a}{-3} = 42$ -126
23. $4x + 2 = -10$ -3
24. $3(7m - 2) = 4(-7m + 2)$ $\frac{2}{7}$
25. $\frac{y - 3}{6} = \frac{2}{3}$ 7
26. $\frac{4}{9} = \frac{8}{r}$ 18
27. What is 35% of 70? 24.5
28. State an inequality for the graph. $x > -2$

Solve each inequality.

29. $y - 9 > 4$ $y > 13$
30. $-\frac{3}{8}x \geq 6$ $x \leq -16$
31. $7 + 3t \leq 14$ $t \leq \frac{7}{3}$
32. $4(a + 6) > 7a - 8$ $a < \frac{32}{3}$
33. Replace __?__ with <, >, or = to make the sentence $-\frac{7}{8}$ __?__ $-\frac{28}{32}$ true. =

Solve each inequality and graph the solution set.

34. $x - 4 < 0$ and $2x + 5 > 1$ $x < 4$ and $x > -2$
35. $|7 - x| \geq 2$ $x \leq 5$ or $x \geq 9$

Problem Solving

Solve each problem.

36. The difference of two integers is -117. One of the integers is -68. What is the other integer? 49 or -185
37. Two meters of copper wire weigh 0.4 kg. How much will 50 meters of wire weigh? 10 kg
38. The sales tax on a \$44 purchase is \$2.31. What is the rate of sales tax? $5\frac{1}{4}\%$
39. Connie earns \$17,000 per year in salary and a 7% commission on her sales. How much must her sales be for her income to be at least \$25,000? at least \$114,285.71

Standardized Test Practice Questions

The test questions on this page deal with fraction concepts. The information at the right may help you with some of the questions. Study the examples.

Directions: Choose the best answer. Write A, B, C, or D.

1. In a class of 27 students, six are honor students. What part of the class are *not* honor students?

   (A) $\frac{7}{11}$ (B) $\frac{7}{9}$ (C) $\frac{2}{7}$ (D) $\frac{2}{9}$

2. Which of the following fractions is less than $\frac{1}{5}$?

   (A) $\frac{3}{14}$ (B) $\frac{21}{100}$ (C) $\frac{2}{11}$ (D) $\frac{101}{501}$

3. Which fraction is greater than $\frac{1}{4}$ but less than $\frac{1}{3}$?

   (A) $\frac{5}{12}$ (B) $\frac{1}{5}$ (C) $\frac{4}{13}$ (D) $\frac{3}{4}$

4. Which of the following is the least?

   (A) 0.77 (B) $\frac{7}{9}$ (C) $\frac{8}{11}$ (D) $\frac{3}{4}$

5. Which of the following is the greatest?

   (A) $\frac{1}{2}$ (B) $\frac{7}{13}$ (C) $\frac{4}{9}$ (D) $\frac{8}{15}$

6. The months from April through December, inclusive, are what fractional part of a year?

   (A) $\frac{7}{12}$ (B) $\frac{3}{4}$ (C) $\frac{5}{12}$ (D) $\frac{2}{3}$

7. What fractional part of a week is one-sixth of a day?

   (A) $\frac{1}{42}$ (B) $\frac{1}{30}$ (C) $\frac{4}{7}$ (D) $\frac{7}{168}$

8. The difference between $42\frac{3}{8}$ minutes and $41\frac{2}{3}$ minutes is approximately how many seconds?

   (A) 18 (B) 63 (C) $22\frac{1}{2}$ (D) 43

1. The fraction $\frac{a}{b}$ means $a \div b$. The *numerator* is a and the *denominator* is b.
2. As the numerator of a positive fraction increases, the value of the fraction increases.

   Example: $\frac{1}{5} < \frac{2}{5} < \frac{3}{5} < \frac{10}{5} < \frac{13}{5}$
3. As the denominator of a positive fraction increases, the value of the fraction decreases.

   Example: $\frac{3}{2} > \frac{3}{3} > \frac{3}{4} > \frac{3}{8} > \frac{3}{11}$
4. A fraction may be used to indicate part of a group.

   $\frac{\text{part}}{\text{whole}} = \text{fractional part}$

   Example: If there are 4 boys in a class of 9 students, then boys represent $\frac{4}{9}$ of the class.

9. Which group of numbers is arranged in descending order?

   (A) $\frac{5}{7}, \frac{7}{12}, \frac{6}{11}, \frac{3}{13}$ (B) $\frac{7}{12}, \frac{5}{7}, \frac{6}{11}, \frac{3}{13}$

   (C) $\frac{5}{7}, \frac{6}{11}, \frac{7}{12}, \frac{3}{13}$ (D) $\frac{3}{13}, \frac{7}{12}, \frac{6}{11}, \frac{5}{7}$

10. Which group of numbers is arranged from greatest to least?

    (A) $3, \frac{1}{4}, -1, -0.5$ (B) $-1, -0.5, \frac{1}{4}, 3$

    (C) $3, \frac{1}{4}, -0.5, -1$ (D) $-0.5, \frac{1}{4}, -1, 3$

11. A car was driven twice as many miles in July as in each of the other 11 months of the year. What fraction of the total mileage for the year occurred in July?

    (A) $\frac{2}{11}$ (B) $\frac{2}{13}$ (C) $\frac{1}{6}$ (D) $\frac{1}{7}$

Using the Standardized Test Practice Questions

The questions on this page may be used to help students prepare for pre-college tests. These questions usually require careful analysis and a thorough understanding of the concepts.

This page could be used as an overnight assignment for better students. Other students may benefit most if only a few questions are assigned at a time.

After students have chosen an answer for each problem, you should discuss how each problem can be solved, or provide copies of the solutions from the *Algebra One Solutions Manual*.

Other practice questions appear in this text on pages 73, 209, 269, 335, 403, 467, and 527. Also see pages 50-58 in the *Teacher Resource Book*.

About the Photo

Expressing the numbers mentioned using powers of ten is called scientific notation. This is an application of powers that is widely used in the scientific community since scientists quite often work with very large and very small numbers. Scientific notation provides a much more concise way of writing such numbers. For example, the number 1,500,000,000,000,000,000 is 1.5×10^{18} in scientific notation.

Teacher Resource Book Materials for Chapter 5
See pages 139-158.

Chapter Test, Form I (Multiple Choice)	139-141
Chapter Test, Form II (Free Response)	142-143
Quizzes	144
Extra Practice	145-146
Application: Using Scientific Notation (Use after 5-4.)	147-148
Application: Using Equations and Inequalities (Use after 5-7.)	149-151
Extension: The Metric System and Scientific Notation (Use after 5-4.)	152
Problem Solving	153
Cumulative Review	154
Answers	155-158

CHAPTER 5

Powers

We live in the Milky Way galaxy. The galaxy closest to us is the Large Magellanic Cloud. It is about 160,000 light years or 1,500,000,000,000,000,000 kilometers from Earth. The number 1,500,000,000,000,000,000 can be expressed more concisely by using powers of ten. You will learn more about powers in this chapter.

Objective 5-1: Multiply monomials.

5-1 Multiplying Monomials

A **monomial** is a number, a variable, or a product of numbers and variables. Some examples of monomials are $-9, y, 7a, 3y^3$, and $\frac{1}{2}abc^5$. Monomials such as -9 and $\frac{5}{3}$, which do not contain variables, are called **constant monomials**, or **constants**. ❶

The following expressions are *not* monomials.

$$m + n \qquad \frac{x}{y} \qquad 3a - 4ab \qquad \frac{1}{x^2}$$

Why is each expression not a monomial?

Consider each of the following products.

$$4 \cdot 16 = 64 \qquad 8 \cdot 16 = 128 \qquad 8 \cdot 32 = 256$$

Replace each factor by a power of 2.

$$2^2 \cdot 2^4 = 2^6 \qquad 2^3 \cdot 2^4 = 2^7 \qquad 2^3 \cdot 2^5 = 2^8$$

Recall that an expression of the form x^n is a power. The base is x and the exponent is n.

Consider the exponents only. How do you obtain the exponent 6 from the exponents 2 and 4? How do you obtain 7 from 3 and 4? How do you obtain 8 from 3 and 5?

Think about $a^2 \cdot a^3$.

a^2 **means** $a^1 \cdot a^1$ and a^3 **means** $a^1 \cdot a^1 \cdot a^1$.

$a^2 \cdot a^3$ **means** $(a^1 \cdot a^1) \cdot (a^1 \cdot a^1 \cdot a^1)$.

$$\underbrace{a^1 \cdot a^1}_{2} \cdot \underbrace{a^1 \cdot a^1 \cdot a^1}_{+\ 3} = a^5 \quad (= 5)$$

$$a^2 \cdot a^3 = a^{2+3} = a^5$$

These and many similar examples suggest that you can multiply powers that have the same base by adding exponents.

For any number a and positive integers m and n, $a^m \cdot a^n = a^{m+n}$.

Product of Powers

Examples

1 Simplify: $x^3 \cdot x^4$

$$x^3 \cdot x^4 = x^{3+4}$$
$$= x^7$$

To multiply powers that have the same base, add the exponents.

2 Simplify: $(3a^2)(4a)$

$$(3a^2)(4a) = 3(4)(a^2)(a)$$
$$= 12a^{2+1}$$
$$= 12a^3$$

Use the commutative and associative properties. Recall $a = a^1$.

5-1

Lesson Notes, p. T33

Readiness Activity

Which of the following expressions have the same value?

$3^2 \cdot 3^3$	$10^2 \cdot 10^2$
5^3	2^6
10^4	$5^2 \cdot 5$
$2^4 \cdot 2^2$	3^5

$3^2 \cdot 3^3$ and 3^5
5^3 and $5^2 \cdot 5$
10^4 and $10^2 \cdot 10^2$
$2^4 \cdot 2^2$ and 2^6

Discussion Questions

If $x^m \cdot x^n = x^{12}$, what are some possible values for m and n? All positive integers m and n such that $m + n = 12$.

Chalkboard Exercises

Simplify.

1a. $y^6 \cdot y^5$ y^{11}
1b. $v^3 \cdot v^3 \cdot v^3$ v^9
2a. $(4b^3)(5b^2)$ $20b^5$
2b. $(6m^4)(2m)$ $12m^5$

❶ Monomial is pronounced mah NOH mee ul .

Chalkboard Exercises

Simplify.

3a. $(xy^4)(x^5y^2)$ x^6y^6

3b. $(r^2s)(r^4s^3)$ r^6s^4

4a. $(2a^4)(2a^3b^2)(-3ab^3)$ $-12a^8b^5$

4b. $(4m^3n^2)(-m^4n^5)(2mn)$ $-8m^8n^8$

4c. $(-3x^2y)(4x^5y^2)(-2x)$ $24x^8y^3$

Suggested Daily Assignments

Enriched
- Written: 2–36 evens, 37–45

Average
- Exp: 7–18
- Written: 1–36

Fundamental
- Exp: 1–18
- Written: 2–36 evens

Examples

3 **Simplify: $(a^3b^2)(a^2b^4)$**

$$(a^3b^2)(a^2b^4) = (a^3 \cdot a^2)(b^2 \cdot b^4)$$
$$= a^{3+2}b^{2+4}$$
$$= a^5b^6$$

4 **Simplify: $(-5x^2)(3x^3y^2)\left(\frac{2}{5}xy^4\right)$**

$$(-5x^2)(3x^3y^2)\left(\tfrac{2}{5}xy^4\right)$$
$$= \left(-5 \cdot 3 \cdot \tfrac{2}{5}\right)(x^2 \cdot x^3 \cdot x)(y^2 \cdot y^4)$$
$$= -6x^6y^6$$

The variables in a monomial are usually arranged in alphabetical order.

Exploratory Exercises

Verify each product by multiplication.

1. $3^1 \cdot 3^2 = 3^3$ $3 \cdot 9 = 27$ **2.** $2^3 \cdot 2^2 = 2^5$ $8 \cdot 4 = 32$ **3.** $4^2 \cdot 4^2 = 4^4$ $16 \cdot 16 = 256$

4. $2^5 \cdot 2^2 = 2^7$ $32 \cdot 4 = 128$ **5.** $5^2 \cdot 5 = 5^3$ $25 \cdot 5 = 125$ **6.** $3^2 \cdot 3^4 = 3^6$ $9 \cdot 81 = 729$

Simplify.

7. $x^3 \cdot x^5$ x^8 **8.** $y^7 \cdot y^7$ y^{14} **9.** $a^4 \cdot a^4$ a^8 **10.** $b^5 \cdot b^2$ b^7 **11.** $m \cdot m^3$ m^4 **12.** $d \cdot d^6$ d^7

13. $(3a^2)(4a^3)$ $12a^5$ **14.** $(-5x^3)(4x^4)$ $-20x^7$ **15.** $(-10x^3y)(2x^2)$ $-20x^5y$

16. $(3y^3z)(7y^4)$ $21y^7z$ **17.** $(y^3z^4)(y^2)$ y^5z^4 **18.** $m^4(m^3b^2)$ m^7b^2

Written Exercises

Simplify.

1. $a^2(a^5)$ a^7 **2.** $b^4 \cdot b^2$ b^6 **3.** $m^5 \cdot m$ m^6

4. $m^2(m^2)$ m^4 **5.** $t^2(t^2)(t^2)$ t^6 **6.** $r^2(r^3)(r^3)$ r^8

7. $a^5(a)(a^7)$ a^{13} **8.** $(b^4)(b^4)(b)$ b^9 **9.** $(a^2b)(ab^4)$ a^3b^5

10. $(x^3y)(xy^3)$ x^4y^4 **11.** $(m^3n)(mn^2)$ m^4n^3 **12.** $(r^3t^4)(r^4t^4)$ r^7t^8

13. $(3a^2b)(2ab^5)$ $6a^3b^6$ **14.** $(4x^2y^3)(2xy^6)$ $8x^3y^9$ **15.** $(3x^4)(-2x^4y^3)$ $-6x^8y^3$

16. $(-8x^3y)(2x^4)$ $-16x^7y$ **17.** $(3x^4y)(4x^2y^2)$ $12x^6y^3$ **18.** $(-2x^2y)(-6x^4y^7)$ $12x^6y^8$

19. $(-3x^5y)(2x^4)$ $-6x^9y$ **20.** $(-2n^4y^3)(3ny^4)$ $-6n^5y^7$ **21.** $(3x^2y^2z)(2x^2y^2z^3)$ $6x^4y^4z^4$

22. $(r^2xy)(-2r^3x)$ $-2r^5x^2y$ **23.** $(5a^2b^2c)(-7a^3)$ $-35a^5b^2c$ **24.** $(2am^3n)(-3am^4)$ $-6a^2m^7n$

25. $(2am)(3a^2m^2n)$ $6a^3m^3n$ **26.** $(4r^2st)(-6s^2t^2)$ $-24r^2s^3t^3$ **27.** $(ab)(ac)(bc)$ $a^2b^2c^2$

28. $(m^2n)(am)(an^2)$ $a^2m^3n^3$ **29.** $\frac{3}{4}a(12b^2)$ $9ab^2$ **30.** $-\frac{5}{6}c(12a^3)$ $-10a^3c$

31. $\left(\frac{1}{2}a^2\right)(6ab^2)$ $3a^3b^2$ **32.** $(-27ay^3)\left(-\frac{1}{3}ay^3\right)$ $9a^2y^6$ **33.** $ab\left(\frac{1}{2}a\right)\left(\frac{1}{2}b\right)\left(\frac{1}{2}c\right)$ $\frac{1}{8}a^2b^2c$

34. $(bc)\left(\frac{2}{3}b\right)\left(\frac{2}{3}a\right)$ $\frac{4}{9}ab^2c$ **35.** $\left(-\frac{1}{8}a\right)\left(-\frac{1}{6}\right)(b)(48c)$ abc **36.** $\left(-\frac{1}{3}c^2b^3a\right)(18a^2b^2c^3)$ $-6a^3b^5c^5$

Challenge

Simplify. Consider all exponents to be positive integers.

37. $y^2 \cdot y^b$ y^{2+b} **38.** $3^x \cdot 3^2$ 3^{x+2} **39.** $a^{x-2} \cdot a^5$ a^{x+3}

40. $2^{4a} \cdot 2^{5a}$ 2^{9a} **41.** $x^{2a} \cdot x^{4a}$ x^{6a} **42.** $(2^{7x+6})(2^{3x-4})$ 2^{10x+2}

43. $(y^{3a+1})(y^{a-6})$ y^{4a-5} **44.** $(x+3)^a \cdot (x+3)^6$ $(x+3)^{a+6}$ **45.** $(x+3)^{2a} \cdot (x+3)^{b-a}$ $(x+3)^{a+b}$

Problem Solving — Look for a Pattern

Another problem-solving strategy is to look for a pattern. When using this strategy, it is important to organize information about the problem. Study the following example.

Example: **How many diagonals can be drawn for a polygon with *n* sides, where *n* is less than or equal to 12?**

Try several cases. That is, draw several polygons and see how many diagonals you can draw for each polygon. Use a table to record the number of sides and the number of diagonals for each polygon. Then study the pattern formed by the numbers in the table.

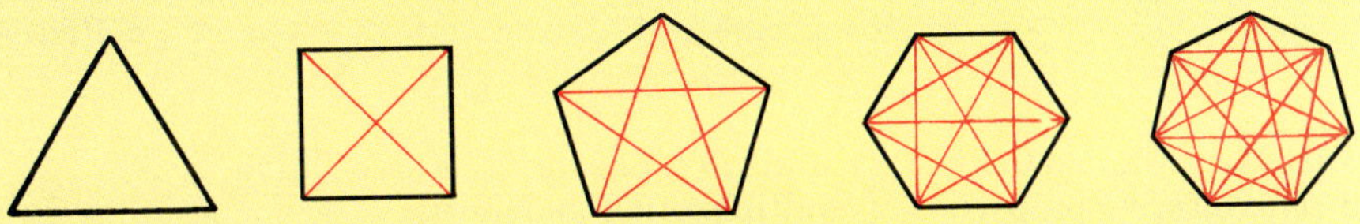

Number of Sides	3	4	5	6	7	8	9	10	11	12
Number of Diagonals	0	2	5	9	14					

+2 +3 +4 +5

Continue the pattern to find the number of diagonals for a polygon with 8, 9, 10, 11, or 12 sides. Thus, the rest of the numbers that belong in the table above are 20, 27, 35, 44, and 54. Do you see why?

Exercises

Solve each problem.

1. There were 10 people at a party. Each person shook hands with each of the other people exactly once. How many handshakes occurred? 45
2. How many squares are shown at the right? (Hint: There are more than 16.) 30

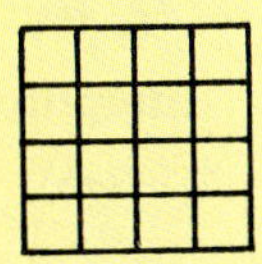

3. How many rectangles are shown below? 15

4. How many triangles are shown at the right? 27

5. Imagine 25 lockers, all closed, and 25 people. Suppose the first person opens every locker. Then the second person closes every second locker. Next the third person changes the state of every third locker. (If it's open, she closes it. If it's closed, she opens it.) Suppose this procedure is continued until the 25th person changes the state of the 25th locker. Which lockers will be open at the end of the procedure? 1, 4, 9, 16, 25

About Problem Solving

Recognizing patterns can make seemingly impossible problems quite simple. Students should enjoy the puzzle type approach to solving these problems.

For enrichment, suggest that students try to discover a formula for the number of diagonals of a polygon with n sides.

$$\text{number of diagonals} = \frac{n(n-3)}{2}$$

Many more problems can be generated by adapting the exercises given. As an extension of Exercise 2, suggest that students find the number of squares on an $n \times n$ checkerboard. Have them find the number of squares for n = 1, 2, 3, 4, and 5. (Exercise 2 shows the case when $n = 4$.) For enrichment, suggest that they try to discover a general formula for the number of squares on an $n \times n$ checkerboard.

n	1	2	3	4	5
Total Squares	1	5	14	30	55

$$\text{number of squares} = 1^2 + 2^2 + 3^2 + \ldots + n^2 = \frac{n(n+1)(2n+1)}{6}$$

5-2

Lesson Notes, p. T34

Discussion Question
If $(a^m)^n = a^{12}$, what are some possible values for m and n? All integers m and n such that $mn = 12$.

Chalkboard Exercises
Simplify.
1a. $(m^2)^4$ m^8
1b. $(y^5)^2$ y^{10}
1c. $(4xy)^3$ $64x^3y^3$
1d. $(2mn)^5$ $32m^5n^5$
2a. $(x^4y^5)^2$ x^8y^{10}
2b. $(a^2b^4)^3$ a^6b^{12}

❶ Clarify the difference between $(a^m)^n$ and $a^m a^n$.

Objective 5-2: Simplify expressions involving powers of monomials.

5-2 Powers of Monomials

Mary wanted to rewrite the expressions $(5^2)^4$ and $(x^6)^2$ using only a single exponent. She wrote the expressions correctly as follows.

$$(5^2)^4 = (5^2)(5^2)(5^2)(5^2) = 5^{2+2+2+2} = 5^8 \qquad (x^6)^2 = (x^6)(x^6) = x^{6+6} = x^{12}$$

How can you obtain the exponent 8 from the exponents 2 and 4? How can you obtain 12 from 6 and 2? These and many similar examples suggest the following rule.

For any number a and any integers m and n,
$(a^m)^n = a^{mn}$.

❶ ***Power of a Power***

Next Mary wanted to find $(xy)^3$ and $(ab)^4$. She used the associative and commutative properties of multiplication as follows.

$$(xy)^3 = (xy)(xy)(xy) = (x \cdot x \cdot x)(y \cdot y \cdot y) = x^3y^3 \qquad (ab)^4 = (ab)(ab)(ab)(ab) = (a \cdot a \cdot a \cdot a)(b \cdot b \cdot b \cdot b) = a^4b^4$$

These examples suggest that the power of a product is the product of the powers.

For all numbers a and b and any integer m,
$(ab)^m = a^mb^m$.

Power of a Product

Examples

1 **Simplify: $(-3ab)^4$**

$$(-3ab)^4 = (-3)^4a^4b^4 = 81a^4b^4$$

2 **Simplify: $(x^2y^4)^3$**

$$(x^2y^4)^3 = (x^2)^3 \cdot (y^4)^3 = x^{2\cdot3}y^{4\cdot3} = x^6y^{12}$$

Example 2 shows the combined use of the rules for the power of a power and the power of a product. This can be stated as follows.

For all numbers a and b and any integers m, n, and p,
$(a^mb^n)^p = a^{mp}b^{np}$.

Power of a Monomial

Examples

3 **Simplify:** $\left(-\frac{2}{3}a^2x^3\right)^3$

$$\left(-\frac{2}{3}a^2x^3\right)^3 = \left(-\frac{2}{3}\right)^3(a^2)^3(x^3)^3$$
$$= -\frac{8}{27}a^6x^9$$

4 **Simplify:** $(9b^4y)^2[(-b)^2]^3$

$$(9b^4y)^2[(-b)^2]^3 = 9^2(b^4)^2y^2(b^2)^3$$
$$= 81b^8y^2b^6$$
$$= 81b^{14}y^2$$

Exploratory Exercises

Simplify.

1. $(m^2)^4$ m^8 **2.** $(n^4)^3$ n^{12} **3.** $(3y)^2$ $9y^2$ **4.** $(4x)^3$ $64x^3$ **5.** $[(-x)y]^4$ x^4y^4 **6.** $[(-p)q]^2$ p^2q^2

7. $[(-a)b^2]^3$ $-a^3b^6$ **8.** $[(-x)y^2]^5$ $-x^5y^{10}$ **9.** $(x^2y^5)^2$ x^4y^{10} **10.** $(n^3r^2)^4$ $n^{12}r^8$ **11.** $\left(\frac{2}{3}a^2\right)^3$ $\frac{8}{27}a^6$ **12.** $\left(-\frac{1}{2}a^4\right)^3$ $-\frac{1}{8}a^{12}$

Written Exercises

Simplify.

1. $(5^3)^3$ 5^9 or 1,953,125 **2.** $(10^2)^2$ 10^4 or 10,000 **3.** $[(-5)^2]^3$ $(-5)^6$ or 15,625 **4.** $[(-4)^2]^2$ $(-4)^4$ or 256

5. $(x^4)^3$ x^{12} **6.** $(m^2)^5$ m^{10} **7.** $[(-y)^3]^6$ y^{18} **8.** $[(-m)^5]^2$ m^{10}

9. $(5c)^3$ $125c^3$ **10.** $(10y)^2$ $100y^2$ **11.** $(-7z)^3$ $-343z^3$ **12.** $(-3n)^4$ $81n^4$

13. $\left(\frac{1}{2}c\right)^2$ $\frac{1}{4}c^2$ **14.** $\left(\frac{2}{5}d\right)^2$ $\frac{4}{25}d^2$ **15.** $(0.4d)^2$ $0.16d^2$ **16.** $(0.6d)^3$ $0.216d^3$

17. $(ab^2)^3$ a^3b^6 **18.** $(a^3x^2)^4$ $a^{12}x^8$ **19.** $(2a^2b)^2$ $4a^4b^2$ **20.** $(3x^2y^3)^3$ $27x^6y^9$

21. $4(a^2b^3)^3$ $4a^6b^9$ **22.** $-3(ax^3y)^2$ $-3a^2x^6y^2$ **23.** $(-5x^3y)^3$ $-125x^9y^3$ **24.** $(-3a^2b^5)^2$ $9a^4b^{10}$

25. $\left(\frac{1}{2}xy^2\right)^3$ $\frac{1}{8}x^3y^6$ **26.** $\left(\frac{3}{4}x^2y^5\right)^2$ $\frac{9}{16}x^4y^{10}$ **27.** $(0.2a^2)^3$ $0.008a^6$ **28.** $(0.3x^3y^2)^2$

29. $(-0.1x^2)^2$ $0.01x^4$ **30.** $(-0.2x^3y)^3$ $-0.008x^9y^3$ **31.** $10^2 \cdot 10^3$ 10^5 **32.** $4^2 \cdot 4^3 \cdot 4^4$ 4^9

33. $(4xy)^2(-3x)$ $-48x^3y^2$ **34.** $(2a)^2(3y)$ $12a^2y$ **35.** $(2a)^3(-3b)$ $-24a^3b$ **36.** $(-3ab)^3(2b^2)$

37. $(4x^2y^3)^3$ $64x^6y^9$ **38.** $(3ab^4)^3$ $27a^3b^{12}$ **39.** $(-2a^2b^3)^4$ $16a^8b^{12}$ **40.** $(-6a^3x^5)^2$

41. $(2x^2)^2\left(\frac{1}{2}y^2\right)^2$ x^4y^4 **42.** $\left(\frac{3}{10}y^2\right)^2(10y^2)^3$ $90y^{10}$ **43.** $(4ab)^2(a^3b)^4$ $16a^{14}b^6$ **44.** $(-3x^2y)^2(2x)^3$

45. $(3a^2)^3 + (5a^2)^3$ $152a^6$ **46.** $(5y)^2 + (3y)(7y)$ $46y^2$ **47.** $(-3x^2y)^3(x^3) - 3(x^2y)^2(x^5y)$

48. $(3ab)^2(2a^2b) + 5a^3(ab^3)$ $23a^4b^3$ **49.** $(a^2)^3 + (10a^3)^2 + (5ab)^6$ $101a^6 + 15{,}625a^6b^6$ **50.** $(-3a^2b)^3 + (5a^6b^3)^2 + 10a^6b^3$ $-17a^6b^3 + 25a^{12}b^6$

Challenge

Evaluate if $x = 2$, $y = 3$, and $z = 4$.

51. x^x 4 **52.** y^x 9 **53.** z^y 64 **54.** $(2x)^2(3)^y$ 432 **55.** $(4x)^y(2x)^y$ 32,768 **56.** $(y + x)^z$ 625

Using Calculators — Powers of Numbers

Some calculators have a power key, labeled y^x. The following example shows how you can use this key to find 7^8.

ENTER: 7 [y^x] 8 [=] The display shows 5764801. So, $7^8 = 5{,}764{,}801$.

Exercises Use a calculator to simplify each expression.

1. $(40.8)^3$ 67,917.312 **2.** $(0.018)^4$ 0.0000001 **3.** $(4^2)^2$ 256 **4.** $(9^2)^3$ 531,441 **5.** $[(2^3)^2]^4$ 16,777,216 **6.** $[(3^2)^4]^2$ 43,046,721

For more exercises on multiplying monomials, see exercises 1-9 on page 557.
For more exercises on evaluating expressions, see exercises 37-48 on page 554.

Chalkboard Exercises

Simplify.

3a. $(-xy)^4$ x^4y^4
3b. $(-xy)^5$ $-x^5y^5$
3c. $(2x^3y^4)^2$ $4x^6y^8$
3d. $\left(-\frac{1}{2}x^3y\right)^4$ $\frac{1}{16}x^{12}y^4$
4a. $(6ab^2)^3(5a)^2$ $5400a^5b^6$
4b. $(3m^3n^2)^4(-n^2)^5$ $-81m^{12}n^{18}$
4c. $(3r^2t)^3(3r^2)^4$ $2187r^{14}t^3$
4d. $5(-x^2)^3(2x^3y^2)^3$ $-40x^{15}y^6$

Suggested Daily Assignments

Enriched
Written: 2–50 evens, 51–56

Average
Exp: 2–12 evens
Written: 1–49 odds
p. 146, 37, 38

Fundamental
Exp: 1–12
Written: 2–44 evens
p. 146, 1–35 odds

Answers to Written

28. $0.09x^6y^4$
36. $-54a^3b^5$
40. $36a^6x^{10}$
44. $72x^7y^2$
47. $-30x^9y^3$

About Using Calculators

You may want to mention that many calculators can not compute a negative number raised to a power. For example, on such calculations, entering $(-2)^3$ will result in an error message.
Some calculators without an x^2-key will find the square of a number as follows:
Enter the number, then [×] [=].

5-3

Lesson Notes, p. T34

Readiness Activity
Simplify.

$\frac{56}{-7}$ -8 $\qquad$ $\frac{-144}{-2}$ 72

$\frac{-81}{9}$ -9 $\qquad$ $\frac{13b}{-39}$ $\frac{b}{-3}$

$\frac{-8c}{64}$ $\frac{-c}{8}$ $\qquad$ $\frac{-5}{-35d}$ $\frac{1}{7d}$

Discussion Question
Why is 0^0 undefined?
0^0 is derived from an expression such as
$0^{x-x} = \frac{0^x}{0^x}$ which is not defined.

Chalkboard Exercises
Simplify.

1a. $\frac{y^9}{y^7}$ $\quad y^2$

1b. $\frac{m^6}{m}$ $\quad m^5$

2a. $\frac{m^7n^5}{mn^2}$ $\quad m^6n^3$

2b. $\frac{x^3y^7}{x^2y^6}$ $\quad xy$

2c. $\frac{3^4a^8b^3}{3^2a^5b}$ $\quad 3^2a^3b^2$ or $9a^3b^2$

Objective 5-3A: Simplify expressions involving quotients of monomials.
Objective 5-3B: Simplify expressions containing negative exponents.

5-3 Dividing Monomials

Consider quotients **a**, **b**, and **c** shown below.

a. $\frac{81}{27} = 3$ $\qquad$ **b.** $\frac{27}{3} = 9$ $\qquad$ **c.** $\frac{243}{9} = 27$

A power of 3 can be substituted for each number.

a. $\frac{3^4}{3^3} = 3^1$ $\qquad$ **b.** $\frac{3^3}{3^1} = 3^2$ $\qquad$ **c.** $\frac{3^5}{3^2} = 3^3$

Consider the exponents only. In example **a**, how do you obtain the exponent 1 from the exponents 4 and 3? Look at examples **b** and **c**. How do you obtain 2 from 3 and 1? How do you obtain 3 from 5 and 2?

You can use your knowledge of exponents to simplify $\frac{b^5}{b^2}$, $b \neq 0$.

$$\frac{b^5}{b^2} = \frac{\cancel{b} \cdot \cancel{b} \cdot b \cdot b \cdot b}{\cancel{b} \cdot \cancel{b}}$$ *Notice that $\frac{b \cdot b}{b \cdot b}$ is equal to 1.*

$$= b \cdot b \cdot b$$ *The quotient has (5 − 2) or 3 factors.*

$$= b^3$$

Notice that you can divide powers that have the same base by subtracting exponents.

Dividing Powers

> **For all integers m and n, and any nonzero number a,**
> $$\frac{a^m}{a^n} = a^{m-n}.$$

Why is 0 not an acceptable replacement for a?

Examples

1 **Simplify:** $\frac{x^8}{x^5}$

$$\frac{x^8}{x^5} = x^{8-5}$$
$$= x^3$$

2 **Simplify:** $\frac{a^4b^3}{ab^2}$

$$\frac{a^4b^3}{ab^2} = \left(\frac{a^4}{a^1}\right)\left(\frac{b^3}{b^2}\right)$$ *Group the powers that have the same base.*

$$= a^{4-1}b^{3-2}$$
$$= a^3b$$

Study the two ways to simplify $\frac{a^3}{a^3}$, $a \neq 0$, shown below.

$$\frac{a^3}{a^3} = \frac{a \cdot a \cdot a}{a \cdot a \cdot a} = 1 \qquad\qquad \frac{a^3}{a^3} = a^{3-3} = a^0$$ *Use the rule for dividing powers.*

Since $\frac{a^3}{a^3}$ cannot have two different values, you can conclude that a^0 is equal to 1. In general, *any* nonzero number raised to the zero power is equal to 1.

0^0 is not defined.

For any nonzero number a, $a^0 = 1$.

Zero Exponent ❶

Study the two ways to simplify $\frac{k^2}{k^7}$, $k \neq 0$, shown below.

$$\frac{k^2}{k^7} = \frac{\cancel{k} \cdot \cancel{k}}{\cancel{k} \cdot \cancel{k} \cdot k \cdot k \cdot k \cdot k \cdot k} = \frac{1}{k \cdot k \cdot k \cdot k \cdot k} = \frac{1}{k^5}$$

$$\frac{k^2}{k^7} = k^{2-7} = k^{-5}$$

Use the rule for dividing powers.

Since $\frac{k^2}{k^7}$ cannot have two different values, you can conclude that k^{-5} is equal to $\frac{1}{k^5}$. This example suggests the following rule.

For any nonzero number a and any integer n, $a^{-n} = \frac{1}{a^n}$.

Negative Exponents ❷

To simplify a quotient of monomials, write an equivalent expression that has positive exponents and no powers of powers. Also, each base should appear only once and all fractions should be in simplest form.

Examples

3 **Simplify:** $\frac{-2r^3s^5}{6r^7s^5}$

$$\frac{-2r^3s^5}{6r^7s^5} = \left(-\frac{2}{6}\right)\left(\frac{r^3}{r^7}\right)\left(\frac{s^5}{s^5}\right) = -\frac{1}{3}r^{-4}s^0 = \left(-\frac{1}{3}\right)\left(\frac{1}{r^4}\right)(1) = -\frac{1}{3r^4}$$ ❸

4 **Simplify:** $\frac{20b^2c^{-4}}{15a^{-2}b^7c^{-3}}$

$$\frac{20b^2c^{-4}}{15a^{-2}b^7c^{-3}} = \left(\frac{20}{15}\right)\left(\frac{1}{a^{-2}}\right)\left(\frac{b^2}{b^7}\right)\left(\frac{c^{-4}}{c^{-3}}\right) = \left(\frac{4}{3}\right)(a^2b^{-5}c^{-1}) = \frac{4a^2}{3b^5c}$$

Note: $\frac{1}{a^{-2}} = \frac{1}{\frac{1}{a^2}} = 1 \cdot a^2 = a^2$

Chalkboard Exercises

Simplify.

3a. $\frac{4x^5}{12x^5}$ $\frac{1}{3}$

3b. $\frac{-15xy^4}{3xy^7}$ $-\frac{5}{y^3}$

3c. $\frac{-3v^4w^9}{-9v^8w^9}$ $\frac{1}{3v^4}$

3d. $\frac{5a^3b^4}{-10a^3b^9}$ $-\frac{1}{2b^5}$

4a. $\frac{144x^5y^{-3}z^5}{12x^6y^3z}$ $\frac{12z^4}{xy^6}$

4b. $\frac{225r^5s^9t^{-2}}{15r^5s^8t^7}$ $\frac{15s}{t^9}$

4c. $\frac{72x^5y^{13}z}{27x^8yz^{-2}}$ $\frac{8y^{12}z^3}{3x^3}$

4d. $\frac{30x^{-2}y^5z^{-3}}{20x^2y^{-3}z^2}$ $\frac{3y^8}{2x^4z^5}$

4e. $\frac{21a^{-4}b^4c^{-2}}{14a^{-2}b^3}$ $\frac{3b}{2a^2c^2}$

❶ Reinforce the zero exponent concept with numerical examples, such as the ones below.

$\frac{2^5}{2^5} = 1 = 2^{5-5} = 2^0$

$\frac{4^2}{4^2} = 1 = 4^{2-2} = 4^0$

❷ Reinforce the negative exponent concept with numerical examples, such as the ones below.

$\frac{5^2}{5^4} = 5^{-2}$ and $\frac{5^2}{5^4} = \frac{1}{5^2}$,

so $5^{-2} = \frac{1}{5^2}$

❸ If you wish, present the following rule for simplifying fractions: If $a \neq 0$ and $y \neq 0$, then $\frac{ax}{ay} = \frac{x}{y}$.

Suggested Daily Assignments

Enriched
- Written: 2–44 evens, 45–54

Average
- Exp: 1–23 odds
- Written: 2–44 evens, 51, 52

Fundamental
- Exp: 1–24
- Written: 1–24
  - p. 149, 1–43 odds

Exploratory Exercises

Simplify. Remember to express the results with positive exponents. Assume no denominator is equal to zero.

1. k^0 1
2. b^0 1
3. y^{-3} $\frac{1}{y^3}$
4. m^{-8} $\frac{1}{m^8}$
5. r^0s^4 s^4
6. $m^{-5}n^0$ $\frac{1}{m^5}$
7. a^0c^{-7} $\frac{1}{c^7}$
8. f^6g^0 f^6
9. $\frac{1}{2}x^3y^{-6}$ $\frac{x^3}{2y^6}$
10. $\frac{2}{3}a^{-7}b^{-4}$ $\frac{2}{3a^7b^4}$
11. $c^0d^{-2}e^{-1}$ $\frac{1}{d^2e}$
12. $x^5y^0z^{-5}$ $\frac{x^5}{z^5}$
13. $\frac{k^9}{k^4}$ k^5
14. $\frac{r^5}{r^2}$ r^3
15. $\frac{x^2}{x^3}$ $\frac{1}{x}$
16. $\frac{c^4}{5c^5}$ $\frac{1}{5c}$
17. $\frac{5n^5}{n^8}$ $\frac{5}{n^3}$
18. $\frac{w^2}{w^9}$ $\frac{1}{w^7}$
19. $\frac{a^2b^7}{a^4}$ $\frac{b^7}{a^2}$
20. $\frac{b^9}{b^4c^3}$ $\frac{b^5}{c^3}$
21. $\frac{1}{x^{-1}}$ x
22. $\frac{1}{r^{-4}}$ r^4
23. $\frac{a^{-4}}{b^{-3}}$ $\frac{b^3}{a^4}$
24. $\frac{r^{-5}}{k^{-1}}$ $\frac{k}{r^5}$

Written Exercises

Simplify. Assume no denominator is equal to zero.

1. $\frac{n^8}{n^5}$ n^3
2. $\frac{w^9}{w^2}$ w^7
3. $\frac{x^2}{x^3}$ $\frac{1}{x}$
4. $\frac{b^6}{b^7}$ $\frac{1}{b}$
5. $\frac{a^0}{a^{-2}}$ a^2
6. $\frac{1}{r^{-3}}$ r^3
7. $\frac{k^{-2}}{k^4}$ $\frac{1}{k^6}$
8. $\frac{m^2}{m^{-4}}$ m^6
9. $\frac{an^6}{n^5}$ an
10. $\frac{xy^7}{y^4}$ xy^3
11. $\frac{an^3}{n^5}$ $\frac{a}{n^2}$
12. $\frac{kn^2}{n^4}$ $\frac{k}{n^2}$
13. $\frac{b^6c^5}{b^3c^2}$ b^3c^3
14. $\frac{(-a)^4b^8}{a^4b^7}$ b
15. $\frac{(-x)^3y^3}{x^3y^6}$ $-\frac{1}{y^3}$
16. $\frac{a^2b^2}{a^4b^5}$ $\frac{1}{a^2b^3}$
17. $\frac{12b^5}{4b^4}$ $3b$
18. $\frac{48a^8}{12a}$ $4a^7$
19. $\frac{12b^4}{60b}$ $\frac{b^3}{5}$
20. $\frac{10m^4}{30m}$ $\frac{m^3}{3}$
21. $\frac{x^3y^6}{x^3y^3}$ y^3
22. $\frac{a^6b^3}{a^2b^9}$ $\frac{a^4}{b^6}$
23. $\frac{a^3b^4}{a^2b^2}$ ab^2
24. $\frac{b^6c^5}{b^{14}c^2}$ $\frac{c^3}{b^8}$
25. $\frac{w^5t^7}{w^3t^{12}}$ $\frac{w^2}{t^5}$
26. $\frac{(-r)^5s^8}{r^5s^2}$ $-s^6$
27. $\frac{30x^4y^7}{-6x^{13}y^2}$ $-\frac{5y^5}{x^9}$
28. $\frac{24a^3b^6}{-2a^2b^2}$ $-12ab^4$
29. $\frac{16b^4c}{-4bc^3}$ $-\frac{4b^3}{c^2}$
30. $\frac{-8a^3b^7}{a^2b^6}$ $-8ab$
31. $\frac{22a^2b^5c^7}{-11abc^2}$ $-2ab^4c^5$
32. $\frac{24x^2y^7z^3}{-6x^2y^3z^1}$ $-4y^4z^2$
33. $\frac{9xyz^5}{x^4}$ $\frac{9yz^5}{x^3}$
34. $\frac{ab^5c}{ac}$ b^5
35. $\frac{7x^3z^5}{4z^{15}}$ $\frac{7x^3}{4z^{10}}$
36. $\frac{27a^4b^6c^9}{15a^3c^{15}}$ $\frac{9ab^6}{5c^6}$
37. $\frac{a^7b^2}{a^{-2}b}$ a^9b
38. $\frac{r^{-5}s^{-2}}{r^2s^5}$ $\frac{1}{r^7s^7}$
39. $\frac{5r^{-1}s}{s^2}$ $\frac{5}{rs}$
40. $\frac{r^{-4}k^2}{5k^2}$ $\frac{1}{5r^4}$
41. $\frac{3m^2n^2}{6m^{-1}k}$ $\frac{m^3n^2}{2k}$
42. $\frac{-b^{-1}c}{4a^{-1}c^2}$ $-\frac{a}{4bc}$
43. $\frac{2xy^{-2}z^4}{3xyz^{-1}}$ $\frac{2z^5}{3y^3}$
44. $\frac{7m^{-1}n^3}{n^2r^{-2}}$ $\frac{7nr^2}{m}$

Challenge

Simplify.

45. $\frac{x^{y+2}}{x^{y-3}}$ x^5
46. $\frac{y^{x-3}}{y^{x+4}}$ $\frac{1}{y^7}$
47. $\frac{(a^{x+2})^2}{(a^{x-3})^2}$ a^{10}
48. $\frac{a^b}{a^{a-b}}$ a^{2b-a}

Solve each of the following for *k*.

49. $x^{k-12} = (x^2)^{-k-3}$ 2
50. $m^k \cdot m^{-15} = (m^3)^{k+2}$ $-\frac{21}{2}$
51. $x^{2k} \cdot x^{3k} = x^{15}$ 3
52. $x^{2k+1} = (x^5)^{k-4}$ 7
53. $a^k \cdot a^3 = \frac{1}{a^{2k}}$ -1
54. $4^k \cdot (4^2)^k = 4^6$ 2

For more exercises on dividing monomials, see exercises 31-36 on page 557.

Objective 5-4A: Express numbers in scientific notation and decimal notation.
Objective 5-4B: Find products and quotients of numbers expressed in scientific notation.

5-4 Scientific Notation

Astronomers use large numbers when measuring the distance from a star to earth. Sometimes it is not desirable to record these numbers in decimal notation.

For example, the earth is about 93,000,000 miles away from the sun.

$93{,}000{,}000 = 9.3 \times 10{,}000{,}000$ *The decimal point was moved 7 places to the left, and placed after the first nonzero digit.*

$= 9.3 \times 10^7$ *$10{,}000{,}000 = 10^7$.*

This form of writing numbers is called **scientific notation**. To write a number in scientific notation, express it as the product of a number between 1 and 10 and a power of 10.

A number is expressed in scientific notation when it is in the form

$$a \times 10^n$$

where $1 \leq a < 10$ and n is an integer.

Definition of Scientific Notation

Examples

1 Express 38,245 in scientific notation.

38,245 *The decimal point must be moved 4 places to the left.*

$38{,}245 = 3.8245 \times 10^4$

2 Express 5,093.4 in scientific notation.

5,093.4 *The decimal point must be moved 3 places to the left.*

$5{,}093.4 = 5.0934 \times 10^3$

3 Express 2.6×10^5 in decimal notation.

$2.6 \times 10^5 = 2.6 \times 100{,}000$

$= 260{,}000$ *Notice that the decimal point was moved 5 places to the right. Why?*

5-4

Lesson Notes, p. T34

Readiness Activity

Evaluate each of the following expressions.

10^4 10,000 10^{-3} $\frac{1}{1000}$

10^{-5} $\frac{1}{100{,}000}$

4.3×10^8 430,000,000 5.68×10^{-4} 0.000568

Discussion Question

What are some examples of adding or subtracting numbers in scientific notation? What rule do these examples suggest?

Answers will vary. A typical answer is

$$\begin{array}{r} 2000 \\ +6000 \\ \hline 8000 \end{array} \quad \text{or} \quad \begin{array}{r} 2 \times 10^3 \\ +6 \times 10^3 \\ \hline 8 \times 10^3 \end{array}.$$

As long as the exponent of ten is the same, numbers can be added or subtracted by adding or subtracting the first term.

Chalkboard Exercises

Express each number in scientific notation.

1. 381,677 3.81677×10^5
2. 96,444.5 9.64445×10^4
3. Express 4.58×10^6 in decimal notation. 4,580,000

Chalkboard Exercises

Express each number in scientific notation.

4a. 0.0000000902
9.02×10^{-8}

4b. 0.09072
9.072×10^{-2}

5. Express 1.8×10^{-4} in decimal notation.
0.00018

Evaluate. Express each result in scientific notation and decimal notation.

6a. $(3 \times 10^{-5})(3 \times 10^{3})$
9×10^{-2} or 0.09

6b. $(5 \times 10^{4})(6 \times 10^{-2})$
3.0×10^{3} or 3000

6c. $\dfrac{4.5 \times 10^{-5}}{1.6 \times 10^{-2}}$
2.8125×10^{-3} or 0.0028125

6d. $\dfrac{7.2 \times 10^{-9}}{1.2 \times 10^{-2}}$
6×10^{-7} or 0.0000006

7a. (0.0042)(330,000)
1.386×10^{3} or 1386

7b. (3,600,000,000)(0.0023)
8.28×10^{6} or 8,280,000

Computer Notes

The use of scientific notation in BASIC is discussed on page 540 of the appendix. Exercises are included on page 542. For enrichment, suggest that students try writing a program for expressing numbers in scientific notation.

About Using Calculators

Students can use the calculator to check if they are adjusting the decimal point correctly on those exercises where the first term is greater than 10 or less than 1. For example, did they write 230×10^{15} as 2.3×10^{17} or 2.3×10^{13}?

Scientific notation is also used to express very small numbers.

$$\begin{aligned} 0.000034 &= 3.4 \times 0.00001 && \textit{The decimal point was moved 5 places to the right.} \\ &= 3.4 \times \frac{1}{100{,}000} \\ &= 3.4 \times \frac{1}{10^5} \\ &= 3.4 \times 10^{-5} && \textit{The exponent of 10 is } -5. \end{aligned}$$

When numbers between zero and one are written in scientific notation, the exponent of 10 is negative.

Examples

4 **Express 0.00319 in scientific notation.**

0.00319 *The number is less than 1. The decimal point must be moved 3 places to the right.*

$0.00319 = 3.19 \times 10^{-3}$

5 **Express 3.2×10^{-7} in decimal notation.**

$$\begin{aligned} 3.2 \times 10^{-7} &= 3.2 \times 0.0000001 \\ &= 0.00000032 && \textit{Notice that the decimal point was moved 7 places to the left.} \end{aligned}$$

You can find products or quotients of numbers that are expressed in scientific notation.

Examples

6 **Evaluate $\dfrac{5.6 \times 10^{-8}}{8.0 \times 10^{-3}}$. Express the result in scientific notation and decimal notation.**

$$\begin{aligned} \frac{5.6 \times 10^{-8}}{8.0 \times 10^{-3}} &= \frac{5.6}{8.0} \times \frac{10^{-8}}{10^{-3}} && \textit{Note } -8-(-3) = -8+3 = -5 \\ &= 0.7 \times 10^{-5} && \textit{Change } 0.7 \times 10^{-5} \textit{ to scientific notation.} \\ &= 7 \times 10^{-6} \text{ or } && 0.000007 \end{aligned}$$

7 **Use scientific notation to find the product of 0.000008 and 3,500,000,000. Express the result in scientific notation and decimal notation.**

$$\begin{aligned} (0.000008)(3{,}500{,}000{,}000) &= (8 \times 10^{-6})(3.5 \times 10^{9}) \\ &= (8)(3.5)(10^{-6})(10^{9}) && \textit{Use the commutative and associative properties.} \\ &= 28(10^{3}) \\ &= 2.8 \times 10^{4} \text{ or } 28{,}000 \end{aligned}$$

Exploratory Exercises

Express each number in the second column in decimal notation. Express each number in the third column in scientific notation.

	Planet	Diameter (km)	From sun (km)	❶
1.	Mercury	5.0×10^3	57,900,000	5000; 5.79×10^7
2.	Venus	1.218×10^4	108,230,000	12,180; 1.0823×10^8
3.	Earth	1.276×10^4	149,590,000	12,760; 1.4959×10^8
4.	Mars	6.76×10^3	227,720,000	6760; 2.2772×10^8
5.	Jupiter	1.427×10^5	778,120,000	142,700; 7.7812×10^8
6.	Saturn	1.21×10^5	1,428,300,000	121,100; 1.4283×10^9

Written Exercises

Express each number in scientific notation.

1. 4293 4.293×10^3 **2.** 5280 5.28×10^3 **3.** 240,000 2.4×10^5 **4.** 389,500 3.895×10^5
5. 0.000319 3.19×10^{-4} **6.** 0.004296 4.296×10^{-3} **7.** 0.000000092 9.2×10^{-8} **8.** 0.00000000317 3.17×10^{-9}
9. 32×10^5 3.2×10^6 **10.** 284×10^3 2.84×10^5 **11.** 0.76×10^7 7.6×10^6 **12.** 0.0031×10^3 3.1×10^0

Evaluate. Express each result in scientific notation and decimal notation.

13. $\frac{4.8 \times 10^3}{1.6 \times 10^1}$ 3×10^2; 300 **14.** $\frac{5.2 \times 10^5}{1.3 \times 10^2}$ 4×10^3; 4000 **15.** $\frac{7.8 \times 10^{-5}}{1.3 \times 10^{-7}}$ 6×10^2; 600
16. $\frac{8.1 \times 10^2}{2.7 \times 10^{-3}}$ 3×10^5; 300,000 **17.** $\frac{1.32 \times 10^{-6}}{2.4 \times 10^2}$ **18.** $\frac{2.31 \times 10^{-2}}{3.3 \times 10^{-9}}$
19. $(2 \times 10^5)(3 \times 10^{-8})$ **20.** $(4 \times 10^2)(1.5 \times 10^6)$ **21.** $(3.1 \times 10^{-2})(2.1 \times 10^5)$
22. $(3.1 \times 10^4)(4.2 \times 10^{-3})$ **23.** $(78 \times 10^6)(0.01 \times 10^3)$ **24.** $(0.2 \times 10^5)(31 \times 10^{-6})$
25. (0.000003)(70,000) 2.1×10^{-1}; 0.21 **26.** (86,000,000)(0.005) 4.3×10^5; 430,000
27. 24,000 ÷ 0.00006 4×10^8; 400,000,000 **28.** 0.0000039 ÷ 650,000 6×10^{-12}; 0.000000000006

Using Calculators — Scientific Notation

Scientific notation can be used to enter very large or very small numbers into many calculators. These numbers are entered using the key labeled [EE] or [Exp]. Use the following steps to enter 230×10^{15}.

ENTER: 230 [EE↓] 15 [=]
DISPLAY: 230 230.00 230.15 2.3 17

Notice that the numeral on the far right is the exponent of 10. That is, 2.3 17 means 2.3×10^{17}. The calculator has expressed 230×10^{15} in scientific notation as 2.3×10^{17}.

Exercises Express each of the following numbers in scientific notation.

1. 234.6×10^9 2.346×10^{11} **2.** 723.4×10^{13} 7.234×10^{15} **3.** 8061×10^{15} 8.061×10^{18} **4.** 9382×10^{10} 9.382×10^{13}
5. 0.0083×10^{21} 8.3×10^{18} **6.** 0.0642×10^{11} 6.42×10^9 **7.** 0.03×10^{19} 3×10^{17} **8.** 0.458×10^{25} 4.58×10^{24}

Suggested Daily Assignments

Enriched
Exp: 1–6
Written: 2–28 evens

Average
Exp: 1–6
Written: 1–28

Fundamental
Exp: 1–6
Written: 1–12, 14–28 evens
p. 152, 25–40

❶ The distances given in the chart are average distances. They vary considerably due to the elliptical orbits of the planets.

Answers to Written

7. 9.2×10^{-8}
17. 5.5×10^{-9}; 0.0000000055
18. 7×10^6; 7,000,000
19. 6×10^{-3}; 0.006
20. 6×10^8; 600,000,000
21. 6.51×10^3; 6510
22. 1.302×10^2; 130.2
23. 7.8×10^8; 780,000,000
24. 6.2×10^{-1}; 0.62

Quiz
Lessons 5-1 to 5-4
Simplify.

1. $(-15x^2y^4)(-4x^6y^9)$ $60x^8y^{13}$
2. $\left(\frac{1}{6}b\right)\left(\frac{4}{9}b\right)\left(-\frac{3}{8}b^3\right)$ $-\frac{1}{36}b^5$
3. $(x^3)^2$ x^6
4. $(3c^2d^2)^3 - 2(c^3d^3)^2 - 4(c^2d^3)(c^4d^3)$ $21c^6d^6$
5. $\frac{y^0}{y^5}$ $\frac{1}{y^5}$
6. $\frac{42x^2z^3}{7x^5z^2}$ $\frac{6z}{x^3}$

Express each of the following in scientific notation.

7. 5,236,000 5.236×10^6
8. $(6 \times 10^2)(-5.4 \times 10^{-8})$ 3.24×10^{-5}

Express the result of each of the following in decimal notation.

9. $\frac{3.92 \times 10^{-4}}{4.9 \times 10^3}$ 0.00000008
10. (2,300,000)(0.0009) 2,070

5-5

Lesson Notes, p. T34

Readiness Activity

Suppose Shirley is 14 years old. Answer the following questions.

(a) How old was Shirley 10 years ago? 4

(b) How old will she be 3 years from now? 17

(c) How old is Mark if he is twice as old as Shirley? 28

Discussion Question

Can a problem contain both extra information and not enough information? yes

Chalkboard Exercises

Solve each problem. If there is not enough information given, write *not enough information*.

1a. Pat is 2 years younger than Lee, and Chris is 5 years older than Lee. The sum of Lee's age and Chris's age is 27. How old is Lee?
Let x = Lee's age
$x + (x + 5) = 27$
$x = 11$
Lee is 11 years old.

1b. Chris is 3 years older than David. David is twice as old as Rick. The sum of Rick's age and David's age is 81. How old is Rick?
Let x = Rick's age
$x + 2x = 81$
$x = 27$
Rick is 27 years old.

2. Paul is 4 years older than his sister. How old is Paul?
not enough information

Objective 5-5A: Solve verbal problems (age problems) that contain extra information.
Objective 5-5B: Recognize problems which do not contain enough information.

5-5 Problem Solving: Age Problems

To solve verbal problems you must analyze each sentence carefully. Decide what is asked and explore how the given facts are related. Many problems contain extra information. Other problems do not contain enough information. Study the following problems.

Examples

1 **Maria is 8 years older than Jose. Carla is 2 years younger than Maria. The sum of Maria's age and Jose's age is 38. How old is Jose?**

Explore This problem asks for Jose's age. What facts are given?

Maria is 8 years older than Jose.
The sum of Maria's age and Jose's age is 38.

You do not need the information about Carla's age to solve this problem.

Define variables, then use an equation to solve the problem.

Let a = Jose's age now.
$a + 8$ = Maria's age now.

Plan

$$\underbrace{\text{Jose's age}}_{a} + \underbrace{\text{Maria's age}}_{(a+8)} = \underbrace{38}_{38}$$

Solve

$$\begin{aligned} a + (a + 8) &= 38 \\ 2a + 8 &= 38 \\ 2a + 8 - 8 &= 38 - 8 \\ 2a &= 30 \\ a &= 15 \end{aligned}$$

Examine If Jose is 15 years old, then Maria must be 23 years old. Since the sum of their ages is 38, the solution is correct.

2 **Luanne is 7 years younger than Jolene. Jolene is 2 years older than Beth. How old is Beth?**

Explore Let b = Beth's age now.
$b + 2$ = Jolene's age now.
$(b + 2) - 7$ = Luanne's age now.

Plan There is no other information given about how the ages are related. You cannot write an equation.

This problem does not contain enough information.

Example

3 **Six years ago, Mr. Winters was five times as old as his son Mark. How old is Mark now if his age is one-third of his father's present age?**

Explore

Let a = Mark's age now.
$3a$ = Mr. Winters' age now. *Why?*
$a - 6$ = Mark's age 6 years ago.
$3a - 6$ = Mr. Winters' age 6 years ago.

Plan

$$\underbrace{\text{Mr. Winters' age}}_{3a - 6} = \underbrace{\text{5 times}}_{5 \times} \underbrace{\text{Mark's age.}}_{(a - 6)}$$

Solve

$$3a - 6 = 5(a - 6)$$
$$3a - 6 = 5a - 30$$
$$-2a = -24$$
$$a = 12$$

Mark is now 12 years old. *Examine this solution.*

Exploratory Exercises

State an expression for each of the following.

1. Marilyn is now n years old. Represent her age:
a. 5 years ago. $n - 5$ **b.** in 11 years. $n + 11$ **c.** x years ago. $n - x$

2. Don is now x years old. Represent his age:
a. 7 years ago. $x - 7$ **b.** in 12 years. $x + 12$ **c.** in d years. $x + d$

3. Paul is now $(n + 7)$ years old. Represent his age:
a. in 7 years. $n + 14$ **b.** 4 years ago. $n + 3$ **c.** n years ago. 7
d. in $(n + 2)$ years. $2n + 9$ **e.** $(n - 4)$ years ago. 11 **f.** in $5n$ years. $6n + 7$

4. Helene is now $(2n + 8)$ years old. Represent her age:
a. 5 years ago. $2n + 3$ **b.** in 6 years. $2n + 14$ **c.** n years ago. $n + 8$
d. in $(n - 3)$ years. $3n + 5$ **e.** $(n + 1)$ years ago. $n + 7$ **f.** in $3n$ years. $5n + 8$

5. Mary is m years old. Represent John's age if he is:
a. twice as old as Mary. $2m$ **b.** one-third as old as Mary. $\frac{m}{3}$
c. 5 years younger than Mary. $m - 5$ **d.** 8 years older than Mary. $m + 8$

Written Exercises

Use an equation to solve each problem. If there is not enough information given, write *not enough information*.

1. Abe is 3 years older than Mindy. The sum of their ages is 39. What are their ages now?
Abe, 21; Mindy, 18

2. Charlie is 14 years younger than Jack. The sum of their ages is 74. What are their ages now?
Jack, 44; Charlie, 30

Chalkboard Exercises

Solve each problem.

3a. Five years ago, Sue was twice as old as her brother Jason. Sue is five years older than Jason. How old is Jason now?

Let x = Jason's age now.
$x + 5$ = Sue's age now
$x - 5$ = Jason's age 5 years ago.
$(x + 5) - 5 = x$ = Sue's age 5 years ago
$x = 2(x - 5)$
$10 = x$
Jason is 10 years old.

3b. Eighteen years ago Alan was twice as old as Tony. Alan is eight years older than Tony. Find Tony's age now.

Let x = Tony's age now.
$x + 8$ = Alan's age now.
$x - 18$ = Tony's age 18 years ago.
$(x + 8) - 18 = x - 10$ = Alan's age 18 years ago.
$x - 10 = 2(x - 18)$
$x = 26$
Tony is 26 years old.

Suggested Daily Assignments

Enriched
Exp: 1–5
Written: 2–16 evens, 17–22
Mini-Review: 1–5

Average
Exp: 1–5
Written: 1–10
Mini-Review: 1–5

Fundamental
Exp: 1–5
Written: 1–10
p. 155, 13–27 odds
Mini-Review: 1–5

3. Nica is five times as old as his sister. The sum of their ages is 12. How old is Nica's sister? 2

4. Joshua is 7 years younger than his brother, and 30 years younger than his father. How old is Joshua? not enough information

5. Erin is 25 years younger than her father. The sum of their ages is 75. How old is Erin? 25

6. Matt is 5 years older than his sister, and 23 years younger than his mother. If the sum of Matt's age and his mother's age is 35, how old is Matt? 6

7. Lisa is 6 years older than her brother Tom. Their father's age is twice the sum of their ages. How old is Lisa if her father is 32? 11

8. Todd is 12 years older than Delores. In 2 years, Todd will be twice as old as Delores. How old is Todd now? 22

9. Twelve years ago, Thea was 7 years old. In 8 years, Thea will be three times as old as Nancy. Nancy is two years younger than Cindy. How old is Nancy now? 1

10. The sum of David's age and Ben's age is 40. Ann is 3 years older than David. David is 4 years older than Ben. How old is David? 22

11. Barney is 54 years old and Hugh is 38. How many years ago was Barney three times as old as Hugh? 30

12. Hector is now 28 years old and Henry is 8 years old. In how many years will Hector be twice as old as Henry? 12

13. Ramona is 5 years older than Alex. Six years ago, Alex was twice as old as Maria. How old is Alex now? not enough information

14. Felicia is 23 years younger than her mother and 25 years younger than her father. How old is Felicia? not enough information

15. Eight years ago, the sum of Pete's age and Matt's age was 26. Matt is 4 years older than Pete. How old is Pete now? 19

16. Dana is 3 years younger than Natalie. In 7 years, the sum of their ages will be 63. What are their ages now? Dana, 23; Natalie, 26

Challenge

Use an equation to solve each problem.

17. Shannon is 8 years younger than Susie and in 6 years will be four-fifths as old as Susie. What are their ages now? Susie, 34; Shannon, 26

18. Mr. Dunn is 7 times as old as his son, Jeff. In 5 years, he will be four times as old as Jeff. What are their ages now? Mr. Dunn, 35; Jeff, 5

19. Erica is $\frac{2}{3}$ as old as Gerda. In 6 years, she will be $\frac{3}{4}$ as old as Gerda. What are their ages now? Gerda, 18; Erica, 12

20. Violet is $\frac{5}{6}$ as old as Lynne. In 20 years, Violet will be $\frac{10}{11}$ as old as Lynne. What are their ages now? Lynne, 24; Violet, 20

21. Harvey is now one-half as old as his mother. Harvey is 12 years older than his brother Henry. Henry is one-fourth as old as his mother. How old is Harvey? 24

22. The sum of Arlie's age and Bert's age is 39. In 7 years, Arlie will be 3 years older than Bert was 5 years ago. How old is Arlie? 15

mini-review

Graph the set of numbers on a number line.

1. $\{\ldots -3, -2, -1, 0, 1, 2\}$

Solve.

2. $r - (-3) = -15$ -18
3. $\frac{s + 5}{s + 1} = \frac{7}{6}$ 23
4. $6t - 14 < 10$ $t < 4$
5. $-0.7m > 4.2$ $m < -6$

Answer to Mini-Review

1.

Objective 5-6A: Solve problems involving percent of increase or percent of decrease.
Objective 5-6B: Solve problems involving discount or sales tax.

5-6 Problem Solving: Percents

A jacket that cost \$50 last year is now priced at \$60. The price increased by \$10 since last year. You can write a ratio that compares the amount of increase to the price last year. The ratio can be changed to a percent.

$$\frac{\text{amount of increase}}{\text{price last year}} \quad \frac{10}{50} = \frac{20}{100} \text{ or } 20\%$$

A percent is a ratio of a number to 100.

The amount of increase is 20% of the price last year. Therefore, you could say that the price of the jacket *increased by 20%* since last year. The **percent of increase** was 20%.

The **percent of decrease** can be found in a similar manner, as shown in the following example.

Example

1 A sweater that originally cost \$35 is now on sale for \$28. Find the percent of decrease.

Explore The price decreased from \$35 to \$28. The amount of decrease was \$7.

Plan You need to write a ratio that compares the amount of decrease with the original price. Then, express the ratio as a percent.

$$\frac{\text{amount of decrease}}{\text{original price}} \quad \frac{7}{35} = \frac{x}{100}$$

To express $\frac{7}{35}$ as a percent, change it to the ratio of some number to 100.

Solve

$$\frac{7}{35} = \frac{x}{100}$$
$$700 = 35x$$
$$\frac{700}{35} = x$$
$$20 = x$$

❶

The percent of decrease was 20%.

Examine Check the solution with the words of the problem. The original price of the sweater was \$35. Find 20% of \$35 and subtract the result from \$35.

$$20\% \text{ of } 35 = \frac{20}{100}(35)$$
$$= \frac{1}{5}(35)$$
$$= 7$$

The amount of decrease was \$7.

Since \$35 − \$7 = \$28, the solution is correct.

Sometimes an increase or decrease is given as a percent, and you must find the amount. The following problem involves a discount. A discount of 35% means that the price is *decreased* by 35%.

In advertisements, discounts are often stated as %-off.

5-6

Lesson Notes, p. T35
Enrichment, p. T35

Readiness Activity
Change each fraction to a percent.

$\frac{11}{100}$ 11% $\quad \frac{2}{25}$ 8%
$\frac{3}{4}$ 75% $\quad \frac{1}{5}$ 20%
$\frac{1}{8}$ 12.5% $\quad \frac{3}{2}$ 150%

Discussion Question
An amount is increased by 10%. The result is decreased by 10%. Is the final result equal to the original amount? Try some specific amounts.
No

Chalkboard Exercises
Solve each problem.

1a. A retailer increased the price of a ballpoint pen from 40¢ to 45¢. Find the percent of increase.
12.5%

1b. A car salesman reduced the price of a car from \$6000 to \$5850. Find the percent of decrease.
2.5%

❶ Another way to solve this problem is shown below. 7 is what percent of 35?

$$7 = \frac{x}{100} \cdot 35$$
$$700 = 35x$$
$$20 = x$$

Chalkboard Exercises

Solve each problem.

2a. Fred bought a stereo that ordinarily sells for \$220. He received a 20% discount. What was the discount price?
\$176

2b. Tami received a 5% discount on a calculator. The original price was \$15. What is the sale price?
\$14.25

3. Tim bought a pair of running shoes for \$35. There was a 4% sales tax. Find the tax and the total price.
Tax = (35.00)(.04) = \$1.40
Total = 35.00 + 1.40 = \$36.40

❷ An equation which corresponds to this problem is: $495.50 - (0.35)(495.50) = x$ Better students may solve the problem by computing 65% of \$495.50. Ask them to explain why the same result is obtained by either method.

❸ Better students may solve this problem by computing 105% of \$31.78. Ask them to explain why the same result is obtained.

Example 2

Amy bought a television set that had an original price of \$495.50. She received a 35% discount. What was the discount price?

Explore The original price was \$495.50, and the discount was 35%.

Plan You need to find the amount of discount, then subtract that amount from \$495.50. The result is the discount price.

Solve 35% of 495.50 = 0.35(495.50) *Note 35% = $\frac{35}{100}$ or 0.35.*
= 173.425 *Round 173.425 to 173.43.*

❷ The amount of discount is \$173.43. Subtract this amount from the original price.

495.50 − 173.43 = 322.07

The discount price was \$322.07.

Examine Here is another way to solve the problem. The discount was 35%, so the discount price was 65% of the original price. Find 65% of \$495.50.

0.65 (495.50) = 322.075

This method produces the same discount price, \$322.07.

The sales tax on a purchase is a *percent* of the purchase price. To find the total price, you must calculate the *amount* of sales tax and add it to the purchase price.

Example 3

Frank Orfanello purchased a new tennis racket for \$31.78. He also had to pay a sales tax of 5%. Find the amount of tax and the total price.

Explore The price is \$31.78 and the tax rate is 5%.

Plan First find 5% of \$31.78. Then add the result to \$31.78.

Solve 5% of 31.78 = 0.05(31.78) *Note 5% = 0.05.*
= 1.589 *Round 1.589 to 1.59.*

The amount of tax is \$1.59. Now find the total price.

❸
$$\underbrace{\text{purchase price}}_{31.78} + \underbrace{\text{amount of tax}}_{1.59} = \underbrace{\text{total price}}_{33.37}$$

The total price is \$33.37.

Examine Since 100 + 5 = 105, the total price is 105% of the purchase price. Find 105% of \$31.78 and compare the result to \$33.37.

1.05(31.78) = 33.369 The total price of \$33.37 is correct.

Sometimes it is helpful to use equations to solve percent problems.

Example

4 **Jane Shriver works as a sales person in a department store. One of her benefits is a 20% discount on all items she buys for herself. She paid $54 for a new dress. What was the price before the discount?**

Explore Let x = price of dress before discount.
Then $0.20x$ = amount of discount. *Note 20% = 0.20.*

Plan $\underbrace{\text{price}}_{x} - \underbrace{\text{discount}}_{0.20x} = \underbrace{\text{discount price}}_{54}$

Solve

$$x - 0.20x = 54$$
$$(1 - 0.20)x = 54 \quad \textit{Distributive Property}$$
$$0.80x = 54$$
$$x = 67.5$$

❹

The price before the discount was $67.50.

Examine If the price before the discount was $67.50 and Jane paid $54 for the dress, the amount of discount was $13.50. Since $13.50 is 20% of $67.50, the solution is correct.

Exploratory Exercises

Solve each problem.

1. What is 50% of 20? 10
2. What is 50% of 350? 175
3. What is 25% of 32? 8
4. What is 25% of 60? 15
5. What is 75% of 60? 45
6. What is 75% of 120? 90
7. 18 is 50% of what number? 36
8. 92 is 50% of what number? 184
9. 10 is 25% of what number? 40
10. 30 is 75% of what number? 40

Written Exercises

Solve each problem.

1. A number x increased by 40% is equal to 14. 10
2. A number x decreased by 20% is equal to 16. 20
3. 14 is 50% less than n. 28
4. 55 is 10% more than n. 50
5. The price in dollars (p) plus 5% tax is equal to $3.15. $3.00
6. The price in dollars (p) minus a 15% discount is $3.40. $4.00

Chalkboard Exercises

Solve each problem.

4a. Patty bought a dress on sale at a department store. If she paid $45 for the dress, and there was a 20% discount, what was the original price?
x = original price
$x - .20x = 45$
$x = 56.25$
original price = $56.25

4b. A person spent a total of $86.92 shopping. Part of that total was a 6% sales tax. What was the price without the tax?
x = price before tax
$x + .06x = 86.92$
$x = 82.00$
price before tax = $82.00

4c. A retailer marks his goods 25% above the wholesale price. If the retail price of a jacket is $68.75, what is the wholesale price?
x = the wholesale price
$x + .25x = 68.75$
$x = 55.00$
wholesale price = $55

❹ Better students may realize that a discount of 20% is equivalent to paying 80% of full price. 80% of the full price is $54, which leads to the equation $0.80x = 54$.

Suggested Daily Assignments

Enriched
- Exp: 1–9 odds
- Written: 1–12, 14–32 evens, 33–37

Average
- Exp: 1–10
- Written: 1–28
  - p. 158, 13–17

Fundamental
- Exp: 1–10
- Written: 1–26
  - p. 158, 11–16

Calculator Notes
Some calculators can easily compute answers for discount and tax problems, such as Exercises 21-26, as follows:
(number) − (discount rate)% + (tax rate)% = customer price.

Copy and complete the following chart. The first line is given as a sample.

	Earlier Amount	*Later Amount*	*Did the amount increase (I) or decrease (D)?*	*Amount of Increase or Decrease*	*Percent of Increase or Decrease*
	$50	$70	I	$20	40%
7.	$100	$94	**a.** D	**b.** $6	**c.** 6%
8.	$100	$108	**a.** I	**b.** $8	**c.** 8%
9.	$200	**b.** $172	D	**a.** $28	14%
10.	$300	**b.** $396	I	**a.** $96	32%
11.	**a.** $48	$60	I	$12	**b.** 25%
12.	**a.** $72	$36	D	$36	**b.** 50%

Notice that 20 is 40% of 50.

Solve each problem.

13. What is 30% more than 30? 39

14. What is 75% less than 80? 20

15. What percent of 11 is 6? $54\frac{6}{11}$%

16. What percent of 80 is 60? 75%

17. A price decreased from $50 to $40. Find the percent of decrease. 20%

18. A price increased from $40 to $50. Find the percent of increase. 25%

19. An item priced $36 has a 25% discount. Find the discount price. $27

20. Sales tax of 6% is added to a purchase of $11. Find the total price. $11.66

Find the customer price for each of the following. When there is a discount and sales tax, compute the discount price first.

21. Stereo Set: $345.00
Discount: 12% $303.60

22. Ten-speed Bike: $148.00
Discount: 18% $121.36

23. Clothing: $74.00
Sales Tax: 6.5% $78.81

24. Books: $38.50
Sales Tax: 6% $40.81

25. Shoes: $44.00
Discount: 10%
Sales Tax: 4% $41.18

26. Auto Tires: $154.00
Discount: 20%
Sales Tax: 5% $129.36

Solve each problem.

27. Wilma paid $92.04 for new school clothes. This included 4% sales tax. What was the cost of the clothes before taxes? $88.50

28. Millie paid $45.10, including $7\frac{1}{2}$% tax, for a pair of jeans. What was the cost of the jeans before taxes? $41.95

29. Jason paid $13.96 for a new shirt, after receiving a 20% discount. What was the price before the discount? $17.45

30. Ben received a discount of $4.35 on a new radio. The discount price was $24.65. What was the percent of discount? 15%

31. A group of 25 people share equally in the profit from a bake sale. What percent does each person receive? 4%

32. The original selling price of a stove was $550. This price was increased by 20%. The increased price was then discounted by 10%. What was the new selling price? $594

33. Forty percent of the students who auditioned for the school play were cut in the first week. Twenty percent of the remaining students were chosen for the play. If 12 students were chosen, how many students were in the original audition? 100

34. If a is 180% of b, then b is what percent of a? $55\frac{5}{9}$% or 55.6%

Challenge

Solve each problem.

35. Which is better, a discount of 30%, or successive discounts of 15% and 15%? 30% discount

36. Which is better, successive discounts of 10%, 5%, and 5%, or successive discounts of 5%, 5%, and 10%? no difference

37. A coat that sells for $80 is discounted 20%. The sales tax is 6%. Is it better for the buyer to have the discount applied before the sales tax is added or after the sales tax is added? no difference

Using Calculators — The Percent Key

Most calculators have a key labeled [%]. This key can be used to find percents of numbers. The way this key is used varies for different types of calculators. Two possible key sequences are shown in the following example.

Example: Find 18% of 46.

ENTER:	46	[×]	18	[%]	
DISPLAY:	46		18	8.28	

or

ENTER:	46	[×]	18	[%]	[=]
DISPLAY:	46		18	0.18	8.28

The correct answer is 8.28.

Try to obtain the same result using the percent key on your calculator. Refer to the instruction manual if necessary.

Exercises

Use a calculator to find the value of each of the following.

1. 20% of 96 19.2
2. 38% of 600 228
3. 32.7% of 91 29.757
4. 123% of 60 73.8
5. 0.7% of 90 0.63
6. 88% of 93 81.84
7. 3.75% of 509 19.0875
8. 0.92% of 92 0.8464
9. 0.002% of 80 0.0016

About Using Calculators
You may want to caution students that the calculator can convert percents to decimals, but not decimals to percents.

About Applications
This feature will provide reinforcement in solving percent of increase problems. Ask students if they think the CPI will ever be less than \$100. Have them explain their answer.

Applications in Business — Consumer Price Index

The Consumer Price Index is used to keep track of the change in the prices of goods and services bought by most people in the United States. Current prices are compared to average prices during 1967. For example, an index of 310 points means that goods and services that cost \$100 in 1967 would now cost \$310. Monthly changes in the CPI are often expressed as percents.

Example: **The CPI was 308.8 in April, 1984 and 309.7 in May, 1984. Find the monthly percent increase.**

$309.7 - 308.8 = 0.9$ *Subtract the earlier value from the later value.*

$\frac{0.9}{308.8} = \frac{x}{100}$ *Express the difference as a percent of the earlier value.*

$90 = 308.8x$

$\frac{90}{308.8} = x$

$0.29 = x$ *Round to the nearest tenth.*

The monthly percent increase was 0.3%

Exercises

Compute the monthly percent increase in the CPI to the nearest tenth.

1. March 245.1
   April 247.8 1.1%
2. Jan. 179.3
   Feb. 180.7 0.8%
3. June 234.8
   July 236.4 0.7%
4. August 294.3
   Sept. 295.4 0.4%
5. Oct. 301.4
   Nov. 306.1 1.6%
6. April 280.6
   May 281.8 0.4%

Compute the yearly percent increase in the CPI to the nearest tenth. The CPI for two successive years is given.

7. Jan. 195.3
   Jan. 209.4 7.2%
8. July 223.7
   July 251.4 12.4%
9. Dec. 245.1
   Dec. 283.8 15.8%
10. Aug. 291.4
    Aug. 310.8 6.7%
11. May 263.1
    May 295.6 12.4%
12. Nov. 236.4
    Nov. 275.2 16.4%

Objective 5-7: Solve mixture problems by organizing the information in a chart.

5-7 Problem Solving: Mixtures

Sue Murphy sold tickets for the basketball game. Each adult ticket costs \$2.50 and each student ticket costs \$1.10. The total number of tickets sold was 105 and the total income was \$202.30. How many of each kind of ticket were sold?

Equations can be used to solve problems such as this. It is often helpful to use a chart to organize the information.

Example

1 Use an equation to solve the problem above.

Explore Let a = the number of adult tickets sold.
$105 - a$ = the number of student tickets sold.

Plan

	Number of Tickets	*Price per Ticket*	*Total Price*
Adult	a	2.50	$2.50a$
Student	$105 - a$	1.10	$1.10(105 - a)$

$$\underbrace{\text{Income from Adult Tickets}}_{2.50a} + \underbrace{\text{Income from Student Tickets}}_{1.10(105 - a)} = \underbrace{\text{Total Income}}_{202.30}$$

Solve

$$\begin{aligned} 2.50a + 1.10(105 - a) &= 202.30 \\ 2.50a + 115.50 - 1.10a &= 202.30 \\ 2.50a - 1.10a &= 202.30 - 115.50 \\ 1.40a &= 86.80 \\ a &= 62 \end{aligned}$$

There were 62 adult tickets sold. There were 105 − 62, or 43 student tickets sold.

Examine If 62 adult tickets were sold, the total income from adult tickets was \$155. If 43 student tickets were sold, the total income from student tickets was \$47.30.
Since 155 + 47.30 = \$202.30, the solution is correct.

Some problems involve mixtures of solutions. When solving such problems, compare the amounts of the substances in the solutions. Sometimes these amounts are given as percents.

5-7

Lesson Notes, p. T35
Enrichment, p. T35

Discussion Question
How many different combinations of coins have a total value of .50¢?
Many, depending on how many and what coins are used.

Chalkboard Exercises
Solve each problem.

1a. Judy bought five records for \$30.75. Some cost \$4.95 each and others cost \$6.95 each. How many of each did she buy?

	Number	Value
\$4.95 records	x	$4.95x$
\$6.95 records	$5 - x$	$6.95(5 - x)$

$$4.95x + 6.95(5 - x) = 30.75$$
$$x = 2$$

She bought 2 records at \$4.95 and 3 records at \$6.95.

1b. Jeff has \$8.35 in dimes and quarters. If he has 40 coins in all, how many dimes does he have?

	Number	Value
dimes	x	$10x$
quarters	$40 - x$	$25(40 - x)$

$$10x + 25(40 - x) = 835$$
$$x = 11 \text{ dimes}$$

Chalkboard Exercises

Solve the problem.

2. Molly has 48 mL of a solution which is 25% acid. How much water should she add to obtain a solution which is 15% acid?

	Amount (mL)	Acid
25% solution	48	0.25(48)
water	x	0
15% solution	$48 + x$	$0.15 \cdot (48 + x)$

$0.25(48) = 0.15(48 + x)$

$32 = x$

Molly should add 32 mL of water.

Suggested Daily Assignments

Enriched
- Exp: 1–4
- Written: 2–12 evens, 13–16
- Mini-Review: 1–5

Average
- Exp: 1–4
- Written: 1–12
  - p. 162, 29–34
- Mini-Review: 1–5

Fundamental
- Exp: 1–4
- Written: 1–12
  - p. 162, 27–32
- Mini-Review: 1–5

Example 2

Hal is doing a chemistry experiment that calls for a 30% solution of copper sulfate. Hal has 40 mL of 25% solution. How many milliliters of a 60% solution should Hal add to obtain the required 30% solution?

Explore Let x = amount of 60% solution to be added.

Plan

	Amount of Solution (mL)	*Amount of Copper Sulfate*
25% solution	40	0.25(40)
60% solution	x	$0.60x$
30% solution (mixture)	$40 + x$	$0.30(40 + x)$

Amount of Copper Sulfate in 25% Solution + Amount of Copper Sulfate in 60% Solution = Amount of Copper Sulfate in Mixture

$$0.25(40) + 0.60x = 0.30(40 + x)$$

Solve

$$0.25(40) + 0.60x = 0.30(40 + x)$$
$$10 + 0.6x = 12 + 0.3x$$
$$0.3x = 2$$
$$x = 6.7$$ *Round to the nearest tenth.*

Hal should add 6.7 mL of the 60% solution to the 40 mL of 25% solution. *Examine this solution.*

Exploratory Exercises

Copy and complete the chart for each problem. Then state an equation.

1. Bonnie bought 16 paperback books for \$10.95. Some costs 60¢ each and the rest cost 75¢ each. How many of each did she buy?

	Number	*Total Price*
60¢ books	x	$60x$
75¢ books	$16 - x$	$75(16 - x)$

$60x + 75(16 - x) = 1095$

2. Ralph has \$2.55 in dimes and quarters. He has eight more dimes than quarters. How many quarters does he have?

	Number	*Total Value*
Quarters	x	$25x$
Dimes	$x + 8$	$10(x + 8)$

$25x + 10(x + 8) = 255$

3. Peanuts sell for \$3.00 a pound. Cashews sell for \$6.00 a pound. How many pounds of cashews should be mixed with 12 pounds of peanuts to obtain a mixture selling for \$4.20 a pound?

	Pounds	*Total Price*
\$3.00 peanuts	12	36
\$6.00 cashews	x	$6x$
\$4.20 mixture	$12 + x$	$4.20(12 + x)$

$6x + 36 = 4.20(12 + x)$

4. How much whipping cream (9% butterfat) should be added to 1 gallon of milk (4% butterfat) to obtain a 6% butterfat mixture?

	Gallons	*Amount of Butterfat*
9% butterfat	x	$0.09x$
4% butterfat	1	0.04(1)
6% butterfat	$x + 1$	$0.06(x + 1)$

$0.09x + 0.04 = 0.06(x + 1)$

Written Exercises

Solve each problem.

1. A chemist has 2.5 liters of a solution which is 70% acid. How much water should be added to obtain a 50% acid solution? 1 liter

2. How much pure copper must be added to 50 kg of an alloy containing 12% copper to raise the copper content to 21%? 5.7 kg

3. Matt has 500 mL of a 60% solution of silver nitrate. How many milliliters of a 30% silver nitrate solution should be added to obtain a 50% solution? 250 mL

4. A liter of cream has 9.2% butterfat. How much skim milk containing 2% butterfat should be added to the cream to obtain a mixture with 6.4% butterfat? 0.64 liter

5. Adrienne has 5 more dimes than nickels. In all, she has \$2.30. How many nickels does she have? 12 nickels

6. Lois has 27 coins in nickels and dimes. In all she has \$1.90. How many of each does she have? 16 nickels; 11 dimes

7. How much coffee costing \$3 a pound should be mixed with 5 pounds of coffee costing \$3.50 a pound to obtain a mixture costing \$3.25 a pound? 5 pounds

8. Theatre tickets cost \$2.50 for children and \$3.50 for adults. The price for 8 tickets was \$23.00. How many adult tickets were purchased? 3 adult tickets

9. On the first day of school, 264 notebooks were sold. Some sold for 95¢ each and the rest sold for \$1.25 each. How many of each were sold if the total sales were \$297.00? 110 at 95¢; 154 at \$1.25

10. Ground chuck sells for \$1.75 a pound. How many pounds of ground round selling for \$2.45 a pound should be mixed with 20 pounds of ground chuck to obtain a mixture that sells for \$2.05 a pound? 15 lb

11. Drew has twice as many quarters as nickels and three more dimes than nickels. He has \$4.20 in all. How many of each does he have? 6 nickels, 12 quarters, 9 dimes

12. Norbert has 42 coins in nickels, dimes, and quarters. If he has 8 more nickels than dimes and has \$7.15 in all, how many of each does he have? 5 dimes, 13 nickels, 24 quarters

Challenge

13. A 60 mL mixture of alcohol and iodine is 2% iodine. How much alcohol must be evaporated away to leave a mixture that is 8% iodine? 45 mL

14. Don has twice as much money in nickels as in dimes. He has 30 coins in all. How many of each coin does he have? 24 nickels, 6 dimes

15. Jackie's car radiator has a 9-liter capacity. The radiator contains a solution that is 20% antifreeze. To be safe, she would like the solution to contain 25% antifreeze. How much of the solution should she drain and replace with 100% antifreeze? 0.5625 L

16. A storage battery contains 1000 mL of a solution of 18% acid. The battery should contain a 21% solution for best efficiency. How much of the 18% solution should be drained and replaced with 100% acid? 36.6 mL

mini-review

Find each sum.

1. $-84 + 17$ -67
2. $-\frac{7}{8} + \frac{3}{5}$ $-\frac{11}{40}$
3. $41ab + (-13cd) + (-12ab) + 56cd$ $29ab + 43cd$

Graph the solution set.

4. $5r \geq 4r + 6$ and $3r < 30$
5. $|m - 3| \leq 5$

Answers to Mini-Review

4. 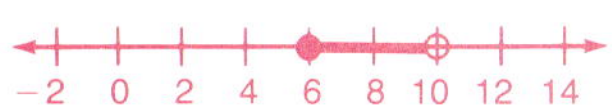

5. 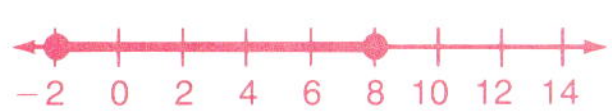

About Using Computers
This feature provides extra practice in using scientific notation and order of operations. You may want to have students write a program to compare the results of evaluating $(2ab^2)^3$ and $2^3a^3b^6$.

Answers to Exercises

1. 10 READ A, B, C
   20 DATA 4, 6, 8
   30 PRINT A ↑ 2 * B ↑ 3 * C ↑ 4
   40 END

For exercises 2–8, change line 30 in the program for exercise 1 as indicated below.

2. 30 PRINT (−2 * A) ↑ 2 * (4 * B) ↑ 3
3. 30 PRINT (4 * A ↑ 2 * B ↑ 4) ↑ 3
4. 30 PRINT (−6 * A * B ↑ 3 * C ↑ 2) ↑ 4
5. 30 PRINT (.5 * A * B) ↑ 2 * (2 * C) ↑ 2
6. 30 PRINT (−5 * A ↑ 2 * C ↑ 3) ↑ 2 * (2 * B ↑ 2) ↑ 2
7. 30 PRINT (A * C) ↑ 3 + (3 * B) ↑ 2
8. 30 PRINT (2 * A ↑ 3) ↑ 2 + (A * B ↑ 2) ↑ 2
9. PRINT (4.2E + 08)/(2.3E + 15)
10. PRINT (3.4E − 10)/(1.2E + 10)
11. PRINT (4.24E − 12)/(3.1E − 09)
12. PRINT (5E + 19) * (6E − 08)
13. PRINT (2E + 12) * (1.3E + 07)
14. PRINT (1.4E − 10) * (2.7E − 08)

Using Computers

Powers and E Notation

In the BASIC language, powers are written using the symbols ↑ or ∧. For example, $(2ab^2)^3$ can be written (2*A*B ↑ 2) ↑ 3. The following program evaluates this expression if $a = 9$ and $b = 10$.

```
10 READ A,B
20 DATA 9,10
30 PRINT (2 * A * B↑2)↑3
40 END

RUN
5.832E + 09
```

Lines 10 and 20 assign the variables A the value 9 and B the value 10.

Notice that the output for this program is in E notation. This is the computer equivalent of scientific notation. Thus, 5.832E + 09 means 5.832×10^9. The computer used E notation because there were more than six significant digits in the output. (Some computers allow nine significant digits.)

Expressions such as $\dfrac{3.8 \times 10^{-9}}{1.3 \times 10^{-4}}$ can be evaluated using E notation.

This can be done by entering the following command into a computer.

```
PRINT (3.8E-09)/(1.3E-04)
```

The answer will be printed.

```
2.92307692 E-05
```

Exercises

Write a program that will evaluate each of the following expressions if $a = 4$, $b = 6$, and $c = 8$.

1. $a^2b^3c^4$ **2.** $(-2a)^2(4b)^3$ **3.** $(4a^2b^4)^3$ **4.** $(-6ab^3c^2)^4$

5. $\left(\frac{1}{2}ab\right)^2(2c)^2$ **6.** $(-5a^2c^3)^2(2b^2)^2$ **7.** $(ac)^3 + (3b)^2$ **8.** $(2a^3)^2 + (ab^2)^2$

Write PRINT commands that will evaluate each of the following expressions.

9. $\dfrac{4.2 \times 10^8}{2.3 \times 10^{15}}$ **10.** $\dfrac{3.4 \times 10^{-10}}{1.2 \times 10^{10}}$ **11.** $\dfrac{4.24 \times 10^{-12}}{3.1 \times 10^{-9}}$

12. $(5 \times 10^{19})(6 \times 10^{-8})$ **13.** $(2 \times 10^{12})(1.3 \times 10^7)$ **14.** $(1.4 \times 10^{-10})(2.7 \times 10^{-8})$

Vocabulary

monomial (145)
constant (145)
scientific notation (153)
percent of increase (159)
percent of decrease (159)

Chapter Summary

1. A monomial is a number, a variable, or a product of numbers and variables. (145)
2. Monomials that do not contain variables are called constants. (145)
3. Product of Powers Rule: For any number a and positive integers m and n,
$$a^m \cdot a^n = a^{m+n}. \quad (145)$$
4. Power of a Power Rule: For any number a and any integers m and n,
$$(a^m)^n = a^{mn}. \quad (148)$$
5. Power of a Product Rule: For all numbers a and b and any integer m,
$$(ab)^m = a^m b^m. \quad (148)$$
6. Power of a Monomial Rule: For all numbers a and b and any integers m, n, and p,
$$(a^m b^n)^p = a^{mp} b^{np}. \quad (148)$$
7. Dividing Powers: For all integers m and n and any nonzero number a,
$$\frac{a^m}{a^n} = a^{m-n}. \quad (150)$$
8. Zero Exponent: For any nonzero number a, $a^0 = 1$. (151)
9. Negative Exponents: For any nonzero number a and any integer n,
$$a^{-n} = \frac{1}{a^n}. \quad (151)$$
10. Scientific Notation: A number is expressed in scientific notation when it is in the form $a \times 10^n$ where $1 \le a < 10$ and n is an integer. (153)

Quiz
Lessons 5-5 to 5-7
Solve each problem.

1. Joan is three times as old as Charlotte. In eight years the sum of their ages will be 100. What are their ages now?
c = Charlotte's age now
$(c + 8) + (3c + 8) = 100$
$c = 21$
$3c = 63$
Charlotte 21, Joan 63
2. A camping tent is regularly priced at \$79.99. A 20% discount is offered. What is the price after the discount? \$63.99
3. A price increased from \$85 to \$110. Find the percent of increase. 29.4%
4. Hi Price Clothes has a new suit priced at \$280. If there is a $6\frac{1}{2}$% tax, find the final price. \$298.20
5. A chemist has 800 cm^3 of a solution which is 25% acid. How much 100% acid must be added to obtain a 40% acid solution?
$0.25(800) + 1.00x = 0.40(800 + x)$
$x = 200$
200 cm^3 of 100% acid must be added.

Chapter Review

5-1 **Simplify.**

1. $b^2 \cdot b^7$ b^9
2. $y^3 \cdot y^3 \cdot y$ y^7
3. $(a^2b)(a^2b^2)$ a^4b^3
4. $(a^2)(a^2b)$ a^4b
5. $(3ab)(-4a^2b^3)$ $-12a^3b^4$
6. $(-4a^2x)(-5a^3x^4)$ $20a^5x^5$

5-2 **Simplify.**

7. $(7a)^2$ $49a^2$
8. $(4a^2b)^3$ $64a^6b^3$
9. $\left(\frac{1}{3}b^2\right)^2$ $\frac{1}{9}b^4$
10. $(-0.5)^3a^4$ $-0.125a^4$
11. $(-3xy)^2(4x)^3$ $576x^5y^2$
12. $(5a^2)^3 + 7(a^6)$ $132a^6$

5-3 **Simplify. Assume no denominator is equal to zero.**

13. $\dfrac{y^{10}}{y^6}$ y^4
14. $\dfrac{(3y)^0}{6a}$ $\dfrac{1}{6a}$
15. $\dfrac{42b^7}{14b^4}$ $3b^3$
16. $\dfrac{27b^{-2}}{14b^{-3}}$ $\dfrac{27b}{14}$
17. $\dfrac{3a^3bc^2}{18a^2b^3c^4}$ $\dfrac{a}{6b^2c^2}$
18. $\dfrac{-16a^3b^2c^4d}{-48a^4bcd^3}$ $\dfrac{bc^3}{3ad^2}$

5-4 **Express each number in scientific notation.**

19. 240,000 2.4×10^5
20. 0.000314 3.14×10^{-4}

Evaluate. Express each result in scientific notation.

21. $(2 \times 10^5)(3 \times 10^6)$ 6×10^{11}
22. $(3 \times 10^3)(1.5 \times 10^6)$ 4.5×10^9
23. $\dfrac{5.4 \times 10^{-3}}{0.9 \times 10^4}$ 6×10^{-7}
24. $\dfrac{8.4 \times 10^{-6}}{1.4 \times 10^{-9}}$ 6×10^3

5-5 **Use an equation to solve each problem.**

25. Joe is 10 years older than Jim. The sum of their ages is 64. What are their ages? Jim, 27; Joe, 37
26. Mary is 20 years old. Her sister, Sara, is 5 years old. In how many years will Mary be twice as old as Sara? 10 years

5-6 **Solve each problem.**

27. What is 80% less than 90? 18
28. Gasoline prices increased from \$1.10 to \$1.20 per gallon. Find the percent of increase. 9.1%
29. Find the total price if $5\frac{1}{4}$% sales tax is added to a purchase of \$21.00. \$22.11
30. Lee received a discount of \$2.40 on an item marked \$48.00. What was the percent of discount? 5%

5-7 **Solve each problem.**

31. Joe has 32 coins in nickels and dimes. He has 10 more nickels than dimes and has \$2.15 in all. How many of each kind of coin does he have? 11 dimes, 21 nickels
32. How many pounds of nuts costing \$2.00 per pound should be mixed with 10 pounds of nuts costing \$3.00 per pound to obtain a mixture costing \$2.25 per pound? 30 pounds

Chapter Test

Simplify. Assume no denominator is equal to zero.

1. $y^2 \cdot y^{13}$ y^{15}

2. $a^2 \cdot a^3 \cdot b^4 \cdot b^5$ a^5b^9

3. $(a^2b)(a^3b^2)$ a^5b^3

4. $(-12abc)(4a^2b^3)$ $-48a^3b^4c$

5. $(9a)^2$ $81a^2$

6. $(-3a)^4(a^5b)$ $81a^9b$

7. $\left(\frac{1}{5}r^2\right)^2$ $\frac{1}{25}r^4$

8. $(-0.3a)^3$ $-0.027a^3$

9. $(-5a^2)(-6b)^2$ $-180a^2b^2$

10. $(5a)^2b + 7a^2b$ $32a^2b$

11. $\frac{y^{11}}{y^6}$ y^5

12. $\frac{y^3x}{yx}$ y^2

13. $\frac{63a^2bc}{9abc}$ $7a$

14. $\frac{49a^2bc}{21ab^3c^2}$ $\frac{7a}{3b^2c}$

15. $\frac{14ab^{-3}}{21a^2b^{-5}}$ $\frac{2b^2}{3a}$

16. $\frac{10a^2bc}{20a^{-1}b^{-1}c}$ $\frac{a^3b^2}{2}$

Express in scientific notation.

17. 5280 5.28×10^3

18. 0.00378 3.78×10^{-3}

Evaluate. Express each result in scientific notation.

19. $(3 \times 10^3)(2 \times 10^4)$ 6×10^7

20. $(4 \times 10^{-3})(2 \times 10^{16})$ 8×10^{13}

21. $\frac{25 \times 10^3}{5 \times 10^{-3}}$ 5×10^6

22. $\frac{91 \times 10^{18}}{13 \times 10^{14}}$ 7×10^4

Solve each problem. Les, 50; Jim, 55

23. Jim is 5 years older than Les. The sum of their ages is 105. What are their ages?

24. What is 0.5% of 234? 1.17

25. What percent of 200 is 86? 43%

26. 42 is 20% of what number? 210

27. A tire listed for \$63.00 is sold for \$49.00. What is the percent of discount? $22\frac{2}{9}\%$

28. Don has 42 coins in dimes, nickels, and quarters. He has twice as many dimes as quarters and two more nickels than quarters. In all he has \$5.10. How many of each kind of coin does Don have?

10 quarters, 20 dimes, 12 nickels

About the Application
Other *Applications in Probability* features can be found on pages 108 and 242. *Applications in Statistics* features can be found on pages 302 and 368.

Applications in Probability — Independent Events

Kyong Song wishes to travel from Chatsworth to Northridge to Granada Hills. She can travel from Chatsworth to Northridge on Lassen Street, Devonshire Street, or Plummer Street. She can travel from Northridge to Granada Hills on Henderson Street or Rinaldi Street. Her possible routes can be shown on a **tree diagram** as follows.

Lassen Street — Henderson Street
Lassen Street — Rinaldi Street
Devonshire Street — Henderson Street
Devonshire Street — Rinaldi Street
Plummer Street — Henderson Street
Plummer Street — Rinaldi Street

A set of all of the possible outcomes, called a **sample space**, is {Lassen-Henderson, Lassen-Rinaldi, Devonshire-Henderson, Devonshire-Rinaldi, Plummer-Henderson, Plummer-Rinaldi}. Notice that there are six possible routes and that six is the product of the number of routes from Chatsworth to Northridge (3), and from Northridge to Granada Hills (2). Assuming that each route is equally likely, you can use the tree diagram or the sample space to find the following probabilities.

$$P(\text{Devonshire as the first street}) = \frac{1}{3}.$$

$$P(\text{Rinaldi as the second street}) = \frac{3}{6} = \frac{1}{2}.$$

$$P(\text{Devonshire, Rinaldi}) = \frac{1}{6}.$$

$$\begin{aligned} P(\text{Devonshire, Rinaldi}) &= P(\text{Devonshire as first Street}) \cdot P(\text{Rinaldi as second street}) \\ &= \frac{1}{3} \cdot \frac{1}{2} \\ &= \frac{1}{6} \end{aligned}$$

In this example, Kyong's choice of the second street is not affected by her choice of the first street. When the outcome of the second event is not affected by the outcome of the first, the events are **independent**.

Probability of Two Independent Events

The probability of two independent events, A and B, occurring is the product of their individual probabilities.

$$P(A \text{ and } B) = P(A) \cdot P(B)$$

Example 1: **A bag contains 4 red marbles and 3 white marbles. One marble is selected, returned to the bag, and then a second marble is selected. Find the probability of each of the following.**

a. P**(both red)**

$$P(\text{both red}) = P(\text{red}) \cdot P(\text{red})$$
$$= \frac{4}{7} \cdot \frac{4}{7} = \frac{16}{49}$$

b. P**(red and white)** *Select red followed by white.*

$$P(\text{red and white}) = P(\text{red}) \cdot P(\text{white})$$
$$= \frac{4}{7} \cdot \frac{3}{7} = \frac{12}{49}$$

Exercises

State whether the following events are independent.

1. flip a coin then flip it again independent

2. toss a die and flip a coin independent

3. draw a marble from a bag and then draw a second without replacement dependent

4. draw a card from a deck, replace it and draw again independent

A nickel is flipped and then a dime is flipped. Find the probability of each of the following outcomes in the indicated order.

5. P(head and tail) $\frac{1}{4}$

6. P(both tails) $\frac{1}{4}$

7. P(both heads) $\frac{1}{4}$

8. P(tail and head) $\frac{1}{4}$

A bag contains 5 red, 3 blue, and 7 white marbles. A marble is selected, returned to the bag, and a second marble is selected. Find the probability of each of the following outcomes in the indicated order.

9. P(both red) $\frac{1}{9}$

10. P(white and red) $\frac{7}{45}$

11. P(both blue) $\frac{1}{25}$

12. P(red and blue) $\frac{1}{15}$

13. P(blue and red) $\frac{1}{15}$

14. P(both white) $\frac{49}{225}$

Suppose three marbles are selected and returned to the bag of marbles in Exercises 9-14. Find the probability of each of the following outcomes in the indicated order.

15. P(all three red) $\frac{1}{27}$

16. P(red, white, blue) $\frac{7}{225}$

17. P(all three blue) $\frac{1}{125}$

18. P(white, blue, blue) $\frac{7}{375}$

19. P(red, blue, red) $\frac{1}{45}$

20. P(all three white) $\frac{343}{3375}$

Theo and Luke Mallinson play on the Cherry Creek High School basketball team. Theo can make 3 out of 5 free throws and Luke can make 4 out of 7 free throws. What is the probability of each of the following?

21. Luke will make his next two free throws. $\frac{16}{49}$

22. Theo will make his next two free throws. $\frac{9}{25}$

23. Luke and Theo (in that order) will both make their next free throws. $\frac{12}{35}$

24. Theo will miss his next three free throws. $\frac{8}{125}$

About the Photo

Numerous terms contain the prefix poly-, which signifies plurality. Suggest that students check a dictionary for terms which begin with this prefix. Some examples are polycrystalline, polyurethane, and polysulfide. Discuss the math-related terms mentioned, polygon and polynomial.

You may want to point out that a figure is a polygon if and only if it is formed by three or more coplanar (lie on the same plane) line segments that are also noncollinear (do not lie on the same line). Each segment intersects exactly two other segments only at their endpoints. The line segments are called sides and the endpoints are called vertices.

Algebra is concerned mainly with polynomials. Emphasis on performing the basic operations with polynomials is essential. Skills learned in this chapter assure success later in solving algebraic sentences and other problems.

Teacher Resource Book Materials for Chapter 6 See pages 159-180.

Chapter Test, Form I (Multiple Choice)	159-161
Chapter Test, Form II (Free Response)	162-163
Quizzes	164-165
Extra Practice	166-168
Application: Using Equations (Use after 6-9.)	169-171
Extension: Special Polynomial Products (Use after 6-5.)	172
Extension: Compound Interest (Use after 6-8.)	173
Problem Solving	174
Cumulative Review	175
Answers	176-180

CHAPTER 6

Polynomials

Poly-, a prefix meaning many or several, is used in some mathematical words. There are many *poly*gons, plane figures with many angles, in the photograph above. In algebra, expressions consisting of one or more monomials are called *poly*nomials. In this chapter you will learn to add, subtract, and multiply polynomials.

Objective 6-1A: Find the degree of a polynomial.
Objective 6-1B: Arrange the terms of a polynomial so that the powers of a certain variable are in ascending or descending order.

6-1 Polynomials

A **polynomial** is a monomial or a sum of monomials. Recall that a monomial is a number, a variable, or a product of numbers and variables. A **binomial** is a polynomial with two terms, and a **trinomial** is a polynomial with three terms.

A monomial has one term.

Examples of each of these types of polynomials are given in the following chart.

monomial	binomial	trinomial
$5x^2$	$3x + 2$	$5x^2 - 2x + 7$
$4abc$	$4x + 5y$	$a^2 + 2ab + b^2$
-7	$3x^2 - 8xy$	$4a + 2b^2 - 3c$

The expression $x + \frac{3}{b}$ is not a polynomial because $\frac{3}{b}$ is not a monomial.

Example 1

State whether each expression is a polynomial. If the expression is a polynomial, identify it as either a monomial, binomial, or trinomial.

a. $8x^2 - 3xy$

The expression $8x^2 - 3xy$ can be written as $8x^2 + (-3xy)$. Therefore, $8x^2 - 3xy$ is a polynomial because it can be written as the sum of two monomials, $8x^2$ and $-3xy$. Since it has two terms, $8x^2 - 3xy$ is a binomial.

b. $\frac{5}{2y^2} + 7y + 6$

The expression $\frac{5}{2y^2} + 7y + 6$ is not a polynomial because $\frac{5}{2y^2}$ is not a monomial.

c. $3x^2 + 2x + 4$

The expression $3x^2 + 2x + 4$ is a polynomial because it is the sum of three monomials, $3x^2$, $2x$, and 4. Since it has three terms, $3x^2 + 2x + 4$ is a trinomial.

The **degree** of a monomial is the sum of the exponents of its variables.

monomial	degree
$5x^2$	2
$4ab^3c^4$	$1 + 3 + 4 = 8$
-9	0

❶ ❷

What is the degree of $-\frac{1}{2}x$?

6-1

Lesson Notes, p. T35

Readiness Activity
What is the exponent in each of the following expressions?
$4x^2$ 2 $7y^3$ 3 $8a$ 1
$3c^5$ 5 $56m^8$ 8 $25r^4$ 4

Discussion Questions
(Use after page 175)
1. Simplify $x^2 \cdot x^3$. What is the degree of the first factor? the second factor? the product? x^5; 2, 3, 5
2. Simplify $5 \cdot 2x^4$. What is the degree of the first factor? the second factor? the product? $10x^4$; 0, 4, 4
3. Simplify $0 \cdot 7x^6$. What is the degree of the first factor? the second factor? the product? 0; no degree, 6, no degree

Chalkboard Exercises
State whether each expression is a polynomial. If the expression is a polynomial, identify it as either a monomial, binomial, or trinomial.
1a. $7y^3 - 4y^2 + y$ trinomial
1b. $10x^3y^2z$ monomial
1c. $4u^2 + 3v + \frac{1}{w}$ not a polynomial
1d. $9t^2 + 8t$ binomial

❶ Emphasize that only a *nonzero* constant has a degree of 0. The number 0 and expressions such as $0x^2$ and $0a^2b^2$, which represent 0, have *no* degree.

❷ Explain to students that -9 can be expressed as $-9x^0$. Therefore, the degree of -9 is 0.

Chalkboard Exercises

Find the degree of each polynomial.

2a. $3x^2 + 5$ 2
2b. $7y^3 + 2y$ 3
2c. $ay^2 + z$ 3
2d. $a^2 + 2ab + b^4$ 4
2e. $xy^2 + x^3y^2 + xy$ 5
2f. $p^5 + p^3m^3 + 4m$ 6
2g. $y^7 + y^6 + 3x^4m^4$ 8

❸ The polynomials in the first row are written so that the powers of a are in ascending or descending order.

Suggested Daily Assignments

Enriched
Written: 1–37 odds
Mini-Review: 1–5

Average
Exp: 2–38 evens
Written: 1–24, 25–37 odds
Mini-Review: 1–5

Fundamental
Exp: 1–39
Written; 1–30
Mini-Review: 1–5

To find the degree of a polynomial, first find the degree of each of its terms. The degree of the polynomial is the greatest of the degrees of its terms.

Example

2 **State the degree of each polynomial.**

a. $8x^3 - 2x^2 + 7$

The polynomial $8x^3 - 2x^2 + 7$ has three terms, $8x^3$, $-2x^2$, and 7. Their degrees are 3, 2, and 0. Therefore, the degree of $8x^3 - 2x^2 + 7$ is 3.

b. $6x^2y + 5x^3y^2z - x + x^2y^2$

The polynomial $6x^2y + 5x^3y^2z - x + x^2y^2$ has four terms, $6x^2y$, $5x^3y^2z$, $-x$, and x^2y^2. Their degrees are 3, 6, 1, and 4. Therefore, the degree of $6x^2y + 5x^3y^2z - x + x^2y^2$ is 6.

The terms of a polynomial are usually arranged so that the powers of one variable are in either ascending or descending order.

Ascending Order	Descending Order
$3 + 5a - 8a^2 + a^3$	$a^3 - 8a^2 + 5a + 3$ ❸
(in x) $5xy + x^3y^2 - x^4 + x^5y^2$	(in x) $x^5y^2 - x^4 + x^3y^2 + 5xy$
(in y) $x^3 - 3x^2y + 4x^2y^2 - y^3$	(in y) $-y^3 + 4x^2y^2 - 3x^2y + x^3$

Exploratory Exercises

State whether each expression is a polynomial. If the expression is a polynomial, identify it as either a monomial, binomial, or trinomial.

1. $5x^2y + 3xy + 7$ trinomial
2. $\frac{5}{k} - k^2y$ not a polynomial
3. 17 monomial
4. $4t^2 + 3t$ binomial
5. $\frac{3}{a^3}$ not a polynomial
6. $5a^2b^3$ monomial
7. $k^3 - \frac{2}{k}$ not a polynomial
8. $\frac{a^3}{3}$ monomial
9. $3a^2x - 5a$ binomial
10. $4x^2 + 3y^2z$ binomial
11. $4b^2c - 8abc + b^3$ trinomial
12. -12 monomial
13. 0 monomial
14. $x^2 - \frac{x}{2} + \frac{1}{3}$ trinomial
15. $5a^2b - \frac{1}{2}ab^2 + \frac{1}{b^3}$ not a polynomial

State the degree of each monomial.

16. $100x$ 1
17. -18 0
18. $29xyz$ 3
19. $8x^2$ 2
20. $17x^2y$ 3
21. $-14x^3z$ 4
22. 0 no degree
23. $51x^5yz$ 7
24. $28xst^3$ 5
25. $14m^2n^3$ 5
26. $4x$ 1
27. $-36xw^4$ 5

State the degree of each polynomial.

28. $12s + 21t$ 1
29. 19 0
30. $14x + 3y$ 1
31. $22x + 5y^4$ 4
32. $-5m^2 + 9m^5$ 5
33. $7x^3 + 4xy + 3xz^3$ 4
34. $17r^2t + 3r + t^2$ 3
35. $29x + x^{29}$ 29
36. $17x^3y^4 - 11xy^5$ 7
37. $22m^4n^2 - 14mn^6$ 7
38. $-2a^2b^{10} + 3a^9b$ 12
39. $13s^2t^2 + 4st^2 - 5s^5t$ 6

Written Exercises

Find the degree of each polynomial.

1. $11x^2$ 2
2. $5r - 3s + 7t$ 1
3. $x + 2y^2 + 3z^3$ 3
4. $n - 2p^2$ 2
5. $27x^4 - 3x^3yz$ 5
6. $12x^4y^5 + xy^6$ 9
7. $29n^2 + 17n^2t^2$ 4
8. $32xyz + 11x^2y + 17xz^2$ 3
9. $4xy + 9xz^2 + 17rs^3$ 4
10. $11wxyz - 9w^4$ 4
11. $2xy^2z + 5xyz^5 + x^4$ 7
12. $5x^2 - 2x^5$ 5
13. $n^2 + r^3 + s^{14} + z^2$ 14
14. $-4yzw^4 + 10x^4z^2w - 2z^2w^3$ 7
15. $6mn^4t - 3m^4nt^2 + mn^6$ 7
16. $2x^3yz - 5xy^3z + 11z^5$ 5
17. 2^2r^2 2
18. $3^3xy + 5x$ 2

Arrange the terms of each polynomial so that the powers of x are in ascending order.

19. $3 + x^4 + 2x^2$ $3 + 2x^2 + x^4$
20. $1 + x^3 + x^5 + x^2$ $1 + x^2 + x^3 + x^5$
21. $2x^2 + 5ax + a^3$ $a^3 + 5ax + 2x^2$
22. $-2x^2y + 3xy^3 + x^3$ $3xy^3 - 2x^2y + x^3$
23. $17bx^2 + 11b^2x - x^3$ $11b^2x + 17bx^2 - x^3$
24. $21p^2x + 3px^3 + p^4$ $p^4 + 21p^2x + 3px^3$

Arrange the terms of each polynomial so that the powers of x are in descending order.

25. $-7 + x^2 + 4x$ $x^2 + 4x - 7$
26. $-6x + x^5 + 4x^3 - 20$ $x^5 + 4x^3 - 6x - 20$
27. $5x^2 - 3x^3 + 2x + 7$ $-3x^3 + 5x^2 + 2x + 7$
28. $a^2x + x^3 - 3$ $x^3 + a^2x - 3$
29. $5b + \frac{2}{3}bx + b^3x^2$ $b^3x^2 + \frac{2}{3}bx + 5b$
30. $\frac{3}{4}x^3y - x^2 + 4 + \frac{2}{3}x$ $\frac{3}{4}x^3y - x^2 + \frac{2}{3}x + 4$
31. $a + x$ $x + a$
32. $11x^2 + 7ax^3 - 3x + 2a$
33. $4x^3y + 3xy^4 - x^2y^3 + y^4$
34. $7a^3x - 8a^3x^3 + \frac{1}{5}x^5 + \frac{2}{3}x^2$
35. $\frac{1}{3}s^2x^3 + 4x^4 - \frac{2}{5}s^4x^2 + \frac{1}{4}x$
36. $0.2mx^4 - 1.3x^5 + 0.4m^2 + 2.6x^3$
37. $1.7t^3x + 2.4tx^5 - 0.3tx^2 + 5.1x^3$

mini-review

Simplify.

1. $x^2 + y^2 + 3xy + 2x^2 - y^2$ $3x^2 + 3xy$
2. $\dfrac{-18}{-\frac{2}{3}}$ 27
3. $y^4 \cdot y^3 \cdot y$ y^8
4. $(2ab^2c^3)^4$ $16a^4b^8c^{12}$
5. $\dfrac{5x^4yz^2}{20x^2y^3z^2}$ $\dfrac{x^2}{4y^2}$

Answers to Written

32. $7ax^3 + 11x^2 - 3x + 2a$
33. $4x^3y - x^2y^3 + 3xy^4 + y^4$
34. $\frac{1}{5}x^5 - 8a^3x^3 + \frac{2}{3}x^2 + 7a^3x$
35. $4x^4 + \frac{1}{3}s^2x^3 - \frac{2}{5}s^4x^2 + \frac{1}{4}x$
36. $-1.3x^5 + 0.2mx^4 + 2.6x^3 + 0.4m^2$
37. $2.4tx^5 + 5.1x^3 - 0.3tx^2 + 1.7t^3x$

6-2

Lesson Notes, p. T36

Discussion Questions

1. Is addition of polynomials commutative? Yes
2. Is subtraction of polynomials commutative? No

Readiness Activity

Name the like terms in each of the following pairs of polynomials.

$4y + 2w$, $3y - 5w$

$4y, 3y; 2w, -5w$

$8t^2 + 4t - 2$, $7t^2 - 3t + 1$

$8t^2, 7t^2; 4t, -3t; -2, 1$

$3xy + 4y^2 - 2x$, $3y^2 - 4xy + x$

$3xy, -4xy; 4y^2, 3y^2; -2x, x$

$10m^2 + 5m - 1$, $5m^2 + 3$

$10m^2, 5m^2; -1, 3$

Chalkboard Exercises

Add.

1. $(3a + 7b) + (6a - 3b)$
   $9a + 4b$

2a. $(10y^2 + 7y + 5) + (y^2 - 2y - 4)$
   $11y^2 + 5y + 1$

2b. $(10m^2 - 4mn + n^2) + (2m^2 + 4mn - 7n^2)$
   $12m^2 - 6n^2$

2c. $(9v^2 + 7w - 2) + (4v^2 - 9)$
   $13v^2 + 7w - 11$

Add in column form.

3. $(9y - 7x + 15a) + (-3y + 8x - 8a)$

$$\begin{array}{r} 9y - 7x + 15a \\ + \; -3y + 8x - \;\; 8a \\ \hline 6y + \;\; x + \;\; 7a \end{array}$$

4. $(3a^2 + 3ab - b^2) + (4ab + 6b^2)$

$$\begin{array}{r} 3a^2 + 3ab - b^2 \\ + \qquad 4ab + 6b^2 \\ \hline 3a^2 + 7ab + 5b^2 \end{array}$$

❶ If students add monomials such as $5x^2$ and $4x^2$ incorrectly to get $9x^4$, emphasize proper use of the distributive property to simplify like terms. Also review exponent rules.

Objective 6-2: Add and subtract polynomials.

6-2 Adding and Subtracting Polynomials

You have used properties to simplify expressions.

$4a + 3a - 2 = (4 + 3)a - 2$	*Distributive Property*
$= 7a - 2$	*Substitution Property of Equality*
$5x^2 + 3y + 2x^2 - y = 5x^2 + 2x^2 + 3y - y$	*Commutative Property for Addition*
$= (5 + 2)x^2 + (3 - 1)y$	*Distributive Property*
$= 7x^2 + 2y$	*Substitution Property of Equality*

Suppose you want to add the polynomials $(3x + 2y)$ and $(8x + 3y)$. You can use the same properties to find such sums.

Examples

1 Add: $(3x + 2y) + (8x + 3y)$

To add polynomials, add their like terms.

$(3x + 2y) + (8x + 3y) = (3x + 8x) + (2y + 3y)$	*Associative and Commutative Properties for Addition*
$= (3 + 8)x + (2 + 3)y$	*Distributive Property*
$= 11x + 5y$	*Substitution Property of Equality*

2 Add: $(-3x^2 + 2x + 7) + (6x^2 - 5x - 3)$

$$\begin{aligned}(-3x^2 + 2x + 7) + (6x^2 - 5x - 3) &= (-3x^2 + 6x^2) + [2x + (-5x)] + [7 + (-3)] \\ &= (-3 + 6)x^2 + [2 + (-5)]x + [7 + (-3)] \\ &= 3x^2 + (-3x) + 4 \\ &= 3x^2 - 3x + 4\end{aligned}$$

Sometimes it is convenient to find the sum by writing the polynomials in column form.

Examples

3 Add in column form: $(5x^2 - 3xy + 7y^2) + (4x^2 + xy - 3y^2)$

❶
$$\begin{array}{r} 5x^2 - 3xy + 7y^2 \\ 4x^2 + \;\; xy - 3y^2 \\ \hline 9x^2 - 2xy + 4y^2 \end{array}$$

4 Add in column form: $(3y^2 + 5y - 6) + (7y^2 - 9)$

$$\begin{array}{r} 3y^2 + 5y - \;\; 6 \\ 7y^2 \qquad - \;\; 9 \\ \hline 10y^2 + 5y - 15 \end{array}$$

Notice that like terms are aligned.

Recall that you can subtract a rational number by adding its additive inverse or opposite.

$a - b = a + (-b)$

additive inverses

$-3 - (-5) = 2$ $\qquad$ $-3 + 5 = 2$ ❷

same result

Similarly, you can subtract a polynomial by adding its additive inverse. To find the additive inverse of a polynomial, replace each term by its additive inverse.

Another way to find the additive inverse is to multiply each term by -1.

Polynomial	Additive Inverse
$x + 2y$	$-x - 2y$
$2x^2 - 3x + 5$	$-2x^2 + 3x - 5$
$-8x + 5y - 7z$	$8x - 5y + 7z$
$3x^3 - 2x^2 - 5x$	$-3x^3 + 2x^2 + 5x$

❸

Examples

5 **Subtract: $(5x^2 - 2x + 7) - (2x^2 + 4x - 3)$**

Add the additive inverse of $(2x^2 + 4x - 3)$.

$$\begin{aligned}(5x^2 - 2x + 7) - (2x^2 + 4x - 3) &= (5x^2 - 2x + 7) + [-(2x^2 + 4x - 3)]\\ &= (5x^2 - 2x + 7) + (-2x^2 - 4x + 3)\\ &= 5x^2 - 2x + 7 - 2x^2 - 4x + 3\\ &= (5x^2 - 2x^2) + (-2x - 4x) + (7 + 3)\\ &= 3x^2 - 6x + 10\end{aligned}$$

6 **Subtract: $(2x^2 - 3xy + 5y^2) - (x^2 - 3xy - 2y^2)$**

$$\begin{aligned}(2x^2 - 3xy + 5y^2) - (x^2 - 3xy - 2y^2) &= (2x^2 - 3xy + 5y^2) + [-(x^2 - 3xy - 2y^2)]\\ &= (2x^2 - 3xy + 5y^2) + (-x^2 + 3xy + 2y^2)\\ &= (2x^2 - x^2) + (-3xy + 3xy) + (5y^2 + 2y^2)\\ &= x^2 + 7y^2\end{aligned}$$

7 **Subtract in column form: $(4x^2 + 5xy - 3y^2) - (6x^2 + 8xy + 3y^2)$**

$$\begin{array}{rr} & 4x^2 + 5xy - 3y^2\\ (-) & \underline{6x^2 + 8xy + 3y^2}\end{array} \quad -(6x^2 + 8xy + 3y^2) \quad \begin{array}{rr} & 4x^2 + 5xy - 3y^2\\ (+) & \underline{-6x^2 - 8xy - 3y^2}\\ & -2x^2 - 3xy - 6y^2\end{array}$$

additive inverse

Chalkboard Exercises

Subtract.

5a. $(7a - 10b) - (3a + 4b)$ $\quad 4a - 14b$

5b. $(6y - 8z) - 6y$ $\quad -8z$

6a. $\left(\frac{3}{4}a^2 - \frac{1}{3}b^2 + c^2\right) - \left(\frac{1}{4}a^2 + \frac{2}{3}b^2\right)$ $\quad \frac{1}{2}a^2 - b^2 + c^2$

6b. $(14a^2 - 3a + 7) - (6a^2 + 5a + 17)$ $\quad 8a^2 - 8a - 10$

Subtract in column form.

7a. $(3y^2 + 7y + 8) - (2y^2 - 4y + 3)$

$$\begin{array}{rr} & 3y^2 + 7y + 8\\ (+) & \underline{-2y^2 + 4y - 3}\\ & y^2 + 11y + 5\end{array}$$

7b. $(4z + 7 + 9z^2) - (11z^2 - 8z + 6)$

$$\begin{array}{rr} & 9z^2 + 4z + 7\\ (+) & \underline{-11z^2 + 8z - 6}\\ & -2z^2 + 12z + 1\end{array}$$

❷ There is less chance for error if students subtract by adding the additive inverse.

❸ Emphasize that the additive inverse of every term must be found.

Chalkboard Exercise

Subtract and check by addition.

8. $(9ab + 6a^2 + 7m^2) - (4m^2 - 6a^2 + 12ab)$

$12a^2 - 3ab + 3m^2$

Check: $12a^2 - 3ab + 3m^2$

$(+) -6a^2 + 12ab + 4m^2$

$6a^2 + 9ab + 7m^2$

Computer Notes

For students with previous programming experience, suggest the following project: Write a computer program for finding the sum or difference of two polynomials of the form $Ax^3 + Bx^2 + Cx + D$.

Suggested Daily Assignments

Enriched
Written: 6–54 evens

Average
Exp: 2–28 evens
Written: 2–46 evens

Fundamental
Exp: 1–28
Written: 2–18 evens, 33–36
p. 177, 31–37

Answers to Exploratory

9. $-x^2 - 3x - 7$
10. $-x^2 - 7xy - y^2$
11. $-2y^2 + 7y - 12$
12. $-6a^2 + 3ab - b^2$
13. $10r^2 - 4rs + 6s$

Answers to Written

13. $3n^2 + 13n + 11$
14. $6a^2 + 13a + 8$
16. $x^2 - 15x + 10$
17. $x^2 + \frac{4}{3}x + \frac{1}{4}$
18. $2x^2 - \frac{16}{3}x - 3$
19. $7ax^2 + 3a^2x - 5ax + 2x$
20. $\frac{2}{7}a^2 - \frac{5}{4}a + 1$
21. $-3x^2y + 6xy^2 - 2y^2$
22. $m^2 - 6m + 1$

Example

8 **Subtract and check by addition:** $(5a^3 + 2a^2 - 7) - (3a^3 + 8a^2 - 2a + 4)$

$$\begin{array}{r} 5a^3 + 2a^2 \qquad - 7 \\ (-)\underline{3a^3 + 8a^2 - 2a + 4} \end{array} \quad -(3a^3 + 8a^2 - 2a + 4) \rightarrow \quad \begin{array}{r} 5a^3 + 2a^2 \qquad - 7 \\ (+)\underline{-3a^3 - 8a^2 + 2a - 4} \\ 2a^3 - 6a^2 + 2a - 11 \end{array}$$

additive inverse

Check:
$$\begin{array}{rl} 2a^3 - 6a^2 + 2a - 11 & \text{difference} \\ (+)\ \underline{3a^3 + 8a^2 - 2a + 4} & \text{second polynomial} \\ 5a^3 + 2a^2 \qquad - 7 & \text{first polynomial} \end{array}$$

Exploratory Exercises

State the additive inverse of each polynomial.

1. 29 -29
2. $-15x^2y$ $15x^2y$
3. $3x + 2y$ $-3x - 2y$
4. $5a + 9b$ $-5a - 9b$
5. $7a - 6b$ $-7a + 6b$
6. $4x^2 - 5$ $-4x^2 + 5$
7. $-8m + 7n$ $8m - 7n$
8. $-11x + 13y$ $11x - 13y$
9. $x^2 + 3x + 7$
10. $x^2 + 7xy + y^2$
11. $2y^2 - 7y + 12$
12. $6a^2 - 3ab + b^2$
13. $-10r^2 + 4rs - 6s$
14. $-6a + 5b - c$ $6a - 5b + c$
15. $-x^3 - x - 1$ $x^3 + x + 1$
16. $-4h^2 - 5hk - k^2$ $4h^2 + 5hk + k^2$
17. $-3ab^2 + 5a^2b - b^3$ $3ab^2 - 5a^2b + b^3$
18. $x^3 + 5x^2 - 3x - 11$ $-x^3 - 5x^2 + 3x + 11$

Add or subtract.

19. $(2n + 7p) + (8n + 2p)$ $10n + 9p$
20. $(3r - 9s) + (4r + 8s)$ $7r - s$
21. $(2a + 3b) + (5a - 2b)$ $7a + b$
22. $(7m + 3n) - (3m + 5n)$ $4m - 2n$
23. $(5x + 8y) - (-3x + 5y)$ $8x + 3y$
24. $[-5a + (-8b)] + (4a + 8b)$ $-a$
25. $(5n + r) + (7n - 3r)$ $12n - 2r$
26. $(-5r + 2s) - (-3r - 2s)$ $-2r + 4s$
27. $(5r - 3s) - (5r + 3s)$ $-6s$
28. $(7m + 2n) - (-6m + 7n)$ $13m - 5n$

Written Exercises

Add or subtract.

1. $(5x + 6y) + (2x + 8y)$ $7x + 14y$
2. $(4a + 6b) + (2a + 3b)$ $6a + 9b$
3. $(7n + 11m) - (4m + 2n)$ $7m + 5n$
4. $(7y + 9x) - (6x + 5y)$ $3x + 2y$
5. $(3x - 7y) + (3y + 4x)$ $7x - 4y$
6. $(3x - 2y) + (5x + 8y)$ $8x + 6y$
7. $(5a - 6m) - (2a + 5m)$ $3a - 11m$
8. $(3s - 5t) - (8t + 2s)$ $s - 13t$
9. $(5m + 3n) + 8m$ $13m + 3n$
10. $(12x + 7y) + 8y$ $12x + 15y$
11. $(13x + 9y) - 11y$ $13x - 2y$
12. $(9x + 3y) - 9y$ $9x - 6y$
13. $(n^2 + 5n + 3) + (2n^2 + 8n + 8)$
14. $(5a^2 + 7a + 5) + (a^2 + 6a + 3)$
15. $(3 + 2a + a^2) - (5 + 8a + a^2)$ $-2 - 6a$
16. $(3x^2 - 7x + 4) - (2x^2 + 8x - 6)$
17. $\left(\frac{1}{2}x^2 + \frac{5}{3}x - \frac{1}{4}\right) + \left(\frac{1}{2}x^2 - \frac{1}{3}x + \frac{1}{2}\right)$
18. $\left(\frac{3}{4}x^2 + \frac{2}{3}x - 7\right) + \left(\frac{5}{4}x^2 - 6x + 4\right)$
19. $(5ax^2 + 3a^2x - 5x) + (2ax^2 - 5ax + 7x)$
20. $\left(\frac{5}{7}a^2 - \frac{3}{4}a + \frac{1}{2}\right) - \left(\frac{3}{7}a^2 + \frac{1}{2}a - \frac{1}{2}\right)$
21. $(2xy + 6xy^2 + y^2) - (3x^2y + 2xy + 3y^2)$
22. $\left(\frac{3}{8}m^2 - 4m + \frac{2}{3}\right) + \left(\frac{5}{8}m^2 - 2m + \frac{1}{3}\right)$

23. $(3mn^2 + 3mn - n^3) - (5mn^2 + n + 2n^3)$ $-2mn^2 + 3mn - n - 3n^3$

24. $(x^3 - 3x^2y + 4xy^2 + y^3) - (7x^3 + x^2y - 9xy^2 + y^3)$ $-6x^3 - 4x^2y + 13xy^2$

Add.

25. $4a + 5b - 6c + d$
$3a - 7b + 2c + 8d$
$\underline{2a - b \quad + 7d}$ $9a - 3b - 4c + 16d$

26. $2x^2 - 5x + 7$
$5x^2 + 7x - 3$
$\underline{x^2 - x + 11}$ $8x^2 + x + 15$

27. $5ax^2 + 3a^2x - 7a^3$
$\underline{2ax^2 - 8a^2x \quad + 4}$
$7ax^2 - 5a^2x - 7a^3 + 4$

28. $a^3 \quad - b^3$
$\underline{3a^3 + 2a^2b - b^2 + 2b^3}$
$4a^3 + 2a^2b - b^2 + b^3$

Subtract.

29. $6x^2y^2 - 3xy - 7$
$\underline{5x^2y^2 + 2xy + 3}$ $x^2y^2 - 5xy - 10$

30. $5x^2 \quad - 4$
$\underline{3x^2 + 8x + 4}$ $2x^2 - 8x - 8$

31. $11m^2n^2 + 2mn - 11$
$\underline{5m^2n^2 - 6mn + 17}$ $6m^2n^2 + 8mn - 28$

32. $2a - 7$
$\underline{5a^2 + 8a - 11}$ $-5a^2 - 6a + 4$

Subtract and check by addition.

33. $11m^2n^2 + 4mn - 6$
$\underline{5m^2n^2 - 6mn + 17}$
$6m^2n^2 + 10mn - 23$

34. $7z^2 \quad + 4$
$\underline{3z^2 + 2z - 6}$ $4z^2 - 2z + 10$

35. $4y - 11$
$\underline{3y^2 + 4y - 11}$ $-3y^2$

Find the measure of the third side of each triangle. *P* is the measure of the perimeter.

36. $P = 3x + 3y$ $x + y$

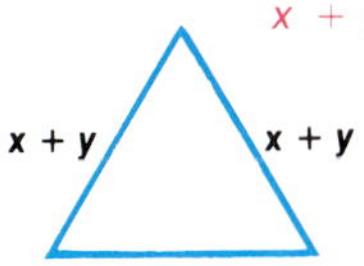

37. $P = 7x + 2y$ $2x + 6y$

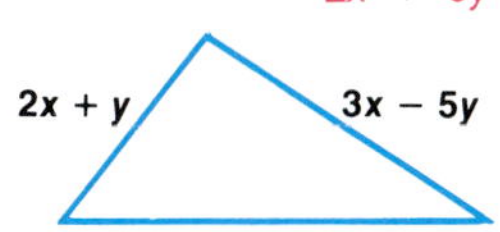

38. $P = 11x^2 - 29x + 10$ $5x^2 - 23x - 23$

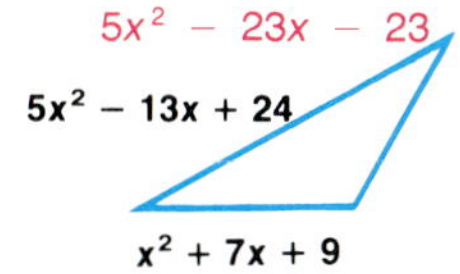

Challenge

Two angles are complementary if the sum of their degree measures is 90. Find the complement of each angle below.

39. 85 5
40. $3x$ $90 - 3x$
41. $x - 2$ $92 - x$
42. $2x + 40$ $50 - 2x$

Two angles are supplementary if the sum of their degree measures is 180. Find the supplement of each angle below.

43. 45 135
44. $2x$ $180 - 2x$
45. $8x - 10$ $190 - 8x$
46. $5x + 130$ $50 - 5x$

The sum of the degree measures of the three angles of a triangle is 180. Find the degree measure of the third angle of each triangle given the degree measures of the other two angles.

47. 80; 70 30
48. $3x$; $x + 2$ $178 - 4x$
49. $x - 1$; $x + 5$ $176 - 2x$
50. $2x - 3$; $3x - 8$ $191 - 5x$
51. $5 - 2x$; $7 + 8x$ $168 - 6x$
52. $x^2 - 8x + 2$; $x^2 - 3x - 1$ $-2x^2 + 11x + 179$
53. $3x^2 - 5$; $4x^2 + 2x + 1$ $-7x^2 - 2x + 184$
54. $4 - 2x$; $x^2 - 1$ $-x^2 + 2x + 177$

Quiz
Lessons 6-1 to 6-2

1. Find the degree of the following polynomial and arrange it in descending order.
$r^3 + 6r + 7r^2 - 2r^4 + 3r^5$
The degree is 5.
$3r^5 - 2r^4 + r^3 + 7r^2 + 6r$

2. Arrange $ry^4 - 2r^2y + 3ry^3 - 7y^5$ so that the powers of y are in ascending order.
$-2r^2y + 3ry^3 + ry^4 - 7y^5$

Simplify each of the following.

3. $(9r^2 - 3s^3 + 8) + (11s^3 - 12 + 5r^2)$
$14r^2 + 8s^3 - 4$

4. $(-16y^4 + 3x^2 - 2z - 6) - (12y^4 - 6x^2 - 3z - 4)$
$-28y^4 + 9x^2 + z - 2$

About Reading Algebra
A thorough understanding of powers and parentheses is essential for the understanding of multiplication of polynomials and the word problems associated with them.

Answers to Exercises
1. four squared
2. three cubed
3. *a* to the fifth power
4. *m* to the ninth power
5. five times *b* squared
6. twelve times *r* the quantity to the fifth power
7. nine times *x* cubed times *y*
8. *x* plus two times *y* the quantity squared
9. four times *m* squared times *n* to the fourth power
10. The quantity, 6*a* squared *b*, raised to the fourth power.
11. *a* minus *b* cubed
12. The quantity, 2*a* plus *b*, raised to the fourth power.

Reading Algebra — Powers

Powers containing the exponent 2 or 3 can be read in two ways.

x^2 is read "*x* squared" or "*x* to the second power."
x^3 is read "*x* cubed" or "*x* to the third power."

Powers containing numerical exponents other than 2 or 3 are usually read as follows.

x^0 is read "*x* to the zero power."
x^6 is read "*x* to the sixth power."
x^n is read "*x* to the *n*th power."

Recall that an exponent indicates the number of times that the base is used as a factor. Suppose you are to write each of the following in symbols.

Words	Symbols
3 times *x* squared	$3x^2$
3 times *x* the quantity squared	$(3x)^2$

In the second expression, parentheses are used to show that the expression $3x$ is used as a factor twice.

$$(3x)^2 = (3x)(3x)$$

The phrase *the quantity* is used to indicate parentheses when reading expressions.

Exercises

State how you would read each of the following expressions.

1. 4^2
2. 3^3
3. a^5
4. m^9
5. $5b^2$
6. $(12r)^5$
7. $9x^3y$
8. $(x + 2y)^2$
9. $4m^2n^4$
10. $(6a^2b)^4$
11. $a - b^3$
12. $(2a + b)^4$

Decide if the expressions are equivalent. Write *yes* or *no*.

13. $12 \cdot x^3$ and $12 \cdot x \cdot x \cdot x$ yes
14. $3xy^5$ and $3(xy)^5$ no
15. $(2a)^3$ and $8a^3$ yes
16. $(ab)^2$ and $a^2 \cdot b^2$ yes
17. xy^3 and x^3y^3 no
18. $4(x^3)^2$ and $4x^6$ yes

Objective 6-3: Multiply a polynomial by a monomial by using the distributive property.

6-3 Multiplying a Polynomial by a Monomial

You can use the distributive property to multiply a polynomial by a monomial.

Examples

1 Multiply: $5a(3a^2 + 4)$

$5a(3a^2 + 4) = 5a(3a^2) + 5a(4)$ *Use the distributive property.*

$= 15a^3 + 20a$

2 Multiply: $2m^2(5m^2 - 7m + 8)$

$2m^2(5m^2 - 7m + 8) = 2m^2(5m^2) + 2m^2(-7m) + 2m^2(8)$ *Use the product of powers rule.*

$= 10m^4 - 14m^3 + 16m^2$

3 Multiply: $-3xy(2x^2y + 3xy^2 - 7y^3)$

❶ $-3xy(2x^2y + 3xy^2 - 7y^3) = -3xy(2x^2y) + (-3xy)(3xy^2) + (-3xy)(-7y^3)$

$= -6x^3y^2 - 9x^2y^3 + 21xy^4$

4 Simplify: $2a(5a^2 + 3a - 2) + 8(3a^2 - 7a + 1)$

$2a(5a^2 + 3a - 2) + 8(3a^2 - 7a + 1)$

$= 2a(5a^2) + 2a(3a) + 2a(-2) + 8(3a^2) + 8(-7a) + 8(1)$

$= 10a^3 + 6a^2 - 4a + 24a^2 - 56a + 8$

$= 10a^3 + 30a^2 - 60a + 8$ *Combine like terms.*

Exploratory Exercises

Multiply.

1. $-5a(12a^2)$ $-60a^3$
2. $4x^2(7x^3)$ $28x^5$
3. $2(5x - 3)$ $10x - 6$
4. $8(3a + 5)$ $24a + 40$
5. $7a(5a + 8)$ $35a^2 + 56a$
6. $3m(8m + 7)$ $24m^2 + 21m$
7. $3ab(5a - 3)$ $15a^2b - 9ab$
8. $2mn(m - 7)$ $2m^2n - 14mn$
9. $-8a^2(3a^2 + 7a)$ $-24a^4 - 56a^3$
10. $-4m^3(5m^2 + 2m)$ $-20m^5 - 8m^4$
11. $3xy(3xy + 2x)$ $9x^2y^2 + 6x^2y$
12. $7ab(5ab^2 + b^2)$ $35a^2b^3 + 7ab^3$

Written Exercises

Multiply.

1. $5(3a + 7)$ $15a + 35$
2. $8(7m + 2)$ $56m + 16$
3. $-3(8x + 5)$ $-24x - 15$
4. $-7(5x + 8)$ $-35x - 56$
5. $\frac{1}{2}x(8x + 6)$ $4x^2 + 3x$
6. $\frac{2}{3}a(6a + 15)$ $4a^2 + 10a$
7. $3b(5b + 8)$ $15b^2 + 24b$
8. $7a(8a + 11)$ $56a^2 + 77a$
9. $-2x(5x + 11)$ $-10x^2 - 22x$
10. $-5y(7y + 12)$ $-35y^2 - 60y$
11. $1.1a(2a + 7)$ $2.2a^2 + 7.7a$
12. $2.6b(5b - 1)$ $13b^2 - 2.6b$
13. $7a(3a^2 - 2a)$ $21a^3 - 14a^2$
14. $5b(8b^2 - 7b)$ $40b^3 - 35b^2$
15. $3st(5s^2 + 2st)$ $15s^3t + 6s^2t^2$
16. $5mn(5m^2 + 3mn)$ $25m^3n + 15m^2n^2$
17. $7xy(5x^2 - y^2)$ $35x^3y - 7xy^3$
18. $4ab(3a^2 - 7b^2)$ $12a^3b - 28ab^3$

6-3

Lesson Notes, p. T36
Enrichment, p. T36

Discussion Questions

1. If a polynomial of degree n is combined with a polynomial of degree m and the result is a polynomial of degree $n + m$, what operation was used to combine the polynomials? multiplication
2. If a polynomial of degree n is added to a polynomial of degree m, what is the degree of the sum? m or n whichever is greater

Readiness Activity

Simplify.

$6a(7a^2)$ $42a^3$

$2mn(5m^3)$ $10m^4n$

$4c^2d(2cd^2)$ $8c^3d^3$

$-2x^3(6x^2y)$ $-12x^5y$

$\frac{1}{2}yz(14z)$ $7yz^2$

$-\frac{3}{4}r^3(2r^2t)$ $-\frac{3}{2}r^5t$

❶ Make sure that students are careful when distributing the negative term.

Chalkboard Exercises

Multiply.

1a. $5(7n - 2)$ $35n - 10$

1b. $\frac{3}{4}a(8a + 12)$ $6a^2 + 9a$

2. $6rs(r^2s - 3)$ $6r^3s^2 - 18rs$

3a. $4m^2(3m + 2n - 4p)$ $12m^3 + 8m^2n - 16m^2p$

3b. $\frac{2}{5}a(10a^2 + 15a + 20)$ $4a^3 + 6a^2 + 8a$

3c. $-\frac{x}{3}(27x^2 - 6x + 12)$ $-9x^3 + 2x^2 - 4x$

Simplify.

4a. $3b(7b^2 + 5b + 4) + 7(2b^2 - b + 6)$ $21b^3 + 29b^2 + 5b + 42$

4b. $3r(9r^2 + 7r - 12) - 4(4r^2 - 3r + 7)$ $27r^3 + 5r^2 - 24r - 28$

Suggested Daily Assignments

Enriched
Written: 8–48 evens, 49, 50
p. 181, 39–53 odds

Average
Exp: 1–11 odds
Written: 1–47 odds

Fundamental
Exp: 1–12
Written: 2–44 evens
p. 180, 19–32

Answers to Written

22. $56c^3d^3 - 8c^2d^3 + 8cd^4$
27. $15x^4y - 35x^3y^2 + 5x^2y^3$
28. $7a^6b^2 - 35a^4b^3 + 42a^2b^4$
29. $36m^4n + 4m^3n - 20m^2n^2$
30. $18x^4y - 14x^3y + 2x^2y^2$
31. $-32x^2y^2 - 56x^2y + 112xy^3$
32. $-7a^2b^2 - 77a^3b^2 + 77ab^3$
33. $-3x^3 - \frac{1}{3}x^2 + \frac{5}{3}x$
34. $4a^3 - 6a^2 + \frac{16}{5}a$
35. $-16m^3n + 6m^2n^2 - 2mn^3$
36. $-14a^3m^3 + 49a^2m^2 - 77am$
37. $-\frac{1}{4}ab^4 + \frac{1}{3}ab^3 - \frac{3}{4}ab^2$
38. $-4x^3y - \frac{8}{3}x^2y^2 + \frac{2}{9}xy^3$
39. $7a^4 + 21a^3 - a + 25$
40. $-m^3 - 66m^2 - 32m + 77$
41. $-3m^3 + 41m^2 - 14m - 16$
42. $53a^3 - 57a^2 + 7a$
43. $21a^3 - 6a^2 - 46a + 28$
44. $-7r^3 + 37r^2 + 105r - 9$

(continued on p. 185)

About the Excursion
For more information about Emmy Noether, see the March 1982 issue of *The Mathematics Teacher*.

19. $2a(5a^3 - 7a^2 + 2)$ $10a^4 - 14a^3 + 4a$
20. $5a(a^2 + 9a - 3)$ $5a^3 + 45a^2 - 15a$
21. $7x^2y(5x^2 - 3xy + y)$ $35x^4y - 21x^3y^2 + 7x^2y^2$
22. $8cd^2(7c^2d - cd + d^2)$
23. $5y(8y^3 + 7y^2 - 3y)$ $40y^4 + 35y^3 - 15y^2$
24. $8v(7v^3 + v^2 - 7v)$ $56v^4 + 8v^3 - 56v^2$
25. $-4x(7x^2 - 4x + 3)$ $-28x^3 + 16x^2 - 12x$
26. $-8a(5a^2 + 8a - 3)$ $-40a^3 - 64a^2 + 24a$
27. $5x^2y(3x^2 - 7xy + y^2)$
28. $7a^2b^2(a^4 - 5a^2b + 6b^2)$
29. $4m^2(9m^2n + mn - 5n^2)$
30. $2x^2(9x^2y - 7xy + y^2)$
31. $-8xy(4xy + 7x - 14y^2)$
32. $-7ab(ab + 11a^2b - 11b^2)$
33. $-\frac{1}{3}x(9x^2 + x - 5)$
34. $\frac{2}{5}a(10a^2 - 15a + 8)$
35. $-2mn(8m^2 - 3mn + n^2)$
36. $-7am(2a^2m^2 - 7am + 11)$
37. $-\frac{3}{4}ab^2\left(\frac{1}{3}b^2 - \frac{4}{9}b + 1\right)$
38. $-\frac{1}{3}xy\left(12x^2 + 8xy - \frac{2}{3}y^2\right)$

Simplify.

39. $2a(a^3 - 2a^2 + 7) + 5(a^4 + 5a^3 - 3a + 5)$
40. $6m(m^2 - 11m + 4) - 7(m^3 + 8m - 11)$
41. $5m^2(m + 7) - 2m(5m^2 - 3m + 7) + 2(m^3 - 8)$
42. $6a^2(3a - 4) + 5a(7a^2 - 6a + 5) - 3(a^2 + 6a)$
43. $3a^2(a - 4) + 6a(3a^2 + a - 7) - 4(a - 7)$
44. $8r^2(r + 8) - 3r(5r^2 - 11) - 9(3r^2 - 8r + 1)$
45. $2.5t(8t - 12) + 5.1(6t^2 + 10t - 20)$
46. $3.2a(5a + 1.1) + 9.3(6a^2 - 1.1a + 3.5)$
47. $\frac{3}{4}m(8m^2 + 12m - 4) + \frac{3}{2}(8m^2 - 9m)$
48. $\frac{1}{2}a(a^2 + 7a - 8) + \frac{1}{4}(5a^3 + 8a^2 - 3a)$

Challenge

49. $\frac{3}{4}a(a^2 + 0.79a + 0.345) - \frac{4}{5}a\left(\frac{1}{4}a^2 + 0.295a + \frac{1}{2}\right)$
50. $\frac{2}{5}x^2(x^3 + 0.86x^2 - 0.49) - 0.67x\left(0.34x^4 - \frac{2}{5}x^3 - 0.631\right)$

Excursions in Algebra — History

Amalie Emmy Noether (1882–1935) was a German mathematician who worked with the structure of algebra. She studied non-commutative algebras. In the period 1930–1933 she was the center of mathematical activity at the University of Göttingen with her research program and her influence on many students.

Noether's strength as a mathematician lay in her ability to operate abstractly with concepts. In her hands the axiomatic method (using axioms or properties) became a powerful tool of mathematical research. Albert Einstein paid her a great tribute in 1935: "In the realm of algebra . . . , she discovered methods which have proved of enormous importance in the development of the present day younger generation of mathematicians."

Objective 6-4A: Use the FOIL method to multiply two binomials.
Objective 6-4B: Multiply any two polynomials by using the distributive property.

6-4 Multiplying Polynomials

The distributive property can be used to multiply polynomials as well as to multiply a monomial by a polynomial. To find the product of two binomials, use the distributive property twice.

Example 1

Multiply: $(2x + 3)(5x + 8)$

$(2x + 3)(5x + 8) = 2x(5x + 8) + 3(5x + 8)$ *Use the distributive property.*

$= 2x(5x) + 2x(8) + 3(5x) + 3(8)$ *Use the distributive property again.*

$= 10x^2 + 16x + 15x + 24$ *Simplify each term.*

$= 10x^2 + 31x + 24$ *Combine like terms.*

Although two binomials can always be multiplied as shown above, the following shortcut called the **FOIL method** is used frequently.

Multiply the **F**irst terms. $2x \cdot 5x$ — $(2x + 3)(5x + 8)$ — $10x^2$

$+$

Multiply the **O**uter terms. $2x \cdot 8$ — $(2x + 3)(5x + 8)$ — $16x$

$+$

Multiply the **I**nner terms. $3 \cdot 5x$ — $(2x + 3)(5x + 8)$ — $15x$

$+$

Multiply the **L**ast terms. $3 \cdot 8$ — $(2x + 3)(5x + 8)$ — 24

$(2x + 3)(5x + 8) = 10x^2 + 16x + 15x + 24$

$= 10x^2 + 31x + 24$ *Combine the like terms $16x$ and $15x$.*

❶ ***FOIL Method for Multiplying Two Binomials***

To multiply two binomials, find the sum of the products of
F the first terms,
O the outer terms,
I the inner terms, and
L the last terms.

6-4

Lesson Notes, p. T36

Discussion Question
If a binomial is multiplied by a trinomial, how many terms can the product contain?
2, 3, 4, 5, or 6

Chalkboard Exercises
Multiply.
1a. $(m - 3)(m + 4)$
$m^2 + m - 12$
1b. $(n + 6)(n + 1)$
$n^2 + 7n + 6$

❶ Remind students that FOIL is a memory device, not a mathematical theorem or rule.

Answers to Written
(continued from p. 184)
45. $50.6t^2 + 21t - 102$
46. $71.8a^2 - 6.71a + 32.55$
47. $6m^3 + 21m^2 - \frac{33}{2}m$
48. $\frac{7}{4}a^3 + \frac{11}{2}a^2 - \frac{19}{4}a$

Answers to Challenge
49. $0.55a^3 + 0.3565a^2 - 0.14125a$
50. $0.1722x^5 + 0.612x^4 - 0.196x^2 + 0.42277x$

Chalkboard Exercises

Multiply.

2. $(2m + 6)(m - 1)$
$2m^2 + 4m - 6$

3. $(5a - 1)(5a + 6)$
$25a^2 + 25a - 6$

4. $\left(2x + \frac{1}{3}\right)\left(3x + \frac{1}{5}\right)$
$6x^2 + \frac{7}{5}x + \frac{1}{15}$

5. $(2x + 3)(x^2 + 3x + 8)$
$2x^3 + 9x^2 + 25x + 24$

6a. $(b^2 + 7b - 2)(4b^2 - 2b + 6)$
$4b^4 + 26b^3 - 16b^2 + 46b - 12$

6b. $(m^2 + 5m - 3)(m^2 + 7m + 4)$
$m^4 + 12m^3 + 36m^2 - m - 12$

❷ When finding the product $(3x - 5)(5x + 2)$ the -5 is distributed in the same way as a negative 5. You may want to rewrite $(3x - 5)(5x + 2)$ as $(3x + -5)(5x + 2)$ and show how to find this product first.

Examples

2 **Multiply: $(x + 5)(x + 7)$**

F L; I O

$$(x + 5)(x + 7) = \overbrace{x \cdot x}^{F} + \overbrace{x \cdot 7}^{O} + \overbrace{5 \cdot x}^{I} + \overbrace{5 \cdot 7}^{L}$$
$$= x^2 + 7x + 5x + 35$$
$$= x^2 + 12x + 35$$

3 **Multiply: $(3x - 5)(5x + 2)$**

❷

$$(3x - 5)(5x + 2) = \overset{F}{3x(5x)} + \overset{O}{3x(2)} + \overset{I}{(-5)(5x)} + \overset{L}{(-5)(2)}$$
$$= 15x^2 + 6x - 25x - 10$$
$$= 15x^2 - 19x - 10$$

4 **Multiply: $\left(\frac{1}{3}a - \frac{2}{3}\right)\left(\frac{1}{2}a + \frac{1}{3}\right)$**

$$\left(\frac{1}{3}a - \frac{2}{3}\right)\left(\frac{1}{2}a + \frac{1}{3}\right) = \left(\frac{1}{3}a\right)\left(\frac{1}{2}a\right) + \left(\frac{1}{3}a\right)\left(\frac{1}{3}\right) + \left(-\frac{2}{3}\right)\left(\frac{1}{2}a\right) + \left(-\frac{2}{3}\right)\left(\frac{1}{3}\right)$$
$$= \frac{1}{6}a^2 + \frac{1}{9}a - \frac{1}{3}a - \frac{2}{9}$$
$$= \frac{1}{6}a^2 + \frac{1}{9}a - \frac{3}{9}a - \frac{2}{9}$$
$$= \frac{1}{6}a^2 - \frac{2}{9}a - \frac{2}{9}$$

The distributive property can be used to multiply any two polynomials.

Examples

5 **Multiply: $(2x + 5)(3x^2 - 5x + 4)$**

$$(2x + 5)(3x^2 - 5x + 4) = (2x + 5)(3x^2) + (2x + 5)(-5x) + (2x + 5)(4)$$
$$= 6x^3 + 15x^2 - 10x^2 - 25x + 8x + 20$$
$$= 6x^3 + 5x^2 - 17x + 20$$

6 **Multiply: $(x^2 - 5x + 4)(2x^2 + x - 7)$**

$$(x^2 - 5x + 4)(2x^2 + x - 7)$$
$$= (x^2 - 5x + 4)(2x^2) + (x^2 - 5x + 4)(x) + (x^2 - 5x + 4)(-7)$$
$$= 2x^4 - 10x^3 + 8x^2 + x^3 - 5x^2 + 4x - 7x^2 + 35x - 28$$
$$= 2x^4 - 10x^3 + x^3 + 8x^2 - 5x^2 - 7x^2 + 4x + 35x - 28$$
$$= 2x^4 - 9x^3 - 4x^2 + 39x - 28$$

Polynomials can also be multiplied in column form.

Multiply by x.	**Multiply by 5.**	**Add.**
$x^2 + 7x - 6$	$x^2 + 7x - 6$	$x^2 + 7x - 6$
$x + 5$	$x + 5$	$x + 5$
$x^3 + 7x^2 - 6x$	$x^3 + 7x^2 - 6x$	$x^3 + 7x^2 - 6x$
	$5x^2 + 35x - 30$	$5x^2 + 35x - 30$
		$x^3 + 12x^2 + 29x - 30$

❸

In the partial products, notice that like terms are aligned.

Examples

7 **Multiply in column form: $(4x + 7)(5x^2 - 3x + 7)$**

$$\begin{array}{rrrr}
 & 5x^2 & -\ 3x & +\ 7 \\
 & & 4x & +\ 7 \\
\hline
20x^3 & -\ 12x^2 & +\ 28x & \\
 & 35x^2 & -\ 21x & +\ 49 \\
\hline
20x^3 & +\ 23x^2 & +\ 7x & +\ 49
\end{array}$$

$20x^3 - 12x^2 + 28x$ ← *This is the product of $4x$ and $5x^2 - 3x + 7$.*

$35x^2 - 21x + 49$ ← *This is the product of 7 and $5x^2 - 3x + 7$.*

8 **Multiply in column form: $(x^3 + 5x - 6)(2x - 9)$**

Since there is no x^2 in $x^3 + 5x - 6$, $0x^2$ is used as a placeholder.

$$\begin{array}{rrrrr}
x^3 & +\ 0x^2 & +\ 5x & -\ 6 & \\
2x & -\ 9 & & & \\
\hline
2x^4 + 0x^3 & +\ 10x^2 & -\ 12x & & \\
-\ 9x^3 & +\ 0x^2 & -\ 45x & +\ 54 & \\
\hline
2x^4 - 9x^3 & +\ 10x^2 & -\ 57x & +\ 54 &
\end{array}$$

$2x^4 + 0x^3 + 10x^2 - 12x$ — *This is the product of $2x$ and $x^3 + 5x - 6$.*

$-9x^3 + 0x^2 - 45x + 54$ — *This is the product of -9 and $x^3 + 5x - 6$.*

Exploratory Exercises

State the sum of the product of the inner terms and the product of the outer terms.

1. $(x + 5)(x + 3)$ $8x$
2. $(r - 3)(r + 7)$ $4r$
3. $(2a + 1)(a + 5)$ $11a$
4. $(3y + 2)(y - 3)$ $-7y$
5. $(x + 5)(5x - 3)$ $22x$
6. $(x + 2)(4x - 7)$ x
7. $(3x + 2)(2x + 3)$ $13x$
8. $(3a - 4)(4a + 1)$ $-13a$
9. $(5b - 3)(2b + 1)$ $-b$
10. $(3z + 5)(5z - 3)$ $16z$
11. $(2m + 4)(m + 5)$ $14m$
12. $(3x + 7)(5x - 1)$ $32x$

Written Exercises

Multiply.

1. $(a + 3)(a + 7)$ $a^2 + 10a + 21$
2. $(c + 2)(c + 8)$ $c^2 + 10c + 16$
3. $(m - 5)(m - 11)$ $m^2 - 16m + 55$
4. $(x - 4)(x - 8)$ $x^2 - 12x + 32$
5. $(x + 11)(x - 4)$ $x^2 + 7x - 44$
6. $(y + 3)(y - 7)$
7. $(2x + 1)(x + 8)$ $2x^2 + 17x + 8$
8. $(5y - 3)(y + 2)$ $5y^2 + 7y - 6$
9. $(4a + 3)(2a - 1)$ $8a^2 + 2a - 3$

6. $y^2 - 4y - 21$

Chalkboard Exercises
Multiply in column form.

7. $(3x - 7)(2x^2 + x + 5)$

$$\begin{array}{rrrr}
 & 2x^2 & +\ x & +\ 5 \\
 & 3x & -\ 7 & \\
\hline
6x^3 & +\ 3x^2 & +\ 15x & \\
+ & -\ 14x^2 & -\ 7x & -\ 35 \\
\hline
6x^3 & -\ 11x^2 & +\ 8x & -\ 35
\end{array}$$

8. $(3x - 5)(x^3 - 2x^2 + 6)$

$$\begin{array}{rrrrr}
x^3 & -\ 2x^2 & +\ 0x & +\ 6 & \\
3x & -\ 5 & & & \\
\hline
3x^4 - 6x^3 & +\ 0x^2 & +\ 18x & & \\
+\ \ -\ 5x^3 & +\ 10x^2 & +\ 0x & -\ 30 & \\
\hline
3x^4 - 11x^3 & +\ 10x^2 & +\ 18x & -\ 30 &
\end{array}$$

❸ Have students practice multiplication in vertical form by using the chalkboard exercises for Examples 5 and 6. Using the same examples will confirm to the students that the methods are equivalent.

Suggested Daily Assignments

Enriched
- Written: 2–58 evens, 59–66
- Mini-Review: 1–5

Average
- Exp: 2–12 evens
- Written: 2–36 evens, 49–52, 59–61
- Mini-Review: 1–5

Fundamental
- Exp: 1–12
- Written: 2–30 evens, 49–52 p. 184, 19–43 odds
- Mini-Review: 1–5

Answers to Written

27. $0.63x^2 + 3.9xy + 6y^2$
28. $0.4x^2 - 0.06x - 0.18$
29. $0.24a^2 - 0.63a - 1.32$
30. $a^3 - 3a^2 - a + 18$
31. $m^3 + 14m^2 + 47m - 14$
32. $2x^3 + 15x^2 - 11x - 9$
33. $5a^3 - 13a^2 + 49a + 22$
34. $6x^3 - 5x^2 + 8x + 55$
35. $12s^3 + 47s^2 + 4s - 45$
36. $12x^3 + 11x^2y - 11xy^2 - 3y^3$
37. $30x^3 - 37x^2y + 55xy^2 - 18y^3$
38. $0.06m^3 + 0.07m^2 - 1.99m + 1.4$
39. $0.15n^3 + 3.07n^2 - 24.35n + 12$
40. $0.42a^3 + 0.23a^2 - 0.76a - 0.55$
41. $0.63t^3 - 0.40t^2 + 1.41t + 0.34$
42. $\frac{1}{2}a^3 + a^2 + \frac{19}{36}a + \frac{1}{9}$
43. $\frac{4}{9}x^3 - \frac{5}{6}x^2y + \frac{13}{24}xy^2 - \frac{1}{8}y^3$
44. $\frac{1}{8}m^3 + \frac{3}{8}m^2 - \frac{3}{2}m - \frac{9}{2}$
45. $\frac{1}{12}a^3 + \frac{5}{12}a^2 + \frac{1}{12}a - \frac{5}{6}$
46. $2x^4 - 17x^3 + 23x^2 + 30x - 24$
47. $a^4 - a^3 - 8a^2 - 29a - 35$
48. $30a^4 + 7a^3b - 7a^2b^2 - 22ab^3 - 8b^4$
49. $20x^4 - 9x^3 + 73x^2 - 39x + 99$
50. $-6x^4 - 5x^3 + 7x^2 - 71x + 40$
51. $-24a^3 - 34a^2 + 19a + 15$
52. $-25x^3 + 20x^2 + 31x - 14$
53. $5x^6 - 25x^5 + 13x^4 + 10x^3 - 5x^2 - 5x + 3$
54. $-35b^5 + 14b^4 - 18b^3 - 19b^2 + 14b - 12$
55. $\frac{1}{4}a^6 + 2a^3 - 9a^2 + 4$
56. $-\frac{10}{3}x^5 + \frac{7}{3}x^4 - 6x^3 + \frac{4}{3}x^2 - x$
57. $0.6x^5 + 9x^4 - 6.06x^3 - 1.5x^2 - 8.4x + 6$
58. $0.5a^4 - 2a^3b + 6.5a^2b^2 - 9ab^3 + 7b^4$
62. $6y^{4a} - 15y^{3a+2b} - 8y^{a+b} + 20y^{3b}$
63. $8x^{3y} - 20x^{6y} + 12x^{4y} - 30x^{7y}$
64. $18z^{3r} + 36z^{5r} - 16z^{7r} - 32z^{9r}$

10. $(5a + 3)(3a + 1)$ $15a^2 + 14a + 3$

11. $(8a + 3)(5a + 4)$ $40a^2 + 47a + 12$

12. $(2x + 3y)(5x + 2y)$ $10x^2 + 19xy + 6y^2$

13. $(2a + 3b)(5a - 2b)$ $10a^2 + 11ab - 6b^2$

14. $(2b + 5d)(4b + 8d)$ $8b^2 + 36bd + 40d^2$

15. $(7y - 1)(2y - 3)$ $14y^2 - 23y + 3$

16. $(5q + 2r)(8q - 3r)$ $40q^2 + qr - 6r^2$

17. $(5r - 7s)(4r + 3s)$ $20r^2 - 13rs - 21s^2$

18. $\left(2x - \frac{1}{2}\right)\left(5x + \frac{1}{2}\right)$ $10x^2 - \frac{3}{2}x - \frac{1}{4}$

19. $\left(4x - \frac{2}{3}\right)\left(5x + \frac{1}{3}\right)$ $20x^2 - 2x - \frac{2}{9}$

20. $\left(3x + \frac{1}{4}\right)\left(6x - \frac{1}{2}\right)$ $18x^2 - \frac{1}{8}$

21. $\left(4x - \frac{1}{2}\right)\left(3x + \frac{3}{4}\right)$ $12x^2 + \frac{3}{2}x - \frac{3}{8}$

22. $\left(\frac{1}{3}x + \frac{1}{2}\right)\left(\frac{2}{3}x - \frac{1}{2}\right)$ $\frac{2}{9}x^2 + \frac{1}{6}x - \frac{1}{4}$

23. $\left(\frac{1}{4}y + \frac{1}{3}x\right)\left(\frac{3}{4}y - \frac{2}{3}x\right)$ $\frac{3}{16}y^2 + \frac{1}{12}xy - \frac{2}{9}x^2$

24. $(2r + 0.1)(5r - 0.3)$ $10r^2 - 0.1r - 0.03$

25. $(2a + 0.7)(3a + 0.8)$ $6a^2 + 3.7a + 0.56$

26. $(0.2x + 0.5y)(0.2x - 0.5y)$ $0.04x^2 - 0.25y^2$

27. $(0.7x + 2y)(0.9x + 3y)$

28. $(0.5x + 0.3)(0.8x - 0.6)$

29. $(0.3a - 1.2)(0.8a + 1.1)$

30. $(a + 2)(a^2 - 5a + 9)$

31. $(m + 7)(m^2 + 7m - 2)$

32. $(2x + 1)(x^2 + 7x - 9)$

33. $(5a + 2)(a^2 - 3a + 11)$

34. $(3x + 5)(2x^2 - 5x + 11)$

35. $(4s + 5)(3s^2 + 8s - 9)$

36. $(4x - 3y)(3x^2 + 5xy + y^2)$

37. $(5x - 2y)(6x^2 - 5xy + 9y^2)$

38. $(0.3m + 2)(0.2m^2 - 1.1m + 0.7)$

39. $(0.3n + 8)(0.5n^2 - 3.1n + 1.5)$

40. $(0.6a + 0.5)(0.7a^2 - 0.2a - 1.1)$

41. $(0.9t + 0.2)(0.7t^2 - 0.6t + 1.7)$

42. $\left(\frac{1}{2}a + \frac{2}{3}\right)\left(a^2 + \frac{2}{3}a + \frac{1}{6}\right)$

43. $\left(\frac{2}{3}x - \frac{1}{2}y\right)\left(\frac{2}{3}x^2 - \frac{3}{4}xy + \frac{1}{4}y^2\right)$

44. $\left(\frac{1}{4}m + \frac{3}{4}\right)\left(\frac{1}{2}m^2 - 6\right)$

45. $\left(\frac{1}{3}a + \frac{2}{3}\right)\left(\frac{1}{4}a^2 + \frac{3}{4}a - \frac{5}{4}\right)$

46. $(x^2 - 7x + 4)(2x^2 - 3x - 6)$

47. $(a^2 + 2a + 5)(a^2 - 3a - 7)$

48. $(6a^2 + 5ab + 4b^2)(5a^2 - 3ab - 2b^2)$

Multiply in column form.

49. $(5x^2 - 6x + 9)(4x^2 + 3x + 11)$

50. $(-2x^2 + 3x - 8)(3x^2 + 7x - 5)$

51. $(3a + 5)(2a - 8a^2 + 3)$

52. $(5x - 2)(7 - 5x^2 + 2x)$

53. $(5x^4 - 2x^2 + 1)(x^2 - 5x + 3)$

54. $(-7b^3 + 2b - 3)(5b^2 - 2b + 4)$

55. $\left(\frac{1}{2}a^3 - 3a + 2\right)\left(\frac{1}{2}a^3 + 3a + 2\right)$

56. $\left(\frac{2}{3}x^3 - \frac{1}{3}x^2 + x\right)(-5x^2 + x - 1)$

57. $(0.2x^2 + 3x - 2)(3x^3 - 0.3x - 3)$

58. $(0.5a^2 - ab + b^2)(a^2 - 2ab + 7b^2)$

Challenge

Multiply.

59. $(3x^a + y)(5x^a + 2y)$ $15x^{2a} + 11x^ay + 2y^2$

60. $(8x^n + x^{2n})(4x^n - x^{2n})$ $32x^{2n} - 4x^{3n} - x^{4n}$

61. $(2a^n + b^m)(3a^n + 5b^m)$ $6a^{2n} + 13a^nb^m + 5b^{2m}$

62. $(2y^a - 5y^{2b})(3y^{3a} - 4y^b)$

63. $(4x^{2y} + 6x^{3y})(2x^y - 5x^{4y})$

64. $(9z^r - 8z^{5r})(2z^{2r} + 4z^{4r})$

65. $(6m^{2x} - 10m^{2x})(2m^{5x} + 6m^{3x})$ $-8m^{7x} - 24m^{5x}$

66. $(7r^{4a} + 2r^{4a})(11r^{6a} - 3r^{4a})$ $99r^{10a} - 27r^{8a}$

mini-review

Solve.

1. $6(x + 3) = 4(x - 1)$ -11

2. $\frac{9}{m} = \frac{30}{20}$ 6

3. $8y - 3 \geq 7y$ $3 \leq y$

4. $\frac{b}{-7} < -8$ $b > 56$

5. $5(m - 1) > 8m + 2$ $m < -\frac{7}{3}$

For more exercises on multiplying polynomials, see exercises 10-30 on page 557.

Objective 6-5: Develop and use the patterns for $(a + b)^2$, $(a - b)^2$, and $(a + b)(a - b)$.

6-5 Some Special Products

Study the following diagram. There are two ways to find the area of the large square.

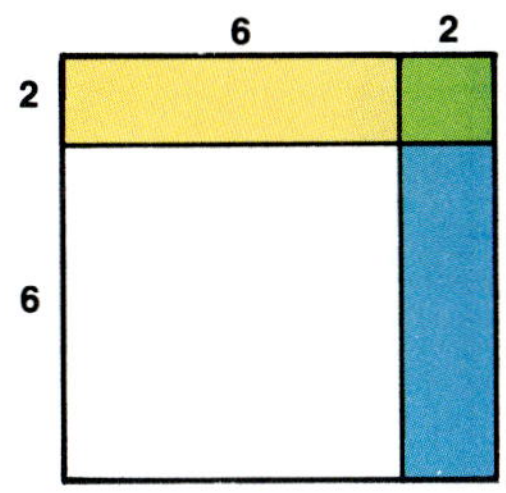

Method 1. The length of each side is 6 + 2 units. Find the square of (6 + 2).

$$(6 + 2)^2 = 8^2$$
$$= 64$$

Method 2. The area of the large square is the sum of the areas of the smaller parts.

$$6^2 + 6 \cdot 2 + 6 \cdot 2 + 2^2$$
$$= 36 + 12 + 12 + 4$$
$$= 64$$

Notice that $(6 + 2)^2 = 6^2 + 6 \cdot 2 + 6 \cdot 2 + 2^2$

Using a similar procedure, the general form $(a + b)^2$ can be simplified as follows.

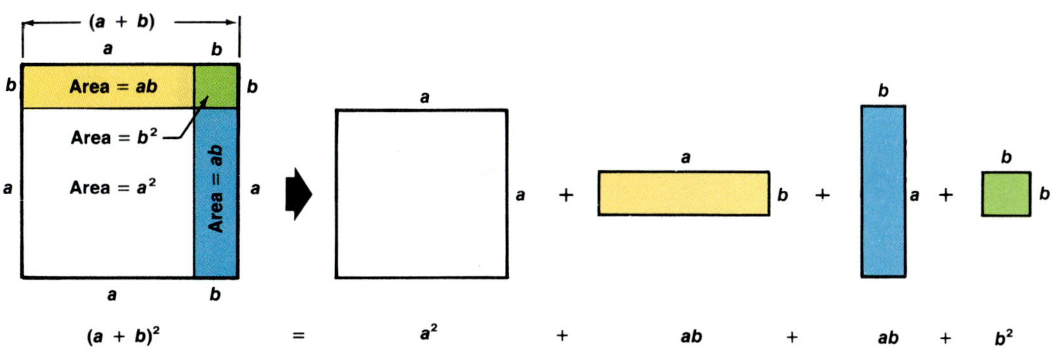

$$(a + b)^2 = a^2 + ab + ab + b^2$$
$$= a^2 + 2ab + b^2$$

Check this result using the FOIL method.

$$(a + b)^2 = a^2 + 2ab + b^2$$

Square of a Sum

6-5

Lesson Notes, p. T36
Enrichment, p. T36

Readiness Activity
Use the FOIL method to find each product.

$(x - 3)^2$	$x^2 - 6x + 9$
$(a + 1)^2$	$a^2 + 2a + 1$
$(2z - 2)^2$	$4z^2 - 8z + 4$
$(m - 5)(m + 5)$	$m^2 - 25$
$(r - 6)(r + 6)$	$r^2 - 36$
$(4q + 3)(4q - 3)$	$16q^2 - 9$

Chalkboard Exercises

Multiply.

1a. $(3a + 2)^2$
$9a^2 + 12a + 4$

1b. $(5b + 7)^2$
$25b^2 + 70b + 49$

2. $(2x + 3y)^2$
$4x^2 + 12xy + 9y^2$

3. $(y - 7)^2$
$y^2 - 14y + 49$

4. $(6m - 4n)^2$
$36m^2 - 48mn + 16n^2$

❶ Emphasize to students that the square of a difference is another case of multiplying binomials by FOIL.

$$\begin{aligned}(a - b)^2 &= (a - b)(a - b)\\ &\quad\ \ \text{F}\quad \text{O}\quad \text{I}\quad \text{L}\\ &= a^2 - ab - ba + b^2\\ &= a^2 - 2ab + b^2\end{aligned}$$

Use the following diagram to show $(a - b)^2$.

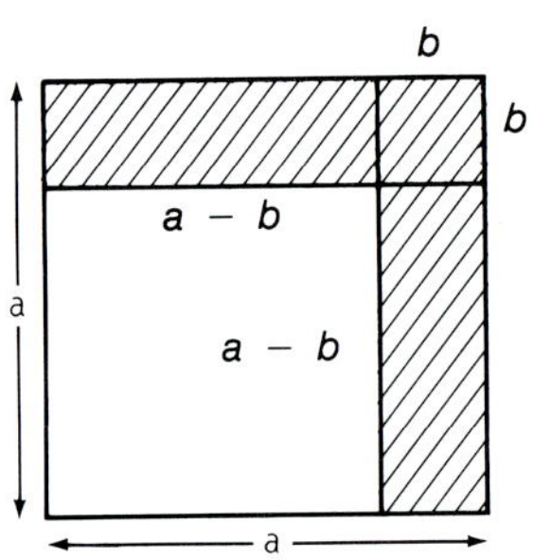

$$\begin{aligned}(a - b)^2 &= a^2 - ab - b(a - b)\\ &= a^2 - ab - ba + b^2\\ &= a^2 - 2ab + b^2\end{aligned}$$

Examples

1 **Multiply: $(x + 3)^2$**

Use the square of a sum, $(a + b)^2 = a^2 + 2ab + b^2$.

$$\begin{aligned}(x + 3)^2 &= x^2 + 2(x)(3) + (3)^2 \\ &= x^2 + 6x + 9\end{aligned}$$

Replace a with x and b with 3.

2 **Multiply: $(5m + 3n)^2$**

Use $(a + b)^2 = a^2 + 2ab + b^2$.

$$\begin{aligned}(5m + 3n)^2 &= (5m)^2 + 2(5m)(3n) + (3n)^2\\ &= 25m^2 + 30mn + 9n^2\end{aligned}$$

To find $(a - b)^2$, express $(a - b)$ as $[a + (-b)]$ and square it as a sum.

$$\begin{aligned}(a - b)^2 &= [a + (-b)]^2\\ &= a^2 + 2(a)(-b) + (-b)^2\\ &= a^2 - 2ab + b^2\end{aligned}$$

❶

$$(a - b)^2 = a^2 - 2ab + b^2$$

Square of a Difference

Compare the square of a sum and the square of a difference. How do they differ?

Examples

3 **Multiply: $(c - 2)^2$**

Use the square of a difference, $(a - b)^2 = a^2 - 2ab + b^2$.

$$\begin{aligned}(c - 2)^2 &= c^2 - 2(c)(2) + (2)^2\\ &= c^2 - 4c + 4\end{aligned}$$

Replace a with c and b with 2.

4 **Multiply: $(3x - 2y)^2$**

Use $(a - b)^2 = a^2 - 2ab + b^2$.

$$\begin{aligned}(3x - 2y)^2 &= (3x)^2 - 2(3x)(2y) + (2y)^2\\ &= 9x^2 - 12xy + 4y^2\end{aligned}$$

You can use the distributive property to check this.

You can use the FOIL method to find the product of a sum and a difference of the same two numbers. For example find $(a + b)(a - b)$.

❷

$$\begin{aligned}(a + b)(a - b) &= a(a) + a(-b) + b(a) + b(-b)\\ &= a^2 - ab + ba - b^2\\ &= a^2 - b^2\end{aligned}$$

This product is called the **difference of squares**.

❸

$$(a + b)(a - b) = a^2 - b^2$$

Product of a Sum and a Difference

Examples

5 **Multiply: $(x + 5)(x - 5)$**

$(a + b)(a - b) = a^2 - b^2$

$(x + 5)(x - 5) = x^2 - 5^2$ *Replace a with x and b with 5.*

$= x^2 - 25$

6 **Multiply: $(5a + 6b)(5a - 6b)$**

$(a + b)(a - b) = a^2 - b^2$

$(5a + 6b)(5a - 6b) = (5a)^2 - (6b)^2$

$= 25a^2 - 36b^2$

Exploratory Exercises

Multiply.

1. $(a + 2b)^2$ $a^2 + 4ab + 4b^2$
2. $(a - 3b)^2$ $a^2 - 6ab + 9b^2$
3. $(2x + y)^2$ $4x^2 + 4xy + y^2$
4. $(3x - 2y)^2$ $9x^2 - 12xy + 4y^2$
5. $(3m^2 + 2n)^2$
6. $(3m - 4x)^2$
7. $(a^2 + b)^2$ $a^4 + 2a^2b + b^2$
8. $(2a + 3)(2a - 3)$ $4a^2 - 9$
9. $(3a - 8b)^2$
10. $(4x - 3)(4x + 3)$ $16x^2 - 9$
11. $(5a - 3b)(5a - 3b)$
12. $(4x^2 + y)^2$
13. $(3x - y)^2$ $9x^2 - 6xy + y^2$
14. $(5m^2 + 7n)^2$
15. $(2x^2 - 3y^3)^2$

Written Exercises

Multiply.

1. $(4x + y)^2$ $16x^2 + 8xy + y^2$
2. $(5x + y)^2$ $25x^2 + 10xy + y^2$
3. $(2a - b)^2$ $4a^2 - 4ab + b^2$
4. $(5a - b)^2$ $25a^2 - 10ab + b^2$
5. $(6m + 2n)^2$ $36m^2 + 24mn + 4n^2$
6. $(7x + 3y)^2$ $49x^2 + 42xy + 9y^2$
7. $(4x - 9y)^2$ $16x^2 - 72xy + 81y^2$
8. $(5r - 7s)^2$ $25r^2 - 70rs + 49s^2$
9. $(5a - 12b)^2$
10. $(6a - 5b)^2$ $36a^2 - 60ab + 25b^2$
11. $(5x + 6y)^2$ $25x^2 + 60xy + 36y^2$
12. $(11m + 7n)^2$
13. $\left(\frac{1}{2}a + b\right)^2$ $\frac{1}{4}a^2 + ab + b^2$
14. $\left(\frac{1}{3}x + y\right)^2$ $\frac{1}{9}x^2 + \frac{2}{3}xy + y^2$
15. $\left(\frac{4}{3}x - 2y\right)^2$
16. $\left(\frac{3}{4}a - 2b\right)^2$ $\frac{9}{16}a^2 - 3ab + 4b^2$
17. $(0.7x - 1.1)^2$
18. $(0.3a - 2.5)^2$
19. $(1.1x + y)^2$ $1.21x^2 + 2.2xy + y^2$
20. $(3.1 + 2y)^2$
21. $(a^2 - 3b^3)^2$
22. $(x^3 - 5y^2)^2$ $x^6 - 10x^3y^2 + 25y^4$
23. $(3x + 5)(3x - 5)$ $9x^2 - 25$
24. $(8a + 2b)(8a - 2b)$ $64a^2 - 4b^2$

Chalkboard Exercises

Multiply.

5a. $(x - 9)(x + 9)$ $x^2 - 81$
5b. $(3x - 8)(3x + 8)$ $9x^2 - 64$
6a. $(4x + 3y)(4x - 3y)$ $16x^2 - 9y^2$
6b. $(3x^2 - 2y)(3x^2 + 2y)$ $(9x^4 - 4y^2)$

❷ A geometric development of the difference of squares is presented in Chapter 7 on page 217.

❸ Have students compare the following expressions:
square of a difference
$(a - b)^2$
difference of squares
$a^2 - b^2$

Suggested Daily Assignments

Enriched
Written: 2–42 evens, 44–52

Average
Exp: 1–15 odds
Written: 2–42 evens
p. 188, 35–43 odds

Fundamental
Exp: 1–15
Written: 2–34 evens
p. 187–8, 1–29 odds, 53–58

Answers to Exploratory

5. $9m^4 + 12m^2n + 4n^2$
6. $9m^2 - 24mx + 16x^2$
9. $9a^2 - 48ab + 64b^2$
11. $25a^2 - 30ab + 9b^2$
12. $16x^4 + 8x^2y + y^2$
14. $25m^4 + 70m^2n + 49n^2$
15. $4x^4 - 12x^2y^3 + 9y^6$

Answers to Written

9. $25a^2 - 120ab + 144b^2$
12. $121m^2 + 154mn + 49n^2$
15. $\frac{16}{9}x^2 - \frac{16}{3}xy + 4y^2$
17. $0.49x^2 - 1.54x + 1.21$
18. $0.09a^2 - 1.5a + 6.25$
20. $9.61 + 12.4y + 4y^2$
21. $a^4 - 6a^2b^3 + 9b^6$

Answers to Written

25. $10.24x^4 - 8.96x^2y + 1.96y^2$
26. $3.24a^4 - 8.28a^2y^2 + 5.29y^4$
27. $\frac{1}{9}a^2 - \frac{9}{4}b^2$
28. $\frac{16}{9}x^4 - y^2$
29. $49a^4 - 14a^2b + b^2$
30. $64x^4 - 48x^2y + 9y^2$
31. $9a^4 + 6a^2b + b^2$
32. $16x^6 - 24x^3y^2 + 9y^4$
33. $36a^6 - 12a^5 + a^4$
34. $\frac{9}{25}x^4 - \frac{12}{25}x^2y + \frac{4}{25}y^2$
35. $\frac{1}{16}a^4 - \frac{1}{6}a^2b^3 + \frac{1}{9}b^6$
36. $0.04a^6 + 0.04a^3b^2 + 0.01b^4$
37. $1.69x^4 + 5.46x^2y^3 + 4.41y^6$
38. $x^{4a} + 2x^{2a}y^{2a} + y^{4a}$
39. $x^{6n} - 2x^{3n}y^{2n} + y^{4n}$

Answers to Challenge

44. $a^4 + 4a^3b + 6a^2b^2 + 4ab^3 + b^4$
45. $16x^4 - 32x^3y + 24x^2y^2 - 8xy^3 + y^4$
46. $81m^4 + 216m^3n + 216m^2n^2 + 96mn^3 + 16n^4$
47. $a^5 + 5a^4b + 10a^3b^2 + 10a^2b^3 + 5ab^4 + b^5$
48. $a^5 - 5a^4b + 10a^3b^2 - 10a^2b^3 + 5ab^4 - b^5$
49. $x^3 - 6x^2y + 12xy^2 - 8y^3$
50. $16x^4 - 96x^3y + 216x^2y^2 - 216xy^3 + 81y^4$
51. $\frac{1}{8}x^3 + \frac{3}{4}x^2y + \frac{3}{2}xy^2 + y^3$
52. $0.0081m^4 - 0.108m^3n + 0.54m^2n^2 - 1.2mn^3 + n^4$

About Using Calculators
Some calculators have different labels for the keys that perform the store and recall functions. Have students bring in different types of calculators and compare the function keys.

25. $(3.2x^2 - 1.4y)^2$ **26.** $(1.8a^2 - 2.3y^2)^2$ **27.** $\left(\frac{1}{3}a+\frac{3}{2}b\right)\left(\frac{1}{3}a-\frac{3}{2}b\right)$

28. $\left(\frac{4}{3}x^2-y\right)\left(\frac{4}{3}x^2+y\right)$ **29.** $(7a^2 - b)^2$ **30.** $(8x^2 - 3y)^2$

31. $(3a^2 + b)^2$ **32.** $(4x^3 - 3y^2)^2$ **33.** $(6a^3 - a^2)^2$

34. $\left(\frac{3}{5}x^2-\frac{2}{5}y\right)^2$ **35.** $\left(\frac{1}{4}a^2-\frac{1}{3}b^3\right)^2$ **36.** $(0.2a^3 + 0.1b^2)^2$

37. $(1.3x^2 + 2.1y^3)^2$ **38.** $(x^{2a} + y^{2a})^2$ **39.** $(x^{3n} - y^{2n})^2$

40. $(a^{2n} - b^{5n})(a^{2n} - b^{5n})$ $a^{4n} - 2a^{2n}b^{5n} + b^{10n}$ **41.** $(x^{2n} + y^n)(x^{2n} - y^n)$ $x^{4n} - y^{2n}$

42. $(a^{3n} - b^{5n})(a^{3n} + b^{5n})$ $a^{6n} - b^{10n}$ **43.** $(x^{3n} + y^{5n})(x^{3n} + y^{5n})$ $x^{6n} + 2x^{3n}y^{5n} + y^{10n}$

Challenge

Multiply.

44. $(a + b)^4$ **45.** $(2x - y)^4$ **46.** $(3m + 2n)^4$

47. $(a + b)^5$ **48.** $(a - b)^5$ **49.** $(x - 2y)^3$

50. $(2x - 3y)^4$ **51.** $\left(\frac{1}{2}x + y\right)^3$ **52.** $(0.3m - n)^4$

Using Calculators — The Store and Recall Keys

Many calculators have keys labeled STO and RCL. The key labeled STO is the store key. When you enter a number and then press this key, the number is *stored* in the memory of the calculator.

The key labeled RCL is the recall key. When you press this key, the calculator retrieves a stored number from the memory and displays it.

The store and recall keys are helpful when evaluating expressions.

Example: Evaluate $5 - 4x^2$ when x is 7.29.

First, store 7.29.

Some calculators have different labels for store. and recall keys.

ENTER: 7.29 [STO]
DISPLAY: 7.29

Then evaluate the expression.

ENTER:	5	[−]	4	[×]	[RCL]	[x^2]	[=]
DISPLAY:	5		4		7.29	53.1441	−207.5764

When x is 7.29, the value of $5 - 4x^2$ is -207.5764.

Exercises

Evaluate each expression for the given value.

1. $2 - 3x^2$, when $x = 1.03$ −1.1827
2. $-6 - 7a$, when $a = 0.358$ −8.506
3. $2a^2 + 5a - 7$, when $a = 3.81$ 41.0822
4. $4c^2 + c - 9$, when $c = -0.006$ −9.005856
5. $(7 - x)^2$, when $x = -0.007$ 49.098049
6. $7 - x^2$, when $x = -0.007$ 6.999951

Objective 6-6: Solve equations that require addition, subtraction or multiplication on one or both sides of the equal sign before the equation can be solved.

6-6 Solving Equations

Many equations contain polynomials that must be added, subtracted, or multiplied before the equation can be solved.

Examples

1 Solve: $-8 - (9 - 4m) = 15$

$$-8 - (9 - 4m) = 15$$
$$-8 - 9 + 4m = 15$$ *Replace each term in the parentheses with its additive inverse.*
$$-17 + 4m = 15$$
$$-17 + 17 + 4m = 15 + 17$$
$$4m = 32$$
$$\frac{4m}{4} = \frac{32}{4}$$
$$m = 8$$

The solution is 8. *Check this result.*

2 Solve: $5(x - 3) + 3x = 8(9 - x) + 21$

$$5(x - 3) + 3x = 8(9 - x) + 21$$
$$5x - 15 + 3x = 72 - 8x + 21$$
$$8x - 15 = 93 - 8x$$
$$8x + 8x - 15 + 15 = 93 + 15 - 8x + 8x$$
$$16x = 108$$
$$\frac{16x}{16} = \frac{108}{16}$$
$$x = 6.75$$

The solution is 6.75. *Check this result.*

3 Solve: $x(x - 3) + 4x - 3 = 8x + 4 + x(3 + x)$

$$x(x - 3) + 4x - 3 = 8x + 4 + x(3 + x)$$
$$x^2 - 3x + 4x - 3 = 8x + 4 + 3x + x^2$$
$$x^2 + x - 3 = x^2 + 11x + 4$$
$$x - 3 = 11x + 4$$ *Subtract x^2 from both sides of the equation.*
$$x - 11x - 3 + 3 = 11x - 11x + 4 + 3$$
$$-10x = 7$$
$$\frac{-10x}{-10} = \frac{7}{-10}$$
$$x = -\frac{7}{10} \text{ or } -0.7$$

The solution is -0.7. *Check this result.*

6-6

Lesson Notes, p. T37

Readiness Activity

Solve and check each equation.

$5n + 2 = 2n + 8$ 2
$4x - 3 = 7 + 2x$ 5
$3(r + 1) = 2(r - 6) + 3$ -12
$10(3m - 5) = -4(6m - 1)$ 1

Discussion Question

Solve the equation $5x - x(x + 1) = x - 2 - x(x - 2)$ by first simplifying both sides of the equation. What happens to the x^2 terms?
-2; They drop out of the equation.

Chalkboard Exercises

Solve each equation.

1. $16 = 8 - (12 - 5x)$ 4
2. $(3y + 2) - 4 = 30 - 4(y + 1)$ 4
3. $4r + 2 + r(r - 1) = r(r + 4) + 5$ -3

Suggested Daily Assignments

Enriched
  Written: 2–46 evens

Average
  Exp: 1–9 odds
  Written: 1–35 odds

Fundamental
  Exp: 1–10
  Written: 1–22
    p. 191, 1–23 odds

Quiz
Lessons 6-3 to 6-6
Simplify each of the following.

1. $-2(5a^4 - 6a^2 - 7a)$
   $-10a^4 + 12a^2 + 14a$
2. $(r + 9t)(r - 4t)$
   $r^2 + 5rt - 36t^2$
3. $(y^2 - 3y - 2)(2y^2 + 4y - 3)$
   $2y^4 - 2y^3 - 19y^2 + y + 6$
4. $(a + 3b)^2$
   $a^2 + 6ab + 9b^2$
5. $(5x + 3y)(5x - 3y)$
   $25x^2 - 9y^2$
6. $2(5a + 6b - 3c) - 7(8a - 4b + 2c)$
   $-46a + 40b - 20c$

Exploratory Exercises

Simplify.

1. $3x + 4x - 5 - 6$ $7x - 11$
2. $12 - 3y + 12y - 23$ $9y - 11$
3. $3y + 5y - 8 - 34$ $8y - 42$
4. $17 - (3a + 4) - 6a$ $13 - 9a$
5. $6y - 8y + 34 - 34y$ $-36y + 34$
6. $8r + 34 + 39r - 8r$ $39r + 34$
7. $12(q + 4) - 2q$ $10q + 48$
8. $5(x - 3) + 22 - 8x$ $-3x + 7$
9. $-3(x - 12) + 5(x - 23)$ $2x - 79$
10. $-9(9 - y) - 8(12 - y) + 23$ $17y - 154$

Written Exercises

Solve each equation.

1. $3x + 8x - 7 = 21x - 5$ $-\frac{1}{5}$
2. $9x - 8 + 4x = 7x + 16$ 4
3. $4y + 8y - 7 = 30y + 19$ $-\frac{13}{9}$
4. $8y + 16 + 8y = 21 - 9y$ $\frac{1}{5}$
5. $-8 + 12a = 14a - 23 + 5a$ $\frac{15}{7}$
6. $7a + 23 - 12a = 34 + 23a$ $-\frac{11}{28}$
7. $-9a + 34 - 2a = 28 - 5a + 27$ $-\frac{7}{2}$
8. $9a - 11 = 23 + 24a - 27a$ $\frac{17}{6}$
9. $11(a - 3) + 5 = 2a + 44$ 8
10. $-3(2a - 12) + 48 = 3a - 3$ $\frac{29}{3}$
11. $26 + 4(3a - 5) = 14a - 15$ $\frac{21}{2}$
12. $-13 + 3(3a + 11) = -33 - 3a$ $-\frac{53}{12}$
13. $57 + 2a = 3a - 5(a - 9)$ -3
14. $29 - 3a = 2(3a - 4) + 3$ $\frac{34}{9}$
15. $3a - 35 = 4(a + 12)$ -83
16. $13 - 3a = 23(2a - 3)$ $\frac{82}{49}$
17. $2(5w - 12) = 6(-2w + 3) + 2$ 2
18. $-11(3a - 4) = -3(-4a - 14)$ $\frac{2}{45}$
19. $-6(12 - 2w) = 7(-2 - 3w)$ $\frac{58}{33}$
20. $15(3a - 4) = -4(3a - 3)$ $\frac{24}{19}$
21. $-5(2x - 13) + 12 = 3(2x - 5)$ $\frac{23}{4}$
22. $7(x - 12) = 13 + 5(3x - 4)$ $-\frac{77}{8}$
23. $\frac{1}{2}(2x - 34) = \frac{2}{3}(6x - 27)$ $\frac{1}{3}$
24. $\frac{3}{4}(8z - 12) = \frac{5}{6}(12z - 18)$ $\frac{3}{2}$
25. $19 - (2y + 3) = 2(y + 3) + y$ 2
26. $2(a + 2) + 3a = 13 - (2a + 2)$ 1
27. $x(x + 2) + 3x = x(x - 3)$ 0
28. $w(w + 12) = w(w + 14) + 12$ -6
29. $a(a - 6) + 2a = 3 + a(a - 2)$ $-\frac{3}{2}$
30. $q(2q + 3) + 20 = 2q(q - 3)$ $-\frac{20}{9}$
31. $x(x + 8) - x(x + 3) - 23 = 3x + 11$ 17
32. $y(y - 12) + y(y + 2) + 25 = 2y(y + 5) - 15$ 2

Challenge

Solve each equation for x or y.

33. $2x = 5a$ $\frac{5a}{2}$
34. $4aby = 8a^2bc$ $2ac$
35. $30b^2c^2d = 10bc^2y$ $3bd$
36. $y - 2a = 0$ $2a$
37. $4n + 3y = 2m$ $\frac{2m - 4n}{3}$
38. $ax + 3 = b$ $\frac{b - 3}{a}$
39. $2x - 7 = 8 + a$ $\frac{15 + a}{2}$
40. $2y + b = 3 - 5y$ $\frac{3 - b}{7}$
41. $7ax - 2d = 3ax + 6d$ [illegible]
42. $4x + 5mn = 7mn + 2x$ mn
43. $by - c = a - d$ $\frac{a - d + c}{b}$
44. $3(y - 2a) = 24a$ $10a$
45. $a(x + b) = 3ab + 5$ $\frac{2ab + 5}{a}$
46. $2(3a - 2x) = 5a + x$ $\frac{a}{5}$
47. $4(y + 3a) = 4a - y$ $-\frac{8a}{5}$

Objective 6-7: Solve perimeter and area problems using equations.

6-7 Problem Solving: Perimeter and Area

Recall that a polygon is a closed plane figure formed by line segments. The **perimeter** of a polygon is the sum of the lengths of its sides. Some common units of length are inches (in.), feet (ft), yards (yd), centimeters (cm), and meters (m).

Example 1

In a certain isosceles triangle the third side is 3 inches shorter than either of the congruent sides. If the perimeter is 69 inches, find the lengths of the sides.

An isosceles triangle has at least two congruent sides.

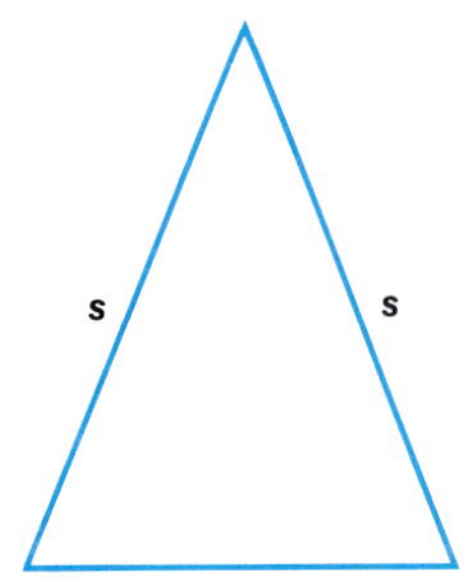

❶

Explore Let the following expressions represent the lengths of the sides.

s *length of each congruent side*

$(s - 3)$ *length of third side*

Plan Recall that the perimeter of a triangle is the sum of the lengths of its sides. To find these lengths, solve the following equation.

First Side + *Second Side* + *Third Side* = *Perimeter of the Triangle*

Solve

$$s + s + (s - 3) = 69$$
$$3s - 3 = 69$$
$$3s = 72$$
$$s = 24$$

Therefore, the length of each congruent side is 24 inches. The length of the third side is 24 − 3 or 21 inches.

Examine The sum of the lengths of the three sides is 24 + 24 + 21 or 69 inches.

The **area** of a polygon is the measurement of the region bounded by the polygon. Area is measured in square units. Some common units of area are square inches, square feet, square yards, square centimeters, and square meters.

6-7

Lesson Notes, p. T37
Enrichment, p. T37

Discussion Question
A rectangular plot of ground is to be enclosed with 48 feet of fencing. What are some of the possible areas that can be enclosed? What is the greatest area that can be enclosed?
Answers will vary; 12 · 12
144 ft^2 is greatest

Chalkboard Exercises
Use an equation to solve each problem.

1a. A triangle has two congruent sides. The third side is 5 inches shorter. The perimeter is 34 inches. What are the lengths of the sides?
13 in., 13 in., 8 in.

1b. A triangle has two congruent sides. The third side is 2 inches longer. The perimeter is 41 inches. What are the lengths of the sides?
13 in., 13 in., 15 in.

❶ Encourage students to draw a diagram as an aid in solving area problems.

Use an equation to solve each problem.

2a. Edith has a stamp which is 0.5 cm longer than it is wide. A border that is 1 cm in width surrounds the stamp. The total area of the border is 11 cm^2. What are the dimensions of the stamp?
1.5 cm by 2.0 cm

2b. Donnie has a rectangular garden 15 feet longer than it is wide. A sidewalk that is 4 feet in width surrounds the garden. The total area of the sidewalk is 632 square feet. What are the dimensions of the garden?
28 feet by 43 feet

Suggested Daily Assignments

Enriched
- Exp: 1–2
- Written: 1–9
  - p. 194, 33–47 odds
- Mini-Review: 1–5

Average
- Exp: 1–2
- Written: 1–8
  - p. 194, 34–38 evens
- Mini-Review: 1–5

Fundamental
- Exp: 1–2
- Written: 1–6
  - p. 194, 23–30
- Mini-Review: 1–5

Example

2 **Maria Coulson has a rectangular garden that is 10 feet longer than it is wide. A sidewalk that is 3 feet in width surrounds the garden. The total area of the sidewalk is 396 square feet. What are the dimensions of the garden?**

Explore

Let x = width of garden.
$(x + 10)$ = length of garden.
$(x + 6)$ = width of garden and sidewalk.
$(x + 16)$ = length of garden and sidewalk.

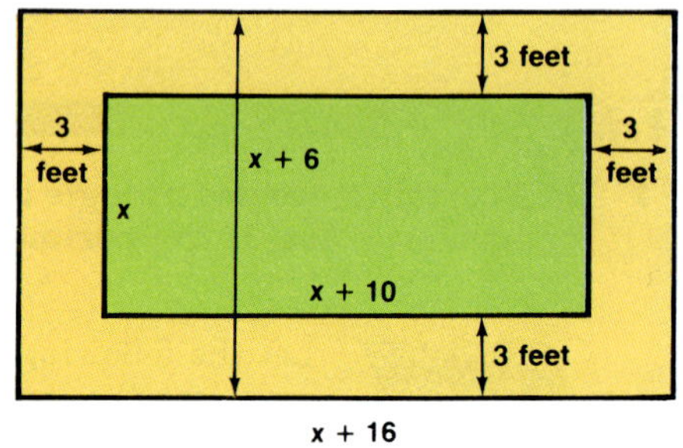

Since the area of a rectangle is the product of its length and width, the following expressions represent area.

$x(x + 10)$ *area of garden*
$(x + 6)(x + 16)$ *area of garden and sidewalk*

Plan

To find the dimensions of the garden, solve the following equation.

area of garden and sidewalk − *area of garden* = *area of sidewalk*

Solve

$$(x + 6)(x + 16) - x(x + 10) = 396$$
$$x^2 + 22x + 96 - x^2 - 10x = 396$$
$$12x + 96 = 396$$
$$12x = 300$$
$$x = 25$$

Therefore, the width is 25 feet. The length is $(x + 10)$ or 35 feet.

Examine

To examine the solution, compute the total area of the garden and sidewalk in two ways and compare the results.

area of garden + *area of sidewalk*
$25 \times 35 + 396 = 1271$

total length × *total width*
$(35 + 6) \times (25 + 6)$
$41 \times 31 = 1271$

Exploratory Exercises

Write an expression that represents each of the following.

1. The length of a rectangle is 5 units more than the width. Find the length if the measure of the width is: **a.** n **b.** $2t$ **c.** $2n + 3$ **d.** $t - 1$
a. $n + 5$ b. $2t + 5$ c. $2n + 8$ d. $t + 4$

2. The length of a rectangle is 2 units less than twice the width. Find the length if the measure of the width is: **a.** t **b.** $2x$ **c.** $3a + 2$ **d.** $2t + 1$
a. $2t - 2$ b. $4x - 2$ c. $6a + 2$ d. $4t$

For more exercises on adding polynomials, see exercises 1-24 on page 556.
For more exercises on subtracting polynomials, see exercises 25-50 on page 556.

Written Exercises

Use an equation to solve each problem.

1. The length of a rectangle is 4 feet more than twice the width. The perimeter is 116 feet. Find the dimensions of the rectangle. 18 ft by 40 ft

2. A certain triangle has two congruent sides. The third side is 17 cm shorter than either of the equal sides. If the perimeter is 91 cm, what is the length of the third side? 19 cm

3. To get a square photograph to fit into a square frame, Linda LaGuardia had to trim a 1-inch strip from each side of the photo as shown below. In all, she trimmed off 40 square inches. What were the original dimensions of the photograph? 11 in. by 11 in.

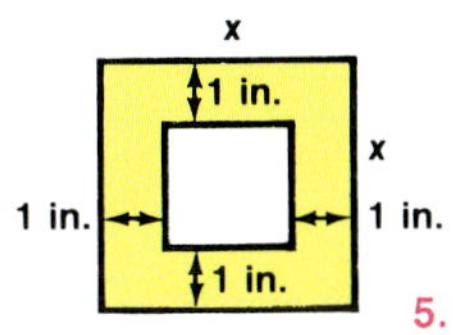

4. A rectangular garden is 5 feet longer than twice its width. It has a sidewalk 3 feet wide on two of its sides, as shown below. The area of the sidewalk is 213 square feet. Find the dimensions of the garden. 21 ft by 47 ft

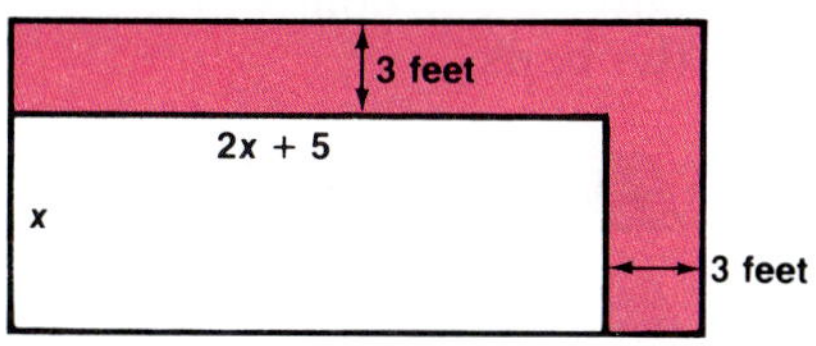

5. 8 cm, 16 cm, 13 cm

5. The second side of a triangle is twice the length of the first. The third side is 3 cm less than the second side. What are the lengths of the sides if the perimeter is 37 cm?

6. The three sides of a triangle have measures that are consecutive odd numbers. What are the lengths of the sides if the perimeter is 87 m? 27 m, 29 m, 31 m

7. The length of a rectangle is 20 yards greater than the width. If the length was decreased by 5 yards, and the width increased by 4 yards, the area would remain unchanged. Find the original dimensions of the rectangle. 60 yd by 80 yd

8. The length of a rectangle is 7 cm less than twice its width. If the length was increased by 11 cm and the width decreased by 6 cm, the area would be decreased by 40 square centimeters. Find the original dimensions of the rectangle. 16 cm by 25 cm

mini-review

Simplify.

1. $\frac{mn^3}{n^6}$ $\frac{m}{n^3}$ **2.** $\left(\frac{1}{4}x^3\right)^2$ $\frac{x^6}{16}$

Solve.

3. What number is 70% less than 80? 24

4. Find three consecutive odd integers whose sum is 261. 85, 87, 89

5. The sum of two consecutive positive even integers is at most 10. What are the integers? 2, 4; 4, 6

Challenge

9. Mr. Herrera had a concrete sidewalk built on three sides of his yard as shown at the right. The yard measures 24 by 42 feet. The longer walk is 3 feet wide. The price of the concrete was $22 per square yard, and the total bill was $902. What is the width of the walk on the remaining two sides? 4.5 ft

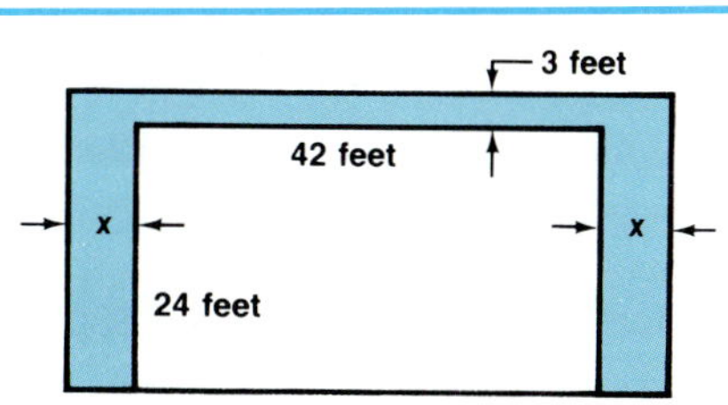

6-8

Lesson Notes, p. T37

Readiness Activity
Write each percent as a decimal.

16% 0.16 10% 0.10
8% 0.08 9% 0.09
$12\frac{1}{2}\%$ 0.125 $14\frac{3}{4}\%$ 0.1475

Discussion Questions

1. Why do banks pay interest on savings accounts? Answers will vary.
2. Why are interest rates on loans higher than interest rates on savings accounts? Because banks need to make a profit.

Chalkboard Exercises
Substitute the following information in Example 1, then solve the new problem.

1. The money was deposited at 8% for 9 months. It earned \$72.00 interest. What was the original deposit? \$1200

Substitute the following in information in Example 2, then solve the new problem.

2. The amount of \$12,000 was invested, part at 12% and part at 13.25%. The interest earned was \$1540. How much money was invested at each rate? \$4000 at 12% \$8000 at 13.25%

❶ Remind students that interest rate is annual. Therefore, time should be expressed in years when working with the simple interest formula.

Objective 6-8: Solve simple interest problems using the formula $I = Prt$.

6-8 Problem Solving: Simple Interest

Anita Seltzer deposited an amount of money in the bank at 7% annual interest. After 6 months, she received \$52.50 interest. How much money had Anita deposited in the bank?

The formula $I = Prt$ is used to solve **simple interest** problems such as the one above. In the formula, I represents interest, P represents the amount of money invested or principal, r represents the annual interest rate, and t represents time in years.

Examples

1 Solve the problem above to find out how much money Anita had deposited in the bank.

Explore Let P = amount of money Anita deposited.

Plan
$I = Prt$
$52.50 = P(0.07)\left(\frac{1}{2}\right)$ *Change 7% to 0.07. Change 6 months to $\frac{1}{2}$ year.*

Solve
$52.50 = 0.035P$
$1500 = P$

Anita had deposited \$1500 in the bank.

Examine When P is 1500, I is $(1500)(0.07)\left(\frac{1}{2}\right)$ or \$52.50.

2 Marilyn Mallinson invested \$15,000, part at 12% annual interest and the balance at 15% annual interest. Last year she earned \$1995 in interest. How much money did she invest at each rate?

Let n = amount of principal invested at 12%.
$(15{,}000 - n)$ = amount of principal invested at 15%.

Since $I = Prt$, the following expressions represent the interest earned on each portion of the investment.

Interest on 12% investment ❶
$I = n \cdot 0.12 \cdot 1$
$= 0.12n$

Interest on 15% investment
$I = (15{,}000 - n) \cdot 0.15 \cdot 1$
$= 2250 - 0.15n$

Use the following equation to find the amount of money invested at each rate.

Interest at 12% + Interest at 15% = Total Interest

$$0.12n + 2250 - 0.15n = 1995$$
$$-0.03n = -255$$
$$n = 8500$$

Therefore, Marilyn invested \$8500 at 12% and $(15{,}000 - n)$ or \$6500 at 15%.

Exploratory Exercises

Use $I = Prt$ to find the missing quantity.

1. Find I, if P = \$8,000, r = 6%, and t = 1 year. \$480
2. Find I, if P = \$5,000, $r = 12\frac{1}{2}\%$, and t = 5 years. \$3125
3. Find t, if I = \$1890, P = \$6000, and r = 9%. $3\frac{1}{2}$ years
4. Find t, if I = \$2160, P = \$6,000, and r = 8%. $4\frac{1}{2}$ years
5. Find r, if I = \$2430, P = \$9,000, and t = 2 years, 3 months. 12%
6. Find r, if I = \$780, P = \$6500, and t = 1 year. 12%
7. Find P, if I = \$756, r = 9%, and $t = 3\frac{1}{2}$ years. \$2400
8. Find P, if I = \$196, r = 10% and t = 7 years. \$280
9. Find r, if I = \$3,487.50, P = \$6,000, and t = 3 years, 9 months. $15\frac{1}{2}\%$
10. Find t, if I = \$3528, P = \$8,400, and $r = 10\frac{1}{2}\%$. 4 years

Written Exercises

1. \$6400 at 8%, \$3600 at 12%
2. \$1200 at 10%, \$6000 at 14%

Solve each problem.

1. Michele Limotta invested \$10,000 for one year, part at 8% annual interest and the balance at 12% annual interest. Her total interest for the year was \$944. How much money did she invest at each rate?
2. Steve Devine invested \$7,200 for one year, part at 10% annual interest and the balance at 14% annual interest. His total interest for the year was \$960. How much money did he invest at each rate?
3. Fred Ferguson invested \$5,000 for one year, part at 9% annual interest and the balance at 12% annual interest. The interest from the investment at 9% was \$198 more than the interest from the investment at 12%. How much money did he invest at 9%? \$3800
4. Angela Raimondi wants to invest \$8500, part at 14% annual interest and part at 12% annual interest. If she wants to earn the same amount of interest from each investment, how much should she invest at 14%? (Answer to the nearest cent.) \$3923.08
5. In one year, Charlotte earned the same interest from an investment at 8% annual interest as an investment at 12% annual interest. She had invested \$1500 more in the 8% account. How much money did she have invested at 12%? \$3000
6. John and Inger Johnson have \$8000 to invest. They want to earn \$1050 interest for the year. Money can be invested at annual interest rates of either 12% or 15%. What is the minimum amount at 15%, with the balance at 12%, that will give \$1050 in interest? \$3000
7. Charlie and Bonnie Hall have invested \$2500 at 10% annual interest. They have \$6000 more to invest. At what rate must they invest the \$6000 to have a total annual interest of \$1120? $14\frac{1}{2}\%$
8. Ken Bauman invested \$7525, part at 16% annual interest and the balance at 11% annual interest. He earned twice as much interest from the 11% investment as from the 16% investment. How much money did he have invested at 11%? \$5600

Suggested Daily Assignments

Enriched
Exp: 2–10 evens
Written: 1–8

Average
Exp: 1–10
Written: 1–8

Fundamental
Exp: 1–10
Written: 1–6
p. 197, 7, 8

About Applications
As an extension of this application, discuss checking accounts which earn interest. Usually, interest is added monthly, based on the average daily balance. Ask students whether the interest should be added to or subtracted from the balance in the check register. (It should be added.)

Applications in Banking

Checking Account

Del Vargas has a checking account. He records his checks and deposits in the check register below. To find his balance, Del adds any deposits to his previous balance and subtracts the checks he has written.

CHECK NO.	DATE	CHECK ISSUED TO	AMOUNT OF DEPOSIT	AMOUNT OF CHECK	BALANCE	
					35.72	←to previous balance
—	11/10	Deposit	115.00		150.72	←add deposits
428	11/12	S.E. Electric		25.50	125.22	←subtract checks
429	11/12	Star Oil Co.		21.35	103.87	
430	11/19	Kelly's Casuals		42.14	61.73	
431	11/23	Photo Journal Inc		6.50	55.23	
432	11/25	Palmer's Market		29.83	25.40	
—	11/25	Deposit	119.75		145.15	
433	11/28	Central Loan		90.00	55.15	

Each month Del receives a bank statement along with his canceled checks and deposit slips. The statement shows which checks and deposits the bank has processed. Unprocessed checks and deposits are said to be outstanding. Parts of the statement are shown below.

ENDING BALANCE
$31.90

CHECK NO.	AMOUNT
CHECKS	
428	25.50
429	21.35
430	42.14
432	29.83
DEPOSITS	
	115.00

Del must bring his statement and register ending balances into agreement. To the bank's ending balance, he adds any outstanding deposits. Then he subtracts any outstanding checks or service charges. This number should equal the ending balance in Del's check register.

Exercises

See above for 1–6.

1-6. In Del's check register, find the balance after each check or deposit is made.

7. What checks and deposits are outstanding? Deposit 119.75; checks 431, 433

8. Bring the ending balance (from Exercise 6) into agreement with the bank balance.

Answers to Exercises

8.
```
   31.90
+ 119.75
  151.65
-   6.50
  145.15
-  90.00
   55.15
```

Objective 6-9: Solve uniform motion problems using the formula $d = rt$.

6-9 Problem Solving: Uniform Motion

When an object moves at a constant speed, or rate, it is said to be in **uniform motion**. The formula $d = rt$ is used to solve uniform motion problems. In the formula, d represents distance, r represents rate, and t represents time. When solving uniform motion problems, it is often helpful to draw a diagram and to organize relevant information in a chart.

In the formula, r can represent an average rate instead of a constant rate.

Examples

1 **Manuel Fernandez rides his bicycle at a speed of 8 mph (miles per hour). How long will it take him to ride 28 miles?**

Explore Let t = the time it takes Manuel to ride 28 miles

Plan
$$d = rt$$
$$28 = 8t$$

Solve
$$3\tfrac{1}{2} = t$$

Manuel will take $3\frac{1}{2}$ hours to ride 28 miles.

Examine When t is $3\frac{1}{2}$, d is $8 \cdot 3\frac{1}{2}$ or 28 miles.

2 **Dan and Donna Wyatt leave their home in Chatsworth at the same time. They travel in opposite directions. Dan travels at 80 km/h (kilometers per hour) and Donna travels at 72 km/h. In how many hours will they be 760 km apart?**

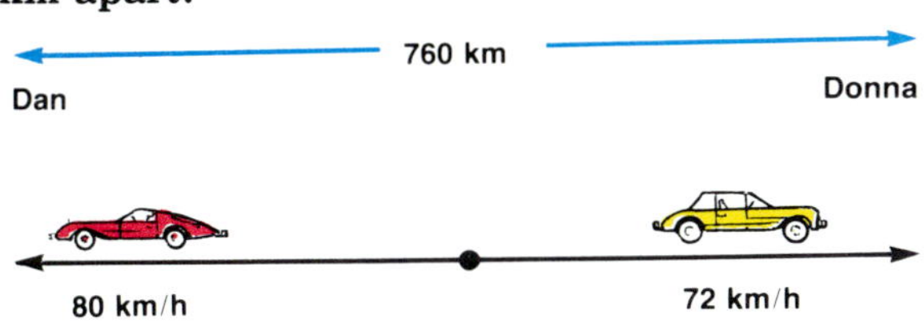

Draw a diagram to help analyze the problem.

Let t represent the number of hours.

Organize the information in a chart.

	r	$\cdot\ t$ =	d
Dan	80	t	$80t$
Donna	72	t	$72t$

Dan travels 80t km.
Donna travels 72t km.

They travel a total of 760 km.

Dan's distance + Donna's distance = Total distance

$$80t + 72t = 760$$
$$152t = 760$$
$$t = 5$$

In 5 hours, Dan and Donna will be 760 km apart.

Examine this solution.

6-9

Lesson Notes, p. T37
Enrichment, p. T37

Discussion Questions

What units of measure do you think should be used for each rate below?

1. The speed of a person running. m per s, ft per s, yd per s
2. The speed of a caterpillar. cm per min, in per min
3. The rate of growth of a child. in per year, cm per year
4. The speed of light. light years per second

Chalkboard Exercises

Solve each problem.

1a. Carla drove her car at a rate of 84 km/h. How far did she drive in 16 hours? 1334 km

1b. Lisa and Michael traveled 2 miles in $\frac{1}{2}$ hour. What was their rate? 4 mph

2a. Suppose, as in Example 2 Dan and Donna leave at the same time, travel in opposite directions, but Dan travels 60 km/h and Donna 110 km/h. In how many hours will they be 935 km apart? 5.5 hours

2b. Suppose Dan and Donna leave at the same time travelling the same direction. Dan drives 70 km/h, Donna drives 85 km/h. How long until they are 90 km apart? 6 hours

Chalkboard Exercises

3a. Solve each problem. Jake left home on his bicycle for a long-distance ride. Abbey left 3 hours later on her motorcycle, carrying his lunch. Her speed was 42 mph and she caught up with him in $1\frac{1}{2}$ hours. How fast was Jake traveling? 14 mph

3b. At 12:00 p.m. a plane leaves Washington for Chicago traveling at a rate of 240 mph. A second plane leaves Washington at 1:00 p.m. traveling at a rate of 400 mph. The planes arrive in Chicago at the same time. What time will they arrive? 2:30 p.m.

Suggested Daily Assignments

Enriched
- Exp: 1–5 odds
- Written: 2–14 evens, 15

Average
- Exp: 1–4
- Written: 1–14

Fundamental
- Exp: 1–6
- Written: 1–11
  - p. 199, 7, 8

Answers to Exploratory

1. a. 160 km b. 480 km c. 80 h km
2. a. 120 miles b. 180 miles c. 10 miles d. $\frac{2}{3}k$ miles
3. a. 40 mph b. $\frac{240}{t}$ mph
4. a. 54 mph b. $\frac{270}{x}$ mph
5. a. 8 hours b. 4 hours
6. a. 9 hours b. 12 hours c. $\frac{360}{x}$ hours

Example 3

At 8:00 A.M. Peggy leaves home driving at 35 mph. A half hour later, Doug discovers that she left her briefcase, so he drives 50 mph to catch up with her. If Doug is delayed 15 minutes with a flat tire, at what time will he catch up to Peggy?

Let x = the time Peggy travels until Doug arrives

	r ·	t =	d
Peggy	35	x	$35x$
Doug	50	$\left(x - \frac{3}{4}\right)$	$50\left(x - \frac{3}{4}\right)$

Peggy travels x hours. Doug starts $\frac{1}{2}$ hour later and is delayed for $\frac{1}{4}$ hour. Doug travels $\left(x - \frac{3}{4}\right)$ hours.

Peggy and Doug travel the same distance.

$$35x = 50\left(x - \frac{3}{4}\right)$$
$$35x = 50x - \frac{75}{2}$$
$$-15x = -\frac{75}{2}$$
$$x = 2\frac{1}{2}$$

Peggy has been traveling for $2\frac{1}{2}$ hours when Doug catches up to her.

Doug catches up to Peggy at 8 A.M. + $2\frac{1}{2}$ hours or 10:30 A.M.

Exploratory Exercises

Use $d = rt$ to answer each question.

1. Pat is driving 80 km/h. How far will she travel: **a.** in 2 hours? **b.** in 6 hours? **c.** in h hours?
2. Bob is driving 40 mph. How far will he travel: **a.** in 3 hours? **b.** in $4\frac{1}{2}$ hours? **c.** in 15 minutes? **d.** in k minutes?
3. Marilyn traveled 240 miles. What was her rate if she made the trip: **a.** in 6 hours? **b.** in t hours?
4. Juan traveled 270 miles. What was his rate if he made the trip: **a.** in 5 hours? **b.** in x hours?
5. Rudy rode his bicycle 72 km. How long did it take him if his rate was: **a.** 9 km/h? **b.** 18 km/h?
6. Patsy traveled 360 miles. How long did it take her if her rate was: **a.** 40 mph? **b.** 30 mph? **c.** x mph?

Written Exercises

Solve each problem. Use charts and diagrams if necessary.

1. Two trains leave Bridgeport at the same time, one traveling north, the other south. The first train travels at 40 mph and the second at 30 mph. In how many hours will the trains be 245 miles apart? $3\frac{1}{2}$ hours
2. Ike and Mike leave home traveling on their bicycles in opposite directions. Mike travels 10 km/h and Ike travels 12 km/h. In how many hours will they be 110 kilometers apart? 5 hours

3. Rosita drives from Boston to Cleveland, a distance of 616 miles. Her rest, gasoline, and food stops amount to 2 hours. What was her rate if the trip took 16 hours? 44 mph

4. At the same time Kris leaves Washington, D.C. for Detroit, Amy leaves Detroit for Washington, D.C. The distance between the cities is 510 miles. Amy drives 5 mph faster than Kris. How fast is Kris driving if they pass each other in 6 hours? 40 mph

5. Two cyclists are traveling in the same direction on the same course. One travels 20 mph and the other 14 mph. After how many hours will they be 15 miles apart? $2\frac{1}{2}$ hours

6. Art leaves at 10:00 A.M., traveling at 50 mph. At 11:30 A.M., Jennifer starts in the same direction at 45 mph. When will they be 100 miles apart? 4:30 PM

7. Boat *A* leaves the pier at 9:00 A.M., at 8 knots (nautical miles per hour). A half hour later, Boat *B* leaves the same pier in the same direction traveling at 10 knots. At what time will Boat *B* overtake Boat *A*? 11:30 AM

8. Bob is driving 40 mph. After Bob has driven 30 miles, Jack starts driving in the same direction. At what rate must Jack drive to catch up to Bob in 5 hours? 46 mph

9. An express train travels 80 km/h from Wheaton to Whitfield. A passenger train, traveling 48 km/h, takes 2 hours longer for the same trip. How far apart are Wheaton and Whitfield? 240 km

10. Farmer Brown drives to town at 36 mph and returns at 48 mph. If his total driving time is $3\frac{1}{2}$ hours, how far is his home from town? 72 miles

11. Two airplanes leave Dallas at the same time and fly in opposite directions. One plane travels 80 mph faster than the other. After three hours, they are 2940 miles apart. What is the rate of each plane? 450 mph, 530 mph

12. Jackson runs a 440 yard run in 55 seconds and Owen runs it in 88 seconds. To have Jackson and Owen finish at the same time, how many yards headstart should Owen be given? 165 yd

13. At 1:30 P.M., a plane leaves Tucson for Baltimore, a distance of 2240 miles. The plane flies 280 mph. A second plane leaves Tucson at 2:15 P.M., and is scheduled to land in Baltimore 15 minutes before the first plane. At what rate must the second plane travel to arrive on schedule? 320 mph

14. Sherry leaves Columbus, Ohio, on a train traveling 70 mph to Dallas. Twelve hours later, Don leaves Columbus by plane traveling 500 mph to Dallas. If Columbus and Dallas are 1050 miles apart, will Don arrive before or after Sherry? Don will arrive before Sherry.

Challenge

15. Two trains are 240 miles apart traveling toward each other on parallel tracks. One travels 35 mph and the other travels 45 mph. At the front of the faster train is a bee that can fly 75 mph. Assume that the bee can change direction instantaneously. It flies from one train to the other until the trains pass each other. When the trains pass, what distance has the bee flown? 225 miles

About Using Computers
The program in this feature demonstrates the power of the iterative process of a computer. As an extension, you may want to have students modify the program to see how many years it would take to triple the money.

Using Computers

Compound Interest

Suppose you have deposited $1000 in a savings account that pays 7.5% interest compounded annually. If you do not withdraw or deposit any additional money to the account, in how many years will your money double?

The formula $I = Prt$ is used to solve this problem:

$$I = Prt$$
$$I = 1000 \cdot 0.075 \cdot 1$$
$$I = 75$$

The interest earned after 1 year is $75. The new principal is $1075. This procedure is now repeated using $1075 as the principal and continues until the principal is greater than $2000.

Certainly you could solve the problem with a calculator, but a computer has the advantage of performing the operations very quickly. Use this BASIC program to determine in how many years your money will double.

```
10   PRINT "ENTER THE PRINCIPAL, THE RATE AS A DECIMAL,
       AND THE TIME IN YEARS"
20   INPUT P,R,T         When interest is compounded annually, T = 1.
30   LET DEP = P         DEP is the amount deposited.
40   LET Y = 0
50   PRINT "YEAR", "PRINCIPAL"
60   LET I = P * R * T        Line 60 calculates the interest.
70   LET P = P + I            Line 70 adds interest to the principal.
80   LET Y = Y + T            Line 80 keeps a count of the years.
90   PRINT Y, "$";P
100  IF P > 2 * DEP THEN 120
110  GOTO 60
120  PRINT "IT TOOK ";Y;" YEARS TO DOUBLE YOUR MONEY"
130  END
```

Exercises

1. If $P = \$1000$, $r = 7.5\%$, and $t = 1$, in how many years will your money double? 10 years
2. If $P = \$1000$, $r = 10\%$, and $t = 1$, in how many years will your money double? 8 years
3. Run the program several times using $P = \$1000$ and $t = 1$, but vary the rate of interest. What is the relationship between the interest rate and the number of years it takes to double the principal? As the interest rate increases, the number of years decreases.
4. If interest is compounded semiannually, $t = 0.5$. In how many years would your money double if $P = 1000$, $r = 10\%$ and $t = 0.5$? $7\frac{1}{2}$ years
5. Run the program several times using $P = \$1000$ and $r = 10\%$, but vary the time, t. What is the relationship between t and the number of years it takes to double the principal? As t decreases, the number of years decreases.

Vocabulary

polynomial (175)
binomial (175)
trinomial (175)
degree (175)
ascending order (176)
descending order (176)
FOIL method (185)
square of a sum (190)
square of a difference (190)
difference of squares (191)
perimeter (195)
area (195)
simple interest (198)
uniform motion (201)

Chapter Summary

1. The degree of a monomial is the sum of the exponents of its variables. The degree of a nonzero constant is 0. (175)
2. Expressions like $8x + y$, $3x^2 + 2x + 4$, and $5 - 3y + 5xy^2$ are called polynomials. (175)
3. A polynomial is an expression that can be written as a sum of monomials. A polynomial with two terms is called a binomial. A polynomial with three terms is called a trinomial. (175)
4. The degree of a polynomial is the greatest of the degrees of its terms. (176)
5. The terms of a polynomial are usually arranged so that the powers of one variable are in ascending or descending order (176)
6. The same properties used to simplify expressions can be used to add polynomials. (178)
7. To subtract a polynomial from a polynomial, add the additive inverse. (179)
8. To multiply two binomials, find the sum of the products of the **F**irst terms, the **O**uter terms, the **I**nner terms, and the **L**ast terms. This method is called FOIL. (185)
9. Square of a Sum:
$(a + b)^2 = (a + b)(a + b) = a^2 + 2ab + b^2$. (190)
10. Square of a Difference:
$(a - b)^2 = (a - b)(a - b) = a^2 - 2ab + b^2$. (190)
11. Product of a Sum and a Difference:
$(a + b)(a - b) = a^2 - b^2$. (191)
12. The formula $I = Prt$ is used to solve simple interest problems. (198)
13. The formula $d = rt$ is used to solve uniform motion problems. (201)

Quiz
Lessons 6-7 to 6-9
Solve each problem.

1. Russ leaves Stanton traveling 50 mph. Ellen leaves 2 hours later traveling 55 mph in the opposite direction. How many hours will Russ have traveled by the time they are 415 miles apart?
Let t represent the number of hours Russ travels. Then, $t - 2$ represents the number of hours Ellen travels.
$50t + 55(t - 2) = 415$
$t = 5$
5 hours
2. The length of a rectangle is 3 cm longer than twice its width. The perimeter is 42 cm. Find the area of the rectangle.
w = width
$2w + 3$ = length
$2[w + (2w + 3)] = 42$
$w = 6$
$2w + 3 = 15$
Area: $6 \cdot 15$ or 90 cm^2
3. Bill and Nancy Walker invested \$1600 at 15% annual interest and \$1400 at 12% annual interest. If they earned \$1224 in interest, how many years was the money invested at those rates?
t represents time in years
$(1600)(0.15)t + (1400)(0.12)t = 1224$
$240t + 168t = 1224$
$t = 3$
The money was invested for 3 years.

Chapter Review

6-1 **Arrange the terms of each polynomial so that the powers of x are in descending order.**

1. $3x^4 - x + x^2 - 5$ $3x^4 + x^2 - x - 5$

2. $ax^2 - 5x^3 + a^2x - a^3$ $-5x^3 + ax^2 + a^2x - a^3$

6-2 **Add or subtract.**

3. $(2x^2 - 5x + 7) - (3x^3 + x^2 + 2)$ $-3x^3 + x^2 - 5x + 5$

4. $(x^2 - 6xy + 7y^2) + (3x^2 + xy - y^2)$ $4x^2 - 5xy + 6y^2$

5. $(x^2 - 5x + 3) - (3x - 4)$ $x^2 - 8x + 7$

6. $\left(-\frac{1}{2}x^2 - \frac{1}{3}x + 2\right) + \left(\frac{3}{2}x^2 + \frac{1}{3}x - 4\right)$ $x^2 - 2$

6-3 **Multiply. Simplify when possible.**

7. $7xy(x^2 + 4xy - 8y^2)$ $7x^3y + 28x^2y^2 - 56xy^3$

8. $x(3x - 5) + 7(x^2 - 2x + 9)$ $10x^2 - 19x + 63$

9. $4x^2(x + 8) - 3x(2x^2 - 8x + 3) + 3(x^3 - 7x + 1)$ $x^3 + 56x^2 - 30x + 3$

6-4 **Multiply.**

10. $(x + 5)(3x - 2)$ $3x^2 + 13x - 10$

11. $(4x - 3)(x + 4)$ $4x^2 + 13x - 12$

12. $(x - 4)(x^2 + 5x - 7)$ $x^3 + x^2 - 27x + 28$

13. $\left(\frac{1}{2}a + 1\right)\left(\frac{3}{4}a^2 - 2a + 4\right)$ $\frac{3}{8}a^3 - \frac{1}{4}a^2 + 4$

6-5 **Multiply.**

14. $(x - 6)(x + 6)$ $x^2 - 36$

15. $(5x - 3y)(5x + 3y)$ $25x^2 - 9y^2$

16. $(4x + 7)^2$ $16x^2 + 56x + 49$

17. $(8x - 5)^2$ $64x^2 - 80x + 25$

6-6 **Solve each equation.**

18. $4a + 9a - 7 = 14a + 6$ -13

19. $4 + 6x = 2(4x - 2)$ 4

Solve each problem.

6-7 **20.** Bill McClure is making a picture frame whose length will be 4 in. greater than its width. The frame will have a uniform width of 2 in. If the area of the frame will be 192 square inches, what will the inside dimensions of the picture frame be? 24 in. by 20 in.

21. The sum of two congruent sides of a certain triangle is 22 in. greater than the base. If the perimeter is 50 inches, what is the length of each congruent side? 18 inches

6-8 **22.** Brenda Smith invested \$8000 for one year, part at 8% and the remainder at 12%. Her total interest for the year was \$744. How much money did she invest at each rate? \$5400 at 8%, \$2600 at 12%

23. Francie earned \$786 interest on a one year investment earning 15% annual interest. How much money did Francie invest? \$5240

6-9 **24.** At 8:00 A.M., Alan drove west at 35 mph. At 9:00 A.M., Winnie drove east from the same point at 42 mph. When will they be 266 miles apart? 12:00 noon

25. Karen leaves home driving 32 mph. A half hour later, her sister Gail drives 40 mph to catch up to her. When will Gail catch up to Karen? 2 hours

Chapter Test

Arrange the terms of each polynomial so that the powers of x are in descending order.

1. $5x^2 - 3 + x^3 + 5x$ $x^3 + 5x^2 + 5x - 3$

2. $5 - xy^3 + x^3y^2 - x^2$ $x^3y^2 - x^2 - xy^3 + 5$

Simplify each of the following.

3. $(a + 5)(a - 5)$ $a^2 - 25$

4. $(2x - 5)(7x + 3)$ $14x^2 - 29x - 15$

5. $(3a^2 + 3)[2a - (-6)]$ $6a^3 + 18a^2 + 6a + 18$

6. $3x^2y^3(2x - xy^2)$ $6x^3y^3 - 3x^3y^5$

7. $-4xy(5x^2 - 6xy^3 + 2y^2)$ $-20x^3y + 24x^2y^4 - 8xy^3$

8. $(4x^2 - y^2)(4x^2 + y^2)$ $16x^4 - y^4$

9. $\left(\frac{1}{2}x + 4\right)\left(\frac{1}{2}x + 3\right)$ $\frac{1}{4}x^2 + \frac{7}{2}x + 12$

10. $(0.2a - 0.4)^2$ $0.04a^2 - 0.16a + 0.16$

11. $0.3b(0.4b^2 - 0.7b + 4)$ $0.12b^3 - 0.21b^2 + 1.2b$

12. $(2a^2b + b^2)^2$ $4a^4b^2 + 4a^2b^3 + b^4$

13. $x^2(x - 8) - 3x(x^2 - 7x + 3) + 5(x^3 - 6x^2)$ $3x^3 - 17x^2 - 9x$

14. $a^2(a + 5) + 7a(a^2 + 8) - 7(a^3 - a + 2)$ $a^3 + 5a^2 + 63a - 14$

Solve each equation.

15. $5y - 8 - 13y = 12y + 6$ $-\frac{7}{10}$

16. $3(4m + 5) = 6m - 9$ -4

Solve each problem.

17. The length of a rectangle is 3 inches less than twice the width. The perimeter is 84 inches. Find the dimensions of the rectangle. 15 in. by 27 in.

18. The length of a rectangle is eight times its width. If the length was decreased by 10 meters and the width was decreased by 2 meters, the area would be decreased by 162 square meters. Find the original dimensions. 7 m by 56 m

19. Peggy invested \$5,000 for one year. Maureen also invested \$5,000 for one year. Maureen's account earned interest at a rate of 10% per year. At the end of the year, Maureen's account earned \$125 more than Peggy's account. What was the rate of interest on Peggy's account? $7\frac{1}{2}\%$

20. The second side of a triangle is twice the length of the first side. The length of the third side is 12 inches more than the length of the first side. If the perimeter is 304 inches, what is the length of each side?

73 in., 146 in., 85 in.

21. Two cyclists start toward each other from two towns that are 96 miles apart. One cyclist rides at 18 mph and the other rides at 14 mph. In how many hours will they meet? 3 hours

22. Joe invested \$5000 for one year at 9% interest. Harry invested some money at the same time at 8% interest. At the end of the year, Joe and Harry together had earned \$810 in interest. How much money did Harry invest? \$4500

Using the Cumulative Review If students have difficulty with a particular topic, have them review the appropriate lesson. You may want to provide copies of the solutions from the *Algebra One Solutions Manual*. This Cumulative Review covers Chapters 1-6.

Other Cumulative Reviews appear on pages 72, 142, 268, 334, 402, 466, and 526.

Answer

18.

Cumulative Review

1. Evaluate $3b^3y$ if $b = -4$ and $y = -\frac{4}{3}$. 256

2. transitive property of equality

State the property shown.

2. If $a = 6 + 1$ and $6 + 1 = b$ then $a = b$.
3. $12x^2 + 4xy = 4x(3x + y)$ distributive
4. $(4 + x) + 6 = 4 + (x + 6)$

4. associative property of addition

Simplify.

5. $4a + 3(a + 4)$ $7a + 12$

Find each sum or difference.

6. $-21 + 13$ -8
7. $\frac{3}{5} + \left(-\frac{3}{10}\right)$ $\frac{3}{10}$
8. $-21 + (-5) + (-18)$ -44
9. $-13.4 - 14.3$ -27.7

Solve.

10. $5 + a = 17$ 12
11. $-13 = -42 - p$ -29

Simplify.

12. $-3(-6) - 6(-5)$ 48
13. $\dfrac{-\frac{3}{5}}{9}$ $-\frac{1}{15}$

Solve.

14. $\frac{3}{5}x = -8$ $-\frac{40}{3}$
15. $7a - 5 = 30$ 5
16. $7 = 3 + 4(m - 1)$ 2
17. 48 is what percent of 144? $33\frac{1}{3}\%$
18. Graph the solution set of the inequality $x \geq -3$ on a number line.

Solve each inequality.

19. $2a < a + 3$ $a < 3$
20. $-16r \geq -48$ $r \leq 3$
21. $10y - 3(y + 4) \leq 0$ $y \leq \frac{12}{7}$
22. Which is a better buy, a gallon of milk for \$1.67 or a half-gallon of milk for \$0.96? a gallon for \$1.67

23. Write the following compound sentence without *and*: $x > 4$ and $x \leq 7$. $4 < x \leq 7$
24. Solve: $|x + 4| \leq 1$ $-5 \leq x \leq -3$

Simplify.

25. $(2r^2s)(8rs^3)$ $16r^3s^4$
26. $(4a^2b^3)^3$ $64a^6b^9$
27. y^{-4} $\frac{1}{y^4}$
28. $\dfrac{52m^8}{14m^5}$ $\dfrac{26m^3}{7}$
29. Express 983,000,000 in scientific notation. 9.83×10^8
30. Find the degree of $27x^2 + 4xy - 13y^3$. 3

Simplify.

31. $(4x^2 - 7xy + 8y^2) - (3x^2 - 2xy - 5y^2)$ $x^2 - 5xy + 13y^2$
32. $\frac{1}{2}y\left(2y + \frac{1}{y}\right)$ $y^2 + \frac{1}{2}$
33. $(2x + 5)(3x - 8)$ $6x^2 - x - 40$
34. $(3m + 4)^2$ $9m^2 + 24m + 16$

Solve.

35. $4z - 2 + 3z = 2(8 - z)$ 2

Problem Solving

Solve each problem.

36. Find three consecutive integers whose sum is 117. 38, 39, 40
37. Six times an integer decreased by 5 is greater than 67. What is the least integer that will satisfy this condition? 13
38. Maggie is 3 years older than Agnes. The sum of their ages is 35. What are their ages now? Maggie, 19; Agnes, 16
39. Mike has 400 mL of 75% solution of silver nitrate. How many milliliters of a 30% solution should be added to obtain a 50% solution? 500 mL
40. Use $I = Prt$ to find t, if $I = \$1192.50$, $P = \$4500$, and $r = 13\frac{1}{4}\%$. 2 years

Standardized Test Practice Questions

The test questions on this page deal with ratios, proportions, and percents. The information at the right may help you with some of the questions.

Directions: Choose the best answer. Write A, B, C, or D.

1. 0.7% is the ratio of 7 to

(A) 100 **(B)** $\frac{1}{10}$
(C) $\frac{1}{1000}$ **(D)** 1000

2. 37.5% of a pound is equivalent to what fractional part of a pound?

(A) $\frac{37}{100}$ **(B)** $\frac{3}{8}$ **(C)** $\frac{37}{99}$ **(D)** $37\frac{1}{2}$

3. Five-sixths is how many sevenths?

(A) $5\frac{5}{7}$ **(B)** $8\frac{2}{5}$ **(C)** $5\frac{5}{6}$ **(D)** $4\frac{2}{7}$

4. How many elevenths is 75%?

(A) $8\frac{1}{4}$ **(B)** $\frac{3}{4}$ **(C)** $6\frac{9}{11}$ **(D)** $\frac{12}{11}$

5. 90% of 270 is 2.7% of

(A) 729 **(B)** 900
(C) 2000 **(D)** 9000

6. A box contains 60 red, blue, and green pens. If 35% are red and 9 are blue, what percent are green?

(A) 40 **(B)** 50 **(C)** 60 **(D)** 65

7. Thirty-six of 90 students in a high school graduating class are boys. Nine boys and 18 girls plan to attend college. What percent of the class are girls who plan to attend college?

(A) 20 **(B)** 40 **(C)** 60 **(D)** $66\frac{2}{3}$

8. 150% of $5c$ is b. What percent of $2b$ is c?

(A) $5\frac{1}{2}$ **(B)** $6\frac{2}{3}$ **(C)** 15 **(D)** 75

9. Two-fifths times five-sevenths is equal to what number times six-elevenths?

(A) $\frac{11}{21}$ **(B)** $\frac{55}{63}$ **(C)** $\frac{20}{77}$ **(D)** $\frac{22}{43}$

---

1. A ratio is an expression that compares two quantities.

Example: If a container holds 4 pens and 7 pencils, the ratio of pencils to pens is 7:4 or $\frac{7}{4}$.

2. Two equivalent ratios form a proportion. The cross products of a proportion are equal. That is, if $\frac{a}{b} = \frac{c}{d}$, then $ad = bc$.

3. *Percent* means *hundredths.*

7% means $\frac{7}{100}$ or 7 out of 100.

---

10. A woman owns $\frac{3}{4}$ of a business. She sells half of her share for \$30,000. What is the total value of the business?

(A) \$45,000 **(B)** \$75,000
(C) \$80,000 **(D)** \$112,000

11. On a map, a line segment 1.75 inches long represents 21 miles. What distance in miles is represented by a segment 0.6 inches long?

(A) 6 **(B)** 7.2 **(C)** 12.6 **(D)** 24.15

12. A cake recipe that serves 6 people calls for 2 cups of flour. How many cups of flour would be needed to bake a smaller cake that serves 4 people?

(A) $\frac{3}{4}$ **(B)** 1 **(C)** $1\frac{1}{4}$ **(D)** $1\frac{1}{3}$

13. The price of an item was reduced by 10%, then later reduced by 20%. The two reductions were equivalent to a single reduction of

(A) 15% **(B)** 28% **(C)** 30% **(D)** 70%

14. The discount price for a \$50 item is \$40. What is the discount price for a \$160 item if the rate of discount is $1\frac{1}{2}$ times the rate of discount for the \$50 item?

(A) 20 **(B)** 48 **(C)** 96 **(D)** 112

Using the Standardized Test Practice Questions

The questions on this page may be used to help students prepare for pre-college tests. These questions usually require careful analysis and a thorough understanding of the concepts.

This page could be used as an overnight assignment for better students. Other students may benefit most if only a few questions are assigned at a time.

After students have chosen an answer for each problem, you should discuss how each problem can be solved, or provide copies of the solutions from the *Algebra One Solutions Manual*.

Other practice questions appear in this text on pages 73, 143, 269, 335, 403, 467, and 527. Also see pages 50-58 in the *Teacher Resource Book*.

About the Photo

Polynomial expressions involving a squared variable often represent quantities such as the measure of the area of a garden. In algebra, such expressions are simplified using the important process called factoring.

Factoring means finding numbers that when multiplied together equal the given product. A given product may have more than one set of factors. For example, the numbers 1 and 12 are factors of 12 but so are 2 and 6, and 3 and 4. However, there is only one prime factorization, namely $2 \cdot 2 \cdot 3$ or $2^2 \cdot 3$.

Algebraic expressions also have factors. For example, the factors of the polynomial $y^2 - 16y + 64$ are $(y - 8)$ and $(y - 8)$.

Teacher Resource Book Materials for Chapter 7 See pages 181-200.

Chapter Test, Form I (Multiple Choice)	181-183
Chapter Test, Form II (Free Response)	184-185
Quizzes	186-187
Extra Practice	188-190
Application: Using Factoring (Use after 7-8.)	191-193
Extension: Finding the GCF by Euclid's Algorithm (Use after 7-1.)	194
Problem Solving	195
Cumulative Review	196
Answers	197-200

CHAPTER 7

Factoring

The measure of the area of a certain square garden can be represented by the polynomial $y^2 - 16y + 64$. What is the measure of one side of the garden? To answer questions such as this one, you can use *factoring*. In this chapter, you will learn to factor certain types of polynomials.

Objective 7-1A: Find the prime factorization of an integer.
Objective 7-1B: Find the greatest common factor (GCF) for a set of numbers or monomials.

7-1 Factors, Greatest Common Factors

If two or more numbers are multiplied, each number is a **factor** of the product. For example, you can express 18 as the product of different pairs of positive integers.

$18 = 1 \cdot 18 \qquad 18 = 2 \cdot 9 \qquad 18 = 3 \cdot 6$

The integers 1, 18, 2, 9, 3, and 6 are factors of 18.

Some integers have only two whole-number factors, the integer itself and 1. These integers are called **prime numbers**. ❶

A prime number is an integer, greater than 1, whose only whole number factors are 1 and itself.

Definition of Prime Number

The prime numbers less than 50 are 2, 3, 5, 7, 11, 13, 17, 19, 23, 29, 31, 37, 41, 43, and 47.

A composite number is any positive integer, except 1, that is not prime.

Definition of Composite Number

The first few composite numbers are 4, 6, 8, 9, 10, and 12.

The number 9 is a factor of 18, but 9 is not a *prime* factor because it is not a prime number. When a composite number is expressed as a product of factors that are all prime, the expression is called the **prime factorization** of the number. The prime factorization of 18 is $2 \cdot 3 \cdot 3$ or $2 \cdot 3^2$. ❷

The number 1 is neither prime nor composite.

Example

1 Express the composite number 84 as a product of prime factors.

Begin by dividing 84 by the least prime number that is a factor. Then use each prime in order as many times as it is a factor.

$84 = 2 \cdot 42$ — *The least prime number that is a factor of 84 is 2.*

$2 \cdot 2 \cdot 21$

$2 \cdot 2 \cdot 3 \cdot 7$ — *When all factors are prime the process stops.*

The expression $2 \cdot 2 \cdot 3 \cdot 7$ or $2^2 \cdot 3 \cdot 7$ is the prime factorization of 84. ❸

A negative integer may be expressed as a product of a positive integer and −1. For example, to find the factors of −525, first write it as the product of −1 and 525. Then find the prime factors of 525.

$$-525 = -1 \cdot 525$$
$$= -1 \cdot 3 \cdot 5^2 \cdot 7$$ ❹

$525 = 3 \cdot 5^2 \cdot 7$

7-1

Lesson Notes, page T38

Readiness Activity

Which of the following numbers are divisible by 2? by 3? by 5?

26	2	40	2, 5
75	3, 5	54	2, 3
30	2, 3, 5	210	2, 3, 5

Discussion Questions

1. Name some numbers that have an odd number of factors. What is similar about these numbers?
   4, 9, 16, 25, . . . ;
   They are all perfect square numbers.
2. Name some pairs of numbers for which the lesser number is the GCF.
   Any pair where the lesser number divides the greater with a remainder of 0.

Chalkboard Exercises

Find the prime factorization of each number.

1a. 52 $2^2 \cdot 13$
1b. 63 $3^2 \cdot 7$
1c. 250 $2 \cdot 5^3$
1d. 625 5^4
1e. 80 $2^4 \cdot 5$
1f. 770 $2 \cdot 5 \cdot 7 \cdot 11$

❶ See the Excursion on Page 213 for a method of finding prime numbers.

❷ $18 = 1 \cdot 18$
$2 \cdot 9$
$2 \cdot 3 \cdot 3$

❸ The order of prime factors in a prime factorization is not critical. Usually, the factors are ordered from least prime factor to greatest. For example, 630 is $2 \cdot 3 \cdot 3 \cdot 5 \cdot 7$ or $2 \cdot 3^2 \cdot 5 \cdot 7$.

❹ −1 is factored out first so that the factoring will be consistent with prime factoring of positive numbers.

A monomial such as $20a^2b$ may be expressed in factored form as a product of prime numbers and variables with no exponent greater than 1.

$$\begin{aligned} 20a^2b &= 2 \cdot 10 \cdot a \cdot a \cdot b \\ &= 2 \cdot 2 \cdot 5 \cdot a \cdot a \cdot b \end{aligned}$$

Two or more integers may have some common factors. Consider the prime factorizations of 90 and 105.

$$90 = 2 \cdot 3 \cdot 3 \cdot 5 \qquad 105 = 3 \cdot 5 \cdot 7$$

The integers 90 and 105 have the common factors 3, 5, and $3 \cdot 5$ or 15. The greatest of these, 15, is called the **greatest common factor (GCF)** of 90 and 105. ⑤

The number 1 is also a common factor.

Definition of Greatest Common Factor.

The greatest common factor of two or more integers is the greatest of the factors common to each integer.

The GCF of two or more monomials is the product of their common factors, when each monomial is in factored form.

Examples

2 **Find the GCF of 54, 63, and 180.**

$$\begin{aligned} 54 &= 2 \cdot (3) \cdot (3) \cdot 3 \\ 63 &= (3) \cdot (3) \cdot 7 \\ 180 &= 2 \cdot 2 \cdot (3) \cdot (3) \cdot 5 \end{aligned}$$

Find the common factors.

The GCF of 54, 63, and 180 is $3 \cdot 3$, or 9.

3 **Find the GCF of $8a^2b$ and $18a^2b^2c$.**

$$\begin{aligned} 8a^2b &= (2) \cdot 2 \cdot 2 \cdot (a) \cdot (a) \cdot (b) \\ 18a^2b^2c &= (2) \cdot 3 \cdot 3 \cdot (a) \cdot (a) \cdot (b) \cdot b \cdot c \end{aligned}$$

The GCF of $8a^2b$ and $18\ a^2b^2c$ is $2a^2b$.

Exploratory Exercises

State whether the second integer is a factor of the first.

1. 16; 8 yes
2. 15; 3 yes
3. 41; 6 no
4. 55; 11 yes
5. 72; 36 yes

State whether each number is prime or composite. If the number is composite, find its prime factorization.

6. 89 prime
7. 39 composite; $3 \cdot 13$
8. 24 composite; $2^3 \cdot 3$
9. 91 composite; $13 \cdot 7$
10. 53 prime

Chalkboard Exercises

Find the GCF for each of the following.

2a. 35, 165 5
2b. 18, 22 2
2c. 42, 125 1
2d. 6, 18, 36 6
3a. $9a^2$, $15a^2b$ $3a^2$
3b. $6x^2y$, $9xy^2$ $3xy$
3c. $3a^2b$, $16ab^2$, $21ab$ ab

⑤ Emphasize that 15 is the greatest number that will divide 90 and 105 evenly.

Computer Notes

A computer program for finding the factors of a number is given on page 542. The program uses the built-in function INT(X), which finds the greatest integer less than or equal to X.
The following enrichment activity could be assigned:
Write a computer program for finding the prime factorization of any whole number.

Suggested Daily Assignments

Enriched
Exp: 8–13
Written: 9–25 odds, 33–55 odds

Average
Exp: 2–20 evens
Written: 2–54 evens

Fundamental
Exp: 1–20
Written: 1–40

Find the GCF of each pair of numbers.

11. 4, 12 4 **12.** 10, 15 5 **13.** 9, 36 9 **14.** 15, 5 5 **15.** 11, 22 11

16. 15, 45 15 **17.** 20, 30 10 **18.** 18, 35 1 **19.** 36, 72 36 **20.** 16, 18 2

Written Exercises

Find the prime factorization of each number.

1. 21 $3 \cdot 7$ **2.** 28 $2^2 \cdot 7$ **3.** 60 $2^2 \cdot 3 \cdot 5$ **4.** 51 $3 \cdot 17$ **5.** 39 $3 \cdot 13$

6. 63 $3^2 \cdot 7$ **7.** 34 $2 \cdot 17$ **8.** 72 $2^3 \cdot 3^2$ **9.** 112 $2^4 \cdot 7$ **10.** 150 $2 \cdot 3 \cdot 5^2$

11. 304 $2^4 \cdot 19$ **12.** 216 $2^3 \cdot 3^3$ **13.** 300 $2^2 \cdot 3 \cdot 5^2$ **14.** 1000 $2^3 \cdot 5^3$ **15.** 1540 $2^2 \cdot 5 \cdot 7 \cdot 11$

Factor each expression. Do not use exponents.

16. -64 **17.** -26 **18.** -240 **19.** -500 **20.** -231

21. $98a^2b$ **22.** $44ab^2c^3$ **23.** $196b^2$ **24.** $756(ab)^3$ **25.** $-102x^3y$

Find the GCF for each of the following.

26. 16, 60 4 **27.** 15, 50 5 **28.** -80, 45 5

29. 29, -58 29 **30.** 305, 55 5 **31.** 252, 126 126

32. 128, 245 1 **33.** 95, 304 19 **34.** $7y^2$, $14y^3$ $7y^2$

35. $17a$, $34a^2$ $17a$ **36.** $-12ab$, $4a^2b^2$ $4ab$ **37.** $4xy$, $-6x$ $2x$

38. 6, 8, 4 2 **39.** 5, 10, 15 5 **40.** 16, 24, 30 2

41. 18, 30, 54 6 **42.** 24, 84, 168 12 **43.** 16, 22, 30 2

44. $50n^4$, $40n^2p^2$ $10n^2$ **45.** $20a^2$, $24ab^5$ $4a$ **46.** $60x^2y^2$, $35xz^3$ $5x$

47. $12an^2$, $40a^4$ $4a$ **48.** $56x^3y$, $49ax^2$ $7x^2$ **49.** $12mn$, $10mn$, $15mn$ mn

50. $6a^2$, $18b^2$, $9b^3$ 3 **51.** $8b^4$, $5c$, $3a^2b$ 1 **52.** $15abc$, $35a^2c$, $105a$ $5a$

53. $14a^2b^2$, $18ab$, $2a^3b^3$ $2ab$ **54.** $18a^2b^2$, $6b$, $42a^2b^3$ $6b$ **55.** $24a^2b$, $28axy$, $36ay$ $4a$

Excursions in Algebra — Sieve of Eratosthenes

Eratosthenes was an early Greek (c. 200 B.C.) who was one of the first to calculate the circumference of the earth. He is also remembered in mathematics for the "sieve of Eratosthenes," a method for finding prime numbers.

To use his method, list the positive integers beginning with 2. Each multiple of a prime is to be crossed out. For example, cross out every second number after 2, every third number after 3, every fifth number after 5, and so on. When the process is finished, the numbers not crossed out are prime. *Begin by crossing out every second number after two.*

2 3 ~~4~~ 5 ~~6~~ 7 ~~8~~ 9 ~~10~~ 11 ~~12~~ 13 ~~14~~ 15 ~~16~~ 17
~~18~~ 19 ~~20~~ 21 ~~22~~ 23 ~~24~~ 25 ~~26~~ 27 ~~28~~ 29 ~~30~~

Exercises

1. Copy and complete the sieve of Eratosthenes above for integers to 30.
2. List the positive integers from 2 to 100. Use the sieve of Eratosthenes to find the prime numbers between 2 and 100.

For more exercises on prime factorization, see exercises 1-16 on page 558.

Answers to Written

16. $-1 \cdot 2 \cdot 2 \cdot 2 \cdot 2 \cdot 2 \cdot 2$
17. $-1 \cdot 2 \cdot 13$
18. $-1 \cdot 2 \cdot 2 \cdot 2 \cdot 2 \cdot 3 \cdot 5$
19. $-1 \cdot 2 \cdot 2 \cdot 5 \cdot 5 \cdot 5$
20. $-1 \cdot 3 \cdot 7 \cdot 11$
21. $2 \cdot 7 \cdot 7 \cdot a \cdot a \cdot b$
22. $2 \cdot 2 \cdot 11 \cdot a \cdot b \cdot b \cdot c \cdot c \cdot c$
23. $2 \cdot 2 \cdot 7 \cdot 7 \cdot b \cdot b$
24. $2 \cdot 2 \cdot 3 \cdot 3 \cdot 3 \cdot 7 \cdot a \cdot a \cdot a \cdot b \cdot b \cdot b$
25. $-1 \cdot 2 \cdot 3 \cdot 17 \cdot x \cdot x \cdot x \cdot y$

About the Excursion

The sieve of Eratosthenes is still a very useful method of determining primes. It requires systematic thinking and technique.

Answers to Excursion

1. prime numbers between 1 and 30: 2, 3, 5, 7, 11, 13, 17, 19, 23, 29
2. prime numbers between 2 and 100: 2, 3, 5, 7, 11, 13, 17, 19, 23, 29, 31, 37, 41, 43, 47, 53, 59, 61, 67, 71, 73, 79, 83, 89, 97

7-2

Lesson Notes, page T38

Discussion Question
If the greatest common factor of x^n and x^m is x^m, what can you say about n and m? $m < n$

Chalkboard Exercises
Factor each expression.

1a. $18a^2 + 4a$
$2a(9a + 2)$
1b. $14a + 7a^3$
$7a(2 + a^2)$
2a. $3a + 12ab$
$3a(1 + 4b)$
2b. $6a^2 + 18a^2b$
$6a^2(1 + 3b)$
3a. $a + a^2b - a^3b$
$a(1 + ab - a^2b)$
3b. $8a^3 + 4a^2 + 2a$
$2a(4a^2 + 2a + 1)$
4a. $3x^5 + 6x^4 + 12x^2 + 18x$
$3x(x^4 + 2x^3 + 4x + 6)$
4b. $14xy + 21xy^2 - 35x^2y$
$7xy(2 + 3y - 5x)$

1 When factoring exercises include fractions, answers may vary. For Exercise 33, other possible answers are $\frac{1}{4}x(2x - a)$ and $x\left(\frac{1}{2}x - \frac{1}{4}a\right)$.

Objective 7-2: Use the GCF and the distributive property to factor polynomials.

7-2 Factoring Using the Distributive Property

The distributive property has been used to multiply a polynomial by a monomial. It can also be used to write a polynomial in factored form.

Multiplying	**Factoring**
$3(a + b) = 3a + 3b$	$3a + 3b = 3(a + b)$
$x(y - z) = xy - xz$	$xy - xz = x(y - z)$
$3y(4x + 2) = 3y(4x) + 3y(2)$	$12xy + 6y = 3y(4x) + 3y(2)$
$= 12xy + 6y$	$= 3y(4x + 2)$

Examples

1 **Use the distributive property to write $10y^2 + 15y$ in factored form.**

First, find the greatest common factor of $10y^2$ and $15y$.

$$10y^2 = 2 \cdot 5 \cdot y \cdot y$$
$$15y = 3 \cdot 5 \cdot y$$

The GCF is 5y.

Then, express each term as a product of the GCF and its remaining factors.

$$10y^2 + 15y = 5y(2y) + 5y(3)$$
$$= 5y(2y + 3)$$ *Use the distributive property.*

2 **Factor: $21ab^2 - 33a^2bc$**

$$21ab^2 = 3 \cdot 7 \cdot a \cdot b \cdot b$$
$$33a^2bc = 3 \cdot 11 \cdot a \cdot a \cdot b \cdot c$$

The GCF is 3ab.

Express the terms as products.

$$21ab^2 - 33a^2bc = 3ab(7b) - 3ab(11ac)$$
$$= 3ab(7b - 11ac)$$ *Use the distributive property.*

3 **Factor: $12a^5b + 8a^3 - 24a^3c$**

$$12a^5b = 2 \cdot 2 \cdot 3 \cdot a \cdot a \cdot a \cdot a \cdot a \cdot b$$
$$8a^3 = 2 \cdot 2 \cdot 2 \cdot a \cdot a \cdot a$$
$$24a^3c = 2 \cdot 2 \cdot 2 \cdot 3 \cdot a \cdot a \cdot a \cdot c$$

The GCF is $4a^3$.

$$12a^5b + 8a^3 - 24a^3c = 4a^3(3a^2b) + 4a^3(2) - 4a^3(6c)$$
$$= 4a^3(3a^2b + 2 - 6c)$$

4 **Factor: $6x^3y^2 + 14x^2y + 2x^2$**

$$6x^3y^2 = 2 \cdot 3 \cdot x \cdot x \cdot x \cdot y \cdot y$$
$$14x^2y = 2 \cdot 7 \cdot x \cdot x \cdot y$$
$$2x^2 = 2 \cdot x \cdot x$$

The GCF is $2x^2$.

$$6x^3y^2 + 14x^2y + 2x^2 = 2x^2(3xy^2 + 7y + 1)$$

Exploratory Exercises

Find the GCF of the terms in each expression.

1. $3y^2 + 12$ 3
2. $4a + 2b$ 2
3. $5y - 9y^2$ y
4. $6a + 3b$ 3
5. $9b + 5c$ 1
6. $3a + 6a$ $3a$
7. $9y^2 + 3y$ $3y$
8. $11a + 10q$ 1
9. $8x^2 - 4x$ $4x$
10. $12a^2b + 6a$ $6a$
11. $7ac - 21a^2$ $7a$
12. $16c - 21b$ 1
13. $21xyz - xy$ xy
14. $14b^2 - 42c$ 14
15. $42a - 13r$ 1

Written Exercises

Factor each expression.

1. $16x + 4y$ $4(4x + y)$
2. $24x^2 + 12y^2$ $12(2x^2 + y^2)$
3. $12xy + 12x^2$ $12x(y + x)$
4. $5a^2b + 10ab$ $5ab(a + 2)$
5. $11x + 44x^2y$ $11x(1 + 4xy)$
6. $16x^2 + 8x$ $8x(2x + 1)$
7. $25a^2b^2 + 30ab^3$ $5ab^2(5a + 6b)$
8. $18xy^2 - 24x^2y$ $6xy(3y - 4x)$
9. $14xy - 18xy^2$ $2xy(7 - 9y)$
10. $27a^2b + 9b^3$ $9b(3a^2 + b^2)$
11. $36a^2b^2 - 12ab$
12. $3a^2b - 6a^2b^2$ $3a^2b(1 - 2b)$
13. $14xy^2 + 2xy$ $2xy(7y + 1)$
14. $15xy^3 + y^4$ $y^3(15x + y)$
15. $x^5y - x$ $x(x^4y - 1)$
16. $29xy - 3x$ $x(29y - 3)$
17. $17a - 41a^3b$ $a(17 - 41a^2b)$
18. $a + a^4b^3$ $a(1 + a^3b^3)$
19. $3x^2y + 9y^2 + 6$ $3(x^2y + 3y^2 + 2)$
20. $5a^2 + 10ab - 15b^2$ $5(a^2 + 2ab - 3b^2)$
21. $2a^3b^2 - 16a^2b^3 + 8ab$ $2ab(a^2b - 8ab^2 + 4)$
22. $3x^3y + 9xy^2 + 36xy$ $3xy(x^2 + 3y + 12)$
23. $24x^2y^2 + 12xy + x$ $x(24xy^2 + 12y + 1)$
24. $28a^2b^2c^2 + 21a^2bc^2 - 14abc$
25. $12ax + 20bx + 32cx$ $4x(3a + 5b + 8c)$
26. $a + a^2b + a^3b^3$ $a(1 + ab + a^2b^3)$
27. $ax^3 + 5bx^3 + 9cx^3$ $x^3(a + 5b + 9c)$
28. $14a^3x + 19a^3y + 11a^3z$
29. $6x^2 - 9xy + 24x^2y^2$ $3x(2x - 3y + 8xy^2)$
30. $42abc - 12a^2b^2 + 3a^2c^2$
31. $24abx + 12ax^2 + 6x^3$ $6x(4ab + 2ax + x^2)$
32. $x^5 + 5x^4 + 3x^2 + 2x$ $x(x^4 + 5x^3 + 3x + 2)$
33. $\frac{1}{2}x^2 - \frac{1}{4}ax$ $\frac{1}{2}x\left(x - \frac{1}{2}a\right)$ ❶
34. $\frac{2}{3}x + \frac{1}{3}y$ $\frac{1}{3}(2x + y)$
35. $\frac{4}{5}x^2y + \frac{3}{5}y^2$ $\frac{1}{5}y(4x^2 + 3y)$
36. $\frac{2}{5}a - \frac{2}{5}b + \frac{4}{5}c$ $\frac{2}{5}(a - b + 2c)$

11. $12ab(3ab - 1)$ 28. $a^3(14x + 19y + 11z)$

Excursions in Algebra — Divisibility Rules

Some simple rules can help you determine if a number is divisible by 2, 3, 5, or 10.

If the ones digit is 0, 2, 4, 6, or 8, the number is divisible by 2.
If the ones digit is 0 or 5, the number is divisible by 5.
If the ones digit is 0, the number is divisible by 10.
If the *sum* of the digits is divisible by 3, the number is divisible by 3.

For example, the sum of the digits of 741 is 7 + 4 + 1 or 12. 741 is divisible by 3 because 12 is divisible by 3.

Exercises

Determine whether each number is divisible by 2, 3, 5, or 10.

1. 44
2. 75
3. 110
4. 123
5. 405
6. 570

Suggested Daily Assignments

Enriched
Exp: 2–14 evens
Written: 8–36 evens

Average
Exp: 1–15 odds
Written: 1–35 odds

Fundamental
Exp: 1–15
Written: 1–35 odds
p. 213, 41–52

Answers to Written

24. $7abc(4abc + 3ac - 2)$
30. $3a(14bc - 4ab^2 + ac^2)$

Answers to Excursion

1. divisible by 2
2. divisible by 3, 5
3. divisible by 2, 5, 10
4. divisible by 3
5. divisible by 3, 5
6. divisible by 2, 3, 5, 10

Quiz
Lessons 7–1 to 7–2

1. Find the prime factorization of 819. $3^2 \cdot 7 \cdot 13$
2. Factor $-84x^3y^2$. $-1 \cdot 2 \cdot 2 \cdot 3 \cdot 7 \cdot x \cdot x \cdot x \cdot y \cdot y$
3. Find the GCF of $34a^2b^3$ and $85ab^2$. $17ab^2$
4. Find the GCF of 20, 90, and 36. 2

Factor each expression.

5. $35x^2 + 5x$ $5x(7x + 1)$
6. $8m^2 + 4am + 16my$ $4m(2m + a + 4y)$
7. $12xy^2 - 8x^2y$ $4xy(3y - 2x)$
8. $r^4 + r^3s + r^2s^2$ $r^2(r^2 + rs + s^2)$

About the Excursion
Have students determine divisibility rules for 4, 6, and 9 by using the divisibility rules that are given.

About Applications
You may want to have some students redraw these diagrams (using centimeters instead of meters) to see if the formulas agree with the measurement of area done piece by piece.

Applications in Geometry

Area

You can simplify equations involving area by using the distributive property and factoring. Consider this example.

A deck 3 meters wide is to be built around a swimming pool that is shown in the diagram at the right. In order to determine the amount of material that is needed to build the deck, it is necessary to calculate the area of the deck.

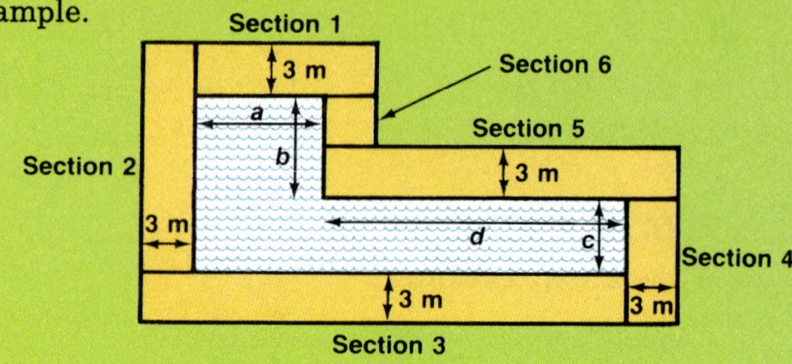

To find the area of the deck, find the sum of the areas of the rectangular regions.

$$\text{Area} = \underbrace{3(a+3)}_{\text{Section 1}} + \underbrace{3(b+c+3)}_{\text{Section 2}} + \underbrace{3(a+d+3)}_{\text{Section 3}} + \underbrace{3(c+3)}_{\text{Section 4}} + \underbrace{3(d+3)}_{\text{Section 5}} + \underbrace{3(b-3)}_{\text{Section 6}}$$

Use the distributive property.

$A = 3a + 9 + 3b + 3c + 9 + 3a + 3d + 9 + 3c + 9 + 3d + 9 + 3b - 9$

Combine like terms.

$A = 6a + 6b + 6c + 6d + 36$

Factor the polynomial.

$A = 6(a + b + c + d + 6)$

The area of the deck can be represented by $A = 6(a + b + c + d + 6)$.

Exercises

For each diagram, write an equation that represents the area of the shaded region.

1. $A = 4(a + b + 4)$

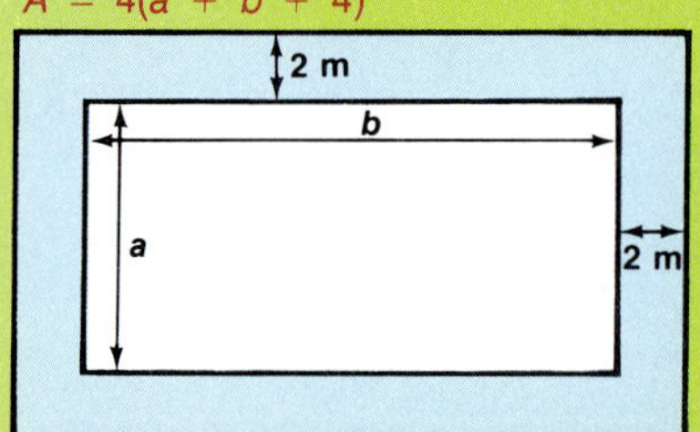

2. $A = 4(a + c + d + 8)$

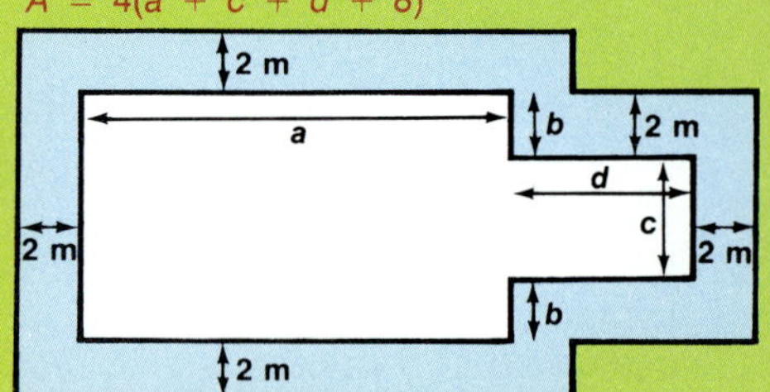

3. $A = 8(a + b + c + 2d + 8)$

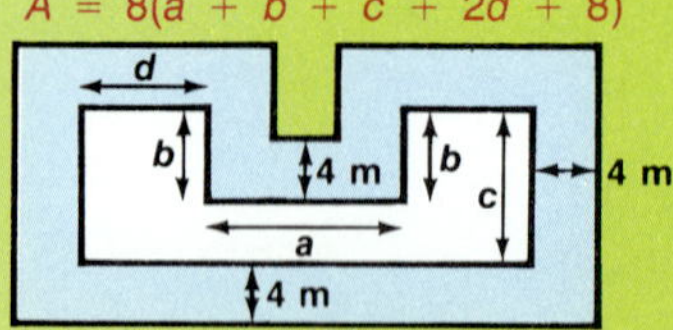

4. $A = 4\pi(a + 1)$

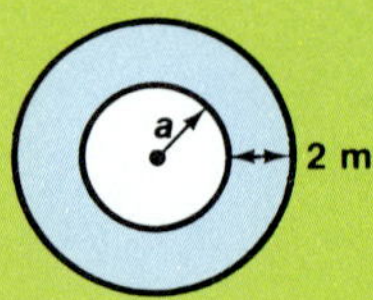

Objective 7-3: Recognize and factor differences of squares.

7-3 Factoring Differences of Squares

Consider the two squares that are shown. The area of the larger square is a^2, and the area of the smaller square is b^2.

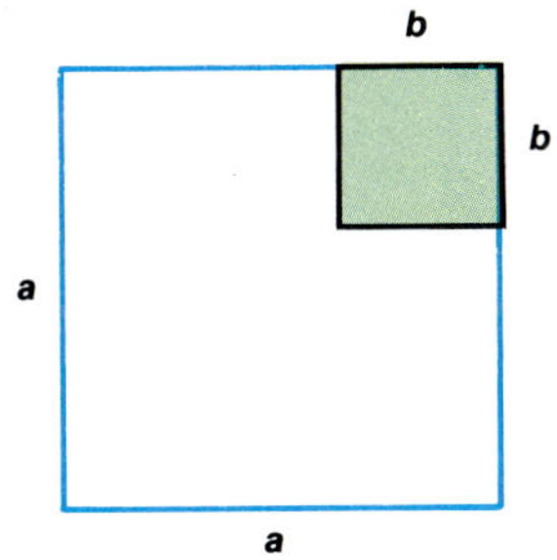

The area $a^2 - b^2$ can be found by subtracting the area of the smaller square from the area of the larger square.

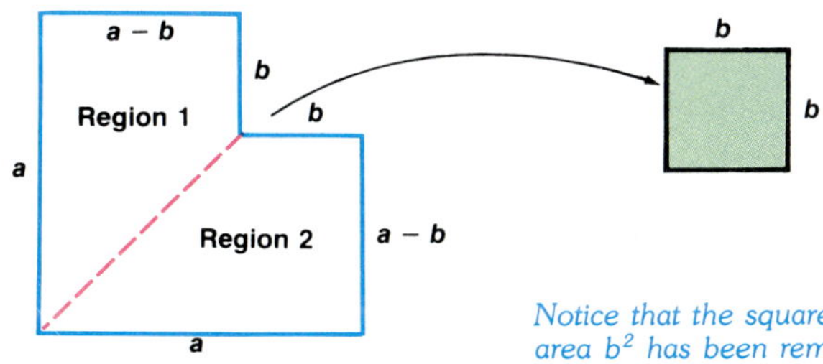

Notice that the square with area b^2 has been removed.

Rearranging these regions as shown below, you can see that $a^2 - b^2$ is equal to the product of $(a - b)$ and $(a + b)$.

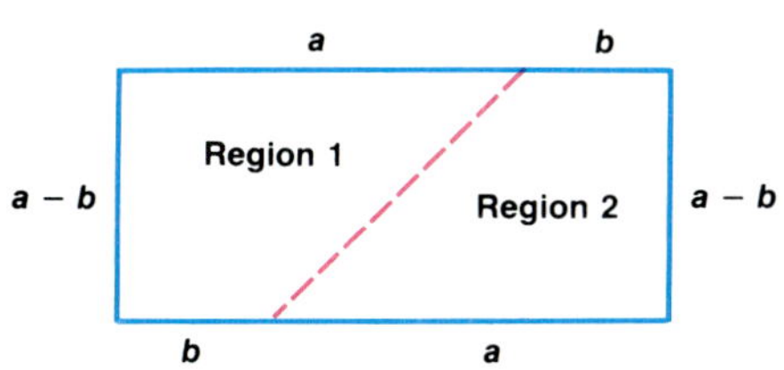

The area of the rectangle is $(a - b)(a + b)$. This is the same as $a^2 - b^2$.

$$a^2 - b^2 = (a - b)(a + b) = (a + b)(a - b)$$

❶ ***Difference of Squares***

You can use the difference of squares to factor binomials of the form $a^2 - b^2$.

7-3

Lesson Notes, page T38
Enrichment, page T39

Readiness Activity

Find the products.

$(a + 4)(a - 4)$	$a^2 - 16$
$(p - 5)(p + 5)$	$p^2 - 25$
$(4m - 1)(4m + 1)$	$16m^2 - 1$
$(3x + 2y)(3x - 2y)$	$9x^2 - 4y^2$
$(a + b)(a - b)$	$a^2 - b^2$

Discussion Question

If $x^n - y^m$ is the difference of two squares, what are some possible values of n and m? any two positive even integers

❶ Students should memorize the form of a difference of squares.

Chalkboard Exercises
Factor each polynomial completely.

1. $m^2 - 4$
$(m - 2)(m + 2)$
2a. $4x^2 - 9y^2$
$(2x - 3y)(2x + 3y)$
2b. $4k^2 - 25m^2$
$(2k - 5m)(2k + 5m)$
3. $\frac{1}{16}v^2 - \frac{9}{25}w^2$
$\left(\frac{1}{4}v - \frac{3}{5}w\right)\left(\frac{1}{4}v + \frac{3}{5}w\right)$
4a. $9r^2 - 36m^2$
$9(r - 2m)(r + 2m)$
4b. $8x^2 - 32y^2$
$8(x - 2y)(x + 2y)$
5. $48a^4 - 3b^4$
$3(2a - b)(2a + b)(4a^2 + b^2)$

❷ Point out that this is an application of the distributive property.

Suggested Daily Assignments

Enriched
Written: 2–44 evens, 46, 47

Average
Exp: 2–12 evens
Written: 2–44 evens

Fundamental
Exp: 1–12
Written: 1–30
p. 215, 2–32 evens

Examples

1 **Factor:** $a^2 - 64$

$a^2 - 64 = (a)^2 - (8)^2$ — *$a \cdot a = a^2$ and $8 \cdot 8 = 64$.*
$= (a - 8)(a + 8)$ — *Use the difference of squares.*

2 **Factor:** $9x^2 - 100y^2$

$9x^2 - 100y^2 = (3x)^2 - (10y)^2$ — *$3x \cdot 3x = 9x^2$ and $10y \cdot 10y = 100y^2$.*
$= (3x - 10y)(3x + 10y)$ — *Use the difference of squares.*

3 **Factor:** $\frac{1}{4}y^2 - \frac{4}{9}p^2$

$\frac{1}{4}y^2 - \frac{4}{9}p^2 = \left(\frac{1}{2}y\right)^2 - \left(\frac{2}{3}p\right)^2$ — *Why?*
$= \left(\frac{1}{2}y - \frac{2}{3}p\right)\left(\frac{1}{2}y + \frac{2}{3}p\right)$ — *Check your answers using FOIL.*

Sometimes the terms of a binomial have common factors. If so, the GCF should always be factored out before factoring the difference of squares. Occasionally, factoring the difference of squares needs to be used more than once in a problem. ❷

Examples

4 **Factor:** $16x^2 - 36y^2$

$16x^2 - 36y^2 = 4(4x^2 - 9y^2)$ — *4 is the GCF of $16x^2$ and $36y^2$.*
$= 4(2x - 3y)(2x + 3y)$ — *$2x \cdot 2x = 4x^2$ and $3y \cdot 3y = 9y^2$.*

5 **Factor:** $162a^4 - 32y^8$

$162a^4 - 32y^8 = 2(81a^4 - 16y^8)$ — *Why?*
$= 2(9a^2 - 4y^4)(9a^2 + 4y^4)$ — *$9a^2 + 4y^4$ cannot be factored.*
$= 2(3a - 2y^2)(3a + 2y^2)(9a^2 + 4y^4)$

Exploratory Exercises

State whether each binomial can be factored as the difference of squares.

1. $x^2 - y^2$ yes
2. $a^2 + b^2$ no
3. $b^2 - 25$ yes
4. $a^2 - 4b^4$ yes
5. $9y^2 + a^2$ no
6. $25a^2 - 16$ yes
7. $9x^2 - 5$ no
8. $a - 9$ no
9. $8c^2 - 7$ no
10. $36a^2 - 49$ yes
11. $\frac{4}{9}a^2 - \frac{1}{4}$ yes
12. $\frac{16}{25}a^2 + 1$ no

Written Exercises

Factor each polynomial completely.

1. $a^2 - 9$ $(a + 3)(a - 3)$
2. $x^2 - 49$ $(x + 7)(x - 7)$
3. $4x^2 - 9y^2$ $(2x + 3y)(2x - 3y)$
4. $x^2 - 36y^2$ $(x + 6y)(x - 6y)$
5. $a^2 - 4b^2$ $(a + 2b)(a - 2b)$
6. $1 - 9y^2$ $(1 + 3y)(1 - 3y)$
7. $16a^2 - 9b^2$
8. $3m^2 - 6n^2$ $3(m^2 - 2n^2)$
9. $16a^2 - 25$ $(4a + 5)(4a - 5)$
10. $2a^2 + 18a$ $2a(a + 9)$
11. $6x^2 + 12$ $6(x^2 + 2)$
12. $8x^2 - 12y^2$ $4(2x^2 - 3y^2)$
13. $2z^2 - 98$ $2(z - 7)(z + 7)$
14. $12a^2 - 48$ $12(a + 2)(a - 2)$
15. $8x^2 - 18$ $2(2x + 3)(2x - 3)$
16. $45x^2 - 20z^2$
17. $25y^2 - 49z^4$
18. $9x^4 - 25y^4$
19. $17 - 68a^2$
20. $144x^2 - 9y^2$
21. $25a^2x^2 - 1$
22. $4x^2 - 64y^2$ $4(x + 4y)(x - 4y)$
23. $12a^2 - 12$ $12(a + 1)(a - 1)$
24. $9x^4 - 16y^2$
25. $36x^2 - 81y^4$
26. $28x^2 - 7$ $7(2x + 1)(2x - 1)$
27. $81 - 9x^2$
28. $-16 + 9x^2$ $(3x + 4)(3x - 4)$
29. $-9 + 4y^2$ $(2y + 3)(2y - 3)$
30. $15x^2 - 60y^2$
31. $x^4 - y^4$
32. $a^4 - b^2$ $(a^2 + b)(a^2 - b)$
33. $x^4 - 1$
34. $\frac{1}{4}x^2 - 16$ $\left(\frac{1}{2}x + 4\right)\left(\frac{1}{2}x - 4\right)$
35. $\frac{9}{2}x^2 - \frac{49}{2}y^2$
36. $\frac{2}{3}x^2 - \frac{8}{3}$ $\frac{2}{3}(x + 2)(x - 2)$
37. $x^8 - 1$
38. $(a + b)^2 - m^2$
39. $(x - y)^2 - y^2$
40. $p^2 - (m + n)^2$
41. $a^2 - (b - c)^2$
42. $y^4 - 81$
43. $x^4 - 16$
44. $162a^5 - 32ab^4$
45. $98x^6 - 128y^8$

43. $(x^2 + 4)(x + 2)(x - 2)$ 44. $2a(9a^2 + 4b^2)(3a + 2b)(3a - 2b)$ 45. $2(7x^3 + 8y^4)(7x^3 - 8y^4)$

Challenge

46. Find positive integers m and n such that $m^2 - n^2 = 11$. $m = 6, n = 5$
47. Find positive integers x and y such that $x^2 - y^2 = 15$. $x = 8, y = 7$ or $x = 4, y = 1$

Using Calculators — Solving Equations

You can use your calculator to solve and check equations such as $0.35x + 6.789x = 15$. First, find the sum of 0.35 and 6.789 and store it.

ENTER:	.35	[+]	6.789	[=]	[STO]
DISPLAY:	0.35		6.789	7.139	7.139

When you enter .35, the display shows 0.35.

Then, divide 15 by 7.139. *You are dividing both sides of the equation by 7.139*

ENTER:	15	[÷]	[RCL]	[=]
DISPLAY:	15		7.139	2.1011346

The solution is about 2.101 *Use your calculator to check this solution.*

Exercises

Solve and check each equation.

1. $0.02y + 3.87y = 45$ 11.568123
2. $6.09x + 0.089x = 9$ 1.4565463
3. $2.5x + 7.3x - 5.06x = -1$
4. $7.8y + 9.2y - 0.06y = -20$ −1.1806375
5. $35 = 56.7a - 60.02a$ −10.542168
6. $75 = 0.008c - 4.2c + 9.01c$ 15.566625

Answers to Written

7. $(4a + 3b)(4a - 3b)$
16. $5(3x + 2z)(3x - 2z)$
17. $(5y - 7z^2)(5y + 7z^2)$
18. $(3x^2 + 5y^2)(3x^2 - 5y^2)$
19. $17(1 + 2a)(1 - 2a)$
20. $9(4x + y)(4x - y)$
21. $(5ax + 1)(5ax - 1)$
24. $(3x^2 + 4y)(3x^2 - 4y)$
25. $9(2x + 3y^2)(2x - 3y^2)$
27. $9(3 + x)(3 - x)$
30. $15(x + 2y)(x - 2y)$
31. $(x - y)(x + y)(x^2 + y^2)$
33. $(x^2 + 1)(x + 1)(x - 1)$
35. $\frac{1}{2}(3x + 7y)(3x - 7y)$
37. $(x^4 + 1)(x^2 + 1)(x + 1)(x - 1)$
38. $(a + b + m)(a + b - m)$
39. $(x - 2y)x$
40. $(p + m + n)(p - m - n)$
41. $(a - b + c)(a + b - c)$
42. $(y^2 + 9)(y + 3)(y - 3)$

About Using Calculators
Have students justify each step. For example, $0.35x + 6.789x = 7.139x$ by the distributive property.

Answer to Using Calculator
3. −0.21097046

About the Excursion
Because converting from one metric unit to another involves powers of 10, you may prefer to approach this feature using scientific notation.

Excursions in Algebra — Mass in the Metric System

In the metric system, a unit of mass is the gram (g). One gram is defined as the mass of one cubic centimeter of water at its maximum density.

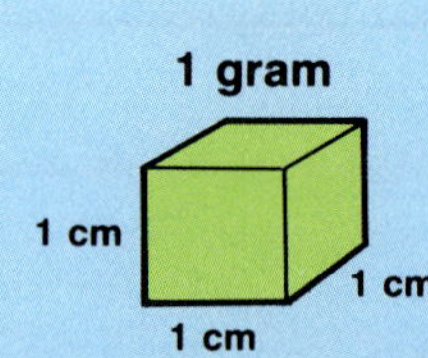

1 cubic centimeter of water at 4°C and 1 atmosphere of pressure has a mass of 1 gram.

You can think of a gram as about the mass of two paper clips.

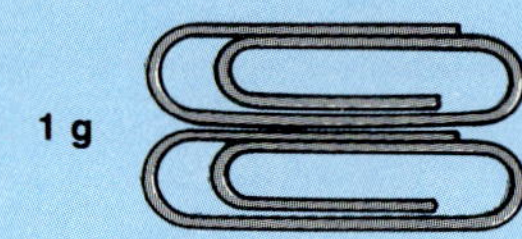

A gram is too small to use conveniently. Therefore, the kilogram (kg) is used as the basic unit for mass. There are 1000 grams in a kilogram. The mass of this textbook is about one kilogram.

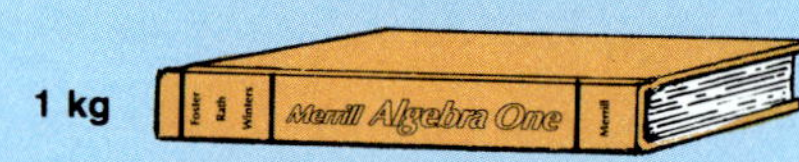

A metric ton (t) equals 1000 kilograms. Shippers measure grain in metric tons. A metric ton is about the mass of a small automobile.

The milligram (mg) is used to measure very small quantities. One gram equals 1000 milligrams.

Exercises

Would you use gram or kilogram to measure the mass of each object?

1. bag of groceries kg
2. pencil g
3. mouse g
4. wrestler kg
5. suitcase kg
6. pepper g

Complete the following.

7. 3 kg = ___ g 3000
8. 8.5 kg = ___ g 8500
9. 5000 g = ___ kg 5
10. 6700 g = ___ kg 6.7
11. 8 g = ___ mg 8000
12. 6.5 g = ___ mg 6500
13. 3000 mg = ___ g 3
14. 3900 mg = ___ g 3.9
15. 9.52 kg = ___ g 9520
16. 3852 g = ___ kg 3.852
17. 3598 mg = ___ g 3.598
18. 6.257 g = ___ mg 6257
19. 38.575 kg = ___ g 38,575
20. 875 g = ___ kg 0.875
21. 575 mg = ___ g 0.575
22. 5 t = ___ kg 5000
23. 5300 kg = ___ t 5.3
24. 88,000 kg = ___ t 88
25. 40 mg = ___ kg 0.00004
26. 260 mg = ___ kg 0.00026
27. 2 kg = ___ mg 2,000,000
28. 1.93 kg = ___ mg 1,930,000
29. 10 mg = ___ t 0.00000001
30. 10 t = ___ mg 10,000,000,000

Objective 7-4: Factor perfect square trinomials.

7-4 Perfect Squares and Factoring

Recall $1^2 = 1, 2^2 = 4, 3^2 = 9, 4^2 = 16$, and so on. Numbers such as 1, 4, 9, and 16 are called **perfect squares**. Products of the form $(a + b)^2$ and $(a - b)^2$ are also called perfect squares and these expansions are called **perfect square trinomials**.

You should learn to recognize these patterns.

$$(a + b)^2 = a^2 + 2ab + b^2$$
$$(a - b)^2 = a^2 - 2ab + b^2$$

❶ ***Perfect Square Trinomials***

These patterns can help you factor trinomials such as $y^2 + 16y + 64$ and $4x^2 - 20xy + 25y^2$.

Finding a Product	Factoring
$(y + 8)^2 = y^2 + 2(8)y + (8)^2$	$y^2 + 16y + 64 = (y)^2 + 2(8y) + (8)^2$
$= y^2 + 16y + 64$	$= (y + 8)^2$
$(2x - 5y)^2$	$4x^2 - 20xy + 25y^2$
$= (2x)^2 - 2(2x)(5y) + (5y)^2$	$= (2x)^2 - 2(2x)(5y) + (5y)^2$
$= 4x^2 - 20xy + 25y^2$	$= (2x - 5y)^2$

To determine if a trinomial can be factored in this way, first decide if it is a perfect square. In other words, decide if it can be written in either the form $a^2 + 2ab + b^2$ or $a^2 - 2ab + b^2$. Study the following examples.

Example 1

Determine if $x^2 + 22x + 121$ is a perfect square. If it is, factor it.

To decide if $x^2 + 22x + 121$ is a perfect square, answer each question below.

❷
a. Is x^2 a perfect square? $x^2 = (x)^2$ *Yes*
b. Is 121 a perfect square? $121 = (11)^2$ *Yes*
c. Is the middle term twice the product of x and 11? $22x = 2(x)(11)$ *Yes*

Since all three answers are "yes," the trinomial $x^2 + 22x + 121$ is a perfect square. It can be factored as follows.

$$x^2 + 22x + 121 = (x)^2 + 2(x)(11) + (11)^2$$
$$= (x + 11)^2$$

7-4

Lesson Notes, page T39

Readiness Activity
Find the products.
$(t - 1)^2$ $t^2 - 2t + 1$
$(r + 3)^2$ $r^2 + 6r + 9$
$(2x + 4)^2$ $4x^2 + 16x + 16$
$(a + b)^2$ $a^2 + 2ab + b^2$
$(a - b)^2$ $a^2 - 2ab + b^2$

Discussion Question
Find two perfect square numbers a^2 and b^2 so that $2ab$ is also a perfect square number. One of a and b is a perfect square; the other is twice a perfect square.

Chalkboard Exercises
Determine whether each trinomial is a perfect square trinomial. If it is, factor it.
1a. $a^2 + 18a + 81$
yes, $(a + 9)^2$
1b. $x^2 + 20x + 100$
yes, $(x + 10)^2$

❶ Caution students against confusing the difference of squares with the square of a difference.
$(a + b)(a - b) = a^2 - b^2$
$(a - b)^2 \neq a^2 - b^2$

❷ Steps a, b, and c give students a system for determining perfect square trinomials.

Chalkboard Exercises

Determine whether each trinomial is a perfect square trinomial. If it is, factor it.

2a. $25m^2 - 70m + 49$ yes, $(5m - 7)^2$

2b. $16x^2 - 40x + 25$ yes, $(4x - 5)^2$

3a. $x^2 - 10x - 20$ no

3b. $y^2 + 10y - 42$ no

4a. $4y^2 - 20y + 36$ no

4b. $9x^2 - 12x + 16$ no

5a. $36q^2 - 84qr + 49r^2$ yes, $(6q - 7r)^2$

5b. $25x^2 - 20xy + 4y^2$ yes, $(5x - 2y)^2$

Examples

2 **Determine if $16a^2 + 72a + 81$ is a perfect square. If it is, factor it.**

a. Is the first term a square? $16a^2 = (4a)^2$ *Yes*

b. Is the last term a square? $81 = (9)^2$ *Yes*

c. Is the middle term twice the product of $4a$ and 9? $72a = 2(4a)(9)$ *Yes*

$16a^2 + 72a + 81$ is a perfect square.

$$16a^2 + 72a + 81 = (4a)^2 + 2(4a)(9) + (9)^2$$
$$= (4a + 9)^2$$

3 **Determine if $4a^2 - 20a + 15$ is a perfect square. If it is, factor it.**

a. Is the first term a square? $4a^2 = (2a)^2$ *Yes*

b. Is the last term a square? $15 = (?)^2$ *No*

$4a^2 - 20a + 15$ is not a perfect square.

4 **Determine if $16x^2 - 26x + 49$ is a perfect square. If it is, factor it.**

a. Is the first term a square? $16x^2 = (4x)^2$ *Yes*

b. Is the last term a square? $49 = (7)^2$ *Yes*

c. Is the middle term 2 times the product of $4x$ and 7? $26x \neq 2(4x)(7)$ *No*

$16x^2 - 26x + 49$ is *not* a perfect square.

5 **Factor $9x^2 - 12xy + 4y^2$.**

$9x^2 - 12xy + 4y^2$ is a perfect square. *Why?*

$$9x^2 - 12xy + 4y^2 = (3x)^2 - 2(3x)(2y) + (2y)^2$$
$$= (3x - 2y)^2$$

Suggested Daily Assignments

Enriched
- Written: 2–46 evens, 47, 48
- Mini-Review: 1–5

Average
- Exp: 1–11 odds
- Written: 1–37 odds, 39-46
- Mini-Review: 1–5

Fundamental
- Exp: 1–12
- Written: 1–45 odds p. 219, 31–40
- Mini-Review: 1–5

Exploratory Exercises

Determine whether each trinomial is a perfect square trinomial. If it is, state its factors.

1. $a^2 + 4a + 4$ yes, $(a + 2)^2$
2. $b^2 + 14b + 49$ yes, $(b + 7)^2$
3. $x^2 - 10x - 100$ no
4. $y^2 + 10y - 100$ no
5. $n^2 - 13n + 36$ no
6. $p^2 + 12p + 36$ yes, $(p +$
7. $y^2 - 8y + 10$ no
8. $x^2 + 17x + 21$ no
9. $4x^2 - 4x + 1$ yes, $(2x -$
10. $9y^2 - 10y + 4$ no
11. $9b^2 - 6b + 1$ yes, $(3b - 1)^2$
12. $4a^2 - 20a + 25$ yes, $(2a$

Written Exercises

Factor, if possible.

1. $a^2 + 12a + 36$ $(a + 6)^2$
2. $b^2 + 10b + 25$ $(b + 5)^2$
3. $x^2 + 16x + 64$ $(x + 8)^2$
4. $y^2 + 14y + 49$ $(y + 7)^2$
5. $4a^2 + 4a + 1$ $(2a + 1)^2$
6. $64b^2 + 16b + 1$ $(8b + 1)^2$
7. $1 - 10a + 25a^2$ $(1 - 5a)^2$
8. $1 - 12y + 36y^2$ $(1 - 6y)^2$
9. $9a^2 + 6a + 2$ not factorable
10. $16c^2 - 8c + 49$ not factorable
11. $25x^2 + 20x + 4$ $(5x + 2)^2$
12. $9a^2 - 24a + 16$ $(3a - 4)^2$
13. $121y^2 + 22y + 1$ $(11y + 1)^2$
14. $81n^2 + 36n + 4$ $(9n + 2)^2$
15. $25b^2 - 30b + 9$ $(5b - 3)^2$
16. $25x^2 - 120x + 144$ $(5x - 12)^2$
17. $64b^2 - 72b + 81$ not factorable
18. $49m^2 - 126m + 81$ $(7m - 9)^2$
19. $m^2 + 16mn + 64n^2$ $(m + 8n)^2$
20. $25a^2 + 40a + 9$ not factorable
21. $9x^2 + 24xy + 16y^2$ $(3x + 4y)^2$
22. $16p^2 - 40pq + 25q^2$ $(4p - 5q)^2$
23. $144p^2 + 72p + 9$ $9(4p + 1)^2$
24. $144n^2 + 168n + 49$ $(12n + 7)^2$
25. $64x^2 - 80x + 25$ $(8x - 5)^2$
26. $2x^2 - 20x + 50$ $2(x - 5)^2$
27. $3x^2 + 36x + 108$ $3(x + 6)^2$
28. $3x^2 + 18x + 48$ $3(x^2 + 6x + 16)$
29. $6x^2 + 18x + 24$ $6(x^2 + 3x + 4)$
30. $3b^2 - 18bc + 27c^2$ $3(b - 3c)^2$
31. $4x^2 + 4xz^2 + z^4$ $(2x + z^2)^2$
32. $4x^2 - 28xy^2 + 49y^4$ $(2x - 7y^2)^2$
33. $\frac{1}{4}a^2 + 3a + 9$ $\left(\frac{1}{2}a + 3\right)^2$
34. $\frac{4}{9}b^2 - \frac{16}{3}b + 16$ $\left(\frac{2}{3}b - 4\right)^2$
35. $m^4 + 12m^2n^2 + 36n^4$ $(m^2 + 6n^2)^2$
36. $16x^2y^2 - 72xy + 81$ $(4xy - 9)^2$
37. $9a^2 + \frac{24}{5}a + \frac{16}{25}$ $\left(3a + \frac{4}{5}\right)^2$
38. $a^2 - \frac{4}{3}a + \frac{16}{36}$ $\left(a - \frac{2}{3}\right)^2$

❸

Find the replacement for c that would make a perfect square.

39. $9x^2 + cxy + 49y^2$ 42 or −42
40. $16m^2 + cmp + 25p^2$ 40 or −40
41. $cx^2 + 28x + 49$ 4
42. $cy^2 + 50y + 25$ 25
43. $4x^2 + 4xy + c$ y^2
44. $9a^2 + 24ab + c$ $16b^2$
45. $9x^2 - 12xy + c$ $4y^2$
46. $64x^2 - 16xy + c$ y^2

Challenge

47. What is the length of a side of a square whose area is represented by the expression $x^2 - 20x + 100$? $x - 10$
48. The area of a square is represented by $y^2 + 40y + 400$. What is its perimeter? $4y + 80$

mini-review

1. Express 0.00000253 in scientific notation. 2.53×10^{-6}

Solve.

2. Les is 6 years younger than twice Jim's age. The sum of their ages is 30. What are their ages? Jim, 12; Les, 18

Simplify.

3. $x(2x + 5) - 7(x^2 + 3x - 2)$ 3. $-5x^2 - 16x + 14$
4. $(2y^2 + 5y - 3) + (7y^2 - 8y + 9)$ 4. $9y^2 - 3y + 6$
5. $(2y + 5)(3y - 7)$ 5. $6y^2 + y - 35$

❸ Written Exercises 39-46 involve completing the square. This topic is formally introduced in Lesson 13-3.

7-5

Lesson Notes, page T39

Readiness Activity
Use the FOIL method to find each product.

$(a + 5)(a - 2)$ $a^2 + 3a - 10$
$(s - 4)(s - 1)$ $s^2 - 5s + 4$
$(2r + 6)(r + 2)$
$2r^2 + 10r + 12$
$(d - 5)(3d + 2)$
$3d^2 - 13d - 10$
$(y + a)(y + b)$
$y^2 + ay + by + ab$

Discussion Questions
(Use after Example 2)

1. List all positive pairs of factors of 24.
   1, 24; 2, 12; 3, 8; 4, 6
2. List all negative pairs of factors of 24.
   −1, −24; −2, −12; −3, −8; −4, −6
3. Why can only negative pairs of factors of 24 be considered when factoring $x^2 - 10x + 24$?
   Because the middle term is negative.
4. Why can only positive pairs of factors of 24 be considered when factoring $x^2 + 11x + 24$?
   Because the middle term is positive.

Chalkboard Exercises
Factor.

1a. $a^2 + 6a + 8$
$(a + 2)(a + 4)$
1b. $x^2 + 20x + 19$
$(x + 1)(x + 19)$

❶ To factor $x^2 + 10x + 21$, students should find the factors of 21 in pairs. Then they should determine the combination that yields the sum of 10.

❷ The pattern $(x + m)(x + n)$ can be used to factor trinomials of the form $ax^2 + bx + c$, where $a = 1$.

Objective 7-5: Factor trinomials of in form $x^2 + bx + c$.

7-5 Factoring $x^2 + bx + c$

Recall that if two numbers are multiplied, each number is a *factor* of the product. Similarly, if two binomials are multiplied, each binomial is a *factor* of the product.

Use the FOIL method to find the product $(x + 3)(x + 7)$.

$$\begin{aligned} & \quad\;\; \text{F} \qquad\; \text{O} \qquad\; \text{I} \qquad\; \text{L} \\ (x + 3)(x + 7) &= x \cdot x + x \cdot 7 + 3 \cdot x + 3 \cdot 7 \\ &= x^2 + 7x + 3x + 3 \cdot 7 \\ &= x^2 + (7 + 3)x + 3 \cdot 7 \\ &= x^2 + 10x + 21 \end{aligned}$$ ❶

21 is the product of 3 and 7.
10 is the sum of 3 and 7.

The binomials $(x + 3)$ and $(x + 7)$ are factors of $x^2 + 10x + 21$.

In general, the product of the binomials $(x + m)$ and $(x + n)$ can be found by using FOIL.

$$\begin{aligned} (x + m)(x + n) &= x^2 + nx + mx + mn \\ &= x^2 + (m + n)x + mn \end{aligned}$$ ❷

This pattern can be used to factor certain types of trinomials. For example, to factor $x^2 + 7x + 10$ into the form $(x + m)(x + n)$, find the factors of 10 whose sum is 7. Consider only positive factors of 10 since $(m + n)$ and mn are both positive.

$mn = 10$
$m + n = 7$

Factors of 10	Sum of Factors
1, 10	1 + 10 = 11
2, 5	2 + 5 = 7

Since the sum must be 7, the factors are 2 and 5.

$$x^2 + 7x + 10 = (x + 2)(x + 5)$$ *Check by using FOIL.*

Example

1 **Factor: $y^2 + 8y + 12$**

$y^2 + 8y + 12 = y^2 + (m + n)y + mn$ $\quad m + n = 8$ and $mn = 12$

Find the factors of 12 whose sum is 8. Consider only positive integers.

Factors of 12	Sum of Factors
1, 12	1 + 12 = 13
2, 6	2 + 6 = 8
3, 4	3 + 4 = 7

The factors are 2 and 6. Therefore, $y^2 + 8y + 12 = (y + 2)(y + 6)$.
Check that $(y + 2)(y + 6) = y^2 + 8y + 12$ by using FOIL.

Example

2 **Factor: $a^2 - 9a + 18$**

$a^2 - 9a + 18 = a^2 + (m + n)a + mn$ $m + n = (-9)$ and $mn = 18$

Since $(m + n)$ is negative and mn is positive, the factors of 18 must *both* be *negative*. Do you see why?

Factors of 18	Sum of Factors
$-1, -18$	$-1 + (-18) = -19$
$-2, -9$	$-2 + (-9) = -11$
$-3, -6$	$-3 + (-6) = -9$

Therefore, $a^2 - 9a + 18 = (a - 3)(a - 6)$. *Check by using FOIL.*

Notice that in the trinomial $x^2 + 5x - 6$ the value of c is *negative*. Since $x^2 + 5x - 6$ can be expressed as $x^2 + 5x + (-6)$, the same pattern can be used to factor trinomials in which c is negative. ❸

Examples

3 **Factor: $x^2 + 5x - 6$**

$x^2 + 5x - 6 = x^2 + (m + n)x + mn$ where $m + n = 5$ and $mn = -6$. Since $(m + n)$ is positive and mn is negative, exactly one factor of each pair of factors below must be negative. Do you see why?

Factors of -6	Sum of Factors
$1, -6$	$1 + (-6) = -5$
$-1, 6$	$-1 + 6 = 5$
$2, -3$	$2 + (-3) = -1$
$-2, 3$	$-2 + 3 = 1$

Select the factors -1 and 6. Therefore, $x^2 + 5x - 6 = (x - 1)(x + 6)$. *Check by using FOIL.*

4 **Factor: $a^2 - 3a - 10$**

$a^2 - 3a - 10 = a^2 + (m + n)a + mn$ $m + n = -3$ and $mn = -10$

$2 + (-5) = -3$ and $2(-5) = -10$.

Therefore, $a^2 - 3a - 10 = (a + 2)(a - 5)$.

When the value of c is negative, the factors of a trinomial are a *difference* and a *sum*. For $x^2 + 5x - 6$, the difference is $(x - 1)$ and the sum is $(x + 6)$. For $a^2 - 3a - 10$ the difference is $(a - 5)$ and the sum is $(a + 2)$.

Chalkboard Exercises

Factor.

2a. $b^2 - 10b + 16$
$(b - 2)(b - 8)$

2b. $y^2 - 16y + 28$
$(y - 2)(y - 14)$

2c. $c^2 - 42c + 41$
$(c - 1)(c - 41)$

3a. $x^2 + 2x - 8$
$(x + 4)(x - 2)$

3b. $x^2 + 10x - 24$
$(x + 12)(x - 2)$

4a. $a^2 - 6a - 16$
$(a - 8)(a + 2)$

4b. $a^2 - 13a - 48$
$(a + 3)(a - 16)$

❸ Show students that for the quadratic $x^2 + bx + c$ where c is negative, the sign of the bx term determines the signs of the factors of c. For example, in $x^2 - 7x - 8$ the factor of 8 with the largest absolute value must be negative since b is negative.

Chalkboard Exercises

Factor each trinomial, if possible. If the trinomial cannot be factored, write *prime*.

5a. $26 - 20v + v^2$
prime

5b. $m^2 + 19m + 77$
prime

Find all values of k so that each trinomial can be factored using integers.

6a. $x^2 + kx + 27$
28, 12, −12, −28

6b. $x^2 + kx + 12$
13, 8, 7, −7, −8, −13

❹ Explain to students that this polynomial need not be rewritten in order to be factored.

❺ Another way of stating this definition is a prime polynomial cannot be written as a product of two polynomials, neither of which is 1.

❻ For Exploratory Exercises 1-12, remind students that the pattern is $(x + m)(x + n) = x^2 + (m + n)x + mnx$.

Suggested Daily Assignments

Enriched
Exp: 10–15
Written: 10–42 evens, 43–54

Average
Exp: 2–20 evens
Written: 1–41 odds, 43–54

Fundamental
Exp: 1–20
Written: 1–30, 43–45
p. 223, 2–30 evens

Answers to Written

2. $(a + 1)(a + 21)$
3. $(c + 3)(c - 1)$
6. $(a - 5)(a - 7)$
11. $(z - 13)(z + 3)$
13. $(r - 12)(r + 2)$
14. $(y - 12)(y + 3)$
15. $(y + 10)(y - 2)$
16. $(n + 15)(n - 3)$
17. $(p + 2)(p + 8)$
18. $(g + 2)(g + 13)$
19. $(b - 4)(b - 7)$
21. $(15 + m)(1 + m)$
22. $(12 + a)(3 + a)$

Example

5 **Factor: $14 + 13x + x^2$**

The polynomial $14 + 13x + x^2$ can be written as $x^2 + 13x + 14$. ❹

$x^2 + 13x + 14 = x^2 + (m + n)x + mn$ $\quad m + n = 13$ and $mn = 14$

Factors of 14	Sum of Factors
1, 14	$1 + 14 = 15$
2, 7	$2 + 7 = 9$

Why should you test only positive factors?

There are no factors m and n such that $m + n = 13$ and $mn = 14$. Therefore, $x^2 + 13x + 14$ cannot be factored using integers.

A polynomial that *cannot* be written as a product of two polynomials is called a **prime polynomial**. In Example 5 the trinomial $x^2 + 13x + 14$ is prime. ❺

Example

6 **Find all values of k so that the trinomial $x^2 + kx + 8$ can be factored using integers.**

For $x^2 + kx + 8$, the value of k is the sum of the factors of 8.

Factors of 8	Sum of Factors
1, 8	$1 + 8 = 9$
2, 4	$2 + 4 = 6$
−1, −8	$-1 + (-8) = -9$
−2, −4	$-2 + (-4) = -6$

The values of k are 9, 6, −9, and −6.

Exploratory Exercises ❻

For each polynomial, find integers m and n whose product is the last term and whose sum is the coefficient of the second term.

1. $x^2 + 15x + 14$ 1, 14
2. $y^2 + 9y + 14$ 2, 7
3. $b^2 - 8b + 12$ −2, −6
4. $x^2 - 19x + 34$ −2, −17
5. $p^2 + 5p - 36$ 9, −4
6. $r^2 - 10r - 24$ 2, −12
7. $y^2 - 7y - 30$ 3, −10
8. $g^2 + 5g - 24$ 8, −3
9. $x^2 + 18x + 45$ 3, 15
10. $t^2 - 19t + 48$ −3, −16
11. $c^2 - 17c - 60$ 3, −20
12. $a^2 + 13a - 48$ 16, −3

Complete the factoring.

13. $p^2 + 9p - 10 = (p + ■)(p - 1)$ 10
14. $y^2 + 3y - 28 = (y + 7)(y ■ 4)$ −
15. $x^2 - 2x - 35 = (x + 5)(x - ■)$ 7
16. $a^2 - 4a - 21 = (a + ■)(a - 7)$ 3
17. $h^2 + 6h - 16 = (h - 2)(h ■■)$ + 8
18. $x^2 - 9x - 36 = (x - 12)(x ■■)$ + 3
19. $a^2 + 4am - 21m^2 = (a - 3m)(a ■■)$ + 7m
20. $p^2 + 16rp + 63r^2 = (p ■■)(p + 9r)$ +

Written Exercises

Factor each trinomial, if possible. At least two cannot be factored using integers. If the trinomial cannot be factored using integers, write *prime*.

1. $y^2 + 12y + 27$ $(y + 3)(y + 9)$
2. $a^2 + 22a + 21$
3. $c^2 + 2c - 3$
4. $x^2 + 2x - 15$ $(x + 5)(x - 3)$
5. $y^2 - 8y + 15$ $(y - 3)(y - 5)$
6. $a^2 - 12a + 35$
7. $m^2 - m - 20$ $(m - 5)(m + 4)$
8. $x^2 - 5x - 24$ $(x - 8)(x + 3)$
9. $h^2 + 5h + 8$ prime
10. $c^2 + 3c + 6$ prime
11. $z^2 - 10z - 39$
12. $y^2 - 7y + 60$ prime
13. $r^2 - 10r - 24$
14. $y^2 - 9y - 36$
15. $y^2 + 8y - 20$
16. $n^2 + 12n - 45$
17. $p^2 + 10p + 16$
18. $g^2 + 15g + 26$
19. $b^2 - 11b + 28$
20. $x^2 - 9x + 14$ $(x - 2)(x - 7)$
21. $15 + 16m + m^2$
22. $36 + 15a + a^2$
23. $y^2 + 11y + 28$ $(y + 7)(y + 4)$
24. $x^2 + 14x + 33$
25. $a^2 - 18a - 40$
26. $b^2 + 13b - 30$
27. $x^2 - 3x - 18$
28. $y^2 - 7y - 30$
29. $36 - 16m + m^2$ prime
30. $36 - 13y + y^2$
31. $m^2 - 6m - 55$
32. $h^2 - 3h - 108$
33. $66 - 17j + j^2$
34. $90 - 19c + c^2$
35. $42 - 23a + a^2$
36. $50 - 27x + x^2$
37. $c^2 - 2cd - 8d^2$
38. $x^2 - 4xy - 5y^2$
39. $a^2 + 2ab - 3b^2$
40. $n^2 + 4an - 32a^2$ $(n + 8a)(n - 4a)$
41. $(a + b)^2 - 5(a + b) - 6$ $(a + b - 6)(a + b + 1)$
42. $(c + q)^2 - 8(c + q) - 9$ $(c + q - 9)(c + q + 1)$

Find all values of k so that each trinomial can be factored using integers.

43. $x^2 + kx + 10$ 7, −7, 11, −11
44. $m^2 + km + 6$ 5, −5, 7, −7
45. $r^2 + kr - 13$ 12, −12
46. $y^2 + ky - 4$ 3, −3, 0
47. $n^2 + kn + 7$ 8, −8
48. $p^2 + kp + 17$ 18, −18
49. $a^2 + ka - 5$ 4, −4
50. $b^2 + kb - 16$
51. $t^2 + kt + 21$
52. $c^2 + kc + 12$ 7, −7, 8, −8, 13, −13
53. $s^2 + ks - 14$ 5, −5, 13, −13
54. $y^2 + ky - 15$ 2, −2, 14, −14

Excursions in Algebra — Factoring Cubes

You know how to factor the difference of two squares. Can you factor the difference of two cubes? Consider $a^3 - b^3$.

$$\begin{aligned} a^3 - b^3 &= a^3 - a^2b + a^2b - b^3 \quad \text{Notice that } -a^2b + a^2b = 0. \\ &= a^2(a - b) + b(a^2 - b^2) \\ &= a^2(a - b) + b(a - b)(a + b) \\ &= (a - b)[a^2 + b(a + b)] \\ &= (a - b)(a^2 + ab + b^2) \end{aligned}$$

Example: **Factor $y^3 - 27$.**

y^3 is the cube of y. 27 is the cube of 3. Now use the pattern above.

$$y^3 - 27 = (y - 3)(y^2 + 3y + 9)$$

Exercises

Factor using the pattern above.

1. $x^3 - 64$
2. $8x^3 - y^3$
3. $27a^3 - 64b^3$
4. Find a pattern for factoring $a^3 + b^3$ by using a method similar to the one above.

Answers to Written (continued)

24. $(x + 11)(x + 3)$
25. $(a - 20)(a + 2)$
26. $(b + 15)(b - 2)$
27. $(x - 6)(x + 3)$
28. $(y - 10)(y + 3)$
30. $(9 - y)(4 - y)$
31. $(m - 11)(m + 5)$
32. $(h - 12)(h + 9)$
33. $(11 - j)(6 - j)$
34. $(10 - c)(9 - c)$
35. $(21 - a)(2 - a)$
36. $(2 - x)(25 - x)$
37. $(c - 4d)(c + 2d)$
38. $(x - 5y)(x + y)$
39. $(a + 3b)(a - b)$
50. 6, −6, 15, −15, 0
51. 10, −10, 22, −22

Quiz
Lessons 7-3 to 7-5

Determine whether each of the following is a perfect square. If it is, factor it.

1. $x^2 - 5x + 6$ no
2. $9x^2 - 24xy + 16y^2$ yes, $(3x - 4y)^2$

Factor each expression, if possible.

3. $b^2 - 144$ $(b - 12)(b + 12)$
4. $9a^2 - 12ab + 4b^2$ $(3a - 2b)^2$

Factor each trinomial, if possible. If the trinomial cannot be factored using integers, write *prime*.

5. $m^2 - 7m + 10$ $(m - 5)(m - 2)$
6. $p^2 + 14p + 49$ $(p + 7)^2$
7. $b^2 + 7b - 6$ prime
8. $x^2 - 8x - 9$ $(x - 9)(x + 1)$
9. $y^2 + 13y - 48$ $(y + 16)(y - 3)$

About the Excursion

Point out that a is the cube root of a^3 and b is the cube root of b^3.

Answers to Excursion

1. $(x - 4)(x^2 + 4x + 16)$
2. $(2x - y)(4x^2 + 2xy + y^2)$
3. $(3a - 4b)(9a^2 + 12ab + 16b^2)$
4. $(a + b)(a^2 - ab + b^2)$

7-6

Lesson Notes, page T39

Readiness Activity
Name all pairs of terms from the following list that have a common monomial factor. Then name the GCF of each pair.
$5x^2y$ $3ab^2$ $25mn$
$16a^3$ $7pq^3$ $49p^2q^2$
$5x^2y$, $25mn$; 5
$3ab^2$, $16a^3$; a
$7pq^3$, $49p^2q^2$; $7pq^2$

Discussion Question
How many ways can you find to factor
$a^4 + a^2b^2 - a^2d^2 - b^2d^2$?
Answer will vary.

Chalkboard Examples
Factor. Check by using FOIL.
1a. $7mn + 14n + 5m + 10$
$(7n + 5)(m + 2)$
1b. $arx + 4am + amx + 4ar$
$(4a + ax)(r + m)$
2a. $36a - 4ab - 63 + 7b$
$(4a - 7)(9 - b)$
2b. $49c - 7cd - 21 + 3d$
$(7c - 3)(7 - d)$
3a. Factor $8b^2c + 16b + 3bc + 6$ in two different ways.
Two typical answers are shown.
a. $(8b^2c + 16b) + (3bc + 6)$
$= 8b(bc + 2) + 3(bc + 2)$
$= (8b + 3)(bc + 2)$
b. $(8b^2c + 3bc) + (16b + 6)$
$= bc(8b + 3) + 2(8b + 3)$
$= (bc + 2)(8b + 3)$

❶ Make sure students realize that if one way of grouping does not succeed, they should try another.

❷ Point out to students that $(3 - a)$ and $(a - 3)$ are additive inverses since $(3 - a) + (a - 3) = 0$.

Objective 7-6: Factor polynomials containing four terms by grouping.

7-6 Factoring by Grouping

Some polynomials have four terms. It may be possible to factor these polynomials by first grouping the terms in pairs and factoring a monomial from each group. Then use the distributive property again to factor the common binomial.

Examples

1 **Factor: $3xy - 21y + 5x - 35$**

$$\begin{aligned} 3xy - 21y + 5x - 35 &= (3xy - 21y) + (5x - 35) && \text{Group pairs of terms that have a common monomial factor.} \\ &= 3y(x - 7) + 5(x - 7) && \text{Factor the GCF from each group.} \\ &= 3y(x - 7) + 5(x - 7) && \text{Notice that } (x-7) \text{ is a common factor.} \\ &= (3y + 5)(x - 7) && \text{Use the distributive property.} \end{aligned}$$

Check by using FOIL.

$$\begin{aligned} (3y + 5)(x - 7) &= \overset{F}{3y(x)} + \overset{O}{3y(-7)} + \overset{I}{5(x)} + \overset{L}{5(-7)} \\ &= 3xy - 21y + 5x - 35 \end{aligned}$$

2 **Factor: $8x^2y - 5x - 24xy + 15$**

$$\begin{aligned} 8x^2y - 5x - 24xy + 15 &= (8x^2y - 5x) + (-24xy + 15) && \text{Why are the terms grouped this way?} \\ &= x(8xy - 5) + (-3)(8xy - 5) && \text{What is the common factor?} \\ &= (x - 3)(8xy - 5) \end{aligned}$$

Check: $(x - 3)(8xy - 5) = x(8xy) + x(-5) + (-3)(8xy) + (-3)(-5)$
$= 8x^2y - 5x - 24xy + 15$

Sometimes you can group the terms in more than one way. The following example shows another way to factor the polynomial in Example 2 above. ❶

Example

3 **Factor: $8x^2y - 5x - 24xy + 15$**

$$\begin{aligned} 8x^2y - 5x - 24xy + 15 &= (8x^2y - 24xy) + (-5x + 15) \\ &= 8xy(x - 3) + (-5)(x - 3) \\ &= (8xy - 5)(x - 3) && \text{The result is the same as in Example 2.} \end{aligned}$$

Recognizing binomials that are additive inverses is often helpful in factoring. For example, the binomials $(3 - a)$ and $(a - 3)$ are additive inverses since $(3 - a)$ and $-1(a - 3)$ are equivalent. What is the additive inverse of $(5 - y)$?

❷ $-1(a-3) = (-1)(a) + (-1)(-3)$
$= -a + 3$
$= 3 - a$

Example

4 **Factor: $15x - 3xy + 4y - 20$**

$$\begin{aligned}15x - 3xy + 4y - 20 &= (15x - 3xy) + (4y - 20)\\ &= 3x(5 - y) + 4(y - 5)\\ &= -3x(y - 5) + 4(y - 5)\\ &= (-3x + 4)(y - 5)\end{aligned}$$

The binomials $(5 - y)$ and $(y - 5)$ are additive inverses.

$(5 - y) = -1(y - 5)$.

Check: $$\begin{aligned}(-3x + 4)(y - 5) &= (-3x)(y) + (-3x)(-5) + 4(y) + 4(-5)\\ &= -3xy + 15x + 4y - 20\\ &= 15x - 3xy + 4y - 20\end{aligned}$$

Chalkboard Exercises

Factor.

4a. $8x - 2xy + 5y - 20$ $(-2x + 5)(y - 4)$

4b. $12y - 4xy + 3x - 9$ $(-x + 3)(4y - 3)$

Exploratory Exercises

Express each polynomial in factored form.

1. $k(r + s) - m(r + s)$ $(k - m)(r + s)$
2. $y(m + n) - 4(m + n)$ $(y - 4)(m + n)$
3. $4b(x - y) + y(x - y)$ $(4b + y)(x - y)$
4. $t(t - s) + s(t - s)$ $(t + s)(t - s)$
5. $7rp\,(r - 3p) - 4q(r - 3p)$ $(7rp - 4q)(r - 3p)$
6. $3ab(a - 4) - 8(a - 4)$ $(3ab - 8)(a - 4)$
7. $8m(x + y) + (x + y)$ $(8m + 1)(x + y)$
8. $p(p - 3r) - (p - 3r)$ $(p - 3r)(p - 1)$
9. $5mp(2m + 3p) - 3bc(2m + 3p)$
10. $5z(5x^2 - 3y^2) + m^2(5x^2 - 3y^2)$
11. $4m(y - 5) + 3p(5 - y)$ $(4m - 3p)(y - 5)$
12. $5y(z - 6) - 3m(6 - z)$ $(5y + 3m)(z - 6)$
13. $7x(a + b) + 3m(a + b) - 4p(b + a)$
14. $3a(x - y) - 4b(y - x) + 5c(x - y)$
15. $a(8r - 3y) + b(8r - 3y) + c(3y - 8r)$ $(a + b - c)(8r - 3y)$
16. $x(4a^2 + 3b^2) - 3y(4a^2 + 3b^2) + z(3b^2 + 4a^2)$ $(x - 3y + z)(4a^2 + 3b^2)$

14. $(3a + 4b + 5c)(x - y)$

Suggested Daily Assignments

Enriched
- Exp: 13–16
- Written: 2–44 evens

Average
- Exp: 1–15 odds
- Written: 1–43 odds

Fundamental
- Exp: 1–16
- Written: 1–12, 14–42 evens
  - p. 227, 31–40, 46–48

Answers to Exploratory

9. $(5mp - 3bc)(2m + 3p)$
10. $(5z + m^2)(5x^2 - 3y^2)$
13. $(a + b)(7x + 3m - 4p)$

Written Exercises

Find the common factor for the grouped binomials.

> **Sample:** $(2a^2 + ab) + (4ac + 2bc)$
> $(2a^2 + ab) + (4ac + 2bc) = a(2a + b) + 2c(2a + b)$
> The common factor is $(2a + b)$.

1. $(bx + by) + (3ax + 3ay)$ $(x + y)$
2. $(3mx + 2my) + (3kx + 2ky)$ $(3x + 2y)$
3. $(a^2 + 3ab) + (2ac + 6bc)$ $(a + 3b)$
4. $(rx + 2ry) + (kx + 2ky)$ $(x + 2y)$
5. $(a^2 - 4ac) + (ab - 4bc)$ $(a - 4c)$
6. $(10x^2 - 6xy) + (15x - 9y)$ $(5x - 3y)$
7. $(4m^2 - 3mp) + (3p - 4m)$ $(4m - 3p)$
8. $(16k^2 - 28kp) + (7p^2 - 4kp)$ $(4k - 7p)$
9. $(6x^3 + 7x^2y) + (6x + 7y)$ $(6x + 7y)$
10. $(5a + 2b) + (10a^2 + 4ab)$ $(5a + 2b)$
11. $(3a^2b + 2ab^3) + (6ab + 4b^3)$ $(3a + 2b^2)$
12. $(15a^2 - 21ab) + (20ab - 28b^2)$ $(5a - 7b)$

Answers to Written

27. $(a - 3b^2)(5a - 4b)$
28. $(2x - 5y)(2a - 7b)$
30. $5(a^2 + ab + cd + c^2)$
33. $(7m + 2n)(p + q)(p - q)$
34. $(5a + 3b)(a + b)(a - b)$
38. $(2a^2 - 5b^2)(a - b)$
39. $(7m - 5y)(a + b)(a - b)$
40. $(3x + 4a)(k + m)(k - m)$
41. $(m - 2)(a + 2)(a + 3)$
42. $(a + 1)(x - 5)(x + 3)$

About the Excursion

When discussing the example in this feature, it may be helpful to point out that $a^2 + 4a + 4 - 9b^2$ may be written in the form $m^2 - n^2$, where $m = a + 2$ and $n = 3b$. Thus, $a^2 + 4a + 4 - 9b^2 = (a^2 + 4a + 4) - 9b^2 = (a + 2)^2 - (3b)^2$.

Answers to Excursion

1. $(x + 2 - 5y)(x + 2 + 5y)$
2. $(y + 4 - 6z)(y + 4 + 6z)$
3. $(x + y + r)(x + y - r)$
4. $(m - k + 3)(m + k - 3)$
5. $(3 - 3y + x)(3 + 3y - x)$
6. $(c - a - 5b)(c + a + 5b)$
7. $(2a + 3b - 4x)(2a + 3b + 4x)$
8. $(2a + x - b)(2a + x + b)$

Factor. Check by using FOIL.

13. $2ax + 6xc + ba + 3bc$ $(2x + b)(a + 3c)$
14. $6mx - 4m + 3rx - 2r$ $(2m + r)(3x - 2)$
15. $ay + 4p + 4a + yp$ $(a + p)(y + 4)$
16. $3my - ab + am - 3by$ $(m - b)(3y + a)$
17. $2my + 7x + 7m + 2xy$ $(m + x)(2y + 7)$
18. $zr + 6q + rq + 6z$ $(r + 6)(z + q)$
19. $ay - ab + cb - cy$ $(a - c)(y - b)$
20. $3ax - 6bx + 8b - 4a$ $(a - 2b)(3x - 4)$
21. $a^2 - 2ab + a - 2b$ $(a + 1)(a - 2b)$
22. $2ab + 2am - b - m$ $(2a - 1)(b + m)$
23. $4ax + 3ay + 4bx + 3by$ $(4x + 3y)(a + b)$
24. $x^2 + 5xy + ax + 5ay$ $(x + 5y)(x + a)$
25. $3m^2 - 5m^2p + 3p^2 - 5p^3$ $(m^2 + p^2)(3 - 5p)$
26. $10x^2 - 14xy + 5xy - 7y^2$ $(5x - 7y)(2x +$
27. $5a^2 - 4ab + 12b^3 - 15ab^2$
28. $4ax - 14bx + 35by - 10ay$
29. $6a^2 - 6ab + 3cb - 3ca$ $3(2a - c)(a - b)$
30. $5a^2 + 5ab + 5cd + 5c^2$
31. $ax + a^2x - a - 2a^2$ $a(x + ax - 1 - 2a)$
32. $5xy + 15x - 6y - 18$ $(5x - 6)(y + 3)$
33. $7mp^2 + 2np^2 - 7mq^2 - 2nq^2$
34. $5a^3 + 3a^2b - 5ab^2 - 3b^3$
35. $x^3 + 2x^2 - x - 2$ $(x + 1)(x - 1)(x + 2)$
36. $y^3 + y^2 - y - 1$ $(y + 1)^2(y - 1)$
37. $a^3 - a^2b + ab^2 - b^3$ $(a^2 + b^2)(a - b)$
38. $2a^3 - 5ab^2 - 2a^2b + 5b^3$
39. $7m(a^2 - b^2) + 5y(b^2 - a^2)$
40. $3x(k^2 - m^2) - 4a(m^2 - k^2)$
41. $a^2(m - 2) + 5a(m - 2) + 6(m - 2)$
42. $x^2(a + 1) - 2x(a + 1) - 15(a + 1)$
43. $2a^2x + 3a^2y - 14ax - 21ay + 24x + 36y$ $(2x + 3y)(a - 3)(a - 4)$
44. $ax^2 - 3bx^2 - 7ax + 21bx - 18a + 54b$ $(a - 3b)(x - 9)(x + 2)$

Excursions in Algebra — Grouping Three Terms

In the following example, the first three terms have been grouped to form a perfect square trinomial.

Factor: $a^2 + 4a + 4 - 9b^2$

$$\begin{aligned} a^2 + 4a + 4 - 9b^2 &= (a^2 + 4a + 4) - 9b^2 && \text{The perfect square trinomial } a^2 + 4a + 4 \text{ can be factored as } (a + 2)^2. \\ &= (a + 2)^2 - 9b^2 \\ &= [(a + 2) - 3b][(a + 2) + 3b] && \text{Use the difference of squares.} \\ &= (a + 2 - 3b)(a + 2 + 3b) \end{aligned}$$

Exercises

Factor each polynomial.

1. $x^2 + 4x + 4 - 25y^2$
2. $y^2 + 8y + 16 - 36z^2$
3. $x^2 + 2xy + y^2 - r^2$
4. $m^2 - k^2 + 6k - 9$
5. $9 - 9y^2 - x^2 + 6xy$
6. $c^2 - 10ab - a^2 - 25b^2$
7. $4a^2 + 12ab + 9b^2 - 16x^2$
8. $4a^2 + 4ax + x^2 - b^2$

Reading Algebra Reading Speed

When you read mathematics, do not expect to read rapidly. Mathematics is a concise language. Try reading aloud each of these statements.

Liquids have no definite shape.
$x^2 + 5x + 4 = 0$

The equation takes longer to read because the mathematical symbols are compact forms of words and concepts. In words, it would be read as *x squared plus five x plus four equals zero*. The following concepts are involved.

x is a variable.
The raised 2 in x^2 means x is used as a factor twice.
$5x$ means 5 multiplied by x.
You must also know the meaning of $+$, $=$, 5, 4, and 0.

For understanding, you need to translate the symbols into words and concepts.

Suppose you read a section in this book. First read it rapidly to get a general idea of the content. Then reread it slowly so that you understand all of the ideas in the section. Turn to page 236 and read section 7-8 rapidly to find the main ideas in that section.

The main ideas in the section are listed here.

The section gives a summary of the methods used to factor polynomials.
If there is a greatest common factor, factor it out first.
Continue factoring until each of the remaining factors is prime.

Now reread section 7-8 carefully to find out how to factor a polynomial completely.

Exercises

Write each expression in words.

1. $3x^2 + 14x + 15$
2. $(3x + 5)(x + 3)$
3. $3xy - 6x + 5y - 10$
4. $(3x + 5)(y - z)$
5. List the concepts contained in exercise 1.
6. List the concepts contained in exercise 2.
7. List the methods of factoring polynomials that are discussed in section 7-8.

About Reading Algebra
It is important to emphasize the value of symbols in mathematics. Many students fail to realize that while symbols save space, the concepts they convey can be quite important.

Answers to Exercises

1. Three x squared plus fourteen x plus fifteen.
2. The quantity of three x plus five times the quantity of x plus three.
3. Three xy minus six x plus five y minus ten.
4. The quantity of three x plus five times the quantity of y minus z.
5. x is a variable; The raised 2 in x^2 means x is used as a factor twice; $3x^2$ means 3 multiplied by x^2 and $14x$ means 14 times x; You must know the meaning of $+$, 3, 14, and 15.
6. x is a variable; $3x$ means 3 multiplied by x; $(3x + 5)(x + 3)$ means $3x + 5$ multiplied by $x + 3$. You must know the meaning of $+$, 3, and 5.
7. GCF, difference of squares, perfect square trinomial, trinomial that has two binomial factors, pairs of terms that have a common monomial factor

7-7

Lesson Notes, page T39

Discussion Question
Can you find a trinomial in the form $ax^2 + bx + c$ in which a, b, and c are all prime numbers, but the trinomial itself is not prime (that is, can be factored)?
Answers will vary. Some typical answers are $2x^2 + 11x + 5$ and $11x^2 + 23x + 2$.

❶ This method eliminates trial and error. It is especially helpful when a and c, of $ax^2 + bx + c$, have a great number of factors.

Objective 7-7: Factor trinomials of the form $ax^2 + bx + c$.

7-7 Factoring $ax^2 + bx + c$

In the trinomial $x^2 - 3x - 18$, the coefficient of x^2 is 1. To factor this trinomial, you find the factors of -18 whose sum is -3.

Consider the trinomial $2y^2 + 7y + 6$. In this trinomial the coefficient of y^2 is not 1. To factor a trinomial such as this one, another method is needed.

What is the coefficient of y^2 in $2y^2 + 7y + 6$?

Study how FOIL is used to multiply $(3x + 4)$ by $(2x + 7)$.

$$(3x + 4)(2x + 7) = \overset{F}{6x^2} + \overset{O}{21x} + \overset{I}{8x} + \overset{L}{28}$$

$$= 6x^2 + (21 + 8)x + 28$$

$21 \cdot 8 = 168$

$6 \cdot 28 = 168$

Notice that the *product* of 6 and 28 is the same as the *product* of 21 and 8. You can use this method to factor the trinomial $2y^2 + 7y + 6$.

$2y^2 + 7y + 6$	The product of 2 and 6 is 12.
$2y^2 + (\square + \square)y + 6$	You need to find two integers whose *product is 12* and whose *sum is 7.*

Factors of 12	Sum of Factors
1, 12	$1 + 12 = 13$
2, 6	$2 + 6 = 8$
3, 4	$3 + 4 = 7$

❶

$2y^2 + (3 + 4)y + 6$	Select the factors 3 and 4.
$2y^2 + 3y + 4y + 6$	
$(2y^2 + 3y) + (4y + 6)$	Group pairs of terms that have a common monomial factor.
$y(2y + 3) + 2(2y + 3)$	Factor the GCF from each group.
$(y + 2)(2y + 3)$	Use the distributive property.

Therefore, $2y^2 + 7y + 6 = (y + 2)(2y + 3)$. *Check by using FOIL.*

Example 1, which follows, illustrates that the order in which the factors are selected does not affect the result.

Examples

1 **Factor: $3x^2 + 11x + 6$**

$3x^2 + 11x + 6$ — *The product of 3 and 6 is 18.*

Factors of 18	Sum of Factors
1, 18	$1 + 18 = 19$
2, 9	$2 + 9 = 11$
3, 6	$3 + 6 = 9$

$3x^2 + (\ \ + \ \ )x + 6$ — *Select the factors 2 and 9.*

$3x^2 + (2 + 9)x + 6$ or $3x^2 + (9 + 2)x + 6$

$3x^2 + 2x + 9x + 6$ — $3x^2 + 9x + 2x + 6$

❷ $(3x^2 + 2x) + (9x + 6)$ — $(3x^2 + 9x) + (2x + 6)$

$x(3x + 2) + 3(3x + 2)$ — $3x(x + 3) + 2(x + 3)$ *Factor the GCF from each group.*

$(x + 3)(3x + 2)$ — $(3x + 2)(x + 3)$ *Use the distributive property.*

Notice that $(x + 3)(3x + 2) = (3x + 2)(x + 3)$.

Therefore, $3x^2 + 11x + 6 = (x + 3)(3x + 2)$. *Check by using FOIL.*

2 **Factor: $5x^2 - 17x + 14$**

$5x^2 - 17x + 14$ — *The product of 5 and 14 is 70.*
Both factors of 70 must be negative. Why?

Factors of 70	Sum of Factors
$-1, -70$	$-1 + (-70) = -71$
$-2, -35$	$-2 + (-35) = -37$
$-5, -14$	$-5 + (-14) = -19$
$-7, -10$	$-7 + (-10) = -17$

$5x^2 + (\ \ + \ \ )x + 14$

$5x^2 + [-10x + (-7)x] + 14$

$(5x^2 - 10x) + (-7x + 14)$

$5x(x - 2) + (-7)(x - 2)$ — *What is the common factor?*

$(5x - 7)(x - 2)$

Therefore, $5x^2 - 17x + 14 = (5x - 7)(x - 2)$. *Check by using FOIL.*

Chalkboard Exercises

Factor.

1a. $2y^2 + 7y + 6$ $(y + 2)(2y + 3)$

1b. $6y^2 + 5y + 1$ $(2y + 1)(3y + 1)$

1c. $4y^2 + 11y + 6$ $(4y + 3)(y + 2)$

2a. $2m^2 - 11m + 15$ $(2m - 5)(m - 3)$

2b. $2x^2 - 11x + 12$ $(2x - 3)(x - 4)$

❷ Show students that the same factors are obtained when $3x^2 + 2x + 9x + 6$ is grouped as follows. $(3x^2 + 9x) + (2x + 6) = 3x(x + 3) + 2(x + 3) = (3x + 2)(x + 3)$

Chalkboard Exercises

Factor.

3a. $3r^2 - 7r - 6$ $(r - 3)(3r + 2)$

3b. $10x^2 - 3x - 4$ $(5x - 4)(2x + 1)$

Example

3 Factor: $8y^2 - 6y - 9$

$8y^2 - 6y - 9$

The product of 8 and −9 is −72. Exactly one factor of −72 in each pair must be negative. Why? Since the sum is negative, the factor with the greater absolute value is negative.

Factors of −72	Sum of Factors
1, −72	1 + (−72) = −71
2, −36	2 + (−36) = −34
3, −24	3 + (−24) = −21
4, −18	4 + (−18) = −14
6, −12	6 + (−12) = −6

You can stop listing factors when you find a pair that works.

$8y^2 + (\blacksquare + \blacksquare)y - 9$

$8y^2 + [6y + (-12)y] - 9$

$(8y^2 + 6y) + (-12y - 9)$

$2y(4y + 3) + (-3)(4y + 3)$

$(2y - 3)(4y + 3)$

Therefore, $8y^2 - 6y - 9 = (2y - 3)(4y + 3)$. *Check by using FOIL.*

Suggested Daily Assignments

Enriched
- Exp: 11–14
- Written: 10–54 evens
- Mini-Review: 1–5

Average
- Exp: 2–14 evens
- Written: 2–54 evens
- Mini-Review: 1–5

Fundamental
- Exp: 1–14
- Written: 1–40
  - p. 230, 13–33 odds
- Mini-Review: 1–5

Exploratory Exercises

For each trinomial $ax^2 + bx + c$, supply the missing numbers so that the sum will equal b and the product will equal ac.

1. $3y^2 + 11y + 6 = 3y^2 + (\blacksquare + \blacksquare)y + 6$
2. $3c^2 + 14c + 8 = 3c^2 + (\blacksquare + \blacksquare)c + 8$
3. $4y^2 + 11y + 6 = 4y^2 + (\blacksquare + \blacksquare)y + 6$
4. $6b^2 + 31b + 5 = 6b^2 + (\blacksquare + \blacksquare)b + 5$
5. $4x^2 - 8x + 3 = 4x^2 + (\blacksquare + \blacksquare)x + 3$
6. $2y^2 - 11y + 15 = 2y^2 + (\blacksquare + \blacksquare)y + 15$
7. $2k^2 - 9k + 10 = 2k^2 + (\blacksquare + \blacksquare)k + 10$
8. $4a^2 - 25a + 6 = 4a^2 + (\blacksquare + \blacksquare)a + 6$
9. $5c^2 + 6c - 8 = 5c^2 + (\blacksquare + \blacksquare)c - 8$
10. $2x^2 + x - 21 = 2x^2 + (\blacksquare + \blacksquare)x - 21$
11. $5r^2 - 13r - 6 = 5r^2 + (\blacksquare + \blacksquare)r - 6$
12. $6a^2 + 7a - 3 = 6a^2 + (\blacksquare + \blacksquare)a - 3$
13. $3m^2 + 11m - 20 = 3m^2 + (\blacksquare + \blacksquare)m - 20$
14. $3d^2 - 5d - 28 = 3d^2 + (\blacksquare + \blacksquare)d - 28$

Answers to Exploratory

1. 2, 9	2. 2, 12
3. 3, 8	4. 1, 30
5. −2, −6	6. −5, −6
7. −4, −5	8. −1, −24
9. −4, 10	10. −6, 7
11. −15, 2	12. 9, −2
13. −4, 15	14. −12, 7

Written Exercises

Factor each trinomial, if possible. At least two of the trinomials cannot be factored using integers.

1. $3y^2 + 8y + 5$ $(3y + 5)(y + 1)$
2. $7a^2 + 22a + 3$ $(7a + 1)(a + 3)$
3. $3x^2 + 8x + 4$ $(3x + 2)(x$
4. $3a^2 + 14a + 15$ $(3a + 5)(a + 3)$
5. $8m^2 - 10m + 3$ $(4m - 3)(2m - 1)$
6. $2y^2 - 7y + 3$ $(2y - 1)(y$
7. $2h^2 - h - 3$ $(2h - 3)(h + 1)$
8. $3y^2 + 5y - 2$ $(3y - 1)(y + 2)$
9. $3k^2 + 7k - 6$ $(3k - 2)(k$
10. $3m^2 - 7m - 6$
11. $6p^2 - p - 2$
12. $4b^2 + 5b - 6$ $(4b - 3)(b$
13. $2a^2 + 3a - 14$
14. $2x^2 + 5x - 12$
15. $6t^2 + 5t - 6$ $(3t - 2)(2t +$
16. $7n^2 - 22n + 3$
17. $2y^2 - 5y + 3$
18. $3x^2 + 4x - 15$ $(3x + 5)(x$

10. $(3m + 2)(m - 3)$ 11. $(3p - 2)(2p + 1)$ 13. $(2a + 7)(a - 2)$
14. $(2x - 3)(x + 4)$ 16. $(7n - 1)(n - 3)$ 17. $(2y - 3)(y - 1)$

19. $2q^2 - 9q - 18$
20. $6y^2 - 11y + 4$
21. $6m^2 + 19m + 10$
22. $4y^2 - 17y - 15$
23. $6x^2 - 19x - 11$
24. $10n^2 - 19n + 7$
25. $12r^2 - 11r + 3$
26. $9k^2 - 12k + 4$
27. $8p^2 - 18p + 9$
28. $5b^2 - 13b - 10$
29. $15p^2 + 14p - 8$
30. $3t^2 - 32t + 20$
31. $6y^2 - 19y + 15$
32. $10k^2 - 11k - 6$
33. $6s^2 + 7s - 20$
34. $12b^2 + 17b + 6$
35. $18x^2 + 55x + 25$
36. $16m^2 + 14m - 15$
37. $18c^2 + 41c - 10$
38. $15y^2 + 17y - 18$
39. $18r^2 - 19r - 12$
40. $15x^2 - 13xy + 2y^2$
41. $8m^2 - 14mn + 3n^2$
42. $3s^2 - 10st - 8t^2$
43. $16x^2 - 16xy - 5y^2$
44. $16a^2 - 38ab - 5b^2$
45. $20p^2 + 11pq - 4q^2$
46. $25r^2 + 25rs + 6s^2$ $(5r + 3s)(5r + 2s)$
47. $9k^2 + 30km + 25m^2$ $(3k + 5m)^2$
48. $36a^2 + 9ab - 10b^2$ $(12a - 5b)(3a + 2b)$
49. $20s^2 + 17st - 24t^2$ $(5s + 8t)(4s - 3t)$
50. $3x^2 - 30xy + 56y^2$ prime
51. $8x^2 - 42xq + 27q^2$ $(4x - 3q)(2x - 9q)$
52. $12r^2 - 16rs - 11s^2$ $(2r + s)(6r - 11s)$
53. $14c^2 + 41cd + 15d^2$ $(2c + 5d)(7c + 3d)$
54. $21x^2 + 52xy + 32y^2$ $(7x + 8y)(3x + 4y)$

mini-review

Find each product.

1. $\left(-\frac{7}{12}\right)\left(\frac{6}{7}\right)\left(-\frac{3}{4}\right)$ $\frac{3}{8}$
2. $(4x^2y)(-3x^3y^2)$ $-12x^5y^3$
3. $8ay(-2ax + 4y^2)$
4. $(2x - 5)^2$ $4x^2 - 20x + 25$
5. $(a + 2)(2a^3 + 7a - 3)$ $2a^4 + 4a^3 + 7a^2 + 11a - 6$

Using Calculators — Checking Solutions

You can use your calculator to check whether $\frac{1}{18}$ is a solution of $243y^2 = \frac{3}{4}$. First enter $\left(\frac{1}{18}\right)^2$ by using the reciprocal and square keys. Then store it.

ENTER:	18	[1/x]	[x^2]	[STO]
DISPLAY:	18	.05555556	.00308642	.00308642

Then, evaluate $243y^2$ for $y = \frac{1}{18}$ and compare it to the value of $\frac{3}{4}$.

ENTER:	243	[×]	[RCL]	[=]
DISPLAY:	243		.00308642	0.75

ENTER:	3	[÷]	4	[=]
DISPLAY:	3		4	0.75

Since $0.75 = 0.75$, the number $\frac{1}{18}$ is a solution of $243y^2 = \frac{3}{4}$.

Exercises

Use your calculator to check whether the given number is a solution of the equation.

1. $128x^2 = 2$; $\frac{1}{8}$ yes
2. $108y^2 = 3$; $\frac{1}{5}$ no
3. $144a^2 - 16 = 0$; $-\frac{1}{3}$ yes
4. $25x^2 - 30x = -9$; $\frac{3}{5}$ yes
5. $49c^2 = -112c - 64$; $-\frac{8}{7}$ yes
6. $4a^2 = -20a - 25$; $\frac{2}{5}$ no

Answers to Written
19. $(2z + 3)(q - 6)$
20. $(3y - 4)(2y - 1)$
21. $(3m + 2)(2m + 5)$
22. $(4y + 3)(y - 5)$
23. $(3x - 11)(2x + 1)$
24. $(5n - 7)(2n - 1)$
25. prime
26. $(3k - 2)^2$
27. $(4p - 3)(2p - 3)$
28. prime
29. $(5p - 2)(3p + 4)$
30. $(3t - 2)(t - 10)$
31. $(2y - 3)(3y - 5)$
32. $(2k - 3)(5k + 2)$
33. $(2s + 5)(3s - 4)$
34. $(4b + 3)(3b + 2)$
35. $(9x + 5)(2x + 5)$
36. $(8m - 5)(2m + 3)$
37. $(9c - 2)(2c + 5)$
38. $(5y + 9)(3y - 2)$
39. $(2r - 3)(9r + 4)$
40. $(5x - y)(3x - 2y)$
41. $(4m - n)(2m - 3n)$
42. $(3s + 2t)(s - 4t)$
43. $(4x - 5y)(4x + y)$
44. $(8a + b)(2a - 5b)$
45. $(4p - q)(5p + 4q)$

Answer to Mini-Review
3. $-16a^2xy + 32ay^3$

About Using Calculators
Have students compare using a calculator to check these answers to checking them manually using fractions. They should find the calculator checks to be easier.

7-8

Lesson Notes, page T40
Enrichment, page T40

Discussion Question
Can a binomial be a perfect square? No

Chalkboard Exercises
Factor completely. Check by using FOIL or the distributive property.

1a. $4x^2 - 64$
$4(x - 4)(x + 4)$
1b. $6x^2 - 54$
$6(x + 3)(x - 3)$
2a. $2a^2 + 12a + 16$
$2(a + 2)(a + 4)$
2b. $12r^2 - 28r - 24$
$4(3r + 2)(r - 3)$
2c. $2a^2 + 4ab + 2b^2$
$2(a + b)^2$
2d. $6m^3 + 12m^2 + 6m$
$6m(m + 1)^2$

❶ The chart shows a practical set of guidelines that students can refer to when factoring.

❷ It is assumed that the directions "factor" always mean "factor completely".

❸ Emphasize checking factorization by FOIL or the distributive property.

Objective 7-8: Review and summarize the various methods of factoring.

7-8 Summary of Factoring

You have used many methods to factor polynomials. This table can help you decide which method to use.

Check for:	*Number of Terms* *Two*	*Three*	*Four or More*
greatest common factor	✓	✓	✓
difference of squares	✓		
perfect square trinomial		✓	
trinomial that has two binomial factors		✓	
pairs of terms that have a common monomial factor			✓

❶

If there is a GCF, factor it out first. Then, check the appropriate factoring methods in the order shown in the table. Using these methods, factor until each of the remaining factors is prime. When you follow this procedure, you are *factoring a polynomial completely*. ❷

Examples

1 **Factor: $3x^2 - 27$**

Since $3x^2 - 27$ has two terms, first check for the GCF. Then check for the difference of squares.

$3x^2 - 27 = 3(x^2 - 9)$ — *3 is the GCF.*
$= 3(x - 3)(x + 3)$ — *$x^2 - 9$ is the difference of squares.*

❸ *Check:* $3(x - 3)(x + 3) = 3(x^2 - 9)$
$= 3x^2 - 27$

The binomial $3x^2 - 27$ is completely factored as $3(x - 3)(x + 3)$.

2 **Factor: $6y^2 - 20y - 16$**

Since $6y^2 - 20y - 16$ has three terms, check for the GCF, a perfect square trinomial, and a trinomial that has two binomial factors.

$6y^2 - 20y - 16 = 2(3y^2 - 10y - 8)$ — *2 is the GCF.*

The trinomial $3y^2 - 10y - 8$ is not a perfect square. Why? Try factoring $3y^2 - 10y - 8$ using two binomial factors.

$6y^2 - 20y - 16 = 2(3y^2 - 10y - 8)$
$= 2(y - 4)(3y + 2)$

The trinomial $6y^2 - 20y - 16$ is completely factored as $2(y - 4)(3y + 2)$.

Example

3 **Factor: $2m^3 + 3m^2n - 8m - 12n$**

Since $2m^3 + 3m^2n - 8m - 12n$ has four terms, first check for the GCF. Then, check for pairs of terms that have a common monomial factor.

$$\begin{aligned} 2m^3 + 3m^2n - 8m - 12n &= (2m^3 + 3m^2n) + (-8m - 12n) \\ &= m^2(2m + 3n) + (-4)(2m + 3n) \\ &= (m^2 - 4)(2m + 3n) \\ &= (m - 2)(m + 2)(2m + 3n) \end{aligned}$$

$m^2 - 4$ is the difference of squares.

The polynomial $2m^3 + 3m^2n - 8m - 12n$ is completely factored as $(m - 2)(m + 2)(2m + 3n)$. *Check by multiplying the factors.*

Exploratory Exercises

Factor. Check by using FOIL or the distributive property.

1. $3x^2 + 15$ $3(x^2 + 5)$
2. $4y^2 - 24$ $4(y^2 - 6)$
3. $5ax + 6ay$ $a(5x + 6y)$
4. $8mn^2 - 13m^2n$ $mn(8n - 13m)$
5. $12ax^2 + 18ay^2$ $6a(2x^2 + 3y^2)$
6. $4c^2d + 16bc^2$ $4c^2(d + 4b)$
7. $5x^2 - 10x^2y$ $5x^2(1 - 2y)$
8. $4m^2 - 20m^2n$ $4m^2(1 - 5n)$
9. $3a^2b + 6ab + 9ab^2$
10. $2x^2y + 8x^2y^2 + 10x^2y^3$ $2x^2y(1 + 4y + 5y^2)$
11. $a^2 - 9b^2$ $(a + 3b)(a - 3b)$
12. $9x^2 - 16y^2$ $(3x + 4y)(3x - 4y)$

Written Exercises

Factor completely. Check by using FOIL or the distributive property.

1. $2a^2 - 72$ $2(a + 6)(a - 6)$
2. $3y^2 - 147$ $3(y + 7)(y - 7)$
3. $m^3 + 6m^2 + 9m$ $m(m + 3)^2$
4. $18y + 12y^2 + 2y^3$ $2y(3 + y)^2$
5. $4a^3 - 36a$ $4a(a + 3)(a - 3)$
6. $3x^3 - 27x$ $3x(x + 3)(x - 3)$
7. $m^4 - p^2$ $(m^2 + p)(m^2 - p)$
8. $b^4 - 16$ $(b^2 + 4)(b + 2)(b - 2)$
9. $2k^2 + 3k + 1$
10. $6r^2 + 13r + 6$
11. $6y^2 - 24x^2$ $6(y + 2x)(y - 2x)$
12. $2m^2 - 32n^2$
13. $3y^2 + 21y - 24$
14. $6x^2 + 27x - 15$
15. $20y^2 + 34y + 6$
16. $5a^2 + 7a + 3$ prime
17. $2b^2 + 6b + 2$ $2(b^2 + 3b + 1)$
18. $3b^2 - 36$ $3(b^2 - 12)$
19. $m^2 + 8mn + 16n^2$ $(m + 4n)^2$
20. $4a^2 + 12ab + 9b^2$ $(2a + 3b)^2$
21. $6p^2 + 9p - 105$
22. $12c^2 + 10c - 42$
23. $3ax^2 + 16ax + 21a$
24. $4y^3 - 12y^2 + 8y$
25. $9a^3 + 66a^2 - 48a$
26. $2m^2 - mb - 10b^2$
27. $3x^2 - 9xy - 30y^2$
28. $a^2b^3 - 25b$
29. $m^3n^2 - 49m$
30. $2a^2 - 4ab - 70b^2$
31. $3x^2 + 24xy - 99y^2$
32. $3y^4 - 48$
33. $4x^4 - 324$
34. $m^2p + m^2 - 36p - 36$
35. $9y^4 + 8y^2 - 1$
36. $27k^2m + 27km^2 - 12m^3$
37. $60r^3 - 54r^2s + 12rs^2$
38. $8y^4 + 14y^2 - 4$
39. $x^2y^2 - z^2 - y^2 + x^2z^2$

37. $6r(5r - 2s)(2r - s)$ 38. $2(2y + 1)(2y - 1)(y^2 + 2)$ 39. $(y^2 + z^2)(x + 1)(x - 1)$

Challenge

Factor completely by first factoring out a common decimal or fraction.

40. $0.3x^2 - 4.8$ $0.3(x + 4)(x - 4)$
41. $0.7y^2 - 6.3$ $0.7(y + 3)(y - 3)$
42. $\frac{1}{4}a^2 + a + 1$ $\frac{1}{4}(a + 2)^2$
43. $\frac{1}{3}b^2 + 2b + 3$ $\frac{1}{3}(b + 3)^2$
44. $y^2 + \frac{5}{12}y - \frac{1}{6}$
45. $\frac{1}{4}r^2 + \frac{3}{2}r + 2$
46. $0.4r^2 + 1.2r + 0.8$ $0.4(r + 2)(r + 1)$
47. $0.7y^2 + 3.5y + 4.2$ $0.7(y + 2)(y + 3)$
48. $m^2 - \frac{5}{12}m - \frac{1}{4}$

For more exercises on factoring, see exercises 17-64 on page 558.

Chalkboard Exercises

Factor completely.

3a. $3r^3 + 5r^2s - 15r - 25s$ $(r^2 - 5)(3r + 5s)$

3b. $6m^3 - 14m^2n - 24m + 56n$ $2(m + 2)(m - 2)(3m - 7n)$

Suggested Daily Assignments

Enriched
Written: 1–39 odds, 40–48

Average
Exp: 2–12 evens
Written: 1–40

Fundamental
Exp: 1–12
Written: 1–30
p. 235, 41–50

Answer to Exploratory

9. $3ab(a + 2 + 3b)$

Answers to Written

9. $(2k + 1)(k + 1)$
10. $(3r + 2)(2r + 3)$
12. $2(m + 4n)(m - 4n)$
13. $3(y + 8)(y - 1)$
14. $3(2x - 1)(x + 5)$
15. $2(5y + 1)(2y + 3)$
21. $3(2p - 7)(p + 5)$
22. $2(3c + 7)(2c - 3)$
23. $a(3x + 7)(x + 3)$
24. $4y(y - 2)(y - 1)$
25. $3a(3a - 2)(a + 8)$
26. $(2m - 5b)(m + 2b)$
27. $3(x - 5y)(x + 2y)$
28. $b(ab + 5)(ab - 5)$
29. $m(mn + 7)(mn - 7)$
30. $2(a - 7b)(a + 5b)$
31. $3(x + 11y)(x - 3y)$
32. $3(y^2 + 4)(y - 2)(y + 2)$
33. $4(x^2 + 9)(x + 3)(x - 3)$
34. $(m + 6)(m - 6)(p + 1)$
35. $(3y + 1)(3y - 1)(y^2 + 1)$
36. $3m(3k - m)(3k + 4m)$

Answers to Challenge

44. $\frac{1}{12}(4y - 1)(3y + 2)$
45. $\frac{1}{4}(r + 4)(r + 2)$
48. $\frac{1}{12}(4m - 3)(3m + 1)$

About Using Computers
It may be worthwhile to review conversion from base two to base ten and back before assigning this section. Students may find it helpful to label the binary numeral 11101 as follows:

2^4	2^3	2^2	2^1	2^0
1	1	1	0	1

Using Computers

Bits and Bytes

The decimal numeration system is a base 10 system. The ten digits 0, 1, 2, 3, 4, 5, 6, 7, 8, and 9 are used to name numbers.

Internally, a digital computer uses only two digits, 0 and 1. These two digits (bits) are used in a base two, or binary, numeration system.

Powers of Two

$2^0 = 1$	$2^4 = 16$
$2^1 = 2$	$2^5 = 32$
$2^2 = 4$	$2^6 = 64$
$2^3 = 8$	$2^7 = 128$

In this system the position of each digit gives its place value as a power of 2. If 0 and 1 are used, then 11101 in base two means 1 sixteen, 1 eight, 1 four, 0 twos, and 1 one.

binary numeral → $11101 = 1 \cdot 2^4 + 1 \cdot 2^3 + 1 \cdot 2^2 + 0 \cdot 2^1 + 1$

$= 16 + 8 + 4 + 0 + 1$

$= 29$ ← *decimal numeral*

To show which base is used, you can write $11101_{two} = 29_{ten}$.

The circuits of a computer store numbers in base 2. The diagram depicts how a computer would store 1010_{two}.

A color spot indicates a circuit that conducts electricity. It represents 1.

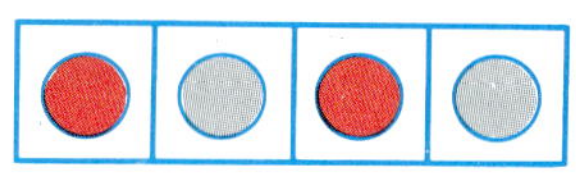

A grey spot indicates a circuit that does not conduct electricity. It represents 0.

Bits are the alphabet of machine language. The pattern formed by a group of bits stands for a number, and that number itself may stand for a character or even a whole word. Usually, eight bits make up one **byte**. Computer size is often stated in bytes of memory.

In mathematics and business 1K = 1000; in a computer, 1K = 1024. A small personal computer may have only 4K bytes of memory, but most have from 16K to 128K bytes. A time sharing computer may have 20 times as much.

Exercises

Change each binary numeral to a decimal numeral.

1. 1101_{two} 13 **2.** 10110_{two} 22 **3.** 11111_{two} 31 **4.** 10101_{two} 21

Draw a diagram that depicts how a computer would store each of the following.

5. 1100_{two} **6.** 1001_{two} **7.** 19_{ten} **8.** 33_{ten}

Answers to Exercises

5.
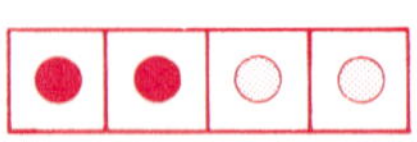

6.
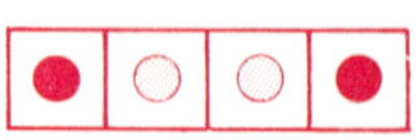

7.
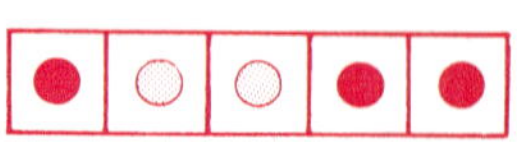

8.
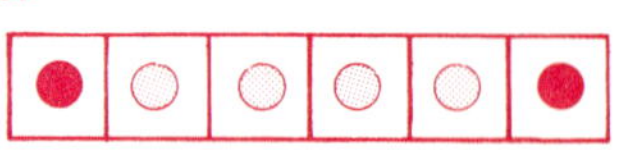

Vocabulary

factor (211)
prime number (211)
composite number (211)
prime factor (211)
prime factorization (211)
greatest common factor, GCF (212)
difference of squares (217)
perfect square trinomial (221)
prime polynomial (226)

Chapter Summary

1. A prime number is an integer, greater than 1, whose only whole number factors are 1 and itself. (211)
2. A composite number is any positive integer, except 1, that is not prime. (211)
3. Expressing a composite number as a product of prime factors is called prime factorization. (211)
4. The greatest common factor (GCF) of two or more integers is the greatest of the factors common to each integer. (212)
5. The distributive property may be used to express a polynomial in factored form. (214)
6. Difference of Squares:
   $a^2 - b^2 = (a - b)(a + b) = (a + b)(a - b)$ (217)
7. The result of squaring an integer is called a perfect square. (221)
8. The square of a binomial is a perfect square trinomial. (221)
9. Factoring Perfect Square Trinomials:
   $a^2 + 2ab + b^2 = (a + b)^2$
   $a^2 - 2ab + b^2 = (a - b)^2$ (221)
10. The pattern $(x + m)(x + n) = x^2 + (m + n)x + mn$ can be used to factor trinomials of the form $x^2 + bx + c$. (224)
11. When the value of c is negative, the factors of $x^2 + bx + c$ are a difference and a sum. (225)
12. A polynomial that cannot be written as a product of two polynomials is called a prime polynomial. (226)

Quiz
Lessons 7-6 to 7-8

Factor.

1. $a(a - 2b) + 1(a - 2b)$
   $(a + 1)(a - 2b)$
2. $3c^2 + 12cb - 2c - 8b$
   $(c + 4b)(3c - 2)$
3. $3b^2 - 2by - 3b + 2y$
   $(b - 1)(3b - 2y)$
4. $3x^2 + 13x + 12$
   $(3x + 4)(x + 3)$
5. $2a^2 - 7a - 15$
   $(2a + 3)(a - 5)$
6. $2m^2 - 8m + 3$
   not possible

Factor completely.

7. $x^4 - y^4$
   $(x - y)(x + y)(x^2 + y^2)$
8. $r^3 + 6r^2s + 8rs^2$
   $r(r + 4s)(r + 2s)$
9. $y^2 - 7y + 10$
   $(y - 2)(y - 5)$

13. To factor a trinomial of the form $ax^2 + bx + c$, first find two factors whose product is $a \cdot c$ and whose sum is b. Then, rename b as the sum of these factors. Continue to factor by grouping the terms in pairs and factoring the GCF from each group. (232).

14. Summary of Factoring (236)

Check for:	*Number of Terms* Two	Three	Four or More
greatest common factor	✓	✓	✓
difference of squares	✓		
perfect square trinomial		✓	
trinomial that has two binomial factors		✓	
pairs of terms that have a common monomial factor			✓

Chapter Review

7-1 Find the prime factorization of each of the following.

1. 42 $2 \cdot 3 \cdot 7$
2. 66 $2 \cdot 3 \cdot 11$
3. 80 $2 \cdot 2 \cdot 2 \cdot 2 \cdot 5$ or $2^4 \cdot$
4. $96y^3x^2$ 4. $2 \cdot 2 \cdot 2 \cdot 2 \cdot 2 \cdot 3 \cdot x \cdot x \cdot y \cdot y \cdot y$
5. $121(ab)^4$ 5. $11 \cdot 11 \cdot a \cdot a \cdot a \cdot a \cdot b \cdot b \cdot b \cdot b$
6. $16a^3b$ 6. $2 \cdot 2 \cdot 2 \cdot 2 \cdot a \cdot a \cdot a \cdot$

Find the GCF of each of the following.

7. 35, 50 5
8. 12, 18, 40 2
9. $16abc$, $30a^2b$ $2ab$

Factor each polynomial completely. If a polynomial cannot be factored using integers, write *prime*.

7-2

10. $13x + 26y$ $13(x + 2y)$
11. $6x^2y + 12xy + 6$ $6(x^2y + 2xy + 1)$
12. $24a^2b^2 - 18ab$ $6ab(4ab$
13. $24ab + 18ac + 32a^2$ $2a(12b + 9c + 16a)$
14. $\frac{3}{5}a - \frac{3}{5}b + \frac{6}{5}c$
15. $\frac{3}{4}x + \frac{1}{4}y$ $\frac{1}{4}(3x + y)$

7-3

16. $b^2 - 16$ $(b + 4)(b - 4)$
17. $25 - 9y^2$ $(5 + 3y)(5 - 3y)$
18. $16a^2 - 81b^4$
19. $2y^2 - 128$ $2(y + 8)(y - 8)$
20. $\frac{1}{4}y^2 - \frac{9}{16}z^2$
21. $x^4 - 1$ $(x^2 + 1)(x + 1)($

7-4

22. $a^2 + 18a + 81$ $(a + 9)^2$
23. $16x^2 - 8x + 1$ $(4x - 1)^2$
24. $9x^2 - 12x + 4$
25. $32x^2 - 80x + 50$ $2(4x - 5)^2$
26. $6a^2 - 24a + 6$
27. $y^2 - \frac{3}{2}y + \frac{9}{16}$ $\left(y - \frac{3}{4}\right)^2$

Answers to Chapter Review

14. $\frac{3}{5}(a - b + 2c)$
18. $(4a + 9b^2)(4a - 9b^2)$
20. $\frac{1}{4}(y + \frac{3}{2}z)(y - \frac{3}{2}z)$
24. $(3x - 2)^2$
26. $6(a^2 - 4a + 1)$

7-5 **28.** $y^2 + 7y + 12$ **29.** $b^2 - 8b + 15$ **30.** $y^2 - 9y - 36$

31. $b^2 + 5b - 6$ **32.** $r^2 + 23r + 132$ **33.** $a^2 - 3a - 28$

34. $x^2 - 13x + 30$ **35.** $s^2 + 10s - 39$ **36.** $x^2 + 11xy + 24y^2$

37. $m^2 - mn - 42n^2$ $(m - 7n)(m + 6n)$ **38.** $a^2 - 10ab + 9b^2$ $(a - b)(a - 9b)$ **39.** $r^2 - 8rs - 65s^2$ $(r - 13s)(r + 5s)$

7-6 **40.** $a^2 - 4ac + ab - 4bc$ $(a + b)(a - 4c)$ **41.** $4m^2 - 3mp + 3p - 4m$

42. $2rs + 6ps + rm + 3mp$ $(2s + m)(r + 3p)$ **43.** $bm + 7r + rm + 7b$ $(m + 7)(b + r)$

44. $10x^2 - 6xy + 15x - 9y$ $(5x - 3y)(2x + 3)$ **45.** $16k^2 - 4kp - 28kp + 7p^2$ $(4k - p)(4k - 7p)$

7-7 **46.** $3y^2 + 14y + 8$ **47.** $4m^2 + 11m + 6$ **48.** $4y^2 - 7y + 3$

49. $2r^2 - 3r - 20$ **50.** $3a^2 - 13a + 14$ **51.** $4m^2 - 8m + 3$

52. $6b^2 + 7b + 3$ prime **53.** $3x^2 + 11x - 20$ $(3x - 4)(x + 5)$ **54.** $12x^2 + 31x + 20$ $(3x + 4)(4x + 5)$

7-8 **55.** $3x^2 - 12$ **56.** $12mx + 3xb + 4my + by$ **57.** $12ay^2 + 31ay + 20a$

58. $2r^3 - 18r^2 + 28r$ $2r(r - 2)(r - 7)$ **59.** $mx^2 + bx^2 - 49m - 49b$ $(x + 7)(x - 7)(m + b)$ **60.** $4y^2 + 2y - 6$ $2(2y + 3)(y - 1)$

Chapter Test

Find the prime factorization of each of the following.

1. 95 $5 \cdot 19$ **2.** 300 $2 \cdot 2 \cdot 3 \cdot 5 \cdot 5$ **3.** $18a^2b$ $2 \cdot 3 \cdot 3 \cdot a \cdot a \cdot b$

Find the GCF of each of the following.

4. 36, 20 4 **5.** $18a^2b$, $28a^3b^2$ $2a^2b$ **6.** $6x^2y^3$, $12x^2y^2z$, $15x^2y$ $3x^2y$

Factor, if possible.

7. $25y^2 - 49w^2$ **8.** $w^2 - 16w + 64$ **9.** $x^2 + 14x + 24$

10. $36a^2 - 48a + 16$ **11.** $y^2 - 8y + 7$ **12.** $a^2 - 11ab + 18b^2$

13. $28m^2 + 18m$ **14.** $m^2 + 7m + 10$ **15.** $5x^2 + 20y^2$

16. $h^2 - 3h - 10$ **17.** $8m^2 - 72n^2$ **18.** $r^2 + 3r - 18$

19. $x^2 + 18x + 81$ **20.** $a^2 - 12a - 13$ **21.** $r^2 - rs - 72s^2$

22. $36m^2 - 24mn + 4n^2$ **23.** $3x^2 + 19x + 28$ **24.** $5x^2 - 19x + 14$

25. $36m^2 + 60mn + 25n^2$ **26.** $192a^2 - 75b^2$ **27.** $6p^2 + 7p - 3$

28. $16a^2b^2 - c^2d^2$ **29.** $2m^2 + 11m + 12$ **30.** $36a^2b^3 - 45ab^4$

31. $4m^2 - 196$ $4(m + 7)(m - 7)$ **32.** $6x^3 + 15x^2 - 9x$ $3x(2x - 1)(x + 3)$

33. $3a^2b - 8a - 12ab - 32$ prime **34.** $4my - 20m + 15p - 3py$

35. $15a + 6b + 10a^2 + 4ab$ $(5a + 2b)(3 + 2a)$ **36.** $y^2m + y^2n - 4m - 4n$ $(y + 2)(y - 2)(m + n)$

Answers to Chapter Review

28. $(y + 3)(y + 4)$
29. $(b - 3)(b - 5)$
30. $(y - 12)(y + 3)$
31. $(b - 1)(b + 6)$
32. $(r + 11)(r + 12)$
33. $(a - 7)(a + 4)$
34. $(x - 3)(x - 10)$
35. $(s + 13)(s - 3)$
36. $(x + 3y)(x + 8y)$
41. $(m - 1)(4m - 3p)$
46. $(3y + 2)(y + 4)$
47. $(4m + 3)(m + 2)$
48. $(4y - 3)(y - 1)$
49. $(2r + 5)(r - 4)$
50. $(3a - 7)(a - 2)$
51. $(2m - 1)(2m - 3)$
55. $3(x + 2)(x - 2)$
56. $(3x + y)(4m + b)$
57. $a(3y + 4)(4y + 5)$

Answers to Chapter Test

7. $(5y + 7w)(5y - 7w)$
8. $(w - 8)^2$
9. $(x + 2)(x + 12)$
10. $4(3a - 2)^2$
11. $(y - 1)(y - 7)$
12. $(a - 2b)(a - 9b)$
13. $2m(14m + 9)$
14. $(m + 2)(m + 5)$
15. $5(x^2 + 4y^2)$
16. $(h - 5)(h + 2)$
17. $8(m + 3n)(m - 3n)$
18. $(r + 6)(r - 3)$
19. $(x + 9)^2$
20. $(a - 13)(a + 1)$
21. $(r - 9s)(r + 8s)$
22. $4(3m - n)^2$
23. $(3x + 7)(x + 4)$
24. $(5x - 14)(x - 1)$
25. $(6m + 5n)^2$
26. $3(8a + 5b)(8a - 5b)$
27. $(2p + 3)(3p - 1)$
28. $(4ab + cd)(4ab - cd)$
29. $(2m + 3)(m + 4)$
30. $9ab^3(4a - 5b)$
34. $(4m - 3p)(y - 5)$

About the Application
Other *Applications in Probability* features can be found on pages 108 and 172. Applications in Statistics features can be found on pages 302 and 368.

Applications in Probability

Dependent Events

Cecile, Tom, Mike, Allison, Carl, and Eric would all like to help plan the senior prom. Two of them will be chosen to help. Mr. O'Donnell decides to choose the two students by writing their names on cards, putting the cards in a hat, and choosing two cards without replacing the first card. What is the probability that the cards he draws will contain the names Allison and Eric in that order?

Since there were six students originally, the probability of selecting Allison is $\frac{1}{6}$. After the first student is selected, only five remain. So, the probability of selecting Eric on the second draw is $\frac{1}{5}$. Thus, the probability of selecting Allison and then Eric is $\frac{1}{6} \cdot \frac{1}{5}$ or $\frac{1}{30}$.

Notice that selecting Eric on the second draw is more likely than selecting Allison on the first draw because the sample space was decreased from 6 to 5 elements. The outcome of the second event is affected by the outcome of the first event. Such events are called **dependent events**.

Probability of Two Dependent Events

The probability of two dependent events, *A* and *B*, occurring in that order is the product of the probability of *A* occurring and the probability of *B* occurring given that *A* has occurred.

$$P(A \text{ and } B) = P(A) \cdot P(B \mid A)$$

$P(B \mid A)$ is read "the probability of B given that A has occurred."

Example 1: **A bag contains 5 blue marbles and 7 red marbles. Two marbles are to be selected, without replacing the first one. What is the probability of selecting a blue marble and then a red marble?**

Since replacement does not occur, the probability of selecting the red marble is altered by selecting the blue marble. The event is dependent.

$$P(\text{blue and red}) = P(\text{blue}) \cdot P(\text{red} \mid \text{blue})$$

$$= \frac{5}{12} \cdot \frac{7}{11} \quad P(A \text{ and } B) = P(A) \cdot P(B \mid A)$$

$$= \frac{35}{132}$$

Example 2: **A bag contains 3 blue, 5 red, and 8 green marbles. If 3 marbles are selected in succession without replacement, what is the probability that they are all blue?**

$$P(\text{3 blue}) = P(\text{blue}) \cdot P(\text{blue} \mid \text{blue}) \cdot P(\text{blue} \mid \text{blue, blue})$$

$$= \frac{3}{16} \cdot \frac{2}{15} \cdot \frac{1}{14}$$

Each time there is one less blue marble.

$$= \frac{6}{3360} \text{ or } \frac{1}{560}$$

Exercises

State whether the following events are dependent or independent.

1. choosing two names from a list of eight without replacement dependent
2. tossing two 6's on a die on two consecutive rolls independent
3. drawing two kings from a deck of cards with replacement between draws independent
4. tossing a 5 then a 4 on a die independent
5. tossing a 1 on a die then drawing a queen from a deck of cards independent
6. drawing two hearts from a deck of cards without replacement between draws dependent

A bag contains 5 red, 8 blue, and 3 green marbles. What is the probability of selecting each of the following in the indicated order?

7. two red with no replacement $\frac{1}{12}$
8. two red with replacement $\frac{25}{256}$
9. blue, then green, with replacement $\frac{3}{32}$
10. blue, then green, with no replacement $\frac{1}{10}$
11. red, then blue, then green, with no replacement $\frac{1}{28}$
12. red, then blue, then green, with replacement $\frac{15}{512}$
13. three green with replacement $\frac{27}{4096}$
14. three green with no replacement $\frac{1}{560}$
15. four blue with replacement $\frac{1}{16}$
16. four green with replacement $\frac{81}{65,536}$

Cathy Bentz wishes to select two other students to help decorate for a party. There are 10 girls and 15 boys from which to choose. What is the probability that the following occurs?

17. Mary and Bill are selected in that order $\frac{1}{600}$
18. a girl and a boy are selected in that order $\frac{1}{4}$
19. two girls are selected $\frac{3}{20}$
20. two boys are selected $\frac{7}{20}$

About the Photo

Factoring can be used to solve problems such as the one presented below the photograph. On page 260, students are introduced to the formula $h = vt - 16t^2$. The formula applies to projectiles that are launched vertically from the earth, and that do not continue to use fuel after blast-off.

When the appropriate values of the problem presented below the photograph are substituted into the formula $h = vt - 16t^2$, the resulting equation is $32{,}000 = 1920t - 16t^2$. Solving the equation results in two reasonable answers, 20 seconds after blast-off and 100 seconds after blast-off. This means that the rocket will have ascended to a height of 32,000 feet after 20 seconds and descended to a height of 32,000 feet after 100 seconds.

The formula has been derived from the more general equation $s = v_0t + \left(\frac{-at^2}{2}\right)$, where s represents the distance from the initial position, v_0 represents initial velocity, a represents acceleration of gravity and t represents time spent traveling. Acceleration due to gravity is 32 feet/sec^2.

Teacher Resource Book Materials for Chapter 8 See pages 201-216.

CHAPTER 8

Applications of Factoring

A rocket travels at an average speed of 1920 feet per second. How many seconds after it is fired will it be 32,000 feet high? Problems such as this can be solved by using equations. In this chapter, you will learn how factoring can be used to solve some types of equations.

Objective 8-1: Use the zero product property to solve equations in factored form.

8-1 Solving Equations Already in Factored Form

Consider the following products.

$3(0) = 0 \qquad -8(0) = 0 \qquad \left[\frac{1}{2} + \left(-\frac{1}{2}\right)\right](0) = 0 \qquad (x + 2)(0) = 0$

The above products all equal zero. Notice that in each case *at least one* of the factors is zero.

For all numbers a and b, if $ab = 0$ then $a = 0$ or $b = 0$. ***Zero Product Property***

You can use this property to solve equations already written in factored form.

Examples

1 Solve and check: $2x(x - 5) = 0$

If $2x(x - 5) = 0$, then $2x = 0$ or $x - 5 = 0$ *Zero Product Property*

$2x = 0$ or $x - 5 = 0$ *Solve both equations.*

$x = 0$ or $x = 5$

Check: $2x(x - 5) = 0$

$2(0)(0 - 5) \stackrel{?}{=} 0$ or $2(5)(5 - 5) \stackrel{?}{=} 0$

$0(-5) \stackrel{?}{=} 0 \qquad 10(0) \stackrel{?}{=} 0$

$0 = 0 \qquad 0 = 0$

The solutions of $2x(x - 5) = 0$ are 0 and 5. The solution set is $\{0, 5\}$.

2 Solve and check: $(y + 2)(3y + 5) = 0$

$(y + 2)(3y + 5) = 0$

$y + 2 = 0$ or $3y + 5 = 0$ *Zero Product Property*

$y = -2 \qquad 3y = -5$

$y = -\frac{5}{3}$

Check: $(y + 2)(3y + 5) = 0$

$(-2 + 2)[3(-2) + 5] \stackrel{?}{=} 0$ or $\left(-\frac{5}{3} + 2\right)\left[3\left(-\frac{5}{3}\right) + 5\right] \stackrel{?}{=} 0$

$0[3(-2) + 5] \stackrel{?}{=} 0 \qquad \left(-\frac{5}{3} + 2\right)(-5 + 5) \stackrel{?}{=} 0$

$0 = 0 \qquad \left(-\frac{5}{3} + 2\right)(0) \stackrel{?}{=} 0$

$0 = 0$

The solution set is $\left\{-2, -\frac{5}{3}\right\}$.

8-1

Lesson Notes, page T40

Readiness Activity

True or false?

$10 \cdot 0 = 0$ true

$2 \cdot 6 = 0$ false

$0 \cdot -5 = 0$ true

$-3 \cdot 7 = 0$ false

What must one of the factors be for a product to equal zero? 0

Discussion Question

If $(a + b)(a - b) = 0$, what can be concluded about a and b?

$a = b$ or $a = -b$

Chalkboard Exercises

Solve and check.

1. $3x(x - 12) = 0$ $\{0, 12\}$

2a. $(y + 2)(y - 5) = 0$ $\{-2, 5\}$

2b. $(y + 2)(3y - 5) = 0$ $\left\{-2, 1\frac{2}{3}\right\}$

2c. $(2y + 1)(3y - 4) = 0$ $\left\{-\frac{1}{2}, 1\frac{1}{3}\right\}$

2d. $(2y - 7)(y + 6) = 0$ $\left\{-6, 3\frac{1}{2}\right\}$

Chalkboard Exercises

Solve each problem. Disregard unreasonable solutions.

3a. The product of 3 times my age and 74 less than 2 times my age is 0. How old am I? 37

3b. The product of 18 more than twice my age and 51 less than 3 times my age is zero. How old am I? 17

Suggested Daily Assignments

Enriched
Written: 1–30

Average
Exp: 2–20 evens
Written: 1–26

Fundamental
Exp: 1–20
Written: 1–20

Answers to Exploratory

1. $x = 0$ or $x + 3 = 0$
2. $y = 0$ or $y - 12 = 0$
3. $3r = 0$ or $r - 4 = 0$
4. $7a = 0$ or $a + 6 = 0$
5. $3t = 0$ or $4t - 32 = 0$
6. $2x = 0$ or $5x - 10 = 0$
7. $x - 6 = 0$ or $x + 4 = 0$
8. $b - 3 = 0$ or $b - 5 = 0$
9. $a + 3 = 0$ or $3a - 12 = 0$
10. $2y + 22 = 0$ or $y - 1 = 0$
11. $2y + 8 = 0$ or $3y + 24 = 0$
12. $4x + 4 = 0$ or $2x + 6 = 0$
13. $x - 3 = 0$
14. $3x - 9 = 0$ or $5x - 15 = 0$
15. $3x + 2 = 0$ or $x - 7 = 0$
16. $x - 8 = 0$ or $2x + 7 = 0$
17. $4x - 7 = 0$ or $3x + 5 = 0$
18. $3x - 5 = 0$ or $4x - 7 = 0$
19. $2x + 3 = 0$ or $x + 7 = 0$
20. $4x - \frac{1}{3} = 0$ or $3x + \frac{1}{8} = 0$

Answers to Written

1. $\{0, -3\}$
2. $\{0, 12\}$
3. $\{0, 4\}$
4. $\{0, -6\}$
5. $\{0, 8\}$
6. $\{0, 2\}$
7. $\{6, -4\}$
8. $\{3, 5\}$
9. $\{-3, 4\}$
10. $\{-11, 1\}$
11. $\{-4, -8\}$
12. $\{-1, -3\}$
13. $\{3\}$
14. $\{3\}$
15. $\left\{-\frac{2}{3}, 7\right\}$
16. $\left\{8, -\frac{7}{2}\right\}$
17. $\left\{\frac{7}{4}, -\frac{5}{3}\right\}$
18. $\left\{\frac{5}{3}, \frac{7}{4}\right\}$
19. $\left\{-\frac{3}{2}, -7\right\}$
20. $\left\{\frac{1}{12}, -\frac{1}{24}\right\}$

Example

3 **Maria gave this puzzle to her friends. "The product of 4 times my age and 45 less than 3 times my age is 0. How old am I?" Find Maria's age now.**

Explore This problem, can be solved by using an equation. Let y = Maria's age now.

Plan $4y(3y - 45) = 0$

Solve $4y = 0$ or $3y - 45 = 0$
$y = 0$ $\quad 3y = 45$
$\quad y = 15$

Examine Although 0 is a solution of the equation $4y(3y - 45) = 0$, Maria cannot be 0 years old. Therefore, disregard 0 as a solution. Maria is 15 years old.

Exploratory Exercises

State the conditions under which each equation will be true.
Sample: $x(x + 7) = 0$ will be true if $x = 0$ or $x + 7 = 0$.

1. $x(x + 3) = 0$
2. $y(y - 12) = 0$
3. $3r(r - 4) = 0$
4. $7a(a + 6) = 0$
5. $3t(4t - 32) = 0$
6. $2x(5x - 10) = 0$
7. $(x - 6)(x + 4) = 0$
8. $(b - 3)(b - 5) = 0$
9. $(a + 3)(3a - 12) = 0$
10. $(2y + 22)(y - 1) = 0$
11. $(2y + 8)(3y + 24) = 0$
12. $(4x + 4)(2x + 6) = 0$
13. $(x - 3)(x - 3) = 0$
14. $(3x - 9)(5x - 15) = 0$
15. $(3x + 2)(x - 7) = 0$
16. $(x - 8)(2x + 7) = 0$
17. $(4x - 7)(3x + 5) = 0$
18. $(3x - 5)(4x - 7) = 0$
19. $(2x + 3)(x + 7) = 0$
20. $\left(4x - \frac{1}{3}\right)\left(3x + \frac{1}{8}\right) = 0$

Written Exercises

1-20. Solve and check each equation in Exploratory Exercises 1-20.

Solve each problem. Disregard unreasonable solutions.

21. Sue told Jennifer, "The product of twice my age decreased by 32 and 5 times my age is zero. How old am I?" Find Sue's age. 16 yr
22. The product of a certain positive number decreased by 5 and the same number increased by 7 is 0. What is the number? 5
23. The product of a certain negative number increased by 2 and the same number decreased by $\frac{3}{4}$ is 0. What is the number? -2
24. The result is 0 when 7 is multiplied by the sum of 6 and a certain number. What is the number? -6

Challenge

Write an equation in factored form that would have each given solution set.

25. $\{-3, 5\}$ $(x + 3)(x - 5) = 0$
26. $\{0, -6\}$ $x(x + 6) = 0$
27. $\{-9, -11\}$
28. $\left\{2, -\frac{2}{3}\right\}$ $(x - 2)\left(x + \frac{2}{3}\right) = 0$
29. $\left\{-\frac{3}{4}, -\frac{5}{6}\right\}$ $\left(x + \frac{3}{4}\right)\left(x + \frac{5}{6}\right) = 0$
30. $\left\{0, -7, -\frac{1}{3}\right\}$

27. $(x + 9)(x + 11) = 0$
30. $x(x + 7)\left(x + \frac{1}{3}\right) = 0$

Problem Solving — Using Tables

A useful problem-solving strategy is to organize relevant information by using a table. Study the following problem.

To mail a small package costs a certain amount for the first ounce and 17¢ for each additional ounce. A 5-ounce package costs 90¢ to mail. How much would a 10-ounce package cost?

You know that a 5-ounce package costs 90¢ to mail. Write this specific case in your table. You also know that it costs 17¢ more for each additional ounce. Use this information to generate more cases.

Number of Ounces	1	2	3	4	5	6	7	8	9	10
Cost					0.90	1.07	1.24	1.41	1.58	1.75

Therefore, a 10-ounce package would cost \$1.75 to mail.

Using the pattern in the table, you could find the cost of mailing a 1-ounce package.

A 4-ounce package costs 0.90 − 0.17 or 0.73 to mail.

Number of Ounces	1	2	3	4	5	6	7	8	9	10
Cost	0.22	0.39	0.56	0.73	0.90	1.07	1.24	1.41	1.58	1.75

A 1-ounce package costs 22¢ to mail. The following expression represents the cost of mailing an n-ounce package.

cost of first ounce + (number of ounces − first ounce) 0.17

$$0.22 + (n - 1)\,0.17$$

Exercises

Use a table to solve each problem.

1. Bob Hartschorn was charged \$2.12 for a 20-minute call to Chicago. If Bob was charged \$2.32 for a 22-minute call, how much should he be charged for an hour call? \$6.12
2. Find the sum of the first three consecutive odd integers. What is the sum of the first five consecutive odd integers? What is the sum of the first fifty consecutive odd integers? 9; 25; 2500
3. Find pairs of positive integers whose sum is 45 and whose difference is less than 9.
4. Find the positive integers less than 50 that have an odd number of factors.
5. To determine a grade point average, four points are given for an A, three for a B, two for a C, one for a D, and zero for an F. If a student has a total of 13 points for 5 classes, what combinations of grades could he have?
6. Joan tore a sheet of paper in half. Then she tore each of the resulting pieces in half. If she continued this process 15 more times, how many pieces of paper would she have at the end? 2^{17}

About Problem Solving
This feature presents yet another strategy that can be used to solve verbal problems. Remind students that they have already used charts to help them solve mixture and uniform-motion problems. Provide help to students who may need to learn how to set up a table and complete it.

Answers to Exercises

3. 19 and 26, 20 and 25, 21 and 24, 22 and 23
4. 1, 4, 9, 16, 25, 36, 49
5. 3A's, 1D, 1F; 2A's, 1B, 1C, 1F; 2A's, 2C's, 1D; 2A's, 1B, 2D's; 1A, 2B's, 1C, 1D; 1A, 3B's, 1F; 1A, 1B, 3C's; 4B's, 1D; 3B's, 2C's;

8-2

Lesson Notes, page T40

Discussion Question

Factor $4x^3 - 6x^2 - 4x + 6$ and state the factoring method used in each step.

a. $2(2x^3 - 3x^2 - 2x + 3)$
$= 2[(2x^3 - 3x^2) + (-2x + 3)]$

b. $= 2[x^2(2x - 3) - 1(2x - 3)]$
$= 2[(x^2 - 1)(2x - 3)]$

c. $= 2[(x - 1)(x + 1)(2x - 3)]$

Method used

a. GCF

b. Grouping

c. Difference of Squares

Chalkboard Exercises

Factor each of the following polynomials.

1. $7x^3 - 28x$
$7x(x + 2)(x - 2)$

2a. $3x^3 + 12x^2 + 12x$
$3x(x + 2)^2$

2b. $4b^2x^2 - 40b^2x + 100b^2$
$4b^2(x - 5)^2$

3a. $2a^3 + 5a^2b - 3ab^2$
$a(2a - b)(a + 3b)$

3b. $8x^4 + 8x^3y - 6x^2y^2$
$2x^2(2x - y)(2x + 3y)$

4. $12ax^2 + 8bx^2 - 3ay^2 - 2by^2$
$(2x - y)(2x + y)(3a + 2b)$

❶ Explain that the terms $-5bx^2$ and $45by^2$ were grouped together and then a -1 was factored out. You may want to add the following step to help clarify this. $(4ax^2 - 36ay^2) + (-5bx^2 + 45by^2) = (4ax^2 - 36ay^2) - (5bx^2 - 45by^2)$

Suggested Daily Assignments

Enriched
Written: 2–46 evens, 47–54
Mini-Review: 1–5

Average
Exp: 1–12
Written: 1–46
Mini-Review: 1–5

Fundamental
Exp: 1–12
Written: 1–45 odds
p. 246, 21–24
Mini-Review: 1–5

Objective 8-2: Review the various methods of factoring.

8-2 Factoring Review

Equations can be easily solved if the equation is factorable. You have used several methods to factor polynomials. A summary of these is given in the following chart.

Check for:	Number of Terms: Two	Three	Four or More
greatest common factor	✓	✓	✓
difference of squares	✓		
perfect square trinomial		✓	
trinomial that has two binomial factors		✓	
pairs of terms that have a common monomial factor			✓

Examples

1 Factor: $2x^3 - 50x$

$2x^3 - 50x = 2x(x^2 - 25)$ — *2x is the GCF.*

❶ $= 2x(x - 5)(x + 5)$ — *$x^2 - 25$ is the difference of two squares.*

2 Factor: $5a^2x^2 - 30ax^2 + 45x^2$

$5a^2x^2 - 30ax^2 + 45x^2 = 5x^2(a^2 - 6a + 9)$ — *$5x^2$ is the GCF.*

$= 5x^2(a - 3)(a - 3)$ — *$a^2 - 6a + 9$ is a perfect square trinomial.*

$= 5x^2(a - 3)^2$

3 Factor: $6m^3 - 11m^2n - 10mn^2$

$6m^3 - 11m^2n - 10mn^2 = m(6m^2 - 11mn - 10n^2)$ — *m is the GCF.*

$= m(2m - 5n)(3m + 2n)$

4 Factor: $4ax^2 - 36ay^2 - 5bx^2 + 45by^2$

$$4ax^2 - 36ay^2 - 5bx^2 + 45by^2 = (4ax^2 - 36ay^2) - (5bx^2 - 45by^2)$$
$$= 4a(x^2 - 9y^2) - 5b(x^2 - 9y^2)$$
$$= (x^2 - 9y^2)(4a - 5b)$$
$$= (x - 3y)(x + 3y)(4a - 5b)$$

Exploratory Exercises

Indicate which method of factoring you would apply first to each polynomial. Use the table from the previous page.

1. $x^3 - 5x^2$ **2.** $64c^2 - 25p^2$ **3.** $9x^2 - 12xy + 4y^2$

4. $16m^4 - 64n^2$ **5.** $35z^2 + 13z - 12$ **6.** $x^2 + y^2$

7. $8x^2 - 72x + 162$ **8.** $3x^2 - xy - 6x + 2y$ **9.** $1 - 49k^2$

10. $4x^4 - 3x^3y - 12x^2 + 9xy$ **11.** $x^2 - 5x + 6$ **12.** $x^2 - (y + z)^2$

Written Exercises

1-12. Factor each polynomial in Exploratory Exercises 1-12.

Factor each of the following polynomials.

13. $6x^2 + 10xy$ $2x(3x + 5y)$ **14.** $35ab - 15b^3$ $5b(7a - 3b^2)$ **15.** $4a^2 - 9b^2$ $(2a - 3b)(2a + 3b)$

16. $25v^4 - 36w^4$ **17.** $x^2 + 12x + 36$ $(x + 6)^2$ **18.** $a^2 - 8a + 16$ $(a - 4)^2$

19. $12y^2 + 19y - 21$ **20.** $20m^2 + 17mn - 24n^2$ **21.** $x^3 + 5x^2 - 9x - 45$

22. $4x^3 + 12x^2 - x - 3$ **23.** $m^4 - m^2n^2$ **24.** $x^6 - x^4y^4$

25. $20a^2c^2 + 60a^2c + 45a^2$ $5a^2(2c + 3)^2$ **26.** $48a^2m^2 - 24a^2m + 3a^2$ $3a^2(4m - 1)^2$

27. $15x^3y - 6x^2y^3 + 3x^2yz$ $3x^2y(5x - 2y^2 + z)$ **28.** $5ab^2c + 5ab^4 - 15a^3b^2$ $5ab^2(c + b^2 - 3a^2)$

29. $8x^3 - 32xy^4$ $8x(x - 2y^2)(x + 2y^2)$ **30.** $48a^5y - 147ay^7$ $3ay(4a^2 - 7y^3)(4a^2 + 7y^3)$

31. $x^3 + 2x^2 + 8x + 16$ $(x^2 + 8)(x + 2)$ **32.** $4a^3 + 3a^2 + 8a + 6$ $(a^2 + 2)(4a + 3)$

33. $12x^2 - 12xy - 72y^2$ $12(x - 3y)(x + 2y)$ **34.** $10ab^2 + 20abc - 80ac^2$

35. $(x + y)^2 - (a - b)^2$ **36.** $(3a + 2b)^2 - 9(x + 3y)^2$

37. $6x^2 + 2xy - 4y^2$ $2(3x - 2y)(x + y)$ **38.** $10x^2 + 29xy + 10y^2$ $(5x + 2y)(2x + 5y)$

39. $\frac{m^2}{4} - \frac{9}{25}$ $\left(\frac{m}{2} - \frac{3}{5}\right)\left(\frac{m}{2} + \frac{3}{5}\right)$ **40.** $\frac{x^4}{9} - 49y^2$ $\left(\frac{x^2}{3} - 7y\right)\left(\frac{x^2}{3} + 7y\right)$

41. $6ax - 21ay + 10bx - 35by$ **42.** $20ax - 4ay - 15bx + 3by$

43. $2x^2 - 7x - 8$ prime **44.** $2x^2 - 7x - 4$ $(2x + 1)(x - 4)$

45. $x^2 + 6x + 9 - y^2$ $(x + 3 - y)(x + 3 + y)$ **46.** $a^4 - b^4 - 2b^2c^2 - c^4$ $(a^2 - b^2 - c^2)(a^2 + b^2 + c^2)$

Challenge

47. $0.01x^2 - 0.06xy + 0.09y^2$ $0.01(x - 3y)^2$

48. $0.15x^4 - 0.019x^2y + 0.0006y^2$

49. $0.12m^3n^2 - 0.001m^2n$ $0.001m^2n(120mn - 1)$

50. $\frac{3}{5}a^2b - \frac{4}{5}a^3b^2$ $\frac{1}{5}a^2b(3 - 4ab)$

51. $\frac{1}{4}x^2 - \frac{3}{5}xy + \frac{9}{25}y^2$ $\left(\frac{1}{2}x - \frac{3}{5}y\right)^2$

52. $\frac{m^2}{3} + \frac{2}{3}m - \frac{35}{3}$ $\frac{1}{3}(m + 7)(m - 5)$

53. $x^4 - 2x^2y + 3x^3 - 6xy$ $x(x + 3)(x^2 - 2y)$

54. $4p^2m - 5p^2n - 8pm + 10pn$ $p(p - 2)(4m - 5n)$

mini-review

Solve.

1. $4x + 9 < 6x$ $x > \frac{9}{2}$

2. $-\frac{1}{6}x \leq \frac{5}{3}$ $x \geq -10$

3. $2r + 3.5 > 7.1$ $r > 1.8$

4. $|m - 1| \leq 3$ $-2 \leq m \leq 4$

Graph the solution set.

5. $x \geq -3$ and $x < 4$

Answers to Exploratory

1. greatest common factor
2. difference of squares
3. perfect square trinomial
4. greatest common factor
5. trinomial that has two binomial factors
6. prime
7. greatest common factor
8. pairs of terms that have a common monomial factor
9. difference of squares
10. greatest common factor
11. trinomial that has two binomial factors
12. difference of squares

Answers to Written

1. $x^2(x - 5)$
2. $(8c - 5p)(8c + 5p)$
3. $(3x - 2y)^2$
4. $16(m^2 - 2n)(m^2 + 2n)$
5. $(5z + 4)(7z - 3)$
6. prime
7. $2(2x - 9)^2$
8. $(x - 2)(3x - y)$
9. $(1 - 7k)(1 + 7k)$
10. $x(x^2 - 3)(4x - 3y)$
11. $(x - 3)(x - 2)$
12. $(x - y - z)(x + y + z)$

16\. $(5v^2 - 6w^2)(5v^2 + 6w^2)$
19\. $(4y - 3)(3y + 7)$
20\. $(5m + 8n)(4m - 3n)$
21\. $(x - 3)(x + 3)(x + 5)$
22\. $(2x - 1)(2x + 1)(x + 3)$
23\. $m^2(m - n)(m + n)$
24\. $x^4(x - y^2)(x + y^2)$
34\. $10a(b + 4c)(b - 2c)$
35\. $(x + y - a + b) \cdot (x + y + a - b)$
36\. $(3a + 2b - 3x - 9y) \cdot (3a + 2b + 3x + 9y)$
41\. $(3a + 5b)(2x - 7y)$
42\. $(4a - 3b)(5x - y)$

Answer to Challenge

48\. $0.0002(15x^2 - y) \cdot (50x^2 - 3y)$

Answer to Mini-Review

5\.

About Applications
You may wish to reinforce the derivation of the formula by repeating the process with numbers from the original example, as shown below. First year:

$500 + 500(0.06)$
$= 500(1 + 0.06)$
$= 500(1.06)$
$= 530$

Second year:
$530 + 530(0.06)$
$= 500(1.06) + 500(1.06)(0.06)$
$= 500(1.06)(1 + 0.06)$
$= 500(1.06)(1.06)$
$= 500(1.06)^2$
$= 561.80$

Third year:
$561.80 + 561.80(0.06)$
$= 500(1.06)^2 + 500(1.06)^2(0.06)$
$= 500(1.06)^2(1 + 0.06)$
$= 500(1.06)^2(1.06)$
$= 500(1.06)^3$
$= 595.51$

Applications in Banking Factoring and Compound Interest

Many formulas are used daily in business and science. One such formula is used in finding **compound interest**. Compound interest is interest that is added to an original investment at specified time intervals. Thus, the interest becomes part of a new principal.

Suppose you deposited \$500 in a savings account for 3 years at a rate of 6% annual interest.

	amount deposited +	*interest*	*= new principal*	
End of first year	\$500 +	\$500(0.06)	= \$530	
End of second year	\$530 +	\$530(0.06)	= \$561.80	
End of third year	\$561.80 +	\$561.80(0.06)	= \$595.51	*This result is approximate.*

To write a general formula, let P represent the amount deposited or principal and let r represent the annual interest rate. Then, $P \cdot r$ represents the amount of interest earned.

End of first year	$P + P \cdot r$	$= P(1 + r)^1$	*Factor out P.*
End of second year	$P(1 + r) + P(1 + r) \cdot r$	$= P(1 + r)(1 + r)$	*Factor out P(1 + r).*
		$= P(1 + r)^2$	
End of third year	$P(1 + r)^2 + P(1 + r)^2 \cdot r$	$= P(1 + r)^2(1 + r)$	*Factor out $P(1 + r)^2$.*
		$= P(1 + r)^3$	
		$\vdots$	
End of t years		$P(1 + r)^t$	

Therefore, at the end of t years the total amount of savings including interest could be represented by $P(1 + r)^t$.

Suppose that interest is compounded biannually for 1 year at an annual interest rate of 6%. The interest would be computed twice per year at 3% each time. Thus the total amount in savings could be represented by:

$$P\left(1 + \frac{0.06}{2}\right)^{2 \cdot 1}$$

exponent $2 \cdot 1$: interest computed twice per year for 1 year
$\frac{0.06}{2}$: interest computed each time at $\frac{0.06}{2}$ or 0.03

Exercises **Use a calculator to solve each problem.**

1. Gloria invested \$750 for 1 year at 8% annual interest. If interest is compounded biannually, how much will Gloria have at the end of the year? \$811.20
2. Ann invested \$600 in a savings account for 2 years. If the annual interest rate is 12% compounded biannually, what is the total amount Ann will have at the end of 2 years? \$757.49
3. For interest compounded quarterly for 2 years at 12% annual interest the expression is $P\left(1 + \frac{0.12}{4}\right)^{4 \cdot 2}$. If Jim invested \$600 in such an account, how much will he have at the end of 2 years? \$760.06
4. In Exercises 2 and 3, Ann's money was compounded biannually and Jim's was compounded quarterly. Who had more money at the end of 2 years? How much more? Jim, \$2.57

Objective 8-3: Solve simple equations by factoring and using the zero-product property. (These equations contain two or three nonzero terms, and can be easily factored.)

8-3 Solving Some Equations by Factoring

If a polynomial equation can be written in the form $ab = 0$, then the zero product property can be applied to solve the equation.

Examples

1 Solve and check: $x^2 - 9 = 0$

$x^2 - 9 = 0$ *$x^2 - 9$ is the difference of two squares.*

$(x - 3)(x + 3) = 0$

$x - 3 = 0$ or $x + 3 = 0$ *Use the zero product property.*

$x = 3$ $\quad x = -3$

Check: $x^2 - 9 \stackrel{?}{=} 0$

$(3)^2 - 9 \stackrel{?}{=} 0$ or $(-3)^2 - 9 \stackrel{?}{=} 0$

$9 - 9 \stackrel{?}{=} 0$ $\quad 9 - 9 \stackrel{?}{=} 0$

$0 = 0$ $\quad 0 = 0$

The solution set is $\{3, -3\}$.

2 Solve and check: $y^2 = -7y$ ❶

To use the zero product property, one side of the equation must be zero.

$y^2 = -7y$

$y^2 + 7y = 0$

$y(y + 7) = 0$ *Use the distributive property.*

$y = 0$ or $y + 7 = 0$

$y = -7$

Check: $y^2 \stackrel{?}{=} -7y$

$(0)^2 \stackrel{?}{=} -7(0)$ or $(-7)^2 \stackrel{?}{=} -7(-7)$

$0 = 0$ $\quad 49 = 49$

The solution set is $\{0, -7\}$.

3 Solve: $\frac{2}{3}x^2 - 3x = 0$

$\frac{2}{3}x^2 - 3x = 0$

$x\left(\frac{2}{3}x - 3\right) = 0$

$x = 0$ or $\frac{2}{3}x - 3 = 0$

$\frac{2}{3}x = 3$

$x = \frac{9}{2}$ *Check the result.*

The solution set is $\left\{0, \frac{9}{2}\right\}$.

8-3

Lesson Notes, page T40
Enrichment, page T41

Readiness Activity

Arrange the terms of each polynomial so that the powers of x are in descending order.

$x + 2 + x^2 + 4x^3$ — $4x^3 + x^2 + x + 2$

$x^2 + 4 - 3x$ — $x^2 - 3x + 4$

$3x^3 + 4 - 2x$ — $3x^3 - 2x + 4$

$x^2 + 4x^4 - 1$ — $4x^4 + x^2 - 1$

Discussion Questions

1. If $(a + b)^2 = 0$, what must be true of a and b? $a = -b$
2. If $(a - b)^2 = 0$, what must be true of a and b? $a = b$

Chalkboard Exercises

Solve and check.

1a. $a^2 - 169 = 0$ $\{13, -13\}$

1b. $y^2 - 144 = 0$ $\{12, -12\}$

2a. $\frac{3}{5}y = \frac{1}{5}y^2$ $\{0, 3\}$

2b. $4x^2 = -8x$ $\{0, -2\}$

3a. $n^2 - 5n = 0$ $\{0, 5\}$

3b. $\frac{4}{3}y^2 - \frac{1}{3}y = 0$ $\left\{0, \frac{1}{4}\right\}$

❶ Emphasize that an expression must be set equal to zero before the zero product property can be applied. Students often attempt to solve an equation such as $x(x - 4) = 3$ by saying that either $x = 3$ or $x - 4 = 3$. Neither 3 nor 7 is a correct solution!

Chalkboard Exercises
Solve and check.

4a. $a^2 - 24a + 144 = 0$ $\{12\}$

4b. $4m^2 + 20m + 25 = 0$ $\left\{-2\frac{1}{2}\right\}$

5a. A certain number decreased by 5 is multiplied by the same number increased by 5. The product is 56. Find the number. 9 or −9

5b. The square of a number is equal to 14 times the number decreased by 49. Find the number. 7

❷ Remind students that they should define each variable carefully and completely.

Suggested Daily Assignments

Enriched
- Written: 1–35 odds, 36, 37
- Mini-Review: 1–5

Average
- Exp: 1–15 odds
- Written: 1–27, 28–34 evens
- Mini-Review: 1–5

Fundamental
- Exp: 1–15
- Written: 1–27
  - p. 249, 2–30 evens
- Mini-Review: 1–5

Answers to Exploratory

4. $9d(d + 4) = 0$
9. $13y(2y - 1) = 0$
11. $(y + 4)(y - 4) = 0$
12. $(x + 2)(x + 2) = 0$

Examples

4 **Solve: $a^2 + 16a + 64 = 0$**

$a^2 + 16a + 64 = 0$ — *$a^2 + 16a + 64$ is a perfect square.*

$(a + 8)^2 = 0$

$(a + 8)(a + 8) = 0$

$a + 8 = 0$ or $a + 8 = 0$

$a = -8$ $\quad$ $a = -8$

The solution set is $\{-8\}$.

5 **A tinsmith has a square piece of tin with an area of 16 m². Find the length of each side of the piece of tin.**

Explore *Use a diagram.*

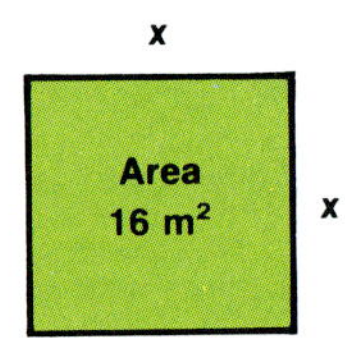

❷

Let x = the length of each side.

Plan $x^2 = 16$ — *The formula for the area of a square is $A = s^2$.*

Solve

$x^2 - 16 = 0$ — *Subtract 16 from both sides.*

$(x - 4)(x + 4) = 0$ — *Factor $x^2 - 16$.*

$x - 4 = 0$ or $x + 4 = 0$

$x = 4$ $\quad$ $x = -4$

Examine -4 m is not a reasonable length for each side of a piece of tin. Therefore, the only possible length for each side is 4 m.

Exploratory Exercises

Express each equation in factored form so that the zero product property can be applied.

1. $n^2 - 3n = 0$ $n(n - 3) = 0$

2. $y^2 - 6y = 0$ $y(y - 6) = 0$

3. $8c^2 + 32c = 0$ $8c(c + 4) =$

4. $9d^2 + 36d = 0$

5. $3x^2 - \frac{3}{4}x = 0$ $3x\left(x - \frac{1}{4}\right) = 0$

6. $x^2 + \frac{5}{3}x = 0$ $x\left(x + \frac{5}{3}\right) = 0$

7. $7y^2 = 14y$ $7y(y - 2) = 0$

8. $8y^2 = -4y$ $4y(2y + 1) = 0$

9. $-13y = -26y^2$

10. $x^2 = 9$ $(x + 3)(x - 3) = 0$

11. $y^2 - 16 = 0$

12. $x^2 + 4x + 4 = 0$

13. $y^2 - 16y + 64 = 0$ $(y - 8)(y - 8) = 0$

14. $x^2 - 8x = -16$ $(x - 4)(x - 4) = 0$

15. $y^2 = 10y - 25$ $(y - 5)(y - 5) =$

Written Exercises

Solve and check.

1. $x^2 - 6x = 0$ {0, 6}
2. $m^2 + 36m = 0$ {0, −36}
3. $2x^2 + 4x = 0$ {0, −2}
4. $4x^2 - 9 = 0$ $\left\{\frac{3}{2}, -\frac{3}{2}\right\}$
5. $a^2 - 81 = 0$ {9, −9}
6. $16x^2 = 64$ {2, −2}
7. $4s^2 = -36s$ {0, −9}
8. $y^2 = 7y$ {0, 7}
9. $x^2 = -8x$ {0, −8}
10. $6y = -3y^2$ {0, −2}
11. $x^2 = 36$ {6, −6}
12. $y^2 = 64$ {8, −8}
13. $y^2 = -5y$ {0, −5}
14. $\frac{1}{2}y^2 - \frac{1}{4}y = 0$ $\left\{0, \frac{1}{2}\right\}$
15. $3y^2 - \frac{4}{3} = 0$ $\left\{\frac{2}{3}, -\frac{2}{3}\right\}$
16. $\frac{2}{3}y = \frac{1}{3}y^2$ {0, 2}
17. $5x^2 = 45$ {3, −3}
18. $2y^2 - 98 = 0$ {7, −7}
19. $4y^2 - 16 = 0$ {2, −2}
20. $m^2 - 24m + 144 = 0$ {12}
21. $y^2 + 10y = -25$ {−5}
22. $25x^2 + 20x = -4$ $\left\{-\frac{2}{5}\right\}$
23. $81n^2 + 36n = -4$ $\left\{-\frac{2}{9}\right\}$
24. $9y^2 = -42y - 49$ $\left\{-\frac{7}{3}\right\}$
25. $y^2 - \frac{16}{81} = 0$ $\left\{\frac{4}{9}, -\frac{4}{9}\right\}$
26. $\frac{5}{6}x^2 = \frac{2}{3}x$ $\left\{0, \frac{4}{5}\right\}$
27. $\frac{3}{4}x^2 - \frac{1}{8}x = 0$ $\left\{0, \frac{1}{6}\right\}$

Use an equation to solve each problem. Disregard any unreasonable solutions.

28. The square of a number decreased by 144 is zero. Find the number. 12 or −12
29. The area of a square tabletop is 4 m^2. Find the length of one side of the tabletop. 2m
30. The square of a number added to 6 times the number is zero. Find the number. 0 or −6
31. The square of a number is equal to 10 times the number decreased by 25. Find the number. 5
32. Jake is as many years less than 30 as Jackie is more than 30. The product of their ages is 884. How old is each?
33. A certain number decreased by 7 is multiplied by the same number increased by 7. The product is 51. Find the number.
34. The square of a number increased by 1070 is 1359. Find the number. 17 or −17
35. If the square of a number is subtracted from 900, the result is 275. Find the number. 25 or −25

Answers to Written
32. Jake is 26. Jackie is 34.
33. −10 or 10

Challenge

36. The height h in meters of a rocket with initial velocity v, t seconds after blast-off, is given by the formula $h = vt - 5t^2$. A toy rocket is launched with an initial velocity of 80 meters/second. How many seconds after blast-off will it hit the ground?
37. A tinsmith has a rectangular piece of tin with a 3-inch square cut from each corner.

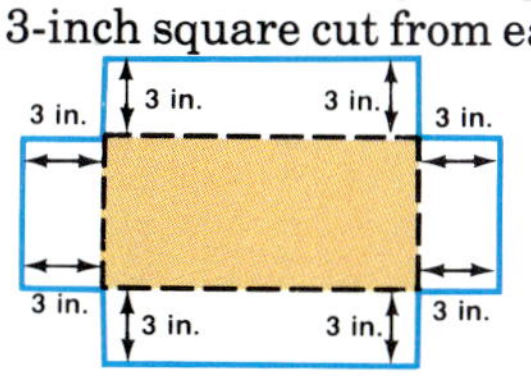

After folding up the sides to form a box, she notices the box is twice as long as it is wide and has a volume of 1350 in^3. Find the length and width of the box.
length is 30 in., width is 15 in.

36. 16 seconds

mini-review

Replace each ? with <, >, or = to make each sentence true.

1. $(-6.3)(0.4)$? 2.5 <
2. $-\frac{6}{11}$? $-\frac{4}{7}$ >

Simplify.

3. $x(3x - 4) - 5(x^2 - 2x + 7)$ 3. $-2x^2 + 6x - 35$
4. $(y^2 + 4y - 3) - (2y^2 - 3y + 8)$ 4. $-y^2 + 7y - 11$

Solve.

5. Karen has 23 coins in dimes and quarters. She has 7 more dimes than quarters and has \$3.50 in all. How many of each coin does she have? 8 quarters, 15 dimes

8-4

Lesson Notes, page T41

Discussion Question
Find some values of c so that $x^2 + c = 0$ has solutions. State the solutions for each value.
Answers will vary.

Chalkboard Exercises
Solve and check each solution.

1a. $y^2 = 18 + 7y$
$\{9, -2\}$

1b. $x^2 + 35 = -12x$
$\{-5, -7\}$

1c. $y^2 + 24 = 10y$
$\{4, 6\}$

2a. $b^3 + 3b^2 = 10b$
$\{0, 2, -5\}$

2b. $x^3 = 14x - 5x^2$
$\{0, 2, -7\}$

2c. $x^3 + 5x^2 = -6x$
$\{0, -2, -3\}$

❶ Remind students that the zero-product property can only be used when an expression is set equal to zero. For example, for the equation $(x - 3)(x + 5) = 10$, it is *not* correct to assume $x - 3 = 10$ or $x + 5 = 10$.

Objective 8-4: Solve equations by factoring and using the zero-product property. (These equations contain three or four nonzero terms.)

8-4 Solving More Equations by Factoring

Some equations of second degree or higher can be written as a product of factors equal to zero. The zero product property can be applied to solve these equations.

Example 1

Solve and check: $2x^2 = 11x - 12$

❶
$2x^2 = 11x - 12$

$2x^2 - 11x + 12 = 0$ — *Rewrite the equation.*

$(2x - 3)(x - 4) = 0$ — *Factor $2x^2 - 11x + 12$.*

$2x - 3 = 0$ or $x - 4 = 0$ — *Use the zero-product property.*

$x = \frac{3}{2}$ or $x = 4$

Check: $2x^2 = 11x - 12$

$2\left(\frac{3}{2}\right)^2 \stackrel{?}{=} 11\left(\frac{3}{2}\right) - 12$ $\qquad$ $2(4)^2 \stackrel{?}{=} 11(4) - 12$

$2\left(\frac{9}{4}\right) \stackrel{?}{=} \frac{33}{2} - \frac{24}{2}$ $\qquad$ $2(16) \stackrel{?}{=} 44 - 12$

$\frac{9}{2} = \frac{9}{2}$ $\qquad$ $32 = 32$

The solution set of $2x^2 = 11x - 12$ is $\left\{\frac{3}{2}, 4\right\}$.

You may have to use more than one method of factoring to solve an equation.

Example 2

Solve and check: $x^3 - 24x = 5x^2$

$x^3 - 24x = 5x^2$

$x^3 - 5x^2 - 24x = 0$ — *Arrange terms in descending order.*

$x(x^2 - 5x - 24) = 0$ — *Factor out the GCF, x.*

$x(x - 8)(x + 3) = 0$ — *Factor $x^2 - 5x - 24$.*

$x = 0$ or $x - 8 = 0$ or $x + 3 = 0$

$x = 0$ or $x = 8$ or $x = -3$

Check: $x^3 - 24x = 5x^2$

$0^3 - 24(0) \stackrel{?}{=} 5(0)^2$ $\qquad$ $8^3 - 24(8) \stackrel{?}{=} 5(8)^2$ $\qquad$ $(-3)^3 - 24(-3) \stackrel{?}{=} 5(-3)^2$

$0 - 0 \stackrel{?}{=} 0$ $\qquad$ $512 - 192 \stackrel{?}{=} 5(64)$ $\qquad$ $-27 + 72 \stackrel{?}{=} 5(9)$

$0 = 0$ $\qquad$ $320 = 320$ $\qquad$ $45 = 45$

The solution set of $x^3 - 24x = 5x^2$ is $\{0, 8, -3\}$.

In Example 2, if you divide both sides of $x(x^2 - 5x - 24) = 0$ by x the result would not be an equivalent equation. This follows since 0 is a solution of the first equation. But 0 is *not* a solution of $x^2 - 5x - 24 = 0$. For this reason, when solving equations do not divide by expressions containing a variable.

Exploratory Exercises

Solve each equation.

1. $(a + 7)(a - 3) = 0$ $\{-7, 3\}$ **2.** $(x - 6)(x + 5) = 0$ $\{6, -5\}$ **3.** $(r - 8)(r - 2) = 0$ $\{8, 2\}$
4. $(b + 12)(b - 3) = 0$ $\{-12, 3\}$ **5.** $(2y + 3)(y + 1) = 0$ **6.** $(3z - 1)(z + 2) = 0$
7. $(3x - 4)(2x - 1) = 0$ $\left\{\frac{4}{3}, \frac{1}{2}\right\}$ **8.** $(2m - 1)(5m + 3) = 0$ **9.** $7(k - 5)(k + 3) = 0$
10. $4(s + 7)(s - 6) = 0$ $\{-7, 6\}$ **11.** $x(2x + 7)(3x + 4) = 0$ $\left\{0, -\frac{7}{2}, -\frac{4}{3}\right\}$ **12.** $z(5z + 4)(3z - 11) = 0$ $\left\{0, -\frac{4}{5}, \frac{11}{3}\right\}$

Written Exercises

Solve and check each equation.

1. $x^2 + 13x + 36 = 0$ $\{-4, -9\}$ **2.** $y^2 + 19y + 34 = 0$ $\{-2, -17\}$ **3.** $a^2 + 4a - 21 = 0$ $\{3, -7\}$
4. $b^2 - 8b - 33 = 0$ $\{-3, 11\}$ **5.** $s^2 + 2s - 63 = 0$ $\{-9, 7\}$ **6.** $c^2 + 17c + 72 = 0$
7. $r^2 - 49 = 0$ $\{-7, 7\}$ **8.** $y^2 - 64 = 0$ $\{-8, 8\}$ **9.** $x^2 - 12x = 0$ $\{0, 12\}$
10. $k^2 - 13k = 0$ $\{0, 13\}$ **11.** $p^2 = 5p + 24$ $\{8, -3\}$ **12.** $r^2 = 18 + 7r$ $\{9, -2\}$
13. $2t^2 + 7t = 15$ $\left\{\frac{3}{2}, -5\right\}$ **14.** $3y^2 - 7y = 20$ $\left\{-\frac{5}{3}, 4\right\}$ **15.** $6z^2 + 5 = -17z$
16. $12m^2 + 3 = -20m$ **17.** $2x^2 + 13x = 24$ $\left\{\frac{3}{2}, -8\right\}$ **18.** $3y^2 + 16y = 35$ $\left\{\frac{5}{3}, -7\right\}$
19. $m^3 - 81m = 0$ $\{0, 9, -9\}$ **20.** $4b^3 - 36b = 0$ $\{0, 3, -3\}$
21. $r^3 - 6r^2 + 8r = 0$ $\{0, 2, 4\}$ **22.** $s^3 + 2s^2 - 35s = 0$ $\{0, -7, 5\}$
23. $5b^3 + 34b^2 = 7b$ $\left\{0, \frac{1}{5}, -7\right\}$ **24.** $2k^3 + 5k^2 = 42k$ $\left\{0, \frac{7}{2}, -6\right\}$
25. $2y^3 - 15y = y^2$ $\left\{0, -\frac{5}{2}, 3\right\}$ **26.** $6x^3 - 7x = 11x^2$ $\left\{0, \frac{7}{3}, -\frac{1}{2}\right\}$
27. $\frac{x^2}{12} - \frac{2}{3}x - 4 = 0$ $\{12, -4\}$ **28.** $x^2 - \frac{1}{6}x - \frac{35}{6} = 0$ $\left\{\frac{5}{2}, -\frac{7}{3}\right\}$
29. $0.03x^2 - x + 3 = 0$ $\left\{\frac{10}{3}, 30\right\}$ **30.** $0.3x^2 - 1.06x + 0.2 = 0$ $\left\{\frac{1}{5}, \frac{10}{3}\right\}$
❷ **31.** $(x + 8)(x + 1) = -12$ $\{-4, -5\}$ **32.** $(r - 1)(r - 1) = 36$ $\{-5, 7\}$
33. $(3y + 2)(y + 3) = y + 14$ $\left\{\frac{2}{3}, -4\right\}$ **34.** $(y + 4)(3y - 2) = -y - 14$ $\left\{-\frac{2}{3}, -3\right\}$
35. $h^3 + h^2 - 4h - 4 = 0$ $\{-2, 2, -1\}$ **36.** $y^3 - y^2 - y + 1 = 0$ $\{-1, 1\}$
37. $9a^3 - 18a^2 - a + 2 = 0$ $\left\{\frac{1}{3}, -\frac{1}{3}, 2\right\}$ **38.** $2m^3 + 5m^2 - 18m - 45 = 0$ $\left\{-\frac{5}{2}, 3, -3\right\}$
39. $y^4 - 8y^2 + 16 = 0$ $\{2, -2\}$ **40.** $m^4 - 2m^2 + 1 = 0$ $\{1, -1\}$
41. $xy + 4x - 3y - 12 = 0$ **42.** $2my + 5m + 8y + 20 = 0$
43. $3rs - 4r + 6s - 8 = 0$ **44.** $4pz - z + 12p - 3 = 0$

Challenge

Write an equation with integral coefficients in the form $ax^2 + bx + c = 0$ that would have the given solutions.

45. $\{2, 3\}$ $x^2 - 5x + 6 = 0$ **46.** $\{-3, 5\}$ $x^2 - 2x - 15 = 0$ **47.** $\left\{\frac{2}{3}, -1\right\}$ $3x^2 + x - 2 = 0$ **48.** $\left\{-3, \frac{3}{4}\right\}$ $4x^2 + 9x - 9 = 0$

Suggested Daily Assignments

Enriched
Exp: 10–12
Written: 2–44 evens, 45–48

Average
Exp: 2–12 evens
Written: 1–44
p. 253, 29–35 odds

Fundamental
Exp: 1–12
Written: 1–30
p. 253, 28–35

Answers to Exploratory

5. $\left\{-\frac{3}{2}, -1\right\}$ **6.** $\left\{\frac{1}{3}, -2\right\}$
8. $\left\{\frac{1}{2}, -\frac{3}{5}\right\}$ **9.** $\{5, -3\}$

❷ Point out that in Written Exercises 31-36 the left side of the equation must first be simplifed before setting the equation equal to zero.

Answers to Written

6. $\{-9, -8\}$
15. $\left\{-\frac{1}{3}, -\frac{5}{2}\right\}$ **16.** $\left\{-\frac{1}{6}, -\frac{3}{2}\right\}$
41. $\{(x, y) | x = 3 \text{ or } y = -4\}$
42. $\left\{(m, y) | m = -4 \text{ or } y = -\frac{5}{2}\right\}$
43. $\left\{(r, s) | r = 2 \text{ or } s = \frac{4}{3}\right\}$
44. $\left\{(p, z) | p = \frac{1}{4} \text{ or } z = -3\right\}$

Quiz
Lessons 8-1 to 8-4
Solve each of the following.
1. $x(2x + 6) = 0$ $\{0, -3\}$
2. $(y - 4)(5y + 1) = 0$ $\left\{4, -\frac{1}{5}\right\}$
3. $a^2 + 6a + 9 = 0$ $\{-3\}$
4. $r^2 - 49 = 0$ $\{-7, 7\}$

Solve and check each equation.
5. $p^2 - 8p + 12 = 0$ $\{2, 6\}$
6. $x^2 - 5x = -6$ $\{2, 3\}$
7. $2t^2 + 5t - 3 = 0$ $\left\{-3, \frac{1}{2}\right\}$

8-5

Lesson Notes, page T41

Readiness Activity

Let x = an integer. What is the next consecutive integer?
$x + 1$

Let x = an odd integer. What are the next two consecutive odd integers?
$x + 2, x + 4$

Let x = an even integer. What are the next three consecutive even integers?
$x + 2, x + 4, x + 6$

Discussion Question

If n = any integer, can you write an expression that must be an even integer? An odd integer?
Answers will vary. Typical answers are given: even, $2n$ and odd, $2n + 1$.

Chalkboard Exercises

Find two consecutive integers whose product is the number given.

1a. 182 — $-13, -14$ or $13, 14$
1b. 90 — $-9, -10$ or $9, 10$

Find two integers whose sum is the first number and whose product is the second number.

2a. 18, 81 — 9, 9
2b. 10, 21 — 3, 7
2c. Find two integers whose difference is 7 and whose product is 144.
9, 16 or $-9, -16$

Computer Notes

Suggest that students write a program for finding the sum and product of pairs of integers between 11 and 20, inclusive (or another specified set of numbers). One possible program is given below.

```
10  FOR X = 11 TO 20
20  FOR Y = 11 TO 20
30  LET S = X + Y
40  LET P = X*Y
50  PRINT X;"+"; Y; "=";S
60  PRINT X; "*"; Y; "=";P
70  NEXT Y
80  NEXT X
90  END
```


Objective 8-5: Solve integer problems by factoring.

8-5 Problem Solving: Integer Problems

Some types of integer problems can be solved by using factoring.

Examples

1 **Find two consecutive integers whose product is 72.**

Explore This problem can be solved by using an equation.
Let x = one integer.
Then $x + 1$ = the next greater integer.

Plan $x(x + 1) = 72$

Solve
$$x(x + 1) = 72$$
$$x^2 + x = 72$$
$$x^2 + x - 72 = 0$$
$$(x + 9)(x - 8) = 0$$
$$x + 9 = 0 \quad \text{or} \quad x - 8 = 0$$
$$x = -9 \quad \text{or} \quad x = 8$$

If $x = -9$, then $x + 1 = -8$.
If $x = 8$, then $x + 1 = 9$.

Examine Since $-9(-8) = 72$ and $8 \cdot 9 = 72$, the consecutive integers are -9 and -8 or 8 and 9.

2 **Find two integers whose sum is 15 and whose product is 54.**

Explore Let n = one integer.
Then $15 - n$ = the other integer.

Plan $n(15 - n) = 54$

Solve
$$n(15 - n) = 54$$
$$15n - n^2 = 54$$
$$-n^2 + 15n - 54 = 0$$
$$n^2 - 15n + 54 = 0 \quad \textit{Multiply both sides by } -1.$$
$$(n - 9)(n - 6) = 0$$
$$n - 9 = 0 \quad \text{or} \quad n - 6 = 0$$
$$n = 9 \quad \text{or} \quad n = 6$$

If $n = 9$, then $15 - n = 6$.
If $n = 6$, then $15 - n = 9$.

Examine Since $6 + 9 = 15$ and $6 \cdot 9 = 54$, the two integers are 6 and 9.

Example

3 **The sum of the squares of two consecutive odd integers is 130. Find the integers.**

Explore Let x = one odd integer.
Then $x + 2$ = the next greater odd integer.

Plan

$$x^2 + (x + 2)^2 = 130$$

Solve

$$x^2 + (x + 2)^2 = 130$$
$$x^2 + x^2 + 4x + 4 = 130$$
$$2x^2 + 4x - 126 = 0$$
$$2(x^2 + 2x - 63) = 0 \quad \text{Factor out the GCF, 2.}$$
$$x^2 + 2x - 63 = 0 \quad \text{Divide both sides by 2.}$$
$$(x + 9)(x - 7) = 0$$
$$x + 9 = 0 \quad \text{or} \quad x - 7 = 0$$
$$x = -9 \quad \text{or} \quad x = 7$$

If $x = -9$, then $x + 2 = -7$.
If $x = 7$, then $x + 2 = 9$.

Examine Since $(-9)^2 + (-7)^2 = 130$ and $(7)^2 + (9)^2 = 130$, the consecutive odd integers are -9 and -7 or 7 and 9.

Exploratory Exercises

Define a variable and state an equation for each sentence.

1. The product of two consecutive integers is 110.
2. The product of two consecutive integers is 156.
3. The product of two consecutive even integers is 168.
4. The product of two consecutive odd integers is 143.
5. The sum of two integers is 15 and their product is 44.
6. The sum of two integers is 22 and their product is 117.
7. The sum of the squares of two consecutive integers is 181.
8. The sum of the squares of two consecutive even integers is 100.
9. The difference of the squares of two consecutive integers is 17.
10. The difference of the squares of two consecutive even integers is 52.

Written Exercises

1-10. Solve each problem in Exploratory Exercises 1-10.

For each problem below, define a variable. Then use an equation to solve the problem.

11. Find two consecutive even integers whose product is 120. 10, 12; -10, -12
12. Find two consecutive even integers whose product is 360. 18, 20; -18, -20

Chalkboard Exercises
For each problem below, define a variable. Then use an equation to solve the problem.

3a. The sum of the squares of two consecutive integers is 221. Find the integers.
10, 11 or -10, -11

3b. The sum of the squares of two consecutive even integers is 100. Find the integers.
6, 8 or -6, -8

3c. The sum of two integers is 16. The sum of their squares is 146. Find the integers.
5, 11

Suggested Daily Assignments

Enriched
Written: 2–32 evens

Average
Exp: 2–10 evens
Written: 1–26

Fundamental
Exp: 1–10
Written: 1–24
p. 255, 31–40

Answers to Exploratory

1. Let x = the smaller integer; $x(x + 1) = 110$
2. Let x = the smaller integer; $x(x + 1) = 156$
3. Let x = the smaller integer; $x(x + 2) = 168$
4. Let x = the smaller integer; $x(x + 2) = 143$
5. Let x = one integer; $x(15 - x) = 44$
6. Let x = one integer; $x(22 - x) = 117$
7. Let x = the smaller integer; $x^2 + (x + 1)^2 = 181$
8. Let x = the smaller integer; $x^2 + (x + 2)^2 = 100$
9. Let x = the smaller integer; $(x + 1)^2 - x^2 = 17$ or $x^2 - (x + 1)^2 = 17$
10. Let x = the smaller integer; $(x + 2)^2 - x^2 = 52$ or $x^2 - (x + 2)^2 = 52$

Answers to Written
See margin on p. 258.

Answers to Written (p. 257)

1. 10, 11; -10, -11
2. 12, 13; -12, -13
3. 12, 14; -12, -14
4. 11, 13; -11, -13
5. 4, 11
6. 9, 13
7. 9, 10; -9, -10
8. 6, 8; -6, -8
9. 8, 9; -8, -9
10. 12, 14; -12, -14

13. Find two consecutive positive odd integers whose product is 195. 13, 15

14. Find two consecutive odd integers whose product is 399. 19, 21; -19, -21

15. Find two consecutive positive integers whose product is 182. 13, 14

16. Find two consecutive positive integers whose product is 272. 16, 17

17. Find two integers whose sum is 11 and whose product is 24. 3, 8

18. Find two integers whose sum is 19 and whose product is 60. 4, 15

19. Find two integers whose difference is 3 and whose product is 88. 8, 11; -8, -11

20. Find two integers whose difference is 23 and whose product is -120. -8, 15; -15, 8

21. The sum of the squares of two consecutive positive odd integers is 202. Find the integers. 9, 11

22. The sum of the squares of two consecutive positive integers is 113. Find the integers.

22. 7, 8 **24.** 11, -8

23. The sum of two integers is 13. The sum of their squares is 97. Find the integers. 4, 9

24. The sum of two integers is 3. The sum of their squares is 185. Find the integers.

25. When one integer is added to the square of the next consecutive integer, the sum is 41. Find the integers. 5, 6; -8, -7

26. When one integer is added to the square of the next consecutive integer, the sum is 55. Find the integers. -9, -8; 6, 7

27. When the square of the second of two consecutive even integers is added to twice the first integer, the sum is 76. Find the integers. 6, 8; -12, -10

28. When the square of the second of two consecutive even integers is added to twice the first integer, the sum is 116. Find the integers. 8, 10; -14, -12

29. Find 3 consecutive odd integers if the difference of the squares of the least and greatest is 120. 13, 15, 17

30. Lyle and Connie Fisher's ages are consecutive odd integers. The product of their ages is 399. If Lyle is older, find their ages.

31. Jim is a year older than Colleen and the difference of the squares of their ages is 97. How old is each? Colleen is 48. Jim is 49.

32. Jennifer said to Peggy, "I am thinking of 2 consecutive even integers. The difference of their squares is 92." What are the numbers? 22, 24

30. Connie is 19. Lyle is 21.

Using Calculators — Special Products

You know that $(x - y)(x + y) = x^2 - y^2$. This fact can be used to multiply numbers mentally. Suppose you want to find $36 \cdot 44$. Notice that $36 = 40 - 4$ and $44 = 40 + 4$.

$$\begin{aligned} 36 \cdot 44 &= (40 - 4)(40 + 4) \\ &= 40^2 - 4^2 \\ &= 1600 - 16 \\ &= 1584 \end{aligned}$$

Let $x = 40$ and $y = 4$ in $(x - y)(x + y) = x^2 - y^2$.

Exercises

Calculate each product mentally and check by using a calculator.

1. $27 \cdot 33$ 891 **2.** $19 \cdot 21$ 399 **3.** $53 \cdot 67$ 3551 **4.** $95 \cdot 105$ 9975

About Using Calculators
You can also utilize the fact that $(x + 1)^2 = x^2 + 2x + 1$ to find the value of expressions like $(101)^2$. For example, let $x = 100$ then $(101)^2 = (100 + 1)^2 = 100^2 + 2(100) + 1 = 10{,}201$.

Excursions in Algebra — Inequalities

When you solve a polynomial equation by factoring, the zero product property is used.

If $ab = 0$, then $a = 0$ or $b = 0$.

Now consider the inequality $ab > 0$. For what values of a and b will this inequality be true? It will be true only when a and b are either both positive or both negative.

If $ab > 0$, then either $a > 0$ and $b > 0$, or $a < 0$ and $b < 0$.

Example: Solve $(x + 6)(x - 2) > 0$ and graph the solution.

$x + 6 > 0$ and $x - 2 > 0$ or $x + 6 < 0$ and $x - 2 < 0$

$x > -6$ and $x > 2$ or $x < -6$ and $x < 2$

$x > 2$ or $x < -6$ *Why?*

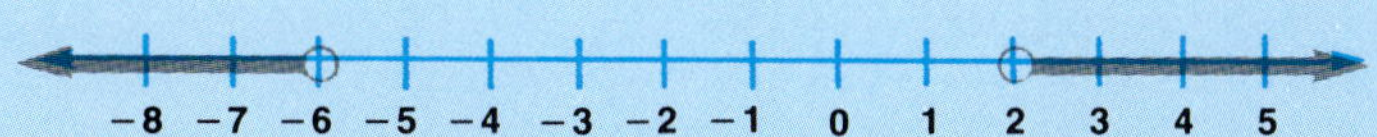

Consider the inequality $ab < 0$. Either a must be positive and b negative or b must be positive and a negative.

If $ab < 0$, then either $a > 0$ and $b < 0$ or $a < 0$ and $b > 0$.

Example: Solve $(x + 7)(x - 8) < 0$ and graph the solution.

$x + 7 > 0$ and $x - 8 < 0$ or $x + 7 < 0$ and $x - 8 > 0$

$x > -7$ and $x < 8$ or $x < -7$ and $x > 8$

This can never be true.

The compound sentence $x > -7$ and $x < 8$ can be written $-7 < x < 8$.

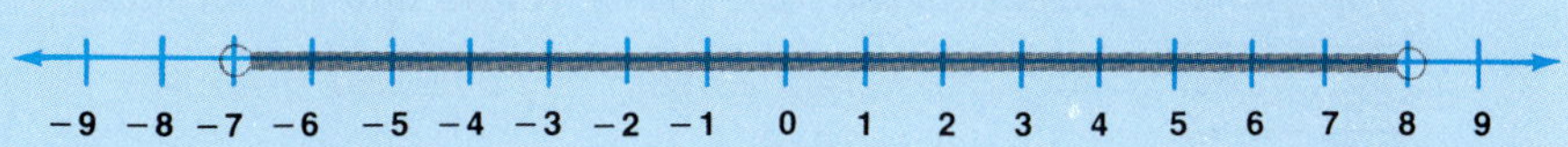

Exercises

Solve each inequality and graph the solution.

1. $x(x + 4) > 0$
2. $x(x - 3) > 0$
3. $(x + 2)(x - 3) \geq 0$
4. $x(x - 5) < 0$
5. $(x + 6)(x - 3) \leq 0$
6. $x^2 + 5x + 6 > 0$
7. $x^2 - 2x - 15 \geq 0$
8. $x^2 + 6x + 5 < 0$
9. $y^2 < 9$
10. $y^2 + 2y > 35$
11. $x^2 > 36$
12. $x^2 - 3x \leq -2$

About the Excursion
You will want to review the meaning of *and* and *or* in conditional statements, as was introduced in Chapter 4. Also review the meaning of using an open circle and a dot when graphing inequalities.

Answers to Exercises

1. $x < -4$ or $x > 0$

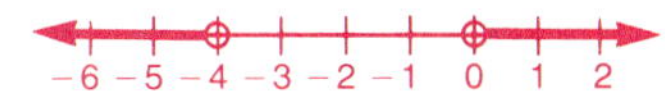

2. $x < 0$ or $x > 3$

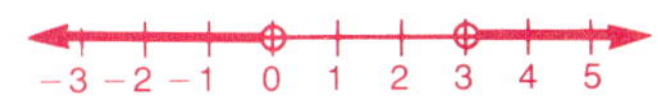

3. $x \leq -2$ or $x \geq 3$

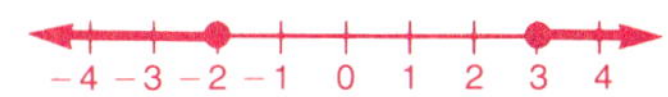

4. $0 < x < 5$

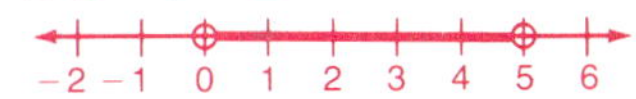

5. $-6 \leq x \leq 3$

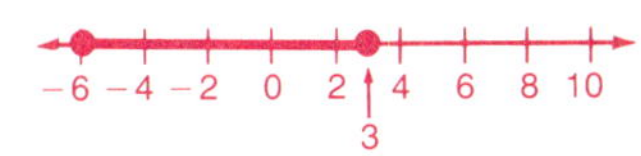

6. $x < -3$ or $x > -2$

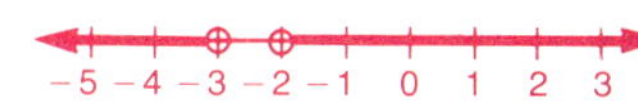

7. $x \leq -3$ or $x \geq 5$

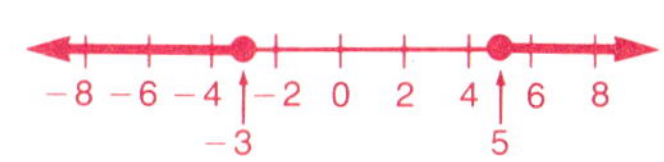

8. $-5 < x < -1$

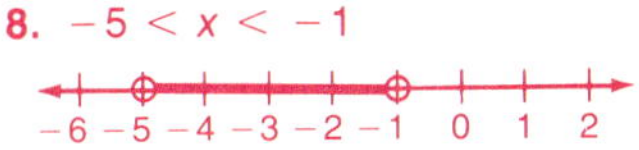

9. $-3 < y < 3$

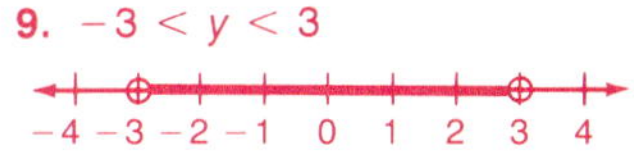

10. $y < -7$ or $y > 5$

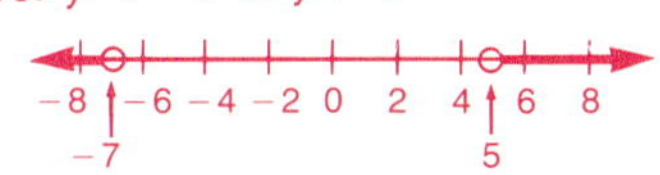

11. $x < -6$ or $x > 6$

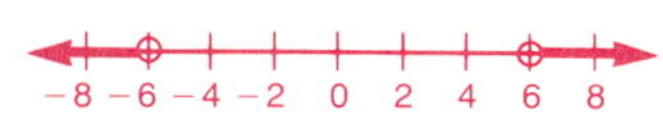

12. $1 \leq x \leq 2$

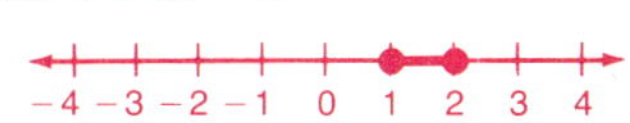

8-6

Lesson Notes, page T41
Enrichment, page T41

Discussion Questions

How high did the flare in Example 1 get? (Hint: Try using a table and finding the height at various times.)
324 feet

Chalkboard Exercises

Use the formula $h = vt - 16t^2$ to solve each of the following problems.

1a. A flare is launched from a life raft with an initial velocity of 160 feet per second. How many seconds will it take for the flare to return to the sea? 10 sec.

1b. When will the flare in the problem above be 336 feet off the ground?
at 3 sec and 7 sec

❶ The constant 16 means 16 feet/sec^2.

❷ The formula $h = vt - 16t^2$ becomes $h = vt - 5t^2$ when height is in meters and initial velocity v is in meters per second. Acceleration due to gravity is 9.8 meters/sec^2. The constant 4.9 is rounded off to the nearest whole number 5.

Objective 8-6: Solve problems that involve velocity and area by factoring.

8-6 Problem Solving: Velocity and Area

If an object is launched into the air its height above the ground after t seconds is given by the formula $h = vt - 16t^2$. In this formula, h represents the height of the object in feet, and v represents the initial velocity in feet per second. Thus, the height of an object with an initial velocity of 144 feet per second is represented by $h = 144t - 16t^2$. ❶ ❷

Example 1

A flare is launched from a life raft with an initial velocity of 144 feet per second. How many seconds will it take for the flare to return to the sea?

Explore Use the formula $h = vt - 16t^2$.

The variable h represents the height of the flare in feet when it returns to the sea. Thus, $h = 0$.

The variable v represents the initial velocity of the flare, in feet per second.
Thus, $v = 144$.

Plan Substitute the appropriate values into the formula.

Solve

$$0 = 144t - 16t^2 \quad \textit{t is the time in seconds.}$$
$$0 = 16t(9 - t)$$
$$16t = 0 \quad \text{or} \quad 9 - t = 0$$
$$t = 0 \quad \text{or} \quad 9 = t$$

Examine The flare returns to the sea in 9 seconds. The answer 0 is not reasonable since it represents the time when the flare is launched.

You can use factoring to solve some area problems. Recall that the area of a rectangle is equal to the product of the length and the width.

Example

2 **Pat Bing has a photograph that is 8 cm long and 6 cm wide. Pat wants to reduce the length and width of the photo by the same amount. She also wants the reduced photo to have half the area of the original photo. By what amount should she reduce the length and width?**

Let x = the amount the length and width should be reduced.

❸

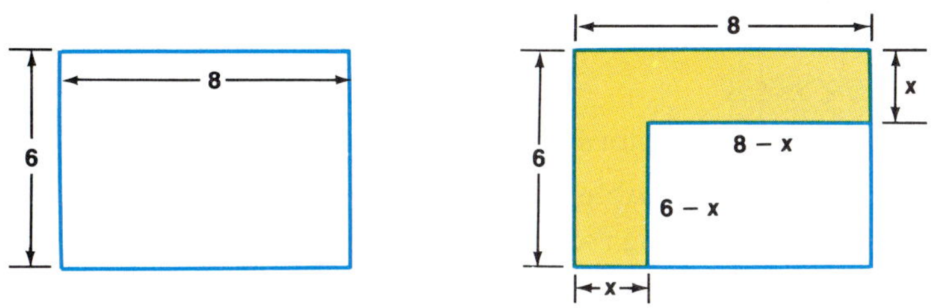

The dimensions of the reduced photo will be $(8 - x)$ and $(6 - x)$.

The area of the photo is $8 \cdot 6$, or 48 cm^2.

The area of the reduced photo will be $\frac{48}{2}$, or 24 cm^2.

$$\text{length} \cdot \text{width} = \text{area}$$

$$(8 - x)(6 - x) = 24 \quad \textit{Substitute the appropriate values into the formula.}$$

$$48 - 8x - 6x + x^2 = 24 \quad \textit{Solve the equation.}$$

$$x^2 - 14x + 48 = 24$$

$$x^2 - 14x + 24 = 0$$

$$(x - 12)(x - 2) = 0$$

$$x - 12 = 0 \quad \text{or} \quad x - 2 = 0$$

$$x = 12 \quad \text{or} \quad x = 2 \quad \textit{Why is 12 cm not a reasonable answer?}$$

Pat should reduce the length and width by 2 cm each.
The dimensions of the reduced photo will be $8 - 2$, or 6 cm and $6 - 2$, or 4 cm.

Exploratory Exercises

Define a variable and state an equation for each problem.

1. The length of a rectangle is 3 feet greater than its width. The area of the rectangle is 108 ft^2. Find the length and width.

2. The length of a rectangle is 4 cm greater than its width. The area of the rectangle is 117 cm^2. Find the length and width.

3. The width of a rectangle is 9 feet less than its length. The area of the rectangle is 90 ft^2. Find the length and width.
Let x = length; $x(x - 9) = 90$

4. The width of a rectangle is 7 cm less than its length. The area of the rectangle is 120 cm^2. Find the length and width.
Let x = length; $x(x - 7) = 120$

Chalkboard Exercises

2a. Rob has a picture that is 12 cm long by 9 cm wide. He wants to reduce the length and width by the same amount to get the area equal to one-half of the original. By what amount must he reduce the length and width? 3 cm

2b. A rectangle is 8 inches wide and 13 inches long. When the length and width are increased by the same amount, the area is increased by 130 sq. in. What are the dimensions of the new rectangle?
13 inches by 18 inches

❸ Encourage students to draw pictures to help them set up equations.

Suggested Daily Assignments

Enriched
Written: 1–8, 10–24 evens

Average
Exp: 1–8
Written: 1–24
p. 258, 27–32

Fundamental
Exp: 1–8
Written: 1–18
p. 258, 25–32

Answers to Exploratory

1. Let x = width; $x(x + 3) = 108$
2. Let x = width; $x(x + 4) = 117$

Answers to Exploratory

5. Let s = length of side of smaller square; $(s + 7)^2 - s^2 = 231$
6. Let s = length of side of smaller square; $s^2 + (3s)^2 = 640$

Answers to Written

1. length, 12 feet; width, 9 feet
2. length, 13 cm; width, 9 cm
3. length, 15 feet; width, 6 feet
4. length, 15 cm; width, 8 cm
5. length of side of smaller square, 13 in., length of side of larger square, 20 in.
6. dimensions of smaller square, 8 ft by 8 ft; dimensions of larger square, 24 ft by 24 ft
7. 11 in.
8. 9 m
9. Let x = width. $x(x + 3) = 40$ The length is 8 m and the width is 5 m.
10. Let x = width. $x(x + 5) = 234$ The length is 18 yd and the width is 13 yd.
11. Let x = amount that length and width are increased. $(4 + x)(7 + x) = 54$ The length is 9 in. and the width is 6 in.
12. Let x = amount that length and width are increased. $(8 + x)(12 + x) = 192$ The length is 16 cm and the width is 12 cm.
13. Let x = width of the strip. $(15 + 2x)(10 + 2x) = 300$ The width of the strip = $2\frac{1}{2}$ feet.
14. Let x = width of the strip. $(120 - 2x)(90 - 2x) = 5400$ The width of the strip = 15 feet.

5. The side of one square is 7 in. longer than the side of a second square. The difference of their areas is 231 in^2. Find the length of their sides.

6. One square has sides three times as long as the sides of a second square. The combined area of the squares is 640 ft^2. Find their dimensions.

7. If the sides of a square are doubled, its area is increased by 363 in^2. What is the original length of the side?
Let s = length of side of square;
$(2s)^2 = s^2 + 363$

8. A rectangle is 24 m long and 16 m wide. By how much must each dimension be decreased in order to decrease the area by 279 m^2?
Let x = amount of decrease;
$(24 - x)(16 - x) = 24(16) - 279$

Written Exercises

1-8. Solve each problem in Exploratory Exercises 1-8.

For each problem, define a variable. Then use an equation to solve the problem.

9. The area of Jane Redfern's living room is 40 m^2. The length of the room is 3 m more than the width. What are its dimensions?

10. The length of Mrs. Boland's garden is 5 yards more than its width. The area of the garden is 234 square yards. What are its dimensions?

11. A rectangle is 4 in. wide and 7 in. long. When the length and width are increased by the same amount, the area is increased by 26 in^2. What are the dimensions of the new rectangle?

12. A photo is 8 cm wide and 12 cm long. The length and width are increased by an equal amount in order to double the area of the photo. What are the dimensions of the new photo?

13. Mr. Steinborn wants to double the area of his garden by adding a strip of uniform width along each of the four sides. The original garden is 10 ft by 15 ft. How wide a strip must be added?

14. A strip of uniform width is plowed along both sides and both ends of a garden 120 ft by 90 ft. How wide is the strip if the garden is half plowed?

Use $h = vt - 16t^2$ to find the missing quantity.

15. Find v, if t = 5 seconds and h = 480 feet. 176 feet per second

16. Find h, if t = 7 seconds and v = 1700 feet per second. 11,116 feet

17. Find h, if t = 2 seconds and v = 110 feet per second. 156 feet

18. Find 2 values for t, if v = 120 feet per second and h = 224 feet. 3.5 seconds, 4 seconds

Use the formula $h = vt - 16t^2$ to solve each of the following problems.

19. A flare is launched from a life raft with an initial velocity of 192 feet per second. How many seconds will it take for the flare to return to the sea? 12 seconds

20. A golf ball is hit into the air with an initial velocity of 64 feet per second. How many seconds will it take for the golf ball to hit the ground? 4 seconds

21. A missile is fired with an initial velocity of 2320 feet per second. When will it be 40,000 feet high? 20 seconds, 125 seconds

22. A rocket is fired with an initial velocity of 1920 feet per second. When will it be 32,000 feet high? 20 seconds, 100 seconds

23. A rocket is fired with an initial velocity of 1640 feet per second. When will it be 816 feet high? $\frac{1}{2}$ second, 102 seconds

24. A flare is launched with an initial velocity of 128 feet per second. How many seconds will it take for the flare to return to the sea? 8 seconds

Objective 8-7: Simplify rational expressions.

8-7 Simplifying Rational Expressions

Factoring polynomials is also a useful tool for simplifying **algebraic fractions.**

Rational numbers and algebraic fractions are similar. Recall that any rational number can be expressed as the quotient of two integers.

Algebraic fractions contain variables. The expressions $\frac{36a^2bc}{24b^2}$, $\frac{2x}{x-5}$, and $\frac{p^2-25}{p^2+4p+1}$ are examples of algebraic fractions.

Assume that the denominator does not equal zero.

To simplify a rational number such as $\frac{6}{21}$, first factor the numerator and denominator. Then eliminate common factors.

$$\frac{6}{21} = \frac{2 \cdot 3}{7 \cdot 3} \qquad \textit{Notice that } \frac{3}{3} = 1.$$

$$= \frac{2 \cdot \cancel{3}}{7 \cdot \cancel{3}} \text{ or } \frac{2}{7} \qquad \textit{Eliminate common factors.}$$

The same procedure can be used to simplify algebraic fractions.

Examples

1 Simplify: $\frac{42y^2}{18xy}$ *Assume the denominator does not equal zero.*

$$\frac{42y^2}{18xy} = \frac{2 \cdot 3 \cdot 7 \cdot y \cdot y}{2 \cdot 3 \cdot 3 \cdot x \cdot y} \qquad \textit{Factor the numerator and denominator.}$$

$$= \frac{\overset{1}{\cancel{2}} \cdot \overset{1}{\cancel{3}} \cdot 7 \cdot y \cdot \overset{1}{\cancel{y}}}{\underset{1}{\cancel{2}} \cdot \underset{1}{\cancel{3}} \cdot 3 \cdot x \cdot \underset{1}{\cancel{y}}} \qquad \textit{Eliminate common factors.}$$

$$= \frac{7y}{3x}$$

2 Simplify: $\frac{y^2 - 9}{y^2 - y - 6}$

$$\frac{y^2 - 9}{y^2 - y - 6} = \frac{(y + 3)(y - 3)}{(y - 3)(y + 2)}$$

$$= \frac{(y + 3)\overset{1}{\cancel{(y - 3)}}}{\underset{1}{\cancel{(y - 3)}}(y + 2)}$$

$$= \frac{y + 3}{y + 2}$$

3 Simplify: $\frac{a + 2}{a^2 - 4}$

$$\frac{a + 2}{a^2 - 4} = \frac{(a + 2) \cdot 1}{(a + 2)(a - 2)}$$

$$= \frac{\overset{1}{\cancel{(a + 2)}} \cdot 1}{\underset{1}{\cancel{(a + 2)}}(a - 2)}$$

$$= \frac{1}{a - 2}$$

8-7

Lesson Notes, page T41

Readiness Activity

Simplify.

$\frac{2}{20}$ $\frac{1}{10}$ $\qquad \frac{4}{14}$ $\frac{2}{7}$

$\frac{-10}{25}$ $\frac{-2}{5}$ $\qquad \frac{8}{-6}$ $\frac{4}{-3}$ or $-1\frac{1}{3}$

$\frac{-12}{-36}$ $\frac{1}{3}$ $\qquad \frac{-15}{-12}$ $\frac{5}{4}$ or $1\frac{1}{4}$

Discussion Question

(Use after Example 2)

Is there any difference between the expressions $\frac{x^2 - 4}{x + 2}$ and $x - 2$?

(Hint: Try substituting $x = -2$ into both expressions.)

Yes, in the first expression $x \neq -2$.

Chalkboard Exercises

Simplify each algebraic fraction. Assume that no denominator is equal to zero.

1a. $\frac{24p^2q}{28pq^2}$ $\quad \frac{6p}{7q}$

1b. $\frac{30rs}{33r^2}$ $\quad \frac{10s}{11r}$

1c. $\frac{42a^2m}{44am^4}$ $\quad \frac{21a}{22m^3}$

2a. $\frac{y^2 - 4}{y^2 + y - 2}$ $\quad \frac{y - 2}{y - 1}$

2b. $\frac{m^2 + m - 30}{m^2 - 25}$ $\quad \frac{m + 6}{m + 5}$

3a. $\frac{r^2 - 1}{r - 1}$ $\quad r + 1$

3b. $\frac{p + 5}{3p^2 - 75}$ $\quad \frac{1}{3p - 15}$

Suggested Daily Assignments

Enriched
Written: 1–45 odds

Average
Exp: 2–12 evens
Written: 1–45

Fundamental
Exp: 1–12
Written: 1–40
p. 262, 19–24

Answers to Written

15. $-\frac{4a}{a^2 + 6ab + 9b}$

16. $\frac{2a^2 - 8}{m^3 + m^2 + m}$

Exploratory Exercises

For each expression, find the greatest common factor (GCF) of the numerator and denominator. Then simplify each expression. Assume that no denominator is equal to zero.

1. $\frac{24}{72}$ 24, $\frac{1}{3}$
2. $\frac{99}{132}$ 33, $\frac{3}{4}$
3. $\frac{42y}{18xy}$ 6y, $\frac{7}{3x}$
4. $\frac{13x}{39x^2}$ 13x, $\frac{1}{3x}$
5. $\frac{38a^2}{42ab}$ 2a, $\frac{19a}{21b}$
6. $\frac{79a^2b}{158a^3bc}$ $79a^2b$, $\frac{1}{2ac}$
7. $\frac{-3x^2y^5}{18x^5y^2}$ $3x^2y^2$, $\frac{-y^3}{6x^3}$
8. $\frac{14y^2z}{49yz^3}$ 7yz, $\frac{2y}{7z^2}$
9. $\frac{x(y+1)}{x(y-2)}$ x, $\frac{y+1}{y-2}$
10. $\frac{4a}{a(a+7)}$ a, $\frac{4}{a+7}$
11. $\frac{m+5}{2(m+5)}$ m + 5, $\frac{1}{2}$
12. $\frac{(a+b)(a-b)}{(a-b)(a-b)}$ $\frac{a+b}{a-b}$, a − b

Written Exercises

Simplify each algebraic fraction. Assume that no denominator is equal to zero.

1. $\frac{y-4}{y^2-16}$ $\frac{1}{y+4}$
2. $\frac{z+3}{z^2-9}$ $\frac{1}{z-3}$
3. $\frac{a+b}{a^2-b^2}$ $\frac{1}{a-b}$
4. $\frac{x+y}{x^2-y^2}$ $\frac{1}{x-y}$
5. $\frac{c^2-4}{(c+2)^2}$ $\frac{c-2}{c+2}$
6. $\frac{y^2-81}{(y-9)^2}$ $\frac{y+9}{y-9}$
7. $\frac{a^2-a}{a-1}$ a
8. $\frac{m^2-2m}{m-2}$ m
9. $\frac{x^2+4}{x^4-16}$ $\frac{1}{x^2-4}$
10. $\frac{y^2+6}{y^4-36}$ $\frac{1}{y^2-6}$
11. $\frac{2x^2+6}{2x+6}$ $\frac{x^2+3}{x+3}$
12. $\frac{3n^2-9}{3n-9}$ $\frac{n^2-3}{n-3}$
13. $\frac{-4y^2}{2y^2-4y^3}$ $-\frac{2}{1-2y}$
14. $\frac{3a^3}{3a^3+6a^2b}$ $\frac{a}{a+2b}$
15. $\frac{-4a^2}{a^3+6a^2b+9ab}$
16. $\frac{2a^3-8a}{am^3+am^2+am}$
17. $\frac{x+y}{x^2+2xy+y^2}$ $\frac{1}{x+y}$
18. $\frac{x+6}{x^2+7x+6}$ $\frac{1}{x+1}$
19. $\frac{x-3}{x^2+x-12}$ $\frac{1}{x+4}$
20. $\frac{x+5}{x^2+7x+10}$ $\frac{1}{x+2}$
21. $\frac{m^2-16}{m^2+8m+16}$ $\frac{m-4}{m+4}$
22. $\frac{n^2-y^2}{n^2-2ny+y^2}$ $\frac{n+y}{n-y}$
23. $\frac{y^2-9}{y^2+6y+9}$ $\frac{y-3}{y+3}$
24. $\frac{y^2+8y-20}{y^2-4}$ $\frac{y+10}{y+2}$
25. $\frac{x^2+8x+15}{x+3}$ x + 5
26. $\frac{2m^2-13m+15}{2m-3}$ m − 5
27. $\frac{6x^2+24x}{x^2+8x+16}$ $\frac{6x}{x+4}$
28. $\frac{8m^2-16m}{m^2-4m+4}$ $\frac{8m}{m-2}$
29. $\frac{4k^2-25}{4k^2-20k+25}$ $\frac{2k+5}{2k-5}$
30. $\frac{x^2-x-20}{x^2+7x+12}$ $\frac{x-5}{x+3}$
31. $\frac{3-x}{6-17x+5x^2}$ $\frac{1}{2-5x}$
32. $\frac{2x-14}{x^2-4x-21}$ $\frac{2}{x+3}$
33. $\frac{5x^2+10x+5}{3x^2+6x+3}$ $\frac{5}{3}$
34. $\frac{3x^2+3x+18}{7x^2+7x+42}$ $\frac{3}{7}$
35. $\frac{m^2-2m-8}{m^2-m-6}$ $\frac{m-4}{m-3}$
36. $\frac{b^2+b-12}{b^2+2b-15}$ $\frac{b+4}{b+5}$
37. $\frac{p^2-4p+4}{p^2+4p-12}$ $\frac{p-2}{p+6}$
38. $\frac{b^2-5b+6}{b^4-13b^2+36}$ $\frac{1}{(b+3)(b+2)}$
39. $\frac{2x^2-5x+3}{3x^2-5x+2}$ $\frac{2x-3}{3x-2}$
40. $\frac{3m^2+8m-3}{6m^2+17m-3}$ $\frac{3m-1}{6m-1}$
41. $\frac{2y^2-18}{2y-6}$ y + 3
42. $\frac{8x^2-x^3}{16-2x}$ $\frac{x^2}{2}$
43. $\frac{2x^2+8x}{x^2-11x}$ $\frac{2x+8}{x-11}$
44. $\frac{6y^3-9y^2}{2y^2+5y-12}$ $\frac{3y^2}{y+4}$
45. $\frac{x^2-x^2y}{x^3-x^3y}$ $\frac{1}{x}$

For more exercises on simplifying algebraic fractions, see exercises 1-6 on page 559.

Using Computers

Sums of Integers

One advantage of using a computer in problem solving is that the computer can perform many operations very quickly. Consider this problem.

> The sum of the first 5 positive even integers is 30. Find the sum of the first 100 positive even integers.

The following BASIC program can be used to find the sum.

```
10  INPUT "HOW MANY EVEN INTEGERS? ";N
20  LET SUM = 0
30  FOR I = 2 TO 2 * N STEP 2
40  LET SUM = SUM + I          Line 40 calculates the sum.
50  NEXT I
60  PRINT "THE SUM OF THE FIRST ";N;" EVEN INTEGERS IS ";SUM
70  END

]RUN
HOW MANY EVEN INTEGERS? 100
THE SUM OF THE FIRST 100 EVEN INTEGERS IS 10100
```

Exercises

1. Run the program several times. Input 1, 2, 3,. . .10 for the value of N. Write the corresponding sum.
2. Modify lines 10, 30, and 60 to find the sum of the first N positive odd integers.
3. For N = 1, 2, 3, . . . 10, write the sum of the first N positive odd integers.
4. Modify the program to find the sum of the first N positive integers. Run the program for N = 1, 2, 3, . . . 10.

1. 2, 6, 12, 20, 30, 42, 56, 72, 90, 110

3. 1, 4, 9, 16, 25, 36, 49, 64, 81, 100

Vocabulary

Zero Product Property (245)   algebraic fractions (263)

Chapter Summary

1. The Zero Product Property:
   For all numbers a and b, if $ab = 0$, then $a = 0$ or $b = 0$. (245)
2. If an equation of second degree or higher can be written as a product of factors equal to zero, the zero product property can be applied to solve the equation. (254)
3. To simplify an algebraic fraction, first factor the numerator and denominator. Then eliminate common factors. (263)

About Using Computers
This demonstrates how the computer can be used to solve certain problems by a straight-forward calculation without using abstract ideas such as induction. You may want to tell students that the formula for finding the sum of the first n positive even integers is $\frac{n}{2}(2 + a_n)$ where n = number of integers and a_n = greatest positive even integer.

Answers to Exercises

2.
```
10  INPUT "HOW MANY ODD INTEGERS";N
30  FOR I = 1 TO 2 * N STEP 2
60  PRINT "THE SUM OF THE FIRST";N;"ODD INTEGERS IS";SUM
```

4.
```
10  INPUT "HOW MANY INTEGERS?";N
30  FOR I = 1 TO N
60  PRINT "THE SUM OF THE FIRST";N;"INTEGERS IS";SUM
```
1, 3, 6, 10, 15, 21, 28, 36, 45, 55

Quiz
Lessons 8–5 to 8–7

For each problem, define a variable, use an equation, and solve.

1. Find two consecutive positive integers such that the sum of their squares is 61.
   $x^2 + (x + 1)^2 = 61$
   $x = 5, x + 1 = 6$
2. The sum of the squares of two consecutive even integers is 100. Find the integers.
   $n^2 + (n + 2)^2 = 100$
   $\{-8, -6\}, \{6, 8\}$
3. The length of a picture is 3 inches more than the width. If the area is 88 in^2, find the length and the width.
   $w(w + 3) = 88$
   width, 8 in.; length, 11 in.
4. A missile is fired with an initial velocity of 3840 ft per sec. How many seconds will it take to return to the initial level?
   $h = vt - 16t^2$
   $0 = 3840t - 16t^2$
   $t = 0$ or $t = 240$
   The missile returns in 240 sec or 4 min.

Simplify.

5. $\dfrac{4x^3y^2z}{20xy^4z}$ $\quad \dfrac{x^2}{5y^2}$
6. $\dfrac{x^2 - 6x - 27}{x^2 + 6x + 9}$ $\quad \dfrac{x - 9}{x + 3}$

Answers to Chapter Review

11. $(3a + 5b)(8m - 3n)$
12. $(3x^2 - 2c)(3x^2 + 2c)\cdot(9x^4 + 4c^2)$

Chapter Review

8-1 **Solve and check each equation.**

1. $y(y + 11) = 0$ $\{0, -11\}$
2. $7a(a - 7) = 0$ $\{0, 7\}$
3. $4t(2t - 10) = 0$ $\{0, 5\}$
4. $(2y - 9)(2y + 9) = 0$ $\left\{\frac{9}{2}, -\frac{9}{2}\right\}$
5. $(3y - 2)(4y + 7) = 0$ $\left\{\frac{2}{3}, -\frac{7}{4}\right\}$
6. $(y - 2)(y + 7) = 0$ $\{2, -7\}$
7. The product of a positive number decreased by 6 and the same number increased by 11 is zero. What is the number? 6

8-2 **Factor.**

8. $50x^2 - 98y^2$ $2(5x - 7y)(5x + 7y)$
9. $6m^3 + m^2 - 15m$ $m(3m + 5)(2m - 3)$
10. $56a^2 - 93a + 27$ $(7a - 9)(8a - 3)$
11. $24am - 9an + 40bm - 15bn$
12. $81x^8 - 16c^4$
13. $49m^2 + 70mn + 25n^2$ $(7m + 5n)^2$
14. $14x^2 + \frac{13}{2}x - 3$ $\frac{1}{2}(7x - 2)(4x + 3)$
15. $8a^2 + 32b^2$ $8(a^2 + 4b^2)$

Solve and check each equation.

8-3

16. $y^2 - 9y = 0$ $\{0, 9\}$
17. $y^2 = -7y$ $\{0, -7\}$
18. $2y^2 - 98 = 0$ $\{7, -7\}$
19. $\frac{3}{4}y = \frac{1}{4}y^2$ $\{0, 3\}$
20. $a^2 + 10a + 25 = 0$ $\{-5\}$
21. $25y^2 + 20y = -4$ $\left\{-\frac{2}{5}\right\}$
22. The square of a number added to seven times the number is zero. What is the number? 0 or -7

8-4

23. $y^2 + 40 = -13y$ $\{-5, -8\}$
24. $x^2 + 12x = -35$ $\{-5, -7\}$
25. $2m^2 + 13m = 24$ $\left\{\frac{3}{2}, -8\right\}$
26. $(x + 6)(x - 1) = 78$ $\{7, -12\}$
27. $6p^2 - 11p + 4 = 0$ $\left\{\frac{4}{3}, \frac{1}{2}\right\}$
28. $6x^3 + 29x^2 + 28x = 0$ $\left\{0, -\frac{4}{3}, -\frac{7}{2}\right\}$

For each problem, define a variable. Then use an equation to solve the problem.

8-5

29. The product of two consecutive odd integers is 143. Find the integers. 11, 13; −11, −
30. The sum of the squares of two consecutive odd integers is 202. Find the integers. 9, 11; −9, −1

8-6

31. A rectangle is 11 inches wide and 13 inches long. When its length and width are decreased by the same amount, its area is decreased by 63 square inches. What are the dimensions of the new rectangle? 8 inches by 10 inches
32. For this exercise, use the formula $h = vt - 16t^2$. A baseball travels upward with an initial velocity of 80 feet per second. How long does it take to return to the ground? 5 seconds

8-7 **Simplify each algebraic fraction.**

33. $\dfrac{3x^2y}{12xy^3z}$ $\quad \dfrac{x}{4y^2z}$
34. $\dfrac{x + y}{x^2 + 3xy + 2y^2}$ $\quad \dfrac{1}{x + 2y}$
35. $\dfrac{x^2 - 9}{x^2 - 6x + 9}$ $\quad \dfrac{x + 3}{x - 3}$
36. $\dfrac{x^2 + 10x + 21}{x^3 + x^2 - 42x}$ $\quad \dfrac{x + 3}{x(x - 6)}$

Chapter Test

Factor.

1. $9x^2 - 36y^4$ $9(x - 2y^2)(x + 2y^2)$
2. $12x^2 + 23x - 24$ $(4x - 3)(3x + 8)$
3. $25a^2 - 20a + 4$ $(5a - 2)^2$
4. $15ac - 21ad + 10bc - 14bd$
5. $3a^2b - 5ab + 11ab^2$ $ab(3a - 5 + 11b)$
6. $50a^2 - 25b^2$ $25(2a^2 - b^2)$
7. $25m^2n^2 + 90mnp + 81p^2$ $(5mn + 9p)^2$
8. $x^3 - 5x^2 - 9x + 45$ $(x - 3)(x + 3)(x - 5)$
9. $x^8 - y^8$ $(x - y)(x + y)(x^2 + y^2)(x^4 + y^4)$
10. $98a^2 - 140ab + 50b^2$ $2(7a - 5b)^2$

Simplify.

11. $\frac{21x^2y}{28ax}$ $\frac{3xy}{4a}$
12. $\frac{x^2 + 7x - 18}{x^2 + 12x + 27}$ $\frac{x - 2}{x + 3}$
13. $\frac{7x^2 - 28}{5x^3 - 20x}$ $\frac{7}{5x}$

Solve each equation.

14. $(3w - 21)(w + 8) = 0$ $\{7, -8\}$
15. $18s^2 + 72s = 0$ $\{0, -4\}$
16. $x^2 - 6x + 9 = 0$ $\{3\}$
17. $6w(4w - 28) = 0$ $\{0, 7\}$
18. $t^2 - 121 = 0$ $\{11, -11\}$
19. $4x^2 = 36$ $\{3, -3\}$
20. $(3m + 5)(2m - 3) = 0$
21. $x^2 - \frac{4}{9} = 0$ $\left\{\frac{2}{3}, -\frac{2}{3}\right\}$
22. $x^2 + 25 = 10x$ $\{5\}$
23. $a^2 - 9a - 52 = 0$ $\{-4, 13\}$
24. $3x^2 - 7x - 20 = 0$ $\left\{-\frac{5}{3}, 4\right\}$
25. $12x^2 - 6 = x$ $\left\{\frac{3}{4}, -\frac{2}{3}\right\}$
26. $10m^2 + 7m = 3$ $\left\{\frac{3}{10}, -1\right\}$
27. $x^3 - 5x^2 - 66x = 0$ $\{-6, 0, 11\}$
28. $x^3 + 7x^2 - 25x - 175 = 0$ $\{-7, -5, 5\}$

For each problem, define a variable. Then use an equation to solve the problem.

29. The product of 5 times Joe's salary and 61,000 less than 4 times Joe's salary is zero. What is Joe's salary? $15,250
30. The sum of two integers is 21 and their product is 104. Find the integers. 8, 13
31. When one integer is added to the square of the next consecutive integer, the sum is 55. Find the integers. 6, 7; −9, −8
32. A photograph is 8 cm wide and 12 cm long. The length and width are increased by an equal amount in order to double the area of the photograph. What are the dimensions of the new photograph? 12 cm by 16 cm

For problems 33 and 34, use $h = vt - 16t^2$.

33. An arrow is shot directly up with an initial velocity of 144 ft/sec. How long does it take to return to the ground? 9 seconds
34. A rocket is fired directly up at an initial velocity of 2240 ft/sec. How high will the rocket be after 70 seconds? 78,400 ft

Answers to Chapter Test

4. $(3a + 2b)(5c - 7d)$

20. $\left\{-\frac{5}{3}, \frac{3}{2}\right\}$

Using the Cumulative Review
If students have difficulty with a particular topic, have them review the appropriate lesson. You may want to provide copies of the solutions from the *Algebra One Solutions Manual*. This Cumulative Review covers Chapters 1-8.

Other Cumulative Reviews appear on pages 72, 142, 208, 334, 402, 466, and 526.

Answers

5.

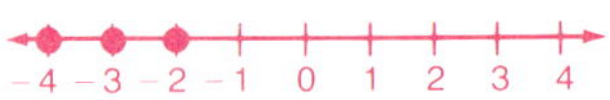

17.

Cumulative Review

1. Write a mathematical expression for 37 increased by twice a number. $37 + 2n$
2. Evaluate: $[4(15 - 3) \div 6 \times 2] + 4^2$ 32
3. State the property shown by $7 \cdot 0 = 0$.
4. Simplify: $\frac{3}{4}(20 - 10x)$ $15 - \frac{15}{2}x$
5. Graph $\{-2, -3, -4\}$ on a number line.
6. Evaluate $b + (-16)$ if $b = 6$. -10

3. multiplicative property of zero

Find each sum or difference.

7. $\frac{3}{4} + \left(-\frac{5}{6}\right) + \frac{3}{2}$ $\frac{17}{12}$
8. $-14 - 21$ -35
9. Solve: $-14 = 6 - z$ 20
10. Simplify: $5(4a - 2a) + 6(7a - 4a)$ $28a$

Solve.

11. $\frac{a + 2}{7} = -10$ -72
12. $7.6x + 11.2 = -13.4x - 9.8$ -1
13. Find 1.8% of 15. 0.27

Solve each inequality.

14. $16 \leq y - 5$ $y \geq 21$
15. $-7p < 84$ $p > -12$
16. $\frac{1}{3}x - 1 \geq -4$ $x \geq -9$
17. Solve the following inequality and graph the solution set:
$2x + 5 \leq 9$ and $3x \geq 2x - 3$ $-3 \leq x \leq 2$

Simplify.

18. $(4a^2b)(-3b^3)$ $-12a^2b^4$
19. $-2(ax^3y)^3$ $-2a^3x^9y^3$
20. $\frac{-64a^3b^5c^4d}{-16a^2b^6c^3d}$ $\frac{4ac}{b}$
21. $7m^2p^3(m^4 - 3mp^5 + 2p^4)$
22. $(2a + 3)(a^2 + 3a + 4)$
23. $(2x + 3y)(2x - 3y)$ $4x^2 - 9y^2$

Solve.

24. $q(q - 1) + 3q = 1 + q(q + 3)$ -1

Factor, if possible.

25. $25x^4 + 5y^2$ $5(5x^4 + y^2)$
26. $a^2 - 16y^2$ $(a + 4y)(a - 4y)$
27. $b^2 - b - 20$ $(b - 5)(b + 4)$
28. $12x^2 + 26x + 12$ $2(3x + 2)(2x + 3)$
29. $2a^2 + 10ab + 3ab + 15b^2$ $(2a + 3b)(a + 5b)$

Solve.

30. $(3y + 2)(y - 5) = 0$ $-\frac{2}{3}, 5$
31. $5m^2 + 5m = 0$ $-1, 0$
32. $a^3 + 6a^2 + 9a = 0$ $-3, 0$
33. $n^2 + n = 2$ $-2, 1$

Problem Solving

Solve each problem.

34. Darlene earned \$340 in 4 days. At that rate how long will it take her to earn \$935? 11 days
35. Alan paid \$96.70 for a special computer program. The price included $7\frac{1}{2}$% sales tax. What was the price without the tax? \$89.95
36. The length of a rectangle is 12 meters greater than the width. If the length was decreased by 5 meters and the width increased by 2 meters, the area would be decreased by 55 square meters. Find the original dimensions of the rectangle.
37. Dylana drives from home to town at an average speed of 30 mph. She returns home at an average speed of 40 mph. If her total driving time is $3\frac{1}{2}$ hours, what is the distance between her home and town? 60 mi
38. When one integer is added to the square of the next consecutive integer, the sum is 155. Find the integers. 11, 12 or -14, -13

36. 23 meters by 35 meters

21. $7m^6p^3 - 21m^3p^8 + 14m^2p^7$ 22. $2a^3 + 9a^2 + 17a + 12$

Standardized Test Practice Questions

The questions on this page involve comparing two quantities, one in Column A and one in Column B. In certain questions, information related to one or both quantities is centered above them. All variables used stand for real numbers.

Directions:

Write A if the quantity in Column A is greater.

Write B if the quantity in Column B is greater.

Write C if the quantities are equal.

Write D if there is not enough information to determine the relationship.

	Column A	Column B
	$x \geq 13$	
1. B	$\frac{37}{3}$	x
2. A	$\frac{0.4}{20}$	0.002
3. A	$5 + 2 \cdot 4 - 2$	$10 \div 5 + 4 - 3$
	$b > c + 1$	
4. A	b	c
	$b < c + 1$	
5. D	b	c
6. B	The number of which 6 is 20%	10% of 310
	$a = b$	
7. C	$-5(a - b)$	$7(3b - 3a)$
8. B	$6\left(\frac{3}{4}\right)0$	$24\left(\frac{1}{6}\right)$
9. C	25% of $\frac{4}{7}$	$\frac{4}{7}$ of $\frac{1}{4}$
10. A	$0.02 \div 0.2$	$0.2 \cdot 0.02$
11. B	The value of $2y + 7$ when $y = -2$.	The value of $2x - 8$ when $x = 6$.

Examples

	Column A	Column B
1.	$\frac{x}{y}$	$\frac{y}{x}$

The answer is D because it is not known whether x and y are positive or negative.

The values of $\frac{x}{y}$ and $\frac{y}{x}$ depend on the choices for x and y.

	Column A	Column B
	$a > 0$	
2.	$4a - 3a$	$4 \cdot 3a$

The answer is B because $12a$ is greater than a for all positive values of a.

	Column A	Column B
12. B	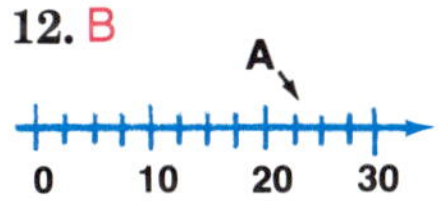 The number indicated by arrow A on the number line above.	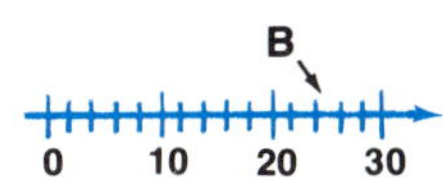 The number indicated by arrow B on the number line above.

13. A The price of a stereo increased by 20%, and then decreased by 20%.

	Column A	Column B
	The original price of the stereo.	The new price of the stereo.
	$a:b = c:d$	
14. C	ad	bc
	$3x - 3y = 24$	
15. A	x	y
	$r < 0 < s$	
16. D	r^2	$\frac{s}{2}$

Using the Standardized Test Practice Questions

The questions on this page may be used to help students prepare for pre-college tests. These questions usually require careful analysis and a thorough understanding of the concepts.

This page could be used as an overnight assignment for better students. Other students may benefit most if only a few questions are assigned at a time.

After students have chosen an answer for each problem, you should discuss how each problem can be solved, or provide copies of the solutions from the *Algebra One Solutions Manual*.

Other practice questions appear in this text on pages 73, 143, 209, 335, 403, 467, and 527. Also see pages 50-58 in the *Teacher Resource Book*.

About the Photo

The photograph illustrates how ordered pairs are related to the coordinates on a map.

The key for a map often gives the location of a city or town as a pair consisting of a letter and a number, which correspond to horizontal and vertical scales. Notice that the letters and numbers along the edges of the map are positioned between, rather than along, the grid lines. The city or town is located in the region which corresponds to the letter and number.

Ordered pairs are used to locate points, rather than regions, in the coordinate plane. The numbers along the scales are positioned on the grid lines rather than between them.

Suggest that students bring maps from home, then discuss the different types of scales used.

Teacher Resource Book Materials for Chapter 9 See pages 217-236.

CHAPTER 9

Functions and Graphs

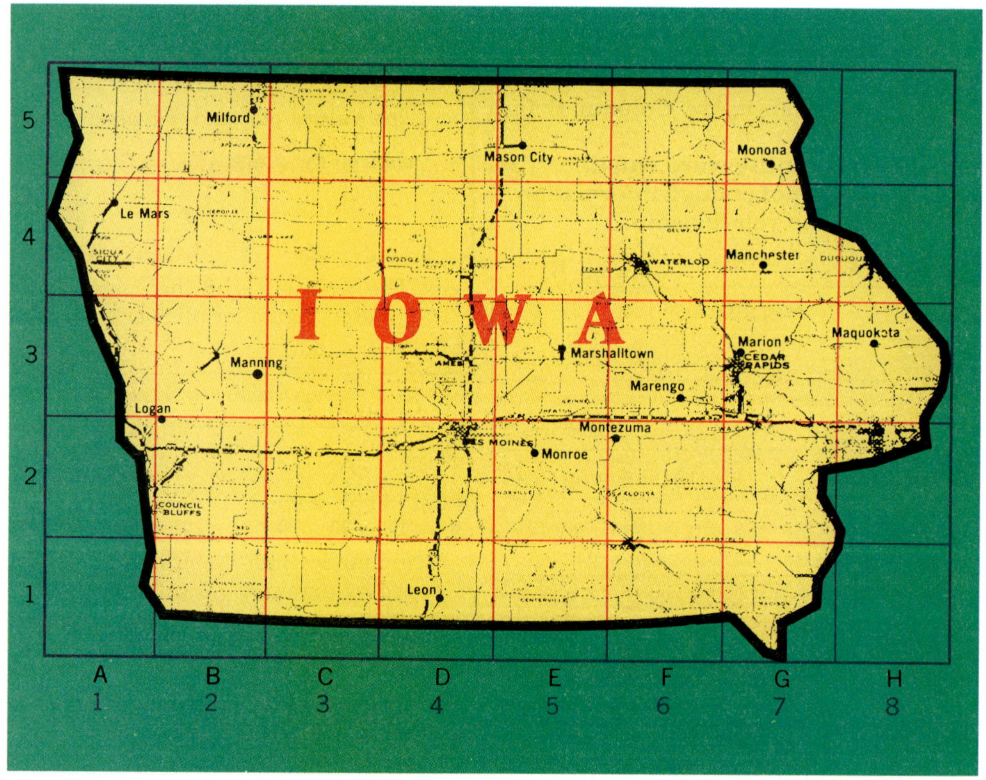

The letters and numbers shown beside the map above are used to identify regions on the map. For example, Manning is in region B–3. This region is above B and across from 3.

Suppose the letters were replaced by numbers as shown. Then the region containing Manning could be identified by 2–3, or by the **ordered pair** (2, 3). Would (3, 2) identify the same region? Why is the order of the numbers important?

In this chapter, you will learn to graph ordered pairs. You will also learn more about functions and their graphs.

Objective 9-1: Graph ordered pairs.

9-1 Ordered Pairs

In mathematics, **ordered pairs** of numbers are used to locate points in the plane. The points are in reference to two perpendicular number lines, as shown at the right. Notice that the lines intersect at their zero points. The point of intersection is called the **origin**. The two number lines are often called the ***x*-axis** and ***y*-axis**. The first component of an ordered pair corresponds to a number on the horizontal, or x-axis. The second component of the ordered pair corresponds to a number on the vertical, or y-axis.

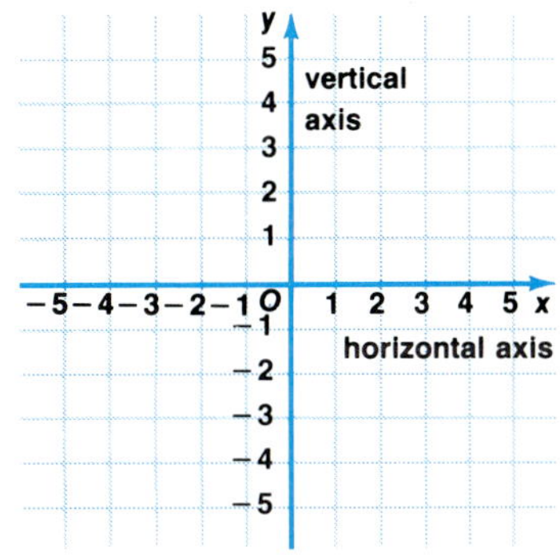

To find the ordered pair that locates point A, think of a horizontal line and vertical line through point A. Notice where these lines intersect the axes. The number on the x-axis that corresponds to A is -3. The number on the y-axis that corresponds to A is 4. Thus, the ordered pair for point A is $(-3, 4)$.

The first component, -3, is called the ***x*-coordinate** of point A. The second component, 4, is called the ***y*-coordinate** of point A. The plane is called the **coordinate plane**.

The first component is sometimes called the abscissa. The second component is sometimes called the ordinate.

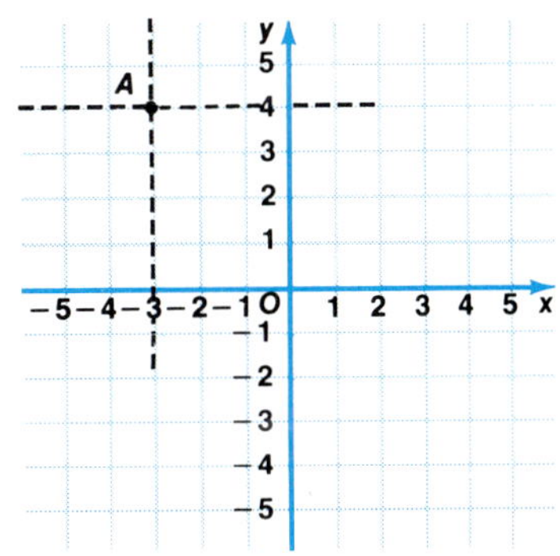

Example

1 Name the ordered pairs that locate points *R*, *S*, *T*, and *U*.

The x-coordinate of point R is 4. The y-coordinate is -3. Thus, the ordered pair for R is $(4, -3)$.

The x-coordinate of point S is 1. The y-coordinate is 4. Thus, the ordered pair for S is $(1, 4)$.

The ordered pair for T is $(-3, -5)$ and the ordered pair for U is $(-5, 3)$. Do you see why?

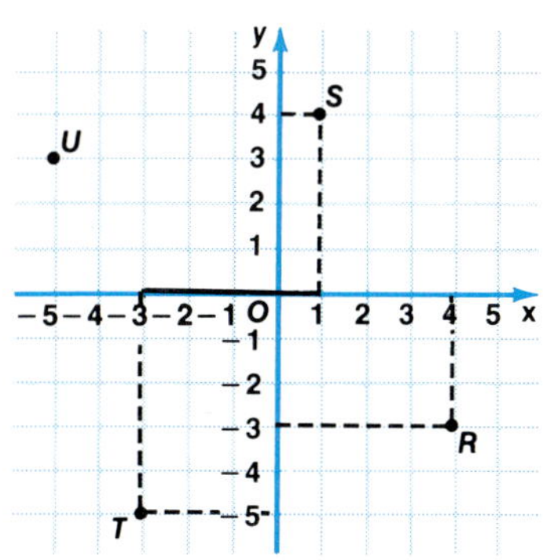

9-1

Lesson Notes, Page T41.

Readiness Activity

Use the number line below to answer each of the following questions.

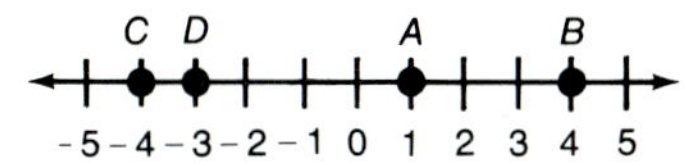

What is the coordinate of point C? -4 of point B? 4

What point is the graph of -3? D of 1? A

Discussion Question

(Use after example 1.)

Graph ordered pairs in which the x-coordinate is equal to the y-coordinate. What pattern do you observe? The points form a line.

Chalkboard Exercises

1. Name the ordered pairs that locate points R, S, T, and U.

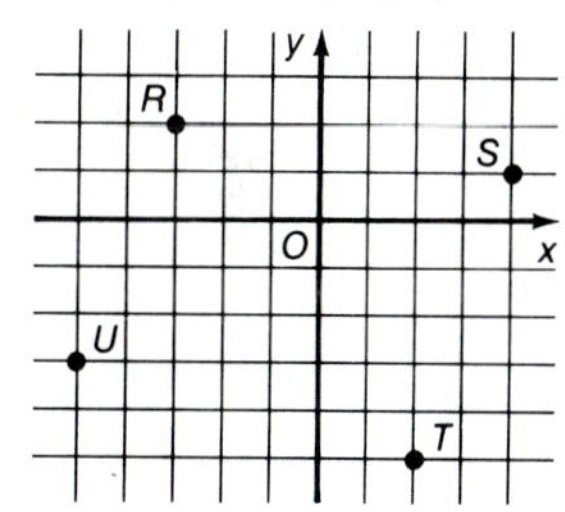

R $(-3, 2)$
S $(4, 1)$
T $(2, -5)$
U $(-5, -3)$

Chalkboard Exercises

2. Graph each point.

$Q(4, 6)$	$R(-4, -6)$
$S(0, 3)$	$T(-3, 0)$
$U(2, -5)$	$V(-2, -3)$
$W(5, 0)$	$X(1, 1)$
$Y(-2, 5)$	$Z(1, -3)$

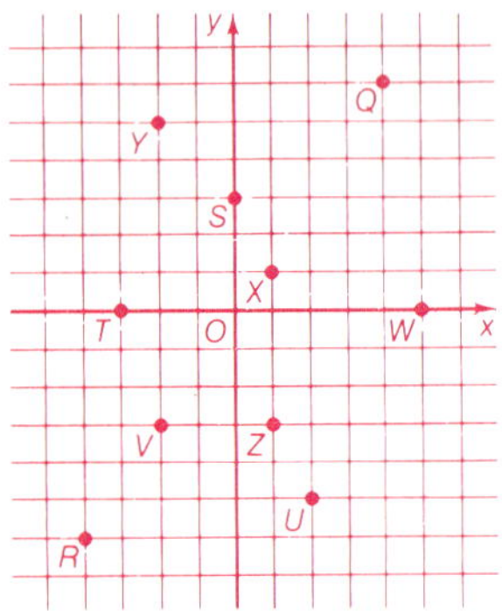

❶ Discuss why the moves for positive numbers are to the right or up, and the moves for negative numbers are to the left or down.

❷ When graphing ordered pairs, students may find it helpful to use their pencils as pointers as they are locating the point for an ordered pair. Emphasize that they should *not* draw a path to the point. In examples 2 and 3, the path to the point is shown for explanatory purposes only.

❸ Point out that the signs of the coordinates in each quadrant remain the same. The following diagram will help clarify this point.

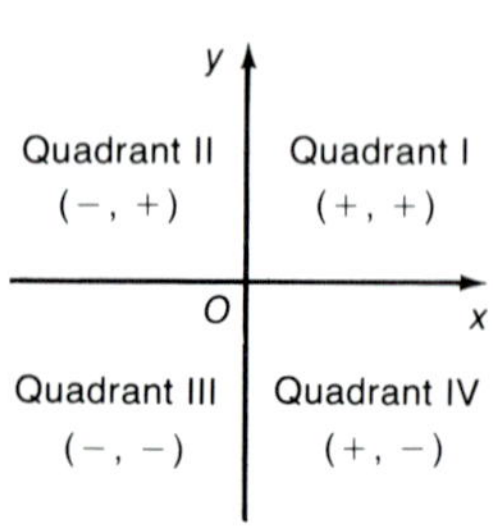

Sometimes a point is named not only by a letter but also by its location. For example, $R(4, 3)$ means that the location of point R is (4, 3).

To **graph** an ordered pair means to place a dot at the point that is located by the ordered pair. This is sometimes called *plotting* the point. When graphing an ordered pair, you should start at the origin. The first component gives the number of units to move right or left. The second component gives the number of units to move up or down. Study these examples. ❶

Examples

2 **Graph point $C(3, 1)$.**

Start at O. Move 3 units to the right. Then move 1 unit up and place a dot. Label the dot with the letter C.

Check to make sure the dot corresponds to 3 on the x-axis and 1 on the y-axis.

3 **Graph point $D(-3, -2)$.**

Start at O. Move 3 units to the left. Do you see why the move was to the left? Then move 2 units down and place a dot. Do you see why the move was down? Label the dot with the letter D.

Check to make sure the dot corresponds to −3 on the x-axis and −2 on the y-axis.

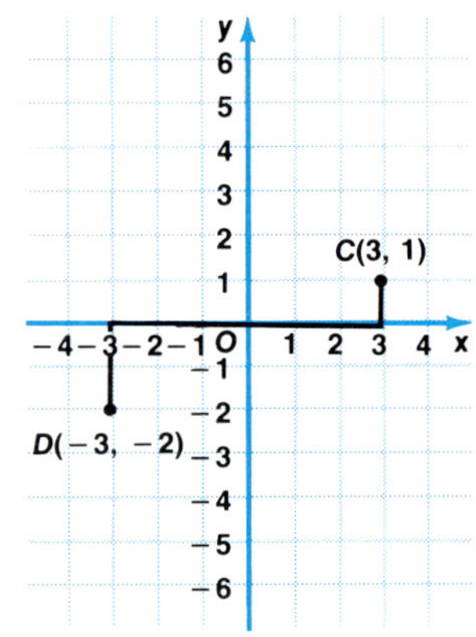

❷

When plotting points, the following is true.

> **1. Exactly one point in the plane is located by a given ordered pair of numbers.**
> **2. Exactly one ordered pair of numbers locates a given point in the plane.**

Completeness Property for Points in the Plane

The x-axis and y-axis separate the plane into four regions, called **quadrants**. The quadrants are numbered as shown at the right. In which quadrant is the graph of $(5, -4)$? In which quadrant is the graph of $(-2, -7)$? In which quadrant is the graph of $(2, 0)$? The axes are not in any quadrant. Can you think of a reason why? ❸

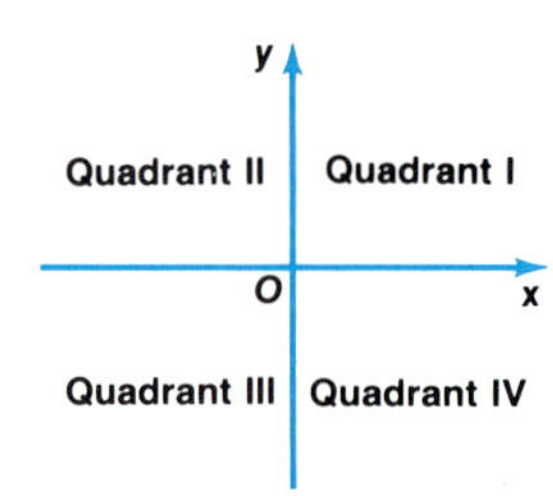

Exploratory Exercises

State the ordered pair for each point.

1. A (1, 4) **2.** B (−1, 1)
3. C (−3, 3) **4.** D (−4, −3)
5. E (−1, −2) **6.** F (3, −2)
7. G (4, 0) **8.** H (4, 2)
9. I (1, 1) **10.** J (0, −1)
11. K (−1, −1) **12.** L (−3, 0)
13. M (−2, 2) **14.** N (2, −3)
15. P (−3, −4) **16.** Q (−4, 1)
17. R (2, 2) **18.** S (5, 1)
19. T (3, −1) **20.** U (0, −5)

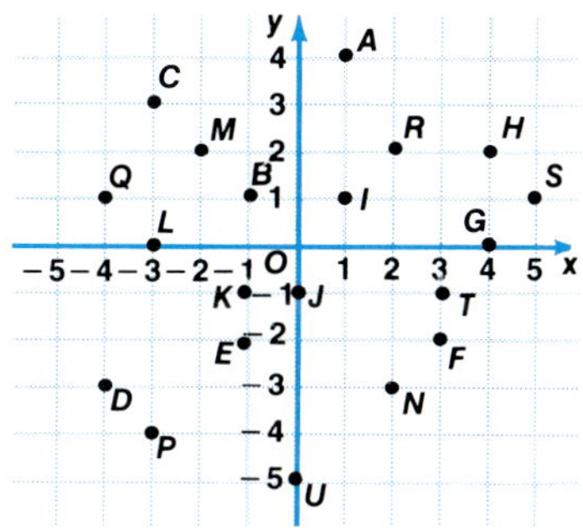

State the quadrant in which the point for each ordered pair is located.

21. (5, 2) I **22.** (−3, −1) III **23.** (−2, 3) II **24.** (6, 0) none
25. (0, −2) none **26.** (−6, −2) III **27.** (4, −3) IV **28.** (5, 1) I
29. (3, −2) IV **30.** (3, 6) I **31.** (−4, 2) II **32.** (−3, 0) none
33. (−3, 1) II **34.** (−4, 0) none **35.** (−1, −3) III **36.** (5, −1) IV

State the quadrant for the graph of (*x*, *y*), given each condition.

37. $x > 0, y < 0$ IV **38.** $x < 0, y < 0$ III
39. $x = 0, y > 0$ none **40.** $x < 0, y > 0$ II
41. $x > 0, y > 0$ I **42.** $x > 0, y = 0$ none
43. $x = -1, y < 0$ III **44.** $x < 0, y = 3$ II

Written Exercises

Graph each point.

1. $A(5, -2)$ **2.** $B(3, 5)$ **3.** $C(-6, 0)$ **4.** $D(-3, 4)$
5. $E(-3, -3)$ **6.** $F(-5, 1)$ **7.** $G(2, -1)$ **8.** $H(4, 0)$
9. $I(3, -4)$ **10.** $J(-3, 0)$ **11.** $K(-5, -2)$ **12.** $L(4, 2)$
13. $M(-5, -4)$ **14.** $N(1, 4)$ **15.** $P(2, -3)$ **16.** $Q(0, -2)$
17. $R(-2, 1)$ **18.** $S(-3, -1)$ **19.** $T(-3, 6)$ **20.** $U(6, 2)$

For each of the following, graph the points named. Then connect the points in alphabetical order and identify the figure.

21. $A(4, 6), B(3, 7), C(2, 6), D(1, 7), E(0, 6), F(1, 8), G(4, 9), H(7, 8), I(8, 6), J(7, 7), K(6, 6), L(5, 7), M(4, 6), N(4, 0), P(5, 0), Q(5, 1)$.

22. $A\left(-3, \frac{1}{2}\right), B(1, 0), C(4, 0), D(7, -4), E(8, -4), F(7, 0), G(10, 0), H(11, -2), I(12, -2), J\left(11\frac{1}{2}, 0\right), K\left(13, \frac{1}{2}\right), L\left(11\frac{1}{2}, 1\right), M(12, 3), N(11, 3), P(10, 1), Q(7, 1), R(8, 5), S(7, 5), T(4, 1), U(1, 1), V\left(-3, \frac{1}{2}\right)$.

Suggested Daily Assignments

Enriched
Exp: 14–36 evens, 37–44
Written: 11–19 odds, 21, 22

Average
Exp: 1–35 odds, 37–44
Written: 1–22

Fundamental
Exp: 1–44
Written: 1–20

Answers to Written 1–20.

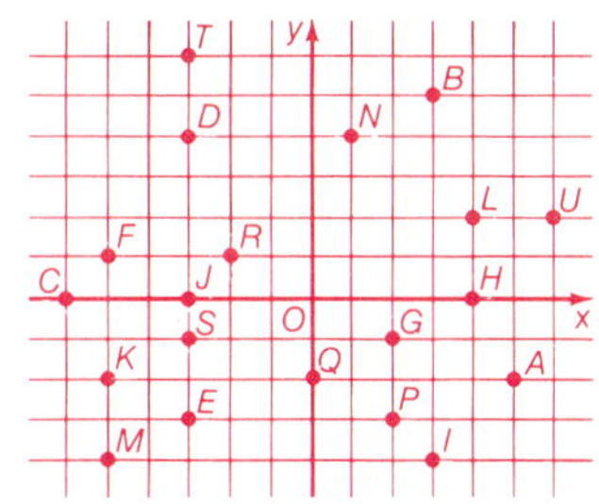

21. umbrella

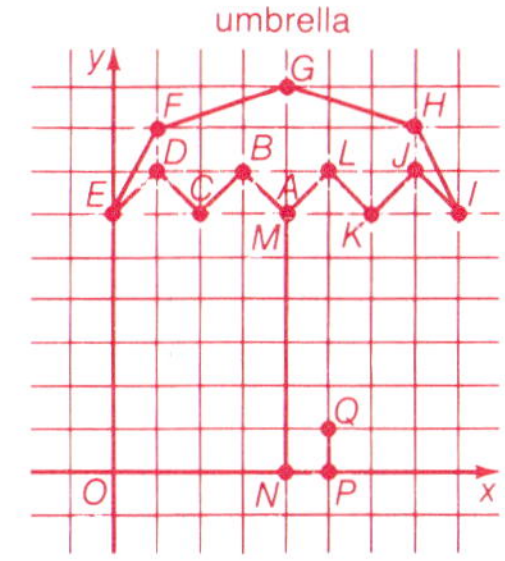

22. airplane or jet

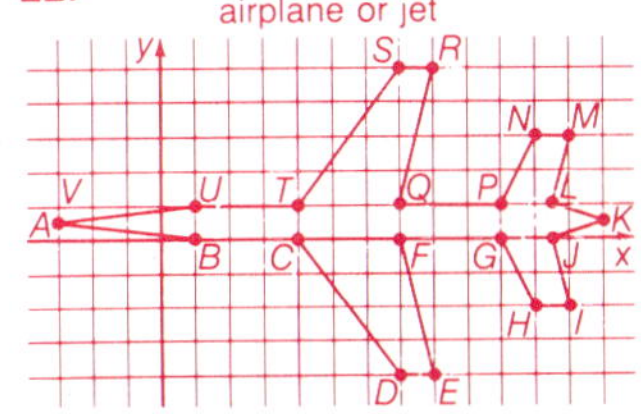

9-2

Lesson Notes, page T42

Discussion Question
(Use after page 274.)
The domain of a relation is {1, 3, 5}, and the range is {2, 4}. If the relation is the set of all possible ordered pairs, how many pairs are there for the relation? 6 pairs

❶ The word *relation* simply indicates a set of pairs of numbers. There may or may not be a reason for the pairing. Stress that a relation can be shown in many different ways. Remind students that the domain is a set of numbers, not a set of ordered pairs. The same is true of the range.

Chalkboard Exercises
Write each relation as a set of ordered pairs. Then state the domain and range of each.

1.

x	y
1	5
2	7
−1	1
−3	−3

{(1, 5), (2, 7), (−1, 1), (−3, −3)}
D = {1, 2, −1, −3,}
R = {5, 7, 1, −3,}

2.

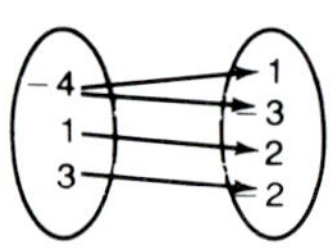

{(−4, 1), (−4, −3), (1, 2), (3, −2)}
D = {−4, 1, 3}
R = {1, −3, 2, −2}

3.

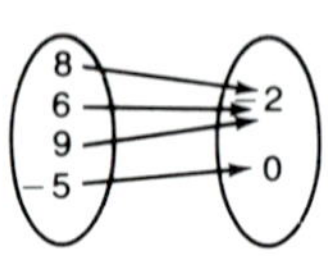

{(8, −2), (6, −2), (9, −2), (−5, 0)}
D = {8, 6, 9, −5}
R = {−2, 0}

Objective 9-2A: Identify the domain and range of a relation.
Objective 9-2B: Show relations as sets of ordered pairs and mappings.

9-2 Relations

Consider the ordered pairs (2, 2), (−2, 3), and (0, −1). In these examples, 2 is paired with 2, −2 is paired with 3, and 0 is paired with −1.

A relation is a set of ordered pairs.

❶ ***Definition of Relation***

A relation can be shown by a set of ordered pairs, a table, a mapping, or a graph.

Set

{(2, 2), (−2, 3), (0, −1)}

Table

x	*y*
2	2
−2	3
0	−1

Mapping

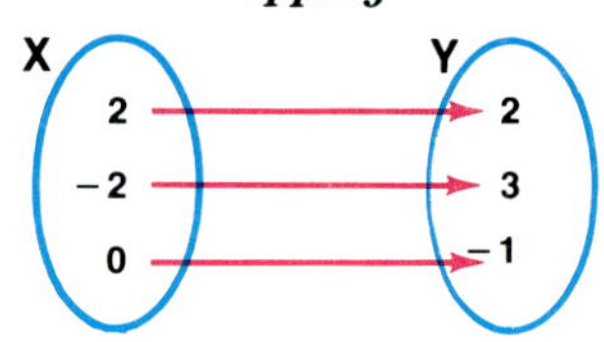

Graph

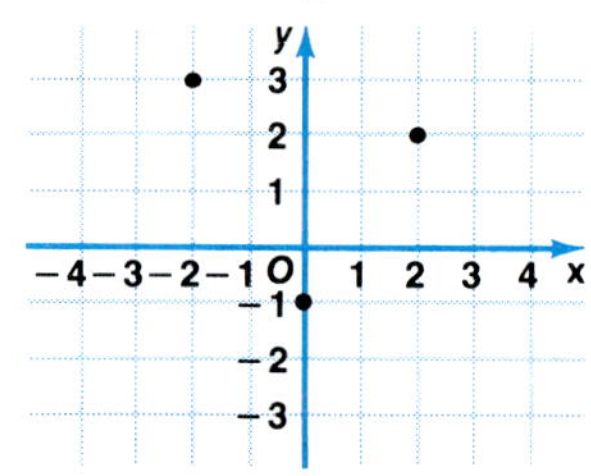

The first components of the above ordered pairs are 2, −2, and 0. The **domain** of the relation is {2, −2, 0}.

The domain of a relation is the set of all *first* components from each ordered pair.

Definition of Domain

The second components are 2, 3, and −1. The **range** of the relation is {2, 3, −1}.

The range of a relation is the set of all *second* components from each ordered pair.

Definition of Range

Examples

1 **State the set of ordered pairs shown by the table. Then state the domain and range of the relation.**

x	y
0	5
2	3
1	-4
-3	3
-1	-2

The set of ordered pairs in the relation is $\{(0, 5), (2, 3), (1, -4), (-3, 3), (-1, -2)\}$.

The domain is the set of first components or $\{0, 2, 1, -3, -1\}$.

The range is the set of second components or $\{5, 3, -4, -2\}$.

2 **State the relation shown by the graph. Then state the domain and range.**

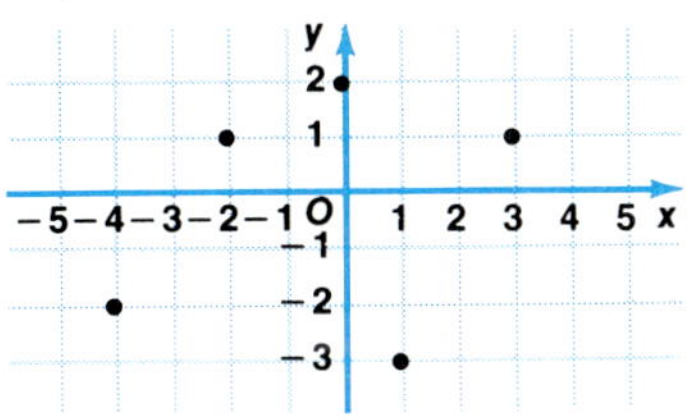

The relation is $\{(3, 1), (0, 2), (-2, 1), (-4, -2), (1, -3)\}$.

The domain is $\{3, 0, -2, -4, 1\}$.

The range is $\{1, 2, -2, -3\}$.

3 **State the relation for the mapping. Then state the domain and range.**

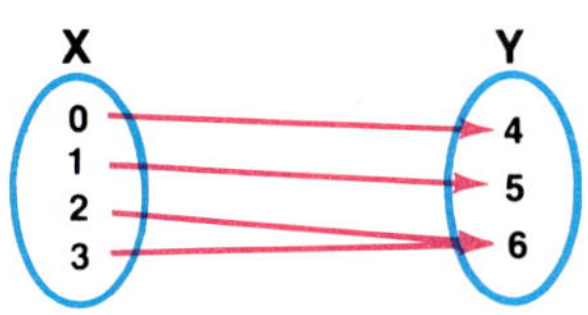

The relation is $\{(0, 4), (1, 5), (2, 6), (3, 6)\}$.

The domain is $\{0, 1, 2, 3\}$.

The range is $\{4, 5, 6\}$.

Exploratory Exercises

State the domain and range of each relation.

1. $\{(0, 2), (1, -2), (2, 4)\}$

2. $\{(-4, 2), (-2, 0), (0, 2), (2, 4)\}$

3. $\{(-3, 1), (-2, 0), (-1, 1), (0, 2)\}$

4. $\{(5, 2), (0, 0), (-9, -1)\}$

5. $\{(7, 5), (-2, -3), (4, 0), (5, -7), (-9, 2)\}$

6. $\{(1, 1), (2, 4), (-2, 4), (3, 9)\}$

7. $\{(-3, 0), (-2, -5), (-2, 6), (3, 7), (0, -17)\}$

8. $\{(-5, 2), (-4, 5), (-2, 1), (-4, 4)\}$

9. $\{(3.1, -1), (-4.7, 3.9), (2.4, -3.6), (-9, 12.12)\}$

10. $\{(4.7, -2.3), (5.1, 2.8), (-3.5, 1)\}$

11. $\left\{\left(\frac{1}{2}, \frac{1}{4}\right), \left(1\frac{1}{2}, -\frac{2}{3}\right), \left(-3, \frac{2}{5}\right), \left(-5\frac{1}{4}, -7\frac{2}{7}\right)\right\}$

12. $\left\{(0, 0), \left(\frac{1}{2}, \frac{1}{8}\right), \left(-\frac{1}{2}, -\frac{1}{8}\right)\right\}$

Suggested Daily Assignments

Enriched
- Exp: 8–16 evens
- Written: 2–28 evens
- Mini-Review: 1–5

Average
- Exp: 2–16 evens
- Written: 1–18, 20–28 evens
- Mini-Review: 1–5

Fundamental
- Exp: 1–16
- Written: 1–18
  - p. 273, 21, 22
- Mini-Review: 1–5

Answers to Exploratory

1. $D = \{0, 1, 2\}$
$R = \{-2, 2, 4\}$

2. $D = \{-4, -2, 0, 2\}$
$R = \{0, 2, 4\}$

3. $D = \{-3, -2, -1, 0\}$
$R = \{1, 0, 2\}$

4. $D = \{5, 0, -9\}$
$R = \{2, 0, -1\}$

5. $D = \{7, -2, 4, 5, -9\}$
$R = \{5, -3, 0, -7, 2\}$

6. $D = \{1, 2, -2, 3\}$
$R = \{1, 4, 9\}$

7. $D = \{-3, -2, 3, 0\}$
$R = \{0, -5, 6, 7, -17\}$

8. $D = \{-5, -4, -2\}$
$R = \{2, 5, 1, 4\}$

9. $D = \{3.1, -4.7, 2.4, -9\}$
$R = \{-1, 3.9, -3.6, 12.12\}$

10. $D = \{4.7, 5.1, -3.5\}$
$R = \{-2.3, 2.8, 1\}$

11. $D = \left\{\frac{1}{2}, 1\frac{1}{2}, -3, -5\frac{1}{4}\right\}$
$R = \left\{\frac{1}{4}, -\frac{2}{3}, \frac{2}{5}, -7\frac{2}{7}\right\}$

12. $D = \left\{0, \frac{1}{2}, -\frac{1}{2}\right\}$
$R = \left\{0, \frac{1}{8}, -\frac{1}{8}\right\}$

Answers to Exploratory

13. {(1, 5), (2, 7), (3, 9), (4, 11)}
$D = \{1, 2, 3, 4\}$
$R = \{5, 7, 9, 11\}$

14. {(1, 3), (2, 2), (4, 9), (6, 5)}
$D = \{1, 2, 4, 6\}$
$R = \{3, 2, 9, 5\}$

15. {(−4, 1), (−2, 3), (0, 1), (2, 3), (4, 1)}
$D = \{-4, -2, 0, 2, 4\}$
$R = \{1, 3\}$

16. {(1, −2), (3, −4), (5, −6), (9, −4), (11, −2)}
$D = \{1, 3, 5, 9, 11\}$
$R = \{-2, -4, -6\}$

Answers to Written

7.
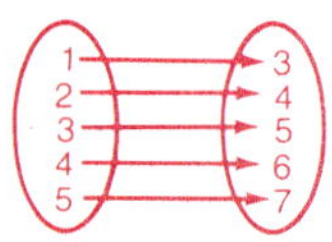

{(1, 3), (2, 4), (3, 5), (4, 6), (5, 7)}

8.
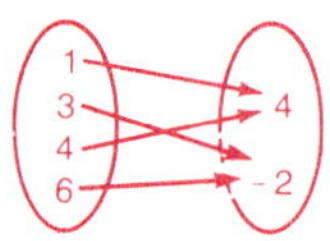

{(1, 4), (3, −2), (4, 4), (6, −2)}

9.
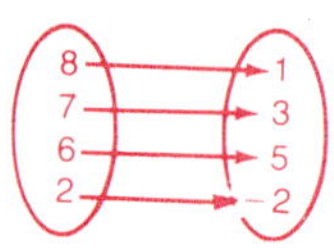

{(8, 1), (7, 3), (6, 5), (2, −2)}

10.
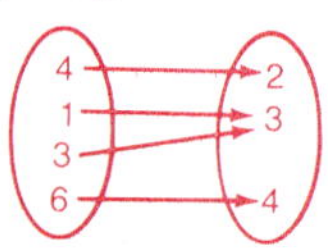

{(4, 2), (1, 3), (3, 3), (6, 4)}

11.
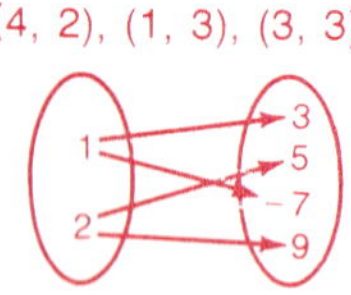

{(1, 3), (2, 5), (1, −7), (2, 9)}

12.
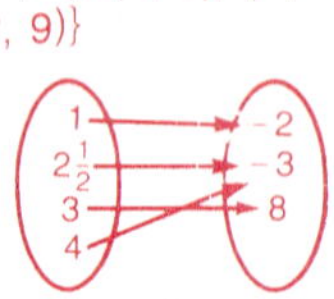

$\left\{(1, -2), \left(2\frac{1}{2}, -3\right), (3, 8), (4, -3)\right\}$

State each relation as a set of ordered pairs. Then state the domain and range of each.

13.

x	y
1	5
2	7
3	9
4	11

14.

x	y
1	3
2	2
4	9
6	5

15.

x	y
−4	1
−2	3
0	1
2	3
4	1

16.

x	y
1	−2
3	−4
5	−6
9	−4
11	−2

Written Exercises

Write the relation for each mapping as a set of ordered pairs.

1.
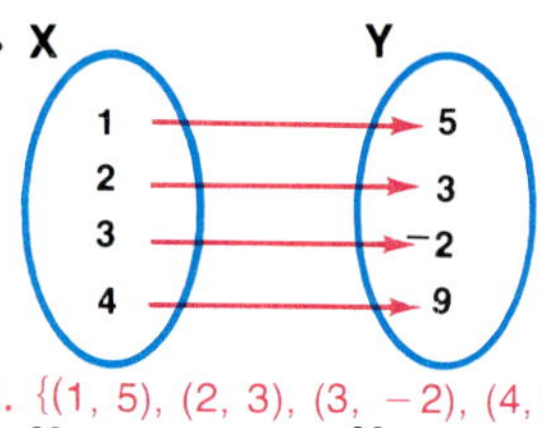

1. {(1, 5), (2, 3), (3, −2), (4, 9)}

2.
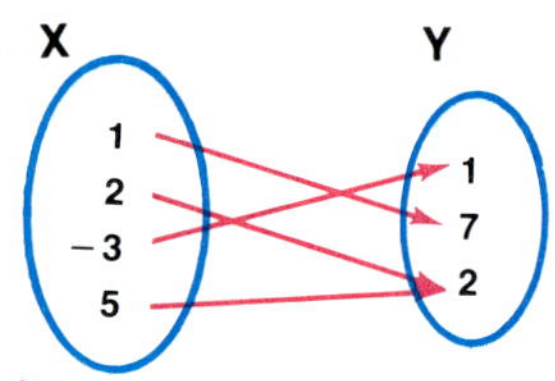

2. {(1, 7), (2, 2), (−3, 1), (5, 2)}

3.
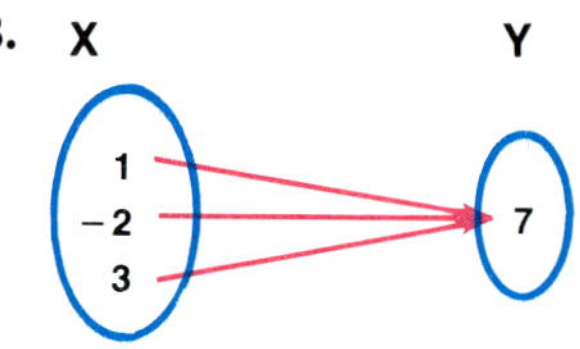

3. {(1, 7), (−2, 7), (3, 7)}

4.
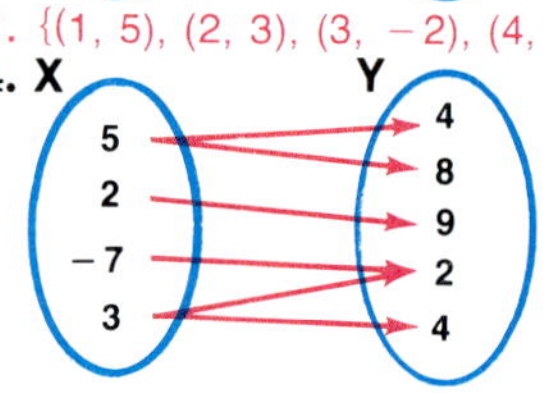

5.
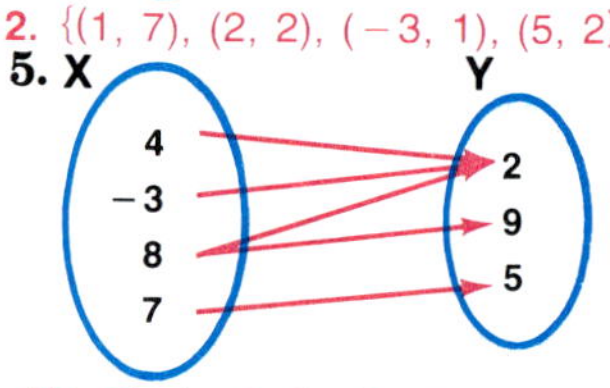

6.
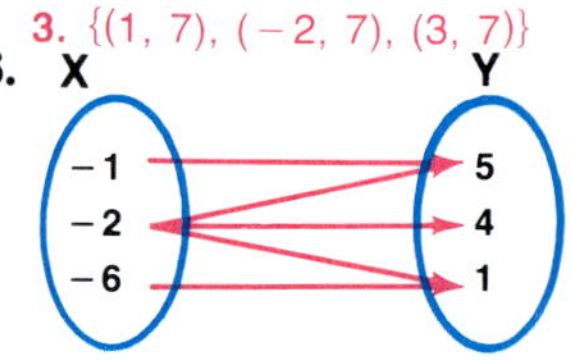

5. {(4, 2), (−3, 2), (8, 2), (8, 9), (7, 5)}

4. {(5, 4), (5, 8), (2, 9), (−7, 2), (3, 2), (3, 4)}

6. {(−1, 5), (−2, 5), (−2, 1), (−2, 4), (−6, 1)}

Draw a mapping for each relation. Then write the relation as a set of ordered pairs.

7.

x	y
1	3
2	4
3	5
4	6
5	7

8.

x	y
1	4
3	−2
4	4
6	−2

9.

x	y
8	1
7	3
6	5
2	−2

10.

x	y
4	2
1	3
3	3
6	4

11.

x	y
1	3
2	5
1	−7
2	9

12.

x	y
1	−2
$2\frac{1}{2}$	−3
3	8
4	−3

Write each relation as a set of ordered pairs. Then state the domain and range of each.

13.
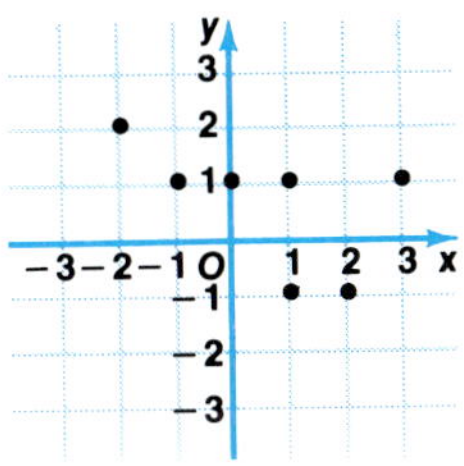

14.
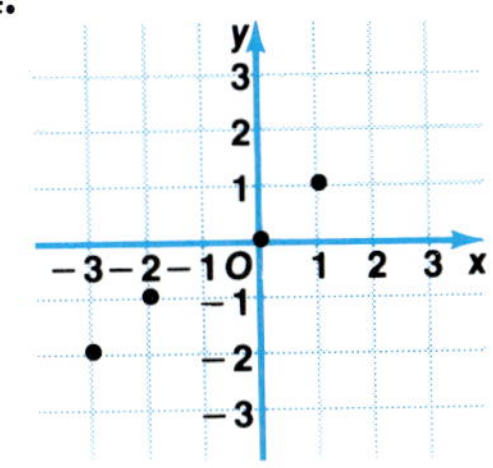

15.
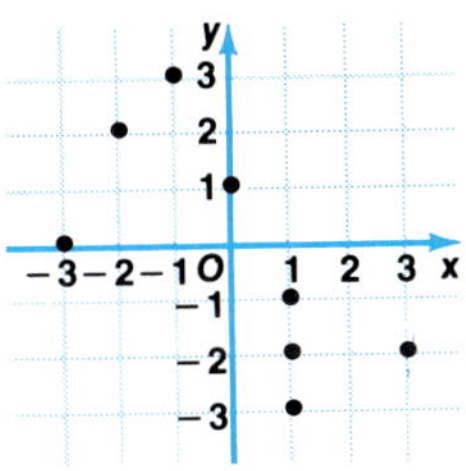

16.
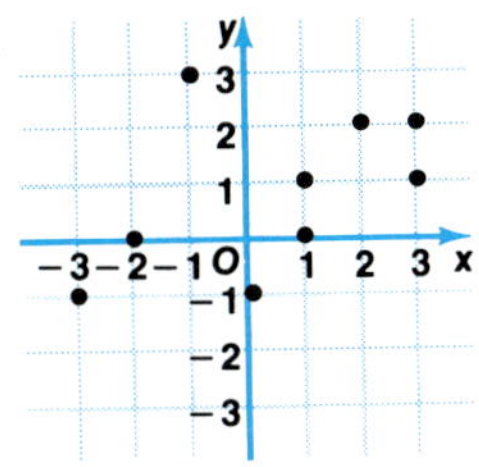

17.
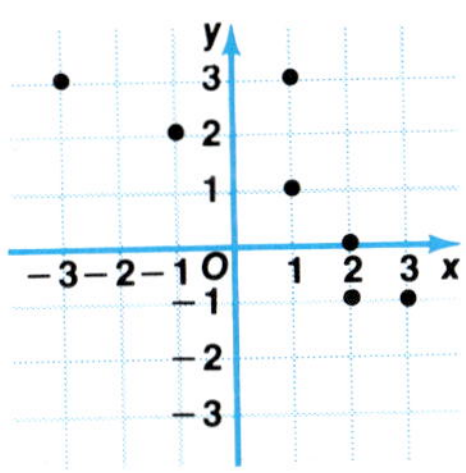

18.
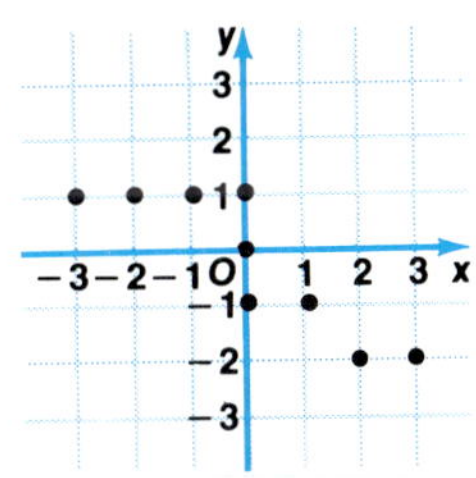

Draw a mapping for each relation.

19. $\{(1, 3), (2, 5), (8, 2), (5, -3)\}$

20. $\{(2, 3), (3, 2)\}$

21. $\{(8, 1), (4, 2), (6, -4), (5, -3), (6, 0)\}$

22. $\{(0, 1)\}$

23. $\{(2, 2), (3, 3), (4, 4)\}$

24. $\{(1, 3), (2, 7), (4, 1), (-3, 3), (3, 3)\}$

25. $\{(-6, 0), (-1, 2), (-3, 4)\}$

26. $\{(-3, 3), (-2, 2), (-1, 1), (0, 0), (1, -1)\}$

27. $\left\{\left(4\frac{1}{2}, \frac{1}{2}\right), \left(2\frac{1}{2}, 1\right), \left(-1\frac{1}{2}, 2\right)\right\}$

28. $\left\{\left(\frac{1}{3}, \frac{3}{4}\right), \left(\frac{2}{3}, -\frac{1}{2}\right), \left(-\frac{3}{4}, \frac{2}{3}\right), \left(\frac{2}{3}, \frac{3}{4}\right)\right\}$

mini-review

Arrange the terms of the polynomial so that the powers of x are in descending order.

1. $8 - xy^3 - 2x^3y^2 - 5x^2$ $\quad -2x^3y^2 - 5x^2 - xy^3 + 8$

Factor.

2. $8m^2 - 72n^2$ $\quad 8(m + 3n)(m - 3n)$

3. $y^2 - \frac{3}{2}y + \frac{9}{16}$ $\quad \frac{1}{16}(4y - 3)^2$

Solve.

4. $(2m + 7)(5m - 3) = 0$ $\quad \left\{-\frac{7}{2}, \frac{3}{5}\right\}$

5. $a^2 + 2a = 63$ $\quad \{7, -9\}$

Answers to Written

13. $\{(-2, 2), (-1, 1), (0, 1), (1, 1), (1, -1), (2, -1), (3, 1)\}$
$D = \{-2, -1, 0, 1, 2, 3\}$
$R = \{2, 1, -1\}$

14. $\{(-3, -2), (-2, -1), (0, 0), (1, 1)\}$
$D = \{-3, -2, 0, 1\}$
$R = \{-2, -1, 0, 1\}$

15. $\{(-3, 0), (-2, 2), (-1, 3), (0, 1), (1, -1), (1, -2), (1, -3), (3, -2)\}$
$D = \{-3, -2, -1, 0, 1, 3\}$
$R = \{0, 2, 3, 1, -1, -2, -3\}$

16. $\{(-3, -1), (-2, 0), (-1, 3), (0, -1), (1, 0), (1, 1), (2, 2), (3, 1), (3, 2)\}$
$D = \{-3, -2, -1, 0, 1, 2, 3\}$
$R = \{-1, 0, 3, 1, 2\}$

17. $\{(-3, 3), (-1, 2), (1, 1), (1, 3), (2, 0), (2, -1), (3, -1)\}$
$D = \{-3, -1, 1, 2, 3\}$
$R = \{3, 2, 1, 0, -1\}$

18. $\{(-3, 1), (-2, 1), (-1, 1), (0, 1), (0, 0), (0, -1), (1, -1), (2, -2), (3, -2)\}$
$D = \{-3, -2, -1, 0, 1, 2, 3\}$
$R = \{1, 0, -1, -2\}$

19.
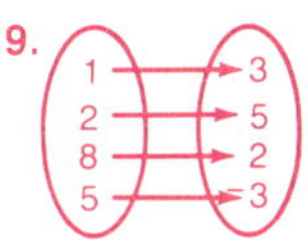

20.
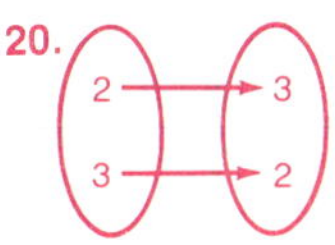

21.
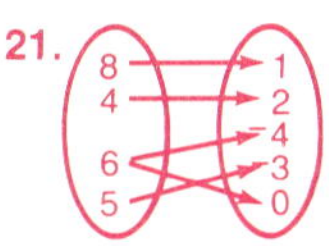

22.
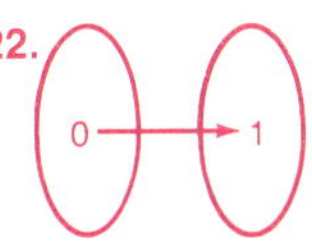

23.
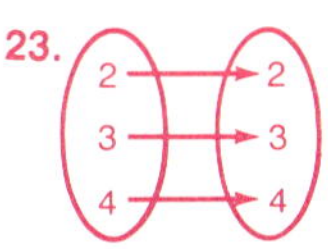

24.
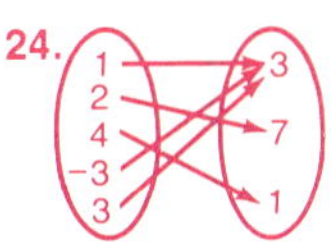

25.
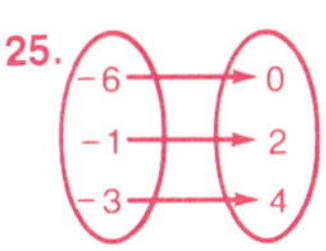

26.
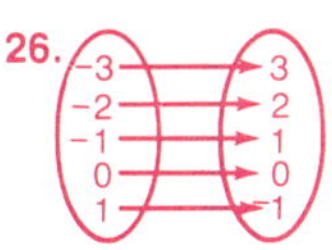

27.
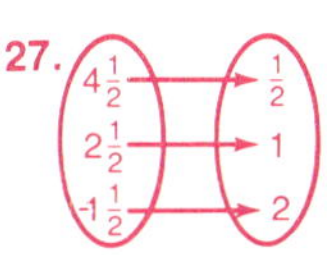

28.
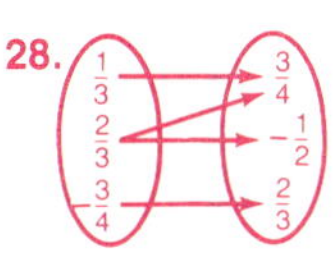

9-3

Lesson Notes, page T42

Readiness Activity

Suppose $y = x + 5$. Find a value for y when

(a) $x = 1$. 6

(b) $x = 3$. 8

(c) $x = -4$. 1

(d) $x = -6$. -1

Discussion Question

(Use after page 278.)

1. Why are the values of x restricted when x represents the number of hours worked per week? Because there are only 168 hours in a week.

2. Suppose x represents the number of hours worked per day. What would the restriction be in this case? $x \leq 24$

Computer Notes

Give students the following problem.

Suppose Anita earns $4 per hour when she works 40 hours or less per week, and $6 per hour for overtime. Write a computer program in BASIC which can be used to determine her pay for x hours when $x \leq 60$.

```
10 FOR X = 1 TO 40
20 LET Y = 4*X
30 PRINT X, Y
40 NEXT X
50 FOR X = 41 TO 60
60 LET Y = 40*4+6*(X-40)
70 PRINT X, Y
80 NEXT X
90 END
```


Objective 9-3A: Solve linear equations for a specific variable.
Objective 9-3B: Solve an equation for a given domain.

9-3 Equations as Relations

Anita works part-time in a bookstore. She earns $4 per hour. She made the following chart to show her earnings for different numbers of hours worked. She let x equal the number of hours she works per week, and y equal her total pay.

x (hours)	$4x$	y (pay)	(x, y)
1	4(1)	4	(1, 4)
5	4(5)	20	(5, 20)
8	4(8)	32	(8, 32)
10	4(10)	40	(10, 40)
12	4(12)	48	(12, 48)
15	4(15)	60	(15, 60)
20	4(20)	80	(20, 80)

The equation $y = 4x$ describes her pay (y) for any number of hours (x). Anita could have selected other values of x. Generally, however, these values are restricted. For example, Anita could not work less than 0 hours or more than 168 hours in a week. The ordered pairs in the chart above are *solutions* to the equation $y = 4x$.

If a true statement results when an ordered pair is substituted into an equation in two variables, the ordered pair is a solution of the equation.

Definition of the Solution of an Equation in Two Variables

Since the solutions to an equation in two variables are ordered pairs, such an equation describes a relation. The set of values of x is the domain of the relation. The set of corresponding values of y is the range.

Examples

1 **Solve $y = 2x + 3$ if the domain is $\{-5, -3, -1, 0, 1, 3, 5, 7, 9\}$.**

Make a table. The values of x come from the domain.

x	$2x + 3$	y	(x, y)
-5	$2(-5) + 3$	-7	$(-5, -7)$
-3	$2(-3) + 3$	-3	$(-3, -3)$
-1	$2(-1) + 3$	1	$(-1, 1)$
0	$2(0) + 3$	3	$(0, 3)$
1	$2(1) + 3$	5	$(1, 5)$
3	$2(3) + 3$	9	$(3, 9)$
5	$2(5) + 3$	13	$(5, 13)$
7	$2(7) + 3$	17	$(7, 17)$
9	$2(9) + 3$	21	$(9, 21)$

The solution set is
$\{(-5, -7), (-3, -3), (-1, 1), (0, 3), (1, 5), (3, 9), (5, 13), (7, 17), (9, 21)\}$.

2 **Solve $2y + 4x = 8$ if the domain is $\{-4, -3, -2, 2, 3, 4\}$.** ❶

Values for y are usually easier to determine if the equation is first solved for y in terms of x.

$$2y + 4x = 8$$
$$2y = 8 - 4x$$
$$\frac{2y}{2} = \frac{8 - 4x}{2}$$
$$y = 4 - 2x$$

Chalkboard Exercises

Solve each equation if the domain is $\{-5, -3, -1, 0, 1, 3, 5, 7, 9\}$.

1a. $y = 5x$
$\{(-5, -25), (-3, -15), (-1, -5), (0, 0), (1, 5), (3, 15), (5, 25), (7, 35), (9, 45)\}$

1b. $y = 3x - 1$
$\{(-5, -16), (-3, -10), (-1, -4), (0, -1), (1, 2), (3, 8), (5, 14), (7, 20), (9, 26)\}$

1c. $y = 4x - 3$
$\{(-5, -23), (-3, -15), (-1, -7), (0, -3), (1, 1) (3, 9), (5, 17), (7, 25), (9, 33)\}$

1d. $y = -3x + 5$
$\{(-5, 20), (-3, 14), (-1, 8), (0, 5), (1, 2), (3, -4), (5, -10), (7, -16), (9, -22)\}$

Solve each equation if the domain is $\{-4, -3, -2, 2, 3, 4\}$.

2a. $y + 2x = 3$ $y = 3 - 2x$
$\{(-4, 11), (-3, 9), (-2, 7), (2, -1), (3, -3), (4, -5)\}$

2b. $4y - 8x = 20$ $y = 5 + 2x$
$\{(-4, -3), (-3, -1), (-2, 1), (2, 9), (3, 11), (4, 13)\}$

2c. $3y + x = 15$ $y = 5 - \frac{1}{3}x$
$\left\{\left(-4, 6\frac{1}{3}\right), (-3, 6), \left(-2, 5\frac{2}{3}\right), \left(2, 4\frac{1}{3}\right), (3, 4), \left(4, 3\frac{2}{3}\right)\right\}$

❶ Transforming equations gives good practice in applying the properties of solving equations.

Chalkboard Exercises
Solve each equation if the domain is $\{-3, 0, 1, 2, 5\}$.

3a. $3m + 6p = 12$ $p = 2 - \frac{1}{2}m$
$\left\{\left(-3, 3\frac{1}{2}\right), (0, 2), \left(1, 1\frac{1}{2}\right), (2, 1), \left(5, -\frac{1}{2}\right)\right\}$

3b. $m + 3k = 14$ $m = 14 - 3k$
$\{(-3, 23), (0, 14), (1, 11), (2, 8), (5, -1)\}$

❷ Make sure students realize that the domain and range are not x's and y's. The domain and range are sets of numbers from which values for a variable are chosen. Letters other than x and y can represent the variables.

❸ Set-builder notation can be used to avoid confusion as to which variable represents the domain and which one represents the range. In this example, the relation can be written as follows:
$\{(r, s)\colon\ 3r + 2s = 11,\ r \in \{-3, 0, 1, 2, 5\}\}$.

Computer Notes
Suggest that students write a computer program for solving an equation, given a specific domain.

Suggested Daily Assignments

Enriched
Exp: 1–3
Written: 1–18, 20–36 evens

Average
Exp: 1–3
Written: 1–30

Fundamental
Exp: 1–3
Written: 1–24
p. 277, 19–28

Now substitute the given values of x to determine the corresponding values of y.

x	$4 - 2x$	y	(x, y)
-4	$4 - 2(-4)$	12	$(-4, 12)$
-3	$4 - 2(-3)$	10	$(-3, 10)$
-2	$4 - 2(-2)$	8	$(-2, 8)$
2	$4 - 2(2)$	0	$(2, 0)$
3	$4 - 2(3)$	-2	$(3, -2)$
4	$4 - 2(4)$	-4	$(4, -4)$

The solution set is $\{(-4, 12), (-3, 10), (-2, 8), (2, 0), (3, -2), (4, -4)\}$.

When variables other than x and y are used, assume that the values of the letters that come first in the alphabet are from the domain. ❷

Example

3 **Solve $3r + 2s = 11$ if the domain is $\{-3, 0, 1, 2, 5\}$.** ❸

The values for r come from the domain. Therefore, first solve the equation so that s is expressed in terms of r.

$$3r + 2s = 11$$
$$2s = 11 - 3r$$
$$\frac{2s}{2} = \frac{11 - 3r}{2}$$
$$s = \frac{11 - 3r}{2}$$

Now substitute the given values of r to determine the corresponding values of s.

r	$\frac{11 - 3r}{2}$	s	(r, s)
-3	$\frac{11 - 3(-3)}{2}$	10	$(-3, 10)$
0	$\frac{11 - 3(0)}{2}$	$\frac{11}{2}$	$\left(0, \frac{11}{2}\right)$
1	$\frac{11 - 3(1)}{2}$	4	$(1, 4)$
2	$\frac{11 - 3(2)}{2}$	$\frac{5}{2}$	$\left(2, \frac{5}{2}\right)$
5	$\frac{11 - 3(5)}{2}$	-2	$(5, -2)$

The solution set is $\left\{(-3, 10), \left(0, \frac{11}{2}\right), (1, 4), \left(2, \frac{5}{2}\right), (5, -2)\right\}$.

Exploratory Exercises

Copy each table. Then find the solutions for each equation for the domain indicated.

1. $y = 3x$

x	y	(x, y)
-4	-12	$(-4, -12)$
-2	-6	$(-2, -6)$
0	0	$(0, 0)$
1	3	$(1, 3)$
2	6	$(2, 6)$
3	9	$(3, 9)$

2. $y = 4x - 3$

x	y	(x, y)
-3	-15	$(-3, -15)$
-2	-11	$(-2, -11)$
-1	-7	$(-1, -7)$
0	-3	$(0, -3)$
2	5	$(2, 5)$
4	13	$(4, 13)$

3. $n = \dfrac{2m + 5}{3}$

m	n	(m, n)
-4	-1	$(-4, -1)$
-2	$\frac{1}{3}$	$\left(-2, \frac{1}{3}\right)$
0	$\frac{5}{3}$	$\left(0, \frac{5}{3}\right)$
1	$\frac{7}{3}$	$\left(1, \frac{7}{3}\right)$
3	$\frac{11}{3}$	$\left(3, \frac{11}{3}\right)$

Written Exercises

Solve each equation for the variable indicated.

1. $x + y = 5$, for y $y = 5 - x$
2. $3x + y = 7$, for y $y = 7 - 3x$
3. $b - 5a = 3$, for b $b = 3 + 5a$
4. $4m + n = 7$, for n $n = 7 - 4m$
5. $8x + 2y = 6$, for y $y = 3 - 4x$
6. $6x + 3y = 12$, for y $y = 4 - 2x$
7. $4a + 3b = 7$, for b $b = \dfrac{7 - 4a}{3}$
8. $6r + 5s = 2$, for s $s = \dfrac{2 - 6r}{5}$
9. $6x = 3y + 2$, for y $y = \dfrac{6x - 2}{3}$
10. $3a = 7b + 8$, for b $b = \dfrac{3a - 8}{7}$
11. $4p = 7 - 2q$, for q $q = \dfrac{4p - 7}{-2}$
12. $7q = 4 + 5m$, for q $q = \dfrac{4 + 5m}{7}$

State which of the ordered pairs given are solutions of the equation.

13. $3x + y = 8$ (**a.**) $(2, 2)$ **b.** $(3, 1)$ (**c.**) $(4, -4)$ **d.** $(8, 0)$
14. $2x + 3y = 11$ **a.** $(3, 1)$ (**b.**) $(1, 3)$ (**c.**) $(-2, 5)$ **d.** $(4, -1)$
15. $2m - 5n = 1$ (**a.**) $(-2, -1)$ **b.** $(2, 1)$ **c.** $\left(-\frac{3}{2}, -\frac{1}{2}\right)$ (**d.**) $(-7, -3)$
16. $3r = 8s - 4$ (**a.**) $\left(\frac{2}{3}, \frac{3}{4}\right)$ (**b.**) $\left(0, \frac{1}{2}\right)$ (**c.**) $(4, 2)$ **d.** $(2, 4)$
17. $3y = x + 7$ (**a.**) $(-1, 2)$ **b.** $(2, -1)$ (**c.**) $(2, 3)$ **d.** $(2, 4)$
18. $4x = 8 - 2y$ (**a.**) $(2, 0)$ **b.** $(0, 2)$ **c.** $\left(\frac{1}{2}, -3\right)$ **d.** $(1, -2)$

Solve each equation if the domain is $\{-2, -1, 0, 2, 5\}$.

19. $y = 2x + 1$
20. $y = 5x - 3$
21. $x + y = 7$
22. $x - y = 4$
23. $2x + y = 7$
24. $5x + y = 4$
25. $2a + 3b = 13$
26. $4r + 3s = 16$
27. $3x = 5 + 2y$
28. $4y = 3 + 2x$
29. $5a = 8 - 4b$
30. $2t = 3 - 5s$
31. $a - 6b = -32$
32. $5x - y = -3$
33. $6x + 5y = 11$
34. $4m + 3n = 10$
35. $5y - 3x = 8$
36. $3z - 6w = 1$

Answers to Written

19. $\{(-2, -3), (-1, -1), (0, 1), (2, 5), (5, 11)\}$
20. $\{(-2, -13), (-1, -8), (0, -3), (2, 7), (5, 22)\}$
21. $\{(-2, 9), (-1, 8), (0, 7), (2, 5), (5, 2)\}$
22. $\{(-2, -6), (-1, -5), (0, -4), (2, -2), (5, 1)\}$
23. $\{(-2, 11), (-1, 9), (0, 7), (2, 3), (5, -3)\}$
24. $\{(-2, 14), (-1, 9), (0, 4), (2, -6), (5, -21)\}$
25. $\left\{\left(-2, \frac{17}{3}\right), (-1, 5), \left(0, \frac{13}{3}\right), (2, 3), (5, 1)\right\}$
26. $\left\{(-2, 8), \left(-1, \frac{20}{3}\right), \left(0, \frac{16}{3}\right), \left(2, \frac{8}{3}\right), \left(5, -\frac{4}{3}\right)\right\}$
27. $\left\{\left(-2, -\frac{11}{2}\right), (-1, -4), \left(0, -\frac{5}{2}\right), \left(2, \frac{1}{2}\right), (5, 5)\right\}$
28. $\left\{\left(-2, -\frac{1}{4}\right), \left(-1, \frac{1}{4}\right), \left(0, \frac{3}{4}\right), \left(2, \frac{7}{4}\right), \left(5, \frac{13}{4}\right)\right\}$
29. $\left\{\left(-2, \frac{9}{2}\right), \left(-1, \frac{13}{4}\right), (0, 2), \left(2, -\frac{1}{2}\right), \left(5, -\frac{17}{4}\right)\right\}$
30. $\left\{\left(-2, \frac{13}{2}\right), (-1, 4), \left(0, \frac{3}{2}\right), \left(2, -\frac{7}{2}\right), (5, -11)\right\}$
31. $\left\{(-2, 5), \left(-1, \frac{31}{6}\right), \left(0, \frac{16}{3}\right), \left(2, \frac{17}{3}\right), \left(5, \frac{37}{6}\right)\right\}$
32. $\{(-2, -7), (-1, -2), (0, 3), (2, 13), (5, 28)\}$
33. $\left\{\left(-2, \frac{23}{5}\right), \left(-1, \frac{17}{5}\right), \left(0, \frac{11}{5}\right), \left(2, -\frac{1}{5}\right), \left(5, -\frac{19}{5}\right)\right\}$
34. $\left\{(-2, 6), \left(-1, \frac{14}{3}\right), \left(0, \frac{10}{3}\right), \left(2, \frac{2}{3}\right), \left(5, -\frac{10}{3}\right)\right\}$
35. $\left\{\left(-2, \frac{2}{5}\right), (-1, 1), \left(0, \frac{8}{5}\right), \left(2, \frac{14}{5}\right), \left(5, \frac{23}{5}\right)\right\}$
36. $\left\{\left(-2, -\frac{11}{3}\right), \left(-1, -\frac{5}{3}\right), \left(0, \frac{1}{3}\right), \left(2, \frac{13}{3}\right), \left(5, \frac{31}{3}\right)\right\}$

About Reading Algebra
Point out that the horizontal and vertical scales on a graph can be different.

You may want to have students bring in graphs from newspapers and magazines. Then have them answer the three questions at the top of this page about their graphs.

Reading Algebra — Graphs

Graphs are often used to help visualize the relationship between two variables. To fully understand the information presented in the graph, ask yourself the following three questions.

1. What does the title indicate will be represented in the graph?
2. What variable is represented along each axis?
3. What units are used along each axis?

Apply each of the three questions to the graph below. The bar graph shows information from a survey.

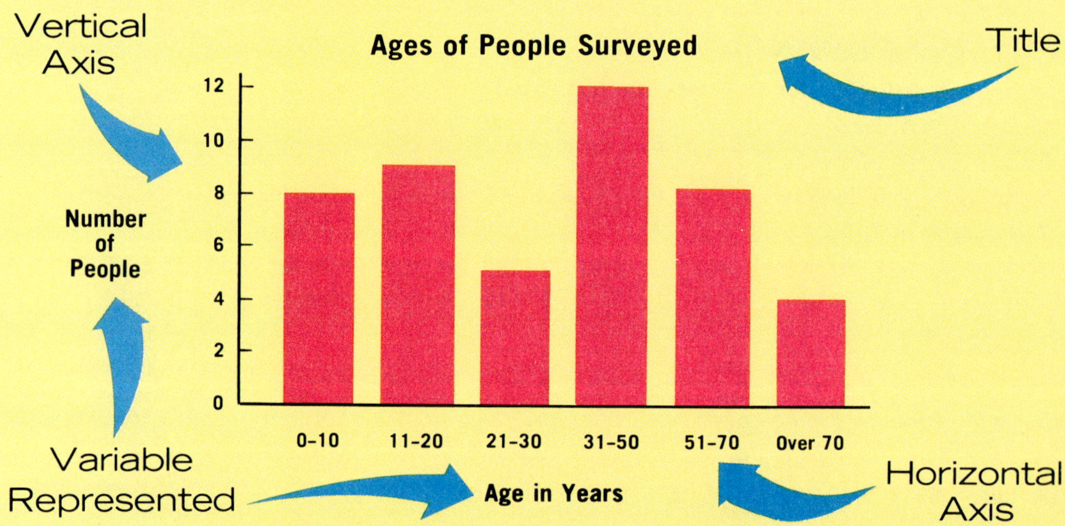

On the graph, the title indicates that the graph shows the number of people of each age who were surveyed. The vertical axis shows the number of people in each age group. The horizontal axis shows age groups in years. Each unit on the vertical axis represents two people. The horizontal axis is labeled to show what age groups were used. Notice that the units used for each axis do not have to be the same.

Exercises

Turn to page 293, *Misleading Graphs*. Answer the three questions at the top of this page about the graphs on page 293. 1. The title indicates that the graph displays information about algebra
2. The horizontal axis represents the chronological order in which the exams were taken. The vertical axis re[p]
the respective score Greg received on each test. 3. Left graph: 0, 10, 20, 30, 40,
right graph: 70, 71, 72, . . . 80; both graphs: 1, 2, 3, 4, 5.

Objective 9-4: Graph linear equations.

9-4 Graphing Relations

Some of the solutions of $y = 2x - 1$ are shown in the table. The solutions can also be shown by a graph.

x	$2x - 1$	y	(x, y)
-2	$2(-2) - 1$	-5	$(-2, -5)$
-1	$2(-1) - 1$	-3	$(-1, -3)$
0	$2(0) - 1$	-1	$(0, -1)$
1	$2(1) - 1$	1	$(1, 1)$
2	$2(2) - 1$	3	$(2, 3)$

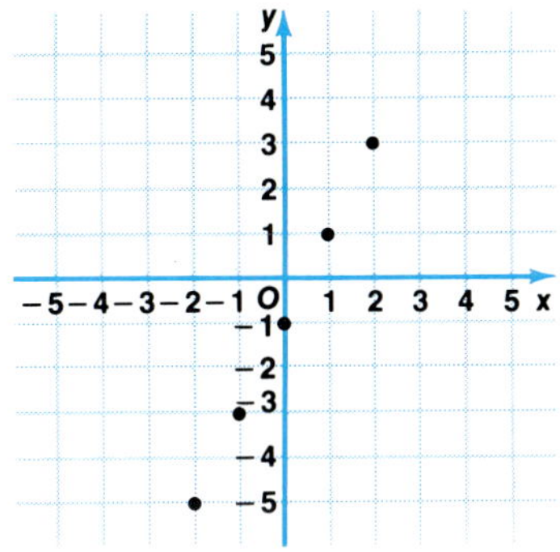

The domain of $y = 2x - 1$ is the set of all numbers. Therefore, an infinite number of ordered pairs are solutions to the equation. ❶

Suppose you draw a line connecting the points in the graph above. All points for the solutions of $y = 2x - 1$ lie on this line. And the coordinates of every point on the line satisfy the equation. Hence, the line is called the graph of $y = 2x - 1$.

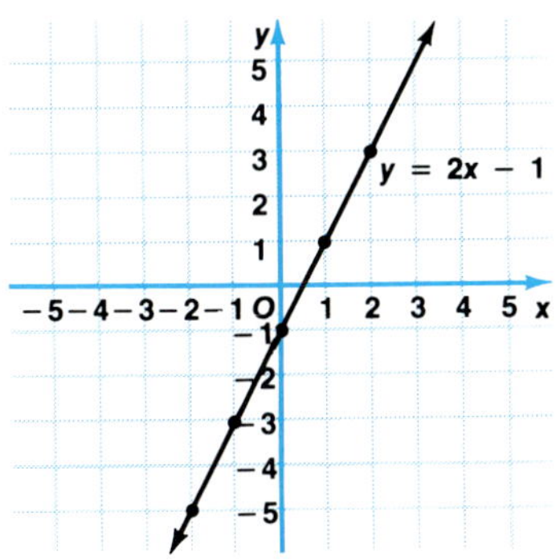

The equation $y = 2x - 1$ can be changed to equivalent forms such as $-y = 1 - 2x$ or $2x - y = 1$. All of these forms will have the same graph. An equation whose graph is a straight line is called a **linear equation**.

A linear equation is an equation that can be written in the form $Ax + By = C$, where A, B, and C are any numbers, and A and B are not both 0.

❷ ***Definition of Linear Equation***

9-4

Lesson Notes, page T42

Discussion Question

The following is a graph of $y = x + 1$. Equivalent equations have the same graph. Can you write an equation that is equivalent to $y = x + 1$?

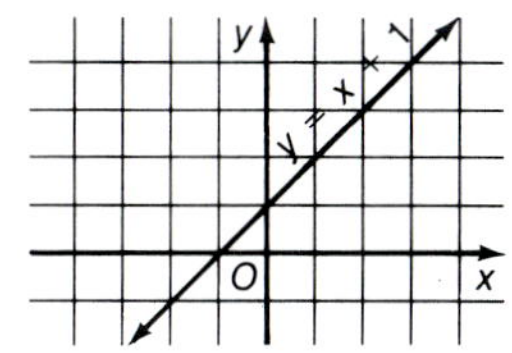

Answers will vary. Some typical answers are:
$y - 1 = x$,
$y - x = 1$, and
$-1 = x - y$.

❶ The text states that the domain of $y = 2x - 1$ is the set of all numbers. More specifically, the domain is the set of *real* numbers. The statement given in the text should not be confusing to students since they have not been introduced to any numbers which are not real. At this point in their studies, students have learned about integers and rational numbers. Irrational numbers, which complete the reals, will be introduced in Chapter 12.

❷ Make sure students understand the definition of a linear equation. The phrase "can be written" does not mean the equation has to appear in that form. However, the equation must be equivalent to an equation in the form $Ax + By = C$. All such equations have degree 1.

Chalkboard Exercises

Graph each equation.

1a. $x - 5y = 5$

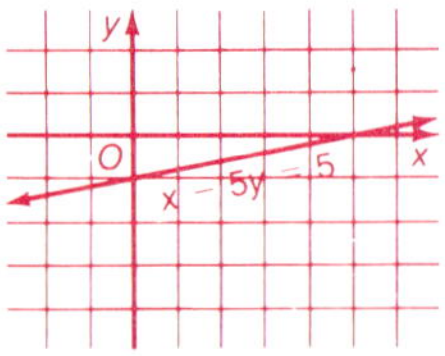

1b. $x + 3y = 3$

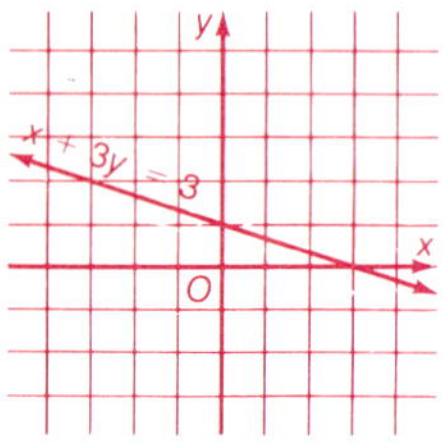

2a. $2x + 5y = 8$

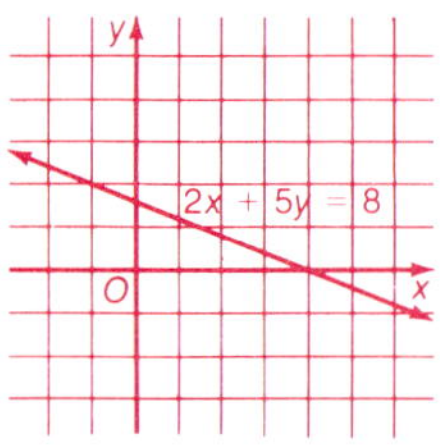

2b. $3x - 4y = 6$

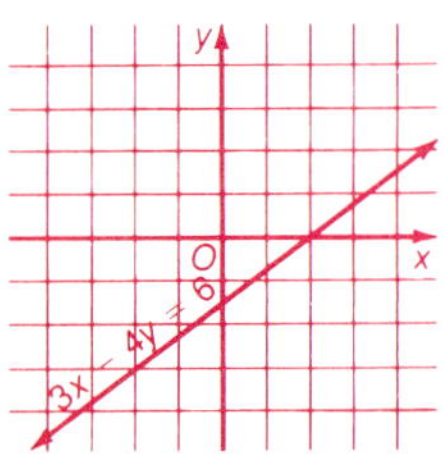

Suggested Daily Assignments

Enriched
- Written: 2–36 evens, 37–45
- Mini-Review: 1–5

Average
- Exp: 1–17 odds
- Written: 1–12, 14–36 evens
- Mini-Review: 1–5

Fundamental
- Exp: 1–18
- Written: 1–12, 13–35 odds p. 281, 25–30
- Mini-Review: 1–5

The equations $4x + 3y = 7$, $2x = 8 + y$, $5m - n = \frac{1}{2}$, $x = 3$, and $y = 7$ are linear equations. The equations $3x + y^2 = 7$ and $\frac{1}{x} + y = 4$ are *not* linear equations. Why?

Examples

1 **Draw the graph of $3x - y = 3$.**

Transform the equation to an equivalent one that expresses y in terms of x.

$$3x - y = 3$$
$$-y = 3 - 3x$$
$$y = 3x - 3$$

Now set up a table of values for x and y. Graph the ordered pairs and connect them with a line.

x	$3x - 3$	y	(x, y)
-2	$3(-2) - 3$	-9	$(-2, -9)$
-1	$3(-1) - 3$	-6	$(-1, -6)$
0	$3(0) - 3$	-3	$(0, -3)$
1	$3(1) - 3$	0	$(1, 0)$
2	$3(2) - 3$	3	$(2, 3)$

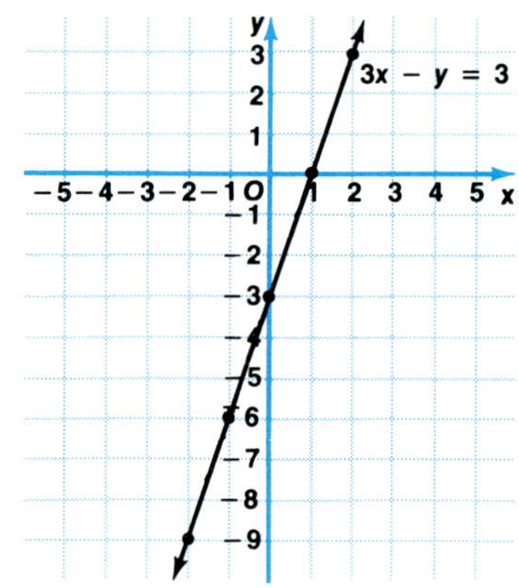

2 **Draw the graph of $3x + 2y = 4$.**

$$3x + 2y = 4$$
$$2y = 4 - 3x$$
$$y = \frac{4 - 3x}{2}$$

x	$\frac{4 - 3x}{2}$	y	(x, y)
-2	$\frac{4 - 3(-2)}{2}$	5	$(-2, 5)$
-1	$\frac{4 - 3(-1)}{2}$	$\frac{7}{2}$	$\left(-1, \frac{7}{2}\right)$
0	$\frac{4 - 3(0)}{2}$	2	$(0, 2)$
1	$\frac{4 - 3(1)}{2}$	$\frac{1}{2}$	$\left(1, \frac{1}{2}\right)$
2	$\frac{4 - 3(2)}{2}$	-1	$(2, -1)$

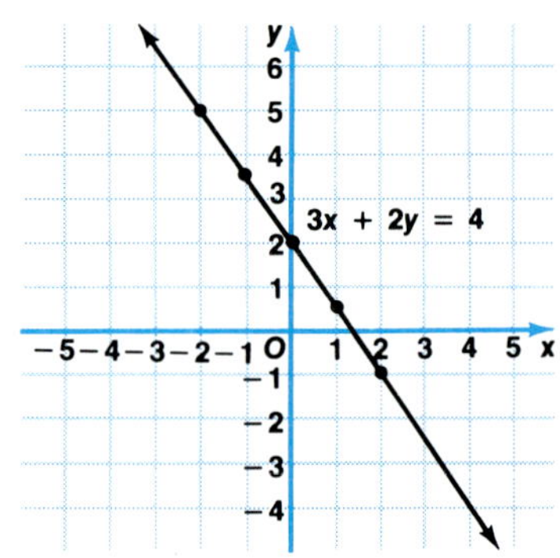

Exploratory Exercises

State whether each equation is a linear equation.

1. $5x + 2y = 7$ yes
2. $3x^2 + 2y = 4$ no
3. $x + \frac{1}{y} = 7$ no
4. $\frac{1}{x} - \frac{1}{y} = \frac{1}{2}$ no
5. $\frac{3}{4}x + 2y = 3$ yes
6. $\frac{3}{5}x - \frac{2}{3}y = 5$ yes
7. $x = y^2$ no
8. $x = y$ yes
9. $\frac{3}{5x} = y$ no
10. $\frac{3}{5}x = y$ yes
11. $\frac{1}{3}x + \frac{1}{3y} = 4$ no
12. $5x - y = 4$ yes
13. $xy^2 = y$ no
14. $xy = 4$ no
15. $\frac{y}{2} = 2$ yes
16. $5m + 2n = 8$ yes
17. $2w = 3z - 1$ yes
18. $\frac{5}{3}x + \frac{2}{3}y = 7$ yes

Written Exercises

State whether each equation is a linear equation.

1. $xy = 2$ no
2. $3x = 2y$ yes
3. $x = y^2$ no
4. $y^2 = x^2$ no
5. $3m - 2n = 8$ yes
6. $3mn^2 = 1$ no
7. $2x - 3 = y^2$ no
8. $5x - 7 = 3y$ yes
9. $8a - 7b = 2a$ yes
10. $3m - 2n = 0$ yes
11. $4x^2 - 3x = y$ no
12. $2m + 5m = 7n$ yes

Graph each equation. See the Teacher's Answer Key.

13. $3m + n = 4$
14. $2x - y = 8$
15. $b = 5a - 7$
16. $y = 3x + 1$
17. $4x + 3y = 12$
18. $5x + 2y = 10$
19. $5x + 3y = 8$
20. $2x + 7y = 9$
21. $3x - 2y = 12$
22. $4x - 3y = 24$
23. $x = 4$
24. $y = -2$
25. $\frac{1}{2}x + y = 8$
26. $x + \frac{1}{3}y = 6$
27. $\frac{2}{3}x - \frac{1}{3}y = 2$
28. $\frac{3}{4}x + \frac{1}{2}y = 6$
29. $x = -\frac{5}{2}$
30. $y = \frac{4}{3}$
31. $\frac{2}{5}x - \frac{3}{5}y = 1$
32. $\frac{3}{4}x - \frac{2}{3}y = 7$
33. $\frac{3}{5}x = 6$
34. $\frac{3}{4}y = 6$
35. $\frac{5}{2}x + \frac{2}{3}y = 1$
36. $\frac{4}{3}x - \frac{3}{4}y = 1$

Challenge

See the Teacher's Answer Key.

37. $y = x^2$
38. $y = x^2 + 2$
39. $y = -(x^2) + 3$
40. $x = 2y^2$
41. $y = -(x^2) - 3$
42. $y - 2x^2 = 3$
43. $y = x^3$
44. $y = 2x^3 + 1$
45. $x = y^3 - 1$

mini-review

1. Find the prime factorization of 96.
2. Find the GCF of $6x^2y^3$, $12x^2y^2z$, and $15x^2y$. $3x^2y$

1. $2 \cdot 2 \cdot 2 \cdot 2 \cdot 2 \cdot 3$

Factor.

3. $m^2 + mn - 42n^2$ $(m + 7n)(m - 6n)$
4. $y^2m + y^2n - 4m - 4n$ $(m + n)(y + 2)(y - 2)$
5. $\frac{2}{3}a - \frac{2}{3}b + \frac{4}{3}c$ $\frac{2}{3}(a - b + 2c)$

Answers to Written
For graphs to exercises 13-36, see p. T82.

Answers to Challenge
For graphs to exercises 37-45, see p. T82.

Quiz
Lessons 9-1 to 9-4

State the quadrant in which the point for each ordered pair is located.

1. $(-5, -1)$ III
2. $(3, -2)$ IV

State each relation as a set of ordered pairs. Then state the domain and range of each.

3. 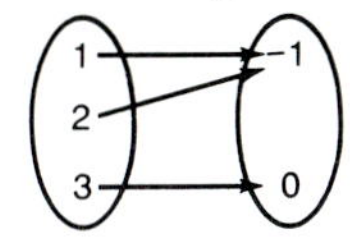

$\{(1, -1), (2, -1), (3, 0)\}$
$D = \{1, 2, 3\}$
$R = \{-1, 0\}$

4. 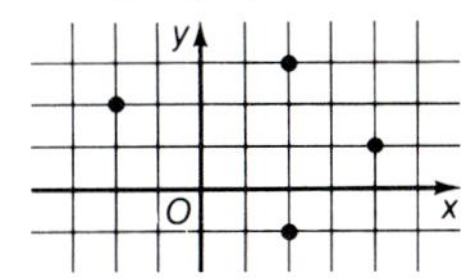

$\{(-2, 2), (2, 3), (2, -1), (4, 1)\}$
$D = \{-2, 2, 4\}$
$R = \{2, 3, -1, 1\}$

Solve each equation for y.

5. $2x - y = 5$
$y = 2x - 5$
6. $3y + 5x = 1$
$y = -\frac{5}{3}x + \frac{1}{3}$

Solve the equation if the domain is $\{-5, 0, 4\}$.

7. $x + 2y = 8$
$\left\{\left(-5, 6\frac{1}{2}\right), (0, 4), (4, 2)\right\}$

State whether or not each equation is a linear equation.

8. $x^2 = 4$ no
9. $y - x = \frac{1}{2}$ yes

Graph.

10. $4x + 3y = 12$

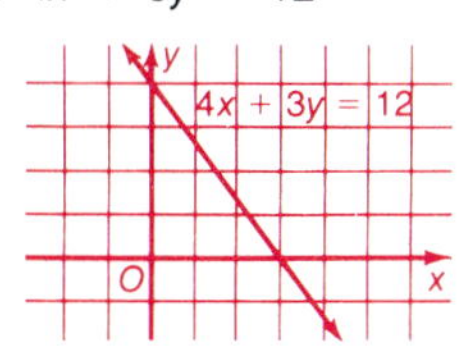

9-5

Lesson Notes, page T42
Enrichment, page T42

Discussion Question
(Use after Example 2.)
The family relationship, *is the daughter of*, does not act like a function, since each name placed before the phrase goes to at least two names after the phrase (mother and father's names). Find a family relationship that acts like a function. is the wife (or husband) of

Chalkboard Exercises
Determine whether each relation is a function.

1a. {(3, 2), (4, 5), (6, 8), (7, 1)} yes
1b. {(3, 2), (4, 2), (5, 2), (6, 2)} yes
2a. {(4, 1), (3, 3), (4, 6), (7, 5)} no
2b. {(−2, −2), (−1, 6), (5, 4), (3, 4), (−1, 5)} no

❶ Encourage students to look at the graphs on page 277 and decide which of the graphs are functions.

Objective 9-5A: Determine whether a given relation is a function.
Objective 9-5B: Calculate functional values for a given function.

9-5 Functions

The following graphs and accompanying tables describe two different relations.

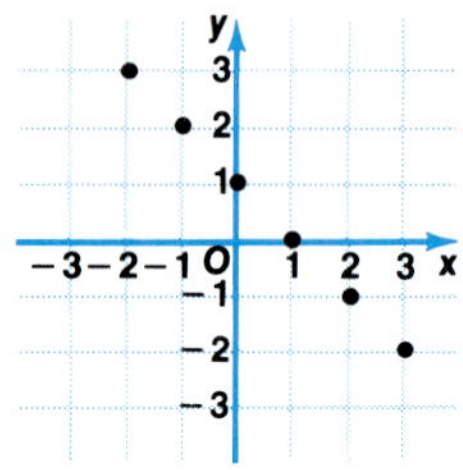

x	y
−2	3
−1	2
0	1
1	0
2	−1
3	−2

For each value of x, there is exactly one value of y.

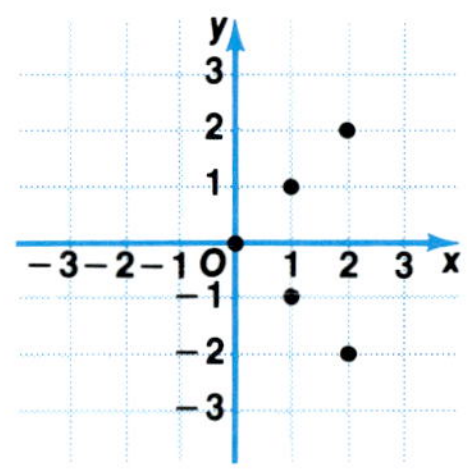

x	y
0	0
1	−1
1	1
2	−2
2	2

For x = 1, there are two values of y, 1 and −1. For x = 2, there are two values, 2 and −2.

The first relation is an example of a special type of relation called a **function**.

A function is a relation in which each element of the domain is paired with exactly one element of the range.

Definition of Function

❶

Examples

1 **Is {(5, −2), (3, 2), (4, −1), (−2, 2)} a function?**

Yes, because each element of the domain is paired with exactly one element of the range.

2 **Is {(−1, 3), (0, 5), (2, 3), (5, −2), (2, 4)} a function?**

No. The element 2 of the domain is paired with more than one element of the range, namely 3 and 4.

Examples

3 **Two relations are described by the following mappings. Which are functions?**

❷

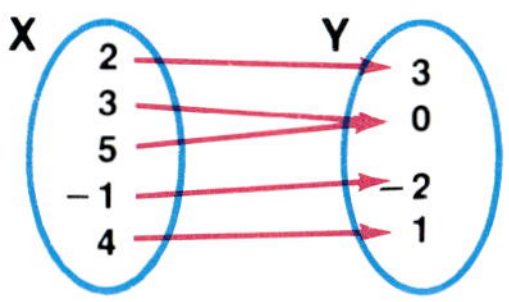

a function

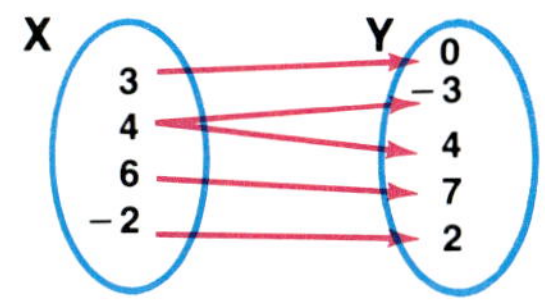

not a function

The relation at the left is a function. For each element in the domain of the relation, there is *only one* corresponding element in the range. Why is the relation at the right not a function?

4 **Is the relation described by $x + 2y = 8$ a function?**

Try substituting a value for x. For example, try 2. What is the corresponding value for y? Can there be more than one value for y when $x = 2$? No. If $x = 2$, the only value for y is 3. Can you find any values for x that will give more than one value for y? No. There are none. Therefore, $x + 2y = 8$ is a function. Specifically, we say that y is a function of x.

Suppose you graph the relation $x + 2y = 8$.

x	2	0	8
y	3	4	0

Now place your pencil at the left of the graph to represent a vertical line. Slowly move the pencil to the right across the graph.

❸

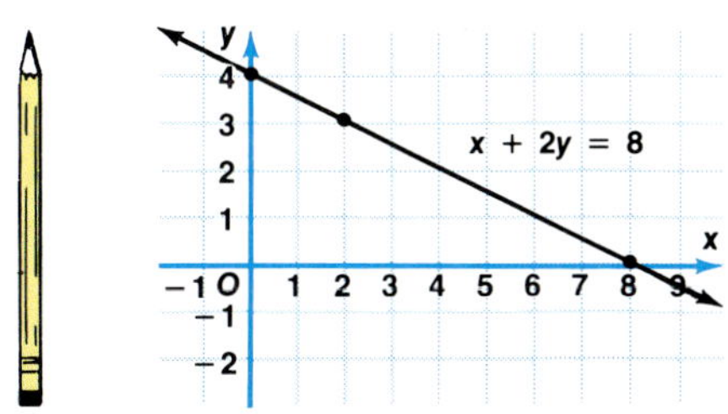

For each value of x, the vertical line passes through no more than one point of the graph. This is true for every function.

If any vertical line passes through no more than one point of the graph of a relation, then the relation is a function.

Vertical Line Test for a Function

Chalkboard Exercises

3. In example 3, suppose the arrowheads were reversed, indicating a mapping from y to x. Which of the mappings would represent functions? The mapping at the right would be a function, but the mapping on the left would not be a function.

Determine whether each relation is a function.

4a. $y = 2x + 1$ yes
4b. $y = x^2 + 1$ yes
4c. $y^2 = x$ no
4d. $y = 4x + 5$ yes

❷ Encourage students to look at the mappings on page 276 and decide which of the mappings are functions.

❸ The vertical line test may be illustrated with graphs on a chalkboard by using a meter stick.

Chalkboard Exercises

5. Use the vertical line test to determine if the relation graphed is a function.

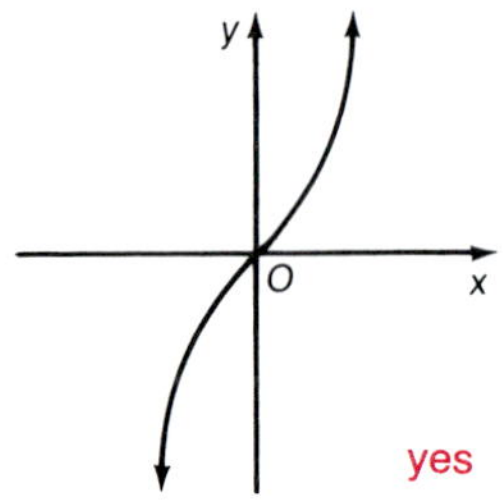

yes

Determine the value for each of the following.

6a. $f(1)$ if $f(x) = 3x - 2$ 1

6b. $f(2)$ if $f(x) = 7 - 4x$ -1

6c. $g(-2)$ if $g(x) = x^2 + 1$ 5

6d. $g(7)$ if $g(x) = 10x - x^2 + 2$ 23

7a. $g(3a)$ if $g(x) = x^2 + 3x + 2$ $9a^2 + 9a + 2$

7b. $g(4a)$ if $g(x) = x^3 - 2x - 3$ $64a^3 - 8a - 3$

8a. $3[f(2)]$ if $f(t) = 8t + 3t^2$ 84

8b. $2[f(-2)]$ if $f(t) = 4t^2 - 6t$ 56

❹ The graph in example 5d may be confusing to students. Point out that the graph consists of a horizontal line *and* the point *P*.

❺ It may be helpful for some students to think of $f(x)$ as "f at x", or the value of f at x.

Example

5 Use the vertical line test to determine if the relation graphed is a function.

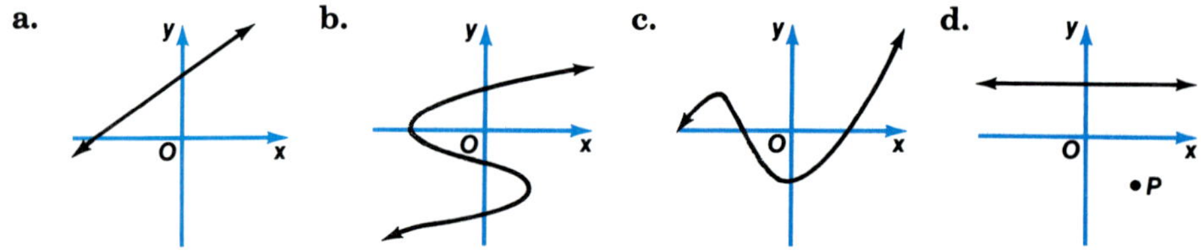

❹

The relations in **a** and **c** are functions since any vertical line passes through no more than 1 point of the graph of the relation. The relation in **b** is *not* a function, since a vertical line near the *y*-axis passes through *three* points. The relation in **d** is *not* a function. A vertical line through point *P* also intersects the graph in another point.

Equations that describe functions are often written in a special way. The equation $y = 2x + 1$ can be written $f(x) = 2x + 1$. The symbol $f(x)$ is read "f of x" and represents the range of a function whose domain is represented by x. Specific values of the domain can then be substituted for x in the equation. For example, if $x = 3$, then $f(3) = 2(3) + 1$ or 7. Note $f(3)$ is a convenient way of referring to the value of y corresponding to $x = 3$. We say $f(3)$ is the **functional value** at $x = 3$. ❺

Letters other than f are sometimes used for names of functions.

Examples

6 Find $f(2)$, $f(5)$, and $f(-3)$ if $f(x) = 2x - 7$.

$$\begin{aligned} f(x) &= 2x - 7 \\ f(2) &= 2(2) - 7 \\ &= 4 - 7 \\ &= -3 \end{aligned} \qquad \begin{aligned} f(x) &= 2x - 7 \\ f(5) &= 2(5) - 7 \\ &= 10 - 7 \\ &= 3 \end{aligned} \qquad \begin{aligned} f(x) &= 2x - 7 \\ f(-3) &= 2(-3) - 7 \\ &= -6 - 7 \\ &= -13 \end{aligned}$$

7 Find $g(2a)$ if $g(x) = x^2 - 2x + 1$.

$$\begin{aligned} g(x) &= x^2 - 2x - 1 \\ g(2a) &= (2a)^2 - 2(2a) - 1 \qquad \textit{Substitute 2a for x.} \\ &= 4a^2 - 4a - 1 \end{aligned}$$

8 Find $4[f(3)]$ if $f(t) = 100t - 5t^2$.

$$\begin{aligned} f(t) &= 100t - 5t^2 \\ 4[f(3)] &= 4[100(3) - 5(3)^2] \\ &= 4[300 - 45] \\ &= 4 \cdot 255 \\ &= 1020 \end{aligned}$$

Exploratory Exercises

State whether each relation is a function.

1. $\{(3, 4), (5, 4), (-2, 3), (5, 3)\}$ no
2. $\{(8, 4), (5, -2), (6, 3), (2, 3)\}$ yes
3. $\{(4, 1), (-2, 3), (0, 5), (5, 0), (3, 3)\}$ yes
4. $\{(5, 1), (5, 2), (5, 3), (5, 4)\}$ no
5. $3x = 7y$ yes
6. $2x - y^2 = 3$ no
7. $x^2 = y - 4$ yes
8. $3m + 5n = 4$ yes
9. $3s^2 + 2t^2 = 7$ no
10. $5a^2 - 7 = b$ yes

Use the vertical line test to determine whether each graph represents a function.

11. no

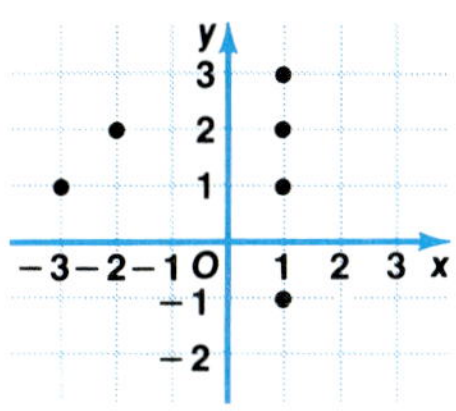

12. yes

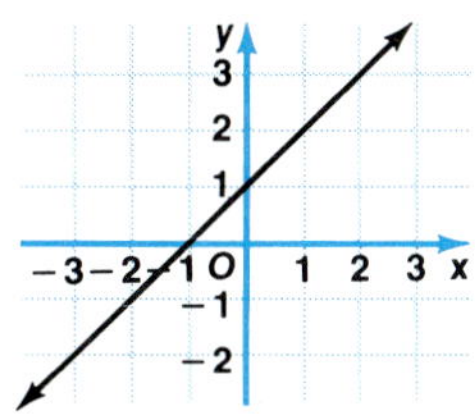

13. no

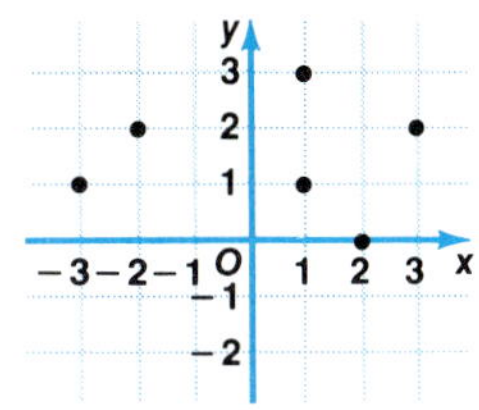

14. yes

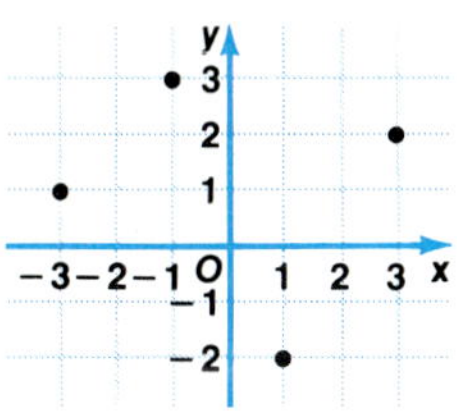

15. yes

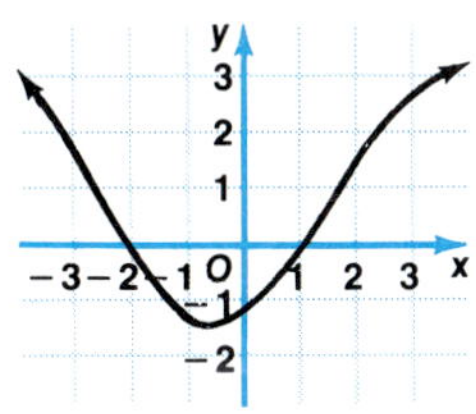

16. no

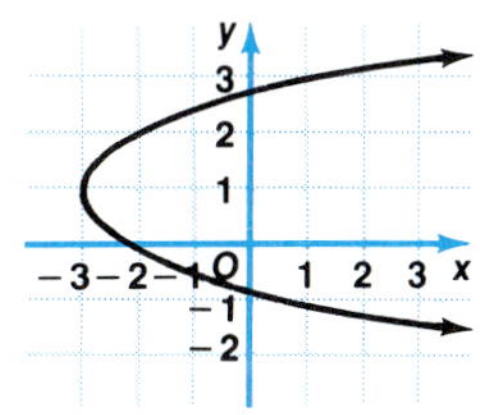

Given $g(x) = 2x - 1$, determine the value for each of the following.

17. $g(2)$ 3
18. $g(-4)$ -9
19. $g(-6)$ -13
20. $g(0)$ -1
21. $g\left(\frac{1}{2}\right)$ 0
22. $g\left(\frac{5}{2}\right)$ 4

Written Exercises

Determine whether each relation is a function.

1. $\{(3, 1), (5, 1), (7, 1)\}$ yes
2. $\{(1, 3), (1, 5), (1, 7)\}$ no
3. $\{(-2, 4), (1, 3), (5, 2), (1, 4)\}$ no
4. $\{(-3, 3), (-2, 2), (-1, 1), (0, 0)\}$ yes
5. $\{(6, -1), (1, 4), (2, 3), (6, 1)\}$ no
6. $\{(5, 4), (6, 1), (-2, 3), (0, 3)\}$ yes
7. $\{(5, 4), (-6, 5), (4, 5), (0, 4)\}$ yes
8. $\{(3, -2), (4, -1), (-2, 5), (4, 5)\}$ no
9. $3x + 5y = 7$ yes
10. $y = 2$ yes
11. $4x - 7y = 3$ yes
12. $x^2 + y = 7$ yes
13. $x + y^2 = 7$ no
14. $x = -3$ no
15. $\frac{1}{x} = y$ yes
16. $3 = x - y$ yes
17. $4x = 5y$ yes
18. $x = -2$ no
19. $\frac{2}{3}x = \frac{3}{2}y + 1$ yes
20. $x^2 - y^2 = 3$ no

Given $f(x) = 3x - 5$ and $g(x) = x^2 - x$, determine the value for each of the following.

21. $f(2)$ 1
22. $f(-3)$ -14
23. $g(-4)$ 20
24. $g(3)$ 6
25. $f\left(\frac{1}{2}\right)$ $-3\frac{1}{2}$
26. $f\left(\frac{2}{3}\right)$ -3
27. $g\left(-\frac{1}{2}\right)$ $\frac{3}{4}$
28. $g\left(\frac{1}{3}\right)$ $-\frac{2}{9}$
29. $f(2a)$ $6a - 5$
30. $g(2b)$ $4b^2 - 2b$
31. $3[f(5)]$ 30
32. $2[g(-2)]$ 12
33. $f(5.5)$ 11.5
34. $1.5[f(-3)]$ -21
35. $f(a + 3)$ $3a + 4$
36. $g(b - 3)$ $b^2 - 7b + 12$

Suggested Daily Assignments

Enriched
- Exp: 2–22 evens
- Written: 2–36 evens

Average
- Exp: 1–22
- Written: 1–20, 21–35 odds

Fundamental
- Exp: 1–22
- Written: 1–24
  - p. 285, 14–36 evens

About the Application
The coordinate method for finding area can be introduced by finding the area of a triangle.

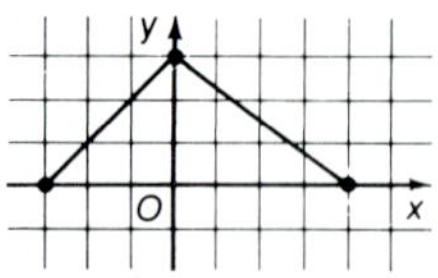

The steps are described below.

Step 1 (0, 3)
(−3, 0)
(4, 0)
(0, 3)

Step 2 $D = (0 \cdot 0) + (-3 \cdot 0) + (4 \cdot 3) = 12$

Step 3 $U = (-3 \cdot 3) + (4 \cdot 0) + (0 \cdot 0) = -9$

Step 4 $A = \frac{1}{2}(D - U) = \frac{1}{2}(12 - (-9)) = 10\frac{1}{2}$

By this method the area is $10\frac{1}{2}$ square units. Students should recognize the formula $A = \frac{1}{2}bh$ for the area of a triangle. This formula can be used to check the result obtained above.

$$A = \frac{1}{2}bh = \frac{1}{2}(7)(3) = 10\frac{1}{2}$$

Discuss with students how to verify the result in the example on this page.

Applications in Surveying

Areas

Many land regions have irregular shapes. Aerial surveyors often use coordinates when finding areas of such regions. The coordinate method described in the steps below can be used to find the area of *any* polygonal region. Study how this method is used to find the area of the region at the right.

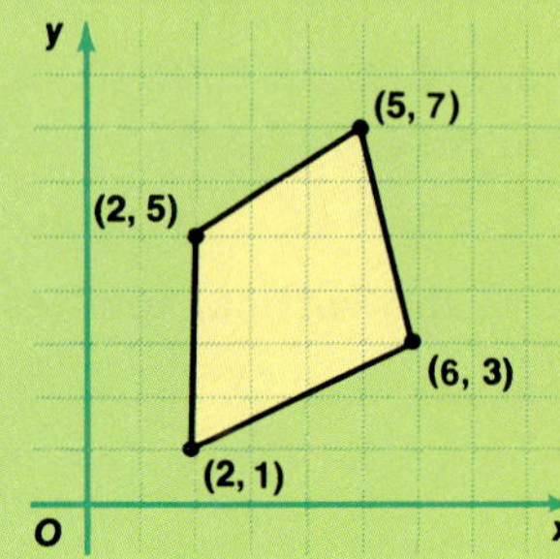

Step 1 List the ordered pairs for the vertices in counter-clockwise order, repeating the first ordered pair at the bottom of the list.

(5, 7)
(2, 5)
(2, 1)
(6, 3)
(5, 7)

Step 2 Find D, the sum of the downward diagonal products (from left to right.)

$$D = (5 \cdot 5) + (2 \cdot 1) + (2 \cdot 3) + (6 \cdot 7) = 25 + 2 + 6 + 42 \text{ or } 75$$

Step 3 Find U, the sum of the upward diagonal products (from left to right.)

$$U = (2 \cdot 7) + (2 \cdot 5) + (6 \cdot 1) + (5 \cdot 3) = 14 + 10 + 6 + 15 \text{ or } 45$$

Step 4 Use the formula $A = \frac{1}{2}(D - U)$ to find the area.

$$A = \frac{1}{2}(D - U) = \frac{1}{2}(75 - 45) = \frac{1}{2}(30) \text{ or } 15$$

The area is 15 square units. Count the number of square units enclosed by the polygon. Does this result seem reasonable?

Exercises

Use the coordinate method to find the area of each region.

1. 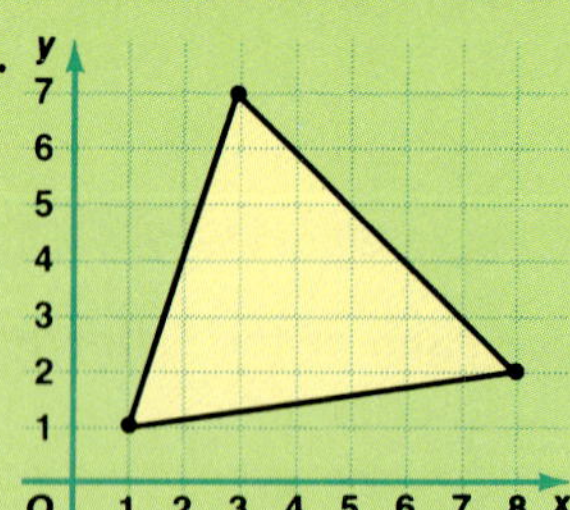

20 square units

2. 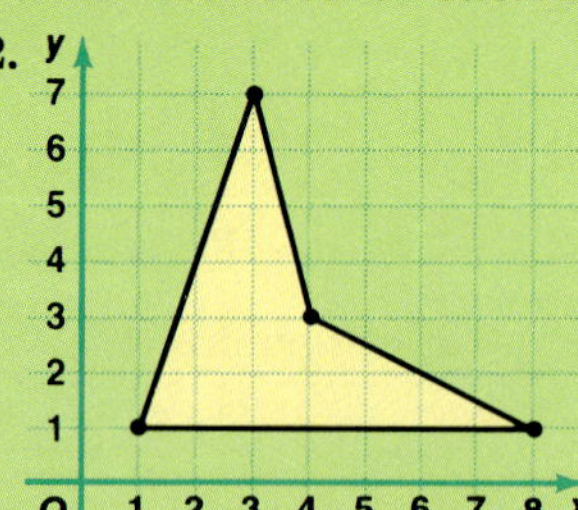

14 square units

3. 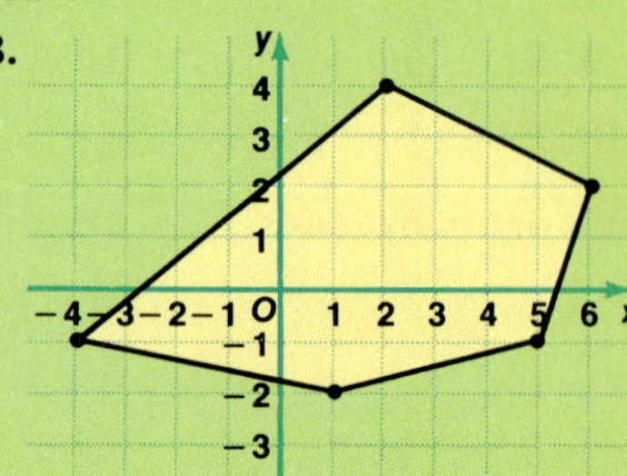

34 square units

Objective 9-6: Graph inequalities in the coordinate plane.

9-6 Graphing Inequalities in Two Variables.

The graph of the equation $x = 3$ is a line that separates the coordinate plane into two regions. One region is shaded blue. The other is shaded yellow. Each region is called a **half-plane**. The line for $x = 3$ is called the **boundary** for each half-plane.

❶

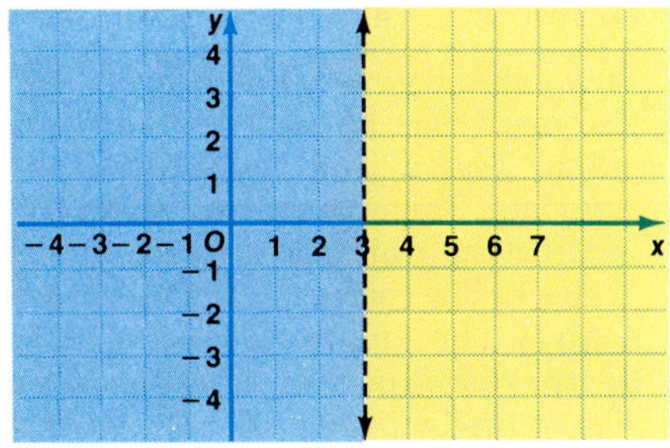

The inequality $x > 3$ describes the yellow region. The x-coordinate of every point in that region is greater than 3. The inequality $x < 3$ describes the blue region. Points on the line for $x = 3$ are not part of either region.

Consider the graphs of $y > x + 1$ and $y \le x + 1$. The boundary line in both is the line for $y = x + 1$.

$y > x + 1$

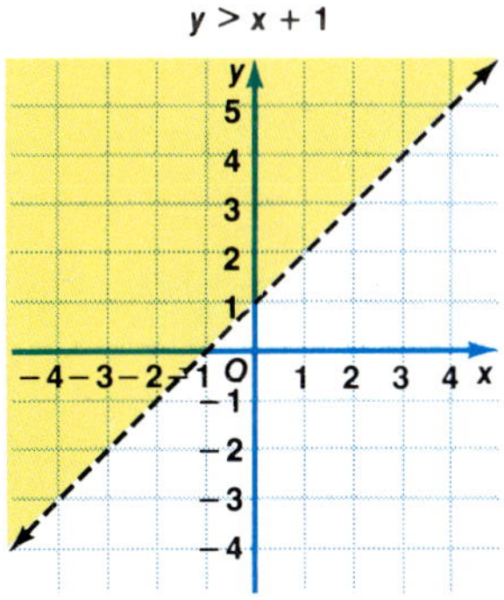

$y \le x + 1$

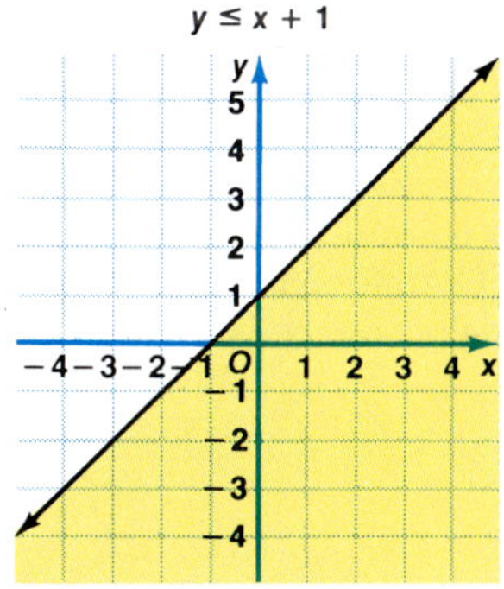

In the graph of $y > x + 1$, the boundary is *not* part of the graph. Therefore the boundary is shown as a broken line. However, all points above the broken line are part of the graph. The graph is called an **open half-plane**.

The inequality $y \le x + 1$ means $y < x + 1$ or $y = x + 1$. In the graph of $y \le x + 1$, the boundary *is* part of the graph and is shown as a solid line. The graph also contains all points below the line. The graph is called a **closed half-plane**. ❷

9-6

Lesson Notes, page T42

Readiness Activity

Graph the solution set of each inequality on a number line.

$x \ge 2$

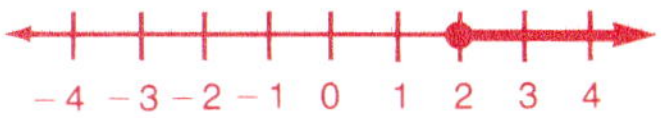

$m < -1$

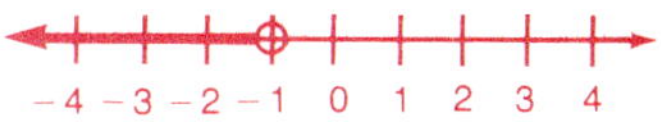

$r \le 0$

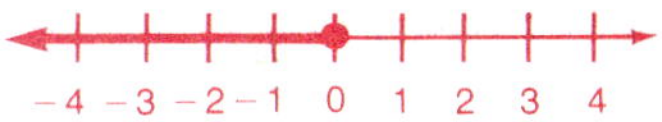

$t > 4$

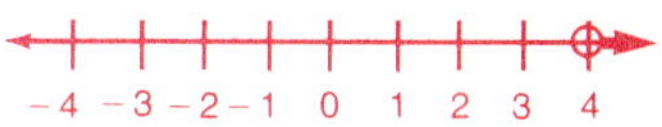

Discussion Questions

(Use after page 291.)

1. Name some ordered pairs that are solutions to the inequality $x + y < 0$. Answers may vary.
2. How would you describe the graph of $x + y < 0$? All points below the line $y = -x$.

❶ Point out to students that the graph shown on this page are graphs of *relations*. Every x-value is paired with many y-values.

❷ Relate the open and closed half-planes used in graphing inequalities in two variables to the open and closed endpoints used in graphing inequalities in one variable.

Chalkboard Exercises

Graph each inequality.

1. $y > x + 1$

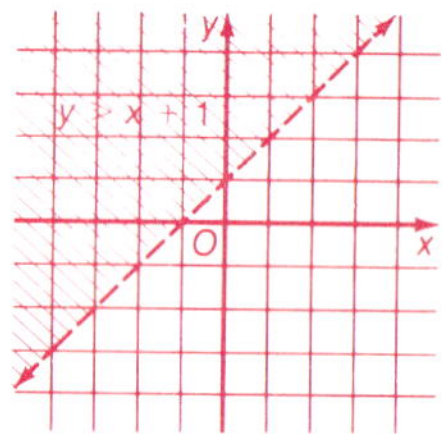

2. $3y - 2x \le 6$

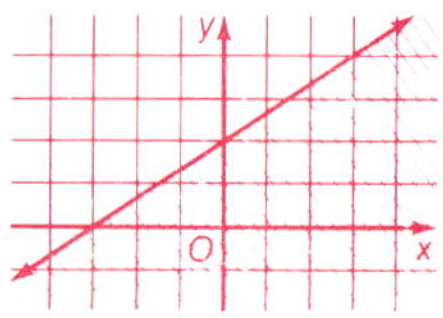

3. $y < 5x$

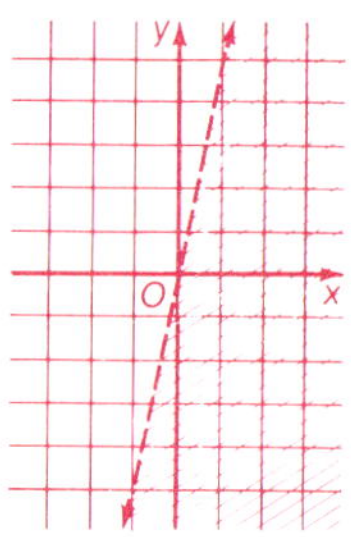

Examples

1 **Graph $y > -4x - 3$.**

First graph $y = -4x - 3$. Draw it as a broken line since this boundary is *not* part of the graph. Test a point on either side of the boundary. This will tell you which half-plane is part of the graph. The origin is an easy point to check. The coordinates of the origin, (0, 0), satisfy the inequality $y > -4x - 3$. So, the coordinates of all the points on that side of the boundary satisfy the inequality.

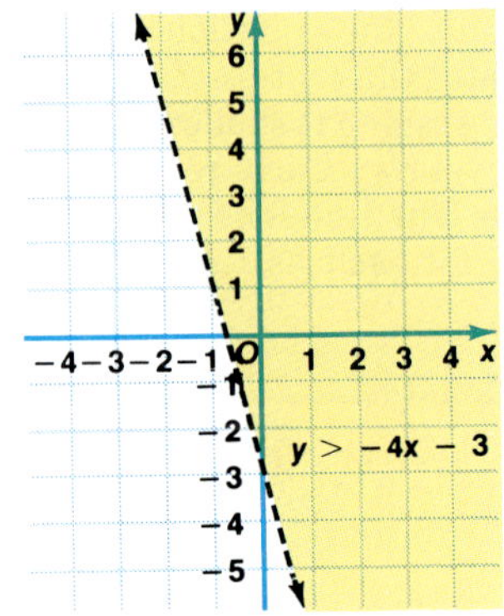

2 **Graph $-2y + 3x \ge 8$.**

First solve for y.

$$-2y + 3x \ge 8$$
$$-2y \ge -3x + 8$$
$$y \le \frac{3}{2}x - 4$$

Then graph $y = \frac{3}{2}x - 4$. Draw it as a solid line since this boundary *is* part of the graph. Then test a point to determine which half-plane is part of the graph. The origin is not part of the graph. Therefore, the graph is the right half-plane including the boundary.

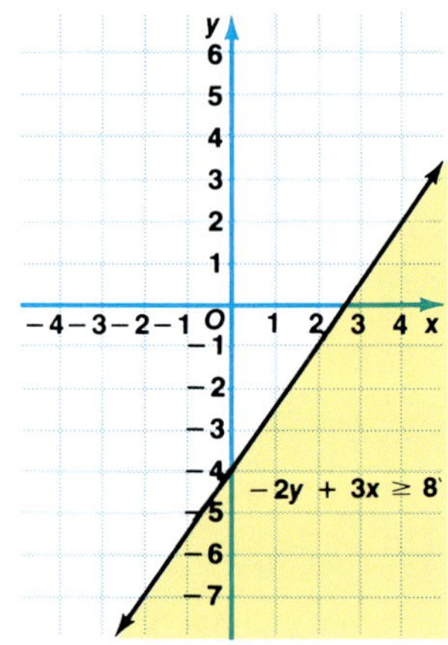

3 **Graph $y \le 3x$.**

First graph $y = 3x$. Draw it as a solid line since this boundary *is* part of the graph. Then test a point to determine which half-plane is part of the graph. The origin is on the boundary, so you must test a point other than the origin. Try (1, 1). Since (1, 1) satisfies the inequality, the graph is the half-plane including (1, 1).

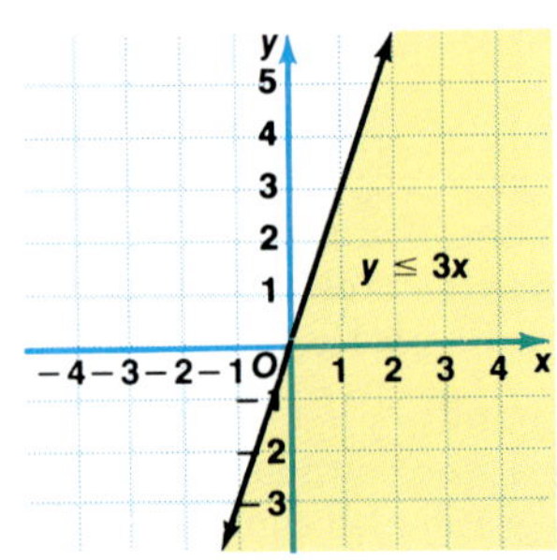

Exploratory Exercises

State which of the ordered pairs given are solutions of the inequality.

1. $x + 2y \geq 3$	**a.** (3, 1)	**b.** (4, −1)	**c.** (−2, 2)
2. $2x - 3y \leq 1$	**a.** (2, 1)	**b.** (5, −1)	**c.** (1, 1)
3. $5x + 2 > 3y$	**a.** (2, 3)	**b.** (4, 7)	**c.** (−2, −3)
4. $4y - 8 \geq 0$	**a.** (0, 2)	**b.** (2, 5)	**c.** (−2, 0)
5. $3x - 2y \leq 6$	**a.** (2, 0)	**b.** (4, −3)	**c.** (0, 3)
6. $-2x < 8 - y$	**a.** (5, 10)	**b.** (3, 6)	**c.** (−4, 0)
7. $3x + 4y < 7$	**a.** (1, 1)	**b.** (2, −1)	**c.** (−2, 4)
8. $y > x - 1$	**a.** (5, 8)	**b.** (3, 3)	**c.** (−2, −3)

Written Exercises See Teacher's Answer Key.

Graph each inequality. Draw a separate set of axes for each.

1. $y > 3$	**2.** $y \leq -2$	**3.** $x \leq -1$	**4.** $x > 4$
5. $x + y > 1$	**6.** $x + y < 2$	**7.** $x + y < -2$	**8.** $x + y > 4$
9. $2x - y < 1$	**10.** $3x + y > 1$	**11.** $2x + 3y \geq -2$	**12.** $2x - 2y \leq 2$
13. $x - 2y < 4$	**14.** $4y + x < 16$	**15.** $2x > 3y$	**16.** $x < y$
17. $-x < -y$	**18.** $-y > x$	**19.** $y - x > 0$	**20.** $2x < -y$
21. $f(x) > x - 1$	**22.** $g(x) \leq x + 1$	**23.** $g(x) \leq 3x - 1$	**24.** $f(x) > 4x - 1$

Challenge

25. $y > \lvert x \rvert$	**26.** $\lvert y \rvert \geq 2$	**27.** $y > 2$ and $x < 3$	**28.** $y > 2$ or $y < 1$

Excursions in Algebra — Misleading Graphs

Greg Friedman had received scores of 72, 74, 75, 77, and 78 on his first five algebra exams. The two graphs below were made to show how his test scores increased. Do they show the same results?

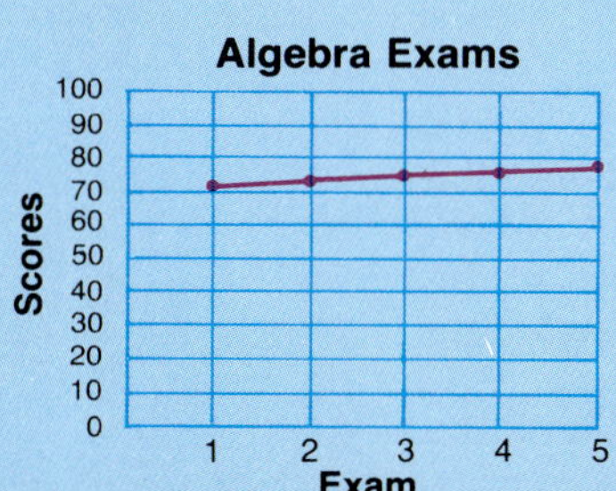

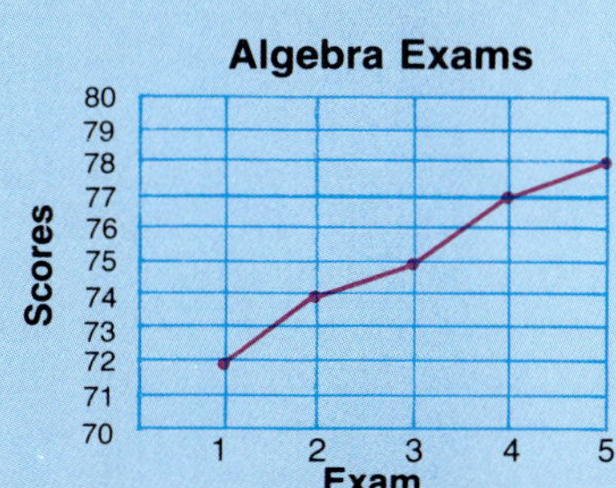

Both graphs use the same data but appear to show different results. Different scales change the appearance of the graphs. Sometimes different scales are used to make the graphs easier to read. Other times scales are chosen to make the data appear to support one point of view. Which graph do you think Greg showed his parents?

Suggested Daily Assignments

Enriched
- Exp: 1–8
- Written: 2–24 evens, 25–28

Average
- Exp: 1–8
- Written: 1–20
  - p. 289, 22–36 evens

Fundamental
- Exp: 1–8
- Written: 1–16
  - p. 289, 25–32

Answers to Written
See p. T83.

Answers to Challenge
See p. T84.

About the Excursion
Discuss the graphs in the Excursion. For the graph at the right, ask students why the results seem distorted.

9-7

Lesson Notes, page T43
Enrichment, page T43

Readiness Activity
Suppose $x = 1$ and $y = 3$. Can you write an equation using x and y such that these values are a solution to the equation? Answers may vary. Answer the question above if $x = 3$ and $y = 1$. Answers may vary.

Discussion Question
Can you find an equation for the relationship between the variables in the following chart?

x	-7	0	$4\frac{1}{2}$	57	50,000
y	3	3	3	3	3

$y = 3$

Objective 9-7: Write an equation for a relation, given a chart of values.

9-7 Finding Equations from Relations

Each package of chewing gum contains five sticks of gum. You can make a chart to show the relationship between the number of packages of gum and the number of sticks. Use x for the number of packages and y for the number of sticks.

x	1	2	3	4	5	6
y	5	10	15	20	25	30

This relation can also be shown as an equation. Since y is always five times x, the equation is $y = 5x$. Another way to discover this relationship is to study the differences between successive values of x and y.

+1 +1 +1 +1 +1

x	1	2	3	4	5	6
y	5	10	15	20	25	30

+5 +5 +5 +5 +5

Notice that the differences of the y-values are exactly five times the differences of the x-values. This suggests the relation $y = 5x$.

The following chart shows a different relationship.

+1 +1 +1 +1 +1

a	1	2	3	4	5	6
b	7	11	15	19	23	27

+4 +4 +4 +4 +4

Notice that the differences of the b-values are four times the differences of the a-values. This suggests the relation $b = 4a$. However, $b = 4a$ does *not* describe the relationship shown in the chart. If you substitute 1 for a in $b = 4a$, then $b = 4(1)$ or 4. According to the chart, the b-value should be 7, not 4. To make $b = 7$ when $a = 1$, you need to add 3. This suggests the relation $b = 4a + 3$. Check the relationship with other values.

If $a = 2$ and $b = 4a + 3$
then $b = 4(2) + 3$
$= 8 + 3$
$= 11$ ✓

If $a = 5$ and $b = 4a + 3$
then $b = 4(5) + 3$
$= 20 + 3$
$= 23$ ✓

Examples

1 **Write an equation for the relation given in the chart.**

x	1	2	3	4	5	6
y	1	4	7	10	13	16

❶

Find the differences.

+1 +1 +1 +1 +1

x	1	2	3	4	5	6
y	1	4	7	10	13	16

+3 +3 +3 +3 +3

The differences suggest $y = 3x$, but this equation does *not* describe the relation. If $x = 1$ then $y = 3$ not 1. You need to subtract 2 from $3x$. Thus, $y = 3x - 2$ will describe the relation.

2 **Write an equation for the relation given in the chart.**

m	1	2	4	5	6	9
n	9	6	0	−3	−6	−15

Find the differences.

+1 +2 +1 +1 +3

m	1	2	4	5	6	9
n	9	6	0	−3	−6	−15

−3 −6 −3 −3 −9

Notice that the values for m are increasing, but the values for n are decreasing. The differences suggest $n = -3m$, but this equation does *not* describe the relation. If $m = 1$ then $n = -3$, not 9. For n to equal 9, you must add 12 to -3. Thus, the equation for the relation is $n = -3m + 12$, or $n = 12 - 3m$.

Exploratory Exercises

Find the missing terms in each sequence.

1. 1, 3, 5, 7, 9, _11_, _13_, _15_
2. 4, 7, 10, 13, _16_, _19_, _22_
3. 12, 9, 6, 3, _0_, _−3_, _−6_
4. −11, −9, −7, −5, _−3_, _−1_, _1_
5. 27, 34, 41, 48, _55_, _62_, _69_
6. 31, 21, 11, _1_, _−9_, _−19_
7. 4, 7, 12, 19, _28_, _39_, _52_
8. 7, 19, 30, 40, 49, _57_, _64_, _70_
9. 1, 4, 9, 16, _25_, _36_, _49_
10. 4, 16, 36, 64, _100_, _144_, _196_
11. 3, 5, 8, 13, 21, _34_, _55_, _89_
12. −2, 4, −8, 16, _−32_, _64_, _−128_

Chalkboard Exercises
Write an equation for the relationship between the variables in each chart.

1a.

x	1	2	3	4	5
y	6	12	18	24	30

$y = 6x$

1b.

x	1	2	3	4	5	6
y	−1	1	3	5	7	9

$y = 2x - 3$

2.

x	1	2	3	4	5	6
y	5	7	9	11	13	15

$y = 2x + 3$

❶ For better students, you could show the set builder notation for each relation. The relation in this chart is equivalent to $\{(x, y): y = 3x - 2, x \in \{1, 2, 3, 4, 5, 6\}\}$.

Suggested Daily Assignments

Enriched
- Exp: 1–11 odds
- Written: 1–12, 14–22 evens, 23–26

Average
- Exp: 1–12
- Written: 1–23
  - p. 293, 21–24

Fundamental
- Exp: 1–12
- Written: 1–22
  - p. 293, 17–24

Written Exercises

Find the missing terms for each sequence.

1. 5, 7, 9, 11, 13, 15, 17
2. 12, 9, 6, 3, 0, −3, −6
3. −1, −4, −7, −10, −13, −16
4. 5, 10, 15, 20, 25, 30, 35
5. 6, 11, 16, 21, 26, 31
6. 1, 2, 4, 8, 16, 32, 64
7. 3, 1, $\frac{1}{3}$, __, __, __ 7. $\frac{1}{9}, \frac{1}{27}, \frac{1}{81}$
8. 4, −2, 1, $-\frac{1}{2}$, __, __, __ 8. $\frac{1}{4}, -\frac{1}{8}, \frac{1}{16}$
9. 1, 0, 1, 0, 1, 0, 1
10. 1, 11, 21, 31, 41, 51, 61
11. 3, 1, −1, −3, −5, −7, −9
12. 17, 14, 11, 8, 5, 2

Write an equation for the relationship between the variables in each chart. Then copy and complete each chart.

13. $y = 4x$

x	1	2	3	4	5	6
y	4	8	12	16	20	24

14. $y = -3x$

x	1	2	3	4	5	6
y	−3	−6	−9	−12	−15	−18

15. $n = 2m + 1$

m	−3	−2	−1	0	1	2	3
n	−5	−3	−1	1	3	5	7

16. $b = 4a + 5$

a	−2	−1	0	1	2	3	4
b	−3	1	5	9	13	17	21

17. $b = 4a + 3$

a	−4	−2	0	2	4	6	8
b	−13	−5	3	11	19	27	35

18. $y = 6x - 1$

x	1	3	5	7	9	11	13
y	5	17	29	41	53	65	77

19. $y = 15 - x$

x	1	2	3	4	5	6	7
y	14	13	12	11	10	9	8

20. $n = 11 - m$

m	−2	−1	0	1	2	3	4
n	13	12	11	10	9	8	7

21. $b = 3 - 5a$

a	−5	−3	−1	1	2	4	7
b	28	18	8	−2	−7	−17	−32

22. $y = 18 - 2x$

x	−4	−2	0	2	4	6	8
y	26	22	18	14	10	6	2

Challenge

Write an equation for each set of ordered pairs.

23. {(−2, 4), (−1, 1), (0, 0), (1, 1), (2, 4), (3, 9)}
24. {(1, 2), (2, 9), (3, 28), (4, 65), (5, 126)}
25. {(−2, 11), (−1, 14), (0, 15), (1, 14), (2, 11), (3, 6)}
26. {(1, 24), (2, 12), (3, 8), (4, 6), (6, 4)}

23. $y = x^2$ 24. $y = x^3 + 1$ 25. $y = 15 - x^2$ 26. $xy = 24$

Using Calculators — Ordered Pairs

Write a linear equation for each set of ordered pairs.

1. {(−2, −5), (−1, −3.5), (−0.5, −2.75), (1, −0.5)} $y = \frac{3}{2}x - 2$
2. {(5, −16.5), (−1.5, 3), (1.5, −6), (10, −31.5)} $y = -3x - 1.5$
3. {(2, −11), (4, −7), (8, 1), (16, 17), (32, 49)} $y = 2x - 15$
4. {(4, 69), (6, 91), (10, 135), (15, 190)} $y = 11x + 25$

About Using Calculators
Encourage students to put the ordered pairs in table form.

Using Computers

FOR/NEXT Statements

An important step in graphing an equation is to select several values for x and then find the corresponding values for y. This means that the same expression is evaluated many times with different values for x. A computer can perform the operations easily using a loop.

FOR/NEXT statements provide the simplest means for establishing a loop.

This BASIC program uses a loop to generate several ordered pairs that could be used to graph the equation $y = 2x - 1$.

```
10  FOR X = -2 TO 2
20  LET Y = 2 * X - 1
30  PRINT "(";X;",";Y;")"
40  NEXT X
50  END
```

When the program is executed, five ordered pairs, $(-2, -5)$, $(-1, -3)$, $(0, -1)$, $(1, 1)$, and $(2, 3)$, are printed.

1. As the coefficient of x increases and decreases the slope increases and decreases respectively. All five lines intersect the y-axis at the point $(0, 2)$.
2. When the coefficient of x is negative the graphs have negative slope. The graphs run from quadrant II to IV. All the graphs intersect the y-axis at the point $(0, 2)$.
3. The graphs of these three equations form three parallel lines. The graphs intersect the y-axis at the points $(0, -3)$, $(0, 0)$, $(0, 4)$ respectively.
4. The graphs are three parallel lines with negative slope. The graphs intersect the y-axis at the points $(0, -3)$, $(0, 0)$, $(0, 4)$ respectively.

Exercises

1. Modify line 20 of the program to generate ordered pairs for $y = x + 2$, $y = 2x + 2$, $y = 3x + 2$, $y = 10x + 2$, and $y = 0.5x + 2$. Graph the five equations on the same axes. What effect does the coefficient of x have on the graph? At what point do the graphs intersect the y-axis?
2. Generate ordered pairs for $y = -x + 2$, $y = -2x + 2$, $y = -3x + 2$, $y = -10x + 2$, and $y = -0.5x + 2$. Graph the five equations on the same axes. If the coefficient of x is negative, describe the graphs. At what point do the graphs intersect the y-axis?
3. Generate ordered pairs for $y = 2x - 3$, $y = 2x$, and $y = 2x + 4$. Graph the three equations on the same axes. Describe the graphs. At what points do the graphs intersect the y-axis?
4. Generate ordered pairs for $y = -2x - 3$, $y = -2x$, and $y = -2x + 4$. Graph the three equations on the same axes. Describe the graphs. At what points do these graphs intersect the y-axis?

About Using Computers
The exercises in this feature will help prepare students for the concepts of slope and y-intercept, which are presented in the next chapter.

Quiz
Lessons 9–5 to 9–7

State whether or not each relation is a function.

1. {(1, 6), (2, 5), (3, 4), (4, 4)} yes
2. $y + 2 = x$ yes

Given $f(x) = x^2 - 3$, determine the value for each of the following.

3. $f(2)$ 1
4. $f(-5)$ 22

Graph each inequality in the coordinate plane.

5. $x - 2y < 4$

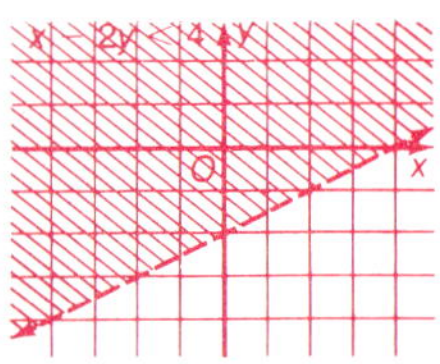

6. $2x + 3y \geq 2$

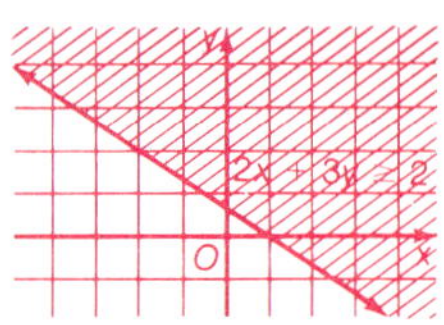

Find the missing terms for each sequence.

7. 3, 7, 11, 15, 19, 23, 27
8. −2, −4, −8, −16, −32, −64, −128

Write an equation for the relationship between the variables in each chart. Then copy and complete each chart.

9.

x	1	2	3	4	5	6
y	−1	−4	−7	−10	−13	−16

$y = -3x + 2$

10.

x	−6	−3	0	3	6	9
y	−2	−1	0	1	2	3

$y = \frac{1}{3}x$

Vocabulary

ordered pairs (271)
origin (271)
x-axis (271)
y-axis (271)
x-coordinate (271)
y-coordinate (271)
coordinate plane (271)
graph (272)
quadrants (272)
relation (274)
domain (274)
range (274)
solution (279)
linear equation (283)
function (286)
functional value (288)
half-plane (291)
boundary (291)
open half-plane (291)
closed half-plane (291)

Chapter Summary

1. Ordered pairs of numbers are used to locate points in the plane. The first component (*x*-coordinate) of an ordered pair corresponds to a number on the horizontal, or *x*-axis. The second component (*y*-coordinate) of the ordered pair corresponds to a number on the vertical, or *y*-axis. (271)
2. Completeness Property for Points in the Plane: Exactly one point in the plane is located by a given ordered pair of numbers. Exactly one ordered pair of numbers locates a given point in the plane. (272)
3. The *x*-axis and *y*-axis separate the plane into four regions, called quadrants. (272)
4. Definition of Relation: A relation is a set of ordered pairs. (274)
5. Definition of Domain: The domain of a relation is the set of all first components from each ordered pair. (274)
6. Definition of Range: The range of a relation is the set of all second components from each ordered pair. (274)
7. The solutions to equations in two variables are ordered pairs. (279)
8. Definition of Linear Equation: A linear equation is an equation that can be written in the form $Ax + By = C$, where A, B, and C are any numbers, and A and B are not both 0. (283)
9. Definition of Function: A function is a relation in which each element of the domain is paired with exactly one element of the range. (286)

10. Vertical Line Test for a Function: If any vertical line passes through no more than one point of the graph of a relation, then the relation is a function. (287)

11. $f(x)$ is read "f of x" and represents the range of a function whose domain is represented by x. (288)

12. When graphing inequalities, an open half-plane is used if the boundary is not part of the solution set. A closed half-plane is used if the boundary is part of the solution set. (291)

Chapter Review

9-1 **State the quadrant in which the point for each ordered pair is located.**

1. (8, −2) IV
2. (−3, −3) III
3. (5, 6) I
4. (−2, 5) II

Graph each point.

5. $A(6, -5)$
6. $B(-4, 0)$
7. $C(-2, -3)$
8. $D(3, 5)$

9-2 **State the domain and range for each relation.**

9. {(3, 5), (2, 6), (5, 7)}
10. {(4, 1), (4, −2), (4, 6), (4, −1)}
11. {(−3, 5), (−3, 6), (4, 5), (4, 6)}
12. {(1, 3), (1, 5), (1, 7)}

State each relation as a set of ordered pairs. Then state the domain and range.

13.

x	y
−1	4
3	4
4	6
0	−4

14.

x	y
−1	1
−2	4
−3	9
−4	6

15.

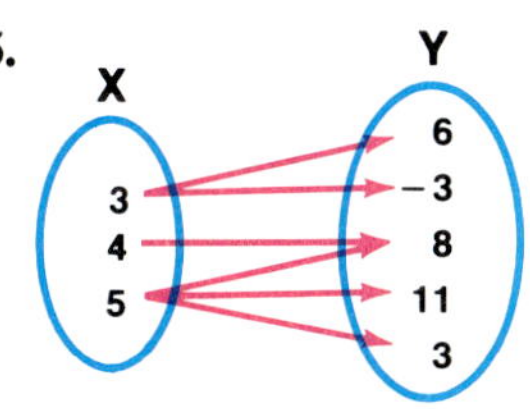

16.

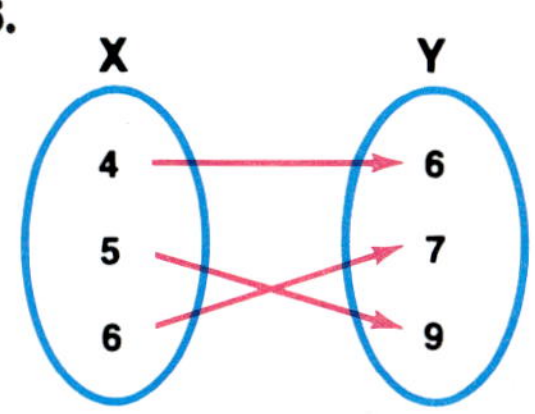

17.

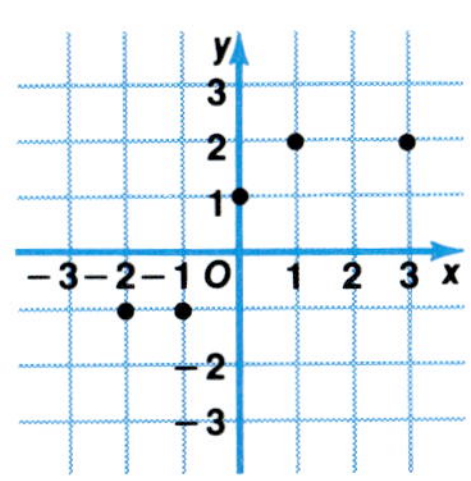

18.

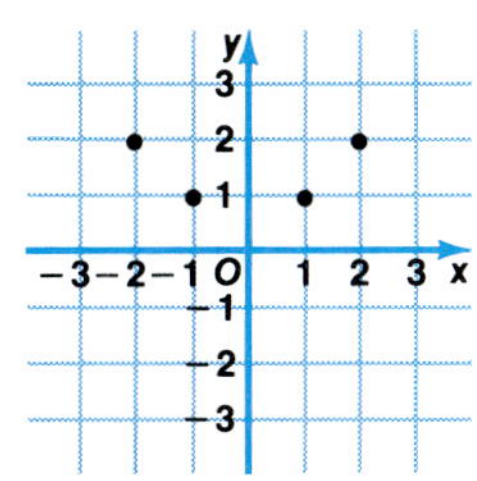

Answers to Chapter Review 5–8.

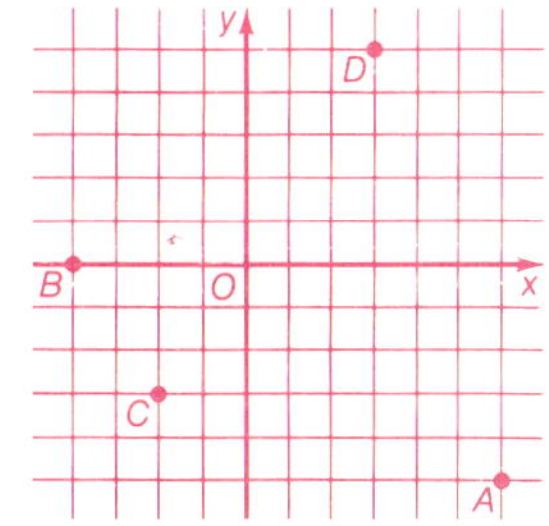

9. D = {3, 2, 5}
   R = {5, 6, 7}
10. D = {4}
    R = {1, −2, 6, −1}
11. D = {−3, 4}
    R = {5, 6}
12. D = {1}
    R = {3, 5, 7}
13. {(−1, 4), (3, 4), (4, 6), (0, −4)}
    D = {−1, 3, 4, 0}
    R = {4, 6, −4}
14. {(−1, 1), (−2, 4), (−3, 9), (−4, 6)}
    D = {−1, −2, −3, −4}
    R = {1, 4, 9, 6}
15. {(3, 6), (3, −3), (4, 8), (5, 8), (5, 11), (5, 3)}
    D = {3, 4, 5}
    R = {6, −3, 8, 11, 3}
16. {(4, 6), (5, 9), (6, 7)}
    D = {4, 5, 6}
    R = {6, 7, 9}
17. {(−2, −1), (−1, −1), (0, 1), (1, 2), (3, 2)}
    D = {−2, −1, 0, 1, 3}
    R = {−1, 1, 2}
18. {(−2, 2), (−1, 1), (1, 1), (2, 2)}
    D = {−2, −1, 1, 2}
    R = {1, 2}

Answers to Chapter Review

31.

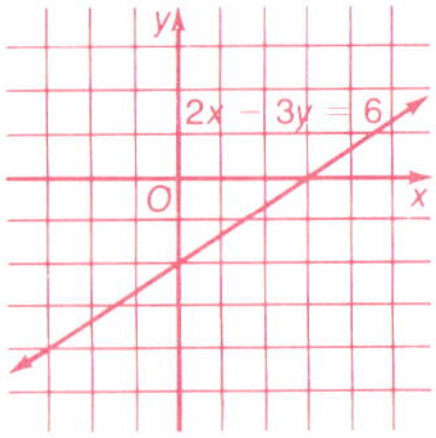

32.

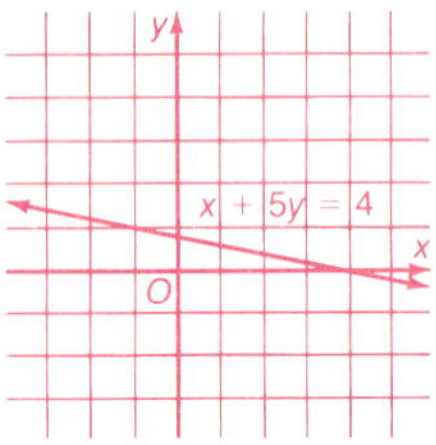

33.

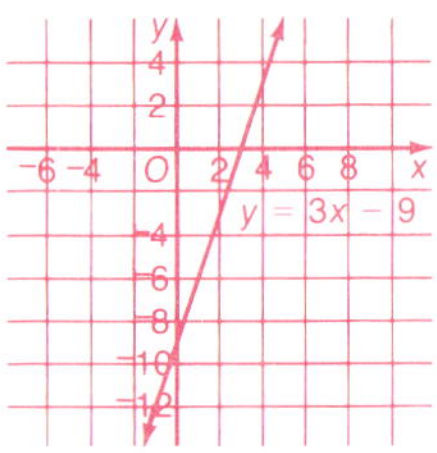

34.

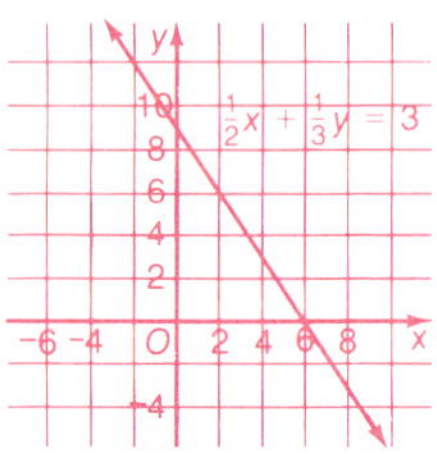

45.

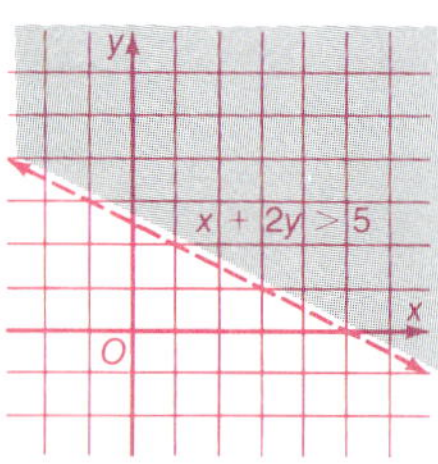

46.

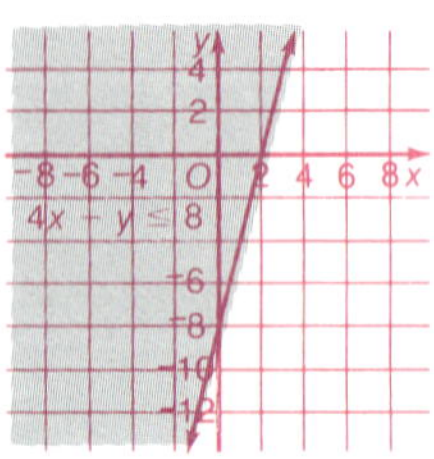

9-3 **Solve each equation for y in terms of x.**

19. $3x + y = 7$ $y = 7 - 3x$

20. $4x - 3y = 9$ $y = \frac{4x - 9}{3}$

State which of the ordered pairs are solutions of the given equation.

21. $2x - 3y = 8$ **a.** (4, 4) **b.** (6, 2) **c.** (4, 0) **d.** (2, 4)

22. $y = 3x - 6$ **a.** (0, 2) **b.** (2, 0) **c.** (6, 6) **d.** (−6, 0)

23. $\{(-4, -11), (-2, -3), (0, 5), (2, 13), (4, 21)\}$ 24. $\{(-4, -13), (-2, -11), (0, -9), (2, -7), (4,$

Solve each equation if the domain is $\{-4, -2, 0, 2, 4\}$.

23. $y = 4x + 5$

24. $x - y = 9$

25. $3x + 2y = 9$

26. $4x - 3y = 0$

25. $\left\{\left(-4, \frac{21}{2}\right), \left(-2, \frac{15}{2}\right), \left(0, \frac{9}{2}\right), \left(2, \frac{3}{2}\right), \left(4, -\frac{3}{2}\right)\right\}$ 26. $\left\{\left(-4, -\frac{16}{3}\right), \left(-2, -\frac{8}{3}\right), (0, 0), \left(2, \frac{8}{3}\right), (4,\right.$

9-4 **State whether each equation is a linear equation.**

27. $3x - y^2 = 6$ no

28. $xy = 8$ no

29. $y = 6$ yes

30. $2x - 8y = 5$ yes

Graph each equation.

31. $2x - 3y = 6$

32. $x + 5y = 4$

33. $y = 3x - 9$

34. $\frac{1}{2}x + \frac{1}{3}y = 3$

9-5 **State whether each relation is a function.**

35. $\{(3, 2), (5, 3), (4, 3), (5, 2)\}$ no

36. $\{(3, 8), (9, 3), (-3, 8), (5, 3)\}$ yes

37. $3x - 4y = 7$ yes

38. $x^2 - y = 4$ yes

Given $g(x) = x^2 - x + 1$, determine the value for each of the following.

39. $g(2)$ 3 **40.** $g(-2)$ 7 **41.** $g(0)$ 1 **42.** $g\left(\frac{1}{2}\right)$ $\frac{3}{4}$ **43.** $g(-1)$ 3 **44.** $g(2a)$ $4a^2 - 2a$

9-6 **Graph each inequality in the coordinate plane. Draw a separate set of axes for each.**

45. $x + 2y > 5$

46. $4x - y \le 8$

47. $3x - 2y < 6$

48. $\frac{1}{2}y \ge x + 4$

9-7 **Write an equation for the relationship between the variables in each chart.**

$y = 3x + 5$

49.

x	0	1	2	3	4
y	5	8	11	14	17

$y = x - 4$

50.

x	2	4	5	7	10
y	−2	0	1	3	6

Chapter Test

State the quadrant in which the point for each ordered pair is located.

1. $(-3, 5)$ II
2. $(-4, -1)$ III
3. $(5, -9)$ IV

State each relation as a set of ordered pairs. Then state the domain and range.

4.

x	y
1	3
2	7
3	−3
5	−2

5.

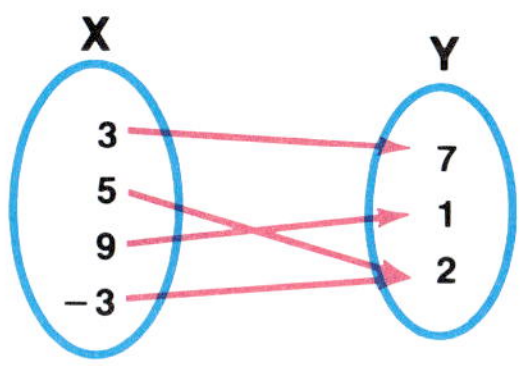

6.

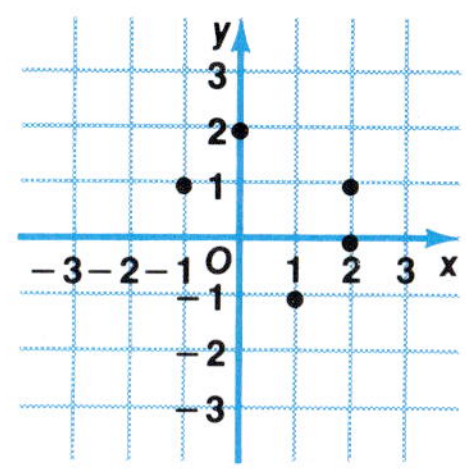

Solve each equation if the domain is $\{-2, -1, 0, 1, 3\}$.

7. $y = 3x + 10$
8. $2x - 5y = 4$
9. $2y - x = 8$

State whether each equation is a linear equation.

10. $5x = 17 - 4y$ yes
11. $y = x^2 - y$ no
12. $y = \frac{1}{x}$ no

Graph each equation.

13. $x - 2y = 8$
14. $5x - 2y = 8$
15. $\frac{2}{3}x + \frac{3}{4}y = 1$

State whether each relation is a function.

16. $\{(2, 4), (4, 2), (5, 4), (4, 6)\}$ no
17. $\{(-6, 9), (4, 5), (8, -2), (0, 0)\}$ yes
18. $8x - 3y = 7$ yes
19. $2x = 9$ no

Given $f(x) = 2x - 3$, determine the value for each of the following.

20. $f(-3)$ -9
21. $f(7)$ 11
22. $f(0)$ -3

Graph each of the following in the coordinate plane. Draw a separate set of axes for each.

23. $4x + 3y = 12$
24. $3x - 2y < 6$
25. $y \geq 5x + 1$
26. $4x + 2y = 9$

Write an equation for the relationship between the variables in the chart.

$y = 5x - 2$

27.

x	1	2	3	4	7
y	3	8	13	18	33

Answers to Chapter Review (continued)

47.

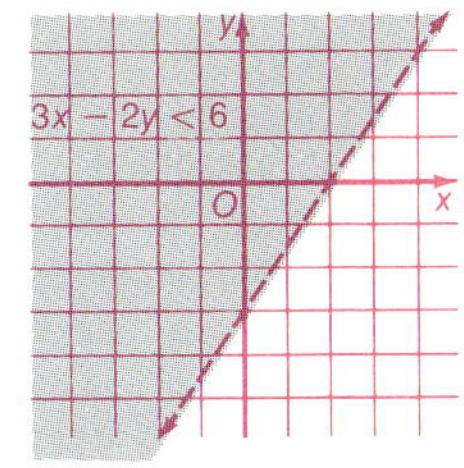

48.

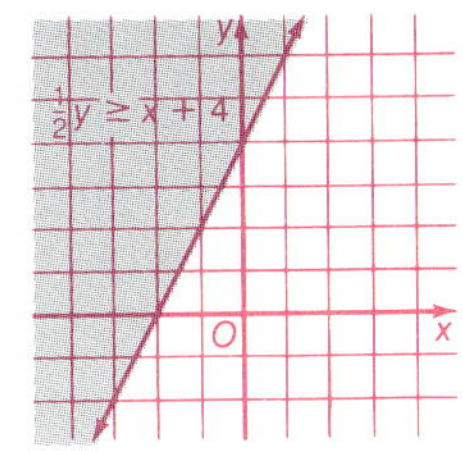

Answers to Chapter Test

4. $\{(1, 3), (2, 7), (3, -3), (5, -2)\}$
   $D = \{1, 2, 3, 5\}$
   $R = \{3, 7, -3, -2\}$
5. $\{(3, 7), (5, 2), (9, 1), (-3, 2)\}$
   $D = \{3, 5, 9, -3\}$
   $R = \{7, 1, 2\}$
6. $\{(-1, 1), (0, 2), (1, -1), (2, 0), (2, 1)\}$
   $D = \{-1, 0, 1, 2\}$
   $R = \{1, -1, 2, 0\}$
7. $\{(-2, 4), (-1, 7), (0, 10), (1, 13), (3, 19)\}$
8. $\left\{\left(-2, -\frac{8}{5}\right), \left(-1, -\frac{6}{5}\right), \left(0, -\frac{4}{5}\right), \left(1, -\frac{2}{5}\right), \left(3, \frac{2}{5}\right)\right\}$
9. $\left\{(-2, 3), \left(-1, \frac{7}{2}\right), (0, 4), \left(1, \frac{9}{2}\right), \left(3, \frac{11}{2}\right)\right\}$

(continued in the Teacher's Answer Key, p. T84)

About the Application
Another *Applications in Statistics* feature can be found on page 368. *Applications in Probability* features can be found on pages 108, 172, and 242. Students should be familiar with these graphs from previous math courses. Encourage them to find examples of both bar and line graphs in newspapers and magazines.

Applications in Statistics

Recording Data

Statistics provide techniques for collecting, organizing, analyzing, and interpreting numerical information called **data**. Organized data is easier to read and interpret. One way to organize data is by using frequency tables. Consider the following example.

Suppose a farmer recorded the number of eggs collected daily for 20 days. He recorded the following numbers:

8, 5, 6, 8, 7, 9, 8, 6, 4, 5, 8, 9, 11, 7, 6, 9, 4, 7, 8, 9

The table below was constructed to organize this data and to show the frequency distribution for the data.

Number of Eggs	Tally	Frequency
4	||	2
5	||	2
6	|||	3
7	|||	3
8	~~||||~~	5
9	||||	4
10		0
11	|	1
	Total	20

Note that the number in the frequency column corresponds to the number of tally marks.

This same data can be organized in intervals.

Number of Eggs	Frequency
4 – 5	4
6 – 7	6
8 – 9	9
10 – 11	1

Graphs are also used to present data and show relationships. There are several ways of presenting the data in the following table.

Students Enrolled in Graham College

Year	1960	1965	1970	1975	1980	1985
Enrollment	2870	3050	5000	7900	8550	10,400

This graph is called a **bar graph**. It shows how specific quantities compare.

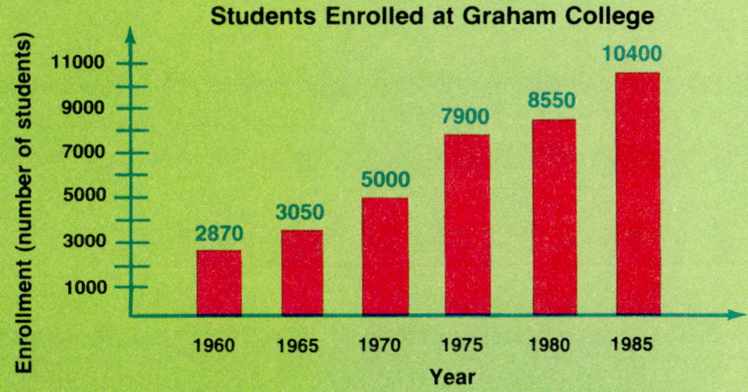

This graph is called a **line graph**. It is helpful for showing trends or changes.

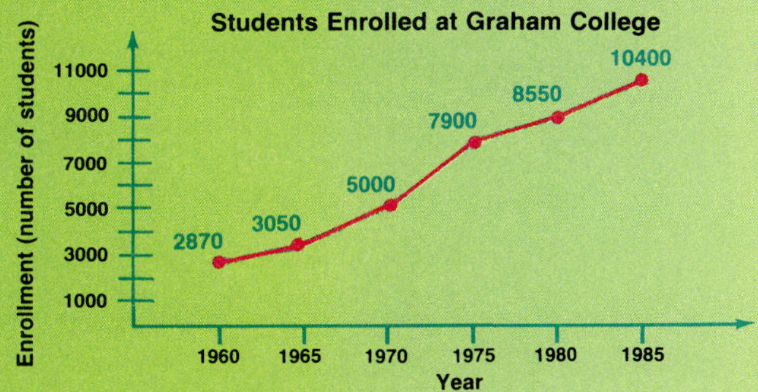

Exercises

1. Make a frequency table from the following data. Organize the data using the following intervals: 60–65, 66–70, 71–75, 76–80, 81–85, 86–90, 91–95, and 96–100.

Test scores on a mathematics exam:

64	82	93	75	96	89	68	71	78	80
99	85	70	90	95	77	86	65	88	91
66	78	76	98	73	84	94	96	77	90

Draw a bar graph and a line graph for each of the following tables.

2. Number of Students at Broadview High School

Grade	9th	10th	11th	12th
Number of Students	250	400	325	230

3. Percent of 18 Year Olds with High School Diplomas

Year	1940	1950	1955	1960	1965	1970	1975	1980
Percent	49.3	60.0	62.5	70.0	73.9	75.6	74.8	74.5

4. Postal Rates for First Class Mail

Years	1966	1969	1972	1975	1978	1981	1985
Cost/ounce	5¢	6¢	8¢	13¢	15¢	20¢	22¢

Answers to Exercises

1.

Score	Tally	Frequency
60–65	||	2
66–70	|||	3
71–75	|||	3
76–80	卌 |	6
81–85	|||	3
86–90	卌	5
91–95	||||	4
96–100	||||	4

For answers to exercises 2-4, see p. T85.

About the Photo

Slope is encountered not only on roadways as illustrated in the photograph, but also on train and amusement park rides. An incline railroad takes tourists up and down Lookout Mountain near Chattanooga, Tennessee. Roller coaster rides in amusement parks begin on an incline. Ramps are commonplace in sidewalks and buildings for convenience of the handicapped and elderly. Every incline has slope because an incline makes an oblique angle with the plane at the horizon.

One method of mining coal is called slope mining. An opening in the coal mine is made by a bore that goes down to the coal seams on a gradual incline or a gentle slope. The coal is removed from the mine by a belt conveyor that moves coal up the slope to the surface.

Emphasize that slope is the ratio of rise to run. It will be evident that parallel lines have the same slope. For perpendicular lines, the slope of one line must be the negative reciprocal of the other.

Teacher Resource Book Materials for Chapter 10 See pages 237-260.

Chapter Test, Form I (Multiple Choice)	237-239
Chapter Test, Form II (Free Response)	240-242
Quizzes	243-244
Extra Practice	245-247
Application: Equations in Slope-Intercept Form (Use after 10-3.)	248-249
Application: Olympic Women's Shot Put Winners (Use after 10-5.)	250
Application: Celsius and Kelvin Temperatures (Use after 10-5.)	251
Extension: A Scatter Graph (Use after 10-5.)	252
Extension: Formulas for a Line (Use after 10-5.)	253
Problem Solving	254
Cumulative Review	255
Answers	256-260

CHAPTER 10

Lines and Slopes

The road sign shows the steepness or slope of the hill. This information aids drivers in gauging their speed. Slope is also used in mathematics to give information about the graphs of equations.

Objective 10-1: Find the slope of a line, given the coordinates of two points on the line.

10-1 Slope

Jack Ferguson is a truck driver. When driving he sees many road signs similar to the one shown on the preceding page. Suppose that the slope of a hill is 6%. This means that for every 100 feet of *run* (horizontal distance), there is a *rise* (vertical distance) of 6 feet. The ratio of rise to run is called **slope**. The slope of a line describes its steepness or rate of change. ❶

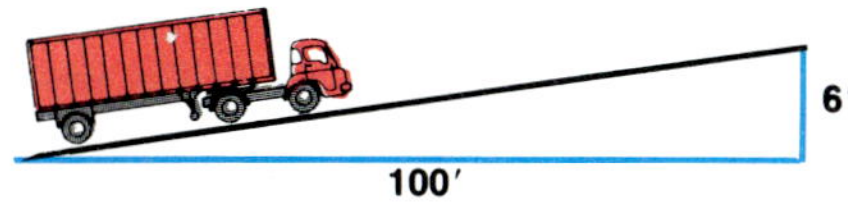

On the graph below, the line passes through the origin, (0, 0), and (4, 3). The rise (change in y-coordinates) is 3, while the run, (change in the x-coordinates) is 4. Thus, the slope of this line is $\frac{3}{4}$.

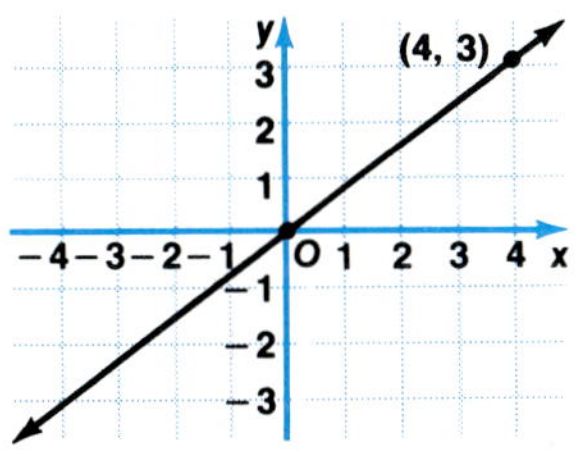

The slope of a line is the ratio of the change in y to the corresponding change in x.

$$\text{Slope} = \frac{\text{change in } y}{\text{change in } x}$$

Definition of Slope ❷

Example

1 **Determine the slope of each line.**

a.

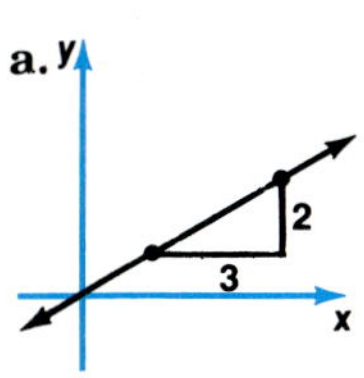

$$\frac{\text{change in } y}{\text{change in } x} = \frac{2}{3}$$

b.

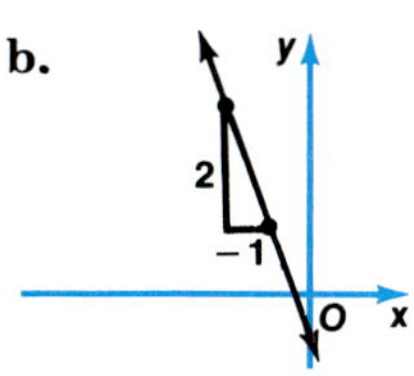

$$\frac{\text{change in } y}{\text{change in } x} = \frac{2}{-1} = -2$$

c.

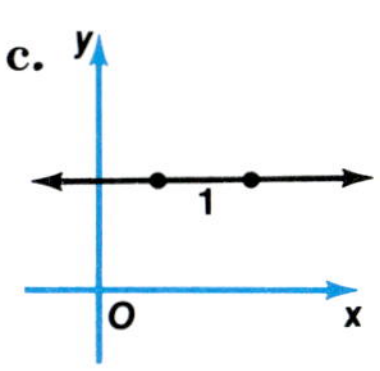

$$\frac{\text{change in } y}{\text{change in } x} = \frac{0}{1} = 0$$

d.

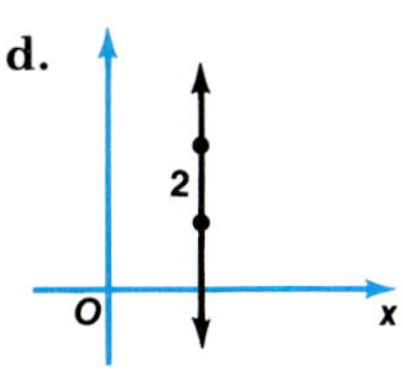

$$\frac{\text{change in } y}{\text{change in } x} = \frac{2}{0}$$ (crossed out)

The slope is undefined.

10-1

Lesson Notes, page T43

Discussion Question
(Use after Example 1.)
What happens to the slope of the line suggested by the minute hand on a clock between 5 p.m. and 6 p.m.? It starts out having no slope. From 5:00 to 5:15 it has every positive slope possible. At 5:15, it has a slope of zero. From 5:15 to 5:30, it has every negative slope possible. At 5:30, it again has no slope. This same pattern repeats from 5:30 to 6:00.

Chalkboard Exercises

1. What is the slope of the line through the origin and the point named?

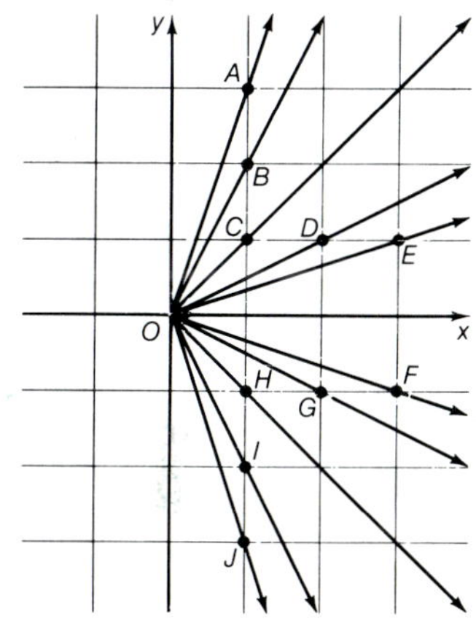

A	3	F	$-\frac{1}{3}$
B	2	G	$-\frac{1}{2}$
C	1	H	−1
D	$\frac{1}{2}$	I	−2
E	$\frac{1}{3}$	J	−3

❶ Discuss the introductory example.

$$\text{slope} = \frac{\text{rise}}{\text{run}} = \frac{6'}{100'} = \frac{6}{100} = \frac{3}{50}$$

❷ In advanced courses, slope is often defined as $\frac{\Delta y}{\Delta x}$. The Greek letter delta (Δ) is used to represent differences.

Chalkboard Exercises

Determine the slope of the line containing the points listed.

2a.

x	y
−3	−1
0	0
3	1
6	2
9	3

$\frac{1}{3}$

2b.

x	y
−2	−3
0	0
2	3
4	6
6	9

$\frac{3}{2}$

Determine the slope of the line passing each pair of points named below.

3a. (1, −2), (6, 2) $\frac{4}{5}$

3b. (−3, 4), (1, −2) $-\frac{3}{2}$

3c. (2, 5), (6, 2) $-\frac{3}{4}$

3d. (−3, 4), (2, 5) $\frac{1}{5}$

❸ A vertical line is said to have no slope.

❹ Stress that the slope of a line is the same between any two points on the line.

❺ Emphasize how to read expressions such as x_1 and y_2. Explain that the subscripts tell us that, for example, x_1 and x_2 both represent x-coordinates, but for different points.

Notice that the line extending from lower left to upper right has a positive slope. The line extending from upper left to lower right has a negative slope. The slope of the horizontal line is 0.

For a vertical line the change in x would be zero. Since division by zero is not defined we say that *the slope of a vertical line is undefined.* ❸

Example 2

Determine the slope of the line containing the points listed.

x	y
−2	−3
0	0
2	3
4	6
6	9

(Each step: x +2, y +3)

Notice that y increases 3 units for each 2 units that x increases.

$$\text{slope} = \frac{\text{change in } y}{\text{change in } x} = \frac{3}{2}$$

From the above example and many other similar examples it is clear that the slope of a line can be determined from the coordinates of any two points on the line. ❹

Determining Slope Given Two Points

Given the coordinates of two points on a line, (x_1, y_1) and (x_2, y_2), the slope, m, can be found as follows.

$$m = \frac{y_2 - y_1}{x_2 - x_1} \text{ where } x_2 \neq x_1$$

Example 3

Determine the slope of the line passing through (3, −9) and (4, −12).

$$m = \frac{y_2 - y_1}{x_2 - x_1}$$

y_2 is read "y sub 2." The 2 is a subscript. ❺

$$= \frac{-12 - (-9)}{4 - 3} = \frac{-3}{1} = -3$$

The difference of y-coordinates was expressed as $-12 - (-9)$. Suppose $-9 - (-12)$ had been used instead. The corresponding difference of the x-coordinates would have been $3 - 4$.

Notice that $\frac{-9 - (-12)}{3 - 4}$ is also equal to −3.

Example

4 **Determine the value of r so the line through $(r, 4)$ and $(9, -2)$ has a slope of $-\frac{3}{2}$.**

$$m = \frac{y_2 - y_1}{x_2 - x_1}$$

$$-\frac{3}{2} = \frac{-2 - 4}{9 - r}$$

$$-\frac{3}{2} = \frac{-6}{9 - r}$$

$$-3(9 - r) = -6(2) \quad \textit{Means-Extremes Property}$$

$$-27 + 3r = -12 \quad \textit{Solve for r.}$$

$$3r = 15$$

$$r = 5$$

Chalkboard Exercises

Determine the value of r so that the line through each pair of points has the given slope.

4a. $(3, 2), (-1, r)$; $m = 2$

$2 = \frac{r - 2}{-1 - 3}$

$r = -6$

4b. $(0, 4), (r, -2)$; $m = 3$

$3 = \frac{-2 - 4}{r - 0}$

$r = -2$

4c. $(r, 2), (1, 5)$; $m = -\frac{1}{3}$

$-\frac{1}{3} = \frac{5 - 2}{1 - r}$

$r = 10$

4d. $(-1, r), (3, -4)$; $m = \frac{1}{4}$

$\frac{1}{4} = \frac{-4 - r}{3 - (-1)}$

$r = -5$

Suggested Daily Assignments

Enriched
- Exp: 2–14 evens
- Written: 1–12, 14–36 evens
- Mini-Review: 1–5

Average
- Exp: 1–15
- Written: 1–36
- Mini-Review: 1–5

Fundamental
- Exp: 1–15
- Written: 1–24, 26–36 evens
- Mini-Review: 1–5

Exploratory Exercises

State the slope of each line.

1.

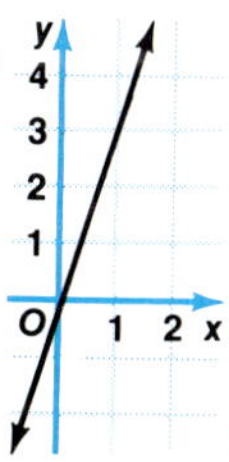

3

2.

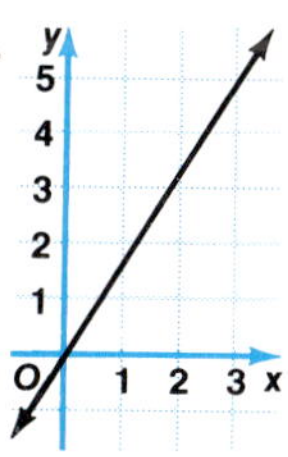

$\frac{5}{3}$

3.

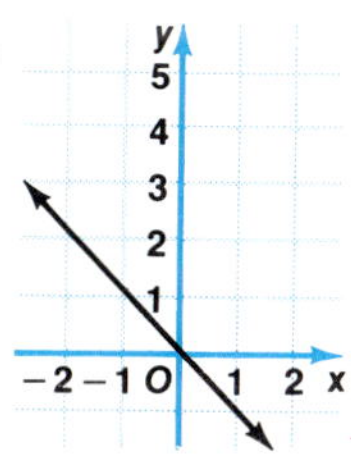

-1

4.

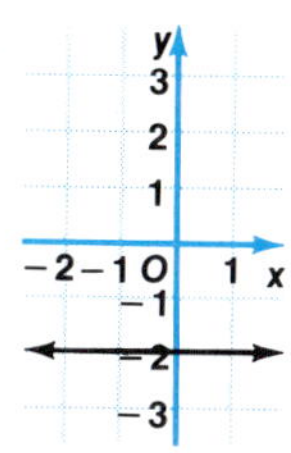

0

For each table, state the change in y and the change in x. Then determine the slope of the line passing through those points.

5.

x	y
0	0
1	2
2	4
3	6
4	8

$\frac{2}{1} = 2$

6.

x	y
0	0
1	1
2	2
3	3
4	4

$\frac{1}{1} = 1$

7.

x	y
−2	2
−1	1
0	0
1	−1
2	−2

$\frac{-1}{1} = -1$

8.

x	y
−2	−8
−1	−4
0	0
1	4
2	8

$\frac{4}{1} = 4$

9.

x	y
−6	8
−3	4
0	0
3	−4
6	−8

$\frac{-4}{3} = -\frac{4}{3}$

Determine the slope of the line passing through each pair of points named below.

10. $(3, 1), (5, 4)$ $\frac{3}{2}$

11. $(8, -1), (7, -2)$ 1

12. $(6, 1), (6, 8)$ undefined

13. $(1, 4), (-2, -3)$ $\frac{7}{3}$

14. $(0, 0), (5, -2)$ $-\frac{2}{5}$

15. $(6, 4), (-6, -4)$ $\frac{2}{3}$

Written Exercises

Determine the slope of each line named below.

1. a 3
2. b -1
3. c $\frac{4}{3}$
4. d 0
5. e undefined
6. f $-\frac{1}{5}$
7. g $\frac{1}{3}$
8. h -2

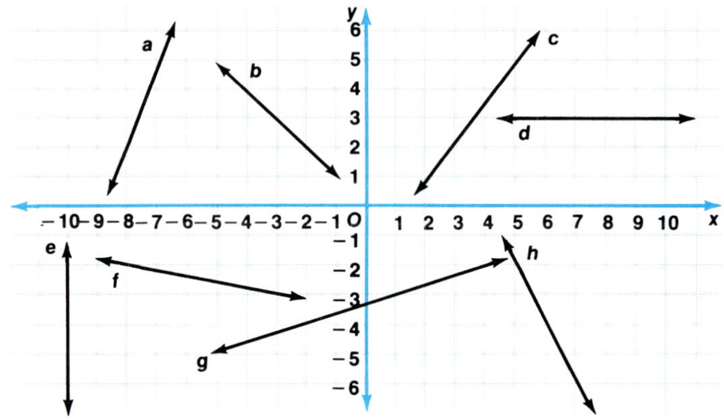

9\. lines go from lower left to upper right

9. Look at lines a, c, and g above. These lines have positive slopes. What are the general directions of lines with positive slopes?

10\. lines go from upper left to lower right

10. Look at lines b, f, and h above. These lines have negative slopes. What are the general directions of lines with negative slopes?

11. What is the slope of a horizontal line? zero

12. What is the slope of a vertical line? undefined

Determine the slope of the line passing through each pair of points named below.

13. (3, 4), (4, 6) 2
14. (5, 2), (7, 6) 2
15. $(-3, 6), (-5, 9)$ $-\frac{3}{2}$
16. $(-2, 1), (-6, 3)$ $-\frac{1}{2}$
17. $(-3, 6), (-8, 4)$ $\frac{2}{5}$
18. $(-1, 11), (-5, 4)$ $\frac{7}{4}$
19. $(7, -4), (9, -1)$ $\frac{3}{2}$
20. $(11, -1), (14, -6)$ $-\frac{5}{3}$
21. $(18, -4), (6, -10)$ $\frac{1}{2}$
22. $(12, -3), (14, -7)$ -2
23. $(14, 3), (-11, 3)$ 0
24. $(-4, -6), (-3, -8)$ -2
25. $(0, 0), \left(\frac{1}{3}, \frac{1}{6}\right)$ $\frac{1}{2}$
26. $\left(\frac{3}{4}, \frac{1}{2}\right), \left(\frac{1}{2}, \frac{3}{4}\right)$ -1
27. $\left(\frac{5}{8}, \frac{1}{3}\right), \left(\frac{1}{8}, \frac{2}{3}\right)$ $-\frac{2}{3}$
28. $\left(\frac{3}{4}, 1\right), \left(\frac{3}{4}, -1\right)$ undefined
29. $\left(\frac{2}{3}, 5\right), \left(\frac{1}{3}, 2\right)$ 9
30. $\left(3\frac{1}{2}, 5\frac{1}{4}\right), \left(2\frac{1}{2}, 6\right)$ $-\frac{3}{4}$

Determine the value of r so that the line through each pair of points has the given slope.

31. $(9, r), (6, 3), m = -\frac{1}{3}$ 2
32. $(r, 4), (7, 3), m = \frac{3}{4}$ $\frac{25}{3}$
33. $(4, -7), (-2, r), m = \frac{8}{3}$ -23
34. $(6, -2), (r, -6), m = -4$ 7
35. $(r, 7), (11, r), m = -\frac{1}{5}$ 6
36. $(4, r), (r, 2), m = -\frac{5}{3}$ 7

mini-review

Solve.

1. $\frac{3}{4}x = -36$ $x = -48$
2. $\frac{r - 6}{-3} = 5$ $r = -9$
3. $\frac{3}{4} = \frac{x + 5}{x + 7}$ $x = 1$
4. $3y - 0.5 \ge 3.1$ $y \ge 1.2$
5. The length of a rectangle is 5 feet more than 3 times the width. The perimeter is 250 feet. Find the dimensions of the rectangle. 30 feet × 95 feet

Objective 10-2A: Write a linear equation in standard form given the coordinates of a point on the line and the slope of the line.

Objective 10-2B: Write a linear equation in standard form given the coordinates of two points on the line.

10-2 Equations of Lines in Point-Slope and Standard Form

The graph at the right shows the nonvertical line, l, and point R having coordinates (x_1, y_1). The slope of line l is represented by m.

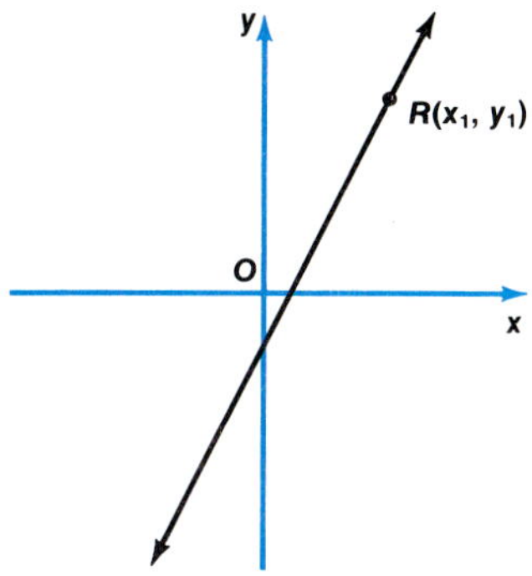

❶

Let (x, y) represent any other point on line l. Since m is the slope of the line, all points (x, y) on the line satisfy the following equation.

$$\frac{y - y_1}{x - x_1} = m \qquad \textit{Assume } x - x_1 \neq 0.$$

If both sides of the equation are multiplied by $(x - x_1)$ the result is the equation $y - y_1 = m(x - x_1)$. This form of an equation of a line is called the **point-slope form**.

For a given point (x_1, y_1) on a nonvertical line, the point-slope form of a linear equation is

$$y - y_1 = m(x - x_1).$$

❷ ***Point-Slope Form***

If you know the slope of a line and the coordinates of one point on the line, you can write the equation of the line.

Example 1

Write the equation of the line passing through (2, −4) and having a slope of $\frac{2}{3}$.

Substitute in the point-slope form. Let $(x_1, y_1) = (2, -4)$ and $m = \frac{2}{3}$.

$$y - y_1 = m(x - x_1)$$
$$y - (-4) = \tfrac{2}{3}(x - 2)$$
$$y + 4 = \tfrac{2}{3}(x - 2)$$

The equation of the line is $y + 4 = \frac{2}{3}(x - 2)$.

10-2

Lesson Notes, Page T44

Discussion Question

A line passes through (0, 0) and has a slope of 2. Name the coordinates of some other points on this line.

Any point whose coordinates are in the form $(a, 2a)$.

Chalkboard Exercises

Write the point-slope form of the equation for the line passing through the given point and having the given slope.

1a. (3, 5), $\frac{4}{3}$

$y - 5 = \frac{4}{3}(x - 3)$

1b. (−2, 5), $-\frac{3}{2}$

$y - 5 = -\frac{3}{2}(x + 2)$

1c. (−3, 2), $-\frac{1}{2}$

$y - 2 = -\frac{1}{2}(x + 3)$

1d. (0, 5), −3

$y - 5 = -3x$

❶ Point R represents a *specific* point. Since (x, y) represents *any* other point on the line, it is not shown on the line.

❷ The point-slope form of a linear equation is not unique. This form is used as an intermediate step when finding the standard form or slope-intercept form.

Chalkboard Exercises

Write each equation in standard form.

2a. $y - 6 = -\frac{3}{2}(x + 1)$

$3x + 2y = 9$

2b. $y + 1 = -\frac{1}{2}(x + 4)$

$x + 2y = -6$

Write the standard form of the equation for the line through each pair of points.

3a. $(-1, 2), (5, 5)$
$x - 2y = -5$

3b. $(1, 4), (6, 1)$
$3x + 5y = 23$

3c. $(3, 2), (5, 5)$
$3x - 2y = 5$

3d. $(-2, 5), (5, 2)$
$3x + 7y = 29$

Recall that any linear equation can be expressed in the form $Ax + By = C$. This is called the **standard form**. An equation that is written in point-slope form can be easily changed to standard form.

Example 2

Write $y + 4 = \frac{2}{3}(x - 2)$ in standard form.

$$y + 4 = \frac{2}{3}(x - 2)$$

$$3(y + 4) = 2(x - 2) \quad \textit{Multiply both sides by 3.}$$

$$3y + 12 = 2x - 4$$

$$-2x + 3y = -16$$

$$2x - 3y = 16 \quad \textit{Multiply both sides by } -1 \textit{ to obtain a positive coefficient for } x.$$

$2x - 3y = 16$ is in standard form.

You can also find the equation of a line if you know the coordinates of two points on the line.

Example 3

Write the standard form of the equation for the line passing through $(5, -2)$ and $(-3, 8)$.

First determine the slope.

$$m = \frac{y_2 - y_1}{x_2 - x_1}$$

$$= \frac{8 - (-2)}{-3 - 5} \quad \textit{Let } y_2 = 8,\ y_1 = -2,\ x_2 = -3, \textit{ and } x_1 = 5.$$

$$m = \frac{10}{-8} \text{ or } -\frac{5}{4}$$

Use the point-slope form. The slope is $-\frac{5}{4}$. The point (x_1, y_1) equals either $(5, -2)$ or $(-3, 8)$.

$$y - y_1 = m(x - x_1) \quad \textit{Let } (x_1, y_1) = (5, -2)$$

$$y - (-2) = -\frac{5}{4}(x - 5)$$

$$y + 2 = -\frac{5}{4}(x - 5) \quad \textit{This equation is in point-slope form.}$$

Now write the equation in standard form.

$$y + 2 = -\frac{5}{4}(x - 5)$$

$$4(y + 2) = -5(x - 5)$$

$$4y + 8 = -5x + 25$$

$$5x + 4y = 17 \quad \textit{This equation is in standard form.}$$

Check to see that you get the same result using the point $(-3, 8)$.

A horizontal line has a slope of zero. The point-slope form can be used to write the equation of horizontal lines. For example, the equation of a line through $(5, 2)$ and $(-9, 2)$ is $(y - 2) = 0(x - 5)$ or $y = 2$.

The slope of a vertical line is undefined. Therefore the point-slope form cannot be used for such lines. The equation of a line through $(3, 5)$ and $(3, -9)$ is $x = 3$ since the x-coordinate of every point on the line is equal to 3. Note that the equation $x = 3$ is in standard form.

Exploratory Exercises

State the slope and the point through which the line passes for each linear equation.

1. $y - 2 = 3(x - 5)$ 3; (5, 2)
2. $y - 3 = 4(x - 6)$ 4; (6, 3)
3. $y - (-5) = -2(x + 1)$ −2; (−1, −5)
4. $y + 4 = -2[x - (-2)]$ −2; (−2, −4)
5. $y + 6 = -\frac{3}{2}(x + 5)$
6. $y + 7 = -\frac{3}{4}(x + 4)$
7. $2(x - 3) = y + \frac{3}{2}$ 2; $\left(3, -\frac{3}{2}\right)$
8. $-\frac{2}{3}(x + 7) = y - \frac{3}{4}$
9. $-\frac{3}{5}(x + 6) = y + \frac{3}{8}$
10. $y = 3$ 0; (0, 3)
11. $y = -2$ 0; (0, −2)
12. $x = 1$ undefined; (1, 0)

Express each equation in standard form.

13. $y - 3 = 2\left(x + \frac{3}{2}\right)$
14. $y + 5 = -3\left(x - \frac{1}{3}\right)$
15. $y + 1 = \frac{2}{3}(x + 2)$
16. $y + \frac{3}{2} = \frac{1}{2}(x + 4)$
17. $y - \frac{2}{3} = \frac{5}{3}(x + 7)$
18. $y - 1 = \frac{5}{6}\left(x + \frac{3}{5}\right)$

Written Exercises

Write the standard form of the equation for the line passing through the given point and having the given slope.

1. $(5, 4), -\frac{2}{3}$ $2x + 3y = 22$
2. $(-6, -3), -\frac{1}{2}$ $x + 2y = -12$
3. $(9, 1), \frac{2}{3}$ $2x - 3y = 15$
4. $(8, 2), \frac{3}{4}$ $3x - 4y = 16$
5. $(4, -3), 2$ $2x - y = 11$
6. $(-2, 4), -3$ $3x + y = -2$
7. $(-6, 1), \frac{3}{2}$ $3x - 2y = -20$
8. $(6, -2), \frac{4}{3}$ $4x - 3y = 30$
9. $(5, 7), 0$ $y = 7$
10. $(-2, 6), 0$ $y = 6$
11. $(1, 3)$, undefined $x = 1$
12. $(-2, 1)$, undefined $x = -2$

Write the standard form of the equation for the line through each pair of points.

13. $(5, 4), (6, 3)$ $x + y = 9$
14. $(9, 1), (8, 2)$ $x + y = 10$
15. $(6, 1), (7, -4)$ $5x + y = 31$
16. $(8, 3), (5, -1)$ $4x - 3y = 23$
17. $(4, -2), (8, -3)$ $x + 4y = -4$
18. $(6, -1), (4, -2)$
19. $(-6, 1), (-8, 2)$
20. $(-5, 1), (6, -2)$
21. $(-8, 2), (-1, -2)$
22. $(4, -2), (4, 8)$ $x = 4$
23. $(5, -4), (5, 5)$ $x = 5$
24. $(5, 3), (-6, 3)$ $y = 3$
25. $\left(\frac{3}{4}, 1\right), \left(2, \frac{1}{2}\right)$ $4x + 10y = 13$
26. $\left(\frac{1}{2}, \frac{3}{4}\right), \left(\frac{2}{3}, \frac{4}{5}\right)$ $3x - 10y = -6$
27. $\left(2\frac{1}{2}, \frac{1}{3}\right), \left(\frac{3}{4}, 1\frac{1}{2}\right)$
28. $(4, 2), (-7, 2)$ $y = 2$
29. $\left(-8, \frac{1}{2}\right), \left(9, \frac{1}{2}\right)$ $2y = 1$
30. $(3, -2), (3, 5)$ $x = 3$
31. $(-2, 7), \left(-2, \frac{16}{3}\right)$ $x = -2$
32. $\left(-\frac{2}{9}, 3\right), \left(\frac{8}{9}, 3\right)$ $y = 3$
33. $\left(-2, \frac{2}{3}\right), \left(-2, \frac{2}{7}\right)$ $x = -2$

Suggested Daily Assignments

Enriched
Exp: 7–14
Written: 2–12 evens, 22–32 evens

Average
Exp: 1–17 odds
Written: 1–33 odds

Fundamental
Exp: 1–18
Written: 1–21
p. 308, 25–35 odds

Answers to Exploratory

5. $-\frac{3}{2}$; $(-5, -6)$
6. $-\frac{3}{4}$; $(-4, -7)$
8. $-\frac{2}{3}$; $\left(-7, \frac{3}{4}\right)$
9. $-\frac{3}{5}$; $\left(-6, -\frac{3}{8}\right)$
13. $2x - y = -6$
14. $3x + y = -4$
15. $2x - 3y = -1$
16. $x - 2y = -1$
17. $5x - 3y = -37$
18. $5x - 6y = -9$

Answers to Written

18. $x - 2y = 8$
19. $x + 2y = -4$
20. $3x + 11y = -4$
21. $4x + 7y = -18$
27. $2x + 3y = 6$

10-3

Lesson Notes, page T44
Enrichment, page T44

Readiness Activity
Graph $y = 2x - 1$.
What are the coordinates of the point where the line crosses the y-axis?

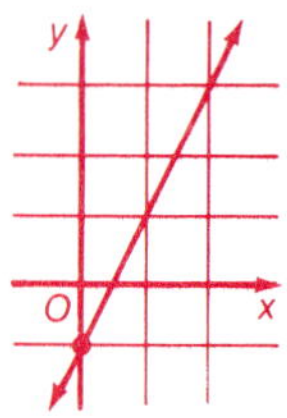

(0, −1)

Repeat the above activity for the equation $y = \frac{4}{3}x + 3$.

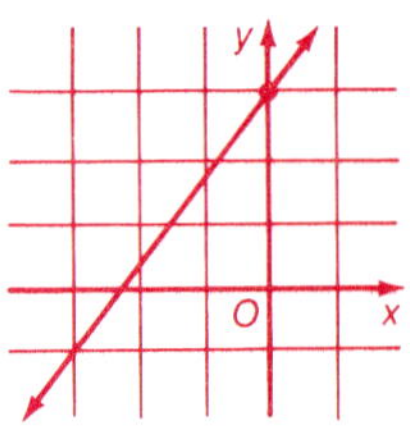

(0, 3)

Discussion Question
(Use after definition of slope-intercept form.)
The equations of vertical lines cannot be written in slope-intercept form. Why?
Vertical lines have no slope or y-intercepts.

Chalkboard Exercises
State the slope and y-intercept of the graph of each equation.

1a. $y = 2x + 7$ 2, 7
1b. $y = -x + 5$ −1, 5
1c. $y = \frac{1}{3}x + 6$ $\frac{1}{3}$, 6

Objective 10-3A: Write an equation in slope-intercept form given the slope and y-intercept of the line.
Objective 10-3B: Determine the slope and y-intercept of the graph of a given equation.
Objective 10-3C: Determine the x- and y-intercepts of the graph of a given equation.

10-3 Slope-Intercept Form

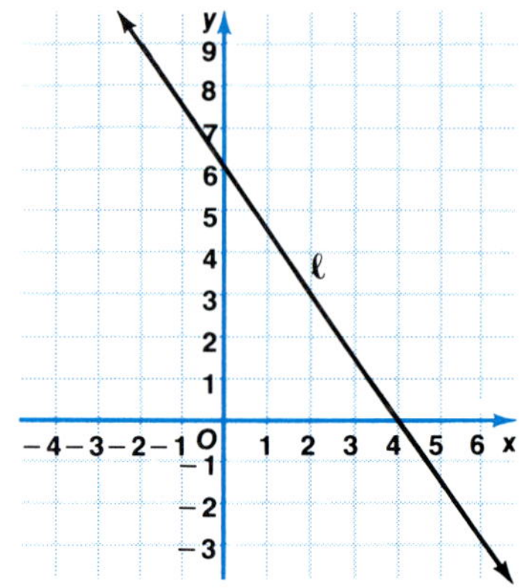

The x-coordinate of the point where a line crosses the x-axis is called the **x-intercept** of the line. Line l crosses the x-axis at (4, 0). Therefore, the x-intercept is 4. Note that the corresponding y-coordinate is 0.

Similarly, the y-coordinate of the point where the line crosses the y-axis is called the **y-intercept** of the line. Line l crosses the y-axis at (0, 6). Therefore, the y-intercept is 6. Note that the corresponding x-coordinate is 0.

Consider the graph at the right. The line with slope m crosses the y-axis at $(0, b)$. You can write an equation for this line using the point-slope form. Let $(x_1, y_1) = (0, b)$.

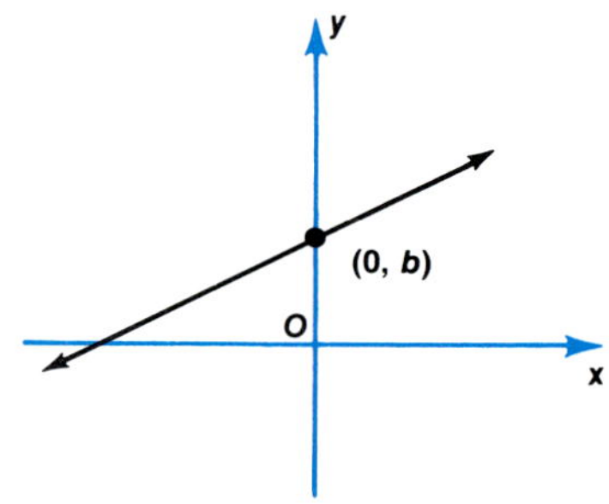

$$y - y_1 = m(x - x_1)$$
Replace y_1 by b.
Replace x_1 by 0.
$$y - b = m(x - 0)$$
$$y = mx + b$$
↑ *slope* ↑ *y-intercept*

An equation in the form $y = mx + b$ is said to be in slope-intercept form.

Slope-Intercept Form

Example

1 **State the slope and y-intercept of the graph of $y = 5x - 3$.**

$$y = mx + b$$
$$y = 5x + (-3)$$

Since $m = 5$, the slope is 5.
Since $b = -3$, the y-intercept is -3.

Compare the slope-intercept form of a linear equation with the standard form, $Ax + By = C$.

Solve for y.

$$Ax + By = C$$
$$By = -Ax + C \qquad \text{Assume } B \neq 0.$$
$$y = -\frac{A}{B}x + \frac{C}{B} \qquad \text{This equation is in slope-intercept form.}$$

($-\frac{A}{B}$ ↑ slope; $\frac{C}{B}$ ↑ y-intercept)

If the equation is given in standard form and B is *not* zero, the slope of the line can be determined by computing $-\frac{A}{B}$ and the y-intercept can be determined by computing $\frac{C}{B}$.

Example 2

State the slope and y-intercept of the graph of $3x + 2y = 6$.

In the equation $3x + 2y = 6$, $A = 3$, $B = 2$, and $C = 6$.

Slope: $-\frac{A}{B} = -\frac{3}{2}$ **y-intercept:** $\frac{C}{B} = \frac{6}{2}$ or 3

Check: Write the equation in slope-intercept form.

$$3x + 2y = 6$$
$$2y = -3x + 6$$
$$y = -\frac{3}{2}x + 3 \qquad \text{The slope is } -\frac{3}{2} \text{ and the } y\text{-intercept is } 3.$$

Recall that the x-coordinate of the y-intercept is 0 and the y-coordinate of the x-intercept is 0. You can use these facts to find the x- and y-intercepts of the graph of a linear equation.

Example 3

Determine the x- and y-intercepts of the graph of $3x - 6y = 12$.

To find the x-intercept, let $y = 0$.

$$3x - 6y = 12$$
$$3x - 6(0) = 12$$
$$3x = 12$$
$$x = 4$$

The x-intercept is 4. The graph crosses the x-axis at (4, 0).

To find the y-intercept, let $x = 0$.

$$3x - 6y = 12$$
$$3(0) - 6y = 12$$
$$-6y = 12$$
$$y = -2$$

The y-intercept is -2. The graph crosses the y-axis at $(0, -2)$.

Chalkboard Exercises

State the slope and y-intercept of the graph of each equation.

2a. $3x + 4y = 12$ $-\frac{3}{4}, 3$

2b. $-2x + 6y = 9$ $\frac{1}{3}, \frac{3}{2}$

2c. $\frac{1}{2}x + y = 6$ $-\frac{1}{2}, 6$

Determine the x- and y-intercepts for the graph of each equation.

3a. $2x + 3y = 5$ $\frac{5}{2}, \frac{5}{3}$

3b. $4x - 2y = 7$ $\frac{7}{4}, -\frac{7}{2}$

3c. $2y = 3x + 4$ $-\frac{4}{3}, 2$

3d. $2x = 4y - 3$ $-\frac{3}{2}, \frac{3}{4}$

3e. $7x - 3y = 12$ $\frac{12}{7}, -4$

Suggested Daily Assignments

Enriched
Exp: 14–20
Written: 2–38 evens, 40–43

Average
Exp: 1–19 odds, 20
Written: 1–21, 22–38 evens

Fundamental
Exp: 1–20
Written: 1–37 odds
p. 311, 22–33

Answers to Written

22. $m = -\frac{2}{5}$, $b = 2$; $y = -\frac{2}{5}x + 2$

23. $m = -\frac{3}{4}$, $b = 3$; $y = -\frac{3}{4}x + 3$

24. $m = 5$, $b = -15$; $y = 5x - 15$

25. $m = 3$, $b = -9$; $y = 3x - 9$

26. $m = -\frac{2}{5}$, $b = \frac{8}{5}$; $y = -\frac{2}{5}x + \frac{8}{5}$

27. $m = -\frac{7}{4}$, $b = 2$; $y = -\frac{7}{4}x + 2$

28. $m = \frac{5}{4}$, $b = -\frac{11}{4}$; $y = \frac{5}{4}x - \frac{11}{4}$

29. $m = \frac{7}{3}$, $b = -\frac{10}{3}$; $y = \frac{7}{3}x - \frac{10}{3}$

30. $m = -6$, $b = 15$; $y = -6x + 15$

31. $m = 12$, $b = -24$; $y = 12x - 24$

32. $m = -2$, $b = 12$; $y = -2x + 12$

33. $m = -4$, $b = 12$; $y = -4x + 12$

For answers to exercises 34-39, see p. T85.

The y-intercept of an equation in the form $Ax + By = C$ is $\frac{C}{B}$. To find a formula for the x-intercept, let $y = 0$.

$$\begin{aligned} Ax + By &= C \\ Ax + B(0) &= C \\ Ax &= C \\ x &= \frac{C}{A} \end{aligned}$$

The x-intercept is $\frac{C}{A}$, $A \neq 0$.

For the equation $2x - 3y = 8$, the x-intercept is $\frac{8}{2}$ or 4.

Exploratory Exercises

State the slope and y-intercept of the graph of each equation.

1. $y = 5x + 3$ 5, 3
2. $y = 2x + 1$ 2, 1
3. $y = 3x - 7$ 3, −7
4. $y = 6x - 8$ 6, −8
5. $y = \frac{1}{3}x$ $\frac{1}{3}$, 0
6. $y = \frac{3}{4}x + 1$ $\frac{3}{4}$, 1
7. $y = \frac{3}{5}x - \frac{1}{4}$ $\frac{3}{5}$, $-\frac{1}{4}$
8. $y = \frac{2}{3}x$ $\frac{2}{3}$, 0
9. $y = \frac{3}{5}x + \frac{1}{4}$ $\frac{3}{5}$, $\frac{1}{4}$
10. $2x + 3y = 5$ $-\frac{2}{3}$, $\frac{5}{3}$
11. $-x + 4y = 3$ $\frac{1}{4}$, $\frac{3}{4}$
12. $2x - y = 1$ 2, −1
13. $y - 6x = 5$ 6, 5
14. $3y - 8x = 2$ $\frac{8}{3}$, $\frac{2}{3}$
15. $2y - 3x = 7$ $\frac{3}{2}$, $\frac{7}{2}$
16. $5y = -8x - 2$ $-\frac{8}{5}$, $-\frac{2}{5}$
17. $4y = 2x + 7$ $\frac{1}{2}$, $\frac{7}{4}$
18. $3y = -2x + 11$ $-\frac{2}{3}$, $\frac{11}{3}$
19. State an equation for a line that has an undefined slope. $x = c$ (c is any number)
20. State an equation for a line that has no y-intercept. $x = c$ (c is any number)

Written Exercises

Write a linear equation in slope-intercept form for the line with the given slope and y-intercept.

1. $m = 3$, $b = 1$ $y = 3x + 1$
2. $m = 2$, $b = 3$ $y = 2x + 3$
3. $m = -3$, $b = 5$ $y = -3x$
4. $m = -4$, $b = 3$ $y = -4x + 3$
5. $m = 4$, $b = -2$ $y = 4x - 2$
6. $m = 2$, $b = -5$ $y = 2x$
7. $m = \frac{1}{2}$, $b = 5$ $y = \frac{1}{2}x + 5$
8. $m = \frac{2}{3}$, $b = 6$ $y = \frac{2}{3}x + 6$
9. $m = -\frac{5}{4}$, $b = 3$ $y = -\frac{5}{4}x$

Determine the x- and y-intercepts for the graph of each equation.

10. $3x + 2y = 6$ 2, 3
11. $5x + 3y = 15$ 3, 5
12. $5x + y = 10$ 2, 10
13. $2x + y = 6$ 3, 6
14. $3x + 4y = 24$ 8, 6
15. $2x - 7y = 28$ 14, −4
16. $3x - 2y = -5$ $-\frac{5}{3}$, $\frac{5}{2}$
17. $2x + 5y = -11$ $-\frac{11}{2}$, $-\frac{11}{5}$
18. $8x + y = 4$ $\frac{1}{2}$, 4
19. $6x + 2y = 3$ $\frac{1}{2}$, $\frac{3}{2}$
20. $3x - 7y = 9$ 3, $-\frac{9}{7}$
21. $-8x - 3y = 8$ −1, $-\frac{8}{3}$

Determine the slope and y-intercept for the graph of each equation. Then write each equation in slope-intercept form.

22. $2x + 5y = 10$ **23.** $3x + 4y = 12$ **24.** $5x - y = 15$

25. $3x - y = 9$ **26.** $2x + 5y = 8$ **27.** $7x + 4y = 8$

28. $5x - 4y = 11$ **29.** $7x - 3y = 10$ **30.** $2x + \frac{1}{3}y = 5$

31. $3x - \frac{1}{4}y = 6$ **32.** $\frac{1}{2}x + \frac{1}{4}y = 3$ **33.** $\frac{2}{3}x + \frac{1}{6}y = 2$

34. $3x = 2y - 7$ **35.** $5x = 8 - 2y$ **36.** $3y = 2x - 7$

37. $8y = 4x + 12$ **38.** $1.1x - 0.2y = 3.2$ **39.** $0.3x - 0.5y = 1.8$

Challenge

Solve each problem.

40. Find the coordinates of a point on the graph of $3x - 4y = -20$, if the y-coordinate is twice the x-coordinate. (4, 8)

41. Find the coordinates of a point on the graph of $4x - y = -2$, if the y-coordinate is three times the x-coordinate. (−2, −6)

42. Find the coordinates of a point on the graph of $x + 2y = 11$, if the y-coordinate is 5 less than the x-coordinate. (7, 2)

43. Find the coordinates of a point on the graph of $7x + 3y = 2$, if the y-coordinate is 4 more than the x-coordinate. (−1, 3)

Excursions in Algebra — Rise Over Run

Slope is sometimes defined as $\frac{\text{rise}}{\text{run}}$. In the coordinate plane, this is just another way of saying $\frac{\text{change in } y}{\text{change in } x}$.

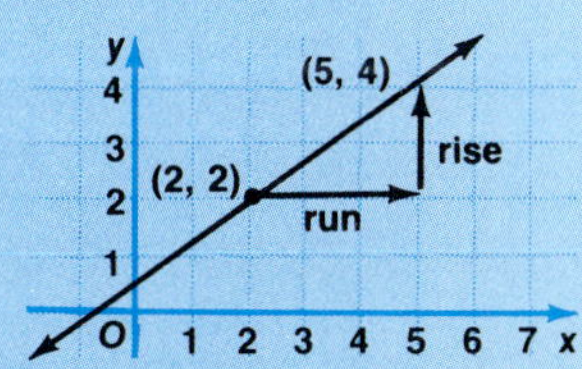

When building a stairway, a carpenter considers the ratio of riser to tread. This ratio is equivalent to $\frac{\text{rise}}{\text{run}}$ and describes the steepness of the stairs.

$$\frac{\text{riser}}{\text{tread}} = \frac{\text{rise}}{\text{run}} = \frac{8}{12} = \frac{2}{3}$$

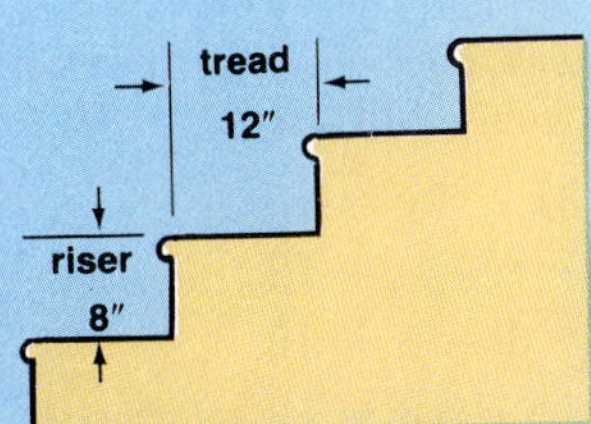

Exercises

Write the $\frac{\text{riser}}{\text{tread}}$ ratio for each of the following.

	Riser	*Tread*	
1.	9″	12″	$\frac{3}{4}$
2.	6″	$7\frac{1}{2}$″	$\frac{4}{5}$
3.	7″	9″	$\frac{7}{9}$

4. For exercises 1-3, which stairs are the steepest? 2

Quiz
Lessons 10–1 to 10–3

Determine the slope of the line passing through each pair of points named.

1. (−1, 3), (4, −2) −1

2. (3, 5), (−3, 1) $\frac{2}{3}$

Find A or B if the graph of the given equation has the indicated slope.

3. $Ax + 4y = 7; -\frac{1}{2}$

$-\frac{A}{4} = -\frac{1}{2}$ so $A = 2$

4. $6x + By = -5; 2$

$-\frac{6}{B} = 2$ so $B = -3$

Write an equation in standard form for the line through each pair of points.

5. (−2, 5), (0, 2)
$3x + 2y = 4$

6. (1, 7), (3, −2)
$9x + 2y = 23$

Write each of the following in slope-intercept form, then find the slope and y-intercept for the graph of each.

7. $2x - 7y = 1$
$y = \frac{2}{7}x - \frac{1}{7}; m = \frac{2}{7}$, $b = -\frac{1}{7}$

8. $6x - y - 12 = 0$
$y = 6x - 12; m = 6$, $b = -12$

Write an equation in slope-intercept form for the line with the given slope and y-intercept.

9. $m = \frac{1}{3}, b = -7$
$y = \frac{1}{3}x - 7$

10. $m = -4, b = \frac{2}{3}$
$y = -4x + \frac{2}{3}$

About the Excursion
This excursion presents an excellent application of slope. As an extention of this excursion, you may want to have students measure the riser and tread of stairways in their homes. Then ask them to compare their ratios with one another.

10-4

Lesson Notes, page T44

Discussion Question
State an equation of a line that has equal x- and y-intercepts. What is the slope of the line? Answers for linear equation will vary. slope $= -1$

Chalkboard Exercises
Graph each equation using the x- and y-intercepts.

1a. $3x - y = 9$

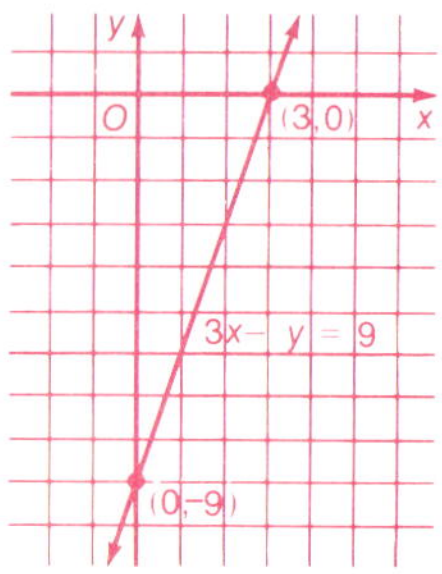

1b. $2x + 5y = 20$

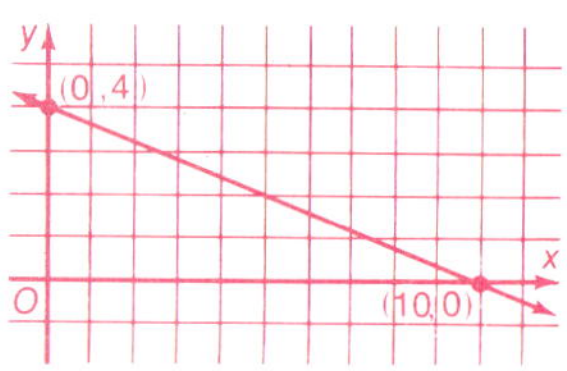

1c. $-3x + 5y = 15$

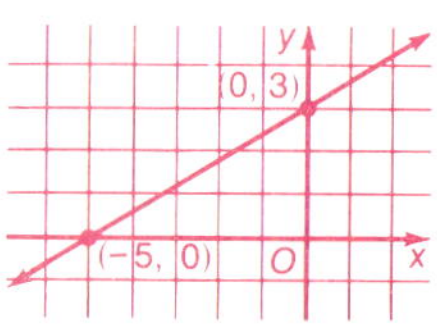

❶ The method shown in this example for checking graphs should be used only when a point on the graph appears to have integer coordinates. Point out that if the coordinates are not actually integers, the equation probably will not check. The best way to check such equations is to find another ordered pair that satisfies the equation, and see if the point for the ordered pair lies on the line.

Objective 10-4: Graph linear equations using the x- and y-intercepts or the slope and y-intercept.

10-4 Graphing Linear Equations

Previously you have graphed a linear equation by finding any two ordered pairs that satisfied the equation.

However, there are other methods that can be used. One such method is to use the x- and y-intercepts.

Example 1

Graph $3x - 2y = 12$ by using the x- and y-intercepts.

To find the x-intercept, let $y = 0$.

$$3x - 2y = 12$$
$$3x - 2(0) = 12$$
$$3x = 12$$
$$x = 4 \qquad \frac{C}{A} = 4$$

To find the y-intercept, let $x = 0$.

$$3x - 2y = 12$$
$$3(0) - 2y = 12$$
$$-2y = 12$$
$$y = -6 \qquad \frac{C}{B} = -6$$

Graph (4, 0) and (0, −6). Then draw the line that passes through these points.

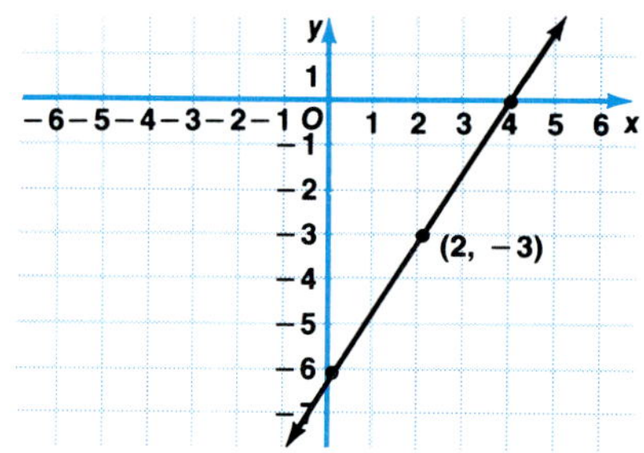

❶ To check, choose some other point on the line and see if it satisfies the equation. In this case, you could use (2, −3).

$$3x - 2y = 12$$
$$3(2) - 2(-3) \stackrel{?}{=} 12$$
$$6 + 6 \stackrel{?}{=} 12$$
$$12 = 12 \quad \textit{It checks!}$$

If an equation is in standard form and A, B, or C is zero, the graph of the equation has only one intercept. In such cases, you must find some other ordered pairs that satisfy the equation.

When $A = 0$, the graph is horizontal. When $B = 0$, the graph is vertical. When $C = 0$, the graph passes through the origin.

If a linear equation is in slope-intercept form, then it is convenient to use the slope and y-intercept to draw the graph of the equation.

Example

2 **Graph $y = 3x + 2$ by using the slope and y-intercept.**

The y-intercept is 2 so the graph passes through (0, 2). Since the slope is 3, we know

$$3 = \frac{3}{1} = \frac{\text{change in } y}{\text{change in } x}.$$

Starting at (0, 2) go up 3 units and to the right 1 unit to (1, 5). Then draw the line through (0, 2) and (1, 5).

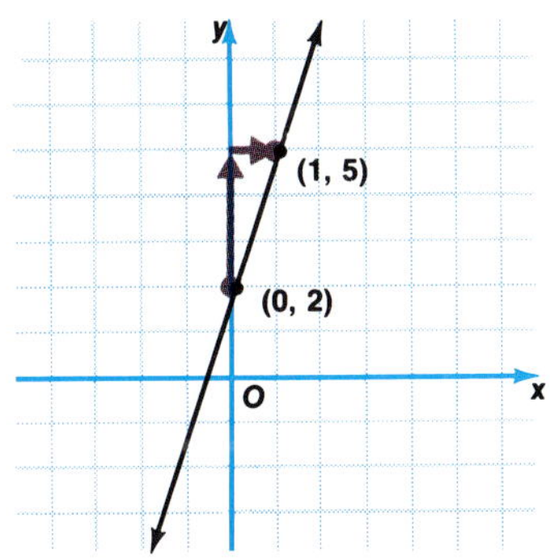

To check, you should find at least one other ordered pair whose graph should be on the line. Two such ordered pairs are (2, 8) and (−1, −1). Do you see how these ordered pairs can be obtained?

If the equation is in standard form, it is sometimes convenient to use the slope and y-intercept to draw the graph. Recall that for an equation in standard form, $-\frac{A}{B}$ is the slope of the line and $\frac{C}{B}$ is the y-intercept.

Consider the line $-4x + 3y = 12$. The slope, $-\frac{A}{B}$, is $\frac{4}{3}$. The y-intercept, $\frac{C}{B}$, is $\frac{12}{3}$ or 4. To draw the graph, use the slope and y-intercept, as in Example 2, and check the graph with at least one other ordered pair on the line.

Chalkboard Exercises

Graph each equation using the slope-intercept method.

2a. $y = 4x - 1$

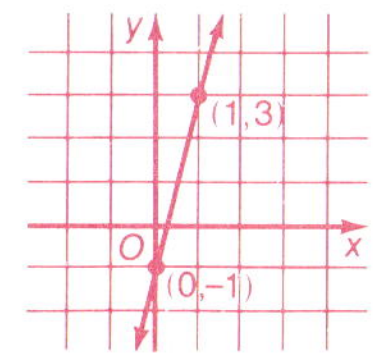

2b. $y = \frac{1}{2}x + 3$

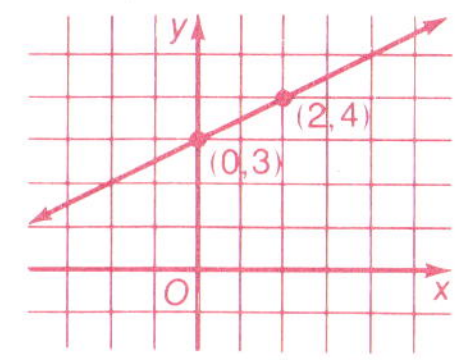

2c. $y = -\frac{1}{4}x - 4$

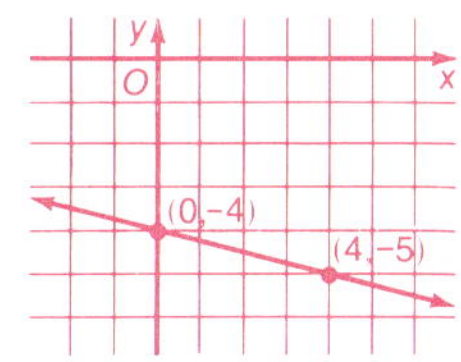

2d. $3x - 6y = 12$

$y = \frac{1}{2}x - 2$

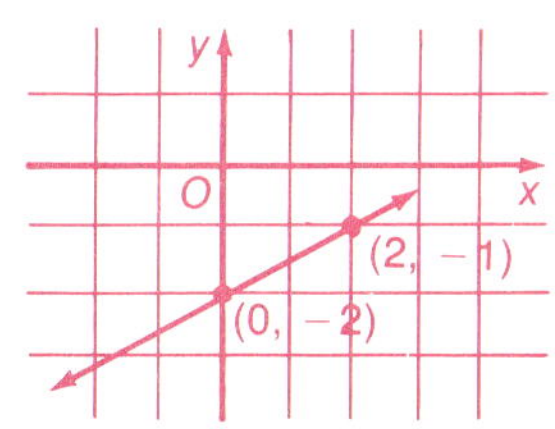

Suggested Daily Assignments

Enriched
- Exp: 8–24 evens
- Written: 2–24 evens
- Mini-Review: 1–5

Average
- Exp: 1–24
- Written: 1–24
- Mini-Review: 1–5

Fundamental
- Exp: 1–24
- Written: 2–24 evens
  - pp. 314-5, 2-36 evens
- Mini-Review: 1–5

Answers to Written

See the Teacher's Answer Key, p. T86, for graphs to exercises 1-24.

Exploratory Exercises

State the x-intercept and y-intercept for each equation.

1. $4x + y = 8$ 2, 8
2. $6x + y = 6$ 1, 6
3. $3x + 4y = 6$ $2, \frac{3}{2}$
4. $5x + 2y = 20$ 4, 10
5. $2x - y = 8$ 4, −8
6. $3x - 2y = 6$ 2, −3
7. $7x + 2y = 10$ $\frac{10}{7}, 5$
8. $5x + 6y = 8$ $\frac{8}{5}, \frac{4}{3}$
9. $5x - \frac{1}{2}y = 2$ $\frac{2}{5}, -4$
10. $3x + \frac{1}{3}y = 2$ $\frac{2}{3}, 6$
11. $y = 3x - 1$ $\frac{1}{3}, -1$
12. $y = 4x + 3$ $-\frac{3}{4}, 3$
13. $2x = 5 - 3y$ $\frac{5}{2}, \frac{5}{3}$
14. $3x = 4y + 1$ $\frac{1}{3}, -\frac{1}{4}$
15. $\frac{1}{2}x + \frac{1}{3}y = 0$ 0, 0

In each of the following the slope and a point on the line are given. Name two other points on each line. Answers may vary. Possible answers are given.

16. $m = \frac{4}{3}$, (0, 1) (3, 5), (6, 9)
17. $m = 2$, (0, 1) (1, 3), (2, 5)
18. $m = \frac{5}{2}$, (0, 3) (2, 8), (4, 13)
19. $m = -7$, (0, 4) (1, −3), (2, −10)
20. $m = -\frac{1}{4}$, (6, 1) (2, 2), (−2, 3)
21. $m = \frac{1}{2}$, (5, 2) (7, 3), (9, 4)
22. $m = -\frac{2}{3}$, (−3, 2) (0, 0), (3, −2)
23. $m = 8$, (−3, 6) (−2, 14), (−4, −2)
24. $m = 0$, (3, 2) (4, 2), (5, 2)

Written Exercises

See Teacher's Answer Key for these graphs.

Graph each equation using the x- and y-intercepts.

1. $6x - 3y = 6$
2. $4x + 5y = 20$
3. $5x - y = -10$
4. $2x + 10y = 5$
5. $2x + 5y = -10$
6. $7x - 2y = -7$
7. $y = 6x - 9$
8. $x + \frac{1}{2}y = 4$
9. $\frac{1}{2}x + \frac{2}{3}y = -3$
10. $x = 8y - 4$
11. $x = \frac{3}{4}y + 6$
12. $\frac{2}{3}y = \frac{1}{2}x + 6$

Graph each equation using the slope-intercept method.

13. $y = \frac{2}{3}x + 3$
14. $y = \frac{3}{4}x + 4$
15. $y = -\frac{3}{5}x - 1$
16. $y = -\frac{3}{4}x + 1$
17. $y = \frac{3}{2}x - 5$
18. $y = \frac{1}{2}x + 4$
19. $-4x + y = 6$
20. $-3x + 2y = 12$
21. $-2x + y = 3$
22. $3y - 7 = 2x$
23. $5x + 2 = 7y$
24. $\frac{3}{4}x + \frac{1}{2}y = 4$

mini-review

Simplify.

1. $(4xy^3)(-5x^3yz)$
2. $(-2a^2b^3)^4$
3. $\frac{5a^3b^2c^3}{-20a^4b^2}$
4. $\frac{x^2-5x-24}{2x^2-4x-30}$
5. $2x(x^3-3x^2+7) + 5(x^4+5x^3-3x+5)$

Answers:
1. $-20x^4y^4z$
2. $16a^8b^{12}$
3. $\frac{c^3}{-4a}$
4. $\frac{x-8}{2(x-5)}$
5. $7x^4 + 19x^3 - x + 25$

Problem Solving

Graphs and Estimation

The freezing point of water on the Celsius and Fahrenheit scales can be represented by the ordered pair (0, 32), where 0 is the Celsius temperature and 32 is the Fahrenheit temperature. Similarly, the boiling point of water can be represented by (100, 212). With these two points you can draw the graph of the line showing the relationship between Celsius and Fahrenheit.

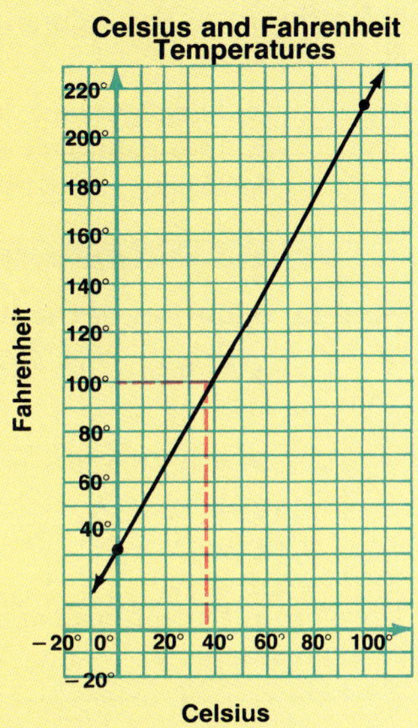

The graph above can be used to quickly estimate equivalent temperatures. For a Fahrenheit temperature of 100, for example, you can estimate the corresponding Celsius temperature by first locating 100 on the Fahrenheit scale. Follow the value horizontally until you reach the line, then drop vertically to the Celsius scale. The corresponding Celsius temperature is between 35° and 40°.

Many other relationships can be shown graphically. For example, the speed of sound through water is 1454 m/s, or about 1.5 km/s. The graph at the right shows the relationship between the time in seconds and the distance in kilometers. Note that the ordered pairs (4, 6) and (6, 9) were used to draw the graph.

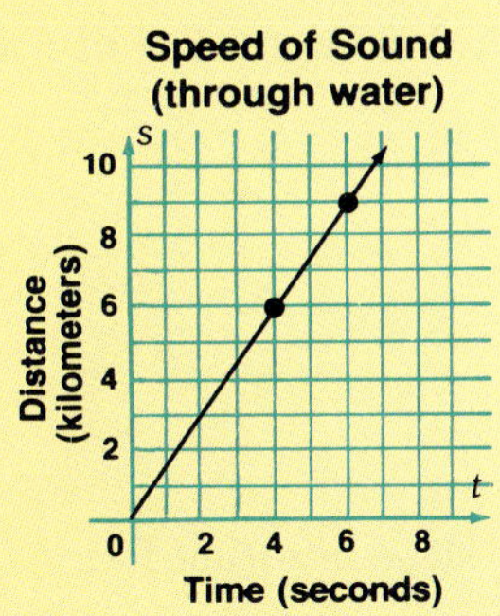

Exercises Estimates may vary.

Use the graph at the top of this page to estimate the Celsius or Fahrenheit equivalents.

1. 80°F 30°C **2.** 65°F 20°C **3.** 20°C 65°F **4.** 50°C 120°F

Use the graph to solve each problem.

5. Using sonar, a ship finds that it takes sound waves 2.5 seconds to reach the ocean floor. Estimate the depth of the ocean at this point. 3.5 km

6. Estimate how many seconds it would take sound waves to reach a depth of 5 km. 3.2 s

About Problem Solving
In this feature, graphs are presented as tools for estimating certain quantities. Notice that the examples given involve relationships that are linear in nature. Ask students if they can think of other linear relationships such as these. You may want to have them draw a graph for their relationships.

10-5

Lesson Notes, page T44
Enrichment, page T44

Readiness Activity
Write a linear equation in slope-intercept form for the line with the given slope and y-intercept.

$m = 5, b = -1$ $y = 5x - 1$
$m = -3, b = 2$ $y = -3x + 2$
$m = 1, b = -6$ $y = x - 6$
$m = \frac{1}{3}, b = 5$ $y = \frac{1}{3}x + 5$

Discussion Questions
Is it possible to write the equation of a line given its graph? How? What difficulties might occur?
yes; find the coordinates of two points on the line; the graph could be inaccurate, or difficult to interpret exactly.

Chalkboard Exercises
Write an equation in slope-intercept form for the line having the given slope and passing through the indicated point.

1a. 3; (1, 4)
$y = 3x + 1$

1b. $\frac{1}{2}$; (−2, 3)
$y = \frac{1}{2}x + 4$

1c. −2; (−2, −1)
$y = -2x - 5$

❶ An alternate way to solve this example is to find the slope and then use the point-slope form, using the coordinates of either point, as shown below.

$$(y - 1) = -1(x - 5)$$
$$y - 1 = -x + 5$$
$$y = -x + 6$$
or
$$[y - (-2)] = -1(x - 8)$$
$$y + 2 = -x + 8$$
$$y = -x + 6$$

Objective 10-5A: Write a linear equation in slope-intercept form given the slope and the coordinates of a point on the line.
Objective 10-5B: Write a linear equation in slope-intercept form given the coordinates of two points of the line.

10-5 Writing Slope-Intercept Equations of Lines

If you have enough information about a line, you can write an equation for the line. For example, you might know any of the following:

1. The slope and a point on the line.
2. Two points on the line.
3. The x- and y-intercepts.

Example 1

Find an equation of the line whose slope is 3 and that passes through (4, −2).

$y = mx + b$ *Use slope-intercept form.*
$y = 3x + b$ *The slope m is 3.*
$-2 = 3(4) + b$ *Substitute 4 for x and −2 for y.*
$-2 = 12 + b$ *Now solve for b.*
$-14 = b$

The slope-intercept form of the equation for the line is $y = 3x + (-14)$. This can also be expressed as $y = 3x - 14$. *In standard form the equation is $3x - y = 14$.*

The following procedure can be used to find an equation when two points on the line are known.

❶

Example 2

Find an equation of the line that passes through (5, 1) and (8, −2).

$$m = \frac{y_2 - y_1}{x_2 - x_1}$$ *First determine the slope.*
$$= \frac{-2 - 1}{8 - 5}$$
$$= \frac{-3}{3}$$
$$= -1$$

In slope-intercept form $y = -1x + b$ or $y = -x + b$.

Now substitute the coordinates of either point into the equation and solve for b.

Using (5, 1)	or	Using (8, − 2)
$y = -x + b$		$y = -x + b$
$1 = -(5) + b$		$-2 = -(8) + b$
$6 = b$		$6 = b$

The slope-intercept form of the equation is $y = -x + 6$. *The standard form is $x + y = 6$.*

You can also use the slope and y-intercept obtained from the graph to find the equation of a line.

Example

3 **Find an equation for line PQ whose graph is shown below.**

First determine the slope. Start at P. The y-coordinate *decreases by 4* as you move from P to Q. The x-coordinate *increases by 3* as you move from P to Q.

$$\text{slope} = \frac{\text{change in } y}{\text{change in } x} = \frac{-4}{3} = -\frac{4}{3}$$

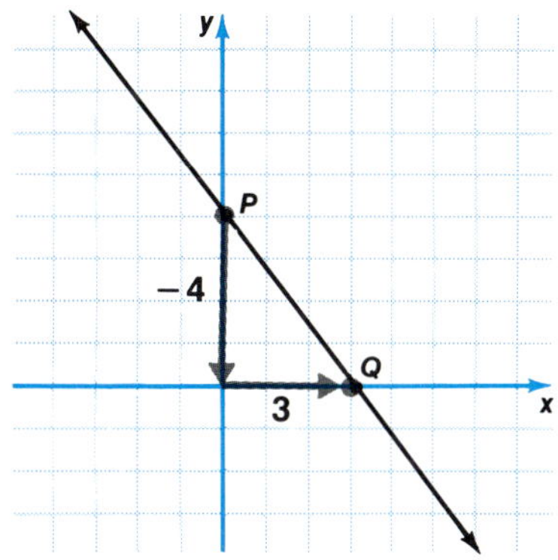

The line intersects the y-axis at (0, 4) so the y-intercept is 4.

Now substitute these values into the slope-intercept form.

$$y = mx + 4$$

$$y = -\tfrac{4}{3}x + 4$$

The equation for line PQ is $y = -\frac{4}{3}x + 4$. *In standard form the equation is $4x + 3y = 12$.*

Exploratory Exercises

In each of the following a point on the line is given. Determine the y-intercept, b.

1. (2, 1), $y = 3x + b$ -5 **2.** (6, 2), $y = -2x + b$ 14 **3.** (4, 5), $y = -x + b$ 9

4. (2, 2), $y = -\frac{3}{2}x + b$ 5 **5.** (8, 1), $y = \frac{3}{4}x + b$ -5 **6.** $(-6, 5)$, $y = -\frac{2}{3}x + b$ 1

7. $(1, -2)$, $y = -\frac{5}{3}x + b$ $-\frac{1}{3}$ **8.** $(3, -1)$, $y = \frac{5}{6}x + b$ $-\frac{7}{2}$ **9.** (2, 3), $y = \frac{4}{3}x + b$ $\frac{1}{3}$

State the slope and y-intercept for each line. Then express the equation of the line in slope-intercept form.

10. a
11. b
12. c
13. d
14. e
15. f

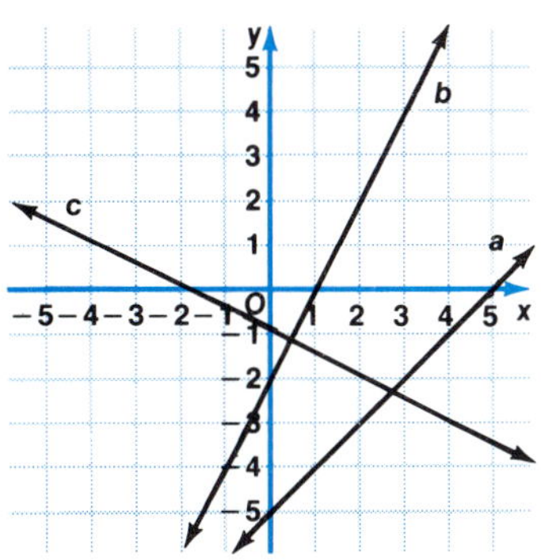

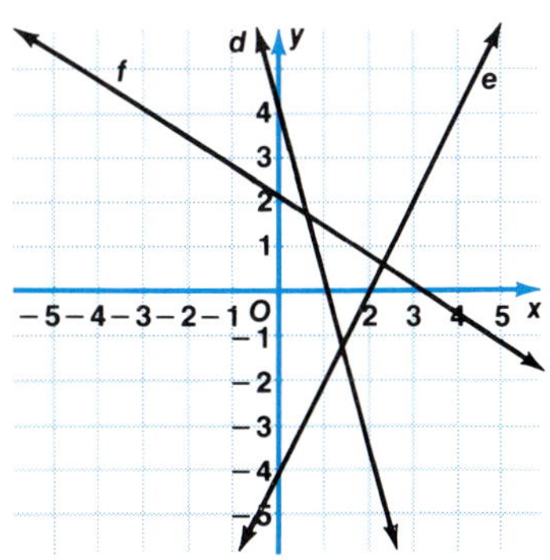

Chalkboard Exercises

Write an equation in slope-intercept form for the line that passes through each pair of points.

2a. (1, 2), (5, 6)
$y = x + 1$

2b. $(-3, 2)$, $(-1, 1)$
$y = -\frac{1}{2}x + \frac{1}{2}$

2c. $(-1, -1)$, (2, 5)
$y = 2x + 1$

2d. $(1, -3)$, $(2, -1)$
$y = 2x - 5$

Find an equation for each line whose graph is shown below.

3a.

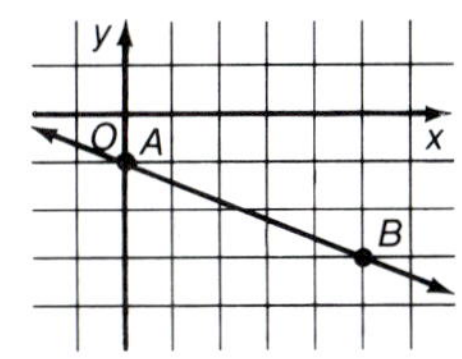

$y = -\frac{2}{5}x - 1$

3b.

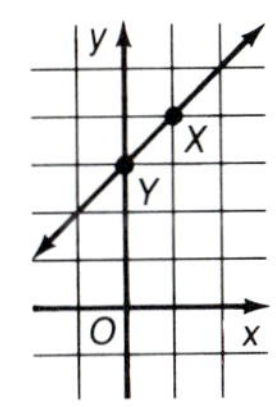

$y = x + 3$

Suggested Daily Assignments

Enriched
Exp: 7–15
Written: 1–21 odds

Average
Exp: 1–15
Written: 1–21

Fundamental
Exp: 1–15
Written: 1–21 odds
p. 318, 1–23 odds

Answers to Exploratory

10. 1, -5, $y = x - 5$
11. 2, -2, $y = 2x - 2$
12. $-\frac{1}{2}$, -1, $y = -\frac{1}{2}x - 1$
13. -4, 4, $y = -4x + 4$
14. 2, -4, $y = 2x - 4$
15. $-\frac{2}{3}$, 2, $y = -\frac{2}{3}x + 2$

Quiz
Lessons 10–4 to 10–5
Graph the following equation by using the x- and y-intercepts.
1. $x - 2y = 8$

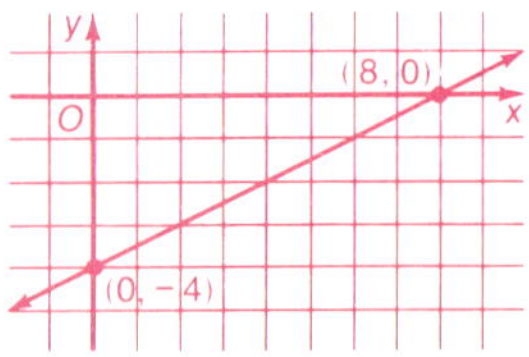

Graph the following equation by using the slope-intercept method.
2. $y = 2x - 3$

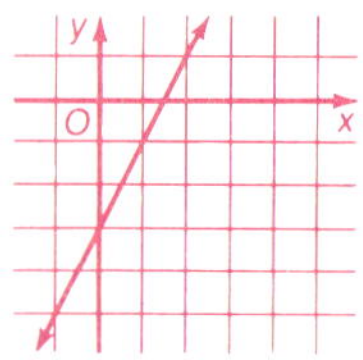

Write an equation in standard form for the line having the given slope and passing through the indicated point.
3. 2; (3, 8) $2x - y = -2$

Write an equation in slope-intercept form for the line satisfying the given conditions.
4. $m = \frac{1}{2}$ and passes through (2, 4) $y = \frac{1}{2}x + 3$
5. passes through (1, 5) and (3, −1) $y = -3x + 8$
6. x-intercept, 1; y-intercept, −4 $y = 4x - 4$

About Using Calculators
This feature shows how a calculator can be used to help write a linear equation in slope-intercept form when the coordinates of two points on the line are given. Notice that the coordinates in this feature are either decimals or very large or very small integers.
Point out to students the slopes and y-intercepts that are found in the exercises are approximate.

Written Exercises

Write an equation in slope-intercept form for the line having the given slope and passing through the indicated point.

1. 3; (5, −2) $y = 3x - 17$
2. 2; (4, −2) $y = 2x - 10$
3. $\frac{2}{3}$; (−1, 0) $y = \frac{2}{3}x + \frac{2}{3}$
4. $\frac{1}{2}$; (5, 3) $y = \frac{1}{2}x + \frac{1}{2}$
5. −5; (5, 4) $y = -5x + 29$
6. −3; (3, 7) $y = -3x + 16$
7. $\frac{3}{4}$; (−2, −4) $y = \frac{3}{4}x - \frac{5}{2}$
8. $-\frac{5}{3}$; (−3, −5) $y = -\frac{5}{3}x - 10$
9. $\frac{1}{4}$; (0, 8) $y = \frac{1}{4}x + 8$

Write an equation in slope-intercept form for the line that passes through each pair of points.

10. (−1, 7), (8, −2) $y = -x + 6$
11. (5, 2), (−7, −4) $y = \frac{1}{2}x - \frac{1}{2}$
12. (6, 0), (0, 4) $y = -\frac{2}{3}x + 4$
13. (4, 1), (5, 2) $y = x - 3$
14. (8, −1), (7, −1) $y = -1$
15. (1, 0), (0, 1) $y = -x + 1$
16. (6, 3), (−2, 4) $y = -\frac{1}{8}x + \frac{15}{4}$
17. (−4, 3), (6, −5)
18. (8, 0), (2, 5) $y = -\frac{5}{6}x + \frac{20}{3}$
19. (5, 7), (−1, 6) $y = \frac{1}{6}x + \frac{37}{6}$
20. (−6, 2), (3, −5)
21. (5, 7), (0, 6) $y = \frac{1}{5}x + 6$

17. $y = -\frac{4}{5}x - \frac{1}{5}$ 20. $y = -\frac{7}{9}x - \frac{8}{3}$

Using Calculators — Writing Equations

Example: Determine slope-intercept form of the line passing through (7.6, 10.8) and (12.2, 93.7). Round values to the nearest hundredth.

First determine the slope, $m = \frac{y_2 - y_1}{x_2 - x_1}$

ENTER: [(] 93.7 [−] 10.8 [)] [÷] [(] 12.2 [−] 7.6 [)] [=]

The display shows 18.021739. Rounded to the nearest hundredth, the slope is 18.02.

Recall that $y = mx + b$. To determine the y-intercept, solve for b. Thus, $b = -mx + y$. Let $(x, y) = (7.6, 10.8)$.

ENTER: 18.02 [+/−] [×] 7.6 [+] 10.8 [=]

The display shows −126.152. Rounded to the nearest hundredth, the y-intercept is −126.15.

Therefore, the slope intercept form is $y = 18.02x - 126.15$.

Exercises

Write an equation in slope-intercept form for the line that passes through each pair of points. Round values to the nearest hundredth.

1. (5, −285), (10, −225) $y = 12x - 345$
2. (100, 293), (125, 368) $y = 3x - 7$
3. (5, −225), (16, −885) $y = -60x + 75$
4. (7, 12.7), (−1, 3.9) $y = 1.1x + 5$
5. (100, 257.5), (200, 507.5)
6. (1.3, 5.1), (2.7, 14.9) $y = 7x - 4$
7. (11.4, 25.8), (−2.4, −17.5)
8. (18.9, −3.4), (−15.3, 5.9) $y = -0.27x + 1.70$

5. $y = 2.5x + 7.5$ 7. $y = 3.14x - 10.00$

For more exercises on linear equations, see exercises 1-36 on page 562.

Applications in Aeronautics — Airplane Flying Time

A light twin-engine aircraft has a useful carrying load of 2061 pounds. Of this a maximum of 1164 pounds of fuel can be carried. The remaining weight may be made up of people and cargo. If the plane is to carry more cargo or people, less fuel is carried and there is less flying time.

Shown below is a graph of the amount of fuel in pounds needed to fly a given number of hours.

Suppose a plane carries 519 pounds of cargo and 6 people weighing a total of 960 pounds. How many hours of flying time does the plane have?

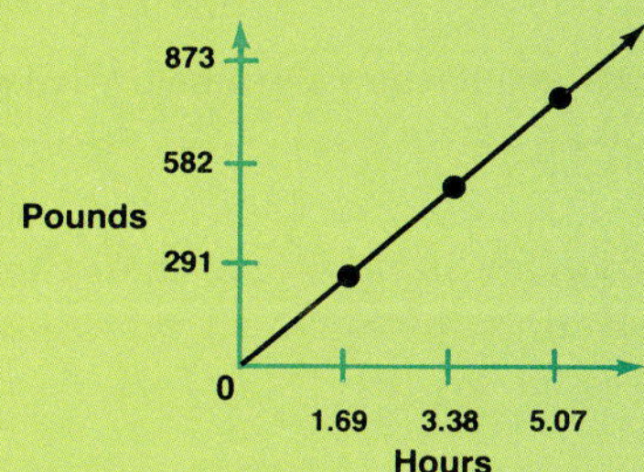

First, determine the amount of fuel the plane can carry. Then check the graph for the number of flying hours.

$$\begin{array}{r} 519 \\ +960 \\ \hline 1479 \end{array}$$ *total pounds of cargo and people*

$$\begin{array}{r} 2061 \\ -1479 \\ \hline 582 \end{array}$$ *total pounds allowed*; *pounds of fuel that can be carried*

582 pounds of fuel permit about 3.38 hours of flying time.

Exercises

Solve.

1. How many hours of flying time does a plane have if it is carrying 1000 pounds of cargo and 4 people weighing a total of 770 pounds? 1.69 hours
2. Approximately how many hours of flying time does a plane have if it is carrying 2 people weighing a total of 305 pounds and 906 pounds of cargo? 4.9 hours
3. How many pounds of cargo can be carried if a plane must fly 3.38 hours to reach its destination? The pilot and copilot each weigh 145 pounds. 1189 pounds

About the Application
This application shows students how linear graphs are used to solve real-life problems. Point out to students that some of the answers obtained from the graph will be approximate. Refer students to the feature Problem Solving: Graphs and Estimation on page 319 for further explanation as to why some answers will be approximate.

10-6

Lesson Notes, page T45

Readiness Activity
State whether each of the following pairs of lines are parallel or perpendicular.

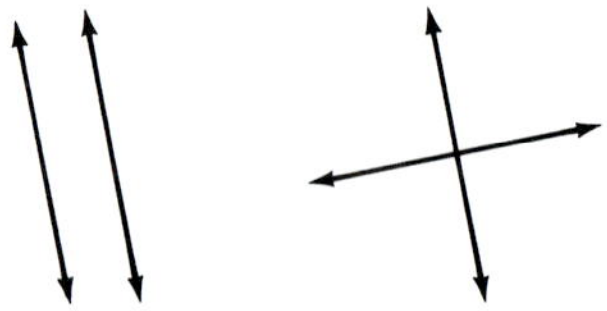

parallel perpendicular

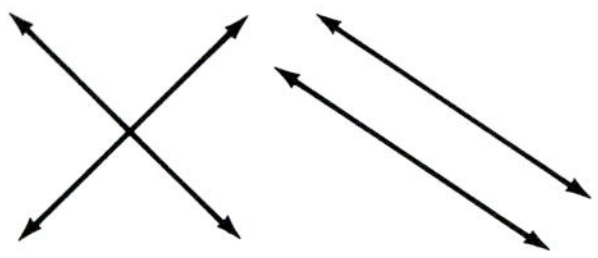

perpendicular parallel

Discussion Questions

1. Write some equations of lines that are perpendicular to a line which has an undefined slope.
   $y = b$, b is a constant.
2. Write some equations of lines that are parallel to a line which has an undefined slope.
   $x = c$, c is a constant.

Chalkboard Exercises
Write an equation for the line that is parallel to the graph of each equation and that passes through the indicated point. Use slope-intercept form.

1a. $x - 2y = 4$; $(1, 5)$
$y = \frac{1}{2}x + \frac{9}{2}$

1b. $3x + 2y = 4$; $(3, 2)$
$y = -\frac{3}{2}x + \frac{13}{2}$

1c. $2x - 4y = -3$; $(-1, 4)$
$y = \frac{1}{2}x + \frac{9}{2}$

1d. $4x - y = -2$; $(-2, 0)$
$y = 4x + 8$

Objective 10-6: Write an equation of a line that is parallel or perpendicular to the graph of a given equation and that passes through a given point.

10-6 Parallel and Perpendicular Lines

The graphs below show parallel lines.

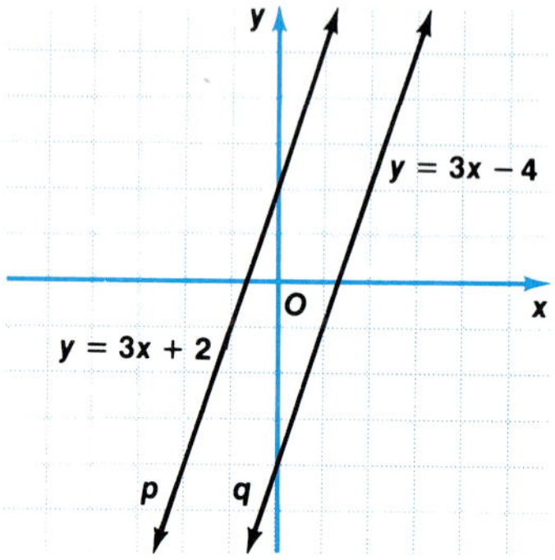

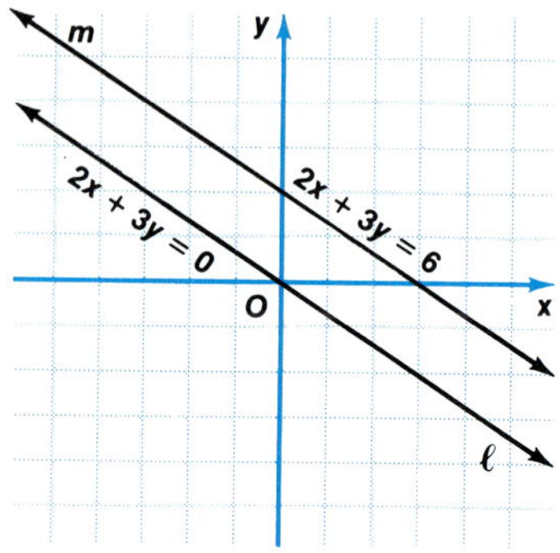

What is the slope of line p? Line q?

What is the slope of line l? Line m?

What is your guess about the relationship between the slopes of parallel lines?

> **If two lines have the same slope, then they are parallel. All vertical lines are parallel.**

Definition of Parallel Lines

Example 1

Find the equation of the line that is parallel to the graph of $5x - 2y = 6$ and that passes through $(4, -2)$. Use slope-intercept form.

The slope of the graph of $5x - 2y = 6$ is $\frac{5}{2}$.

Therefore the slope-intercept form of an equation whose graph is parallel to the graph of $5x - 2y = 6$ is

$$y = \frac{5}{2}x + b.$$

Now substitute $(4, -2)$ into that equation and solve for b.

$$-2 = \frac{5}{2}(4) + b$$
$$-2 = 10 + b$$
$$-12 = b$$

The equation of the line is $y = \frac{5}{2}x - 12$. *In standard form the equation is $5x - 2y = 24$.*

Line l is shown in the figure on the left below. Suppose the axes and the graph are rotated counterclockwise 90°. After the rotation, line l will coincide with line l' of the figure shown in color.

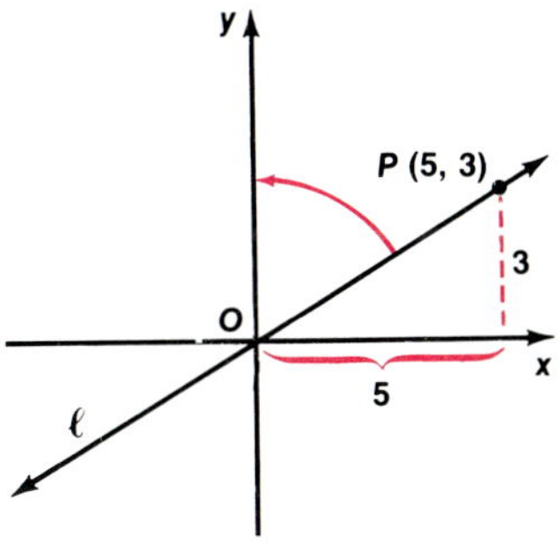

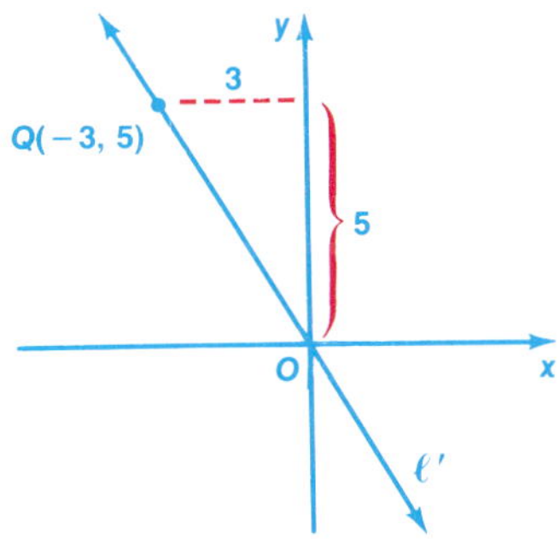

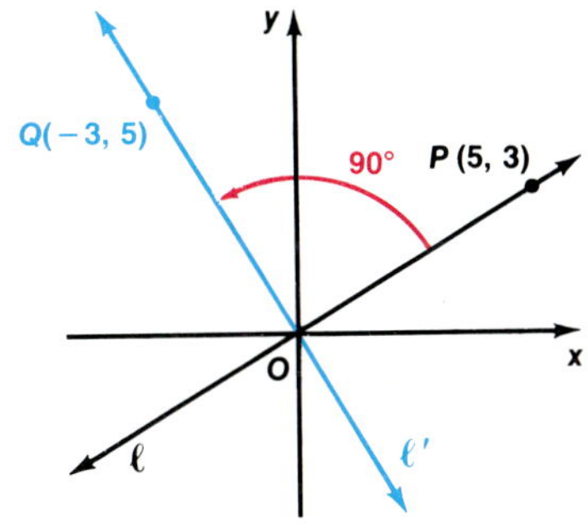

The figure at the right above shows lines l and l'. Note that line l' is perpendicular to line l. Study their slopes.

Slope of l

$$\frac{y_2 - y_1}{x_2 - x_1} = \frac{0 - 3}{0 - 5} = \frac{-3}{-5} = \frac{3}{5}$$

Slope of l'

$$\frac{y_2 - y_1}{x_2 - x_1} = \frac{0 - 5}{0 - (-3)} = \frac{-5}{3} = -\frac{5}{3}$$

Compare the slopes. What is their product?

❶

If the product of the slopes of two lines is −1, then the lines are perpendicular. In a plane, vertical lines are perpendicular to horizontal lines.

Definition of Perpendicular Lines

Example

2 Show that the lines for $7x + 3y = 4$ and $3x - 7y = 1$ are perpendicular.

The slope of the line for $7x + 3y = 4$ is $-\frac{7}{3}$.

The slope of the line for $3x - 7y = 1$ is $\frac{3}{7}$.

Since $-\frac{7}{3} \cdot \frac{3}{7} = -1$, the lines are perpendicular.

Chalkboard Exercises

Determine whether the graphs of the following equations are perpendicular.

2a. $3x + 4y = 2$ and $6x - 8y = 7$

Slopes: $-\frac{3}{4}$ and $\frac{3}{4}$ no

2b. $2x + 3y = 7$ and $3x - 2y = 1$

Slopes: $-\frac{2}{3}$ and $\frac{3}{2}$ yes

2c. $4x + y = 5$ and $x - 4y = 3$

Slopes: -4 and $\frac{1}{4}$ yes

2d. $y = \frac{2}{3}x + 7$ and $2x + 3y = 5$

Slopes: $\frac{2}{3}$ and $-\frac{2}{3}$ no

❶ You may wish to introduce the term *negative reciprocal.* If you find the negative reciprocal of the slope of a given line, you will obtain the slope of a line perpendicular to the given line. The negative reciprocal of n is $-\frac{1}{n}$.

Chalkboard Exercises
Write an equation for the line that is perpendicular to the graph of each equation and that passes through the indicated point. Use slope-intercept form.

3a. $2x + 3y = 9$; $(5, 2)$
$y = \frac{3}{2}x - \frac{11}{2}$

3b. $x - 2y = 4$; $(-1, 4)$
$y = -2x + 2$

3c. $3x + 5y = 6$; $(3, 6)$
$y = \frac{5}{3}x + 1$

Suggested Daily Assignments

Enriched
Exp: 10–15
Written: 1–18

Average
Exp: 2–14 evens
Written: 1–18

Fundamental
Exp: 1–15
Written: 1–6, 10–15
p. 322, 2–20 evens

Answers to Written

1. $y = -\frac{3}{5}x - 1$
2. $y = -6x - 9$
3. $y = \frac{3}{4}x$
4. $y = -\frac{2}{3}x + \frac{14}{3}$
5. $y = \frac{5}{2}x - 4$
6. $y = \frac{4}{3}x - \frac{16}{3}$
7. $y = \frac{3}{4}x + \frac{1}{2}$
8. $y = -\frac{1}{3}x - \frac{13}{3}$
9. $y = x - 9$
10. $y = -\frac{3}{5}x + \frac{14}{5}$
11. $y = \frac{8}{3}x + 4$
12. $y = -\frac{1}{3}x + 1$
13. $y = \frac{1}{3}x + 2$
14. $y = -\frac{1}{5}x - 1$
15. $y = -\frac{3}{2}x - \frac{9}{2}$
16. $y = \frac{7}{3}x - \frac{16}{3}$
17. $y = \frac{9}{5}x$
18. $y = -\frac{1}{2}x - 4$

Example

3 **Find an equation of the line that is perpendicular to the line for $2x - 5y = 3$ and that passes through $(-2, 7)$. Use slope-intercept form.**

The slope of the line for $2x - 5y = 3$ is $\frac{2}{5}$. The slope of the line perpendicular to that line is $-\frac{5}{2}$. The equation in slope-intercept form for the line perpendicular to the line for $2x - 5y = 3$ is

$$y = -\frac{5}{2}x + b.$$

Now substitute $(-2, 7)$ into the equation.

$$7 = -\frac{5}{2}(-2) + b$$
$$7 = 5 + b$$
$$2 = b$$

The equation in slope-intercept form is $y = -\frac{5}{2}x + 2$.
In standard form the equation is $5x + 2y = 4$.

Exploratory Exercises

State the slopes of the lines parallel and perpendicular to the graphs of each equation.

1. $5x - y = 7$ $5, -\frac{1}{5}$
2. $3x + 4y = 2$ $-\frac{3}{4}, \frac{4}{3}$
3. $2x - y = 7$ $2, -\frac{1}{2}$
4. $6x + 3y = 4$ $-2, \frac{1}{2}$
5. $2x - 3y = 7$ $\frac{2}{3}, -\frac{3}{2}$
6. $7x + y = 4$ $-7, \frac{1}{7}$
7. $x = 7$ undefined, 0
8. $2x + 7y = 8$ $-\frac{2}{7}, \frac{7}{2}$
9. $5x - 3y = 4$ $\frac{5}{3}, -\frac{3}{5}$
10. $y = -4$ 0, undefined
11. $y = 4x + 2$ $4, -\frac{1}{4}$
12. $y = 3x - 7$ $3, -\frac{1}{3}$
13. $3y = 2x + 5$ $\frac{2}{3}, -\frac{3}{2}$
14. $4x = 2y - 7$ $2, -\frac{1}{2}$
15. $3x = 4 - 3y$ $-1, 1$

Written Exercises

Write an equation for the line that is parallel to the graph of each equation and that passes through the indicated point. Use slope-intercept form.

1. $y = -\frac{3}{5}x + 4$; $(0, -1)$
2. $6x + y = 4$; $(-2, 3)$
3. $y = \frac{3}{4}x - 1$; $(0, 0)$
4. $2x + 3y = 1$; $(4, 2)$
5. $5x - 2y = 7$; $(0, -4)$
6. $4x - 3y = 2$; $(4, 0)$
7. $y = \frac{3}{4}x - 1$; $(2, 2)$
8. $y = -\frac{1}{3}x + 7$; $(2, -5)$
9. $x = y$; $(7, -2)$

Write an equation for the line that is perpendicular to the graph of each equation and that passes through the indicated point. Use slope-intercept form.

10. $5x - 3y = 7$; $(8, -2)$
11. $3x + 8y = 4$; $(0, 4)$
12. $y = 3x - 2$; $(6, -1)$
13. $y = -3x + 7$; $(-3, 1)$
14. $y = 5x - 3$; $(0, -1)$
15. $y = \frac{2}{3}x + 1$; $(-3, 0)$
16. $3x + 7y = 4$; $(1, -3)$
17. $5x + 9y = 3$; $(0, 0)$
18. $y = 2x - 7$; $(4, -6)$

Objective 10-7: Find the coordinates of the midpoint of a line segment in the coordinate plane given the coordinates of the endpoints.

10-7 Midpoint of a Line Segment

The coordinate of the midpoint, P, of line segment AB shown at the right can be found as follows.

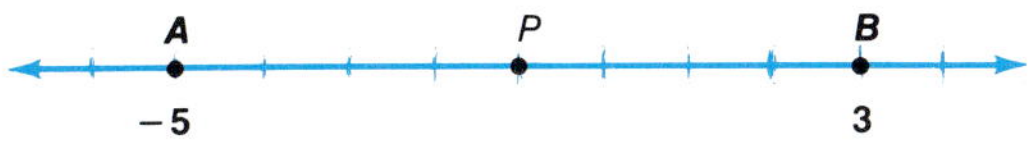

Find the distance from A to B on the number line. $|-5 - 3| = 8$

The distance from A to P on the number line is $\frac{1}{2}(8)$ or 4.

Add the distance, 4, to the coordinate of A, -5, to obtain the coordinate of P. Thus, the coordinate of P is -1.

A simpler approach is to find the average of the coordinates.

$$P = \frac{x_1 + x_2}{2} = \frac{-5 + 3}{2} = \frac{-2}{2} = -1$$

The coordinate of the midpoint, P, of two points, x_1, and x_2, on a number line is

$$P = \frac{x_1 + x_2}{2}.$$

Midpoint on a Number Line

This method can be extended to find the coordinates of the midpoint of a line segment in the coordinate plane.

The coordinates of the midpoint, (x, y), of a line segment whose endpoints are (x_1, y_1) and (x_2, y_2) are

$$(x, y) = \left(\frac{x_1 + x_2}{2}, \frac{y_1 + y_2}{2}\right).$$

Midpoint of a Line Segment

Example

1 Find the coordinates of the midpoint of the line segment from (−2, 4) to (8, 6).

$$(x, y) = \left(\frac{x_1 + x_2}{2}, \frac{y_1 + y_2}{2}\right)$$

$$= \left(\frac{-2 + 8}{2}, \frac{4 + 6}{2}\right)$$

$$= (3, 5)$$

10-7

Lesson Notes, page T45
Enrichment, page T45

Readiness Activity
State whether point M is the midpoint of each of the following line segments.

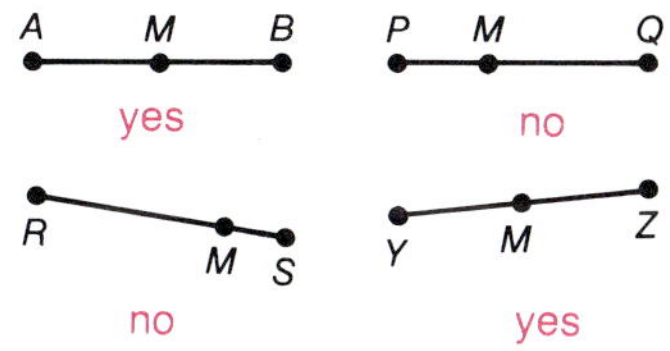

Discussion Question
Find the general form of the coordinates of the endpoints of line segments whose midpoint is the origin. (a, b) and $(-a, -b)$

Chalkboard Exercises
Write the coordinates of the midpoint of the line segment whose endpoints are given.

1a. (1, 2), (7, 6) (4, 4)
1b. (−3, 4), (1, 2) (−1, 3)
1c. (−3, 4), (7, 6) (2, 5)
1d. (1, 2), (4, 10) $\left(2\frac{1}{2}, 6\right)$
1e. (4, 10), (7, 6) $\left(5\frac{1}{2}, 8\right)$

Chalkboard Exercises

If P is the midpoint of line segment AB, find the coordinates of the missing point, A or B.

2a. $A(3, 2)$, $P(6, 3)$ $B(9, 4)$
2b. $A(6, 3)$, $P(3, 2)$ $B(0, 1)$
2c. $A(3, 2)$, $P(1, 5)$ $B(-1, 8)$
2d. $P(6, 3)$, $B(1, 5)$ $A(11, 1)$
2e. $P(1, 5)$, $B(6, 3)$ $A(-4, 7)$

❶ Some students may discover an easier method to solve problems similar to this one that are related to finding points on a line when given the slope.

The coordinates of the other endpoint can be found by studying the change in x and the change in y, as shown below.

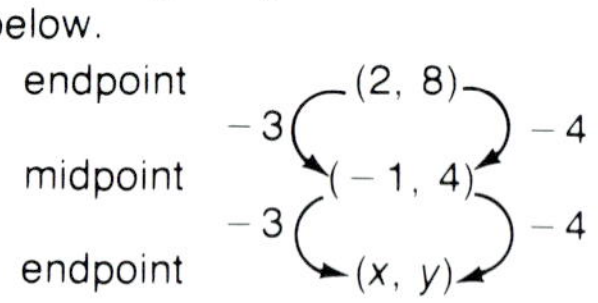

The coordinates of the other endpoint are $(-1-3, 4-4)$ or $(-4, 0)$.

Suggested Daily Assignments

Enriched
- Exp: 10–20 evens
- Written: 8–34 evens, 35–52

Average
- Exp: 1–21 odds
- Written: 2–22 evens, 23–44

Fundamental
- Exp: 1–21
- Written: 1–40
  - p. 326, 7–9, 16–18

❶

Example

2 **If one endpoint of a line segment is (2, 8) and the midpoint is (−1, 4), find the coordinates of the other endpoint.**

$$(x, y) = \left(\frac{x_1 + x_2}{2}, \frac{y_1 + y_2}{2}\right)$$

$$(-1, 4) = \left(\frac{2 + x_2}{2}, \frac{8 + y_2}{2}\right)$$ *Substitute $(-1, 4)$ for (x, y). Substitute $(2, 8)$ for (x_1, y_1).*

Now separate this last expression into two equations.

$-1 = \frac{2 + x_2}{2}$ *Solve for x_2.*

$-2 = 2 + x_2$

$-4 = x_2$

$4 = \frac{8 + y_2}{2}$ *Solve for y_2.*

$8 = 8 + y_2$

$0 = y_2$

The coordinates of the endpoint are $(-4, 0)$.

Exploratory Exercises

State the coordinate of the point midway between each pair of points on the number line.

1. 4 and 8 6
2. 3 and 1 2
3. 10 and 18 14
4. −2 and 6 2
5. −3 and 13 5
6. −4 and −10 −7
7. −3 and 6 $\frac{3}{2}$
8. 5 and −6 $-\frac{1}{2}$
9. −10 and 15 $\frac{5}{2}$

State the coordinates of the midpoint of the line segment whose endpoints are given.

10. (5, 1), (9, 3) (7, 2)
11. (2, 10), (8, −4) (5, 3)
12. (6, 5), (8, −3) (7, 1)
13. (15, −4), (3, −12) (9, −8)
14. (−11, 3), (3, −15) (−4, −6)
15. (2, 0), (0, 12) (1, 6)
16. (9, −4), (1, 14) (5, 5)
17. (5, −6), (−7, 3) $\left(-1, -\frac{3}{2}\right)$
18. (2, 5), (11, 2) $\left(\frac{13}{2}, \frac{7}{2}\right)$
19. (5, −4), (6, −7) $\left(\frac{11}{2}, -\frac{11}{2}\right)$
20. (x, y), (a, b) $\left(\frac{x + a}{2}, \frac{y + b}{2}\right)$
21. $(2x, 3y)$, $(6x, y)$ $(4x, 2y)$

Written Exercises

Write the coordinates of the midpoint of the line segment whose endpoints are given.

1. (8, 4), (12, 2) (10, 3)
2. (9, 5), (17, 3) (13, 4)
3. (11, 4), (9, 2) (10, 3)
4. (17, 9), (11, −3) (14, 3)
5. (14, 4), (2, 0) (8, 2)
6. (19, −3), (11, 5) (15, 1)
7. (4, 2), (8, −6) (6, −2)
8. (−6, 5), (8, −11) (1, −3)
9. (−6, 5), (8, −11) (1, −3)
10. (5, −2), (7, 3) $\left(6, \frac{1}{2}\right)$
11. (8, 2), (12, −5) $\left(10, -\frac{3}{2}\right)$
12. (−11, 6), (13, 4) (1, 5)
13. (12, −3), (5, 8) $\left(\frac{17}{2}, \frac{5}{2}\right)$
14. (4, 7), (8, 0) $\left(6, \frac{7}{2}\right)$
15. (2, 5), (4, 1) (3, 3)
16. (9, 10), (−8, 4) $\left(\frac{1}{2}, 7\right)$
17. (−3, 6), (4, 5) $\left(\frac{1}{2}, \frac{11}{2}\right)$
18. (−8, −5), (−2, 8) $\left(-5, \frac{3}{2}\right)$

19. $\left(\frac{1}{2}, 3\right)$, $(1, -2)$ $\left(\frac{3}{4}, \frac{1}{2}\right)$

20. $\left(\frac{5}{6}, \frac{1}{3}\right)$, $\left(\frac{1}{6}, \frac{1}{3}\right)$ $\left(\frac{1}{2}, \frac{1}{3}\right)$

21. $\left(3\frac{1}{2}, 2\frac{1}{4}\right)$, $\left(5\frac{1}{3}, 1\frac{1}{2}\right)$ $\left(4\frac{5}{12}, 1\frac{7}{8}\right)$

22. $\left(2\frac{1}{5}, 3\frac{1}{3}\right)$, $\left(5\frac{3}{5}, 2\frac{1}{3}\right)$ $\left(3\frac{9}{10}, 2\frac{5}{6}\right)$

If *P* is the midpoint of line segment *AB*, find the coordinates of the missing point *A*, *B*, or *P*.

23. $A(3, 5)$, $P(11, 7)$ $B(19, 9)$
24. $A(3, 5)$, $P(5, -7)$ $B(7, -19)$
25. $A(5, 9)$, $B(-7, 3)$ $P(-1, 6)$
26. $B(11, -4)$, $P(3, 8)$ $A(-5, 20)$
27. $B(5, 3)$, $P(9, 7)$ $A(13, 11)$
28. $A(11, -6)$, $B(5, -9)$ $P\left(8, -\frac{15}{2}\right)$
29. $P(5, -9)$, $A(4, -11)$ $B(6, -7)$
30. $P(3, 9)$, $B(-4, 1)$ $A(10, 17)$
31. $A(4, -7)$, $B(-8, 1)$ $P(-2, -3)$
32. $A(7, 4)$, $P(9, -3)$ $B(11, -10)$
33. $P(3, -5)$, $A(-3, 8)$ $B(9, -18)$
34. $P(5, 6)$, $B(5, 7)$ $A(5, 5)$
35. The center of a circle is $(3, -2)$ and one endpoint of a diameter is $(8, 3)$. Find the other endpoint of the diameter. $(-2, -7)$
36. The two endpoints of the diameter of a circle are $(8, -2)$ and $(4, -6)$. Find the coordinates of the center. $(6, -4)$

The coordinates on a map for certain cities are given below.

Los Angeles $(-6, -1)$	**Dallas $(1, -3)$**	**Chicago $(3, 3)$**
Atlanta $(5, -1)$	**Miami $(6, -4)$**	**Boston $(7, 5)$**

Find the coordinates of the point on the map midway between each pair of cities.

37. Los Angeles and Boston $\left(\frac{1}{2}, 2\right)$
38. Atlanta and Dallas $(3, -2)$
39. Chicago and Miami $\left(\frac{9}{2}, -\frac{1}{2}\right)$
40. Boston and Atlanta $(6, 2)$
41. Dallas and Boston $(4, 1)$
42. Miami and Dallas $\left(\frac{7}{2}, -\frac{7}{2}\right)$
43. Chicago and Los Angeles $\left(-\frac{3}{2}, 1\right)$
44. Boston and Chicago $(5, 4)$

Challenge

Find the coordinates of *P* on line segment *AB* if *P* is one-fourth of the distance from *A* to *B*.

45. $A(8, 4)$, $B(12, 12)$ $(9, 6)$
46. $A(-3, 9)$, $B(5, 1)$ $(-1, 7)$
47. $A(-3, 2)$, $B(5, 4)$ $\left(-1, \frac{5}{2}\right)$
48. $A(2, -6)$, $B(9, 5)$ $\left(\frac{15}{4}, -\frac{13}{4}\right)$

For quadrilateral *ABCD*, determine whether the diagonals of *ABCD* bisect each other.

49. $A(8, 6)$, $B(5, 5)$, $C(4, 2)$, $D(7, 3)$ yes
50. $A(-2, 6)$, $B(2, 11)$, $C(3, 8)$, $D(-1, 3)$ yes
51. $A(11, 6)$, $B(1, -2)$, $C(-2, 4)$, $D(3, 8)$ no
52. $A(8, -2)$, $B(3, -5)$, $C(-3, 5)$, $D(2, 8)$ yes

About Using Computers
The program given can be used to determine if 3 given points are collinear. Point out to students that it would take them much longer than it does the computer to identify 3 collinear points. Explain that to complete exercise 5, only one line needs to be added to the program. Explain to students that REM statements are an important way of telling someone who is using a program what it does. In very long programs, many REM statements are used to describe the purpose of the different parts of the program.

Using Computers

Testing Points On A Line

Each of the two graphs has 3 points plotted. Are the points on the same line?

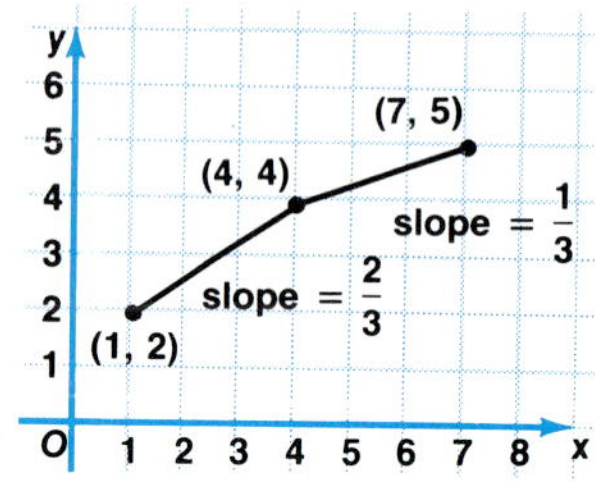

The points are not on the same line. The segments have different slopes.

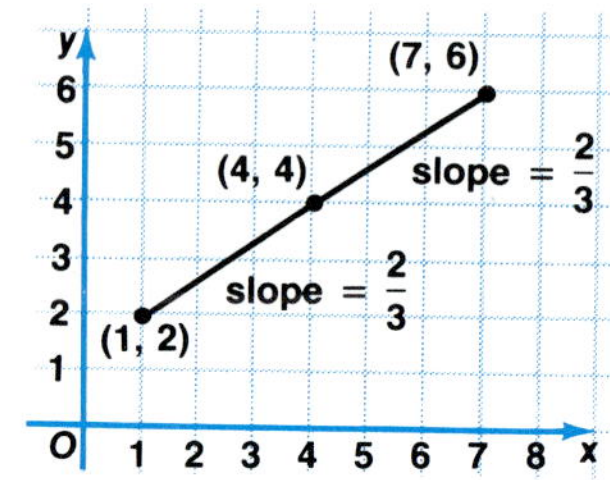

The points are on the same line. The segments have the same slope and have a common point between.

You can write a computer program that tests if any 3 points are on the same line. Let the coordinates of the points to be tested be (A, B), (C, D), and (E, F).

```
10 REM THIS PROGRAM DETERMINES IF 3 POINTS
     LIE ON THE SAME LINE
20 INPUT A, B, C, D, E, F
30 IF (D-B)/(C-A)=(F-D)/(E-C) THEN 60
40 PRINT A;",";B,C;",";D,E;",";F,
   "POINTS ARE NOT ON SAME LINE"
50 GO TO 70
60 PRINT A;",";B,C;",";D,E;",";F,
     "POINTS ARE ON THE SAME LINE"
70 END
```

REM statements often tell how the program works.

Three points on the same line are called collinear.

When the ordered pairs (1, 2), (13, 23), and (5, 9) are used, the output is as follows.

```
1, 2     13, 23     5, 9
POINTS ARE ON THE SAME LINE
```

When the ordered pairs (−5, 2), (−5, 7), (3, 8) are used, a division-by-zero error occurs in line 30. Add the following lines to the program to eliminate the error.

```
25 IF C - A=0 OR E - C=0 THEN PRINT "AT LEAST ONE LINE HAS AN
   UNDEFINED SLOPE": GOTO 70
```

Exercises Use the computer program to determine if each set of points lie on one line.

1. (1, 5), (16, 14), (−4, 2) yes
2. (−2, −3), (2, 1), (5, 6) no
3. (−459, −80), (865, 163), (54, 1) no
4. (7, 14), (−5, 10), (−5, 8) no
5. Modify the program to print the slope for each segment. See student's work.

Vocabulary

slope (305)
point-slope form (309)
standard form (310)
x-intercept (312)
y-intercept (312)
slope-intercept form (312)
slopes of parallel lines (324)
slopes of perpendicular lines (325)
midpoint of a line segment (327)

Chapter Summary

1. The slope of a line is the ratio of the change in y to the corresponding change in x. Slope $= \frac{\text{change in } y}{\text{change in } x}$. (305)
2. Given the coordinates of two points on a line, (x_1, y_1) and (x_2, y_2), the slope m can be found as follows.
$$m = \frac{y_2 - y_1}{x_2 - x_1} \quad (306)$$
3. For a given point (x_1, y_1) on a nonvertical line, the point-slope form of a linear equation is
$$y - y_1 = m(x - x_1). \quad (309)$$
4. The standard form of a linear equation is $Ax + By = C$ where A, B, and C are integers and A and B are not both zero. (310)
5. The x-coordinate of the point where a line crosses the x-axis is called the x-intercept of the line. (312)
6. The y-coordinate of the point where a line crosses the y-axis is called the y-intercept of the line. (312)
7. An equation of the form $y = mx + b$ is said to be in slope-intercept form. (312)
8. If two lines have the same slope, then they are parallel. All vertical lines are parallel. (324)
9. If the product of the slopes of two lines is -1, then the lines are perpendicular. In a plane, vertical lines are perpendicular to horizontal lines. (325)
10. The coordinate of the midpoint, P, of two points, x_1 and x_2, on a number line is
$$P = \frac{x_1 + x_2}{2}. \quad (327)$$
11. The coordinates of the midpoint, (x, y), of a line segment whose endpoints are (x_1, y_1) and (x_2, y_2) are
$$(x, y) = \left(\frac{x_1 + x_2}{2}, \frac{y_1 + y_2}{2}\right). \quad (327)$$

Quiz
Lessons 10-6 to 10-7

Write an equation in slope-intercept form for the line satisfying the given conditions.

1. parallel to $2x - y = -5$ and passes through $(-4, 2)$ $y = 2x + 10$
2. perpendicular to $3x + 5y = 8$ and passes through $(1, 0)$ $y = \frac{5}{3}x - \frac{5}{3}$

Find the coordinates of the midpoint of the line segment that has the given endpoints.

3. $(3, -6)$, $(5, 2)$ $(4, -2)$
4. $(0, 5)$, $(-7, 7)$ $\left(-3\frac{1}{2}, 6\right)$

Answers to Chapter Review

5. $y - 1 = -\frac{4}{11}(x - 8)$; $4x + 11y = 43$

6. $y - 5 = 0$; $y = 5$

7. $y - 5 = \frac{5}{2}x$; $5x - 2y = -10$

8. $y = -2(x + 3)$; $2x + y = -6$

15.

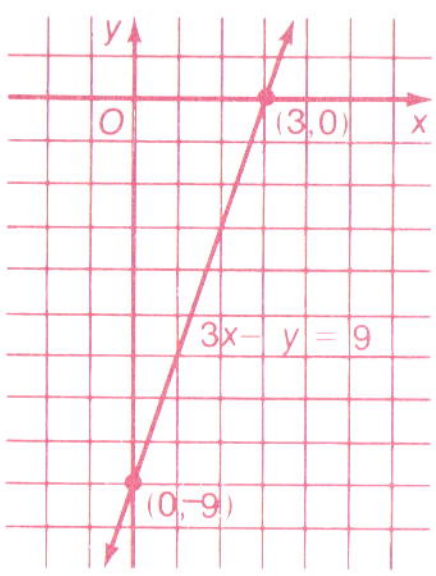

16.

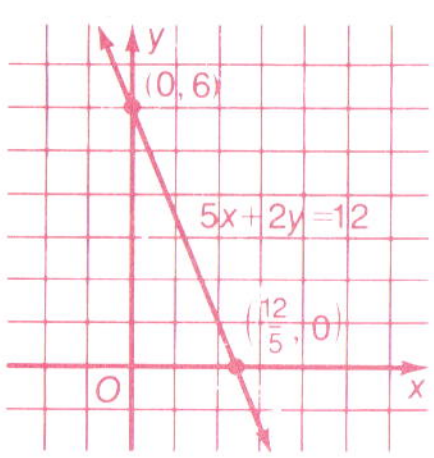

17.

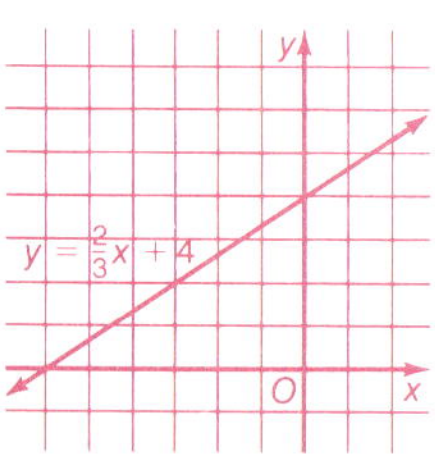

18.

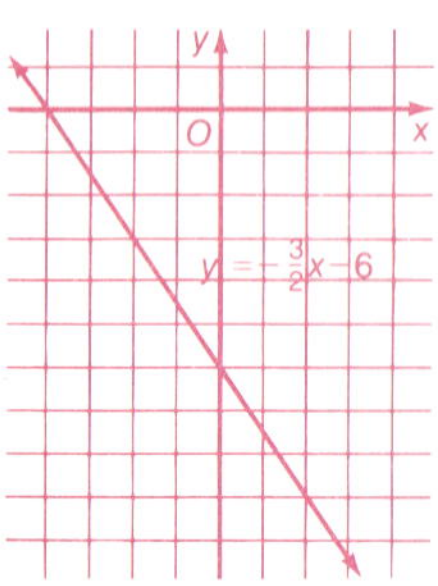

Chapter Review

10-1 Determine the slope of the line passing through each pair of points named.

1. $(-3, 5), (4, 5)$ 0
2. $(8, 3), (2, 5)$ $-\frac{1}{3}$
3. $(-2, 5), (-2, 9)$ undefined
4. $(-3, -5), (9, -1)$ $\frac{1}{3}$

10-2 Write an equation in point-slope form and standard form for the line through each pair of points.

5. $(8, 1), (-3, 5)$
6. $(-2, 5), (9, 5)$
7. $(0, 5), (-2, 0)$
8. $(-3, 0), (0, -6)$

10-3 Determine the slope and y-intercept of the graph of each equation.

9. $3x - 2y = 7$ $\frac{3}{2}, -\frac{7}{2}$
10. $y = \frac{1}{4}x + 3$ $\frac{1}{4}, 3$
11. $8x + y = 4$ $-8, 4$
12. $x = 2y - 7$ $\frac{1}{2}, \frac{7}{2}$

Determine the x- and y-intercepts for the graph of each equation.

13. $2x + 5y = 10$ $(5, 0), (0, 2)$
14. $3x + 4y = 15$ $(5, 0), \left(0, \frac{15}{4}\right)$

10-4 Graph each equation using the x- and y-intercepts.

15. $3x - y = 9$
16. $5x + 2y = 12$

Graph each equation using the slope-intercept method.

17. $y = \frac{2}{3}x + 4$
18. $y = -\frac{3}{2}x - 6$

10-5 Write an equation for the line satisfying the given conditions. Use slope-intercept form.

19. $m = 4$ and passes through $(6, -2)$ $y = 4x - 26$
20. Passes through $(9, 5)$ and $(-3, -4)$ $y = \frac{3}{4}x - \frac{7}{4}$
21. Passes through $(2, 2)$ and has y-intercept of 7 $y = -\frac{5}{2}x + 7$
22. Slope $= -\frac{3}{5}$ and y-intercept $= 3$ $y = -\frac{3}{5}x + 3$

10-6 Write an equation for the line that is parallel to the graph of each equation and that passes through the indicated point. Use slope-intercept form.

23. $4x - y = 7$; $(2, -1)$ $y = 4x - 9$
24. $3x + 9y = 1$; $(3, 0)$ $y = -\frac{1}{3}x + 1$

Write an equation for the line that is perpendicular to the graph of each equation and that passes through the indicated point. Use slope-intercept form.

25. $2x - 7y = 1$; $(-4, 0)$ $y = -\frac{7}{2}x - 14$
26. $8x - 3y = 7$; $(4, 5)$ $y = -\frac{3}{8}x + \frac{13}{2}$

10-7 Find the coordinates of the midpoint of $\overline{AB}$.

27. $A(3, 5), B(9, -3)$ $(6, 1)$
28. $A(2, 7), B(8, 4)$ $\left(5, \frac{11}{2}\right)$

Chapter Test

Determine the slope of the line passing through each pair of points.

1. (9, 2), (3, −4) 1 **2.** (8, 3), (8, 1) undefined **3.** (4, −5), (−2, −5) 0

Determine the slope and *y*-intercept for each equation.

4. $x - 8y = 3$ $\frac{1}{8}, -\frac{3}{8}$

5. $3x - 2y = 9$ $\frac{3}{2}, -\frac{9}{2}$

6. $\frac{1}{2}x + \frac{3}{4}y = 2$ $-\frac{2}{3}, \frac{8}{3}$

7. $y = 7$ 0, 7

Write an equation in standard form for the line satisfying the given conditions.

8. Passes through (2, 5) and (8, −3). $4x + 3y = 23$

9. Passes through (−2, −1) and (6, −4). $3x + 8y = -14$

10. Has slope of 2 and *y*-intercept = 3. $2x - y = -3$

11. Has *y*-intercept −4 and passes through (5, −3). $x - 5y = 20$

12. Slope = $\frac{3}{4}$ and passes through (6, −2). $3x - 4y = 26$

Write an equation in slope-intercept form for the line satisfying the given conditions.

13. Passes through (4, −2) and the origin. $y = -\frac{1}{2}x$

14. Passes through (−2, −5) and (8, −3). $y = \frac{1}{5}x - \frac{23}{5}$

15. Passes through (6, 4) with *y*-intercept = −2. $y = x - 2$

16. Slope = $-\frac{2}{3}$ and *y*-intercept = 5. $y = -\frac{2}{3}x + 5$

17. Slope = 6 and passes through (−3, −4). $y = 6x + 14$

Write the equation of the line satisfying the given conditions. Use slope-intercept form.

18. Parallel to $6x - y = 7$ and passes through (−2, 8). $y = 6x + 20$

19. Parallel to $3x + 7y = 4$ and passes through (5, −2). $y = -\frac{3}{7}x + \frac{1}{7}$

20. Perpendicular to $5x - 3y = 9$ and passes through the origin. $y = -\frac{3}{5}x$

21. Perpendicular to $x + 3y = 7$ and passes through (5, 2). $y = 3x - 13$

Find the coordinates of the midpoint of the segment whose endpoints are named.

22. (9, 3), (3, 6) $\left(6, \frac{9}{2}\right)$

23. (−2, −7), (6, −5) (2, −6)

Graph each equation.

24. $4x - 3y = 24$

25. $2x + 7y = 16$

Answers to Chapter Test

24.

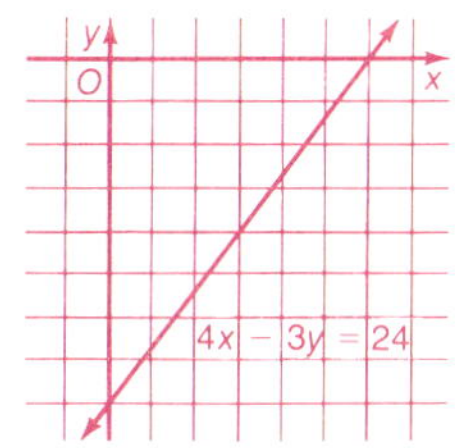

25.

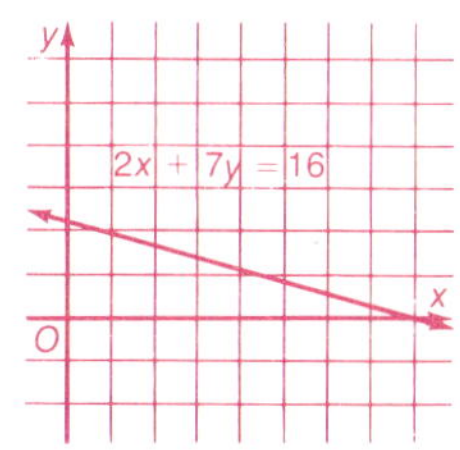

Using the Cumulative Review
If students have difficulty with a particular topic, have them review the appropriate lesson. You may want to provide copies of the solutions from the *Algebra One Solutions Manual*. This Cumulative Review covers Chapters 1-10.

Other Cumulative Reviews appear on pages 72, 142, 208, 268, 402, 466, and 526.

Answers

10.

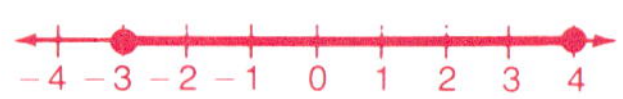

27.

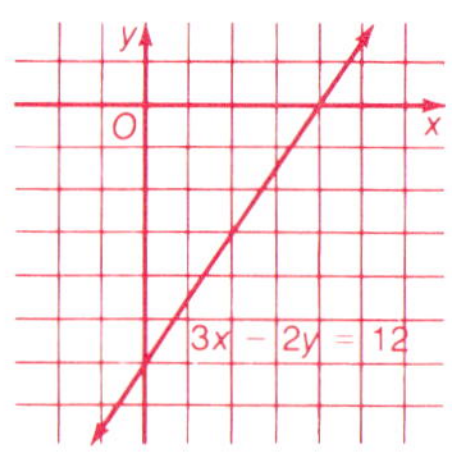

32.

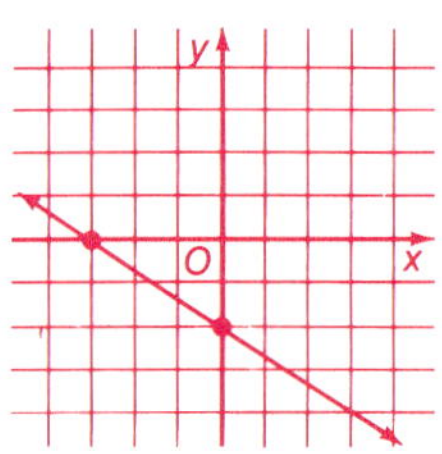

Cumulative Review

1. Solve: $5\frac{2}{3} + 1\frac{1}{2} = x$ $7\frac{1}{6}$
2. Simplify: $14a^2 + 13b^2 - 9a^2$ $5a^2 + 13b^2$
3. Find the sum: $-19.72 + (-13.81)$ -33.53
4. Solve: $r + (-6) = -27$ -21
5. Simplify: $\frac{-27}{-\frac{1}{3}}$ 81

Solve.

6. $4y = -32$ -8
7. $7 - \frac{1}{2}(x + 2) = 10$ -8
8. State an inequality for the graph. $x \neq 3$

9. Replace ? with <, >, or = to make the sentence $-\frac{3}{5}$? $-\frac{5}{7}$ true. >
10. Solve $|2m - 1| \leq 7$ and graph its solution set. $-3 \leq m \leq 4$
11. Simplify: $(-7z^2)^3$ $-343z^6$
12. Express the following product in scientific notation: $(0.00000004)(800)$ 3.2×10^{-5}

Simplify. 14. $63y^4 + 49y^3 - 21y^2 + 28y$

13. $(4m^2 + mp - 3p^2) - (-3m^2 + 4mp + 5p^2)$ $7m^2 - 3mp - 8p^2$
14. $7y(9y^3 + 7y^2 - 3y + 4)$
15. $(3x + 4)^2$ $9x^2 + 24x + 16$
16. Find the prime factorization of 230. $2 \cdot 5 \cdot 23$

Factor, if possible.

17. $4a^2 - 8ab + 12ab^2$ $4a(a - 2b + 3b^2)$
18. $b^2 - 25$ $(b - 5)(b + 5)$
19. $m^2 - 6m + 9$ $(m - 3)^2$
20. $12a^2 + 29a + 15$ $(4a + 3)(3a + 5)$
21. $8x^2y - 3x - 16xy + 6$ $(8xy - 3)(x - 2)$

Solve.

22. $5x(2x - 3) = 0$ $0, \frac{3}{2}$
23. $3m^2 - 108 = 0$ $-6, 6$
24. Simplify: $\frac{p^2 + p - 2}{p^2 - p - 6}$ $\frac{p - 1}{p - 3}$
25. State the quadrant in which the point for the ordered pair $(-3, 5)$ is located. II
26. Solve the equation $3x - y = 2$ if the domain is $\{-1, 0, 3, 5\}$. $\{(-1, -5), (0, -2), (3, 7), (5, 13)\}$
27. Graph: $3x - 2y = 12$
28. Given $f(x) = 5x^2 - 3x + 2$, find $f(-3)$. 56
29. Write an equation for the relationship between the variables in the following chart. $y = -\frac{3}{2}x - 6$

x	0	2	4	6
y	-6	-9	-12	-15

30. Determine the slope of the line passing through $(4, -2)$ and $(8, 1)$. $\frac{3}{4}$
31. Write an equation in slope-intercept form for the line with a slope of -4 and a y-intercept of 3. $y = -4x + 3$
32. Graph $2x + 3y = -6$ using the slope-intercept method.
33. Find the coordinates of the midpoint of the line segment whose endpoints are $(-6, 4)$ and $(2, 0)$. $(-2, 2)$

Problem Solving

Solve each problem.

34. Five times a number decreased by 3 more than twice the number is 153. Find the number. 52
35. Find two integers whose sum is 12 and whose product is 32. 4, 8
36. The area of a rectangle is 64 in^2. The length is 4 more than 3 times the width. Find the length and width. 4 inches by 16 in
37. A person invests \$4000, part at 6% annual interest and the rest at 8% annual interest. If the investment earns \$270 in interest at the end of one year, how much was invested at each rate? \$2500 at 6% \$1500 at 8%

Standardized Test Practice Questions

The test questions on this page deal with expressions and equations. The information at the right may help you with some of the questions.

Directions: Choose the best answer. Write A, B, C, or D.

1. If $(-9)(-9)(-9) = (-9)(-9)(-9)n$, then $n =$

   (A) 1 (B) 0 (C) -9 (D) 9

2. If $(1 - 2 + 3 - 4 + 5 - 6) = -3$, then $2(-1 + 2 - 3 + 4 - 5 + 6) =$

   (A) -6 (B) 6 (C) 3 (D) -3

3. If $48k(5) = 20$, then $k =$

   (A) $\frac{5}{48}$ (B) $\frac{1}{12}$ (C) $\frac{1}{24}$ (D) $\frac{20}{48}$

4. If $23(66 + x) = 2300$, then $x =$

   (A) 1,185 (B) 782
   (C) 34 (D) 44

5. If $3y + 2$ is an odd integer, what is the next consecutive odd integer?

   (A) $3y + 4$ (B) $5y + 2$
   (C) $5y + 4$ (D) $y + 2$

6. $\frac{3(1.8 - 2.6) - (1.8 - 2.6)}{2} =$

   (A) 0.8 (B) -0.8
   (C) -1.5 (D) -3.4

7. $-3(a - b) =$

   (A) $3(b - a)$ (B) $3(a + b)$
   (C) $3(-b) + (-a)$ (D) $3ab$

8. If $3y = 7$, the value of $12y - 5$ is

   (A) 43 (B) 7 (C) 21 (D) 23

9. If $\frac{x}{8} + 3 = 1$, the value of $\frac{x}{2}$ is

   (A) -32 (B) -16 (C) -8 (D) 4

> 1. Many problems can be solved without much calculating if the basic mathematical concepts are understood. Always look carefully at what is asked, and think of possible shortcuts for solving the problem.
> 2. Check your solutions by substituting values for the variables.

10. If $x = 1$, $y = -2$, and $z = 2$, then $\frac{x^2y}{(x - z)^2} =$

    (A) -2 (B) 2 (C) $-\frac{1}{3}$ (D) $\frac{1}{3}$

11. If $1 + \frac{c}{12} = 2\frac{3}{4}$, then $c =$

    (A) 33 (B) 32 (C) 21 (D) 12

12. A number added to one-third of itself results in a sum of 40. What is the number?

    (A) 32 (B) 30 (C) 10 (D) 27

13. How many dollars do you have if you have n nickels, d dimes, and k quarters?

    (A) $\frac{5n + 25k + 10d}{100}$

    (B) $\frac{25n + 10d + 4k}{100}$

    (C) $20n + 10d + 4k$

    (D) $\frac{20n + 10d + 4k}{100}$

14. Dave is twice as old as Bob, who is 3 years older than Steve. If Steve is 7 years old, how old is Dave?

    (A) 14 (B) 18 (C) 20 (D) 22

15. If $\frac{x}{5} = 2$, then $\frac{x}{2} =$

    (A) 10 (B) 5 (C) $\frac{1}{2}$ (D) $\frac{1}{5}$

16. If $20x$ cartons fill $\frac{x}{5}$ trucks, how many trucks are needed to hold 400 cartons?

    (A) 20 (B) 8,000 (C) 4 (D) 100

Using the Standardized Test Practice Questions

The questions on this page may be used to help students prepare for pre-college tests. These questions usually require careful analysis and a thorough understanding of the concepts.

This page could be used as an overnight assignment for better students. Other students may benefit most if only a few questions are assigned at a time.

After students have chosen an answer for each problem, you should discuss how each problem can be solved, or provide copies of the solutions from the *Algebra One Solutions Manual*.

Other practice questions appear in this text on pages 73, 143, 209, 269, 403, 467, and 527. Also see pages 50-58 in the *Teacher Resource Book*.

About the Photo

A system of linear equations can be used to solve motion problems such as the one presented on this page. Many digit and mixture problems can also be solved using systems of equations.

Two equations in a system place two conditions on the variables simultaneously. For this reason, systems of equations are sometimes called simultaneous equations. An ordered pair that satisfies both equations in the system is called a solution of the system. A system has no solutions when there is no ordered pair that will satisfy both equations simultaneously.

Graphical solutions provide visual explanation of these concepts. The graph of a system of linear equations with one solution consists of intersecting lines. The graph of a system with no solutions consists of parallel lines.

Upon completion of this chapter, challenge better students to solve the problem presented on this page. The speed of the wind is about 37 mph.

Teacher Resource Book Materials for Chapter 11 See pages 261-284.

Chapter Test, Form I (Multiple Choice)	261-263
Chapter Test, Form II (Free Response)	264-266
Quizzes	267-268
Extra Practice	269-271
Application: Keeping Nature's Balance (Use after 11-3.)	272-273
Application: Misc. Equations (Use after 11-8.)	274-275
Extension: Absolute Value Inequalities (Use after 11-6.)	276
Extension: A Shortcut (Use after 11-7.)	277
Problem Solving	278
Cumulative Review	279
Answers	280-284

CHAPTER 11

Systems of Open Sentences

An airplane flies from New York to Los Angeles in 4 hours 20 minutes. The return trip from Los Angeles to New York takes 3 hours 50 minutes. The airplane can fly 600 mph in still air. What is the speed of the air current? In this chapter, you will learn how to solve problems such as this by using systems of equations.

Objective 11-1: Solve systems of equations by graphing.

11-1 Graphing Systems of Equations

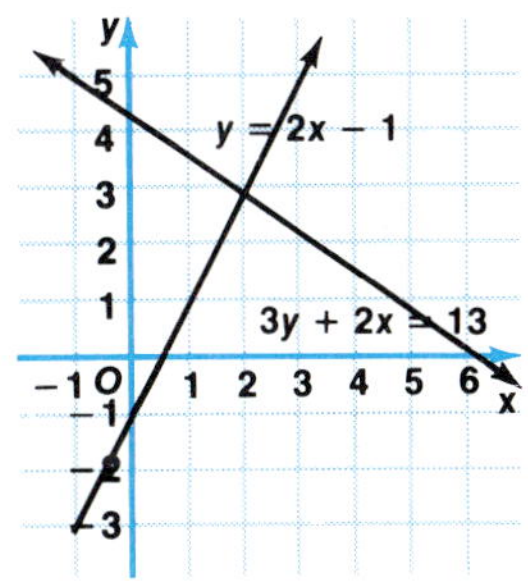

The graphs of $y = 2x - 1$ and $3y + 2x = 13$ are shown at the right.

Notice that the graphs intersect at the point (2, 3). Since this point lies on the graph of each equation, it follows that the ordered pair (2, 3) satisfies both equations.

This can be checked by substituting (2, 3) into each equation.

$$\begin{aligned} y &= 2x - 1 \\ 3 &\stackrel{?}{=} 2(2) - 1 \\ 3 &= 3 \end{aligned} \qquad \begin{aligned} 3y + 2x &= 13 \\ 3(3) + 2(2) &\stackrel{?}{=} 13 \\ 13 &= 13 \end{aligned}$$

The equations $y = 2x - 1$ and $3y + 2x = 13$ together are called a **system of equations**. The solution of this system of equations is (2, 3). ❶

Examples

1 **Graph the equations $x + y = 6$ and $y = 2x$. Then find the solution of the system of equations.**

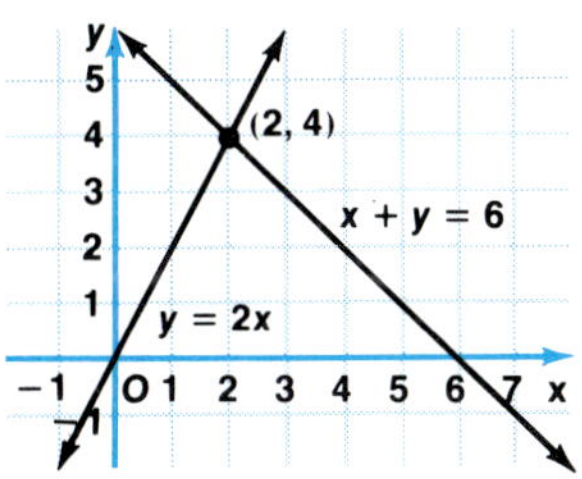

The graphs intersect at (2, 4). Therefore, (2, 4) is the solution of the system of equations $x + y = 6$ and $y = 2x$.

Check:

$$\begin{aligned} x + y &= 6 \\ 2 + 4 &\stackrel{?}{=} 6 \\ 6 &= 6 \end{aligned} \qquad \begin{aligned} y &= 2x \\ 4 &\stackrel{?}{=} 2(2) \\ 4 &= 4 \end{aligned}$$

2 **Graph the equations $y = x - 4$ and $x + \frac{1}{2}y = \frac{5}{2}$. Then find the solution of the system of equations.**

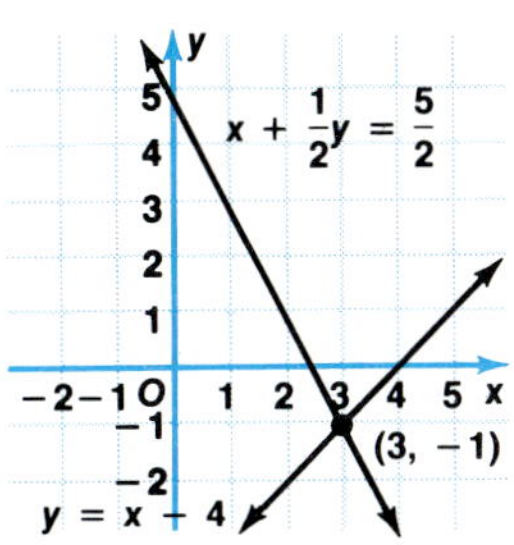

The graphs intersect at (3, −1). Therefore, (3, −1) is the solution of the system of equations $y = x - 4$ and $x + \frac{1}{2}y = \frac{5}{2}$.

Check:

$$\begin{aligned} y &= x - 4 \\ -1 &\stackrel{?}{=} 3 - 4 \\ -1 &= -1 \end{aligned} \qquad \begin{aligned} x + \tfrac{1}{2}y &= \tfrac{5}{2} \\ 3 + \tfrac{1}{2}(-1) &\stackrel{?}{=} \tfrac{5}{2} \\ \tfrac{6}{2} - \tfrac{1}{2} &\stackrel{?}{=} \tfrac{5}{2} \\ \tfrac{5}{2} &= \tfrac{5}{2} \end{aligned}$$

11-1

Lesson Notes, page T45

Readiness Activity

Graph the equations $y = 2x - 1$ and $y = 3x + 1$ on the same set of axes. What are the coordinates of the point where the graphs intersect?

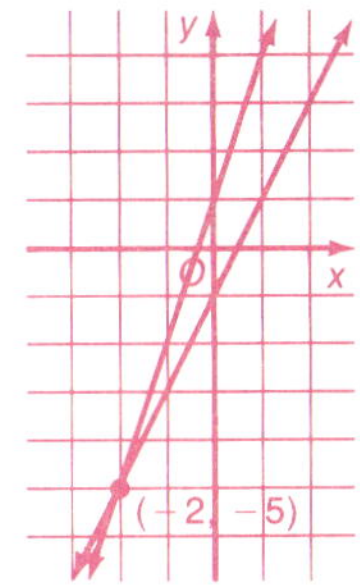

Discussion Questions

(Use after example 2.)

State the solution of each system of equations.

1. $x = 3$, $y = 2$ (3, 2)
2. $x = 5$, $y = -1$ (5, −1)
3. $x = 0$, $y = -4$ (0, −4)
4. $x = -10$, $y = 3$ (−10, 3)
5. Can you suggest a shortcut for solving such systems? All answers are of the form (x, y).

Chalkboard Exercises

Graph each pair of equations. Then find the solution of each system of equations.

1a. $y = 2x$; $x + y = 3$ (1, 2)

1b. $y = x$; $x + y = 6$ (3, 3)

1c. $y = -x$; $x - y = 2$ (1, −1)

2a. $y = 3x + 1$; $x - \frac{1}{2}y = -\frac{1}{2}$ (0, 1)

2b. $y = -4x + 20$; $x - \frac{2}{3}y = \frac{4}{3}$ (4, 4)

❶ Point out that since there are two variables, the solution, if it exists, will be an ordered pair or ordered pairs.

Suggested Daily Assignments

Enriched
Exp: 2–10 evens, 11–14
Written: 2–24 evens

Average
Exp: 1–14
Written: 1–24

Fundamental
Exp: 1–14
Written: 1–15

Answers to Written
See the Teacher's Answer Key, p. T87, for graphs to exercises 1–24.

Exploratory Exercises

State the ordered pair for the intersection of each pair of lines.

1. a and b (6, 3)
2. a and c (−3, 3)
3. a and d (1, 3)
4. a and e (6, 3)
5. b and c (0, 0)
6. b and d (2, 1)
7. b and e (6, 3)
8. c and d (5, −5)
9. c and e (3, −3)
10. d and e $\left(3\frac{1}{2}, -2\right)$

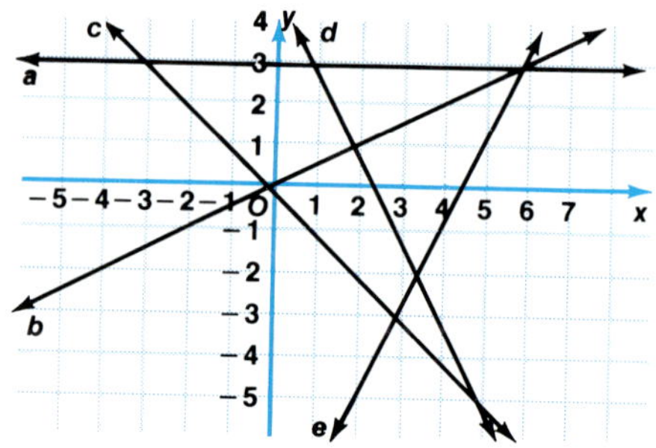

State which of the ordered pairs is a solution for the equation given.

11. $x + 2y = 7$ *a, c, d* **a.** (3, 2) **b.** (6, 1) **c.** (9, −1) **d.** (−1, 4)
12. $3x - 2y = 5$ *c, d* **a.** (1, 1) **b.** (−1, 1) **c.** (1, −1) **d.** (5, 5)
13. $1.5x + 0.3y = 3$ *a, d* **a.** (0, 10) **b.** (2, 1) **c.** (−3, 5) **d.** (4, −10)
14. $\frac{1}{2}x + \frac{2}{3}y = 3$ *a* **a.** (2, 3) **b.** (2, 1) **c.** (4, 5) **d.** (3, 2)

Written Exercises

See Teacher's Answer Key for graphs.

Graph each pair of equations. Then state the solution of each system of equations.

1. $x - y = 2$, $2x + 3y = 9$ (3, 1)
2. $3x - 2y = 10$, $x + y = 0$ (2, −2)
3. $x + y = 8$, $x - y = 2$ (5, 3)
4. $5x - 3y = 12$, $2x - 3y = 3$ (3, 1)
5. $y = x$, $x + y = 4$ (2, 2)
6. $y = -3x$, $4x + y = 2$ (2, −6)
7. $x + y = 5$, $x - 2y = -4$ (2, 3)
8. $y = x - 1$, $y + x = 11$ (6, 5)
9. $y = -x$, $y = 2x$ (0, 0)
10. $x + y = 3$, $y = x + 3$ (0, 3)
11. $y = x + 3$, $3y + x = 5$ (−1, 2)
12. $\frac{1}{2}x + y = -\frac{9}{2}$, $x - y = 6$ (1, −5)
13. $2x + 3y = -17$, $y = x - 4$ (−1, −5)
14. $y = x - 3$, $2x - y = 8$ (5, 2)
15. $\frac{1}{3}x + 2y = 1$, $\frac{3}{4}x + \frac{1}{4}y = -2$ (−3, 1)
16. $0.1x + 0.2y = 0.8$, $0.5x - 0.5y = -0.5$ (2, 3)
17. $0.4x + 0.3y = 2.4$, $1.2x - 1.8y = -3.6$ (3, 4)
18. $3x + y = -11$, $\frac{1}{2}x + 2y = 0$ (−4, 1)
19. $\frac{1}{2}x + \frac{2}{3}y = -3$, $4x - \frac{1}{3}y = -7$ (−2, −3)
20. $\frac{1}{3}x - \frac{1}{4}y = 2$, $9x + \frac{3}{4}y = 24$ (3, −4)
21. $0.2a = 0.3b$, $0.4a - 0.2b = 0.2$ (0.75, 0.5)
22. $3.4x + 6.3y = 4.4$, $2.1x + 3.7y = 3.1$ (5, −2)
23. $2x - 8y = 64$, $\frac{1}{4}x + \frac{1}{7}y = 0$ (4, −7)
24. $x + 2y = 0$, $\frac{1}{3}x + \frac{1}{3}y = -1$ (−6, 3)

Applications in Business Break-Even Analysis

The Murphy Company, a leading gadget maker, wanted to know its break-even volume. This is the volume (number of gadgets) at which revenue (income) equals cost.

Jan Martinez, an accountant, made the following graph for the Murphy Company. Revenue is described by the equation $y = 5x$. Cost is described by the equation $y = 4x + 200$. Jan graphed these equations and found the point of intersection.

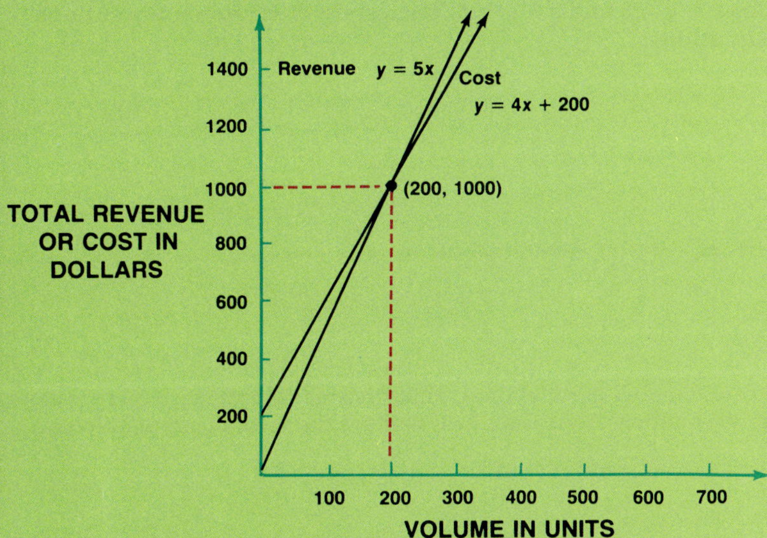

The graphs intersect at (200, 1000). This tells Jan that the Murphy Company must sell 200 gadgets to break even. At 200 units, the Murphy Company will incur \$1000 of costs and receive \$1000 in revenue. To make a profit, the Murphy Company must sell more than 200 gadgets.

Exercises

Make a break-even graph for each company. State the number of units necessary to break even and the revenue at that point.

1. Block Company revenue: $y = 7x$ cost: $y = x + 300$ 50, \$350

2. Zwik Company revenue: $y = 4x$ cost: $y = 2x + 150$ 75, \$300

3. Wu Company revenue: $y = 11x$ cost: $y = 3x + 240$ 30, \$330

4. Carilli Company revenue: $y = 42x$ cost: $y = 3x + 780$ 20, \$840

About the Application
Explain to students that revenue means total income. Point out that in this example, the slopes of the lines are almost the same. Thus, the revenue or income rises just a little faster than the cost. This means it will take large-volume selling before a large profit is made.

Answers to Exercises

1.
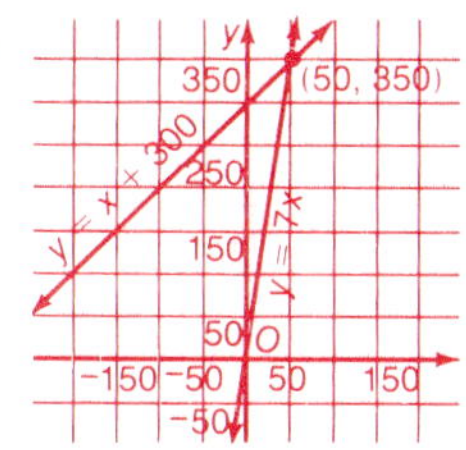

2.
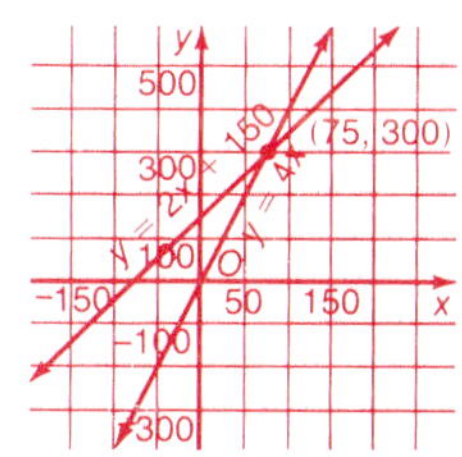

3.
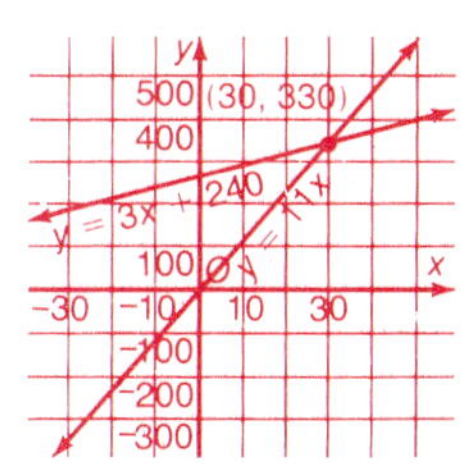

4.
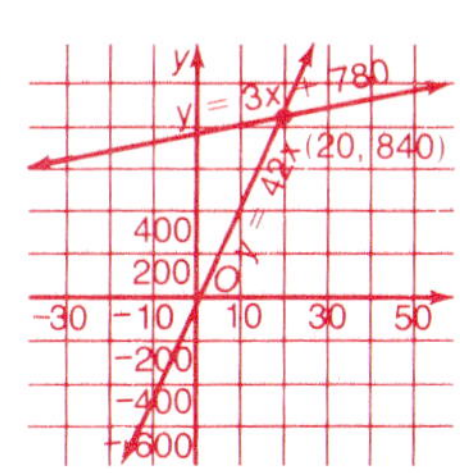

11-2

Lesson Notes, page T45

Readiness Activity
If two lines are parallel, where do they intersect? How many solutions do you think a system of equations whose graphs are parallel lines has? They do not intersect.; none
Suppose you graph two equations and you get the same line. How many solutions do you think this system has? an infinite number

Discussion Question
(Use after example 2.)
A system of equations cannot be both inconsistent and dependent. Why?
A system that is inconsistent has no solution, whereas a system that is dependent is consistent and has infinitely many solutions.

Chalkboard Exercises
Graph each pair of equations. Then state the number of solutions of the system of equations.

1. $2y - x = -4$
   $2y - x = 2$

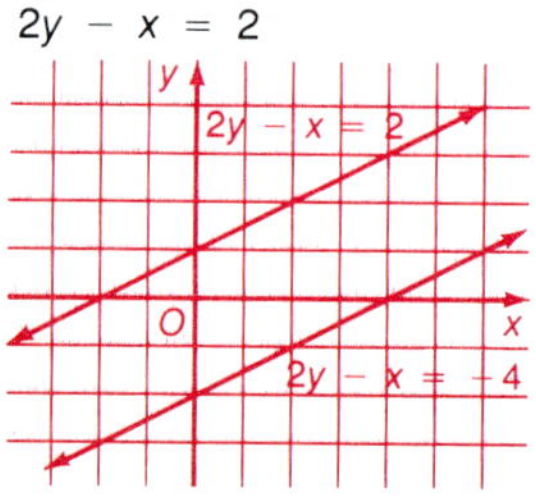

no solution

2. $3x - y = 2$
   $12x - 4y = 8$

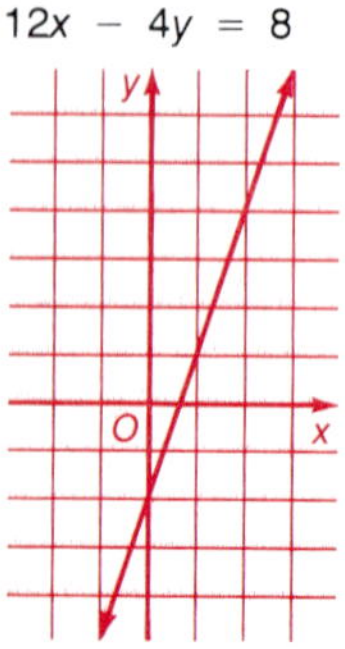

infinitely many solutions

Objective 11-2: Determine whether a system of equations has one solution, no solutions, or infinitely many solutions by graphing.

11-2 Systems of Equations

Not every system of equations has one ordered pair as its solution. Some systems have no solution. Others have infinitely many solutions.

A system of equations that has exactly one ordered pair as its solution is said to be consistent and independent.

Examples

1 **Graph the equations $x + y = 3$ and $x + y = 4$. Then state the number of solutions of the system of equations.**

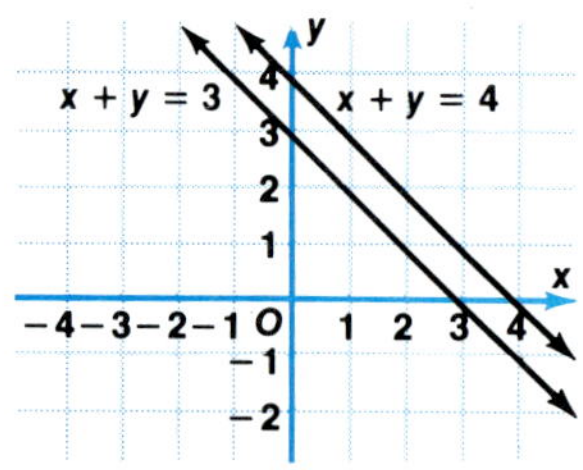

A system that has no solution is said to be inconsistent.

The graphs of the equations are parallel lines. Since they do not intersect, there is no solution to the system of equations. Note that the lines have the same slope but different x- and y-intercepts.

2 **Graph the equations $2x + y = 3$ and $4x + 2y = 6$. Then state the number of solutions of the system of equations.**

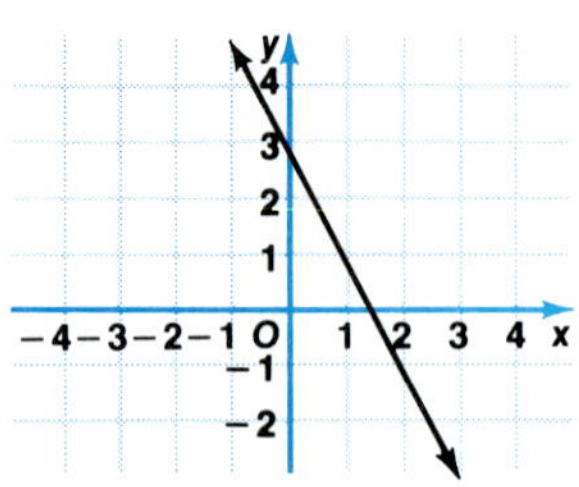

A system that has infinitely many solutions is said to be consistent and dependent.

The graphs of the equations have the same slope and intercepts. Any ordered pair on the graph satisfies both equations. So, there are infinitely many solutions to the system of equations.

Exploratory Exercises ❶

State the slope and y-intercept for the graph of each equation. Then state whether the system of equations has one solution, no solution, or infinitely many solutions.

1. $x + y = 4$ $m = -1, b = 4$
$2x + 3y = 9$ $m = -\frac{2}{3}, b = 3$
one

2. $x + y = 6$ $m = -1, b = 6$
$x - y = 2$ $m = 1, b = -2$
one

3. $x + y = 6$
$3x + 3y = 3$

4. $x + y = 1$ $m = -1, b = 1$
$2x - 2y = 8$ $m = 1, b = -4$
one

5. $x + 2y = 5$
$3x - 15 = -6y$

6. $2x + 4y = 8$
$x + 2y = 4$

7. $y = -3x$
$6y - x = -38$

8. $x + y = 1$ $m = -1, b = 1$
$3x + 5y = 7$ $m = -\frac{3}{5}, b = \frac{7}{5}$
one

9. $3x + 6 = 7y$
$x + 2y = 11$

10. $x + 5y = 10$ $m = -\frac{1}{5}, b = 2$
$x + 5y = 15$ $m = -\frac{1}{5}, b = 3$
none

11. $3x - 8y = 4$
$6x - 42 = 16y$

12. $2x + 3y = 5$
$-6x - 9y = -15$

12. $m = -\frac{2}{3}, b = \frac{5}{3}$
$m = -\frac{2}{3}, b = \frac{5}{3}$
infinitely many

Written Exercises See Teacher's Answer Key for graphs.

Graph each pair of equations. Then state whether the system has one solution, no solution, or infinitely many solutions. If the system has one solution, state it.

1. $3x + y = 6$ (2, 0)
$y + 2 = x$

2. $3x - y = 4$ (0, −4)
$6x + 2y = -8$

3. $x + 2y = 5$ none
$2x + 4y = 1$

4. $2x - y = -4$ (13, 30)
$-3x + y = -9$

5. $2x + 5 = 3y$ (2, 3)
$x = 2$

6. $x + y = 4$ none
$x + y = 9$

7. $y = 5$ none
$y = 7$

8. $y = 6$ (9, 6)
$x = 9$

9. $2x + 3y = 4$
$-4x - 6y = -8$
infinitely many

10. $9x - 2 = 4y$
$x + y = 6$ (2, 4)

11. $x + y = 4$ (7, −3)
$x - y = 10$

12. $x + 2y = 6$ none
$x + 2y = 8$

13. $\frac{1}{2}x + y = 3$ none
$\frac{1}{2}y + \frac{1}{4}x = 2$

14. $\frac{1}{2}x + \frac{1}{3}y = 6$
$y = \frac{1}{2}x + 2$ (8, 6)

15. $\frac{2}{3}x + \frac{1}{4}y = 4$ infinitely many
$\frac{1}{3}x + \frac{1}{8}y = 2$

16. $\frac{1}{4}x - \frac{1}{3}y = -1$
$\frac{1}{3}x - \frac{1}{2}y = -2$
(12, 12)

17. $0.3y = 0.4x - 1.2$
$1.2x + 0.9y = -3.6$
(0, −4)

18. $0.2x + 0.1y = 1$ none
$x = 1 - 0.5y$

mini-review

1. State the quadrant in which $(-2, 5)$ is located. II
2. State the domain and range for $\{(-4, 1), (-2, 3), (0, 1), (2, 3), (4, 1)\}$.
$D = \{-4, -2, 0, 2, 4\}$; $R = \{1, 3\}$

Solve.

3. $x^2 + 12x = -35$ 4. $\frac{1}{2}y^2 - \frac{1}{4}y = 0$
5. The sum of the squares of two consecutive odd integers is 130. Find the integers. −9, −7 or 7, 9

3. $\{-5, -7\}$ 4. $\{0, \frac{1}{2}\}$

Suggested Daily Assignments

Enriched
Exp: 1–11 odds
Written: 1–17 odds
Mini-Review: 1–5

Average
Exp: 1–12
Written: 1–18
Mini-Review: 1–5

Fundamental
Exp: 1-12
Written: 1–12
p. 338, 16–21
Mini-Review: 1–5

Answers to Exploratory

3. $m = -1, b = 6$
$m = -1, b = 1$
no solutions

5. $m = -\frac{1}{2}, b = \frac{5}{2}$
$m = -\frac{1}{2}, b = \frac{5}{2}$
infinitely many

6. $m = -\frac{1}{2}, b = 2$
$m = -\frac{1}{2}, b = 2$
infinitely many

7. $m = -3, b = 0$
$m = \frac{1}{6}, b = -\frac{19}{3}$
one

9. $m = \frac{3}{7}, b = \frac{6}{7}$
$m = -\frac{1}{2}, b = \frac{11}{2}$
one

11. $m = \frac{3}{8}, b = -\frac{1}{2}$
$m = \frac{3}{8}, b = -\frac{21}{8}$
none

❶ For systems of equations in the Exploratory Exercises that have one solution, you may wish to have students state the solution. The solutions are given below.

1. (3, 1)   **2.** (4, 2)   **4.** $\left(\frac{5}{2}, -\frac{3}{2}\right)$
7. (2, −6)   **8.** (−1, 2)   **9.** (5, 3)

Answers to Written
See the Teacher's Answer Key, p. T88, for graphs to exercises 1–18.

11-3

Lesson Notes, page T45

Readiness Activity

Let $y = 3$. Find the value of x in each of the following equations by substituting 3 for y and solving for x.

$y - x = 6$ -3

$4x - y = 9$ 3

$2y - 3x = 6$ 0

$3x + 2y = 8$ $\frac{2}{3}$

Discussion Question

(Use after example 1.)
Suppose you want to use the substitution method to solve the following system of equations.

$2x = 6 - 3y$

$3y = 2x + 12$

What shortcuts can you use?
Substitute $6 - 3y$ for $2x$ or $2x + 12$ for $3y$.

Chalkboard Exercises

Use substitution to solve each system of equations.

1a. $y = 2x$
$2x + 5y = -12$ $(-1, -2)$

1b. $2y = -3x$
$4x + y = 5$ $(2, -3)$

1c. $x + y = -4$
$2x + y = -6$ $(-2, -2)$

1d. $3x + y = 7$
$10x - 3y = -2$ $(1, 4)$

1e. $x + 4y = 1$
$3x + 7y = 8$ $(5, -1)$

1f. $2x - 3y = -6$
$-y + 4x = 8$ $(3, 4)$

1g. $x - 2y = 4$
$3x + 4y = 2$ $(2, -1)$

1h. $x + y = 6$
$3x + y = 15$ $\left(\frac{9}{2}, \frac{3}{2}\right)$

❶ Be sure students know that the answer to this question is yes. You may want to have them do the substitution to emphasize that they will still get $y = 2$.

❷ When solving systems of equations, encourage students to check their answers by substituting the ordered pairs into the original equations.

Objective 11-3: Solve systems of equations by the substitution method.

11-3 Substitution

Most systems of equations are easier to solve by algebraic methods than by graphing. One such algebraic method is called **substitution**.

Consider the system of equations $y = 2x$ and $3x + 4y = 11$. In the first equation, y is equal to $2x$. Since y must have the same value in the second equation, you can substitute $2x$ for y in $3x + 4y = 11$.

$$\begin{aligned} 3x + 4y &= 11 \\ 3x + 4(2x) &= 11 \\ 3x + 8x &= 11 \\ 11x &= 11 \\ x &= 1 \end{aligned}$$

Substitute $2x$ for y. Now the equation has only one variable, x. Solve to find the value of x.

Now find the value of y by substituting 1 for x in $y = 2x$.

$$\begin{aligned} y &= 2x \\ y &= 2(1) \\ y &= 2 \end{aligned}$$

Could you also substitute 1 for x in $3x + 4y = 11$? ❶

❷ *Check:*

$y = 2x$	$3x + 4y = 11$
$2 \stackrel{?}{=} 2(1)$	$3(1) + 4(2) \stackrel{?}{=} 11$
$2 = 2$	$3 + 8 \stackrel{?}{=} 11$
	$11 = 11$

The solution of the system of equations is $(1, 2)$.

Example 1

Use substitution to solve the system of equations $x + 6y = 1$ and $3x - 10y = 17$.

Solve the first equation for x.

$$\begin{aligned} x + 6y &= 1 \\ x &= 1 - 6y \end{aligned}$$

Then substitute $1 - 6y$ for x in the second equation and find the value of y.

$$\begin{aligned} 3x - 10y &= 17 \\ 3(1 - 6y) - 10y &= 17 \\ 3 - 18y - 10y &= 17 \\ -28y &= 14 \\ y &= -\tfrac{1}{2} \end{aligned}$$

Substitute $-\frac{1}{2}$ for y in one of the equations to find the value of x.

$$\begin{aligned} x + 6y &= 1 \\ x + 6\left(-\tfrac{1}{2}\right) &= 1 \\ x - 3 &= 1 \\ x &= 4 \end{aligned}$$

The solution to this system is $\left(4, -\frac{1}{2}\right)$.   *Check this result.*

Examples

2 **Use substitution to solve the system of equations $\frac{3}{2}x + y = 3$ and $3x + 2y = 12$.**

Solve the first equation for y.

$$\frac{3}{2}x + y = 3$$
$$y = 3 - \frac{3}{2}x$$

Substitute $3 - \frac{3}{2}x$ for y in the second equation.

$$3x + 2y = 12$$
$$3x + 2\left(3 - \frac{3}{2}x\right) = 12$$
$$3x + 6 - 3x = 12$$
$$6 = 12$$

A false statement, $6 = 12$, occurs. When you graph the equations you will find that they have the same slope but different intercepts. The graphs are parallel lines. Therefore, no solution exists. ③

3 **Use substitution to solve the system of equations $\frac{1}{3}x + \frac{1}{2}y = 6$ and $\frac{2}{3}x + y = 12$.**

Solve the second equation for y.

④
$$\frac{2}{3}x + y = 12$$
$$y = 12 - \frac{2}{3}x$$

Substitute $12 - \frac{2}{3}x$ for y in the first equation.

$$\frac{1}{3}x + \frac{1}{2}y = 6$$
$$\frac{1}{3}x + \frac{1}{2}\left(12 - \frac{2}{3}x\right) = 6$$
$$\frac{1}{3}x + 6 - \frac{1}{3}x = 6$$
$$6 = 6$$

A true statement, $6 = 6$, occurs. When you graph the equations you will find they have the same slope and the same intercepts. The graphs coincide. Therefore, any solution to the first equation is also a solution to the second equation. The system has infinitely many solutions. ⑤

Exploratory Exercises

Solve each equation for x. Then, solve each equation for y.

1. $y + 1 = x$ **2.** $y + 2x = 3$ **3.** $2x + 3y = 6$

4. $y + 13 = \frac{1}{2}x$ **5.** $3y - \frac{1}{2}x = 7$ **6.** $x + y = 5$

7. $2x + 2y = -10$ **8.** $3x + 5y = 8$ **9.** $\frac{2}{3}x - \frac{1}{2}y = 10$

10. $1.6x + 2.3y = 3.2$ **11.** $0.75x + 6 = -0.8y$ **12.** $\frac{2}{3}x - \frac{4}{5}y = 3$

Chalkboard Exercises

Use substitution to solve each system of equations.

2a. $\frac{5}{2}x + 3y = 7$, $5x + 6y = 9$ — no solutions

2b. $3x + 4y = 7$, $\frac{3}{2}x + 2y = 11$ — no solutions

3a. $3x + y = 7$, $6x + 2y = 14$ — infinitely many

3b. $\frac{1}{4}x + y = 5$, $2y - 10 = -\frac{1}{2}x$ — infinitely many

③ Emphasize that there are no ordered pairs that satisfy the system of equations. You can say that the solution set is the empty set, $\emptyset$. The system is inconsistent and independent.

④ When an equation contains fractions, some students may find it easier to eliminate the fractions first. This can be done by multiplying both sides of the equation by the LCM of the denominators. For example, the equation $\frac{1}{3}x + \frac{1}{2}y = 6$ can be multiplied by 6 to obtain $2x + 3y = 36$.

⑤ Emphasize that there are an infinite number of ordered pairs that satisfy the system. You can say that the solution set is $\{(x, y) \mid \frac{1}{3}x + \frac{1}{2}y = 6\}$. The system is consistent and dependent.

Suggested Daily Assignments

Enriched
Written: 1–18

Average
Exp: 2–12 evens
Written: 1–18

Fundamental
Exp: 1–12
Written: 2–18 evens
p. 341, 13–18

Answers to Exploratory

1. $x = y + 1$
   $y = x - 1$
2. $x = \frac{1}{2}(3 - y)$
   $y = 3 - 2x$
3. $x = \frac{1}{2}(6 - 3y)$
   $y = \frac{1}{3}(6 - 2x)$
4. $x = 2(y + 13)$
   $y = \frac{1}{2}x - 13$
5. $x = 2(3y - 7)$
   $y = \frac{1}{3}\left(7 + \frac{1}{2}x\right)$

(continued in the Teacher's Answer Key, p. T88)

Quiz
Lessons 11-1 to 11-3

1. Graph $2x + 3y = -8$ and $x - 2y = 3$.

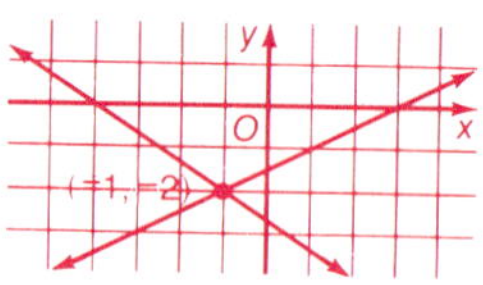

2. Graph $3x - 8 = 2y$ and $6x = 4y - 2$. State the number of solutions.

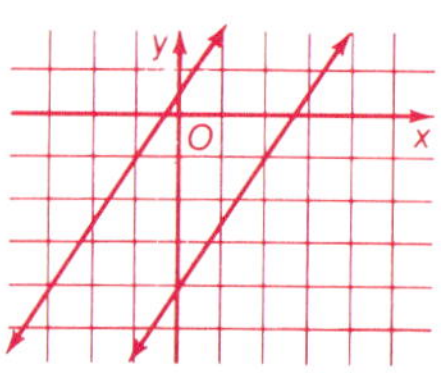

There are no solutions.

3. Solve the following system of equations by substitution.
   $4x + y = -2$
   $7x = 2y + 4$ (0, −2)

About Using Calculators
You may want to extend the exercises by having students check their answers to Written Exercises 1–18. Emphasize that the solution must work in *both* equations to be correct.

Written Exercises

Use substitution to solve each system of equations. State whether the system has no solution, one solution, or infinitely many solutions. If the system has one solution, state it.

1. $y = 2x$; $x + 2y = 8$ $\left(\frac{8}{5}, \frac{16}{5}\right)$

2. $y = 3x$; $x + 2y = -21$ $(-3, -9)$

3. $x = 2y$; $4x + 2y = 15$ $\left(3, \frac{3}{2}\right)$

4. $2x + 3y = 5$; $4x - 9y = 9$ $\left(\frac{12}{5}, \frac{1}{15}\right)$

5. $x = 3 - 2y$; $2x + 4y = 6$ infinitely many

6. $x - 3y = 3$; $2x + 9y = 11$ $\left(4, \frac{1}{3}\right)$

7. $3x + 2y = 14$; $x + \frac{2}{3}y = 4$ none

8. $3x + y = 2$; $2x - y = \frac{1}{2}$ $\left(\frac{1}{2}, \frac{1}{2}\right)$

9. $3x + 5y = 2x$; $x + 3y = y$ $(0, 0)$

10. $x + \frac{1}{2}y = 4$; $y = -2x + 8$ infinitely many

11. $x - 2y = 5$; $3x - 5y = 8$ $(-9, -7)$

12. $2x - y = 7$; $\frac{3}{2}x = \frac{3}{4}y + 5$ none

13. $\frac{1}{4}x - 2y = -3$; $8x + 6y = 44$ $(4, 2)$

14. $2x - y = 3x$; $2x + y = 3y$ $(0, 0)$

15. $\frac{1}{2}x + \frac{1}{3}y = 3$; $-\frac{1}{4}x - \frac{2}{3}y = -3$ $(4, 3)$

16. $-\frac{2}{5}x + \frac{2}{3}y = -8$; $\frac{1}{2}x + \frac{1}{2}y = 2$ $(10, -6)$

17. $0.3x + 0.2y = 0.5$; $0.5x - 0.3y = 0.2$ $(1, 1)$

18. $0.1x + 0.1y = 2$; $0.1x - 0.1y = 0$ $(10, 10)$

Using Calculators — Checking Solutions

You can use a calculator to determine whether an ordered pair is a solution of a system of equations.

For example, follow the steps below to determine whether $(-2, -1)$ is the solution of the system $2x - 5y = 1$ and $3x - 4y = -2$.

ENTER: 2 [×] 2 [+/−] [−] 5 [×] 1 [+/−] [=]

The display shows 1. Therefore $(-2, -1)$ is a solution of the first equation. Now check the second equation.

ENTER: 3 [×] 2 [+/−] [−] 4 [×] 1 [+/−] [=]

The display shows −2. Therefore, $(-2, -1)$ is a solution of the system of equations.

Exercises

Use a calculator to determine whether the given ordered pair is a solution of each system of equations. If the ordered pair is not a solution, find the solution.

1. $5m + 2n = -8$; $4m + 3n = 2$ $(-4, 6)$ yes

2. $0.2a = 0.3b$; $0.4a - 0.2b = 0.2$ $(0.75, 0.5)$ yes

3. $x + y = 40$; $0.2x + 0.45y = 10.5$ $(20, 20)$ no $(30, 10)$

4. $4x + 2y = 10.5$; $2x + 3y = 10.75$ $(1.25, 2.75)$ yes

5. $108x + 537y = -1395$; $-214x - 321y = 535$ $(2, 3)$ no $(2, -3)$

6. $x + 5y = 20.25$; $x + 3y = 13.05$ $(2.25, 3.6)$ yes

Objective 11-4: Solve systems of equations by the elimination method using addition.

11-4 Elimination Using Addition

In some systems of equations the coefficients of either the x or y terms are additive inverses of each other. The simplest way to solve these systems is to add the equations. Because one of the variables is eliminated, this method is called elimination.

Examples

1 Use elimination to solve the system of equations $x - 2y = 5$ and $2x + 2y = 7$.

❶ Add the two equations.

$$\begin{aligned} x - 2y &= 5 \\ 2x + 2y &= 7 \\ \hline 3x \quad &= 12 \\ x &= 4 \end{aligned}$$

The y-coefficients, 2 and −2, are additive inverses.
Add like terms.
Notice the variable y is eliminated.

Substitute 4 for x in either equation and solve for y.

$$\begin{aligned} 2x + 2y &= 7 \\ 2(4) + 2y &= 7 \\ 8 + 2y &= 7 \\ 2y &= -1 \\ y &= -\tfrac{1}{2} \end{aligned}$$

The solution of the system is $\left(4, -\frac{1}{2}\right)$. *Check this result.*

2 The sum of two numbers is 42. Their difference is 6. Find the numbers.

Explore Let x = the greater number.
Let y = the lesser number.

Plan
$x + y = 42$ *sum*
$x - y = 6$ *difference*

Solve

$$\begin{aligned} x + y &= 42 \\ x - y &= 6 \\ \hline 2x \quad &= 48 \\ x &= 24 \\ x + y &= 42 \\ 24 + y &= 42 \\ y &= 18 \end{aligned}$$

The numbers are 18 and 24. *Check the result.*

11-4

Lesson Notes, page T46
Enrichment, page T46

Discussion Question
(Use after example 1.)
Can you give an example of a system of equations in which both variables would be eliminated by addition? What is the solution of your system of equations?
The system must be either inconsistent or consistent and dependent. The solution will either be no solution (for inconsistent systems) or infinitely many solutions (for consistent and dependent systems).

Chalkboard Exercises
Solve each system by elimination.

1a. $4x - 3y = 11$
$2x + 3y = 13$ $\left(4, \frac{5}{3}\right)$

1b. $2x + 5y = 9$
$3x - 5y = 1$ $(2, 1)$

1c. $-3x + 4y = 12$
$3x - 6y = 18$
$(-24, -15)$

1d. $\frac{1}{4}x + \frac{1}{5}y = -20$
$\frac{1}{4}x - \frac{1}{5}y = 20$
$(0, -100)$

Use a system of equations to solve each problem.

2a. The sum of two numbers is 27. Their difference is 5. Find the numbers. 11, 16

2b. The sum of two numbers is 128. Their difference is 114. Find the numbers. 7, 121

❶ To add two equations, first add the left sides. Then add the right sides. The sums are equal. That is, if $a = b$ and $c = d$ then $a + c = b + d$.

Suggested Daily Assignments

Enriched
  Written: 1–25 odds

Average
  Exp: 7–9
  Written: 1–21

Fundamental
  Exp: 1–9
  Written: 1–19
    p. 344, 1–13 odds

❷ You may want to have students solve the problems in the Exploratory Exercises. The solutions are given below.

1. $\left(\frac{11}{3}, \frac{2}{3}\right)$
2. (2, 1)
3. (1, 3)
4. $\left(\frac{5}{2}, \frac{9}{2}\right)$
5. $\left(\frac{1}{5}, 8\right)$
6. (5, 2)
7. $\left(\frac{9}{2}, -3\right)$
8. $\left(-\frac{8}{7}, -\frac{90}{7}\right)$
9. (8, −2)

Calculator Notes

The elimination method for solving systems of equations, using pencil and paper, is practical only when the coefficients are numbers between −15 and 15. With greater coefficients, it is much more practical to use a calculator.

Exploratory Exercises ❷

State whether addition can be used to solve each system.

1. $2x + y = 8$, $x - y = 3$ yes
2. $m + 3n = 5$, $n = 3 - m$ no
3. $3a + b = 6$, $4a + b = 7$ no
4. $y + 3x = 12$, $3y - 3x = 6$ yes
5. $y + 5x = 9$, $y = 7 + 5x$ yes
6. $x = 5$, $x + y = 7$ no
7. $y = -3$, $2x + y = 6$ no
8. $\frac{1}{2}x - \frac{1}{5}y = 2$, $3x + \frac{1}{5}y = -6$ yes
9. $\frac{1}{2}x + \frac{1}{2}y = 3$, $\frac{1}{4}x - \frac{1}{2}y = 3$ yes

Written Exercises

Solve each system by elimination.

1. $x + y = 7$, $x - y = 9$ (8, −1)
2. $r - s = -5$, $r + s = 25$ (10, 15)
3. $-y + x = 6$, $x + y = 5$ $\left(\frac{11}{2}, -\frac{1}{2}\right)$
4. $m - n = 3$, $n + m = 3$ (3, 0)
5. $3x = 13 - y$, $2x - y = 2$ (3, 4)
6. $x + y = 8$, $2x - y = 6$ $\left(\frac{14}{3}, \frac{10}{3}\right)$
7. $2x - 3y = -4$, $x = 7 - 3y$ (1, 2)
8. $5s + 4t = 12$, $3s = 4 + 4t$ $\left(2, \frac{1}{2}\right)$
9. $0.6m - 0.2n = 0.9$, $0.3m = 0.9 - 0.2n$ (2, 1.5)
10. $x + y = 0$, $\frac{1}{3}x = y$ (0, 0)
11. $8x - 3y = 5$, $\frac{1}{3}x + 3y = 20$ $\left(3, \frac{19}{3}\right)$
12. $9x + 2y = 8$, $\frac{3}{2}x - 2y = 6$ $\left(\frac{4}{3}, -2\right)$
13. $4x - \frac{1}{3}y = 8$, $5x + \frac{1}{3}y = 6$ $\left(\frac{14}{9}, -\frac{16}{3}\right)$
14. $\frac{1}{2}x - \frac{1}{2}y = 10$, $\frac{3}{4}x + \frac{1}{2}y = 20$ (24, 4)
15. $\frac{1}{3}x + \frac{1}{5}y = 5$, $\frac{1}{3}x - \frac{1}{5}y = -5$ (0, 25)

Use a system of equations to solve each problem.

16. The sum of two numbers is 45. Their difference is 7. Find the numbers. 19, 26
17. The sum of two numbers is 48. Their difference is 24. Find the numbers. 12, 36
18. The sum of two numbers is 64. Their difference is 42. Find the numbers. 11, 53
19. The sum of two numbers is 101. Their difference is 25. Find the numbers. 38, 63
20. A room is 12 feet longer than it is wide. Half the perimeter of the rectangular floor is 72 feet. Find the length and width. 42 ft, 30 ft
21. Half the perimeter of a rectangular garden is 100 m. The garden is 10 m longer than it is wide. Find its length and width. 55 m, 45 m
22. The difference between twice one number and a lesser number is 21. The sum of the lesser number and twice the greater number is 27. Find the numbers. 3, 12
23. The difference between twice one number and a lesser number is 32. The sum of the lesser number and twice the greater number is 60. Find the numbers. 14, 23
24. Sara McMenemy's farm is 82 acres larger than Jim Underwood's farm. Together the farms contain 276 acres. How large is each farm? 179 acres, 97 acres
25. At a meeting for mathematics editors, there were 7 more women than men. There were 43 editors present. How many men and how many women were there? 18 men, 25 women

Objective 11-5: Solve systems of equations by the elimination method, using multiplication along with addition.

11-5 Elimination Using Multiplication

Some systems of equations cannot be solved directly by addition. Consider the following system.

$$\begin{aligned} 6x + 12y &= -5 \\ 6x + 9y &= -3 \end{aligned}$$

Would a variable be eliminated by addition?

If either equation is multiplied by -1, the system can be solved by addition.

❶ $6x + 12y = -5$ **Multiply by -1.** $-6x - 12y = 5$ *Then, add.*

$6x + 9y = -3$ $\quad 6x + 9y = -3$

$$\begin{aligned} -6x - 12y &= 5 \\ 6x + 9y &= -3 \\ \hline -3y &= 2 \\ y &= -\tfrac{2}{3} \end{aligned}$$

Substitute $-\frac{2}{3}$ for y in either of the original equations and solve for x.

$$\begin{aligned} 6x + 12y &= -5 \\ 6x + 12\left(-\tfrac{2}{3}\right) &= -5 \\ 6x - 8 &= -5 \\ 6x &= 3 \\ x &= \tfrac{1}{2} \end{aligned}$$

The solution of the system is $\left(\frac{1}{2}, -\frac{2}{3}\right)$. *Check this result.*

Example

1 Use elimination to solve the system of equations $\frac{1}{3}x + 2y = 6$ and $x + 3y = -6$.

Multiply the first equation by -3.

$\frac{1}{3}x + 2y = 6$ **Multiply by -3.** $-x - 6y = -18$ *Then, add.*

$x + 3y = -6$

$$\begin{aligned} -x - 6y &= -18 \\ x + 3y &= -6 \\ \hline -3y &= -24 \\ y &= 8 \end{aligned}$$

Substitute 8 for y in either of the original equations and solve for x.

$$\begin{aligned} x + 3y &= -6 \\ x + 3(8) &= -6 \\ x + 24 &= -6 \\ x &= -30 \end{aligned}$$

The solution to the system of equations is $(-30, 8)$. *Check this result.*

11-5

Lesson Notes, page T46

Readiness Activity

State the integer by which you would multiply the first integer in each of the following pairs to make it the additive inverse of the second integer.

$-5, 10$	2	$1, -6$	6
4, 4	-1	3, 6	-2
$-3, 15$	5	2, 8	-4

Discussion Question

If the graph of one of the equations in a system is a vertical or horizontal line, what method would you use to solve the system?

Answers will vary; however, substitution is easiest.

Chalkboard Exercises

Use elimination to solve each system of equations.

1a. $x + y = 8$
$2x - 3y = -9$ (3, 5)

1b. $x - y = 6$
$2x - 2y = 12$
An identity occurs. The system has infinitely many solutions.

1c. $3x + 2y = 6$
$-6x - 4y = 3$
A false sentence occurs. There are no solutions.

1d. $2x + y = 5$
$3x - 2y = -3$ (1, 3)

❶ Another approach is to subtract the bottom equation from the top equation.

Chalkboard Exercises

Use elimination to solve each system of equations.

2a. $3x + 5y = 11$
$2x + 3y = 7$ (2, 1)

2b. $4x + 3y = 8$
$3x - 5y = -23$ (−1, 4)

2c. $2x - 3y = 8$
$-5x + 2y = 13$ (−5, −6)

❷ Emphasize that there are many other combinations of multipliers that can be used. You may want to have students suggest other multipliers.

Suggested Daily Assignments

Enriched
Written: 11–41 odds, 42–45
Mini-Review: 1–5

Average
Exp: 2–12 evens
Written: 1–27 odds, 28–37
p. 346, 22–25
Mini-Review: 1–5

Fundamental
Exp: 1–12
Written: 1–20, 28–33
p. 346, 20–25
Mini-Review: 1–5

In some cases it is necessary to multiply *each* equation by some number in order to solve the system by addition. Consider the following example.

Example

2 Use elimination to solve the system of equations $3a + 4b = -25$ and $2a - 3b = 6$.

Multiply the first equation by 2 and the second equation by −3 to eliminate *a*.

$3a + 4b = -25$ → Multiply by 2. → $6a + 8b = -50$ ❷

$2a - 3b = 6$ → Multiply by −3. → $-6a + 9b = -18$ *Add.*

$$17b = -68$$
$$b = -4$$

Substitute −4 for *b* in either of the equations and solve for *a*.

$$\begin{aligned} 3a + 4b &= -25 \\ 3a + 4(-4) &= -25 \\ 3a - 16 &= -25 \\ 3a &= -9 \\ a &= -3 \end{aligned}$$

The solution to the system of equations is $(-3, -4)$.

The above system of equations can also be solved by eliminating *b*. Multiply the first equation by 3 and the second equation by 4.

$3a + 4b = -25$ → Multiply by 3. → $9a + 12b = -75$ ❷

$2a - 3b = 6$ → Multiply by 4. → $8a - 12b = 24$

$$17a = -51$$
$$a = -3$$

Substitute −3 for *a* in either of the equations and solve for *b*.

$$\begin{aligned} 2a - 3b &= 6 \\ 2(-3) - 3b &= 6 \\ -6 - 3b &= 6 \\ -3b &= 12 \\ b &= -4 \end{aligned}$$

The solution to the system of equations is $(-3, -4)$. Notice that the same result is obtained.

Exploratory Exercises

Explain how to eliminate the variable x in each system. Then, do the same for the variable y.

1. $x + 2y = 5$
   $3x + y = 7$
2. $4x + y = 8$
   $x - 7y = 2$
3. $x + 8y = 3$
   $4x - 2y = 7$
4. $y + x = 9$
   $2y - x = 1$
5. $4x - y = 4$
   $x + 2y = 3$
6. $y - 4x = 11$
   $2y + x = 6$
7. $3y - 8x = 9$
   $y - x = 2$
8. $0.6y + 0.9x = 1.1$
   $y + x = 2$
9. $\frac{3}{2}x + y = \frac{3}{2}$
   $\frac{3}{4}x + y = \frac{1}{4}$
10. $2x + y = 6$
    $3x - 7y = 9$
11. $1.2x + 1.4y = 3.6$
    $x - 2.8y = 7.3$
12. $3x + \frac{2}{3}y = 4$
    $4x + \frac{2}{3}y = 6$

Written Exercises

Use elimination to solve each system of equations.

1. $x - y = 6$, $x + y = 5$ $\left(\frac{11}{2}, -\frac{1}{2}\right)$
2. $x + y = 8$, $2x - y = -6$ $\left(\frac{2}{3}, \frac{22}{3}\right)$
3. $y = 2x$, $2x + y = 10$ $\left(\frac{5}{2}, 5\right)$
4. $x + y = 1$, $y = x$ $\left(\frac{1}{2}, \frac{1}{2}\right)$
5. $3x + 3y = 6$, $2x - y = 1$ $(1, 1)$
6. $3x + 4y = 7$, $3x - 4y = 8$ $\left(\frac{5}{2}, -\frac{1}{8}\right)$
7. $x + 2y = 8$, $3x + 2y = 6$ $\left(-1, \frac{9}{2}\right)$
8. $3x + 7y = -1$, $6x + 7y = 0$ $\left(\frac{1}{3}, -\frac{2}{7}\right)$
9. $3x + 0.2y = 7$, $3x = 0.4y + 6$ $\left(\frac{20}{9}, \frac{5}{3}\right)$
10. $9x + 8y = 7$, $18x - 14 = 16y$ $\left(\frac{7}{9}, 0\right)$
11. $x - 5y = 0$, $2x - 3y = 7$ $(5, 1)$
12. $5x + 3y = 12$, $4x - 3y = 15$ $(3, -1)$
13. $12x - 9y = 114$, $12x + 7y = 82$ $(8, -2)$
14. $\frac{1}{4}x + y = \frac{7}{2}$, $\frac{1}{2}x - \frac{1}{4}y = 1$ $\left(\frac{10}{3}, \frac{8}{3}\right)$
15. $\frac{1}{3}x - y = -1$, $\frac{1}{5}x - \frac{2}{5}y = -1$ $(-9, -2)$
16. $\frac{2}{5}x - \frac{1}{2}y = 1$, $\frac{1}{5}x + \frac{1}{2}y = -1$ $(0, -2)$
17. $\frac{1}{8}(x + y) = 1$, $x - y = 4$ $(6, 2)$
18. $2x - y = 36$, $3x - 0.5y = 26$ $(4, -28)$
19. $3x + \frac{1}{3}y = 10$, $2x - 5 = \frac{1}{3}y$ $(3, 3)$
20. $\frac{1}{2}x - \frac{2}{3}y = 2\frac{1}{3}$, $\frac{3}{2}x + 2y = -25$ $(-6, -8)$
21. $\frac{2x + y}{3} = 15$, $\frac{3x - y}{5} = 1$ $(10, 25)$
22. $x + y = 6000$, $0.06x + 0.08y = 460$ $(1000, 5000)$
23. $x + y = 20$, $0.4x + 0.15y = 4$ $(4, 16)$
24. $1.6x + 0.4y = 1$, $0.4x - 0.1y = 1$ $\left(\frac{25}{16}, -\frac{15}{4}\right)$
25. $3.2y + 7.3x = 0$, $y = -3.5x$ $(0, 0)$
26. $0.25(x + 4y) = 3.5$, $0.5x - 0.25y = 1$ $\left(\frac{10}{3}, \frac{8}{3}\right)$
27. $0.4x - 0.5y = 1$, $0.4x + y = -2$ $(0, -2)$

Answers to Exploratory

1. x: multiply first equation by 3; subtract.
   y: multiply second equation by 2; subtract.
2. x: multiply second equation by 4; subtract.
   y: multiply first equation by 7; add.
3. x: multiply first equation by 4; subtract.
   y: multiply second equation by 4; add.
4. x: add.
   y: multiply first equation by 2; subtract.
5. x: multiply second equation by 4; subtract.
   y: multiply first equation by 2; add.
6. x: multiply second equation by 4; add.
   y: multiply first equation by 2; subtract.
7. x: multiply second equation by 8; subtract.
   y: multiply second equation by 3; subtract.
8. x: multiply second equation by 0.9; subtract.
   y: multiply second equation by 0.6; subtract.
9. x: multiply second equation by 2; subtract.
   y: subtract.
10. x: multiply first equation by 3; second equation by 2; subtract.
    y: multiply first equation by 7; add.
11. x: multiply second equation by 1.2; subtract.
    y: multiply first equation by 2; add.
12. x: multiply first equation by 4; second equation by 3; subtract.
    y: subtract.

For each problem, define two variables. Then, use a system of equations to solve the problem.

28. The sum of two numbers is 57 and their difference is 7. Find the two numbers. 32, 25

29. The difference between two numbers is 20 and their sum is 48. Find the two numbers. 14, 34

30. The sum of two numbers is 22. Five times one of the numbers equals six times the other number. Find the numbers. 12, 10

31. The difference between two numbers is four. Twice the larger number equals three times the sum of the smaller number and 2. Find the numbers. 6, 2

32. Mr. Abbott's farm is 32 acres smaller than Mr. Collins' farm. The two farms together contain 276 acres. How large is each farm? Mr. Abbott's is 122 acres. Mr. Collins' is 154 acres.

33. A 140-meter rope is cut into two pieces. One piece is four times as long as the other. How long is each piece of rope? 28 m, 112 m

34. The difference between the length and width of a rectangle is 7 cm. The perimeter of the rectangle is 50 cm. Find the length and width of the rectangle. 16 cm, 9 cm

35. Layla is three times as old as Diana. In 10 years, Layla will be twice as old as Diana. Find their present ages. Layla is 30. Diana is 10.

36. A father is three times the age of his son. In six years, the father will be $2\frac{1}{2}$ times the age of the son. Find their present ages. The son is 18. The father is 54.

37. A rectangle has a perimeter of 40 cm. The length of the rectangle is 1 cm less than twice the width. Find the length and width of the rectangle. 13 cm, 7 cm

38. The greater of two numbers is twice the lesser. If the greater is increased by 18, the result is 4 less than 4 times the lesser. Find the numbers. 22, 11

39. The greater of two numbers is 1 less than 4 times the lesser. Three times the lesser number is 4 less than the greater. Find the numbers. 19, 5

40. The perimeter of a rectangle is 86 cm. Twice the width exceeds the length by 2 cm. Find the dimensions of the rectangle. 15 cm × 28 cm

41. The sum of Kari's age and her mother's age is 52. Kari's mother is 20 years older than Kari. How old is each? Kari is 16. Her mother is 36.

Challenge

Use elimination to solve each of the following systems.

42. $\frac{1}{x} + \frac{1}{y} = 7$ $\left(\frac{1}{5}, \frac{1}{2}\right)$
$\frac{2}{x} + \frac{3}{y} = 16$

43. $\frac{1}{x + y} = 2$ $\left(\frac{1}{3}, \frac{1}{6}\right)$
$\frac{1}{x - y} = \frac{1}{y}$

44. Explain how the elimination method shows that a system is inconsistent.

45. Explain how the elimination method shows that a system is dependent.

mini-review

1. State whether $xy = 10$ is a linear equation. no

Name the property shown by each of the following.

2. $5(a + b) = 5a + 5b$ 2. distributive

3. If $a = 7$ and $7 = 9 - 2$, then $a = 9 - 2$. 3. transitive

4. $(3x)y = 3(xy)$ 4. associative property for multiplication

5. $a + (-a) = 0$ 5. additive inverse

For more exercises on solving systems of equations, see exercises 37-52 on page 562.

Answers to Challenge

44. The system is inconsistent if a false statement, such as $2 = 4$, occurs.

45. The system is dependent if a true statement, such as $0 = 0$, occurs.

Quiz
Lessons 11-4 to 11-5

1. Solve the following system of equations by elimination.
$5x + 6y = 3$
$11x + 15 = -6y$ $(-3, 3)$

2. Solve the following system of equations by elimination.
$\frac{1}{4}m - \frac{3}{4}n = -5$
$m + \frac{1}{2}n = 8$ $(4, 8)$

Solve the following system of equations.

3. $\frac{1}{3}x - \frac{5}{6}y = -6$
$4x + 7y = 30$ $(-3, 6)$

Excursions in Algebra — Systems with Three Variables

A system of three equations in three variables can be solved by elimination. Consider the following system of equations.

$$\begin{aligned} x + y + z &= 6 \\ x + 2y - z &= 2 \\ 2x - 3y - z &= -7 \end{aligned}$$

First, eliminate one of the variables. In this case, eliminate z by adding the first equation to each of the other equations.

$$\begin{array}{r} x + y + z = 6 \\ \underline{x + 2y - z = 2} \\ 2x + 3y \quad = 8 \end{array} \qquad \begin{array}{r} x + y + z = 6 \\ \underline{2x - 3y - z = -7} \\ 3x - 2y \quad = -1 \end{array}$$

In this way, a system of two equations in two variables is obtained. Then, solve the new system by elimination.

$2x + 3y = 8$ **Multiply by 2.** $4x + 6y = 16$

$3x - 2y = -1$ **Multiply by 3.** $\underline{9x - 6y = -3}$

$$13x = 13$$
$$x = 1$$

Substitute 1 for x in either of the two equations to solve for y.

$$\begin{aligned} 2x + 3y &= 8 \\ 2(1) + 3y &= 8 \\ 3y &= 6 \\ y &= 2 \end{aligned}$$

Finally, substitute 1 for x and 2 for y in any of the original equations to solve for z.

$$\begin{aligned} x + y + z &= 6 \\ 1 + 2 + z &= 6 \\ z &= 3 \end{aligned}$$

The solution of the system of equations is the ordered triple (1, 2, 3). This can be checked by substituting these values into the original system of equations.

Exercises

Solve each system of equations.

1. $x - y + z = 5$, $2x + y + z = 13$, $4x + y - 2z = 12$

2. $-3x + y - z = 6$, $3x - y + 2z = -7$, $x + y + z = 0$

3. $x + 2y - z = 12$, $x - 2y + z = -8$, $2x - y + z = -3$

4. $2x - 2y - z = -1$, $x + 2y + 2z = 9$, $x + y - z = 8$

5. $x + y + z = 5$, $2x + y - z = -1$, $3x - y + z = 1$

6. $5x + 3y + z = -4$, $3x + 2y - z = -4$, $x - 3y + 5z = 4$

7. $a - 2b + c = -9$, $2b + 3c = 16$, $4b = 8$

8. $a + b + c = 0$, $2a + b - c = 2$, $2a + 2b + c = 5$

9. $m + n + p = 3$, $m - p = 1$, $n - p = -4$

About the Excursion

Discuss how to eliminate one of the variables in a system of three equations. Point out that students should choose a pair of the given equations and eliminate a variable. Then they should choose another pair of the given equations and eliminate the same variable. Ask students to suggest reasons the variable z was eliminated first in the example shown. Explain that it does not matter which variable is eliminated first.

Students should look at the equations and see which variable seems easiest to eliminate. The resulting two equations can then be solved as a system, using the methods introduced in this chapter.

Discuss the term *ordered triple*. Point out that solutions to a system of equations in three variables are written as ordered triples. The first component is the x-coordinate, the second component is the y-coordinate, and the third component is the z-coordinate.

Answers to Exercises

1. (4, 2, 3) **2.** (−1, 2, −1) **3.** (2, 3, −4) **4.** (3, 4, −1) **5.** (0, 2, 3) **6.** (−1, 0, 1) **7.** (−9, 2, 4) **8.** (−8, 13, −5) **9.** (3, −2, 2)

11-6

Lesson Notes, page T46
Enrichment, page T46

Readiness Activity
Graph each inequality.

$y \geq x - 2$

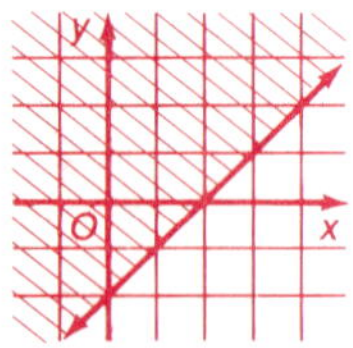

$y < -\frac{x}{3}$

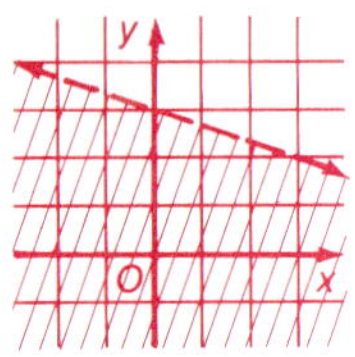

Discussion Question
(Use right before example 1.) Name an ordered pair in the solution set of the system of equations shown at the top of this page. Prove that the point is in the solution set.

This proof can be done by substituting the chosen ordered pair into both inequalities. The ordered pair should make both inequalities into true statements.

Chalkboard Exercises
Solve the system by graphing.

1. $y < -2x + 4$
   $y > 3x - 4$

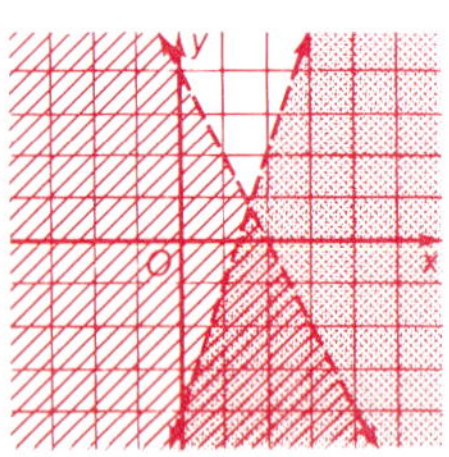

The solution set is represented by the intersection of the shaded regions.

❶ Systems of inequalities are an important aspect of linear programming, which will be introduced in *Merrill Algebra Two.*

Objective 11-6: Solve systems of inequalities by graphing.

11-6 Graphing Systems of Inequalities

Consider the following system of inequalities. ❶

$$y \geq x + 2$$
$$y \leq -2x - 1$$

The solution of this system is the set of all ordered pairs that satisfy both inequalities. To find the solution of this system, graph each inequality. The intersection of the graphs represents the solution of the system.

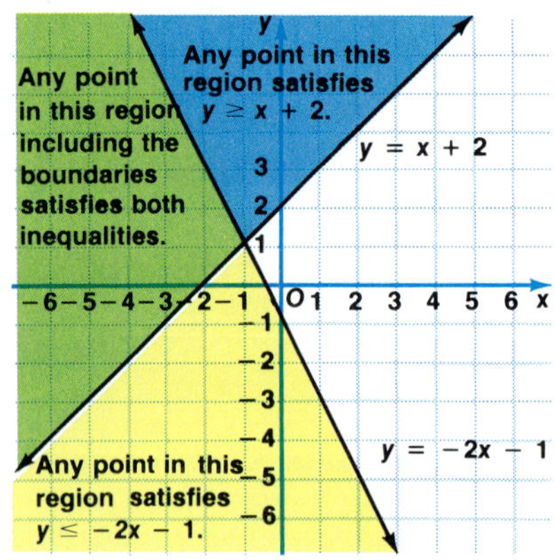

The graphs of $y = x + 2$ and $y = -2x - 1$ are the boundaries of the region. The solution of the system of inequalities is a region that contains an infinite number of ordered pairs.

Example

1 **Solve the system $y > x - 3$ and $y \leq -1$ by graphing.**

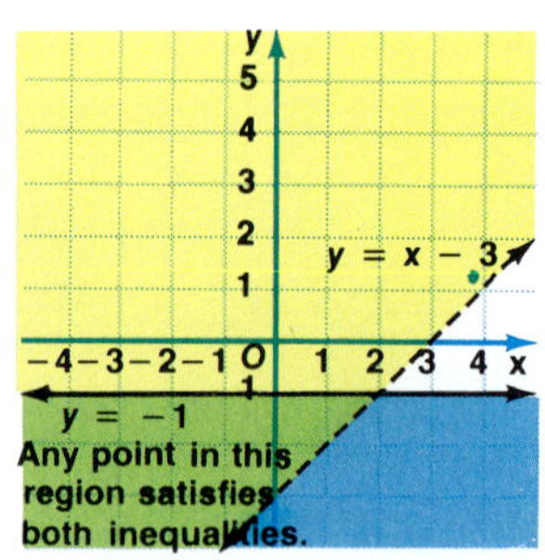

The graphs of $y = -1$ and $y = x - 3$ are boundaries of the region. The graph of $y = x - 3$ is a broken line and is not included in the solution of the system.

A system of inequalities may have no solution. This is illustrated by the next example.

Example

2 **Solve the system $y > x + 2$ and $y < x - 3$ by graphing.**

The graphs of the inequalities do not intersect. Thus, this system has no solutions.

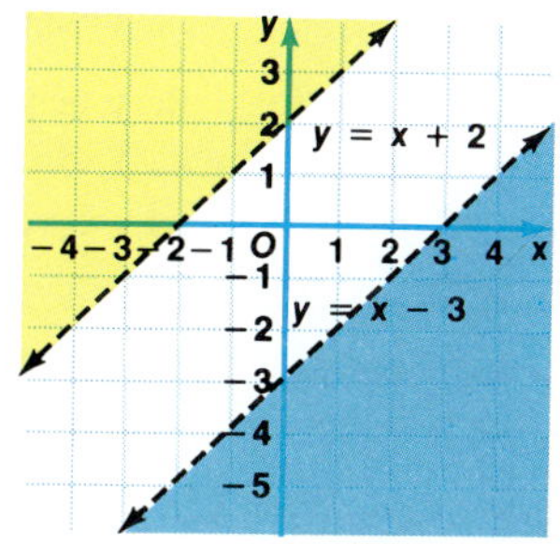

❷

An inequality containing an absolute value expression can be graphed by graphing the equivalent system of inequalities.

❸

Example

3 **Solve the inequality $|x| \leq y$ by graphing.**

The equivalent system of inequalities is $x \leq y$ and $x \geq -y$.
The solution set is represented by the intersection of the shaded regions including the boundaries.

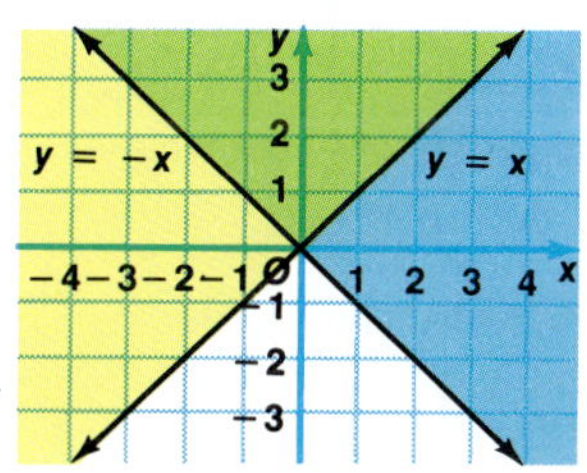

Exploratory Exercises

State whether each ordered pair is a solution of the system $x \leq 3$ and $y > 6$.

1. (3, 7) yes **2.** (6, 2) no **3.** (0, 8) yes **4.** (−3, −3) no

State which region on the graph is the solution of the system of inequalities.

5. $y \geq 2x + 2$ B
$y \leq -x - 1$

6. $y \geq 2x + 2$ A
$y \geq -x - 1$

7. $y \leq 2x + 2$ C
$y \leq -x - 1$

8. $y \leq 2x + 2$ D
$y \geq -x - 1$

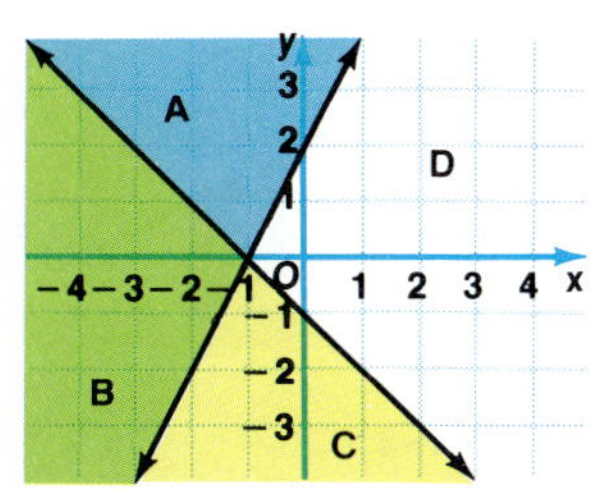

Chalkboard Exercises

Solve each system by graphing.

2. $x + y \geq 5$
$x + y \leq 1$

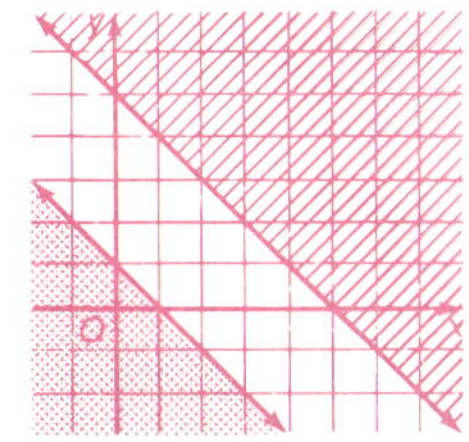

Since the graphs do not intersect, there are no solutions.

3. $|x + 1| < -y$

This is equivalent to the following system:
$y < x + 1$ and $y < -x - 1$

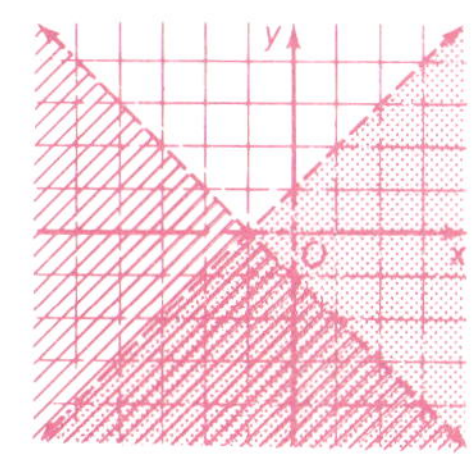

❷ Point out that the upper left region represents the solution set of $y > x + 2$. The lower right region represents the solution of $y < x - 3$. Since the shaded regions do not overlap, the system has no solutions. That is, the solution set is the empty set.

❸ Before introducing this example, review the meaning of open sentences containing absolute value.

Suggested Daily Assignments

Enriched
Exp: 1–8
Written: 2–16 evens, 17–28

Average
Exp: 1–8
Written: 1–26
p. 350, 38–41

Fundamental
Exp: 1–8
Written: 1–20
p. 350, 34–39

Answers to Written
See the Teacher's Answer Key, p. T89, for graphs to exercises 1–20.

Written Exercises

Solve each system by graphing.

1. $x > 3$, $y < 6$
2. $y < -2$, $x < -3$
3. $y > 0$, $x \leq 0$
4. $x > y$, $y < 4$
5. $y > 2$, $y > -x + 2$
6. $y \geq 2x + 1$, $y \leq -x + 1$
7. $y < -2$, $y - x > 1$
8. $x \leq 2$, $y - 3 \geq 5$
9. $y \leq x + 3$, $y \geq x + 2$
10. $x \geq 1$, $y + x \leq 3$
11. $y \geq 3x$, $3y \leq 5x$
12. $y \geq x - 3$, $y \geq -x - 1$
13. $y > x + 1$, $y < x + 3$
14. $y - x < 1$, $y - x > 3$
15. $2y + x < 4$, $3x - y > 6$
16. $y + 2 < x$, $2y - 3 > 2x$

Solve each inequality by graphing.

17. $|y| < x$
18. $|x| \geq y$
19. $|y| + 1 < x$
20. $|x - 1| < y$

Write a system of inequalities for each graph.

21. $x \leq 2$ $y \geq 2$

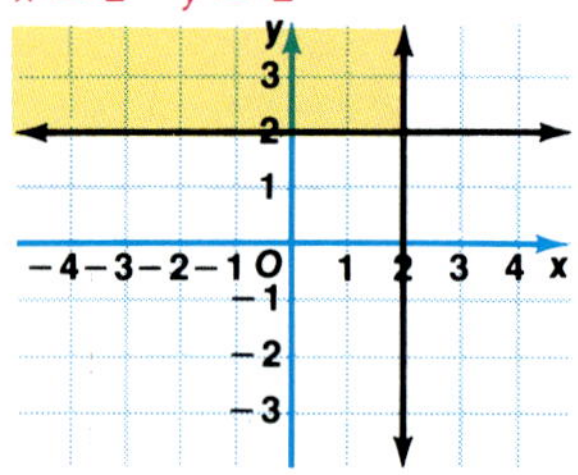

22. $x \leq -1$ $y \geq x$

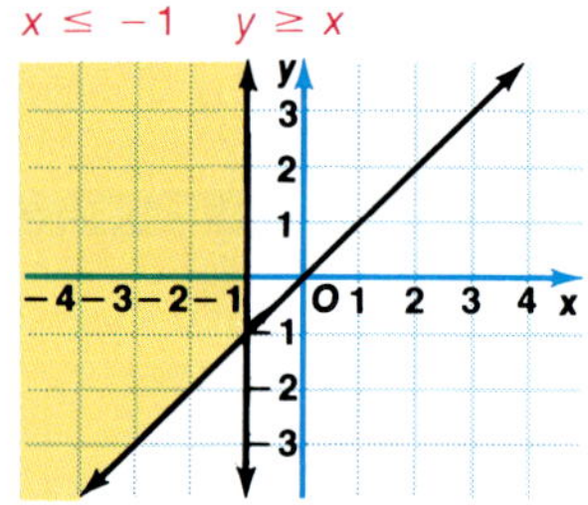

23. $y \geq -x$ $y < 3 - x$

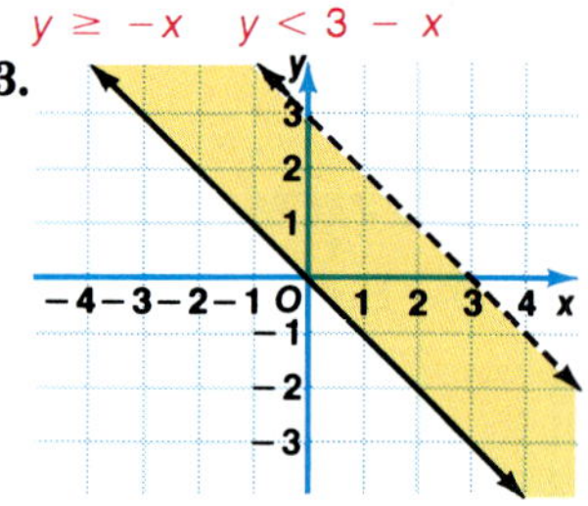

24. 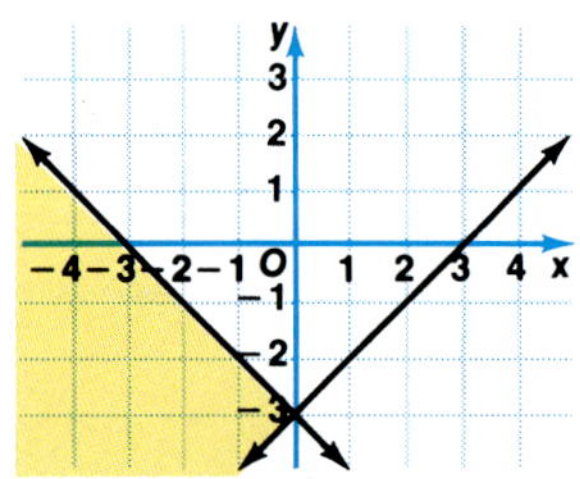

$y \geq x - 3$ $y \leq -x - 3$

25. 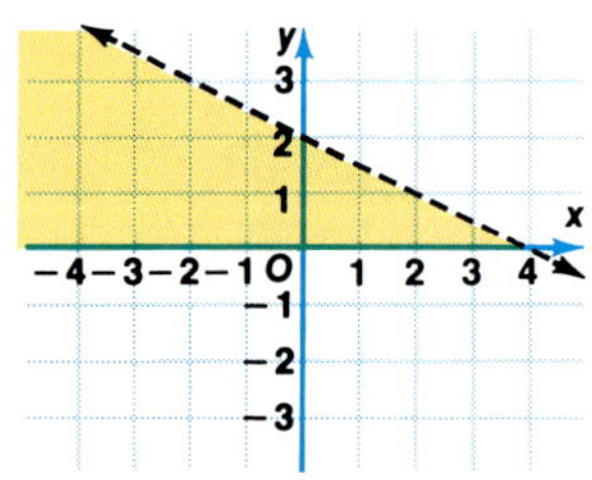

$y < -\frac{1}{2}x + 2$ $y \geq 0$

26. 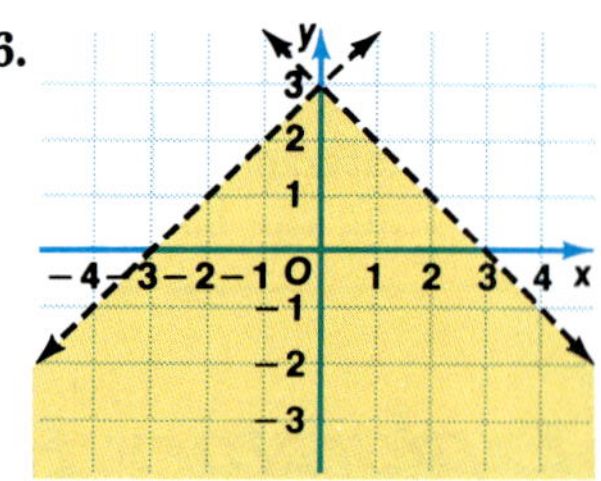

$y < 3 - x$ $y < 3 + x$

Challenge

Write a system of inequalities for each graph.

27. 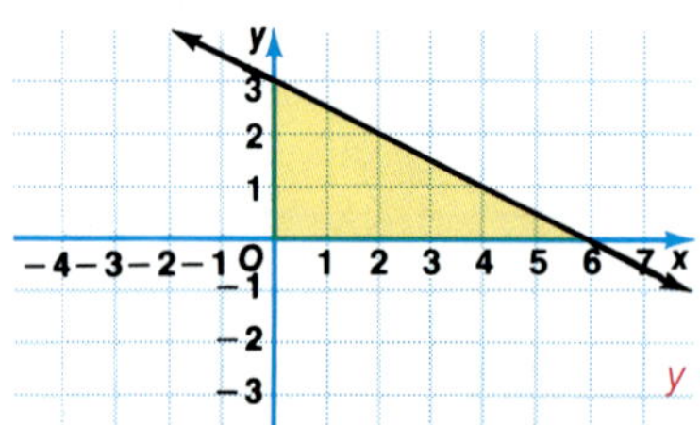

$y \leq -\frac{1}{2}x + 3$
$y \geq 0$
$x \geq 0$

28. 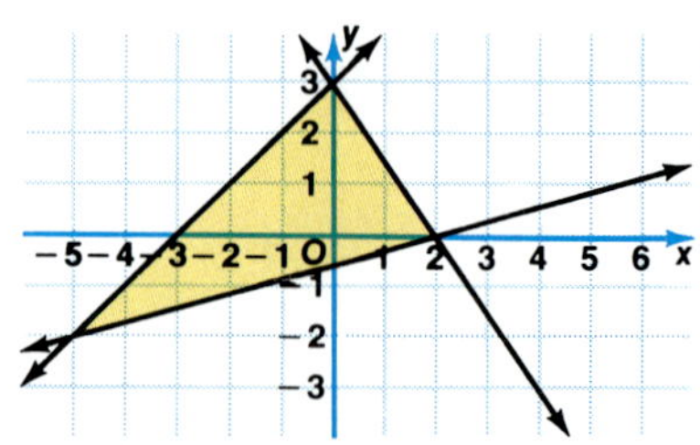

$y \leq x +$
$y \geq \frac{2}{7}x$
$y \leq -\frac{3}{2}$

Problem Solving

Using a Chart

Organizing information into a chart can be a helpful tool in problem solving. Consider this problem.

Mrs. Ruben, a grocer, mixes peanuts that sell for \$1.65 a pound and almonds that sell for \$2.10 a pound. She wants to make 30 pounds of the mixture that sells for \$1.83 a pound. How many pounds of each should Mrs. Ruben include in the mixture?

Explore First identify the variables.
Let x = the number of pounds of peanuts to use.
Let y = the number of pounds of almonds to use.

Plan Organize the variables and the information from the problem into a chart.

	Number of Pounds ×	*Cost per Pound* =	*Total Cost*
Peanuts	x	1.65	$1.65x$
Almonds	y	2.10	$2.10y$
Mixture	30	1.83	1.83(30)

Note that the quantities in red were obtained by multiplying the number of pounds by the cost per pound.

Using the first and last columns of the chart, write two equations.

$$x + y = 30$$
$$1.65x + 2.10y = 1.83(30)$$

The mixture weighs 30 pounds. The cost of the peanuts plus the cost of the almonds equals the cost of the mixture.

Solve the system of equations to find the answer.

Exercises

Complete each chart and write two equations for each problem.

1. Carla invested \$7000, part at 6% and the rest at 10%. The total yearly interest was \$580. How much did she invest at each rate?

	Principal ×	*Rate* =	*Interest*
First part	x	.06	$.06x$
Second part	y	.10	$.10y$
Total	7000		580

$x + y = 7000$

$.06x + .10y = 580$

2. A barge travels 32 miles up a river in 4 hours. The return trip takes 2 hours. Find the rate of the barge in still water and the rate of the current.

	Rate ×	*Time* =	*Distance*
Up River	$x - y$	4	32
Down River	$x + y$	2	32

$4(x - y) = 32$

$2(x + y) = 32$

About Problem Solving
Remind students that they have used charts similar to these in previous lessons. Notice that students are asked only to fill in the charts and then to write the system of equations. You may want to encourage your better students to solve the systems of equations. However, problems similar to the exercises are presented in Lesson 11–8, at which point all students are asked to solve the systems. Thus, this feature is an excellent lead-in to Lesson 11–8.

11-7

Lesson Notes, page T46

Discussion Question
A palindromic number is one that is unchanged if its digits are reversed. Find a palindromic number for which the sum of the digits is 5. Answers will vary. Typical answers are 212, 10301, 1011101.

Chalkboard Exercises
Use a system of equations to solve each problem.

1a. The sum of the digits of a two-digit number is 12. The number is 13 times the tens digit. Find the number. 39

1b. The sum of the digits of a two-digit number is 8. The number is 11 times the tens digit. Find the number. 44

2a. The tens digit of a two-digit number is twice the units digit. If the digits are reversed, the new number is 18 less than the original number. Find the original number. 42

2b. The tens digit of a two-digit number is half of the units digit. If the digits are reversed, the new number is 27 more than the original. Find the original number. 36

❶ A digit is one of the numerals 0 through 9. A two-digit number means a whole number that can be named using two digits.

Objective 11-7: Use systems of equations to solve digit problems.

11-7 Problem Solving: Digit Problems

Many problems relating to the digits of a number can be solved by using systems of equations. ❶

Consider the two-digit number 86.

$$86 = 10(8) + 6$$

Let t = the tens digit.
Let u = the units digit.

Then a two-digit number can be represented as

$$10(t) + u.$$

Examples

1 **The sum of the digits of a two-digit number is 9. The number is 6 times the units digit. Find the number.**

Any two-digit number can be represented by $10t + u$.

Explore Let t = the tens digit.
Let u = the units digit.

Plan $t + u = 9$ *Write a system of equations.*
$10t + u = 6u$

Solve Solve $t + u = 9$ for u and use substitution.

$$t + u = 9 \qquad u = 9 - t$$

$$\begin{aligned} 10t + u &= 6u \\ 10t + (9 - t) &= 6(9 - t) \\ 10t + 9 - t &= 54 - 6t \\ 15t &= 45 \\ t &= 3 \end{aligned}$$

Solve for u.

$$\begin{aligned} t + u &= 9 \\ 3 + u &= 9 \\ u &= 6 \end{aligned}$$

The number is $10(3) + 6$, or 36.

Examine Notice that the sum of the digits is 9 and that 36 is 6 times the units digit, 6.

2 **The tens digit of a two-digit number is twice the units digit. If the digits are reversed, the new number is 36 less than the original number. Find the original number.**

Explore Let t = the tens digit.
Let u = the units digit.
The original number can be represented by $10t + u$.
The new number can be represented by $10u + t$.

Plan $t = 2u$ — *The tens digit is twice the units digit.*

$10t + u - 36 = 10u + t$ — *The new number is 36 less than the original number.*

Solve Use substitution.

$$\begin{aligned} 10t + u - 36 &= 10u + t \\ 10(2u) + u - 36 &= 10u + 2u \\ 20u + u - 36 &= 12u \\ 21u - 36 &= 12u \\ 9u &= 36 \\ u &= 4 \end{aligned}$$

Solve for t.

$$\begin{aligned} t &= 2u \\ t &= 2(4) \\ t &= 8 \end{aligned}$$

The original number is $10(8) + 4$, or 84.

Examine Notice that the tens digit is twice the units digit. If the digits are reversed the new number is 48, and 48 is 36 less than 84.

Written Exercises

Use a system of equations to solve each problem.

1. A two-digit number is 6 times its units digit. The sum of the digits is 6. Find the number. 24

2. The sum of the digits of a two-digit number is 13. Twice the tens digit is two less than 5 times the units digit. Find the number. 94

3. The tens digit of a two-digit number is 6 more than the units digit. The number is 2 more than 8 times the sum of the digits. Find the number. 82

4. The sum of the digits of a two-digit number is 12. If the digits are reversed, the new number is 18 less than the original number. Find the original number. 75

5. The sum of the digits of a two-digit number is 7. If the digits are reversed, the new number is 3 less than 4 times the original number. Find the original number. 16

6. A bank teller reversed the digits in the amount of a check and overpaid a customer by $9. The sum of the digits in the two-digit amount was 9. Find the amount of the check. $45

7. The sum of the digits of a two-digit number is 12. The units digit is twice the tens digit. Find the number. 48

8. The units digit of a two-digit number exceeds twice the tens digit by one. The sum of the digits is 7. Find the number. 25

Suggested Daily Assignments

Enriched
  Written: 2–26 evens

Average
  Written: 1–20

Fundamental
  Written: 1–18
    p. 354, 21–26

9. The sum of the digits of a two-digit number is 6. If the digits are reversed, the new number is three times the original tens digit. Find the original number. 51

10. The tens digit in a two-digit number exceeds twice its units digit by 1. If the digits are reversed, the number is 4 more than 3 times their sum. Find the original number. 52

11. A two-digit number is equal to 7 times the units digit. If 18 is added to the number, its digits are reversed. Find the number. 35

12. The tens digit of a two-digit number exceeds the units digit by 4. If the digits are reversed, the sum of the new number and the original number is 154. Find the original number. 95

13. The tens digit of a two-digit number exceeds twice the units digit by 1. If the digits are reversed, the sum of the new number and the original number is 143. Find the original number. 94

14. The ratio of the tens digit to the units digit of a two-digit number is 1 to 4. The number formed by reversing the digits is 2 less than 3 times the original number. Find the original number. 28

15. The sum of the digits of a two-digit number is 9. If 45 is subtracted from the number, the digits of the number are reversed. Find the number. 72

16. The sum of the digits of a three-digit number is 11. The tens digit is 3 times the hundreds digit and twice the units digit. Find the number. 263

17. The units digit of a three-digit number is 3. The sum of its digits is 9. If the units and hundreds digits are reversed, the sum of the new number and the original number is 909. Find the original number. 603

18. The units digit of a three-digit number is 5. The sum of its digits is 11. If the units and hundreds digits are reversed, the sum of the new number and the original number is 787. Find the original number. 245

19. The tens digit of a three-digit number is 7. The sum of its digits is 10. If the tens and hundreds digits are reversed, the sum of the new number and the original number is 1100. Find the original number. 370

20. The tens digit of a three-digit number is 3. The sum of its digits is 18. If the tens and hundreds digits are reversed, the sum of the new number and the original number is 1008. Find the original number. 639

21. The units digit of a two-digit number divided by the tens digit is 3. The sum of the digits is 12. Find the number. 39

22. The units digit of a two-digit number divides the tens digit twice with a remainder of 1. The sum of the digits is 10. Find the number. 73

23. Suppose that in a certain two-digit positive integer, the digits are reversed. The result is 36 less than the original integer. Find all integers for which this is true. 40, 51, 62, 73, 84, 95

24. If the digits of a two-digit positive integer are reversed, the result is 6 less than twice the original. Find all integers for which this is true. 24

25. The numerator of a fraction is a positive two-digit integer. The denominator is found by reversing the digits of the numerator. If the value of the fraction is $\frac{7}{4}$, find all fractions that satisfy these conditions. $\frac{21}{12}, \frac{42}{24}, \frac{63}{36}, \frac{84}{48}$

26. The sum of the digits of a three-digit number is 12. The tens digit is three less than the hundreds digit and the units digit is three times the hundreds digit. Find the number. 309

Objective 11-8: Use systems of equations to solve uniform-motion problems and mixture problems.

11-8 Problem Solving: Using Systems of Equations

Systems of equations can be used to solve many different problems. Study these examples. ❶

Example 1

A barge travels 24 miles up a river in 3 hours. The return trip takes 2 hours. Find the rate of the barge in still water and the rate of the current.

Explore Let b = the rate of the barge in still water.
Let c = the rate of the current.

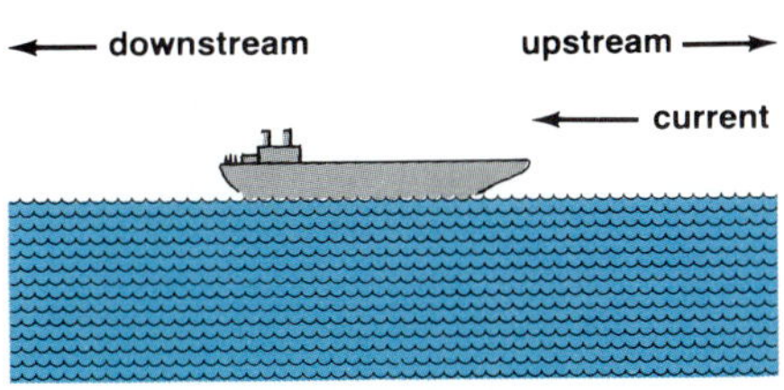

Because of the current, the barge travels slower upstream than downstream.

Use the formula distance = rate × time.

❷

	r	$\times t$	$= d$
Upstream	$b - c$	3	24
Downstream	$b + c$	2	24

Use a chart to organize the information.

Plan

$$3(b - c) = 24 \rightarrow 3b - 3c = 24$$
$$2(b + c) = 24 \rightarrow 2b + 2c = 24$$

Solve

$3b - 3c = 24$ Multiply by 2. $6b - 6c = 48$

$2b + 2c = 24$ Multiply by 3. $6b + 6c = 72$ *Now add.*

$$12b = 120$$
$$b = 10$$

Solve for c.

$$3b - 3c = 24$$
$$3(10) - 3c = 24$$
$$30 - 3c = 24$$
$$-3c = -6$$
$$c = 2$$

Examine The rate of the barge in still water is 10 mph.
The rate of the current is 2 mph.

11-8

Lesson Notes, page T47

Discussion Question

Two girls ran a race and one beat the other by 2 seconds. What other information is needed to determine the times of the two girls? Their rates.

Chalkboard Exercises

Use a system of equations to solve each problem.

1a. A barge travels 32 miles up a river in 4 hours. The return trip takes 2 hours. Find the rate of the barge in still water and the rate of the current. 12 mph, 4 mph

1b. A barge travels 45 miles up a river in 5 hours. The return trip takes 3 hours. Find the rate of the barge in still water and the rate of the current. 12 mph, 3 mph

❶ Encourage students to use charts when solving the problems in this lesson.

❷ Discuss this chart. What is the rate of the barge when it travels upstream? Why is the rate $b - c$? What is the rate of the barge when it travels downstream? Why is the rate $b + c$?

Chalkboard Exercises

Use a system of equations to solve each problem.

2a. Solve example 2 for the same data except that the alloy is to be 35% copper instead of 45%. 600 g, 400 g

2b. Solve example 2 for the same data except that the alloy is to be 38% copper. 480 g, 520 g

2c. Solve example 2 for the same data except that the alloy is to be 44% copper. 240 g, 760 g

2d. Solve example 2 for the same data except that the alloy is to be 32% copper. 720g, 280 g

Suggested Daily Assignments

Enriched
Written: 2–18 evens

Average
Written: 1–18
p. 358, 21–26

Fundamental
Written: 1–12
p. 358, 19–24

Example

2 **A metal alloy is 25% copper. Another alloy is 50% copper. How much of each alloy should be used to make 1000 grams of an alloy that is 45% copper?**

Explore Let a = the number of grams of 25% copper alloy.
Let b = the number of grams of 50% copper alloy.

	25% Copper	*50% Copper*	*45% Copper*
Grams	a	b	1000
Percent	$0.25a$	$0.50b$	0.45(1000)

Plan

$$a + b = 1000$$
$$0.25a + 0.50b = 0.45(1000)$$

Solve Use substitution.

$$a + b = 1000 \qquad 0.25a + 0.50b = 0.45(1000)$$
$$a = 1000 - b \qquad 0.25(1000 - b) + 0.50b = 450$$
$$250 - 0.25b + 0.50b = 450$$
$$0.25b = 200$$
$$b = 800$$

Find the value of a.

$$a + b = 1000$$
$$a + 800 = 1000$$
$$a = 200$$

200 grams of the 25% copper alloy and 800 grams of the 50% copper alloy should be used. *Check this result.*

Written Exercises

Use a system of equations to solve each problem.

1. A boat is rowed 10 miles downstream in two hours, then rowed the same distance upstream in $3\frac{1}{3}$ hours. Find the rate of the boat in still water and the rate of the current. 4 mph, 1 mph

2. While traveling with the wind, a plane flies 300 miles between Chicago and Columbus in 40 minutes. It returns against the wind in 45 minutes. Find the air speed of the plane and the rate of the wind. 425 mph, 25 mph

3. A box contains two different types of candy, weighs 10 pounds, and costs $14.55. One type of candy costs $1.50 a pound. The other type costs $1.35 a pound. How many pounds of each kind are there? 7 lbs., 3 lbs.

4. How many pounds of candy that sells for 80 cents a pound should be mixed with candy that sells for $1.50 a pound to make 20 pounds of a mixture to sell at $1.01 a pound? 14 lbs., 6 lbs.

5. Joe sold 30 peaches from his tree for a total of $7.50. He sold the small ones for 20 cents each and the large ones for 35 cents each. How many of each kind did he sell? 20, 10

6. Louise walks from her home to the city in 4 hours. She can travel the same distance on her bicycle in one hour. If she rides 6 mph faster than she walks, what is her speed on the bicycle? 8 mph

7. An airplane travels 1,800 miles in 3 hours flying with the wind. On the return trip, flying against the wind it takes 4 hours to travel 2,000 miles. Find the rate of the wind and the rate of the plane in still air. 550 mph, 50 mph

8. A gas station attendant has some antifreeze that is 40% alcohol and another type of antifreeze that is 60% alcohol. He wishes to make 1,000 gallons of antifreeze that is 48% alcohol. How much of each kind should he use? 600 gal. of 40% 400 gal. of 60%

9. A man invests $4000, part of it at 10% annual interest and the rest at 12% annual interest. If he receives $460 in interest at the end of one year, how much did he invest at each rate? $1000, $3000

10. Jennifer broke open her "piggy bank" and found 83 coins in nickels and dimes. If she had $6.95 in all, how many coins of each has she? 56 dimes, 27 nickels

11. While driving to Northridge, Mrs. Winters averages 40 mph. On the return trip she averages 56 mph and saves two hours of traveling time. How far from Northridge does she live? 280 mi.

12. Two boys are 15 miles apart. If they walk towards each other they meet in 3 hours. If they both walk in the same direction one boy overtakes the other in 8 hours. How fast does each boy walk? $3\frac{7}{16}$ mph, $1\frac{9}{16}$ mph

13. Two trains start toward each other on parallel tracks at the same time from towns 450 miles apart. One train travels 6 mph faster than the other train. What is the rate of each train if they meet in 5 hours? 42 mph, 48 mph

14. A car traveled due east from a certain point. One hour later another car started from the same point and traveled due west. Five hours after the first car started, they were 392 miles apart. The eastbound car traveled 10 mph faster than the westbound car. What is the speed of each car? 38 mph, 48 mph

15. A riverboat travels 48 miles downstream in the same time that it takes to travel 32 miles upstream. The speed of the boat in still water is 16 miles per hour greater than the speed of the current. Find the speed of the current. 4 mph

16. In still water, a motorboat can travel five times as fast as the current in the river. A trip upriver and back totaling 96 miles can be done in 5 hours. Find the rate of the current. 4 mph

17. When George was 2 miles upstream from the starting point on a canoe trip, he passed a log floating downstream with the current. He paddled upstream for one more hour and returned to the starting point just as the log arrived. What was the rate of the current? 1 mph

18. The Rent-A-Car company rents a compact car for a fixed amount each day plus a fixed amount for each mile driven. Joe spent $134 to rent a car for 5 days and drove it 450 miles. Patti spent $106 to rent the same kind of car for 4 days and drove 350 miles. Find the daily charge and the charge per mile for the compact car. $16, 12c

Quiz
Lessons 11–6 to 11–8

Solve the following system of inequalities by graphing.

1. $x + y < 5$ and $y \geq x - 4$

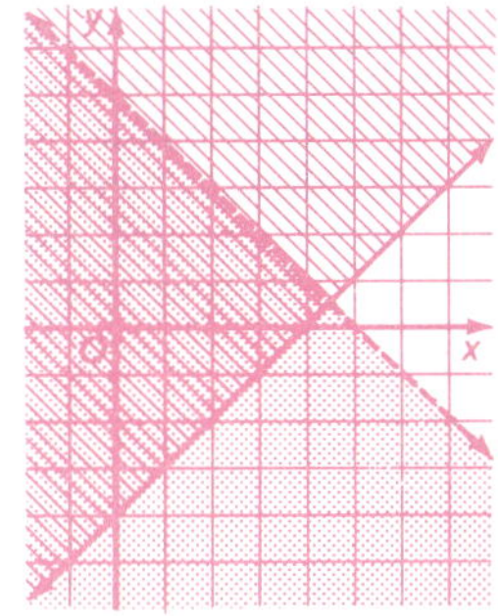

The solution is represented by the intersection of the shaded regions.

Use a system of equations to solve each problem.

2. Alice starts to school at 7:30 a.m. walking at the rate of 3 mph. Her brother starts to school at 7:40 a.m. riding his bike at 6 mph. When will he overtake Alice? 7:50 a.m.
3. The sum of the digits of a two digit number is 9. If the digits are reversed and the resulting number is subtracted from the original number, the result is 27. Find the original number. 63
4. A solution is 25% acid. A chemist wants to make it a 40% acid solution by adding 100% acid. If there are 800 mL of the original solution, how much 100% acid must be added? 200 mL of 100% acid

About the Excursion
Emphasize that in order to use Cramer's Rule, the equations must be in standard form, that is in the form $ax + by = c$. Students may be interested to know that this method of solving systems of equations is a method often used to solve systems using computers. Refer them to Using Computers on page 364.

Excursions in Algebra — Cramer's Rule

A square array of numbers, called a **determinant**, has a number associated with it. For a 2 × 2 array, the value is found by using the diagonals as shown.

$$\begin{vmatrix} 2 & 1 \\ 3 & 4 \end{vmatrix} = 2(4) - 3(1) = 8 - 3 = 5$$

The determinant is denoted by vertical bars.

In general the determinant of a 2 × 2 array is found as follows.

$$\begin{vmatrix} a & b \\ c & d \end{vmatrix} = ad - bc$$ ***Value of a Determinant***

Determinants can be used to solve a system of linear equations. Consider solving the following system of equations.

$$ax + by = c$$
$$dx + ey = f$$

To solve for the variable x in this system, the variable y is eliminated.

$ax + by = c$ — Multiply by e. — $(ae)x + (be)y = ce$ — *Then, add.*

$dx + ey = f$ — Multiply by $-b$. — $-(bd)x - (be)y = -bf$

$$(ae - bd)x = ce - bf$$
$$x = \frac{ce - bf}{ae - bd}$$

To solve for y, eliminate the variable x.

$ax + by = c$ — Multiply by d. — $(ad)x + (bd)y = cd$ — *Then, add.*

$dx + ey = f$ — Multiply by $-a$. — $-(ad)x - (ae)y = -af$

$$(bd - ae)y = cd - af$$
$$y = \frac{cd - af}{bd - ae} = \frac{af - cd}{ae - bd}$$

The general solution to this system of equations is $\left(\frac{ce - bf}{ae - bd}, \frac{af - cd}{ae - bd}\right)$.

Notice that the denominator of each fraction is the same. It can be written as a determinant.

$$ae - bd = \begin{vmatrix} a & b \\ d & e \end{vmatrix}$$

Each numerator also may be written as a determinant.

$$ce - bf = \begin{vmatrix} c & b \\ f & e \end{vmatrix} \text{ and } af - cd = \begin{vmatrix} a & c \\ d & f \end{vmatrix}$$

Therefore, the solution to a system of equations in two variables can be found using determinants. This method is known as **Cramer's Rule**.

The solution to the system of equations $\begin{matrix} ax + by = c \\ dx + ey = f \end{matrix}$ is (x, y)

where $x = \dfrac{\begin{vmatrix} c & b \\ f & e \end{vmatrix}}{\begin{vmatrix} a & b \\ d & e \end{vmatrix}}$ **and** $y = \dfrac{\begin{vmatrix} a & c \\ d & f \end{vmatrix}}{\begin{vmatrix} a & b \\ d & e \end{vmatrix}}$ **and** $\begin{vmatrix} a & b \\ d & e \end{vmatrix} \neq 0$.

Cramer's Rule

Example:

Use Cramer's rule to solve the system $3x - 5y = -7$ and $x + 2y = 16$.

$$x = \frac{\begin{vmatrix} -7 & -5 \\ 16 & 2 \end{vmatrix}}{\begin{vmatrix} 3 & -5 \\ 1 & 2 \end{vmatrix}} = \frac{(-7)(2) - (16)(-5)}{(3)(2) - (1)(-5)} = \frac{66}{11} = 6$$

$$y = \frac{\begin{vmatrix} 3 & -7 \\ 1 & 16 \end{vmatrix}}{\begin{vmatrix} 3 & -5 \\ 1 & 2 \end{vmatrix}} = \frac{(3)(16) - (1)(-7)}{(3)(2) - (1)(-5)} = \frac{55}{11} = 5$$

The solution is (6, 5).

Exercises

Solve each system by using Cramer's Rule.

1. $2x + 2y = 6$, $3x - 4y = 2$ (2, 1)

2. $x - y = -3$, $4x + 3y = -5$ (−2, 1)

3. $3x + 2y = 10$, $x + y = 6$ (−2, 8)

4. $x - y = 4$, $x - 2y = 8$ (0, −4)

5. $x = 3 - 2y$, $y = 2x - 3$ $\left(\frac{9}{5}, \frac{3}{5}\right)$

6. $3x + 2y = 7$, $4x - y = 2$ (1, 2)

About Using Computers
Point out to students that Cramer's Rule is used in this program. Refer them to the Excursion on Cramer's Rule on pages 362 and 363.

Ask students to explain why line 20 checks whether the denominator is zero.

You may want to have students use their modified program from exercise 4 to check their answers to the exercises from the excursion on page 363.

Answers to Exercises

4.
```
60 GOTO 110
70 IF C * E - B * F = 0 THEN 100
80 PRINT "NO SOLUTION"
90 GOTO 110
100 PRINT "INFINITE NUMBER OF SOLUTIONS"
110 END
```
Now running the program for 1-3:
1. NO SOLUTION
2. INFINITE NUMBER OF SOLUTIONS
3. $(-9, -7)$

Answers to Chapter Review

1.
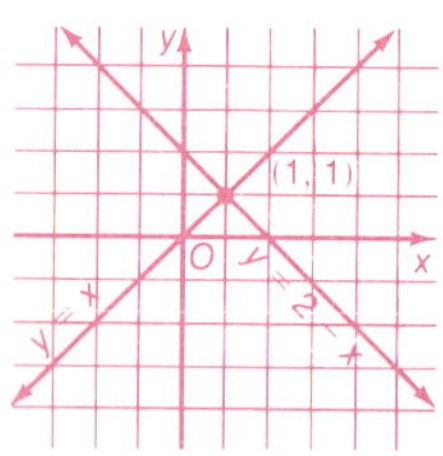

2.
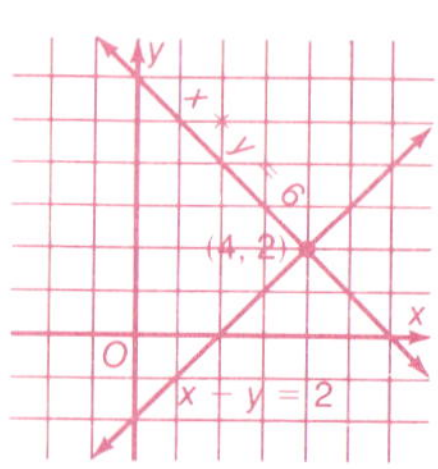

Using Computers

Solving Systems of Equations

Computers are often used as a tool for solving systems of equations. Consider the following system of equations.

$$ax + by = c$$
$$dx + ey = f$$

The general solution to the system of equations is $\frac{ce - bf}{ae - bd}, \frac{af - cd}{ae - bd}$.

This BASIC program will print the solution to a system of equations.

```
10 INPUT A,B,C,D,E,F
20 IF A * E - B * D = 0 THEN 70
30 LET X = (C * E - B * F) / (A * E - B * D)
40 LET Y = (A * F - C * D) / (A * E - B * D)
50 PRINT "(";X;",";Y;") IS A SOLUTION"
60 GOTO 80
70 PRINT "NO UNIQUE SOLUTION"
80 END
```

Line 20 checks whether the denominator is zero.

Notice that line 70 makes no distinction between systems that have no solution and systems that have an infinite number of solutions. Study the following graphs.

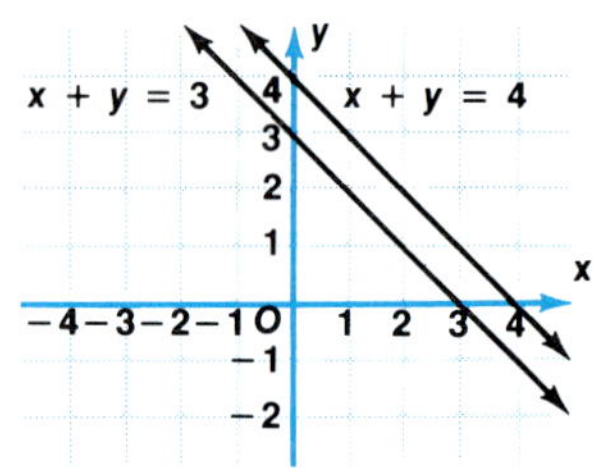

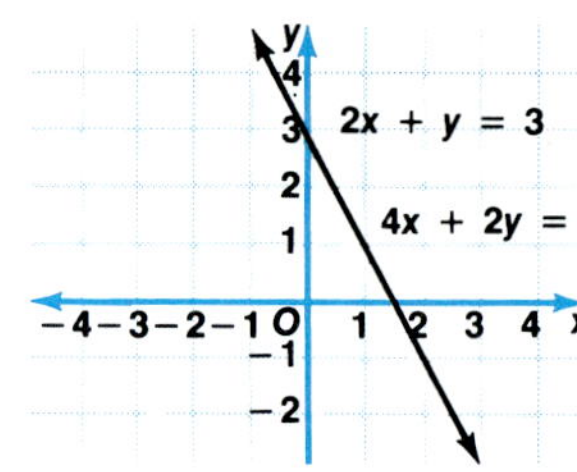

$x + y = 4$ $a = 1, b = 1, c = 4$
$x + y = 3$ $d = 1, e = 1, f = 3$

$ae - bd = 1 \cdot 1 - 1 \cdot 1 = 0$
$ce - bf = 4 \cdot 1 - 1 \cdot 3 = 1$
$af - cd = 1 \cdot 3 - 1 \cdot 4 = -1$

$2x + y = 3$ $a = 2, b = 1, c = 3$
$4x + 2y = 6$ $d = 4, e = 2, f = 6$

$ae - bd = 2 \cdot 2 - 4 \cdot 1 = 0$
$ce - bf = 3 \cdot 2 - 1 \cdot 6 = 0$
$af - cd = 2 \cdot 6 - 3 \cdot 4 = 0$

These and many other examples suggest that if only $ae - bd = 0$, the lines are parallel and the system has no solution. But if $ce - bf$ is also zero, the lines coincide and the system has an infinite number of solutions.

Exercises Solve each system by using the computer program.

1. $5x + 5y = 16$ no unique solution
   $2x + 2y = 5$
2. $7x - 3y = 5$ no unique solution
   $14x - 6y = 10$
3. $x - 2y = 5$ $(-9, -7)$
   $3x - 5y = 8$
4. Modify the computer program so that it will check whether the system has no solution or an infinite number of solutions.

Vocabulary

system of equations (337)
substitution (342)
elimination (345)
system of inequalities (352)

Chapter Summary

1. A system of equations can be solved by graphing. (337)
2. A system of equations may have either no solution, one solution, or infinitely many solutions. (340)
3. Algebraic methods for solving systems of equations are as follows.
   substitution (342)
   elimination (345)
4. A system of inequalities can be solved by graphing. (352)
5. Systems of equations can be used to solve digit problems (356), uniform motion problems (359), and mixture problems. (360)

Chapter Review

11-1 Solve each system of equations by graphing.

1. $y = x$, $y = 2 - x$ (1, 1)
2. $x + y = 6$, $x - y = 2$ (4, 2)
3. $y = x - 1$, $x + y = 11$ (6, 5)

11-2 Graph each pair of equations. Then state whether the system has one solution, no solution, or infinitely many solutions.

4. $3x + y = -4$, $6x + 2y = -8$ infinitely many
5. $y = 5$, $y = 7$ none
6. $x - y = 10$, $x + y = 11$ one
7. $9x + 2 = 4y$, $x + y = 6$ one

11-3 Solve each system of equations by substitution.

8. $y = x - 1$, $4x - y = 19$ (6, 5)
9. $2m + n = 1$, $m - n = 8$ (3, −5)
10. $x = 2y$, $x + y = 6$ (4, 2)
11. $5x = -10y$, $8x - 4y = 40$ (4, −2)
12. $3m - 2n = -4$, $3m + n = 2$ (0, 2)
13. $3x - y = 1$, $2x + 2y = 2$ $\left(\frac{1}{2}, \frac{1}{2}\right)$

Answers to Chapter Review
(continued)

3.
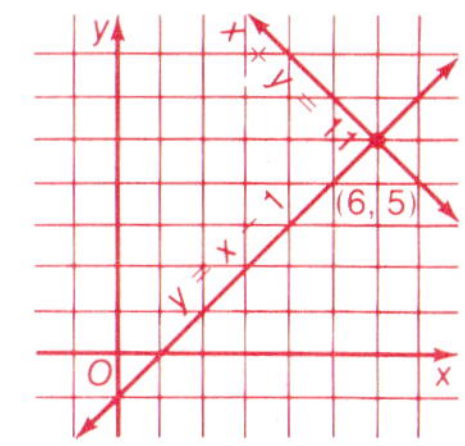

4.
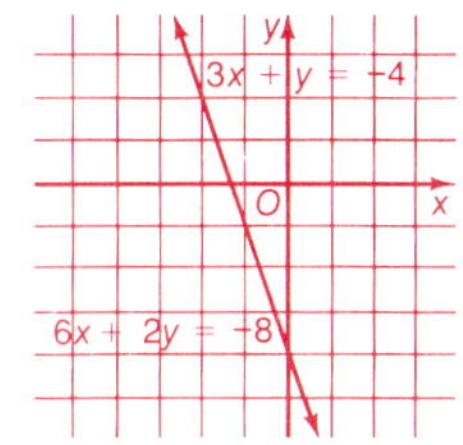

5.
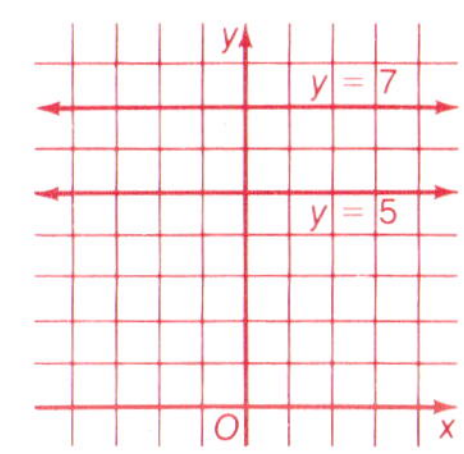

6.
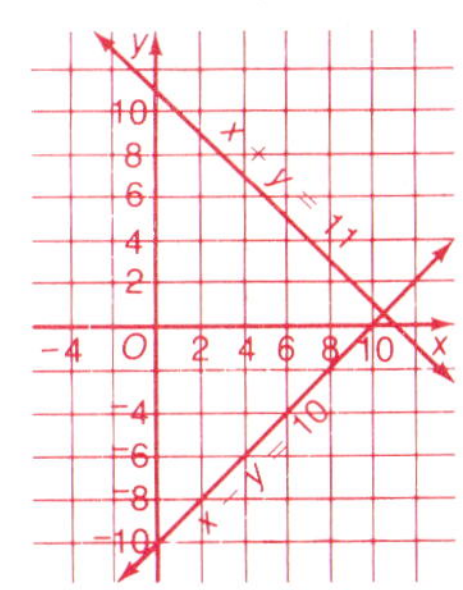

7.
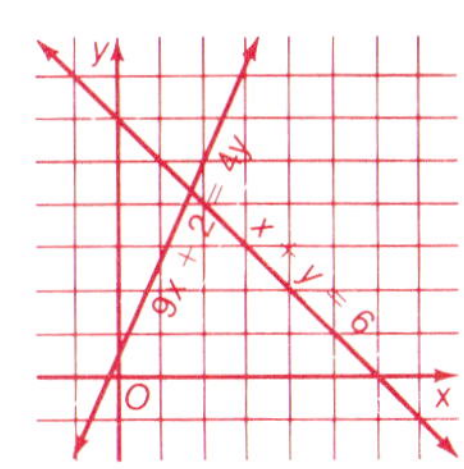

Answers to Chapter Review

The solution is represented by the intersection of the shaded regions.

26.

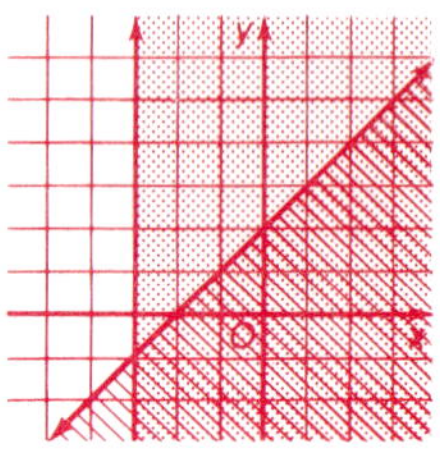

27.

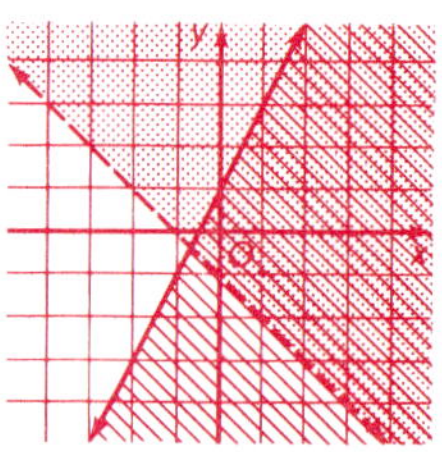

11-4 **Solve each system of equations by elimination using addition.**

14. $x + y = 8$; $x - y = 8$ (8, 0)

15. $2m - n = 4$; $m + n = 2$ (2, 0)

16. $x + 2y = 6$; $-x + 3y = 4$ (2, 2)

17. $2m - n = 5$; $2m + n = 3$ (2, −1)

18. $3x - y = 11$; $x + y = 5$ (4, 1)

19. $6r + 2s = 32$; $-6r + 9s = -21$ (5, 1

11-5 **Solve each system of equations by elimination.**

20. $3x + 3y = 6$; $2x - y = 1$ (1, 1)

21. $x - 2y = 5$; $3x - 5y = 8$ (−9, −7)

22. $2x + 3y = 8$; $x - y = 2$ $\left(\frac{14}{5}, \frac{4}{5}\right)$

23. $6x + 8y = 4$; $2x - y = 5$ (2, −1)

24. $2r + s = 9$; $r + 11s = -6$ (5, −1)

25. $5m + 2n = -8$; $4m + 3n = 2$ (−4, 6

11-6 **Solve each system of inequalities by graphing.**

26. $x \geq -3$; $y \leq x + 2$

27. $y > -x - 1$; $y \leq 2x + 1$

11-7 **Use a system of equations to solve each problem.**

28. The sum of two numbers is 42. Their difference is 18. What are the numbers? 30, 12

29. The difference between the length and width of a rectangle is 7 cm. The perimeter is 50 cm. Find the length and width. 16 cm, 9 cm

30. A two-digit number is 6 times its units digit. The sum of its digits is 3. Find the number. 12

31. A two-digit number is 7 times its units digit. If 18 is added to the number, its digits are reversed. Find the original number. 35

11-8 **32.** Bill mixes candy costing 45¢ per pound with candy costing 65¢ a pound. A 7-pound box of the mix costs \$3.65. How much of each kind of candy should Bill use? $4\frac{1}{2}$ lbs., $2\frac{1}{2}$ lbs.

33. A speedboat travels 60 miles with the current in $1\frac{1}{2}$ hours. The return trip takes 2 hours. What is the rate of the current? 5 mph

Chapter Test

Solve each system of equations by graphing.

1. $y = 3x$ (1, 3)
$x + y = 4$

2. $y = x + 2$ (−5, −3)
$y = 2x + 7$

Graph each pair of equations. Then state whether the system has one solution, no solution, or infinitely many solutions.

3. $3x + y = 5$ infinitely many
$6x + 2y = 10$

4. $x + y = 11$ none
$x + y = 14$

5. $x + y = 5$ one
$x - y = 1$

6. $x = 2y$ one
$x + y = 3$

Solve each system of equations by substitution.

7. $y = 7 - x$ (2, 5)
$y - x = 3$

8. $x + y = 8$ (5, 3)
$x - y = 2$

9. $\frac{1}{8}(x + y) = 1$ (6, 2)
$x - y = 4$

Solve each system of equations by elimination.

10. $-3x - y = -10$ (4, −2)
$3x - 2y = 16$

11. $5x - 3y = 12$ (3, 1)
$2x - 3y = 3$

12. $2x + 5y = 12$ (1, 2)
$x - 6y = -11$

13. $x + y = 6$ $\left(\frac{21}{4}, \frac{3}{4}\right)$
$x - y = 4\frac{1}{2}$

14. $3x + \frac{1}{3}y = 10$ (5, −15)
$2x + \frac{1}{3}y = 5$

15. $8x - 6y = 14$ (1, −1)
$-6x + 9y = -15$

Use a system of equations to solve each problem.

16. The sum of the digits of a two-digit number is 10. If the digits are reversed, the new number is 18 less than the original number. Find the original number. 64

17. Mr. Salvatore mixes nuts costing \$0.90 per pound and nuts costing \$1.30 per pound. He wishes to make 50 pounds that cost \$1.20 per pound. How many pounds of each should he use? $12\frac{1}{2}$ lbs., $37\frac{1}{2}$ lbs.

18. The units digit of a two-digit number exceeds twice the tens digit by 1. The sum of the digits is 10. Find the number. 37

19. Jodi rode her bicycle against the wind for a distance of 15 km in 1 hour. The return trip took 36 minutes. What was the rate of the wind? 5 mph

20. Solve the system of inequalities $y > -x + 2$ and $y \le -3$ by graphing.

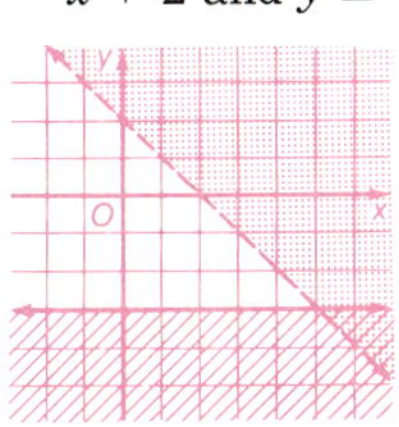

Answers to Chapter Test

1.
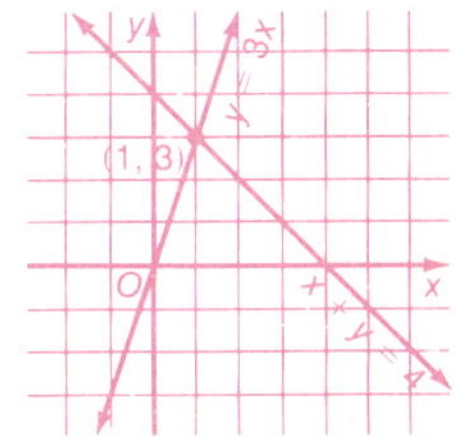

2.
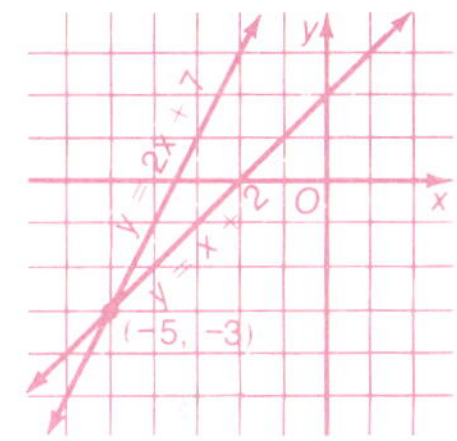

3.
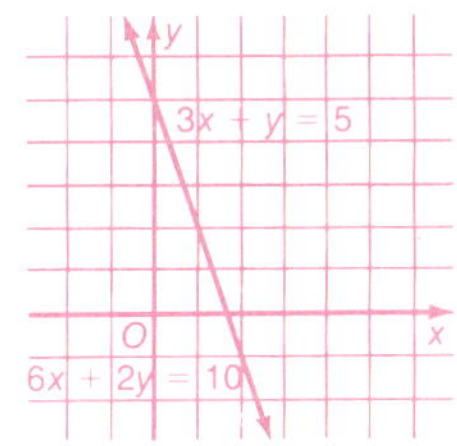

4.
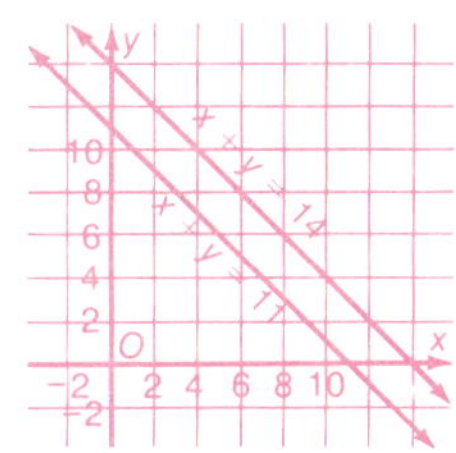

5.
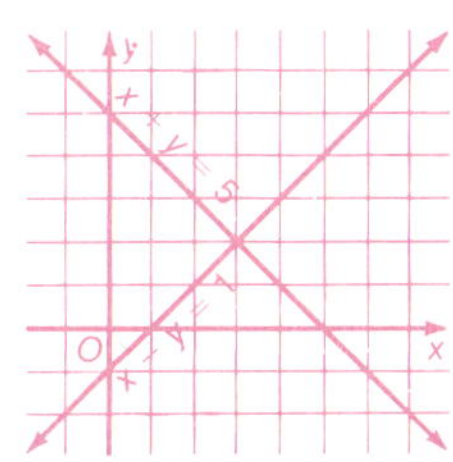

6.
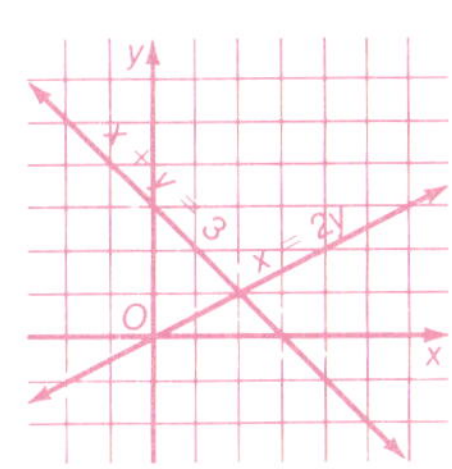

About the Application
Another *Applications in Statistics* feature can be found on page 302. *Applications in Probability* features can be found on pages 108, 172, and 242.

Point out to students that the mean of a set of data is also called an average.

Applications in Statistics — Central Tendencies

A measure of **central tendency** is a number that represents a set of data. Measures of central tendency may be used to *summarize* a set of data or to *compare* one set of data with another. Three measures of central tendency are the **median**, **mode**, and **mean**. They are defined as follows.

Definition of Mean, Median, and Mode

The mean of a set of data is the sum of all the values divided by the number of values.

The median of a set of data is the middle value. If there are two middle values, it is the value halfway between them.

The mode of a set of data is the value that appears more frequently than any other. Some sets of data have multiple modes.

Example: **Find the mean, median, and mode of the salaries of the 10 employees of the Rond Company. Three workers earn \$15,500 each, two earn \$13,200 each, and five earn \$17,050 each.**

To find the mean, add all 10 values. You can use multiplication to shorten the addition.

$$\text{mean} = \frac{5(17{,}050) + 3(15{,}500) + 2(13{,}200)}{10}$$
$$= \frac{85{,}250 + 46{,}500 + 26{,}400}{10}$$
$$= \frac{158{,}150}{10} \text{ or } 15{,}815$$

The mean is \$15,815.

To find the median, first arrange the salaries in descending order. The median is the middle value.

Notice that in this example there are two middle values, the 5th value, \$17,050, and the 6th value, \$15,500. The median is the mean of the two values. Therefore, the median is $\frac{17{,}050 + 15{,}500}{2}$ or \$16,275.

17,050
17,050
17,050
17,050
17,050 ← median
15,500 ← median
15,500
15,500
13,200
13,200

More workers make \$17,050 than any other salary. So it is the most frequently occurring value. Thus, the mode is \$17,050.

The median is \$16,275, the mode is \$17,050, and the mean is \$15,815.

The value of every item in a set of data affects the value of the mean. Thus, when extreme values are included, the mean may become less representative of the set. The values of the median and the mode are not affected by extreme values.

Example: **Find the mean of {1, 3, 6, 86} and {33, 34, 34, 35}.**

{1, 3, 6, 86}

$$\text{mean} = \frac{1 + 3 + 6 + 86}{4} = \frac{96}{4} \text{ or } 24$$

The mean is not close to any one of the four values in this set. In this case, it is not a particularly representative value.

{33, 34, 34, 35}

$$\text{mean} = \frac{33 + 34 + 34 + 35}{4} = \frac{136}{4} \text{ or } 34$$

There are no extreme values in this set. In this case, the mean is a representative value.

Exercises

Find the median, mode, and mean for each set of data.

1. {1, 2, 3, 4, 5} 3, none, 3
2. {4, 6, 8, 10, 12} 8, none, 8
3. {3, 3, 5, 6, 1} 3, 3, 3.6
4. {8, 8, 8, 8, 8, 8} 8, 8, 8
5. {7, 56, 8, 49, 3, 0, 10, 23} 9, none, 19.5
6. {1, 6, 4, 2, 10, 12, 13, 12} 8, 12, 7.5
7. {2.4, 8.1, 2.4, 5.7, 6.5} 5.7, 2.4, 5.02
8. {3.3, 1.6, 8.2, 4.9, 6.7} 4.9, none, 4.94

9. Two dice were tossed 25 times. Find the median, mode, and mean for the following results. 7, 7, 7

11	8	10	9	2
7	10	6	7	9
5	7	7	6	12
9	8	4	4	11
2	3	5	7	6

One of Isaac Newton's most famous quotes is, "If I have seen a little farther than others it is because I have stood on the shoulders of giants."

10. Make a frequency distribution chart for the number of letters in each word of the quote.
11. Find the median for the number of letters in each word. 4
12. Find the mode for the number of letters in each word. 2
13. Find the mean for the number of letters in each word. 3.9

Answer

10.

Number of Letters	Tally	Frequency
1					3
2	𝍸	5
3			1
4						4
5			1
6					3
7				2
8		0
9			1

About the Photo

Square roots are found in many formulas used in many disciplines. The formula given here is used in physics.

The g in the formula is constant for a given location. Acceleration due to gravity is 9.8 m/s^2.

Teacher Resource Book Materials for Chapter 12
See pages 285-302.

Chapter Test, Form I (Multiple Choice)	285-286
Chapter Test, Form II (Free Response)	287-288
Quizzes	289-290
Extra Practice	291-293
Application: Radical Equations (Use after 12-8.)	294-295
Extension: A Space Saving Method (Use after 12-8.)	296
Problem Solving	297
Cumulative Review	298
Answers	299-302

CHAPTER 12

Radical Expressions

When a sky diver jumps from an airplane, the time it takes to reach the ground can be calculated by using the following equation.

$$t = \sqrt{\frac{2s_v}{g}}$$

In this equation s_v is the vertical distance the sky diver falls and g is the acceleration of gravity. Notice that it is necessary to find the square root of $\frac{2s_v}{g}$. In this chapter, you will learn how to find and use square roots.

Objective 12-1: Simplify radicals by using the product and quotient properties of square roots.

12-1 Square Roots

Squaring a number means using that number as a factor two times.

$8^2 = 8 \cdot 8 = 64$ *8 is used as a factor two times.*

❶ $(-8)^2 = (-8)(-8) = 64$ *−8 is used as a factor two times.*

The inverse of squaring is finding a **square root**. To find a square root of 64, you must find *two equal factors* whose product is 64. ❷

$$x^2 = x \cdot x = 64$$

Since 8 times 8 is 64, one square root of 64 is 8. Since −8 times −8 is also 64, another square root of 64 is −8.

Definition of Square Root

If $x^2 = y$, then x is a square root of y.

An expression like $\sqrt{64}$ is called a **radical expression**. The symbol $\sqrt{}$ is a **radical sign**. It indicates the nonnegative or **principal square root** of the expression under the radical sign. This expression is called the **radicand**.

The square root of a negative number is not defined. ❸

radical sign ⟶ $\sqrt{64}$ ⟵ radicand

$\sqrt{64} = 8$ $\sqrt{64}$ indicates the *principal* square root of 64.

$-\sqrt{64} = -8$ $-\sqrt{64}$ indicates the *negative* square root of 64.

$\pm\sqrt{64} = \pm 8$ $\pm\sqrt{64}$ indicates *both* square roots of 64.

± means positive or negative.

Examples

1 Find $\sqrt{81}$.

Since $9^2 = 81$, we know that $\sqrt{81} = 9$.

2 Find $-\sqrt{36}$.

Since $(6)^2 = 36$, we know that $-\sqrt{36} = -6$.

3 Find $\pm\sqrt{0.09}$.

Since $(0.3)^2 = 0.09$, we know that $\pm\sqrt{0.09} = \pm 0.3$.

12-1

Lesson Notes, page T47

Readiness Activity

Evaluate each expression.

4^2	16	$(-4)^2$	16
10^2	100	$(-10)^2$	100
0.5^2	0.25	$(-0.5)^2$	0.25

Discussion Question

(Use after Definition of Square Root)

Are there any numbers that equal their own square roots? 0, 1

Chalkboard Exercises

Simplify.

1a.	$\sqrt{49}$	7
1b.	$\sqrt{121}$	11
1c.	$\sqrt{144}$	12
2a.	$-\sqrt{25}$	−5
2b.	$-\sqrt{100}$	−10
2c.	$-\sqrt{225}$	−15
3a.	$\pm\sqrt{0.49}$	±0.7
3b.	$\pm\sqrt{0.04}$	±0.2
3c.	$\pm\sqrt{1.69}$	±1.3

❶ Make sure students realize that $-x^2$ means $-(x^2)$, not $(-x)^2$. Thus, parentheses are necessary for indicating the square of a negative number.

❷ The inverse of squaring a positive number is finding the principal square root. The inverse of squaring a negative number is finding the negative square root.

❸ One way to clarify the meaning of a radical expression is to state the following:

$\sqrt{a} = b$ if and only if

(i) $b^2 = a$

(ii) $a \geq 0$

(iii) $b \geq 0$

Chalkboard Exercises

Simplify.

4a. $\sqrt{441}$ 21

4b. $\sqrt{1024}$ 32

5a. $\sqrt{\frac{484}{400}}$ $\frac{22}{20}$ or $\frac{11}{10}$

5b. $\sqrt{\frac{0.16}{0.25}}$ $\frac{0.4}{0.5}$ or $\frac{4}{5}$

5c. $\sqrt{\frac{25}{81}}$ $\frac{5}{9}$

6a. $-\sqrt{\frac{49}{289}}$ $-\frac{7}{17}$

6b. $-\sqrt{\frac{36}{361}}$ $-\frac{6}{19}$

4 Encourage students to check square roots by squaring.

5 Before introducing this example, have students find square roots of numbers when given the prime factorizations. Point out that to find the square root of a power, divide the exponent by 2.

6 Emphasize that $(2^3)^2 = 2^6$ and therefore $\sqrt{2^6} = 2^3$.

7 Emphasize that both the product and quotient properties of square roots are true for *real* numbers.

To simplify a radical expression, find the square root of any factors of the radicand that are perfect squares. For example, to simplify $\sqrt{196}$, find any factors of 196 that are perfect squares. You can use the prime factorization of 196 as shown.

$$\begin{aligned}\sqrt{196} &= \sqrt{2 \cdot 2 \cdot 7 \cdot 7}\\ &= \sqrt{2^2 \cdot 7^2}\\ &= \sqrt{2^2} \cdot \sqrt{7^2}\\ &= 2 \cdot 7 \quad \text{④}\\ &= 14 \qquad \textit{Check: } 14^2 = 196\end{aligned}$$

The property used to simplify $\sqrt{196}$ is stated below.

For any numbers a and b, where $a \geq 0$ and $b \geq 0$, $\sqrt{ab} = \sqrt{a} \cdot \sqrt{b}$.

Product Property of Square Roots

Example ⑤

4 Simplify: $\sqrt{576}$

$$\begin{aligned}\sqrt{576} &= \sqrt{2 \cdot 2 \cdot 2 \cdot 2 \cdot 2 \cdot 2 \cdot 3 \cdot 3} && \textit{Find the prime factorization of 576.}\\ &= \sqrt{2^6 \cdot 3^2} && \textit{Express the radicand using exponents.}\\ &= \sqrt{2^6} \cdot \sqrt{3^2} \quad \text{⑥} && \textit{Use the product property of square roots.}\\ &= 2^3 \cdot 3 && \textit{Simplify each radical.}\\ &= 24 \qquad \textit{Check: } 24^2 = 576\end{aligned}$$

A similar property for quotients can also be used to simplify square roots.

For any numbers a and b, where $a \geq 0$ and $b > 0$, $\sqrt{\frac{a}{b}} = \frac{\sqrt{a}}{\sqrt{b}}$.

Quotient Property of Square Roots ⑦

Examples

5 Simplify: $\sqrt{\frac{9}{16}}$

$$\begin{aligned}\sqrt{\frac{9}{16}} &= \frac{\sqrt{9}}{\sqrt{16}} && \textit{Use the quotient property of square roots.}\\ &= \frac{3}{4}\end{aligned}$$

Check: $\left(\frac{3}{4}\right)^2 = \frac{9}{16}$

6 Simplify: $-\sqrt{\frac{81}{121}}$

$$\begin{aligned}-\sqrt{\frac{81}{121}} &= -\frac{\sqrt{81}}{\sqrt{121}}\\ &= -\frac{9}{11}\end{aligned}$$

Check: $\left(-\frac{9}{11}\right)^2 = \frac{81}{121}$

Exploratory Exercises

State the square of each number.

1. 10 100 **2.** 12 144 **3.** -7 49 **4.** -20 400 **5.** 0.3 0.09

6. 0.04 0.0016 **7.** $\frac{1}{2}$ $\frac{1}{4}$ **8.** $\frac{4}{7}$ $\frac{16}{49}$ **9.** $-\frac{7}{8}$ $\frac{49}{64}$ **10.** $-\frac{11}{4}$ $\frac{121}{16}$

Simplify.

11. $\sqrt{121}$ 11 **12.** $\pm\sqrt{36}$ ± 6 **13.** $-\sqrt{81}$ -9 **14.** $\sqrt{25}$ 5 **15.** $\sqrt{\frac{4}{9}}$ $\frac{2}{3}$

16. $\sqrt{\frac{81}{64}}$ $\frac{9}{8}$ **17.** $\pm\sqrt{\frac{49}{121}}$ $\pm\frac{7}{11}$ **18.** $-\sqrt{\frac{9}{100}}$ $-\frac{3}{10}$ **19.** $\sqrt{0.0016}$ 0.04 **20.** $\sqrt{0.09}$ 0.3

Written Exercises

Find the principal square root of each number.

1. 49 7 **2.** 64 8 **3.** 16 4 **4.** 169 13

5. $\frac{25}{36}$ $\frac{5}{6}$ **6.** $\frac{81}{121}$ $\frac{9}{11}$ **7.** $\frac{36}{196}$ $\frac{6}{14}$ or $\frac{3}{7}$ **8.** $\frac{25}{400}$ $\frac{5}{20}$ or $\frac{1}{4}$

9. 0.0009 0.03 **10.** 0.0025 0.05 **11.** 0.0036 0.06 **12.** 0.16 0.4

Simplify.

13. $\sqrt{36}$ 6 **14.** $\sqrt{9}$ 3 **15.** $-\sqrt{100}$ -10 **16.** $\pm\sqrt{144}$ ± 12

17. $\pm\sqrt{25}$ ± 5 **18.** $-\sqrt{169}$ -13 **19.** $\sqrt{0.36}$ 0.6 **20.** $\sqrt{0.0081}$ 0.09

21. $\sqrt{529}$ 23 **22.** $\sqrt{225}$ 15 **23.** $\sqrt{676}$ 26 **24.** $\sqrt{256}$ 16

25. $-\sqrt{441}$ -21 **26.** $-\sqrt{484}$ -22 **27.** $\pm\sqrt{1024}$ ± 32 **28.** $\pm\sqrt{289}$ ± 17

29. $\sqrt{729}$ 27 **30.** $\sqrt{961}$ 31 **31.** $\sqrt{1764}$ 42 **32.** $\sqrt{2025}$ 45

33. $-\sqrt{\frac{289}{100}}$ $-\frac{17}{10}$ **34.** $-\sqrt{\frac{169}{121}}$ $-\frac{13}{11}$ **35.** $\sqrt{\frac{225}{25}}$ 3 **36.** $\sqrt{\frac{144}{196}}$ $\frac{6}{7}$

37. $\sqrt{\frac{256}{361}}$ $\frac{16}{19}$ **38.** $\sqrt{\frac{196}{289}}$ $\frac{14}{17}$ **39.** $\pm\sqrt{\frac{961}{729}}$ $\pm\frac{31}{27}$ **40.** $\pm\sqrt{\frac{484}{1024}}$ $\pm\frac{11}{16}$

41. $\sqrt{\frac{576}{729}}$ $\frac{8}{9}$ **42.** $\sqrt{\frac{441}{529}}$ $\frac{21}{23}$ **43.** $\sqrt{\frac{0.09}{0.16}}$ $\frac{3}{4}$ **44.** $\sqrt{\frac{0.0025}{0.0036}}$ $\frac{5}{6}$

Excursions in Algebra — Radical Sign

Ancient mathematicians commonly wrote the word for root to indicate square roots. Late medieval Latin writers commonly used ℞, a contraction of radix (root), to indicate a square root. The symbol $\sqrt{}$ first appeared in print in Rudolff's *Coss* (1525). It is uncertain whether Rudolff used the symbol because it resembled a small *r* for radix or whether he invented a new symbol.

Suggested Daily Assignments

Enriched
- Exp: 2–20 evens
- Written: 2–44 evens

Average
- Exp: 1–20 evens
- Written: 2–44 evens

Fundamental
- Exp: 1–20
- Written: 1–36

About the Excursion
Students may be interested in finding how other mathematical symbols were developed.

12-2

Lesson Notes, page T47

Discussion Questions
(Use after discussing real numbers.)

1. Are all integers real numbers? yes
2. Are all rational numbers real numbers? yes
3. Are all real numbers irrational numbers? no
4. Are any integers irrational numbers? no

Chalkboard Exercises
Name the set or sets of numbers to which each number belongs. Use N for natural numbers, W for whole numbers, Z for integers, Q for rational numbers, and I for irrational numbers.

1a. -9 Z, Q
1b. 0.368421 . . . I
1c. $\sqrt{81}$ N, W, Z, Q
1d. 0.333333 . . . Q

❶ Point out that irrational numbers are *real* numbers that cannot be expressed in the form $\frac{a}{b}$.

Objective 12-2: Identify irrational numbers.

12-2 Irrational Numbers and Real Numbers

You have encountered natural numbers, whole numbers, integers, and rational numbers. The capital letters **N**, **W**, **Z**, and **Q** are often used to denote these sets of numbers.

Counting or Natural Numbers, **N** $\{1, 2, 3, 4, \ldots\}$

Whole Numbers, **W** $\{0, 1, 2, 3, 4, \ldots\}$

Integers, **Z** $\{\ldots, -3, -2, -1, 0, 1, 2, 3, \ldots\}$

Rational Numbers, **Q** {all numbers that can be expressed in the form $\frac{a}{b}$, where a and b are integers and $b \neq 0$}

Recall that repeating or terminating decimals name rational numbers, since they can be represented as quotients of integers. The square roots of perfect squares also name rational numbers. For example, $\sqrt{0.16}$ is rational since it is equivalent to 0.4, a rational number. Consider the square roots of numbers that are *not* perfect squares.

$\sqrt{2} = 1.414213\ldots$ $\sqrt{3} = 1.732050\ldots$ $\sqrt{7} = 2.645751\ldots$

Notice that none of these decimals terminate or repeat.

These numbers are *not* rational numbers. They are **irrational numbers**. The set of irrational numbers is often denoted by the capital letter **I**.

Irrational numbers are numbers that cannot be expressed in the form $\frac{a}{b}$, where a and b are integers and $b \neq 0$.

❶ ***Definition of Irrational Numbers***

Example

1 **Name the set or sets of numbers to which each number belongs.**

a. -7 — -7 is an integer. Since it can be expressed as $\frac{-7}{1}$, it is also a rational number.

b. **0.8888 . . .** — This repeating decimal is equivalent to $\frac{8}{9}$. Thus, it names a rational number. This number can also be expressed as $0.\overline{8}$. The bar indicates repeating digits.

c. **0.101101110 . . .** — This decimal does not terminate or repeat. Thus, it names an irrational number.

d. $\sqrt{9}$ — Notice $\sqrt{9} = 3$. The number 3 is a natural number, a whole number, an integer, and a rational number.

For review of forms of real numbers, see the exercises on page 552.

Irrational numbers together with rational numbers form the set of **real numbers, R**. You have graphed some rational numbers on the number line. Yet, if you graphed *all* rational numbers, the number line still would not be complete. Irrational numbers complete the number line. Therefore, the graph of all real numbers is the entire number line.

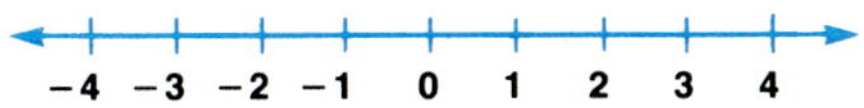

❷

Each real number corresponds to exactly one point on the number line. Each point on the number line corresponds to exactly one real number.

Completeness Property for Points on the Number Line

You can approximate the graph of irrational numbers. The table of square roots on page 544 gives decimal approximations for square roots of integers from 1 to 100.

You can use a calculator or a table of square roots to approximate some irrational numbers.

Example

2 **Use the table on page 544 to find an approximate value for $\sqrt{54}$. Then graph $\sqrt{54}$ on the number line.**

number → n; *square of number* → n^2; *square root of number* → $\sqrt{n}$

n	n^2	$\sqrt{n}$
51	2601	7.141
52	2704	7.211
53	2809	7.280
54	2916	7.348
55	3025	7.416

Locate 54 in the column marked n.

Locate the value for $\sqrt{54}$ in the column marked $\sqrt{n}$.

An approximate value for $\sqrt{54}$ is 7.348.

To check this result, compare $(7.348)^2$ to 54.

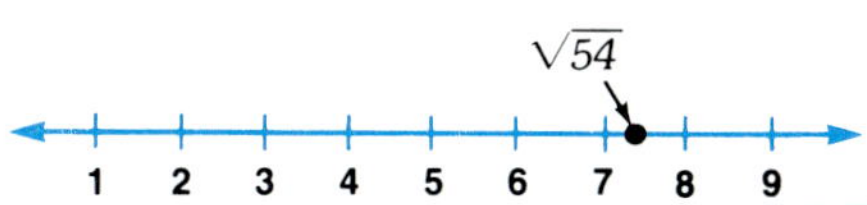

Notice that the table on page 544 also has a column labeled n^2 that lists squares of integers from 1 to 100. You can use this column to determine if an integer between 1 and 10,000 is a perfect square. Any integer that is *not* listed in the n^2 column is *not* a perfect square and has an irrational square root. You can also use the table to find the integers whose values are nearest to the irrational square root.

Chalkboard Exercises

Use the table on page 544 to find an approximate value for each expression.

2a. $\sqrt{83}$ 9.110

2b. $\sqrt{95}$ 9.747

❷ Stress that there are no "holes" in the real number line. It is continuous.

Chalkboard Exercises

Use the table on page 544 to determine whether each square root is rational or irrational. Then state the integers between which the value of the square root lies.

3a. $\sqrt{576}$ rational, 24

3b. $\sqrt{480}$ irrational, between 21 and 22

3c. $\sqrt{500}$ irrational, between 22 and 23

❸ You may want to start out with examples such as $\sqrt{81}$ and $\sqrt{144}$ and progress to $\sqrt{9216}$.

❹ Point out to students that $\sqrt{8556}$ *does* correspond to a unique point on the number line, even though it cannot be represented by a terminating or repeating decimal. An *infinite* decimal corresponds to $\sqrt{8556}$.

Calculator Note

Written Exercises 15–26 can be done using calculators. If the radicand is not a perfect square, the square root must be irrational.

Example

3 **Use the table on page 544 to determine whether each square root is rational or irrational. Then state the two integers between which the value of the square root lies.**

❸ **a.** $\sqrt{9216}$

Look for 9216 in the column marked n^2. Since the number is listed, it is a perfect square. Thus, $\sqrt{9216}$ is a rational number. The corresponding value of n is 96. Thus, the value of $\sqrt{9216}$ is *exactly* 96.

n	n^2	$\sqrt{n}$
91	8281	9.539
92	8464	9.592
93	8649	9.644
94	8836	9.695
95	9025	9.747
96	9216	9.798

❹ **b.** $\sqrt{8556}$

Notice that 8556 is *not* listed in the n^2 column. Thus, $\sqrt{8556}$ is irrational. Since 8556 is greater than 8464 and less than 8649, the value of $\sqrt{8556}$ is between 92 and 93.

Exploratory Exercises

Name the set or sets of numbers to which each number belongs. Use N for natural numbers, W for whole numbers, Z for integers, Q for rational numbers, and I for irrational numbers.

1. -5 Z, Q **2.** $-\frac{1}{2}$ Q **3.** $\frac{6}{3}$ N, W, Z, Q **4.** 16 N, W, Z, Q

5. 0.3333 . . . Q **6.** $\sqrt{11}$ I **7.** $\sqrt{36}$ N, W, Z, Q **8.** 0.125 Q

9. 0.53694 . . . I **10.** $0.\overline{35}$ Q **11.** 0.6125 Q **12.** 3.1416 . . . I

Use the table on page 544 to find an approximate value for each expression.

13. $\sqrt{2}$ 1.414 **14.** $\sqrt{11}$ 3.317 **15.** $\sqrt{20}$ 4.472 **16.** $\sqrt{91}$ 9.539 **17.** 16^2 256 **18.** 28^2 784

19. $\sqrt{31}$ 5.568 **20.** $\sqrt{40}$ 6.325 **21.** $\sqrt{89}$ 9.434 **22.** $\sqrt{29}$ 5.385 **23.** 66^2 4,356 **24.** $(-43)^2$ 18

Written Exercises

Graph each number on the number line.

1. 3 **2.** -1.5 **3.** $1\frac{2}{3}$ **4.** $\sqrt{7}$ **5.** $-\sqrt{5}$ **6.** $\sqrt{20}$

State whether each decimal represents a rational number or an irrational number.

7. 0.10010001 . . . I **8.** 1.23123412 . . . I **9.** $0.\overline{571428}$ Q **10.** $0.\overline{37}$ Q

11. 0.4444 . . . Q **12.** 4.34334333 . . . I **13.** 7.6567876 . . . I **14.** 0.777 . . . Q

Use the table on page 544 to determine if each number is rational or irrational. Then state the integers between which the value of the square root lies.

15. $\sqrt{6724}$ Q; 82 **16.** $\sqrt{350}$ I; 18-19 **17.** $\sqrt{3800}$ I; 61-62 **18.** $\sqrt{1156}$ Q; 34

19. $\sqrt{888}$ I; 29-30 **20.** $\sqrt{289}$ Q; 17 **21.** $\sqrt{9025}$ Q; 95 **22.** $\sqrt{3640}$ I; 60-61

23. $\sqrt{7166}$ I; 84-85 **24.** $\sqrt{7511}$ I; 86-87 **25.** $\sqrt{1444}$ Q; 38 **26.** $\sqrt{3249}$ Q; 57

Suggested Daily Assignments

Enriched
- Exp: 1–12, 14–24 evens
- Written: 1–26

Average
- Exp: 1–24
- Written: 1–26

Fundamental
- Exp: 1–24
- Written: 1–22
  - p. 373, 37–44

Answers to Written

1–6.
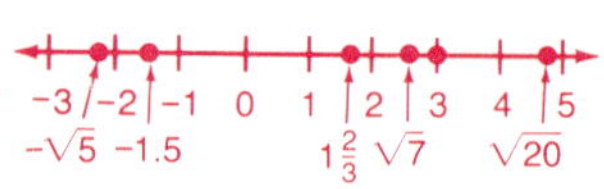

Objective 12-3: Use the divide-and-average method to compute square roots.

12-3 Approximating Square Roots

One method for approximating square roots is called the **divide-and-average method**. This method is used below to approximate $\sqrt{88}$.

Step 1 Isolate 88 between two perfect squares. Choose 9 as an approximation for $\sqrt{88}$ because 88 is closer to 81 than 100.

$$81 < 88 < 100$$
$$9^2 < 88 < 10^2 \quad ❶$$
$$9 < \sqrt{88} < 10$$

Step 2 Divide 88 by 9. Carry the quotient to twice as many digits as in the divisor.

❷ $9\overline{)88.0}$ = 9.7 ← *two-digit quotient*; 9 ↑ *one digit divisor*

Step 3 Average the quotient, 9.7, and the divisor, 9. Carry the average to the same number of digits as the quotient. Do not round.

$$\frac{9.7 + 9}{2} = \frac{18.7}{2} \approx 9.3$$

The symbol ≈ means is approximately equal to. ❸

Step 4 Use the average, 9.3, as the new divisor. Repeat steps 2 and 3.

Divide. $9.3_{\wedge}\overline{)88.0_{\wedge}000}$ = 9.462

Average. $\frac{9.3 + 9.462}{2} = \frac{18.762}{2} \approx 9.381$ ❹

An approximation for $\sqrt{88}$ is 9.381. More accurate approximations may be made by continuing the divide-and-average method.

Example

1 Approximate $\sqrt{19.9}$. Use the divide-and-average method twice.

$$16 < 19.9 < 25$$
$$4^2 < 19.9 < 5^2$$
$$4 < \sqrt{19.9} < 5$$

Choose 4 as the first approximation.

Divide. $4\overline{)19.9}$ = 4.9

Average. $\frac{4.9 + 4}{2} = \frac{8.9}{2} \approx 4.4$

12-3

Lesson Notes, page T48

Discussion Question
On a calculator, press any positive number then press the $\sqrt{\ }$ key repeatedly. What happens? Why?
Converges to 1. As the radicand gets smaller, the square root gets smaller.

Chalkboard Exercises
Use the divide-and-average method twice to approximate each square root.

1a. $\sqrt{60}$ 7.746
1b. $\sqrt{93}$ 9.644
1c. $\sqrt{48}$ 6.928
1d. $\sqrt{302}$ 17.378
1e. $\sqrt{38.6}$ 6.213

❶ Students could use a table of squares to choose the first approximation.

❷ Students may think they can improve accuracy by carrying out the division to many decimal places. Point out that it is no less accurate to stop dividing when a couple of digits in the quotient differ from the divisor. Encourage students to follow the steps given.

❸ Some students may not be familiar with the symbol ≈, which means *is approximately equal to*.

❹ After step 4, have students compare the computed value to the value in the table on page 544.

Calculator Note
You may want to let your students use a calculator when using the divide-and-average method.

Computer Notes
A computer program for finding squares and square roots of numbers is given on page 541 of the appendix. The program uses the internal function SQR(X).

Suggested Daily Assignments

Enriched
Exp: 2–18 evens
Written: 1–12, 14–24 evens, 25–28

Average
Exp: 1–18
Written: 1–25

Fundamental
Exp: 1–18
Written: 1–18
p. 376, 23–26

Quiz
Lessons 12-1 to 12-3
Simplify.
1. $-\sqrt{64}$ -8
2. $\pm\sqrt{\frac{49}{144}}$ $\pm\frac{7}{12}$
3. $\sqrt{441}$ 21
4. $\sqrt{16a^5x^2}$ $4a^2|x|\sqrt{a}$
5. $\sqrt{98c^2d^3}$ $7|c|d\sqrt{2d}$
6. $\sqrt{5}(2\sqrt{5}+\sqrt{3})$ $10+\sqrt{15}$

Approximate using the divide and average method twice.
7. $\sqrt{6}$ 2.45
8. Name the set or sets of numbers to which $\sqrt{7}$ belongs.
I, irrational numbers
R, real numbers

About Using Calculators
The square root key on some calculators is labeled $\sqrt{\ }$ rather than $\sqrt{x}$.

Use 4.4 as the second approximation.

Divide. $4.4_\wedge\overline{)19.9_\wedge000}$ = 4.522

Average. $\frac{4.522 + 4.4}{2} = \frac{8.922}{2}$ or 4.461

An approximation for $\sqrt{19.9}$ is 4.461. *To check this result, compare $(4.461)^2$ to 19.9.*

Exploratory Exercises

State the perfect square integer closest to each number.

1. 14 16
2. 3.2 4
3. 48 49
4. 79 81
5. 63 64
6. 41 36
7. 56 49
8. 27 25
9. 7.8 9
10. 29.3 25
11. 84 81
12. 69.7 64

State a first approximation for each of the following.

13. $\sqrt{13}$ 4
14. $\sqrt{10}$ 3
15. $\sqrt{31}$ 6
16. $\sqrt{35}$ 6
17. $\sqrt{27}$ 5
18. $\sqrt{53}$ 7

Written Exercises

State a first approximation for each of the following.

1. $\sqrt{68}$ 8
2. $\sqrt{78}$ 9
3. $\sqrt{42}$ 6
4. $\sqrt{91}$ 10
5. $\sqrt{95}$ 10
6. $\sqrt{103}$ 10
7. $\sqrt{150}$ 12
8. $\sqrt{300}$ 17
9. $\sqrt{176}$ 13
10. $\sqrt{138}$ 12
11. $\sqrt{147}$ 12
12. $\sqrt{200}$ 14

Use the divide-and-average method twice to approximate each square root.

13. $\sqrt{85}$ 9.219
14. $\sqrt{71}$ 8.426
15. $\sqrt{60.3}$ 7.765
16. $\sqrt{93.5}$ 9.669
17. $\sqrt{149}$ 12.207
18. $\sqrt{185}$ 13.601
19. $\sqrt{193}$ 13.892
20. $\sqrt{206}$ 14.353
21. $\sqrt{131.4}$ 11.463
22. $\sqrt{155.1}$ 12.454
23. $\sqrt{175.6}$ 13.251
24. $\sqrt{115.7}$ 10.756
25. $\sqrt{0.61}$ 0.781
26. $\sqrt{2.314}$ 1.521
27. $\sqrt{0.00462}$ 0.070
28. $\sqrt{0.00932}$ 0.0965

Using Calculators — Square Root Key

The key labeled $\sqrt{x}$ on your calculator is the **square root key**. When this key is pressed, the calculator replaces the number in the display with its principal square root. If a square root is irrational, the calculator display will probably show as many decimal places as it can. The result can be rounded to the nearest hundredth or thousandth.

Exercises

1-24. Use a calculator to check your solutions for Written Exercises 1-24 above.

25. Choose any number and enter it on your calculator. Then press the square root key, followed by the x^2 key. What is the result? Why? Inverse operations result in the original n

26. Choose a negative number and enter it on your calculator. Then press the square root key. What happens? Try some other negative numbers. Can you find any negative numbers that have square roots? Error message is given. No, negative numbers do not have square roots in the field of real numbers.

Objective 12-4A: Simplify irrational square roots.
Objective 12-4B: Simplify radical expressions that contain variables.

12-4 Simplifying Square Roots

The square root of a positive integer is in *simplest form* if the radicand has no perfect square factor other than 1.

The product property of square roots and prime factorization can be used as shown in these examples to simplify radical expressions in which the radicand is *not* a perfect square.

Examples

1 **Simplify: $\sqrt{72}$**

$\sqrt{72} = \sqrt{2 \cdot 2 \cdot 2 \cdot 3 \cdot 3}$ *Prime Factorization*

$= \sqrt{2} \cdot \sqrt{2^2} \cdot \sqrt{3^2}$ *Product Property*

$= \sqrt{2} \cdot 2 \cdot 3$

$= 6\sqrt{2}$

2 **Simplify: $\sqrt{150}$**

$\sqrt{150} = \sqrt{2 \cdot 3 \cdot 5 \cdot 5}$

$= \sqrt{2} \cdot \sqrt{3} \cdot \sqrt{5^2}$

$= \sqrt{2 \cdot 3} \cdot 5$

$= 5\sqrt{6}$

When finding the positive square root of an expression containing variables, you must be sure that the result is *not* negative. Consider that $5^2 = 25$ and $(-5)^2 = 25$. When you find $\sqrt{25}$, however, you want *only* the principal square root. Therefore, absolute values are used as needed to ensure nonnegative results.

$$\sqrt{x^2} = |x|$$

Examples

3 **Simplify: $\sqrt{81y^2}$**

$\sqrt{81y^2} = \sqrt{81} \cdot \sqrt{y^2}$ *Use the product property of square roots.*

$= 9|y|$ *The absolute value of y ensures a nonnegative result.*

4 **Simplify: $\sqrt{200a^4b^3}$**

$\sqrt{200a^4b^3} = \sqrt{2 \cdot 2 \cdot 2 \cdot 5 \cdot 5 \cdot a \cdot a \cdot a \cdot a \cdot b \cdot b \cdot b}$

$= \sqrt{2} \cdot \sqrt{2^2} \cdot \sqrt{5^2} \cdot \sqrt{a^4} \cdot \sqrt{b^2} \cdot \sqrt{b}$ *Product Property of Square Roots*

$= \sqrt{2} \cdot 2 \cdot 5 \cdot a^2 \cdot b \cdot \sqrt{b}$

$= 10a^2b\sqrt{2b}$ *Note that absolute value is not used because a^2 is always nonnegative and b is nonnegative in this case. If b were negative then b^3 would be negative and $\sqrt{200a^4b^3}$ would not be defined.*

The product property can be used to multiply square roots.

12-4

Lesson Notes, page T48

Readiness Activity

Find the prime factorization of each number.

27	3^3	48	$2^4 \cdot 3$
98	$2 \cdot 7^2$	75	$3 \cdot 5^2$
66	$2 \cdot 3 \cdot 11$	130	$2 \cdot 5 \cdot 13$

Discussion Question

(Use after discussing absolute value.)

Find values of *a* and *b* so that it is necessary to use absolute value in the expression $\sqrt{x^a} = |x^b|$.

$a = 2, b = 1$; $a = 6, b = 3$; $a = 10, b = 5$; and so on.

Chalkboard Exercises

Simplify. Use absolute value symbols when necessary.

1a. $\sqrt{147}$ $7\sqrt{3}$
1b. $\sqrt{605}$ $11\sqrt{5}$
2a. $\sqrt{450}$ $15\sqrt{2}$
2b. $\sqrt{350}$ $5\sqrt{14}$
3. $\sqrt{x^6}$ $|x^3|$
4a. $\sqrt{49x^2}$ $7|x|$
4b. $\sqrt{44y^2}$ $2|y|\sqrt{11}$
4c. $\sqrt{120a^4b^2}$ $2a^2|b|\sqrt{30}$
4d. $\sqrt{300x^9y^5}$ $10x^4y^2\sqrt{3xy}$

Chalkboard Exercises

Simplify. Use absolute value symbols when necessary.

5a. $\sqrt{6} \cdot \sqrt{10}$ $2\sqrt{15}$

5b. $2\sqrt{14} \cdot \sqrt{21}$ $14\sqrt{6}$

6a. $\sqrt{5}(\sqrt{5} + \sqrt{11})$ $5 + \sqrt{55}$

6b. $\sqrt{8}(\sqrt{2} + 5\sqrt{6})$ $4 + 20\sqrt{3}$

❶ Emphasize that $7 + \sqrt{35}$ is not the same as $7\sqrt{35}$.

Suggested Daily Assignments

Enriched
- Exp: 2–26 evens
- Written: 12–54 evens

Average
- Exp: 1–25 odds
- Written: 1–54

Fundamental
- Exp: 1–26
- Written: 2–32 evens, 33–54
  p. 378, 23, 24

Answers to Written

29. $8m^2|n^3|\sqrt{5}$
30. $2|x^5y^5|\sqrt{22}$
31. $|x|\sqrt{21y}$
32. $\sqrt{42xy}$
33. $4\sqrt{3}$
34. $10\sqrt{3}$
35. 10
36. 8
37. $14\sqrt{15}$
38. $18\sqrt{14}$
39. 150
40. 336
41. $60\sqrt{3}$
42. $84\sqrt{5}$
43. $3 + 3\sqrt{2}$
44. $7\sqrt{2} + 7$
45. $\sqrt{42} - 3\sqrt{2}$
46. $4 - 2\sqrt{2}$
47. $5\sqrt{2} - \sqrt{10}$
48. $10 - 2\sqrt{5}$
49. $\sqrt{15} + 9$
50. $\sqrt{15} + 25$
51. $12 + 4\sqrt{21}$
52. $4\sqrt{6} + 20\sqrt{3}$

Examples

5 **Simplify:** $\sqrt{10} \cdot \sqrt{20}$

$$\begin{aligned}\sqrt{10} \cdot \sqrt{20} &= \sqrt{10 \cdot 20} \\ &= \sqrt{10 \cdot 10 \cdot 2} \\ &= 10\sqrt{2}\end{aligned}$$

Notice that $\sqrt{10 \cdot 20} = \sqrt{2} \cdot \sqrt{2^2} \cdot \sqrt{5^2}$.

6 **Simplify:** $\sqrt{7}(\sqrt{7} + \sqrt{5})$

$$\begin{aligned}\sqrt{7}(\sqrt{7} + \sqrt{5}) &= \sqrt{7} \cdot \sqrt{7} + \sqrt{7} \cdot \sqrt{5} \\ &= \sqrt{49} + \sqrt{35} \\ &= 7 + \sqrt{35} \quad ❶\end{aligned}$$

Use the distributive property.

Notice that 35 contains no perfect square factor other than 1 and therefore cannot be simplified.

Exploratory Exercises

Simplify.

1. $\sqrt{8}$ $2\sqrt{2}$ **2.** $\sqrt{12}$ $2\sqrt{3}$ **3.** $\sqrt{20}$ $2\sqrt{5}$ **4.** $\sqrt{18}$ $3\sqrt{2}$ **5.** $\sqrt{24}$ $2\sqrt{6}$ **6.** $\sqrt{32}$ $4\sqrt{2}$
7. $\sqrt{48}$ $4\sqrt{3}$ **8.** $\sqrt{19}$ $\sqrt{19}$ **9.** $\sqrt{m^2}$ $|m|$ **10.** $\sqrt{y^6}$ $|y^3|$ **11.** $\sqrt{x^5}$ $x^2\sqrt{x}$ **12.** $\sqrt{a^3}$ $a\sqrt{a}$
13. $\sqrt{8a^3}$ $2a\sqrt{2a}$ **14.** $\sqrt{9a^4}$ $3a^2$ **15.** $\sqrt{a^3b^3}$ **16.** $\sqrt{x^4y^4}$ x^2y^2 **17.** $\sqrt{4} \cdot \sqrt{9}$ 6 **18.** $\sqrt{8} \cdot \sqrt{3}$
19. $\sqrt{5} \cdot \sqrt{10}$ $5\sqrt{2}$ **20.** $\sqrt{6} \cdot \sqrt{9}$ $3\sqrt{6}$ **21.** $\sqrt{11} \cdot \sqrt{11}$ 11 **22.** $\sqrt{10} \cdot \sqrt{10}$ 10
23. $\sqrt{3}(\sqrt{3} + \sqrt{2})$ **24.** $\sqrt{5}(\sqrt{3} + \sqrt{5})$ **25.** $\sqrt{7}(\sqrt{5} - \sqrt{7})$ **26.** $\sqrt{10}(\sqrt{10} - \sqrt{3})$

23. $3 + \sqrt{6}$ **24.** $\sqrt{15} + 5$ **25.** $\sqrt{35} - 7$ **26.** $10 - \sqrt{30}$

15. $ab\sqrt{ab}$

Written Exercises

Simplify. Use absolute value symbols when necessary.

1. $\sqrt{27}$ $3\sqrt{3}$ **2.** $\sqrt{75}$ $5\sqrt{3}$ **3.** $\sqrt{45}$ $3\sqrt{5}$ **4.** $\sqrt{58}$ $\sqrt{58}$
5. $\sqrt{72}$ $6\sqrt{2}$ **6.** $\sqrt{80}$ $4\sqrt{5}$ **7.** $\sqrt{90}$ $3\sqrt{10}$ **8.** $\sqrt{98}$ $7\sqrt{2}$
9. $\sqrt{128}$ $8\sqrt{2}$ **10.** $\sqrt{280}$ $2\sqrt{70}$ **11.** $\sqrt{500}$ $10\sqrt{5}$ **12.** $\sqrt{1000}$ $10\sqrt{10}$
13. $\sqrt{720}$ $12\sqrt{5}$ **14.** $\sqrt{784}$ 28 **15.** $\sqrt{5184}$ 72 **16.** $\sqrt{2916}$ 54
17. $\sqrt{32x^2}$ $4|x|\sqrt{2}$ **18.** $\sqrt{20a^2}$ $2|a|\sqrt{5}$ **19.** $\sqrt{40b^4}$ $2b^2\sqrt{10}$ **20.** $\sqrt{48m^4}$ $4m^2\sqrt{3}$
21. $\sqrt{36a^2b^2}$ $6 \cdot |a| \cdot |b|$ **22.** $\sqrt{80a^2b^3}$ $4|a|b\sqrt{5b}$ **23.** $\sqrt{120a^3b}$ $2|a|\sqrt{30ab}$ **24.** $\sqrt{44m^4n}$ $2m^2\sqrt{11}$
25. $\sqrt{60x^2y^4}$ $2|x|y^2\sqrt{15}$ **26.** $\sqrt{54a^2b^2}$ $3|a| \cdot |b|\sqrt{6}$ **27.** $\sqrt{20m^2n^7}$ $2|m|n^3\sqrt{5n}$ **28.** $\sqrt{147x^5y^4}$ $7x^2y^2\sqrt{}$
29. $\sqrt{320m^4n^6}$ **30.** $\sqrt{88x^{10}y^{10}}$ **31.** $\sqrt{21x^2y}$ **32.** $\sqrt{42xy}$
33. $\sqrt{6} \cdot \sqrt{8}$ **34.** $\sqrt{10} \cdot \sqrt{30}$ **35.** $2\sqrt{5} \cdot \sqrt{5}$ **36.** $4\sqrt{2} \cdot \sqrt{2}$
37. $2\sqrt{3} \cdot 7\sqrt{5}$ **38.** $3\sqrt{7} \cdot 6\sqrt{2}$ **39.** $5\sqrt{10} \cdot 3\sqrt{10}$ **40.** $6\sqrt{8} \cdot 7\sqrt{8}$
41. $4\sqrt{5} \cdot 3\sqrt{15}$ **42.** $7\sqrt{30} \cdot 2\sqrt{6}$ **43.** $\sqrt{3}(\sqrt{3} + \sqrt{6})$ **44.** $\sqrt{7}(\sqrt{14} + \sqrt{7})$
45. $\sqrt{6}(\sqrt{7} - \sqrt{3})$ **46.** $\sqrt{2}(\sqrt{8} - \sqrt{4})$ **47.** $\sqrt{5}(\sqrt{10} - \sqrt{2})$ **48.** $\sqrt{10}(\sqrt{10} - \sqrt{2})$
49. $\sqrt{3}(\sqrt{5} + \sqrt{27})$ **50.** $\sqrt{5}(\sqrt{3} + \sqrt{125})$ **51.** $\sqrt{3}(2\sqrt{12} + 4\sqrt{7})$ **52.** $\sqrt{8}(2\sqrt{3} + 5\sqrt{6})$

53. Find the area of a rectangle whose length is $\sqrt{343}$ cm and whose width is $\sqrt{7}$ cm. 49 cm^2

54. Find the area of a triangle whose base is $\sqrt{6}$ in. and whose altitude is $2\sqrt{6}$ in. 6 in.^2

Reading Algebra — Equivalent Expressions

Many expressions can be written in several different ways. Consider the expression $\frac{1}{xy}$.

$$\frac{1}{xy} = \frac{1}{x \cdot y} = 1/(xy) = 1 \div (xy)$$

All four of these expressions are equivalent because they name the same number.

Notice the importance of the parentheses in the expressions $1/(xy)$ and $1 \div (xy)$. Without the parentheses, the order of operations dictates that the division would come *before* the multiplication. As a result, $1/xy$ and $1 \div xy$ would *not* be equivalent to $\frac{1}{xy}$.

Sometimes parentheses are *not* important when writing equivalent expressions. For example, both $\sqrt{3}^2$ and $(\sqrt{3})^2$ indicate that $\sqrt{3}$ is to be squared.

But sometimes the placement of the parentheses in an expression is *very* important, as shown in the examples below.

$3\sqrt{5^3}$ is equivalent to $3(\sqrt{5^3})$.

$3\sqrt{5^3}$ is *not* equivalent to $(3\sqrt{5})^3$

$\sqrt{7} \cdot \sqrt{3} + \sqrt{2}$ is equivalent to $(\sqrt{7} \cdot \sqrt{3}) + \sqrt{2}$.

$\sqrt{7} \cdot \sqrt{3} + \sqrt{2}$ is *not* equivalent to $\sqrt{7}(\sqrt{3} + \sqrt{2})$.

Exercises

Decide if the expressions are equivalent. Write *yes* or *no*.

1. $\frac{2}{3a}$ and $\frac{2}{3 \cdot a}$ yes

2. $\frac{4}{a + b}$ and $4/a + b$ no

3. ab^2 and $(ab)^2$ no

4. $(7 + \sqrt{2}) \div 8$ and $\frac{7 + \sqrt{2}}{8}$ yes

5. $a + b^2$ and $(a + b)^2$ no

6. $2 + \sqrt{3} \cdot \sqrt{4}$ and $(2 + \sqrt{3})\sqrt{4}$ no

Choose the expressions that are equivalent to the first expression. There may be more than one correct answer.

7. $\frac{m + 2n}{4}$ (**a.**) $(m + 2n)/4$ (**b.**) $(m + 2n) \div 4$ **c.** $m + (2n/4)$

8. $(2\sqrt{4})^6$ **a.** $2(\sqrt{4})^6$ (**b.**) $(2 \cdot \sqrt{4})^6$ **c.** $2\sqrt{4}^6$

9. $2(\sqrt{6} + \sqrt{3})$ **a.** $2 \cdot \sqrt{6} + \sqrt{3}$ (**b.**) $2 \cdot (\sqrt{6} + \sqrt{3})$ **c.** $2\sqrt{6} + \sqrt{3}$

10. $6\sqrt{3} \cdot 4\sqrt{7}$ (**a.**) $6\sqrt{3}(4\sqrt{7})$ (**b.**) $(6\sqrt{3})(4\sqrt{7})$ (**c.**) $(6\sqrt{3} \cdot 4\sqrt{7})$

About Reading Algebra
Understanding the effect of parentheses on expressions is very important. Have students study the examples very carefully before attempting to complete the exercises.

12-5

Lesson Notes, page T48

Discussion Question
(Use after Example 1.)
Find values for m and n so that m and n are not perfect squares, but $\frac{\sqrt{m}}{\sqrt{n}}$ is a perfect square. Answers will vary. Some typical answers are $m = 50$ and $n = 2$ and $m = 12$ and $n = 3$.

Chalkboard Exercises
Simplify.

1a. $\frac{\sqrt{108}}{\sqrt{3}}$ 6

1b. $\frac{\sqrt{98}}{\sqrt{2}}$ 7

2a. $\frac{\sqrt{50}}{\sqrt{5}}$ $\sqrt{10}$

2b. $\frac{2\sqrt{7}}{\sqrt{2}}$ $\sqrt{14}$

3a. $\sqrt{\frac{5}{2}}$ $\frac{\sqrt{10}}{2}$

3b. $\sqrt{\frac{49}{12}}$ $\frac{7\sqrt{3}}{6}$

❶ Point out the different situations in Examples 1 and 2. In Example 1, the denominator is a factor of the numerator; in Example 2, it is not.

❷ Emphasize that multiplication by $\frac{\sqrt{3}}{\sqrt{3}}$ results in an equivalent expression since $\frac{\sqrt{3}}{\sqrt{3}} = 1$.

❸ Explain that the goal in multiplying by $\frac{\sqrt{3}}{\sqrt{3}}$ was to make the radicand in the denominator a perfect square.

Objective 12-5: Simplify expressions involving division of radical expressions.

12-5 Dividing Radical Expressions and Rationalizing Denominators

The quotient property of square roots can be used to simplify expressions and to divide radicals.

Examples

1 Simplify: $\frac{\sqrt{48}}{\sqrt{3}}$

$\frac{\sqrt{48}}{\sqrt{3}} = \sqrt{\frac{48}{3}}$ *Quotient Property of Square Roots*

$= \sqrt{16}$ *Notice that 16 is a perfect square.*

$= 4$ ❶

2 Simplify: $\frac{\sqrt{32}}{\sqrt{3}}$

$\frac{\sqrt{32}}{\sqrt{3}} = \frac{\sqrt{32}}{\sqrt{3}} \cdot \frac{\sqrt{3}}{\sqrt{3}}$ *Notice that $\frac{\sqrt{3}}{\sqrt{3}} = 1$.* ❷

$= \frac{\sqrt{32 \cdot 3}}{\sqrt{3 \cdot 3}}$

$= \frac{\sqrt{96}}{\sqrt{9}}$

$= \frac{\sqrt{2^4} \cdot \sqrt{2} \cdot \sqrt{3}}{3}$ *Now do you see why $\frac{\sqrt{3}}{\sqrt{3}}$ was used?* ❸

$= \frac{4\sqrt{6}}{3}$

The method used to simplify $\frac{\sqrt{32}}{\sqrt{3}}$ in the above example is called **rationalizing the denominator**. Notice that the denominator becomes a rational number.

Example

3 Simplify: $\sqrt{\frac{25}{18}}$

$\sqrt{\frac{25}{18}} = \frac{\sqrt{25}}{\sqrt{18}}$ *Quotient Property of Square Roots*

$= \frac{5}{\sqrt{3^2} \cdot \sqrt{2}}$ *Product Property of Square Roots*

$= \frac{5}{3\sqrt{2}} \cdot \frac{\sqrt{2}}{\sqrt{2}}$ *Rationalize the denominator.*

$= \frac{5\sqrt{2}}{6}$

Some quotients may have binomial expressions, such as $8 + \sqrt{2}$, as denominators. Multiplying $8 + \sqrt{2}$ by $8 - \sqrt{2}$ results in a rational number.

$$\begin{aligned}(8 + \sqrt{2})(8 - \sqrt{2}) &= 8^2 - (\sqrt{2})^2 \\ &= 64 - 2 \\ &= 62\end{aligned}$$

Use the pattern $(a - b)(a + b) = a^2 - b^2$ to simplify.

Binomials of the form $a\sqrt{b} + c\sqrt{d}$ and $a\sqrt{b} - c\sqrt{d}$ are **conjugates** of each other. ❹

Examples

4 **Simplify:** $\dfrac{3}{3 - \sqrt{5}}$

To rationalize the denominator, multiply both numerator and denominator by the conjugate of $3 - \sqrt{5}$, which is $3 + \sqrt{5}$.

$$\begin{aligned}\frac{3}{3 - \sqrt{5}} &= \frac{3}{3 - \sqrt{5}} \cdot \frac{3 + \sqrt{5}}{3 + \sqrt{5}} \\ &= \frac{3^2 + 3\sqrt{5}}{3^2 - (\sqrt{5})^2} \\ &= \frac{9 + 3\sqrt{5}}{9 - 5} \\ &= \frac{9 + 3\sqrt{5}}{4}\end{aligned}$$

Notice that $\dfrac{3 + \sqrt{5}}{3 + \sqrt{5}} = 1$.

Use the distributive property to multiply the numerators.

5 **Simplify:** $\dfrac{2\sqrt{3}}{3\sqrt{6} + 4\sqrt{2}}$

$$\begin{aligned}\frac{2\sqrt{3}}{3\sqrt{6} + 4\sqrt{2}} &= \frac{2\sqrt{3}}{3\sqrt{6} + 4\sqrt{2}} \cdot \frac{3\sqrt{6} - 4\sqrt{2}}{3\sqrt{6} - 4\sqrt{2}} \\ &= \frac{2\sqrt{3} \cdot 3\sqrt{6} - 2\sqrt{3} \cdot 4\sqrt{2}}{(3\sqrt{6})^2 - (4\sqrt{2})^2} \\ &= \frac{6\sqrt{18} - 8\sqrt{6}}{54 - 32} \\ &= \frac{18\sqrt{2} - 8\sqrt{6}}{22} \\ &= \frac{2(9\sqrt{2} - 4\sqrt{6})}{22} \\ &= \frac{9\sqrt{2} - 4\sqrt{6}}{11}\end{aligned}$$

The conjugate of $3\sqrt{6} + 4\sqrt{2}$ is $3\sqrt{6} - 4\sqrt{2}$.

$6\sqrt{18} = 6\sqrt{2 \cdot 3^2} = 18\sqrt{2}$

The GCF of the numerator is 2.

A radical expression is in simplest form when the following conditions are met.

1. No radicands have perfect square factors other than 1.
2. No radicands contain fractions.
3. No radical appears in the denominator of a fraction.

Chalkboard Exercises

Simplify.

4a. $\dfrac{2}{3 + \sqrt{5}}$ $\quad \dfrac{6 - 2\sqrt{5}}{4}$

4b. $\dfrac{1}{4 - \sqrt{2}}$ $\quad \dfrac{4 + \sqrt{2}}{14}$

4c. $\dfrac{3}{5 - \sqrt{3}}$ $\quad \dfrac{15 + 3\sqrt{3}}{22}$

5a. $\dfrac{3\sqrt{5}}{2\sqrt{5} - 3\sqrt{6}}$ $\quad \dfrac{30 + 9\sqrt{30}}{-34}$

5b. $\dfrac{4\sqrt{5}}{3\sqrt{2} + 2\sqrt{3}}$ $\quad \dfrac{6\sqrt{10} - 4\sqrt{15}}{3}$

❹ Some students may need more examples of conjugate expressions. Two expressions containing radicals are conjugates when their product is a rational number. The pairs of expressions below are conjugates. Have students verify this by finding the products.

a. $(3 + \sqrt{2})(3 - \sqrt{2})$ $\quad 7$

b. $(3 + \sqrt{2})(-3 + \sqrt{2})$ $\quad -7$

c. $(3 + \sqrt{2})(6 - 2\sqrt{2})$ $\quad 14$

d. $(\sqrt{6} + \sqrt{3})(\sqrt{6} - \sqrt{3})$ $\quad 3$

e. $(\sqrt{3} + \sqrt{6})(\sqrt{3} - \sqrt{6})$ $\quad -3$

Notice that the conjugate of an expression is not unique. However, the easiest way to write the conjugate of an expression with two terms is to change the sign of the second term.

Suggested Daily Assignments

Enriched
- Exp: 2–12 evens
- Written: 2–36 evens
- Mini-Review: 1–5

Average
- Exp: 1–20
- Written: 1–36
- Mini-Review: 1–5

Fundamental
- Exp: 1–20
- Written: 1–24
  - p. 380, 1–31 odds
- Mini-Review: 1–5

Answers to Written

25. $6\sqrt{3} - 6\sqrt{2}$
27. $\dfrac{20a + 10a\sqrt{a}}{4 - a}$
28. $\dfrac{54b - 9b\sqrt{b}}{36 - b}$
29. $\dfrac{-6\sqrt{5} - 2\sqrt{30}}{3}$
30. $\dfrac{6\sqrt{5} - 3\sqrt{30}}{-2}$
32. $-10\sqrt{6} + 20$
33. $\dfrac{3\sqrt{35} - 5\sqrt{21}}{-15}$
34. $\dfrac{21\sqrt{22} + 18\sqrt{55}}{-82}$
35. $\dfrac{12\sqrt{10} + 3\sqrt{35}}{25}$
36. $\dfrac{12\sqrt{133} - 32\sqrt{57}}{-129}$

Exploratory Exercises

Simplify.

1. $(3 - \sqrt{7})(3 + \sqrt{7})$ 2
2. $(\sqrt{6} - 5)(\sqrt{6} + 5)$ -19
3. $(4 + 2\sqrt{2})(4 - 2\sqrt{2})$ 8

State the conjugate of each expression. Then multiply the expression by its conjugate.

4. $3 + \sqrt{2}$ $3 - \sqrt{2}, 7$
5. $\sqrt{3} + 4$ $\sqrt{3} - 4, -13$
6. $7 - \sqrt{5}$ $7 + \sqrt{5}, 44$
7. $6 + \sqrt{8}$ $6 - \sqrt{8}, 28$
8. $9 - \sqrt{3}$ $9 + \sqrt{3}, 78$
9. $\sqrt{2} + \sqrt{5}$ $\sqrt{2} - \sqrt{5}, -3$
10. $\sqrt{3} - \sqrt{7}$ $\sqrt{3} + \sqrt{7}, -4$
11. $2\sqrt{5} - \sqrt{6}$ $2\sqrt{5} + \sqrt{6}, 14$
12. $2\sqrt{8} + 3\sqrt{5}$ $2\sqrt{8} - 3\sqrt{5},$

State the fraction that each expression should be multiplied by to rationalize the denominator.

13. $\dfrac{3}{\sqrt{5}}$ $\dfrac{\sqrt{5}}{\sqrt{5}}$
14. $\dfrac{7}{\sqrt{11}}$ $\dfrac{\sqrt{11}}{\sqrt{11}}$
15. $\dfrac{3\sqrt{2}}{\sqrt{6}}$ $\dfrac{\sqrt{6}}{\sqrt{6}}$
16. $\dfrac{2\sqrt{3}}{\sqrt{8}}$ $\dfrac{\sqrt{2}}{\sqrt{2}}$
17. $\sqrt{\dfrac{3}{5}}$ $\dfrac{\sqrt{5}}{\sqrt{5}}$
18. $\sqrt{\dfrac{8}{7}}$ $\dfrac{\sqrt{7}}{\sqrt{7}}$
19. $\dfrac{1}{3 + \sqrt{7}}$ $\dfrac{3 - \sqrt{7}}{3 - \sqrt{7}}$
20. $\dfrac{2\sqrt{5}}{4 - \sqrt{3}}$ $\dfrac{4 + \sqrt{3}}{4 + \sqrt{3}}$

Written Exercises

Simplify.

1. $\dfrac{\sqrt{42}}{\sqrt{6}}$ $\sqrt{7}$
2. $\dfrac{\sqrt{20}}{\sqrt{5}}$ 2
3. $\dfrac{\sqrt{10}}{\sqrt{7}}$ $\dfrac{\sqrt{70}}{7}$
4. $\dfrac{\sqrt{7}}{\sqrt{3}}$ $\dfrac{\sqrt{21}}{3}$
5. $\dfrac{\sqrt{6}}{\sqrt{18}}$ $\dfrac{\sqrt{3}}{3}$
6. $\dfrac{\sqrt{5}}{\sqrt{10}}$ $\dfrac{\sqrt{2}}{2}$
7. $\sqrt{\dfrac{3}{7}}$ $\dfrac{\sqrt{21}}{7}$
8. $\sqrt{\dfrac{5}{2}}$ $\dfrac{\sqrt{10}}{2}$
9. $\sqrt{\dfrac{7}{20}}$ $\dfrac{\sqrt{35}}{10}$
10. $\sqrt{\dfrac{11}{32}}$ $\dfrac{\sqrt{22}}{8}$
11. $\sqrt{\dfrac{2}{3}} \cdot \sqrt{\dfrac{5}{2}}$ $\dfrac{\sqrt{15}}{3}$
12. $\sqrt{\dfrac{7}{11}} \cdot \sqrt{\dfrac{10}{7}}$ $\dfrac{\sqrt{110}}{11}$
13. $\sqrt{\dfrac{a}{3}}$ $\dfrac{\sqrt{3a}}{3}$
14. $\sqrt{\dfrac{b}{6}}$ $\dfrac{\sqrt{6b}}{6}$
15. $\sqrt{\dfrac{a^2}{5}}$ $\dfrac{|a|\sqrt{5}}{5}$
16. $\sqrt{\dfrac{m^4}{11}}$ $\dfrac{m^2\sqrt{11}}{11}$
17. $\sqrt{\dfrac{27}{b^2}}$ $\dfrac{3\sqrt{3}}{|b|}$
18. $\sqrt{\dfrac{54}{r^2}}$ $\dfrac{3\sqrt{6}}{|r|}$
19. $\sqrt{\dfrac{5n^5}{4m^5}}$ $\dfrac{n^2\sqrt{5mn}}{2|m^3|}$
20. $\sqrt{\dfrac{11a^3}{10b^3}}$ $\dfrac{|a|\sqrt{110ab}}{10b^2}$
21. $\dfrac{1}{7 - \sqrt{3}}$ $\dfrac{7 + \sqrt{3}}{46}$
22. $\dfrac{1}{6 + \sqrt{3}}$ $\dfrac{6 - \sqrt{3}}{33}$
23. $\dfrac{11}{\sqrt{2} + 5}$ $\dfrac{11\sqrt{2} - 55}{-23}$
24. $\dfrac{10}{\sqrt{5} - 9}$ $\dfrac{5\sqrt{5} + 45}{-38}$
25. $\dfrac{6}{\sqrt{3} + \sqrt{2}}$
26. $\dfrac{12}{\sqrt{6} - \sqrt{5}}$ $12\sqrt{6} + 12\sqrt{5}$
27. $\dfrac{10a}{2 - \sqrt{a}}$
28. $\dfrac{9b}{6 + \sqrt{b}}$
29. $\dfrac{2\sqrt{5}}{-3 + \sqrt{6}}$
30. $\dfrac{-3\sqrt{5}}{-2 - \sqrt{6}}$
31. $\dfrac{-9\sqrt{2}}{-4 + \sqrt{8}}$ $\dfrac{9\sqrt{2} + 9}{2}$
32. $\dfrac{-10\sqrt{6}}{3 + \sqrt{6}}$
33. $\dfrac{2\sqrt{7}}{3\sqrt{5} + 5\sqrt{3}}$
34. $\dfrac{3\sqrt{11}}{7\sqrt{2} - 6\sqrt{5}}$
35. $\dfrac{6\sqrt{5}}{4\sqrt{8} - 2\sqrt{7}}$
36. $\dfrac{4\sqrt{19}}{3\sqrt{7} + 4\sqrt{12}}$

mini-review

1. Given $f(x) = 2x^2 + 6$, find $f(3)$. 24
2. Determine the slope of the line passing through the points with coordinates $(4, -2)$ and $(-6, 8)$. -1
3. Solve $25y^2 + 20y = -4$. $-\frac{2}{5}$
4. Solve $(x + 6)(x - 1) = 78$. $-12, 7$
5. The sum of two numbers is 132. Their difference is 44. Find the numbers. 88, 44

Objective 12-6: Simplify expressions involving addition and subtraction of radical expressions.

12-6 Adding and Subtracting Radical Expressions

Radical expressions in which the radicands are alike can be added or subtracted in the same way that monomials are added or subtracted.

$$3x + 2x = (3 + 2)x = 5x$$
$$3\sqrt{2} + 2\sqrt{2} = (3 + 2)\sqrt{2} = 5\sqrt{2}$$
$$7y - 4y = (7 - 4)y = 3y$$
$$7\sqrt{5} - 4\sqrt{5} = (7 - 4)\sqrt{5} = 3\sqrt{5}$$

Notice that the distributive property was used to simplify each of the radical expressions. ❶

Examples

1 Simplify: $3\sqrt{11} + 6\sqrt{11} - 2\sqrt{11}$

$3\sqrt{11} + 6\sqrt{11} - 2\sqrt{11} = (3 + 6 - 2)\sqrt{11}$ *Distributive Property*

$= 7\sqrt{11}$

2 Simplify: $9\sqrt{7} - 4\sqrt{2} + 3\sqrt{2} - 5\sqrt{7}$

$9\sqrt{7} - 4\sqrt{2} + 3\sqrt{2} - 5\sqrt{7} = 9\sqrt{7} - 5\sqrt{7} - 4\sqrt{2} + 3\sqrt{2}$ *Commutative Property*

$= (9 - 5)\sqrt{7} + (-4 + 3)\sqrt{2}$ *Distributive Property*

$= 4\sqrt{7} - \sqrt{2}$

If the radical expressions are *not* in simplest form, first simplify. Then, if possible, use the distributive property to further simplify the expression. Some expressions, such as $3\sqrt{5} - 2\sqrt{7}$ and $4\sqrt{2} + \sqrt{17}$, *cannot* be simplified further because the radicands are different and there are no common factors.

Example

3 Simplify: $6\sqrt{8} + 11\sqrt{18}$

$6\sqrt{8} + 11\sqrt{18} = 6\sqrt{2^2 \cdot 2} + 11\sqrt{3^2 \cdot 2}$ *Simplify each term.*

$= 6(2\sqrt{2}) + 11(3\sqrt{2})$

$= 12\sqrt{2} + 33\sqrt{2}$

$= 45\sqrt{2}$ *Distributive Property.*

12-6

Lesson Notes, page T48
Enrichment, page T48

Discussion Question

(Use after example 3.)
Find vaues for a and b so that $\sqrt{a} + \sqrt{b} = 5\sqrt{2}$.
$a = 2$ and $b = 32$,
$a = 8$ and $b = 18$,
$a = 0$ and $b = 50$.

Chalkboard Exercises

Simplify.

1. $3\sqrt{5} + 4\sqrt{5} - 2\sqrt{5}$ $5\sqrt{5}$
2. $6\sqrt{7} - 2\sqrt{7} - \sqrt{7} + 4\sqrt{7}$ $7\sqrt{7}$

3a. $3\sqrt{8} + \sqrt{98} + \sqrt{50}$ $18\sqrt{2}$

3b. $\sqrt{20} + \sqrt{45} + 12\sqrt{5}$ $17\sqrt{5}$

❶ If students can use the distributive property with ease, they should have no trouble adding and subtracting expressions with like radicands.

Chalkboard Exercises

Simplify.

4a. $4\sqrt{27} - 2\sqrt{48} + 2\sqrt{20}$
$4\sqrt{3} + 4\sqrt{5}$

4b. $6\sqrt{50} + 3\sqrt{32} - \sqrt{48}$
$42\sqrt{2} - 4\sqrt{3}$

5a. $3\sqrt{3} - 4\sqrt{\frac{1}{3}}$ $\frac{5\sqrt{3}}{3}$

5b. $6\sqrt{5} + \sqrt{125} - 2\sqrt{\frac{1}{5}}$
$\frac{53\sqrt{5}}{5}$

Examples

4 **Simplify:** $7\sqrt{98} + 5\sqrt{32} - 2\sqrt{75}$

$$\begin{aligned} 7\sqrt{98} + 5\sqrt{32} - 2\sqrt{75} &= 7\sqrt{7^2 \cdot 2} + 5\sqrt{2^4 \cdot 2} - 2\sqrt{5^2 \cdot 3} \\ &= 7(7\sqrt{2}) + 5(4\sqrt{2}) - 2(5\sqrt{3}) \\ &= 49\sqrt{2} + 20\sqrt{2} - 10\sqrt{3} \\ &= 69\sqrt{2} - 10\sqrt{3} \end{aligned}$$

5 **Simplify:** $4\sqrt{7} - 5\sqrt{28} + 5\sqrt{\frac{1}{7}}$

$$\begin{aligned} 4\sqrt{7} - 5\sqrt{28} + 5\sqrt{\tfrac{1}{7}} &= 4\sqrt{7} - 5\sqrt{2^2 \cdot 7} + 5\left(\frac{\sqrt{1}}{\sqrt{7}} \cdot \frac{\sqrt{7}}{\sqrt{7}}\right) \\ &= 4\sqrt{7} - 5(2\sqrt{7}) + 5\left(\frac{\sqrt{7}}{7}\right) \\ &= 4\sqrt{7} - 10\sqrt{7} + \frac{5\sqrt{7}}{7} \end{aligned}$$

Rationalize the denominator of $5\sqrt{\frac{1}{7}}$.

$$\begin{aligned} 4\sqrt{7} - 5\sqrt{28} + 5\sqrt{\tfrac{1}{7}} &= \frac{28\sqrt{7} - 70\sqrt{7} + 5\sqrt{7}}{7} \\ &= -\frac{37\sqrt{7}}{7} \end{aligned}$$

The least common denominator is 7.

Exploratory Exercises

Name the expressions in each group that have the same radicand.

1. $5\sqrt{3}, 4\sqrt{6}, 3\sqrt{3}$ $5\sqrt{3}, 3\sqrt{3}$
2. $5\sqrt{14}, -3\sqrt{7}, 4\sqrt{7}$
3. $3\sqrt{12}, 2\sqrt{6}, 5\sqrt{12}$ $3\sqrt{12},$
4. $4\sqrt{2}, 3\sqrt{2}, 2\sqrt{3}$ $4\sqrt{2}, 3\sqrt{2}$
5. $-3\sqrt{3}, 2\sqrt{6}, 12\sqrt{3}$
6. $3\sqrt{3}, 2\sqrt{3}, 3\sqrt{6}$ $3\sqrt{3}, 2\sqrt{}$
7. $2\sqrt{10}, -6\sqrt{10}, 4\sqrt{2}, 7\sqrt{10}$
8. $3\sqrt{2}, 5\sqrt{6}, 2\sqrt{3}, 4\sqrt{7}$ none
9. $2\sqrt{5}, -\sqrt{10}, 3\sqrt{5}, -5\sqrt{5}$

7. $2\sqrt{10}, -6\sqrt{10}, 7\sqrt{10}$ **2.** $-3\sqrt{7}, 4\sqrt{7}$ **5.** $-3\sqrt{3}, 12\sqrt{3}$ **9.** $2\sqrt{5}, 3\sqrt{5}, -5\sqrt{5}$

Simplify.

10. $3\sqrt{5} + 2\sqrt{5}$ $5\sqrt{5}$
11. $4\sqrt{3} - 7\sqrt{3}$ $-3\sqrt{3}$
12. $4\sqrt{7} + 11\sqrt{7}$ $15\sqrt{7}$
13. $8\sqrt{6} + 3\sqrt{6}$ $11\sqrt{6}$
14. $4\sqrt{x} - 5\sqrt{x}$ $-\sqrt{x}$
15. $5\sqrt{y} + 6\sqrt{y}$ $11\sqrt{y}$
16. $2\sqrt{11} + 3\sqrt{5}$ $2\sqrt{11} + 3\sqrt{5}$
17. $3\sqrt{15} - 2\sqrt{5}$ $3\sqrt{15} - 2\sqrt{5}$
18. $10\sqrt{6} - 3\sqrt{6}$ $7\sqrt{6}$
19. $25\sqrt{13} + \sqrt{13}$ $26\sqrt{13}$
20. $15\sqrt{6} - \sqrt{6}$ $14\sqrt{6}$
21. $21\sqrt{19} + 19\sqrt{19}$ $40\sqrt{19}$
22. $18\sqrt{2x} + 3\sqrt{2x}$ $21\sqrt{2x}$
23. $2\sqrt{3a} - 7\sqrt{3a}$ $-5\sqrt{3a}$
24. $3\sqrt{5m} - 5\sqrt{5m}$ $-2\sqrt{5m}$

Written Exercises

Simplify.

1. $4\sqrt{11} + 7\sqrt{11}$ $11\sqrt{11}$
2. $10\sqrt{21} - 8\sqrt{21}$ $2\sqrt{21}$
3. $3\sqrt{13} - 7\sqrt{13}$ $-4\sqrt{13}$
4. $-7\sqrt{7} + 6\sqrt{7}$ $-\sqrt{7}$
5. $5\sqrt{7} + 12\sqrt{7}$ $17\sqrt{7}$
6. $2\sqrt{11} + 4\sqrt{11}$ $6\sqrt{11}$

For more exercises on simplifying radicals, see exercises 1-62 on page 561.

Suggested Daily Assignments

Enriched
- Exp: 2–24 evens
- Written: 2–40 evens

Average
- Exp: 1–24
- Written: 1–40

Fundamental
- Exp: 1–24
- Written: 2–40 evens
  - p. 384, 25–36

7. $4\sqrt{3} + 7\sqrt{3} - 2\sqrt{3}$ $9\sqrt{3}$

8. $5\sqrt{2} + 4\sqrt{2} + 7\sqrt{2}$ $16\sqrt{2}$

9. $8\sqrt{7} - 10\sqrt{7} + 3\sqrt{7}$ $\sqrt{7}$

10. $2\sqrt{11} - 6\sqrt{11} - 3\sqrt{11}$ $-7\sqrt{11}$

11. $5\sqrt{5} + 3\sqrt{5} - 18\sqrt{5}$ $-10\sqrt{5}$

12. $2\sqrt{10} + 2\sqrt{10} + 2\sqrt{10}$ $6\sqrt{10}$

13. $8\sqrt{3} - 2\sqrt{2} + 3\sqrt{2} + 5\sqrt{3}$ $13\sqrt{3} + \sqrt{2}$

14. $-3\sqrt{5} + 2\sqrt{5} + 5\sqrt{2} + 3\sqrt{2}$ $-\sqrt{5} + 8\sqrt{2}$

15. $4\sqrt{6} + \sqrt{7} - 6\sqrt{2} + 4\sqrt{7}$

16. $10\sqrt{10} + 4\sqrt{10} - \sqrt{3} + 6\sqrt{10}$

17. $\sqrt{3} + \sqrt{5}$ $\sqrt{3} + \sqrt{5}$

18. $4\sqrt{6} + 2\sqrt{2}$ $4\sqrt{6} + 2\sqrt{2}$

19. $2\sqrt{3} + \sqrt{12}$ $4\sqrt{3}$

20. $3\sqrt{8} + 7\sqrt{2}$ $13\sqrt{2}$

21. $3\sqrt{7} - 2\sqrt{28}$ $-\sqrt{7}$

22. $3\sqrt{50} - 4\sqrt{5}$ $15\sqrt{2} - 4\sqrt{5}$

23. $4\sqrt{20} + 9\sqrt{45}$ $35\sqrt{5}$

24. $3\sqrt{27} + 5\sqrt{48}$ $29\sqrt{3}$

25. $3\sqrt{28} - 8\sqrt{63}$ $-18\sqrt{7}$

26. $2\sqrt{50} - 3\sqrt{32}$ $-2\sqrt{2}$

27. $\sqrt{18} + \sqrt{108} + \sqrt{50}$ $8\sqrt{2} + 6\sqrt{3}$

28. $\sqrt{48} - \sqrt{12} + \sqrt{300}$ $12\sqrt{3}$

29. $8\sqrt{72} + 2\sqrt{20} - 3\sqrt{5}$ $48\sqrt{2} + \sqrt{5}$

30. $2\sqrt{20} - 3\sqrt{24} - \sqrt{180}$ $-2\sqrt{5} - 6\sqrt{6}$

31. $2\sqrt{108} - \sqrt{27} + \sqrt{363}$ $20\sqrt{3}$

32. $8\sqrt{50} + 5\sqrt{72} - 2\sqrt{98}$ $56\sqrt{2}$

33. $\sqrt{7} + \sqrt{\frac{1}{7}}$ $\frac{8}{7}\sqrt{7}$

34. $\sqrt{6} + \sqrt{\frac{2}{3}}$ $\frac{4}{3}\sqrt{6}$

35. $\sqrt{3} - \sqrt{\frac{1}{3}}$ $\frac{2}{3}\sqrt{3}$

36. $\sqrt{10} - \sqrt{\frac{2}{5}}$ $\frac{4}{5}\sqrt{10}$

37. $3\sqrt{3} - \sqrt{48} + 3\sqrt{\frac{1}{3}}$ 0

38. $14\sqrt{\frac{3}{2}} + 9\sqrt{\frac{2}{3}} + \sqrt{24}$ $12\sqrt{6}$

39. $3\sqrt{\frac{7}{4}} - 10\sqrt{\frac{1}{7}} + 3\sqrt{28}$ $\frac{85}{14}\sqrt{7}$

40. $6\sqrt{\frac{3}{5}} - 3\sqrt{60} + 2\sqrt{\frac{3}{5}}$ $-\frac{22}{5}\sqrt{15}$

Using Calculators — Verifying Answers

You can use a calculator to verify answers to exercises such as the ones above.

Example: Use a calculator to verify the answer to Example 5 on page 386.

In Example 5 on page 386, $4\sqrt{7} - 5\sqrt{28} + 5\sqrt{\frac{1}{7}}$ is simplified to $\frac{-37\sqrt{7}}{7}$.

To verify this answer, first evaluate $4\sqrt{7} - 5\sqrt{28} + 5\sqrt{\frac{1}{7}}$.

ENTER: 4 [×] 7 [√] [−] 5 [×] 28 [√] [+] 5 [×] [(] 1 [÷] 7 [)] [√] [=]

The display shows −13.984686.

Then evaluate $\frac{-37\sqrt{7}}{7}$.

ENTER: [(] 37 [×] 7 [√] [)] [÷] 7 [=] [+/−]

The display again shows −13.984686.
Therefore, the answer has veen verified.

Exercises

1-10. Use a calculator to verify your answers to Written Exercises 31-40.
See students' work.

Quiz
Lessons 12-4 to 12-6
Simplify.

1. $\frac{5}{\sqrt{7}}$ $\frac{5\sqrt{7}}{7}$
2. $\sqrt{3}(\sqrt{6} + 2\sqrt{3})$ $3\sqrt{2} + 6$
3. $\frac{\sqrt{21}}{\sqrt{3}}$ $\sqrt{7}$
4. $\frac{4\sqrt{2}}{3\sqrt{2} - 2}$ $\frac{12 + 4\sqrt{2}}{7}$
5. $3\sqrt{5} + 2\sqrt{20}$ $7\sqrt{5}$
6. $3\sqrt{18} - \sqrt{8} + 2\sqrt{32}$ $15\sqrt{2}$

Answers to Written
15. $4\sqrt{6} - 6\sqrt{2} + 5\sqrt{7}$
16. $20\sqrt{10} - \sqrt{3}$

About Using Calculators
You may want to have students check the answers to all of the Written Exercises.

About Applications
The exercises in the Application are challenging. Exceptional students could find v for other planets. Point out that the gravitational constant is given in N-m^2/kg^2. The abbreviation N stands for newton. One newton of force acting on a mass of one kilogram causes acceleration of one meter per second per second.

You may want to show students why the answer is labeled in m/s.

G is in $\frac{Nm^2}{(kg)^2}$ with

N in $\frac{kg(m)}{s^2}$.

Thus G is in $\frac{m^3}{kg(s^2)}$.

M is in kg.
r is in m.

Therefore, $\frac{2GM}{r}$ is in

$$\frac{\frac{m^3}{kg(s^2)} \cdot kg}{m} \text{ or } \frac{m^2}{s^2}$$

and $\sqrt{\frac{2GM}{r}} = $ m/s.

Applications in Physics

Escape Velocity

The ability to use and compute square roots is often a necessary tool in physics. For example, square roots are used to compute the minimum velocity a spacecraft must have to escape the gravitational force of a planet. The escape velocity, v, can be calculated using the following formula.

$$v = \sqrt{\frac{2GM}{r}}$$

In this formula, G is a gravitational constant, M is the mass of the planet, and r is the radius of the planet.

Example: Compute the escape velocity for earth if its mass is 5.98×10^{24} kg, its radius is 6.37×10^6 m, and $G = 6.67 \times 10^{-11}$ N-m^2/kg^2. The abbreviation N represents newton, a unit of force in the metric system. Using these constants the answers will be in m/s.

$1 \text{ N} = \frac{1 \text{ kg} \cdot \text{m}}{s^2}$

$$v = \sqrt{\frac{2GM}{r}} \qquad M = 5.98 \times 10^{24},\ r = 6.37 \times 10^6,\ G = 6.67 \times 10^{-11}$$

$$= \sqrt{\frac{2(6.67 \times 10^{-11})(5.98 \times 10^{24})}{6.37 \times 10^6}}$$

$$= \sqrt{\frac{7.97732 \times 10^{14}}{6.37 \times 10^6}}$$

$$= \sqrt{1.2523265 \times 10^8} \approx 1.12 \times 10^4$$

The escape velocity for earth is approximately 1.12×10^4 m/s.

Exercises

Compute the escape velocity for the indicated planet using the formula above. Use $G = 6.67 \times 10^{-11}$ N-m^2/kg^2.

1. Mars: Mass 6.46×10^{23} kg; radius 3.39×10^6 m 5.04×10^3 m/s
2. Jupiter: Mass 1.90×10^{27} kg; radius 7.15×10^7 m 5.95×10^4 m/s
3. Mercury: Mass 3.35×10^{23} kg; radius 2.44×10^6 m 4.28×10^3 m/s
4. Neptune: Mass 1.03×10^{26} kg; radius 2.25×10^7 m 2.47×10^4 m/s
5. Venus: Mass 4.90×10^{24} kg; radius 6.06×10^6 m 1.04×10^4 m/s
6. Uranus: Mass 8.73×10^{25} kg; radius 2.35×10^7 m 2.23×10^4 m/s

Objective 12-7: Solve simple radical equations in which only one radical contains a variable.

12-7 Radical Equations

Equations containing radicals with variables in the radicand are called **radical equations**. To solve such equations, first isolate the radical on one side of the equation. Then square both sides of the equation to eliminate the radical. ❶

Examples

1 Solve and check: $\sqrt{y} = 11$

$\sqrt{y} = 11$

$(\sqrt{y})^2 = (11)^2$ *Square both sides to eliminate the radical.*

$y = 121$

Check: $\sqrt{y} = 11$

$\sqrt{121} \stackrel{?}{=} 11$

$11 = 11$

The solution is 121.

2 Solve and check: $\sqrt{3y - 5} - 4 = 0$

$\sqrt{3y - 5} - 4 = 0$

$\sqrt{3y - 5} = 4$ *Isolate the radical by adding 4 to both sides.*

$3y - 5 = 16$ *Square both sides.*

$3y = 21$

$y = 7$

Check: $\sqrt{3y - 5} - 4 = 0$

$\sqrt{3 \cdot 7 - 5} - 4 \stackrel{?}{=} 0$

$\sqrt{16} - 4 \stackrel{?}{=} 0$

$4 - 4 \stackrel{?}{=} 0$

$0 = 0$

The solution is 7.

Consider this equation.

$x = 2$ *The solution is 2.*

Now square both sides.

$x^2 = 4$ *The solutions are 2 and −2.*

Notice that squaring both sides of an equation does not necessarily produce results that satisfy the *original* equation. Therefore you must check *all* solutions when solving radical equations.

12-7

Lesson Notes, page T48

Readiness Activity

Square both sides of each of the following equations.

$\sqrt{x} = 4$ $x = 16$

$\sqrt{m} = 3$ $m = 9$

$\sqrt{2y} = -8$ $2y = 64$

$\sqrt{a + 4} = 2$ $a + 4 = 4$

$\sqrt{\frac{2d}{3}} = -5$ $\frac{2d}{3} = 25$

Describe the results. Squaring both sides eliminates the radical.

Discussion Question

For what values of a is $\sqrt{|a|} = |a|$? 0, 1, −1

Chalkboard Exercises

Solve and check.

1a. $\sqrt{y} = 10$ 100

1b. $\sqrt{m} = 3$ 9

2a. $3 + \sqrt{m} = 13$ 100

2b. $\sqrt{4m + 1} - 5 = 0$ 6

2c. $\sqrt{6m - 3} - 3 = 0$ 2

❶ Emphasize the importance of isolating the radical by showing an example of what happens when students do not.

$\sqrt{y} + 3 = 0$

$(\sqrt{y} + 3)^2 = 0$

$y + 6\sqrt{y} + 9 = 0$

The radical still remains.

Chalkboard Exercises

Solve and check.

3a. $\sqrt{y+2} = y - 4$ 7

3b. $\sqrt{3x-5} = x - 5$ 10

❷ Stress that squaring both sides of an equation often produces results which do not check. Thus, it is important to check *all* possible solutions.

A result that does not check is called an *extraneous value*. In Example 3, the extraneous value is 10.

❸ Point out that some exercises may have no solutions.

Suggested Daily Assignments

Enriched
Exp: 1–8
Written: 2–34 evens

Average
Exp: 1–8
Written: 1–34

Fundamental
Exp: 1–8
Written: 1–20
p. 387, 7–31 odds

Answers to Written

25. $\pm\sqrt{22}$

26. $\pm\frac{4}{3}\sqrt{3}$

Example

3 **Solve and check:** $\sqrt{3x - 14} = 6 - x$

$$\sqrt{3x-14} = 6 - x$$

$$3x - 14 = (6 - x)^2$$ *Square both sides.*

$$3x - 14 = 36 - 12x + x^2$$ *Recall that $(a - b)^2 = a^2 - 2ab + b^2$.*

$$0 = x^2 - 15x + 50$$

$$0 = (x - 5)(x - 10)$$ *Factor.*

$$x - 5 = 0 \quad \text{or} \quad x - 10 = 0$$ *Set each factor equal to 0.*

$$x = 5 \quad \text{or} \quad x = 10$$

❷ ***Check:*** $\sqrt{3x - 14} = 6 - x$

$$\sqrt{3 \cdot 5 - 14} \stackrel{?}{=} 6 - 5 \quad \text{or} \quad \sqrt{3 \cdot 10 - 14} \stackrel{?}{=} 6 - 10$$

$$\sqrt{15 - 14} \stackrel{?}{=} 1 \qquad \sqrt{30 - 14} \stackrel{?}{=} -4$$

$$\sqrt{1} \stackrel{?}{=} 1 \qquad \sqrt{16} \stackrel{?}{=} -4$$

$$1 = 1 \qquad 4 \neq -4$$

Notice that 10 satisfies $3x - 14 = (6 - x)^2$, but does not satisfy the *original* equation. Therefore, 5 is the *only* solution of $\sqrt{3x - 14} = 6 - x$.

Exploratory Exercises ❸

State the solution of each equation.

1. $\sqrt{y} = 3$ 9
2. $\sqrt{a} = 4$ 16
3. $\sqrt{m} = -2$ none
4. $\sqrt{s} = -5$ none
5. $-\sqrt{r} = -2$ 4
6. $-\sqrt{z} = -3$ 9
7. $-\sqrt{y} = 4$ none
8. $\sqrt{x - 3} = 6$ 39

Written Exercises

Solve and check.

1. $\sqrt{r} = 5$ 25
2. $\sqrt{s} = 9$ 81
3. $\sqrt{3x} = 3$ 3
4. $\sqrt{2m} = 4$ 8
5. $\sqrt{3m} = -9$ none
6. $\sqrt{4a} = -1$ none
7. $\sqrt{b} - 5 = 0$ 25
8. $\sqrt{s} + 3 = 0$ none
9. $\sqrt{2d} + 1 = 0$ none
10. $4 - \sqrt{3a} = 0$ $\frac{16}{3}$
11. $5 + \sqrt{2x} = 8$ $\frac{9}{2}$
12. $2 + \sqrt{m} = 13$ 1
13. $\sqrt{4x + 1} = 3$ 2
14. $\sqrt{2x + 7} = 5$ 9
15. $\sqrt{8s + 1} - 5 = 0$ 3
16. $\sqrt{3b - 5} = -4$
17. $\sqrt{\frac{x}{4}} = 6$ 144
18. $\sqrt{\frac{x}{4}} = 10$ 400
19. $\sqrt{\frac{4a}{3}} - 2 = 0$ 3
20. $\sqrt{\frac{9s}{2}} - 6 = 0$ 8
21. $\sqrt{\frac{5x}{4}} - 8 = 2$ 80
22. $\sqrt{\frac{k}{7}} - 10 = -3$ 343
23. $\sqrt{r} = 3\sqrt{5}$ 45
24. $3\sqrt{7} = \sqrt{m}$ 63
25. $5\sqrt{2n^2 - 28} = 20$
26. $4\sqrt{3m^2 - 15} = 4$
27. $\sqrt{2z^2 - 121} = z$ 11
28. $\sqrt{5x^2 - 7} = 2x$
29. $\sqrt{x + 2} = x - 4$ 7
30. $\sqrt{1 - 2x} = 1 + x$ 0
31. $4 + \sqrt{x - 2} = x$ 6
32. $\sqrt{x^2 + 3} = 2x$ 1
33. The square root of the product of 4 and a number is 26. Find the number. 169
34. The square root of the product of 9 and a number is 27. Find the number. 81

For more exercises on solving radical equations, see exercises 63-73 on page 561.

Objective 12-8A: Use the Pythagorean Theorem to find the length of the third side of a right triangle given the lengths of the other two sides.

Objective 12-8B: Determine whether a triangle is a right triangle given the length of its sides.

12-8 The Pythagorean Theorem

On a baseball diamond the distance from one base to the next is 90 feet. How far from home plate is second base? To answer this question, notice that a baseball diamond can be divided into two right triangles.

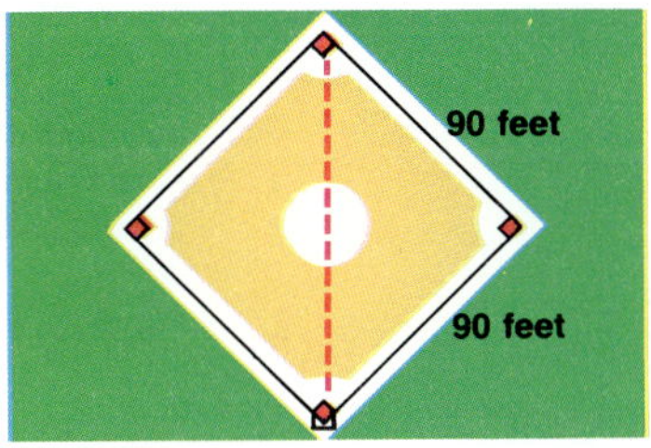

The side opposite the right angle in a right triangle is called the **hypotenuse**. This side is *always* the longest side of a right triangle. The other two sides are called the **legs** of the triangle.

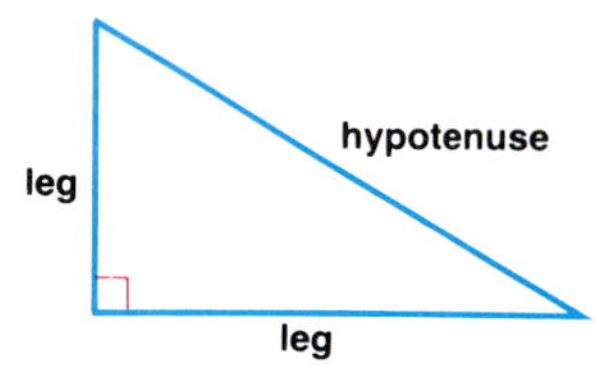

To find the length of the hypotenuse, given the lengths of the legs, you can use a formula proposed by the Greek mathematician Pythagoras.

In a right triangle if a and b are the measures of the legs, and c is the measure of the hypotenuse, then $c^2 = a^2 + b^2$.

❶ ***The Pythagorean Theorem***

If c is the measure of the longest side of a triangle and $c^2 \neq a^2 + b^2$, then the triangle is not a right triangle.

For the baseball diamond, $a = 90$ and $b = 90$. Find the distance from home plate to second base, that is, the length of the hypotenuse.

$$c^2 = a^2 + b^2$$
$$c^2 = 90^2 + 90^2$$
$$c^2 = 8100 + 8100$$
$$c^2 = 2 \cdot 8100$$
$$c = \sqrt{8100 \cdot 2} \quad ❷$$
$$c = 90\sqrt{2} \approx 90(1.414)$$
$$\approx 127$$

$\sqrt{2} \approx 1.414$

An approximation of $90\sqrt{2}$ is 127. So, the distance from home plate to second base is approximately 127 feet.

12-8

Lesson Notes, page T49
Enrichment, page T49

Readiness Activity
State whether each of the following triangles is a right triangle.

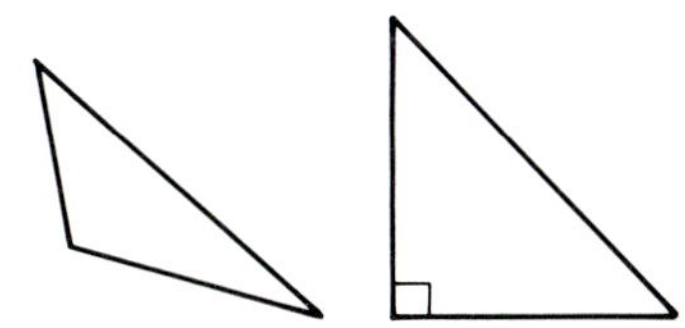

no yes

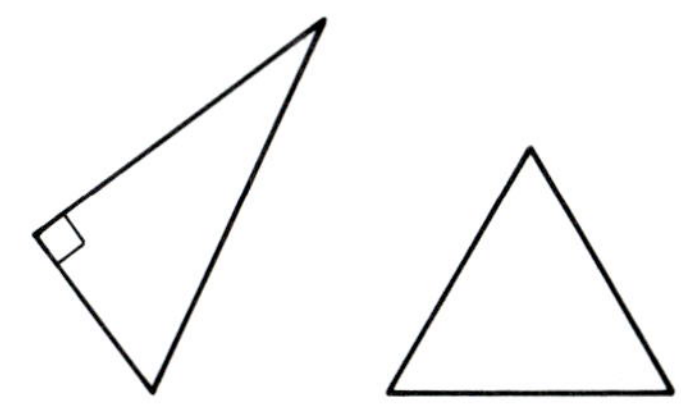

yes no

Discussion Questions

1. An isosceles triangle has two sides of equal length. Can an isosceles triangle be a right triangle? yes
2. An equilateral triangle has three sides of equal length. Can an equilateral triangle be a right triangle? no

❶ Point out that the Pythogorean Theorem can be used even when the sides of the right triangle are given as radicals.

❷ When the Pythagorean Theorem is used to find the length of the hypotenuse of a triangle, the negative square root does not need to be considered. Ask students to explain why.
The length of a side of a triangle is always positive.

Chalkboard Exercises

In the following exercises, c is the measure of the hypotenuse of a right triangle. Find each missing measure. Use the table on page 544.

1a. $a = 12, b = 16, c = ?$ 20

1b. $a = 11, b = \sqrt{23}, c = ?$ 12

1c. $a = \sqrt{5}, b = \sqrt{7}, c = ?$ $2\sqrt{3}$ or 3.464

2a. $a = 4, c = 10, b = ?$ $2\sqrt{21}$ or 9.165

2b. $a = 5, c = 6, b = ?$ $\sqrt{11}$ or 3.317

2c. $c = 10, b = 7, a = ?$ $\sqrt{51}$ or 7.141

❸ Remind students that the result, $4\sqrt{10}$, is an irrational number.

The Pythagorean Theorem can be illustrated geometrically as shown below.

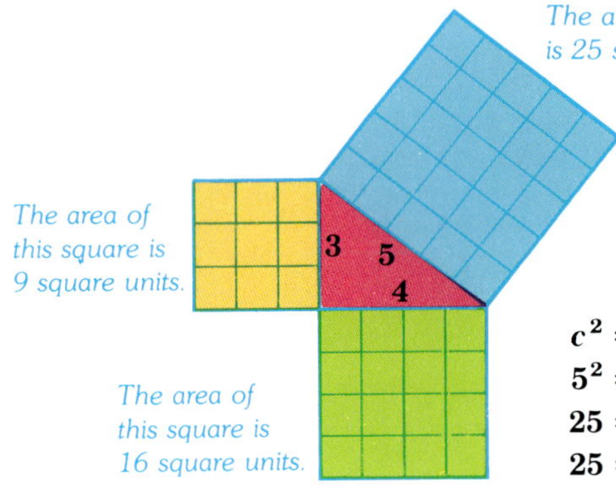

$c^2 = a^2 + b^2$
$5^2 = 3^2 + 4^2$
$25 = 9 + 16$
$25 = 25$

The Pythagorean Theorem can be used to find the length of a side of a right triangle when the lengths of the other two sides are known.

Examples

1 Find the length of the hypotenuse of a right triangle if $a = 15$ and $b = 8$.

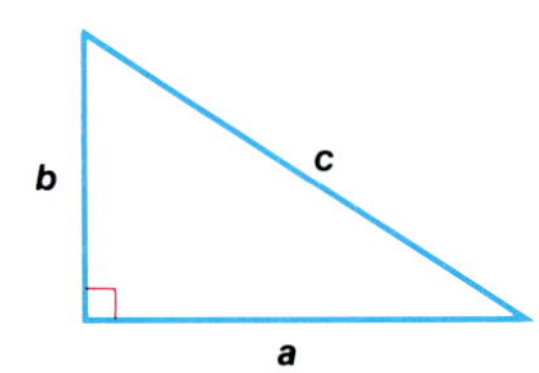

$$c^2 = 15^2 + 8^2$$
$$c^2 = 225 + 64$$
$$c^2 = 289$$
$$c = 17$$

The length of the hypotenuse is 17 units.

2 Find the length of a leg of a right triangle if $a = 6$ and $c = 14$.

$$c^2 = a^2 + b^2$$
$$14^2 = 6^2 + b^2$$
$$196 = 36 + b^2$$
$$160 = b^2$$
$$\sqrt{160} = b$$
$$b = 4\sqrt{10} \approx 4(3.162) \qquad \sqrt{10} \approx 3.162$$
$$\approx 12.648$$

The length of the leg is $4\sqrt{10}$ or approximately 12.648 units. ❸

Example 3

The measures of three sides of a triangle are 5, 7, and 9. Determine whether this triangle is a right triangle.

Since the measure of the longest side is 9, let $c = 9$, $a = 5$, and $b = 7$. Then check to see if $c^2 = a^2 + b^2$.

$$c^2 = a^2 + b^2$$
$$9^2 \stackrel{?}{=} 5^2 + 7^2$$
$$81 \stackrel{?}{=} 25 + 49$$
$$81 \neq 74$$

The numbers do not satisfy the Pythagorean Theorem. Therefore, they are *not* the measures of the sides of a right triangle.

Chalkboard Exercises

The measures of three sides of a triangle are given in each exercise below. Determine whether each triangle is a right triangle.

3a. $a = 3$, $b = 8$, $c = \sqrt{73}$ yes

3b. $a = 8$, $b = 9$, $c = 1$ no

3c. $a = 13$, $b = 84$, $c = 85$ yes

Exploratory Exercises

State the value of the variable in each sentence.

1. $3^2 + 4^2 = c^2$ 5 **2.** $6^2 + 8^2 = c^2$ 10 **3.** $5^2 + 12^2 = c^2$ 13 **4.** $a^2 + 12^2 = 13^2$ 5

5. $a^2 + 15^2 = 17^2$ 8 **6.** $a^2 + 24^2 = 25^2$ 7 **7.** $6^2 + b^2 = 10^2$ 8 **8.** $3^2 + b^2 = 5^2$ 4

State whether each sentence is true or false.

9. $3^2 + 4^2 = 5^2$ true **10.** $9^2 + 10^2 = 11^2$ false **11.** $6^2 + 8^2 = 9^2$ false **12.** $5^2 + 12^2 = 13^2$ true

Written Exercises

In exercises 1-12 below, *c* is the measure of the hypotenuse of a right triangle. Find each missing measure. Use the table on page 544.

1. $a = 5$, $b = 12$, $c = ?$ 13 **2.** $a = 6$, $b = 3$, $c = ?$ $3\sqrt{5}$ or 6.708 **3.** $a = 4$, $b = \sqrt{11}$, $c = ?$ $3\sqrt{3}$ or 5.196

4. $a = \sqrt{7}$, $b = 9$, $c = ?$ **5.** $b = 12$, $c = 15$, $a = ?$ 9 **6.** $b = 10$, $c = 11$, $a = ?$

7. $b = 30$, $c = 34$, $a = ?$ 16 **8.** $a = 11$, $c = 61$, $b = ?$ 60 **9.** $a = 20$, $c = 29$, $b = ?$ 21

10. $a = \sqrt{5}$, $c = \sqrt{30}$, $b = ?$ 5 **11.** $a = \sqrt{7}$, $b = \sqrt{9}$, $c = ?$ 4 **12.** $a = \sqrt{11}$, $c = \sqrt{47}$, $b = ?$ 6

The measures of three sides of a triangle are given in each of the following. Determine whether each triangle is a right triangle.

13. 9, 16, 20 no **14.** 9, 40, 41 yes **15.** 45, 60, 75 yes **16.** 11, 12, 16 no

For each problem, make a drawing. Then use an equation to solve the problem. Use the table on page 544.

17. Find the length of the diagonal of a rectangle whose length is 8 meters and whose width is 5 meters. $\sqrt{89}$ or 9.434 m

18. Justin hikes 7 miles due east and then 3 miles due north. How far is he from the starting point? $\sqrt{58}$ or 7.616 mi

19. The diagonal of a rectangular wall measures 14 meters. One side of the wall is 10 meters long. What is the length of the other side? $4\sqrt{6}$ or 9.798 m

20. A rope from the top of a mast on a sailboat is attached to a point 2 meters from the base of the mast. The rope is 8 meters long. How high is the mast? $2\sqrt{15}$ or 7.746 m

Suggested Daily Assignments

Enriched
Exp: 2–12 evens
Written: 1–20

Average
Exp: 1–12
Written: 1–20

Fundamental
Exp: 1–12
Written: 1–16
p. 390, 21–34

Answers to Written

4. $2\sqrt{22}$ or 9.381

6. $\sqrt{21}$ or 4.583

12-9

Lesson Notes, page T49
Enrichment, page T49

Readiness Activity
How many units long are each of the following line segments?

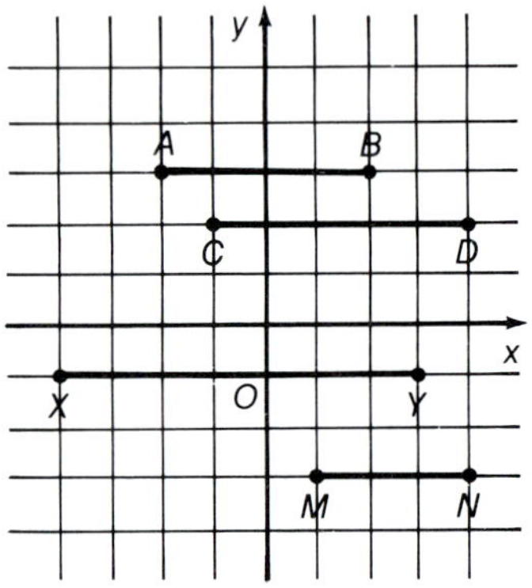

$AB = 4$, $CD = 5$, $XY = 7$, $MN = 3$

Discussion Question
Why is it easy to find the distance between two points in the coordinate plane that form either a vertical or a horizontal line segment?
You can just subtract the coordinates that are different.

Chalkboard Exercises
Find the distance between each pair of points whose coordinates are given.

1a. (2, 0); (3, 5) $\sqrt{26}$ or 5.099
1b. (1, 1); (4, 4) $3\sqrt{2}$ or 4.243
1c. (−1, −2); (−3, 0) $2\sqrt{2}$ or 2.828
1d. (1, −2); (7, 8) $2\sqrt{34}$ or 11.662
1e. (−2, −3); (−9, −10) $7\sqrt{2}$ or 9.899

Objective 12-9: Find the distance between two points in the coordinate plane.

12-9 The Distance Formula

The Pythagorean Theorem can be used to find the distance between any two points in the coordinate plane. For example, find the distance between (−5, 2) and (4, 5).

In the diagram at the right, the distance to be found is the length of segment d. Notice that a right triangle can be formed by drawing lines parallel to the axes from (−5, 2) and (4, 5). The coordinates of the point where these lines meet are (4, 2).

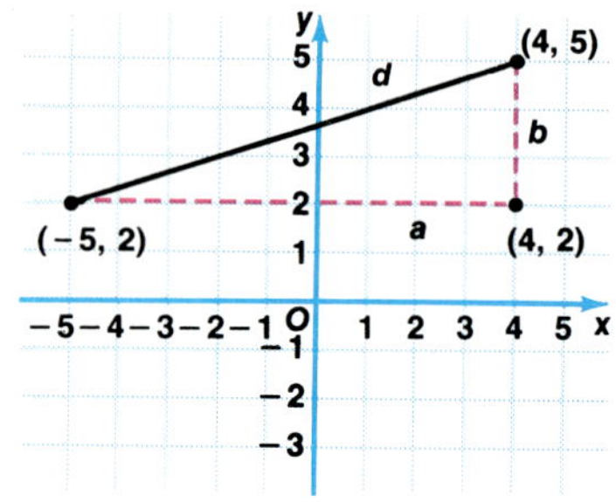

The length of side a of the triangle is the difference of the x-coordinates, $4 - (-5)$, or 9 units. The length of side b of the triangle is the difference of the y-coordinates, $5 - 2$, or 3 units.

Now the Pythagorean Theorem can be used to find the length of d, that is, the distance between (−5, 2) and (4, 5).

$$d^2 = 9^2 + 3^2$$
$$d^2 = 81 + 9$$
$$d^2 = 90$$
$$d = \sqrt{90}$$
$$d = 3\sqrt{10}$$

The distance between (−5, 2) and (4, 5) is $3\sqrt{10}$ units. The method for finding the distance between any two points can be generalized as the following formula.

The distance between any two points with coordinates (x_1, y_1) and (x_2, y_2) is given by the following formula:

$$d = \sqrt{(x_2 - x_1)^2 + (y_2 - y_1)^2}$$

The Distance Formula

Example

1 **Find the distance between the points with coordinates (8, −2) and (5, 3).**

$$d = \sqrt{(x_2 - x_1)^2 + (y_2 - y_1)^2}$$ *Let $x_2 = 5$, $x_1 = 8$, $y_2 = 3$, and $y_1 = -2$.*

$$d = \sqrt{(5 - 8)^2 + [3 - (-2)]^2}$$
$$d = \sqrt{(-3)^2 + 5^2}$$
$$d = \sqrt{9 + 25}$$
$$d = \sqrt{34}$$ *This is approximately 5.831 units.*

Example

2 **Find y if the distance between the points with coordinates (5, y) and (7, −3) is $\sqrt{85}$ units.**

$$d = \sqrt{(x_2 - x_1)^2 + (y_2 - y_1)^2} \quad \textit{Let } x_2 = 7,\ x_1 = 5,\ y_2 = -3, \textit{ and } y_1 = y.$$
$$\sqrt{85} = \sqrt{(7 - 5)^2 + (-3 - y)^2}$$
$$\sqrt{85} = \sqrt{2^2 + (-3 - y)^2}$$
$$\sqrt{85} = \sqrt{4 + 9 + 6y + y^2}$$
$$\sqrt{85} = \sqrt{13 + 6y + y^2}$$
$$85 = 13 + 6y + y^2 \quad \textit{Square both sides.}$$
$$0 = y^2 + 6y - 72$$
$$0 = (y + 12)(y - 6) \quad \textit{Factor.}$$
$$y + 12 = 0 \quad \text{or} \quad y - 6 = 0$$
$$y = -12 \qquad y = 6$$

The value of y is −12 or 6. ❶ *These answers can be checked by substituting −12 and 6 for y in the equation $\sqrt{85} = \sqrt{(7 - 5)^2 + (-3 - y)^2}$.*

Exploratory Exercises

State the values of x_1, x_2, y_1, and y_2 for each pair of points.

1. (3, 4), (6, 8) 3, 6, 4, 8
2. (5, −1), (11, 7) 5, 11, −1, 7
3. (−4, 2), (4, 17) −4, 4, 2, 17
4. (−2, 8), (3, 20) −2, 3, 8, 20
5. (−3, 5), (2, 7) −3, 2, 5, 7
6. (5, 4), (−3, 8) 5, −3, 4, 8
7. (−8, −4), (−3, 8) −8, −3, −4, 8
8. (2, 7), (10, −4) 2, 10, 7, −4
9. (3, 7), (−2, −5) 3, −2, 7, −5
10. (3, 2), (0, 5) 3, 0, 2, 5
11. (2, 2), (5, −1) 2, 5, 2, −1
12. (−8, −7), (0, 8) −8, 0, −7, 8

Written Exercises

Find the distance between each pair of points whose coordinates are given.

1. (−4, 2), (4, 17) 17
2. (5, −1), (11, 7) 10
3. (−3, 5), (2, 7) $\sqrt{29}$ or 5.385
4. (5, 4), (−3, 8) $4\sqrt{5}$ or 8.944
5. (−8, −4), (−3, 8) 13
6. (2, 7), (10, −4)
7. (7, −9), (4, −3) $3\sqrt{5}$ or 6.708
8. (9, −2), (3, −6) $2\sqrt{13}$ or 7.211
9. (10, 8), (2, −3)
10. (11, −2), (−4, 5)
11. (−2, 5), $\left(-\frac{1}{2}, 3\right)$ $\frac{5}{2}$
12. (4, 2), $\left(6, -\frac{2}{3}\right)$ $\frac{10}{3}$
13. $\left(\frac{2}{3}, -4\right)$, (3, −2) $\frac{\sqrt{85}}{3}$
14. $\left(6, -\frac{2}{7}\right)$, (5, −1) $\frac{\sqrt{74}}{7}$
15. $\left(\frac{4}{5}, -1\right)$, $\left(2, -\frac{1}{2}\right)$ $\frac{13}{10}$

The coordinates of a pair of points are given in each of the following. Find two possible values for a if the points are the given distance apart.

16. (4, 7), (a, 3); d = 5 7, 1
17. (3, a), (−4, 2); $d = \sqrt{170}$ 13, −9
18. (8, 1), (5, a); d = 5 5, −3
19. (−3, a), (5, 2); d = 17 17, −13
20. (a, 5), (−7, 3); $d = \sqrt{29}$ −2, −12
21. (5, 9), (a, −3); d = 13 10, 0
22. (a, −4), (2, −3); $d = \sqrt{65}$ 10, −6
23. (−6, −5), (−3, a); $d = \sqrt{13}$ −7, −3
24. (4, −7), (7, a); $d = \sqrt{34}$ −12, −2
25. (−5, a), (4, −2); $d = \sqrt{130}$ −9, 5

Chalkboard Exercises

The coordinates of a pair of points are given in each of the following. Find two possible values for k if the points are the given distance apart.

2a. (3, 5), (−3, k); d = 10
−3, 13

2b. (1, k), (−3, 8); $d = \sqrt{17}$
7, 9

2c. (k, 2), (4, 5); $d = \sqrt{34}$
−1, 9

2d. (2, 0), (k, 5); d = 13
−10, 14

❶ Encourage students to check the results. Point out that the negative result does *not* need to be omitted. It represents the y-coordinate of a point in the plane, not a distance.

Suggested Daily Assignments

Enriched
- Written: 2–14 evens, 16–28
- Mini-Review: 1–5

Average
- Exp: 1–9 odds
- Written: 1–25
- Mini-Review: 1–5

Fundamental
- Exp: 1–9
- Written: 1–25
  - p. 393, 17–20
- Mini-Review: 1–5

Answers to Written

6. $\sqrt{185}$ or 13.601
9. $\sqrt{185}$ or 13.601
10. $\sqrt{274}$ or 16.553

Answers to Challenge

26. 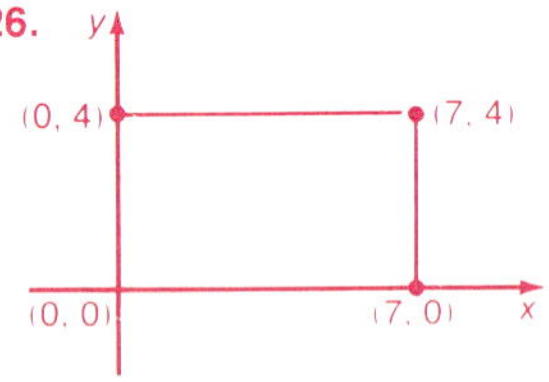

The opposite sides measure 7 and 4. The diagonals measure $\sqrt{65}$.

27. The distance from (2, 6) to (−1, 2) is 5. The distance from (2, 6) to (5, 10) is 5. This distance from (−1, 2) to (5, 10) is 10.

About the Excursion
Point out that a circle does *not* have a unique equation—many different equivalent equations can be written. One equation can be obtained using the distance formula as shown in the Excursion.

Challenge

Use the distance formula to solve each of the following.

26. Show that the points with the coordinates (0, 0), (7, 0), (7, 4), and (0, 4) are the vertices of a rectangle.

27. Show that the point with coordinates (2, 6) is the midpoint of the segment joining the points with coordinates (−1, 2) and (5, 10).

28. Find the distance between $A(\sqrt{8}, \sqrt{3})$ and $B(\sqrt{3}, -\sqrt{8})$. $\sqrt{22}$ or 4.690

mini-review

Write *true* or *false*.

1. The equation $2y = -x + 1$ is in slope-intercept form. false
2. The relation $4x + y = 2$ is a function. true
3. The ordered pair (4, −6) is a solution to the equation $3y - 2 = x$. false
4. The slope of the line with the equation $2y + x = 4$ is $-\frac{1}{2}$. true
5. Parallel lines have the same slope. true

Excursions in Algebra — Equations of Circles

A circle is the set of all points in the coordinate plane that are the same distance from a given point, called the *center.* All points of the circle shown at the right are 4 units from the point with coordinates (0, 0), the center. Choose any point of the circle. The distance from a point (x, y) to the center is 4 units. Now apply the distance formula.

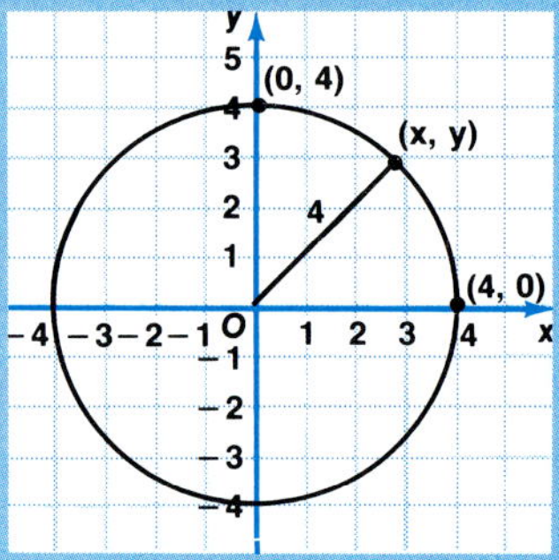

$$d = \sqrt{(x_2 - x_1)^2 + (y_2 - y_1)^2}$$
$$4 = \sqrt{(x - 0)^2 + (y - 0)^2}$$
$$4 = \sqrt{x^2 + y^2}$$
$$16 = x^2 + y^2 \quad \textit{Square both sides.}$$

Thus, $16 = x^2 + y^2$ is the equation for a circle with center at (0, 0) and radius of 4 units.

The equation of a circle with center at (0, 0) and a radius of r units is $x^2 + y^2 = r^2$. ***Equation of a Circle***

Example: **Find the equation of a circle with center at (0, 0) and a radius of 7 units.**

$$x^2 + y^2 = r^2$$
$$x^2 + y^2 = 7^2 \quad \textit{Substitute 7 for r.}$$
$$x^2 + y^2 = 49$$

Exercises **State the radius of each circle whose equation is given below.**

1. $x^2 + y^2 = 100$ 10
2. $x^2 + y^2 = 81$ 9
3. $x^2 + y^2 = 8$ $2\sqrt{2}$
4. $x^2 + y^2 = 10$ $\sqrt{10}$
5. $x^2 + y^2 = \frac{1}{4}$ $\frac{1}{2}$
6. $x^2 + y^2 = \frac{4}{9}$ $\frac{2}{3}$
7. $x^2 + y^2 = 256$ 16
8. $x^2 + y^2 = 576$ 24
9. $x^2 + y^2 = 0.09$ 0.3
10. $x^2 + y^2 = 0.16$ 0.4
11. $x^2 + y^2 = \frac{100}{121}$ $\frac{10}{11}$
12. $x^2 + y^2 = \frac{196}{625}$ $\frac{14}{25}$

Using Computers

Computing Square Roots

Recall that one method for computing a square root, called the divide-and-average method, involves first choosing any approximation for the square root and then calculating more accurate approximations.

Another method for computing $\sqrt{N}$ is to substitute any approximation, A, into the expression $\frac{A^2 + N}{2A}$. Then repeat this process until successive values of A are close together. This method is demonstrated in the following computer program.

```
10   PRINT "FOR WHAT NUMBER DO YOU WANT
     TO FIND THE SQUARE ROOT?"
20   INPUT N
30   PRINT "CHOOSE ANY APPROXIMATION FOR THE SQUARE ROOT"
40   INPUT A
50   LET L = A
60   PRINT A
70   LET A = (A ↑ 2 + N) / (2 * A)
80   IF ABS (A - L) > .000001 THEN 50
90   PRINT "AN APPROXIMATION OF THE SQUARE ROOT OF ";
       N;" IS ";A
100  END
```

The ABS function finds the absolute value.

Notice that the variable L is used to store the old value of A. Then, in line 60, the next value of A is computed. Finally, in line 70, the values of A and L are compared. If the values are close together (in this case, if their difference is less than .000001), then the program ends. If the difference is not less than .000001, the loop is executed again to obtain a closer approximation.

Exercises

Use the program above to approximate each of the following square roots.

1. $\sqrt{10}$ 3.16228 **2.** $\sqrt{19}$ 4.35890 **3.** $\sqrt{23}$ 4.79583 **4.** $\sqrt{92}$ 9.59166 **5.** $\sqrt{198}$ 14.07125

6. $\sqrt{147}$ 12.12436 **7.** $\sqrt{323}$ 17.97220 **8.** $\sqrt{582}$ 24.12468 **9.** $\sqrt{1901}$ 43.60046 **10.** $\sqrt{5460}$ 73.89181

11. In the BASIC language, SQR(A) can be used to find an approximation of the square root of *A*. Write an additional line for the program above to compute and print an approximation using the SQR function. Run the program to compare the two approximations for exercises 1-10.
95 PRINT "ANOTHER APPROXIMATION OF THIS SQUARE ROOT IS "; SQR(N)

About Using Computers
Point out to students that they can use the expression $\frac{A^2 + N}{2A}$ to manually approximate a square root. Although the computer will find an approximation much faster, this method is still a valuable tool.

You may want to explain how the expression $\frac{A^2 + N}{2A}$ was derived. Suppose A is the first approximation for $\sqrt{N}$. By the divide-and-average method, the new approximation would be $\frac{(N \div A) + A}{2}$ or $\frac{A^2 + N}{2A}$. This expression produces the same results as the divide-and-average method.

Quiz
Lessons 12–7 to 12–9

Solve.

1. $\sqrt{x + 1} = 2$ 3
2. $3\sqrt{t + 4} - 2 = 7$ 5
3. Find the length, to 3 decimal places, of the hypotenuse of a right triangle if $a = 5$ and $b = 7$. 8.602
4. Find the length, to 3 decimal places, of a leg of a right triangle if $a = 24$ and $c = 25$. 7
5. Find the distance between $(-6, 4)$ and $(-2, -3)$. $\sqrt{65}$ or 8.062 units
6. Find x if the distance between $(-5, 6)$ and $(x, -4)$ is $\sqrt{164}$. 3 or -13

Vocabulary

square root (371)
radical expression (371)
radical sign (371)
radicand (371)
principal square root (371)
product property of square roots (372)
quotient property of square roots (372)
irrational numbers (374)
real numbers (375)
completeness property for points on the number line (375)
divide-and-average method (377)
rationalizing the denominator (382)
conjugate (383)
radical equations (389)
hypotenuse (391)
legs (391)
Pythagorean Theorem (391)
distance formula (394)

Chapter Summary

1. Definition of Square Root: If $x^2 = y$ then x is a square root of y. (371)
2. The symbol $\sqrt{}$, called a radical sign, indicates a nonnegative square root. The expression under the radical sign is called the radicand. (371)
3. Product Property of Square Roots: For any numbers a and b, if $a \geq 0$ and $b \geq 0$, then $\sqrt{ab} = \sqrt{a} \cdot \sqrt{b}$. (372)
4. Quotient Property of Square Roots: For any numbers a and b, if $a \geq 0$ and $b > 0$, then $\sqrt{\frac{a}{b}} = \frac{\sqrt{a}}{\sqrt{b}}$. (372)
5. The square roots of numbers such as 2, 3, 5, and 7 are irrational numbers. Irrational numbers are numbers that cannot be expressed in the form $\frac{a}{b}$, where a and b are integers, $b \neq 0$. (374)
6. Irrational numbers together with the rational numbers form the set of real numbers. Each real number corresponds to exactly one point on the number line. Each point on the number line corresponds to exactly one real number. (375)
7. One method for approximating square roots is the divide-and-average method. (377)
8. Rationalizing the denominator changes the denominator to a rational number. (382)
9. Binomials of the form $a\sqrt{b} + c\sqrt{d}$ and $a\sqrt{b} - c\sqrt{d}$ are called conjugates of each other. They can be used to rationalize some denominators. (383)

10. A radical expression is in simplest form when the following conditions are met.
   1. No radicands have perfect square factors other than 1.
   2. No radicands contain fractions.
   3. No radical appears in the denominator of a fraction. (383)
11. Square roots having like radicands can be added or subtracted. (385)
12. To solve radical equations, first isolate the radical on one side of the equation. Then square both sides of the equation to eliminate the radical. (389)
13. The Pythagorean Theorem: In a right triangle if a and b are the measures of the legs and c is the measure of the hypotenuse, then $a^2 + b^2 = c^2$. (391)
14. The Distance Formula: The distance between any two points with coordinates (x_1, y_1) and (x_2, y_2) is given by the following formula.

$$d = \sqrt{(x_2 - x_1)^2 + (y_2 - y_1)^2} \quad (394).$$

Chapter Review

12-1 Simplify.

1. $\sqrt{121}$ 11
2. $-\sqrt{64}$ −8
3. $\pm\sqrt{\frac{4}{81}}$ $\pm\frac{2}{9}$
4. $-\sqrt{\frac{100}{225}}$ $-\frac{2}{3}$

12-2 Graph each number on the number line.

5. $-\sqrt{5}$
6. $\sqrt{15}$

State whether each decimal represents a rational number or an irrational number.

7. $0.4\overline{6}$ rational
8. 4.2302300230002 . . . irrational

Use a table to determine the integers between which the value of the square root lies.

9. $\sqrt{250}$ 15 and 16
10. $\sqrt{490}$ 22 and 23

12-3 Approximate each square root. Use the divide-and-average method twice for each exercise.

11. $\sqrt{19}$ 4.359
12. $\sqrt{77}$ 8.775
13. $\sqrt{61.7}$ 7.855

Answers to Chapter Review

5-6.

12-4 **Simplify. Use absolute value symbols when necessary.**

14. $\sqrt{18}$ $3\sqrt{2}$ **15.** $\sqrt{108}$ $6\sqrt{3}$ **16.** $\sqrt{720}$ $12\sqrt{5}$ **17.** $\sqrt{2916}$ 54

18. $\sqrt{4b^2}$ $2|b|$ **19.** $\sqrt{\frac{60}{x^2}}$ $\frac{2\sqrt{15}}{|x|}$ **20.** $\sqrt{44a^2b}$ $2|a|\sqrt{11b}$ **21.** $\sqrt{6} \cdot \sqrt{8}$ $4\sqrt{3}$

22. $5\sqrt{3} \cdot \sqrt{3}$ 15 **23.** $\sqrt{3}(\sqrt{3} + \sqrt{6})$ $3 + 3\sqrt{2}$ **24.** $\sqrt{5}(\sqrt{10} - \sqrt{3})$ $5\sqrt{2}$ −

12-5 **Simplify.**

25. $\frac{\sqrt{35}}{\sqrt{5}}$ $\sqrt{7}$ **26.** $\frac{\sqrt{20}}{\sqrt{7}}$ $\frac{2\sqrt{35}}{7}$ **27.** $\frac{7}{2 - \sqrt{3}}$ $14 + 7\sqrt{3}$

28. $\frac{9}{3 + \sqrt{2}}$ $\frac{27 - 9\sqrt{2}}{7}$ **29.** $\frac{5\sqrt{2}}{7\sqrt{3} + 6\sqrt{5}}$ $\frac{35\sqrt{6} - 30\sqrt{10}}{-33}$ **30.** $\frac{3\sqrt{7}}{4\sqrt{6} - 7\sqrt{7}}$ $\frac{12\sqrt{42} + 147}{-247}$

12-6 **Simplify.**

31. $2\sqrt{13} + 3\sqrt{13}$ $5\sqrt{13}$ **32.** $8\sqrt{15} - 3\sqrt{15}$ $5\sqrt{15}$ **33.** $3\sqrt{7} - 5\sqrt{28}$ $-7\sqrt{7}$

34. $4\sqrt{27} + 6\sqrt{48}$ $36\sqrt{3}$ **35.** $\sqrt{8} + \sqrt{\frac{1}{8}}$ $\frac{9\sqrt{2}}{4}$ **36.** $\sqrt{3} - \sqrt{\frac{1}{3}}$ $\frac{2\sqrt{3}}{3}$

12-7 **Solve and check.**

37. $\sqrt{3x} = 6$ 12 **38.** $\sqrt{3x + 5} = 4$ $\frac{11}{3}$ **39.** $\sqrt{7x - 1} - 5 = 0$ $\frac{26}{7}$

40. $\sqrt{w} = 5\sqrt{6}$ 150 **41.** $\sqrt{x + 4} = x - 8$ 12 **42.** $\sqrt{\frac{4a}{3}} - 2 = 0$ 3

12-8 **Use the Pythagorean Theorem to find each missing measure.**

43. $a = 6, b = 10, c = ?$

44. $a = 10, c = 15, b = ?$

45. $b = 6, c = 12, a = ?$

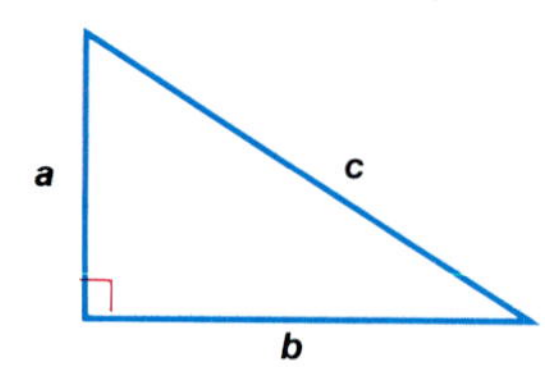

43. $2\sqrt{34}$ or 11.662

44. $5\sqrt{5}$ or 11.180

45. $6\sqrt{3}$ or 10.392

46. Find the length of the diagonal of a rectangle whose length is 15 meters and whose width is 8 meters. 17 m

12-9 **Find the distance between each pair of points whose coordinates are given.**

47. (4, 2), (7, −9) $\sqrt{130}$ or 11.402 **48.** (5, −2), (−8, −3) $\sqrt{170}$ or 13.038

Chapter Test

Simplify.

1. $\sqrt{3} \cdot \sqrt{15}$ $3\sqrt{5}$
2. $\sqrt{40}$ $2\sqrt{10}$
3. $\dfrac{5}{\sqrt{2} + \sqrt{3}}$ $5\sqrt{3} - 5\sqrt{2}$
4. $\sqrt{\dfrac{a^2}{7}}$ $\dfrac{|a|\sqrt{7}}{7}$
5. $\sqrt{54x^4y}$ $3x^2\sqrt{6y}$
6. $(4 + \sqrt{3})(4 - \sqrt{3})$ 13
7. $\dfrac{7}{7 + \sqrt{5}}$ $\dfrac{49 - 7\sqrt{5}}{44}$
8. $2\sqrt{27} + \sqrt{48} - 3\sqrt{3}$ $7\sqrt{3}$
9. $3\sqrt{50} - 2\sqrt{8}$ $11\sqrt{2}$
10. $\sqrt{\dfrac{3x^2}{4n^3}}$ $\dfrac{|x|\sqrt{3n}}{2n^2}$
11. $\sqrt{3xy^3}$ $|y|\sqrt{3xy}$
12. $15\sqrt{6} - \sqrt{72}$ $15\sqrt{6} - 6\sqrt{2}$
13. $\sqrt{\dfrac{32}{9}}$ $\frac{4}{3}\sqrt{2}$
14. $\sqrt{\dfrac{64x^4}{8x^2}}$ $2|x|\sqrt{2}$
15. $\dfrac{6\sqrt{5}}{2\sqrt{2} + 5\sqrt{3}}$ $\dfrac{12\sqrt{10} - 30\sqrt{15}}{-67}$
16. $\sqrt{2}(\sqrt{2} + 4\sqrt{3})$ $2 + 4\sqrt{6}$
17. $\sqrt{6} + \sqrt{\dfrac{2}{3}}$ $\frac{4}{3}\sqrt{6}$
18. $\sqrt{96} \cdot \sqrt{48}$ $48\sqrt{2}$

Solve and check.

19. $\sqrt{t} + 5 = 3$ no solution
20. $\sqrt{5x^2 - 9} = 2x$ 3
21. $\sqrt{4x + 1} = 5$ 6
22. $\sqrt{4x - 3} = 6 - x$ 3

Use the Pythagorean Theorem to find the missing measures.

23. Find c if $a = 8$ and $b = 10$.
24. Find b if $a = 2$ and $c = 8$.
25. Find a if $a = b$ and $c = 12$.

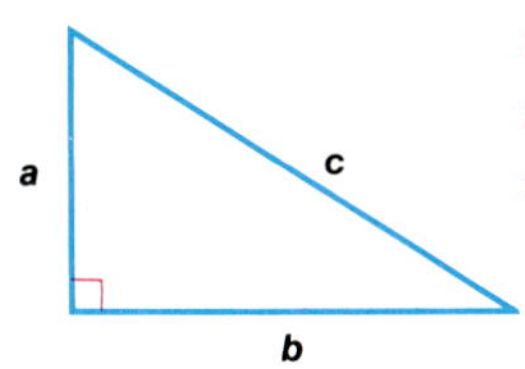

23. $2\sqrt{41}$ or 12.806
24. $2\sqrt{15}$ or 7.746
25. $6\sqrt{2}$ or 8.485

Find the distance between each pair of points whose coordinates are given.

26. (4, 3), (6, −3) $2\sqrt{10}$ or 6.235
27. (−8, 2), (4, −3) 13

Using the Cumulative Review
If students have difficulty with a particular topic, have them review the appropriate lesson. You may want to provide copies of the solutions from the *Algebra One Solutions Manual*. This Cumulative Review covers Chapters 1-12.

Other Cumulative Reviews appear on pages 72, 142, 208, 268, 334, 466, and 526.

Answers

16. 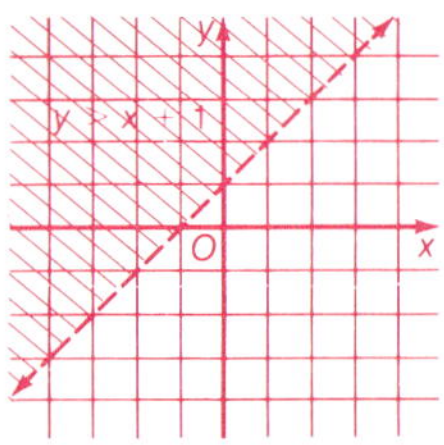

21. 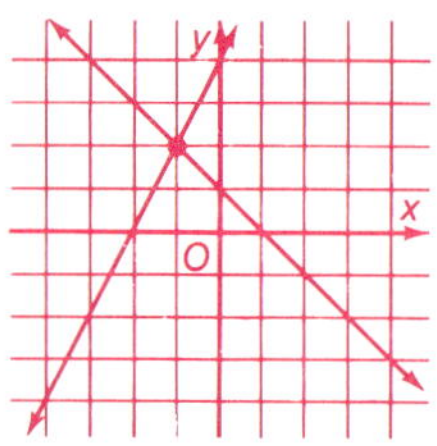

Cumulative Review

1. Evaluate $a^2 - 3c - y$ if $a = 3$, $c = \frac{2}{3}$, and $y = -\frac{4}{3}$. $\frac{25}{3}$
2. Find the sum: $-9.8 + 4.3 + 3.7$ -1.8

Solve.

3. $b - (-11) = -11$ -22
4. $\frac{3}{4}a - 2 = 7$ 12
5. $0.2(x + 3) = 0.3(x - 0.4)$ 7.2
6. $7p > -63$ $p > -9$
7. $5 - 3k \leq k$ $k \geq \frac{5}{4}$

Simplify.

8. $(3a^2)(4a^3)(2a)$ $24a^6$
9. $\frac{5a^{-2}b}{6a^2b^{-4}}$ $\frac{5b^5}{6a^4}$
10. $(y + 3)(2y - 1)$ $2y^2 + 5y - 3$

Factor, if possible.

11. $y^2 + 2y - 63$ $(y + 9)(y - 7)$
12. $18x^2 + 45xy + 18y^2$ $9(2x + y)(x + 2y)$
13. Solve: $6x^2 - 17x + 12 = 0$ $\frac{3}{2}, \frac{4}{3}$
14. Write the following relation as a set of ordered pairs. Then state the domain and range.

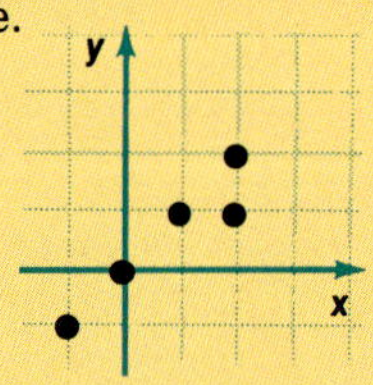

15. Is the relation in Exercise 14 a function? no
16. Graph: $y > x + 1$
17. Write an equation in standard form for the line through $(2, -1)$ and $(4, 1)$. $x - y = 3$
18. Write an equation in slope-intercept form for the line having a slope of $\frac{2}{3}$ and passing through $(5, 1)$. $y = \frac{2}{3}x - \frac{7}{3}$
19. Write an equation for the line that is parallel to the graph of $x + 2y = -1$ and passes through $(1, 4)$. Use slope-intercept form. $y = -\frac{1}{2}x + \frac{9}{2}$
20. If the midpoint of line segment AB is $(7, 2)$, find the coordinates of point A if the coordinates of point B are $(2, -5)$. $(12, 9)$
21. Graph the pair of equations $x + y = 1$ and $-2x + y = 4$. Then state the solution of the system. $(-1, 2)$
22. Use substitution to solve the system of equations $2x - 3y = 5$ and $x - y = 2$. $(1, -1)$
23. Use elimination to solve the system of equations $x - 2y = 3$ and $2x - 3y = 4$. $(-1, -2)$

Simplify.

24. $\pm\sqrt{196}$ ± 14
25. $\sqrt{64x^2y^4}$ $8|x|y^2$
26. $\frac{1}{2 - \sqrt{2}}$ $\frac{2 + \sqrt{2}}{2}$
27. $3\sqrt{8} - 4\sqrt{2} + \sqrt{12}$ $2\sqrt{2} + 2\sqrt{3}$

Solve.

28. $\sqrt{3x - 5} = 5$ 10

Problem Solving

Solve each problem.

29. Find the customer price on a video game tape if there is a 15% discount on the marked price of \$27.95. \$23.76
30. Two cyclists start toward each other from two towns that are 112 miles apart. One cyclist rides at 16 mph and the other at 12 mph. In how many hours will they meet? 4
31. The square of a number added to 4 times the number equals 45. Find the number.
32. A two-digit number is 5 times the sum of its digits. The units digit is one more than the tens digit. Find the number. 45
33. An airplane travels 620 miles in 2 hours flying with the wind. Flying against the wind, it takes 4 hours to travel 1160 miles. Find the rate of the wind and the rate of the plane in still air. plane, 300 mph wind, 10 mph

14. The relation is $\{(-1, -1), (0, 0), (1, 1), (2, 1), (2, 2)\}$. The domain is $\{-1, 0, 1, 2\}$. The range is $\{-1, 0, 1, 2\}$.

31. -9 or 5

Standardized Test Practice Questions

The test questions on this page deal with coordinates and geometry. The information at the right may help you with some of the questions.

Directions: Choose the best answer. Write A, B, C, or D.

1. The midpoint of $\overline{AB}$ is M. If the coordinates of A are $(-3, 2)$ and the coordinates of M are $(-1, 5)$, what are the coordinates of B?

   (A) (1, 10) (B) (1, 8)
   (C) (0, 7) (D) (−5, 8)

2. 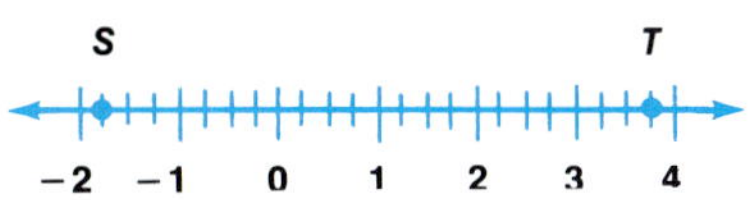

   The length of $\overline{ST}$ is

   (A) 5 (B) $4\frac{1}{2}$ (C) $5\frac{1}{2}$ (D) 6

3. What is the area of the shaded triangle in square units?

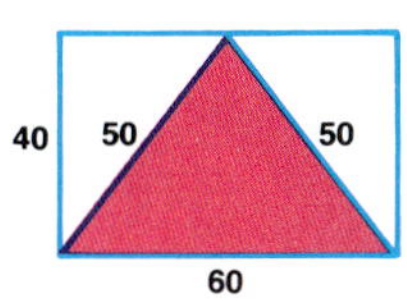

   (A) 1600 (B) 1200 (C) 1500 (D) 2400

4. Seven squares of the same size form a rectangle when placed side-by-side. The perimeter of the rectangle is 496. What is the area of each square?

   (A) 72 ft^2 (B) 324 ft^2
   (C) 900 ft^2 (D) 961 ft^2

5. What is the total length of fencing needed to enclose a rectangular area 46 feet by 34 feet?

   (A) 26 yards 1 foot (B) $26\frac{2}{3}$ yards
   (C) 52 yards 2 feet (D) $53\frac{1}{3}$ yards

> 1. The diameter of a circle is twice the length of the radius.
> 2. The area of a circle with radius r is πr^2.
> 3. The area of a triangle is the product of $\frac{1}{2}$ the base and the height.

6. The number of degrees in the smaller angle formed by the hands of a clock at 12:15 is

   (A) 120 (B) $82\frac{1}{2}$ (C) $92\frac{1}{2}$ (D) 90

7. The length of each side of a square is $\frac{3x}{5} + 1$. The perimeter of the square is

   (A) $\frac{12x + 20}{5}$ (B) $\frac{12x + 4}{5}$
   (C) $\frac{3x + 4}{5}$ (D) $\frac{3x}{5} + 16$

8. If a line passes through point (0, 2) and has a slope of 4, what is the equation of the line?

   (A) $x = 2y + 4$ (B) $x = 4y + 2$
   (C) $y = 4x + 2$ (D) $y = 2x + 4$

9. The larger circle has a diameter of b. The area of the shaded ring in square units is

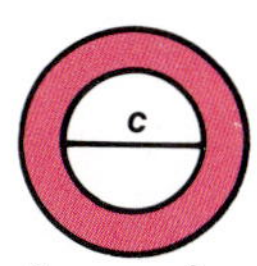

   (A) $b^2 - c^2$ (B) $\pi b^2 - \pi c^2$
   (C) $\frac{1}{4}\pi(b^2 - c^2)$ (D) $\frac{1}{2}\pi(b^2 - c^2)$

10. If a circle of radius 10 meters has its radius decreased by 2 meters, by what percent is its area decreased?

    (A) 20% (B) 40% (C) 80% (D) 36%

11. What is the value of x if the area of the triangle is $\frac{1}{4}$ the area of the square?

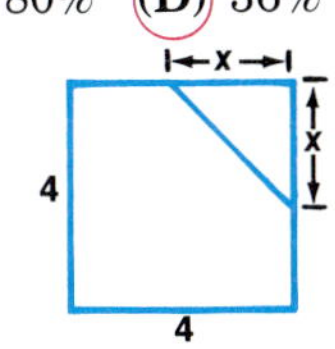

    (A) $\sqrt{2}$ (B) $2\sqrt{2}$ (C) 4 (D) 8

Using the Standardized Test Practice Questions

The questions on this page may be used to help students prepare for pre-college tests. These questions usually require careful analysis and a thorough understanding of the concepts.

This page could be used as an overnight assignment for better students. Other students may benefit most if only a few questions are assigned at a time.

After students have chosen an answer for each problem, you should discuss how each problem can be solved, or provide copies of the solutions from the *Algebra One Solutions Manual*.

Other practice questions appear in this text on pages 73, 143, 209, 269, 335, 467, and 527. Also see pages 50-58 in the *Teacher Resource Book*.

About the Photo

Mars Station is a tracking station complex located in Madrid, Spain. It is used to track and obtain data from Viking spacecraft and the planet Mars.

The tracking antenna in the photograph is a reflecting paraboloid of revolution illuminated by a source of radiation placed at its focus. See the Application on page 412.

A radar antenna operates on the principle of detecting pulses of energy which have been reflected from an object. The antenna system in the photograph is capable of transmitting the primary signal as well as receiving the reflected echo.

The distance from the antenna to the object is calculated by measuring the time required for the pulses of energy to travel to and from the object. The pulses of energy are radio waves which travel at a velocity of 186,000 miles per second. Distance measurements obtained using radar are extremely accurate even though the time required for a pulse to travel to an object one mile away and return to the transmitter is only 10.7 microseconds (10.7 millionths of a second).

Teacher Resource Book Materials for Chapter 13 See pages 303-320.

CHAPTER 13

Quadratics

Mars Station is the site of a tracking antenna for spacecraft. It is capable of tracking spacecraft to the edge of the solar system.

The reflector of this antenna is 64 meters in diameter, and has a shape whose cross section may be described by a quadratic equation. In this chapter, you will learn about quadratic equations and quadratic functions.

404

Objective 13-1A: Find the equation of the axis of symmetry and the coordinates of the maximum or minimum point of a quadratic function.

Objective 13-1B: Graph a quadratic function.

13-1 Graphing Quadratic Functions

An equation such as $y = x^2 - 4x + 1$ describes a type of function known as a **quadratic function**.

A quadratic function is a function that can be described by an equation of the form $y = ax^2 + bx + c$, where $a \neq 0$.

Definition of Quadratic Function

Graphs of quadratic functions have certain common characteristics. For instance, they all have a general shape called a **parabola**.

The table and graph below illustrate some other common characteristics of quadratic functions.

x	$x^2 - 4x + 1$	y
-1	$(-1)^2 - 4(-1) + 1$	6
0	$0^2 - 4(0) + 1$	1
1	$1^2 - 4(1) + 1$	-2
2	$2^2 - 4(2) + 1$	-3
3	$3^2 - 4(3) + 1$	-2
4	$4^2 - 4(4) + 1$	1
5	$5^2 - 4(5) + 1$	6

Notice the matching values in the y-column.

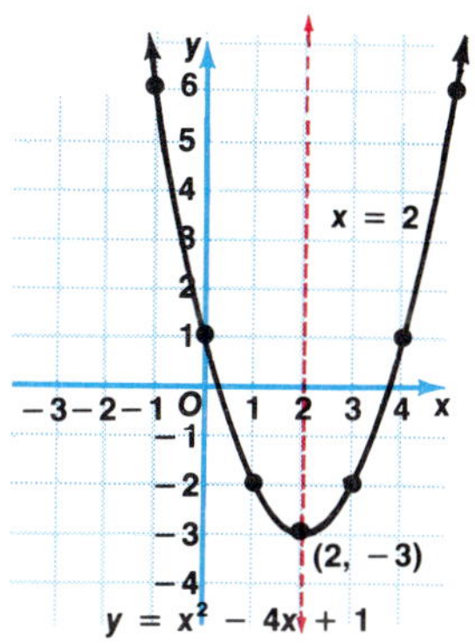

Notice that -3 does *not* have a matching value in the y-column of the table. Also, -3 is the y-coordinate of the lowest point on the graph of $y = x^2 - 4x + 1$. For the graph of $y = x^2 - 4x + 1$, the *lowest point*, called the **minimum point**, has coordinates $(2, -3)$. ❶

The vertical line containing the minimum point is the **axis of symmetry**. The equation of the axis of symmetry for the graph above is $x = 2$.

If the graph of any quadratic function is folded along the axis of symmetry, the two halves coincide. In other words, the two halves of the parabola are *symmetric*.

13-1

Lesson Notes, page T49

Discussion Question
(Use after page 405.)
Must a quadratic function cross the x-axis? The y-axis? no; yes

❶ The minimum or maximum point of a parabola is often called the *vertex*.

Chalkboard Exercises

Graph each of the following.

1a. $y = -x^2 - 2x + 5$

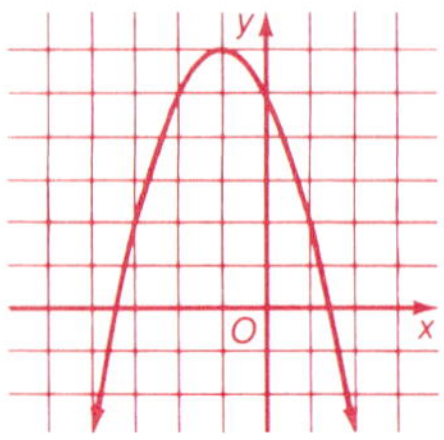

1b. $y = -x^2 - 6x - 6$

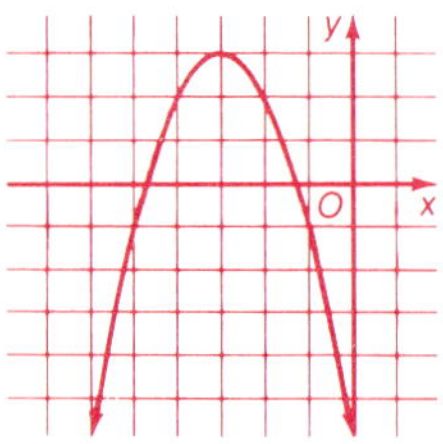

Example

1 **Graph:** $y = -x^2 + 2x + 3$

x	$-x^2 + 2x + 3$	y
-2	$-(-2)^2 + 2(-2) + 3$	-5
-1	$-(-1)^2 + 2(-1) + 3$	0
0	$-0^2 + 2(0) + 3$	3
1	$-(1)^2 + 2(1) + 3$	4
2	$-(2)^2 + 2(2) + 3$	3
3	$-(3)^2 + 2(3) + 3$	0
4	$-(4)^2 + 2(4) + 3$	-5

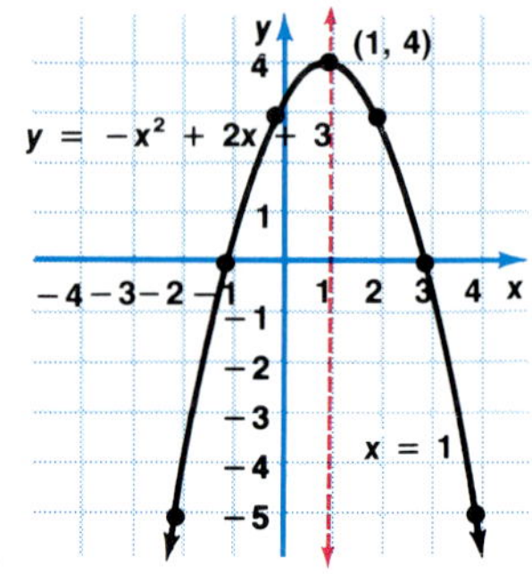

The graph of $y = -x^2 + 2x + 3$ opens *downward*. The equation of the axis of symmetry is $x = 1$. The graph has a *highest point,* or **maximum point**, at (1, 4).

In general, a parabola will open upward and have a minimum point when the coefficient of x^2 is positive. It will open downward and have a maximum point when the coefficient of x^2 is negative. The maximum or minimum point of the graph *always* lies on the axis of symmetry.

Notice that in Example 1 the axis of symmetry is halfway between any two points having the same y-coordinate. Consider the points on the graph whose coordinates are $(-1, 0)$ and $(3, 0)$.

From these coordinates, the equation of the axis of symmetry may be found as shown below.

$$x = \frac{-1 + 3}{2} \quad \text{Add the x-coordinates and divide by 2.}$$
$$= 1$$

The equation of the axis of symmetry is $x = 1$.

In general, the equation of the axis of symmetry for the graph of a quadratic function can be found by using the following rule.

Equation of Axis of Symmetry

The equation of the axis of symmetry for $y = ax^2 + bx + c$, where $a \neq 0$, is

$$x = -\frac{b}{2a}.$$

Example

2 Find the equation of the axis of symmetry and the coordinates of the minimum point for the graph of $y = x^2 - x - 6$. Then use the information to draw the graph.

First, find the equation of the axis of symmetry.

$$x = -\frac{b}{2a}$$
$$= -\frac{-1}{2 \cdot 1} \qquad a = 1, b = -1$$
$$= \frac{1}{2}$$

The equation of the axis of symmetry is $x = \frac{1}{2}$.

Next, find the coordinates of the minimum point. The minimum point lies on the axis of symmetry. Here the x-coordinate of the minimum point will be $\frac{1}{2}$. Find the y-coordinate by substituting $\frac{1}{2}$ for x in $y = x^2 - x - 6$.

$$y = x^2 - x - 6$$
$$= \left(\frac{1}{2}\right)^2 - \frac{1}{2} - 6$$
$$= \frac{1}{4} - \frac{1}{2} - 6$$
$$= -\frac{25}{4}$$

The coordinates of the minimum point are $\left(\frac{1}{2}, -\frac{25}{4}\right)$.

Then, construct a table. For the values of x, choose some integers greater than $\frac{1}{2}$, and some less than $\frac{1}{2}$. This insures that points on both sides of the axis of symmetry are plotted. Use this information to draw the graph.

x	$x^2 - x - 6$	y
-2	$(-2)^2 - (-2) - 6$	0
-1	$(-1)^2 - (-1) - 6$	-4
0	$0^2 - 0 - 6$	-6
1	$1^2 - 1 - 6$	-6
2	$2^2 - 2 - 6$	-4
3	$3^2 - 3 - 6$	0

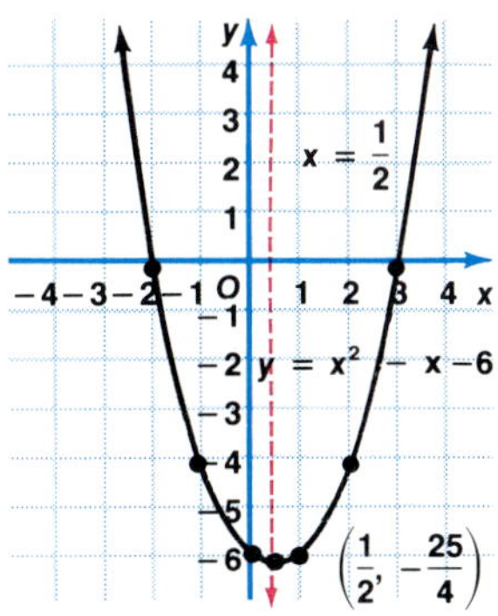

Chalkboard Exercises

Find the equation of the axis of symmetry and the coordinates of the maximum or minimum point of each quadratic function. Then draw the graph.

2a. $y = x^2 - 8x + 9$ $\quad x = 4$; min. $(4, -7)$

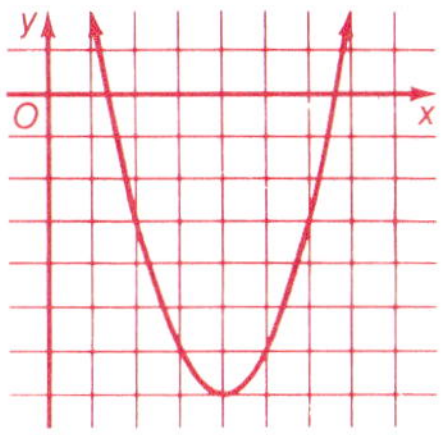

2b. $y = x^2 - 6x + 5$ $\quad x = 3$; min. $(3, -4)$

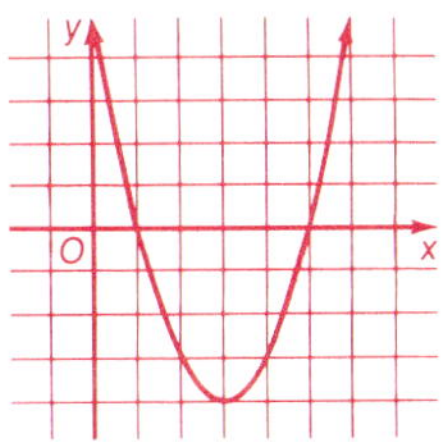

Chalkboard Exercises

Find the equation of the axis of symmetry and the coordinates of the maximum or minimum point for the graph of each quadratic function.

3a. $y = x^2 - 2$ $x = 0$; min $(0, -2)$

3b. $y = -x^2 + 2x + 1$ $x = 1$; max $(1, 2)$

3c. $y = 3x^2$ $x = 0$; min $(0, 0)$

3d. $y = x^2 - 2x - 3$ $x = 1$; min $(1, -4)$

3e. $y = -x^2 + 4x - 1$ $x = 2$; max $(2, 3)$

3f. $y = 4x^2 + 12x + 9$ $x = -\frac{3}{2}$; min $\left(-\frac{3}{2}, 0\right)$

Suggested Daily Assignments

Enriched
- Exp: 1–9
- Written: 2–30 evens

Average
- Exp: 1–9
- Written: 1–18, 20–30 evens

Fundamental
- Exp: 1–18
- Written: 1–24

Answers to Written

1. $x = \frac{5}{2}; \left(\frac{5}{2}, \frac{49}{4}\right)$
2. $x = \frac{5}{2}; \left(\frac{5}{2}, -\frac{31}{4}\right)$
3. $x = 2; (2, 9)$
4. $x = -1; (-1, -1)$
5. $x = \frac{3}{2}; \left(\frac{3}{2}, \frac{137}{4}\right)$
6. $x = 0; (0, 4)$
7. $x = -1; (-1, -20)$
8. $x = -4; (-4, 32)$
9. $x = 0; (0, -9)$
10. $x = -1; (-1, 8)$
11. $x = -1; (-1, 1)$
12. $x = -1; (-1, -16)$
13. $x = 1; (1, 22)$
14. $x = 1; (1, 17)$
15. $x = 0; (0, 4)$
16. $x = -3; (-3, -35)$
17. $x = 1; (1, 42)$
18. $x = \frac{3}{2}; \left(\frac{3}{2}, \frac{29}{2}\right)$

(continued on p. 409 margin)

Example

3 **Find the coordinates of the maximum point for the graph of $y = -2x^2 - 8x + 9$.**

$x = -\frac{b}{2a}$ First, find the equation of the axis of symmetry.

$= -\frac{-8}{2(-2)}$ *$a = -2, b = -8$*

$= -2$ The equation of the axis of symmetry is $x = -2$.

$y = -2(-2)^2 - 8(-2) + 9$ *Since the maximum point lies on the axis of symmetry, substitute -2 for x in $y = -2x^2 - 8x + 9$.*

$= -8 + 16 + 9$

$= 17$

The coordinates of the maximum point are $(-2, 17)$.

Exploratory Exercises

State whether the graph of each quadratic function opens upward or downward.

1. $y = x^2 - 1$ up
2. $y = -x^2 + 3x$ down
3. $y = -x^2 + x + 1$ down
4. $y = x^2 + x + 3$ up
5. $y = -x^2 + 4x + 5$ down
6. $y = 2x^2 - 5x + 2$ up
7. $y = 3x^2 + 9x - 1$ up
8. $y = 2x^2 - 8x + 1$ up
9. $y = -5x^2 + 3x + 2$ down

Find the equation of the axis of symmetry for the graph of each quadratic function.

10. $y = x^2 + x + 3$ $x = -\frac{1}{2}$
11. $y = -x^2 + 4x + 5$ $x = 2$
12. $y = -x^2 + 3x$ $x = \frac{3}{2}$
13. $y = x^2 + 6x + 8$ $x = -3$
14. $y = 3x^2 + 6x + 16$ $x = -1$
15. $y = 5x^2 + 20x + 37$ $x =$
16. $y = 8x + 3 + x^2$ $x = -4$
17. $y = \frac{3}{2} + 7x + 5x^2$ $x = -\frac{7}{10}$
18. $y = -4x^2 + 4x + \frac{5}{2}$ $x =$

Written Exercises

Find the equation of the axis of symmetry and the coordinates of the maximum or minimum point for the graph of each quadratic function.

1. $y = -x^2 + 5x + 6$
2. $y = -x^2 + 5x - 14$
3. $y = x^2 - 4x + 13$
4. $y = x^2 + 2x$
5. $y = -5x^2 + 15x + 23$
6. $y = -3x^2 + 4$
7. $y = 3x^2 + 6x - 17$
8. $y = 3x^2 + 24x + 80$
9. $y = -2x^2 - 9$
10. $y = -3x^2 - 6x + 5$
11. $y = 5x^2 + 10x + 6$
12. $y = 7x^2 + 14x - 9$
13. $y = -7x^2 + 14x + 15$
14. $y = -4x^2 + 8x + 13$
15. $y = 3x^2 + 4$
16. $y = 2x^2 + 12x - 17$
17. $y = -5x^2 + 10x + 37$
18. $y = 2x^2 - 6x + 19$

Find the equation of the axis of symmetry and the coordinates of the maximum or minimum point for the graph of each quadratic function. Then draw the graph.

19. $y = x^2 - 4x - 5$
20. $y = -x^2 + 4x + 5$
21. $y = -x^2 + 6x + 5$
22. $y = x^2 - x - 6$
23. $y = x^2 - 3$
24. $y = -x^2 + 7$
25. $y = 2x^2 + 3$
26. $y = x^2 - 2x - 8$
27. $y = x^2 - x - 12$
28. $y = \frac{1}{4}x^2 - 4x + 3\frac{3}{4}$
29. $y = \frac{1}{2}x^2 + 3x + \frac{9}{2}$
30. $y = -3x^2 - 6x + 4$

For more exercises on finding the maximum or minimum point and the axis of symmetry for the graph of a quadratic function, see exercises 53-61 on page 562.

Applications in Business — Maximum Profit

Many times businesses will raise the prices of their goods or services to increase their profit. However, when they raise their prices, they usually lose some customers. In such situations, the price at which the profit would be a maximum needs to be found.

Example

An auditorium has seats for 1,200 people. For the past several days, the auditorium has been filled to capacity for each show. Tickets currently cost $5.00 and the owner wants to increase the ticket prices. He estimates that for each $0.50 increase in price, 100 fewer people will attend. What ticket price will maximize the profit?

Let x = number of $0.50 price increases. Thus $5.00 + 0.50x$ represents the single ticket price and $1200 - 100x$ represents the number of tickets sold.

$$\begin{aligned} \text{Income} &= \text{number of tickets sold} \cdot \text{ticket price} \\ &= (1200 - 100x) \cdot (5.00 + 0.50x) \\ &= 6000 + 100x - 50x^2 \end{aligned}$$

Notice that the result is a quadratic equation. The graph of the related function, $y = -50x^2 + 100x + 6000$, opens downward and thus has a maximum point. Since this is a maximum point, the x-coordinate gives the number of price increases needed to maximize the profit.

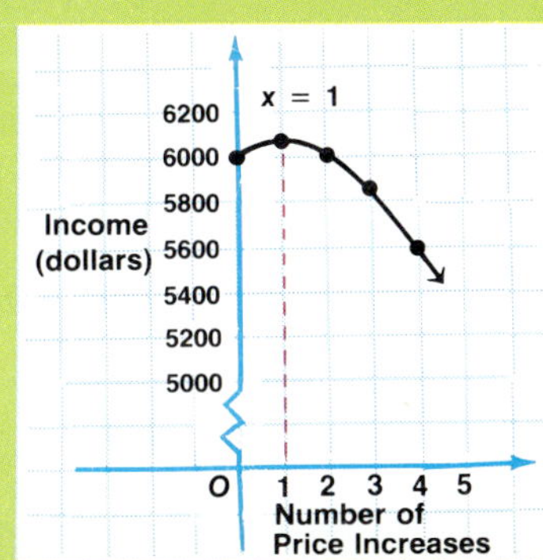

Recall that the x-coordinate of the maximum point is given by the equation of the axis of symmetry.

$$\begin{aligned} x &= -\frac{b}{2a} \\ &= -\frac{100}{2(-50)} \qquad a = -50,\ b = 100 \\ &= 1 \end{aligned}$$

The equation of the axis of symmetry is $x = 1$.

Thus, the profit is maximized when the owner makes one $0.50 price increase. So, the price of one ticket should be $5.50.

Exercises **Solve each problem.**

1. A grocer sells 50 loaves of bread a day. The cost is $0.65 a loaf. The grocer estimates that for each $0.05 price increase, 2 fewer loaves of bread will be sold. What cost will maximize the profit? $0.95 a loaf

2. A bus company transports 500 people a day between Morse Rd. and High St. A one-way fare is $0.50. The owners estimate that for each $0.10 price increase 50 passengers will be lost. What price will maximize their profit? $0.75

Answers to Written (continued)

See the Teacher's Answer Key, p. T89-T90, for graphs to exercises 19-30.

19. $x = 2$; $(2, -9)$
20. $x = 2$; $(2, 9)$
21. $x = 3$; $(3, 14)$
22. $x = \frac{1}{2}$; $\left(\frac{1}{2}, -6\frac{1}{4}\right)$
23. $x = 0$; $(0, -3)$
24. $x = 0$; $(0, 7)$
25. $x = 0$; $(0, 3)$
26. $x = 1$; $(1, -9)$
27. $x = \frac{1}{2}$; $\left(\frac{1}{2}, -12\frac{1}{4}\right)$
28. $x = 8$; $\left(8, -12\frac{1}{4}\right)$
29. $x = -3$; $(-3, 0)$
30. $x = -1$; $(-1, 7)$

About Applications
This is an excellent real-world example that provides reinforcement in such areas as problem solving, multiplying polynomials, and finding maximum points.

13-2

Lesson Notes, page T50

Readiness Activity
Graph each of the following quadratic functions. Then state the x-intercepts of each graph.
$y = x^2 - 3x + 2$

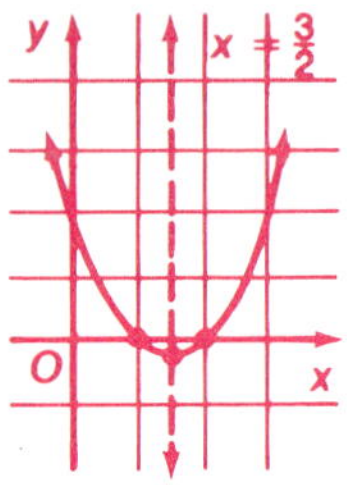

The x-intercepts are 1 and 2.

$y = x^2 - x - 6$

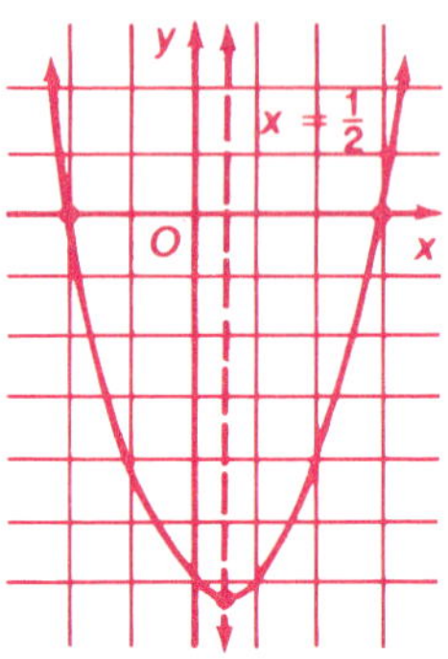

The x-intercepts are -2 and 3.

❶ Explain that the x-coordinates of the x-intercepts are called the zeros of the function.

❷ The axis of symmetry is helpful for determining which elements of the domain to choose when graphing a function.

Objective 13-2: Locate the roots of a quadratic equation by graphing the related function.

13-2 Solving Quadratic Equations by Graphing

Recall that many quadratic equations can be solved by factoring. For example, solve $x^2 + x - 6 = 0$.

$$x^2 + x - 6 = 0 \qquad \textit{Find the factors of } x^2 + x - 6.$$
$$(x + 3)(x - 2) = 0$$
$$x + 3 = 0 \quad \text{or} \quad x - 2 = 0 \qquad \textit{Zero Product Property}$$
$$x = -3 \qquad\qquad x = 2$$

The solutions of an equation are called the **roots** of the equation. The roots of $x^2 + x - 6 = 0$ are -3 and 2. Notice that the roots of $x^2 + x - 6 = 0$ are the x-intercepts of the graph of the related function, $y = x^2 + x - 6$.

x	$x^2 + x - 6$	y
-3	$(-3)^2 + (-3) - 6$	0
-2	$(-2)^2 + (-2) - 6$	-4
-1	$(-1)^2 + (-1) - 6$	-6
0	$0^2 + 0 - 6$	-6
1	$1^2 + 1 - 6$	-4
2	$2^2 + 2 - 6$	0

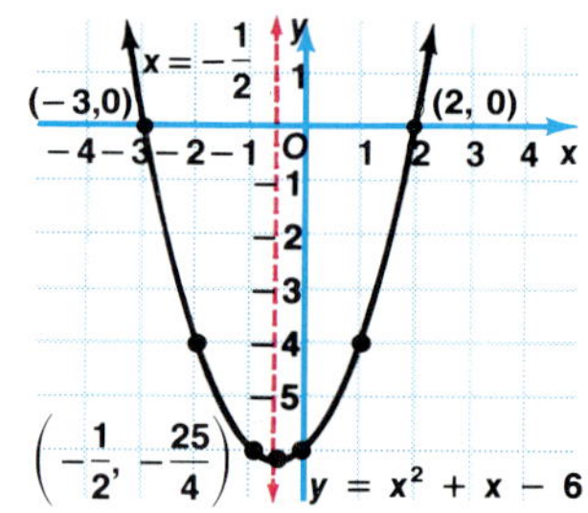

❶

In general, the roots of any quadratic equation of the form $ax^2 + bx + c = 0$ are the x-intercepts of the graph of the related function $y = ax^2 + bx + c$.

Many times, only approximations of roots can be found by graphing. In the following example, the consecutive integers between which the roots lie are found.

Examples

1 Locate the roots of $x^2 - 6x + 6 = 0$ by graphing the related function.

The graph of the related function, $y = x^2 - 6x + 6$, is a parabola that opens upward.

❷ **axis of symmetry**

$$x = -\frac{b}{2a}$$
$$= -\frac{-6}{2 \cdot 1} \text{ or } 3$$

minimum point

Substitute 3 for x in $y = x^2 - 6x + 6$.

$$y = 3^2 - 6(3) + 6$$
$$= -3$$

The equation of the axis of symmetry is $x = 3$ and the coordinates of the minimum point are $(3, -3)$.

x	$x^2 - 6x + 6$	y
1	$1^2 - 6(1) + 6$	1
2	$2^2 - 6(2) + 6$	−2
3	$3^2 - 6(3) + 6$	−3
4	$4^2 - 6(4) + 6$	−2
5	$5^2 - 6(5) + 6$	1

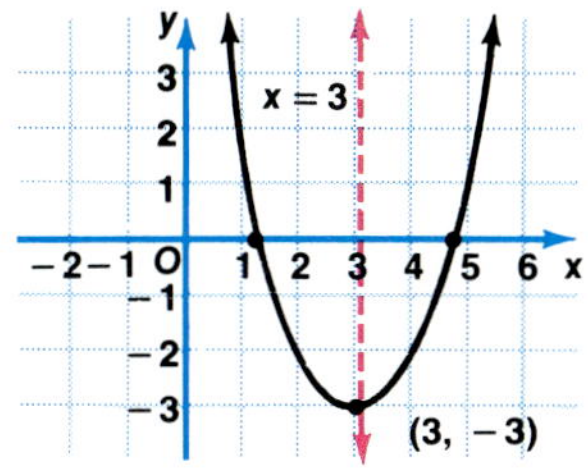

From the graph it can be seen that one root lies between 1 and 2. The other root is between 4 and 5.

2 **Locate the roots of $x^2 - 4x + 5 = 0$ by graphing the related function.**

The graph of the related function $y = x^2 - 4x + 5$ opens upward and has as its axis of symmetry $x = 2$. The coordinates of its minimum point are (2, 1).

x	$x^2 - 4x + 5$	y
0	$0^2 - 4(0) + 5$	5
1	$1^2 - 4(1) + 5$	2
2	$2^2 - 4(2) + 5$	1
3	$3^2 - 4(3) + 5$	2
4	$4^2 - 4(4) + 5$	5

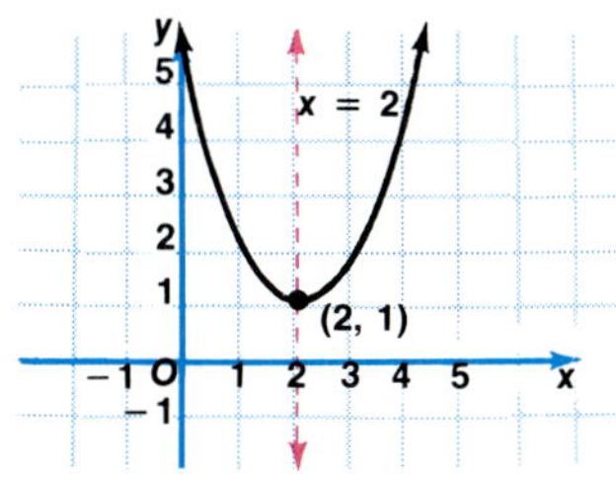

The graph has no x-intercepts since it does not cross the x-axis. Therefore, the equation $x^2 - 4x + 5 = 0$ has no real roots.

Exploratory Exercises

State the roots of each quadratic equation whose related function is graphed below.

1. 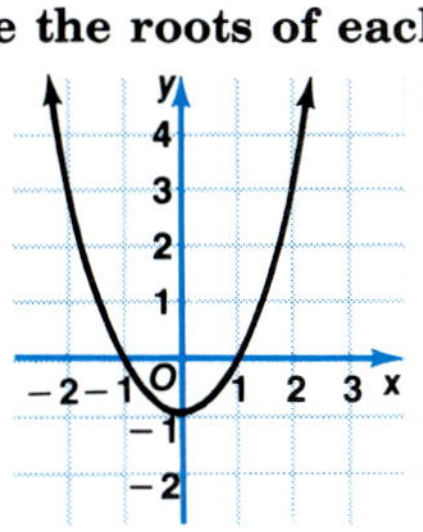

−1, 1

2. 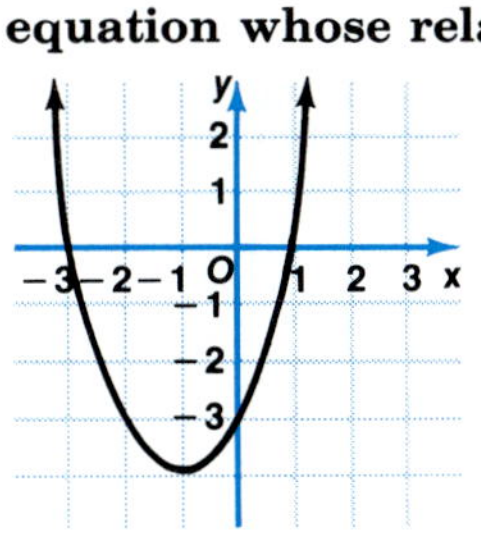

−3, 1

3. 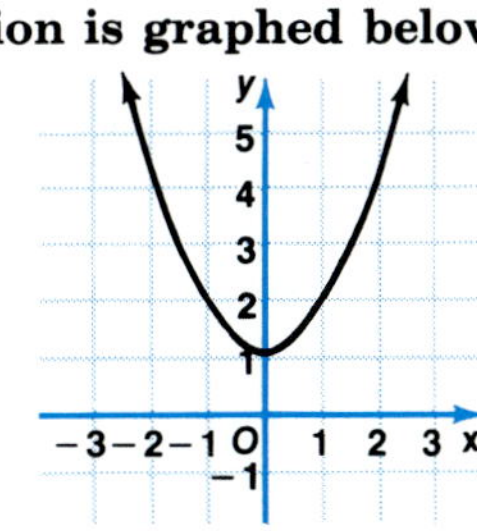

no real roots

Discussion Question
(Use after example 1.)
Suppose the roots of a quadratic equation are opposites of one another. Write the equation of the related function.
Any equations of the form $y = ax^2 - b$ or $y = -ax^2 + b$ where $a > 0$ and $b > 0$.

Chalkboard Exercises
Locate the roots of each equation by graphing the related function.
See students' graphs.

1a. $4x^2 - 4x - 3 = 0$
One root is between 0 and −1. The other root is between 1 and 2.

1b. $x^2 - 2x + 1 = 0$
The only root is 1.

2. $5x^2 - 4x + 1 = 0$
There are no roots.

Suggested Daily Assignments

Enriched
Exp: 1–6
Written: 2–24 evens

Average
Exp: 1–6
Written: 1–24

Fundamental
Exp: 1–6
Written: 1–15
p. 408, 25–30

Answers to Written
For graphs to exercises 1-24, see pages T90-T91.

10. no real roots
11. $3 < x < 4$, $0 < x < 1$
12. $0 < x < 1$, $5 < x < 6$
15. $-2 < x < -1$, $0 < x < 1$
16. $-1 < x < 0$, 3
19. $-1 < x < 0$, $2 < x < 3$
20. $-5 < x < -4$, $-2 < x < -1$

About the Excursion
Ask students to suggest other applications of parabolas. For example, the path of a projectile forms a parabola.

State the roots of each quadratic equation whose related function is graphed below.

4.
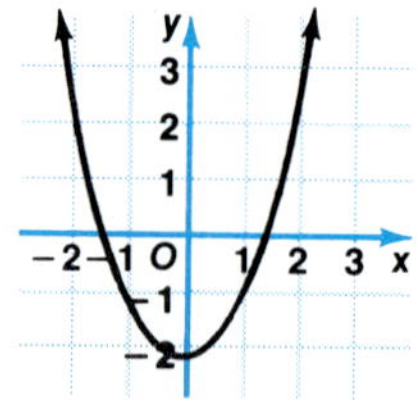

$-2 < x < -1$, $1 < x < 2$

5.
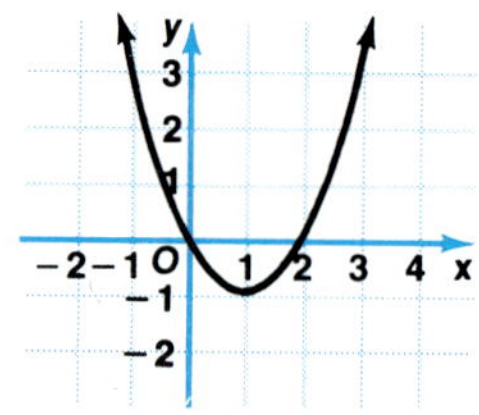

0, 2

6.
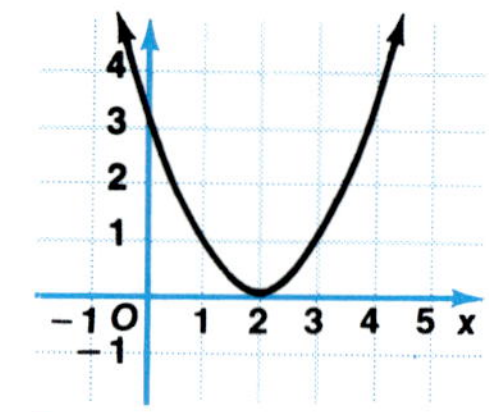

2

Written Exercises

Locate the roots of each equation by graphing the related function.

1. $x^2 - x - 12 = 0$ −3, 4
2. $x^2 + 7x + 12 = 0$ −3, −4
3. $x^2 - 5x - 6 = 0$ 6, −1
4. $x^2 + 3x - 18 = 0$ −6, 3
5. $x^2 - 4 = 0$ −2, 2
6. $x^2 - 9 = 0$ −3, 3
7. $x^2 - 10x = -21$ 3, 7
8. $x^2 + 4x = 12$ −6, 2
9. $x^2 - 2x + 2 = 0$ no real roots
10. $x^2 + 6x + 10 = 0$
11. $x^2 - 4x + 1 = 0$
12. $x^2 - 6x + 4 = 0$
13. $x^2 - 8x + 16 = 0$ 4
14. $x^2 + 6x + 9 = 0$ −3
15. $3x^2 + 4x - 1 = 0$
16. $6x^2 - 13x - 15 = 0$
17. $x^2 - 8x + 18 = 0$ no real roots
18. $x^2 + 2x + 4 = 0$ no real roots
19. $x^2 - 2x - 1 = 0$
20. $x^2 + 6x + 7 = 0$
21. $x^2 + 5x + 9 = 0$ no real roots
22. $x^2 - 3x + 7 = 0$ no real roots
23. $4x^2 - 12x + 3 = 0$ $0 < x < 1$, $2 < x < 3$
24. $4x^2 + 4x - 35 = 0$ $2 < x < 3$, $-4 < x < -3$

Excursions in Algebra — Parabolic Surfaces

Suppose a parabola is rotated about its axis of symmetry. The surface described by this rotation is known as a paraboloid of revolution, or a parabolic surface.

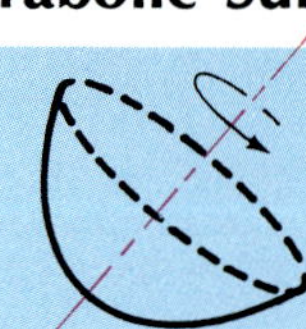

Parabolic surfaces are used in automobile headlamps, radar antennas, solar collectors, and other devices. This is because a parabolic surface can focus incoming energy at a single point, or reflect the energy from a point and send it in one direction.

This point is known as the focus. The focus lies on the axis of symmetry.

The focus for a parabola with equation $y = ax^2 + bx + c$ has the following coordinates.

$$\left(-\frac{b}{2a}, \frac{4ac - b^2 + 1}{4a}\right)$$

Exercises

Find the coordinates of the focus of each parabola whose equation is given.

1. $y = x^2 + 6x + 9$ $\left(-3, \frac{1}{4}\right)$
2. $y = 2x^2 + 4x + 7$ $\left(-1, \frac{41}{8}\right)$
3. $y = \frac{1}{4}x^2 - 2x + 3$ (4, 0)

Objective 13-3: Solve quadratic equations by completing the square.

13-3 Completing the Square

An equation like $x^2 - 36 = 0$ can be solved in the following way.

❶
$$x^2 - 36 = 0$$
$x^2 = 36$ *Add 36 to both sides.*

$\sqrt{x^2} = \sqrt{36}$ *Find the square root of both sides.*

$|x| = 6$ *Recall that $|x| = \sqrt{x^2}$.*

$x = \pm 6$ *$x = \pm 6$ since $|+6| = |-6| = 6$.*

This same method can be used to solve the equation $x^2 - 4x + 4 = 3$.

Example 1

Solve: $x^2 - 4x + 4 = 3$

$x^2 - 4x + 4 = 3$

$(x - 2)^2 = 3$ *Factor $x^2 - 4x + 4$.*

$\sqrt{(x-2)^2} = \sqrt{3}$ *Find the square root of both sides.*

$|x - 2| = \sqrt{3}$

$x - 2 = \pm\sqrt{3}$ *Why is this so?*

$x = 2 \pm \sqrt{3}$ *Add 2 to both sides.*

The solution set is $\{2 - \sqrt{3}, 2 + \sqrt{3}\}$.

The quadratic expression must be a perfect square in order to use the above method. If it is not a perfect square, then a method called **completing the square** may be used.

Consider the pattern for squaring a binomial.

$$(x + 6)^2 = x^2 + 2(6)(x) + 6^2$$
$$= x^2 + 12x + 36$$
$$\downarrow \qquad \uparrow$$
$$\left(\frac{12}{2}\right) \rightarrow 6^2$$
Notice that 36 is 6^2 and 6 is one-half of 12.

To complete the square for an expression of the form $x^2 + bx$, follow the steps listed below.

Step 1 Find one-half of b, the coefficient of x.

Step 2 Square the result of **Step 1**.

Step 3 Add the result of **Step 2** to $x^2 + bx$.

13-3

Lesson Notes, page T50

Discussion Question

How many roots does

$x^2 + bx + \left(\frac{b}{2}\right)^2 = 0$ have?

(Hint: Try some values for b.) 1

Readiness Activity

Factor.

$x^2 + 4x + 4$	$(x + 2)^2$
$x^2 - 6x + 9$	$(x - 3)^2$
$n^2 + 10n + 25$	$(n + 5)^2$
$y^2 - 2y + 1$	$(y - 1)^2$

❶ The equation in the introductory example, $x^2 - 36 = 0$, can be easily solved by factoring. However, its purpose is to show that square roots can be helpful in solving equations.

Chalkboard Exercises

Solve.

1a. $x^2 - 10x + 25 = 1$

$(x - 5)^2 = 1$

$|x - 5| = \sqrt{1}$

$x - 5 = \pm 1$

$x = 5 \pm 1$

The solution set is $\{4, 6\}$.

1b. $x^2 + 6x + 9 = 5$

$(x + 3)^2 = 5$

$|x + 3| = \sqrt{5}$

$x + 3 = \pm\sqrt{5}$

$x = -3 \pm \sqrt{5}$

The solution set is $\{-3 + \sqrt{5}, -3 - \sqrt{5}\}$.

Chalkboard Exercises

Find the value of c that makes each trinomial a perfect square.

2a. $x^2 - 20x + c$ 100

2b. $y^2 + y + c$ $\frac{1}{4}$

2c. $m^2 - 3m + c$ $\frac{9}{4}$

Solve by completing the square. Leave irrational roots in simplest radical form.

3a. $x^2 - 6x + 8 = 0$ $\{2, 4\}$

3b. $x^2 - 6x + 5 = 0$ $\{1, 5\}$

3c. $m^2 + 4m - 9 = 0$ $\{-2 + \sqrt{13}, -2 - \sqrt{13}\}$

3d. $x^2 - 6x + 7 = 0$ $\{3 + \sqrt{2}, 3 - \sqrt{2}\}$

4a. $2m^2 + 7m + 3 = 0$ $\left\{-\frac{1}{2}, -3\right\}$

4b. $2r^2 + r - 3 = 0$ $\left\{1, -\frac{3}{2}\right\}$

4c. $3x^2 + 7x - 6 = 0$ $\left\{\frac{2}{3}, -3\right\}$

❷ Discuss this example thoroughly. Make sure students understand the third step of the solution. The number 9 is added because it completes the square for $x^2 + 6x$. That is, $\left(\frac{6}{2}\right)^2 = 9$.

❸ Encourage students to check the solutions to quadratic equations. A calculator is helpful for checking solutions which are irrational. For example 4, suggest that students evaluate $2x^2 - 9x + 8$ when $x = 3.281$ and when $x = 1.219$. The value of the expression should be very close to zero.

Examples

2 **Find the value of c that makes $x^2 + 14x + c$ a perfect square.**

Step 1 Find one-half of 14. $\frac{14}{2} = 7$

Step 2 Square the result of **Step 1.** $7^2 = 49$

Step 3 Add the result of **Step 2** to $x^2 + 14x$. $x^2 + 14x + 49$

Thus, $c = 49$. Notice that $x^2 + 14x + 49$ is equal to $(x + 7)^2$, which is a perfect square.

3 **Solve $x^2 + 6x - 16 = 0$ by completing the square.** ❷

$x^2 + 6x - 16 = 0$ *Notice that $x^2 + 6x - 16$ is not a perfect square.*

$x^2 + 6x = 16$ *Add 16 to both sides. Then complete the square.*

$x^2 + 6x + 9 = 16 + 9$ *$\left(\frac{6}{2}\right)^2 = 9$, so add 9 to both sides of the equation.*

$(x + 3)^2 = 25$ *Factor $x^2 + 6x + 9$.*

$x + 3 = \pm 5$ *Find the square root of both sides.*

$x = \pm 5 - 3$ *Subtract 3 from both sides.*

$x = 5 - 3$ or $x = -5 - 3$

$x = 2$ $\quad$ $x = -8$

The roots of $x^2 + 6x - 16 = 0$ are 2 and -8. *Check this result.*

Notice that the roots can be used to factor $x^2 + 6x - 16$.

$$x^2 + 6x - 16 = (x - 2)(x + 8)$$

Example

4 **Solve $2x^2 - 9x + 8 = 0$ by completing the square.**

$2x^2 - 9x + 8 = 0$ *To complete the square, the coefficient of x^2 must be 1. Divide both sides of the equation by 2.*

$x^2 - \frac{9}{2}x + 4 = 0$

$x^2 - \frac{9}{2}x = -4$ *Subtract 4 from both sides. Then complete the square.*

$x^2 - \frac{9}{2}x + \frac{81}{16} = -4 + \frac{81}{16}$ *Why was $\frac{81}{16}$ added to both sides?*

$\left(x - \frac{9}{4}\right)^2 = \frac{17}{16}$

$x - \frac{9}{4} = \pm\frac{\sqrt{17}}{4}$

$x = \frac{9 \pm \sqrt{17}}{4}$

The roots of $2x^2 - 9x + 8 = 0$ are $\frac{9 + \sqrt{17}}{4}$ and $\frac{9 - \sqrt{17}}{4}$. In decimal form the roots are approximately $\frac{9 + 4.123}{4}$, or 3.281, and $\frac{9 - 4.123}{4}$, or 1.219. ❸

Exploratory Exercises

State whether each trinomial is a perfect square.

1. $y^2 + 8y + 7$ no **2.** $b^2 + 4b + 3$ no **3.** $m^2 - 10m + 25$ yes

4. $r^2 - 8r + 16$ yes **5.** $k^2 + 10k + 9$ no **6.** $z^2 - 18z + 81$ yes

7. $t^2 + 8t + 15$ no **8.** $d^2 + 12d + 27$ no **9.** $h^2 - 13h + \frac{169}{4}$ yes

Find the value of c that makes each trinomial a perfect square.

10. $x^2 + 8x + c$ 16 **11.** $x^2 + 4x + c$ 4 **12.** $m^2 - 10m + c$ 25

13. $x^2 - 6x + c$ 9 **14.** $z^2 + 2z + c$ 1 **15.** $x^2 + 14x + c$ 49

16. $x^2 - 12x + c$ 36 **17.** $a^2 + 5a + c$ $\frac{25}{4}$ **18.** $x^2 + 7x + c$ $\frac{49}{4}$

19. $x^2 - 7x + c$ $\frac{49}{4}$ **20.** $y^2 + 16y + c$ 64 **21.** $x^2 - 13x + c$ $\frac{169}{4}$

Written Exercises

Solve by completing the square. Leave irrational roots in simplest radical form.

1. $y^2 + 4y + 3 = 0$ $-1, -3$ **2.** $n^2 + 8n + 7 = 0$ $-1, -7$ **3.** $t^2 - 4t = 21$ $7, -3$

4. $a^2 - 10a - 24 = 0$ $12, -2$ **5.** $a^2 - 6a + 8 = 0$ $4, 2$ **6.** $b^2 - 3b = 10$ $5, -2$

7. $z^2 - 4z = 2$ $2 \pm \sqrt{6}$ **8.** $p^2 + 2p - 3 = 0$ $1, -3$ **9.** $y^2 - 8y = 4$ $4 \pm 2\sqrt{5}$

10. $p^2 - p - 6 = 0$ $3, -2$ **11.** $x^2 - 8x = 12$ $4 \pm 2\sqrt{7}$ **12.** $s^2 + 8s + 15 = 0$ $-3, -5$

13. $r^2 + 14r - 10 = 5$ $1, -15$ **14.** $y^2 + 7y + 10 = -2$ $-3, -4$ **15.** $n^2 + 5n + 6 = 0$ $-2, -3$

16. $x^2 - 5x + 2 = -2$ $4, 1$ **17.** $n^2 - 11n - 12 = 0$ $12, -1$ **18.** $x^2 - 4x + 1 = 0$ $2 \pm \sqrt{3}$

19. $x^2 - 6x + 7 = 0$ $3 \pm \sqrt{2}$ **20.** $b^2 - 6b + 4 = 0$ $3 \pm \sqrt{5}$ **21.** $\frac{1}{2}t^2 - 2t - \frac{3}{2} = 0$ $2 \pm \sqrt{7}$

22. $4x^2 - 20x + 25 = 0$ $\frac{5}{2}$ **23.** $2b^2 - b - 14 = 7$ $\frac{7}{2}, -3$ **24.** $2d^2 + 3d - 20 = 0$ $\frac{5}{2}, -4$

25. $x^2 - \frac{7}{2}x + \frac{3}{2} = 0$ $\frac{1}{2}, 3$ **26.** $a^2 - \frac{1}{2}a - \frac{3}{2} = 0$ $-1, \frac{3}{2}$ **27.** $3y^2 - y - 10 = 0$ $-\frac{5}{3}, 2$

28. $0.3x^2 + 0.1x - 0.2 = 0$ $\frac{2}{3}, -1$ **29.** $0.3n^2 - 0.2n = 0.1$ $-\frac{1}{3}, 1$ **30.** $\frac{1}{2}q^2 - \frac{5}{4}q - 3 = 0$ $-\frac{3}{2}, 4$

Challenge

31. $x^2 + 4x + c = 0$ **32.** $x^2 - bx + 8 = 0$ **33.** $x^2 + bx + c = 0$

34. $ax^2 + bx + c = 0$ **35.** $x^2 + xy - 2y^2 = 0$ **36.** $x^2 + 4bx + b^2 = 0$

Using Calculators — Irrational Roots

Solve by completing the square. Then use a calculator to approximate the roots to three decimal places.

1. $x^2 + 2x - 10 = 0$ **2.** $r^2 + 0.25r - 0.5 = 0$ **3.** $2m^2 - 5m + 1 = 0$

4. $a^2 - 6a + 6 = 0$ **5.** $x^2 - 10x = 23$ **6.** $y^2 - 8y = 13$

7. $2x^2 - 6x - 5 = 0$ **8.** $4y^2 - 2y = 1$ **9.** $3x^2 - 7x - 3 = 0$

10. $x^2 - \frac{7}{3}x - \frac{5}{3} = 0$ **11.** $y^2 - \frac{8}{5}y - \frac{2}{5} = 0$ **12.** $t^2 - 3t = -\frac{5}{3}$

Suggested Daily Assignments

Enriched
- Exp: 7–21 odds
- Written: 2–30 evens, 31–36

Average
- Exp: 1–21
- Written: 1–30

Fundamental
- Exp: 1–21
- Written: 1–21
  - p. 412, 16–20

Answers to Challenge

31. $-2 \pm \sqrt{4 - c}$

32. $\frac{b \pm \sqrt{b^2 - 32}}{2}$

33. $\frac{-b \pm \sqrt{b^2 - 4c}}{2}$

34. $\frac{-b \pm \sqrt{b^2 - 4ac}}{2a}$

35. $y, -2y$

36. $b(-2 \pm \sqrt{3})$

About Using Calculators

Problems such as these help emphasize the power and the limitations of calculators. Students should realize that the calculator can only approximate irrational numbers.

Answers to Using Calculators

1. 2.317, −4.317
2. 0.593, −0.843
3. 2.281, 0.219
4. 4.732, 1.268
5. 11.928, −1.928
6. 9.385, −1.385
7. 3.679, −0.679
8. 0.809, −0.309
9. 2.703, −0.370
10. 2.907, −0.573
11. 1.820, −0.220
12. 2.264, 0.736

13-4

Lesson Notes, page T50
Enrichment, page T50

Readiness Activity
Recall that the general form of a quadratic equation is $ax^2 + bx + c = 0$. Name a, b, and c for each of the following quadratic equations.

$2x^2 - 4x + 3 = 0$
2, −4, 3
$x^2 + x - 5 = 0$
1, 1, −5
$7x^2 - 10x + 3 = 0$
7, −10, 3
$-3x^2 + 4x - 12 = 0$
−3, 4, −12

Discussion Question
Find values for r, s, and t so that $\frac{r \pm s\sqrt{x}}{t}$ can be simplified. *r, s, and t must have a common factor other than one.*

❶ Encourage students to memorize the quadratic formula. Emphasize that it can be used to solve any quadratic equation.

Objective 13-4: Solve quadratic equations by using the quadratic formula.

13-4 The Quadratic Formula

The method of completing the square can be used to develop a general formula for solving any quadratic equation. Begin with the general form of a quadratic equation, $ax^2 + bx + c = 0$, where $a \neq 0$.

$$ax^2 + bx + c = 0$$

$$x^2 + \frac{b}{a}x + \frac{c}{a} = 0$$ *Divide by a so the coefficient of x^2 becomes 1.*

$$x^2 + \frac{b}{a}x = -\frac{c}{a}$$ *Subtract $\frac{c}{a}$ from both sides.*

$$x^2 + \frac{b}{a}x + \left(\frac{b}{2a}\right)^2 = -\frac{c}{a} + \left(\frac{b}{2a}\right)^2$$ *Complete the square.*

$$\left(x + \frac{b}{2a}\right)^2 = -\frac{c}{a} + \frac{b^2}{4a^2}$$ *Factor the left side.*

$$\left(x + \frac{b}{2a}\right)^2 = \frac{b^2 - 4ac}{4a^2}$$ *Simplify the right side.*

$$x + \frac{b}{2a} = \pm\sqrt{\frac{b^2 - 4ac}{4a^2}}$$ *Find the square root of both sides.*

$$x + \frac{b}{2a} = \frac{\pm\sqrt{b^2 - 4ac}}{2a}$$ *Simplify the square root on the right side.*

$$x = \frac{\pm\sqrt{b^2 - 4ac}}{2a} - \frac{b}{2a}$$ *Subtract $\frac{b}{2a}$ from both sides.*

$$x = \frac{-b \pm\sqrt{b^2 - 4ac}}{2a}$$ *The result is an expression for x.*

This result is called the **quadratic formula** and can be used to solve any quadratic equation.

The roots of a quadratic equation of the form $ax^2 + bx + c = 0$, where $a \neq 0$, are given by:

$$x = \frac{-b \pm\sqrt{b^2 - 4ac}}{2a}$$

❶ ***The Quadratic Formula***

In order to find a real value for $\sqrt{b^2 - 4ac}$, the value of $b^2 - 4ac$ must be non-negative. If $b^2 - 4ac$ is negative, the equation has no real roots.

Examples

1 **Use the quadratic formula to solve $x^2 - 6x - 7 = 0$.**

$$x = \frac{-b \pm \sqrt{b^2 - 4ac}}{2a}$$

$$= \frac{-(-6) \pm \sqrt{(-6)^2 - 4(1)(-7)}}{2(1)} \quad a = 1,\ b = -6,\ c = -7$$

$$= \frac{6 \pm \sqrt{36 + 28}}{2}$$

$$= \frac{6 \pm \sqrt{64}}{2}$$

$$= \frac{6 \pm 8}{2}$$

$$x = \frac{6 + 8}{2} \text{ or } x = \frac{6 - 8}{2} \quad \textit{Check this result.}$$

$$x = 7 \qquad x = -1$$

The roots are 7 and -1.

2 **Use the quadratic formula to solve $\frac{4}{3}x^2 - 2x = -\frac{1}{3}$.**

First, change the equation to the general form, $ax^2 + bx + c = 0$.

$$\frac{4}{3}x^2 - 2x = -\frac{1}{3}$$

$$\frac{4}{3}x^2 - 2x + \frac{1}{3} = 0 \quad \textit{Add } \tfrac{1}{3} \textit{ to both sides.}$$

$$4x^2 - 6x + 1 = 0 \quad \textit{Multiply both sides by 3.}$$

$$x = \frac{-b \pm \sqrt{b^2 - 4ac}}{2a}$$

$$= \frac{-(-6) \pm \sqrt{(-6)^2 - 4(4)(1)}}{2(4)} \quad a = 4,\ b = -6,\ c = 1$$

$$= \frac{6 \pm \sqrt{36 - 16}}{8}$$

$$= \frac{6 \pm \sqrt{20}}{8}$$

$$= \frac{6 \pm \sqrt{4 \cdot 5}}{8}$$

$$= \frac{6 \pm 2\sqrt{5}}{8}$$

$$= \frac{3 \pm \sqrt{5}}{4} \quad \textit{Check this result.}$$

The roots are $\frac{3 + \sqrt{5}}{4}$ and $\frac{3 - \sqrt{5}}{4}$.

Chalkboard Exercises

Use the quadratic formula to solve each equation. Leave irrational roots in simplest radical form.

1a. $x^2 + 6x + 8 = 0$ $\{-2, -4\}$

1b. $x^2 - 2x - 15 = 0$ $\{-3, 5\}$

1c. $2x^2 + 7x - 15 = 0$ $\left\{-5, \frac{3}{2}\right\}$

1d. $2x^2 + 11x + 5 = 0$ $\left\{-\frac{1}{2}, -5\right\}$

2a. $2x^2 - 2x = 1$ $\left\{\frac{1 \pm \sqrt{3}}{2}\right\}$

2b. $4x^2 = 8x - 1$ $\left\{\frac{2 \pm \sqrt{3}}{2}\right\}$

2c. $-3x + 1 = -x^2$ $\left\{\frac{3 \pm \sqrt{5}}{2}\right\}$

2d. $x^2 + 2x = -\frac{2}{3}$ $\left\{\frac{-3 \pm \sqrt{3}}{3}\right\}$

Suggested Daily Assignments

Enriched
- Exp: 10–15
- Written: 2–42 evens, 43–46
- Mini-Review: 1–5

Average
- Exp: 2–24 evens
- Written: 1–24, 26–42 evens
- Mini-Review: 1–5

Fundamental
- Exp: 1–24
- Written: 2–42 evens p. 415, 22–27
- Mini-Review: 1–5

Answers to Written

31. $-3 \pm 2\sqrt{3}$
32. 0.6, −0.25
33. 1.5, −0.4
34. $3 \pm 2\sqrt{2}$
35. $\frac{2 \pm \sqrt{7}}{2}$
36. $\frac{-3 \pm \sqrt{5}}{2}$
37. $\frac{-5 \pm \sqrt{2}}{2}$
38. $\frac{-2 \pm \sqrt{5}}{2}$
39. $\frac{4 \pm \sqrt{29}}{2}$
40. $\frac{5}{6}, -\frac{3}{4}$
41. $\frac{3}{7}, -\frac{2}{3}$
42. $\frac{3}{5}, -\frac{2}{7}$

Exploratory Exercises

State the values of a, b, and c for each quadratic equation.

1. $x^2 + 7x + 6 = 0$ 1, 7, 6
2. $y^2 + 8y + 15 = 0$ 1, 8, 15
3. $m^2 + 4m + 3 = 0$ 1, 4, 3
4. $2t^2 - t = 15$ 2, −1, −15
5. $4x^2 + 8x = -3$ 4, 8, 3
6. $2y^2 + 3 = -7y$ 2, 7, 3
7. $y^2 - 25 = 0$ 1, 0, −25
8. $2y^2 = 98$ 2, 0, −98
9. $2x^2 + 8x = 0$ 2, 8, 0
10. $3n^2 - 18n = 0$ 3, −18, 0
11. $3m^2 + 2 = -5m$ 3, 5, 2
12. $3k^2 + 11k = 4$ 3, 11, −4

State the value of $b^2 - 4ac$ for each quadratic equation.

13. $x^2 + 5x - 6 = 0$ 49
14. $r^2 + 10r + 9 = 0$ 64
15. $y^2 - 7y - 8 = 0$ 81
16. $z^2 - 13z + 36 = 0$ 25
17. $m^2 - 2m = 8$ 36
18. $y^2 - 2y = 35$ 144
19. $y^2 + y = 12$ 49
20. $4n^2 - 20n = 0$ 400
21. $5t^2 = 125$ 2500
22. $2x^2 - x = 3$ 25
23. $3x^2 + 14x = 5$ 256
24. $3x^2 + 23x + 14 = 0$ 361

Written Exercises

Use the quadratic formula to solve each equation. Leave irrational roots in simplest radical form.

1. $x^2 + 7x + 6 = 0$ −1, −6
2. $y^2 + 8y + 15 = 0$ −3, −5
3. $m^2 + 4m + 3 = 0$ −1, −3
4. $p^2 + 5p + 4 = 0$ −1, −4
5. $2r^2 + r - 15 = 0$ $\frac{5}{2}, -3$
6. $8t^2 + 10t + 3 = 0$ $-\frac{1}{2}, -\frac{3}{4}$
7. $2t^2 - t = 15$ $-\frac{5}{2}, 3$
8. $-4x^2 + 8x = -3$ $\frac{2 \pm \sqrt{7}}{2}$
9. $2y^2 + 3 = -7y$ $-\frac{1}{2}, -3$
10. $y^2 - 25 = 0$ ±5
11. $2y^2 = 98$ ±7
12. $-2x^2 + 8x + 3 = 3$ 0, 4
13. $3n^2 - 18n = 0$ 6, 0
14. $3m^2 + 2 = -5m$ $-1, -\frac{2}{3}$
15. $3k^2 + 11k = 4$ $\frac{1}{3}, -4$
16. $-x^2 + 5x - 6 = 0$ 2, 3
17. $r^2 + 10r + 9 = 0$ −1, −9
18. $y^2 - 7y - 8 = 0$ 8, −1
19. $z^2 - 13z + 36 = 0$ 9, 4
20. $m^2 - 2m - 4 = 4$ 4, −2
21. $y^2 - 2y = 35$ 7, −5
22. $y^2 + y = 12$ 3, −4
23. $4n^2 - 20n = 0$ 5, 0
24. $5t^2 = 125$ ±5
25. $2x^2 - x = 3$ $\frac{3}{2}, -1$
26. $3x^2 + 14x = 5$ $\frac{1}{3}, -5$
27. $3x^2 + 23x + 10 = -4$ $-\frac{2}{3}, -7$
28. $3n^2 - 2n = 1$ $1, -\frac{1}{3}$
29. $y^2 - \frac{3}{5}y + \frac{2}{25} = 0$ $\frac{2}{5}, \frac{1}{5}$
30. $3x^2 - \frac{5}{4}x - \frac{1}{2} = 0$ $\frac{2}{3}, -\frac{1}{4}$
31. $-r^2 - 6r + 3 = 0$
32. $2x^2 - 0.7x - 0.3 = 0$
33. $x^2 - 1.1x - 0.6 = 0$
34. $k^2 - 6k + 1 = 0$
35. $4x^2 - 8x + 3 = 6$
36. $a^2 + 3a + 1 = 0$
37. $4b^2 + 20b + 23 = 0$
38. $4x^2 + 8x - 1 = 0$
39. $-4y^2 + 16y + 13 = 0$
40. $24x^2 - 2x - 15 = 0$
41. $21x^2 + 5x - 6 = 0$
42. $35x^2 - 11x - 6 = 0$

Challenge

Solve each equation by factoring, if possible. Then, solve each equation using the quadratic formula.

43. $y^2 - 2y - 15 = 0$ 5, −3
44. $x^2 - 13x + 42 = 0$ 7, 6
45. $x^2 + 2x - 2 = 0$ $-1 \pm \sqrt{3}$
46. If a quadratic equation can be factored, what can you say about the value of $b^2 - 4ac$? It is a perfect square.

mini-review

Factor.

1. $r^2 - 10r - 24$ $(r - 12)(r + 2)$
2. $6p^2 - 16p - 6$ $2(3p + 1)(p - 3)$

Simplify.

3. $\pm\sqrt{81}$ ±9
4. $\sqrt{75} \cdot \sqrt{3}$ 15
5. $\frac{\sqrt{2}}{\sqrt{3}}$ $\frac{\sqrt{6}}{3}$

For more exercises on using the quadratic formula to solve equations, see exercises 74-82 on page 561.

Problem Solving

List Possibilities

Another strategy for solving problems is to list possibilities. When making a list, use a systematic approach so you do not omit important items. This strategy is often helpful when you need to find the *number* of solutions to a problem. Study the following examples.

Example 1: **Which positive numbers less than 30 are divisible by both 3 and 4?**

List the numbers divisible by 3: 3, 6, 9, 12, 15, 18, 21, 24, 27
List the numbers divisible by 4: 4, 8, 12, 16, 20, 24, 28

Which numbers are in both lists? Since 12 and 24 are in both lists, they are divisible by both 3 and 4.

Example 2: **How many ways can you receive change for a quarter if at least one coin is a dime?**

List the possibilities. List ways which use the fewest number of coins first.

1. dime, dime, nickel *This is the same as dime, nickel, dime.*
2. dime, dime, 5 pennies
3. dime, nickel, nickel, nickel
4. dime, nickel, nickel, 5 pennies
5. dime, nickel, 10 pennies
6. dime, 15 pennies *Are there any other solutions?*

Thus, there are 6 ways to receive change for a quarter if at least one coin is a dime.

Exercises

Solve each problem.

1. Which positive numbers less than 50 are divisible by both 8 and 12? 24, 48
2. Which positive numbers less than 100 are divisible by both 3 and 11? 33, 66, 99
3. How many ways can you receive change for a half-dollar if you receive at least one dime and one quarter? 6
4. How many different whole numbers can be written using the digits 3, 6, 6, and 7? Each number must use all four digits. 12
5. An ice cream shop makes chocolate, butterscotch, or strawberry sundaes. Any sundae can be served with whipped cream, nuts, neither, or both. How many ways can a sundae be served? 12
6. The president, vice-president, secretary, and treasurer of a club are to be seated in four chairs in the front of a meeting room. How many seating arrangements are possible? 24
7. In how many ways can you write 45 as the sum of positive consecutive integers? 5
8. How many positive whole numbers less than 50 can be written as the sum of 2 squares? 17

About Problem Solving
Many other problems can be generated by adapting the ones given in the exercises. Suggest that each student write one or two problems. The extra problems can be used for class discussion, or to share with other members of the class.

Point out that the solutions to divisibility problems, such as exercises 1 and 2, should be whole numbers.

Quiz
Lessons 13-1 to 13-4
Find the equation of the axis of symmetry and the maximum or minimum point for the graph of the quadratic function.

1. $y = 3x^2 - 6x - 4$
$x = 1$; minimum $(1, -7)$

Locate the roots of each of the following quadratic equations.

2. $x^2 + 8x + 16 = 0$ -4
3. $x^2 - 3x - 28 = 0$ $7, -4$

Solve each equation by completing the square. Leave irrational roots in simplest radical form.

4. $a^2 - 8a + 11 = 0$ $4 \pm \sqrt{5}$
5. $3y^2 - 11y = 4$ $4, -\frac{1}{3}$

State the value of $b^2 - 4ac$ for the following equation.

6. $-4t^2 - 3t + 6 = 0$ 105

Use the quadratic formula to solve each equation. Leave irrational roots in simplest radical form.

7. $x^2 - 3x - 28 = 0$ $7, -4$
8. $5y^2 = 8y + 5$ $\frac{4 \pm \sqrt{41}}{5}$

13-5

Lesson Notes, page T50
Enrichment, page T50

Discussion Question
(Use after Example 1.)
Show that if a and c have different signs, then $ax^2 + bx + c = 0$ ($a \neq 0$) has two real roots.
Since $ac < 0$, then $-4ac > 0$. Thus, $b^2 - 4ac > 0$ and $ax^2 + bx + c = 0$ has two real roots.

❶ Emphasize that the discriminant is $b^2 - 4ac$, not $\sqrt{b^2 - 4ac}$.

❷ Remind students that the number of real roots of a quadratic equation is equal to the number of times that the graph of the related function intersects the x-axis.

Objective 13-5: Find the value of the discriminant and use it to determine the nature of the roots of a quadratic equation.

13-5 The Discriminant

In the quadratic formula, the expression $b^2 - 4ac$ is called the **discriminant**. The discriminant can give information about the nature of the roots of a quadratic equation. ❶

In particular, the discriminant is used to determine how many real roots there are. Real roots are roots that are real numbers, that is, *either* rational numbers *or* irrational numbers. Recall that the real roots of an equation are represented by the intersection of the x-axis and the graph of the related function.

Solve $x^2 + 3x - 2 = 0$.

$$x = \frac{-b \pm \sqrt{b^2 - 4ac}}{2a}$$

$$= \frac{-3 \pm \sqrt{3^2 - 4(1)(-2)}}{2(1)} \qquad a = 1, b = 3, c = -2$$

$$= \frac{-3 \pm \sqrt{9 + 8}}{2}$$

$$= \frac{-3 \pm \sqrt{17}}{2}$$

The roots are approximately −3.56 and 0.56.

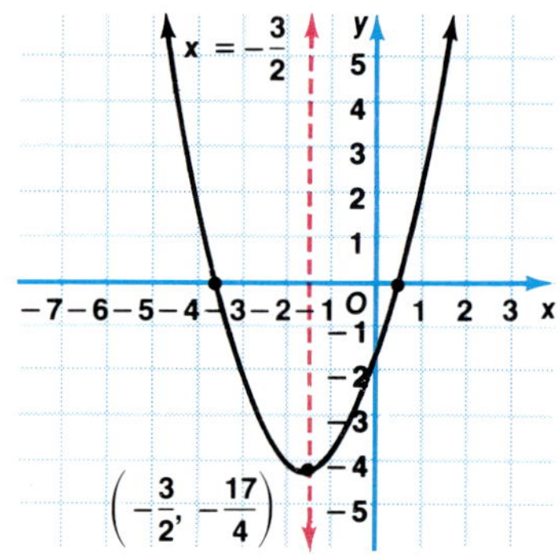

❷ *The graph of the corresponding function intersects the x-axis at two different points.*

Notice that $b^2 - 4ac > 0$ and there are two real roots.

Solve $x^2 - 8x + 16 = 0$.

$$x = \frac{-b \pm \sqrt{b^2 - 4ac}}{2a}$$

$$= \frac{-(-8) \pm \sqrt{(-8)^2 - 4(1)(16)}}{2(1)}$$

$$= \frac{8 \pm \sqrt{64 - 64}}{2}$$

$$= \frac{8 \pm 0}{2} \text{ or } 4$$

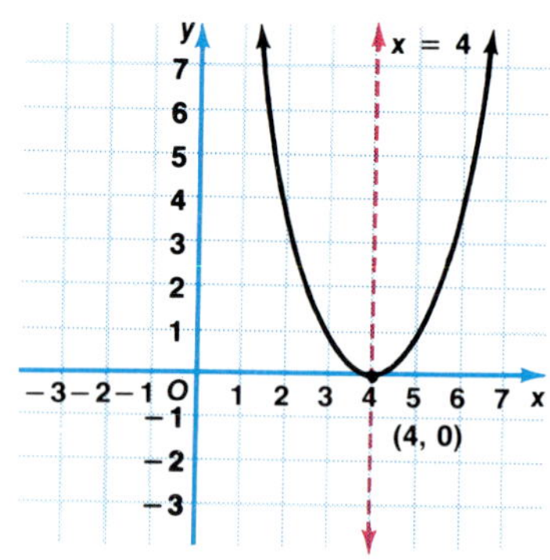

The graph of the corresponding function intersects the x-axis at one point.

Notice that $b^2 - 4ac = 0$ and there is one real root.

Solve $x^2 + 6x + 10 = 0$.

$$x = \frac{-b \pm \sqrt{b^2 - 4ac}}{2a}$$
$$= \frac{-6 \pm \sqrt{6^2 - 4(1)(10)}}{2(1)}$$
$$= \frac{-6 \pm \sqrt{36 - 40}}{2}$$
$$= \frac{-6 \pm \sqrt{-4}}{2}$$

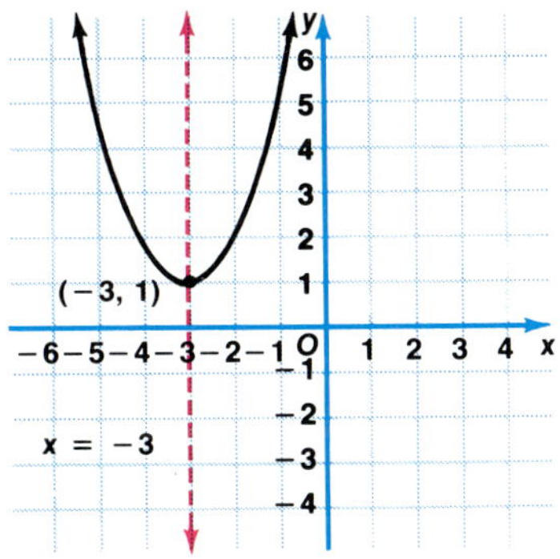

The graph of the corresponding function does not intersect the x-axis

Notice that $b^2 - 4ac < 0$ and there are no real roots since *no* real number can be the square root of a negative number.

The relationship between the value of the discriminant and the nature of the roots of an equation can be summarized as follows.

Nature of Roots of a Quadratic Equation

Discriminant	***Nature of Roots***
$b^2 - 4ac > 0$	**two real roots**
$b^2 - 4ac = 0$	**one real root**
$b^2 - 4ac < 0$	**no real roots**

Examples

1 **State the value of the discriminant of $2x^2 - 10x + 11 = 0$. Then determine the nature of the roots of the equation.**

$b^2 - 4ac = (-10)^2 - 4(2)(11)$ *$a = 2, b = -10, c = 11$*
$= 100 - 88$
$= 12$

Since $b^2 - 4ac > 0$, then $2x^2 - 10x + 11 = 0$ has two real roots.

2 **State the value of the discriminant of $4x^2 - 12x + 9 = 0$. Then determine the nature of the roots of the equation.**

❸ $b^2 - 4ac = (-12)^2 - 4(4)(9)$ *$a = 4, b = -12, c = 9$*
$= 144 - 144$
$= 0$

Since $b^2 - 4ac = 0$, then $4x^2 - 12x + 9 = 0$ has one real root.

Chalkboard Exercises

State the value of the discriminant for each equation. Then determine the nature of the roots of the equation.

1a. $2x^2 - 11x + 13 = 0$
17, two real roots

1b. $3x^2 + 8x - 11 = 0$
196, two real roots

1c. $3x^2 + 7x - 26 = 0$
361, two real roots

2a. $x^2 - 6x + 9 = 0$
0, one real root

2b. $x^2 - 10x + 25 = 0$
0, one real root

2c. $16x^2 + 24x + 9 = 0$
0, one real root

❸ The equation in this example has one distinct real root. Notice that the left side of the equation is a perfect square trinomial.

$$4x^2 - 12x + 9 = 0$$
$$(2x - 3)(2x - 3) = 0$$
$$x = \frac{3}{2} \qquad x = \frac{3}{2}$$

The root is a double root.

Chalkboard Exercises

State the value of the discriminant for each equation. Then determine the nature of the roots of the equation.

3a. $x^2 - x + 1 = 0$ -3, no real roots

3b. $x^2 - 4x + 9 = 0$ -20, no real roots

3c. $7x^2 + 5x + 36 = 0$ -983, no real roots

4 Students may wonder if this equation has any roots. Explain that it does have 2 complex roots. Refer students to the feature on imaginary roots on page 423.

Suggested Daily Assignments

Enriched
- Exp: 10–20 evens
- Written: 11–39 odds

Average
- Exp: 1–15
- Written: 2–38 evens

Fundamental
- Exp: 1–21
- Written: 1–18
  - p. 418, 1–23 odds

Answers to Written

22. $\frac{1 \pm \sqrt{17}}{4}$ 23. $\frac{5 \pm \sqrt{193}}{12}$

24. 7, 2 25. $\frac{-1 \pm \sqrt{41}}{4}$

26. $-\frac{2}{3}, -2$ 27. $1, -\frac{1}{3}$

28. $\frac{-5 \pm \sqrt{10}}{3}$ 29. $\frac{-7 \pm \sqrt{33}}{4}$

30. $\frac{3 \pm \sqrt{37}}{14}$ 31. $\frac{-7 \pm \sqrt{17}}{16}$

32. $1, \frac{2}{3}$ 33. $\frac{-5 \pm \sqrt{133}}{6}$

34. $\frac{4 \pm \sqrt{6}}{2}$ 35. $\frac{3 \pm \sqrt{14}}{5}$

36. $\frac{4 \pm \sqrt{30}}{2}$ 37. $1, \frac{2}{3}$

38. none 39. $\frac{7 \pm \sqrt{69}}{10}$

Example

3 **State the value of the discriminant of $3x^2 + 4x + 2 = 0$. Then determine the nature of the roots of the equation.**

$$b^2 - 4ac = 4^2 - 4(3)(2) \quad a = 3,\ b = 4,\ c = 2$$
$$= 16 - 24$$
$$= -8$$

Since $b^2 - 4ac < 0$, then $3x^2 + 4x + 2 = 0$ has no real roots. **4**

Exploratory Exercises

State the value of the discriminant for each equation. Then determine the nature of the roots of the equation.

1. $x^2 + 3x - 4 = 0$ 25; 2 real
2. $y^2 + 3y + 1 = 0$ 5; 2 real
3. $m^2 + 5m - 6 = 0$ 49; 2 re
4. $r^2 + 3r + 4 = 0$ -7; no real
5. $n^2 + 12n + 5 = 0$ 124; 2 real
6. $t^2 - 12t + 36 = 0$ 0; 1 real
7. $s^2 + 8s + 16 = 0$ 0; 1 real
8. $2k^2 - 7k + 6 = 0$ 1; 2 real
9. $2p^2 - p - 3 = 0$ 25; 2 real
10. $2x^2 + 1 = 7x$ 41; 2 real
11. $h^2 = 6 - h$ 25; 2 real
12. $b^2 - 10b + 12 = 0$ 52; 2 r
13. $x^2 - 1.2x = 0$ 1.44; 2 real
14. $2z^2 + 7z + 50 = 0$ -351; no real
15. $y^2 - 4y - 8 = 0$ 48; 2 real
16. $3x^2 + x + 1 = 0$ -11; no real
17. $3x^2 + 7x - 2 = 0$ 73; 2 real
18. $2x^2 - 2x - 1 = 0$ 12; 2 rea
19. $\frac{4}{3}x^2 + 4x + 3 = 0$ 0; 1 real
20. $\frac{3}{2}m^2 + m = -\frac{7}{2}$ -20; no real
21. $4a^2 + 10a + 6.25 = 0$ 0; 1 real

Written Exercises

Determine the nature of the roots of each equation by using the discriminant. Find all real roots. Leave irrational roots in simplest radical form.

1. $x^2 + 5x + 3 = 0$ $\frac{-5 \pm \sqrt{13}}{2}$
2. $y^2 - 4y + 1 = 0$ $2 \pm \sqrt{3}$
3. $r^2 + 4r - 12 = 0$ 2, -6
4. $z^2 + 8z - 5 = 0$ $-4 \pm \sqrt{21}$
5. $m^2 + 7m + 6 = 0$ $-1, -6$
6. $p^2 - 7p + 6 = 0$ 6, 1
7. $k^2 + 6k + 10 = 0$ no real roots
8. $a^2 + 8a - 12 = 0$ $-4 \pm 2\sqrt{7}$
9. $d^2 + 4d + 7 = 0$ no real
10. $2x^2 + 3x + 1 = 0$ $-\frac{1}{2}, -1$
11. $2x^2 - 3x + 1 = 0$ $1, \frac{1}{2}$
12. $3y^2 + y - 1 = 0$ $\frac{-1 \pm \sqrt{1}}{6}$
13. $m^2 - 14m + 49 = 0$ 7
14. $3n^2 - n - 5 = 0$ $\frac{1 \pm \sqrt{61}}{6}$
15. $15a^2 + 2a + 16 = 0$ no real roots
16. $h^2 - 16h + 64 = 0$ 8
17. $y^2 - 4y - 32 = 0$ 8, -4
18. $11z^2 - z - 3 = 0$ $\frac{1 \pm \sqrt{1}}{22}$
19. $3p^2 - 4p - 1 = 0$ $\frac{2 \pm \sqrt{7}}{3}$
20. $9y^2 - 6y + 1 = 0$ $\frac{1}{3}$
21. $3g^2 - 4g + 1 = 0$ $1, \frac{1}{3}$
22. $2x^2 - x - 2 = 0$
23. $6r^2 - 5r = 7$
24. $-9m + m^2 = -14$
25. $2a^2 + a = 5$
26. $0.3a^2 + 0.8a + 0.4 = 0$
27. $3y^2 - 2y = 1$
28. $3z^2 + 10z + 5 = 0$
29. $2b^2 + 7b + 2 = 0$
30. $7r^2 - 3r - 1 = 0$
31. $8p^2 + 7p + 1 = 0$
32. $3a^2 + 2 = 5a$
33. $0.6v^2 + v = 1.8$
34. $2z^2 + 5 = 8z$
35. $5b^2 = 1 + 6b$
36. $2x^2 - 8x = 7$
37. $x^2 - \frac{5}{3}x = -\frac{2}{3}$
38. $\frac{1}{3}x^2 - 4x + 13\frac{1}{3} = 0$
39. $5c^2 - 7c = 1$

Excursions in Algebra — Imaginary Roots

The equation $x^2 = -1$ has *no* solution among the real numbers. This is because the square of a real number is positive.

We define a new number to be a solution to $x^2 = -1$. This number is designated by the letter i and is called the **imaginary unit.** The imaginary unit i is *not* a real number.

$$i^2 = -1$$

Using i as you would any constant, you can rewrite square roots of negative numbers.

$$i^2 = -1 \quad \text{so} \quad \sqrt{-1} = i$$
$$(2i)^2 = 2^2 i^2 \text{ or } -4 \quad \text{so} \quad \sqrt{-4} = 2i$$
$$(i\sqrt{3})^2 = i^2(\sqrt{3})^2 \text{ or } -3 \quad \text{so} \quad \sqrt{-3} = i\sqrt{3}$$

If the value of the discriminant of an equation is less than zero, the roots of the equation are *not* real numbers. However, the roots can be expressed using the imaginary unit.

Consider the equation $x^2 + 6x + 10 = 0$. The discriminant of this equation is a negative number.

$$\begin{aligned} b^2 - 4ac &= 6^2 - 4(1)(10) \\ &= 36 - 40 \\ &= -4 \end{aligned}$$

Even though the discriminant is less than zero, the quadratic formula can still be used to find the roots of the equation.

$$\begin{aligned} x &= \frac{-b \pm \sqrt{b^2 - 4ac}}{2a} \\ &= \frac{-6 \pm \sqrt{-4}}{2} \\ &= \frac{-6 \pm 2i}{2} \text{ or } -3 \pm i \end{aligned}$$

The roots of $x^2 + 6x + 10 = 0$ are $-3 + i$ and $-3 - i$. Roots that contain the imaginary unit are known as **imaginary roots.**

Exercises

Find the roots of each equation.

1. $x^2 + x + 1 = 0$
2. $x^2 + 2x + 2 = 0$
3. $x^2 - 4x + 5 = 0$
4. $x^2 + 5x + 7 = 0$
5. $2x^2 - 3x + 4 = 0$
6. $3x^2 + x + 1 = 0$
7. $2z^2 + 2z + 3 = 0$
8. $2x^2 + 3x + 3 = 0$
9. $6y^2 + 8y + 5 = 0$
10. $r^2 - 4r + 10 = 0$
11. $5x^2 - 2x + 8 = 0$
12. $3k^2 + 3k + 2 = 0$

About the Excursion

If the discriminant of an equation is less than zero, the roots can be expressed using the imaginary unit, *i*. These roots are called *imaginary roots*. That is, they are from the set of imaginary numbers. Taken together, the set of real numbers and the set of imaginary numbers form the set of *complex numbers*. Complex numbers are numbers of the form $a + bi$, where a and b are real numbers and i is the imaginary unit.

The roots of $x^2 + 6x + 10 = 0$ were shown to be $-3 + i$ and $-3 - i$. Explain to students that these roots can be checked by substitution. Check $-3 + i$.

$$(-3 + i)^2 + 6(-3 + i) + 10 \stackrel{?}{=} 0$$
$$9 - 6i + i^2 - 18 + 6i + 10 \stackrel{?}{=} 0$$
$$i^2 + 1 \stackrel{?}{=} 0$$
Recall that $i^2 = -1$
$$-1 + 1 = 0$$

The root is correct.
Suggest that students check the root $-3 - i$.

Answers

1. $\frac{-1 \pm i\sqrt{3}}{2}$
2. $-1 \pm i$
3. $2 \pm i$
4. $\frac{-5 \pm i\sqrt{3}}{2}$
5. $\frac{3 \pm i\sqrt{23}}{4}$
6. $\frac{-1 \pm i\sqrt{11}}{6}$
7. $\frac{-1 \pm i\sqrt{5}}{2}$
8. $\frac{-3 \pm i\sqrt{15}}{4}$
9. $\frac{-4 \pm i\sqrt{14}}{6}$
10. $2 \pm i\sqrt{6}$
11. $\frac{1 \pm i\sqrt{39}}{5}$
12. $\frac{-3 \pm i\sqrt{15}}{6}$

13-6

Lesson Notes, page T50

Readiness Activity
Quadratic equations can be solved by many methods. Describe the disadvantages of the following methods.

graphing not always exact

factoring sometimes factors cannot easily be determined

completing the square b should be even in equations of the form $x^2 + bx + c = 0$ or this process is not efficient.

Discussion Question
Which method would you use to solve an equation of the form $ax^2 + bx + c = 0$ if a, b, and c are all odd? factoring or quadratic formula

Chalkboard Exercises
Solve each quadratic equation by an appropriate method.

1a. $y^2 - \frac{3}{5}y + \frac{2}{25} = 0$

$25y^2 - 15y + 2 = 0$
$(5y - 1)(5y - 2) = 0$
$\left\{\frac{2}{5}, \frac{1}{5}\right\}$

1b. $2a^2 - 8a + \frac{15}{2} = 0$

$4a^2 - 16a + 15 = 0$
$(2a - 3)(2a - 5) = 0$
$\left\{\frac{5}{2}, \frac{3}{2}\right\}$

1c. $x^2 - x + \frac{3}{16} = 0$

$16x^2 - 16x + 3 = 0$
$(4x - 1)(4x - 3) = 0$
$\left\{\frac{3}{4}, \frac{1}{4}\right\}$

Objective 13-6: Use a variety of methods to solve quadratic equations.

13-6 Solving Quadratic Equations

You have studied a variety of methods for solving quadratic equations. The table summarizes these methods.

Method	*Can be Used*	*Comments*
graphing	always	Not always exact; use only when a picture of function is needed.
factoring	sometimes	Use if constant term is 0 or factors are easily determined.
completing the square	always	Useful for equations of form $x^2 + bx + c = 0$ where b is even.
quadratic formula	always	Other methods may be easier, but this method *always* works.

Use the information in this table to help you decide how to solve a quadratic equation.

To solve:	Try:
$x^2 + 9x + 20 = 0$	factoring; quadratic formula
$2m^2 + 7m + 5 = 0$	quadratic formula
$x^2 - 2x - 2 = 0$	factoring; completing the square; quadratic formula
$x^2 + 6x - 315 = 0$	completing the square; quadratic formula
$6n^2 - 5n = 0$	factoring

To solve an equation with fractional or decimal coefficients, it is sometimes easier to first change the coefficients to integers by multiplication.

Example

1 **Solve: $y^2 - \frac{5}{12}y - \frac{1}{4} = 0$**

$$y^2 - \frac{5}{12}y - \frac{1}{4} = 0$$ *The LCM of 12 and 4 is 12.*

$$12\left(y^2 - \frac{5}{12}y - \frac{1}{4}\right) = 12(0)$$ *Multiply both sides by 12.*

$$12y^2 - 5y - 3 = 0$$

$$(3y + 1)(4y - 3) = 0$$ *Factor or use the quadratic formula.*

$$3y + 1 = 0 \quad \text{or} \quad 4y - 3 = 0$$

$$y = -\frac{1}{3} \quad \text{or} \quad y = \frac{3}{4}$$ *Check this result.*

The solutions are $-\frac{1}{3}$ and $\frac{3}{4}$.

Example

2 **Solve: $x^2 + 0.4x - 3.2 = 0$**

$$x^2 + 0.4x - 3.2 = 0$$

$$10(x^2 + 0.4x - 3.2) = 10(0)$$ *Multiply both sides by 10.* ❶

$$10x^2 + 4x - 32 = 0$$

$$x = \frac{-4 \pm \sqrt{4^2 - 4(10)(-32)}}{2(10)}$$ *Use the quadratic formula or try factoring.*

$$= \frac{-4 \pm \sqrt{16 + 1{,}280}}{20}$$

$$= \frac{-4 \pm 36}{20}$$

$x = -2$ or $x = 1.6$ *Check this result.*

The solutions are -2 and 1.6.

Exploratory Exercises

State the method that seems easiest for solving each equation.

1. $x^2 - 12x + 27 = 0$
2. $y^2 - 19y = -84$
3. $2m^2 + 19m + 9 = 0$
4. $a^2 - 12a - 4 = 0$
5. $3r^2 - 7r - 5 = 0$
6. $z^2 - 2z - 120 = 0$
7. $t^2 - 2t - 15 = 0$
8. $y^2 + 4y = 9$
9. $2b^2 + 1 = 6b$
10. $m^2 - 12m - 30 = 0$
11. $3k^2 - 11k - 7 = 0$
12. $3x^2 - 2x - 5 = 0$

State the number that both sides of each equation should be multiplied by before solving.

13. $x^2 - \frac{1}{2}x - \frac{1}{9} = 0$ 18
14. $x^2 - \frac{7}{6}x - \frac{1}{2} = 0$ 6
15. $3x^2 - \frac{5}{4}x - \frac{1}{2} = 0$ 4
16. $x^2 - 1.3x - 0.3 = 0$ 10
17. $2x^2 - 0.7x - 0.3 = 0$ 10
18. $0.2x^2 - 0.33x - 0.35 = 0$ 100

Written Exercises

Solve each quadratic equation by an appropriate method.

1. $x^2 - 9x + 20 = 0$ 4, 5
2. $y^2 + 10y - 2 = 0$ $-5 \pm 3\sqrt{3}$
3. $2x^2 + 4x + 1 = 0$
4. $3z^2 - 7z - 3 = 0$ $\frac{7 \pm \sqrt{85}}{6}$
5. $x^2 + 3x = -2$ $-2, -1$
6. $r^2 + 13r = -42$ $-7, -6$
7. $2y^2 - 5y + 2 = 0$ $\frac{1}{2}, 2$
8. $3x^2 - 7x - 6 = 0$ $-\frac{2}{3}, 3$
9. $4x^2 - 7x - 2 = 0$ $-\frac{1}{4}, 2$
10. $2k^2 + k - 5 = 0$ $\frac{-1 \pm \sqrt{41}}{4}$
11. $3h^2 - 5h - 2 = 0$ $-\frac{1}{3}, 2$
12. $x^2 - 5x - 7 = 0$
13. $r^2 + 4r + 1 = 0$ $-2 \pm \sqrt{3}$
14. $m^2 + 2m + 8 = 0$ no real roots
15. $y^2 - 3y + 3 = 0$
16. $3z^2 = 5z - 1$ $\frac{5 \pm \sqrt{13}}{6}$
17. $9b = -5b^2 - 3$ $\frac{-9 \pm \sqrt{21}}{10}$
18. $a^2 - 15a = -52$
19. $2z^2 + 4z = 5$
20. $x^2 = 4x + 2$ $2 \pm \sqrt{6}$
21. $-2x - 2 = -x^2$ $1 \pm \sqrt{3}$
22. $x^2 - x + \frac{3}{16} = 0$ $\frac{1}{4}, \frac{3}{4}$
23. $2a^2 - 8a + \frac{15}{2} = 0$ $\frac{3}{2}, \frac{5}{2}$
24. $r^2 + r + \frac{2}{9} = 0$ $-\frac{2}{3}, -\frac{1}{3}$
25. $x^2 - \frac{17}{20}x + \frac{3}{20} = 0$ $\frac{1}{4}, \frac{3}{5}$
26. $y^2 - \frac{3}{5}y + \frac{2}{25} = 0$ $\frac{1}{5}, \frac{2}{5}$
27. $x^2 - 1.1x - 0.6 = 0$
28. $2x^2 - 0.7x - 0.6 = 0$ $-\frac{2}{5}, \frac{3}{4}$
29. $5x^2 - 0.5x - 0.3 = 0$ $-\frac{1}{5}, \frac{3}{10}$
30. $0.7a^2 - 2.8a = 7$

Chalkboard Exercises

Solve each quadratic equation by an appropriate method.

2a. $0.2x^2 - 0.7x - 6 = 0$

$2x^2 - 7x - 60 = 0$

$x = \frac{7 \pm \sqrt{49 + 480}}{4}$

$= \frac{7 \pm 23}{4}$ $\{7.5, -4\}$

2b. $5x^2 - 0.5x - 0.3 = 0$

$50x^2 - 5x - 3 = 0$

$x = \frac{5 \pm \sqrt{25 + 600}}{100}$

$= \frac{5 \pm 25}{100}$ $\{0.3, -0.2\}$

❶ It may not be desirable to multiply the equation by 10 before applying the quadratic formula, especially if calculators are being used. However, some students make fewer mistakes when working with whole numbers.

Suggested Daily Assignments

Enriched
- Exp: 2–18 evens
- Written: 2–30 evens

Average
- Exp: 1–18
- Written: 1–30

Fundamental
- Exp: 1–18
- Written: 1–21
  - p. 422, 19–25

Answers to Exploratory

1. factoring 2. factoring
3. factoring or formula
4. formula or completing the square 5. formula
6. factoring 7. factoring
8. formula or completing the square 9. formula
10. formula or completing the square
11. formula 12. formula

Answers to Written

3. $\frac{-2 \pm \sqrt{2}}{2}$ 12. $\frac{5 \pm \sqrt{53}}{2}$

15. no real roots 18. $\frac{15 \pm \sqrt{17}}{2}$

19. $\frac{-2 \pm \sqrt{14}}{2}$ 27. $-\frac{2}{5}, \frac{3}{2}$

30. $2 \pm \sqrt{14}$

13-7

Lesson Notes, page T51

Discussion Question

Show that if A is the area of a square with perimeter P, then $A = \frac{P^2}{16}$.

Since $P = 4s$, $s = \frac{P}{4}$. Since $A = s^2$, by substitution, $A = \left(\frac{P}{4}\right)^2$. Thus, $A = \frac{P^2}{16}$.

Chalkboard Exercises

Solve each of the following problems.

1a. Find the dimensions of a rectangle if the area is 63 cm^2 and the perimeter is 32 cm.
$63 = w(16 - w)$
7 cm, 9 cm

1b. Find the dimensions of a rectangle if the area is 42 cm^2 and the perimeter is 26 cm.
$42 = w(13 - w)$
6 cm, 7 cm

2. A box is to be formed by cutting squares measuring 4 inches on a side from a square piece of cardboard and then folding the sides. If the volume of the box is to be 576 in^2, what is the original size of the cardboard?
$576 = (x - 8)(x - 8)(4)$ The length of each side is 20 in.

❶ The length of each side of the piece of sheet metal was found to be 18 cm. Thus, the dimensions of the pan would be 14 × 14 × 2. (Discuss how this was obtained.)
Does this pan have a volume of 392 cm^3?
$(14)(14)(2) = 392$

Objective 13-7: Solve verbal problems using quadratic equations.

13-7 Problem Solving: Quadratic Equations

You can use the methods for solving quadratic equations to solve some types of verbal problems.

Examples

1 **A rectangle has a perimeter of 19 centimeters. Its area is 21 square centimeters. Find its dimensions.**

Explore Let l = the measure of the length.
Let w = the measure of the width.

Plan Use the formula $P = 2l + 2w$.

$19 = 2l + 2w$ *Substitute 19 for P.*
$9.5 = l + w$ *Solve for l.*
$l = 9.5 - w$

Now write an equation using the formula $A = lw$.

$21 = (9.5 - w)w$ *Substitute 21 for A and 9.5 − w for l.*

Solve

$21 = 9.5w - w^2$
$w^2 - 9.5w + 21 = 0$ *Subtract $9.5w - w^2$ from both sides.*
$2w^2 - 19w + 42 = 0$ *Multiply both sides by 2.*
$(2w - 7)(w - 6) = 0$ *Factor.*
$w = 3.5$ or $w = 6$

If $w = 3.5$ then $l = 9.5 - 3.5$ or 6.
If $w = 6$, then $l = 9.5 - 6$ or 3.5.
The dimensions are 3.5 cm and 6 cm.

Examine The perimeter of a rectangle with dimensions 3.5 cm and 6 cm is 2(3.5) + 2(6) or 19 cm. The area of this rectangle is (3.5)(6) or 21 sq cm.

2 **A pan is to be formed by cutting squares measuring 2 cm on a side from a square piece of sheet metal and then folding the sides. If the volume of the pan is to be 392 cm^3, what is the original size of the sheet metal?**

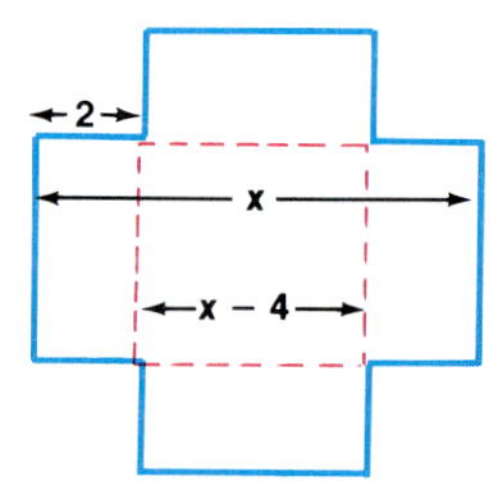

Let x = the measure of each side.
Use the formula $V = lwh$ to write an equation.

$392 = (x - 4)(x - 4)(2)$ *$V = 392$, $l = x - 4$, $w = x - 4$, $h = 2$*

$392 = 2x^2 - 16x + 32$
$2x^2 - 16x - 360 = 0$ *Subtract 392 from both sides.*
$x^2 - 8x - 180 = 0$ *Divide both sides by 2.*
$(x - 18)(x + 10) = 0$ *Factor.*
$x = 18$ or $x = -10$

❶ The length of each side of the sheet metal must be 18 cm since −10 cm is not an appropriate answer. *Examine this solution.*

Written Exercises

Find the dimensions of each rectangle described below. Disregard inappropriate solutions.

1. The perimeter is 30 m. The area is 56 m^2. 7 m, 8 m
2. The perimeter is 34 in. The area is 72 in^2. 8 in., 9 in.
3. The perimeter is 46 yd. The area is 130 yd^2. 10 yd, 13 yd
4. The perimeter is 70 cm. The area is 294 cm^2. 14 cm, 21 cm
5. The length is 2 m more than 3 times the width. The area is 56 m^2. 4 m, 14 m
6. The width is 7 cm less than the length. The area is 78 cm^2. 6 cm, 13 cm
7. The perimeter is 60 in. The area is 200 in^2. 10 in., 20 in.
8. The perimeter is 37 in. The area is 78 in^2.
9. The length is 4 in. more than the width. The area is 45 in^2. 5 in., 9 in.
10. The width is 3 ft less than the length. The area is 54 ft^2. 6 ft, 9 ft
11. The length is $\frac{8}{5}$ times the width. The area is 160 m^2. 10 m, 16 m
12. The width is 0.5 m less than one-half the length. The area is 21 m^2. 3 m, 7 m

8. $6\frac{1}{2}$ in., 12 in.

Solve each of the following problems. Approximate irrational roots and disregard inappropriate solutions.

13. Find two integers whose sum is 14 and whose product is 48. 6, 8
14. Find two integers whose sum is 13 and whose product is 42. 6, 7
15. Find two integers whose difference is 6 and whose product is 135. −15, −9; 9, 15
16. Find two integers whose sum is 12 and whose squares differ by 24. 5, 7
17. A rectangular piece of sheet metal is 3 times as long as it is wide. Squares measuring 2 cm on a side are cut from each corner and the sides are folded to form a pan. If the volume of the pan is 512 cm^3, what are the dimensions of the sheet metal. 12 cm, 36 cm
18. A rectangular piece of sheet metal is twice as long as it is wide. Squares measuring 5 inches on a side are cut from each corner and the sides are folded to form a box. If the volume of the box is 1,760 $in.^3$, what are the dimensions of the sheet metal? 21 in., 42 in.
19. A rectangular piece of glass is twice as long as it is wide. If the length and width are both reduced by 1 cm, the area of the glass becomes 10cm^2. What are the original dimensions of the glass? 3 cm, 6 cm
20. A rectangular piece of sheet metal is 3 in. longer than it is wide. If the length and width are both increased by 2 in. the area increases by 34 in^2. What are the original dimensions of the sheet metal? 6 in., 9 in.
21. The perimeter of a rectangle is 8 meters and its area is 3.84 square meters. Find its dimensions. 1.6 m, 2.4 m
22. A rectangle has a perimeter of 15.4 cm and an area of 14.4 cm^2. Find the dimensions of the rectangle. 3.2 cm, 4.5 cm
23. The length of a rectangle is $\frac{8}{5}$ times its width. The area is 56 square meters. Find the dimensions of the rectangle. 5.92 m, 9.47 m
24. The width of a rectangle is one-half meter less than one-half its length. The area is 21 m^2. Find its dimensions. 3 m, 7 m

Suggested Daily Assignments

Enriched
- Written: 2–10 evens, 13–26
- Mini-Review: 1–5

Average
- Written: 1–20
- Mini-Review: 1–5

Fundamental
- Written: 1–18
  - p. 425, 22–25
- Mini-Review: 1–5

25. Dan Kurtz has a rectangular flower garden that measures 15 m by 20 m. He wishes to place a concrete walk of uniform width around the garden. His budget allows him to cover 74 m^2. How wide can the walk be? 1 m

26. A picture has a square frame that is 5 cm wide. The area of the picture is two-thirds of the total area of the picture and the frame. What are the dimensions of the frame? 54.5 cm, 54.5 cm

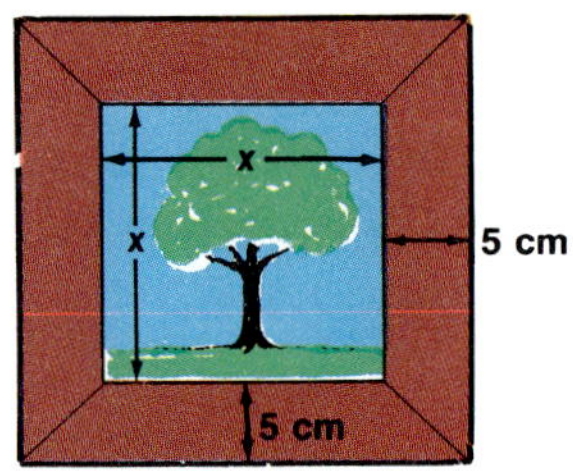

mini-review

Simplify.

1. $\sqrt{3}(\sqrt{3} - \sqrt{6})$ $3 - 3\sqrt{2}$
2. $\frac{12}{2 - \sqrt{5}}$ $-24 - 12\sqrt{5}$
3. $4\sqrt{3} + 5\sqrt{2} - 2\sqrt{18} - 4\sqrt{12}$ $-4\sqrt{3} - \sqrt{2}$
4. Solve $\sqrt{3x + 6} = 3$ 1
5. Find the length of the hypotenuse of a right triangle if its legs measure 6 and 12. $6\sqrt{5}$

Using Calculators — Checking Roots

After finding roots of an equation, verify them by substitution. To see if −4 and 3 are the roots of $2x^2 + 5x - 12 = 0$, replace x with each number and see if the equation holds.

$$2x^2 + 5x - 12 = 0$$

$$2(-4)^2 + 5(-4) - 12 \stackrel{?}{=} 0 \qquad 2(3)^2 + 5(3) - 12 \stackrel{?}{=} 0$$

A calculator may also be used to aid in checking the roots of an equation.

ENTER:	2	×	4	+/−	x^2	+	5	×	4	+/−	−	12	=
DISPLAY:	2	2	4	−4	16	32	5	5	4	−4	12	12	0

ENTER:	2	×	3	x^2	+	5	×	3	−	12	=
DISPLAY:	2	2	3	9	18	5	5	3	33	12	21

By substitution, it is found that −4 is a root of the equation $2x^2 + 5x - 12 = 0$, but that 3 is not.

Exercises

Use a calculator to determine if the given numbers are roots of each equation.

1. −6, −5; $x^2 + x - 30 = 0$

2. 2, −1; $2x^2 - x - 3 = 0$

3. 0.5, 0.6; $10x^2 - 2x - 3 = 0$

4. 0.5, 1.5; $8x^2 - 16x + 6 = 0$

1. −6 is a root; −5 is not a root
2. 2 is not a root; −1 is a root
3. 0.5 is not a root; 0.6 is not a root
4. 0.5 is a root; 1.5 is a root

About Using Calculators
This feature introduces a quick way to use a calculator to check the roots of a quadratic equation. Emphasize the use of the [+/−] key. You may want to have students use this method to check their answers to the Written Exercises on page 425.

Objective 13-8A: State the sum and product of the roots of a quadratic equation.

Objective 13-8B: Write a quadratic equation given its roots.

13-8 The Sum and Product of Roots

An engineer or scientist often must find an equation to describe a certain situation. For example, suppose the roots of a quadratic equation are known to be -3 and 8.

If $x = -3$, then $x + 3 = 0$. If $x = 8$, then $x - 8 = 0$.

$(x + 3)(x - 8) = 0$ *Why is this so?*

$x^2 - 5x - 24 = 0$

The quadratic equation $x^2 - 5x - 24 = 0$ has roots -3 and 8.

Suppose you find the sum and the product of the roots.

$$\text{sum} = -3 + 8 = 5$$
$$\text{product} = -3 \cdot 8 = -24$$

Now look at the equation.

$$x^2 \underbrace{- 5x}_{\text{opposite of sum of roots}} \overbrace{- 24}^{\text{product of roots}} = 0$$

The product of the roots is the constant term.

The opposite of the sum of the roots is the coefficient of x.

This *always* works. Suppose the two roots are r_1 and r_2. Then $(x - r_1)(x - r_2) = 0$. Multiplying the binomials results in the equation $x^2 - (r_1 + r_2)x + r_1r_2 = 0$. Notice that the coefficient of x, $-(r_1 + r_2)$, is the opposite of the sum of the roots. The constant term, r_1r_2, is the product of the roots.

To find a quadratic equation of the general form $x^2 + bx + c = 0$ given its roots:
1. **The coefficient of x is the opposite of the sum of the roots.**
2. **The constant term is the product of the roots.**

Finding Quadratic Equations

Example 1

Find a quadratic equation whose roots are -4 and 7.

$$\text{opposite of sum of roots} = -(-4 + 7) = -3$$
$$\text{product of roots} = -4 \cdot 7 = -28$$

The opposite of the sum of the roots, -3, is the coefficient of x. The product of the roots, -28, is the constant term.

Thus, a quadratic equation whose roots are -4 and 7 is $x^2 - 3x - 28 = 0$.

13-8

Lesson Notes, page T51

Readiness Activity

Solve each of the following quadratic equations. Then find the product and the sum of the roots. Describe the results.

$x^2 + 4x + 3 = 0$ $-3, -1$; product $= 3$, sum $= -4$; product $= c$, sum $= -b$

$x^2 - 6x - 16 = 0$ $-2, 8$; product $= -16$, sum $= 6$; product $= c$, sum $= -b$

$x^2 - 8x + 15 = 0$ $3, 5$; product $= 15$, sum $= 8$; product $= c$, sum $= -b$

Discussion Question

(Use after Example 3.)

Find k such that 3 is a root of the equation $x^2 + kx - 21 = 0$.

Let r = the second root of the equation. Then, since the product of the roots is $\frac{c}{a}$, $3r = \frac{-21}{1}$. So $r = -7$. Since the sum of the roots is $-\frac{b}{a}$, $3 + (-7) = -\frac{k}{1}$. Thus, $k = 4$.

Chalkboard Exercises

Find a quadratic equation having the given roots.

1a. 6, 1 $x^2 - 7x + 6 = 0$

1b. -2, 5 $x^2 - 3x - 10 = 0$

1c. 3, 4 $x^2 - 7x + 12 = 0$

1d. $\sqrt{3}$, $-\sqrt{3}$ $x^2 - 3 = 0$

Chalkboard Exercises

Find a quadratic equation having the given roots.

2a. $\frac{5}{2}$, 1 $2x^2 - 7x + 5 = 0$

2b. -2, $\frac{1}{3}$ $3x^2 + 5x - 2 = 0$

2c. $\frac{3}{5}$, 4 $5x^2 - 23x + 12 = 0$

2d. $\frac{1}{2}$, $-\frac{1}{4}$ $8x^2 - 2x - 1 = 0$

State whether the following numbers are roots of the given equation.

3a. 2, 3; $x^2 + 5x + 6 = 0$ no

3b. 5, 5; $x^2 - 10x + 25 = 0$ yes

3c. $\frac{1}{2}$, $-\frac{1}{2}$; $4x^2 - 1 = 0$ yes

3d. -4, $\frac{1}{3}$; $3x^2 + 11x - 4 = 0$ yes

Suggested Daily Assignments

Enriched
- Exp: 2–26 evens
- Written: 2–28 evens, 29–39

Average
- Exp: 1–31
- Written: 1–36
  - p. 427, 21–26

Fundamental
- Exp: 1–31
- Written: 1–12, 19–30
  - p. 427, 19–25

Example

2 Find a quadratic equation whose roots are $\frac{4}{3}$ and -2.

$$\text{opposite of sum of roots} = -\left[\frac{4}{3} + (-2)\right] = \frac{2}{3}$$

$$\text{product of roots} = \frac{4}{3}(-2) = -\frac{8}{3}$$

The opposite of the sum of the roots, $\frac{2}{3}$, is the coefficient of x. The product of the roots, $-\frac{8}{3}$, is the constant term. Thus, a quadratic equation is $x^2 + \frac{2}{3}x - \frac{8}{3} = 0$. This equation can also be written as $3x^2 + 2x - 8 = 0$.

By studying Example 2, the following rule may be discovered.

Sum and Product of Roots

Given the quadratic equation $ax^2 + bx + c = 0$, where $a \neq 0$, the sum of the roots of the equation is $-\frac{b}{a}$ and the product of the roots of the equation is $\frac{c}{a}$.

Example

3 Are $\frac{3}{2}$ and $-\frac{4}{3}$ roots of $6x^2 - x - 12 = 0$?

Sum of Roots

$$\frac{3}{2} + \left(-\frac{4}{3}\right) = \frac{9}{6} + \left(-\frac{8}{6}\right)$$

$$= \frac{1}{6}$$ *Is this equal to $-\frac{b}{a}$?*

Product of Roots

$$\frac{3}{2}\left(-\frac{4}{3}\right) = -2$$ *Is this equal to $\frac{c}{a}$?*

Since $\frac{-b}{a} = \frac{1}{6}$ and $\frac{c}{a} = -2$, $\frac{3}{2}$ and $-\frac{4}{3}$ are roots of the equation.

Exploratory Exercises

State the sum and product of the roots of each equation.

1. $x^2 - 5x + 6 = 0$ 5, 6
2. $y^2 - 8y - 20 = 0$ 8, −20
3. $z^2 + 12z - 28 = 0$ −12, −28
4. $3m^2 + 6m - 3 = 0$ −2, −1
5. $4k^2 + 20k - 16 = 0$ −5, −4
6. $4t^2 + 8t + 3 = 0$ $-2, \frac{3}{4}$
7. $4x^2 + 4x = 35$ $-1, -\frac{35}{4}$
8. $6a^2 - 13a = 15$ $\frac{13}{6}, -\frac{15}{6}$
9. $6b^2 - 5b = 21$ $\frac{5}{6}, -\frac{21}{6}$

State whether the following numbers are roots of the given equation.

10. $-4, 6; x^2 - 2x - 24 = 0$ yes
11. $-6, 3; y^2 + 3y - 18 = 0$ yes
12. $3, 7; a^2 - 10a + 21 = 0$ yes
13. $2, 3; n^2 + 5n - 6 = 0$ no
14. $-1, 7; b^2 - 8b + 7 = 0$ no
15. $-2, 4; r^2 + 2r + 8 = 0$ no
16. $-3, 4; x^2 + x - 12 = 0$ no
17. $-6, 5; x^2 - x - 30 = 0$ no
18. $-\frac{1}{3}, \frac{1}{2}; x^2 - \frac{1}{6}x - \frac{1}{6} = 0$ yes
19. $-\frac{1}{4}, \frac{1}{6}; x^2 + \frac{5}{12}x + \frac{1}{24} = 0$ no

Find a quadratic equation having the given roots.

20. 4, 1 $x^2 - 5x + 4 = 0$
21. 5, 2
22. -4, 5
23. 3, 8
24. 0, -5 $x^2 + 5x = 0$
25. 1, -6
26. $\frac{3}{2}$, 6
27. $\frac{1}{2}, \frac{-3}{2}$
28. $\frac{2}{3}$, 7 $3x^2 - 23x + 14 = 0$
29. 1.4, 2.2
30. 0.3, -0.6
31. $\frac{2}{5}, \frac{-3}{5}$

Written Exercises

State the sum and product of the roots of each equation.

1. $y^2 + 15y + 54 = 0$ $-15, 54$
2. $a^2 - 5a - 24 = 0$ $5, -24$
3. $m^2 - m - 6 = 0$ $1, -6$
4. $b^2 + 12b - 28 = 0$ $-12, -28$
5. $6x^2 + 31x + 35 = 0$ $-\frac{31}{6}, \frac{35}{6}$
6. $21r^2 + 2r - 8 = 0$
7. $k^2 + 6k - 1 = 0$ $-6, -1$
8. $4y^2 + 4y + 1 = 0$ $-1, \frac{1}{4}$
9. $7h^2 - 33h - 10 = 0$
10. $n^2 - 10n + 23 = 0$ $10, 23$
11. $\frac{1}{2}m^2 - \frac{3}{2}m + 4 = 0$ $3, 8$
12. $z^2 + \frac{13}{2}z - \frac{9}{4} = 0$
13. $m^2 - 1.2m + 0.27 = 0$ $1.2, 0.27$
14. $6k^2 - 0.4k - 0.02 = 0$
15. $2c^2 - \frac{2}{3}c = \frac{1}{6}$ $\frac{1}{3}, -\frac{1}{12}$
16. $12a^2 - \frac{7}{2}a + \frac{1}{4} = 0$ $\frac{7}{24}, \frac{1}{48}$
17. $a^2 + 4a\sqrt{3} + 9 = 0$ $-4\sqrt{3}, 9$
18. $2y^2 + y\sqrt{2} - 6 = 0$

Find a quadratic equation having the given roots.

19. 5, 1 $x^2 - 6x + 5 = 0$
20. 4, 7 $x^2 - 11x + 28 = 0$
21. 6, -5 $x^2 - x - 30 = 0$
22. 7, -8 $x^2 + x - 56 = 0$
23. 1, -10 $x^2 + 9x - 10 = 0$
24. 8, -3 $x^2 - 5x - 24 = 0$
25. 5, -9 $x^2 + 4x - 45 = 0$
26. -2, -17 $x^2 + 19x + 34 = 0$
27. 16, -5 $x^2 - 11x - 80 = 0$
28. $\frac{5}{2}$, 2 $2x^2 - 9x + 10 = 0$
29. $\frac{7}{3}$, -3 $3x^2 + 2x - 21 = 0$
30. $-\frac{3}{4}$, 8 $4x^2 - 29x - 24 = 0$
31. $\frac{2}{3}, \frac{-3}{2}$ $6x^2 + 5x - 6 = 0$
32. $-\frac{4}{5}, \frac{5}{4}$ $20x^2 - 9x - 20 = 0$
33. 6, $-\frac{1}{5}$ $5x^2 - 29x - 6 = 0$
34. $\sqrt{2}, -\sqrt{2}$ $x^2 - 2 = 0$
35. $\sqrt{3}, \sqrt{3}$ $x^2 - 2\sqrt{3}x + 3 = 0$
36. $2 + \sqrt{3}, 2 - \sqrt{3}$ $x^2 - 4x + 1 = 0$
37. $5 - \sqrt{2}, 5 + \sqrt{2}$ $x^2 - 10x + 23 = 0$
38. $\frac{1 + \sqrt{7}}{2}, \frac{1 - \sqrt{7}}{2}$ $2x^2 - 2x - 3 = 0$
39. Find the general form of the quadratic equation with roots q and r. $x^2 - (q + r)x + qr = 0$

Using Calculators — Decimal Coefficients

Use the quadratic formula and a calculator to solve the following equations. Approximate the roots to three decimal places. 1. $-1.784, 1.641$ 3. $0.253, -2.444$

1. $1.4x^2 + 0.2x - 4.1 = 0$
2. $0.2x^2 - 1.3x + 0.6 = 0$ 0.5, 6
3. $2.1y^2 + 4.6y - 1.3 = 0$
4. $3.8m^2 - 2.1m - 1.7 = 0$ $-0.447, 1$
5. $0.4r^2 + 5.2r - 2.3 = 0$ $-13.428, 0.428$
6. $7.9z^2 - 5.1z + 0.6 = 0$ 0.491, 0.155

Answers to Exploratory

21. $x^2 - 7x + 10 = 0$
22. $x^2 - x - 20 = 0$
23. $x^2 - 11x + 24 = 0$
25. $x^2 + 5x - 6 = 0$
26. $2x^2 - 15x + 18 = 0$
27. $4x^2 + 4x - 3 = 0$
29. $x^2 - 3.6x + 3.08 = 0$
30. $x^2 + 0.3x - 0.18 = 0$
31. $25x^2 + 5x - 6 = 0$

Answers to Written

6. $-\frac{2}{21}, -\frac{8}{21}$
9. $\frac{33}{7}, -\frac{10}{7}$
12. $-\frac{13}{2}, -\frac{9}{4}$
14. $\frac{1}{15}, -\frac{1}{300}$
18. $-\frac{\sqrt{2}}{2}, -3$

About Using Calculators

You may want to go through the first exercise with students to make sure they understand how to use a calculator to solve the equation. Emphasize that the roots will only be approximate.

About Using Computers

(p.432)

The program in this feature can be used to help students graph quadratic functions. Emphasize the meaning of the term "zeros of a quadratic function." Explain that the word zero is used because the number or numbers computed are the x-intercepts of the graph.

Point out that the FOR/NEXT loop in lines 90–110 computes the coordinates of 9 pairs of points on the graph. You may want to ask students to modify the program to identify the coordinates of the maximum or minimum point of the graph.

Answers to Exercises

1. (−3, 9) (−2, 2) (−1, −3) (0, −6) (1, −7) (2, −6) (3, −3) (4, 2) (5, 9) Zeros 3.646, −1.646
2. (−4.5, 12.75) (−3.5, 5.75) (−2.5, 0.75) (−1.5, −2.25) (−0.5, −3.25) (0.5, 2.25) (1.5, 0.75) (2.5, 5.75) (3.5, 12.75) Zeros 1.303, −2.303
3. (−3, 35) (−2, 21) (−1, 11) (0, 5) (1, 3) (2, 5) (3, 11) (4, 21) (5, 35) No real zeros
4. (−5, 43) (−4, 22) (−3, 7) (−2, −2) (−1, −5) (0, −2) (1, 7) (2, 22) (3, 43) Zeros 0.291, −2.291
5. (−2, −16) (−1, −9) (0, −4) (1, −1) (2, 0) (3, −1) (4, −4) (5, −9) (6, −16) Zero 2
6. (−2, −30) (−1, −16) (0, −6) (1, 0) (2, 2) (3, 0) (4, −6) (5, −16) (6, −30) Zeros 1, 3

(continued on p. 433)

Using Computers

Graphing Quadratic Functions

The following BASIC program can be used as an aid for graphing quadratic functions.

```
10   PRINT "FOR THE QUADRATIC FUNCTION Y=A*X↑2+B*X+C,
       ENTER A,B, AND C"
20   INPUT A,B,C
30   LET S = -B / (2 * A)
40   PRINT "THE EQUATION OF THE AXIS OF SYMMETRY IS X = ";S
50   LET Y = A * S ↑ 2 + B * S + C
60   IF A > 0 THEN PRINT "THE MINIMUM POINT IS (";S;",";Y;")"
70   IF A < 0 THEN PRINT "THE MAXIMUM POINT IS (";S;",";Y;")"
80   IF A = 0 THEN PRINT "A MUST NOT BE ZERO"
90   FOR X = S - 4 TO S + 4
95   LET Y = A * X ↑ 2 + B * X + C
100  PRINT "(";X;",";Y;")"
110  NEXT X
120  END
```

Line 30 determines the axis of symmetry.

The zeros of a quadratic function are the roots of the related quadratic equation. The program above can be extended to compute the zeros of the function. To do this, delete line 120 and add the lines shown at the right. Notice that the discriminant is used in lines 130 and 140 to check for the number of real zeros. If there are one or two real zeros, the quadratic formula is used to compute the zeros.

```
120  LET D = B ↑ 2 - 4 * A * C
130  IF D > 0 THEN 170
140  IF D = 0 THEN 200
150  PRINT "NO REAL ZEROS"
160  GOTO 220
170  PRINT "THE ZEROS ARE"
180  PRINT ( - B + SQR (D)) /
       (2 * A);" AND ";( - B -
         SQR (D)) / (2 * A)
190  GOTO 220
200  PRINT "THE ZERO IS"
210  PRINT - B / (2 * A)
220  END
```

Exercises

Use the program above to compute the coordinates of several points of the graphs of the following quadratic functions. Also use lines 120-220 above to compute the zeros for each function.

1. $y = x^2 - 2x - 6$ **2.** $y = x^2 + x - 3$ **3.** $y = 2x^2 - 4x + 5$

4. $y = 3x^2 + 6x - 2$ **5.** $y = -x^2 + 4x - 4$ **6.** $y = -2x^2 + 8x - 6$

7. Graph the following equations on one coordinate axis:

$y = 2x^2 + x - 6$, $y = x^2 + x - 6$, and $y = \frac{1}{2}x^2 + x - 6$

8. What effect does the coefficient of x^2 have on the shape of the graph?

Vocabulary

quadratic function (405)
parabola (405)
minimum point (405)
axis of symmetry (405)
maximum point (406)
roots (410)
completing the square (413)
quadratic formula (416)
discriminant (420)

Chapter Summary

1. Definition of Quadratic Function: A quadratic function is a function described by an equation of the form $y = ax^2 + bx + c$, where $a \neq 0$. (405)
2. The graph of a quadratic function has a general shape called a parabola. (405)
3. The minimum point is the lowest point on the graph of a parabola that opens upward. The maximum point is the highest point on the graph of a parabola that opens downward. (405–406)
4. If the graph of a parabola is folded along the axis of symmetry, the two halves of the graph coincide. (405)
5. Equation of Axis of Symmetry: The equation of the axis of symmetry for the graph of $y = ax^2 + bx + c$, where $a \neq 0$, is given by $x = -\frac{b}{2a}$. (406)
6. The roots of a quadratic equation are the x-coordinates of the points where the graph of the corresponding quadratic function crosses the x-axis. (410)
7. To complete the square for $x^2 + bx$, follow these steps.
   Step 1 Find one-half of b, the coefficient of x.
   Step 2 Square the result of **Step 1**.
   Step 3 Add the result of **Step 2** to $x^2 + bx$. (410)
8. The Quadratic Formula: The roots of a quadratic equation of the form $ax^2 + bx + c = 0$, where $a \neq 0$, are given by
$$x = \frac{-b \pm \sqrt{b^2 - 4ac}}{2a}. \quad (416)$$
9. Nature of Roots of a Quadratic Equation: The discriminant, $b^2 - 4ac$, gives the following information about the roots of a quadratic equation.
   1. $b^2 - 4ac > 0$, two real roots
   2. $b^2 - 4ac = 0$, one real root
   3. $b^2 - 4ac < 0$, no real roots (421)
10. The table on page 424 summarizes the methods for solving quadratic equations. (424)
11. To find a quadratic equation of the form $x^2 + bx + c = 0$ given its roots:
    1. The coefficient of x is the opposite of the sum of the roots.
    2. The constant term is the product of the roots. (429)
12. Sum and Product of Roots: Given the quadratic equation $ax^2 + bx + c = 0$, where $a \neq 0$, the sum of the roots of the equation is $-\frac{b}{a}$ and the product of the roots of the equation is $\frac{c}{a}$. (430)

Answers (continued)

7.

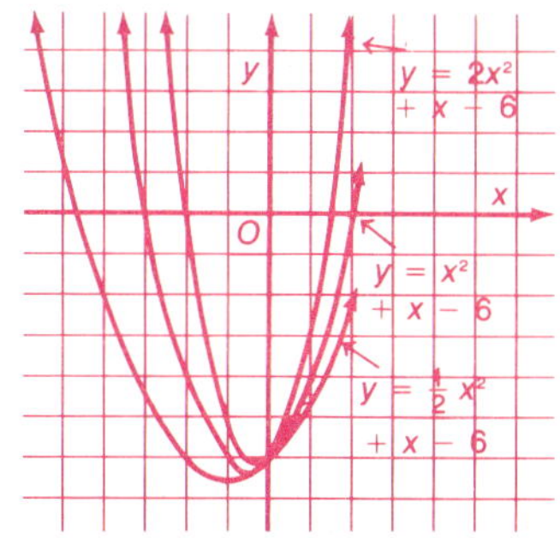

8. The coefficient of x^2 affects the narrowness or width of the parabola.

Quiz
Lessons 13-5 to 13-8

Use the discriminant to determine the number of real roots of each equation.

1. $x^2 + 6x + 10 = 0$
   Since $b^2 - 4ac = -4$, there are no real roots.
2. $4y^2 = -20y - 25$
   Since $b^2 - 4ac = 0$, there is one real root.
3. $2a^2 - 5a = -3$
   Since $b^2 - 4ac = 1$, there are two real roots.

Solve the following problem.

4. An ice skating rink is designed to be 30 m by 60 m. Budget cuts require that the area is reduced to 1000 m². A strip is to be removed from one end and a strip of the same width is to be removed from one side. Find the width of the strips.
   $(30 - w)(60 - w) = 1000$
   $w = 10$
   The width is 10 meters.

State the sum and product of the roots of each equation.

5. $3x^2 - 16x - 12 = 0$
   $\frac{16}{3}, -4$
6. $x^2 = -2x + 35$ $\quad -2, -35$

Write a quadratic equation having the given roots.

7. $\frac{2}{3}, -1$ $\quad 3x^2 + x - 2 = 0$

Answers to Chapter Review

4.

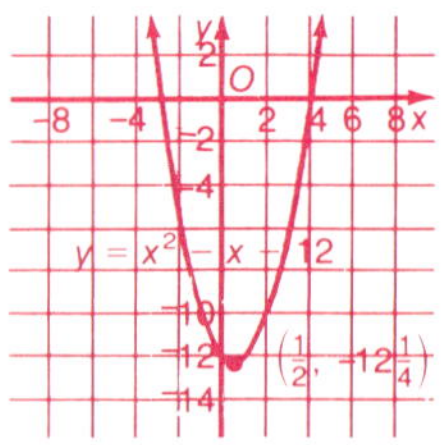

5.

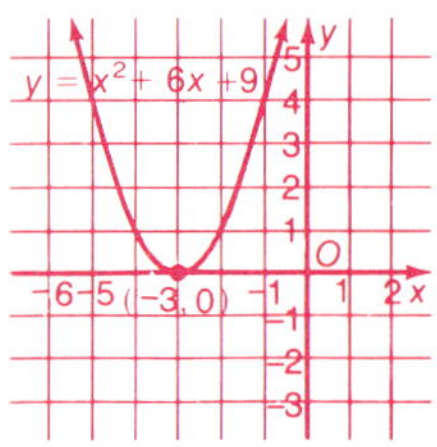

6.

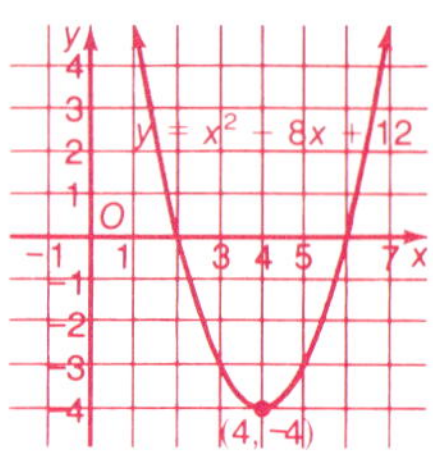

Chapter Review

13-1 **Find the equation of the axis of symmetry and the coordinates of the maximum or minimum point for the graph of each quadratic function.**

1. $y = x^2 - 3x - 4$ $x = \frac{3}{2}; \left(\frac{3}{2}, -\frac{25}{4}\right)$
2. $y = -x^2 + 6x + 16$ $x = 3; (3, 25)$
3. $y = 2x^2 + 9x + 9$ $x = -\frac{9}{4}; \left(-\frac{9}{4}, -\frac{9}{8}\right)$

13-2 **Locate the roots of each equation by graphing the related function.**

4. $x^2 - x - 12 = 0$ $-3, 4$
5. $x^2 + 6x + 9 = 0$ -3
6. $x^2 - 8x + 12 = 0$ $2, 6$

13-3 **Find the value of c that makes each trinomial a perfect square.**

7. $h^2 + 12h + c$ 36
8. $m^2 + 8m + c$ 16
9. $r^2 - 5r + c$ $\frac{25}{4}$

Solve by completing the square.

10. $x^2 - 16x + 32 = 0$ $8 \pm 4\sqrt{2}$
11. $y^2 + 6y + 4 = 0$ $-3 \pm \sqrt{5}$
12. $x^2 - 7x - 5 = 0$ $\frac{7 \pm \sqrt{69}}{2}$
13. $4a^2 + 16a + 15 = 0$ $-\frac{3}{2}, -\frac{5}{2}$

13-4 **Use the quadratic formula to solve each equation.**

14. $x^2 - 8x = 20$ $10, -2$
15. $2x^2 + 7x - 15 = 0$ $\frac{3}{2}, -5$
16. $2m^2 + 3 = 7m$ $3, \frac{1}{2}$
17. $9k^2 - 12k - 1 = 0$ $\frac{2 \pm \sqrt{5}}{3}$
18. $3s^2 - 7s - 2 = 0$ $\frac{7 \pm \sqrt{73}}{6}$
19. $5b^2 + 9b + 3 = 0$ $\frac{-9 \pm \sqrt{21}}{10}$

13-5 **Use the discriminant to determine the nature of the roots for each equation.**

20. $3m^2 - 8m - 40 = 0$ 2 real roots
21. $7x^2 - 6x + 5 = 0$ no real roots
22. $4p^2 + 4p = 15$ 2 real roots
23. $9k^2 - 13k + 4 = 0$ 2 real roots

13-6 **State the method that seems easiest for solving each equation. Then use it to solve the equation.**

24. $r^2 - 12r = -27$ factoring; $9, 3$
25. $m^2 + 10m - 7 = 0$ formula; $-5 \pm 4\sqrt{}$
26. $3a^2 - 11a + 10 = 0$ factoring; $2, \frac{5}{3}$
27. $9x^2 + 11 = 20x$ formula; $\frac{11}{9}, 1$
28. $2x^2 - \frac{17}{6}x + 1 = 0$ formula; $\frac{3}{4}; \frac{2}{3}$
29. $x^2 - \frac{27}{20}x + \frac{3}{5} = 0$ formula; no real roots
30. $x^2 - x + 0.21 = 0$ factoring; $0.3, 0.7$
31. $x^2 - 2.3x + 0.6 = 0$ factoring; $0.3, 2$

13-7 **Solve each problem.**

32. A rectangle has a perimeter of 38 inches. Its area is 84 square inches. Find its dimensions. 7 in., 12 in.

33. Find two integers whose sum is 21 and whose product is 90. 6, 15

13-8 **State the sum and product of the roots of each equation.**

34. $y^2 + 8y - 14 = 0$ $-8, -14$
35. $4a^2 - 6a + 11 = 0$ $\frac{3}{2}, \frac{11}{4}$

Write a quadratic equation having the given roots.

36. $1, -8$ $x^2 + 7x - 8 = 0$
37. $\frac{3}{2}, -4$ $2x^2 + 5x - 12 = 0$
38. $3 + \sqrt{5}, 3 - \sqrt{5}$ $x^2 - 6x + 4 = 0$

Chapter Test

Find the equation of the axis of symmetry and the coordinates of the maximum or minimum point for the graph of each quadratic function.

1. $y = 4x^2 - 8x - 17$ $x = 1; (1, -21)$

2. $y = -3x^2 + 12x + 34$ $x = 2; (2, 46)$

Locate the roots of each equation by graphing the related function.

3. $x^2 + x - 2 = 0$ $-2, 1$

4. $x^2 - 8x + 15 = 0$ $3, 5$

Find the value of *c* that makes each trinomial a perfect square.

5. $x^2 + 14x + c$ 49

6. $x^2 - 21x + c$ $\frac{441}{4}$

Solve by completing the square.

7. $k^2 - 8k - 4 = 0$ $4 \pm 2\sqrt{5}$

8. $m^2 - 6m + 6 = 0$ $3 \pm \sqrt{3}$

Use the quadratic formula to solve each equation.

9. $2x^2 - 5x - 12 = 0$ $4, -\frac{3}{2}$

10. $2m^2 - 9m + 8 = 0$ $\frac{9 \pm \sqrt{17}}{4}$

11. $3y^2 - 2y - 5 = 0$ $\frac{5}{3}, -1$

12. $2y^2 + 3y - 20 = 0$ $\frac{5}{2}, -4$

Use the discriminant to determine the nature of the roots for each equation.

13. $3y^2 - y - 10 = 0$ 2 real roots

14. $4y^2 + 12y + 9 = 0$ 1 real root

15. $y^2 + \sqrt{3}y - 5 = 0$ 2 real roots

16. $x^2 + 2x + 2 = 0$ no real roots

Solve the following equations.

17. $4x^2 - 5x + 1 = 0$ $\frac{1}{4}, 1$

18. $m^2 + 18m + 75 = 0$ $-9 \pm \sqrt{6}$

19. $3x^2 + 2x = 5$ $1, -\frac{5}{3}$

20. $2x^2 - 3x = 10$ $\frac{3 \pm \sqrt{89}}{4}$

21. $7x^2 - \frac{23}{3}x + 2 = 0$ $\frac{3}{7}, \frac{2}{3}$

22. $x^2 - 4.4x + 4.2 = 0$ $\frac{7}{5}, 3$

Solve each problem.

23. A rectangle has a perimeter of 44 centimeters. Its area is 105 square centimeters. Find its dimensions. 7 cm, 15 cm

24. Find two integers whose sum is 22 and whose product is 72. 4, 18

Find a quadratic equation having the given roots.

25. $-2, 5$ $x^2 - 3x - 10 = 0$

26. $6 + \sqrt{3}, 6 - \sqrt{3}$ $x^2 - 12x + 33 = 0$

State the sum and product of the roots of each equation.

27. $m^2 + 15m + 41 = 0$ $-15, 41$

28. $2x^2 - x - 6 = 0$ $\frac{1}{2}, -3$

Answers to Chapter Test

3.

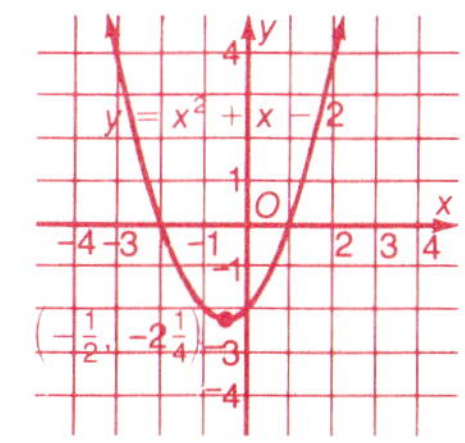

4.

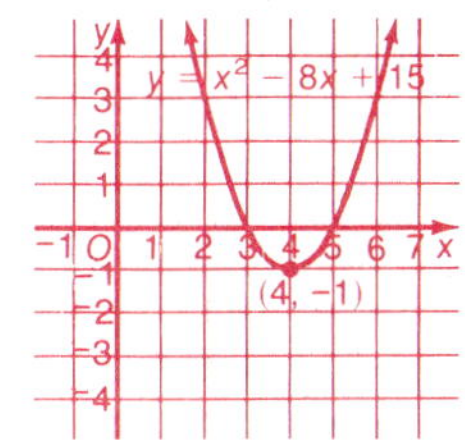

About the Photo

In arithmetic, fractions indicate division. Fractions name rational numbers. Algebraic expressions are also written as fractions to indicate division. A *rational expression* is the quotient of two polynomials. Since division by zero is undefined, all replacements for variables that yield a denominator of zero are excluded.

The analogy presented on this page motivates students into the study of rational expressions. Properties and operations for rational expressions are related to basic principles and previously learned concepts. Students extend their understanding of fractions to include rational algebraic expressions.

Teacher Resource Book Materials for Chapter 14 See pages 321-338.

Chapter Test, Form I (Multiple Choice)	321-323
Chapter Test, Form II (Free Response)	324-325
Quizzes	326
Extra Practice	327-328
Application: Using Rational Expressions and Equations (Use after 14-4.)	329-330
Extension: Synthetic Division (Use after 14-4.)	331
Extension: Continued Fractions (Use after 14-7.)	332
Problem Solving	333
Cumulative Review	334
Answers	335-338

CHAPTER 14

Rational Expressions

Many musical instruments are used in a marching band. Even though the instruments are different, the same music scale applies to each instrument. Similarly, the properties and operations you have applied to common fractions can also be applied to rational expressions.

Objective 14-1: Simplify algebraic fractions.

14-1 Simplifying Algebraic Fractions

Algebraic fractions contain variables. The expressions $\frac{5x + 3}{y}$, $\frac{2}{x}$, and $\frac{a - 2}{a^2 + 4}$ are examples of algebraic fractions.

A fraction indicates division. Zero cannot be used as a denominator because division by zero is undefined. Therefore, any value assigned to a variable that results in a denominator of zero must be excluded from the domain of the variable.

For $\frac{5}{x}$, exclude $x = 0$.

For $\frac{3x + 7}{x + 4}$, exclude $x = -4$.

For $\frac{y^2 - 5}{x^2 - 5x + 6}$, exclude $x = 2$ and $x = 3$. *Factor $x^2 - 5x + 6$ to see why.*

Example 1

For each fraction state the values of the variable that must be excluded.

a. $\frac{5x}{x + 7}$

Exclude the values for which $x + 7 = 0$.

$x + 7 = 0$

$x = -7$

Therefore, x cannot equal -7.

b. $\frac{2a - 3}{a^2 - a - 12}$

Exclude the values for which $a^2 - a - 12 = 0$.

$a^2 - a - 12 = 0$

$(a - 4)(a + 3) = 0$ *Factor $a^2 - a - 12 = 0$.*

$a = 4$ or $a = -3$ *Use the zero product property.*

Therefore, a cannot equal 4 or -3.

Recall that to simplify an algebraic fraction such as $\frac{14a^2bc}{42abc^2}$, first factor the numerator and denominator. Then eliminate common factors.

$$\frac{14a^2bc}{42abc^2} = \frac{2 \cdot 7 \cdot a \cdot a \cdot b \cdot c}{2 \cdot 3 \cdot 7 \cdot a \cdot b \cdot c \cdot c}$$ *$a \neq 0, b \neq 0, c \neq 0$*

$$= \frac{\overset{1}{\cancel{2}} \cdot \overset{1}{\cancel{7}} \cdot \overset{1}{\cancel{a}} \cdot a \cdot \overset{1}{\cancel{b}} \cdot \overset{1}{\cancel{c}}}{\underset{1}{\cancel{2}} \cdot 3 \cdot \underset{1}{\cancel{7}} \cdot \underset{1}{\cancel{a}} \cdot \underset{1}{\cancel{b}} \cdot \underset{1}{\cancel{c}} \cdot c} \text{ or } \frac{a}{3c}$$ *The GCF is 14abc.*

14-1

Lesson Notes, page T51

Discussion Question
Draw a flowchart for factoring polynomials. Use the Summary of Factoring on page 236 as a resource. Test your flowchart with examples.
Answers will vary.

Chalkboard Exercises
For each fraction state the values that must be excluded.

1a. $\frac{3r}{6 - r}$ 6

1b. $\frac{m + 5}{m^2 - 4m + 3}$ 3, 1

1c. $\frac{15k}{k^2 - 64}$ 8, −8

1d. $\frac{p - 3}{p^2 + 9}$ none

1e. $\frac{n + 7}{n^2 - 49}$ 7, −7

Chalkboard Exercises
Simplify each algebraic fraction. State the excluded values of the variables.

1a. $\frac{2x^2 - 50}{x - 5}$

$2x + 10, x \neq 5$

1b. $\frac{3m^2 - 48}{3(m + 4)}$

$m - 4, m \neq -4$

1c. $\frac{y^2 - 4}{y^2 + y - 2}$

$\frac{y - 2}{y - 1}, y \neq -2, 1$

2. $\frac{b - 2m}{12m^2 - 3b^2}$

$-\frac{1}{6m + 3b}, b \neq \pm 2m$

❶ Another way to state the excluded values is x cannot equal y or $-y$.

Suggested Daily Assignments

Enriched
Exp: 9–12
Written: 2–44 evens

Average
Exp: 2–12 evens
Written: 1–43 odds

Fundamental
Exp: 1–12
Written: 1–30

Answers to Exploratory

1. $\frac{1}{y - 4}; y \neq -4, 4$
2. $\frac{1}{a + 1}; a \neq -1, 7$
3. $r; r \neq -3$
4. $m; m \neq 1$
5. $\frac{x}{x^2 - 4}; x \neq 0, 2, -2$
6. $\frac{-3z}{z^2 - 5}; z \neq 0, \pm\sqrt{5}$
7. $\frac{a + 4}{a - 2}; a \neq 2, 4$
8. $\frac{c + 6}{c - 2}; c \neq 2$
9. $(t - 2)(t + 3); t \neq -2, 3$
10. $\frac{x - 6}{(x - 5)(x^2 - 6)}; x \neq -5, 5, \pm\sqrt{6}$
11. $-\frac{1}{w + 4}; w \neq \frac{2}{3}, -4$
12. $\frac{a(b - 7)}{(b + 1)(b + 3)}; b \neq -1, -2, -3$

The same procedure can be used to simplify algebraic fractions having polynomials in the numerator and denominator.

Examples

2 **Simplify $\frac{a^2 - 1}{a^2 + 7a + 6}$. State the excluded values of a.**

$$\frac{a^2 - 1}{a^2 + 7a + 6} = \frac{(a + 1)(a - 1)}{(a + 6)(a + 1)}$$ *Factor $a^2 - 1$. Factor $a^2 + 7a + 6$.*

$$= \frac{\overset{1}{\cancel{(a + 1)}}(a - 1)}{(a + 6)\underset{1}{\cancel{(a + 1)}}}$$ *The GCF is $(a + 1)$.*

$$= \frac{a - 1}{a + 6}$$

The excluded values of a are any values for which $a^2 + 7a + 6 = 0$.

$$a^2 + 7a + 6 = 0$$
$$(a + 6)(a + 1) = 0$$
$$a = -6 \quad \text{or} \quad a = -1$$

Therefore, a cannot equal -6 or -1.

3 **Simplify $\frac{2x - 2y}{y^2 - x^2}$. State the excluded values of x and y.**

$$\frac{2x - 2y}{y^2 - x^2} = \frac{2(x - y)}{(y - x)(y + x)}$$ *Use the distributive property.*

$$= \frac{2(-1)(y - x)}{(y - x)(y + x)}$$ *Notice that $x - y = -1(y - x)$.*

$$= \frac{-2\overset{1}{\cancel{(y - x)}}}{\underset{1}{\cancel{(y - x)}}(y + x)}$$ *Eliminate common factors.*

$$= \frac{-2}{y + x} \quad \text{or} \quad -\frac{2}{y + x}$$

The excluded values of x and y are any values for which $y^2 - x^2 = 0$.

$$y^2 - x^2 = 0$$
$$(y - x)(y + x) = 0$$
$$y = x \quad \text{or} \quad y = -x$$

Therefore, y cannot equal x or $-x$. ❶

Exploratory Exercises

Simplify each algebraic fraction. State the excluded values of the variables.

1. $\dfrac{y + 4}{(y - 4)(y + 4)}$
2. $\dfrac{a - 7}{(a + 1)(a - 7)}$
3. $\dfrac{r(r + 3)}{r + 3}$
4. $\dfrac{m(m - 1)}{m - 1}$
5. $\dfrac{2x^3}{2x^2(x^2 - 4)}$
6. $\dfrac{-3z^2}{z(z^2 - 5)}$
7. $\dfrac{(a - 4)(a + 4)}{(a - 2)(a - 4)}$
8. $\dfrac{(c + 6)(c - 2)}{(c - 2)(c - 2)}$
9. $\dfrac{(t + 2)(t - 2)(t + 3)(t - 3)}{(t + 2)(t - 3)}$
10. $\dfrac{(x + 5)(x - 6)}{(x + 5)(x - 5)(x^2 - 6)}$
11. $\dfrac{-1(3w - 2)}{(3w - 2)(w + 4)}$
12. $\dfrac{a(b + 2)(b - 7)}{(b + 1)(b + 2)(b + 3)}$

Written Exercises

Simplify each algebraic fraction. State the excluded values of the variables.

1. $\dfrac{y - 3}{y^2 - 9}$
2. $\dfrac{s + 6}{s^2 - 36}$
3. $\dfrac{a^2 - 25}{a^2 + 3a - 10}$
4. $\dfrac{x^2 - 49}{x^2 - 2x - 35}$
5. $\dfrac{r^3 - r^2}{r - 1}$
6. $\dfrac{z^2 - 3z}{z - 3}$
7. $\dfrac{4n^2 - 8}{4n - 4}$
8. $\dfrac{6y^3 - 12y^2}{12y^2 - 18}$
9. $\dfrac{3m^3}{6m^2 - 3m}$
10. $\dfrac{7a^3b^2}{21a^2b + 49ab^3}$
11. $\dfrac{x + 3}{x^2 + 6x + 9}$
12. $\dfrac{c - 6}{c^2 - 12c + 36}$
13. $\dfrac{g^2 + g - 2}{g^2 - 3g + 2}$
14. $\dfrac{r^2 - r - 20}{r^2 + 9r + 20}$
15. $\dfrac{m^2 - 36}{m^2 + 5m - 6}$
16. $\dfrac{a^2 - 9}{a^2 + 6a - 27}$
17. $\dfrac{2y - 4}{y^2 + 3y - 10}$
18. $\dfrac{4x + 8}{x^2 + 6x + 8}$
19. $\dfrac{k^2 - 1}{k^2 + 2k + 1}$
20. $\dfrac{b^2 - 9}{b^2 + 6b + 9}$
21. $\dfrac{9 - a^2}{a^2 - a - 6}$
22. $\dfrac{25 - x^2}{x^2 + x - 30}$
23. $\dfrac{-x^2 + 6x - 9}{x^2 - 6x + 9}$
24. $\dfrac{a^2 - 2a + 1}{-a^2 + 2a - 1}$
25. $\dfrac{4y^2 + 7y - 2}{8y^2 + 15y - 2}$
26. $\dfrac{4x^2 - 6x - 4}{2x^2 - 8x + 8}$
27. $\dfrac{3m^2 + 9m + 6}{4m^2 + 12m + 8}$
28. $\dfrac{6r^2 + 12r - 48}{5r^2 - 5r - 10}$
29. $\dfrac{2t^2 - t - 21}{28 - 15t + 2t^2}$
30. $\dfrac{3z^2 + 5z - 2}{4 - 13z + 3z^2}$
31. $\dfrac{b^2 + 2b - 8}{b^4 - 20b^2 + 64}$
32. $\dfrac{a^2 + 2a - 3}{a^4 - 10a^2 + 9}$
33. $\dfrac{2x^2 + 7x - 4}{4x^2 - 4x + 1}$
34. $\dfrac{6y^2 + 7y + 2}{6y^2 + 5y + 1}$
35. $\dfrac{x^4 - 1}{x^4 - 5x^2 + 4}$
36. $\dfrac{x^4 - 16}{x^4 - 8x^2 + 16}$
37. $\dfrac{6s^2 + 17s - 14}{3s^2 - 20s + 12}$
38. $\dfrac{8t^2 - 14t - 15}{12t^2 - 19t - 21}$
39. $\dfrac{c^2 - c - 20}{c^3 + 10c^2 + 24c}$
40. $\dfrac{n^2 - 8n + 12}{n^3 - 12n^2 + 36n}$
41. $\dfrac{a^4 - 5a^2 + 4}{a^2 - a - 2}$
42. $\dfrac{y^4 - 13y^2 + 36}{y^2 + 5y + 6}$
43. $\dfrac{12x^3 + 12x^2 - 9x}{12x^3 + 18x^2 - 12x}$
44. $\dfrac{16a^3 - 24a^2 - 160a}{8a^4 - 36a^3 + 16a^2}$

Answers to Written

1. $\dfrac{1}{y + 3}$; $y \neq -3, 3$
2. $\dfrac{1}{s - 6}$; $s \neq -6, 6$
3. $\dfrac{a - 5}{a - 2}$; $a \neq -5, 2$
4. $\dfrac{x + 7}{x + 5}$; $x \neq -5, 7$
5. r^2; $r \neq 1$
6. z; $z \neq 3$
7. $\dfrac{n^2 - 2}{n - 1}$; $n \neq 1$
8. $\dfrac{y^2(y - 2)}{2y^2 - 3}$; $y \neq \pm\dfrac{\sqrt{6}}{2}$
9. $\dfrac{m^2}{2m - 1}$; $m \neq 0, \dfrac{1}{2}$
10. $\dfrac{a^2b}{3a + 7b^2}$; $a \neq 0, -\dfrac{7b^2}{3}$
11. $\dfrac{1}{x + 3}$; $x \neq -3$
12. $\dfrac{1}{c - 6}$; $c \neq 6$
13. $\dfrac{g + 2}{g - 2}$; $g \neq 1, 2$
14. $\dfrac{r - 5}{r + 5}$; $r \neq -5, -4$
15. $\dfrac{m - 6}{m - 1}$; $m \neq -6, 1$
16. $\dfrac{a + 3}{a + 9}$; $a \neq -9, 3$
17. $\dfrac{2}{y + 5}$; $y \neq -5, 2$
18. $\dfrac{4}{x + 4}$; $x \neq -4, -2$
19. $\dfrac{k - 1}{k + 1}$; $k \neq -1$
20. $\dfrac{b - 3}{b + 3}$; $b \neq -3$
21. $-\dfrac{a + 3}{a + 2}$; $a \neq -2, 3$
22. $-\dfrac{x + 5}{x + 6}$; $x \neq -6, 5$
23. -1; $x \neq 3$
24. -1; $a \neq 1$
25. $\dfrac{4y - 1}{8y - 1}$; $y \neq -2, \dfrac{1}{8}$
26. $\dfrac{2x + 1}{x - 2}$; $x \neq 2$
27. $\dfrac{3}{4}$, $m \neq -2, -1$
28. $\dfrac{6(r + 4)}{5(r + 1)}$; $r \neq -1, 2$
29. $\dfrac{t + 3}{t - 4}$; $t \neq 4, \dfrac{7}{2}$
30. $\dfrac{z + 2}{z - 4}$; $z \neq \dfrac{1}{3}, 4$
31. $\dfrac{1}{(b + 2)(b - 4)}$; $b \neq \pm 4, \pm 2$
32. $\dfrac{1}{(a + 1)(a - 3)}$; $a \neq \pm 1, \pm 3$

Answers (continued)

33. $\frac{x+4}{2x-1}$; $x \neq \frac{1}{2}$
34. $\frac{3y+2}{3y+1}$; $y \neq -\frac{1}{2}, -\frac{1}{3}$
35. $\frac{x^2+1}{(x-2)(x+2)}$; $x \neq \pm 2, \pm 1$
36. $\frac{x^2+4}{x^2-4}$; $x \neq \pm 2$
37. $\frac{2s+7}{s-6}$; $s \neq \frac{2}{3}, 6$
38. $\frac{2t-5}{3t-7}$; $t \neq -\frac{3}{4}, \frac{7}{3}$
39. $\frac{c-5}{c(c+6)}$; $c \neq 0, -4, -6$
40. $\frac{n-2}{n(n-6)}$; $n \neq 0, 6$
41. $(a-1)(a+2)$; $a \neq -1, 2$
42. $(y-2)(y-3)$; $y \neq -3, -2$
43. $\frac{2x+3}{2(x+2)}$; $x \neq 0, \frac{1}{2}, -2$
44. $\frac{2(2a+5)}{a(2a-1)}$; $a \neq 0, \frac{1}{2}, 4$

About Problem Solving
Many students become overwhelmed with problem solving because they do not break the problems into manageable parts. Encourage students to analyze the given information carefully and think about how the facts are related. This should help them identify one or more subgoals to attempt. If they are still not sure how to proceed, they should use trial-and-error to find some examples or counterexamples.

Discuss the conclusion for Example 2. It is a "proof" that no other pairs of unit fractions have a sum of $\frac{1}{2}$.

Many other problems can be generated by adapting the ones given in the exercises. Suggest that each student write one or two problems. The extra problems can be used for class discussion, or to share with other members of the class.

Problem Solving

Identify Subgoals

Sometimes the solution to a problem involves several steps. An important strategy for solving such problems is to identify subgoals. This involves taking steps that will either produce part of the solution, or will make the problem easier to solve. Study the following examples.

Example 1: Find all whole numbers less than 100 whose digits have a sum of 10.

Choose a subgoal. Find all pairs of digits that have a sum of 10.

1, 9   2, 8   3, 7   4, 6   5, 5

Then use each pair of digits to write whole numbers. The solutions are 19, 91, 28, 82, 37, 73, 46, 64, and 55.

Example 2: How many pairs of unit fractions have a sum of $\frac{1}{2}$?

The unit fractions are $\left\{\frac{1}{2}, \frac{1}{3}, \frac{1}{4}, \frac{1}{5}, \ldots\right\}$.

Suppose one of the fractions is $\frac{1}{3}$.

Use subtraction to find the other fraction. $\frac{1}{2} - \frac{1}{3} = \frac{1}{6}$

Thus, $\frac{1}{3} + \frac{1}{6} = \frac{1}{2}$. This is one solution to the problem.

Suppose one of the fractions is $\frac{1}{4}$. $\frac{1}{2} - \frac{1}{4} = \frac{1}{4}$

Thus, $\frac{1}{4} + \frac{1}{4} = \frac{1}{2}$. This is a second solution to the problem.

Suppose one of the fractions is $\frac{1}{5}$. $\frac{1}{2} - \frac{1}{5} = \frac{3}{10}$

Notice $\frac{3}{10}$ is *not* a unit fraction. There is no unit fraction which can be added to $\frac{1}{5}$ to get $\frac{1}{2}$.

Are there any other pairs of unit fractions whose sum is $\frac{1}{2}$? Of the two unit fractions, one would have to be greater than $\frac{1}{4}$ and the other would have to be less than $\frac{1}{4}$. Why? But $\frac{1}{3}$ and $\frac{1}{2}$ are the *only* unit fractions greater than $\frac{1}{4}$. Thus, there are only two pairs of unit fractions that have a sum of $\frac{1}{2}$.

Exercises

Solve each problem.

1. How many pairs of unit fractions have a sum of $\frac{1}{6}$. 5 pairs
2. How many whole numbers less than 1000 have digits whose sum is 10? 63
3. Find all whole numbers between 10 and 1000 that stay the same when the digits are written in reverse order. For example, 686 has this property. 99 whole numbers
4. Suppose the scoring in football is simplified to 7 points for a touchdown and 3 points for a field goal. What scores are impossible to achieve? 1, 2, 4, 5, 8, 11

Objective 14-2: Multiply algebraic fractions.

14-2 Multiplying Fractions

To multiply fractions, you multiply the numerators and multiply the denominators.

$$\frac{3}{5} \cdot \frac{4}{7} = \frac{3 \cdot 4}{5 \cdot 7}$$
$$= \frac{12}{35}$$

This method can be generalized as follows.

For all rational numbers $\frac{a}{b}$ and $\frac{c}{d}$, where $b \neq 0$ and $d \neq 0$,

$$\frac{a}{b} \cdot \frac{c}{d} = \frac{ac}{bd}.$$

Multiplying Fractions

The same method can be used to multiply algebraic fractions.

Examples

1 Simplify $\frac{5}{a} \cdot \frac{b}{7}$. State any excluded values.

$\frac{5}{a} \cdot \frac{b}{7} = \frac{5 \cdot b}{a \cdot 7}$ *Multiply the numerators. Multiply the denominators.*

$= \frac{5b}{7a}$

Since $7a$ cannot equal 0, $a \neq 0$.

2 Simplify $\frac{2a^2d}{3bc} \cdot \frac{9b^2c}{16ad^2}$. State any excluded values.

$\frac{2a^2d}{3bc} \cdot \frac{9b^2c}{16ad^2} = \frac{18a^2b^2cd}{48abcd^2}$ *The GCF is 6abcd.*

$= \frac{3ab}{8d}$ *Change the fraction to simplest form.*

Since bc and ad^2 cannot equal 0, $a \neq 0$, $b \neq 0$, $c \neq 0$, and $d \neq 0$.

From this point on, it will be assumed that all replacements for variables in algebraic fractions that result in denominators equal to zero will be excluded.

You may have used the shortcut shown below for simplifying and finding products at the same time.

$$\frac{3}{4} \cdot \frac{16}{21} = \frac{\overset{1}{\cancel{3}}}{\underset{1}{\cancel{4}}} \cdot \frac{\overset{4}{\cancel{16}}}{\underset{7}{\cancel{21}}} = \frac{4}{7}$$

The same method can be used with algebraic fractions.

14-2

Lesson Notes, page T51

Readiness Activity

Find each product in simplest form.

$\frac{1}{2} \cdot \frac{4}{5}$ $\frac{2}{5}$ $\frac{2}{7} \cdot \frac{49}{50}$ $\frac{7}{25}$ $\frac{2}{3} \cdot \frac{15}{21}$ $\frac{10}{21}$

$\frac{3}{8} \cdot \frac{20}{12}$ $\frac{5}{8}$ $\frac{12}{15} \cdot \frac{25}{16}$ $\frac{5}{4}$ $\frac{10}{18} \cdot \frac{27}{30}$ $\frac{1}{2}$

Discussion Questions

1. Find a value for b so that $\frac{a+b}{b+c} = \frac{a}{c}$. $b = 0$

2. Find values for b so that $\frac{ab}{bc} = \frac{a}{c}$. b = any real number except 0

Chalkboard Exercises

Find each product in simplest form.

1a. $\frac{7}{m} \cdot \frac{r}{p}$

$\frac{7r}{mp}$; $m \neq 0$, $p \neq 0$

1b. $\frac{2b}{3a} \cdot \frac{5b}{7a}$

$\frac{10b^2}{21a^2}$; $a \neq 0$

2a. $\frac{6rst^2}{5s^2t} \cdot \frac{4r^2s}{3st^2}$

$\frac{8r^3}{5st}$; $s \neq 0$, $t \neq 0$

2b. $\frac{3m}{4x} \cdot \frac{x^3}{5am^2}$

$\frac{3x}{20am}$; $a \neq 0$, $m \neq 0$, $x \neq 0$

Chalkboard Examples

Find each product in simplest form.

3a. $\frac{x+6}{2x^2} \cdot \frac{8x^4}{x^2-36}$ $\frac{4x^2}{x-6}$

3b. $\frac{x^2-7x+12}{25x^5} \cdot \frac{5x^3}{x-3}$ $\frac{x-4}{5x^2}$

3c. $\frac{y^2+6y+5}{y^2} \cdot \frac{5y^3}{y+5}$ $5y+5$

4a. $\frac{x^2-9}{3x-9} \cdot \frac{6x-12}{x+3}$ $2x-4$

4b. $\frac{4a-4}{a^2+2a-8} \cdot \frac{a+4}{14a-14}$ $\frac{2}{7a-14}$

5a. $\frac{m^2+19m+84}{3-m} \cdot \frac{m^2-9}{m^2+15m+36}$ $-m-7$

5b. $\frac{x^2-6x+8}{-4+4x-x^2} \cdot \frac{x^2+x-6}{x^2-2x-15}$ $-\frac{x-4}{x-5}$

❶ A common error that students make in simplifying algebraic fractions is shown below.

$$\frac{\overset{1}{\cancel{x}} + \overset{2}{\cancel{4}}}{\underset{1\ 1}{\cancel{2}\cancel{x}}} = 2$$

To help avoid this error, remind students that only common *factors* can be simplified.

Examples

3 **Simplify:** $\frac{x+5}{3x} \cdot \frac{12x^2}{x^2+7x+10}$

$$\frac{x+5}{3x} \cdot \frac{12x^2}{x^2+7x+10} = \frac{x+5}{3 \cdot x} \cdot \frac{2 \cdot 2 \cdot 3 \cdot x \cdot x}{(x+5)(x+2)}$$ *$x \neq 0$, $x \neq -5$, or $x \neq -2$*

$$= \frac{\overset{1}{\cancel{x+5}}}{\underset{1}{\cancel{3}} \cdot \underset{1}{\cancel{x}}} \cdot \frac{2 \cdot 2 \cdot \overset{1}{\cancel{3}} \cdot \overset{1}{\cancel{x}} \cdot x}{\underset{1}{\cancel{(x+5)}}(x+2)}$$ *Eliminate common factors.* **❶**

$$= \frac{4x}{x+2}$$

4 **Simplify:** $\frac{4a+8}{a^2-25} \cdot \frac{a-5}{5a+10}$

$$\frac{4a+8}{a^2-25} \cdot \frac{a-5}{5a+10} = \frac{4(a+2)}{(a-5)(a+5)} \cdot \frac{a-5}{5(a+2)}$$ *$a \neq 5$, $a \neq -5$, or $a \neq -2$*

$$= \frac{4\overset{1}{\cancel{(a+2)}}}{\underset{1}{\cancel{(a-5)}}(a+5)} \cdot \frac{\overset{1}{\cancel{a-5}}}{5\underset{1}{\cancel{(a+2)}}}$$ *Eliminate common factors.*

$$= \frac{4}{5(a+5)}$$

$$= \frac{4}{5a+25}$$

5 **Simplify:** $\frac{x^2-x-6}{9-x^2} \cdot \frac{x^2+7x+12}{x^2+4x+4}$

$$\frac{x^2-x-6}{9-x^2} \cdot \frac{x^2+7x+12}{x^2+4x+4} = \frac{(x-3)(x+2)}{(3-x)(3+x)} \cdot \frac{(x+3)(x+4)}{(x+2)(x+2)}$$ *$x \neq -3$, $x \neq 3$, or $x \neq -2$*

$$= \frac{(x-3)(x+2)}{-1(x-3)(x+3)} \cdot \frac{(x+3)(x+4)}{(x+2)(x+2)}$$ *Notice that $3 - x = -1(x-3)$.*

$$= \frac{\overset{1}{\cancel{(x-3)}}\overset{1}{\cancel{(x+2)}}}{-1\underset{1}{\cancel{(x-3)}}\underset{1}{\cancel{(x+3)}}} \cdot \frac{\overset{1}{\cancel{(x+3)}}(x+4)}{\underset{1}{\cancel{(x+2)}}(x+2)}$$ *Eliminate common factors.*

$$= \frac{x+4}{-1(x+2)}$$

$$= -\frac{x+4}{x+2}$$

Exploratory Exercises

Find each product.

1. $\frac{1}{3} \cdot \frac{5}{8}$ $\frac{5}{24}$
2. $\frac{3}{4} \cdot \frac{5}{7}$ $\frac{15}{28}$
3. $-\frac{5}{6} \cdot \frac{7}{8}$ $-\frac{35}{48}$
4. $\frac{2}{3}\left(-\frac{5}{9}\right)$ $-\frac{10}{27}$
5. $-\frac{7}{8}\left(-\frac{5}{9}\right)$ $\frac{35}{72}$
6. $\frac{3}{a} \cdot \frac{b}{4}$ $\frac{3b}{4a}$
7. $\frac{a}{3} \cdot \frac{a}{5}$ $\frac{a^2}{15}$
8. $\frac{3a}{5} \cdot \frac{2x}{y}$ $\frac{6ax}{5y}$
9. $\frac{2}{3}\left(\frac{1}{2}\right)$ $\frac{1}{3}$
10. $\frac{4}{5} \cdot \frac{5}{8}$ $\frac{1}{2}$
11. $\left(-\frac{5}{9}\right)\left(\frac{3}{10}\right)$ $-\frac{1}{6}$
12. $\left(-\frac{4}{5}\right)\left(-\frac{3}{8}\right)$ $\frac{3}{10}$
13. $\left(-\frac{4}{9}\right)\left(\frac{3}{8}\right)$ $-\frac{1}{6}$
14. $\frac{5}{12} \cdot \frac{4}{9}$ $\frac{5}{27}$
15. $\left(\frac{4}{7}\right)\left(\frac{11}{16}\right)$ $\frac{11}{28}$

Written Exercises

Find each product in simplest form.

1. $\frac{2}{9} \cdot \frac{3}{5}$ $\frac{2}{15}$
2. $\frac{4}{9} \cdot \frac{1}{3}$ $\frac{4}{27}$
3. $\frac{16}{75} \cdot \frac{5}{8}$ $\frac{2}{15}$
4. $\frac{32}{7} \cdot \frac{35}{8}$ 20
5. $\left(\frac{1}{4}\right)^2$ $\frac{1}{16}$
6. $\left(\frac{2}{3}\right)^2$ $\frac{4}{9}$
7. $\left(\frac{4}{9}\right)^3$ $\frac{64}{729}$
8. $\left(\frac{8}{5}\right)^3$ $\frac{512}{125}$
9. $\frac{ab}{ac} \cdot \frac{c}{d}$ $\frac{b}{d}$
10. $\frac{a^2b}{b^2c} \cdot \frac{c}{d}$ $\frac{a^2}{bd}$
11. $\frac{6a^2n}{8n^2} \cdot \frac{12n}{9a}$ a
12. $\frac{10n^3}{6x^3} \cdot \frac{12n^2x^4}{25n^2x^2}$ $\frac{4n^3}{5x}$
13. $\frac{8}{m^2}\left(\frac{m^2}{2c}\right)^2$ $\frac{2m^2}{c^2}$
14. $\left(\frac{2a}{b}\right)^2 \frac{5c}{6a}$ $\frac{10ac}{3b^2}$
15. $\frac{6m^3n}{10a^2} \cdot \frac{4a^2m}{9n^3}$ $\frac{4m^4}{15n^2}$
16. $\frac{7xy^3}{11z^2} \cdot \frac{44z^3}{21x^2y}$ $\frac{4y^2z}{3x}$
17. $\frac{y-3}{7} \cdot \frac{14}{y-3}$ 2
18. $\frac{5n-5}{3} \cdot \frac{9}{n-1}$ 15
19. $\frac{3a-3b}{a} \cdot \frac{a^2}{a-b}$ $3a$
20. $\frac{-(2a+7c)}{6} \cdot \frac{36}{-7c-2a}$ 6
21. $\frac{2a+4b}{5} \cdot \frac{25}{6a+8b}$ $\frac{5a+10b}{3a+4b}$
22. $\frac{3x+30}{2x} \cdot \frac{4x}{4x+40}$ $\frac{3}{2}$
23. $\frac{3}{x-y} \cdot \frac{(x-y)^2}{6}$ $\frac{x-y}{2}$
24. $\frac{a^2-b^2}{4} \cdot \frac{16}{a+b}$ $4a-4b$
25. $\frac{m^2-4}{2} \cdot \frac{4}{m-2}$ $2m+4$
26. $\frac{9}{m-3} \cdot \frac{m^2-9}{12}$ $\frac{3m+9}{4}$
27. $\frac{r^2}{r-s} \cdot \frac{r^2-s^2}{s^2}$ $\frac{r^3+r^2s}{s^2}$
28. $\frac{a^2-b^2}{a-b} \cdot \frac{7}{a+b}$ 7
29. $\frac{x^2-16}{9} \cdot \frac{x+4}{x-4}$ $\frac{x^2+8x+16}{9}$
30. $\frac{y^2-4}{y^2-1} \cdot \frac{y+1}{y+2}$ $\frac{y-2}{y-1}$
31. $\frac{x^2-y^2}{x^2-1} \cdot \frac{x-1}{x-y}$ $\frac{x+y}{x+1}$
32. $\frac{r^2+s^2}{r^2-s^2} \cdot \frac{r-s}{r+s}$ $\frac{r^2+s^2}{r^2+2rs+s^2}$
33. $\frac{m^2+16}{m^2-16} \cdot \frac{m-4}{m+4}$ $\frac{m^2+16}{m^2+8m+16}$
34. $\frac{3k+9}{k} \cdot \frac{k^2}{k^2-9}$ $\frac{3k}{k-3}$
35. $\frac{3a-6}{a^2-9} \cdot \frac{a+3}{a^2-2a}$ $\frac{3}{a^2-3a}$
36. $\frac{y^2-x^2}{y} \cdot \frac{x}{x-y}$ $\frac{-xy-x^2}{y}$
37. $\frac{b+a}{b-a} \cdot \frac{a^2-b^2}{a}$
38. $\frac{3mn^2-3m}{n} \cdot \frac{3m}{n^2-1}$ $\frac{9m^2}{n}$
39. $\frac{x+3}{x+4} \cdot \frac{x}{x^2+7x+12}$ $\frac{x}{x^2+8x+16}$
40. $\frac{x}{x^2+8x+15} \cdot \frac{2x+10}{x^2}$

Suggested Daily Assignments

Enriched
- Written: 2–50 evens, 51–53
- Mini-Review: 1–5

Average
- Exp: 1–15
- Written: 1–40
- Mini-Review: 1–5

Fundamental
- Exp: 1–15
- Written: 1–30
  - p. 439, 31–40
- Mini-Review: 1–5

Answers to Written

37. $\frac{-a^2-2ab-b^2}{a}$

40. $\frac{2}{x^2+3x}$

41. $\dfrac{1}{x^2 + x - 12} \cdot \dfrac{x - 3}{x + 5}$ $\dfrac{1}{x^2 + 9x + 20}$

42. $\dfrac{x - 5}{x^2 - 7x + 10} \cdot \dfrac{x - 2}{3}$ $\dfrac{1}{3}$

43. $\dfrac{b^2 + 20b + 99}{b + 9} \cdot \dfrac{b + 7}{b^2 + 12b + 11}$ $\dfrac{b + 7}{b + 1}$

44. $\dfrac{z^2 - 15z + 50}{z^2 - 9z + 20} \cdot \dfrac{z^2 - 11z + 24}{z^2 - 18z + 80}$ $\dfrac{z - 3}{z - 4}$

45. $\dfrac{b^2 + 19b + 84}{b - 3} \cdot \dfrac{b^2 - 9}{b^2 + 15b + 36}$ $b + 7$

46. $\dfrac{z^2 + 16z + 39}{z^2 + 9z + 18} \cdot \dfrac{z + 5}{z^2 + 18z + 65}$ $\dfrac{1}{z + 6}$

47. $\dfrac{y^2 + 3y^3}{y^2 - 4} \cdot \dfrac{2y + y^2}{y + 4y^2 + 3y^3}$ $\dfrac{y^2}{y^2 - y - 2}$

48. $\dfrac{2m^2 - 9m + 9}{3m^2 + 19m - 14} \cdot \dfrac{m^2 + 14m + 49}{9 - 6m + m^2}$ $\dfrac{2m^2 + 11m - 21}{3m^2 - 11m + 6}$

49. $\dfrac{6y^2 - 5y - 6}{3y^2 - 20y - 7} \cdot \dfrac{y^2 - 49}{12y^3 + 23y^2 + 10y}$

49. $\dfrac{2y^2 + 11y - 21}{12y^3 + 19y^2 + 5y}$

50. $\dfrac{3t^3 - 14t^2 + 8t}{2t^2 - 3t - 20} \cdot \dfrac{16t^2 + 34t - 15}{24t^2 - 25t + 6}$ t

mini-review

Write *true* or *false*.

1. (2, −1) is a solution to the equation $3x + y = 5$. True
2. (4, 5) is a solution to the system of equations $x + y = 9$ and $x - y = 1$. False
3. $\dfrac{\sqrt{16}}{\sqrt{5}} = \dfrac{4}{5}$ False
4. The roots of $x^2 + 3x + 2 = 0$ are −1 and −2. True
5. The sum of the roots of $x^2 - 2x - 8 = 0$ is −2. False

Challenge

Find each product in simplest form.

51. $\dfrac{a}{a^2 + 6a + 9} \cdot \dfrac{a + 3}{a - 5} \cdot \dfrac{a^2 - 25}{5a + 25}$ $\dfrac{a}{5a + 15}$

52. $\dfrac{x^2y}{x^2 + 4xy + 4y^2} \cdot \dfrac{x^2 + 2xy}{xy} \cdot \dfrac{y}{x^4 - 9x^2}$

53. $\dfrac{a^2x - b^2x}{y} \cdot \dfrac{y^2 + y}{a - 2} \cdot \dfrac{4 - 2a}{axy - bxy}$

52. $\dfrac{y}{x^3 + 2x^2y - 18y - 9x}$ 53. $\dfrac{-2(a + b)(y + 1)}{y}$

Excursions in Algebra — Does 2 = 1?

Find the fallacy in the following "proof" of "2 = 1."

$a = b$

$a \cdot a = b \cdot a$ *Multiply both sides by a.*

$a^2 = ab$

$a^2 - b^2 = ab - b^2$ *Subtract b^2 from both sides.*

$(a - b)(a + b) = b(a - b)$ *Factor.*

$\dfrac{(a - b)(a + b)}{(a - b)} = \dfrac{b(a - b)}{(a - b)}$ *Divide both sides by (a − b).* $(a - b) = 0$

$a + b = b$

$b + b = b$ *Substitute b for a.*

$2b = b$

$\dfrac{2b}{b} = \dfrac{b}{b}$ *Divide both sides by b.*

$2 = 1$

About the Excursion
Ask students who have difficulty finding the fallacy what $a - b$ is equal to. If they continue to experience difficulty, ask them what happens when both sides of an equation are divided by zero.

Objective 14-3: Divide algebraic fractions.

14-3 Dividing Fractions

Consider the following products. What is the pattern of the factors in each example?

$$\frac{5}{1}\cdot\frac{1}{5} = 1 \qquad \frac{1}{4}\cdot\frac{4}{1} = 1 \qquad \frac{2}{3}\cdot\frac{3}{2} = 1 \qquad \frac{x}{y}\cdot\frac{y}{x} = 1$$

Remember that all replacements for variables that yield a denominator of zero are excluded.

You should recall that two numbers whose product is 1 are called **multiplicative inverses** or **reciprocals**.

To find the quotient of two fractions, you multiply by the reciprocal of the second fraction.

$$\frac{2}{3} \div \frac{3}{4} = \frac{2}{3}\cdot\frac{4}{3}$$

The reciprocal of $\frac{3}{4}$ is $\frac{4}{3}$.

$$= \frac{8}{9}$$

This method can be generalized as follows.

For all rational numbers $\frac{a}{b}$ and $\frac{c}{d}$, where $b \neq 0$, $c \neq 0$, and $d \neq 0$,

$$\frac{a}{b} \div \frac{c}{d} = \frac{a}{b}\cdot\frac{d}{c}.$$

Dividing Fractions

The same method is used to divide algebraic fractions.

Examples

1 Find the quotient in simplest form: $\frac{5}{x} \div \frac{y}{z}$

$$\frac{5}{x} \div \frac{y}{z} = \frac{5}{x}\cdot\frac{z}{y}$$

The reciprocal of $\frac{y}{z}$ is $\frac{z}{y}$.

$$= \frac{5z}{xy}$$

2 Find the quotient in simplest form: $\frac{2x}{x+1} \div (x-1)$

$$\frac{2x}{x+1} \div (x-1) = \frac{2x}{x+1}\cdot\frac{1}{x-1}$$

The reciprocal of $(x-1)$ is $\frac{1}{x-1}$.

$$= \frac{2x}{(x+1)(x-1)}$$

$$= \frac{2x}{x^2-1}$$

14-3

Lesson Notes, page T52
Enrichment, page T52

Readiness Activity

Find each quotient in simplest form.

$\frac{2}{3} \div \frac{8}{9}$ $\frac{3}{4}$ $\qquad \frac{3}{4} \div \frac{5}{6}$ $\frac{9}{10}$

$\frac{10}{21} \div \frac{25}{42}$ $\frac{4}{5}$ $\qquad \frac{5}{8} \div \frac{1}{60}$ $37\frac{1}{2}$

$\frac{1}{6} \div \frac{2}{15}$ $1\frac{1}{4}$ $\qquad \frac{15}{32} \div \frac{35}{16}$ $\frac{3}{14}$

Discussion Question

For what values of x is $\frac{x^2 - x - 6}{x^2 - 7x + 6}$ not the reciprocal of $\frac{x^2 - 7x + 6}{x^2 - x - 6}$? Why? when $x = 1, -2, 3$, and 6; because for these values of x, either $x^2 - 7x + 6 = 0$ or $x^2 - x - 6 = 0$.

Chalkboard Exercises

Find each quotient in simplest form.

1a. $\frac{5}{a} \div \frac{3}{a}$ $\quad \frac{5}{3}$ or $1\frac{2}{3}$

1b. $\frac{6}{a^2} \div \frac{6}{a}$ $\quad \frac{1}{a}$

1c. $\frac{y^3}{4} \div \frac{y}{16}$ $\quad 4y^2$

1d. $\frac{a^2}{b} \div \frac{a^3}{b^2}$ $\quad \frac{b}{a}$

2a. $\frac{3a}{a+2} \div (a-2)$ $\quad \frac{3a}{a^2-4}$

2b. $\frac{4x^2}{x-4} \div (x-5)$ $\quad \frac{4x^2}{x^2-9x+20}$

Chalkboard Exercises

Find each quotient in simplest form.

3a. $\frac{a - 2}{a^2 - 4} \div \frac{a + 2}{a^2 + a - 2}$ $\frac{a - 1}{a + 2}$

3b. $\frac{y + 5}{y^2 - 4y - 5} \div \frac{y^2 + 4y - 5}{y + 1}$ $\frac{1}{y^2 - 6y + 5}$

Suggested Daily Assignments

Enriched
 Written: 2–34 evens, 35–39

Average
 Exp: 2–24 evens
 Written: 1–20, 22–34 evens

Fundamental
 Exp: 1–24
 Written: 1–20
 p. 443, 31–40

Example

3 **Find the quotient in simplest form:** $\frac{x - y}{x^2 - y^2} \div \frac{x + y}{x^2 + 2xy + y^2}$

$$\frac{x - y}{x^2 - y^2} \div \frac{x + y}{x^2 + 2xy + y^2} = \frac{x - y}{x^2 - y^2} \cdot \frac{x^2 + 2xy + y^2}{x + y}$$

$$= \frac{x - y}{(x + y)(x - y)} \cdot \frac{(x + y)(x + y)}{x + y}$$

$$= \frac{\overset{1}{\cancel{x - y}}}{\underset{1}{\cancel{(x + y)}}\underset{1}{\cancel{(x - y)}}} \cdot \frac{\overset{1}{\cancel{(x + y)}}\overset{1}{\cancel{(x + y)}}}{\underset{1}{\cancel{x + y}}}$$

$$= 1$$

Exploratory Exercises

State the reciprocal of each of the following.

1. $\frac{3}{4}$ $\frac{4}{3}$

2. $\frac{-5}{8}$ $-\frac{8}{5}$

3. $\frac{8}{-3}$ $-\frac{3}{8}$

4. $-\frac{9}{10}$ $-\frac{10}{9}$

5. $\frac{m}{2}$ $\frac{2}{m}$

6. $\frac{x^2}{4}$ $\frac{4}{x^2}$

7. $\frac{5}{2p}$ $\frac{2p}{5}$

8. $\frac{-8}{3n}$ $-\frac{3n}{8}$

9. 6 $\frac{1}{6}$

10. x $\frac{1}{x}$

11. a^2 $\frac{1}{a^2}$

12. $\frac{1}{3}a$ $\frac{3}{a}$

13. $2bc$ $\frac{1}{2bc}$

14. $\frac{2}{5}m^2$ $\frac{5}{2m^2}$

15. $\frac{x + y}{x - y}$ $\frac{x - y}{x + y}$

16. $\frac{a^2 + b}{a - b}$ $\frac{a - b}{a^2 + b}$

17. $\frac{3}{5}a^2b^2$ $\frac{5}{3a^2b^2}$

18. $x - 2$ $\frac{1}{x - 2}$

State a multiplication expression for each of the following.

19. $\frac{3}{8} \div \frac{1}{4}$ $\frac{3}{8} \cdot \frac{4}{1}$

20. $\frac{1}{7} \div \frac{-1}{49}$ $\frac{1}{7} \cdot \left(-\frac{49}{1}\right)$

21. $\frac{1}{3} \div -6$ $\frac{1}{3} \cdot \left(-\frac{1}{6}\right)$

22. $\frac{x}{y} \div \frac{y^2}{x}$ $\frac{x}{y} \cdot \frac{x}{y^2}$

23. $\frac{2x}{4 - 2a} \div \frac{a^2}{b - 2}$ $\frac{2x}{4 - 2a} \cdot \frac{b - 2}{a^2}$

24. $\frac{b - a}{a - b} \div (a + b)$ $\frac{b - a}{a - b} \cdot \frac{1}{a + b}$

Written Exercises

Find each quotient in simplest form.

1. $\frac{5}{6} \div \frac{2}{3}$ $\frac{5}{4}$

2. $\frac{3}{8} \div \frac{-1}{2}$ $-\frac{3}{4}$

3. $-\frac{5}{6} \div \frac{1}{3}$ $-\frac{5}{2}$

4. $\frac{5}{8} \div 5$ $\frac{1}{8}$

5. $\frac{a^2}{b^2} \div \frac{b^2}{a^2}$ $\frac{a^4}{b^4}$

6. $\frac{a}{b} \div \frac{b^3}{a^4}$ $\frac{a^5}{b^4}$

7. $\frac{y^2}{x^2} \div \frac{a^2}{x^2}$ Answer: $\frac{y^2}{a^2}$

8. $\frac{a^2}{b} \div \frac{a^2}{b^2}$ Answer: b

9. $\frac{(-a)^2}{b} \div \frac{a}{b}$ Answer: a

10. $\frac{p^3}{2q} \div \frac{-(p^2)}{4q}$ Answer: $-2p$

11. $\frac{3m}{m + 1} \div (m - 2)$ Answer: $\frac{3m}{m^2 - m - 2}$

12. $\frac{n^2}{n - 3} \div (n + 4)$ Answer: $\frac{n^2}{n^2 + n - 12}$

13. $\frac{b^2 - 9}{4b} \div (b - 3)$ Answer: $\frac{b + 3}{4b}$

14. $\frac{y^2 + 8y + 16}{y^2} \div (y + 4)$ Answer: $\frac{y + 4}{y^2}$

15. $\frac{y^2}{x + 2} \div \frac{y}{x + 2}$ Answer: y

16. $\frac{2a^3}{a + 1} \div \frac{a^2}{a + 1}$ Answer: $2a$

17. $\frac{p^2}{y^2 - 4} \div \frac{p}{2 - y}$ Answer: $\frac{-p}{y + 2}$

18. $\frac{q}{y^2 - 4} \div \frac{q^2}{y + 2}$ Answer: $\frac{1}{qy - 2q}$

19. $\frac{x^2 - 16}{16 - x^2} \div \frac{7}{x}$ Answer: $-\frac{x}{7}$

20. $\frac{y}{5} \div \frac{y^2 - 25}{5 - y}$ Answer: $\frac{-y}{5y + 25}$

21. $\frac{m^2 + 2m + 1}{2} \div \frac{m + 1}{m - 1}$ Answer: $\frac{m^2 - 1}{2}$

22. $\frac{x^2 - 4x + 4}{3} \div \frac{x^2 - 4}{2x}$ Answer: $\frac{2x^2 - 4x}{3x + 6}$

23. $\frac{m^2 + 2mn + n^2}{3m} \div \frac{m^2 - n^2}{2}$ Answer: $\frac{2m + 2n}{3m^2 - 3mn}$

24. $\frac{a^2 + 2ab + b^2}{2x} \div \frac{a + b}{x^2}$ Answer: $\frac{ax + bx}{2}$

25. $\frac{y^2 - 16}{y^2 - 64} \div \frac{y + 4}{y - 8}$ Answer: $\frac{y - 4}{y + 8}$

26. $\frac{k^2 - 81}{k^2 - 36} \div \frac{k - 9}{k + 6}$ Answer: $\frac{k + 9}{k - 6}$

27. $\frac{t^2 + 8t + 16}{w^2 - 6w + 9} \div \frac{2t + 8}{3w - 9}$ Answer: $\frac{3t + 12}{2w - 6}$

28. $\frac{k + 2}{m^2 + 4m + 4} \div \frac{4k + 8}{m + 4}$ Answer: $\frac{m + 4}{4m^2 + 16m + 16}$

29. $\frac{x}{x + 2} \div \frac{x^2}{x^2 + 5x + 6}$ Answer: $\frac{x + 3}{x}$

30. $\frac{x^2 + x - 2}{x^2 + 5x + 6} \div \frac{x^2 + 2x - 3}{x^2 + 7x + 12}$ Answer: $\frac{x + 4}{x + 3}$

31. $\frac{x^2 + 2x - 15}{x^2 - x - 30} \div \frac{x^2 - 3x - 18}{x^2 - 2x - 24}$ Answer: $\frac{x^2 + x - 12}{x^2 - 3x - 18}$

32. $\frac{2m^2 + 7m - 15}{m + 5} \div \frac{9m^2 - 4}{3m + 2}$ Answer: $\frac{2m - 3}{3m - 2}$

33. $\frac{a^2 + 3a - 10}{a^2 + 8a + 15} \div \frac{a^2 - 6a + 8}{12 + a - a^2}$ Answer: -1

34. $\frac{2x^2 - x - 15}{x^2 - 2x - 3} \div \frac{2x^2 + 3x - 5}{1 - x^2}$ Answer: -1

Challenge

Perform the indicated operations.

35. $\frac{x^2 + 5x + 6}{x^2 - x - 12} \cdot \frac{x - 4}{x^2 + 11x + 18} \div \frac{x + 7}{x^2 + 14x + 45}$ Answer: $\frac{x + 5}{x + 7}$

36. $\frac{2x - 3}{2x^2 - 7x + 6} \cdot \frac{x^2 + 3x - 10}{5x + 1} \div \frac{3x^2 + 14x - 5}{3x^2 + 2x - 1}$ Answer: $\frac{x + 1}{5x + 1}$

37. $\frac{x^2 + 2x - 3}{6x^2 - 5x + 1} \cdot \frac{2x^2 + 9x + 4}{2x^2 + 7x + 3} \div \frac{x^2 - 5x + 4}{6x^2 - x - 1}$ Answer: $\frac{3x^2 + 13x + 4}{3x^2 - 13x + 4}$

38. $\frac{4x^2 - 6x - 4}{8x^2 - 2} \div \frac{8x^2 - 8x + 2}{4x^2 - 10x - 6} \cdot \frac{4x^2 - 4x + 1}{x^2 - 5x + 6}$ Answer: $\frac{2x + 1}{2x - 1}$

39. $\frac{x^2 - 1}{2x^2 + 14x + 12} \div \frac{2x^2 - 3x + 1}{8x^2 + 36x - 72} \cdot \frac{x^2 + 3x - 4}{x^2 + 5x - 6}$ Answer: $\frac{4x^2 + 10x - 24}{2x^2 + 11x - 6}$

For more exercises on finding products or quotients of algebraic fractions, see exercises 7-20 on page 559.

14-4

Lesson Notes, page T52
Enrichment, page T52

Discussion Question
(Use after Example 1.)
Is it true that the degree of the quotient is equal to the degree of the dividend minus the degree of the divisor? Yes

Chalkboard Exercises
Find each quotient.

1a. $(x^2 + 12x + 36) \div (x + 6)$ $x + 6$

1b. $(a^2 + 3a + 1) \div (a + 2)$ $a + 1 - \frac{1}{a + 2}$

1c. $(a^3 + a^2 + a + 2) \div (a + 1)$ $a^2 + 1 + \frac{1}{a + 1}$

1d. $(b^2 + 2b - 35) \div (b + 7)$ $b - 5$

1e. $(6y^2 + 7y - 20) \div (2y + 5)$ $3y - 4$

❶ You may prefer to have students keep like terms in vertical columns as shown below.

$$
\begin{array}{r}
x + 3 \\
x + 5 \overline{) x^2 + 8x + 15} \\
\underline{x^2 + 5x } \\
3x + 15 \\
\underline{3x + 15} \\
0
\end{array}
$$

448

Objective 14-4: Divide a polynomial by a binomial.

14-4 Dividing Polynomials

You can use a long division process similar to that used for numbers to divide a polynomial by a polynomial. For example, you can divide $x^2 + 8x + 15$ by $x + 5$ as shown below.

Step 1
To find the first term of the quotient, divide the first term of the dividend (x^2) by the first term of the divisor (x).

$$
\begin{array}{r}
x \\
x + 5 \overline{) x^2 + 8x + 15} \\
\underline{x^2 + 5x } \\
3x
\end{array}
$$

When you divide x^2 by x the result is x.
Multiply $x(x + 5)$.
Subtract.

Step 2
To find the next term of the quotient, divide the first term of the partial dividend ($3x$) by the first term of the divisor (x).

$$
\begin{array}{r}
x + 3 \\
x + 5 \overline{) x^2 + 8x + 15} \\
\underline{x^2 + 5x } \\
3x + 15 \\
\underline{3x + 15} \\
0
\end{array}
$$

❶ *When you divide $3x$ by x the result is 3.*
Multiply $3(x + 5)$.
Subtract.

Therefore, $x^2 + 8x + 15$ divided by $x + 5$ is $x + 3$. Since the remainder is 0, the divisor is a factor of the dividend. This means that $(x + 5)(x + 3) = x^2 + 8x + 15$.

If the divisor is **not** a factor of the dividend, there will be a nonzero remainder. When there is a nonzero remainder the quotient can be expressed as follows:

$$\text{quotient} = \text{partial quotient} + \frac{\text{remainder}}{\text{divisor}}$$

Example

1 **Find the quotient: $(2x^2 - 11x - 20) \div (2x + 3)$**

$$
\begin{array}{rl}
x - 7 & \\
2x + 3 \overline{) 2x^2 - 11x - 20} & \\
\underline{2x^2 + 3x } & \longleftarrow \textit{Multiply } x(2x + 3). \\
-14x - 20 & \longleftarrow \textit{Subtract, and bring down } -20. \\
\underline{-14x - 21} & \longleftarrow \textit{Multiply } -7(2x + 3). \\
1 & \longleftarrow \textit{Subtract. The remainder is 1.}
\end{array}
$$

Therefore, the quotient is $x - 7$ with remainder 1.

Thus, $(2x^2 - 11x - 20) \div (2x + 3) = x - 7 + \frac{1}{2x + 3}$. ⟵ *remainder* ⟵ *divisor*

Notice that in an expression such as $s^3 + 9$ there is no s^2 term and no s term. In such situations, the expression can be renamed using 0 as the coefficient of these terms.

$$s^3 + 9 = s^3 + 0s^2 + 0s + 9$$

Example

2 **Find the quotient:** $\frac{s^3 + 9}{s - 3}$

$$\begin{array}{r} s^2 + 3s + 9 \\ s - 3\overline{)s^3 + 0s^2 + 0s + 9} \\ \underline{s^3 - 3s^2} \\ 3s^2 + 0s \\ \underline{3s^2 - 9s} \\ 9s + 9 \\ \underline{9s - 27} \\ 36 \end{array}$$

Insert $0s^2$ and $0s$. Why is this permissible?

←remainder

Therefore, $\frac{s^3 + 9}{s - 3} = s^2 + 3s + 9 + \frac{36}{s - 3}$.

Exploratory Exercises

For each of the following, state the first term of the quotient.

1. $\frac{a^2 + 3a + 2}{a + 1}$ a
2. $\frac{b^2 + 8b - 20}{b - 2}$ b
3. $\frac{8m^3 + 27}{2m + 3}$ $4m^2$
4. $\frac{2x^2 + 3x - 2}{2x - 1}$ x
5. $\frac{x^3 + 2x^2 - 5x + 12}{x + 4}$ x^2
6. $\frac{2x^3 - 5x^2 + 22x + 51}{2x + 3}$ x^2

Written Exercises

Find each quotient.

1. $(x^2 + 7x + 12) \div (x + 3)$ $x + 4$
2. $(x^2 + 9x + 20) \div (x + 5)$ $x + 4$
3. $(a^2 - 2a - 35) \div (a - 7)$ $a + 5$
4. $(x^2 + 6x - 16) \div (x - 2)$ $x + 8$
5. $(c^2 + 12c + 36) \div (c + 9)$ $c + 3$ R9
6. $(y^2 - 2y - 30) \div (y + 7)$ $y - 9$ R33
7. $(2r^2 - 3r - 35) \div (2r + 7)$ $r - 5$
8. $(3t^2 - 14t - 24) \div (3t + 4)$ $t - 6$
9. $\frac{10x^2 + 29x + 21}{5x + 7}$ $2x + 3$
10. $\frac{12n^2 + 36n + 15}{6n + 3}$ $2n + 5$
11. $\frac{x^3 - 7x + 6}{x - 2}$ $x^2 + 2x - 3$
12. $\frac{4m^3 + 5m - 21}{2m - 3}$ $2m^2 + 3m + 7$
13. $\frac{4t^3 + 17t^2 - 1}{4t + 1}$ $t^2 + 4t - 1$
14. $\frac{2a^3 + 9a^2 + 5a - 12}{a + 3}$
15. $\frac{27c^2 - 24c + 8}{9c - 2}$
16. $\frac{48b^2 + 8b + 7}{12b - 1}$
17. $\frac{6n^3 + 5n^2 + 12}{2n + 3}$
18. $\frac{t^3 - 19t + 9}{t - 4}$
19. $\frac{3s^3 + 8s^2 + s - 7}{s + 2}$
20. $\frac{9d^3 + 5d - 8}{3d - 2}$
21. $\frac{20t^3 - 27t^2 + t + 6}{4t - 3}$
22. $\frac{6x^3 - 9x^2 - 4x + 6}{2x - 3}$
23. $\frac{56x^3 + 32x^2 - 63x - 36}{7x + 4}$

For more exercises on dividing polynomials, see exercises 37-54 on page 557.

Chalkboard Exercises

Find each quotient.

2a. $\frac{m^3 + 2m - 7}{m + 3}$

$m^2 - 3m + 11 - \frac{40}{m + 3}$

2b. $\frac{a^3 - 2a^2 - 12}{a - 5}$

$a^2 + 3a + 15 + \frac{63}{a - 5}$

Suggested Daily Assignments

Enriched
Written: 1–23 odds, 24-28

Average
Exp: 1–6
Written: 1–23

Fundamental
Exp: 1–6
Written: 1–17
p. 447, 22–30 evens

Answers to Written

14. $2a^2 + 3a - 4$
15. $3c - 2 + \frac{4}{9c - 2}$
16. $4b + 1 + \frac{8}{12b - 1}$
17. $3n^2 - 2n + 3 + \frac{3}{2n + 3}$
18. $t^2 + 4t - 3 - \frac{3}{t - 4}$
19. $3s^2 + 2s - 3 - \frac{1}{s + 2}$
20. $3d^2 + 2d + 3 - \frac{2}{3d - 2}$
21. $5t^2 - 3t - 2$
22. $3x^2 - 2$
23. $8x^2 - 9$

About Using Calculators
For better students, you may want to present the following theorems which are introduced in *Merrill Algebra Two.*
The Remainder Theorem: If a polynomial $f(x)$ is divided by $(x - a)$, the remainder is a constant, $f(a)$, and
$f(x) = q(x) \cdot (x - a) + f(a)$
where $q(x)$ is a polynomial with degree one less than the degree of $f(x)$.
A special case of the remainder theorem, called the factor theorem, occurs when $f(a)$ is 0.
The Factor Theorem:
The binomial $(x - a)$ is a factor of the polynomial $f(x)$ if and only if $f(a) = 0$.
Therefore, by the factor theorem, if the result is 0 when $x^3 + 3x^2 - 6x - 8$ is evaluated for $x = 2$, then $(x - 2)$ is a factor of $x^3 + 3x^2 - 6x - 8$.

Quiz
Lessons 14–1 to 14–4
Simplify each of the following.

1. $\dfrac{r^2 + 3r}{r + 3}$ $\quad r$
2. $\dfrac{y^2 - 9}{y^2 + 6y + 9}$ $\quad \dfrac{y - 3}{y + 3}$
3. $\dfrac{a^2 + 8a - 20}{a^2 - 4}$ $\quad \dfrac{a + 10}{a + 2}$

Multiply each of the following.

4. $\left(\dfrac{5a}{c}\right)^2 \cdot \dfrac{c}{10a^2}$ $\quad \dfrac{5}{2c}$
5. $\dfrac{3a^2b^3c}{4ab^2} \cdot \dfrac{6c}{1}$ $\quad \dfrac{9abc^2}{2}$
6. $\dfrac{2x^2 - 10x}{x - 5} \cdot \dfrac{1}{x^2 - 1}$ $\quad \dfrac{2x}{x^2 - 1}$

Divide each of the following.

7. $\dfrac{3d^3c}{a^4} \div \dfrac{-6dc}{a^5}$ $\quad \dfrac{-ad^2}{2}$
8. $\left(\dfrac{x^2}{x^2 - 25y^2}\right) \div \left(\dfrac{x}{x + 5y}\right)$ $\quad \dfrac{x}{x - 5y}$
9. Find the quotient.

$\dfrac{m^2 - 12m - 45}{m + 3}$

$m - 15$

Challenge

Find each quotient.

24. $\dfrac{2x^4 + 3x^3 - 6x^2 + 7x - 6}{x^2 + 2x - 3}$ $\quad 2x^2 - x + 2$

25. $\dfrac{6x^4 + 11x^3 - 15x^2 + 16x - 6}{2x^2 + 5x - 3}$ $\quad 3x^2 - 2x + 2$

26. Find the value of k if $(x + 2)$ is a factor of $x^3 + 7x^2 + 7x + k$. $\quad -6$

27. Find the value of k if $(2m - 5)$ is a factor of $2m^3 - 5m^2 + 8m + k$. $\quad -20$

28. When $x^3 - 7x^2 + 4x + k$ is divided by $(x - 2)$, the remainder is 15. Find k. $\quad 27$

Using Calculators — Finding Factors

Is $(x - 2)$ a factor of $x^3 + 3x^2 - 6x - 8$? One way to determine this is to divide $x^3 + 3x^2 - 6x - 8$ by $(x - 2)$ using the long division process. If the remainder is 0, then $(x - 2)$ is a factor of $x^3 + 3x^2 - 6x - 8$.

Another method can be used to determine whether $(x - 2)$ is a factor. If $(x - 2)$ is a factor, then when $x^3 + 3x^2 - 6x - 8 = 0$, $x - 2 = 0$ and $x = 2$. Substitute 2 for x and evaluate $x^3 + 3x^2 - 6x - 8$. If the result is 0, then $(x - 2)$ is a factor.

You can use your calculator to evaluate $x^3 + 3x^2 - 6x - 8$ when $x = 2$. First rename $x^3 + 3x^2 - 6x - 8$ as a **nested polynomial**.

$x^3 + 3x^2 - 6x - 8 = [x^2 + 3x - 6]x - 8$ — *Write the polynomial in descending order. Factor x out of the first three terms.*

$= [(x + 3)x - 6]x - 8$ — *Within the brackets, factor x out of the first two terms.*

The nested polynomial $[(x + 3)x - 6]x - 8$ is equivalent to $x^3 + 3x^2 - 6x - 8$. Notice that in the nested polynomial no exponent is greater than 1. Since you do *not* have to raise x to a power, the evaluation process is simplified.

Example: Evaluate $[(x + 3)x - 6]x - 8$ when $x = 2$.

ENTER:	2	STO	+	3	=	×	RCL	−	6	=	×	RCL	−	8	=
DISPLAY:	2	2	2	3	5	5	2	10	6	4	4	2	8	8	0

Since the result is 0, $(x - 2)$ is a factor of $x^3 + 3x^2 - 6x - 8$.

Example: Rename $x^4 + 2x^3 + 3x^2 + 4x - 5$ as a nested polynomial.

$x^4 + 2x^3 + 3x^2 + 4x - 5 = [x^3 + 2x^2 + 3x + 4]x - 5$ — *Factor x out of the first 4 terms.*

$= [(x^2 + 2x + 3)x + 4]x - 5$ — *Continue this process until no exponent is greater than 1.*

$= [[(x + 2)x + 3]x + 4]x - 5$

Exercises

Rename each of the following as a nested polynomial. Then use your calculaor to determine whether the given binomial is a factor.

1. $x^3 + 2x^2 - 22x + 21$; $(x - 3)$

2. $x^3 - 6x^2 + 14x - 12$; $(x - 2)$

3. $x^3 - 8x^2 + 13x - 16$; $(x - 8)$

4. $x^4 + 3x^3 + 2x^2 - x + 6$; $(x - 2)$

1. $[(x + 2)x - 22]x + 21$; $(x - 3)$ is a factor. **2.** $[(x - 6)x + 14]x - 12$; $(x - 2)$ is a factor.
3. $[(x - 8)x + 13]x - 16$; $(x - 8)$ is not a factor. **4.** $[[(x + 3)x + 2]x - 1]x + 6$; $(x - 2)$ is not a fa

Objective 14-5: Add and subtract algebraic fractions with like denominators.

14-5 Adding and Subtracting Fractions with Like Denominators

To add or subtract fractions with *like* denominators, you simply add or subtract the numerators. Then write this sum or difference over the common denominator.

$$\frac{3}{7} + \frac{2}{7} = \frac{5}{7} \qquad \frac{7}{9} - \frac{2}{9} = \frac{5}{9}$$

Sometimes the result needs to be changed to simplest form. ❶

$$\frac{3}{8} + \frac{1}{8} = \frac{4}{8} = \frac{1}{2}$$ *The GCF of 4 and 8 is 4.*

$$\frac{9}{16} - \frac{3}{16} = \frac{6}{16} = \frac{3}{8}$$ *The GCF of 6 and 16 is 2.*

These methods can be generalized as follows.

For all rational numbers $\frac{a}{c}$ and $\frac{b}{c}$, where $c \neq 0$,

$$\frac{a}{c} + \frac{b}{c} = \frac{a+b}{c} \text{ and } \frac{a}{c} - \frac{b}{c} = \frac{a-b}{c}.$$

Adding or Subtracting Fractions with Like Denominators

These methods can be used to add or subtract algebraic fractions.

Examples

1 Add and simplify: $\frac{3}{x+2} + \frac{1}{x+2}$.

$$\frac{3}{x+2} + \frac{1}{x+2} = \frac{3+1}{x+2} = \frac{4}{x+2}$$

Since the denominators are both $x + 2$, add the numerators.

2 Subtract and simplify: $\frac{3a+2}{a-7} - \frac{a-3}{a-7}$.

$$\frac{3a+2}{a-7} - \frac{a-3}{a-7} = \frac{(3a+2)-(a-3)}{a-7} = \frac{3a+2-a+3}{a-7} = \frac{2a+5}{a-7}$$

The parentheses indicate that the quantity $(a - 3)$ is subtracted from the quantity $(3a + 2)$. Remember to change the signs of the second polynomial.

14-5

Lesson Notes, page T52

Readiness Activity

Find each sum or difference in simplest form.

$\frac{1}{4} + \frac{2}{4}$ $\frac{3}{4}$ $\frac{1}{6} + \frac{4}{6}$ $\frac{5}{6}$ $\frac{1}{2} + \frac{1}{2}$ 1

$\frac{2}{3} - \frac{1}{3}$ $\frac{1}{3}$ $\frac{4}{5} - \frac{2}{5}$ $\frac{2}{5}$ $\frac{6}{8} - \frac{2}{8}$ $\frac{1}{2}$

❶ Remind students that a fraction is in simplest form when the greatest common factor of its numerator and denominator is 1.

Discussion Question

Find values for a, b, and c so that $\frac{a}{c} + \frac{b}{c} = 1$.

What is the relationship among a, b, and c?

Answers will vary. $a + b = c$, $c \neq 0$

Chalkboard Exercises

Find each sum or difference in simplest form.

1a. $\frac{4}{y+6} + \frac{y}{y+6}$ $\frac{4+y}{y+6}$

1b. $\frac{r}{r+s} + \frac{s}{r+s}$ 1

2a. $\frac{4}{b+5} - \frac{b}{b+5}$ $\frac{4-b}{b+5}$

2b. $\frac{8}{m+7} - \frac{m-1}{m+7}$ $\frac{9-m}{m+7}$

Chalkboard Exercises
Find each sum or difference in simplest form.

3a. $\frac{r}{r^2 - s^2} + \frac{s}{r^2 - s^2}$ $\frac{1}{r - s}$

3b. $\frac{y^2}{y - 5} - \frac{25}{y - 5}$ $y + 5$

3c. $\frac{r^2 - 8}{r + 5} + \frac{2r - 7}{r + 5}$ $r - 3$

3d. $\frac{k^2 + k}{k^2 - 9} + \frac{k - 3}{k^2 - 9}$ $\frac{k - 1}{k - 3}$

4a. $\frac{r - 1}{r - 3} + \frac{9}{3 - r}$ $\frac{r - 10}{r - 3}$

4b. $\frac{m}{m - p} + \frac{2m - 1}{p - m}$ $\frac{-m + 1}{m - p}$

Suggested Daily Assignments

Enriched
- Exp: 2–16 evens
- Written: 2–44 evens
- Mini-Review: 1–5

Average:
- Exp: 1–16
- Written: 1–30, 32–44 evens
- Mini-Review: 1–5

Fundamental
- Exp: 1–16
- Written: 1–24
  - p. 449, 18–20
- Mini-Review: 1–5

Example

3 **Subtract and simplify:** $\frac{8n + 3}{3n + 4} - \frac{2n - 5}{3n + 4}$

$$\frac{8n + 3}{3n + 4} - \frac{2n - 5}{3n + 4} = \frac{(8n + 3) - (2n - 5)}{3n + 4}$$

$$= \frac{8n + 3 - 2n + 5}{3n + 4} \quad \textit{Why?}$$

$$= \frac{6n + 8}{3n + 4}$$

$$= \frac{2\cancel{(3n + 4)}}{\cancel{(3n + 4)}} \quad \textit{Factor and simplify.}$$

$$= 2$$

Sometimes the denominators are additive inverses.

Example

4 **Subtract and simplify:** $\frac{x}{x - 2} - \frac{x + 1}{2 - x}$

$$\frac{x}{x - 2} - \frac{x + 1}{2 - x} = \frac{x}{x - 2} - \frac{x + 1}{-(x - 2)} \quad \textit{Notice that } 2 - x = -(x - 2).$$

$$= \frac{x}{x - 2} - \left(-\frac{x + 1}{x - 2}\right)$$

$$= \frac{x}{x - 2} + \frac{x + 1}{x - 2}$$

$$= \frac{2x + 1}{x - 2}$$

Exploratory Exercises

Find each sum or difference in simplest form.

1. $\frac{5}{8} + \frac{2}{8}$ $\frac{7}{8}$
2. $\frac{9}{16} + \frac{5}{16}$ $\frac{7}{8}$
3. $\frac{4}{a} + \frac{3}{a}$ $\frac{7}{a}$
4. $\frac{6}{b} + \frac{7}{b}$ $\frac{13}{b}$
5. $\frac{b}{x} + \frac{2}{x}$ $\frac{b + 2}{x}$
6. $\frac{5}{2z} + \frac{-7}{2z}$ $-\frac{1}{z}$
7. $\frac{-4k}{t} + \frac{6k}{t}$ $\frac{2k}{t}$
8. $\frac{2n}{m^2} + \frac{3n}{m^2}$ $\frac{5n}{m^2}$
9. $\frac{3}{11} - \frac{2}{11}$ $\frac{1}{11}$
10. $\frac{14}{16} - \frac{15}{16}$ $-\frac{1}{16}$
11. $\frac{a}{5} - \frac{b}{5}$ $\frac{a - b}{5}$
12. $\frac{7}{a} - \frac{4}{a}$ $\frac{3}{a}$
13. $\frac{4}{x} - \frac{6}{x}$ $-\frac{2}{x}$
14. $\frac{7}{a} - \frac{c}{a}$ $\frac{7 - c}{a}$
15. $\frac{8k}{5m} - \frac{3k}{5m}$ $\frac{k}{m}$
16. $\frac{r}{y^2} - \frac{s}{y^2}$ $\frac{r - s}{y^2}$

Written Exercises

Find each sum or difference in simplest form.

26. $\frac{x^2 - 2xy - y^2}{x - y}$

1. $\frac{y}{2} + \frac{y}{2}$ y
2. $\frac{a}{12} + \frac{2a}{12}$ $\frac{a}{4}$
3. $\frac{8}{x} + \frac{2}{x}$ $\frac{10}{x}$
4. $\frac{3}{a} + \frac{6}{a}$ $\frac{9}{a}$
5. $\frac{2y}{b} - \frac{y}{b}$ $\frac{y}{b}$
6. $\frac{3}{x} - \frac{7}{x}$ $-\frac{4}{x}$
7. $\frac{12a}{7} - \frac{3a}{7}$ $\frac{9a}{7}$
8. $\frac{y}{14} - \frac{3y}{14}$ $-\frac{y}{7}$
9. $\frac{y}{2} + \frac{y - 6}{2}$ $y - 3$
10. $\frac{m + 4}{5} + \frac{m - 1}{5}$ $\frac{2m + 3}{5}$
11. $\frac{a + 2}{6} - \frac{a + 3}{6}$ $-\frac{1}{6}$
12. $\frac{b + 3}{7} - \frac{b + 1}{7}$ $\frac{2}{7}$
13. $\frac{x}{x - 1} + \frac{1}{x - 1}$ $\frac{x + 1}{x - 1}$
14. $\frac{x}{x + 1} + \frac{1}{x + 1}$ 1
15. $\frac{8}{y - 2} - \frac{6}{y - 2}$ $\frac{2}{y - 2}$
16. $\frac{y}{b + 6} - \frac{2y}{b + 6}$ $\frac{-y}{b + 6}$
17. $\frac{2n}{2n - 5} + \frac{5}{5 - 2n}$ 1
18. $\frac{x + y}{y - 2} + \frac{x - y}{2 - y}$ $\frac{2y}{y - 2}$
19. $\frac{y}{a + 1} - \frac{y}{a + 1}$ 0
20. $\frac{y}{y - 1} - \frac{1}{y - 1}$ 1
21. $\frac{a + b}{x - 3} + \frac{a + b}{3 - x}$ 0
22. $\frac{a + b}{x - 3} - \frac{a + b}{3 - x}$ $\frac{2a + 2b}{x - 3}$
23. $\frac{r^2}{r - s} + \frac{s^2}{r - s}$ $\frac{r^2 + s^2}{r - s}$
24. $\frac{x^2}{x - y} - \frac{y^2}{x - y}$ $x + y$
25. $\frac{m^2}{m + n} + \frac{2mn + n^2}{m + n}$ $m + n$
26. $\frac{x^2}{x - y} - \frac{2xy + y^2}{x - y}$
27. $\frac{12n}{3n + 2} + \frac{8}{3n + 2}$ 4
28. $\frac{6x}{x + y} + \frac{6y}{x + y}$ 6
29. $\frac{12n}{3n - 2} - \frac{8}{3n - 2}$ 4
30. $\frac{a}{2a - 2b} - \frac{b}{2a - 2b}$ $\frac{1}{2}$
31. $\frac{a^2}{a - b} + \frac{-(b^2)}{a - b}$ $a + b$
32. $\frac{a^2}{a^2 - b^2} + \frac{-(b^2)}{a^2 - b^2}$ 1
33. $\frac{r^2}{r - 3} + \frac{9}{3 - r}$ $r + 3$
34. $\frac{m^2}{4 + m} - \frac{16}{m + 4}$ $m - 4$
35. $\frac{x^2}{x^2 - 1} + \frac{2x + 1}{x^2 - 1}$ $\frac{x + 1}{x - 1}$
36. $\frac{2x + 1}{(x + 1)^2} + \frac{x^2}{(x + 1)^2}$ 1
37. $\frac{x - 1}{(x + 1)^2} - \frac{x - 1}{(x + 1)^2}$ 0
38. $\frac{25}{k + 5} - \frac{k^2}{k + 5}$ $5 - k$
39. $\frac{x}{x^2 + 2x + 1} + \frac{1}{x^2 + 2x + 1}$ $\frac{1}{x + 1}$
40. $\frac{x}{x^2 - 6x + 9} - \frac{3}{x^2 - 6x + 9}$ $\frac{1}{x - 3}$
41. $\frac{3m}{m^2 + 2m + 1} + \frac{3}{m^2 + 2m + 1}$ $\frac{3}{m + 1}$
42. $\frac{8y}{4y^2 + 12y + 9} + \frac{12}{4y^2 + 12y + 9}$ $\frac{4}{2y + 3}$
43. $\frac{2}{t^2 - t - 2} - \frac{t}{t^2 - t - 2}$ $-\frac{1}{t + 1}$
44. $\frac{4}{3r^2 - r - 4} - \frac{3r}{3r^2 - r - 4}$ $-\frac{1}{r + 1}$
45. $\frac{b^2 + 2b}{b^2 + b - 6} + \frac{3b + 6}{b^2 + b - 6}$ $\frac{b + 2}{b - 2}$

mini-review

Simplify.

1. $\frac{2}{3 - \sqrt{3}}$
2. $\sqrt{5} - \sqrt{\frac{1}{5}}$ $\frac{4\sqrt{5}}{5}$

1. $\frac{3 + \sqrt{3}}{3}$

Solve.

3. $y^2 - 4y - 2 = 0$ $2 \pm \sqrt{6}$
4. $3m^2 - 5m = 6$ $\frac{5 \pm \sqrt{97}}{6}$
5. $a^2 + \frac{1}{2}a - \frac{3}{2} = 0$ $-\frac{3}{2}, 1$

14-6

Lesson Notes, page T52

Readiness Activity
Find each sum or difference in simplest form.

$\frac{1}{2} + \frac{1}{4}$ $\frac{3}{4}$ $\frac{2}{3} + \frac{1}{6}$ $\frac{5}{6}$

$\frac{3}{4} - \frac{1}{2}$ $\frac{1}{4}$ $\frac{7}{8} - \frac{1}{4}$ $\frac{5}{8}$

$\frac{1}{8} + \frac{1}{2}$ $\frac{5}{8}$ $\frac{1}{2} - \frac{1}{6}$ $\frac{1}{3}$

Discussion Question
Can the LCD of two rational expressions be equal to the denominator of one of the rational expressions?
Yes, if one denominator is a factor of the other.

Computer Notes
The computer feature on pages 462 and 463 discusses a program for adding two fractions with unlike denominators.

Objective 14-6: Add and subtract algebraic fractions with unlike denominators.

14-6 Adding and Subtracting Fractions with Unlike Denominators

When you add or subtract fractions with unlike denominators, you must first name the fractions so that the denominators are alike. Any common denominator could be used. However, the computation is easier if the **least common denominator** (LCD) is used. Recall that the least common denominator is the **least common multiple** (LCM) of the denominators. Consider the following example.

$\frac{1}{6} + \frac{3}{8}$ *To add these fractions, first find the LCM of 6 and 8.*

The set of multiples of 6 and 8 can be found by multiplying 6 and 8 by each positive integer.

The multiples of 6 are: $1 \cdot 6, 2 \cdot 6, 3 \cdot 6, 4 \cdot 6, 5 \cdot 6, 6 \cdot 6, \ldots$
$6, \quad 12, \quad 18, \quad 24, \quad 30, \quad 36, \ldots$

The multiples of 8 are: $1 \cdot 8, 2 \cdot 8, 3 \cdot 8, 4 \cdot 8, 5 \cdot 8, 6 \cdot 8, \ldots$
$8, \quad 16, \quad 24, \quad 32, \quad 40, \quad 48, \ldots$

The least number that is common to both sets of multiples is 24. Thus, 24 is the LCM of 6 and 8.

Prime factorization can also be used to find the LCM. First, find the prime factorization of each number.

$$6 = 2 \cdot 3$$
$$8 = 2 \cdot 2 \cdot 2$$

Then, to find the LCM, use each prime factor the greatest number of times it appears in either of the factorizations. Notice that the greatest number of times 2 appears as a factor is three times. The greatest number of times 3 appears is once. Therefore, the LCM of 6 and 8 is $2 \cdot 2 \cdot 2 \cdot 3$, or 24.

Thus, the LCD of $\frac{1}{6}$ and $\frac{3}{8}$ is 24. Rename $\frac{1}{6}$ and $\frac{3}{8}$ so that they have denominators of 24.

$$\frac{1}{6} + \frac{3}{8} = \frac{1}{6} \cdot \frac{4}{4} + \frac{3}{8} \cdot \frac{3}{3}$$
$$= \frac{4}{24} + \frac{9}{24}$$
$$= \frac{13}{24}$$

Notice that $\frac{4}{4} = 1$ and $\frac{3}{3} = 1$. Also, $6 \cdot 4 = 24$ and $8 \cdot 3 = 24$.

Similarly to add or subtract algebraic fractions with different denominators, find the LCM of the denominators.

Examples

1 **Find the LCM of $12x^2y$ and $15x^2y^2$.**

$12x^2y = 2 \cdot 2 \cdot 3 \cdot x \cdot x \cdot y$ *Factor each expression.*

$15x^2y^2 = 3 \cdot 5 \cdot x \cdot x \cdot y \cdot y$

$\text{LCM} = 2 \cdot 2 \cdot 3 \cdot 5 \cdot x \cdot x \cdot y \cdot y$ *Use each factor the greatest number of times it appears in either factorization.*

$= 60x^2y^2$

2 **Find the LCM of $x^2 + x - 2$ and $x^2 + 5x - 6$.**

$x^2 + x - 2 = (x - 1)(x + 2)$

$x^2 + 5x - 6 = (x + 6)(x - 1)$

$\text{LCM} = (x - 1)(x + 2)(x + 6)$

After finding the LCM, rename the fractions so that the denominators are alike. Then, add or subtract the numerators.

Examples

3 **Add and simplify: $\frac{6}{5x} + \frac{7}{10x^2}$**

$5x = 5 \cdot x$ *Use each factor the greatest number of times it appears.*

$10x^2 = 2 \cdot 5 \cdot x \cdot x$

The LCD for $\frac{6}{5x}$ and $\frac{7}{10x^2}$ is $2 \cdot 5 \cdot x \cdot x$ or $10x^2$. Since the denominator of $\frac{7}{10x^2}$ is already $10x^2$, only $\frac{6}{5x}$ needs to be renamed.

$$\frac{6}{5x} + \frac{7}{10x^2} = \frac{6}{5x} \cdot \frac{2x}{2x} + \frac{7}{10x^2}$$ *Why do you multiply $\frac{6}{5x}$ by $\frac{2x}{2x}$?*

$$= \frac{12x}{10x^2} + \frac{7}{10x^2}$$

$$= \frac{12x + 7}{10x^2}$$

4 **Subtract and simplify: $\frac{a}{a^2 - 4} - \frac{a}{a + 2}$**

Since $a^2 - 4 = (a - 2)(a + 2)$ and $a + 2 = (a + 2)$, the LCD for $\frac{a}{a^2 - 4}$ and $\frac{4}{a + 2}$ is $(a - 2)(a + 2)$ or $a^2 - 4$.

$$\frac{a}{a^2 - 4} - \frac{4}{a + 2} = \frac{a}{(a - 2)(a + 2)} - \frac{4}{(a + 2)} \cdot \frac{(a - 2)}{(a - 2)}$$

$$= \frac{a - 4(a - 2)}{(a - 2)(a + 2)}$$ *$a - 4(a - 2) = a - 4a + 8$*

$$= \frac{-3a + 8}{a^2 - 4}$$

Chalkboard Exercises

Find the LCM of each of the following.

1a. $6x^2y, 9xy^2$
$18x^2y^2$

1b. $3a^2b, 16ab^2, 21ab$
$336a^2b^2$

2a. $x^2 - 5x + 4, x^2 - 3x - 4$
$(x - 4)(x - 1)(x + 1)$

2b. $x^2 - 9, x^2 + 6x + 9$
$(x - 3)(x + 3)^2$

Find each sum or difference in simplest form.

3a. $\frac{2}{3y^2} - \frac{1}{12y^3}$
$\frac{8y - 1}{12y^3}$

3b. $\frac{4a}{mn} + \frac{2a}{np}$
$\frac{4ap + 2am}{mnp}$

4a. $\frac{r}{r^2 - 16} - \frac{5}{r + 4}$
$-\frac{4r + 20}{r^2 - 16}$

4b. $\frac{3}{y - 3} + \frac{2y}{y^2 - y - 6}$
$\frac{5y + 6}{y^2 - y - 6}$

Chalkboard Exercises

Find each sum or difference in simplest form.

5a. $\frac{y}{(1-y)(y+1)} + \frac{y+2}{(y+1)^2}$ $\frac{-2}{(y-1)(y+1)^2}$

5b. $\frac{a-1}{(a+3)^2} - \frac{a+1}{(3-a)(a+3)}$ $\frac{2a^2+6}{(a-3)(a+3)^2}$

Suggested Daily Assignments

Enriched
- Exp: 1–15 odds
- Written: 1–55 odds

Average
- Exp: 1–15
- Written: 2–54 evens

Fundamental
- Exp: 1–15
- Written: 2–40 evens
  p. 453, 26–38 evens

Example

5 **Add and simplify:** $\frac{x+4}{(2-x)(x+3)} + \frac{x-5}{(x-2)^2}$.

Multiply the first fraction by $\frac{-1}{-1}$ to change $(2-x)$ to $(x-2)$.

$$\frac{x+4}{(2-x)(x+3)} + \frac{x-5}{(x-2)^2} = \frac{-1}{-1}\cdot\frac{(x+4)}{(2-x)(x+3)} + \frac{x-5}{(x-2)^2}$$
$$= \frac{-(x+4)}{(x-2)(x+3)} + \frac{x-5}{(x-2)^2}$$

The LCD for $(x-2)(x+3)$ and $(x-2)^2$ is $(x+3)(x-2)(x-2)$. *Why?*

$$\frac{-(x+4)}{(x-2)(x+3)} + \frac{x-5}{(x-2)^2} = \frac{-(x+4)}{(x-2)(x+3)}\cdot\frac{(x-2)}{(x-2)} + \frac{(x-5)}{(x-2)^2}\cdot\frac{(x+3)}{(x+3)}$$
$$= \frac{-x^2-2x+8+x^2-2x-15}{(x+3)(x-2)(x-2)}$$
$$= \frac{-4x-7}{(x+3)(x-2)^2}$$

Exploratory Exercises

State the LCD for each pair of fractions.

1. $\frac{3}{8}, \frac{5}{12}$ 24 **2.** $\frac{6}{a}, \frac{7}{13}$ 13a **3.** $\frac{6}{11a}, \frac{2}{5}$ 55a **4.** $\frac{2}{a^2}, \frac{5}{7}$ $7a^2$

5. $\frac{2}{a}, \frac{3}{b}$ ab **6.** $\frac{4}{a^2}, \frac{5}{a}$ a^2 **7.** $\frac{6}{b^3}, \frac{7}{ab}$ ab^3 **8.** $\frac{4}{b^4}, \frac{5}{b^5}$ b^5

9. $\frac{3}{20a^2}, \frac{1}{24ab^3}$ $120a^2b^3$ **10.** $\frac{7}{60x^2y^2}, \frac{6}{35xz^3}$ $420x^2y^2z^3$ **11.** $\frac{1}{12an^2}, \frac{3}{40a^4}$ $120a^4n^2$ **12.** $\frac{11}{56x^3y}, \frac{10}{49ax^2}$ $392ax^3y$

13. $\frac{7}{a+5}, \frac{a}{a-3}$ $(a+5)(a-3)$ **14.** $\frac{m}{m+n}, \frac{6}{n}$ $n(m+n)$ **15.** $\frac{x+5}{3x-6}, \frac{x-3}{x-2}$ $3(x-2)$

Written Exercises

Find each sum or difference in simplest form.

1. $\frac{3}{4}+\frac{1}{8}$ $\frac{7}{8}$ **2.** $\frac{5}{7}+\frac{3}{14}$ $\frac{13}{14}$ **3.** $\frac{3}{4}+\frac{2}{9}$ $\frac{35}{36}$ **4.** $\frac{2}{3}+\frac{-7}{8}$ $-\frac{5}{24}$

5. $\frac{-1}{4}-\frac{1}{8}$ $-\frac{3}{8}$ **6.** $\frac{7}{10}-\frac{3}{20}$ $\frac{11}{20}$ **7.** $\frac{6}{7}-\frac{2}{11}$ $\frac{52}{77}$ **8.** $\frac{-3}{12}-\frac{-5}{18}$ $\frac{1}{36}$

9. $\frac{8}{11}+\frac{y}{4}$ $\frac{32+11y}{44}$ **10.** $\frac{t}{3}+\frac{2t}{7}$ $\frac{13t}{21}$ **11.** $\frac{x}{7}-\frac{4}{5}$ $\frac{5x-28}{35}$ **12.** $\frac{2n}{5}-\frac{3m}{4}$ $\frac{8n-15m}{20}$

13. $\frac{5}{2a}+\frac{-3}{6a}$ $\frac{2}{a}$ **14.** $\frac{6}{5}+\frac{a}{5b}$ $\frac{6b+a}{5b}$ **15.** $\frac{7}{3a}-\frac{3}{6a^2}$ $\frac{14a-3}{6a^2}$ **16.** $\frac{7}{a}-\frac{x+1}{2a}$ $\frac{13-x}{2a}$

17. $\frac{5b}{7x} + \frac{3a}{21x^2}$ $\frac{5bx + a}{7x^2}$

18. $\frac{5}{xy} + \frac{6}{yz}$ $\frac{5z + 6x}{xyz}$

19. $\frac{3z}{7w^2} - \frac{2z}{w}$ $\frac{3z - 14wz}{7w^2}$

20. $\frac{6}{x} - \frac{5}{x^2}$ $\frac{6x - 5}{x^2}$

21. $\frac{2}{t} + \frac{t + 3}{s}$

22. $\frac{7}{a} - \frac{x - 1}{b}$

23. $\frac{a}{a + b} + \frac{6}{b}$

24. $\frac{m}{m - n} - \frac{5}{m}$

25. $\frac{4a}{2a + 6} + \frac{3}{a + 3}$ $\frac{2a + 3}{a + 3}$

26. $\frac{5}{3x - 9} + \frac{3}{x - 3}$ $\frac{14}{3x - 9}$

27. $\frac{2x}{x^2 - 5x} - \frac{-3x}{x - 5}$ $\frac{3x + 2}{x - 5}$

28. $\frac{-3}{a - 5} + \frac{-6}{a^2 - 5a}$ $\frac{-3a - 6}{a^2 - 5a}$

29. $\frac{b + 8}{b^2 - 64} + \frac{1}{b - 8}$ $\frac{2}{b - 8}$

30. $\frac{2y}{y^2 - 25} + \frac{y + 5}{y - 5}$

31. $\frac{m - n}{m + n} - \frac{1}{m^2 - n^2}$

32. $\frac{3a + 2}{3a - 6} - \frac{a + 2}{a^2 - 4}$

33. $\frac{a^2}{a^2 - b^2} + \frac{a}{(a - b)^2}$

34. $\frac{x^2}{4x^2 - 9} + \frac{x}{(2x + 3)^2}$ $\frac{2x^3 + 5x^2 - 3x}{(2x - 3)(2x + 3)^2}$

35. $\frac{x}{x^2 + 2x + 1} + \frac{1}{x + 1}$ $\frac{2x + 1}{x^2 + 2x + 1}$

36. $\frac{y}{y^2 - 2y + 1} - \frac{1}{y - 1}$ $\frac{1}{y^2 - 2y + 1}$

37. $\frac{x^2 - 1}{x + 1} + \frac{x^2 + 1}{x - 1}$ $\frac{2(x^3 + 1)}{x^2 - 1}$ or $\frac{2(x^2 - x + 1)}{x - 1}$

38. $\frac{a}{a - b} + \frac{b}{2b + 3a}$ $\frac{3a^2 + 3ab - b^2}{(a - b)(2b + 3a)}$

39. $\frac{k}{2k + 1} - \frac{2}{k + 2}$ $\frac{k^2 - 2k - 2}{(2k + 1)(k + 2)}$

40. $\frac{m - 1}{m + 1} + \frac{4}{2m + 5}$ $\frac{2m^2 + 7m - 1}{(m + 1)(2m + 5)}$

41. $\frac{-18}{y^2 - 9} + \frac{7}{3 - y}$ $\frac{-7y - 39}{y^2 - 9}$

42. $\frac{-3}{5 - a} + \frac{5}{a^2 - 25}$ $\frac{3a + 20}{a^2 - 25}$

43. $\frac{a - 1}{4ab} + \frac{a^2 - a}{16b^2}$ $\frac{(4b + a^2)(a - 1)}{16ab^2}$

44. $\frac{x - 1}{3xy} - \frac{x^2 - x}{9y^2}$

45. $\frac{6}{a^2 - 2ab + b^2} - \frac{6}{a - b}$

46. $\frac{x^2 + 4x - 5}{x^2 - 2x - 3} + \frac{2}{x + 1}$ $\frac{x^2 + 6x - 11}{x^2 - 2x - 3}$

47. $\frac{3}{m - 2} + \frac{2}{2 - m}$ $\frac{1}{m - 2}$

48. $\frac{4}{5 - p} - \frac{3}{p - 5}$ $\frac{7}{5 - p}$

49. $\frac{m - n}{m^2 + 2mn + n^2} - \frac{m + n}{m - n}$

50. $\frac{a - 2}{a^2 + 4a + 4} + \frac{a + 2}{a - 2}$ $\frac{a^3 + 7a^2 + 8a + 12}{(a + 2)^2(a - 2)}$

51. $\frac{3m}{m^2 + 3m + 2} - \frac{3m - 6}{m^2 + 4m + 4}$

52. $\frac{4a}{6a^2 - a - 2} - \frac{5a + 1}{2 - 3a}$ $\frac{10a^2 + 11a + 1}{(3a - 2)(2a + 1)}$

53. $\frac{2x + 1}{(x - 1)^2} + \frac{x - 2}{(1 - x)(x + 4)}$ $\frac{x^2 + 12x + 2}{(x - 1)^2(x + 4)}$

54. $\frac{a + 3}{3a^2 - 10a - 8} + \frac{2a}{a^2 - 8a + 16}$ $\frac{7a^2 + 3a - 12}{(3a + 2)(a - 4)^2}$

55. $\frac{a + 2}{a^2 - 9} - \frac{2a}{6a^2 - 17a - 3}$

49. $\frac{-m^3 + m^2 - 3m^2n - 3mn^2 - 2nm + n^2 - n^3}{(m + n)^2(m - n)}$

Answers to Written

21. $\frac{2s + t^2 + 3t}{st}$

22. $\frac{7b - ax + a}{ab}$

23. $\frac{ab + 6a + 6b}{b(a + b)}$

24. $\frac{m^2 - 5m + 5n}{m(m - n)}$

30. $\frac{y^2 + 12y + 25}{y^2 - 25}$

31. $\frac{m^2 - 2mn + n^2 - 1}{m^2 - n^2}$

32. $\frac{3a - 1}{3a - 6}$

33. $\frac{a^3 - a^2b + a^2 + ab}{(a + b)(a - b)^2}$

44. $\frac{-x^3 + x^2 + 3xy - 3y}{9xy^2}$ or $\frac{(3y - x^2)(x - 1)}{9xy^2}$

45. $\frac{6 - 6a + 6b}{a^2 - 2ab + b^2}$ or $\frac{6(1 - a + b)}{(a - b)^2}$

51. $\frac{9m + 6}{(m + 2)^2(m + 1)}$

55. $\frac{4a^2 + 7a + 2}{(a + 3)(a - 3)(6a + 1)}$

Using Calculators — Evaluating Sums and Differences

Exercises

Use a calculator to evaluate each sum or difference for the values given.

1. $\frac{3}{x - 2} + \frac{4}{x^2 - 4}$; $x = 2.1$
   39.756098

2. $\frac{5y}{4 - w^2} - \frac{3y}{w}$; $y = 0.2$, $w = -0.1$
   6.2506266

3. $\frac{m^2 - 1}{m + 1} + \frac{m^2 + 1}{m - 1}$; $m = 0.5$
   −3

4. $\frac{r - s}{r^2 + 4rs + 4} - \frac{r + 2s}{r - s}$; $r = -0.2$, $s = -0.1$
   −4.0242718

For more exercises on adding and subtracting algebraic fractions, see exercises 21-34 on page 559.

About Using Calculators
Remind students to use parentheses when necessary. For example, for exercise 3, enter the following.

[(] 0.5 [x^2] [−] 1 [)] [÷] [(] 0.5 [+] 1 [)] [+] [(] 0.5 [x^2] [+] 1 [)] [÷] [(] 0.5 [−] 1 [)] [=]

About the Application
An increasing number of people are becoming interested in photography. This application helps students understand a little about how mathematics is used by photographers, both amateur and professional, to determine the amount of light that enters a camera. You may want to explain to students that to determine the correct f-stop and shutter speed for a photograph, the film speed (ASA) is also considered.

Some of your better students may be challenged by the following problem.

The *f*-stops on a camera relate to the diameter of the opening. The *f*-stops are 1.4, 2, 2.8, 4, 5.6, 8, 11, and 16. Show why each successive *f*-stop allows half as much light to enter the camera. (It has to do with area of the lens opening. If each *f*-stop number is squared, the results will form a linear relation.)

Applications in Photography

F-Stops

To take good pictures, photographers must make sure the correct amount of light enters the camera. The light enters the camera through an adjustable opening. The sizes of the opening are called *f*-stops. The *f*-stops are determined by dividing the focal length of the lens by the diameter of the opening. For example, suppose a camera lens has a focal length of 50 mm. The diameter of the opening is 6.25 mm.

$$f\text{-stop} = \frac{\text{focal length}}{\text{diameter of opening}}$$
$$= \frac{50 \text{ mm}}{6.25 \text{ mm}}$$
$$= 8 \qquad \text{The } f\text{-stop is 8.}$$

The dial that shows *f*-stops has the numbers 1.4, 2, 2.8, 4, 5.6, 8, 11, and 16. Each *f*-stop lets twice as much light into the camera as the next *f*-stop. For example, $f/4$ lets in *twice* as much light as $f/5.6$.

Another dial marked 1, 2, 4, 8, 15, 30, 60, 125, 250, 500, and 1000 shows the shutter speed. The shutter speed, in seconds, is the multiplicative inverse of the number on the dial. For example, a shutter speed marked 30 means light enters the camera for $\frac{1}{30}$ of a second. Each shutter speed lets light into the camera for twice as long as the next setting.

Suppose you want to photograph a parked sports car on a sunny day. The setting $f/11$ at $\frac{1}{60}$ would allow the correct amount of light into the camera. For a moving car, you need a faster shutter speed, such as $\frac{1}{125}$. At $\frac{1}{125}$, only half the amount of light is allowed into the camera. For proper exposure, the *f*-stop must be increased to $f/8$.

Exercises

A lens has a focal length of 200 mm. Find the *f*-stop for each opening whose diameter is given.

1. 50 mm 4
2. 25 mm 8
3. 12.5 mm 16
4. 36 mm $5\frac{5}{9}$ or about

For each exposure combination, state two others that give the same exposure.

5. $f/8$ at $\frac{1}{60}$ $f/5.6$ at $\frac{1}{125}$, $f/11$ at $\frac{1}{30}$.
6. $f/5.6$ at $\frac{1}{60}$ $f/8$ at $\frac{1}{30}$, $f/4$ at $\frac{1}{125}$
7. $f/2.8$ at $\frac{1}{125}$ $f/4$ at $\frac{1}{60}$, $f/2$ at $\frac{1}{250}$
8. $f/4$ at $\frac{1}{30}$ $f/2.8$ at $\frac{1}{60}$, $f/5.6$ at $\frac{1}{15}$

Objective 14-7: Simplify mixed expressions and complex fractions.

14-7 Mixed Expressions and Complex Fractions

Algebraic expressions such as $a + \frac{b}{c}$, and $5 + \frac{x - y}{x + 3}$ are called **mixed expressions.** Changing mixed expressions to algebraic fractions is similar to changing mixed numerals to improper fractions.

Mixed expressions contain monomials and algebraic fractions.

Mixed Numeral to Improper Fraction	Mixed Expression to Algebraic Fraction
$3\frac{2}{5} = \frac{3(5) + 2}{5}$	$a + \frac{a^2 + b}{a - b} = \frac{a(a - b) + (a^2 + b)}{a - b}$
$= \frac{15 + 2}{5}$	$= \frac{a^2 - ab + a^2 + b}{a - b}$
$= \frac{17}{5}$	$= \frac{2a^2 - ab + b}{a - b}$

Example

1 Add and simplify: $8 + \frac{x^2 - y^2}{x^2 + y^2}$.

Notice that $8 + \frac{x^2 - y^2}{x^2 + y^2}$ is a mixed expression. Therefore, use the method shown above to find the sum.

$$8 + \frac{x^2 - y^2}{x^2 + y^2} = \frac{8(x^2 + y^2) + (x^2 - y^2)}{x^2 + y^2}$$

$$= \frac{8x^2 + 8y^2 + x^2 - y^2}{x^2 + y^2}$$

$$= \frac{9x^2 + 7y^2}{x^2 + y^2}$$

If a fraction has one or more fractions in the numerator *or* denominator, it is called a **complex fraction**. Some complex fractions are shown below.

$$\frac{3\frac{1}{2}}{5\frac{2}{3}} \qquad \frac{8}{\frac{a}{b}} \qquad \frac{\frac{a + b}{a}}{\frac{a - b}{b}} \qquad \frac{\frac{1}{x} - \frac{1}{y}}{\frac{1}{x} + \frac{1}{y}}$$

14-7

Lesson Notes, page T52

Discussion Question
(Use after page 459.)
Is it possible for a complex fraction to be its own reciprocal? Yes, if it equals 1 or −1.

Chalkboard Exercises
Simplify.

1a. $4 + \frac{x - 3}{x + 7}$ $\quad \frac{5x + 25}{x + 7}$

1b. $3 + \frac{r - 4}{r^2 - 4}$ $\quad \frac{3r^2 + r - 16}{r^2 - 4}$

1c. $m + \frac{3}{m + 4}$ $\quad \frac{m^2 + 4m + 3}{m + 4}$

1 When simplifying complex fractions, the following shortcut may be applied.

$$\frac{\frac{a}{b}}{\frac{c}{d}} = \frac{ad}{bc} \quad \frac{\text{outside product}}{\text{inside product}}$$

Divide the outside product by the inside product.

2 Another way to simplify this expression is to multiply the numerator and denominator by their LCD.

$$\frac{\frac{1}{x}+\frac{1}{y}}{\frac{1}{x}-\frac{1}{y}} = \frac{xy\left(\frac{1}{x}+\frac{1}{y}\right)}{xy\left(\frac{1}{x}-\frac{1}{y}\right)}$$
$$= \frac{y+x}{y-x}$$

Chalkboard Exercises

Simplify.

2a. $\dfrac{\frac{1}{a}-\frac{1}{b}}{\frac{1}{a}+\frac{1}{b}}$ $\quad \dfrac{b-a}{b+a}$

2b. $\dfrac{\frac{2}{m^2}-\frac{1}{m}}{\frac{4}{m}+\frac{1}{m^2}}$ $\quad \dfrac{2-m}{4m+1}$

Consider the complex fraction $\dfrac{\frac{3}{5}}{\frac{7}{8}}$.

To simplify this fraction, rewrite it as $\frac{3}{5} \div \frac{7}{8}$ and proceed as follows.

$$\frac{3}{5} \div \frac{7}{8} = \frac{3}{5} \cdot \frac{8}{7} \text{ or } \frac{24}{35}$$

Recall that to find the quotient, you multiply by $\frac{8}{7}$, the reciprocal of $\frac{7}{8}$.

Similarly, to simplify $\dfrac{\frac{a}{b}}{\frac{c}{d}}$, rewrite it as $\frac{a}{b} \div \frac{c}{d}$ and proceed as follows.

$$\frac{a}{b} \div \frac{c}{d} = \frac{a}{b} \cdot \frac{d}{c} \text{ or } \frac{ad}{bc}$$ **1**

The reciprocal of $\frac{c}{d}$ is $\frac{d}{c}$.

Simplifying Complex Fractions

Any complex fraction $\dfrac{\frac{a}{b}}{\frac{c}{d}}$, where $b \neq 0$, $c \neq 0$, and $d \neq 0$, may be expressed as $\dfrac{ad}{bc}$.

Example

2 **Simplify:** $\dfrac{\frac{1}{x}+\frac{1}{y}}{\frac{1}{x}-\frac{1}{y}}$ **2**

$$\frac{\frac{1}{x}+\frac{1}{y}}{\frac{1}{x}-\frac{1}{y}} = \frac{\frac{y}{xy}+\frac{x}{xy}}{\frac{y}{xy}-\frac{x}{xy}}$$

The LCD of the numerator and the denominator is xy.

$$= \frac{\frac{y+x}{xy}}{\frac{y-x}{xy}}$$

Add to simplify the numerator.

Subtract to simplify the denominator.

$$= \frac{(y+x)}{xy} \cdot \frac{xy}{(y-x)}$$

The reciprocal of $\frac{(y-x)}{xy}$ is $\frac{xy}{(y-x)}$.

$$= \frac{(y+x)}{\cancel{xy}_{1}} \cdot \frac{\cancel{xy}^{1}}{(y-x)}$$

Eliminate common factors.

$$= \frac{y+x}{y-x}$$

Examples

3 Simplify: $\dfrac{x + 4 - \dfrac{1}{x + 4}}{x + 11 + \dfrac{48}{x - 3}}$

$$\frac{x + 4 - \frac{1}{x + 4}}{x + 11 + \frac{48}{x - 3}}$$

$$= \frac{\frac{(x + 4)(x + 4) - 1}{x + 4}}{\frac{(x + 11)(x - 3) + 48}{x - 3}}$$

$$= \frac{\frac{x^2 + 8x + 16 - 1}{x + 4}}{\frac{x^2 + 8x - 33 + 48}{x - 3}}$$

$$= \frac{\frac{x^2 + 8x + 15}{x + 4}}{\frac{x^2 + 8x + 15}{x - 3}}$$

$$= \frac{\overset{1}{\cancel{x^2 + 8x + 15}}}{x + 4} \cdot \frac{x - 3}{\underset{1}{\cancel{x^2 + 8x + 15}}}$$

$$= \frac{x - 3}{x + 4}$$

4 Simplify: $\dfrac{x - \dfrac{x + 4}{x + 1}}{x - 2}$

$$\frac{x - \frac{x + 4}{x + 1}}{x - 2}$$

$$= \frac{\frac{x(x + 1) - (x + 4)}{x + 1}}{x - 2}$$

$$= \frac{\frac{x^2 + x - x - 4}{x + 1}}{x - 2}$$

$$= \frac{\frac{x^2 - 4}{x + 1}}{\frac{x - 2}{1}}$$

$$= \frac{x^2 - 4}{x + 1} \cdot \frac{1}{x - 2}$$

$$= \frac{(x + 2)\overset{1}{\cancel{(x - 2)}}}{(x + 1)\underset{1}{\cancel{(x - 2)}}}$$

$$= \frac{x + 2}{x + 1}$$

Exploratory Exercises

Add and simplify.

1. $4 + \frac{2}{x}$ $\frac{4x + 2}{x}$
2. $8 + \frac{5}{3y}$ $\frac{24y + 5}{3y}$
3. $x + \frac{x}{y}$ $\frac{xy + x}{y}$
4. $z + \frac{2z}{w}$ $\frac{zw + 2z}{w}$
5. $2m + \frac{4 + m}{m}$ $\frac{2m^2 + m + 4}{m}$
6. $3a + \frac{a + 1}{2a}$ $\frac{6a^2 + a + 1}{2a}$
7. $b^2 + \frac{2}{b - 2}$ $\frac{b^3 - 2b^2 + 2}{b - 2}$
8. $3r^2 + \frac{4}{2r + 1}$ $\frac{6r^3 + 3r^2 + 4}{2r + 1}$

Written Exercises

Simplify.

1. $\dfrac{3\frac{1}{2}}{4\frac{3}{4}}$ $\frac{14}{19}$
2. $\dfrac{\frac{x^2}{y}}{\frac{y}{x^3}}$ $\frac{x^5}{y^2}$
3. $\dfrac{\frac{x + 4}{y - 2}}{\frac{x^2}{y^3}}$ $\frac{y^3(x + 4)}{x^2(y - 2)}$
4. $\dfrac{\frac{x^3}{y^2}}{\frac{x + y}{x - y}}$ $\frac{x^3(x - y)}{y^2(x + y)}$

Chalkboard Exercises

Simplify.

3. $\dfrac{x - 2 + \frac{3}{x + 2}}{x + 1 - \frac{10}{x + 4}}$ $\frac{(x + 1)(x + 4)}{(x + 2)(x + 6)}$
4. $\dfrac{r - \frac{r + 5}{r - 3}}{r + 1}$ $\frac{r - 5}{r - 3}$

Suggested Daily Assignments

Enriched
- Exp: 4–8 evens
- Written: 1–23

Average
- Exp: 1–8
- Written: 1–23

Fundamental
- Exp. 1–8
- Written: 1–16
- pp. 456–7, 1-23 odds

Answers to Written

20. $\dfrac{x+4}{x-2}$

23. $\dfrac{x+3}{x+1}$

5. $\dfrac{\frac{x+y}{a+b}}{\frac{x^2-y^2}{a^2-b^2}}$ $\dfrac{a-b}{x-y}$

6. $\dfrac{\frac{x-y}{x+y}}{\frac{x+y}{x-y}}$ $\dfrac{(x-y)^2}{(x+y)^2}$

7. $\dfrac{\frac{1}{x}+\frac{1}{y}}{\frac{1}{y}-\frac{1}{x}}$ $\dfrac{x+y}{x-y}$

8. $\dfrac{\frac{a+b}{x}}{\frac{a-b}{y}}$ $\dfrac{y(a+b)}{x(a-b)}$

9. $\dfrac{\frac{x^2+8x+15}{x^2+x-6}}{\frac{x^2+2x-15}{x^2-2x-3}}$ $\dfrac{x+1}{x-2}$

10. $\dfrac{\frac{a^2-6a+5}{a^2+13a+42}}{\frac{a^2-4a+3}{a^2+3a-18}}$ $\dfrac{a-5}{a+7}$

11. $\dfrac{\frac{y^2-1}{y^2+3y-4}}{y+1}$ $\dfrac{1}{y+4}$

12. $\dfrac{\frac{a^2-2a-3}{a^2-1}}{a-3}$ $\dfrac{1}{a-1}$

13. $\dfrac{\frac{a^2+2a}{a^2+9a+18}}{\frac{a^2-5a}{a^2+a-30}}$ $\dfrac{a+2}{a+3}$

14. $\dfrac{\frac{x^2+4x-21}{x^2-9x+18}}{\frac{x^2+3x-28}{x^2-10x+24}}$ 1

15. $\dfrac{x-\frac{15}{x-2}}{x-\frac{20}{x-1}}$ $\dfrac{(x+3)(x-1)}{(x-2)(x+4)}$

16. $\dfrac{m+\frac{35}{m+12}}{m-\frac{63}{m-2}}$ $\dfrac{(m+5)(m-2)}{(m+12)(m-9)}$

17. $7+\dfrac{x^2+y^2}{x^2-4y^2}$

17. $\dfrac{8x^2-27y^2}{x^2-4y^2}$

18. $5+\dfrac{a^2+11}{a^2-1}$ $\dfrac{6a^2+6}{a^2-1}$

19. $\dfrac{x+2+\frac{2}{x+5}}{x+6+\frac{6}{x+1}}$ $\dfrac{x+1}{x+5}$

20. $\dfrac{x+5+\frac{3}{x+1}}{x-1-\frac{3}{x+1}}$

21. $\dfrac{\frac{a^2-a-1}{a-1}}{a-\frac{1}{a-1}}$ 1

22. $\dfrac{x-1-\frac{15}{x+1}}{x-4}$ $\dfrac{x+4}{x+1}$

23. $\dfrac{x+5+\frac{4}{x+1}}{x+3}$

Using Computers

Adding Fractions

You do not have to write special programs to tell a computer how to add integers and decimals. However, this is not the case with addition of fractions. The following program tells the computer how to add two fractions. The program uses the definition $\frac{a}{b}+\frac{c}{d}=\frac{ad+bc}{bd}$.

```
10  PRINT "ENTER THE NUMERATOR AND THE
      DENOMINATOR OF THE FIRST FRACTION."
20  PRINT "ENTER THE NUMERATOR AND THE
      DENOMINATOR OF THE SECOND FRACTION."
30  INPUT A,B,C,D
40  LET N = A * D + B * C
50  LET X = B * D
60  PRINT "THE SUM IS ";N;"/";X
70  END
```

About Using Computers
Have students substitute many different numerical values for *a*, *b*, *c*, and *d* in the formula $\frac{a}{b}+\frac{c}{d}=\frac{ad+bc}{bd}$. Ask them why the sum $\frac{ad+bc}{bd}$ is not necessarily in simplest form. Explain that lines 60-170 can be used to obtain a sum that is in simplest form. You may want to have students who are knowledgable in BASIC programming explain exactly what is going on in lines 60-170.

However, the sum given by this program may not always be in simplest form. The following lines can be added to the program so that the sum is given in simplest form.

```
60  IF N < X THEN 90
70  LET Y = X
80  GOTO 100
90  LET Y = N
100 FOR I = Y TO 1 STEP - 1
110 IF INT (N / I) < > N / I THEN 130
120 IF INT (X / I) = X / I THEN 140
130 NEXT I
140 LET N = N / I
150 LET X = X / I
160 PRINT "THE SUM IS ";N;"/";X
170 END
```

Exercises

Use the program to find the following sums in simplest form.

1. $\frac{2}{3} + \frac{4}{9}$ $\frac{10}{9}$ **2.** $\frac{8}{11} + \frac{1}{22}$ $\frac{17}{22}$ **3.** $\frac{4}{6} + \frac{1}{4}$ $\frac{11}{12}$ **4.** $\frac{7}{16} + \frac{9}{10}$ $\frac{107}{80}$ **5.** $\frac{5}{12} + \frac{7}{18}$ $\frac{29}{36}$

6. Write a program similar to the one above to subtract two fractions.

changes: 40 LET N = A * D − B * C
160 PRINT "THE DIFFERENCE IS";N; "/"; X

Vocabulary

algebraic fractions (437)
multiplicative inverse (445)
reciprocal (445)
least common denominator (454)
least common multiple (454)
mixed expression (459)
complex fraction (459)

Chapter Summary

1. Algebraic fractions contain variables. (437)
2. Zero cannot be used as a denominator because division by zero is undefined. (437)
3. To simplify an algebraic fraction, first factor the numerator and denominator. Then eliminate common fractors. (437)
4. Multiplying Fractions: For all rational numbers $\frac{a}{b}$ and $\frac{c}{d}$, where $b \neq 0$ and $d \neq 0$, $\frac{a}{b} \cdot \frac{c}{d} = \frac{ac}{bd}$. (441)
5. Dividing Fractions: For all rational numbers $\frac{a}{b}$ and $\frac{c}{d}$, where $b \neq 0$, $c \neq 0$, and $d \neq 0$, $\frac{a}{b} \div \frac{c}{d} = \frac{a}{b} \cdot \frac{d}{c}$. (445)

Quiz
Lessons 14–5 to 14–7

Add or subtract.

1. $\frac{x}{3x(x-1)} - \frac{1}{3x^2 - 3x}$ $\frac{1}{3x}$

2. $\frac{3x}{x-y} + \frac{4x}{y-x}$

$\frac{-x}{x-y}$ or $\frac{x}{y-x}$

3. $\frac{k}{k^2 + 2k + 1} - \frac{k+2}{k+1}$

$\frac{-(k^2 + 2k + 2)}{(k+1)^2}$ or $\frac{-k^2 - 2k - 2}{k^2 + 2k + 1}$

Simplify each of the following.

4. $\dfrac{\frac{x^2 - y^2}{2}}{\frac{x-y}{4}}$

$2(x + y)$ or $2x + 2y$

5. $\dfrac{\frac{5a^2 - 20}{2a + 2}}{\frac{10a - 20}{4a}}$

$\frac{a(a+2)}{a+1}$ or $\frac{a^2 + 2a}{a+1}$

Answers

15. $m^2 - 9m - 3 - \frac{40}{m - 4}$

16. $\frac{1}{x - 1}$

17. $\frac{7 + a}{x^2}$

18. 2

19. $\frac{2 - x}{x - y}$

6. Adding or Subtracting Fractions with Like Denominators: For all rational numbers $\frac{a}{c}$ and $\frac{b}{c}$, where $c \neq 0$,
$$\frac{a}{c} + \frac{b}{c} = \frac{a + b}{c} \text{ and } \frac{a}{c} - \frac{b}{c} = \frac{a - b}{c}. \quad (451)$$
7. To add or subtract fractions that have unlike denominators, rename the fractions by using the least common denominator. (454)
8. Simplifying Complex Fractions: Any complex fraction $\frac{\frac{a}{b}}{\frac{c}{d}}$, where $b \neq 0, c \neq 0$, and $d \neq 0$, may be expressed as $\frac{ad}{bc}$. (460)

Chapter Review

14-1 Simplify each algebraic fraction. State the excluded values of the variables.

1. $\frac{3x^2y}{12xy^3z}$ $\frac{x}{4y^2z}$; $x \neq 0, y \neq 0, z \neq 0$
2. $\frac{x + y}{x^2 + 3xy + 2y^2}$ $\frac{1}{x + 2y}$; $x \neq -y, -2y$
3. $\frac{x^2 - 9}{x^2 - 6x + 9}$ $\frac{x + 3}{x - 3}$; $x \neq 3$
4. $\frac{x^2 + 10x + 21}{x^3 + x^2 - 42x}$ $\frac{x + 3}{x(x - 6)}$; $x \neq 0, 6, -7$

14-2 Find each product in simplest form.

5. $\frac{7}{9} \cdot \frac{a^2}{b}$ $\frac{7a^2}{9b}$
6. $\frac{5x^2y}{8ab} \cdot \frac{12a^2b}{25x}$ $\frac{3axy}{10}$
7. $\frac{2x + 5}{x^2 - 7x + 12} \cdot \frac{3x^2 - 10x - 8}{6x^2 + 19x + 10}$ $\frac{1}{x - 3}$
8. $\frac{x^2 + x - 12}{x + 2} \cdot \frac{x + 4}{x^2 - x - 6}$ $\frac{(x + 4)^2}{(x + 2)^2}$

14-3 Find each quotient in simplest form.

9. $\frac{7x^2y}{10a^3} \div \frac{14x^3}{25a^2b}$ $\frac{5by}{4ax}$
10. $\frac{2x^2 + x - 15}{x^2 + x - 6} \div \frac{x + 3}{2x^2 - 9x + 10}$ $\frac{(2x - 5)^2}{x + 3}$
11. $\frac{7a^2b}{x^2 + x - 30} \div \frac{3a}{x^2 + 15x + 54}$ $\frac{7ab(x + 9)}{3(x - 5)}$
12. $\frac{m^2 + 4m - 21}{m^2 + 8m + 15} \div \frac{m^2 - 9}{m^2 + 12m + 35}$ $\frac{(m + 7)^2}{(m + 3)^2}$

14-4 Find each quotient.

13. $x + 3\overline{)x^3 + 7x^2 + 10x - 6}$ $x^2 + 4x - 2$
14. $2a - 5\overline{)6a^3 - 19a^2 + 2a + 15}$ $3a^2 - 2a - 4 - \frac{5}{2a - 5}$
15. $m - 4\overline{)m^3 - 13m^2 + 33m - 28}$

14-5 Find each sum or difference in simplest form.

16. $\frac{x}{x^2 - 1} + \frac{1}{x^2 - 1}$
17. $\frac{7}{x^2} + \frac{a}{x^2}$
18. $\frac{2x}{x - 3} - \frac{6}{x - 3}$
19. $\frac{2}{x - y} + \frac{x}{y - x}$

14-6 Find each sum or difference in simplest form.

20. $\frac{5a}{3x} - \frac{2}{4x^2y}$ $\frac{10axy - 3}{6x^2y}$
21. $\frac{x}{x + 3} - \frac{5}{x - 2}$ $\frac{x^2 - 7x - 15}{(x + 3)(x - 2)}$
22. $\frac{2x + 3}{x^2 - 4} + \frac{6}{x + 2}$ $\frac{8x - 9}{x^2 - 4}$
23. $\frac{3x}{x^2 + 3x - 10} + \frac{7}{x^2 - 5x + 6}$ $\frac{3x^2 - 2x + 35}{(x + 5)(x - 2)(x - 3)}$

14-7 Simplify.

24. $\dfrac{\dfrac{x^2}{y^3}}{\dfrac{3x}{9y^2}}$ $\dfrac{3x}{y}$

25. $\dfrac{\dfrac{x-3}{x+5}}{\dfrac{x+5}{x}}$ $\dfrac{x^2-3x}{(x+5)^2}$

26. $\dfrac{\dfrac{a^2-13a+40}{a^2-4a-32}}{\dfrac{a-5}{a+7}}$ $\dfrac{a+7}{a+4}$

27. $\dfrac{x-\dfrac{35}{x+2}}{x+\dfrac{42}{x+13}}$ $\dfrac{(x-5)(x+13)}{(x+2)(x+6)}$

Chapter Test

Simplify.

1. $\dfrac{\frac{5}{9}}{\frac{2}{3}}$ $\dfrac{5}{6}$

2. $\dfrac{21x^2y}{28ax}$ $\dfrac{3xy}{4a}$

3. $\dfrac{x^2+7x-18}{x^2+12x+27}$ $\dfrac{x-2}{x+3}$

4. $\dfrac{x^2-x-56}{x^2+x-42}$ $\dfrac{x-8}{x-6}$

5. $\dfrac{7x^2-28}{5x^3-20x}$ $\dfrac{7}{5x}$

6. $\dfrac{2x^2-5x-3}{x^2+2x-15}$ $\dfrac{2x+1}{x+5}$

Perform the indicated operations and simplify.

7. $\dfrac{3x}{x+3}+\dfrac{5x}{x+3}$ $\dfrac{8x}{x+3}$

8. $\dfrac{2x}{x-7}-\dfrac{14}{x-7}$ 2

9. $\dfrac{2x}{x+7}+\dfrac{4}{x+4}$ $\dfrac{2x^2+12x+28}{(x+7)(x+4)}$

10. $\dfrac{2a+1}{2a-3}+\dfrac{a-3}{3a+2}$ $\dfrac{8a^2-2a+11}{(2a-3)(3a+2)}$

11. $\dfrac{x+5}{x+2}+6$ $\dfrac{7x+17}{x+2}$

12. $\dfrac{x-2}{x-8}+x+5$ $\dfrac{x^2-2x-42}{x-8}$

13. $\dfrac{3x+2}{4x+1}+\dfrac{7}{x}$ $\dfrac{3x^2+30x+7}{4x^2+x}$

14. $\dfrac{3x-8}{x+4}+\dfrac{9}{x+1}$ $\dfrac{3x^2+4x+28}{x^2+5x+4}$

15. $\dfrac{x^2+4x-32}{x+5}\cdot\dfrac{x-3}{x^2-7x+12}$ $\dfrac{x+8}{x+5}$

16. $\dfrac{3x^2+2x-8}{x^2-4}\div\dfrac{6x^2+13x-28}{2x^2-3x-35}$ $\dfrac{x-5}{x-2}$

17. $\dfrac{4x^2+11x+6}{x^2-x-6}\div\dfrac{x^2+8x+16}{x^2+x-12}$ $\dfrac{4x+3}{x+4}$

18. $\dfrac{3x^2+5x-28}{x^2-3x-28}\cdot\dfrac{x^2-8x+7}{3x-7}$ $x-1$

19. $\dfrac{x-\dfrac{24}{x+5}}{x-\dfrac{72}{x-1}}$ $\dfrac{(x-3)(x-1)}{(x-9)(x+5)}$

20. $\dfrac{\dfrac{x+5}{x-2}}{\dfrac{x^2}{x-3}}$ $\dfrac{(x-3)(x+5)}{x^2(x-2)}$

21. $\dfrac{\dfrac{x^2-x-6}{x^2+2x-15}}{\dfrac{x^2-2x-8}{x^2+x-20}}$ 1

22. $\dfrac{\dfrac{2}{3m}+\dfrac{3}{m^2}}{\dfrac{2}{5m}+\dfrac{5}{m}}$ $\dfrac{5(2m+9)}{81m}$

Using the Cumulative Review
If students have difficulty with a particular topic, have them review the appropriate lesson. You may want to provide copies of the solutions from the *Algebra One Solutions Manual*. This Cumulative Review covers Chapters 1-14.

Other Cumulative Reviews appear on pages 72, 142, 208, 268, 334, 402, and 526.

Answers

4. 

10. $\{(-3, -\frac{2}{5}), (-1, \frac{1}{5}), (0, \frac{1}{2}), (4, \frac{17}{10}), (6, \frac{23}{10})\}$

13. slope $= \frac{3}{4}$; y-intercept $= 4$; $y = \frac{3}{4}x + 4$

14. 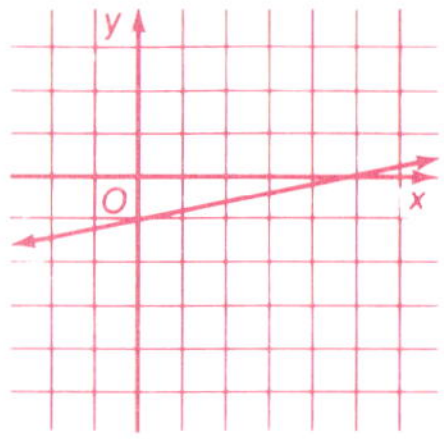

16. $(2, \frac{1}{2})$

17. 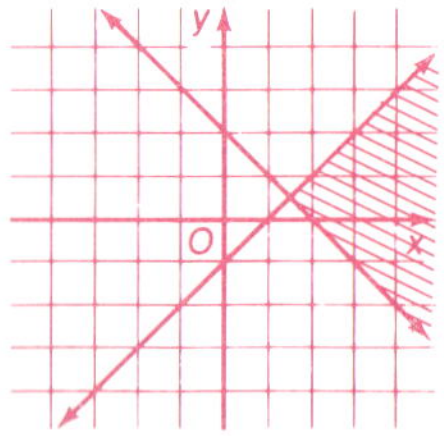

21. axis of symmetry: $x = -1$
minimum point: $(-1, 0)$

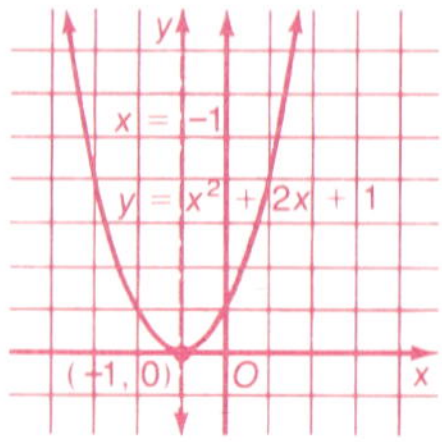

Cumulative Review

1. State the property shown by $(a + b)(b - a) = (b + a)(b - a)$. commutative property for addition
2. Find the difference: $-\frac{4}{5} - \left(-\frac{3}{8}\right)$ $-\frac{17}{40}$
3. Solve: $5(a + 21) = -2(-7a - 4)$ $\frac{97}{9}$
4. Graph the solution set of the inequality $y \geq 5$ on a number line.
5. Simplify: $(4a^2b^4)^3$ $64a^6b^{12}$
6. Solve: $24 - (3c + 1) = 4(c - 2) + 2c$ $\frac{31}{9}$

Factor, if possible.

7. $9x^2 - 16y^4$ $(3x - 4y^2)(3x + 4y^2)$
8. $12a^2 + 53an + 30n^2$ $(3a + 2n)(4a + 15n)$
9. Solve: $(x + 5)(x - 4) = 0$ $-5, 4$
10. Solve the equation $3x - 10y = -5$ if the domain is $\{-3, -1, 0, 4, 6\}$.
11. Write an equation for the relationship between the variables in the following chart. $b = a - 6$

a	1	2	3	4
b	−5	−4	−3	−2

12. Determine the slope of the line passing through $(8, -1)$ and $(0, 0)$. $-\frac{1}{8}$
13. Find the slope and y-intercept for the graph of $-3x + 4y = 16$. Then write the equation in slope-intercept form.
14. Graph $x - 5y = 5$ using the x- and y-intercepts.
15. Graph the system $-x - 2y = -8$ and $-3x - 6y = 6$. Then state whether the system has one solution, no solutions, or an infinite number of solutions. If the system has one solution, state it. no solution
16. Use elimination to solve the system of equations $2x = 5 - 2y$ and $2x - 2y = 3$.
17. Solve the system $y \leq x - 1$ and $y \geq -x + 2$ by graphing.
18. State the integers between which the value of $\sqrt{5894}$ lies. between 76 and 77
19. Approximate $\sqrt{85}$. Use the divide-and-average method twice. 9.220
20. Use the Pythagorean Theorem to find the measure of the hypotenuse of a right triangle if the measures of the legs are 7 and 8. $\sqrt{113}$
21. Find the equation of the axis of symmetry and the coordinates of the maximum or minimum point for the graph of $y = x^2 + 2x + 1$. Then use the information to draw the graph.
22. Use the quadratic formula to solve $n^2 - 2n = 56$. $1 \pm \sqrt{57}$
23. Find a quadratic equation having the roots 3 and -4. $x^2 + x - 12 = 0$

Simplify.

24. $\frac{x^2 - 9}{x^2 - 6x + 9}$ $\frac{x + 3}{x - 3}$
25. $\frac{x - y}{x + y} \cdot \frac{x^2 - y^2}{x + y}$ $\frac{(x - y)^2}{x + y}$
26. $\frac{2a}{a^2 - b - 2b^2} + \frac{2b}{a + b}$ $\frac{2a + 2ab - 4b^2}{(a + b)(a - 2b)}$

Problem Solving

Solve each problem.

27. Reggie is 10 years older than Lois. Five years ago, Reggie was three times as old as Lois. How old is Lois now? 10 years old
28. A certain triangle has two congruent sides. The third side is 7 in. shorter than either of the equal sides. If the perimeter is 31 in., what is the length of the third side? $5\frac{2}{3}$ inches
29. The sum of the squares of two consecutive even positive integers is 164. Find the integers. 8, 10
30. A rectangle is 2 cm longer than it is wide. If the length and width are both increased by 3 cm, the area becomes 288 cm^2. Find the original dimensions of the rectangle. 13 cm by 15 cm

Standardized Test Practice Questions

The test questions on this page deal with averages and rational expressions. The information at the right may help you with some of the questions.

Directions: Choose the best answer. Write A, B, C, or D.

1. $\frac{68 + 68 + 68 + 68}{4} =$

   (A) 17 (B) 68 (C) 136 (D) 272

2. Carrie's bowling scores for four games are $b + 2$, $b + 3$, $b - 2$, and $b - 1$. What must her score be on her fifth game to average $b + 2$?

   (A) $b + 8$ (B) b
   (C) $b - 2$ (D) $b + 5$

3. Which of the following *cannot* be the average of 10, 7, 13, 2, and x, if $x > 4$?

   (A) 7 (B) 9 (C) 13 (D) 258

4. The average of eight integers can be

   (A) 127.6 (B) 130.8
   (C) 131.3 (D) 135.5

5. The sum of six integers is what percent of the average of six integers?

   (A) 0.001% (B) 2% (C) 10% (D) 600%

6. What is the average of $70 - c$, $70 + 2c$, and $45 - c$?

   (A) 75 (B) $61\frac{2}{3} + \frac{2}{3}c$ (C) $61\frac{2}{3}$ (D) 75

7. What is the average of $2b - 4$, $b + 5$, and $3b + 8$?

   (A) $b + 3$ (B) $2b$
   (C) $6b + 9$ (D) $2b + 3$

8. If $\frac{x}{2}$, $\frac{x}{5}$, and $\frac{x}{7}$ are whole numbers, x could be

   (A) 20 (B) 35 (C) 50 (D) 70

> 1. To find the average of a group of numbers, first add the numbers. Then divide by the quantity of numbers added.
> 2. The average of a group of numbers cannot be greater than the greatest number in the group, nor less than the least number.
> 3. Most questions involving rational expressions can be solved by writing the expression in a different form.

9. If $abc = 4$ and $b = c$, then $a =$

   (A) c^2 (B) $\frac{1}{4c}$ (C) $\frac{4}{c^2}$ (D) $\frac{1}{c^2}$

10. If $\frac{1}{b - d} = 4$, then $d =$

    (A) $b - \frac{1}{4}$ (B) $4b - 1$
    (C) $\frac{b + 1}{4}$ (D) $b + 4$

11. If $\frac{2a}{5b} = 6$, then $\frac{2a - 5b}{5b} =$

    (A) 6 (B) $\frac{2}{5}$ (C) 15 (D) 5

12. If $3\frac{1}{5}c = 2\frac{1}{2}b$ and $c \neq 0$, then $\frac{b}{c} =$

    (A) $\frac{25}{32}$ (B) $\frac{32}{25}$ (C) $\frac{7}{8}$ (D) $\frac{11}{10}$

13. If the product of a number and b is increased by y, the result is t. Find the number in terms of b, y, and t.

    (A) $\frac{t - y}{b}$ (B) $\frac{y - t}{b}$
    (C) $\frac{b + y}{t}$ (D) $t - by$

14. Mark can type 60 words per minute and there is an average of 360 words per page. At this rate, how many *hours* would it take him to type k pages?

    (A) $\frac{k}{6}$ (B) $\frac{k}{60}$ (C) $\frac{k}{10}$ (D) $\frac{10}{k}$

Using the Standardized Test Practice Questions

The questions on this page may be used to help students prepare for pre-college tests. These questions usually require careful analysis and a thorough understanding of the concepts.

This page could be used as an overnight assignment for better students. Other students may benefit most if only a few questions are assigned at a time.

After students have chosen an answer for each problem, you should discuss how each problem can be solved, or provide copies of the solutions from the *Algebra One Solutions Manual*.

Other practice questions appear in this text on pages 73, 143, 209, 269, 335, 403, and 527. Also see pages 50-58 in the *Teacher Resource Book*.

About the Photo

Variables are inversely proportional when the product of two variables equals a constant. As one variable increases or decreases in value, the other variable must decrease or increase proportionately. Two forms of the equation describing inverse variation are $xy = k$ and $y = \frac{k}{x}$.

In the chapter, rational expressions are also used to solve problems involving direct variations. In direct variation, when an increase or decrease in the value of one variable occurs, the other variable also increases or decreases proportionately. Two forms of the equation describing direct variation are $y = kx$ and $k = \frac{y}{x}$.

Teacher Resource Book Materials for Chapter 15 See pages 339-358.

CHAPTER 15

Applications of Rational Expressions

The construction of a hot air balloon requires an understanding of the property of gases. The volume of the balloon varies inversely with the applied pressure. That is, the volume decreases as the pressure increases. As a balloon rises, changes in air pressure occur. How does a balloonist compensate for changing air pressure?

Rational expressions are used to solve problems involving variations. In this chapter, you will learn more about using rational expressions.

Objective 15-1: Solve equations containing rational expressions.

15-1 Solving Fractional Equations

Recall that you can solve equations containing fractions by using the least common denominator of all the fractions in the equation. The fractions are eliminated by multiplying both sides of the equation by the common denominator. This method can also be used with equations containing rational expressions.

Examples

1 Solve: $\frac{x-4}{4} + \frac{x}{3} = 6$

$$\frac{x-4}{4} + \frac{x}{3} = 6$$ *The LCD of the fractions is 12.*

$$12\left(\frac{x-4}{4} + \frac{x}{3}\right) = 6 \cdot 12$$ *Multiply both sides of the equation by 12.*

$$3(x-4) + 4(x) = 72$$ *The fractions are eliminated.*

$$3x - 12 + 4x = 72$$
$$7x = 84$$
$$x = 12$$

Check: $\frac{x-4}{4} + \frac{x}{3} = 6$

$$\frac{12-4}{4} + \frac{12}{3} \stackrel{?}{=} 6$$
$$2 + 4 \stackrel{?}{=} 6$$
$$6 = 6$$

The solution is 12.

2 Solve: $\frac{3}{2x} - \frac{2x}{x+1} = -2$

$$\frac{3}{2x} - \frac{2x}{x+1} = -2$$ *Note that $x \neq -1$, $x \neq 0$. The LCD of the rational expressions is $2x(x+1)$.*

$$2x(x+1)\left(\frac{3}{2x} - \frac{2x}{x+1}\right) = 2x(x+1)(-2)$$ *Multiply both sides of the equation by $2x(x+1)$.*

$$3(x+1) - 2x(2x) = -4x^2 - 4x$$
$$3x + 3 - 4x^2 = -4x^2 - 4x$$
$$7x = -3$$
$$x = -\frac{3}{7}$$ *Check this result.*

The solution is $-\frac{3}{7}$.

15-1

Lesson Notes, page T52

Readiness Activity

State the LCD for each set of fractions.

$\frac{2}{3}, \frac{1}{2}$ 6 $\frac{5}{8}, \frac{1}{12}$ 24

$\frac{6}{7}, \frac{5}{21}$ 21 $\frac{1}{8}, \frac{5}{32}, \frac{3}{16}$ 32

$\frac{1}{2}, \frac{2}{9}, \frac{5}{12}$ 36

Discussion Question

Is it true that

$$\frac{a}{a+1} - \frac{b}{b+1} = 0$$

whenever $a = b$?

Yes, as long as a and b do not equal -1.

Chalkboard Exercises

Solve each equation.

1a. $\frac{k+1}{3} - \frac{k}{5} = 3$ 20

1b. $\frac{2-r}{4} + \frac{3r}{5} = \frac{r+2}{3}$ 10

1c. $\frac{2a+7}{4} + \frac{3a}{8} = \frac{7}{8}$ -1

2a. $\frac{b^2-6}{9} + \frac{b}{2} = \frac{11}{6}$ $3, -\frac{15}{2}$

2b. $\frac{6}{x} - \frac{9}{x-1} = \frac{1}{4}$ $-3, -8$

Chalkboard Exercises

Solve each equation.

3a. $\frac{5}{x+1} = \frac{6x}{1-x} - 1$ $-2, \frac{3}{7}$

3b. $2 - \frac{4}{1-x} = \frac{4}{x^2 - x}$ -2

4a. $-\frac{a}{a+1} + \frac{a+1}{4-a} = \frac{-5}{a^2 - 3a - 4}$ 2

4b. $\frac{1}{x-4} - \frac{2}{16 - x^2} = \frac{3}{x+4}$ 9

❶ If you wish, have students name the restrictions on the domain of an equation before solving it. In Example 3, the domain does not include 3. In Example 4, the domain does not include 1 or -1.

Suggested Daily Assignments

Enriched
- Exp: 10–12
- Written: 12–30 evens, 31–36
- Mini-Review: 1–5

Average
- Exp: 2–12 evens
- Written: 1–30
- Mini-Review: 1–5

Fundamental
- Exp: 1–12
- Written: 1–24, 26–36 evens
- Mini-Review: 1–5

❶

Examples

3 **Solve:** $x - \frac{2}{x-3} = \frac{x-1}{3-x}$

$x - \frac{2}{x-3} = \frac{x-1}{3-x}$ *$x \neq 3$ Note that $3 - x = -(x - 3)$.*

$x - \frac{2}{x-3} = -\frac{x-1}{(x-3)}$ *The LCD is $(x - 3)$.*

$(x-3)\left(x - \frac{2}{x-3}\right) = -\left(\frac{x-1}{x-3}\right)(x-3)$ *Multiply both sides of the equation by $(x - 3)$.*

$$x(x-3) - 2 = -(x-1)$$
$$x^2 - 3x - 2 = -x + 1$$
$$x^2 - 2x - 3 = 0$$

$(x-3)(x+1) = 0$ *Factor.*

$x - 3 = 0$ or $x + 1 = 0$ *Zero Product Property.*

$x = 3$ or $x = -1$

Since x cannot equal 3, the only solution is -1.

4 **Solve:** $\frac{2m}{m-1} + \frac{m-5}{m^2-1} = 1$

$\frac{2m}{m-1} + \frac{m-5}{m^2-1} = 1$ *What values for m are excluded?*

$\frac{2m}{m-1} + \frac{m-5}{(m+1)(m-1)} = 1$ *Factor $m^2 - 1$. The LCD is $(m + 1)(m - 1)$.*

$$(m+1)(m-1)\left(\frac{2m}{m-1} + \frac{m-5}{(m+1)(m-1)}\right) = (m+1)(m-1) \cdot 1$$
$$2m(m+1) + (m-5) = m^2 - 1$$
$$2m^2 + 2m + m - 5 = m^2 - 1$$
$$m^2 + 3m - 4 = 0$$

$(m+4)(m-1) = 0$ *Factor.*

$m + 4 = 0$ or $m - 1 = 0$ *Zero Product Property*

$m = -4$ or $m = 1$

The solution is -4. Why is 1 not a solution?

Exploratory Exercises

State the LCD for each pair of fractions.

1. $\frac{m}{2}, \frac{m}{3}$ 6

2. $\frac{x}{5}, \frac{2x}{3}$ 15

3. $\frac{3b}{4}, \frac{5b}{8}$ 8

4. $\frac{1}{x}, \frac{5x}{x+1}$ $x(x+1)$

5. $\frac{r}{r^2-1}, \frac{5}{r-1}$ $r^2 - 1$

6. $\frac{m}{2m^2+3m-35}, \frac{8}{2m-7}$ $(2m-7)(m+5)$

State the LCD for each set of fractions.

7. $\frac{6}{x}, \frac{7}{x-1}, \frac{1}{4}$ $4x(x-1)$

8. $\frac{5}{k-1}, \frac{7}{k}, \frac{1}{k+1}$ $k(k+1)(k-1)$

9. $\frac{5k}{k+5}, \frac{k^2}{k+3}, \frac{1}{k+3}$ $(k+5)(k+3)$

10. $\frac{7}{h+1}, \frac{1}{2}, \frac{2h+5}{h-1}$ $2(h+1)(h-1)$

11. $\frac{3x}{x^2-1}, \frac{1}{x+1}, \frac{1}{x-1}$ $(x+1)(x-1)$

12. $\frac{2x+1}{4x^2-1}, \frac{1}{3}, \frac{1}{2x+1}$ $3(2x+1)(2x-1)$

Written Exercises

Solve each equation.

1. $\frac{2a-3}{6} = \frac{2a}{3} + \frac{1}{2}$ -3
2. $\frac{3x}{5} + \frac{3}{2} = \frac{7x}{10}$ 15
3. $\frac{3a}{2} + \frac{5}{4} = \frac{5a}{2}$ $\frac{5}{4}$
4. $\frac{2b-3}{7} - \frac{b}{2} = \frac{b+3}{14}$ $-\frac{9}{4}$
5. $\frac{x+1}{x} + \frac{x+4}{x} = 6$ $\frac{5}{4}$
6. $\frac{18}{b} = \frac{3}{b} + 3$ 5
7. $\frac{3}{5x} + \frac{7}{2x} = 1$ $\frac{41}{10}$
8. $\frac{11}{2x} - \frac{2}{3x} = \frac{1}{6}$ 29
9. $\frac{5x}{x+1} + \frac{1}{x} = 5$ $\frac{1}{4}$
10. $\frac{5k}{k+2} + \frac{2}{k} = 5$ $\frac{1}{2}$
11. $\frac{2}{3r} - \frac{3r}{r-2} = -3$ $-\frac{1}{4}$
12. $\frac{3n}{n+2} - \frac{5}{7} = 4$ $-\frac{11}{2}$
13. $\frac{m}{m+1} + \frac{5}{m-1} = 1$ $-\frac{3}{2}$
14. $\frac{r-1}{r+1} - \frac{2r}{r-1} = -1$ 0
15. $\frac{4x}{2x+3} - \frac{2x}{2x-3} = 1$ $\frac{1}{2}$
16. $\frac{4x}{3x-2} + \frac{2x}{3x+2} = 2$ -2
17. $\frac{c}{c-4} - \frac{6}{4-c} = c$ $6, -1$
18. $\frac{5}{z-3} - \frac{z}{3-z} = z$ $-1, 5$
19. $\frac{5}{5-p} - \frac{p^2}{5-p} = -2$ $-5, 3$
20. $\frac{r^2}{r-7} + \frac{50}{7-r} = 14$ $8, 6$
21. $\frac{14}{b-6} = \frac{1}{2} + \frac{6}{b-8}$ $20, 10$
22. $\frac{2a-3}{a-3} - 2 = \frac{12}{a+3}$ 5
23. $\frac{r}{3r+6} - \frac{r}{5r+10} = \frac{2}{5}$ -3
24. $\frac{x-2}{x} - \frac{x-3}{x-6} = \frac{1}{x}$ 3
25. $\frac{2b-5}{b-2} - 2 = \frac{3}{b+2}$ 1
26. $\frac{z+3}{z-1} + \frac{z+1}{z-3} = 2$ 2
27. $\frac{7}{k-3} - \frac{1}{2} = \frac{3}{k-4}$ $5, 10$
28. $\frac{x+2}{x-2} - \frac{2}{x+2} = \frac{-7}{3}$ $-1, \frac{2}{5}$
29. $\frac{3w}{w^2-5w+4} = \frac{2}{w-4} + \frac{3}{w-1}$ 7
30. $\frac{3k}{k^2-5k+6} = \frac{2}{k-3} + \frac{3}{k-2}$ $\frac{13}{2}$
31. $\frac{7}{x^2-5x} + \frac{3}{5-x} = \frac{4}{x}$ $\frac{27}{7}$
32. $\frac{9}{b^2-7b+12} = \frac{5}{b-3} + \frac{2}{b-4}$ 5
33. $\frac{6}{z+2} + \frac{3}{z^2-4} = \frac{2z-7}{z-2}$ $-\frac{1}{2}, 5$
34. $\frac{m+3}{m+5} + \frac{2}{m-9} = \frac{-20}{m^2-4m-45}$ $3, 1$
35. $\frac{4}{k^2-8k+12} = \frac{k}{k-2} + \frac{1}{k-6}$ -1
36. $\frac{h^2-7h-8}{3h^2+2h-8} + \frac{1}{h+2} = 0$ 6

mini-review

Solve.

1. $12r - 11 \geq 6r + 13$ $r \geq 4$
2. $x^2 - 5x = 0$ $0, 5$
3. $\sqrt{x-2} = x - 2$ $3, 2$

Simplify.

4. $\frac{x^2-9}{x^2-2x} \cdot \frac{x}{x-3}$ $\frac{x+3}{x-2}$
5. $\frac{x^2+6x+8}{x^2+2x-3} \div \frac{x^2+3x-4}{x^2+7x+12}$ $\frac{(x+4)(x+2)}{(x-1)^2}$

For more exercises on solving equations involving algebraic fractions, see exercises 40-52 on page 560.

Computer Notes

The solutions to fractional equations can be checked with a computer. For example, the following program can be used to check the solutions to Written Exercise 18.

```
10 READ Z
20 DATA -1, 5
30 IF 5/(Z - 3) -
     (-Z/(3 - Z)) = Z
     THEN 60
40 PRINT Z;"IS NOT A
     SOLUTION"
50 GO TO 10
60 PRINT Z;"IS A SOLUTION"
70 GO TO 10
80 END
```

When the computer has used all of the data in line 20, it will print an OUT OF DATA error message.

15-2

Lesson Notes, page T53

Discussion Question

(Use after Example 1.)

Working alone, it usually takes Marshall $1\frac{1}{2}$ hours to clean the basement. His brother Marvin can clean the basement in $1\frac{1}{4}$ hours. The last time they cleaned the basement together it took them 1 hour. Their dad claimed they slowed each other down. Is their dad correct?

Yes, their time working together should be about 41 minutes.

❶ Discuss Example 1 thoroughly. Katina's rate of work is $\frac{1}{3}$ of the job per hour. In t hours, she can complete $\frac{t}{3}$ of the job. Toshio's rate is $\frac{1}{4}$ of the job per hour. In t hours, she can complete $\frac{t}{4}$ of the job. If they work together for t hours, they can complete one job. Therefore, the equation is $\frac{t}{3} + \frac{t}{4} = 1$.

Chalkboard Exercises

Solve each problem. Use $rt = w$ and a chart.

1a. Jan can paint a chair in 45 minutes. Jeff can do the same job in 35 minutes. If they work together, how long will it take them to paint 6 chairs?

	$r \cdot t = w$		
Jan	$\frac{1}{45}$	t	$\frac{t}{45}$
Jeff	$\frac{1}{35}$	t	$\frac{t}{35}$

$$\frac{t}{45} + \frac{t}{35} = 6$$

$$t \approx 118$$

It will take 118 minutes.

1b. A swimming pool can be filled by one pipe in 3 hours and by another pipe in 10 hours. How long will it take

Objective 15-2A: Solve work problems using the formula $rt = w$.
Objective 15-2B: Solve uniform-motion problems involving current or wind speed using the formula $rt = d$.

15-2 Problem Solving: Work and Uniform Motion

Katina Marsh can wash and wax cars at the rate of 1 every 3 hours. Working at this rate, in 1 hour she can complete $\frac{1}{3}$ of the job. In 2 hours she can complete $\frac{1}{3} \cdot 2$ or $\frac{2}{3}$ of the job. In t hours, she can complete $\frac{1}{3} \cdot t$ or $\frac{t}{3}$ of the job.

This and many other similar examples suggest the following formula.

$$(\text{rate of working}) \cdot (\text{time}) = (\text{work done})$$
$$r \cdot t = w$$

Suppose Toshio Meko washes and waxes 1 car in 4 hours. If Katina and Toshio work together, how long will it take them to wash and wax one car? This problem can be solved as follows.

Example 1

Solve the problem above to find how long it will take Katina and Toshio to wash and wax one car.

❶

Explore Let t = time in hours for Katina and Toshio to wash and wax one car.

Plan In t hours, Katina can do $\frac{1}{3} \cdot t$ or $\frac{t}{3}$ of the job.

In t hours, Toshio can do $\frac{1}{4} \cdot t$ or $\frac{t}{4}$ of the job.

	r	$\cdot\ t$	$= w$
Katina	$\frac{1}{3}$	t	$\frac{t}{3}$
Toshio	$\frac{1}{4}$	t	$\frac{t}{4}$

$\frac{t}{3} + \frac{t}{4} = 1$ *Together they complete 1 job.*

Solve $4t + 3t = 12$ *Multiply both sides of the equation by the LCD, 12.*

$7t = 12$

$t = \frac{12}{7}$

Katina and Toshio can do the job in $\frac{12}{7}$, or $1\frac{5}{7}$ hours.

This is about 1 hour and 43 minutes.

Examine Replace t by $\frac{12}{7}$. $\frac{\frac{12}{7}}{3} + \frac{\frac{12}{7}}{4} = \frac{4}{7} + \frac{3}{7} = \frac{7}{7}$ or 1

Recall that uniform motion problems can be solved by using a formula similar to the one used to solve work problems.

$$(\text{rate}) \cdot (\text{time}) = (\text{distance})$$
$$r \cdot t = d$$

Example

2 **While on a fishing trip, Sally and her brother rented a boat. The top speed of the boat in still water was 3 miles per hour. At this rate a 9-mile trip downstream with the current took the same amount of time as a 3-mile trip upstream against the current. What was the rate of the current?**

Explore Let c = the rate of the current.
The rate of the boat when traveling downstream, or *with the current*, is 3 mph *plus* the rate of the current. That is, $3 + c$. The rate when traveling upstream, or *against the current*, is 3 mph *minus* the rate of the current. That is, $3 - c$. **2**

Plan

	r ·	t	$= d$
downstream	$3 + c$	$\frac{9}{3+c}$	9
upstream	$3 - c$	$\frac{3}{3-c}$	3

Solving $rt = d$ for t, $t = \frac{d}{r}$.

Solve

$$\frac{9}{3+c} = \frac{3}{3-c}$$ *time downstream = time upstream*

$$9(3 - c) = 3(3 + c)$$ *Multiply both sides of the equation by the LCD, $(3 + c)(3 - c)$.* **3**

$$27 - 9c = 9 + 3c$$

$$-12c = -18$$

$$c = \frac{3}{2}$$

The rate of the current is $\frac{3}{2}$ or $1\frac{1}{2}$ miles per hour. *Examine this solution.*

Exploratory Exercises

Answer each question.

1. Luisa can paint her house in 8 days. What part of it will she paint: **a.** in 1 day? **b.** in 3 days? **c.** in x days? $\frac{1}{8}, \frac{3}{8}, \frac{x}{8}$

2. Drew can build a garage in n days. What part of it will he build: **a.** in 1 day? **b.** in 4 days? **c.** in x days? $\frac{1}{n}, \frac{4}{n}, \frac{x}{n}$

3. Gabe can do a job alone in 8 days. Stephen can do the same job alone in 10 days. What part of the job can Gabe do: **a.** in 1 day? **b.** in x days? What part of the job can Stephen do: **c.** in 1 day? **d.** in x days? What part of the job can they do together: **e.** in 1 day? **f.** in x days? $\frac{1}{8}, \frac{x}{8}, \frac{1}{10}, \frac{x}{10}, \frac{9}{40}, \frac{9x}{40}$

4. Pat can do a job alone in 4 days. Yoki can do the same job alone in 12 days. What part of the job can Pat do: **a.** in 1 day? **b.** in x days? What part of the job can Yoki do: **c.** in 1 day? **d.** in x days? What part of the job can they do together: **e.** in 1 day? **f.** in x days? $\frac{1}{4}, \frac{x}{4}, \frac{1}{12}, \frac{x}{12}, \frac{1}{3}, \frac{x}{3}$

to fill the pool if the water flows through both pipes?

	r ·	t	$= w$
Pipe 1	$\frac{1}{3}$	t	$\frac{t}{3}$
Pipe 2	$\frac{1}{10}$	t	$\frac{t}{10}$

$\frac{t}{3} + \frac{t}{10} = 1$ $\quad t = 2\frac{4}{13}$

It will take $2\frac{4}{13}$ hr or about 2 hours and 18 minutes.

Solve the problem. Use $rt = d$ and a chart.

2. An airplane can fly at a rate of 500 mph in calm air. It can fly 2575 miles with the wind in the same time it can fly 2425 miles against the wind. Find the speed of the wind.

	r ·	t	$= d$
with wind	$500 + w$	$\frac{2575}{500+w}$	2575
Against wind	$500 - w$	$\frac{2425}{500-w}$	2425

$$\frac{2575}{500+w} = \frac{2425}{500-w}$$

15 mph

2 Emphasize the difference between traveling downstream and upstream. Also emphasize the fact that the times are equal.

3 Another way to solve this equation is to cross multiply.

Suggested Daily Assignments

Enriched
Exp: 1–5
Written: 1–12

Average
Exp: 1–5
Written: 1–12

Fundamental
Exp: 1–5
Written: 1–8
p. 471, 25–35 odds

5. The top flying speed of an open cockpit biplane is 120 mph. At this speed, a 420-mile trip flying with the wind takes the same amount of time as a 300-mile trip flying against the wind. What is the speed of the wind?

   a. Let s = the speed of the wind. Copy and fill in each entry in the chart at the right.

   b. Write an equation to solve this problem. $\frac{420}{120 + s} = \frac{300}{120 - s}$

	r	$\cdot$ t	$=$ d
with the wind	$120 + s$	$\frac{420}{120 + s}$	420
against the wind	$120 - s$	$\frac{300}{120 - s}$	300

Written Exercises

Solve each problem. Use $rt = w$ or $rt = d$ and a chart.

1. Jane can wash the windows of a building in 4 hours. Jim can do the same job in 6 hours. If they work together, how long will it take them to wash the windows? $2\frac{2}{5}$ hours

2. Mark can clean the garage in 6 hours. Rosetta can do the same job in 8 hours. If they work together, how long will it take them to clean the garage? $3\frac{3}{7}$ hours

3. Helena can do a job in 5 days. Jefferson can do the same job in 8 days. If they work together, how long will it take them to complete the job? $3\frac{1}{13}$ days

4. Frank can do a job in 10 hours. Keith can do the same job in 15 hours. If they work together, how long will it take them to complete the job? 6 hours

5. A swimming pool can be filled by one pipe in 12 hours and by another pipe in 4 hours. How long will it take to fill the pool if the water flows through both pipes? 3 hours

6. A swimming pool can be filled by one pipe in 10 hours. The drain pipe can empty the pool in 15 hours. If both pipes are open, how long will it take to fill the pool? 30 hc

7. John and Denise together can mow a lawn in 12 minutes. It takes John 20 minutes to do the job alone. How long would it take Denise to do the job alone? 30 minutes

8. Cindy and Erica together can do a job in $3\frac{3}{5}$ hours. Cindy can do the job alone in 6 hours. How many hours will it take Erica to do the job alone? 9 hours

9. A long distance cyclist pedaling at a steady rate travels 30 miles with the wind. He can travel only 18 miles against the wind in the same amount of time. If the rate of the wind is 3 mph, what is the cyclist's rate without the wind? 12 mph

10. A tugboat pushing a barge up the Ohio River takes 1 hour longer to travel 36 miles up the river than to travel the same distance down the river. If the rate of the current is 3 mph, find the speed of the tugboat and barge in still water. 15 mph

11. An airplane can fly at a rate of 600 mph in calm air. It can fly 2520 miles with the wind in the same time it can fly 2280 miles against the wind. Find the speed of the wind. 30 mph

12. A motorboat takes $\frac{2}{3}$ as much time to travel 10 miles downstream as it does to travel the same distance upstream. If the rate of the current is 5 mph, find the speed of the motorboat in still water. 25 mph

Problem Solving

Using Subgoals

Another problem-solving strategy is to use subgoals. Study the following example.

Example: **Three workers can make three lamps in three days. How many lamps can seven workers working at the same rate make in twenty-one days?**

To solve this problem, first find out how many lamps each worker can make in three days. To do this, divide 3 lamps by 3 workers.

$$3 \div 3 = 1$$

So, each worker can make 1 lamp in 3 days. Use this information to find how many lamps each worker can make in 21 days. To do this, divide 21 by 3 since each worker takes 3 days to make a lamp.

$$21 \div 3 = 7$$

So, each worker can make 7 lamps in 21 days. Now determine how many lamps 7 workers can make by multiplying 7, the number of lamps each worker can make, by 7, the number of workers.

$$7 \times 7 = 49$$

Thus, 7 workers can make 49 lamps in 21 days.

Exercises

Solve each problem.

1. Five farmers can plow five fields in five hours. How many fields can eight farmers plow in fifteen hours? 24 fields
2. Four people can pick eight baskets of apples in two hours. How many baskets of apples can 12 people pick in $\frac{1}{2}$ hour? 6 baskets
3. A flower shop has 112 orders for corsages. Three designers can make 12 corsages in 2 hours. How many designers are needed to complete the orders in 8 hours? 7 designers
4. A pipe 250 cm long needs to be cut into pieces 25 cm long. How many minutes will it take to cut the pipe if each cut takes two minutes? 18 minutes

About Problem Solving
This feature shows students how to use simpler problems to solve a more difficult problem. Point out that determining the number of lamps each worker can make in three days enables us to determine the number of lamps each worker can make in twenty-one days. The answer is then easily determined by multiplying by seven workers.

15-3

Lesson Notes, page T53

Readiness Activity
Solve for x.

$x + a = 3b$ $\quad x = 3b - a$

$r - 2x = 7y$ $\quad x = \frac{7y - r}{-2}$

$ax - ay = z$ $\quad x = \frac{z + ay}{a}$

$\frac{3}{x} + 1 = y$ $\quad x = \frac{3}{y - 1}$

Discussion Question
(Use after Example 2.)
Find values for f, a, and b so that $\frac{1}{f} = \frac{1}{a} + \frac{1}{b}$.
Choose any nonzero values for a and b, with $a \neq -b$.
Then $f = \frac{ab}{b + a}$.

Chalkboard Exercises
Solve each formula for the variable indicated.

1a. $P = 2w + 2\ell$, for ℓ

$\ell = \frac{P - 2w}{2}$

1b. $I = Prt$, for P $\quad P = \frac{I}{rt}$

1c. $A = \frac{1}{2}h(a + b)$, for b

$b = \frac{2A}{h} - a$

1d. $S = \frac{a_1}{1 - r}$, for r

$r = 1 - \frac{a_1}{S}$

1e. $a_n = a_1 + (n - 1)d$, for d

$d = \frac{a_n - a_1}{n - 1}$

2a. $\frac{a}{b} + n = 4n$, for b $\quad b = \frac{a}{3n}$

2b. $\frac{3}{n} = \frac{2}{a} - \frac{3}{b}$, for n

$n = \frac{3ab}{2b - 3a}$

❶ This formula is used to find the sum of the first n terms of an arithmetic series, where A is the first term and t is the nth term.

Quiz, Lessons 15-1 to 15-3
(See margin on p. 479.)

Objective 15-3: Solve formulas for a specified variable.

15-3 Using Formulas

Expressions and equations involving rational expressions often contain more than one variable. Sometimes it is useful to solve the equations or formulas for one of the variables. This is often true with formulas. Study the following example.

Example 1 Solve for n: $S = \frac{n}{2}(A + t)$ ❶

$$S = \frac{n}{2}(A + t)$$

$$2S = 2\left[\frac{n}{2}(A + t)\right]$$

$$2S = n(A + t)$$

$$\frac{2S}{A + t} = n$$

The formula below applies to camera and lens systems. In the formula, f is the focal length of the lens, a is the distance from the object to the lens, and b is the distance from the image to the lens.

$$\frac{1}{f} = \frac{1}{a} + \frac{1}{b}$$

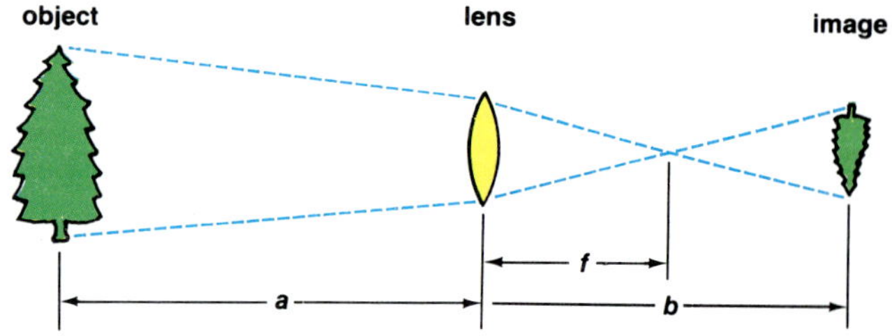

Example 2 Solve for f: $\frac{1}{f} = \frac{1}{a} + \frac{1}{b}$

$$\frac{1}{f} = \frac{1}{a} + \frac{1}{b}$$

$$abf\left(\frac{1}{f}\right) = abf\left(\frac{1}{a} + \frac{1}{b}\right) \quad \textit{The LCD is abf.}$$

$$ab = bf + af$$

$$ab = (b + a)f \quad \textit{Factor } bf + af.$$

$$\frac{ab}{b + a} = f$$

$$f = \frac{ab}{b + a} \quad \textit{Symmetric Property of Equality.}$$

Exploratory Exercises

Answer each question.

1. In Example 1, what steps would you take to solve for A?
2. In Example 1, what steps would you take to solve for t?
3. In Example 2, what steps would you take to solve for a?
4. In Example 2, what steps would you take to solve for b?

Written Exercises

Solve each motion formula for the variable indicated.

1. $a = \frac{v}{t}$, for t $t = \frac{v}{a}$
2. $v = r + at$, for a $a = \frac{v - r}{t}$
3. $s = vt + \frac{1}{2}at^2$, for v $v = \frac{2s - at^2}{2t}$
4. $s = vt + \frac{1}{2}at^2$, for a $a = \frac{2s - 2vt}{t^2}$
5. $F = G\left(\frac{Mm}{d^2}\right)$, for M $M = \frac{Fd^2}{Gm}$
6. $F = G\left(\frac{Mm}{d^2}\right)$, for d $d = \pm\sqrt{\frac{GMm}{F}}$
7. $f = \frac{W}{g} \cdot \frac{V^2}{R}$, for V $V = \pm\sqrt{\frac{fgR}{W}}$
8. $f = \frac{W}{g} \cdot \frac{V^2}{R}$, for R $R = \frac{WV^2}{fg}$

Solve each business formula for the variable indicated.

9. $A = P + Prt$, for P $P = \frac{A}{1 + rt}$
10. $Prt = I$, for r $r = \frac{I}{Pt}$
11. $I = \left(\frac{100 - P}{P}\right)\frac{365}{R}$, for P $P = \frac{36{,}500}{IR + 365}$
12. $I = \frac{365d}{360 - dr}$, for d $d = \frac{360I}{Ir + 365}$
13. $a = \frac{r}{2y} - 0.25$, for y $y = \frac{r}{2a + 0.5}$
14. $c = \frac{P - 100}{P}$, for P $P = \frac{100}{1 - c}$

Solve each electronics formula for the variable indicated.

15. $H = (0.24)I^2Rt$, for R
16. $P = \frac{E^2}{R}$, for E
17. $\frac{1}{R_T} = \frac{1}{R_1} + \frac{1}{R_2}$, for R_1
18. $I = \frac{E}{r + R}$, for R
19. $I = \frac{nE}{nr + R}$, for n
20. $I = \frac{E}{\frac{r}{n} + R}$, for r

Solve each mathematical formula for the variable indicated.

21. $y = mx + b$, for m
22. $A = \frac{1}{2}h(a + b)$, for h
23. $m = \frac{y_2 - y_1}{x_2 - x_1}$, for y_2
24. $m = \frac{y_2 - y_1}{x_2 - x_1}$, for x_1
25. $\frac{P}{D} = Q + \frac{R}{D}$, for R
26. $\frac{P}{D} = Q + \frac{R}{D}$, for D

Solve each equation for n.

27. $\frac{1}{2}n + b = n$
28. $\frac{n}{x} = \frac{y}{r}$
29. $\frac{n}{a} + \frac{b}{c} = d$
30. $\frac{a}{c} = n + bn$
31. $\frac{n}{a + b} = c - n$
32. $\frac{1}{n} = \frac{1}{a} + \frac{1}{b}$
33. $\frac{n + 2}{b} = \frac{n + b}{c}$
34. $r = \frac{16M}{a(n + 1)}$
35. $\frac{a}{n} = \frac{n}{b}$ $n = \pm\sqrt{ab}$
36. $\frac{a}{n} + \frac{a}{b} = \frac{b}{n}$ $n = \frac{b^2 - ab}{a}$
37. $\frac{r}{n} - \frac{2}{k} = \frac{k}{n}$ $n = \frac{rk - k^2}{2}$
38. $\frac{1}{2n} = \frac{n}{2x}$ $n = \pm\sqrt{x}$

Suggested Daily Assignments

Enriched
Written: 2–38 evens

Average
Exp: 1–4
Written: 2–26 evens, 27–38

Fundamental
Exp: 1–4
Written: 1–8, 28–38 evens
p. 474, 9–12

Answers to Exploratory

1. Multiply by 2, divide by n, then subtract t.
2. Multiply by 2, divide by n, then subtract A.
3. Multiply by abf, subtract af, then divide by $b - f$.
4. Multiply by abf, subtract bf, then divide by $a - f$.

Answers to Written

15. $R = \frac{H}{0.24I^2t}$
16. $E = \pm\sqrt{PR}$
17. $R_1 = \frac{R_TR_2}{R_2 - R_T}$
18. $R = \frac{E - Ir}{I}$
19. $n = \frac{IR}{E - Ir}$
20. $r = \frac{En - IRn}{I}$
21. $m = \frac{y - b}{x}$
22. $h = \frac{2A}{a + b}$
23. $y_2 = mx_2 - mx_1 + y_1$
24. $x_1 = \frac{mx_2 - y_2 + y_1}{m}$
25. $R = P - DQ$
26. $D = \frac{P - R}{Q}$
27. $n = 2b$
28. $n = \frac{xy}{r}$
29. $n = \frac{acd - ab}{c}$
30. $n = \frac{a}{c + cb}$
31. $n = \frac{ac + bc}{a + b + 1}$
32. $n = \frac{ab}{a + b}$
33. $n = \frac{b^2 - 2c}{c - b}$
34. $n = \frac{16M - ar}{ar}$

15-4

Lesson Notes, page T53
Enrichment, page T53

Readiness Activity
Solve each proportion.

$\frac{2}{3} = \frac{x}{9}$ 6 $\frac{y}{4} = \frac{5}{2}$ 10

$\frac{10}{y} = \frac{3}{7}$ $23\frac{1}{3}$ $\frac{25}{12} = \frac{5}{x}$ $2\frac{2}{5}$

Discussion Question
(Use after Example 1.)
Is it true that a person's height varies directly as his or her age? No

Chalkboard Exercises
Solve each problem.

1a. A car uses 8 gallons of gasoline to travel 290 miles. How much gasoline will the car use to travel 400 miles? 11 gallons

1b. If 4 pounds of peanuts cost \$7.50, how much will $2\frac{1}{2}$ pounds cost? \$4.69

❶ Point out that the phrase "*y* varies directly as *x*" can also be phrased "*y* is directly proportional to *x*." Both of these phrases mean that as *y* increases or decreases in value, *x* increases or decreases in value.

❷ *k* is also referred to as the *constant of proportionality*.

Objective 15-4: Solve problems involving direct variations.

15-4 Direct Variation

Diego delivers newspapers. He is paid 3¢ for each newspaper he delivers. The following table relates the number of newspapers Diego delivers (x) and his income (y).

x	20	50	100	120	150
y	\$0.60	\$1.50	\$3.00	\$3.60	\$4.50

Diego's income depends *directly* on the number of newspapers he delivers. The relationship between newspapers delivered and income is shown by the equation $y = 0.03x$. Such an equation is called a **direct variation**. We say that *y varies directly as x*. ❶

Definition of Direct Variation

A direct variation is described by an equation of the form $y = kx$, where k is not zero.

In a direct variation, k is called the **constant of variation**. To find ❷ the constant of variation, divide both sides of $y = kx$ by x.

$$\frac{y}{x} = k \quad \text{or} \quad k = \frac{y}{x}$$

Example 1

The weight of an object on the moon varies directly as its weight on the earth. A certain astronaut weighs 168 pounds on earth and 28 pounds on the moon. Kristina weighs 108 pounds on earth. What would she weigh on the moon?

Let x = weight on the moon.
Let y = weight on the earth.

Then $y = kx$. Find the value of k.

$k = \frac{y}{x}$

$k = \frac{168}{28}$ *Substitute the astronaut's weights for x and y.*

$k = 6$ *The constant of variation is 6.*

Next find Kristina's weight on the moon. Find x when y is 108.

$y = kx$

$108 = 6x$ *Substitute 108 for y and 6 for k.*

$18 = x$

Thus, Kristina would weigh 18 pounds on the moon.

Recall that an equation of the form $y = mx + b$ is a linear equation. Notice that the equation for direct variation, $y = kx$, is also a linear equation, where $m = k$ and $b = 0$.

Direct variations are also related to proportions. From the table on Diego's newspapers and income, many true proportions can be formed. Two proportions are shown below.

number of papers → $\frac{20}{60} = \frac{50}{150}$ ← income (number of papers, income)

number of papers → $\frac{20}{50} = \frac{60}{150}$ ← income

Two general forms for proportions such as these can be derived from the equation $y = kx$. Let (x_1, y_1) be a solution for $y = kx$. Let a second solution be (x_2, y_2). Then $y_1 = kx_1$ and $y_2 = kx_2$. A proportion can be written using x_1, x_2, y_1, and y_2. Study the following.

$y_1 = kx_1$ *This equation describes a direct variation.*

$\frac{y_1}{y_2} = \frac{kx_1}{kx_2}$ *Use the division property of equality. Since y_2 and kx_2 are equivalent, you can divide the left side by y_2 and the right side by kx_2.*

❸ $\frac{y_1}{y_2} = \frac{x_1}{x_2}$ *Simplify.*

Study how another proportion can be formed from this proportion.

$\frac{y_1}{y_2} = \frac{x_1}{x_2}$

$x_2y_1 = x_1y_2$ *Find the cross products.*

$\frac{x_2y_1}{y_1y_2} = \frac{x_1y_2}{y_1y_2}$ *Divide both sides by y_1y_2.*

$\frac{x_2}{y_2} = \frac{x_1}{y_1}$ *Simplify.*

Be consistent when using proportions: compare x's and y's; or compare the first two elements and the second two elements.

You can use proportions to solve problems involving direct variation.

Example 2

If y varies directly as x and $y = 27$ when $x = 6$, find x when $y = 45$.

Use the following proportion to solve the problem.

$\frac{y_1}{y_2} = \frac{x_1}{x_2}$

Since y is 27 when x is 6, let $y_1 = 27$ and $x_1 = 6$. Let $y_2 = 45$ and solve for x_2.

$\frac{27}{45} = \frac{6}{x_2}$ *Substitute.*

$27x_2 = 6(45)$ *Find the cross products.*

$27x_2 = 270$

$x_2 = 10$ Thus, $x = 10$ when $y = 45$.

Chalkboard Exercises

Solve each problem. Assume that y varies directly as x.

2a. If $y = 6$ when $x = 8$, find y when $x = 12$. 9

2b. If $y = 8$ when $x = 20$, find x when $y = 6$. 15

2c. If $y = 3$ when $x = 9$, find y when $x = 40.5$. 13.5

2d. If $x = -5$ when $y = 6$, find y when $x = -8$. 9.6

2e. If $y = \frac{2}{3}$ when $x = \frac{3}{4}$, find y when $x = \frac{5}{8}$. $\frac{5}{9}$

❸ The highlighted forms of direct variations are very useful in many applications. These forms are used in the study of similar triangles and proportional line segments.

Quiz

Lessons 15-1 to 15-3

Solve each equation.

1. $\frac{1 + x}{x} = \frac{5}{x}$ 4

2. $1 + \frac{2}{x - 1} = \frac{2}{x^2 - x}$ −2

Solve for y.

3. $\frac{a}{y} + \frac{x}{a} = \frac{x}{y}$

$y = \frac{ax - a^2}{x}$ or $a - \frac{a^2}{x}$

Solve each problem.

4. Jean and John together can do a job in 2 weeks. Jean can do the job alone in 3 weeks. How long will it take John to do the job alone?

$\frac{2}{3} + \frac{2}{t} = 1$

$t = 6$

It will take 6 weeks.

5. Gary can go 15 miles upstream in his boat in the same amount of time he can go 30 miles downstream. If the speed of the current is 3 mph, find the speed of the boat.

9 mph

Suggested Daily Assignments

Enriched
- Exp: 1–8
- Written: 1–12, 19–26

Average
- Exp: 1–8
- Written: 1–26

Fundamental
- Exp: 1–8
- Written: 1–9, 20–26 evens
  - p. 477, 21–24, 27–37 odds

❹ You may prefer to have students solve Written Exercises 10–18 before solving 1–9.

Answers to Written

10. 4, $y = 4x$
11. -4, $y = -4x$
12. $\frac{1}{5}$, $y = \frac{1}{5}x$
13. $\frac{1}{2}$, $y = \frac{1}{2}x$
14. $-\frac{2}{3}$, $y = -\frac{2}{3}x$
15. $\frac{8}{3}$, $y = \frac{8}{3}x$
16. $\frac{4}{5}$, $y = \frac{4}{5}x$
17. $\frac{17}{22}$, $y = \frac{17}{22}x$
18. $\frac{17}{26}$, $y = \frac{17}{26}x$

About Using Calculators

After using a calculator to solve these problems, you may want to have students solve some of the problems in the Written Exercises using a calculator.

Exploratory Exercises

State which of the following are direct variations. For each direct variation, state the constant of variation.

1. $y = 3x$ yes, 3
2. $ab = 8$ no
3. $5x^2 = 6y$ no
4. $-3a = b$ yes, -3
5. $d = 7t$ yes, 7
6. $r = \frac{1}{3}m$ yes, $\frac{1}{3}$
7. $\frac{1}{6}y = \frac{5}{3}$ no
8. $z = \frac{3}{p}$ no

Written Exercises ❹

Solve each problem. Assume that y varies directly as x.

1. If $y = 12$ when $x = 3$, find y when $x = 7$. 28
2. If $y = -8$ when $x = 2$, find y when $x = 10$. -40
3. If $y = 3$ when $x = 15$, find y when $x = -25$. -5
4. If $y = -7$ when $x = -14$, find y when $x = 20$. 10
5. If $y = -6$ when $x = 9$, find x when $y = -4$. 6
6. If $y = -8$ when $x = -3$, find x when $y = 6$. $2\frac{1}{4}$
7. If $y = 12$ when $x = 15$, find x when $y = 21$. $26\frac{1}{4}$
8. If $y = 17$ when $x = 22$, find x when $y = 13$. $16\frac{14}{17}$
9. If $y = 1.7$ when $x = 2.6$, find x when $y = 3.4$. 5.2

10-18. For exercises 1-9, find the constant of variation. Then write an equation of the form $y = kx$ for each variation.

Solve each problem.

19. If 6 pounds of sugar cost \$2.00, how much will 40 pounds cost? \$13.33
20. If 2 m of copper wire weigh 0.3 kg, how much will 75 m weigh? 11.25 kg
21. Ken's wages vary directly as the time he works. If his wages for 4 days are \$110, how much will they be for 17 days? \$467.50
22. Deidre's wages vary directly as the time she works. If her wages for six days are \$121, what are her wages for 20 days? \$403.
23. A car uses 5 gallons of gasoline to travel 143 miles. How much gasoline will the car use to travel 200 miles? about 7 gallons
24. A car uses 9 liters of gasoline to travel 88 km. How much gasoline will the car use to travel 215 km? about 22 liters
25. Charles' Law says that the volume of a gas is directly proportional to its temperature. If the volume of a gas is 2.5 cubic feet at 150° (absolute temperature), what is the volume of the same gas when the temperature is 200° (absolute temperature)? $3\frac{1}{3}$ ft^3
26. In an electrical transformer, voltage is directly proportional to the number of turns on the coil. If 110 volts comes from 55 turns, what would be the voltage produced by 66 turns? 132 volts

Using Calculators — Variations

Use a calculator to help solve each problem. Assume that y varies directly as x.

1. If $y = 5.8$ when $x = 6.5$, find x when $y = 8.7$. 9.75
2. If $y = 4.6$ when $x = 2.4$, find y when $x = 6.3$. 12.075

Objective 15-5: Solve problems involving inverse variations.

15-5 Inverse Variation

Anne Marie plans to drive on a trip of 600 miles. Her time, in hours, for the trip will be determined by her average rate in miles per hour. The following table shows the time (t) for various rates (r).

r	30	40	45	50	60
t	20	15	$13\frac{1}{3}$	12	10

The formula $d = rt$ is used to find values for t.

Notice that as the rate *increases,* the time to make the trip *decreases*. As the rate *decreases,* the time *increases*. We say that the *rate varies inversely as the time.*

An inverse variation is described by an equation of the form $xy = k$, where k is not zero.

Definition of Inverse Variation

Sometimes an inverse variation is written in the form $y = \frac{k}{x}$. We say that y *varies inversely as* x.

Example

1 If y varies inversely as x and $y = 6$ when $x = 18$, find x when $y = -3$.

First find the value of k.

$xy = k$

$18(6) = k$ *Substitute 18 for x and 6 for y.*

$108 = k$ *The constant of variation is 108.*

Then find x when $y = -3$.

$xy = k$

$x(-3) = 108$ *Substitute −3 for y and 108 for k.*

$x = -36$

Thus, $x = -36$ when $y = -3$.

Consider again the table of rates and times for Anne Marie's trip. There are many ways to form true mathematical statements from the data. Compare the following mathematical statements.

$$50 \cdot 12 = 60 \cdot 10 \qquad \frac{50}{60} = \frac{10}{12}$$

Two general forms for mathematical statements such as these can be derived from the equation $xy = k$.

15-5

Lesson Notes, page T53
Enrichment, page T53

Discussion Question
(Use after Example 1.)
Is it true that the percent grade on a test varies inversely as the number of incorrect answers? Yes

Chalkboard Exercises
Solve each problem. Assume that y varies inversely as x.

1a. If $y = 6$ when $x = 12$, find x when $y = 9$. 8

1b. If $y = 4$ when $x = 6$, find x when $y = 12$. 2

1c. If $y = 21$ when $x = 3$, find y when $x = 9$. 7

Chalkboard Exercises

Solve each problem. Assume that y varies inversely as x.

2a. If $y = 6$ when $x = 2$, find x when $y = 4$. 3

2b. If $y = 15$ when $x = 21$, find x when $y = 27$. $11\frac{2}{3}$

2c. If $y = 12$ when $x = 15$, find x when $y = 18$. 10

❶ Emphasize the difference between the proportions below.

Direct Proportion	Inverse Proportion
$\frac{x_1}{x_2} = \frac{y_1}{y_2}$	$\frac{x_1}{x_2} = \frac{y_2}{y_1}$

Suggested Daily Assignments

Enriched
- Exp: 2–12 evens
- Written: 1–12, 21–30

Average
- Exp: 1–12
- Written 1–30

Fundamental
- Exp: 1–12
- Written: 1–10, 22–30 evens p. 480, 10–12, 19–25 odds

Let (x_1, y_1) be a solution of an inverse variation, $xy = k$. Let (x_2, y_2) be a second solution. Then $x_1y_1 = k$ and $x_2y_2 = k$. Study the following.

$x_1y_1 = k$

$x_1y_1 = x_2y_2$ *You can substitute x_2y_2 for k because $x_2y_2 = k$.*

The equation $x_1y_1 = x_2y_2$ is called the *product rule for inverse variations*. Study how it can be used to form a proportion.

$x_1y_1 = x_2y_2$

$\frac{x_1y_1}{x_2y_1} = \frac{x_2y_2}{x_2y_1}$ *Divide both sides by x_2y_1.*

$\frac{x_1}{x_2} = \frac{y_2}{y_1}$ *Notice that this proportion is different from the proportions for direct variation on page 479.* ❶

The product rule or the proportion above can be used to solve problems involving inverse variation.

Example

2 **If y varies inversely as x and $y = 3$ when $x = 12$, find x when $y = 4$.**

Let $x_1 = 12$, $y_1 = 3$, *and* $y_2 = 4$. Solve for x_2.

a. Use the product rule.

$x_1y_1 = x_2y_2$

$12 \cdot 3 = x_2 \cdot 4$

$\frac{36}{4} = x_2$

$9 = x_2$

b. Use the proportion.

$\frac{x_1}{x_2} = \frac{y_2}{y_1}$

$\frac{12}{x_2} = \frac{4}{3}$

$36 = 4 \cdot x_2$

$9 = x_2$

Thus, $x = 9$ when $y = 4$.

Exploratory Exercises

State which of the following are inverse variations. For each inverse variation, state the constant of variation.

1. $ab = 6$ yes, 6
2. $c = 3.14d$ no
3. $\frac{50}{y} = x$ yes, 50
4. $\frac{1}{5}a = d$ no
5. $\frac{-13}{a} = b$ yes, −13
6. $14 = ab$ yes, 14
7. $xy = 1$ yes, 1
8. $y = \frac{1}{x}$ yes, 1
9. $bh = 40$ yes, 40
10. $a = \frac{7}{b}$ yes, 7
11. $d = 4t^2$ no
12. $s = 3t$ no

Written Exercises ❷

Solve each problem. Assume that y varies inversely as x.

1. If $y = 24$ when $x = 8$, find y when $x = 4$. 48

2. If $y = -6$ when $x = -2$, find y when $x = 5$. $2\frac{2}{5}$

3. If $y = \frac{1}{3}$ when $x = 5$, find y when $x = \frac{1}{4}$. $6\frac{2}{3}$

4. If $y = 7$ when $x = \frac{2}{3}$, find y when $x = 7$. $\frac{2}{3}$

5. If $y = -8$ when $x = 2$, find x when $y = 7$. $-2\frac{2}{7}$

6. If $y = 99$ when $x = 11$, find x when $y = 11$. 99

7. If $x = 2.8$ when $y = 5.6$, find y when $x = 3.5$. 4.48

8. If $x = 2.7$ when $y = 8.1$, find y when $x = 3.6$. 6.075

9. If $x = 1.2$ when $y = 1.8$, find x when $y = 1.5$. 1.44

10. If $x = 6.9$ when $y = 4.3$, find x when $y = 6.45$. 4.6

11-20. For exercises 1-10, find the constant of variation, k. Write an equation of the form $xy = k$ for each variation.

Solve each problem.

21. Susan drove 3 hours at a rate of 50 mph. How long would the same distance take Susan if she drove at 45 mph? $3\frac{1}{3}$ hours

22. Gary drove 4 hours at a rate of 80 km/h. How long would the same distance take Gary if he drove at 75 km/h? $4\frac{4}{15}$ hours

Boyle's law states that the volume of a gas (V) varies inversely with applied pressure (P). This is shown by the formula $P_1V_1 = P_2V_2$. Use this formula to solve each problem.

23. Pressure acting on 60 m^3 of a gas is raised from 1 atmosphere to 2 atmospheres. What new volume does the gas occupy? 30 m^3

24. Pressure acting on 8.0 m^3 of a gas is 20 atmospheres. The pressure is reduced until the volume is 20 m^3. What is the new pressure acting on the gas? 8 atm

25. A helium-filled balloon has a volume of 16 m^3 at sea level. The pressure at sea level is 1 atmosphere. The balloon rises to a point in the atmosphere where the pressure is 0.75 atmosphere. What is its volume? $21\frac{1}{3}$ m^3

26. A helium-filled balloon has a volume of 2 m^3 at sea level. The pressure at sea level is 1 atmosphere. The balloon rises to a point in the atmosphere where its volume is 6 m^3. What is the pressure at this height? $\frac{1}{3}$ atm

Suppose the area of a triangle remains constant, while the lengths of the base and the altitude vary. The altitude is 6 inches when the base is 10 inches. Use this information to solve each problem.

27. Find the altitude when the base is 15 inches. 4 in.

28. Find the base when the altitude is 12 inches. 5 in.

In sound and harmonics, the frequency of a vibrating string is inversely proportional to its length. Use this information to solve each problem.

29. A violin string 10 inches long vibrates at a frequency of 512 cycles per second. Find the frequency of an 8-inch string. 640 cycles per second

30. A piano string 36 inches long vibrates at a frequency of 480 cycles per second. Find the frequency of the string if it were shortened to 24 inches. 720 cycles per second

❷ You may prefer to have students solve Written Exercises 11–20 before solving 1–10.

Answers to Written

11. 192; $xy = 192$
12. 12; $xy = 12$
13. $\frac{5}{3}$; $xy = \frac{5}{3}$
14. $\frac{14}{3}$; $xy = \frac{14}{3}$
15. -16; $xy = -16$
16. 1089; $xy = 1089$
17. 15.68; $xy = 15.68$
18. 21.87; $xy = 21.87$
19. 2.16; $xy = 2.16$
20. 29.67; $xy = 29.67$

15-6

Lesson Notes, page T54

Discussion Question
(Use after Example 2.)
Is the balance point of a pencil closer to the point or the eraser? Why? Closer to the eraser, because that end weighs more.

Chalkboard Exercises
Solve each problem.

1. A 120-pound weight is located 8 feet from the fulcrum of a lever. How much weight at a distance of 10 feet on the opposite side of the fulcrum would balance it? 96 lb
2. A 200-pound weight is located 5 feet from the fulcrum. How far away should 125 pounds be placed to balance the lever? 8 ft

❶ The result of this example can be approximated in feet and inches.

$6\frac{6}{11}$ ft ≈ 6 ft 7 in.

❷ Encourage students to check the results of this example.

Objective 15-6: Solve lever problems by applying the product rule for inverse variations.

15-6 Lever Problems

Laura and Jason are on a seesaw. They want the seesaw to balance. Laura weighs 132 pounds and Jason weighs 108 pounds. Which person should sit closer to the fulcrum (the pivot point)?

If you have observed people on a seesaw, you may have noticed that the heavier person must sit closer to the fulcrum to balance the seesaw. This is an example of an inverse variation. A seesaw is a type of lever. All lever problems involve inverse variation.

For all lever problems in this section, assume that the lever is weightless.

Suppose weights w_1 and w_2 are placed on a lever at a distance d_1 and d_2, respectively, from the fulcrum. The lever is balanced when $w_1d_1 = w_2d_2$.

Property of Levers

The property of levers is illustrated at the right.

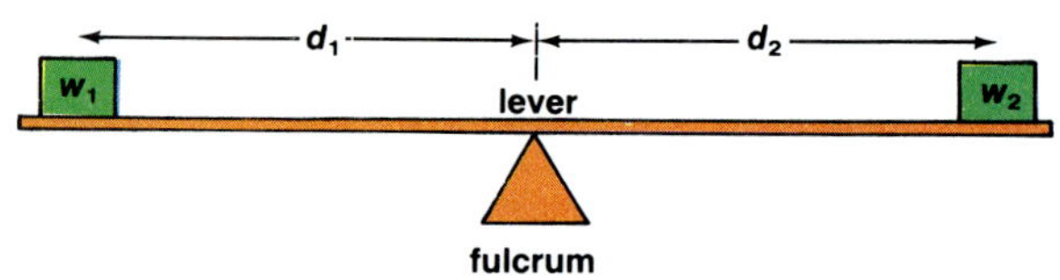

Examples

1 **The fulcrum of a 16-ft seesaw is placed in the middle. Jason, who weighs 108 pounds, is seated 8 feet from the fulcrum. How far from the fulcrum should Laura sit if she weighs 132 pounds?**

Use the property of levers, $w_1d_1 = w_2d_2$.

Let $w_1 = 108$, $d_1 = 8$, and $w_2 = 132$. Solve for d_2.

$$w_1d_1 = w_2d_2$$
$$108 \cdot 8 = 132 \cdot d_2$$
$$864 = 132d_2$$
$$6\tfrac{6}{11} = d_2$$

Laura should sit $6\frac{6}{11}$ feet from the fulcrum. ❶

2 **An 8-ounce weight is placed at one end of a yardstick. A 10-ounce weight is placed at the other end. Where should the fulcrum be placed to have the yardstick balanced?**

Let d = distance from fulcrum to 8-ounce weight, in inches.
Then $36 - d$ = distance from fulcrum to 10-ounce weight, in inches.

$$8d = 10(36 - d)$$
$$8d = 360 - 10d$$
$$18d = 360$$
$$d = 20$$

The fulcrum should be placed 20 inches from the 8-ounce weight. ❷

Examples

3 **Dean Cadwell wants to lift a large rock with the use of a long crowbar. The short end of the crowbar is 6 inches from the fulcrum and the long end is 5 feet from the fulcrum. What is the maximum weight that Dean can lift using this method if he weighs 165 pounds?** ❸

The distances from the fulcrum must be expressed with the same unit of measure.

5 feet = 60 inches

Dean's weight should be at the long end of the crowbar, so let $w_1 = 165$, $d_1 = 60$, and $d_2 = 6$.

$$\begin{aligned} w_1 d_1 &= w_2 d_2 \\ 165 \cdot 60 &= w_2 \cdot 6 \\ 9900 &= 6w_2 \\ 1650 &= w_2 \end{aligned}$$

Dean can lift 1650 pounds with the crowbar.

4 **Patti and Cathy are seated on the same side of a seesaw. Patti is 6 feet from the fulcrum and weighs 115 pounds. Cathy is 8 feet from the fulcrum and weighs 120 pounds. Jack is seated on the other side of the seesaw, 10 feet from the fulcrum. If the seesaw is balanced, how much does Jack weigh?** ❹

Draw a diagram.

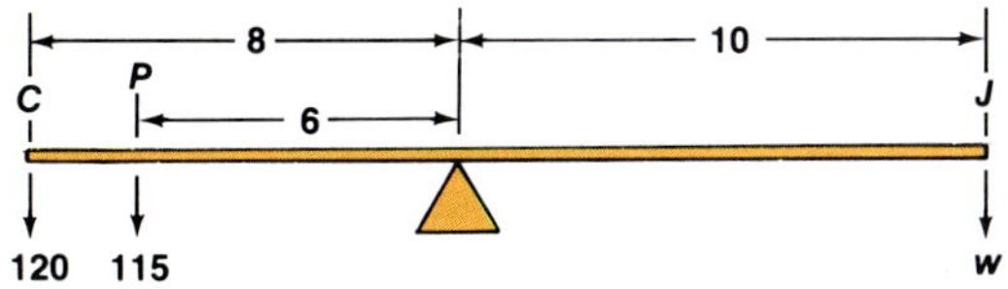

Let w = Jack's weight.

$$\begin{aligned} 120(8) + 115(6) &= 10w \\ 960 + 690 &= 10w \\ 1650 &= 10w \\ 165 &= w \end{aligned}$$

Thus, Jack weighs 165 pounds.

Exploratory Exercises

For each of the following, suppose the two people are on a seesaw. For the seesaw to balance, which person must sit closer to the fulcrum?

1. Jorge, 168 pounds
   Emilio, 220 pounds Emilio
2. Shawn, 114 pounds
   Shannon, 97 pounds Shawn
3. Juanita, 49 pounds
   Lucia, 49 pounds same distance
4. Ryna, 52 kg
   Helen, 55 kg Helen ❺
5. Jack, 50 kg
   Betty, 58 kg Betty
6. Jimmy, 72 pounds
   Jeff, 68 pounds Jimmy

Chalkboard Exercises

Solve each problem.

3. For Example 3, what is the maximum weight that Dean can lift if he weighs 200 pounds? 2000 pounds
4. For Example 4, find Jack's weight if he is seated 9 feet from the fulcrum.
   $183\frac{1}{3}$ pounds

❸ Discuss the use of levers to lift weights. Sometimes the fulcrum can be shifted so that more weight can be lifted. Also point out that holding the end of a lever or crowbar shortens the length, *d*, slightly.

❹ In this example two persons are seated on the same side of a seesaw, and a third person is seated on the other side. Point out that the formula to use in this case is $w_1d_1 + w_2d_2 = w_3d_3$. In general, when two or more weights are placed on the same side of a lever, find the product of *w* and *d* for each weight, then add.

❺ To theoretical physicists, a kilogram is a unit of mass, not a unit of weight (force). In everyday usage, however, such a distinction is overlooked.

Suggested Daily Assignments

Enriched
- Exp: 1–6
- Written: 1–8
- Mini-Review: 1–5

Average
- Exp: 1–6
- Written: 1–8
- Mini-Review: 1–5

Fundamental
- Exp: 1–6
- Written: 1–6
  - p. 483, 11–14, 21–29 odds
- Mini-Review: 1–5

Written Exercises

Solve each problem.

1. Mary Jo weighs 120 pounds and Doug weighs 160 pounds. They are seated at opposite ends of a seesaw. Doug and Mary Jo are 14 feet apart, and the seesaw is balanced. How far is Mary Jo from the fulcrum? 8 feet

2. Grant, who weighs 150 pounds, is seated 8 feet from the fulcrum of a seesaw. Mariel is seated 10 feet from the fulcrum. If the seesaw is balanced, how much does Mariel weigh? 120 pounds

3. Weights of 100 pounds and 115 pounds are placed on a lever. The two weights are 15 feet apart, and the lever is balanced. How far from the fulcrum is the 100-pound weight? 8.02 feet

4. A lever has a 140-pound weight on one end and a 160-pound weight on the other end. The lever is balanced, and the 140-pound weight is exactly one foot farther from the fulcrum than the 160-pound weight. How far from the fulcrum is the 160-pound weight? 7 feet

5. Amy, who weighs 108 pounds, is seated 5 feet from the fulcrum of a seesaw. Barbara is seated on the same side of the seesaw, two feet farther from the fulcrum than Amy. Barbara weighs 96 pounds. The seesaw is balanced when Sue, who weighs 101 pounds, sits on the other side. How far is Sue from the fulcrum? 12 feet

6. Christy, who weighs 112 pounds, is seated 6 feet from the fulcrum of a seesaw. Sooyun, who weighs 124 pounds, is seated on the same side of the seesaw, 8 feet from the fulcrum. The other side of the seesaw is 9 feet long. Is it possible for Seiju, who weighs 180 pounds, to balance the seesaw? no

7. John Kofmehl is using a crowbar to move a boxcar in a railyard. The total length of the bar is 6 feet, and the fulcrum is 4 inches from one end. If John weighs 180 pounds, how much force can he exert against the wheel? 3060 pounds

8. To move a log, Norma Barone places a rock 2 feet from the log to use as a fulcrum. She then uses a 12-foot plank as a lever. If Norma weighs 140 pounds, will she be able to lift the log that weighs 600 pounds? yes

mini-review

Write *True* or *false*.

1. $(3a^2b)^3 = 9a^5b^3$ false
2. $(2m^2n + mn^2) - (m^2n - 2mn^2) = m^2n - mn^2$ false
3. $x^3 + 2x^2 + x + 2 = (x^2 + 1)(x + 2)$ true
4. $\dfrac{6x^2 - x - 12}{2x + 3} = 3x - 4$ false
5. $\dfrac{y}{y - 1} - \dfrac{y + 1}{y + 2} = \dfrac{1}{(y - 1)(y + 2)}$ false

Applications in Physics

Universal Gravitation

Legend has it that while watching an apple fall from a tree, Sir Isaac Newton recognized that the apple fell because an unbalanced force acted on it. This force was gravity. Newton wondered if this special force was peculiar to the earth. Did other bodies also have it? Perhaps every body exerts a gravitational force on every other body.

Newton assumed that the gravitational force between any two bodies acts in the same way as the force between a planet and the sun. From this assumption, he wrote the **law of universal gravitation**. This law states that *every body in the universe attracts every other body in the universe with a force that varies directly with the product of the masses and inversely with the square of the distance between the centers of the two masses*. This law is written as follows.

$$F = G\frac{m_1 m_2}{r^2}$$

In the equation, m_1 and m_2 are the masses of the two bodies in kilograms, r is the distance in meters between the centers of the masses, and G is a universal constant. The value of G is 6.67×10^{-11} N · m²/kg². Using these units, the force, F, will be in newtons (N).

Example: **Two freight cars, each of mass 3.0×10^5 kg, are located on adjacent tracks. Their centers are 9.0 m apart. What gravitational force exists between them?**

$$\begin{aligned} F &= G\frac{m_1 m_2}{r^2} \\ &= (6.67 \times 10^{-11})\frac{(3.0 \times 10^5)(3.0 \times 10^5)}{(9.0)^2} \\ &= (6.67 \times 10^{-11})\frac{(9.0 \times 10^{10})}{81} \\ &= 0.074 \end{aligned}$$

The force between them is 0.074 N.

Exercises

Solve each problem. Assume the distance, r, is the distance between the centers of the two masses.

1. Two people are standing 2.0 m apart. One has a mass of 80 kg. The other has a mass of 60 kg. What is the gravitational force between them? 8.004×10^{-8} N
2. Two ships are docked next to each other. Their centers of gravity are 40 m apart. One ship weighs 9.8×10^7 N. The other ship weighs 1.96×10^8 N. What gravitational force exists between them? 8.007×10^2 N

About the Application
Emphasize the units of measure that are associated with each quantity in the equation. You may want to have students show why the force (E) will be in newtons (N).

15-7

Lesson Notes, page T54
Enrichment, page T54

Readiness Activity

The formula for the area (A) of a trapezoid is $A = \frac{1}{2}h(a + b)$ where h is the altitude, and a and b are the lengths of the bases.

Find the area of each trapezoid for which h, a, and b are given below.

$h = 2, a = 5, b = 1$ 6
$h = 3, a = 4, b = 6$ 15
$h = 5, a = 3, b = 1$ 10
$h = 1, a = 2, b = 1$ $1\frac{1}{2}$

Discussion Question

Are there any values for x and y so that $x + y = \frac{1}{x} + \frac{1}{y}$?

Yes, if x and y are reciprocals.

Chalkboard Exercises

R_1 and R_2 are connected in series. Find R_T.

1a. $R_1 = 5$ ohms, $R_2 = 11$ ohms 16 ohms
1b. $R_1 = 8.2$ ohms, $R_2 = 4.6$ ohms 12.8 ohms
1c. $R_1 = 3.7$ ohms, $R_2 = 4.9$ ohms 8.6 ohms

❶ Point out that when more than two resistances are connected in series, the total resistance is the sum of the individual resistances.

$R_T = R_1 + R_2 + R_3 + \ldots$

When more than two resistors are connected in parallel, the following formula is used for the total resistance.

$$\frac{1}{R_T} = \frac{1}{R_1} + \frac{1}{R_2} + \frac{1}{R_3} + \ldots$$

Objective 15-7: Calculate resistances of circuits by applying knowledge of rational expressions.

15-7 Rational Expressions in Science

Many scientific formulas, such as the ones for electrical resistance, contain rational expressions.

Electricity can be described as the flow of electrons through a conductor, such as a copper wire. Electricity flows more freely through some conductors than others. The force opposing the flow is called *resistance*. The unit of resistance commonly used is the *ohm*.

Resistances can occur one after another, in *series*. They can also occur in branches of the conductor going in the same direction, in *parallel*. Study the diagrams below. Formulas for the total resistance, R_T, are given under the diagrams.

Series

$R_T = R_1 + R_2$

Parallel

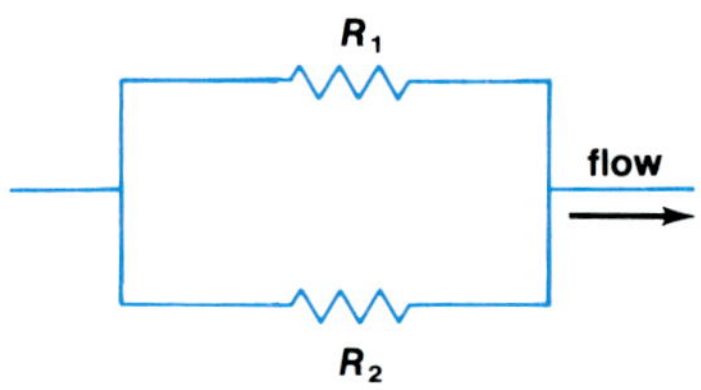

$$\frac{1}{R_T} = \frac{1}{R_1} + \frac{1}{R_2}$$

❶

Example

1 **Assume that $R_1 = 4$ ohms and $R_2 = 3$ ohms. Compute the total resistance of the conductor when the resistances are in series.**

$R_T = R_1 + R_2$
$R_T = 4 + 3$
$R_T = 7$

Thus, the total resistance is 7 ohms.

Example 2

Assume that R_1 = 5 ohms and R_2 = 6 ohms. Compute the total resistance of the conductor when the resistances are in parallel.

$$\frac{1}{R_T} = \frac{1}{R_1} + \frac{1}{R_2}$$

$$\frac{1}{R_T} = \frac{1}{5} + \frac{1}{6}$$

$$\frac{1}{R_T} = \frac{11}{30}$$

$$30 \cdot 1 = R_T \cdot 11$$ *Find the cross products.*

$$\frac{30}{11} = R_T$$

Thus, the total resistance is $\frac{30}{11}$, or $2\frac{8}{11}$ ohms.

A *circuit,* or path for the flow of electrons, often has some resistances connected in series and others in parallel.

Example 3

A parallel circuit has one branch in series as shown at the right. Given that the total resistance is $2\frac{1}{4}$ ohms, R_1 = 3 ohms, and R_2 = 4 ohms, find R_3.

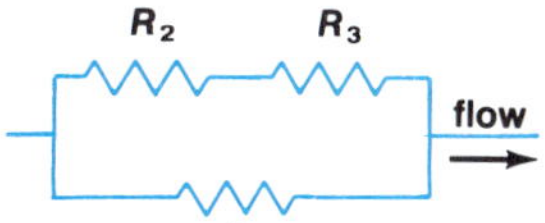

$$\frac{1}{R_T} = \frac{1}{R_1} + \frac{1}{R_2 + R_3}$$ *The total resistance of the branch in series is $R_2 + R_3$.*

$$\frac{1}{2\frac{1}{4}} = \frac{1}{3} + \frac{1}{4 + R_3}$$

$$\frac{4}{9} = \frac{1}{3} + \frac{1}{4 + R_3}$$ *Note that $\frac{1}{2\frac{1}{4}} = \frac{1}{\frac{9}{4}} = 1 \div \frac{9}{4} = 1 \cdot \frac{4}{9} = \frac{4}{9}$.*

$$\frac{4}{9} - \frac{1}{3} = \frac{1}{4 + R_3}$$

$$\frac{1}{9} = \frac{1}{4 + R_3}$$

$$4 + R_3 = 9$$ *Find the cross products.*

$$R_3 = 5$$

Thus, the third resistance is 5 ohms.

Chalkboard Exercises

R_1 and R_2 are connected in parallel.

2a. Find R_T if R_1 = 6 ohms and R_2 = 9 ohms. $\frac{18}{5}$ or $3\frac{3}{5}$ ohms

2b. Find R_T if R_1 = 4 ohms and R_2 = 10 ohms. $\frac{20}{7}$ or $2\frac{6}{7}$ ohms

2c. Find R_1 if R_T = 4 ohms and R_2 = 8 ohms. 8 ohms

2d. Find R_2 if R_T = 3 ohms and R_1 = 4 ohms. 12 ohms

Solve each problem for the circuit shown in Example 3.

3a. Find R_3 if $R_T = \frac{14}{3}$ ohms, R_1 = 6 ohms, and R_2 = 9 ohms. 12 ohms

3b. Find R_1 if R_T = 10 ohms, R_2 = 12 ohms, and R_3 = 8 ohms. 20 ohms

Suggested Daily Assignments

Enriched
Written: 1–17

Average
Written: 1–15

Fundamental
Written: 1–12
p. 486, 7–8

Written Exercises

Problems 1-6 relate to the diagram below. Solve each problem.

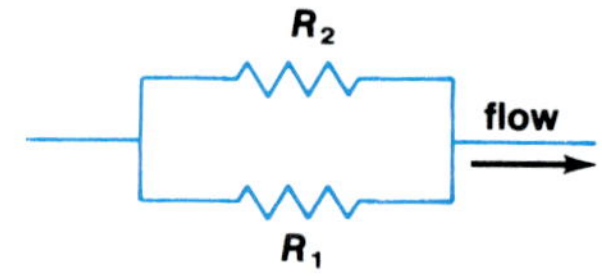

1. Find the total resistance, R_T, given that $R_1 = 8$ ohms and $R_2 = 6$ ohms. $3\frac{3}{7}$ ohms
2. Find the total resistance, R_T, given that $R_1 = 4.5$ ohms and $R_2 = 3.5$ ohms. $1\frac{31}{32}$ ohms
3. Find R_1, given that total resistance is $2\frac{2}{9}$ ohms and $R_2 = 5$ ohms. 4 ohms
4. Find R_1, given that total resistance is $3\frac{3}{7}$ ohms and $R_2 = 8$ ohms. 6 ohms
5. Find R_1 and R_2, given that total resistance is $2\frac{2}{3}$ ohms and R_1 is two times as great as R_2. 8 ohms, 4 ohms
6. Find R_1 and R_2, given that total resistance is $2\frac{1}{4}$ ohms and R_1 is three times as great as R_2. 9 ohms, 3 ohms

Solve each problem.

7. Resistances of 3 ohms, 6 ohms, and 9 ohms are connected in series. What is the total resistance? 18 ohms
8. Eight lights on a Christmas tree are connected in series. Each has a resistance of 12 ohms. What is the total resistance? 96 ohms
9. Three coils with resistances of 3 ohms, 4 ohms, and 6 ohms are connected in parallel. What is the total resistance? $1\frac{1}{3}$ ohms
10. Three coils with resistances of 4 ohms, 6 ohms, and 15 ohms are connected in parallel. What is the total resistance? $2\frac{2}{29}$
11. Three appliances are connected in parallel: a lamp of resistance 120 ohms, a toaster of resistance 20 ohms, and an iron of resistance 12 ohms. Find the total resistance. $7\frac{1}{17}$ ohms
12. Three appliances are connected in parallel: a lamp of resistance 60 ohms, an iron of resistance 20 ohms, and a heating coil of resistance 80 ohms. Find the total resistance. $12\frac{12}{19}$ ohms

Problems 13-15 relate to the diagram below. Solve each problem.

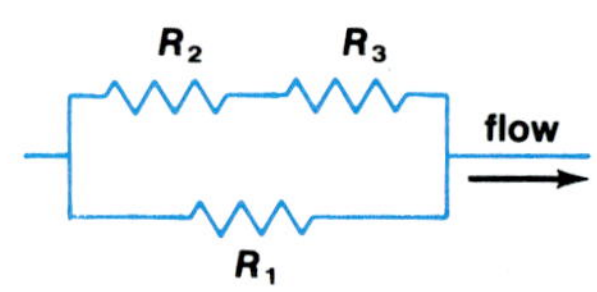

13. Find the total resistance, given that $R_1 =$ 5 ohms, $R_2 = 4$ ohms, and $R_3 = 3$ ohms. $2\frac{11}{12}$ ohms
14. Find R_1, given that the total resistance is $2\frac{10}{13}$ ohms, $R_2 = 3$ ohms, and $R_3 = 6$ ohms. 4 ohms
15. Find R_2, given that the total resistance is $3\frac{1}{2}$ ohms, $R_1 = 5$ ohms, and $R_3 = 4$ ohms. $7\frac{2}{3}$ ohms

Challenge

Solve each problem.

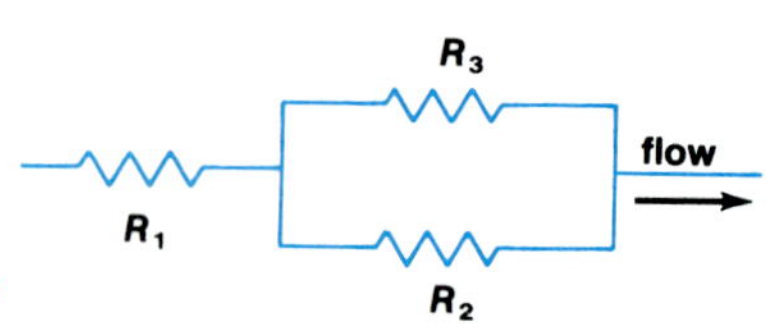

16. Write an expression for the total resistance for the diagram at the right.
17. Find the total resistance, given that $R_1 = 5$ ohms, $R_2 = 4$ ohms, and $R_3 = 6$ ohms. $7\frac{2}{5}$ ohms

16. $R_T = R_1 + \dfrac{R_2R_3}{R_3 + R_2}$

Objective 15-8: Solve problems involving joint variations, quadratic direct variations, or combined variations.

15-8 Other Variations

Consider the formula for the area of a triangle, $A = \frac{1}{2}bh$. Notice that the area depends directly on *both* the base and the height. We say that the area *varies jointly* as the base and the height. Such a variation is known as a **joint variation**.

Definition of Joint Variation

A joint variation is described by an equation of the form $z = kxy$, where k is not zero.

A joint variation can also be expressed as a proportion. One such proportion that can be derived from the equation $z = kxy$ is shown below.

$$\frac{z_1}{x_1 y_1} = \frac{z_2}{x_2 y_2}$$

Example 1

Assume z varies jointly as x and y, and $z = 42$ when $x = 18$ and $y = 7$. Find z when $x = \frac{3}{4}$ and $y = 8$.

Use the proportion $\frac{z_1}{x_1 y_1} = \frac{z_2}{x_2 y_2}$ to solve the problem. Let $z_1 = 42$, $x_1 = 18$, $y_1 = 7$, $x_2 = \frac{3}{4}$, and $y_2 = 8$.

$$\frac{42}{18 \cdot 7} = \frac{z_2}{\frac{3}{4} \cdot 8} \quad \textit{Substitute.}$$

$$\frac{42}{126} = \frac{z_2}{6} \quad \textit{Find the cross products.}$$

$$252 = 126z_2$$

$$2 = z_2$$

Thus, $z = 2$ when $x = \frac{3}{4}$ and $y = 8$.

Consider the formula for the area of a circle, $A = \pi r^2$. Notice that the area depends directly on the *square* of the radius. We say that the area varies directly as the square of the radius. Such a variation is known as a **quadratic direct variation**.

15-8

Lesson Notes, page T54

Readiness Activity
Solve each problem. If y varies directly as x and $y = 3$ when $x = 2$, find x when $y = 12$. 8
If y varies inversely as x and $y = 6$ when $x = 14$, find x when $y = -1$. -84

Discussion Questions
(Use after Example 3.)

1. Is it true that rate and time vary jointly as distance? Yes, $d = rt$ where $k = 1$.
2. Is it true that work varies directly as time and inversely as the rate? No, $w = rt$ is a joint variation.

Chalkboard Exercises
Solve each problem. Assume that z varies jointly as x and y.

1a. If $z = 6$ when $x = 10$ and $y = 1$, find x when $z = 12$ and $y = 2$. 10

1b. If $z = 9$ when $x = 2$ and $y = 15$, find y when $x = \frac{2}{3}$ and $z = 3$. 15

Chalkboard Exercises

Solve each problem. Assume that y varies directly as the square of x.

2a. If $y = 12$ when $x = 3$, find y when $x = 2$. $5\frac{1}{3}$

2b. If $y = 2$ when $x = 4$, find x when $y = 18$. 12

❶ Combined variation involves both direct and indirect variation.

A quadratic direct variation is described by an equation of the form $y = kx^2$, where k is not zero.

Definition of Quadratic Direct Variation

A quadratic direct variation can also be expressed as a proportion of the form $\frac{y_1}{{x_1}^2} = \frac{y_2}{{x_2}^2}$.

Example

2 **If y varies directly as x^2 and $y = 10$ when $x = 2$, find y when $x = 5$.**

Use the proportion $\frac{y_1}{{x_1}^2} = \frac{y_2}{{x_2}^2}$ to solve the problem. Let $y_1 = 10$, $x_1 = 2$, and $x_2 = 5$.

$\frac{10}{2^2} = \frac{y_2}{5^2}$ *Substitute.*

$250 = 4y_2$ *Find the cross products.*

$62\frac{1}{2} = y_2$

Thus, $y = 62\frac{1}{2}$ when $x = 5$.

When a slide projector illuminates a screen, the brightness of the picture depends on the intensity of the bulb used in the projector. However, as the screen is moved farther away from the projector, the same amount of light must cover a larger area on the screen. As a result, the picture becomes less bright. The brightness of the picture varies directly as the intensity of the bulb and inversely as the square of the distance from the bulb. Such a variation is known as a **combined variation**.

❶

A combined variation is described by an equation of the form $z = \frac{kx}{y}$, where k is not zero.

Definition of Combined Variation

A combined variation can also be expressed as a proportion of the form $\frac{z_1 y_1}{x_1} = \frac{z_2 y_2}{x_2}$.

Example 3

If z varies directly as x and inversely as the square of y, and $z = 1.8$ when $x = 100$ and $y = 0.4$, find z when $x = 150$ and $y = 0.3$.

Use the proportion $\frac{z_1(y_1)^2}{x_1} = \frac{z_2(y_2)^2}{x_2}$ to solve the problem.

Let $z_1 = 1.8$, $x_1 = 100$, $y_1 = 0.4$, $x_2 = 150$, and $y_2 = 0.3$.

$$\frac{1.8(0.4)^2}{100} = \frac{z_2(0.3)^2}{150} \quad \textit{Substitute.}$$

$$\frac{0.288}{100} = \frac{0.09z_2}{150} \quad \textit{Find the cross products.}$$

$$43.2 = 9z_2$$

$$4.8 = z_2$$

Thus, $z = 4.8$ when $x = 150$ and $y = 0.3$.

Exploratory Exercises

State whether each of the following is joint variation, a quadratic direct variation, or a combined variation.

1. $a = 4bc$ joint
2. $0.2x = yz$ joint
3. $r = \frac{4s}{t}$ combined
4. $q = \frac{2}{3}mn$ joint
5. $w = 8z^2$ quadratic
6. $\frac{2}{3}a^2 = b$ quadratic
7. $xy = \frac{8}{9}w$ joint or combined
8. $\frac{0.46c}{d} = a$ combined or combined

Written Exercises

Solve each problem. Assume that z varies jointly as x and y.

1. If $z = 4$ when $x = 2$ and $y = 4$, find z when $x = 1$ and $y = 4$. 2
2. If $z = -6$ when $x = -1$ and $y = -2$, find x when $z = -3$ and $y = 2$. $\frac{1}{2}$
3. If $z = 3$ when $x = 15$ and $y = 10$, find y when $x = -25$ and $z = 5$. -10
4. If $z = 18$ when $x = 22$ and $y = 12$, find y when $x = 10$ and $z = -3$. $-4\frac{2}{5}$

Solve each problem. Assume that y varies directly as the square of x.

5. If $y = 2$ when $x = 2$, find y when $x = 4$. 8
6. If $y = -8$ when $x = -4$, find y when $x = 6$. -18
7. If $y = 0.1$ when $x = 0.3$, find x when $y = 0.009$. ± 0.09
8. If $y = -2$ when $x = -3$, find x when $y = -4$. $\pm 3\sqrt{2}$

Solve each problem. Assume that z varies directly as x and inversely as y.

9. If $z = 4$ when $x = 1$ and $y = 2$, find z when $x = 10$ and $y = 20$. 4
10. If $z = -5$ when $x = -12$ and $y = -2$, find x when $z = -4$ and $y = -1$. -4.8
11. If $z = 14$ when $x = -2$ and $y = 14$, find y when $z = -15$ and $x = 6$. 39.2
12. If $z = 0.6$ when $x = 2$ and $y = 0.4$, find z when $x = 9$ and $y = 0.1$. 10.8

The power (P) of an electrical circuit varies directly as the square of the voltage (V) and inversely as the resistance (R). Use this information to solve each problem.

13. If $P = 8$ when $V = 12$ and $R = 6$, find R when $P = 6$ and $V = 15$. 12.5
14. If $P = 2$ when $V = 8$ and $R = 4$, find R when $P = 5$ and $V = 10$. 2.5

Chalkboard Exercises

Solve each problem. Assume that z varies directly as x and inversely as y.

3a. If $z = -1$ when $x = 4$ and $y = 15$, find x when $z = -2$ and $y = 5$. $2\frac{2}{3}$

3b. If $z = 0.2$ when $x = 4$ and $y = 10$, find y when $z = 0.5$ and $x = 4$. 4

Suggested Daily Assignments

Enriched
- Exp: 1–8
- Written: 1–14

Average
- Exp: 1–8
- Written: 1–14

Fundamental
- Exp: 1–8
- Written: 1–14
  - p. 490, 13–15

About Using Computers
Explain to students that data such as 999999 are often called "dummy data" because they are not used in computations. Emphasize that values used for dummy data are unreasonable real data. Point out that dummy data are often used in READ-DATA statements to avoid an OUT OF DATA error message. You may want to have students write a program using READ-DATA statements and dummy data.

Using Computers

Variations

A computer can be used to solve problems involving variations. The following program is for problems involving combined variations. It uses the proportion $\frac{z_1y_1}{x_1} = \frac{z_2y_2}{x_2}$.

```
10   PRINT "ENTER VALUES FOR Z1,Y1,X1,Z2,Y2,AND X2."
20   PRINT "(ENTER 999999 FOR THE VARIABLE WHOSE VALUE
       IS UNKNOWN)"
30   INPUT Z1,Y1,X1,Z2,Y2,X2
40   IF Z2 = 999999 THEN 90
50   IF Y2 = 999999 THEN 120
60   LET X2 = (Z2 * Y2 * X1) / (Z1 * Y1)
70   PRINT "X = ";X2;" WHEN Z = ";Z2;" AND Y = ";Y2
80   GOTO 140
90   LET Z2 = (Z1 * Y1 * X2) / (Y2 * X1)
100  PRINT "Z = ";Z2;" WHEN X = ";X2;" AND Y = ";Y2
110  GOTO 140
120  LET Y2 = (Z1 * Y1 * X2) / (Z2 * X1)
130  PRINT "Y = ";Y2;" WHEN X = ";X2;" AND Z = ";Z2
140  END
```

Line 60 is executed when X2 = 999999.

Notice that the number 999999 is not used in any computations in the program. This number tells the computer which formula to use to solve the problem. For example, in line 40, if $z_2 = 999999$ then the computer goes to line 90. In line 90, the proportion $\frac{z_1y_1}{x_1} = \frac{z_2y_2}{x_2}$ has been solved for z_2, the variable whose value is unknown.

Exercises

1. Use the program above to verify your answers to Written Exercises 9-12 on page 493.
2. Write a program similar to the one above to solve problems involving joint variations. Use the program to verify your answers to Written Exercises 1-4 on page 493.
3. Write a program similar to the one above to solve problems involving quadratic direct variations. Use the program to verify your answers to Written Exercises 5-8 on page 493.

1. See student's work.
2. Change lines 60, 90, and 120 as shown below.

```
60   LET X2 = (Z2*X1*Y1)/(Z1*Y2)
90   LET Z2 = (Z1*X2*Y2)/(X1*Y1)
120  LET Y2 = (Z2*X1*Y1)/(Z1*X2)
```

See students' work.

Answers to Exercises

3.

```
10  PRINT "ENTER VALUES FOR
      Y1, X1, Y2, AND X2."
20  PRINT "(ENTER 999999 FOR
      THE VARIABLE WHOSE
      VALUE IS UNKNOWN.)"
30  INPUT Y1, X1, Y2, X2
40  IF Y2 = 999999 THEN 80
50  LET X2 = SQR((Y2 *
      X1 ↑ 2)/Y1)
60  PRINT "X = ";X2;"WHEN Y =
      ";Y2
70  GOTO 100
80  LET Y2 = (Y1 * X2 ↑ 2)/X1 ↑ 2
90  PRINT "Y = ";Y2;"WHEN X =
      ";X2
100 END
```

Vocabulary

direct variation (478)
constant of variation (478)
inverse variation (481)
property of levers (484)
joint variation (491)
quadratic direct variation (491)
combined variation (492)

Chapter Summary

1. Equations containing rational expressions can be solved by first multiplying every term by the least common denominator. (469)
2. The formula $rt = w$ is used to solve work problems. (472)
3. Definition of Direct Variation: A direct variation is described by an equation of the form $y = kx$, where k is not zero. (478)
4. Direct variation can be expressed in terms of two proportions. Let (x_1, y_1) and (x_2, y_2) be solutions of a direct variation. Then $\frac{x_1}{y_1} = \frac{x_2}{y_2}$ and $\frac{x_1}{x_2} = \frac{y_1}{y_2}$. The cross product can also be used: $x_1y_2 = x_2y_1$. (479)
5. Definition of Inverse Variation: An inverse variation is described by an equation of the form $xy = k$, where k is not zero. (481)
6. Inverse variation can be expressed in terms of a proportion. Let (x_1, y_1) and (x_2, y_2) be solutions of an inverse variation. Then $\frac{x_1}{x_2} = \frac{y_2}{y_1}$. The product rule for inverse variation is: $x_1y_1 = x_2y_2$. (482)
7. Property of Levers: Suppose weights w_1 and w_2 are placed on a lever at distances d_1 and d_2, respectively, from the fulcrum. The lever is balanced when $w_1d_1 = w_2d_2$. (484)
8. Definition of Joint Variation: A joint variation is described by an equation of the form $z = kxy$, where k is not zero. (492)
9. Definition of Quadratic Direct Variation: A quadratic direct variation is described by an equation of the form $y = kx^2$, where k is not zero. (492)
10. Definition of Combined Variation: A combined variation is described by an equation of the form $z = \frac{kx}{y}$, where k is not zero. (492)

Quiz
Lessons 15-4 to 15-8
Solve each problem.

1. If $x = 5$ when $y = 8$, find y when $x = 2.5$. Assume that y varies inversely as x. 16
2. If $y = \frac{1}{2}$ when $x = 3$, find y when $x = \frac{1}{6}$. Assume that y varies inversely as x. 9
3. Weights of 60 pounds and 40 pounds are placed on a lever. The weights are 50 inches apart, and the lever is in balance. How far from the fulcrum is the 60-pound weight? 20 inches
4. Resistances of 5 ohms, 8 ohms, and 6 ohms are connected in series. Find the total resistance. 19 ohms
5. Resistances of 4 ohms and 8 ohms are connected in parallel. Find the total resistance. $2\frac{2}{3}$ ohms
6. If $z = 7$ when $y = 2$ and $x = 3$ find z when $x = 5$ and $y = 1.2$. Assume that z varies jointly as x and y. 7

Chapter Review

15-1 Solve each equation.

1. $\frac{4x}{3} + \frac{7}{2} = \frac{7x}{12}$ $-\frac{14}{3}$

2. $\frac{y+2}{y} - \frac{y-3}{y} = \frac{3}{4}$ $\frac{20}{3}$

3. $\frac{3}{x} + \frac{1}{x-5} = \frac{1}{2x}$ $\frac{25}{7}$

4. $\frac{1}{h+1} + \frac{2}{3} = \frac{2h+5}{h-1}$ $-2, -\frac{5}{2}$

5. $\frac{3x+2}{x^2+7x+6} = \frac{1}{x+6} + \frac{4}{x+1}$ $-\frac{23}{2}$

6. $\frac{3m-2}{2m^2-5m-3} - \frac{2}{2m+1} = \frac{4}{m-3}$ 0

15-2 Solve each problem.

7. Bill can paint the outside of a house in 40 hours. Roberta can do the job in 48 hours. If they work together, how long will it take them to complete the job? $21\frac{9}{11}$ hours

8. An airplane can fly at a rate of 525 mph in calm air. It can fly 1605 miles with the wind in the same amount of time it can fly 1545 miles against the wind. Find the speed of the wind. 10 mph

15-3 Solve each formula for the variable indicated.

9. $\frac{1}{f} = \frac{1}{a} + \frac{1}{b}$, for a $a = \frac{bf}{b-f}$

10. $S = \frac{a}{1-r}$, for r $r = \frac{S-a}{S}$

15-4 Solve each problem.

11. If y varies directly as x and $y = 9$ when $x = 3$, find y when $x = 4$. 12

12. If y varies directly as x and $y = 12.5$ when $x = 5$, find y when $x = 6$. 15

13. Bill uses 16 gallons of gasoline every 3 days while driving to work. How much gasoline will he use in 14 days of driving? $74\frac{2}{3}$ gallons

15-5 Solve each problem.

14. If y varies inversely as x and $y = 2$ when $x = 5$, find y when $x = 20$. $\frac{1}{2}$

15. If y varies inversely as x and $y = 6.9$ when $x = 1.7$, find y when $x = 5.1$. 2.3

16. If the volume of a gas is inversely proportional to applied pressure, and the volume is 80 m^3 at 2 atmospheres of pressure, what will be the volume at $\frac{1}{4}$ atmosphere of pressure? 640 m^3

15-6 Solve each problem.

17. A meter stick rests on a fulcrum at the 50 centimeter mark. A 10-gram weight is placed on the meter stick 25 cm from the fulcrum. Where should a 5-gram weight be placed in order to balance the stick? On the other end of the meter stick, 50 cm from the fulcrum.

18. A fulcrum is placed 1 foot from the end of an 8-foot lever. Will a 200-pound man on the other end of the lever be able to lift a 1500-pound stone? no

15-7 Solve each problem.

19. Assume that $R_1 = 4$ ohms and $R_2 = 6$ ohms. What is the total resistance of the conductor if R_1 and R_2 are: **a.** connected in series? **b.** connected in parallel? 10 ohms, $2\frac{2}{5}$ ohms

20. A parallel circuit has one branch in series. The total resistance is expressed by the formula: $\frac{1}{R_T} = \frac{1}{R_1} + \frac{1}{R_2 + R_3}$. Find R_1 if $R_T = 3$ ohms, $R_2 = 3$ ohms, and $R_3 = 9$ ohms. 4 ohms

15-8 **Solve each problem.**

21. If z varies jointly as x and y, and $z = 12$ when $x = 4$ and $y = 6$, find z when $x = 3$ and $y = 2$. 3

22. If y varies directly as x^2 and $y = 30$ when $x = 5$, find y when $x = 10$. 120

23. If z varies directly as x and inversely as y, and $z = 0.3$ when $x = 10$ and $y = 0.2$, find z when $x = 20$ and $y = 0.6$. 0.2

Chapter Test

Solve each equation.

1. $\frac{y+3}{6} = \frac{y+2}{12} - \frac{2}{5}$ $-\frac{44}{5}$

2. $\frac{x+1}{x} + \frac{6}{x} = x + 7$ 1, −7

3. $\frac{4m}{m-3} + \frac{6}{3-m} = m$ 6, 1

4. $\frac{-2b-9}{b^2+7b+12} = \frac{b}{b+3} + \frac{2}{b+4}$ −5

5. $\frac{1}{y-4} - \frac{2}{y-8} = \frac{-1}{y+6}$ $\frac{16}{9}$

6. $\frac{m+3}{m-1} + \frac{m+1}{m-3} = \frac{22}{3}$ $\frac{3}{2}$, 4

Solve each formula for the variable indicated.

7. $F = G\left(\frac{Mm}{d^2}\right)$, for G $G = \frac{Fd^2}{Mm}$

8. $\frac{1}{R_T} = \frac{1}{R_1} + \frac{1}{R_2}$, for R_2 $R_2 = \frac{R_1R_T}{R_1 - R_T}$

Solve each problem.

9. Willie can do a job in 6 days. Myra can do the same job in $4\frac{1}{2}$ days. If they work together how long will it take to complete the job? $2\frac{4}{7}$ days

10. The top speed of a boat in still water is 5 mph. At this speed, a 21-mile trip downstream took the same amount of time as a 9-mile trip upstream. Find the rate of the current. 2 mph

11. If y varies directly as x and $y = 7$ when $x = 3$, find y when $x = 5$. $11\frac{2}{3}$

12. If y varies inversely as x and $y = 4$ when $x = 9$, find y when $x = 5$. $7\frac{1}{5}$

13. Pressure varies inversely with volume of a confined gas. If the volume of a gas is 400 cm^3 at a pressure of 4 atmospheres, what is the pressure when the volume is 300 cm^3? $5\frac{1}{3}$ atm

14. Jimmie and Anne Marie are on a seesaw. Jimmie weighs 60 pounds and is seated 8 feet from the center. How far from the center should Anne Marie sit if she weighs 75 pounds? $6\frac{2}{5}$ ft

15. Given the formula $\frac{1}{R_T} = \frac{1}{R_1} + \frac{1}{R_2}$ for the resistance of a circuit, determine the resistance R_1 if $R_T = 1.2$ ohms and $R_2 = 4$ ohms. $1\frac{5}{7}$ ohms

16. If z varies directly as x and inversely as y, and $z = -6$ when $x = -1$ and $y = 2$, find z when $x = 4$ and $y = -3$. −16

About the Photo

The most popular billiards game is pocket billiards, or pool. The game begins when fifteen numbered balls are arranged in a triangular formation near the foot of a rectangular table. The object of the game is to drive the balls into the six pockets of the table. The first player stands at the head of the table and uses a cue (long stick) to drive the cue ball into the formation to break it apart. The photograph is a timed exposure showing the 15 balls before and after the break shot.

An understanding of angles is important in playing billiards. Suppose a ball hits a point on the side of the table. Imagine a line drawn through the point and perpendicular to the side. The angle formed by the path of the ball and this line is the angle of incidence. The corresponding angle made by the rebounding ball is the angle of reflection. These two angles always have equal measures.

Teacher Resource Book Materials for Chapter 16 See pages 359-374.

Chapter Test, Form I (Multiple Choice)	359-361
Chapter Test, Form II (Free Response)	362-363
Quizzes	364
Extra Practice	365-366
Application: Good Tennis Serves (Use after 16-4.)	367
Application: Solving a Mystery (Use after 16-7.)	368
Extension: The Law of Sines (Use after 16-6.)	369
Problem Solving	370
Cumulative Review	371
Answers	372-374

CHAPTER 16

Trigonometry

In playing pool, similar triangles are used when banking a shot. Pool is one of many games that involve triangles. Trigonometry is a branch of mathematics that involves triangles.

Objective 16-1A: Find the complement and supplement of an angle.
Objective 16-1B: Find the measure of the third angle of a triangle given the measures of the other two angles.

16-1 Angles

A protractor can be used to measure angles as shown below.

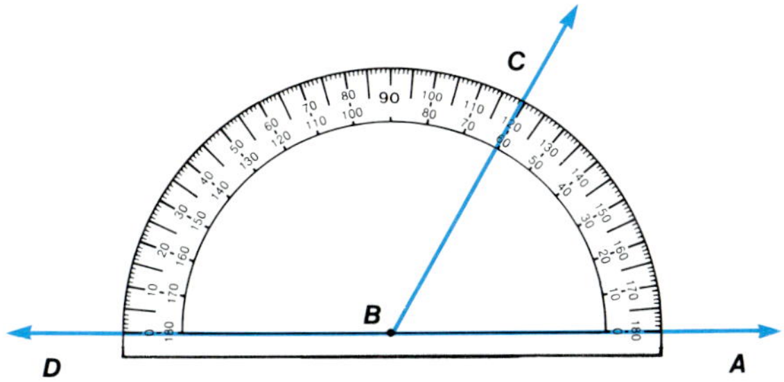

❶

Angle ABC measures 60°. However, where ray BC cuts the protractor, there are two readings—60° and 120°. The 120° is the measure of angle DBC. What is the sum of the measures of angles ABC and DBC?

Two angles are supplementary if the sum of their measures is 180°.

Supplementary Angles

Example 1

The measure of an angle is three times its supplement. Find the measure of each angle.

$$x + 3x = 180$$
$$4x = 180$$
$$x = 45$$

Let x be the measure of the lesser angle and 3x the measure of the greater angle.

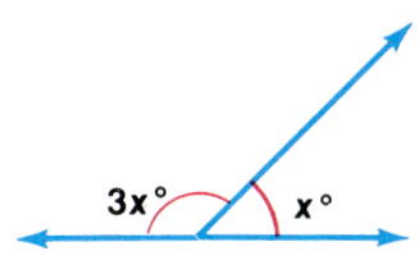

The measures are 45° and 3 · 45°, or 135°.

Two angles are complementary if the sum of their measures is 90°.

Complementary Angles

Example 2

An angle is 16° greater than its complement. Find the measure of each angle.

$$x + (x + 16) = 90$$
$$2x + 16 = 90$$
$$2x = 74$$
$$x = 37$$

Let x be the measure of the lesser angle and x + 16 the measure of the greater angle.

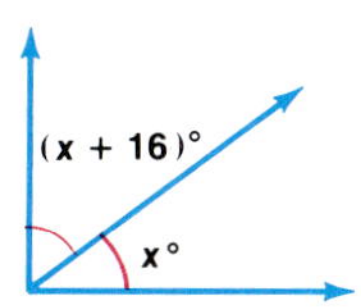

The measures are 37° and 37° + 16°, or 53°.

16-1

Lesson Notes, page T54
Enrichment, page T54

Readiness Activity
Name each of the following angles in two ways.

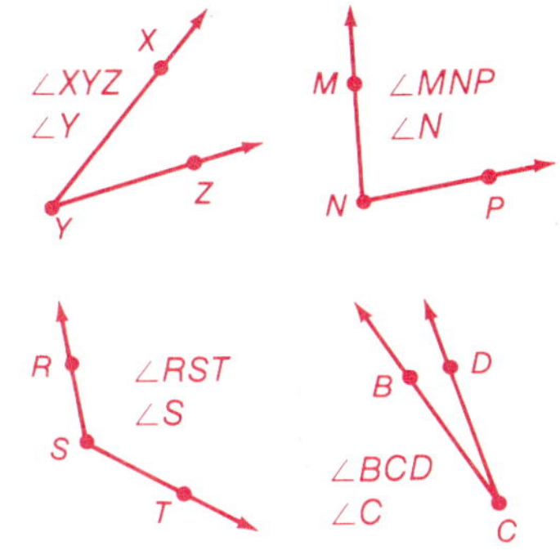

Discussion Question
(Use after Example 4.)
Can a triangle contain two right angles? Why or why not?
No, because if a triangle did contain two right angles, the sum of their measures would be 180°. Thus the measure of the third angle would have to be 0° and this is impossible.

Chalkboard Exercises
Solve each of the following.

1. The measure of an angle is 20° less than 3 times its supplement. Find the measure of each angle.
130°, 50°
2. An angle measures 42° less than its complement. Find the measure of each angle.
24°, 66°

❶ When discussing the introduction to this lesson, point out to students that angle ABC is the supplement of angle DBC and vice versa.

Chalkboard Exercises

Solve each of the following.

3a. Find the measure of the third angle of a triangle if the other two angles measure 32° and 82°. 66°

3b. Find the measure of the third angle of a triangle if the other two angles measure 15° and 113°. 52°

4a. Find the measure of the third angle of a right triangle if the measure of one acute angle is 30°. 60°

4b. Find the measure of the third angle of a right triangle if the measure of one acute angle is 18°. 72°

❷ The following activity can be used to reinforce the concept of the sum of angle measures of a triangle. Have each student make a paper triangle and tear off the corners. If the corners are placed side-by-side, touching at the vertices, they will form a 180° angle (straight angle), as shown below.

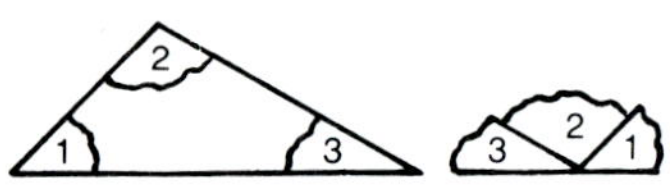

What is the sum of the measures of the three angles of a triangle? Use a protractor to measure the angles of each triangle below. Then find the sum of the angle measures of each triangle.

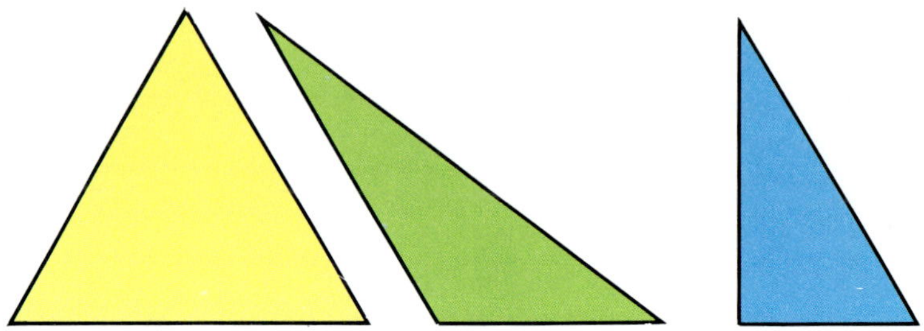

What did you discover? In each case, your sum should be approximately 180°.

❷

The sum of the measures of the angles in any triangle is 180°.

Sum of the Angles of a Triangle

Examples

3 **What are the measures of the angles of an equilateral triangle?**

In an equilateral triangle, the sides are congruent and the angles have equal measures.

$$x + x + x = 180 \quad \text{Let } x \text{ be the measure of each angle.}$$
$$3x = 180$$
$$x = 60$$

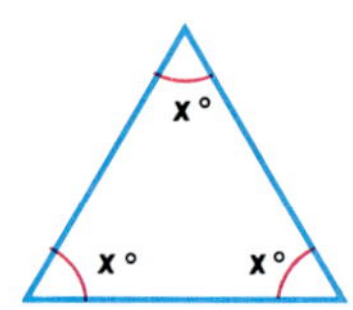

Each angle measures 60°.

4 **What are the measures of the angles of an isosceles right triangle?**

An isosceles right triangle contains a right angle and two congruent angles.

$$x + x + 90 = 180$$
$$2x + 90 = 180$$
$$2x = 90$$
$$x = 45$$

What does x represent?
How do you know one angle is 90°?

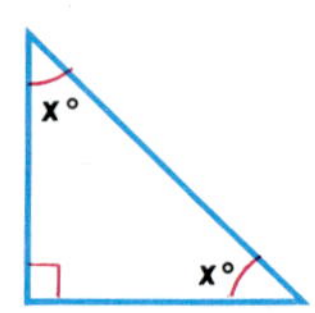

The measures are 45°, 45°, and 90°.

Exploratory Exercises

Find the complement of each angle whose measure is listed below.

1. 85° 5° **2.** 42° 48° **3.** 13° 77° **4.** 45° 45°

5. 55° 35° **6.** $x°$ $(90 - x)°$ **7.** $(2x + 40)°$ $(50 - 2x)°$ **8.** $(x - 7)°$ $(97 - x)°$

Find the supplement of each angle whose measure is listed below.

9. 130° 50° **10.** 65° 115° **11.** 127° 53° **12.** 87° 93°

13. 90° 90° **14.** $y°$ $(180 - y)°$ **15.** $(3x + 5)°$ $(175 - 3x)°$ **16.** $(x - 20)°$ $(200 - x)°$

Written Exercises

Find both the complement and the supplement for each angle whose measure is listed below.

1. 42° 48°, 138° **2.** 87° 3°, 93° **3.** 125° none, 55° **4.** 160° none, 20°

5. 90° 0; 90° **6.** 68° 22°, 112° **7.** 21° 69°, 159° **8.** 174° none, 6°

9. 99° none, 81° **10.** $x°$ **11.** $3y°$ **12.** $(x + 30)°$

13. $(x - 38)°$ $(128 - x)°, (218 - x)°$ **14.** $5x°$ $(90 - 5x)°, (180 - 5x)°$ **15.** $(90 - x)°$ $x°, (90 + x)°$ **16.** $(180 - y)°$ $(y - 90)°, y°$

In each of the following, the measures of two angles of a triangle are given. Find the measure of the third angle.

17. 16°, 42° 122° **18.** 40°, 70° 70° **19.** 50°, 45° 85° **20.** 90°, 30° 60°

21. 89°, 90° 1° **22.** $x°, y°$ $(180 - x - y)°$ **23.** $x°, (x + 20)°$ $(160 - 2x)°$ **24.** $y°, (y - 10)°$ $(190 - 2y)°$

Solve each of the following.

25. One of the congruent angles of an isosceles triangle measures 37°. Find the measures of the other angles. 37°, 106°

26. The measures of the angles of a certain triangle are consecutive even integers. Find their measures. 58°, 60°, 62°

27. An angle measures 38° less than its complement. Find the measures of the two angles. 26°, 64°

28. One angle of a triangle measures 53°. Another angle has measure 37°. What is the measure of the third angle? 90°

29. One angle of a triangle measures 10° more than the second. The measure of the third angle is twice the sum of the first two angles. Find the measure of each angle.

30. One of two complementary angles measures 30° more than 3 times the other. Find the measure of each of the angles. 15°, 75°

31. Find the measure of an angle that measures 10° more than its complement. 50°

32. Find the measure of an angle that is 30° less than its supplement. 75°

33. Find the measure of an angle that is one-half the measure of its complement. 30°

34. Find the measure of an angle that is one-half the measure of its supplement. 60°

mini-review

Simplify.

1. $-\sqrt{81}$ -9 **2.** $5\sqrt{3} \cdot \sqrt{3}$ 15

3. $4\sqrt{27} + 6\sqrt{48}$ $36\sqrt{3}$ **4.** $\frac{\sqrt{35}}{\sqrt{7}}$ $\sqrt{5}$

Solve.

5. $\sqrt{x} = 5\sqrt{6}$ 150

Suggested Daily Assignments

Enriched
- Exp: 5–8, 13–16
- Written: 2–24 evens, 25–34
- Mini-Review: 1–5

Average
- Exp: 2–16 evens
- Written: 1–34
- Mini-Review: 1–5

Fundamental
- Exp: 1–16
- Written: 1–24, 26–34 evens
- Mini-Review: 1–5

Answers to Exploratory

10. $(90 - x)°, (180 - x)°$
11. $(90 - 3y)°, (180 - 3y)°$
12. $(60 - x)°, (150 - x)°$
29. 25°, 35°, 120°

501

16-2

Lesson Notes, page T55

Readiness Activity
Name all of the sides and angles in each of the following triangles

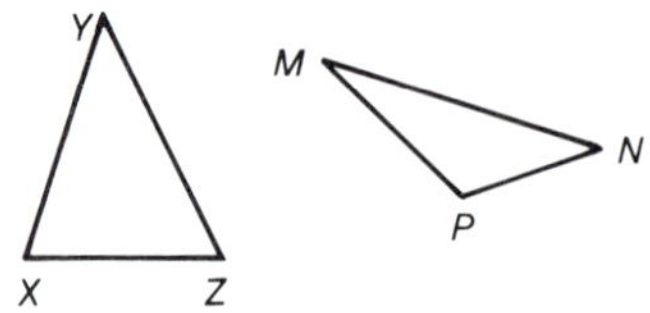

$\angle X$, $\angle Y$, $\angle Z$, $\overline{XY}$, $\overline{YZ}$, $\overline{XZ}$ $\angle M$, $\angle N$, $\angle P$, $\overline{MN}$, $\overline{NP}$, $\overline{PM}$

Discussion Question
Can the lengths of the sides of a 30°-60° right triangle all be whole numbers? Why or why not? No, because the measure of the side opposite the 60° angle is $\sqrt{3}$ times the measure of the leg adjacent to the 60° angle.

Chalkboard Exercises
Find the length of $\overline{BC}$ in triangle ABC.

1a.

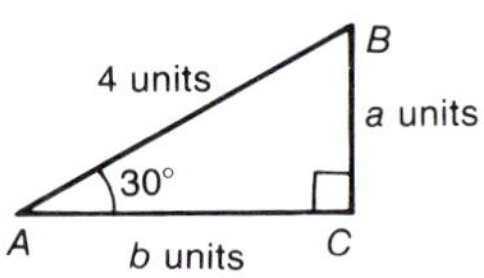

2 units

1b.

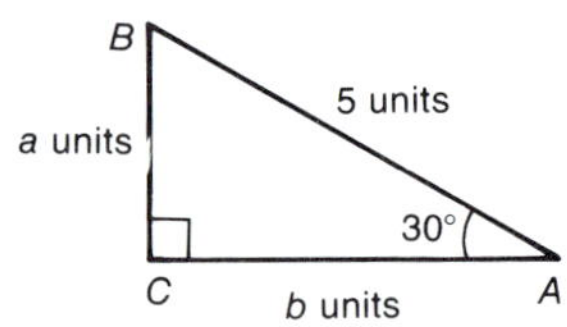

$2\frac{1}{2}$ units

❶ Emphasize that $\overline{AB}$ denotes line segment AB. Also point out that lower case letters are used to denote the measure of the side opposite a given angle.

❷ Review with students the meaning of *bisect*.

Objective 16-2: Find the measures of the sides of a 30°-60° right triangle given the measure of one side.

16-2 30° – 60° Right Triangles

Suppose the measure of one acute angle of a right triangle is 30°. The measure of the other acute angle is 90° – 30° or 60°. Such triangles are called **30° – 60° right triangles**.

Examine the equilateral triangle shown at the right. What is the measure of angle B?

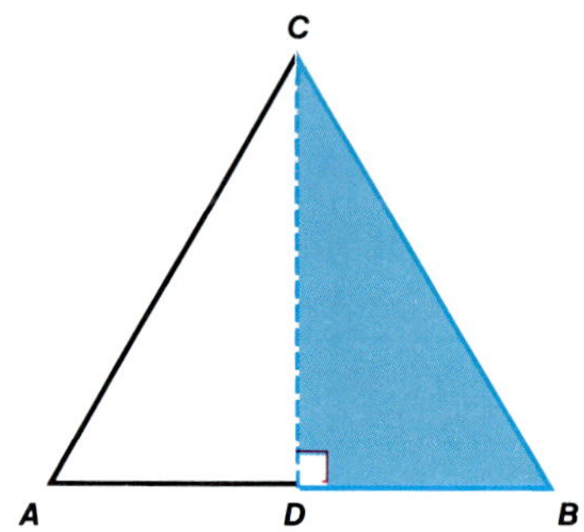

❶ Line segment CD (denoted $\overline{CD}$) can be drawn perpendicular to $\overline{AB}$. What kind of right triangle is triangle CDB?

❷ $\overline{CD}$ bisects $\overline{AB}$. Since the measure of $\overline{DB}$ is one-half the measure of $\overline{AB}$, it is also one-half the measure of the hypotenuse, $\overline{BC}$.

In any 30° – 60° right triangle, the measure of the side opposite the 30° angle is one-half the measure of the hypotenuse.

30° – 60° Right Triangles

In calculations, lower case letters are used to designate the measures of the sides of a triangle. For example, the measure of the side opposite angle R is r.

Example 1

Find the length of $\overline{BC}$ in triangle ABC.

$c = 2a$ *The measure of the hypotenuse is c.*
$8 = 2a$
$4 = a$

Thus, $\overline{BC}$ is 4 units long.

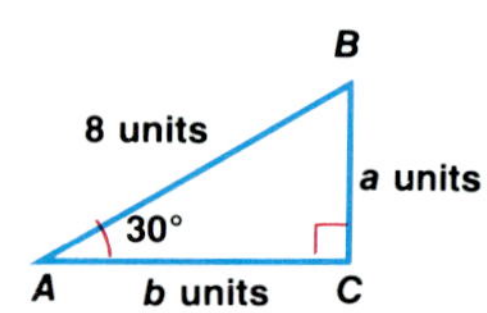

Example

2 **Find the length of $\overline{PQ}$ in triangle PQR.**

$r = 2q$ *The measure of the side opposite the 30° angle is q.*

$= 2 \cdot 6$

$= 12$

Thus, $\overline{PQ}$ is 12 units long.

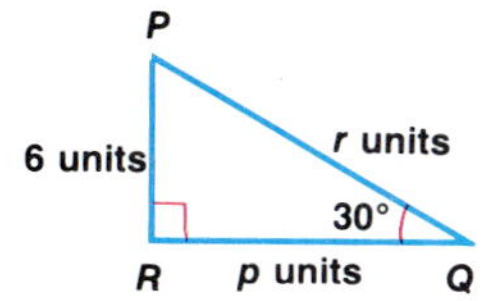

Suppose a represents the measure of the side opposite the 30° angle. The Pythagorean Theorem is used to find the length of the side opposite the 60° angle in a 30°–60° right triangle.

$a^2 + b^2 = c^2$ *Pythagorean Theorem*

$a^2 + b^2 = (2a)^2$ *The measure of the hypotenuse is twice the measure of the side opposite the 30° angle.*

$a^2 + b^2 = 4a^2$

$b^2 = 4a^2 - a^2$

$b^2 = 3a^2$

$b = a\sqrt{3}$ *How is this obtained?*

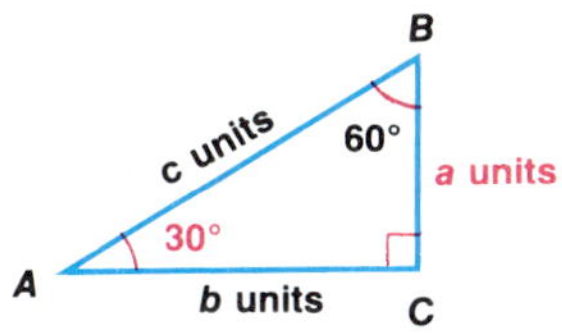

In a 30°–60° right triangle, if a is the measure of the side opposite the 30° angle, then $a\sqrt{3}$ is the measure of the side opposite the 60° angle.

This relationship should be memorized.

Examples

3 **Find the length of $\overline{AC}$ in triangle ABC.**

$b = a\sqrt{3}$ *The measure of the side opposite the 30° angle is a.*

$= 5\sqrt{3}$

$\approx 5 \cdot 1.732$

≈ 8.660

Thus, $\overline{AC}$ is approximately 8.660 units long.

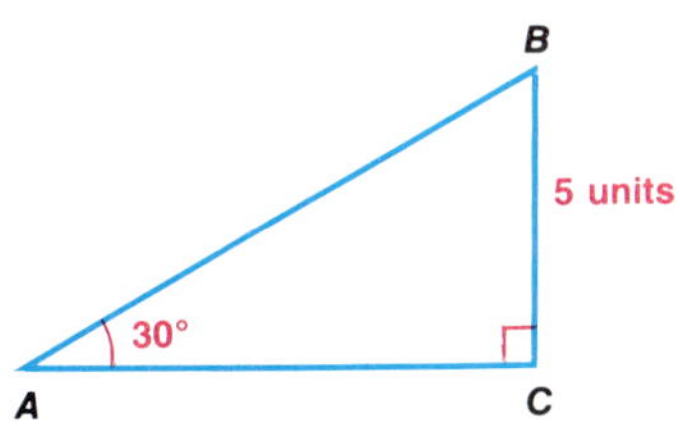

4 **Find the length of $\overline{XY}$ and $\overline{XZ}$ in triangle XYZ.**

$x = z\sqrt{3}$

$8\sqrt{3} = z\sqrt{3}$

$8 = z$

Since $y = 2z$, it is clear that $y = 2 \cdot 8$, or 16.

Thus, $\overline{XY}$ is 8 units long and $\overline{XZ}$ is 16 units long.

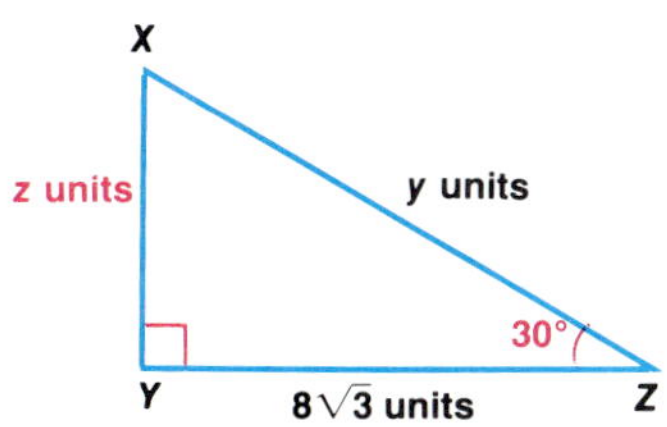

Chalkboard Exercises

Find the length of $\overline{PQ}$ in triangle PQR.

2a.

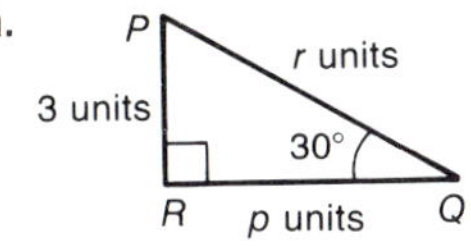

6 units

2b.

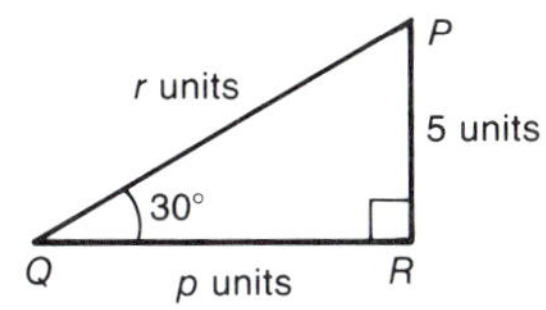

10 units

Find the length of $\overline{AC}$ in triangle ABC.

3a.

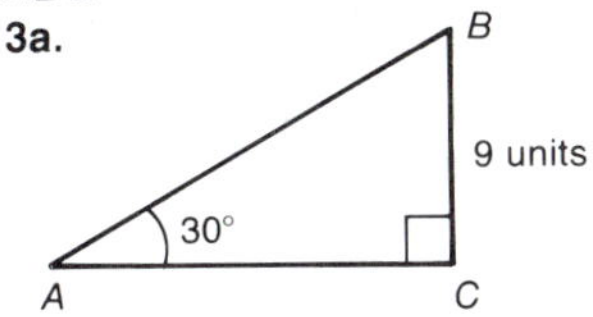

15.588 units

3b.

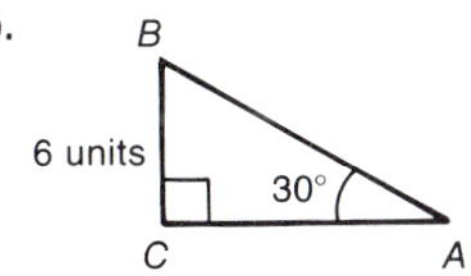

10.392 units

Find the length of $\overline{XY}$ and $\overline{XZ}$ in triangle XYZ.

4a.

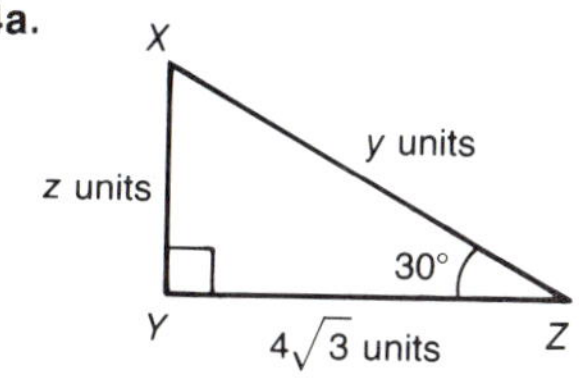

$XY = 4$ units
$XZ = 8$ units

4b.

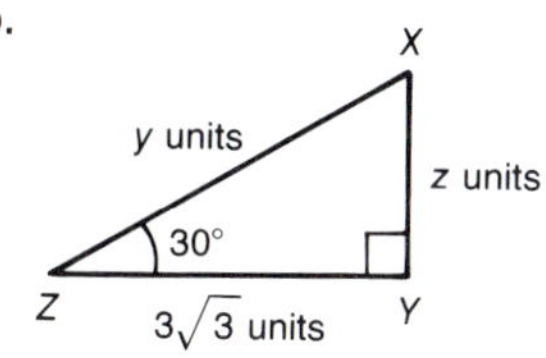

$XY = 3$ units
$XZ = 6$ units

Suggested Daily Assignments

Enriched
Exp: 2–22 evens
Written: 1–12

Average
Exp: 1–21 odds
Written: 1–12

Fundamental
Exp: 1–22
Written: 1–8
p. 501, 25–33 odds

Exploratory Exercises

Each of the following is the length of the hypotenuse of a 30° – 60° right triangle. For each triangle, find the length of the side opposite the 30° angle.

1. 8 m 4 m **2.** 16 cm 8 cm **3.** 13 mm 6.5 mm **4.** 9 mi 4.5 mi

5. $4\frac{1}{2}$ in. $2\frac{1}{4}$ in. **6.** $3\frac{3}{8}$ in. $1\frac{11}{16}$ in. **7.** 16.36 m 8.18 m **8.** 4.63 cm 2.315 cm

Each of the following is the length of the side opposite the 30° angle in a 30° – 60° right triangle. Find the length of the hypotenuse in each triangle.

9. 7 m 14 m **10.** 6.2 cm 12.4 cm **11.** 4.35 mm 8.70 mm **12.** $4\frac{1}{2}$ mi 9 mi

13. $6\frac{3}{8}$ in. $12\frac{3}{4}$ in. **14.** 13 m 26 m **15.** 3.86 cm 7.72 cm **16.** $7\frac{3}{4}$ in. $15\frac{1}{2}$ in.

Each of the following is the length of the side opposite the 60° angle in a 30° – 60° right triangle. Find the length of the other two sides in each triangle.

17. $4\sqrt{3}$ ft 4 ft, 8 ft **18.** $2\sqrt{3}$ cm 2 cm, 4 cm **19.** $8\sqrt{3}$ m 8 m, 16 m

20. $\sqrt{3}$ mi 1 mi, 2 mi **21.** $7\sqrt{3}$ yd 7 yd, 14 yd **22.** $9\sqrt{3}$ mm 9 mm, 18 mm

Written Exercises

Find the length of the third side in each triangle. (Use $\sqrt{3} \approx 1.732$.)

1.

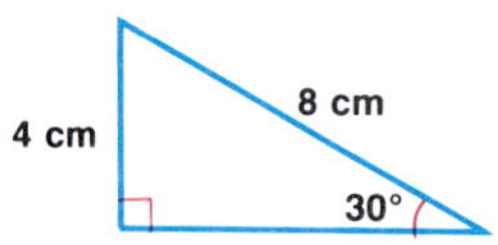

$4\sqrt{3}$ or 6.928 cm

2.

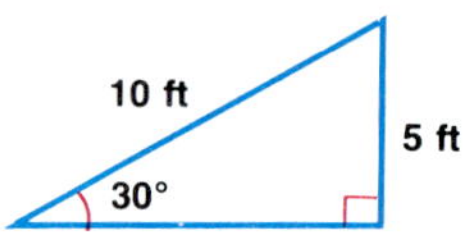

$5\sqrt{3}$ or 8.660 ft

3. 10 units

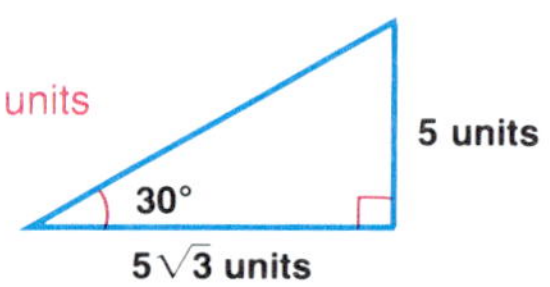

4. 7 units

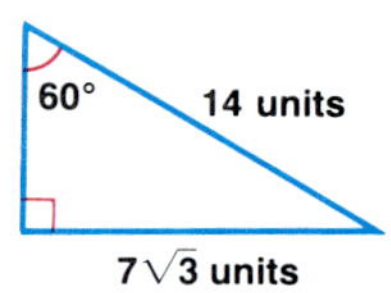

Find the missing lengths for each of the 30° – 60° right triangles described below. (Use $\sqrt{3} \approx 1.732$.)

	Hypotenuse	*Side Opposite 30° Angle*	*Side Opposite 60° Angle*
5.	6 m	3 m	$3\sqrt{3}$ or 5.196 m
6.	4.75 mm	2.375 mm	$2.375\sqrt{3}$ or 4.114 mm
7.	$3\frac{1}{2}$ in.	$1\frac{3}{4}$ in.	$\frac{7}{4}\sqrt{3}$ or 3.031 in.
8.	16 cm	8 cm	$8\sqrt{3}$ or 13.856 cm
9.	13 m	6.5 m	$6.5\sqrt{3}$ or 11.258 m
10.	$6\frac{1}{2}$ in.	$3\frac{1}{4}$ in.	$\frac{13}{4}\sqrt{3}$ or 5.629 in.
11.	4 m	2 m	$\sqrt{12}$ m
12.	7 cm	3.5 cm	$3.5\sqrt{3}$ cm

Objective 16-3: Use proportions to find the missing measures for similar triangles.

16-3 Similar Triangles ❶

Are the triangles below the same size? Do they have the same shape?

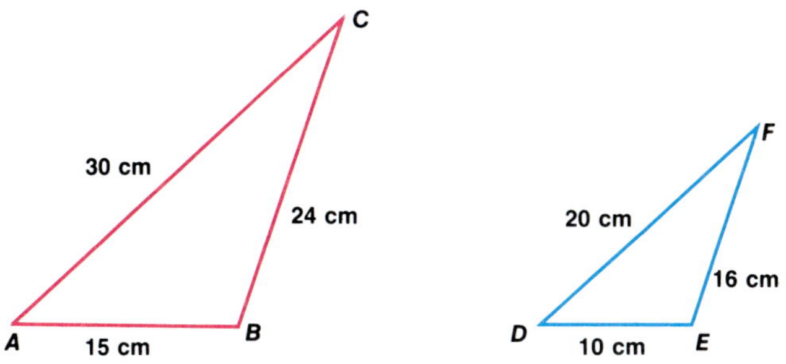

Measure the angles of triangle *ABC*. Compare these with the measures of triangle *DEF*. What do you discover? ❷

Triangles that have three congruent angles are called **similar triangles**. The two triangles above are similar triangles. Similar triangles have the same shape, but not necessarily the same size.

If two triangles are similar, the congruent angles are called **corresponding angles**. Also, the sides opposite congruent angles are called **corresponding sides**.

Let's compare the measures of the corresponding sides. Throughout the remainder of this chapter, *AB* will mean the measure of $\overline{AB}$, *BC* will mean the measure of $\overline{BC}$, and so on. ❸

$$\frac{AB}{DE} = \frac{15}{10} \qquad \frac{BC}{EF} = \frac{24}{16} \qquad \frac{AC}{DF} = \frac{30}{20}$$

$$= \frac{3}{2} \qquad\qquad = \frac{3}{2} \qquad\qquad = \frac{3}{2}$$

What do you discover?

If two triangles are similar, the measures of their corresponding angles are equal and the measures of their corresponding sides are proportional.

Similar Triangles

This means that for the triangles above you can write the following proportions.

$$\frac{AB}{DE} = \frac{BC}{EF} \qquad \frac{AB}{DE} = \frac{AC}{DF} \qquad \frac{BC}{EF} = \frac{AC}{DF}$$

16-3

Lesson Notes, page T55

Readiness Activity

Solve each proportion

$\frac{3}{4} = \frac{x}{16}$ 12 $\qquad \frac{3}{12} = \frac{1}{y}$ 4

$\frac{15}{a} = \frac{25}{2}$ $\frac{6}{5}$ $\qquad \frac{m}{24} = \frac{5}{16}$ $7\frac{1}{2}$

Discussion Questions

(Use after page 505.)

1. Are all isosceles triangles similar? No
2. Are all right triangles similar? No
3. Are all isosceles right triangles similar? Yes

❶ You may wish to review ratios and proportions before introducing this lesson. Similar triangles, along with ratios and proportions, are very important in the study of geometry.

❷ The angles in triangle *ABC* have the same measures as the angles in triangle *DEF*.

❸ Emphasize that *AB* is used to denote the measure of $\overline{AB}$.

Proportions can be used to find the missing measures of similar triangles.

Example

1 If a tree 6 feet tall casts a shadow 4 feet long, how high is a flagpole that casts a shadow 18 feet long?

Triangle JKL is similar to triangle PQR.

$$\frac{JK}{PQ} = \frac{KL}{QR}$$

$$\frac{6}{x} = \frac{4}{18}$$

$$4x = 6 \cdot 18 \quad \textit{Cross multiply.}$$

$$4x = 108$$

$$x = 27$$

The flagpole is 27 feet high.

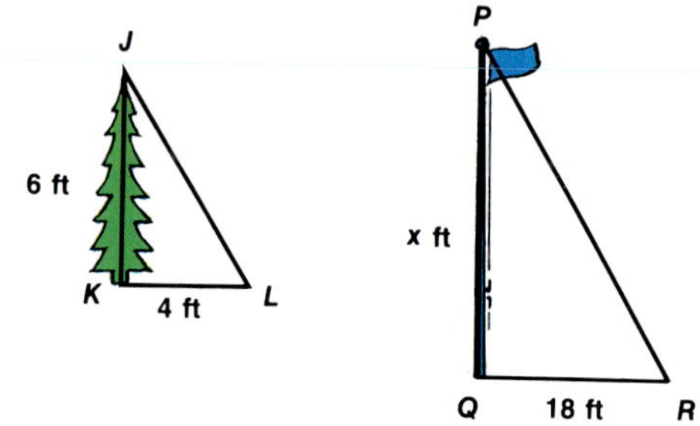

Tell how you could use the method shown above to find the height of a building.

Example

2 Find the distance, UV, across the pond shown below.

Triangle STU is similar to triangle WVU.

$$\frac{TU}{VU} = \frac{ST}{WV}$$

$$\frac{70}{x} = \frac{50}{75}$$

$$50x = 70 \cdot 75$$

$$50x = 5250$$

$$x = 105$$

❹

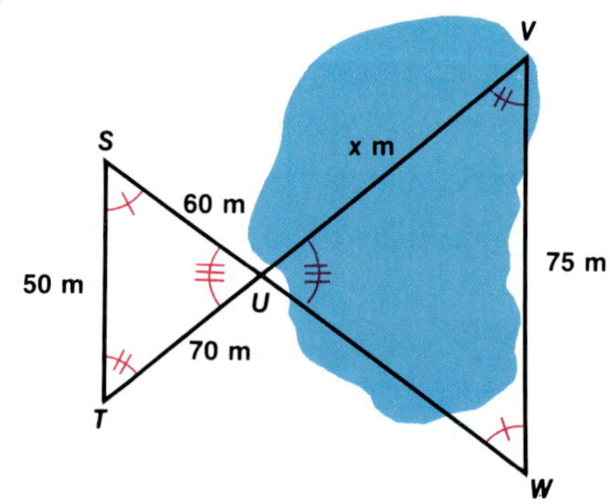

The pond is 105 meters across.

Can you name some other instances where you can use similar triangles as above to find distances?

Chalkboard Exercises

Use similar triangles to solve each of the following.

1a. A tree 9 ft tall casts a shadow 5 ft long. How high is a flagpole that casts a shadow 15 ft long? 27 ft

1b. A tree 5 ft tall casts a shadow 2 ft long. How tall is a flagpole that casts a shadow 9 ft long? $22\frac{1}{2}$ ft

2a. In Example 2, suppose $\overline{VW}$ is 90 meters long. Find x. 126 m

2b. In Example 2, suppose $\overline{ST}$ is 40 meters long. Find x. 131.25 m.

❹ Point out that the marks across the angle symbols indicate congruent angles. Have students identify the pairs of congruent angles in the diagram.

Suggested Daily Assignments

Enriched
- Exp: 1–4
- Written: 1–12

Average
- Exp: 1–4
- Written: 1–12

Fundamental
- Exp: 1–4
- Written: 1–8
  - p. 504, 9–12

Exploratory Exercises

For each pair of similar triangles, list the corresponding angles and the corresponding sides.

1.

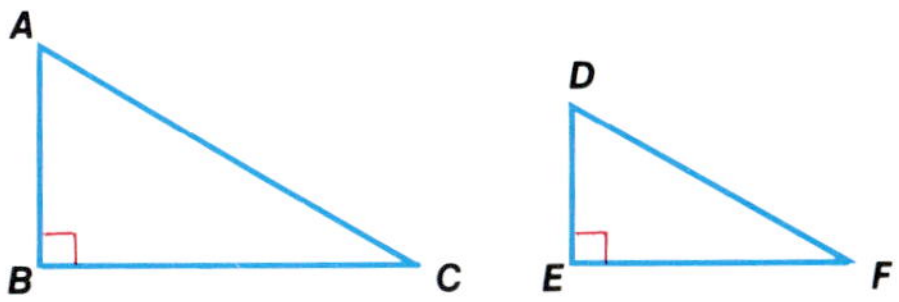

2.

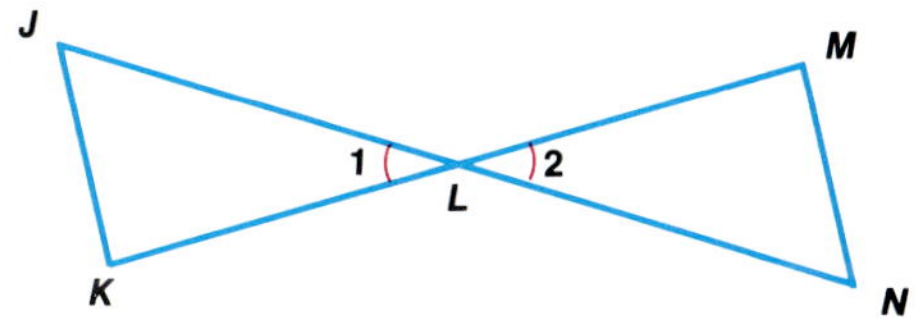

3.

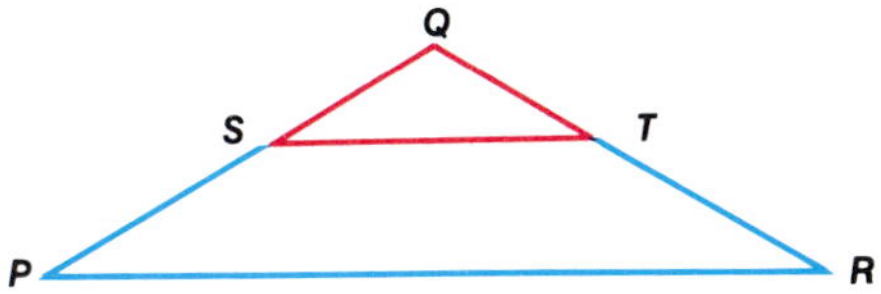

4.

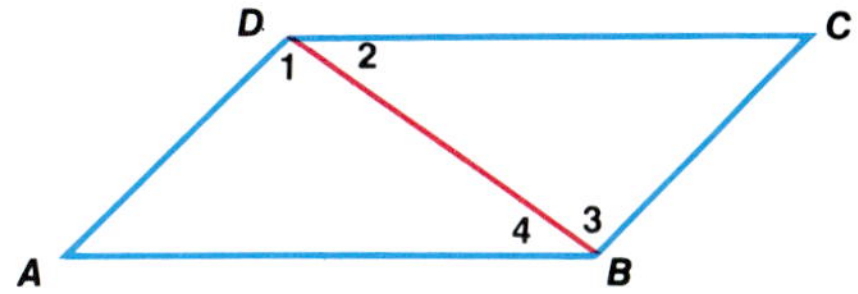

Written Exercises

Triangles *ABC* and *DEF* are similar. For each set of measures, find the measure of the missing sides.

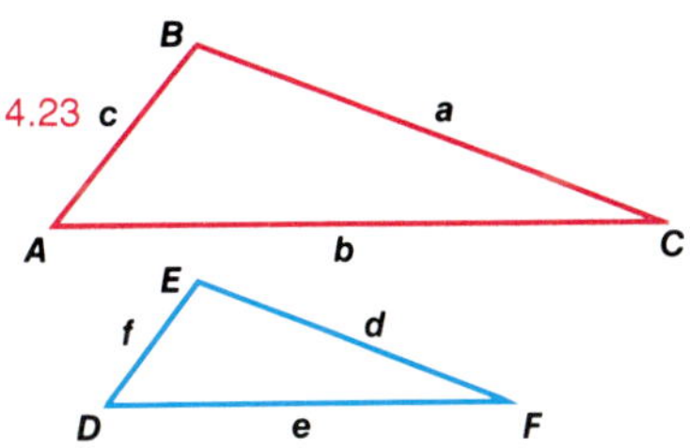

1. $a = 5, d = 7, f = 6, e = 5$ $b = \frac{25}{7}, c = \frac{30}{7}$
2. $c = 11, f = 6, d = 5, e = 4$ $a = \frac{55}{6}, b = \frac{22}{3}$
3. $b = 4.5, d = 2.1, e = 3.4, f = 3.2$ $a \approx 2.78, c \approx 4.23$
4. $a = 16, c = 12, b = 13, e = 7$ $d = \frac{112}{13}, f = \frac{84}{13}$
5. $a = 17, b = 15, c = 10, f = 6$ $d = 10.2, e = 9$
6. $c = 18, f = 12, d = 18, e = 16$ $a = 27, b = 24$
7. $a = 4\frac{1}{4}, b = 5\frac{1}{2}, e = 2\frac{3}{4}, f = 1\frac{3}{4}$ $c = \frac{7}{2}, d = \frac{17}{8}$
8. $c = 7\frac{1}{2}, f = 5, a = 10\frac{1}{2}, b = 15$ $d = 7, e = 10$

Use similar triangles to solve each of the following.

9. In building a roof, a 5-foot support is to be placed at point *B* as shown on the diagram. Find the length of the support that is to be placed at point *A*. $\frac{125}{9}$ ft

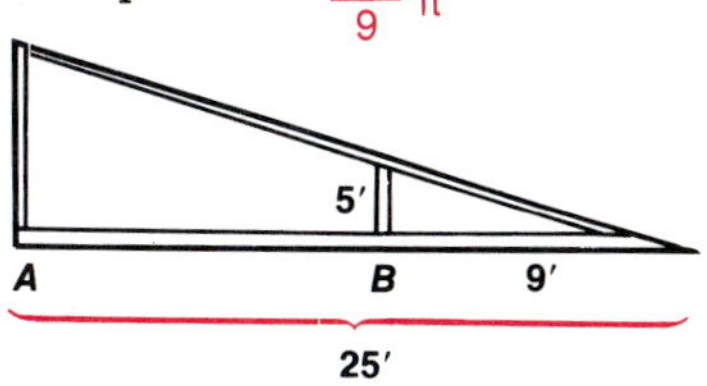

10. Triangle *ABC* is similar to triangle *EDC*. Find the distance across the lake from point *A* to point *B*. 105 m

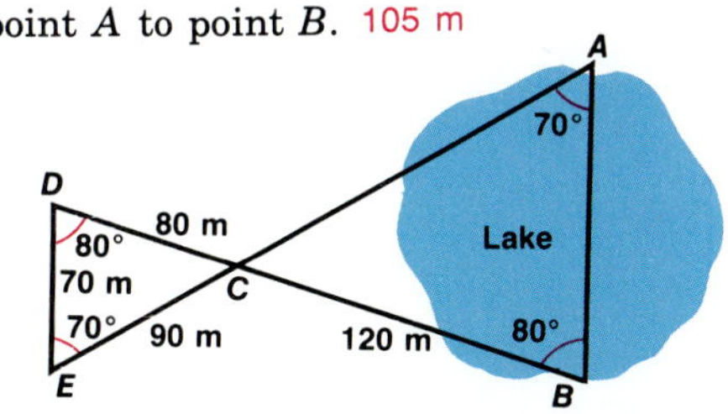

11. A triangle has sides of 9, 15, and 18 cm. The longest side of a similar triangle is 22 cm in length. Find the length of the shortest side of that triangle. 11 cm

12. A fence post one meter high casts a shadow 170 cm long. Find the height of a flagpole whose shadow is 8000 cm in length. 4706 cm

Answers to Exploratory

1. $\angle A$ and $\angle D$; $\angle B$ and $\angle E$; $\angle C$ and $\angle F$; AC and DF; AB and DE; BC and EF
2. $\angle 1$ and $\angle 2$; $\angle J$ and $\angle N$; $\angle K$ and $\angle M$; JK and NM; JL and NL; KL and ML
3. $\angle Q$ and $\angle Q$; $\angle S$ and $\angle P$; $\angle T$ and $\angle R$; QS and QP; QT and QR; ST and PR
4. $\angle 1$ and $\angle 3$; $\angle 4$ and $\angle 2$; $\angle A$ and $\angle C$; AD and CB; AB and CD; DB and BD

Quiz
Lessons 16-1 to 16-3

Find the complement and the supplement for each angle whose measure is given.

1. $120°$ no complement; $60°$
2. $(50 - x)°$ $(40 + x)°$, $(130 + x)°$

Solve each problem.

3. Two angles of a triangle measure $(2x)°$ and $(2x + 10)°$; find the measure of the third angle. $(170 - 4x)°$
4. Find the measure of the third side of the triangle. (Use $\sqrt{3} = 1.732$.)

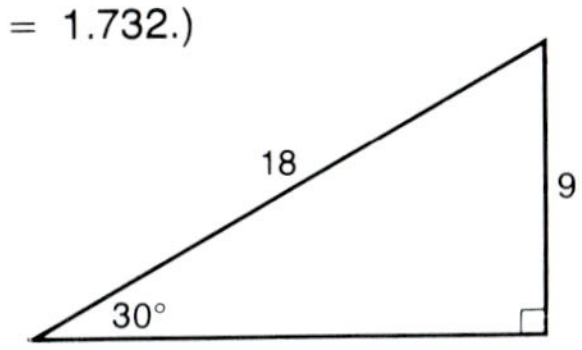

$9\sqrt{3}$ or 15.588

5. When Joe Ellwanger stands 2 ft from a lamp post, his shadow is 3 ft long. If Joe is 6 ft tall, how tall is the lamp post? 10ft

16-4

Lesson Notes, page T55

Readiness Activity

Are all right triangles similar? No

Are all right triangles whose acute angles measure 30° and 60° similar? Yes

Are all right triangles whose two acute angles are congruent similar? Yes

What must be true for two right triangles to be similar? Their corresponding acute angles must be congruent.

Discussion Question

(Use after page 508.)

What is the relationship of the tangents of two complementary angles? They are reciprocals of one another.

❶ Point out to students that the trigonometric ratios are useful only because right triangles with the same angle measures are similar. For example, all 30°-60° right triangles are similar. Students could verify this by drawing a pair of similar triangles, then measuring the sides and computing the trigonometric ratios. The ratios for each triangle should be the same. (They may vary slightly due to measurement error.)

508

Objective 16-4: Compute the sine, cosine, and tangent of an angle of a right triangle given the measures of the sides.

16-4 Trigonometric Ratios

For similar triangles, ratios of the measures of corresponding sides can be written. If enough of these measures are known, these ratios can be used to find the measures of the remaining sides.

For every right triangle, certain ratios can be set up. These ratios, called **trigonometric ratios**, involve not only the measures of the sides but the measures of the acute angles as well. Again, if enough is known, these ratios can be used to find the measures of the remaining parts of the triangle. ❶

A typical right triangle is shown below.

Side BC is *opposite* angle A.

Side AC is *adjacent* to angle A.

Side AB is the *hypotenuse* and is opposite the right angle C.

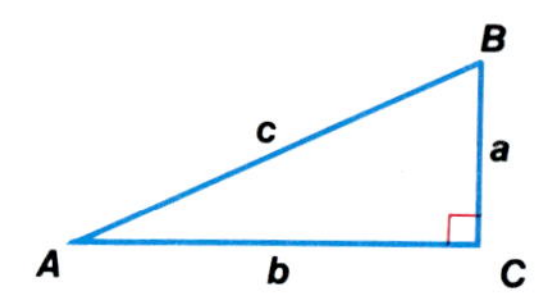

Notice that in this triangle

a denotes the measure of the side opposite angle A.

b denotes the measure of the side opposite angle B.

c denotes the measure of the side opposite angle C.

Three common trigonometric ratios are defined as follows.

Definition of Trigonometric Ratios

$$\textbf{sine of angle } A = \frac{\textbf{measure of side opposite angle } A}{\textbf{measure of hypotenuse}}$$

$$\sin A = \frac{a}{c}$$

$$\textbf{cosine of angle } A = \frac{\textbf{measure of side adjacent to angle } A}{\textbf{measure of hypotenuse}}$$

$$\cos A = \frac{b}{c}$$

$$\textbf{tangent of angle } A = \frac{\textbf{measure of side opposite angle } A}{\textbf{measure of side adjacent to angle } A}$$

$$\tan A = \frac{a}{b}$$

Example 1

Find the sine of angle D in the $30°-60°$ right triangle below.

$\sin D = \frac{EF}{DE}$

Since side EF is opposite the $30°$ angle, its measure is one-half the measure of hypotenuse DE.

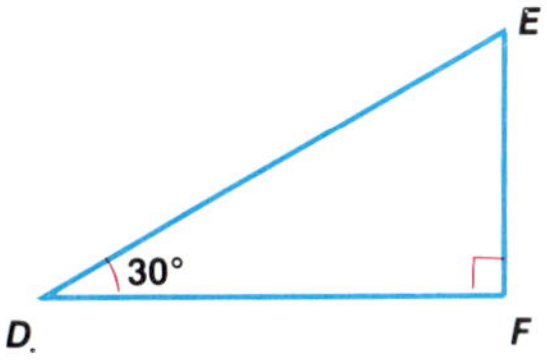

Therefore, $\sin D = \frac{1}{2}$ or $\sin 30° = \frac{1}{2}$.

Are all $30°-60°$ right triangles the same size?

Regardless of the size, the measure of the side opposite the $30°$ angle is one-half the measure of the hypotenuse. Therefore, we see that a trigonometric ratio like sine $30°$ will always be the same regardless of the triangle's size.

Example 2

For triangle ABC, express $\sin A$, $\cos A$, $\tan A$, $\sin B$, $\cos B$, and $\tan B$ to 3 decimal places.

Notice $a = 7$
$b = 24$
$c = 25$

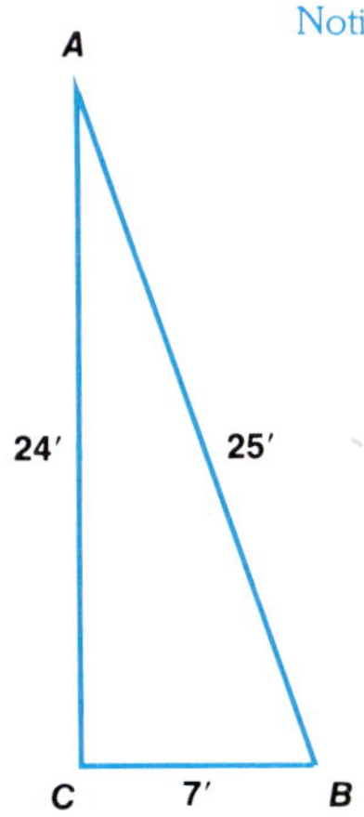

$\sin A = \frac{a}{c} = \frac{7}{25} = 0.280$

$\cos A = \frac{b}{c} = \frac{24}{25} = 0.960$

$\tan A = \frac{a}{b} = \frac{7}{24} \approx 0.292$

$\sin B = \frac{b}{c} = \frac{24}{25} = 0.960$

$\cos B = \frac{a}{c} = \frac{7}{25} = 0.280$

$\tan B = \frac{b}{a} = \frac{24}{7} \approx 3.429$

Chalkboard Exercises

1a. Find the sine of angle E in the triangle in Example 1.
$\frac{\sqrt{3}}{2}$

1b. Find the cosine of angle D in the triangle in Example 1.
$\frac{\sqrt{3}}{2}$

1c. Find the cosine of angle E in the triangle in Example 1.
$\frac{1}{2}$

For each exercise, the measures of the sides of triangle ABC are given. Express the sine, cosine, and tangent of angle A to 3 decimal places.

2a. $AC = 12$, $BC = 16$, $AB = 20$

$\sin A = \frac{16}{20} = 0.8$

$\cos A = \frac{12}{20} = 0.6$

$\tan A = \frac{16}{12} = 1.333$

2b. $AC = 8$, $BC = 6$, $AB = 10$

$\sin A = \frac{6}{10} = 0.6$

$\cos A = \frac{8}{10} = 0.8$

$\tan A = \frac{6}{8} = 0.75$

2c. $AC = 5$, $BC = 12$, $AB = 13$

$\sin A = \frac{12}{13} \approx 0.923$

$\cos A = \frac{5}{13} \approx 0.385$

$\tan A = \frac{12}{5} = 2.4$

Calculator Notes
You may want to allow students to use a calculator for the Written Exercises.

Suggested Daily Assignments

Enriched
Exp: 2–18 evens, 19–21
Written: 1–9

Average
Exp: 1–21
Written: 1–9

Fundamental
Exp: 1–21
Written: 1–6
p. 507, 9–12

Answers to Written

1. $\sin A \approx 0.690$, $\sin B \approx 0.724$
$\cos A \approx 0.724$, $\cos B \approx 0.690$
$\tan A \approx 0.952$, $\tan B \approx 1.050$
2. $\sin D \approx 0.969$, $\sin E \approx 0.246$
$\cos D \approx 0.246$, $\cos E \approx 0.969$
$\tan D \approx 3.938$, $\tan E \approx 0.254$
3. $\sin R \approx 0.324$, $\sin T \approx 0.946$
$\cos R \approx 0.946$, $\cos T \approx 0.324$
$\tan R \approx 0.343$, $\tan T \approx 2.917$
4. $\sin P \approx 0.882$, $\sin Q \approx 0.471$
$\cos P \approx 0.471$, $\cos Q \approx 0.882$
$\tan P \approx 1.875$, $\tan Q \approx 0.533$
5. $\sin J \approx 0.899$, $\sin K \approx 0.438$
$\cos J \approx 0.438$, $\cos K \approx 0.899$
$\tan J \approx 2.051$, $\tan K \approx 0.488$
6. $\sin A \approx 0.753$, $\sin B \approx 0.658$
$\cos A \approx 0.658$, $\cos B \approx 0.753$
$\tan A \approx 1.146$, $\tan B \approx 0.873$
7. $\sin J = 0.800$, $\sin L = 0.600$
$\cos J = 0.600$, $\cos L = 0.800$
$\tan J \approx 1.333$, $\tan L = 0.750$
8. $\sin X \approx 0.709$, $\sin Z \approx 0.709$
$\cos X \approx 0.709$, $\cos Z \approx 0.709$
$\tan X = 1.000$, $\tan Z = 1.000$
9. $\sin E \approx 0.220$, $\sin F \approx 0.976$
$\cos E \approx 0.976$, $\cos F \approx 0.220$
$\tan E \approx 0.225$, $\tan F \approx 4.444$

Exploratory Exercises

For the three triangles below, express each of the following trigonometric ratios as a fraction in simplest form.

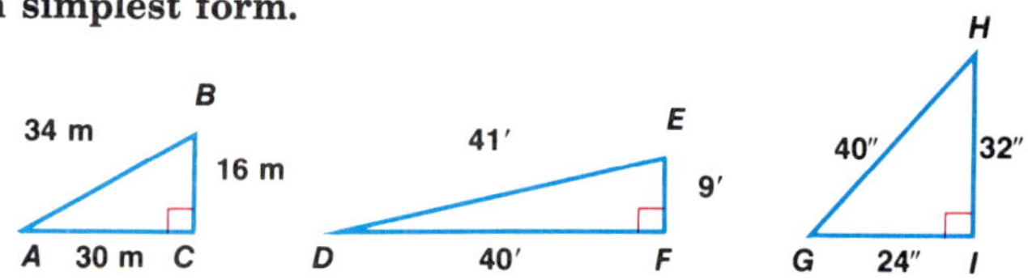

1. $\sin A$ $\frac{8}{17}$
2. $\cos A$ $\frac{15}{17}$
3. $\tan A$ $\frac{8}{15}$
4. $\cos B$ $\frac{8}{17}$
5. $\sin B$ $\frac{15}{17}$
6. $\tan B$ $\frac{15}{8}$
7. $\sin D$ $\frac{9}{41}$
8. $\cos D$ $\frac{40}{41}$
9. $\tan D$ $\frac{9}{40}$
10. $\cos E$ $\frac{9}{41}$
11. $\sin E$ $\frac{40}{41}$
12. $\tan E$ $\frac{40}{9}$
13. $\sin G$ $\frac{4}{5}$
14. $\cos G$ $\frac{3}{5}$
15. $\tan G$ $\frac{4}{3}$
16. $\cos H$ $\frac{4}{5}$
17. $\sin H$ $\frac{3}{5}$
18. $\tan H$ $\frac{3}{4}$
19. Are angles A and B complementary angles? Is $\sin A = \cos B$? yes; yes
20. Are angles D and E complementary angles? Is $\cos D = \sin E$? yes; yes
21. Based on your answers for problems 19-20, what is the relationship of the sine and cosine of complementary angles? They are equal.

Written Exercises

Express the sine, cosine, and tangent of each acute angle below to 3 decimal places.

1.
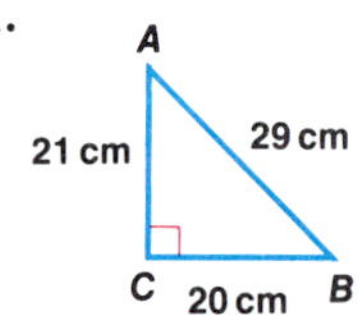

2.
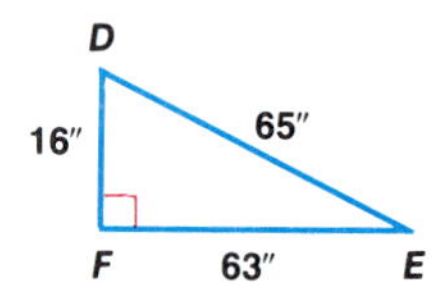

3.
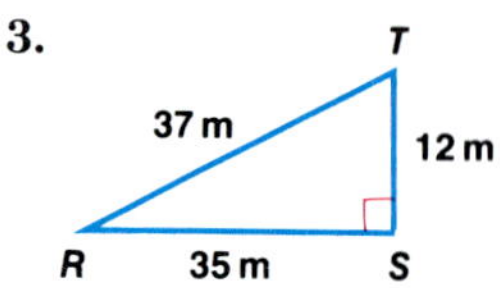

4.
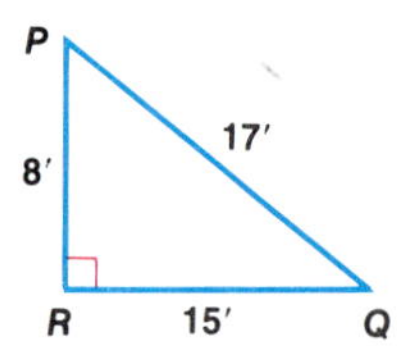

5.
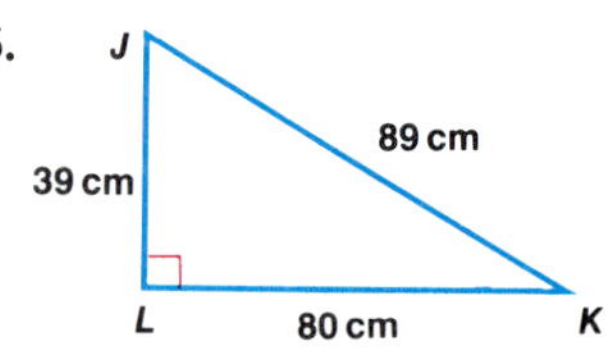

6.
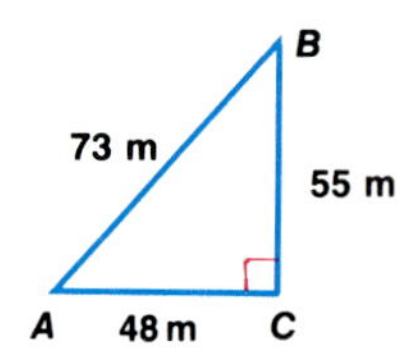

7.
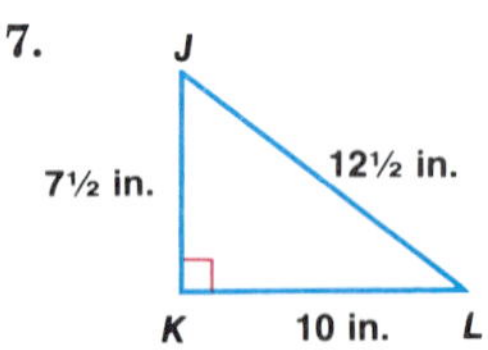

8.
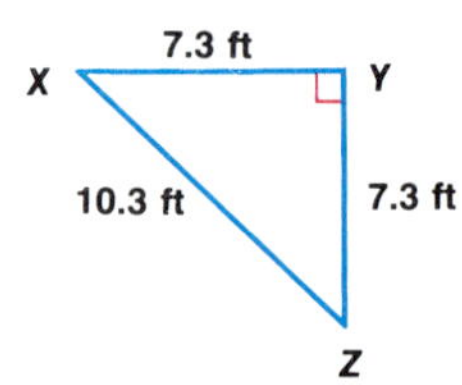

9.
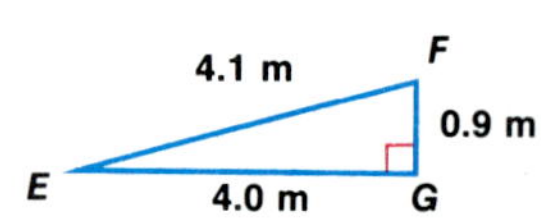

Reading Algebra

Greek Letters

As shown below, letters from the English alphabet can be used to convey mathematical concepts.

$$x^2 + 7 = 0 \qquad A = \frac{1}{2}bh \qquad \overline{AB} \qquad \sin A$$

In higher mathematics, letters from the Greek alphabet are often used in the same way. The table below shows upper case and lower case letters of the Greek alphabet.

GREEK ALPHABET

Greek Letter	*Greek Name*	*English Equivalent*	*Greek Letter*	*Greek Name*	*English Equivalent*
Α α	Alpha	a	Ν ν	Nu	n
Β β	Beta	b	Ξ ξ	Xi	x
Γ γ	Gamma	g	Ο ο	Omicron	ŏ
Δ δ	Delta	d	Π π	Pi	p
Ε ε	Epsilon	ĕ	Ρ ρ	Rho	r
Ζ ζ	Zeta	z	Σ σ	Sigma	s
Η η	Eta	ē	Τ τ	Tau	t
Θ θ	Theta	th	Υ υ	Upsilon	u
Ι ι	Iota	i	Φ φ	Phi	ph
Κ κ	Kappa	k	Χ χ	Chi	ch
Λ λ	Lambda	l	Ψ ψ	Psi	ps
Μ μ	Mu	m	Ω ω	Omega	ō

The most commonly used Greek letter in mathematics is π. This letter is used to represent the ratio of the circumference of a circle to its diameter. An approximate value for π is 3.14.

Other Greek letters are commonly used in higher mathematics. For example, Δ is commonly used to represent the difference between two measures, θ is often used to represent the measure of an angle, and Σ stands for the sum of a set of numbers.

Exercises

How would you say each of the following?

1. $\sin \theta$ sine of theta
2. $\cos \phi$ cosine of phi
3. $\tan \psi$ tangent of psi
4. Δx delta *x*
5. $r = 0.7\ \Omega$ *r* equals 0.7 omega
6. $\alpha - \lambda$ alpha minus lambda
7. $\sin(\alpha + \beta)$ sine of alpha plus beta
8. $|x - y| < \gamma$ The absolute value of *x* minus *y* is less than gamma

About Reading Algebra
Point out to students that Greek letters are used in higher mathematics because some of mathematics had its origins in Greece. Be sure students can recognize and pronounce the Greek letters in the exercises.

16-5

Lesson Notes, page T55
Enrichment, page T55

Discussion Question
Why are no sine or cosine values greater than one?
Because the measure of the hypotenuse is the denominator of both of these ratios and the hypotenuse is always the longest side of a right triangle.

Chalkboard Exercises
Use the table on page 545 to find a value for each of the following ratios.

1a. sin 32° 0.5299
1b. tan 76° 4.0108
1c. cos 47° 0.6820

Use the table to find the measure of each angle to the nearest degree.

2a. $\sin A = \frac{1}{2}$ 30°
3a. $\cos A = 0.9500$ 18°
3b. $\sin B = 0.8999$ 64°
3c. $\tan C = 1.5000$ 56°

512

Objective 16-5: Read trigonometric tables.

16-5 Using Tables

If you know the measures of the sides of a right triangle, you can calculate the sine, cosine, and tangent of any of its angles.

Tables such as those on page 545 list approximations of these ratios in decimal form.

Examples

1 Find tan 28°.

Find 28° in the angle column.

Find the corresponding reading in the tan column.

Angle	sin	cos	tan
26°	0.4384	0.8988	0.4877
27°	0.4540	0.8910	0.5095
28°	0.4695	0.8829	0.5317
29°	0.4848	0.8746	0.5543
30°	0.5000	0.8660	0.5774

$$\tan 28° \approx 0.5317$$

2 If $\cos A = \frac{1}{2}$, find the measure of angle A.

First express the ratio as a decimal.

$$\frac{1}{2} = 0.5000$$

Find this decimal in the cos column.

Angle	sin	cos	tan
56°	0.8290	0.5592	1.4826
57°	0.8387	0.5446	1.5399
58°	0.8480	0.5299	1.6003
59°	0.8572	0.5150	1.6643
60°	0.8660	0.5000	1.7321

The corresponding angle is 60°, so angle $A = 60°$.

3 Find the measure of angle A if $\sin A = 0.9711$.

The measure of angle A must be between 76° and 77° since 0.9711 is between 0.9703 and 0.9744.

Angle	sin	cos	tan
76°	0.9703	0.2419	4.0108
77°	0.9744	0.2250	4.3315
78°	0.9781	0.2079	4.7046
79°	0.9816	0.1908	5.1446

Because 0.9711 is closer to 0.9703 than it is to 0.9744, the measure of angle A is closer to 76° than 77°. Hence, the measure of angle $A \approx 76°$.

Exploratory Exercises

Use the table on page 545 to answer the following.

1. At what angle is the sine at its maximum (greatest value)? At its minimum (least value)? 90°; 0°
2. At what angle is the cosine at its minimum? At its maximum? 90°, 0°
3. For what angle are the sine and cosine equal? 45°
4. What is the tangent of the angle at which the sine and cosine are equal? 1
5. What are the minimum and maximum values of the tangent? no maximum or minimum values

Written Exercises ❶

Use the table on page 545 to find a value for each of the following ratios.

1. cos 25° 0.9063	**2.** tan 31° 0.6009	**3.** sin 89° 0.9998	**4.** tan 14° 0.2493
5. cos 76° 0.2419	**6.** sin 22° 0.3746	**7.** cos 83° 0.1219	**8.** tan 24° 0.4452
9. sin 68° 0.9272	**10.** tan 9° 0.1584	**11.** sin 27° 0.4540	**12.** cos 18° 0.9511
13. cos 42° 0.7431	**14.** tan 50° 1.1918	**15.** sin 45° 0.7071	**16.** cos 30° 0.8660
17. sin 30° 0.5000	**18.** tan 30° 0.5774	**19.** sin 60° 0.8660	**20.** cos 60° 0.5000

Use the table to find the measure of each angle to the nearest degree.

21. $\sin A = 0.4384$ 26°	**22.** $\cos B = 0.4848$ 61°	**23.** $\tan D = 1.3250$ 53°
24. $\cos \angle 1 = 0.9781$ 12°	**25.** $\tan \angle ABC = 5.1446$ 79°	**26.** $\sin A = 0.9620$ 74°
27. $\cos B = 0.3900$ 67°	**28.** $\tan \angle 2 = 1.7321$ 60°	**29.** $\sin D = 0.2756$ 16°
30. $\cos A = 0.3746$ 68°	**31.** $\cos B = 0.7660$ 40°	**32.** $\sin \angle DEF = 0.9848$ 80°
33. $\tan C = 57.2900$ 89°	**34.** $\sin X = 0.0520$ 3°	**35.** $\cos Z = 0.9986$ 3°
36. $\tan A = 0.0524$ 3°	**37.** $\sin B = 0.5000$ 30°	**38.** $\cos B = 0.8660$ 30°

Find values for each of the following.

39. sin 30° − cos 60° 0	**40.** tan 45° + sin 0° 1	**41.** sin 90° + cos 0° 2
42. tan 0° + sin 90° 1	**43.** sin 0° − cos 60° −0.5	**44.** sin 90° − tan 0° 1
45. sin 30° − cos 30° −0.3660	**46.** cos 60° − sin 30° 0	**47.** tan 45° − tan 45° 0

Using Calculators — Trigonometric Functions

You can use a calculator to find the value for cos 25° as follows.

ENTER: 25 [COS]

The display shows 0.963078. Thus, rounded to ten-thousandths, cos 25° = 0.9631.
You can also find the measure of the angle for $\sin A = 0.4384$ as follows.

ENTER: 0.4384 [INV] [SIN]

The display shows 26.001839. Thus, to the nearest degree, angle $A \approx 26°$.

Exercises

Use a calculator to check your answers for Written Exercises 1-38 above.

Suggested Daily Assignments

Enriched
Exp: 1–5
Written: 2–46 evens

Average
Exp: 1–5
Written: 2–46 evens

Fundamental
Exp: 1–5
Written: 1–38, 40–46 evens
p. 510, 7–9

❶ The values given in Written Exercises 23, 26, 27, and 34 are not listed in the table. These angles are rounded to the nearest degree.

About Using Calculators
Explain to students that the tan key on a calculator can be used in the same way the sin and cos keys are used.

16-6

Lesson Notes, page T55

Discussion Questions

(Use after Example 1.)

1. Can a right triangle be solved if the lengths of two sides are given? Yes
2. Can a right triangle be solved if the measures of both acute angles are given? No
3. Can a right triangle be solved if the length of one side and the measure of one acute angle are given? Yes

Chalkboard Exercises

1. Find the measures of angles A and C below.

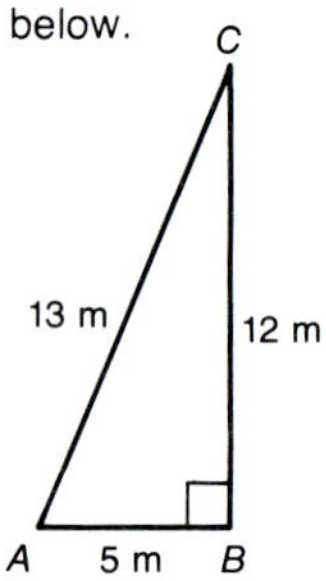

$\sin A = \frac{12}{13} \approx 0.9231;$

$m\angle A \approx 67,\ m\angle C \approx 23$

2. Find the length of $\overline{DE}$ in triangle DEF.

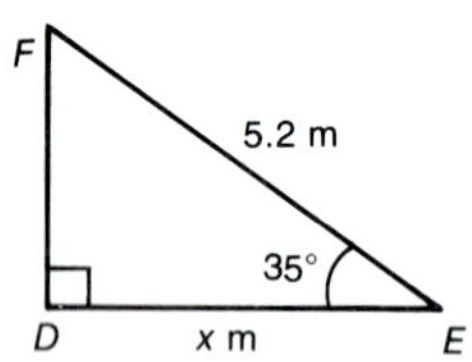

$\cos 35° = \frac{x}{5.2}$

$0.8192 \approx \frac{x}{5.2}$

$4.3 \approx x$

$4.3m$

Objective 16-6: Use trigonometric ratios to solve right triangles.

16-6 Solving Right Triangles

To solve a triangle means to find all the missing measures of the triangle. The trigonometric ratios can be used to solve a triangle. The ratio used depends upon what measures are given and what measures are missing. Sometimes, more than one ratio can be used.

Example 1

Find the measures of angles A and C below.

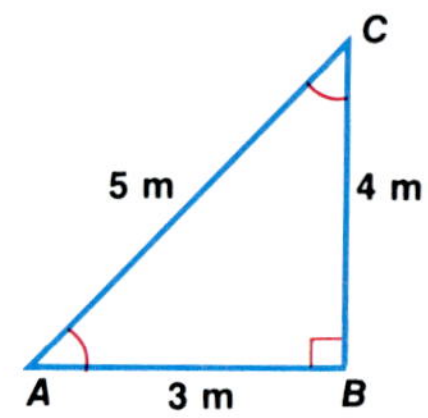

Suppose you use the sine ratio.

$$\sin A = \frac{4}{5}$$
$$= 0.8000$$

Using the table or a calculator, we see that angle A is approximately 53°. Thus, angle C is about $90° - 53°$, or 37°.

How would you use the cosine ratio to find the measure of angle A? Tell how you could use the tangent ratio to find the measure of angle A.

Example 2

Find the length of $\overline{DE}$ in triangle DEF.

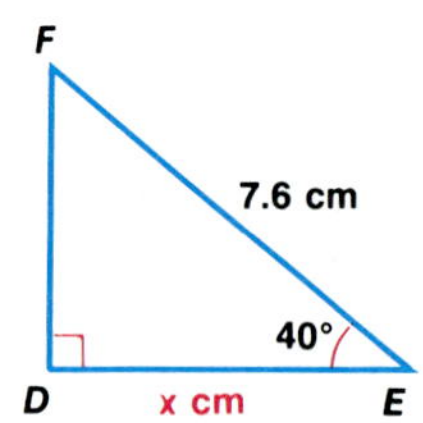

In this case, use the cosine ratio.

$$\cos 40° = \frac{x}{7.6}$$
$$0.7660 \approx \frac{x}{7.6}$$
$$(0.7660)7.6 \approx x$$
$$5.8 \approx x$$

Thus, $\overline{DE}$ is approximately 5.8 cm long.

Try using $\sin 50° = \frac{x}{7.6}$ to solve for the measure of $\overline{DE}$. Did you get the same result as in Example 2?

Examples

3 **Find the length of $\overline{JK}$ in triangle JKL.**

In this case, use the tangent ratio.

$$\tan 48° = \frac{5.8}{x}$$

$$1.1106 \approx \frac{5.8}{x}$$

$$1.1106x \approx 5.8$$

$$x \approx 5.2$$

Thus, $\overline{JK}$ is about 5.2 mm long.

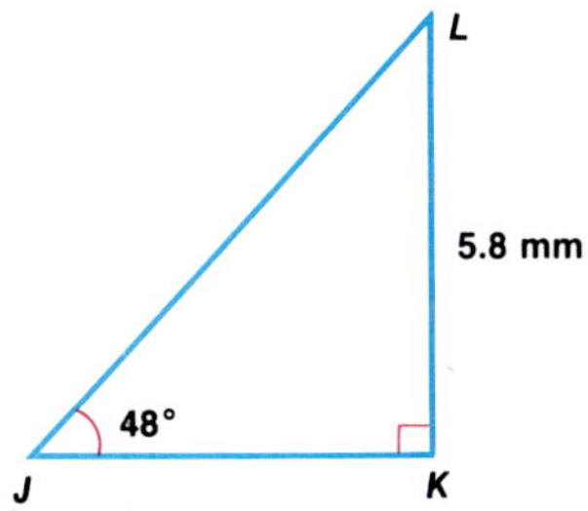

4 **Find the missing measures in the triangle below.**

Angle R = 90° − 36°, or 54°

$$\sin 36° = \frac{x}{18}$$

$$0.5878 \approx \frac{x}{18}$$

$$10.6 \approx x$$

Thus, $\overline{QR}$ is about 10.6 inches long.

$$\cos 36° = \frac{y}{18}$$

$$0.8090 \approx \frac{y}{18}$$

$$14.56 \approx y$$

Thus, $\overline{PQ}$ is about 14.56 inches long.

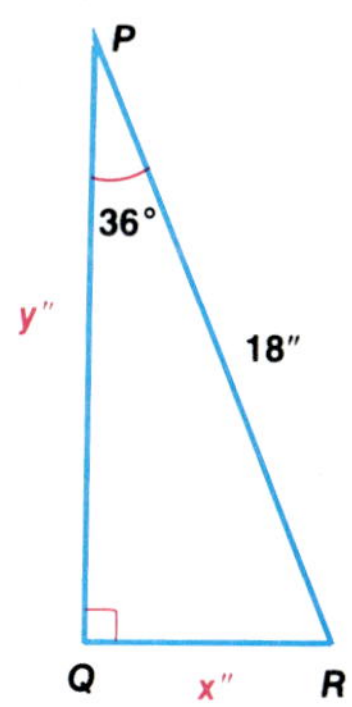

Exploratory Exercises

State which trigonometric ratios you would use to find the missing measures in each of the following.

1. $\sin 46° = \frac{b}{15}$, $\cos 46° = \frac{a}{15}$

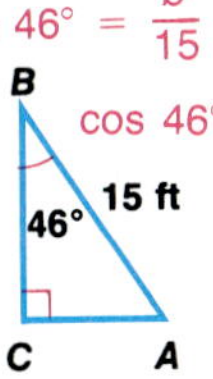

2. $\cos 39° = \frac{12}{y}$, $\tan 39° = \frac{x}{12}$

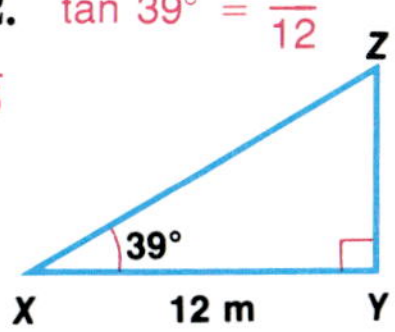

3. $\sin 42° = \frac{6}{c}$, $\tan 42° = \frac{6}{a}$

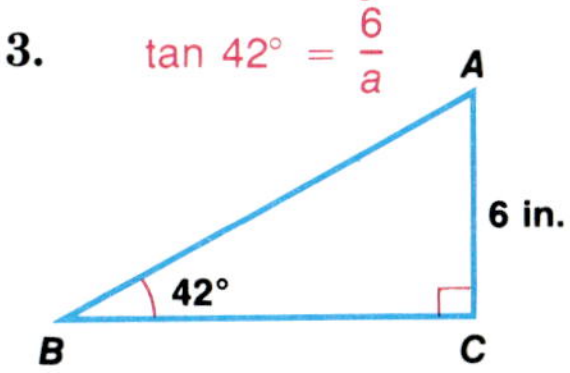

4. $\sin 71° = \frac{l}{12}$, $\cos 71° = \frac{n}{12}$

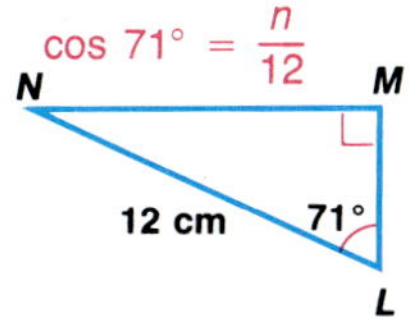

Chalkboard Exercises

3. Find the length of $\overline{JK}$ in triangle JKL.

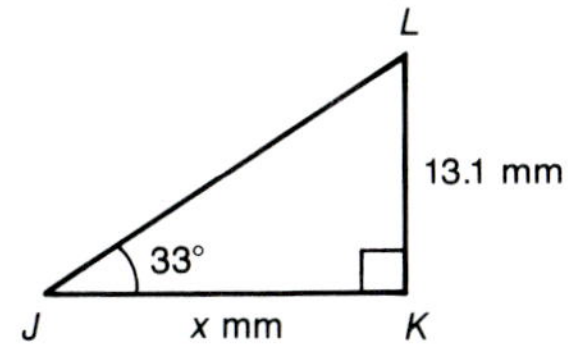

$\tan 33° = \frac{13.1}{x}$

$0.6494 \approx \frac{13.1}{x}$

$x \approx 20.2$

20.2 mm

4. Find the missing measures in the triangle below.

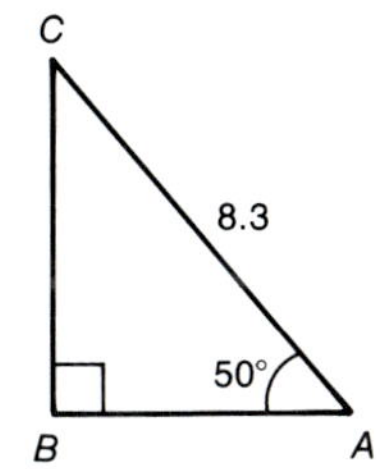

$AB \approx 5.3$

$CB \approx 6.4$

$m\angle C = 40$

Suggested Daily Assignments

Enriched
- Written: 1–26
- Mini-Review: 1–5

Average
- Exp: 2–8 evens
- Written: 1–26
- Mini-Review: 1–5

Fundamental
- Exp: 1–8
- Written: 1–18
  - p. 513, 39–47 odds
- Mini-Review: 1–5

Answers to Written

1. $\angle A = 44°$; $BC = 10.4$ ft.; $AC = 10.8$ ft.
2. $\angle Z = 51°$; $XZ = 15.4$ m; $ZY = 9.7$ m
3. $\angle A = 48°$; $BC = 6.7$ in.; $AB = 9.0$ in.
4. $\angle N = 19°$; $ML = 3.9$ cm; $MN = 11.3$ cm
5. $\angle A = 26°$; $\angle B = 64°$; $AC = 8.1$ cm
6. $\angle A = 23°$; $\angle B = 67°$; $AC = 12$ ft.
7. $\angle A = 52°$; $\angle B = 38°$; $BC = 10.2$ mm
8. $\angle A = 53°$; $\angle B = 37°$; $AB = 15$ in.
17. $\angle B = 59°$; $AC = 10$ m; $AB = 11.6$ m
18. $\angle A = 37°$; $\angle B = 53°$; $AC = 8$ in.
19. $\angle A = 48°$; $BC = 7.4$ in.; $AC = 6.7$ in.
20. $\angle A = 39°$; $\angle B = 51°$; $AB = 6.4$ ft.
21. $\angle A = 38°$; $\angle B = 52°$; $BC = 8.7$ ft.
22. $\angle A = 57°$; $\angle B = 33°$; $BC = 9.2$ m
23. $\angle A = 50°$; $BC = 7.2$ cm; $AB = 9.3$ cm
24. $\angle A = 62°$; $AC = 8.5$ cm; $AB = 18.1$ cm
25. $\angle B = 45°$; $AC = 1$ ft.; $BC = 1$ ft.
26. $\angle B = 15°$; $BC = 11.2$ km; $AB = 11.6$ km

How would you find the measures of angles A and B below?

5.

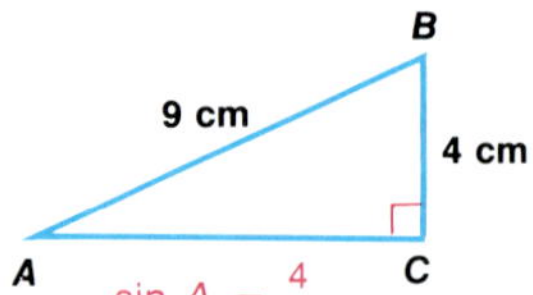

$\sin A = \frac{4}{9}$

$\cos B = \frac{4}{9}$

6.

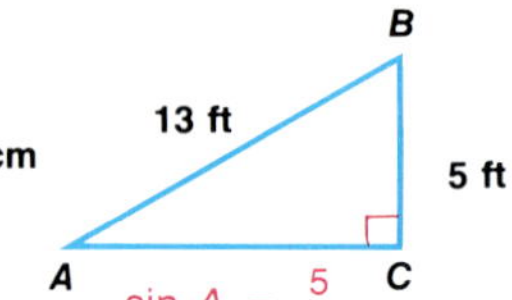

$\sin A = \frac{5}{13}$

$\cos B = \frac{5}{13}$

7.

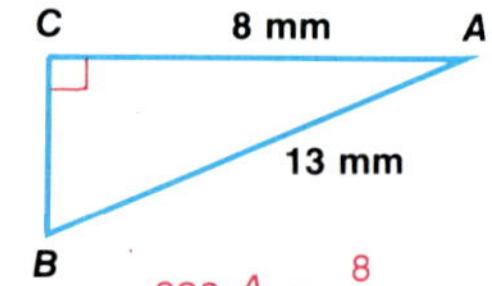

$\cos A = \frac{8}{13}$

$\sin B = \frac{8}{13}$

8.

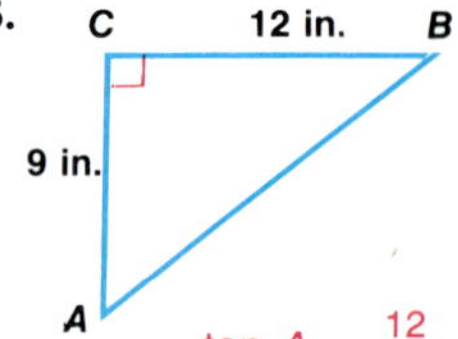

$\tan A = \frac{12}{9}$

$\tan B = \frac{9}{12}$

Written Exercises

1-8. Solve each triangle in the Exploratory Exercises.

Solve each triangle below.

9.

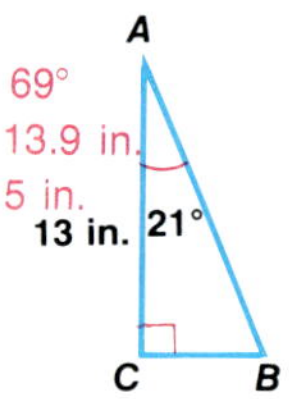

$\angle B = 69°$
$AB = 13.9$ in.
$BC = 5$ in.

10.

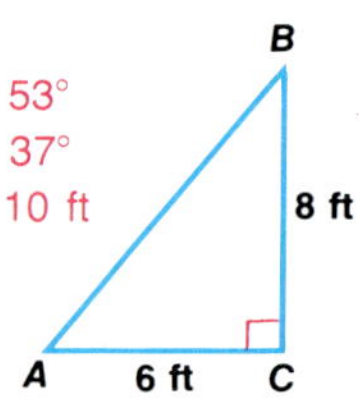

$\angle A = 53°$
$\angle B = 37°$
$AB = 10$ ft

11.

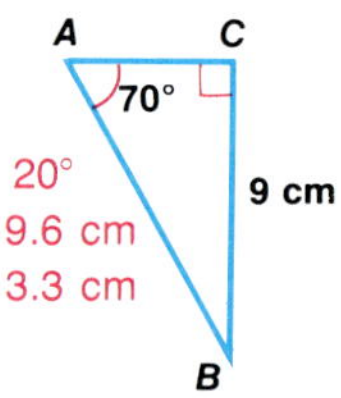

$\angle B = 20°$
$AB = 9.6$ cm
$AC = 3.3$ cm

12.

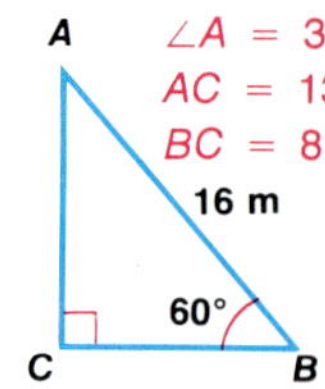

$\angle A = 30°$
$AC = 13.$
$BC = 8$ m

13.

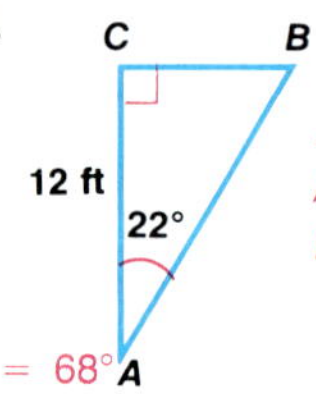

$\angle B = 68°$
$AB = 12.9$ ft
$BC = 4.8$ ft

14.

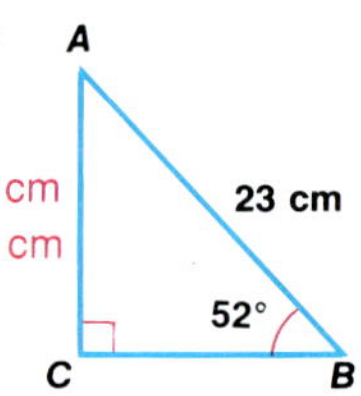

$\angle A = 38°$
$AC = 18.1$ cm
$BC = 14.2$ cm

15.

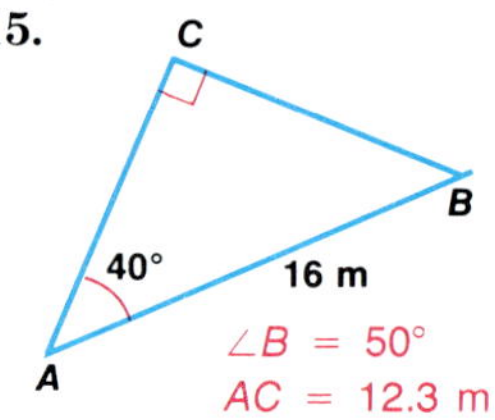

$\angle B = 50°$
$AC = 12.3$ m
$BC = 10.3$ m

16.

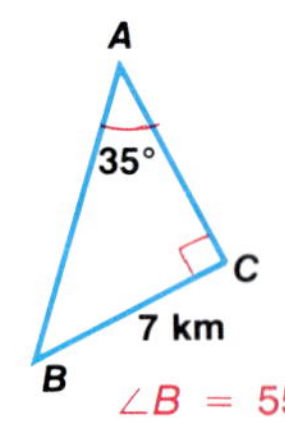

$\angle B = 55°$
$AB = 12.2$
$AC = 10$ k

Solve each of the following right triangles. (Angle C is a right angle.)

17. angle $A = 31°$, $a = 6$ m
18. $a = 6$ in., $c = 10$ in.
19. angle $B = 42°$, $c = 10$ in.
20. $b = 5$ ft, $a = 4$ ft
21. $c = 14$ ft, $b = 11$ ft
22. $c = 11$ m, $b = 6$ m
23. angle $B = 40°$, $b = 6$ cm
24. angle $B = 28°$, $a = 16$ cm
25. angle $A = 45°$, $c = \sqrt{2}$ ft
26. angle $A = 75°$, $b = 3$ km

mini-review

Multiply or divide.

1. $\frac{2a + 8}{a^2 - 25} \cdot \frac{a - 5}{5a + 10}$ 1. $\frac{2(a + 4)}{5(a + 5)(a + 2)}$

2. $\frac{y^2}{x^2} \div \frac{a^2}{x^3}$ $\frac{xy^2}{a^2}$

Add or subtract.

3. $\frac{2x}{x - 7} - \frac{14}{x - 7}$ 2

4. $\frac{2}{x - y} + \frac{y}{y - x}$ $\frac{2 - y}{x - y}$

Solve.

5. $\frac{3}{x} + \frac{1}{x - 5} = \frac{1}{2x}$ $\frac{25}{7}$

Objective 16-7: Use trigonometric ratios to solve verbal problems.

16-7 Problem Solving: Using Trigonometry

In order to use the trigonometric ratios to solve problems, it is helpful to understand the meaning of angles of **elevation** and **depression**.

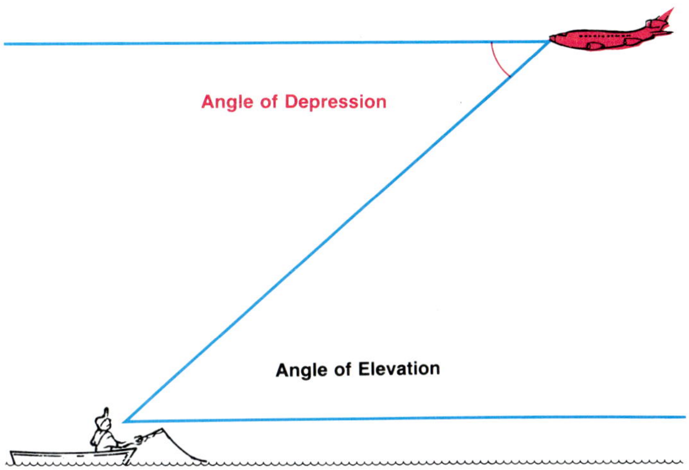

The angle of elevation or depression is the angle formed by the line of sight to an object and a horizontal line.

Example 1

A radio tower casts a shadow 120 meters long when the angle of elevation of the sun is 41°. How tall is the tower?

Explore It would be helpful to first draw a picture for the problem.

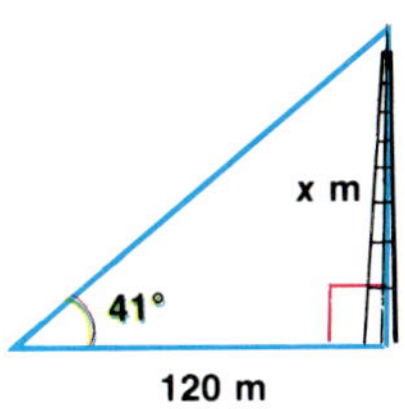

Note that the angle of elevation opens upward.

Plan $\tan 41° = \frac{x}{120}$

Solve

$$0.8693 \approx \frac{x}{120}$$

$$(0.8693)120 \approx x$$

$$104.3 \approx x$$

The height of the tower is approximately 104.3 meters. *Examine the solution.*

16-7

Lesson Notes, page T55

Discussion Question

(Use after Example 1.)
A boy is watching a friend climb an observation tower. Does the measure of the angle of elevation to the friend vary directly as the height of the friend from the ground?

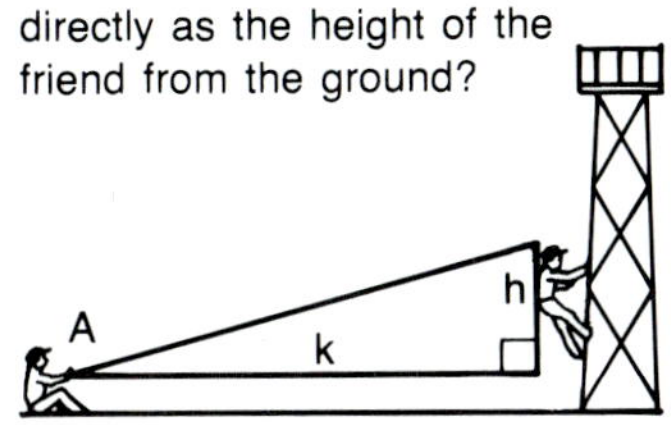

No, the measure of the angle does not vary directly as the height. But the tangent of the angle of elevation varies directly as the height.

$\tan A = \frac{h}{k} = \frac{1}{k} \cdot h$,

$\frac{1}{k}$ is constant.

Chalkboard Exercises

Use trigonometric ratios and the table on page 545 to solve each of the following problems.

1a. A tower casts a shadow of 110 m. The angle of elevation of the sun is 47°. How tall is the tower? 118 m

1b. A tower 93.7 m tall casts a shadow of 150 m. Find the angle of elevation to the nearest degree. 32°

Chalkboard Exercises

Use trigonometric ratios and the table on page 545 to solve each of the following problems.

2a. A boat is sighted from a tower 40 m high. The angle of depression is 24°. How far is the boat from the base of the tower? 89.8 m

2b. Cables are attached to the top of a tower 35 m above the base. The cables form an angle of depression of 65°. How far from the base is the tower anchored? How long are the cables? 16.3 m; 38.6 m

Suggested Daily Assignments

Enriched
- Exp: 1–4
- Written: 1–14

Average
- Exp: 1–4
- Written: 1–12

Fundamental
- Exp: 1–4
- Written: 1–12
  - p. 516, 20–26 evens

Example

2 **From the top of an observation tower 50 meters high, a forest ranger spotted a deer at an angle of depression of 28°. How far was the deer from the base of the tower?**

First make a drawing.

Notice that the angle of depression opens downward.

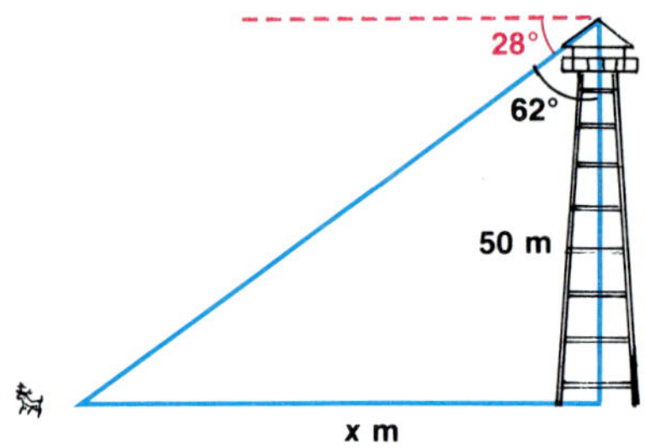

$$\tan 62° = \frac{x}{50}$$

$$1.8807 \approx \frac{x}{50}$$

$$(1.8807)50 \approx x$$

$$94 \approx x$$

The deer is about 94 meters from the tower.

Exploratory Exercises

Name the angles of elevation and depression in each of the following.

1.

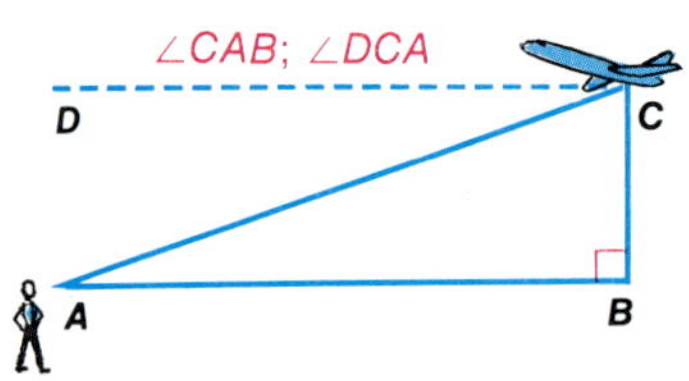

2.

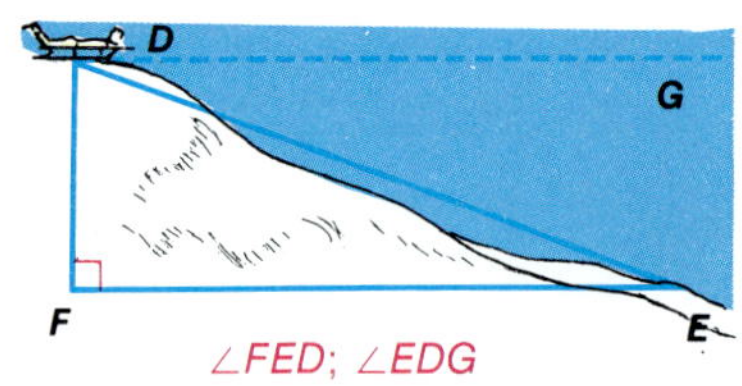

3.

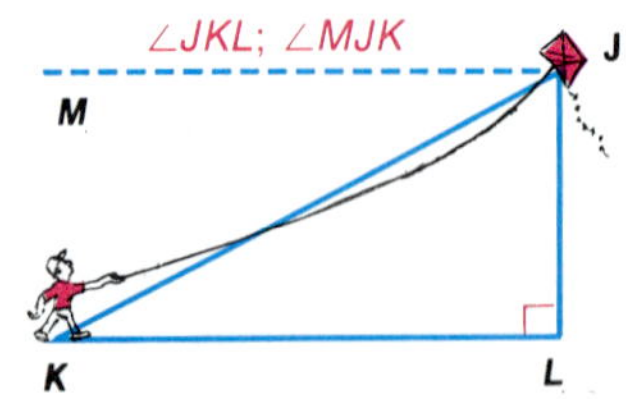

4.

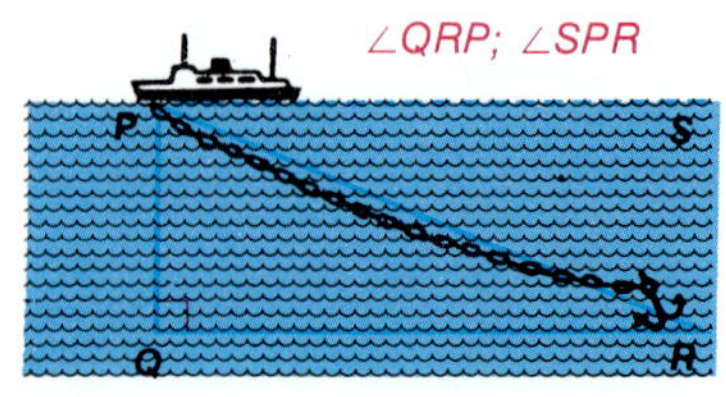

Written Exercises ❶

Use trigonometric ratios and the table on page 545 to solve each problem.

1. A road rises 38 feet vertically over a horizontal distance of 540 feet. What is the angle of elevation of the road? 4°

2. At a point 200 feet from the base of a flagpole, the angle of elevation is 62°. Find the height of the flagpole and the distance from the point to the top. 376 ft; 426 ft

3. A chimney casts a shadow 75 feet long when the angle of elevation of the sun is 41°. How tall is the chimney? 65 ft

4. A train in the mountains rises 8 feet for every 200 feet it moves along the track. Find the angle of elevation of the tracks. 2.3°

5. At a point 210 feet from the base of a building, the angle of elevation to the top of the building is 55°. How tall is the building? 300 ft

6. How far will a submarine travel when going to a depth of 300 feet if its course has an angle of depression of 25°? 709.9 ft

7. How long of a guy wire will be needed for a TV tower if the wire is fastened to the tower 40 feet above the ground and forms an angle of 52° with the tower? 65 ft

8. A roof is constructed as shown in the diagram below. Find the pitch (angle of elevation) of the roof. 29°

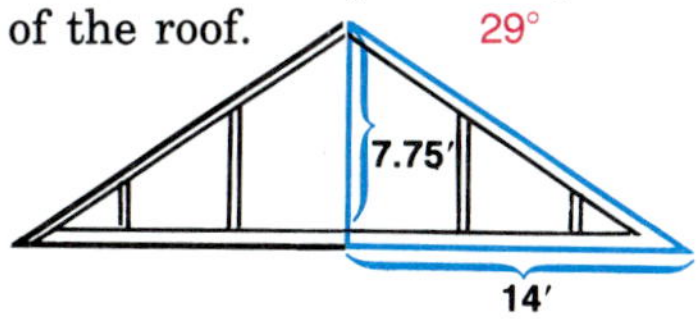

9. Find the area of a right triangle in which one acute angle is 25° and the leg opposite that angle is 40 cm in length. 1716 cm^2

10. From the top of a 70-meter lighthouse, an airplane was observed that was directly over a ship. The angle of elevation of the plane was 18°, while the angle of depression of the ship was 25°. Find the distance from the ship to the foot of the lighthouse and the height of the plane. 150.1 m; 118.8 m

11. A camera has a field of vision of 50° as shown below. The camera is being used to photograph a ship that is 1300 feet long. How far from the ship must the camera be held to photograph the entire ship? 1,393.9 ft

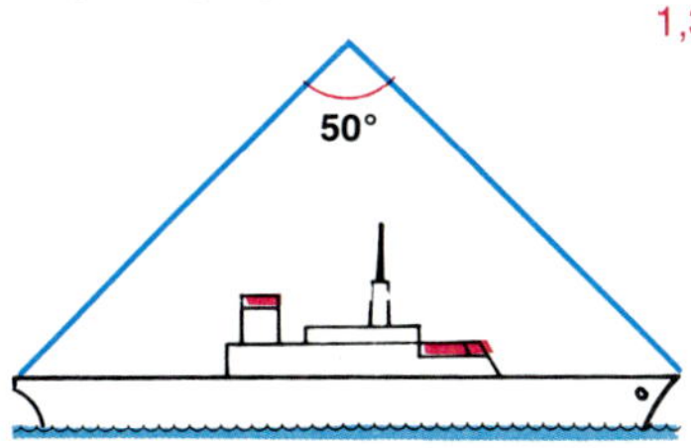

12. An airplane, at an altitude of 2000 feet, is directly over a power plant. The navigator finds the angle of depression of the airport to be 19°. How far is the plane from the airport? How far is it from the power plant to the airport? 6143 ft; 5808 ft

Challenge

Solve the following triangles.

13.

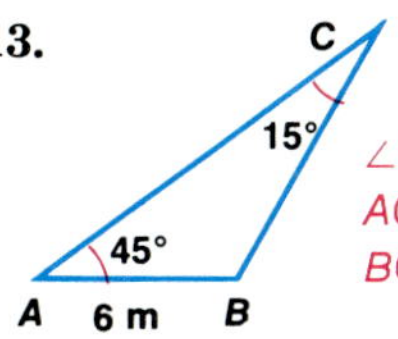

$\angle B = 120°$
$AC = 20.1$ m
$BC = 16.4$ m

14.

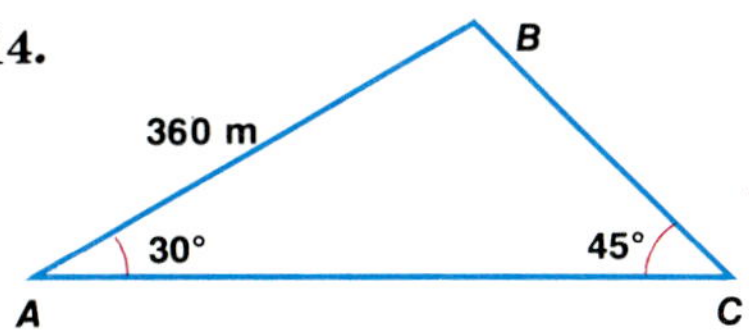

$\angle B = 105°$
$BC = 254.6$ m
$AC = 491.8$ m

❶ Encourage students to draw diagrams to help them solve these problems.

About the Application
An important factor which surveyors must consider is elevation. Sometimes the surveyor is not at the same elevation as the object being measured, and measurements must be adjusted accordingly. The example and exercises are based on the assumption that the instrument and object are at level ground.

Applications in Surveying

Angle of Elevation

Glenn Hatfield is a licensed surveyor. In addition to measuring land, he uses his surveying instruments to check building construction. Glenn frequently uses trigonometry in his work.

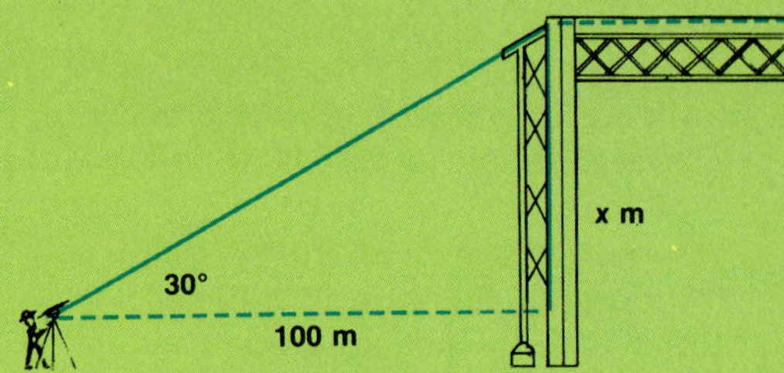

As shown above, Glenn has used his instrument to measure the angle of elevation and find the height from the top of the instrument to the top of the bridge.

$$\tan 30° = \frac{x}{100}$$

$$0.5774 \approx \frac{x}{100}$$

$$57.74 \approx x$$

If the instrument height is 1.54 m, what is the height of the bridge? 59.28 m

Use trigonometry to find the heights of various objects for the following measurements.

	Angle of Elevation	*Distance from Instrument*	*Instrument Height*	
1.	35°	100 m	1.45 m	**1.** 71.47 m
2.	26°	62.56 m	1.73 m	**2.** 32.24 m
3.	11°	131.37 m	1.68 m	**3.** 27.22 m
4.	3°	895 m	1.54 m	**4.** 48.44 m

Using Computers

Trigonometric Functions

Trigonometric functions are used in many applications of mathematics. A computer can calculate values of trigonometric functions very rapidly. However, the values are found for angle measures in *radians*, not for angle measures in degrees. For an angle whose measure is D degrees, the measure in radians, R, is $D \cdot \pi/180$.

Trigonometric functions usually available in BASIC include SIN (sine), COS (cosine), and TAN (tangent). The following program uses the SIN function to print a table of sines.

```
 5 PRINT "ANGLE" , "SINE"
10 FOR D = 0 TO 360 STEP 30
20 LET R = D * 3.1416 / 180
25 LET R1 = INT ( SIN (R) * 1000 + .5)/1000
30 PRINT D,R1
40 NEXT D
50 END
```

ANGLE	SINE
0	0
30	.5
60	.866
90	1
120	.866
150	.5
180	0
210	-.5
240	-.866
270	-1
300	-.866
330	-.5
360	0

The sine function generates ordered pairs. To graph the sine function, use the horizontal axis for the degree values. Use the vertical axis for the sine. After plotting the points, complete the graph by connecting the plotted points with a smooth continuous curve as shown.

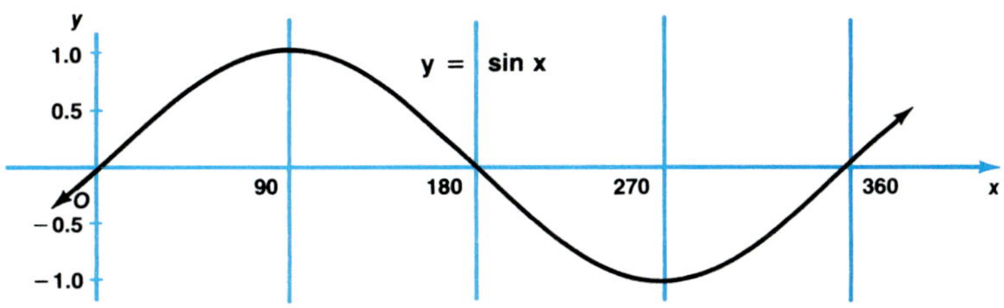

1\. 5 PRINT "ANGLE", "COSINE"
25 LET R1 = INT(COS(R) * 1000 + .5)/1000

Exercises

1. Modify the program to print a table of cosines and then graph the cosine function.

State whether the value of each of the following is positive or negative.

2. sin 30° positive
3. cos 150° negative
4. sin 330° negative
5. cos 60° positive

State which is greater.

6. cos 30° or cos 90° cos 30°
7. sin 0° or sin 90° sin 90°

Find the values of x for which each of the following is true.

8. cos x = 1 0°, 360°
9. sin x = 1 90°
10. sin x = 0 0°, 180°, 360°
11. cos x = −1 180°
12. How are the graphs of the sine function and cosine function related?
They have the same shape, but are located in different positions on the x-axis.

About Using Computers
Explain to students that computers can be programmed to draw the graph mentioned in this feature. You may want to discuss the graphs only with your more advanced students.

Quiz
Lessons 16–4 to 16–7

1. Express the sine, cosine, and tangent of angle B to 3 decimal places.

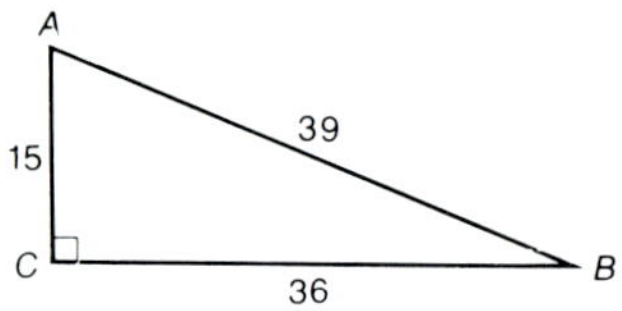

$\sin B = \frac{15}{39} \approx 0.385$

$\cos B = \frac{36}{39} \approx 0.923$

$\tan B = \frac{15}{36} \approx 0.417$

Use the table on page 545 to find a value for each of the following.

2. sin 88° 0.9994
3. cos 33° 0.8387
4. tan 56° 1.4826

For each exercise, find the measure of angle B to the nearest degree.

5. $\cos B = 0.5340$ 58°
6. $\tan B = 0.8956$ 42°

Find the missing measures in each triangle.

7.

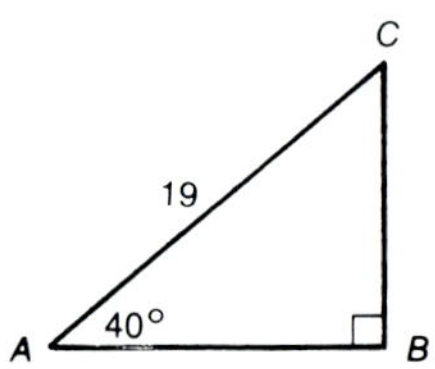

$CB \approx 12.2$
$AB \approx 14.6$
$C = 50°$

8.

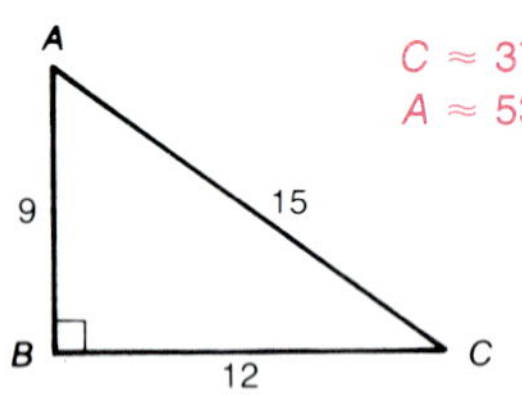

$C \approx 37°$
$A \approx 53°$

Solve each problem.

9. A cross-country slope rises 2 ft for every 60 ft of horizontal distance. Find the angle of elevation. 1.9° or 2°

Vocabulary

supplementary angles (499)
complementary angles (499)
30° – 60° right triangles (502)
similar triangles (505)
corresponding angles (505)
corresponding sides (505)
trigonometric ratios (508)
sine (508)
cosine (508)
tangent (508)
solving triangles (514)
angle of elevation (5107)
angle of depression (517)

Chapter Summary

1. Two angles are supplementary if the sum of their measures is 180°. (499)
2. Two angles are complementary if the sum of their measures is 90°. (499)
3. The sum of the measures of the angles in any triangle is 180°. (500)
4. In any 30° – 60° right triangle, the measure of the side opposite the 30° angle is one half the measure of the hypotenuse. (502)
5. In a 30° – 60° right triangle, if a is the measure of the side opposite the 30° angle, then $a\sqrt{3}$ is the measure of the side opposite the 60° angle. (503)
6. Two triangles are similar if the measures of their corresponding angles are equal. (505)
7. If two triangles are similar, the measures of their corresponding sides are proportional. (505)
8. Sine of an angle:

$$\sin A = \frac{\text{measure of side opposite angle } A}{\text{measure of hypotenuse}} \quad (508)$$

9. Cosine of an angle:

$$\cos A = \frac{\text{measure of side adjacent to angle } A}{\text{measure of hypotenuse}} \quad (508)$$

10. Tangent of an angle:

$$\tan A = \frac{\text{measure of side opposite angle } A}{\text{measure of side adjacent to angle } A} \quad (508)$$

11. The trigonometric ratios can be used to find the missing measures of a triangle. (514)

10. The angle of depression is 22° from the top of a building to a point 300 ft from the base of the building. How tall is the building? 121 ft

Chapter Review

16-1 **Find the complement of each angle whose measure is listed below.**

1. 66° 24°
2. 42° 48°
3. $y°$ $(90 - y)°$

Find the supplement of each angle whose measure is listed below.

4. 62° 118°
5. 148° 32°
6. $m°$ $(180 - m)°$

In each of the following, the measures of two angles of a triangle are given. Find the measure of each third angle.

7. 16°, 72° 92°
8. 42°, 121° 17°
9. $y°$, $x°$ $(180 - x - y)°$

16-2 **Find the missing measures of each of the 30°–60° right triangles described below. (Use $\sqrt{3} \approx 1.732$.)**

	Hypotenuse	Side Opposite 30° Angle	Side Opposite 60° Angle
10.	8 cm	4 cm	$4\sqrt{3}$ or 6.928 cm
11.	4.25 cm	2.125 cm	$2.125\sqrt{3}$ or 3.681 cm
12.	12.4 m	6.2 m	$6.2\sqrt{3}$ or 10.738 m
13.	7 in.	$3\frac{1}{2}$ in.	$3.5\sqrt{3}$ or 6.062 in.
14.	4 in.	2 in.	$2\sqrt{3}$ in.
15.	9 m	4.5 m	$4.5\sqrt{3}$ m

16-3 **Triangles *ABC* and *DEF* are similar.**

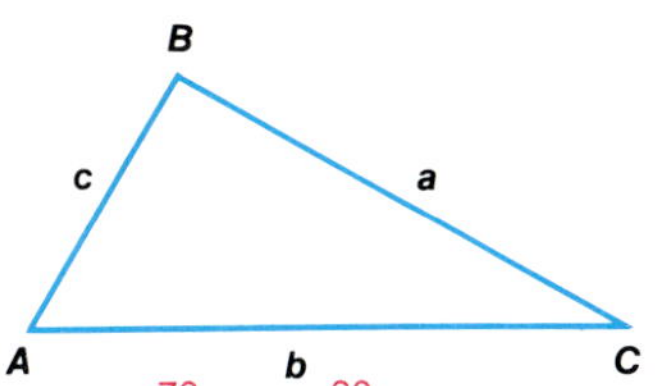

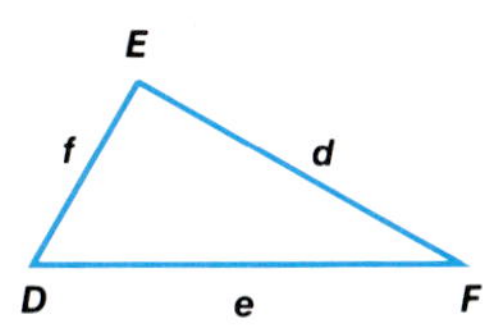

16. $b = \frac{70}{11}$, $c = \frac{30}{11}$

For each set of measures, find the measures of the missing sides.

16. $a = 5$, $d = 11$, $f = 6$, $e = 14$
17. $c = 16$, $b = 12$, $a = 10$, $f = 9$
18. $a = 8$, $c = 10$, $b = 6$, $f = 12$

17. $d = \frac{45}{8}$, $e = \frac{27}{4}$

18. $d = \frac{48}{5}$, $e = \frac{36}{5}$

16-4 **Use the triangle below. Express each of the following ratios as a fraction.**

19. sin A 19. $\frac{28}{53}$
20. cos B 20. $\frac{28}{53}$
21. cos A 21. $\frac{45}{53}$
22. sin B 22. $\frac{45}{53}$
23. tan A 23. $\frac{28}{45}$
24. tan B 24. $\frac{45}{28}$

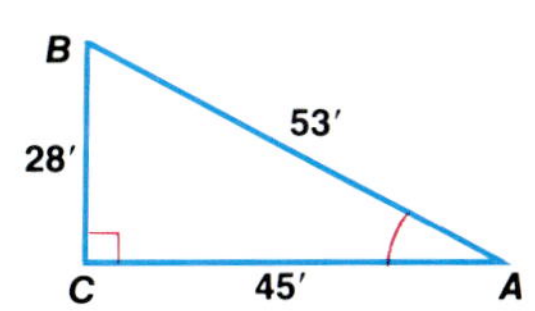

State each of the following correct to 3 decimal places.

25. sin A 0.528

26. cos A 0.849

27. tan B 1.607

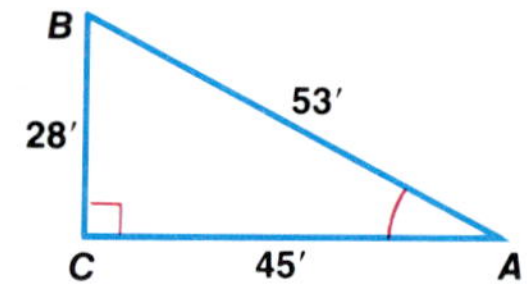

16-5 **Use the table of trigonometric ratios on page 545 or a calculator to find a value for each of the following.**

28. cos 61° 0.4848 **29.** sin 42° 0.6691 **30.** tan 13° 0.2309

31. sin 84° 0.9945 **32.** cos 17° 0.9563 **33.** tan 66° 2.2460

16-6 **Using the triangles below, find each of the following.**

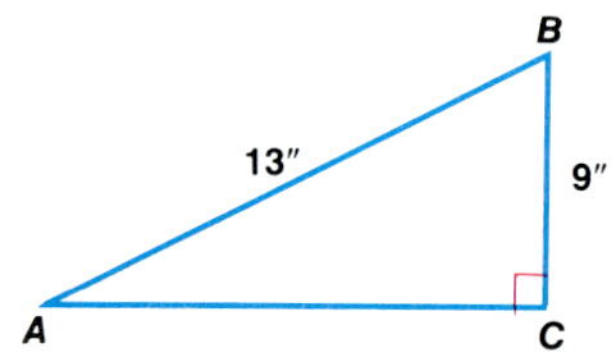

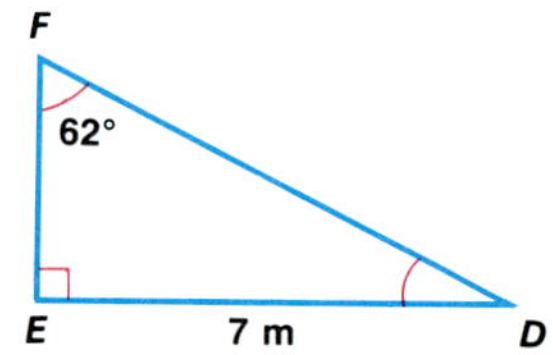

34. angle A 44° **35.** angle D 28°

36. angle B 46° **37.** side EF 3.72 m

38. side AC 9.38″ **39.** side FD 7.93 m

16-7 **Solve each problem.**

40. A cliff is 250 feet above the ocean. From the cliff, the angle of depression of a boat in the water is 10°. How far is the boat from the base of the cliff? 1418 ft

41. A weather balloon is directly above a tree. The angle of elevation is 60° when you are 100 meters from the tree. How high is the balloon? 173.2 m

42. The diagonal of a rectangle is 16 cm long and makes an angle of 50° with a side of the rectangle. Find the length and width of the rectangle. 12.26 cm; 10.28 cm

Chapter Test

Find the complement and supplement of each angle whose measure is listed below.

1. 28° 62°; 152° **2.** 69° 21°; 111° **3.** $(y + 20)°$ $(70 - y)°$; $(160 - y)°$

In each of the following, the measures of two angles of a triangle are given. Find the measure of the third angle.

4. 16°, 47° 117° **5.** 89°, 66° 25° **6.** 45°, 120° 15°

Find the missing measures of each of the 30°–60° right triangles described below. (Use $\sqrt{3} \approx 1.732$.)

	Hypotenuse	Side Opposite 30° Angle	Side Opposite 60° Angle
7.	17 in.	8.5 in.	$8.5\sqrt{3}$ or 14.722 in.
8.	16 ft	8 ft	$8\sqrt{3}$ or 13.856 ft
9.	18 m	9 m	$9\sqrt{3}$ m

Triangles *ABC* and *JKL* are similar. For each set of measures, find the missing side measures.

10. $c = 20, l = 15, k = 16, j = 12$ $a = 16; b = \frac{64}{3}$

11. $c = 12, b = 13, a = 6, l = 10$ $j = 5; k = \frac{65}{6}$

12. $k = 5, c = 6.5, b = 7.5, a = 4.5$ $j = 3; l = \frac{13}{3}$

13. $l = 1\frac{1}{2}, c = 4\frac{1}{2}, k = 2\frac{1}{4}, a = 3$ $j = 1; b = \frac{27}{4}$

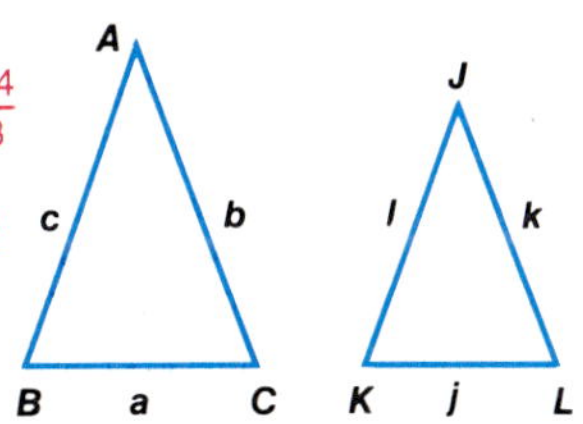

14. $\angle B = 34°$; $b \approx 11.47$; $c \approx 20.5$ **15.** $\angle A = 37°$; $\angle B = 53°$; $c = 20$

Solve each of the following right triangles. (Angle *C* is a right angle.)

14. angle $A = 56°$, $a = 17$ **15.** $a = 12$, $b = 16$

16. angle $B = 42°$, $c = 10$ **17.** $b = 21$, $c = 29$

16. $\angle A = 48°$; $a = 7.43$; $b = 6.69$ **17.** $\angle A = 44°$; $\angle B = 46°$; $a = 20$

Solve each problem.

18. A 6-foot pole casts a 4-foot shadow. How tall is a tree which casts a 50-foot shadow? 75 ft

19. A kite is flying at the end of a 300-foot string. Assuming the string is straight and forms an angle of 58° with the ground, how high is the kite? 254.4 ft

20. A plane is 1000 feet above the ground. The angle of depression of the landing strip is 20°. How far is the plane from the landing strip? 2747 ft

Using the Cumulative Review
If students have difficulty with a particular topic, have them review the appropriate lesson. You may want to provide copies of the solutions from the *Algebra One Solutions Manual*. This Cumulative Review covers Chapters 1-16.

Other Cumulative Reviews appear on pages 72, 142, 208, 268, 334, 402, and 466.

Answer

9. 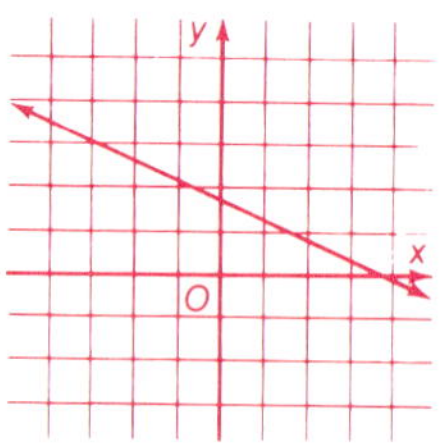

Cumulative Review

1. State the property shown by $4379 \cdot 0 = 0$ multiplicative property of zero
2. Solve: $63 = 11 - x$ -52
3. 40% of what number is 30? 75
4. Solve: $\frac{1}{2}a + 3 \geq -\frac{1}{3}(a + 6)$ $a \geq -6$
5. Simplify: $\frac{-18x^6y^2}{24x^4y^6}$ $-\frac{3x^2}{4y^4}$

Factor, if possible.

6. $5x^2 + 25xy - 30xy^2$ $5x(x + 5y - 6y^2)$
7. $m^2 + 14m + 35$ not factorable
8. Solve: $3y^2 - 13y = -14$ $\frac{7}{3}, 2$
9. Graph: $2x + 5y = 8$
10. Given $f(x) = x^2 - 2x - 4$, find $f\left(\frac{1}{2}\right)$. $-4\frac{3}{4}$
11. Write an equation in standard form for the line passing through (1, 3) and having a slope of -2. $2x + y = 5$
12. Write an equation for the line that is perpendicular to the graph of $5x - 9y = 4$ and passes through (2, 1). Use slope-intercept form. $y = -\frac{9}{5}x + \frac{23}{5}$
13. Use substitution to solve the system of equations $\frac{1}{2}x - y = 6$ and $4x + 2y = 8$. $(4, -4)$
14. Solve: $\sqrt{4x^2 + 4x + 1} = 7$ $-4, 3$
15. Find the distance between $(3, -6)$ and $(-7, -4)$. $2\sqrt{26}$
16. Solve $r^2 - 6r + 4 = 0$ by completing the square. $3 \pm \sqrt{5}$
17. Determine the number of real roots for $6x^2 + x = 1$ by using the discriminant. Find all real roots. 2 real roots; $-\frac{1}{2}, \frac{1}{3}$
18. Solve: $8k = k^2 + 1$ $4 \pm \sqrt{15}$

Find each quotient.

19. $\frac{2m^2 + 5m - 3}{m^2 - 9} \div \frac{4m^2 - 1}{2m^2 - 5m - 3}$ 1
20. $(3t^3 - 11t^2 - 31t + 7) \div (3t + 1)$ $t^2 - 4t - 9 + \frac{16}{3t + 1}$

21. Simplify: $2 + \frac{a^2 + 1}{a - 1}$ $\frac{a^2 + 2a - 1}{a - 1}$
22. Solve: $\frac{4}{t - 2} - \frac{2t - 3}{t^2 - 4} = \frac{5}{t - 2}$ $\frac{1}{3}$
23. If y varies directly as x and $y = 14$ when $x = 3$, find x when $y = 28$. 6
24. If y varies inversely as x and $y = 2$ when $x = 15$, find x when $y = 3$. 10
25. Find the complement and the supplement of an angle whose measure is 32°.
26. The length of the side opposite the 60° angle in a 30°–60° right triangle has a length of $5\sqrt{3}$ cm. Find the length of the other two sides. 5 cm, 10 cm

25. complement = 58°
supplement = 148°

29. 27 m, 9 m

Problem Solving

Solve each problem.

27. 12, 13, 14, 15

27. Find four consecutive integers such that the sum of the first and third is 26.
28. Tom and Carroll live 70 miles apart. At noon, they begin bicycling towards each other. If Tom rides 4 mph faster than Carroll and they meet at 2:30 P.M., how fast does each travel? Tom, 16 mph Carroll, 12 mph
29. The length of a rectangle is 3 times its width. The area of the rectangle is 243 square meters. Find the length and width.
30. The product of the digits of a two-digit number is 18. The units digit is 3 more than the tens digit. Find the number. 36
31. Carlos can do a job in 5 days. Cameron can do the same job in 3 days. If they work together, how long will it take them to complete the job? $1\frac{7}{8}$ days
32. A helicopter flying at an altitude of 3000 feet is directly over a car. How far is the car from the helicopter's landing spot if the angle of depression of the landing spot is 20°? 8242 feet

Standardized Test Practice Questions

The questions on this page involve comparing two quantities, one in Column A and one in Column B. In certain questions, information related to one or both quantities is centered above them. All variables used stand for real numbers.

Directions:

Write A if the quantity in Column A is greater.

Write B if the quantity in Column B is greater.

Write C if the quantities are equal.

Write D if there is not enough information to determine the relationship.

	Column A	Column B
1. B	$(0.64)^2$	$\sqrt{64}$
2. C	$\frac{1}{\sqrt{2}}$	$\frac{1}{2}\sqrt{2}$
3. A	9^2	$\sqrt{900}$
	$x^2 = 25$, $y^2 = 36$	
4. D	x	y
	$x > 1$	
5. A	$\frac{x^3 + x^2}{x}$	$\frac{x^3 + x^2}{x^2}$
6. D	a given chord in a given circle	the radius of the same circle
7. D	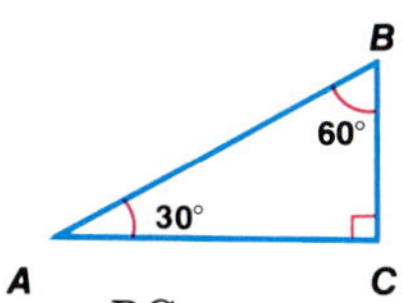 BC	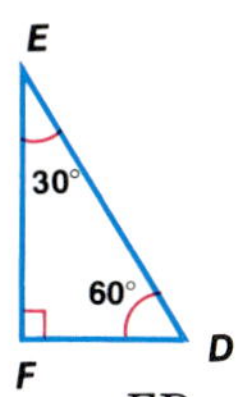 FD
8. A	$\frac{3}{4} + \frac{3}{7}$	$\frac{19}{21} - \frac{3}{7}$
9. A	$(5 + 7)^2$	$5^2 + 7^2$

Examples

Column A	Column B
$\frac{1}{k} < 0$	
1. $-\frac{1}{k}$	k

The answer is A because $-\frac{1}{k}$ is positive, while k is negative.

Column A	Column B
2. $\frac{3^4 + 3^5}{3^4}$	$\frac{3^2 + 3^2}{3^2}$

The answer is C because the simplest form of each quantity is 1 + 3 or 4.

Hint: $\frac{3^4 + 3^5}{3^4} = \frac{3^4}{3^4} + \frac{3^5}{3^4}$

	Column A	Column B
	The distance from City A to City B is 50 miles.	
10. B	The time required to travel from A to B at 55 mph.	60 minutes
	$a > b > c$; $a \neq 0, b \neq 0, c \neq 0$	
11. D	ab	bc
	$a > b > c > 0$	
12. A	ab	bc
13. A	$(x + 2)(x + 3)$	$(x + 1)(x + 4)$
	$r > 7, 0 < t < 6$	
14. B	$\frac{1}{r}$	$\frac{1}{t}$
	$3x + 2 = y$	
15. B	$30x + 15$	$10y$

Using the Standardized Test Practice Questions

The questions on this page may be used to help students prepare for pre-college tests. These questions usually require careful analysis and a thorough understanding of the concepts.

This page could be used as an overnight assignment for better students. Other students may benefit most if only a few questions are assigned at a time.

After students have chosen an answer for each problem, you should discuss how each problem can be solved, or provide copies of the solutions from the *Algebra One Solutions Manual*.

Other practice questions appear in this text on pages 73, 143, 209, 269, 335, 403, and 467. Also see pages 50-58 in the *Teacher Resource Book*.

Overview
The main topics included in this appendix are assignment of variables, FOR-NEXT loops, and IF-THEN statements. Computer systems may vary in several ways such as sign-on, sign-off procedures, panic stop, spacing used in printouts, and so on. The teacher will need to guide the students carefully where their system varies. Make sure students know how to gain access to the particular computer system available to them, and how to correct typing errors when entering a program.

❶ Emphasize that variables may be represented by a letter or a letter followed by a single digit. Some computer systems allow a variable to be represented by several letters, such as AB or SUM.

❷ Some computers perform sign changes as second in the order of operations. That is, the expression −X ↑ 2 is sometimes interpreted as (−X) ↑ 2 instead of −(X ↑ 2) as in algebra. To avoid misinterpretation, use parentheses. Another option is to show −X ↑ 2 as −1 * X ↑ 2. Emphasize that, in many systems, symbols cannot appear in succession without the use of parentheses. Thus, A * −B may need to be entered as A * (−B).

Extra Notes
Some systems use capital and lower case letters. Capital letters are used for BASIC expressions in this text.

Appendix: BASIC

The BASIC Language

BASIC (Beginner's All-Purpose Symbolic Instruction Code) is a computer language. The symbols used in BASIC are similar to those used in algebra. Compare the symbols in the following lists.

Algebra	BASIC	Algebra	BASIC
+	+	>	>
−	−	≥	> =
× or ·	*	<	<
÷	/	≤	< =
=	=	≠	< >
4^3	4↑3 or 4 ^ 3	x	X

A variable in BASIC is represented by a capital letter or a capital letter followed by a numeral. Examples are A, B, X, N1, and M2. ❶

In BASIC, the multiplication symbol may never be omitted. To write A times B, write A * B, not AB as in algebra.

Examples of algebraic and BASIC expressions are shown below.

Algebra	BASIC
$6 + 3$	6 + 3
$a - 4$	A − 4
$7t$, $7 \times t$, or $7 \cdot t$	7 * T
$24 \div x$ or $\frac{24}{x}$	24 / X
$S \geq n^2$	S > = N↑2
$y = 8\frac{1}{2}$	Y = 8.5
$y_2 \neq x_1 + 4$	Y2 < > X1 + 4

BASIC uses parentheses as they are used in algebra for grouping. ❷ Also, the same order of operations is used as in algebra.

1. Do all operations in parentheses, from the innermost parentheses outward.
2. Evaluate all powers from left to right.
3. Then do all multiplications and divisions from left to right.
4. Then do all additions and subtractions from left to right.

Order of Operations in BASIC

Examples

1 **Evaluate: 4 * 5 + 9 + 8**

```
4 * 5 + 9 + 8 = 20 + 9 + 8     First, multiply.
              = 29 + 8         Then, add from left to right.
              = 37
```

2 **Evaluate: 3 + 9 − 2 * (6 − 1)↑2 / 10**

```
3 + 9 − 2 * (6 − 1)↑2 / 10 = 3 + 9 − 2 * (5)↑2 / 10   Do all operations in parentheses.
                           = 3 + 9 − 2 * 25 / 10      Evaluate powers.
                           = 3 + 9 − 50 / 10          Do all multiplications and divisions
                           = 3 + 9 − 5                from left to right. Then, do all
                           = 12 − 5                   additions and subtractions from
                           = 7                        left to right.
```

A **computer program** is a series of statements that gives directions to the computer. The purpose of a program is to get information into the computer (**input**), do the calculations, and then get the results out of the computer (**output**).

A sample program is shown below.

```
10  PRINT 11.2 + 9.3 }
20  END              } statements
↑
line numbers
```

In a computer program, each statement has a **line number**. Numbering by tens permits statements to be inserted later. The computer follows the instructions in numerical order. A program must end with an END statement.

The programmer chooses the line numbers, integers from 1 to 9999, usually.

❸

One way to get information into the computer is to use an INPUT statement. When the computer executes an INPUT statement it prints a question mark and waits until the user types in a number.

Example

3 **Write a program to compute the sum of two numbers.**

```
10  INPUT A        Type a number and press RETURN
20  INPUT B        Type a number and press RETURN
30  PRINT A + B    The computer adds the two numbers and prints the sum.
40  END
```

The program may also be written this way.

```
10  INPUT A, B     Type a number, a comma, and a number.
20  PRINT A + B
30  END
```


❸ Some systems will run programs without END statements. However, encourage students to include the END statement to avoid potential technical problems which could occur by omitting or using more than one END statement.

❹ Point out that the READ statement *must* appear *before* the variables are used. However, the DATA statement can appear anywhere before the END statement. The purpose of READ-DATA statements is to make a program reusable simply by changing the DATA line.

❺ Emphasize that DATA statements require input only in integer or decimal form and that they *cannot include operations* such as 3 + 7, 3/7, 3 * 7, etc. Also emphasize that variables from the READ statement and numbers from the DATA list are assigned in the same order in which both are written.

❻ The example using the READ and DATA statements has been kept simple to illustrate the structure of the program. Point out that the commas in the DATA list are used to separate items. Emphasize that a number such as 61,756 must be entered as 61756 *without* a comma.

❼ Some systems must be given the command BEGIN, rather than RUN, to execute a program.

Extra Notes
Make sure students know how to gain access to the particular computer system available to them, and how to correct typing errors when entering a program.

❽ Check to see that students use parentheses properly in problems similar to exercise 9. Note that 5X + 3/2X − 1 represents $5X + \frac{3}{2X} - 1$. The correct expression is (5X + 3)/(2X − 1).

❾ Point out that there are many programs which can be written to accomplish the same task. There is more than one correct solution for exercises 32-36.

Answers

32.
```
10 READ A, B, C
20 DATA 31, 64, 82
30 PRINT A + B + C
40 END
```
33.
```
10 READ A, B, C
20 DATA 31, 64, 82
30 PRINT A * B * C
40 END
```
34.
```
10 READ A, B
20 DATA 382, 94
30 PRINT A/B
40 END
```
35.
```
10 READ A, B
20 DATA 382, 94
30 PRINT A − B
40 END
```
36.
```
10 READ W, L
20 DATA 3.9, 8.2
30 PRINT 2 * W + 2 * L
40 END
```


Another way to get information into the computer is to use a READ statement and a DATA statement. Each READ statement needs a DATA statement. The DATA statement may appear anywhere in the program but is usually placed after the READ statement. To modify the program in example 3 to include READ and DATA statements, we would replace lines 10 and 20 with ❹ ❺

```
10 READ A, B
20 DATA 22, 37
```
❻ ❼

After the program is entered, the command RUN is entered into the computer. This command instructs the computer to execute, or run, the program.

Commands, such as RUN, do not have line numbers.

If the program in example 3 is entered into the computer followed by RUN, the value of A + B is computed and printed.

Written Exercises

❽

Change each algebraic expression to a BASIC expression.

1. $a + b + c$ A + B + C
2. $6m - 7n + 8$ 6 * M − 7 * N + 8
3. $a + b \cdot 9$ A + B * 9
4. $4(y + 3)$ 4 * (Y + 3)
5. $38 \div y$ 38/Y
6. $3x^2 + 4x + 9$ 3 * X↑2 + 4 * X + 9
7. $a + \frac{5}{3 + a}$ A + 5/(3 + A)
8. $\frac{a}{b} + n$ A/B + N
9. $\frac{5x + 3}{2x - 1}$ (5 * X + 3)/(2 * X − 1)

Evaluate each of the following expressions.

10. 4 + 15 − 3 16
11. 3 * 9 * 4 − 2 106
12. 3↑2 + 5 14
13. (3 + 4)↑(2 + 1) 343
14. 2 * ((4 + 9) * 3) 78
15. 8 * (4 * (−6 + 9) / 2) / 16 3

Copy each of the following BASIC expressions. Put numerals under each operation sign to show the order of operations.

Sample: (6+ A) / 3 + X↑3
1 3 4 2

16. (X − Y) / Z↑3 + X * Y − Z (1 3 2 5 4 6)
17. 4 * (A + B)↑2 + D / E (3 1 2 5 4)
18. (M2 − M1) / (N2 − N1) + C (1 3 2 4)
19. A + B / C − 5 * M + 6 (3 1 4 2 5)

Evaluate each BASIC expression. Let A = 1, B = 2, C = 3, X = 12, Y = 0, Z = 0.5, A1 = 0.25, and A2 = 6.

20. B * C 6
21. C * X 36
22. X − B * C 6
23. A1 * X + A2 9
24. A * C − Y * Z 3
25. (B * B) * C↑2 36
26. B * X / A2 4
27. B * C * (X − Y) 72
28. (X − 2 * C) * (2 * C − B) 2
29. (X − 2 * C) / B 3
30. X * A1 * Z 1.5
31. 4 * (B * C↑2 − X * Z) / B 24

Write a program to do each of the following using either an INPUT statement or READ and DATA statements. ❾

32. Find the sum of 31, 64, and 82.
33. Find the product of 31, 64, and 82.
34. Find the quotient of 382 and 94.
35. Find the difference of 382 and 94.
36. Find the perimeter of a rectangle with a width of 3.9 cm and a length of 8.2 cm.

Assignment Statements

In BASIC, the equals sign, =, is used in a slightly different way than in algebra. In algebra, both sides of an equation may have many terms and variables. In BASIC the left side of an equation may have only a variable. This is shown by the examples below.

Algebra	BASIC
$x + 16 = 7$	X = 7 − 16
$4x + 7y = 5x + 16$	X = 7 * Y − 16
$y - 4 = 3(x - 2)^2$	Y = 3 * (X − 2)↑2 + 4

The LET statement is used to assign values to variables. The general form of a LET statement is shown below.

line number LET *variable* = *expression*

In a LET statement, the equals sign tells the computer to assign the value of the expression on the right to the variable on the left. The LET statement is another way to provide a program with data. ❶

```
10  LET X = 6
20  LET Y = 12
30  LET Z = 20
40  LET W = X + Y + Z
50  PRINT W
60  END
```

In this program 6 is assigned to X, 12 to Y, and 20 to Z. The value of X + Y + Z is computed and assigned to W. The output is 38, the value of W.

READ-DATA statements also assign values to variables.

```
10  READ R
20  DATA 0.05
30  LET I = 1000 * R * 10
40  PRINT R, I
50  END
```

Lines 10 and 20 assign 0.05 to R.

A comma in a PRINT statement causes the output to be printed in columns.

Sometimes a programmer wishes to repeat the same operation in a program. This can be done using a GO TO statement.
The general form of the GO TO statement follows.

Line number GO TO *line number*

❶ Emphasize that in LET statements the equals sign indicates assignment, not equality. The value of the expression on the right is assigned to the variable on the left. All variables on the right should have a previously assigned value. A sentence such as X = X + 1 is meaningful in BASIC although it has no algebraic solution.

Extra Notes
Most systems accept assignment statements without LET. Thus, the two lines below are equivalent in most systems.
10 X = 6 − Y
10 LET X = 6 − Y
However, using LET may help students think of the equals sign as assignment.

❷ On some computers, GO TO must be entered as one word, GOTO.

Extra Notes
A GO TO statement is convenient for setting up a repeated operation (a loop) in a program. The disadvantage, however, is that the types of programs shown here terminate with the OUT OF DATA message.

❸ When a READ statement occurs within a loop, the DATA statement should include multiple sets of data. In Example 1, notice that the READ statement has two variables, L and W. When the computer reads line 10 the first time, it will assign the value 8 to L and the value 5 to W. When the computer reads line 10 the second time, it will assign 7 to L and 2 to W. Discuss what values will be assigned the next two times line 10 is read.

❹ Point out that the punctuation in PRINT statements affects the positioning of the output. A comma or semicolon causes the output to be printed in columns or zones. A semicolon causes the output to be printed close together on one line.

Extra Notes
An INPUT statement can be used instead of a READ statement. When the computer comes to an INPUT statement, it prints a question mark and waits until the data are entered. No DATA statement is necessary. The following program uses an INPUT statement.

```
10  INPUT A, B
20  PRINT A + B
30  END
```

When the computer comes to line 10 it will print a question mark. Then the values for A and B must be entered, separated by a comma. Next the computer

Study the use of the GO TO statement in the following program. ❷

```
10  READ R
20  DATA 0.05, 0.055, 0.06, 0.065
30  I = 1000 * R * 10
40  PRINT R, I
50  GO TO 10
60  END
```

The computer assigns the first value in line 20 to R.

Line 50 instructs the computer to return to line 10. Then the second value in line 20 is assigned to R.

The process of returning to line 10 is continued until all the values in line 20 are used. Then the computer prints an error message such as OUT OF DATA.

Example 1

Write and run a program which finds the areas of four rectangles with dimensions 8 cm and 5 cm, 7 cm and 2 cm, 9 cm and 6.2 cm, and 4.5 cm and 3 cm.

```
10  READ L, W          ❸
20  LET A = L * W
30  PRINT A,           ❹
40  GO TO 10
50  DATA 8, 5, 7, 2, 9, 6.2, 4.5, 3
60  END
RUN
40        14        55.8
13.5
OUT OF DATA IN 10
```

Use A = lw, the formula for the area of a rectangle.

The result of assigning 8 to L and 5 to W is 40. The results from the other values are 14, 55.8, and 13.5.

Compare the two programs below. Notice that the program on the right uses READ-DATA statements in place of LET statements. What advantage do you see in using READ-DATA statements rather than LET statements?

```
10  LET A = 3
11  LET B = 4
12  LET C = 5
13  LET D = 6
14  LET X = 2
20  LET Y = (A + B + C)/(D * X)
30  PRINT Y
40  END
```

```
10  READ A, B, C, D, X
15  DATA 3, 4, 5, 6, 2
20  LET Y = (A + B + C)/(D * X)
30  PRINT Y
40  END
```

Written Exercises

Find the value of X in each statement. Let A = 3, B = 4, and M1 = 16.

1. 190 LET X = 6 * A 18
2. 25 LET X = A * B + 5 17
3. 20 LET X = B + M1 * B 68
4. 30 LET X = M1 / B 4
5. 20 LET X = A↑4 81
6. 40 LET X = M1 − B + 3 * A 21

Each algebraic expression below is followed by a BASIC expression. Rewrite each BASIC expression so that it matches the algebraic expression.

7. $\frac{a+2}{y+3}$ A + 2 / Y + 3 (A + 2)/(Y + 3)
8. $\frac{(3+x)^2}{2r}$ (3 + X)↑2 / 2 * R (3 + X) ↑ 2/(2 * R)
9. $\frac{3a}{y+2}$ 3A / (Y + 2) (3 * A)/(Y + 2)
10. $3a^2 + y$ 3 * A↑(2 + Y) 3 * A ↑ 2 + Y

Correct the error in each BASIC expression or statement below.

11. 7Y + 16 7 * Y + 16
12. (3 + 2 / 5 (3 + 2)/5
13. 4 * Y↑3 + 5Y 4 * Y ↑ 3 + 5 * Y
14. 30 LET Y = 2A 30 LET Y = 2 * A
15. 60 LET 4 * A=B+3+C 60 LET A = (B + 3 + C)/4
16. 90 LET A + B = C 90 LET A = C − B

For each of the following, write a program to find the value of X.

17. A = 2, B = 4, C = −1, X = A + B * (C − 1)
18. A = 5, B = 2, C = 10, X = A / B + C ↑ 2 − A

Write the output for each program below.

19.
```
10   LET A = 4
20   LET B = 5
30   PRINT A * B
40   END
```
20

20.
```
10   READ P, R, T
20   DATA 300, 0.06, 5
30   LET I = P * R * T
40   PRINT P, R, T, I
50   END
```
300 .06 5 90

21.
```
10   READ R
20   LET P = 3.14159
30   LET C = 2 * P * R
40   DATA 2, 4, 7, 9
50   PRINT C
60   GO TO 10
70   END
```
12.56636
25.13272
43.98226
56.54862
OUT OF DATA IN 10

22.
```
10   READ A, B, C
20   LET S = A + B + C
30   LET A1 = S / B
40   PRINT A, B, C, A1
50   DATA 4, 5, 6
60   END
```
4 5 6 3

Write a program to solve each of the following.

23. Find the area of a triangle with a base of 15 units and an altitude of 8 units.
24. Find the volume of a sphere with a radius of 14 cm if $V = \frac{4}{3}\pi r^3$.
25. Find the total surface area of a box with a length of 17 cm, a width of 9 cm, and a height of 4.5 cm. Use $A = 2lw + 2lh + 2hw$.

will go on to line 20 and print the sum. An advantage of INPUT statements is that the program can be run many times, with different data each time. It is helpful to include a PRINT statement before each INPUT statement, to explain what data should be entered. The following line could be added to the program above.

```
15   PRINT "ENTER TWO
       VALUES TO BE ADDED"
```

The computer will print any characters between quotation marks in a PRINT statement.

Answers

Programs may vary. Sample programs are given.

17.
```
10   LET A = 2
20   LET B = 4
30   LET C = −1
40   LET X = A + B *
       (C − 1)
50   PRINT X
60   END
```

18.
```
10   READ A, B, C
20   DATA 5, 2, 10
30   LET X = A/B +
       C ↑ 2 − A
40   PRINT X
50   END
```

23.
```
10   LET B = 15
20   LET H = 8
30   LET A = (1/2) * B * H
40   PRINT A
50   END
```

24.
```
10   READ R
20   DATA 14
30   LET V =
       (4/3) * 3.14159
       * R ↑ 3
40   PRINT V
50   END
```

25.
```
10   READ L, W, H
20   DATA 17, 9, 4.5
30   LET A = 2 * L * W +
       2 * L * H +
       2 * H * W
40   PRINT A
50   END
```

IF-THEN Statements

The IF-THEN statement is used to compare two numbers. It instructs the computer what to do based on the results of the comparison. The general form of the IF-THEN statement is as follows.

line number IF *sentence* THEN *line number* ❶

The sentence uses one of the following symbols.

❷

Symbol	Meaning	Example
=	is equal to	X = 3
>	is greater than	A > B
<	is less than	10 < 21
> =	is greater than or equal to	A > = 16
< =	is less than or equal to	R < = 10
< >	is not equal to	6 < > 10

The IF-THEN statement is used to compare two numbers. If the sentence containing the two numbers is true, then the computer goes to the line whose number follows THEN. If the sentence is false, then the computer simply goes to the next line of the program.

Study the use of IF-THEN statements in the programs below.

```
10  LET X = 5
20  IF X > 10 THEN 40
30  PRINT X
40  END
```

Line 10 instructs the computer to assign the value 5 to X.
In line 20, the sentence X > 10 is false, and the computer goes to line 30.
The output for this program is 5, the value of X.

```
10  LET X = 15
20  IF X > 10 THEN 40
30  GO TO 50
40  PRINT X
50  END
```

Since X = 15, the sentence in line 20 is true, and the computer goes to line 40. What line would the computer go to if the sentence in line 20 were X < 10?

Suppose you wish to find the areas of ten different squares with sides of 1 unit, 2 units, 3 units, and so on up to 10 units. You can use a LET statement to assign the variables. Consider the following.

```
10  LET S = 1
20  LET S = S + 1
```

In algebra, the statement S = S + 1 is nonsense. But in BASIC, it means that S should be assigned a new value equal to 1 more than its previous value.

❶ An IF-THEN statement always includes an algebraic sentence. The computer evaluates the left side and right side of the sentence, then compares the results. Discuss what happens when the sentence is true and what happens when it is false.

❷ In IF-THEN statements the equals sign signifies equality as in algebra. More than one variable can occur on the left side of the equals sign.

Related Exercises

If A = 6 and B = 3, which line does the computer go to after line 10?

1. 10 IF A + B >= 5 THEN 50
   20 PRINT A — 50
2. 10 IF A * B > 19 THEN 50
   20 PRINT B — 20
3. 10 IF A + B↑2 < 12 THEN 70
   20 PRINT A↑2 — 20

Determine the output for the following program.

4. 10 LET X = 2
   20 PRINT X
   30 LET X = X + 2
   40 IF X <= 100 THEN 20
   50 END

The program will print the even integers from 2 to 100.

The program continues as follows.

```
10  LET S = 1
12  LET A = S↑2         Use A = s², the formula for the area of a square.
14  PRINT A
20  LET S = S + 1       In line 20, the value of S is increased by 1.
30  GO TO 12            Line 30 repeats the program for each new value of S.
40  END
```

As the program is written now, it will continue indefinitely. A line is needed to stop the program when S is greater than 10. An IF-THEN statement can be used to do this.

```
25 IF S > 10 THEN 40
```

Study the use of the IF-THEN statement in the program below.

```
10  LET S = 1
12  LET A = S↑2
14  PRINT A
20  LET S = S + 1
25  IF S > 10 THEN 40
      ↓ no          yes
30  GO TO 12
40  END
```

❸

If S is less than or equal to 10, the computer is sent back to line 12. If S is greater than 10, the program ends.

Example

1 Write a program that tells whether each of the following numbers is positive or negative: −3, 6, −11.

❹

```
10  READ N
20  DATA −3, 6, −11, 1000
30  IF N = 1000 THEN 90
40  IF N > 0 THEN 70
50  PRINT N; " IS NEGATIVE"
60  GO TO 10
70  PRINT N; " IS POSITIVE"
80  GO TO 10
90  END
RUN
−3 IS NEGATIVE
6 IS POSITIVE
−11 IS NEGATIVE
```

Line 30 stops the program when 1000 is read in the DATA line.

Line 40 asks. "Is N positive?" If yes, the computer goes to line 70. If no, the computer goes to the next statement, line 50.

The computer prints characters enclosed in quotation marks. A semicolon causes output to be printed close together.

No OUT OF DATA line is printed.

❸ If a variable without a previously assigned value is used, some computers automatically assign the value 0 to the variable. If line 10 was deleted from the program at the top of the page, the value of S after line 20 would be 1.

Extra Notes
In PRINT statements, the computer prints any characters, including punctuation marks, that are enclosed in quotation marks. A computer may or may not require punctuation between a variable and material enclosed by quotation marks.

❹ In Example 1, note that the insertion of the data 1000 allows the programmer to avoid an OUT OF DATA error message.

Written Exercises

In each of the following, state the line number the computer goes to after line 10. Let A = 10, B = 6, and X = 20.

1.
```
10 IF A < 20 THEN 75
20 PRINT A
```
line 75

2.
```
10 IF A > = 20 THEN 80
20 PRINT 3 * A
```
line 20

3.
```
10 IF A <> B THEN 60
20 PRINT "HELLO"
```
line 60

4.
```
10 IF A + X < B THEN 40
20 PRINT A + X
```
line 20

5.
```
10 IF X↑3 > A * B THEN 60
20 X = X + 2
```
line 60

6.
```
10 IF A - X = B - 16 THEN 40
20 PRINT A, B, X
```
line 40

For problems 7-10, tell whether the values of X, Y, or both X and Y will be printed. Give the values to be printed. Use the program below.

```
10 IF X > Y THEN 40
20 LET X = X + 3
30 LET Y = Y + 2
40 IF X > = Y THEN 60
50 PRINT X
60 PRINT Y
70 END
```

7. X = 12, Y = 12 Y is printed; 14
8. X = 14, Y = 19 X, Y are printed; 17, 21
9. X = 9, Y = 10 Y is printed; 12
10. X = 21, Y = 5 Y is printed; 5

Write a BASIC program, using an IF-THEN statement, to solve each of the following problems.

11. Print the integers from 10 to 1 in descending order.
12. Print the squares of the integers from 20 to 40 inclusive.
13. Print the cubes of the odd integers from 7 to 21 inclusive.
14. Read a real number. Print the number, its additive inverse, and its multiplicative inverse if it exists.

Compound statements may be used in IF-THEN statements as shown in the programs in problems 15 and 16. Write the output for each program.

15.
```
10 READ A, B
20 DATA 3, 4, 2, 4
30 IF A = 3 AND B = 4 THEN 60
40 PRINT "HOW ARE YOU?"
50 GO TO 80
60 PRINT "HELLO"
70 GO TO 10
80 END
```
HELLO HOW ARE YOU?

16.
```
10 READ A, B
20 DATA 3, 5, 5, 6
30 IF A = 3 OR B = 4 THEN 60
40 PRINT "HOW ARE YOU?"
50 GO TO 80
60 PRINT "HELLO"
70 GO TO 10
80 END
```
HELLO HOW ARE YOU?

Answers

Sample programs are given below.

11.
```
10 LET I = 10
20 PRINT I
30 LET I = I - 1
40 IF I = 0 THEN 60
50 GOTO 20
60 END
```

12.
```
10 LET I = 20
20 PRINT I, I↑2
30 LET I = I + 1
40 IF I > 40 THEN 60
50 GOTO 20
60 END
```

13.
```
10 LET I = 7
20 PRINT I, I↑3
30 LET I = I + 2
40 IF I > 21 THEN 60
50 GOTO 20
60 END
```

14.
```
10 READ A
20 IF A = 0 THEN 40
30 PRINT A, - A, 1/A
35 GOTO 10
40 PRINT A, -A, "NO
     MULT. INV."
50 GOTO 10
60 DATA 6, -3, 0, 4, -9
70 END
```

FOR-NEXT Statements

FOR-NEXT statements can increase the efficiency of a program. Compare the following programs.

```
10  LET S = 1
12  LET A = S↑2
14  PRINT A
20  LET S = S + 1
25  IF S > 10 THEN 40
30  GO TO 12
40  END
```

```
100  FOR S = 1 TO 10 STEP 1
120  LET A = S↑2
140  PRINT A
300  NEXT S
400  END
```

In a single line in the program at the right, line 100, the value of S is stated at 1 and is increased in steps of 1 until it reaches 10.

100 FOR S = 1	TO 10	STEP 1
LET S = 1	If S > 10 THEN 40	S = S + 1

Line 100 in the program at the right above accomplishes the same thing as lines 10, 20, and 25 do in the program on the left above.

When the computer reads line 100 the first time, S is assigned the value 1. When line 300, NEXT S, is encountered, the computer returns to line 100 and increases the value of S by 1. This process continues until the value of S is greater than 10. The computer then goes to the line following NEXT S.

The general form for FOR-NEXT statements is shown below.

line number FOR *variable* = *number* TO *number* STEP *number* ❶

line number NEXT *variable*

A FOR statement must always be paired with a NEXT statement. The number following STEP may be positive or negative.

Example 1 **Write a program to print the integers from 1 to 10. Use FOR-NEXT statements.**

```
10  FOR I = 1 TO 10      STEP is not needed when increasing by 1.
20  PRINT I
30  NEXT I
40  END
```


❶ Emphasize that if no step is indicated in a FOR statement, there will be increments of 1. Other increments must be specified. Point out that steps can be fractions or decimals such as 1/2 or 0.1.

Extra Notes
The following program is given to illustrate how most computers interpret FOR-NEXT statements.

```
10  FOR X = 2 TO 20 STEP 3
20  PRINT X
30  NEXT X
40  END
```

The first time the computer reads line 10, it assigns the value 2 to X. Each time the computer returns to line 10, it adds 3 to the previous value of X and then compares the result to 20. If X is less than or equal to 20, it continues to the next line. If X is greater than 20, the computer skips to the line following "NEXT X." In this program, the last value assigned to X would be 23 rather than 20. However, the computer will not print 23 since it skips to line 40 after comparing 23 to 20.

Examples

2 **Write a program that prints the even integers from 2 to 100.**

```
10  FOR I = 2 TO 100 STEP 2
20  PRINT I
30  NEXT I
40  END
```

3 **Write a program that prints a table of the integers from 1 to 3 and their squares.**

```
10  PRINT "NUMBER", "SQUARE"
20  FOR I = 1 TO 3
30  PRINT I, I↑2
40  NEXT I
50  END
```

Line 10 prints headings for a table.

NUMBER	SQUARE
1	1
2	4
3	9

This is how the table appears.

FOR-NEXT statements, like GO TO statements and IF-THEN statements, can be used in programs to create loops so that parts of the programs are repeated. There are only two ways these loops can appear. ❷

Nested Loops

```
┌─FOR X
│ ┌─FOR Y
│ └─NEXT Y
└─NEXT X
```

The loops do not cross.

Independent Loops

```
┌─FOR X
└─NEXT X
┌─FOR Y
└─NEXT Y
```

The loops do not cross. They are not nested.

Not Acceptable

```
  ┌─FOR X
┌─┼─FOR Y
│ └─NEXT X
└───NEXT Y
```

These loops cross.

The program below generates all ordered pairs with first element from set A and second element from set B where A = {1, 2, 3, 4} and B = {6, 7, 8}.

```
10  FOR A = 1 TO 4 ←─┐
20  FOR B = 6 TO 8 ←┐│
30  PRINT A; ","; B, ││
40  NEXT B ─────────┘│
50  NEXT A ──────────┘
60  END
```

Note that the loop for B is nested within the loop for A.

❷ Encourage students to experiment with programs that contain loops, such as the ones in Examples 2 and 3. Emphasize that if a program has more than one loop, the loops must not cross.

Answers

Sample programs are given below.

5.
```
10  FOR I = 1 to 25 STEP 2
20  PRINT I
30  NEXT I
40  END
```

6.
```
5   LET S = 0
10  FOR I = 1 TO 5
20  READ N
30  DATA 4, 5, 10, 11, 20
40  LET S = S + N
50  NEXT I
60  PRINT S
70  END
```

7.
```
10  FOR I = 1 TO 25
20  PRINT I, I↑2, I↑3
30  NEXT I
40  END
```

8.
```
10  PRINT "FAHRENHEIT", "CELSIUS"
20  FOR F = 0 TO 100 STEP 2
30  LET C = 5 * (F - 32)/9
40  PRINT F, C
50  NEXT F
60 END
```

The output for the program on the preceding page follows.

```
1, 6    1, 7    1, 8    2, 6    2, 7    2, 8
3, 6    3, 7    3, 8    4, 6    4, 7    4, 8
```

Notice that A equals the same value while B goes through the loop.

Written Exercises

Find the error or errors in each of the following.

1.
```
10 IF I = 1 TO 5
20 LET Y = 3 * I + 4
30 PRINT I, Y
40 NEXT I
50 END
```
line 10 should be: FOR I = 1 TO 5

2.
```
10 FOR I = 1 TO 10 STEP 2
20 LET Y = 3 * I + 10
30 PRINT I, Y
40 NEXT Y
50 END
```
line 40 should be: 40 NEXT I

3.
```
 5 LET S = 0
10 READ M, N
20 FOR J = M + N STEP 3
30 LET S = S + J
40 NEXT J
50 PRINT S
60 END
```
Data line is missing. line 20 should be: 20 FOR J = M TO N STEP 3

4.
```
10 FOR X = 1 TO 5
20 FOR Y = 2 TO 10
30 PRINT X * Y
40 NEXT X
50 NEXT Y
60 END
```
lines 40 and 50 should be: 40 NEXT Y 50 NEXT X

Write a program using FOR-NEXT statements, to do each of the following.

5. Print the odd numbers from 1 to 25.

6. Find the sum of any five numbers. Use READ-DATA statements.

7. Print a table of squares and cubes from 1 to 25.

8. The formula for converting Fahrenheit to Celsius temperature is $C = \frac{5}{9}(F - 32)$. Generate Celsius temperatures for all even Fahrenheit values from 0 to 100.

9. Use the formula in problem 8 to generate Fahrenheit values for all integer Celsius values from -20 to 40.

10. Print an addition table showing all addition facts from $1 + 1 = 2$ to $4 + 4 = 8$.

11. Print a multiplication table showing all multiplication facts from $1 \cdot 1 = 1$ to $10 \cdot 10 = 100$.

12. Print a table of all positive integers and their squares such that none of the squares is greater than 1,000.

13. Find the pairs of integers that satisfy the equation $x^2 + y^2 = 100$.

14. Find the pairs of integers that satisfy the equation $x^2 \cdot y^2 = 400$.

Answers

9.
```
10 PRINT "CELSIUS",
     "FAHRENHEIT"
20 FOR C = -20 TO 40
30 LET F = (9/5) * C +
     32
40 PRINT C, F
50 NEXT C
60 END
```

10.
```
10 FOR I = 1 TO 4
20 FOR J = 1 TO 4
30 PRINT I; "+"; J; "=";
     I + J
40 NEXT J
50 NEXT I
60 END
```

11.
```
10 FOR I = 1 TO 10
20 FOR J = 1 TO 10
30 PRINT I; "*"; J; "=";
     I * J
40 NEXT J
50 NEXT I
60 END
```

12.
```
10 FOR I = 1 TO 40
20 LET Y = I↑2
30 IF Y > 1000 THEN 60
40 PRINT I, Y
50 NEXT I
60 END
```

13.
```
10 FOR X = -10 TO 10
20 FOR Y = -10 TO 10
30 IF X↑2 + Y↑2 = 100
     THEN 50
40 GOTO 60
50 PRINT X, Y
60 NEXT Y
70 NEXT X
80 END
```

14.
```
10 FOR X = -20 TO 20
20 FOR Y = -20 TO 20
30 IF X↑2 * Y↑2 = 400
     THEN 50
40 GOTO 60
50 PRINT X, Y
60 NEXT Y
70 NEXT X
80 END
```

Special Features of BASIC ❷

❶ When the results of computations exceed 6 significant digits, the computer will use E notation. This is the computer equivalent to scientific notation. The E means "times 10 to the given power."

Result of computation	E Notation	Meaning
15,000,000	1.5E + 07	1.5×10^{7}
37,867,275	3.7867275E + 07	3.7867275×10^{7}
0.003629	3.629E − 03	3.629×10^{-3}

Sometimes it is necessary to save the values of variables for use later in a program. Consider the following program in which values of variables are *not* saved.

```
10  READ N
20  DATA 3, 4, 7, 11, -2, 14
30  PRINT N
40  GO TO 10
50  END
```

Each time the computer executes line 10, the previous value of N is erased.

The use of subscripted variables allows you to save values for future use. A subscripted variable is written as a letter A to Z followed by a numeral in parentheses.

```
10  DIM A(15)
20  FOR I = 1 TO 15
30  READ A(I)
35  PRINT A(I)
40  NEXT I
50  DATA 15, 14, 13, 12, 11, 10, 9, 8, 7, 6, 5, 4, 3, 2, 1
60  END
```

A(I) is a subscripted variable where I represents an integer from 1 to 15.

Line 10 uses a DIMENSION statement that tells the computer to reserve 15 spaces for the values of the subscripted variables A(I). The number of spaces reserved must be greater than or equal to the number of spaces needed. Line 30 reads and stores numbers from the data line in subscripted variables. For example, 15 is stored in position A(1), 13 is stored in position A(3), 11 is stored in position A(5), and so on. Line 35 will cause the computer to print all stored values in subscripted variables A(I). ❸

❶ Some computers use a different form of E notation. For example, 3.7E+07 may be shown as 0.37E+08.

❷ Some computers print up to nine digits before using E notation.

❸ Systems may vary concerning when a DIM statement is required, or the maximum allowable size for a subscript. Most BASIC systems automatically provide 10 storage spaces for any subscripted variable. The DIMENSION is usually required when the subscripts are greater than 10.

Extra Notes
Subscripted variables provide a very convenient way to quickly generate many variables for large sets of data.

Algebra	BASIC
$x_1, x_2, \ldots$	X(1), X(2), . . .

Subscripted variables are especially useful within a loop. Also, subscripted variables can be more easily manipulated within a program because the subscript is numerical.

BASIC contains useful internal functions. These are functions that are built into the computer to perform special operations. Some of them are shown below. Examples of their use follow. ❹

ABS(X)	This function finds the absolute value of X.
SQR(X)	This function finds the square root of X.
INT(X)	This function finds the greatest integer less than or equal to X.

Examples

1 Write a program to read a number and print it and its absolute value.

```
10 READ N
20 DATA 6, -3, 0, 4, 0.31, -9, 10000
30 IF N = 10000 THEN 60
40 PRINT N, ABS(N)
50 GO TO 10
60 END
```

The programmer does not need the absolute value of 10000. 10000 is a flag that signals the end of the data.

2 Write a program to print the square and square root of the first 10 positive integers.

```
 5 PRINT "N", "N SQUARED", "SQRT of N"
10 FOR N = 1 TO 10
20 PRINT N,N↑2, SQR(N)
30 NEXT N
40 END
```

Some examples of the greatest integer function are given as follows. ❺

INT(5.72) = 5	INT(−8.9) = −9
INT(−13) = −13	INT(44.99) = 44

Suppose you wish to determine if 2 is a factor of 7. If INT(7 / 2) = 7 / 2, then 7 / 2 must be an integer. Of course, you know that 7 / 2 = 3.5 and INT(7 / 2) = 3. Therefore, 2 is *not* a factor of 7.

In general, if INT(N / X) = N / X, then X is a factor of N. Example 3 illustrates this principle.

❹ Students should realize that many procedures occur so often in programming that they are "built-in" to the computer. A few of these are illustrated in this lesson. Systems may vary in the way an internal function is stated in a program.

❺ The INT function can be used for rounding. The following examples show how expressions can be rounded to the nearest integer.
INT(6.3 + 0.5) = 6
INT(6.6 + 0.5) = 7
In general, INT(X + 0.5) rounds to the nearest integer.

Answers
Sample programs are given below.

13.
```
 5 DIM A(15)
10 FOR I = 1 TO 15
20 READ A(I)
30 DATA 3, 6, 9, 12, 15,
     18, 21, 24, 27, 30,
     33, 36, 39, 42, 45
40 NEXT I
50 PRINT A(3), A(7), A(14)
60 END
```

14.
```
10 DIM A(20)
20 FOR I = 1 TO 20
30 LET A(I) = I
40 NEXT I
50 FOR I = 20 TO 1
     STEP -1
60 PRINT A(I)
70 NEXT I
80 END
```

Answers

15.
```
10 READ N
20 IF N < 0 THEN 40
30 GOTO 10
40 PRINT "THE
     ABSOLUTE VALUE
     OF ";N;" IS ";
     ABS (N)
50 GOTO 10
60 DATA 3, -4, 9, 2, 1,
     -6.8, 0, 4, -9
70 END
```

16.
```
5 PRINT "A", "B",
     "ABS (N)"
10 READ A, B
20 DATA 30, 40, 21, 38,
     42, 26, 39, 46, 0, 0
30 IF A = 0 THEN 70
40 LET X = A - B
50 PRINT A, B, ABS(X)
60 GOTO 10
70 END
```

17.
```
10 PRINT "NUMBER",
     "SQUARE ROOT"
20 FOR I = 1 TO 6
30 READ N
35 DATA 7, -3, 0, -9,
     42, 36
40 IF N < 0 THEN 60
50 PRINT N, SQR(N)
60 NEXT I
70 END
```

18.
```
5 PRINT "NUMBER",
     "SQUARE ROOT"
10 FOR I = 93 TO 121
20 PRINT I, SQR(I)
30 NEXT I
40 END
```

19.
```
5 PRINT "N", "INT(N)"
10 FOR I = 1 TO 6
20 READ N
30 DATA 3.7, 0, 6, 0.31,
     -2.56, -4.01
40 PRINT N, INT(N)
50 NEXT I
60 END
```

(continued on p. 543)

Example

3 **Write a program to find all the factors of 48.**

```
10 FOR I = 1 TO 48
20 IF INT(48/I) = 48/I THEN 40
30 GO TO 50
40 PRINT I
50 NEXT I
60 END
```

Written Exercises

Write each of the following using scientific notation. Then write each in E notation.

1. 8,200,000 8.2×10^7 8.2E + 07

2. 0.00000005108 5.108×10^{-8} 5.108E − 08

3. 27,372,800,000 2.73728×10^{10} 2.73728E + 10

Write the value of each expression in decimal notation.

4. 8.091E + 08 809,100,000

5. 2.2771E − 04 0.00022771

6. 1.461E + 11 146,100,000,000

Write the output of each of the following PRINT statements. Let A = −5.1, B = 9, C = 5.1, and D = 1.

7. PRINT ABS(A) 5.1

8. PRINT ABS(A * D) 5.1

9. PRINT SQR(B) − SQR(D) 2

10. PRINT INT(C) 5

11. PRINT INT(ABS(A)) 5

12. PRINT ABS(INT(A)) 6

Write a program to solve each of the following. Use BASIC functions.

13. Store 15 given numbers in a subscripted variable. Print the 3rd, 7th, and 14th numbers.

14. Store 20 given numbers in a subscripted variable. Print the numbers in reverse order.

15. Read a given number and test if it is negative. If it is, then print its absolute value.

16. Subtract two given numbers and print the absolute value of their difference.

17. Test each of the following to see if it is negative: 7, −3, 0, −9, 42, 36. If not, print the number and its square root.

18. Print the square roots of all integers from 93 to 121 inclusive.

19. Find the greatest integer less than or equal to each of these numbers: 3.7, 0, 6, 0.31, −2.56, −4.01.

20. Use the greatest integer function to determine if nine given numbers are even or odd.

21. Use the greatest integer function to round a given decimal to the nearest tenth.

22. Find all integers between 1 and 100 that are multiples of 6.

Symbols

Symbol	Meaning	Symbol	Meaning
$=$	is equal to	π	pi
$\neq$	is not equal to	$\{\ \}$	set
$>$	is greater than	$\%$	percent
$<$	is less than	$^\circ$	degrees
$\geq$	is greater than or equal to	$a{:}b$	ratio of a to b
$\leq$	is less than or equal to	$f(x)$	f of x, the value of f at x
$\approx$	is approximately equal to	(a, b)	ordered pair a, b
$\cdot$	times	$\overline{AB}$	line segment AB
$-$	negative	AB	measure of $\overline{AB}$
$+$	positive	$\sqrt{\ }$	principal square root
$\pm$	positive or negative	$\cos A$	cosine of A
$-a$	opposite or additive inverse of a	$\sin A$	sine of A
$\vert a \vert$	absolute value of a	$\tan A$	tangent of A
$a \stackrel{?}{=} b$	Does a equal b?		

Metric System

Symbol	Unit	Symbol	Unit
mm	millimeter	h	hour
cm	centimeter	min	minute
m	meter	s	second
km	kilometer	km/h	kilometer per hour
g	gram	m/s	meters per second
kg	kilogram	°C	degrees Celsius
mL	milliliter		
L	liter		

Answers (continued)

20.
```
10 FOR I = 1 TO 9
20 READ N
30 DATA 4, 5, −6, 7, 8,
     −9, 10, 11, −12
40 IF N/2 = INT(N/2)
     THEN 70
50 PRINT N; "IS ODD"
60 GOTO 80
70 PRINT N; "IS EVEN"
80 NEXT I
90 END
```

21.
```
10 READ A
20 DATA 5.461, −8.39,
     1.24, −6.92
30 LET N = INT(A *
     10 + .5)
40 LET R = N/10
50 PRINT A; "ROUNDS
     TO"; R
55 GOTO 10
60 END
```

22.
```
10 FOR I = 1 TO 100
20 IF (I/6) = INT(I/6)
     THEN 40
30 GOTO 50
40 PRINT I
50 NEXT I
60 END
```

SQUARES AND APPROXIMATE SQUARE ROOTS

n	n^2	$\sqrt{n}$	n	n^2	$\sqrt{n}$
1	1	1.000	51	2601	7.141
2	4	1.414	52	2704	7.211
3	9	1.732	53	2809	7.280
4	16	2.000	54	2916	7.348
5	25	2.236	55	3025	7.416
6	36	2.449	56	3136	7.483
7	49	2.646	57	3249	7.550
8	64	2.828	58	3364	7.616
9	81	3.000	59	3481	7.681
10	100	3.162	60	3600	7.746
11	121	3.317	61	3721	7.810
12	144	3.464	62	3844	7.874
13	169	3.606	63	3969	7.937
14	196	3.742	64	4096	8.000
15	225	3.873	65	4225	8.062
16	256	4.000	66	4356	8.124
17	289	4.123	67	4489	8.185
18	324	4.243	68	4624	8.246
19	361	4.359	69	4761	8.307
20	400	4.472	70	4900	8.367
21	441	4.583	71	5041	8.426
22	484	4.690	72	5184	8.485
23	529	4.796	73	5329	8.544
24	576	4.899	74	5476	8.602
25	625	5.000	75	5625	8.660
26	676	5.099	76	5776	8.718
27	729	5.196	77	5929	8.775
28	784	5.292	78	6084	8.832
29	841	5.385	79	6241	8.888
30	900	5.477	80	6400	8.944
31	961	5.568	81	6561	9.000
32	1024	5.657	82	6724	9.055
33	1089	5.745	83	6889	9.110
34	1156	5.831	84	7056	9.165
35	1225	5.916	85	7225	9.220
36	1296	6.000	86	7396	9.274
37	1369	6.083	87	7569	9.327
38	1444	6.164	88	7744	9.381
39	1521	6.245	89	7921	9.434
40	1600	6.325	90	8100	9.487
41	1681	6.403	91	8281	9.539
42	1764	6.481	92	8464	9.592
43	1849	6.557	93	8649	9.644
44	1936	6.633	94	8836	9.695
45	2025	6.708	95	9025	9.747
46	2116	6.782	96	9216	9.798
47	2209	6.856	97	9409	9.849
48	2304	6.928	98	9604	9.899
49	2401	7.000	99	9801	9.950
50	2500	7.071	100	10000	10.000

TRIGONOMETRIC RATIOS

Angle	sin	cos	tan	Angle	sin	cos	tan
0°	0.0000	1.0000	0.0000	45°	0.7071	0.7071	1.0000
1°	0.0175	0.9998	0.0175	46°	0.7193	0.6947	1.0355
2°	0.0349	0.9994	0.0349	47°	0.7314	0.6820	1.0724
3°	0.0523	0.9986	0.0524	48°	0.7431	0.6691	1.1106
4°	0.0698	0.9976	0.0699	49°	0.7547	0.6561	1.1504
5°	0.0872	0.9962	0.0875	50°	0.7660	0.6428	1.1918
6°	0.1045	0.9945	0.1051	51°	0.7771	0.6293	1.2349
7°	0.1219	0.9925	0.1228	52°	0.7880	0.6157	1.2799
8°	0.1392	0.9903	0.1405	53°	0.7986	0.6018	1.3270
9°	0.1564	0.9877	0.1584	54°	0.8090	0.5878	1.3764
10°	0.1736	0.9848	0.1763	55°	0.8192	0.5736	1.4281
11°	0.1908	0.9816	0.1944	56°	0.8290	0.5592	1.4826
12°	0.2079	0.9781	0.2126	57°	0.8387	0.5446	1.5399
13°	0.2250	0.9744	0.2309	58°	0.8480	0.5299	1.6003
14°	0.2419	0.9703	0.2493	59°	0.8572	0.5150	1.6643
15°	0.2588	0.9659	0.2679	60°	0.8660	0.5000	1.7321
16°	0.2756	0.9613	0.2867	61°	0.8746	0.4848	1.8040
17°	0.2924	0.9563	0.3057	62°	0.8829	0.4695	1.8807
18°	0.3090	0.9511	0.3249	63°	0.8910	0.4540	1.9626
19°	0.3256	0.9455	0.3443	64°	0.8988	0.4384	2.0503
20°	0.3420	0.9397	0.3640	65°	0.9063	0.4226	2.1445
21°	0.3584	0.9336	0.3839	66°	0.9135	0.4067	2.2460
22°	0.3746	0.9272	0.4040	67°	0.9205	0.3907	2.3559
23°	0.3907	0.9205	0.4245	68°	0.9272	0.3746	2.4751
24°	0.4067	0.9135	0.4452	69°	0.9336	0.3584	2.6051
25°	0.4226	0.9063	0.4663	70°	0.9397	0.3420	2.7475
26°	0.4384	0.8988	0.4877	71°	0.9455	0.3256	2.9042
27°	0.4540	0.8910	0.5095	72°	0.9511	0.3090	3.0777
28°	0.4695	0.8829	0.5317	73°	0.9563	0.2924	3.2709
29°	0.4848	0.8746	0.5543	74°	0.9613	0.2756	3.4874
30°	0.5000	0.8660	0.5774	75°	0.9659	0.2588	3.7321
31°	0.5150	0.8572	0.6009	76°	0.9703	0.2419	4.0108
32°	0.5299	0.8480	0.6249	77°	0.9744	0.2250	4.3315
33°	0.5446	0.8387	0.6494	78°	0.9781	0.2079	4.7046
34°	0.5592	0.8290	0.6745	79°	0.9816	0.1908	5.1446
35°	0.5736	0.8192	0.7002	80°	0.9848	0.1736	5.6713
36°	0.5878	0.8090	0.7265	81°	0.9877	0.1564	6.3138
37°	0.6018	0.7986	0.7536	82°	0.9903	0.1392	7.1154
38°	0.6157	0.7880	0.7813	83°	0.9925	0.1219	8.1443
39°	0.6293	0.7771	0.8098	84°	0.9945	0.1045	9.5144
40°	0.6428	0.7660	0.8391	85°	0.9962	0.0872	11.4301
41°	0.6561	0.7547	0.8693	86°	0.9976	0.0698	14.3007
42°	0.6691	0.7431	0.9004	87°	0.9986	0.0523	19.0811
43°	0.6820	0.7314	0.9325	88°	0.9994	0.0349	28.6363
44°	0.6947	0.7193	0.9657	89°	0.9998	0.0175	57.2900
45°	0.7071	0.7071	1.0000	90°	1.0000	0.0000	∞

The exercises in the Algebraic Skills Review can be used for remediation, to supplement exercises in the lessons, or as review material at the end of the school year.

On this page, exercises 1-30 provide practice in adding integers. Exercises 31-60 provide practice in subtracting integers. The rest of the exercises provide practice in solving equations.

Algebraic Skills Review

Integer Equations: Addition and Subtraction

Solve each equation.

1. $-5 + (-8) = x$ −13
2. $-7 + 4 = y$ −3
3. $-4 + 8 = t$ 4
4. $9 + (-2) = a$ 7
5. $6 + 6 = b$ 12
6. $3 + (-8) = m$ −5
7. $-7 + (-9) = v$ −16
8. $-5 + 5 = z$ 0
9. $-19 + 43 = c$ 24
10. $51 + (-26) = w$ 25
11. $-37 + (-48) = d$ −85
12. $-93 + 44 = e$ −49
13. $67 + (-82) = n$ −15
14. $28 + 46 = f$ 74
15. $29 + (-37) = s$ −8
16. $-94 + (-58) = g$ −152
17. $-18 + 63 = p$ 45
18. $28 + (-52) = j$ −24
19. $77 + 57 = r$ 134
20. $47 + (-29) = x$ 18
21. $-18 + 26 = a$ 8
22. $-65 + (-75) = k$ −140
23. $21 + (-47) = h$ −26
24. $-15 + 52 = q$ 37
25. $y = -13 + (-98)$ −111
26. $u = -5 + 82$ 77
27. $s = -47 + 26 + (-18)$ −3
28. $a = -71 + (-85) + (-16)$ −172
29. $41 + 57 + (-32) = m$ 66
30. $82 + (-14) + (-35) = c$ 33

31. $-4 - (-2) = t$ −2
32. $5 - (-6) = p$ 11
33. $-9 - 3 = x$ −12
34. $5 - (-5) = k$ 10
35. $-3 - (-8) = m$ 5
36. $3 - 9 = w$ −6
37. $-6 - 8 = w$ −14
38. $0 - 6 = v$ −6
39. $j = -10 - (-4)$ −6
40. $-23 - 45 = a$ −68
41. $28 - (-14) = z$ 42
42. $-53 - (-61) = f$ 8
43. $c = -16 - 47$ −63
44. $90 - 43 = g$ 47
45. $71 - (-47) = q$ 118
46. $-99 - (-26) = s$ −73
47. $38 - (-19) = t$ 57
48. $-20 - (-92) = j$ 72
49. $18 - 47 = y$ −29
50. $h = -15 - (-81)$ 66
51. $-42 - 63 = b$ −105
52. $-84 - 47 = r$ −131
53. $42 - (-47) = d$ 89
54. $y = -19 - (-63)$ 44
55. $16 - (-84) = n$ 100
56. $42 - (-26) = k$ 68
57. $-52 - (-33) = x$ −19
58. $-35 - 86 = a$ −121
59. $v = -8 - (-47)$ 39
60. $33 - 51 = t$ −18

61. $-2 + g = 7$ 9
62. $9 + s = -5$ −14
63. $-7 + k = -2$ 5
64. $-4 + y = -9$ −5
65. $m + 6 = 2$ −4
66. $t + (-4) = 10$ 14
67. $h - (-2) = 6$ 4
68. $v - 7 = -4$ 3
69. $a - (-6) = -5$ −11
70. $r - (-3) = -8$ −11
71. $j - (-8) = 5$ −3
72. $x - 8 = -9$ −1
73. $-2 - x = -8$ 6
74. $14 = -48 + b$ 62
75. $c + (-26) = 45$ 71
76. $z - (-57) = -39$ −96
77. $d + (-44) = -61$ −17
78. $n - 38 = -19$ 19
79. $-77 = w + 23$ −100
80. $e - (-26) = 41$ 15
81. $p - 47 = 22$ 69
82. $-63 - f = -82$ 19
83. $87 = t + (-14)$ 101
84. $q + (-53) = 27$ 80

Integer Equations: Multiplication and Division

Exercises 1-30 provide practice in multiplying integers. Exercises 31-60 provide practice in dividing integers. Exercises 61-90 provide practice in solving equations.

Solve each equation.

1. $x = (-8)(-4)$ 32
2. $(-3)5 = t$ −15
3. $(-7)(-2) = a$ 14
4. $(-9)8 = b$ −72
5. $6(-5) = v$ −30
6. $k = 8(6)$ 48
7. $14(-26) = s$ −364
8. $(-46)(-25) = g$ 1150
9. $(-71)(-20) = y$ 1420
10. $(-42)66 = h$ −2772
11. $(-97)47 = w$ −4559
12. $53(-32) = c$ −1696
13. $19(-46) = x$ −874
14. $(-82)0 = e$ 0
15. $72(43) = m$ 3096
16. $(-18)(-18) = d$ 324
17. $24(-29) = u$ −696
18. $f = (-39)45$ −1755
19. $(-76)(-34) = s$ 2584
20. $(-81)(-18) = q$ 1458
21. $(-65)28 = t$ −1820
22. $71(-38) = p$ −2698
23. $j = 49(-92)$ −4508
24. $36(24) = a$ 864
25. $(-42)78 = z$ −3276
26. $(-54)(-77) = r$ 4158
27. $n = (-6)(-127)(-4)$ −3048
28. $(13)(-12)(95) = w$ −14,820
29. $(-1)(45)(-45) = v$ 2025
30. $(-3)(61)(99) = y$ −18,117

31. $72 \div (-8) = g$ −9
32. $-64 \div 8 = b$ −8
33. $-45 \div (-9) = y$ 5
34. $56 \div (-7) = z$ −8
35. $42 \div 6 = e$ 7
36. $-24 \div (-6) = m$ 4
37. $992 \div (-32) = a$ −31
38. $-4428 \div 54 = k$ −82
39. $x = -600 \div (-24)$ 25
40. $1472 \div (-64) = p$ −23
41. $-564 \div (-47) = h$ 12
42. $-504 \div 14 = j$ −36
43. $-2201 \div 71 = r$ −31
44. $1512 \div (-28) = n$ −54
45. $765 \div (-85) = q$ −9
46. $-1591 \div (-37) = f$ 43
47. $s = -1080 \div 36$ −30
48. $3432 \div (-52) = v$ −66
49. $2730 \div 78 = k$ 35
50. $-3936 \div 96 = c$ −41
51. $-1476 \div 41 = z$ −36
52. $1496 \div (-22) = a$ −68
53. $2646 \div (-63) = t$ −42
54. $w = -4730 \div (-55)$ 86
55. $-1092 \div (-26) = x$ 42
56. $-2700 \div (-75) = e$ 36
57. $1127 \div 49 = y$ 23
58. $d = 1900 \div (-38)$ −50
59. $-845 \div 13 = w$ −65
60. $-1596 \div (-42) = a$ 38

61. $-5p = 35$ −7
62. $7g = -49$ −7
63. $-3x = -24$ 8
64. $a \div (-6) = -2$ 12
65. $m \div (-8) = 8$ −64
66. $q \div 9 = -3$ −27
67. $41j = 1476$ 36
68. $62y = -2356$ −38
69. $b \div (-21) = 13$ −273
70. $-33n = -1815$ 55
71. $k \div 46 = -41$ −1886
72. $w \div 17 = 24$ 408
73. $c \div (-59) = -7$ 413
74. $-56h = 1792$ −32
75. $-42z = 1512$ −36
76. $j \div (-27) = 27$ −729
77. $89s = -712$ −8
78. $-18v = -1044$ 58
79. $d \div (-34) = -43$ 1462
80. $f \div 14 = -63$ −882
81. $45t = 810$ 18
82. $-74w = 1554$ −21
83. $-49e = -2058$ 42
84. $r \div (-16) = -77$ 1232
85. $x \div (-26) = 47$ −1222
86. $-23 = t \div 44$ −1012
87. $-962 = -37g$ 26
88. $-3040 = 95k$ −32
89. $84 = x \div 97$ 8148
90. $-108 = m \div (-12)$ 1296

Exercises 1-24 provide review of adding and subtracting fractions. Exercises 25-48 provide practice in adding and subtracting fractions including negative numbers. Exercises 49-72 provide practice in solving equations.

Fraction Equations: Addition and Subtraction

Solve each equation and express answers in simplest form.

1. $\frac{3}{11} + \frac{6}{11} = x$ $\frac{9}{11}$

2. $\frac{4}{7} + \frac{5}{7} = a$ $\frac{9}{7}$

3. $\frac{5}{9} - \frac{2}{9} = t$ $\frac{1}{3}$

4. $\frac{17}{18} - \frac{5}{18} = w$ $\frac{2}{3}$

5. $\frac{1}{3} + \frac{2}{9} = b$ $\frac{5}{9}$

6. $\frac{1}{2} - \frac{1}{3} = v$ $\frac{1}{6}$

7. $\frac{3}{4} - \frac{9}{16} = s$ $\frac{3}{16}$

8. $\frac{2}{3} + \frac{8}{15} = r$ $1\frac{1}{5}$

9. $\frac{5}{6} - \frac{3}{4} = d$ $\frac{1}{12}$

10. $\frac{4}{9} + \frac{1}{6} = c$ $\frac{11}{18}$

11. $m = \frac{7}{9} + \frac{3}{8}$ $1\frac{11}{72}$

12. $\frac{11}{12} - \frac{7}{10} = j$ $\frac{13}{60}$

13. $\frac{5}{6} - \frac{5}{12} = p$ $\frac{5}{12}$

14. $4\frac{2}{3} + 1\frac{8}{15} = k$ $6\frac{1}{5}$

15. $5\frac{1}{2} - 2\frac{1}{3} = w$ $3\frac{1}{6}$

16. $8\frac{1}{12} - 5\frac{5}{12} = e$ $2\frac{2}{3}$

17. $7 - 1\frac{4}{9} = h$ $5\frac{5}{9}$

18. $n = \frac{3}{16} + \frac{7}{12}$ $\frac{37}{48}$

19. $4\frac{1}{2} - 2\frac{2}{3} = q$ $1\frac{5}{6}$

20. $7\frac{1}{12} - 4\frac{5}{8} = x$ $2\frac{11}{24}$

21. $11\frac{5}{6} + 9\frac{7}{15} = f$ $21\frac{3}{10}$

22. $y = 9\frac{2}{7} - 5\frac{5}{6}$ $3\frac{19}{42}$

23. $\frac{1}{4} + \frac{5}{6} + \frac{7}{12} = c$ $1\frac{2}{3}$

24. $\frac{5}{6} + \frac{2}{9} + \frac{3}{4} = z$ $1\frac{29}{36}$

25. $-\frac{2}{13} + \left(-\frac{3}{13}\right) = t$ $-\frac{5}{13}$

26. $-\frac{11}{18} + \frac{17}{18} = f$ $\frac{1}{3}$

27. $-\frac{9}{10} - \frac{7}{10} = n$ $-1\frac{3}{5}$

28. $-\frac{7}{11} - \left(-\frac{3}{11}\right) = a$ $-\frac{4}{11}$

29. $\frac{1}{12} - \left(-\frac{7}{12}\right) = w$ $\frac{2}{3}$

30. $\frac{17}{21} + \left(-\frac{10}{21}\right) = g$ $\frac{1}{3}$

31. $\frac{1}{4} + \left(-\frac{2}{3}\right) = p$ $-\frac{5}{12}$

32. $b = -\frac{1}{6} - \frac{8}{9}$ $-1\frac{1}{18}$

33. $\frac{1}{3} - \frac{5}{6} = m$ $-\frac{1}{2}$

34. $-\frac{1}{2} + \left(-\frac{3}{5}\right) = a$ $-1\frac{1}{10}$

35. $\frac{3}{7} + (-5) = s$ $-4\frac{4}{7}$

36. $-\frac{5}{9} - 2 = k$ $-2\frac{5}{9}$

37. $t = 1\frac{1}{2} - \left(-\frac{3}{4}\right)$ $2\frac{1}{4}$

38. $-\frac{3}{8} + \frac{4}{7} = c$ $\frac{11}{56}$

39. $\frac{3}{5} - \left(-3\frac{1}{4}\right) = v$ $3\frac{17}{20}$

40. $-8\frac{7}{8} - \left(-4\frac{5}{12}\right) = r$ $-4\frac{11}{24}$

41. $-3\frac{1}{6} + 5\frac{1}{15} = d$ $1\frac{9}{10}$

42. $-1\frac{8}{9} + \left(-5\frac{7}{12}\right) = h$ $-7\frac{17}{36}$

43. $7\frac{5}{6} + \left(-8\frac{7}{8}\right) = j$ $-1\frac{1}{24}$

44. $-3\frac{1}{2} - 4\frac{5}{9} = e$ $-8\frac{1}{18}$

45. $q = \frac{11}{16} - 12$ $-11\frac{5}{16}$

46. $-5\frac{11}{20} + 4\frac{7}{12} = z$ $\frac{-29}{30}$

47. $-1\frac{1}{12} - \left(-\frac{2}{3}\right) = w$ $\frac{-5}{12}$

48. $-4\frac{16}{21} + \left(-7\frac{5}{9}\right) = y$ $-12\frac{20}{63}$

49. $\frac{3}{13} + p = \frac{10}{13}$ $\frac{7}{13}$

50. $e + \frac{4}{15} = \frac{13}{15}$ $\frac{3}{5}$

51. $\frac{2}{5} + n = \frac{2}{3}$ $\frac{4}{15}$

52. $j - \frac{5}{18} = \frac{17}{18}$ $1\frac{2}{9}$

53. $r - \frac{1}{4} = \frac{5}{16}$ $\frac{9}{16}$

54. $b - \frac{1}{2} = \frac{2}{5}$ $\frac{9}{10}$

55. $s + \frac{2}{7} = 2$ $1\frac{5}{7}$

56. $\frac{7}{10} - a = \frac{1}{2}$ $\frac{1}{5}$

57. $1\frac{5}{6} + x = 2\frac{1}{4}$ $\frac{5}{12}$

58. $4\frac{1}{4} = w + 2\frac{1}{3}$ $1\frac{11}{12}$

59. $d - 1\frac{5}{7} = 6\frac{1}{4}$ $7\frac{27}{28}$

60. $h - \frac{3}{4} = 2\frac{5}{8}$ $3\frac{3}{8}$

61. $t - \frac{2}{3} = 1\frac{5}{8}$ $2\frac{7}{24}$

62. $g + \frac{5}{6} = \frac{4}{9}$ $-\frac{7}{18}$

63. $q - \frac{7}{10} = -\frac{11}{15}$ $-\frac{1}{30}$

64. $-\frac{3}{7} + c = \frac{1}{2}$ $\frac{13}{14}$

65. $-\frac{3}{4} = v + \left(-\frac{1}{8}\right)$ $-\frac{5}{8}$

66. $f - \left(-\frac{1}{8}\right) = \frac{3}{10}$ $\frac{7}{40}$

67. $m - \left(-1\frac{3}{8}\right) = -2\frac{1}{2}$ $-3\frac{7}{8}$

68. $-6\frac{5}{6} + y = 7\frac{7}{15}$ $14\frac{3}{10}$

69. $7\frac{1}{6} - z = -5\frac{2}{3}$ $12\frac{5}{6}$

70. $-2\frac{1}{3} + w = -5\frac{5}{6}$ $-3\frac{1}{2}$

71. $-6\frac{1}{7} + k = -\frac{4}{21}$ $5\frac{20}{21}$

72. $-4\frac{5}{12} = t - \left(-10\frac{1}{36}\right)$ $-14\frac{4}{9}$

Fraction Equations: Multiplication and Division

Exercises 1-24 provide review of multiplying and dividing fractions. Exercises 25-48 provide practice in multiplying and dividing fractions including negative numbers. Exercises 49-72 provide practice in solving equations.

Solve each equation and express answers in simplest form.

1. $\frac{1}{7}\left(\frac{1}{3}\right) = x$ $\frac{1}{21}$
2. $\frac{2}{3}\left(\frac{1}{5}\right) = v$ $\frac{2}{15}$
3. $\frac{5}{6}\left(\frac{3}{10}\right) = y$ $\frac{1}{4}$
4. $2 \div \frac{1}{3} = j$ 6
5. $\frac{5}{6} \div \frac{1}{6} = f$ 5
6. $\frac{1}{4} \div \frac{5}{8} = c$ $\frac{2}{5}$
7. $\frac{2}{3}(9) = b$ 6
8. $\frac{5}{18}\left(\frac{3}{10}\right) = r$ $\frac{1}{12}$
9. $\frac{8}{15} \div \frac{1}{10} = k$ $5\frac{1}{3}$
10. $g = \frac{1}{2} \div 8$ $\frac{1}{16}$
11. $\frac{3}{14} \div \frac{2}{7} = y$ $\frac{3}{4}$
12. $\frac{7}{12}\left(\frac{4}{5}\right) = w$ $\frac{7}{15}$
13. $\frac{6}{13} \div \frac{5}{7} = z$ $\frac{42}{65}$
14. $\frac{24}{25}\left(\frac{15}{32}\right) = a$ $\frac{9}{20}$
15. $\frac{7}{10}\left(\frac{5}{28}\right) = w$ $\frac{1}{8}$
16. $2\frac{2}{3}\left(\frac{4}{5}\right) = n$ $2\frac{2}{15}$
17. $t = \frac{7}{8}\left(4\frac{1}{4}\right)$ $3\frac{23}{32}$
18. $1\frac{3}{4} \div \frac{7}{12} = e$ 3
19. $1 \div 2\frac{3}{5} = d$ $\frac{5}{13}$
20. $2\frac{1}{10}\left(4\frac{2}{7}\right) = q$ 9
21. $1\frac{3}{5} \div 11\frac{1}{5} = s$ $\frac{1}{7}$
22. $3\frac{1}{8}\left(2\frac{4}{5}\right)\left(\frac{5}{7}\right) = p$ $6\frac{1}{4}$
23. $3\frac{2}{3}\left(\frac{1}{8}\right)\left(1\frac{1}{11}\right) = h$ $\frac{1}{2}$
24. $m = 3\frac{1}{21} \div 1\frac{21}{35}$ $1\frac{19}{21}$

25. $\frac{1}{5}\left(-\frac{1}{8}\right) = p$ $-\frac{1}{40}$
26. $-\frac{2}{9}\left(\frac{1}{3}\right) = w$ $-\frac{2}{27}$
27. $-\frac{5}{8}\left(-\frac{4}{5}\right) = c$ $\frac{1}{2}$
28. $-3 \div \frac{1}{2} = y$ -6
29. $-\frac{7}{9} \div \left(-\frac{1}{9}\right) = r$ 7
30. $\frac{2}{3} \div \left(-\frac{4}{9}\right) = t$ $-\frac{3}{2}$
31. $-\frac{2}{5}(-10) = m$ 4
32. $\frac{2}{3} \div \left(-\frac{7}{9}\right) = h$ $-\frac{6}{7}$
33. $-\frac{9}{15}\left(\frac{5}{9}\right) = s$ $-\frac{1}{3}$
34. $-\frac{9}{14} \div \left(-\frac{3}{7}\right) = a$ $1\frac{1}{2}$
35. $\frac{7}{16} \div \left(-\frac{7}{11}\right) = z$ $-\frac{11}{16}$
36. $j = -\frac{4}{5}(30)$ -24
37. $-7 \div 4 = q$ $-1\frac{3}{4}$
38. $-4\frac{9}{10}\left(-1\frac{5}{21}\right) = b$ $6\frac{1}{15}$
39. $-5\frac{3}{5} \div 4\frac{1}{5} = g$ $-1\frac{1}{3}$
40. $-5\frac{3}{5} \div \left(-4\frac{1}{5}\right) = e$ $1\frac{1}{3}$
41. $v = 6\frac{1}{4}\left(-1\frac{7}{15}\right)$ $-9\frac{1}{6}$
42. $3\frac{1}{3}\left(-4\frac{1}{2}\right) = w$ -15
43. $-2\left(1\frac{5}{18}\right) = k$ $-2\frac{5}{9}$
44. $4\frac{2}{5} \div \left(-\frac{11}{15}\right) = p$ -6
45. $-2\frac{5}{8} \div 7\frac{1}{2} = x$ $-\frac{7}{20}$
46. $5 \div (-11) = n$ $-\frac{5}{11}$
47. $d = -2\frac{3}{10}\left(-\frac{5}{12}\right)$ $\frac{23}{24}$
48. $-9\frac{1}{3}\left(-3\frac{3}{4}\right) = f$ 35

49. $\frac{1}{3}a = 5$ 15
50. $\frac{4}{7}k = 4$ 7
51. $\frac{2}{5}x = \frac{4}{7}$ $1\frac{3}{7}$
52. $w \div 5 = 3$ 15
53. $w \div \frac{1}{4} = \frac{3}{8}$ $\frac{3}{32}$
54. $c \div \frac{3}{10} = \frac{1}{2}$ $\frac{3}{20}$
55. $\frac{7}{11}t = \frac{4}{5}$ $1\frac{9}{35}$
56. $h \div \frac{1}{8} = \frac{4}{11}$ $\frac{1}{22}$
57. $z \div 6 = \frac{5}{12}$ $2\frac{1}{2}$
58. $1\frac{1}{2}d = \frac{6}{7}$ $\frac{4}{7}$
59. $\frac{10}{33} = b \div 4\frac{2}{5}$ $1\frac{1}{3}$
60. $2\frac{1}{6}j = 5\frac{1}{5}$ $2\frac{2}{5}$
61. $s \div 2\frac{1}{6} = 2\frac{2}{5}$ $5\frac{1}{5}$
62. $1\frac{3}{24}g = 3\frac{1}{8}$ $2\frac{7}{9}$
63. $3 = 1\frac{7}{11}q$ $1\frac{5}{6}$
64. $n \div \frac{2}{3} = -\frac{4}{9}$ $-\frac{8}{27}$
65. $-1\frac{3}{4}p = -\frac{5}{8}$ $\frac{5}{14}$
66. $v \div \left(-\frac{7}{11}\right) = 1\frac{2}{7}$ $-\frac{9}{11}$
67. $-1\frac{3}{5} = e \div \left(-3\frac{1}{5}\right)$ $5\frac{3}{25}$
68. $-\frac{5}{9}r = 7\frac{1}{2}$ $-13\frac{1}{2}$
69. $3\frac{4}{7}x = -3\frac{3}{4}$ $-1\frac{1}{20}$
70. $a \div 3\frac{2}{7} = -8\frac{3}{4}$ $-28\frac{3}{4}$
71. $-2\frac{4}{7}m = -3\frac{3}{8}$ $1\frac{5}{16}$
72. $f \div \left(-3\frac{1}{8}\right) = -3\frac{2}{5}$ $10\frac{5}{8}$

Exercises 1-30 provide review of adding and subtracting decimals. Exercises 31-60 provide practice in adding and subtracting decimals including negative decimals. Exercises 61-90 provide practice in solving equations.

Decimal Equations: Addition and Subtraction

Solve each equation.

1. $0.53 + 0.26 = x$ 0.79
2. $14.756 + 0.185 = k$ 14.941
3. $0.711 - 0.158 = z$ 0.553
4. $12.01 - 0.83 = s$ 11.18
5. $0.4 + 0.86 = n$ 1.26
6. $1.4 - 0.12 = a$ 1.28
7. $57.5 + 7.94 = m$ 65.44
8. $10.04 - 0.18 = f$ 9.86
9. $5 - 1.63 = r$ 3.37
10. $5.92 + 7.3 = b$ 13.22
11. $12 + 9.6 = y$ 21.6
12. $28.05 - 9.95 = c$ 18.1
13. $0.2 + 6.51 + 2.03 = y$ 8.74
14. $4.4 + 30.6 + 11.2 = z$ 46.2
15. $0.007 + 3 + 10.02 = h$ 13
16. $w = 20.13 - 12.5$ 7.63
17. $2.3 - 0.846 = t$ 1.454
18. $11 - 1.1 = p$ 9.9
19. $6.2 + 5.54 + 13.66 = g$ 25.4
20. $a = 412 - 0.007$ 411.993
21. $101.12 + 9.099 = s$ 110.21
22. $66.4 - 5.288 = d$ 61.112
23. $84.083 - 17 = m$ 67.083
24. $q = 0.046 + 5.8 + 11.37$ 17
25. $8 - 3.49 = n$ 4.51
26. $8.77 + 0.3 + 52.9 = x$ 61.97
27. $14.7 - 5.8364 = e$ 8.8636
28. $66.68421 - 18.465 = v$ 48.21921
29. $y = 0.0013 + 2.881$ 2.8823
30. $127.11 + 48 + 0.143 = u$ 17

31. $-0.47 + 0.62 = h$ 0.15
32. $-4.5 + (-12.8) = x$ −17.3
33. $-1.7 + 0.24 = p$ −1.46
34. $-6.831 - (-2.648) = c$ −4.183
35. $-4.23 - 2.47 = b$ −6.7
36. $2.64 - (-5.9) = k$ 8.54
37. $10 + (-0.43) = r$ 9.57
38. $6.7 - (-0.64) = v$ 7.34
39. $-6.71 - (-8) = e$ 1.29
40. $14.14 + (-1.4) = a$ 12.74
41. $1.2 - 6.73 = j$ −5.53
42. $-9.7 + (-0.86) = d$ −10.
43. $-7 - 4.63 = w$ −11.63
44. $-0.17 - (-14.6) = g$ 14.43
45. $m = 1.8 + (-14.14)$ −12.
46. $5.003 + (-0.47) = f$ 4.533
47. $0.88 - 42 = s$ −41.12
48. $-6.2 + (-27.47) = j$ −33
49. $n = -1.4962 + 2.118$ 0.6218
50. $2.4 - (-1.736) = q$ 4.136
51. $4.16 + (-5.909) = t$ −1.7
52. $17 + (-0.45) = w$ 16.55
53. $10 - 13.463 = a$ −3.463
54. $f = -82.007 - 3.218$ −85
55. $-11.264 + (-8.2) = z$ −19.464
56. $-56 + 2.783 = s$ −53.217
57. $-0.682 - (-0.81) = y$ 0.
58. $r = -23 + 4.093$ −18.907
59. $2.08 - (-0.094) = t$ 2.174
60. $-51.34 + (-5.1346) = x$ −56.47

61. $2.2 + a = 11.4$ 9.2
62. $h + 1.83 = 8.42$ 6.59
63. $c + 5.4 = -11.33$ −16.73
64. $m - 0.41 = 0.85$ 1.26
65. $p - 1.1 = 14.9$ 16
66. $r - 0.76 = -3.2$ −2.44
67. $t + (-6.47) = -22.3$ −15.83
68. $-6.11 + b = 14.321$ 20.431
69. $k - (-4) = 7.9$ 3.9
70. $k - 99.7 = -46.88$ 52.82
71. $w + (-17.8) = -5.63$ 12.17
72. $-5 = y - 22.7$ 17.7
73. $13.475 + d = 4.09$ −9.385
74. $-5 - q = 1.19$ −6.19
75. $-3.214 + f = -16.04$ −1
76. $-88.9 = s - 6.21$ −82.69
77. $2 + e = 1.008$ −0.992
78. $n + (-4.361) = 59.78$ 64.1
79. $4.8 - j = -5.834$ 10.634
80. $w - 0.73 = -1.8$ −1.07
81. $-8 = g - (-4.821)$ −12.8
82. $-2.315 + x = -15$ −12.685
83. $m + (-1.4) = 0.07$ 1.47
84. $v - 5.234 = -1.051$ 4.183
85. $7.1 = v - (-0.62)$ 6.48
86. $s + 6.4 = -0.11$ −6.51
87. $t - (-46.1) = -3.673$ −4
88. $k + (-1.604) = -0.45$ 1.154
89. $81.6 + p = -6.73$ −88.33
90. $-0.1448 - z = -2.6$ 2.45

Decimal Equations: Multiplication and Division

Solve each equation.

1. $46(0.5) = e$ 23
2. $108(0.9) = b$ 97.2
3. $g = 6.47(39)$ 252.33
4. $0.04(197) = f$ 7.88
5. $r = 67(5.892)$ 394.764
6. $2.8(4.27) = d$ 11.956
7. $0.061(5.5) = m$ 0.3355
8. $0.62(0.13) = c$ 0.0806
9. $4.007(1.95) = q$ 7.81365
10. $6.25 \div 5 = w$ 1.25
11. $t = 91.8 \div 27$ 3.4
12. $7.31 \div 43 = h$ 0.17
13. $5.91 \div 0.3 = a$ 19.7
14. $167.5 \div 2.5 = k$ 67
15. $4.7208 \div 0.84 = v$ 5.62
16. $p = 278.1 \div 6.18$ 45
17. $30{,}176 \div 9.43 = n$ 3200
18. $0.1001 \div 0.77 = j$ 0.13
19. $2.11(0.059) = w$ 0.12449
20. $s = 0.4484 \div 1.18$ 0.38
21. $0.0062(84.7) = x$ 0.52514
22. $0.03912 \div 1.63 = z$ 0.024
23. $230.4 \div 0.072 = m$ 3200
24. $w = 59.8(100.23)$
25. $v = 432 \div 9.6$ 45
26. $0.008(0.0045) = x$ 0.000036
27. $1.21(0.47)(9.3) = s$
28. $0.0418 \div 0.19 = x$ 0.22
29. $0.032(13)(2.6) = t$ 1.0816
30. $0.0001926 \div 0.00321 = y$ 0.06

31. $-5(0.2) = x$ −1
32. $-1.7(-44) = f$ 74.8
33. $72(1.01) = c$ 72.72
34. $627(-0.14) = a$ −87.78
35. $-2.3(7.81) = n$ −17.963
36. $r = -1.02(-4.4)$ 4.488
37. $57.6 \div (-12) = b$ −4.8
38. $160.8 \div 24 = h$ 6.7
39. $-16.38 \div (-0.7) = t$ 23.4
40. $m = -15.54 \div 2.1$ −7.4
41. $-0.405 \div (-0.27) = a$ 1.5
42. $-598 \div 0.13 = p$ −4600
43. $0.45(-0.0016) = k$ −0.00072
44. $y = -0.002052 \div 0.054$ −0.038
45. $6.7284 \div 1.08 = d$ 6.23
46. $-0.0066(-91.8) = w$ 0.60588
47. $455 \div (-1.82) = q$ −250
48. $-0.905(0.208) = g$
49. $-2.4827 \div (-6.71) = e$ 0.37
50. $0.153 \div (-0.017) = z$ −9
51. $j = -462.1(0.0094)$
52. $56.1(2.3) = y$ 129.03
53. $0.07553 \div 0.0083 = v$ 9.1
54. $-1.7(-0.121) = s$ 0.2057
55. $t = -0.6612 \div (-0.114)$ 5.8
56. $-0.026(45.1) = x$ −1.1726
57. $59(-0.00042) = w$
58. $7.93(-5.036) = c$ −39.93548
59. $9.2397 \div 1.9 = t$ 4.863
60. $-0.000101 \div 0.001 = m$ −0.101

61. $7c = 4.2$ 0.6
62. $37p = 81.4$ 2.2
63. $57k = 0.1824$ 0.0032
64. $1.5m = 9.9$ 6.6
65. $1.296 = 0.48d$ 2.7
66. $0.0022b = 0.1958$ 89
67. $t \div 110 = 2.8$ 308
68. $x \div 71 = 0.33$ 23.43
69. $r \div 0.85 = 10$ 8.5
70. $h \div 1.98 = 6.7$ 13.266
71. $a \div 0.002 = 0.109$ 0.000218
72. $n \div 40.6 = 0.021$ 0.8526
73. $100.8x = 9374.4$ 93
74. $2.61 = f \div 9.5$ 24.795
75. $1.7118 = 0.317e$ 5.4
76. $0.0603g = 0.0043416$ 0.072
77. $w \div 0.0412 = 60$ 2.472
78. $q \div 1.07 = 0.088$ 0.09416
79. $5j = -32.15$ −6.43
80. $-1.2v = 112.8$ −94
81. $-0.013s = -0.00923$ 0.71
82. $w \div (-2) = -2.48$ 4.96
83. $z \div 2.8 = -6.2$ −17.36
84. $a \div (-0.53) = -0.034$
85. $k \div (-0.013) = -0.7$ 0.0091
86. $-4.63t = -125.473$ 27.1
87. $7.9y = 1583.16$ 200.4
88. $6.05p = -1573$ −260
89. $g \div 9.9 = 12$ 118.8
90. $x \div (-0.063) = 0.015$ −0.000945

Exercises 1-30 provide review of multiplying and dividing decimals. Exercises 31-60 provide practice in multiplying and dividing decimals including negative numbers. Exercises 61-90 provide practice in solving equations.

Answers

24. 5993.754
27. 5.28891
48. −0.18824
51. −4.34374
57. −0.02478
84. 0.01802

Forms of Real Numbers

Write each fraction in simplest form.

1. $\frac{13}{26}$ $\frac{1}{2}$	**2.** $\frac{9}{12}$ $\frac{3}{4}$	**3.** $-\frac{36}{42}$ $-\frac{6}{7}$	**4.** $\frac{5}{60}$ $\frac{1}{12}$
5. $-\frac{24}{32}$ $-\frac{3}{4}$	**6.** $-\frac{10}{35}$ $-\frac{2}{7}$	**7.** $\frac{54}{63}$ $\frac{6}{7}$	**8.** $-\frac{45}{60}$ $-\frac{3}{4}$
9. $\frac{48}{84}$ $\frac{4}{7}$	**10.** $-\frac{28}{42}$ $-\frac{2}{3}$	**11.** $-\frac{72}{96}$ $-\frac{3}{4}$	**12.** $\frac{75}{105}$ $\frac{5}{7}$
13. $-\frac{16}{100}$ $-\frac{4}{25}$	**14.** $\frac{24}{60}$ $\frac{2}{5}$	**15.** $\frac{15}{27}$ $\frac{5}{9}$	**16.** $-\frac{99}{111}$ $-\frac{33}{37}$
17. $\frac{126}{700}$ $\frac{9}{50}$	**18.** $-\frac{198}{462}$ $-\frac{3}{7}$	**19.** $-\frac{84}{1080}$ $-\frac{7}{90}$	**20.** $-\frac{525}{1155}$ $-\frac{5}{11}$

Change each fraction to a decimal.

21. $\frac{1}{4}$ 0.25	**22.** $-\frac{3}{10}$ -0.3	**23.** $\frac{1}{50}$ 0.02	**24.** $\frac{2}{3}$ $0.\overline{6}$
25. $-\frac{1}{9}$ $-0.\overline{1}$	**26.** $-\frac{16}{25}$ -0.64	**27.** $-\frac{9}{20}$ -0.45	**28.** $\frac{1}{11}$ $0.\overline{09}$
29. $\frac{5}{9}$ $0.\overline{5}$	**30.** $-\frac{5}{8}$ -0.625	**31.** $\frac{43}{100}$ 0.43	**32.** $-\frac{5}{6}$ $-0.8\overline{3}$
33. $-\frac{7}{11}$ $-0.\overline{63}$	**34.** $\frac{3}{7}$ $0.\overline{428571}$	**35.** $\frac{4}{5}$ 0.8	**36.** $-\frac{7}{12}$ $-0.58\overline{3}$
37. $-\frac{15}{16}$ -0.9375	**38.** $-\frac{8}{15}$ $-0.5\overline{3}$	**39.** $\frac{1}{6}$ $0.1\overline{6}$	**40.** $-\frac{11}{32}$ -0.34375
41. $\frac{9}{11}$ $0.\overline{81}$	**42.** $\frac{11}{16}$ 0.6875	**43.** $-\frac{11}{15}$ $-0.7\overline{3}$	**44.** $\frac{124}{125}$ 0.992

Change each mixed numeral to a decimal.

45. $-5\frac{1}{2}$ -5.5	**46.** $14\frac{17}{100}$ 14.17	**47.** $6\frac{3}{25}$ 6.12	**48.** $-7\frac{1}{3}$ $-7.\overline{3}$
49. $4\frac{3}{25}$ 4.12	**50.** $-20\frac{2}{9}$ $-20.\overline{2}$	**51.** $12\frac{3}{4}$ 12.75	**52.** $-10\frac{5}{6}$ $-10.8\overline{3}$
53. $-1\frac{4}{9}$ $-1.\overline{4}$	**54.** $-9\frac{16}{50}$ -9.32	**55.** $-2\frac{2}{11}$ $-2.\overline{18}$	**56.** $13\frac{13}{40}$ 13.325
57. $3\frac{5}{12}$ $3.41\overline{6}$	**58.** $-8\frac{5}{7}$ $-8.\overline{714285}$	**59.** $2\frac{3}{5}$ 2.6	**60.** $11\frac{1}{12}$ $11.08\overline{3}$
61. $-44\frac{3}{8}$ -44.375	**62.** $19\frac{8}{15}$ $19.5\overline{3}$	**63.** $-67\frac{7}{10}$ -67.7	**64.** $5\frac{3}{16}$ 5.1875
65. $78\frac{2}{9}$ $78.\overline{2}$	**66.** $-108\frac{1}{20}$ -108.05	**67.** $-51\frac{6}{7}$ $-51.\overline{857142}$	**68.** $8\frac{1}{15}$ $8.0\overline{6}$

Change each decimal to a fraction in simplest form.

69. 0.3 $\frac{3}{10}$	**70.** 0.14 $\frac{7}{50}$	**71.** 0.013 $\frac{13}{1000}$	**72.** -1.25 $-1\frac{1}{4}$
73. 4.2 $4\frac{1}{5}$	**74.** -20.05 $-20\frac{1}{20}$	**75.** $0.\overline{3}$ $\frac{1}{3}$	**76.** -14.50 $-14\frac{1}{2}$
77. $-12.\overline{7}$ $-12\frac{7}{9}$	**78.** 6.125 $6\frac{1}{8}$	**79.** -8.6 $-8\frac{3}{5}$	**80.** $-8.\overline{6}$ $-8\frac{2}{3}$
81. 23.15 $23\frac{3}{20}$	**82.** -0.37 $-\frac{37}{100}$	**83.** -33.85 $-33\frac{17}{20}$	**84.** 1.16 $1\frac{4}{25}$
85. -2.27 $-2\frac{27}{100}$	**86.** 16.75 $16\frac{3}{4}$	**87.** -5.375 $-5\frac{3}{8}$	**88.** 4.26 $4\frac{13}{50}$
89. 7.1875 $7\frac{3}{16}$	**90.** -9.45 $-9\frac{9}{20}$	**91.** $5.2\overline{6}$ $5\frac{4}{15}$	**92.** -0.324 $-\frac{81}{250}$

Percents

Write each decimal as a percent.

1. 0.71 71%
2. 0.4 40%
3. 0.835 83.5%
4. 1.05 105%
5. 0.009 0.9%
6. 0.27 27%
7. 2.5 250%
8. 0.706 70.6%

Write each fraction as a percent.

9. $\frac{31}{100}$ 31%
10. $\frac{1}{2}$ 50%
11. $\frac{4}{5}$ 80%
12. $\frac{3}{10}$ 30%
13. $\frac{1}{8}$ 12.5%
14. $\frac{5}{4}$ 125%
15. $\frac{2}{3}$ $66.\overline{6}\%$
16. $\frac{4}{11}$ $36.\overline{36}\%$

Write each percent as a decimal.

17. 14% 0.14
18. 10% 0.1
19. 450% 4.5
20. 6% 0.06
21. 27.5% 0.275
22. 4.2% 0.042
23. 190.5% 1.905
24. 0.3% 0.003

Write each percent as a fraction in simplest form.

25. 17% $\frac{17}{100}$
26. 40% $\frac{2}{5}$
27. 8% $\frac{2}{25}$
28. 75% $\frac{3}{4}$
29. 0.9% $\frac{9}{1000}$
30. 2.5% $\frac{1}{40}$
31. 45.6% $\frac{57}{125}$
32. 1.05% $\frac{21}{2000}$

Solve each of the following.

33. 10% of 70 is __. 7
34. 20% of 35 is __. 7
35. 4% of 250 is __. 10
36. 255% of 160 is __. 408
37. 115% of 24 is __. 27.6
38. 130% of 60 is __. 78
39. __ is 3.7% of 300. 11.1
40. __ is 22.5% of 260. 58.5
41. __ is 52.6% of 150. 78.9
42. 5 is __% of 20. 25
43. 3 is __% of 10. 30
44. 17 is __% of 68. 25
45. __% of 500 is 55. 11
46. __% of 96 is 12. 12.5
47. __% of 81 is 27. $33.\overline{3}$
48. 40% of __ is 12. 30
49. 10% of __ is 16. 160
50. 65% of __ is 26. 40
51. 25 is $33\frac{1}{3}\%$ of __. 75
52. 54 is 108% of __. 50
53. 1.28 is 16% of __. 8
54. 6.3% of 400 is __. 25.2
55. __% of 40 is 25. 62.5
56. 68% of __ is 85. 125
57. 44% of __ is 37.4. 85
58. 53% of 62 is __. 32.86
59. __% of 16 is 56. 350
60. __ is 235% of 270. 634.5
61. __ is 5.8% of 45. 2.61
62. 28 is __% of 21. $133.\overline{3}$
63. 2.5% of __ is 1. 40
64. __% of 20 is 26.2. 131
65. 420% of __ is 336. 80
66. __% of 45 is 30. $66.\overline{6}$
67. $87\frac{1}{2}\%$ of __ is 14. 16
68. $66\frac{2}{3}\%$ of 81 is __. 54
69. __ is 12.4% of 15. 1.86
70. 135 is 675% of __. 20
71. 45 is __% of 36. 125
72. 14.5% of 18 is __. 2.61
73. 180.5% of 200 is __. 361
74. 98.1 is __% of 90. 109
75. __% of 85 is 102. 120
76. 3% of __ is 18. 600
77. 44% of __ is 37.4. 85
78. __% of 170 is 153. 90
79. 738 is 72% of __. 1025
80. $266\frac{2}{3}\%$ of 561 is __. 1496

Evaluating Expressions

Evaluate each expression if $a = 3$, $b = 5$, $c = 12$, and $d = 9$.

1. $8 + c$ 20 **2.** $d - 4$ 5 **3.** $b \cdot d$ 45 **4.** $c \div a$ 4

5. $a + c + d$ 24 **6.** $c - d$ 3 **7.** $a \cdot b \cdot c$ 180 **8.** $d - a$ 6

9. $a \cdot c$ 36 **10.** $\frac{d}{a}$ 3 **11.** $\frac{13 + c}{b}$ 5 **12.** $\frac{a + d}{4}$ 3

Evaluate each expression if $e = 2$, $f = 5$, $g = 6$, and $h = 10$.

13. $8g$ 48 **14.** fh 50 **15.** g^2 36 **16.** e^5 32

17. $7f^2$ 175 **18.** g^2h^3 36,000 **19.** $\frac{h^4}{f^2}$ 400 **20.** $3e^2g$ 72

21. $8g^2f^2$ 7200 **22.** $e^4f^2h^3$ 400,000 **23.** $20e^3f^3g$ 120,000 **24.** $\frac{3e^2f^3}{g}$ 250

Evaluate each expression if $x = 3$, $j = 4$, $k = 9$, and $m = 20$.

25. $k^2 - 4k + 6$ 51 **26.** $(m + j) \div 3$ 8 **27.** $x^3j^2 - 4m$ 352

28. $(xj + k) \div x$ 7 **29.** $5j^2 \div m + k^2$ 85 **30.** $j^3 + mk + 4x^4$ 568

31. $(j^3 + m)k - 4x$ 744 **32.** $(xj)^2 + km^2$ 3744 **33.** $k^3 + m \div j - 5j^2$ 654

34. $(5 + j)^2 \div k + m^2$ 409 **35.** $(m - k)^3 \div (2j + 3)$ 121 **36.** $(x^4 + k)\,m^2 - kx$ 35,973

Evaluate each expression if $n = -1$, $p = 6$, $q = -8$, $r = 15$, and $s = -24$.

37. $pr + 2q$ 74 **38.** $pn^4 + s$ −18 **39.** $r^2 - q + 5s$ 113

40. $pq^2 \div ns$ 16 **41.** $(p + 2q)n - r$ −5 **42.** $(p + q)^5n^5 + s$ 8

43. $pr^2 + ns - 6q$ 1422 **44.** $p(r^2 + ns) - 6q$ 1542 **45.** $(q + r + s)p + n$ −103

46. $\frac{4(p^2 + q^2)}{2q} - s$ −1 **47.** $(p + n)^3 + \left(\frac{s}{p}\right)q$ 157 **48.** $[(r + s)q]p$ 432

The formula for the total surface area of a rectangular solid is $T = 2lw + 2wh + 2lh$ where T is the total surface area of the solid, l is its length, w is its width, and h is its height. Find the total surface area of each rectangular solid.

49. $l = 8$, $w = 5$, $h = 14$. 444 **50.** $l = 4$, $w = 2.5$, $h = 3$ 59 **51.** $l = 7$, $w = 7$, $h = 16$ 546

52. $l = 14$, $w = 17$, $h = 11$ 1158 **53.** $l = 21$, $w = 18$, $h = 6$ 1224 **54.** $l = 3.7$, $w = 1.2$, $h = 3.5$ 43.18

The formula to change Fahrenheit degrees to Celsius degrees is $C = \frac{5}{9}(F - 32)$ where C is the temperature in Celsius degrees and F is the temperature in Fahrenheit degrees. Change each temperature in Fahrenheit degrees to Celsius degrees.

55. 86° F 30°C **56.** 5° F −15°C **57.** −13° F −25°C **58.** 41° F 5°C

59. −40° F −40°C **60.** 23° F −5°C **61.** −58° F −50°C **62.** 374° F 190°C

Inequalities

Replace each ■ with >, <, or = to make each sentence true.

1. 9 ■ 12 $<$ **2.** 14 ■ 7 $>$ **3.** -3 ■ 0 $<$ **4.** -7 ■ -3 $<$

5. -5 ■ 3 $<$ **6.** $7 + 8$ ■ 15 $=$ **7.** $-8 + 5$ ■ -6 $>$ **8.** $-24 \div (-8)$ ■ -3 $>$

9. $-3 - (-9)$ ■ -12 $>$ **10.** -9 ■ $-3 + (-4)$ $<$ **11.** $-6 \cdot 3$ ■ -18 $=$ **12.** 10 ■ $36 \div 4$ $>$

13. 8 ■ $-40 \div 5$ $>$ **14.** $5 - (-4)$ ■ 9 $=$ **15.** 27 ■ $4 \cdot 7$ $<$ **16.** $4 - 7$ ■ 3 $<$

Solve each inequality.

17. $x + 4 < 10$ $x < 6$ **18.** $a + 7 \geq 15$ $a \geq 8$ **19.** $g + 5 > -8$ $g > -13$

20. $c + 9 \leq 3$ $c \leq -6$ **21.** $z - 4 > 20$ $z > 24$ **22.** $h - (-7) > -2$ $h > -9$

23. $m - 14 \leq -9$ $m \leq 5$ **24.** $d - (-3) < 13$ $d < 10$ **25.** $\frac{g}{-8} < 4$ $g > -32$

26. $\frac{w}{3} > -12$ $w > -36$ **27.** $\frac{p}{5} < 8$ $p < 40$ **28.** $\frac{t}{-4} \geq -10$ $t \leq 40$

29. $7b \geq -49$ $b \geq -7$ **30.** $-5j < -60$ $j > 12$ **31.** $-8f < 48$ $f > -6$

32. $-2 + 9n \leq 10n$ $n \geq -2$ **33.** $-5e + 9 > 24$ $e < -3$ **34.** $3y - 4 > -37$ $y > -11$

35. $7s - 12 < 13$ $s < \frac{25}{7}$ **36.** $-6v - 3 \geq -33$ $v \leq 5$ **37.** $-2k + 12 < 30$ $k > -9$

38. $-2x + 1 < 16 - x$ $x > -15$ **39.** $15t - 4 > 11t - 16$ $t > -3$ **40.** $13 - y \leq 29 + 2y$ $y \geq -\frac{16}{3}$

41. $5q + 7 \leq 3(q + 1)$ $q \leq -2$ **42.** $2(w + 4) \geq 7(w - 1)$ $w \leq 3$ **43.** $-4t - 5 > 2t + 13$ $t < -3$

44. $9m + 7 < 2(4m - 1)$ $m < -9$ **45.** $3\left(a + \frac{2}{3}\right) \geq a - 1$ $a \geq -\frac{3}{2}$ **46.** $3(3y + 1) < 13y - 8$ $y > \frac{11}{4}$

47. $2 + x < -5$ or $2 + x > 5$ $x < -7$ or $x > 3$ **48.** $-4 + t > -5$ or $-4 + t < 7$

49. $3 \leq 2g + 7$ and $2g + 7 \leq 15$ $-2 \leq g \leq -4$ **50.** $7 - 3s < 13$ and $7s < 3s + 12$

51. $2x + 1 < -3$ or $3x - 2 > 4$ $x < -2$ or $x > 2$ **52.** $2v - 2 \leq 3v$ and $4v - 1 \geq 3v$

53. $3b - 4 \leq 7b + 12$ and $8b - 7 \leq 25$ $-4 \leq b \leq 4$ **54.** $-9 < 2z + 7 < 10$

55. $5m - 8 \geq 10 - m$ or $5m + 11 < -9$ $m \geq 3$ or $m < -4$ **56.** $12c - 4 \leq 5c + 10$ or $-4c - 1 \leq c + 24$

57. $2h - 2 \leq 3h \leq 4h - 1$ $h \geq 1$ **58.** $3p + 6 < 8 - p$ and $5p + 8 \geq p + 6$

59. $4a + 3 < 3 - 5a$ or $a - 1 \geq -a$ $a < 0$ or $a \geq \frac{1}{2}$ **60.** $d - 4 < 5d + 14 < 3d + 26$

61. $2r + 8 > 16 - 2r$ and $7r + 21 < r - 9$ **62.** $-4j + 3 < j + 22$ and $j - 3 < 2j - 15$

63. $3n \neq 9$ and $6n - 5 \leq 2n + 7$ $n < 3$ **64.** $7e \neq -21$ and $5e + 8 \geq e + 6$

65. $2(q - 4) \leq 3(q + 2)$ or $q - 8 \leq 4 - q$ $-14 \leq q$ or $q \leq 6$ **66.** $\frac{1}{2}w + 5 \geq w + 2 \geq \frac{1}{2}w + 9$

67. $|g + 6| > 8$ **68.** $|t - 5| \leq 3$ **69.** $|a + 5| \geq 0$ **70.** $|y - 9| < 19$

71. $|2m - 5| > 13$ **72.** $|14 - w| \geq 20$ **73.** $|3p + 5| \leq 23$ **74.** $|6b - 12| \leq 36$

75. $|25 - 3x| < 5$ **76.** $|7 + 8x| > 39$ **77.** $|4c + 5| \geq 25$ **78.** $|4 - 5s| > 46$

79. $|8 - 2h| \leq 25$ **80.** $|-8 - 10q| \geq 17$ **81.** $|7r - 5| < 3$ **82.** $|-3n - 12| < 15$

Answers

48. {all numbers}
50. $-2 < s < 3$
52. $v \geq 1$
54. $-8 < z < \frac{3}{2}$
56. {all numbers}
58. $-\frac{1}{2} \leq p < \frac{1}{2}$
60. $-\frac{9}{2} < d < 6$
61. no solutions
62. $-\frac{19}{5} < j < 12$
64. $e \neq 3$ and $e \geq -\frac{1}{2}$
66. no solutions
67. $g > 2$ or $g < -14$
68. $2 \leq t \leq 8$
69. {all numbers}
70. $-10 < y < 28$
71. $m > 9$ or $m < -4$
72. $w \leq -6$ or $w \geq 34$
73. $-\frac{28}{3} \leq p \leq 6$
74. $-4 \leq b \leq 8$
75. $\frac{20}{3} < x < 10$
76. $x > 4$ or $x < -\frac{23}{4}$
77. $c \geq 5$ or $c \leq -\frac{15}{2}$
78. $s < -\frac{42}{5}$ or $s > 10$
79. $-\frac{17}{2} \leq h \leq \frac{33}{2}$
80. $q \leq -\frac{5}{2}$ or $q \geq \frac{9}{10}$
81. $\frac{2}{7} < r < \frac{8}{7}$
82. $-9 < n < 1$

Answers

11. $4a^2 - 6a + 11$
12. $8x + 5y - 13z$
13. $11m + 4mn - 7n$
14. $5x + 14y - 3z$
15. $6 + 8g - 13h - 6j$
16. $-3x^2 + 3x + 20$
17. $-12t^2 - 8ts - 3s^2$
18. $6g + 5h - 9 - 6k$
36. $4v - 20w + 8x$
37. $4a^2 - 12ab - 10b^2$
38. $-6r - 9t - 2s$
39. $9ax^2 - 3ax - 12a$
40. $h^3 + h^2 + 8$
41. $7d - 7e - 3f - 6$
42. $11z^2 - 3x^2 - 8y^2 - 12$
43. $4b + 14$
44. $15j^4k^2 - 15j^2k - 3$
45. $8x^2 + 5xy - 15y^2$
46. $13z^4 - 2z^3 - 5z^2$
47. $4p - 2q + 5t$
48. $-19h + 19j - k + 3$
49. $6a + b - 8x - 6y + 11z$
50. $3m^2 - 7mn + 8n^2 - 5m + 5n$

Polynomials: Addition and Subtraction

Find each sum.

1. $(3x - 4y) + (8x + 6y)$ $11x + 2y$
2. $(12b + 2a) + (7b - 13a)$ $19b - 11a$
3. $(7m - 8n) + (4m - 5n)$ $11m - 13n$
4. $(5x^2 + 3x) + (4x^2 + 2x)$ $9x^2 + 5x$
5. $(-6s - 11t) + (5s - 6t)$ $-s - 17t$
6. $(-14g - h) + (-8g + 5h)$ $-22g + 4h$
7. $(4p - 7q) + (5q - 8p)$ $-4p - 2q$
8. $(5y^2 - 7y) + (7y - 3y^2)$ $2y^2 + 14y$
9. $(9b^3 - 3b^2) + (12b^2 + 4b)$ $9b^3 + 9b^2 + 4b$
10. $(2r + 8s) + (-3s - 9t)$ $2r + 5s - 9t$
11. $(2a^2 + 4a + 5) + (2a^2 - 10a + 6)$
12. $(7x - 2y - 5z) + (x + 7y - 8z)$
13. $(-3m + 9mn - 5n) + (14m - 2n - 5mn)$
14. $(5x + 8y + 3z) + (-6z + 6y)$
15. $(6 - 4g - 9h) + (12g - 4h - 6j)$
16. $(x^2 - 4x + 8) + (12 + 7x - 4x^2)$
17. $(-7t^2 + 4ts - 6s^2) + (3s^2 - 12ts - 5t^2)$
18. $(7g + 8h - 9) + (-g - 3h - 6k)$
19. $(8a^2 - 4ab - 3b^2 + a - 4b) + (3a^2 + 6ab - 9b^2 + 7a + 9b)$ $11a^2 + 2ab - 12b^2 + 8a + 5$
20. $(-3v + 14w - 12x - 13y + 6z - 8) + (16 - 6v + 2x - 5y - 2z)$ $8 - 9v + 14w - 10x - 18y + 4z$
21. $(3y^2 - 7y + 6) + (3 - 2y^2 - 5y) + (y^2 - 8y - 12)$ $2y^2 - 20y - 3$
22. $(4a^2 - 10b^2 + 7c^2) + (2c^2 - 5a^2 + 2b) + (7b^2 - 7c^2 + 7a)$ $-a^2 - 3b^2 + 2c^2 + 2b + 7a$
23. $(5x^2 + 3) + (4 - 7x - 9x^2) + (2x - 3x^2 - 5) + (2x - 6)$ $-7x^2 - 3x - 4$
24. $(9p - 13p^2) + (7p^2 + 5q^2) + (-6p - 12q) + (3q - 8q^2)$ $-6p^2 + 3p - 3q^2 - 9q$

Find each difference.

25. $(5g + 3h) - (2g + 7h)$ $3g - 4h$
26. $(2e - 5f) - (7e - f)$ $-5e - 4f$
27. $(-3m + 8n) - (6m - 4n)$ $-9m + 12n$
28. $(6a^2 - 9a) - (4a^2 + 2a)$ $2a^2 - 11a$
29. $(-11k + 6) - (-6k - 8)$ $-5k + 14$
30. $(9y^2 - 4y) - (-6y^2 - 8y)$ $15y^2 + 4y$
31. $(-r - 3s) - (2s - 5r)$ $4r - 5s$
32. $(13c^2 - 4c) - (5c - 12c^2)$ $25c^2 - 9c$
33. $(g^3 - 2g^2) - (5g^2 - 7)$ $g^3 - 7g^2 + 7$
34. $(7a + 4b) - (7b - 6c)$ $7a - 3b + 6c$
35. $(z^2 + 6z - 8) - (4z^2 - 7z - 5)$ $-3z^2 + 13z - 3$
36. $(6v - 12w - 2x) - (2v + 8w - 10x)$
37. $(6a^2 - 7ab - 4b^2) - (6b^2 + 2a^2 + 5ab)$
38. $(3r - 7t) - (2t + 2s + 9r)$
39. $(7ax^2 + 2ax - 4a) - (5ax - 2ax^2 + 8a)$
40. $(h^3 + 4h^2 - 7h) - (3h^2 - 7h - 8)$
41. $(4d + 3e - 8f) - (-3d + 10e - 5f + 6)$
42. $(-3z^2 + 4x^2 - 8y^2) - (7x^2 - 14z^2 - 12)$
43. $(2b^2 + 7b - 2) - (2b^2 + 3b - 16)$
44. $(15j^4k^2 - 7j^2k + 8) - (8j^2k + 11)$
45. $(9x^2 - 11xy - 3y^2) - (12y^2 + x^2 - 16xy)$
46. $(17z^4 - 5z^2 + 3z) - (4z^4 + 2z^3 + 3z)$
47. $(-4p - 7q - 3t) - (-8t - 5q - 8p)$
48. $(-14h + 16j - 7k) - (-3j + 5h - 6k - 3)$
49. $(14a + 9b - 2x - 11y + 4z) - (8a + 8b + 6x - 5y - 7z)$
50. $(7m^2 - 3mn + 4n^2 - 2m - 8n) - (4m^2 + 3m - 4n^2 - 13n + 4mn)$

Polynomials: Multiplication and Division

Find each product.

1. $t^3 \cdot t^6$ t^9
2. $g^5 \cdot g^9$ g^{14}
3. $(3x^2y)(-5x^3y^8)$ $-15x^5y^9$
4. $(7p^4q^7r^2)(4p^5r^7)$ $28p^9q^7r^9$
5. $(2a^2b)(-b^2c^3)(-8ab^2c^4)$
6. $(e^4f^6g)^5$
7. $(-2m^6n^2)^6$
8. $(-3h^2k^3)^3(5hj^6k^8)^2$
9. $(v^4w)^6(-1v^3w^2)^8$
10. $5y(y^2 - 3y + 6)$
11. $-ab(3b^2 + 4ab - 6a^2)$
12. $4st^2(-4s^2t^3 + 7s^5 - 3st^3)$
13. $(d + 2)(d + 3)$
14. $(z + 7)(z - 4)$
15. $(m - 5)(m - 8)$
16. $(2x - 5)(x + 6)$
17. $(7a - 4)(2a - 5)$
18. $(t + 7)^2$
19. $(q - 4h)^2$
20. $(w - 12)(w + 12)$
21. $(2b + 4d)(2b - 4d)$
22. $(4x + y)(2x - 3y)$
23. $(7v + 3)(v + 4)$
24. $(4e + 3)(4e + 3)$
25. $(7s - 8)(3s - 2)$
26. $(5b - 6)(5b + 6)$
27. $(4g + 3h)(2g - 5h)$
28. $(5c - 2d)^2$
29. $(10x + 11y)(10x - 11y)$
30. $(12r - 4s)(5r + 8s)$

Divide.

31. $\frac{24x^5}{8x^2}$ $3x^3$
32. $\frac{s^7t^4}{s^5}$ s^2t^4
33. $\frac{-9h^2k^4}{18h^5j^3k^4}$ $-\frac{1}{2h^3j^3}$
34. $\frac{3m^7n^2p^4}{9m^2np^3}$ $\frac{m^5np}{3}$
35. $\frac{9a^2b^7c^3}{12a^5b^4c}$ $\frac{3b^3c^2}{4a^3}$
36. $\frac{-15xy^5z^7}{-10x^4y^6z^4}$ $\frac{3z^3}{2x^3y}$
37. $\frac{-5w^4v^2 - 3w^3v}{w^3}$ $-5wv^2 - 3v$
38. $\frac{8g^4h^4 + 4g^3h^3}{2gh^2}$ $4g^3h^2 + 2g^2h$
39. $\frac{x^2 - 2x - 15}{x - 3}$ $x + 1 - \frac{12}{x - 3}$
40. $\frac{q^2 - 10q + 24}{q - 4}$ $q - 6$
41. $\frac{2j^2 + 10j + 12}{j + 3}$ $2j + 4$
42. $\frac{12d^2 + d - 6}{3d - 2}$ $4d + 3$
43. $\frac{4s^3 + 4s^2 - 9s - 18}{2s + 3}$ $2s^2 - s - 3 - \frac{9}{2s + 3}$
44. $\frac{z^3 - 27}{z - 3}$ $z^2 + 3z + 9$
45. $\frac{6n^3 + 7n^2 - 29n + 12}{3n - 4}$ $2n^2 + 5n - 3$
46. $\frac{3e^2 + 3e - 80}{e - 5}$ $3e + 18 + \frac{10}{e - 5}$
47. $\frac{-6k^2 + 3k + 12}{2k + 3}$ $-3k + 6 - \frac{6}{2k + 3}$
48. $\frac{t^3 + 3t}{t - 1}$ $t^2 + t + 4 + \frac{4}{t - 1}$
49. $\frac{4g^3 - 4g - 20}{2g - 4}$
50. $\frac{9a^3 - 3a^2 + 7a + 2}{3a + 1}$
51. $\frac{8c^3 - 2c^2 + 2c - 4}{2c - 2}$
52. $\frac{8y^3 - 1}{2y + 1}$ $4y^2 - 2y + 1 - \frac{2}{2y + 1}$
53. $\frac{12m^3 + m^2 - 20}{4m - 5}$ $3m^2 + 4m + 5 + \frac{5}{4m - 5}$
54. $\frac{5b^3 - 2b^2 + 7b + 4}{5b + 3}$ $b^2 - b + 2 - \frac{2}{5b + 3}$

Answers

5. $16a^3b^5c^7$
6. $e^{20}f^{30}g^5$
7. $64m^{36}n^{12}$
8. $-675h^8j^{12}k^{25}$
9. $v^{48}w^{22}$
10. $5y^3 - 15y^2 + 30y$
11. $-3ab^3 - 4a^2b^2 + 6a^3b$
12. $-16s^3t^5 + 28s^6t^2 - 12s^2t^5$
13. $d^2 + 5d + 6$
14. $z^2 + 3z - 28$
15. $m^2 - 13m + 40$
16. $2x^2 + 7x - 30$
17. $14a^2 - 43a + 20$
18. $t^2 + 14t + 49$
19. $q^2 - 8hq + 16h^2$
20. $w^2 - 144$
21. $4b^2 - 16d^2$
22. $8x^2 - 10xy - 3y^2$
23. $7v^2 + 31v + 12$
24. $16e^2 + 24e + 9$
25. $21s^2 - 38s + 16$
26. $25b^2 - 36$
27. $8g^2 - 14hg - 15h^2$
28. $25c^2 - 20cd + 4d^2$
29. $100x^2 - 121y^2$
30. $60r^2 + 76rs - 32s^2$
49. $2g^2 + 4g + 6 + \frac{4}{2g - 4}$
50. $3a^2 - 2a + 3 - \frac{1}{3a + 1}$
51. $4c^2 + 3c + 4 + \frac{4}{2c - 2}$

Answers
22. $6g^4h^3(g - 2h^3k - 3g^2h^2)$
30. $2b^2(a + c^2)^2$
44. $2(e^2 + f)(g + 2h)$
45. $(x + y)(x - y)(5x - 2y)$
58. $(b^2 + 1)(a - 1)(a + 1)$
61. $(c - 2d)(c + 2d)(a^2 + b^2)$
62. $18mn(2m^2 - 5mn + 2n)$

Factoring

Find the prime factorization of each integer. Write each negative integer as the product of −1 and its prime factors.

1. 35 $5 \cdot 7$
2. 12 $2^2 \cdot 3$
3. 72 $2^3 \cdot 3^2$
4. 64 2^6
5. −75 $-1 \cdot 3 \cdot 5^2$
6. 70 $2 \cdot 5 \cdot 7$
7. 85 $5 \cdot 17$
8. −92 $-1 \cdot 2^2 \cdot 23$
9. −117 $-1 \cdot 3^2 \cdot 13$
10. −114 $-1 \cdot 2 \cdot 3 \cdot 19$
11. 243 3^5
12. −360 $-1 \cdot 2^3 \cdot 3^2 \cdot$
13. 405 $5 \cdot 3^4$
14. 605 $5 \cdot 11^2$
15. −5292 $-1 \cdot 2^2 \cdot 3^3 \cdot 7^2$
16. 5076 $2^2 \cdot 3^3 \cdot 47$

Factor.

17. $10g + 35h$ $5(2g + 7h)$
18. $t^3s^2 - t^2$ $t^2(ts^2 - 1)$
19. $15a^2b - 24a^5b^2$ $3a^2b(5 - 8a^3b)$
20. $18c^4d - 30c^3e$ $6c^3(3cd - 5e)$
21. $36m^4n^2p + 12m^5n^3p^2$ $12m^4n^2p(3 + mnp)$
22. $6g^5h^3 - 12g^4h^6k - 18g^6h^5$
23. $p^2 - q^2$ $(p - q)(p + q)$
24. $144x^2 - 49y^2$ $(12x - 7y)(12x + 7y)$
25. $75r^2 - 48$ $3(5r - 4)(5r + 4)$
26. $64v^2 - 100w^4$ $4(4v - 5w^2)(4v + 5w^2)$
27. $g^2 + 4g + 4$ $(g + 2)^2$
28. $t^2 - 22t + 121$ $(t - 11)^2$
29. $9n^2 - 36nm + 36m^2$ $9(n - 2m)^2$
30. $2a^2b^2 + 4ab^2c^2 + 2b^2c^4$
31. $g^2 - 14g + 48$ $(g - 8)(g - 6)$
32. $z^2 + 15z + 36$ $(z + 12)(z + 3)$
33. $12 - 13b + b^2$ $(12 - b)(1 - b)$
34. $x^2 + 17xy + 16y^2$ $(x + y)(x + 16y)$
35. $g^2 - 4g - 32$ $(g - 8)(g + 4)$
36. $h^2 + 12h - 28$ $(h + 14)(h - 2)$
37. $s^2 - 13st - 30t^2$ $(s - 15t)(s + 2t)$
38. $3a^2 + 11a + 10$ $(3a + 5)(a + 2)$
39. $6y^2 + 2y - 20$ $2(3y - 5)(y + 2)$
40. $12j^2 - 34j - 20$ $2(2j + 1)(3j - 10)$
41. $24a^2 - 57ax + 18x^2$ $3(8a - 3x)(a - 2x)$
42. $2sx - 4tx + 2sy - 4ty$ $2(x + y)(s - 2t)$
43. $8ac - 2ad + 4bc - bd$ $(2a + b)(4c - d)$
44. $2e^2g + 2fg + 4e^2h + 4fh$
45. $5x^3 - 2x^2y - 5xy^2 + 2y^3$
46. $4p^2 + 12pr + 9r^2$ $(2p + 3r)^2$
47. $169 - 16t^2$ $(13 - 4t)(13 + 4t)$
48. $b^2 - 11b - 42$ $(b - 14)(b + 3)$
49. $30g^2h - 15g^3$ $15g^2(2h - g)$
50. $3b^2 - 13bd + 4d^2$ $(3b - d)(b - 4d)$
51. $a^2x - 2a^2y - 5x + 10y$ $(a^2 - 5)(x - 2y)$
52. $s^2 + 30s + 225$ $(s + 15)^2$
53. $18v^2 + 42v + 12$ $6(3v + 1)(v + 2)$
54. $4k^2 + 2k - 12$ $2(2k - 3)(k + 2)$
55. $5z^3 - 8z^2 - 21z$ $z(5z + 7)(z - 3)$
56. $5g^2 - 20h^2$ $5(g - 2h)(g + 2h)$
57. $30x^2 - 125x + 70$ $5(2x - 7)(3x - 2)$
58. $a^2b^2 - b^2 + a^2 - 1$
59. $8t^4 + 56t^3 + 98t^2$ $2t^2(2t + 7)^2$
60. $3p^3q^2 + 27pq^2$ $3pq^2(p^2 + 9)$
61. $a^2c^2 + b^2c^2 - 4a^2d^2 - 4b^2d^2$
62. $36m^3n - 90m^2n^2 + 36mn^2$
63. $4x^2z^2 + 7xyz^2 - 36y^2z^2$ $z^2(4x - 9y)(x + 4y)$
64. $4g^2j^2 - 25h^2j^2 - 4g^2 + 25h^2$ $(j - 1)(j + 1)(2g - 5h)(2g + 5h$

Algebraic Fractions

Simplify.

1. $\dfrac{48a^2b^5c}{32a^7b^2c^3}$ $\dfrac{3b^3}{2a^5c^2}$

2. $\dfrac{-28x^3y^4z^5}{42xyz^2}$ $\dfrac{-2x^2y^3z^3}{3}$

3. $\dfrac{k+3}{4k^2+7k-15}$ $\dfrac{1}{4k-5}$

4. $\dfrac{t^2-s^2}{5t^2-2st-3s^2}$ $\dfrac{t+s}{5t+3s}$

5. $\dfrac{6g^2-19g+15}{12g^2-6g-18}$ $\dfrac{3g-5}{6g+6}$

6. $\dfrac{2d^2+4d-6}{d^4-10d^2+9}$ $\dfrac{2}{(d-3)(d+1)}$

Find each product or quotient in simplest form.

7. $\dfrac{5m^2n}{12a^2}\cdot\dfrac{18an}{30m^4}$ $\dfrac{n^2}{4am^2}$

8. $\dfrac{25g^7h}{28t^3}\cdot\dfrac{42s^2t^3}{5g^5h^2}$ $\dfrac{15g^2s^2}{2h}$

9. $\dfrac{6a+4b}{36}\cdot\dfrac{45}{3a+2b}$ $\dfrac{5}{2}$

10. $\dfrac{x^2y}{18z}\div\dfrac{2yz}{3x^2}$ $\dfrac{x^4}{12z^2}$

11. $\dfrac{p^2}{14qr^3}\div\dfrac{2r^2p}{7q}$ $\dfrac{p}{4r^5}$

12. $\dfrac{3d}{2d^2-3d}\div\dfrac{9}{2d-3}$ $\dfrac{1}{3}$

13. $\dfrac{t^2-2t-15}{t-5}\cdot\dfrac{t+5}{t+3}$ $t+5$

14. $\dfrac{5e-f}{5e+f}\div(25e^2-f^2)$ $\dfrac{1}{(5e+f)^2}$

15. $\dfrac{8}{6c^2+17c+10}\div\dfrac{6}{c+2}$

16. $\dfrac{3v^2-27}{15v}\cdot\dfrac{v^2}{v+3}$ $\dfrac{v(v-3)}{5}$

17. $\dfrac{3k^2-10k+3}{5}\div\dfrac{3k-1}{15k}$ $3k^2-9k$

18. $\dfrac{3g^2+15g}{4}\cdot\dfrac{g^2}{g+5}$ $\dfrac{3g^3}{4}$

19. $\dfrac{x^2-2x-15}{2x^2-7x-15}\cdot\dfrac{4x^2-4x-15}{2x^2+x-15}$ 1

20. $\dfrac{a^2-16}{3a^2-13a+4}\div\dfrac{3a^2+11a-4}{a^3}$ $\dfrac{a^3}{(3a-1)^2}$

Find each sum or difference in simplest form.

21. $\dfrac{j+4}{3}+\dfrac{j-7}{3}$ $\dfrac{2j-3}{3}$

22. $\dfrac{15n}{5n+3}+\dfrac{9}{5n+3}$ 3

23. $\dfrac{2a-3}{b}-\dfrac{a-5}{b}$ $\dfrac{a+2}{b}$

24. $\dfrac{25}{5-g}-\dfrac{g^2}{5-g}$ $5+g$

25. $\dfrac{s}{t^2}-\dfrac{r}{3t}$ $\dfrac{3s-rt}{3t^2}$

26. $\dfrac{7}{ab}+\dfrac{4}{bc}$ $\dfrac{7c+4a}{abc}$

27. $\dfrac{2}{2p+3}+\dfrac{p}{3p+2}$

28. $\dfrac{x}{x+y}-\dfrac{5}{y}$ $\dfrac{xy-5x-5y}{xy+y^2}$

29. $\dfrac{c}{c^2-4c}-\dfrac{5c}{c-4}$ $\dfrac{1-5c}{c-4}$

30. $\dfrac{t+10}{t^2-100}+\dfrac{1}{t-10}$ $\dfrac{2}{t-10}$

31. $\dfrac{1}{g^2-6gh+9h^2}-\dfrac{3}{g-3h}$

32. $\dfrac{x}{x+2}+\dfrac{x^2+3x}{x^2+5x+6}$

33. $\dfrac{3d}{d^2-3d-10}+\dfrac{d+1}{d^2-8d+15}$

34. $\dfrac{k+2}{k^2-8k+16}-\dfrac{k+3}{k^2+k-20}$

Answers

15. $\dfrac{4}{3(6c+5)}$

27. $\dfrac{2p^2+9p+4}{(2p+3)(3p+2)}$

31. $\dfrac{1-3g+9h}{g^2-6gh+9h^2}$

32. $\dfrac{2x}{x+2}$

33. $\dfrac{4d^2-6d+2}{(d-5)(d+2)(d-3)}$

34. $\dfrac{8k+22}{(k-4)(k-4)(k+5)}$

Answers

30. $\left\{-\frac{5}{4}, -5\right\}$

36. $\left\{\frac{4}{3}, -3\right\}$

Solving Equations

Solve each equation.

1. $2x - 5 = 3$ 4
2. $4t + 5 = 37$ 8
3. $7a + 6 = -36$ −6
4. $47 = -8g + 7$ −5
5. $-3c - 9 = -24$ 5
6. $5k - 7 = -52$ −9
7. $5s + 4s = -72$ −8
8. $6(y - 5) = 18$ 8
9. $-21 = 7(p - 10)$ 7
10. $2m + 5 - 6m = 25$ −5
11. $3z - 1 = 23 - 3z$ 4
12. $5b + 12 = 3b - 6$ −9
13. $\frac{e}{5} + 6 = -2$ −40
14. $\frac{d}{4} - 8 = -5$ 12
15. $\frac{p + 10}{3} = 4$ 2
16. $\frac{h - 7}{6} = 1$ 13
17. $\frac{5f + 1}{8} = -3$ −5
18. $\frac{4n - 8}{-2} = 12$ −4
19. $\frac{2a}{7} + 9 = 3$ −21
20. $\frac{-3t - 4}{2} = 8$ $-\frac{20}{3}$
21. $\frac{6v - 9}{3} = v$ 3
22. $|s - 4| = 7$ 11, −3
23. $|5g + 8| = 33$ 5, $-\frac{41}{5}$
24. $|16 - 3b| = 22$ −2, $\frac{38}{3}$
25. $t(t - 4) = 0$ 0, 4
26. $4p(p + 7) = 0$ 0, −7
27. $7m(2m - 12) = 0$ 0, 6
28. $(x - 5)(x + 8) = 0$ 5, −8
29. $(3s + 6)(2s - 7) = 0$ −2, $\frac{7}{2}$
30. $(4g + 5)(2g + 10) = 0$
31. $h^2 + 4h = 0$ 0, −4
32. $3b^2 - 3b = 0$ 0, 1
33. $m^2 - 16 = 0$ 4, −4
34. $w^2 + w - 30 = 0$ −6, 5
35. $2c^2 - 14c + 24 = 0$ 3, 4
36. $6p^2 + 10p - 24 = 0$
37. $5n^2 = 15n$ 0, 3
38. $4x^2 + 20x + 25 = 0$ $-\frac{5}{2}$
39. $3y^2 = 75$ 5, −5
40. $\frac{a}{3} - \frac{a}{4} = 3$ 36
41. $\frac{k}{6} + \frac{2k}{3} = -\frac{5}{2}$ −3
42. $\frac{r + 3}{r} + \frac{r - 12}{r} = 5$ −3
43. $\frac{2y}{y - 4} - \frac{3}{5} = 3$ 9
44. $\frac{2t}{t + 3} + \frac{3}{t} = 2$ 3
45. $\frac{2g}{2g + 1} - \frac{1}{2g - 1} = 1$ 0
46. $\frac{8n^2}{2n^2 - 5n - 3} = 4$ $-\frac{3}{5}$
47. $5 - \frac{5x}{x - 4} = \frac{2}{x^2 - 4x}$ $-\frac{1}{10}$
48. $\frac{2}{e + 1} - \frac{3}{e + 2} = 0$ $-\frac{7}{5}$
49. $\frac{1}{2d - 1} - \frac{12d}{6d^2 + d - 2} = \frac{-4}{3d + 2}$ −2
50. $\frac{4z}{2z + 1} - \frac{6 - z}{2z^2 - 5z - 3} = 2$ 0
51. $\frac{b + 1}{b - 2} - \frac{3b}{3b + 2} = \frac{20}{3b^2 - 4b - 4}$ $\frac{18}{11}$
52. $\frac{2s}{9s^2 - 3s - 2} + \frac{2}{3s + 1} = \frac{4}{3s - 2}$ −2
53. $\frac{m}{20} = \frac{9}{15}$ 12
54. $\frac{12}{21} = \frac{20}{f}$ 35
55. $\frac{4}{9} = \frac{16}{t + 8}$ 28
56. $\frac{4}{14} = \frac{2h - 1}{21}$ 3.5
57. $\frac{2 + c}{c - 5} = \frac{8}{9}$ −58
58. $\frac{2y + 4}{y - 3} = \frac{2}{3}$ $-\frac{9}{2}$

Radicals

Simplify.

1. $\sqrt{25}$ 5 **2.** $-\sqrt{64}$ -8 **3.** $\pm\sqrt{576}$ ± 24 **4.** $\sqrt{900}$ 30

5. $\pm\sqrt{0.01}$ ± 0.1 **6.** $\sqrt{1.44}$ 1.2 **7.** $-\sqrt{0.0016}$ -0.04 **8.** $\sqrt{4.84}$ 2.2

9. $\sqrt{\dfrac{36}{121}}$ $\dfrac{6}{11}$ **10.** $-\sqrt{\dfrac{16}{36}}$ $-\dfrac{2}{3}$ **11.** $\sqrt{\dfrac{81}{49}}$ $\dfrac{9}{7}$ **12.** $\pm\sqrt{\dfrac{64}{100}}$ $\pm\dfrac{4}{5}$

13. $\sqrt{75}$ $5\sqrt{3}$ **14.** $\sqrt{20}$ $2\sqrt{5}$ **15.** $\sqrt{162}$ $9\sqrt{2}$ **16.** $\sqrt{700}$ $10\sqrt{7}$

17. $\sqrt{4x^4y^3}$ $2x^2y\sqrt{y}$ **18.** $\sqrt{12ts^3}$ $2s\sqrt{3ts}$ **19.** $\sqrt{175m^4n^6}$ **20.** $\sqrt{99a^3b^7}$

21. $\sqrt{\dfrac{54}{g^2}}$ $\dfrac{3\sqrt{6}}{|g|}$ **22.** $\sqrt{\dfrac{32c^5}{9d^2}}$ $\dfrac{4c^2\sqrt{2c}}{3|d|}$ **23.** $\sqrt{\dfrac{27p^4}{3p^2}}$ $3|p|$ **24.** $\sqrt{\dfrac{243y^7}{3y^4}}$ $9y\sqrt{y}$

25. $\sqrt{7}\,\sqrt{3}$ $\sqrt{21}$ **26.** $\sqrt{3}\,\sqrt{15}$ $3\sqrt{5}$ **27.** $6\sqrt{2}\,\sqrt{3}$ $6\sqrt{6}$ **28.** $5\sqrt{6}\cdot 2\sqrt{3}$ $30\sqrt{2}$

29. $\sqrt{5}(\sqrt{3} + \sqrt{10})$ **30.** $\sqrt{2}(\sqrt{6} + \sqrt{32})$ **31.** $\sqrt{5}\,\sqrt{t}$ $\sqrt{5t}$ **32.** $\sqrt{18}\,\sqrt{g^3}$ $3g\sqrt{2g}$

33. $\sqrt{12k}\,\sqrt{3k^5}$ $6k^3$ **34.** $\sqrt{15m^2}\,\sqrt{6n^3}$ **35.** $\dfrac{\sqrt{3}}{\sqrt{5}}$ $\dfrac{\sqrt{15}}{5}$ **36.** $\sqrt{\dfrac{2}{7}}$ $\dfrac{\sqrt{14}}{7}$

37. $\dfrac{3\sqrt{6}}{\sqrt{2}}$ $3\sqrt{3}$ **38.** $\sqrt{\dfrac{x}{8}}$ $\dfrac{\sqrt{2x}}{4}$ **39.** $\sqrt{\dfrac{t^2}{3}}$ $\dfrac{|t|\sqrt{3}}{3}$ **40.** $\sqrt{\dfrac{20}{a}}$ $\dfrac{2\sqrt{5a}}{a}$

41. $(5 + \sqrt{3})(5 - \sqrt{3})$ 22 **42.** $(\sqrt{17} + \sqrt{11})(\sqrt{17} - \sqrt{11})$ 6 **43.** $(2\sqrt{5} + \sqrt{7})(2\sqrt{5} - \sqrt{7})$ 13

44. $\dfrac{1}{3 + \sqrt{5}}$ $\dfrac{3 - \sqrt{5}}{4}$ **45.** $\dfrac{2}{\sqrt{3} - 5}$ $\dfrac{\sqrt{3} + 5}{-11}$ **46.** $\dfrac{12}{\sqrt{8} - \sqrt{6}}$ $12\sqrt{2} + 6\sqrt{6}$ **47.** $\dfrac{10}{\sqrt{13} + \sqrt{7}}$

48. $\dfrac{14}{3\sqrt{2} + \sqrt{5}}$ **49.** $\dfrac{\sqrt{3}}{\sqrt{3} - 5}$ $\dfrac{3 + 5\sqrt{3}}{-22}$ **50.** $\dfrac{\sqrt{6}}{7 - 2\sqrt{3}}$ $\dfrac{7\sqrt{6} + 6\sqrt{2}}{37}$ **51.** $\dfrac{5\sqrt{10}}{3\sqrt{3} + 2\sqrt{5}}$

52. $6\sqrt{13} + 7\sqrt{13}$ $13\sqrt{13}$ **53.** $9\sqrt{15} - 4\sqrt{15}$ $5\sqrt{15}$ **54.** $2\sqrt{11} - 8\sqrt{11}$ $-6\sqrt{11}$ **55.** $2\sqrt{12} + 5\sqrt{3}$ $9\sqrt{3}$

56. $2\sqrt{27} - 4\sqrt{12}$ $-2\sqrt{3}$ **57.** $4\sqrt{8} - 3\sqrt{5}$ $8\sqrt{2} - 3\sqrt{5}$ **58.** $8\sqrt{32} + 4\sqrt{50}$ $52\sqrt{2}$ **59.** $6\sqrt{20} + \sqrt{45}$ $15\sqrt{5}$

60. $2\sqrt{63} + 8\sqrt{45} - 6\sqrt{28}$ $24\sqrt{5} - 6\sqrt{7}$ **61.** $10\sqrt{\dfrac{1}{5}} - \sqrt{45} - 12\sqrt{\dfrac{5}{9}}$ $-5\sqrt{5}$ **62.** $3\sqrt{\dfrac{1}{3}} - 9\sqrt{\dfrac{1}{12}} + \sqrt{243}$ $\dfrac{17}{2}\sqrt{3}$

Solve and check.

63. $\sqrt{t} = 10$ 100 **64.** $\sqrt{3g} = 6$ 12 **65.** $\sqrt{y} - 2 = 0$ 4 **66.** $5 + \sqrt{a} = 9$ 16

67. $\sqrt{2k} - 4 = 8$ 72 **68.** $\sqrt{5y + 4} = 7$ 9 **69.** $\sqrt{10x^2 - 5} = 3x$ $\sqrt{5}$ **70.** $\sqrt{2a^2 - 144} = a$ 12

71. $\sqrt{b^2 + 16} + 2b = 5b$ $\sqrt{2}$ **72.** $\sqrt{m + 2} + m = 4$ 2 **73.** $\sqrt{3 - 2c} + 3 = 2c$ $\dfrac{3}{2}$

Use the quadratic formula to solve each equation.

74. $s^2 + 8s + 7 = 0$ $-7, -1$ **75.** $d^2 - 14d + 24 = 0$ 2, 12 **76.** $3h^2 = 27$ $-3, 3$

77. $n^2 - 3n + 1 = 0$ $\dfrac{3 \pm \sqrt{5}}{2}$ **78.** $2z^2 + 5z - 1 = 0$ $\dfrac{-5 \pm \sqrt{33}}{4}$ **79.** $3w^2 - 8w + 2 = 0$

80. $3f^2 + 2f = 6$ $\dfrac{-1 \pm \sqrt{19}}{3}$ **81.** $2r^2 - r - 3 = 0$ $\dfrac{3}{2}, -1$ **82.** $x^2 - 9x = 5$ $\dfrac{9 \pm \sqrt{101}}{2}$

Answers

19. $5m^2|n^3|\sqrt{7}$

20. $3ab^3\sqrt{11ab}$

29. $\sqrt{15} + 5\sqrt{2}$

30. $2\sqrt{3} + 8$

34. $3|m|n\sqrt{10n}$

47. $\dfrac{5\sqrt{13} - 5\sqrt{7}}{3}$

48. $\dfrac{42\sqrt{2} - 14\sqrt{5}}{13}$

51. $\dfrac{15\sqrt{30} - 50\sqrt{2}}{7}$

79. $\dfrac{4 \pm \sqrt{10}}{3}$

Answers
1. $y = 2x + 5$
2. $y = -4x + 1$
3. $y = \frac{1}{2}x - 3$
4. $y = -x - 6$
9. $y = 3x + 2$
10. $y = -6x - 5$
11. $y = 2x + 3$
12. $y = -x - 4$
17. $3x - y = -6$
18. $4x + y = 1$
19. $2x - 3y = 21$
20. $x - 4y = -2$

Equations in Two Variables

Write an equation in slope-intercept form for each given slope and y-intercept.

1. $m = 2, b = 5$ **2.** $m = -4, b = 1$ **3.** $m = \frac{1}{2}, b = -3$ **4.** $m = -1, b = -6$

5. $m = \frac{3}{2}, b = 1$ $y = \frac{3}{2}x + 1$ **6.** $m = 5, b = \frac{1}{4}$ $y = 5x + \frac{1}{4}$ **7.** $m = \frac{2}{5}, b = -4$ $y = \frac{2}{5}x - 4$ **8.** $m = -\frac{3}{4}, b = \frac{1}{2}$ $y = -\frac{3}{4}x + \frac{1}{2}$

Write each equation in slope-intercept form.

9. $-3x + y = 2$ **10.** $6x + y = -5$ **11.** $2x - y = -3$ **12.** $-x - y = 4$

13. $2x + 5y = 10$ $y = -\frac{2}{5}x + 2$ **14.** $x - 4y = -2$ $y = \frac{1}{4}x + \frac{1}{2}$ **15.** $-9x + 3y = -18$ $y = 3x - 6$ **16.** $4x + 7y = 3$ $y = -\frac{4}{7}x + \frac{3}{7}$

Write each equation in standard form.

17. $y = 3x + 6$ **18.** $y = -4x + 1$ **19.** $y = \frac{2}{3}x - 7$ **20.** $y = \frac{1}{4}x + \frac{1}{2}$

21. $y = -\frac{5}{3}x - \frac{1}{3}$ $5x + 3y = -1$ **22.** $y = 2x - \frac{1}{2}$ $4x - 2y = 1$ **23.** $\frac{5}{6}y = \frac{1}{4}x + \frac{2}{3}$ $3x - 10y = -8$ **24.** $\frac{1}{5}x = \frac{7}{10}y - \frac{3}{4}$ $4x - 14y = -15$

Write the equation of the line with each x-intercept and y-intercept. Use slope-intercept form.

25. x-intercept $= 2$; y-intercept $= -1$ $y = \frac{1}{2}x - 1$ **26.** x-intercept $= -3$; y-intercept $= -2$ $y = -\frac{2}{3}x - 2$

27. x-intercept $= 1$; y-intercept $= 5$ $y = -5x + 5$ **28.** x-intercept $= -\frac{1}{2}$; y-intercept $= 3$ $y = 6x + 3$

Write the equation of the line that passes through each pair of points. Use slope-intercept form.

29. $(1, 3), (2, 7)$ $y = 4x - 1$ **30.** $(4, 8), (2, 4)$ $y = 2x$ **31.** $(-2, 3), (3, 1)$ $y = \frac{-2}{5}x + \frac{11}{5}$ **32.** $(0, 0), (-2, -3)$ $y = \frac{3}{2}x$

33. $(5, 1), (3, -2)$ $y = \frac{3}{2}x - \frac{13}{2}$ **34.** $(2, -1), (5, -4)$ $y = -x + 1$ **35.** $(8, 6), (10, 3)$ $y = -\frac{3}{2}x + 18$ **36.** $(-4, -1), (-1, -$ $y = -2x -$

Solve each system of equations.

37. $y = 3x$; $4x + 2y = 30$ $(3, 9)$ **38.** $a = -2b$; $3a + 5b = 21$ $(42, -21)$ **39.** $n = m + 4$; $3m + 2n = 19$ $\left(\frac{11}{5}, \frac{31}{5}\right)$ **40.** $h = k - 7$; $2h - 5k = -2$ $(-11, -4)$

41. $s + 2t = 6$; $3s - 2t = 2$ $(2, 2)$ **42.** $c + 2d = 10$; $-c + d = 2$ $(2, 4)$ **43.** $3v + 5w = -16$; $3v - 2w = -2$ $(-2, -2)$ **44.** $e - 5f = 12$; $3e - 5f = 6$ $(-3, -3)$

45. $-3p + 2q = 10$; $-2p - q = -5$ $(0, 5)$ **46.** $2a + 5b = 13$; $4a - 3b = -13$ $(-1, 3)$ **47.** $5s + 3t = 4$; $-4s + 5t = -18$ $(2, -2)$ **48.** $2g - 7h = 9$; $-3g + 4h = 6$ $(-6, -3)$

49. $2c - 6d = -16$; $5c + 7d = -18$ $(-5, 1)$ **50.** $6m - 3n = -9$; $-8m + 2n = 4$ $\left(\frac{1}{2}, 4\right)$ **51.** $3x - 5y = 8$; $4x - 7y = 10$ $(6, 2)$ **52.** $9a - 3b = 5$; $a + b = 1$ $\left(\frac{2}{3}, \frac{1}{3}\right)$

Find the equation of the axis of symmetry and the maximum or minimum point for the graph of each quadratic function.

53. $y = -x^2 + 2x - 3$ $x = 1, (1, -2)$ **54.** $y = x^2 - 4x - 4$ $x = 2, (2, -8)$ **55.** $y = 3x^2 + 6x + 3$ $x = -1, (-1, 0)$

56. $y = 2x^2 + 12x$ $x = -3, (-3, -18)$ **57.** $y = x^2 - 6x + 5$ $x = 3, (3, -4)$ **58.** $y = 4x^2 - 1$ $x = 0, (0, -1)$

59. $y = -2x^2 - 2x + 4$ $x = -\frac{1}{2}, \left(-\frac{1}{2}, \frac{9}{2}\right)$ **60.** $y = \frac{1}{2}x^2 + 4x + \frac{1}{4}$ $x = -4, \left(-4, -\frac{31}{4}\right)$ **61.** $y = 6x^2 - 12x - 4$ $x = 1, (1, -10)$

Glossary

abscissa (271) The first component of an ordered pair; the x-coordinate.

absolute value (44) The absolute value of a number is the number of units that it is from zero on the number line.

addend (44) In an addition expression of the form $a + b = c$, the addends are a and b.

additive identity (12) The number 0 is the additive identity since the sum of any number and 0 is equal to the number.

additive inverse (53) Two numbers are additive inverses if their sum is zero. The additive inverse, or opposite, of a is $-a$.

algebraic fractions (437) Fractions that contain variables are algebraic fractions.

area of a polygon (195) The measurement of the region bounded by the polygon.

associative property for addition (18) For any numbers a, b, and c,

$$(a + b) + c = a + (b + c).$$

associative property for multiplication (18) For any numbers a, b, and c,

$$(ab)c = a(bc).$$

axis of symmetry (405) A straight line with respect to which a figure is symmetric.

base (3) In an expression of the form x^n, the base is x.

binomial (175) A polynomial with exactly two terms.

boundary (291) A line that separates a graph into half-planes.

coefficient (16) *See* numerical coefficient.

combined variation (492) A combined variation is described by an equation of the form $zy = kx$, where k is not 0.

commutative property for addition (18) For any numbers a and b,

$$a + b = b + a.$$

commutative property for multiplication (18) For any numbers a and b, $ab = ba$.

comparison property (111) For any two numbers a and b, exactly one of the following sentences is true.

$$a < b \qquad a = b \qquad a > b$$

complementary angles (499) Two angles are complementary if the sum of their measures is 90°.

completeness property for points in the plane (272) When plotting points, the following is true.

1. Exactly one point in the plane is located by a given ordered pair of numbers.
2. Exactly one ordered pair of numbers locates a given point in the plane.

complex fraction (459) If a fraction has one or more fractions in the numerator or denominator, it is called a complex fraction.

composite number (211) Any positive integer, except 1, that is not prime.

compound sentence (128) Two sentences connected by *and* or *or*.

conjugates (383) Two binomials of the form $a\sqrt{b} + c\sqrt{d}$ and $a\sqrt{b} - c\sqrt{d}$.

consecutive even integers (93) Numbers given when beginning with an even integer and counting by two's.

consecutive integers (93) Integers in counting order.

consecutive odd integers (93) Numbers given when beginning with an odd integer and counting by two's.

consistent (319) A system of equations is consistent and independent if it has one ordered pair as its solution. A system of equations is consistent and dependent if it has infinitely many ordered pairs as its solution.

constant (145) A monomial that does not contain variables.

coordinate (39) The coordinate of a point is the number that corresponds to it on the number line.

coordinate plane (271) The number plane formed by two perpendicular number lines that intersect at their zero points.

cosine (508) In a right triangle, the cosine of angle A =

$$\frac{\text{measure of side adjacent to angle } A}{\text{measure of hypotenuse}}$$

degree (175) The degree of a monomial is the sum of the exponents of its variables. The degree of a nonzero constant is 0. The degree of a polynomial is the greatest of the degrees of its terms.

direct variation (478) A direct variation is described by an equation of the form $y = kx$, where k is not zero.

discriminant (420) In the quadratic formula, the expression $b^2 - 4ac$ is called the discriminant.

distance formula (394) The distance between any two points (x_1, y_1) and (x_2, y_2) is given by the formula

$$d = \sqrt{(x_2 - x_1)^2 + (y_2 - y_1)^2}.$$

distributive property (15) For any numbers a, b, and c:

1. $a(b + c) = ab + ac$ and $(b + c)a = ba + ca$.
2. $a(b - c) = ab - ac$ and $(b - c)a = ba - ca$.

dividend (3) In the division expression $a \div b = c$, the dividend is a.

divisor (3) In the division expression $a \div b = c$, the divisor is b.

domain (274) The domain of a relation is the set of all first components from each ordered pair.

element (11) One of the members of a set.

elimination method (345) A method for solving systems of equations in which the equations are added or subtracted to eliminate one of the variables. Multiplication of one or both equations may occur before the equations are added or subtracted.

empty set (11) A set with no elements.

equals sign (9) The equals sign between two expressions indicates that if the sentence is true, the expressions name the same number.

equation (9) A mathematical sentence that contains the equals sign.

equivalent equations (57) Equations that have the same solution.

evaluate (6) To find the value of an expression when the values of the variables are known.

exponent (4) A number used to tell how many times a number is used as a factor. In an expression of the form x^n, the exponent is n.

expression (3) An expression consists of one or more numbers and variables with the operations of addition, subtraction, multiplication, and division.

extremes (97) *See* proportion.

factor (3) In a multiplication expression, the numbers multiplied are called factors.

FOIL method for multiplying binomials (185) To multiply two binomials, find the sum of the products of

F the first terms,
O the outer terms,
I the inner terms, and
L the last terms.

formula (22) An equation that states a rule for the relationship between certain quantities.

function (286) A function is a relation in which each element of the domain is paired with exactly one element of the range.

greatest common factor (GCF) (212) The GCF of two or more integers is the greatest factor that is common to each of the integers.

grouping symbols (7) Symbols used to clarify or change the order of operations in an expression. Parentheses, brackets, and the fraction bar are grouping symbols.

half-plane (291) The region of a graph on one side of a boundary.

hypotenuse (391) The side opposite the right angle in a right triangle.

identity (88) An equation that is true for every value of the variable.

inconsistent (319) A system of equations is inconsistent if it has no solution.

inequality (111) Any sentence containing $<$, $>$, $\neq$, $\leq$, or $\geq$.

integers (Z) (39) The set of numbers $\{\ldots, -3, -2, -1, 0, 1, 2, 3, \ldots\}$.

intersection (118) The intersection of two sets consists of all the points that belong to both of the sets.

inverse operations (80) Operations that undo each other, such as addition and subtraction.

inverse variation (481) An inverse variation is described by an equation of the form $xy = k$, where k is not zero.

irrational numbers (I) (374) Numbers that cannot be expressed in the form $\frac{a}{b}$, where a and b are integers, $b \neq 0$.

isosceles triangle (500) A triangle that has two congruent sides.

joint variation (491) A joint variation is described by an equation of the form $z = kxy$, where k is not zero.

least common denominator (LCD) (454) The least common multiple of the denominators of two or more fractions.

least common multiple (LCM) (454) The LCM of two or more integers is the least positive integer that is divisible by each of the integers.

legs (391) The adjacent sides of the right angle of a right triangle.

like terms (16) Terms that contain the same variables, with corresponding variables raised to the same power.

linear equation (283) A linear equation is an equation that may be written in the form $Ax + By = C$, where A, B, and C are any numbers and A and B are not both 0.

maximum point (406) The highest point on the graph of a curve such as a parabola which opens down.

means (97) *See* proportion.

midpoint (327) The midpoint of a line segment is the point that is halfway between the endpoints of the segment.

minimum point (405) The lowest point on the graph of a curve such as a parabola which opens up.

monomial (145) A number, a variable, or a product of numbers and variables.

multiplicative identity (12) The number 1 is the multiplicative identity since the product of any number and 1 is equal to the number.

multiplicative inverse (79) Two numbers are multiplicative inverses if their product is 1. The multiplicative inverse, or reciprocal, of a is $\frac{1}{a}$.

multiplicative property of zero (12) For any number a, $a \cdot 0 = 0$.

natural numbers (N) (374) The set of numbers $\{1, 2, 3, \ldots\}$.

negative number (39) A number that is graphed on the negative side of the number line.

numerical coefficient (16) The numerical part of a term.

open sentence (9) A sentence containing a symbol(s) to be replaced in order to determine if the sentence is true or false.

opposite (53) The opposite of a number is its additive inverse.

order of operations (6)
1. Evaluate all powers.
2. Then do all multiplications and divisions from left to right.
3. Then do all additions and subtractions from left to right.

ordered pair (271) In mathematics, pairs of numbers used to locate points in the plane.

ordinate (271) The second component of an ordered pair; the y-coordinate.

origin (271) The point of intersection of the two axes of the coordinate plane.

parabola (405) The general shape of the graph of a quadratic function.

parallel lines (324) Lines that have the same slope are parallel. All vertical lines are parallel.

percent of decrease (159) The ratio of an amount of decrease to the previous amount, expressed as a percent.

percent of increase (159) The ratio of an amount of increase to the previous amount, expressed as a percent.

percent proportion (102)

$$\frac{\text{Percentage}}{\text{Base}} = \text{Rate or } \frac{P}{B} = \frac{r}{100}.$$

perimeter (195) The perimeter of a polygon is the sum of the lengths of its sides.

perpendicular lines (325) Two lines are perpendicular if the product of their slopes is -1. In a plane, vertical lines are perpendicular to horizontal lines.

polynomial (175) An expression that can be written as a sum of monomials.

point-slope form (309) For a given point (x,y) on a nonvertical line, the point-slope form of a linear equation is $y - y_1 = m(x - x_1)$.

power (4) An expression in the form x^n. The base is x and the exponent is n.

prime factorization (211) The expression of a composite number as the product of its prime factors.

prime number (211) An integer, greater than 1, whose only positive factors are 1 and itself.

prime polynomial (226) A polynomial that cannot be written as a product of two or more polynomials.

principal square root (371) The nonnegative square root of the expression.

product (3) The result of multiplication.

product property of square roots (372) For any numbers a and b, where $a \geq 0$ and $b \geq 0$, $\sqrt{ab} = \sqrt{a} \cdot \sqrt{b}$.

proportion (97) An equation of the form $\frac{a}{b} = \frac{c}{d}$ which states that two ratios are equal. The first and fourth terms (a and d) are called the extremes. The second and third terms (b and c) are called the means.

Pythagorean Theorem (391) In a right triangle if a and b are the measures of the legs, and c is the measure of the hypotenuse, then $c^2 = a^2 + b^2$.

quadrant (271) One of the four regions into which two perpendicular number lines separate the plane.

quadratic formula (416) The solutions of a quadratic equation of the form $ax^2 + bx + c = 0$, where $a \neq 0$, are given by

$$x = \frac{-b \pm \sqrt{b^2 - 4ac}}{2a}.$$

quadratic function (405) A quadratic function is a function described by an equation of the form $y = ax^2 + bx + c$, where $a \neq 0$.

quotient property of square roots (372) For any numbers a and b, where $a \geq 0$ and $b > 0$,

$$\sqrt{\frac{a}{b}} = \frac{\sqrt{a}}{\sqrt{b}}.$$

radical expression (371) An expression of the form $\sqrt{a}$.

radical equations (389) Equations containing radicals with variables in the radicand.

radical sign (371) The symbol $\sqrt{\ }$ indicating the principal or nonnegative square root.

radicand (371) The expression under the radical sign.

range (274) The range of a relation is the set of all second components from each ordered pair.

ratio (97) A comparison of two numbers by division. The ratio of a to b is $\frac{a}{b}$.

rational numbers (Q) (47) Numbers that can be expressed in the form $\frac{a}{b}$, where a and b are integers, $b \neq 0$.

real numbers (R) (375) Irrational numbers together with rational numbers form the set of real numbers.

reciprocal (79) The reciprocal of a number is its multiplicative inverse.

reflexive property of equality (12) For any number a, $a = a$.

relation (274) A set of ordered pairs.

repeating decimal (374) A decimal numeral in which a digit or a group of digits repeat is called a repeating decimal.

replacement set (11) The set of numbers that a variable may represent.

right triangle (391) A triangle that has a 90° angle.

root of an equation (410) A solution of the equation.

scientific notation (153) A number is expressed in scientific notation when it is in the form $a \times 10^n$, where $1 \leq a < 10$ and n is an integer.

set (11) A collection of objects or numbers.

similar triangles (505) If two triangles are similar, the measures of their corresponding angles are equal and the measures of their corresponding sides are proportional.

simplest form of an expression (16) An expression in simplest form has no like terms and no parentheses.

sine (508) In a right triangle, the sine of angle A =

$$\frac{\text{measure of side opposite angle } A}{\text{measure of hypotenuse}}.$$

slope (305) The slope of a line is the ratio of the change in y to the corresponding change in x.

$$\text{slope} = \frac{\text{change in } y}{\text{change in } x}$$

slope-intercept form (312) The slope-intercept form of the equation of a line is $y = mx + b$. The slope of the line is m, and the y-intercept is b.

solution (9) A replacement for the variable in an open sentence which results in a true sentence.

solution set (113) The set of all replacements for the variable in an open sentence which make the sentence true.

solve (9) To solve an open sentence means to find all the solutions.

square root (371) If $x^2 = y$, then x is a square root of y.

standard form of linear equation (310) A linear equation in standard form is $Ax + By = C$ where A, B, and C are integers, and A and B are not both zero.

subset (11) A set that is made from the elements of another set.

substitution method (342) A method for solving systems of equations. One variable is expressed in terms of the other variable in one equation. Then the expression is substituted into the other equation.

substitution property of equality (13) For any numbers a and b, if $a = b$ then a may be replaced by b.

supplementary angles (499) Two angles are supplementary if the sum of their measures is 180°.

symmetric property of equality (12) For any numbers a and b, if $a = b$ then $b = a$.

system of equations (337) A set of equations with the same variables.

tangent (508) In a right triangle the tangent of angle A =

$$\frac{\text{measure of side opposite angle } A}{\text{measure of side adjacent to angle } A}.$$

term (15) A number, a variable, or a *product* or *quotient* of numbers and variables. The terms of an expression are separated by the symbols + and −.

terminating decimal (374) A specific type of repeating decimal in which only a zero repeats.

transitive property of equality (12) For any numbers a, b, and c, if $a = b$ and $b = c$, then $a = c$.

trigonometric ratios (508) Ratios in a right triangle that involve the measures of the sides and the measures of the angles.

trinomial (175) A polynomial having exactly three terms.

union (129) The union of two sets consists of all the points that belong to at least one of the sets.

uniform motion (201) When an object moves at a constant speed, or rate, it is said to be in uniform motion.

variable (3) In a mathematical sentence, a variable is a symbol used to represent an unspecified number.

vertical line test (287) If any vertical line drawn on the graph of a relation passes through no more than one point of its graph, then the relation is a function.

whole numbers (W) (374) The set of numbers $\{0, 1, 2, 3, \ldots\}$.

x-axis (271) The horizontal number line which helps to form the coordinate plane.

x-coordinate (271) The first component of an ordered pair.

x-intercept (312) The value of x when y is 0.

y-axis (271) The vertical number line which helps to form the coordinate plane.

y-coordinate (271) The second component of an ordered pair.

y-intercept (312) The value of y when x is 0.

zero product property (245) For all numbers a and b, if $ab = 0$, then $a = 0$ or $b = 0$.

Selected Answers

CHAPTER 1 THE LANGUAGE OF ALGEBRA

1-1 Page 5

Exploratory **1.** $7x$ **3.** $a + 19$ **5.** b^3 **7.** $2 \div k$ or $\frac{2}{k}$ **9.** 25^2 **11.** $n - 4$ **13.** x squared or x to the second power **15.** 5 cubed or 5 to the third power **17.** n to the first power ***Written*** **1.** 5^3 **3.** $7a^4$ **5.** $\frac{1}{2}a^2b^3$ **7.** 5^3x^2y **9.** $x + 17$ **11.** x^3 **13.** $2x^2$ **15.** $8x - 17$ **17.** $94 + 2x$ **19.** $\frac{3}{4}x^2$ **21.** $x^3 + 7$ **23.** 16 **25.** 125 **27.** 1024

1-2 Page 7

Exploratory **1.** Multiply 2 and 4. Then add 3. **3.** Divide 8 by 4. Then multiply by 2. **5.** Multiply 6 and 2. Then subtract from 12. **7.** Subtract 3 from 9. Then square the result. **9.** Square 9. Square 3. Then subtract. **11.** Square 3. Then multiply by 7. **13.** Subtract 3 from 5. Square the result. Then multiply by 4. **15.** Add 8 and 6. Divide by 2. Then add 2. ***Written*** **1.** 2 **3.** 48 **5.** 48 **7.** 14 **9.** 6 **11.** 8 **13.** 316 **15.** 1 **17.** $\frac{11}{13}$ **19.** $\frac{26}{3}$ or $8\frac{2}{3}$ **21.** 15 **23.** 1.32 **25.** 31 **27.** 75 **29.** 3 **31.** 2 **33.** 13.2 **35.** 3.04 **37.** 413 **39.** $\frac{11}{18}$ **41.** 1920 **43.** $2(a + b)$; 7 **45.** $b^2 + c$; $\frac{1}{4}$

1-3 Page 10

Exploratory **1.** false **3.** false **5.** false **7.** true **9.** false **11.** true **13.** true **15.** false **19.** 2 **21.** Foster, Rath, or Winters ***Written*** **1.** 11 **3.** 11.97 **5.** 9 **7.** 1.45 **9.** 5 **11.** 2 **13.** 11.05 **15.** 6 **17.** $\frac{7}{13}$ **19.** $\frac{7}{8}$ **21.** $\frac{7}{4}$ **23.** $5\frac{5}{6}$

1-4 Page 13

Exploratory **1.** reflexive **3.** symmetric **5.** transitive **7.** substitution **9.** 7 **11.** 1 ***Written*** **1.** symmetric **3.** transitive **5.** multiplicative identity **7.** reflexive **9.** multiplicative property of zero **11.** reflexive **13.** substitution **15.** symmetric **17.** multiplicative property of zero **19.** multiplicative identity

1-5 Page 17

Exploratory **1.** 5 **3.** 0.2 **5.** $\frac{1}{5}$ **7.** 0.5 **9.** $6bc$ and bc **11.** $4xy$ and $5xy$ **13.** $\frac{m^2n}{2}$ and $5m^2n$ **15.** v^3 and $4v^3$ **17.** $2cd^2$ and $3d^2c$ **19.** $3a + 3c$ **21.** $28 - 8a$ **23.** $x(c + d)$ **25.** $r(3 - k)$ ***Written*** **1.** $10a$ **3.** $18a$ **5.** am **7.** $38mn$ **9.** $22x^2$ **11.** $41a$ **13.** $22y^2 + 3$ **15.** $16a + 17b + 6$ **17.** $12a + 5b$ **19.** $2a + 3b$ **21.** $8x - 3y$ **23.** $30a + 6b$ **25.** $2a^2b + 6ab^2$ **27.** $14x + 14$ **29.** $x^2 + \frac{3}{4}x$ **31.** $0.98k^2 + 8.5k$ **33.** $\frac{5}{4}n + 3n^2$

1-6 Page 19

Exploratory **1.** associative property for addition **3.** distributive property **5.** commutative property for multiplication **7.** associative property for multiplication **9.** commutative property for addition **11.** no, no, no **13. a.** commutative property for addition **b.** associative property for addition **c.** distributive property **d.** substitution property of equality ***Written*** **1.** commutative property for addition **3.** associative property for addition **5.** commutative property for multiplication **7.** additive identity property **9.** commutative property for addition **11.** distributive property **13.** multiplicative identity property **15.** $12a + 6b$ **17.** $5x + 10y$ **19.** $\frac{5}{3}x^2 + 5x$ **21.** $3a + 13b + 2c$ **23.** $5 + 9ac + 14b$ **25.** $15x + 10y$ **27.** $\frac{3}{4} + \frac{5}{3}x + \frac{4}{3}y$ **29.** $3.1x + 1.54$ **31.** $12 + 30x + 45y$

1-7 Page 23

Exploratory **1.** $A = s^2$ **3.** $p = 4s$ ***Written*** **1.** 236 square units **3.** 499 square units **5.** 503.68 square units **7.** 129 square units **9.** 384 square units **11.** $\frac{25}{64}$ square units **13.** 63.2 square units **15.** 21 square units **17.** 3740 square units **19.** 8 square units **21.** 191 square units **23.** $2x + y^2 = z$ **25.** $x + a^2 = n$ **27.** $r = (a - b)^3$ **29.** $(abc)^2 = k$ **31.** $y = 2m + n^2$ **33.** $z = 29 - xy$ **35.** 150 miles **37.** 3300 meters

1-8 Page 25

Written **1. a.** \$1 bills **b.** 7 **c.** \$267 **d.** none **e.** \$157 **f.** end of day **g.** $5n$ **3. a.** doesn't say **b.** 7¢ **c.** \$7.18 **d.** more **e.** $(n - 7)$¢ or $(359 - n)$¢ **5. a.** nickels **b.** pennies **c.** 3 **d.** 2 **e.** doesn't say **f.** dimes **7. a.** $\frac{3}{4}$ **b.** $\frac{1}{3}$ **c.** $\frac{n}{4}$ **d.** yes **9. a.** no **b.** $x - 8$ **c.** juniors **d.** juniors **11. a.** 48 **b.** 72 **c.** no

1-9 Page 30

Exploratory **1.** $49 - n$ **3.** $w + 4$ **5.** $2t + 8$ **7.** \$1.59 ÷ 3 **9.** \$5.65n ***Written*** **1.** Let n = the number; $n + 24 = 89$ **3.** Let n = the number; $n - 19 = 83$ **5.** Let a = Tyrone's age; $2a + 17 = 53$ **7.** Let n = Melissa's age now; $(n + 4) + (n + 3 + 4) = 59$ **9.** Let x = Bob's height (in inches); $x + (x + 5) = 137$ **11.** Let x = Bob's age now; $(x - 5) + (x + 27 - 5) = 45$ **13.** Let n = the total; $(134) + (134 - 17) = n$ **15.** Let n = the number of pounds lost each week; $145 - 6n = 125$

CHAPTER REVIEW Page 36

1. a^4 **3.** $2x - 17$ **5.** $8x$ **7.** 320 **9.** 4 **11.** 0.45 **13.** 4 **15.** 2 **17.** 2.2 **19.** additive identity property **21.** multiplicative property of zero **23.** reflexive **25.** $11x$ **27.** $8b + 6$ **29.** commutative property for addition **31.** associative property for addition **33.** $14a + 9b$ **35.** $18 - d^2 = f$ **37.** less **39.** Let x = Carol's weight; $x + (x + 8) = 182$

CHAPTER 2 ADDING AND SUBTRACTING RATIONAL NUMBERS

2-1 Page 40

Exploratory **1.** 3 **3.** -10 **5.** 6 **7.** -7 **9.** -9 **11.** -2 **13.** $\{-3, 1, 2\}$ **15.** $\{-3, -2, -1\}$ **17.** $\{\ldots, -4, -3, -2, -1, 0, 1\}$ **19.** -3 **21.** $+650$ **23.** $+12$ **25.** -450 **27.** $+37$ ***Written*** **1.** $\{-1, 0, 1, 2\}$ **3.** $\{-5, -3, -1\}$ **5.** $\{-3, -2, -1, 0\}$ **7.** $\{-6, -4, -2, 0, 2\}$ **23.** 9, 1, -7 **25.** $-5, -13, -21$ **27.** $-2, -10, -18$

2-2 Page 43

Exploratory **1.** $-3 + 5 = 2$ **3.** $4 + (-5) = -1$ **5.** $-1 + (-4) = -5$ **7.** $-4 + 3 = -1$ **9.** $4 + (-6) = -2$ **11.** In $4 + (-6)$, the parentheses show the sign of the number. In $4 + (5 - 3)$ the parentheses mean to evaluate $5 - 3$ first. ***Written*** **1.** 13 **3.** -16 **5.** -5 **7.** -5 **9.** 0 **11.** 4 **13.** -5 **15.** 5 **17.** 9 **19.** -3 **21.** 0 **23.** 12 **25.** -17 **27.** -26 **29.** -12°C **31.** 25th floor

2-3 Page 45

Exploratory **1.** 8 **3.** 6 **5.** 17 **7.** 21 **9.** 0 **11.** $-$ **13.** $+$ **15.** $+$ **17.** $-$ **19.** $+$ **21.** $+$ **23.** n **25.** n ***Written*** **1.** 16 **3.** -19 **5.** 13 **7.** 26 **9.** 21 **11.** -8 **13.** -29 **15.** 4 **17.** -9 **19.** 16 **21.** 90 **23.** -54 **25.** 8 **27.** 9 **29.** -15 **31.** -22 **33.** 0 **35.** -18 **37.** 6 **39.** 2 **41.** 5 **43.** 3 **45.** -2 **47.** -30 **49.** 613 **51.** -539 **53.** -203 **55.** 500 **57.** 343 **59.** 587 **61.** 74 **63.** 672 **65.** -1074

2-4 Page 48

Exploratory **1.** $\frac{3}{4}$ **3.** 1.76 **5.** $\frac{3}{11}$ **7.** 1.82 **9.** $\frac{3}{82}$ **11.** $+$ **13.** $-$ **15.** $-$ **17.** $+$ **19.** $-$ **21.** $+$ ***Written*** **1.** -2 **3.** $-\frac{1}{11}$ **5.** $-\frac{1}{6}$ **7.** 0.88 **9.** -5.6 **11.** -14.7 **13.** $\frac{1}{6}$ **15.** $-\frac{1}{6}$ **17.** $-\frac{21}{8}$ **19.** 0.622 **21.** -0.665 **23.** -0.3005 **25.** $\frac{7}{30}$ **27.** $\frac{29}{60}$ **29.** $\frac{3}{20}$

2-5 Page 52

Exploratory **1.** 0 **3.** -0.1 **5.** 1 **7.** $4m$ **9.** $4r$ **11.** $3b - 4y$ ***Written*** **1.** -26 **3.** 6 **5.** 80 **7.** $-5a$ **9.** $30b$ **11.** $-38z$ **13.** $-\frac{19}{12}$ **15.** $\frac{5}{14}$ **17.** $\frac{7}{48}$ **19.** -14 **21.** -12 **23.** -8.7 **25.** -7.54 **27.** -17.4 **29.** $-97a + 48k$ **31.** $89mp - 24ps$ **33.** $3a$ **35.** $24w$ **37.** $2.2k$ **39.** $-0.73x$ **41.** -6 **43.** $-\$4.66$ (decrease)

2-6 Page 55

Exploratory **1.** -6 **3.** -5 **5.** 13 **7.** a **9.** 0 **11.** -3.7 **13.** $-\frac{3}{7}$ **15.** $\frac{8}{17}$ **17.** -7 **19.** 13 **21.** 7 **23.** 16 **25.** -56 **27.** -7.1 **29.** $\frac{3}{4}$ **31.** $8 + (-13)$ **33.** $-17 + (-8)$ **35.** $-1.7 + 1.5$ **37.** $9y + (-3y)$ ***Written*** **1.** 8 **3.** 15 **5.** -18 **7.** 40 **9.** 52 **11.** -11 **13.** $\frac{4}{5}$ **15.** $\frac{1}{4}$ **17.** -21 **19.** 11.5 **21.** -28.9 **23.** -153.8 **25.** -17 **27.** -1.4 **29.** $7m$ **31.** $-22p$ **33.** $33b$ **35.** $-35z$ **37.** -2 **39.** 58 **41.** 8.9 **43.** -16.7 **45.** -5.5 **47.** 8 **49.** $-\frac{5}{8}$ **51.** $-\frac{52}{21}$ **53.** $-\frac{67}{34}$

2-7 Page 59

Exploratory **1.** -21 **3.** 5 **5.** 10 **7.** 16 **9.** 5 **11.** -3 ***Written*** **1.** -3 **3.** -19 **5.** -32 **7.** -19 **9.** -19 **11.** -16 **13.** 15 **15.** 52 **17.** 28 **19.** -10 **21.** 8 **23.** -6 **25.** -3.6 **27.** -0.8 **29.** -1.3 **31.** $-\frac{17}{16}$ **33.** 2 **35.** $-\frac{52}{45}$

2-8 Page 61

Exploratory **1.** $m - 8$ **3.** $y + 11$ **5.** $b - 17$ **7.** $n + 23$ **9.** $z - 31$ **11.** $p + 47$ ***Written*** **1.** -17 **3.** -25 **5.** -12 **7.** 20 **9.** -14 **11.** -15 **13.** -3 **15.** -19 **17.** -2 **19.** -12 **21.** 9 **23.** -24 **25.** -36 **27.** -9 **29.** -21 **31.** -9 **33.** 21 **35.** -20 **37.** -2.8 **39.** -8.4 **41.** 2.32 **43.** $\frac{15}{14}$ **45.** $\frac{31}{20}$ **47.** $-\frac{23}{42}$

2-9 Page 64

Exploratory **1. a.** How far downstream was she from her starting point? **b.** 110 km **c.** 320 km **d.** Use a positive number for the distance upstream and a negative number for the distance downstream.

e. Let d = distance from starting point; $110 + (-320) = d$ **3. a.** $35\frac{3}{8}$ in. **b.** $10\frac{1}{8}$ and $12\frac{3}{16}$ **d.** third side **e.** The perimeter is equal to the sum of the lengths of the three sides. **f.** Let x = the measure of the third side; $10\frac{1}{8} + 12\frac{3}{16} + x = 35\frac{3}{8}$ **5.** Let f = floor; $1 + 14 + (-9) = f$ **7.** Let n = number; $n - 13 = -5$ **9.** Let r = runs; $41 - 17 = r$ ***Written*** **1.** -210 (210 km downstream) **3.** $13\frac{1}{16}$ **5.** 6 **7.** 8 **9.** 24 **11.** 63 **13.** $-19°$ **15.** 153 seconds **17.** \$54.44 **19.** 77 **21.** -38 **23.** -43 m (43 m below sea level) **25.** -74 m (74 m below sea level) **27.** -85 m (85 m below cave entrance) **29.** $35\frac{5}{8}$

CHAPTER REVIEW Page 69

1. $\{-3, 1, 2, 5\}$ **3.** $\{3, 4, 5, \ldots\}$ **9.** -1 **11.** -6 **13.** 8 **15.** -4 **17.** -53 **19.** -1 **21.** 2.7 **23.** 3.649 **25.** 66 **27.** $-23b$ **29.** $\frac{23}{12}$ **31.** -3.9 **33.** $-51pq + 53k$ **35.** -22 **37.** 4 **39.** 12.37 **41.** $-6x$ **43.** 17 **45.** -6.7 **47.** $-\frac{11}{14}$ **49.** -8 **51.** -24 **53.** 38 **55.** 53 **57.** -13 **59.** 19 **61.** Let n = the number; $n + (-16) = 39$; 55 **63.** Let c = change in temperature; $15 - (-2) = c$; The temperature rose 17° F.

CHAPTER 3 MULTIPLYING AND DIVIDING RATIONAL NUMBERS

3-1 Page 77

Exploratory **1.** negative **3.** positive **5.** positive **7.** negative **9.** negative **11.** negative **13.** positive **15.** negative **17.** Both a and b are positive or both are negative. **19.** a can be either positive or negative. **21.** positive ***Written*** **1.** -6 **3.** 56 **5.** 30 **7.** -110 **9.** $-\frac{1}{6}$ **11.** $-\frac{1}{2}$ **13.** 24 **15.** $-\frac{6}{5}$ **17.** 60 **19.** $-\frac{12}{35}$ **21.** $-\frac{7}{2}$ **23.** 3 **25.** $-\frac{1}{5}$ **27.** 0 **29.** -26 **31.** -10 **33.** 14 **35.** -5 **37.** $\frac{4}{3}$ or $1\frac{1}{3}$ **39.** $-\frac{5}{8}$ **41.** $7t$ **43.** $-21x$ **45.** $2x$ **47.** $5a - 3b$ **49.** $\frac{1}{6}a - \frac{1}{6}b$

3-2 Page 80

Exploratory **1.** $\frac{1}{3}$ **3.** none **5.** $-\frac{1}{14}$ **7.** $\frac{3}{2}$ **9.** $-\frac{11}{3}$ **11.** $\frac{7}{10}$ **13.** $-\frac{5}{3}$ **15.** $\frac{4}{13}$ **17.** $-\frac{7}{17}$ ***Written*** **1.** 6 **3.** -5 **5.** 14 **7.** -5 **9.** 6 **11.** 4 **13.** -19 **15.** -4 **17.** $6a$ **19.** $-7a$ **21.** 7 **23.** -5 **25.** $-\frac{5}{48}$ **27.** $-\frac{7}{80}$ **29.** $\frac{1}{12}$ **31.** $-\frac{35}{2}$ **33.** $\frac{21}{2}$ **35.** $\frac{27}{2}$ **37.** $a + 3$ **39.** $-a - 5$ **41.** $-10a - 15b$ **43.** $-10a + 5b$

3-3 Page 83

Exploratory **1.** 3 **3.** $\frac{4}{3}$ **5.** $\frac{9}{4}$ **7.** $-\frac{1}{8}$ **9.** 9 **11.** 4 **13.** 4 **15.** -7 **17.** -8 **19.** -6 ***Written*** **1.** 7 **3.** -7 **5.** -7 **7.** 17 **9.** $\frac{40}{9}$ **11.** $-\frac{11}{3}$ **13.** $\frac{14}{3}$ **15.** $\frac{81}{11}$ **17.** -14 **19.** 48 **21.** 70 **23.** -275 **25.** -90 **27.** -1885 **29.** -9 **31.** -25 **33.** 35 **35.** $-\frac{81}{4}$ **37.** $\frac{250}{3}$ **39.** $-\frac{336}{11}$ **41.** $\frac{14}{9}$ **43.** 6 **45.** -8

3-4 Page 85

Exploratory **1.** Add 7 to both sides. Then divide both sides by 3. **3.** Subtract 5. Then divide by 2. **5.** Multiply by 5. Then subtract 2. **7.** Subtract 2. Then multiply by $\frac{11}{3}$. ***Written*** **1.** 3 **3.** 5 **5.** $\frac{6}{5}$ **7.** -9 **9.** $-\frac{25}{3}$ **11.** 4 **13.** -153 **15.** -69 **17.** -81 **19.** -32 **21.** 25 **23.** -104 **25.** 11 **27.** $\frac{65}{7}$ **29.** 28 **31.** 8 **33.** 8 **35.** 16 **37.** -12 **39.** -3

3-5 Page 89

Exploratory **1.** Add 1 to both sides. Subtract $3x$ from both sides. **3.** Add 3 to both sides. Add $5y$ to both sides. Divide both sides by 9. **5.** Multiply $x + 1$ by 3. Subtract 3 from both sides. Divide both sides by 3. **7.** Multiply $2x - 3$ by 6. Add 18 to both sides. Then divide both sides by 12. ***Written*** **1.** 10 **3.** $-\frac{1}{2}$ **5.** no solutions **7.** identity **9.** 2 **11.** 4 **13.** identity **15.** $\frac{13}{5}$ **17.** no solutions **19.** -11 **21.** -2 **23.** no solutions **25.** 2 **27.** identity **29.** 15 **31.** 10 **33.** -30 **35.** 10 **37.** -9 **39.** 4.2 **41.** $\frac{45}{26}$ **43.** $\frac{1}{3}$ **45.** $\frac{3}{5}$ **47.** no solutions **49.** $\frac{5}{3}$

3-6 Page 91

Exploratory **1.** 12, $9x - 84 = 96 + 8x$ **3.** 2, $8t - 14 = 5t - 6$ **5.** 10, $52z = 30 + 17z$ ***Written*** **1.** 16 **3.** $\frac{11}{6}$ **5.** 49 **7.** 44 **9.** $\frac{8}{5}$ **11.** 18 **13.** $\frac{3}{8}$ **15.** -2 **17.** 9.6 **19.** $\frac{21}{2}$ **21.** -1 **23.** $2d - r$ **25.** $ef - d$ **27.** $\frac{3z + 2y}{e}$ **29.** $\frac{c - d}{2}$ **31.** $\frac{b}{a} - 1$ **33.** $\frac{5}{3}(b - a)$

3-7 Page 94

Exploratory **1.** 2, 3, 4 **3.** $-4, -2, 0$ **5.** 13, 15, 17, 19 **7.** $x + (x + 1) = 17$ **9.** $x + (x + 2) = -36$ **11.** $3x + 4 = -11$ ***Written*** **1.** 28, 29, 30 **3.** 31, 32, 33, 34 **5.** no solutions **7.** 31, 33 **9.** 31, 33, 35 **11.** 21 **13.** 46 **15.** 38 **17.** 260 m **19.** 8th, 9th, 10th, 11th **21.** \$64 **23.** 16 **25.** 140 **27.** 30, 32, 34, 36 **29.** 17, 19 **31.** 21, 22 **33.** 50 m, 60 m **35.** 9 m, 7 m

3-8 Page 99

Exploratory **1.** $\frac{3}{11}$ **3.** $\frac{21}{16}$ **5.** $\frac{2}{1}$ **7.** $\frac{24}{7}$ **9.** $\frac{4}{1}$

Written **1.** 6 **3.** 5 **5.** $\frac{28}{3}$ **7.** 6 **9.** 5 **11.** $\frac{8}{5}$ **13.** 20 **15.** 33 gallons **17.** 267.9 km **19.** 27

3-9 Page 103

Exploratory **1.** 31% **3.** 30% **5.** 80% **7.** $37\frac{1}{2}$% **9.** 175% **11.** 40% **13.** 70% **15.** $12\frac{1}{2}$% **17.** 300% **19.** 32 **21.** 68 **23.** 25% ***Written*** **1.** 24 **3.** 60 **5.** 65% **7.** 30.375 **9.** 140 **11.** 25% **13.** $242.80 **15.** $62\frac{1}{2}$% **17.** $6540 **19.** $560 **21.** 160% **23.** 40 **25.** $5 **27.** 540 **29.** 40 **31.** 480 **33.** 30% **35.** $65

CHAPTER REVIEW Page 106

1. -99 **3.** $-\frac{3}{7}$ **5.** $-5a - 12b$ **7.** $-9b$ **9.** 18 **11.** 8 **13.** -16 **15.** 10 **17.** -18 **19.** 69 **21.** 7 **23.** 3 **25.** -6 **27.** $\emptyset$ **29.** $cd - y$ **31.** $\frac{4a + 9b}{5}$ **33.** 25, 27, 29 **35.** 50 **37.** 22 **39.** 60% **41.** 17% **43.** 48 **45.** $87\frac{1}{2}$%

CHAPTER 4 INEQUALITIES

4-1 Page 114

Exploratory **1.** false **3.** true **5.** true **7.** true **9.** false **11.** true **13.** false **15.** {all numbers greater than 3} **17.** {all numbers less than 6} **19.** {all numbers greater than or equal to -4} **21.** yes **23.** no **25.** yes ***Written*** **1.** $x > 3$ **3.** $x \neq -3$ **5.** $x \geq 0$ **7.** $x < -3$ **9.** $x \geq -7$ **11.** $<$ **13.** $=$ **15.** $<$ **17.** $>$ **19.** $=$ **21.** $<$ **23.** $<$ **25.** $>$ **27.** $<$ **29.** $<$ **31.** $<$

4-2 Page 117

Exploratory **1.** -7 **3.** 13 **5.** $-3y$ **7.** $-5z$ **9.** -3 **11.** 7 ***Written*** **1.** $a < 8$ **3.** $r \geq 32$ **5.** $b < 8$ **7.** $m > -19$ **9.** $y \leq -8$ **11.** $n > 11$ **13.** $n > 34$ **15.** $n \leq -9$ **17.** $n < -11$ **19.** $s > -15$ **21.** $t \leq 34.6$ **23.** $m > 6.38$ **25.** $r > 1$ **27.** $x < \frac{5}{6}$ **29.** $-6 > f$ **31.** $n > 6$ **33.** $14 > x$ **35.** $-4 \leq a$ **37.** $r < -6.6$ **39.** $z > 168.93$ **41.** $y \geq -\frac{1}{12}$ **43.** $r < \frac{29}{12}$

4-3 Page 121

Exploratory **1.** 3, no **3.** -8, yes **5.** $\frac{1}{3}$, no **7.** $-\frac{1}{4}$, yes **9.** 6, no **11.** -11, yes **13.** $\frac{3}{4}$, no **15.** $-\frac{3}{5}$, yes ***Written*** **1.** $p < 6$ **3.** $s < -7$ **5.** $r \leq -\frac{35}{2}$ **7.** $s > \frac{17}{4}$ **9.** $h < 450$ **11.** $b \leq -128$ **13.** $r < -0.7$ **15.** $y > \frac{10}{3}$ **17.** $s \leq -6$ **19.** $k < -\frac{8}{7}$ **21.** $m > 0.7$ **23.** $t \leq -1.4$ **25.** $z < 0.08$ **27.** $x > -\frac{4}{5}$ **29.** $w \geq \frac{13}{12}$ **31.** $a < -9$

4-4 Page 123

Exploratory **1.** Add 1 to both sides. Then divide both sides by 3. **3.** Add 7 to both sides. Then divide both sides by 4. **5.** Subtract 32 from both sides. Then divide both sides by 14. **7.** Add 12 to both sides. Then divide both sides by 11. **9.** Subtract 7 from both sides. Then multiply both sides by 4. **11.** Add 5 to both sides. Then multiply both sides by -4 and reverse the direction of the inequality. **13.** Subtract 13 from both sides. Then divide both sides by -2 and reverse the direction of the inequality. **15.** Subtract k from both sides. Subtract 7 from both sides.

Written **1.** $r > 8$ **3.** $p < -\frac{3}{2}$ **5.** $y \geq -1$ **7.** $x > 80$ **9.** $x < -5$ **11.** $n < \frac{21}{5}$ **13.** $y \leq 0$ **15.** $c < 2$ **17.** $x < 40$ **19.** $x > -\frac{1}{2}$ **21.** $x \geq -31$ **23.** $y < 4$

4-5 Page 125

Exploratory **1.** $\frac{4}{5}$ **3.** $\frac{10}{11}$ **5.** $\frac{6}{5}$ **7.** $-\frac{1}{4}$ **9.** $-\frac{9}{7}$ **11.** $-\frac{15}{13}$ **13.** $\frac{0.2}{2}$ **15.** $\frac{0.8}{3}$ **17.** $\frac{459}{10}$

Written **1.** $<$ **3.** $>$ **5.** $<$ **7.** $>$ **9.** $<$ **11.** $=$ **13.** $<$ **15.** $<$ **17.** $\frac{79}{21}$ and $\frac{97}{28}$; second item **19.** $\frac{91}{184}$ and $\frac{189}{340}$; first item **21.** $\frac{179}{2.1}$ and $\frac{169}{1.9}$; first item **23.** $\frac{93}{27}$, $\frac{79}{20}$; first item **25.** $\frac{149}{100}$, $\frac{98}{75}$; second item **27.** Dudley's Market **29.** 26-ounce pie

4-6 Page 130

Exploratory **1.** false **3.** true **5.** false **7.** true **9.** true **11.** true ***Written*** **1.** $0 \leq m < 9$ **3.** $\frac{3}{4} < p \leq \frac{11}{9}$ **5.** $-\frac{4}{5} < z < \frac{2}{3}$ **7.** $-\frac{13}{7} < m < -\frac{6}{5}$ **9.** $-4.9 \leq a \leq -2.4$ **27.** {all numbers} **29.** $m < -1$ **31.** $q < -2$ or $q > -1$ **33.** $x \geq -5$ and $x < 1$ **35.** $x > -1$ and $x < 3$ **37.** $x > -4$ and $x < 1$ **39.** $x < \frac{3}{2}$ **41.** $x \leq -3$ or $x \geq 1$ **43.** $x < -3$ or $x > 3$ **45.** $x \geq -3$ and $x < 5$

4-7 Page 133

Exploratory **1.** $x = 4$ or $x = -4$ **3.** $y > 3$ or $y < -3$ **5.** $y < \frac{5}{2}$ and $y > -\frac{5}{2}$ **7.** $x + 2 > 3$ or $x + 2 < -3$ **9.** $x + 2 < 3$ and $x + 2 > -3$ **11.** $2x - 5 \geq 3$ or $2x - 5 \leq -3$ **13.** $7 - x = 4$ or $7 - x = -4$ **15.** $3x + 1 > 6$ or $3x + 1 < -6$ ***Written*** **1.** $y > 3$ or $y < -5$ **3.** $y < 5$ and $y > -3$ **5.** $y \geq 1$ and $y \leq 3$ **7.** {all numbers} **9.** $y \geq 8$ or $y \leq 2$ **11.** {1} **13.** $x \leq \frac{5}{2}$ and $x \geq \frac{9}{2}$ **15.** $x \leq$

-10 or $x \geq 8$ **17.** $y \leq \frac{1}{2}$ or $y \geq \frac{9}{2}$ **19.** $\left\{\frac{4}{3}\right\}$
21. $t < \frac{4}{3}$ and $t > -1$ **23.** $\{2, -2\}$ **25.** $|x| = 1$
27. $|x| < 3$ **29.** $|x| \geq 2$

4-8 Page 137

Written **1.** The number is greater than or equal to 5 **3.** 7 and 9; 5 and 7; 3 and 5; 1 and 3 **5.** He must deliver 70 or more **7.** 145 kg or greater **9.** His sales were greater than \$150,000 and less than \$250,000 **11.** \$12.10 or less **13.** 3, 4, or 5 **15.** Her score must be 9.7 or greater. **17.** His score must be greater than 9.4. **19.** 38, 40

CHAPTER REVIEW Page 140

1. false **3.** true **5.** $>$ **7.** $<$ **13.** $y \geq 6$ **15.** $n < -11$ **17.** $y < -\frac{1}{6}$ **19.** $h \geq -104$ **21.** $a \leq -4$ **23.** $r < -24$ **25.** $t < -1.4$ **27.** $k > \frac{1}{5}$ **29.** $z > -\frac{7}{10}$ **31.** $r > 1.8$ **33.** $y \leq -\frac{19}{3}$ **35.** $z < -10$ **37.** $<$ **39.** second item **41.** $-3 \leq x \leq 17$ **47.** $m \leq 6$ and $m \geq -4$ **49.** The number is greater than $\frac{7}{8}$ and less than $\frac{17}{8}$.

CHAPTER 5 POWERS

5-1 Page 146

Exploratory **7.** x^8 **9.** a^8 **11.** m^4 **13.** $12a^5$ **15.** $-20x^5y$ **17.** y^5z^4 ***Written*** **1.** a^7 **3.** m^6 **5.** t^6 **7.** a^{13} **9.** a^3b^5 **11.** m^4n^3 **13.** $6a^3b^6$ **15.** $-6x^8y^3$ **17.** $12x^6y^3$ **19.** $-6x^9y$ **21.** $6x^4y^4z^4$ **23.** $-35a^5b^2c$ **25.** $6a^3m^3n$ **27.** $a^2b^2c^2$ **29.** $9ab^2$ **31.** $3a^3b^2$ **33.** $\frac{1}{8}a^2b^2c$ **35.** abc

5-2 Page 149

Exploratory **1.** m^8 **3.** $9y^2$ **5.** x^4y^4 **7.** $-a^3b^6$ **9.** x^4y^{10} **11.** $\frac{8}{27}a^6$ ***Written*** **1.** 5^9 **3.** $(-5)^6$ or 5^6 **5.** x^{12} **7.** y^{18} **9.** $125c^3$ **11.** $-343z^3$ **13.** $\frac{1}{4}c^2$ **15.** $0.16d^2$ **17.** a^3b^6 **19.** $4a^4b^2$ **21.** $4a^6b^9$ **23.** $-125x^9y^3$ **25.** $\frac{1}{8}x^3y^6$ **27.** $0.008a^6$ **29.** $0.01x^4$ **31.** 10^5 **33.** $-48x^3y^2$ **35.** $-24a^3b$ **37.** $64x^6y^9$ **39.** $16a^8b^{12}$ **41.** x^4y^4 **43.** $16a^{14}b^6$ **45.** $152a^6$ **47.** $-30x^9y^3$ **49.** $101a^6 + 15{,}625a^6b^6$

5-3 Page 152

Exploratory **1.** 1 **3.** $\frac{1}{y^3}$ **5.** $\frac{1}{s^4}$ **7.** $\frac{1}{c^7}$ **9.** $\frac{x^3}{2y^6}$ **11.** $\frac{1}{d^2e}$ **13.** k^5 **15.** $\frac{1}{x}$ **17.** $\frac{5}{n^3}$ **19.** $\frac{b^7}{a^2}$ **21.** x **23.** $\frac{b^3}{a^4}$ ***Written*** **1.** n^3 **3.** $\frac{1}{x}$ **5.** a^2 **7.** $\frac{1}{k^6}$ **9.** an **11.** $\frac{a}{n^2}$ **13.** b^3c^3 **15.** $-\frac{1}{y^3}$ **17.** $3b$ **19.** $\frac{b^3}{5}$ **21.** y^3 **23.** ab^2 **25.** $\frac{w^2}{t^5}$ **27.** $-\frac{5y^5}{x^9}$ **29.** $-\frac{4b^3}{c^2}$ **31.** $-2ab^4c^5$ **33.** $\frac{9yz^5}{x^3}$ **35.** $\frac{7x^3}{4z^{10}}$ **37.** a^9b **39.** $\frac{5}{rs}$ **41.** $\frac{m^3n^2}{2k}$ **43.** $\frac{2z^5}{3y^3}$

5-4 Page 155

Exploratory **1.** 5000; 5.79×10^7 **3.** 12,760; 1.4959×10^8 **5.** 142,700; 7.7812×10^8
Written **1.** 4.293×10^3 **3.** 2.4×10^5 **5.** 3.19×10^{-4} **7.** 9.2×10^{-8} **9.** 3.2×10^6 **11.** 7.6×10^6 **13.** 3×10^2; 300 **15.** 6×10^2; 600 **17.** 5.5×10^{-9}; 0.0000000055 **19.** 6×10^{-3}; 0.006 **21.** 6.51×10^3; 6510 **23.** 7.8×10^8; 780,000,000 **25.** 2.1×10^{-1}; 0.21 **27.** 4×10^8; 400,000,000

5-5 Page 157

Exploratory **1. a.** $n - 5$ **b.** $n + 11$ **c.** $n - x$ **3. a.** $n + 14$ **b.** $n + 3$ **c.** 7 **d.** $2n + 9$ **e.** 11 **f.** $6n + 7$ **5. a.** $2m$ **b.** $\frac{m}{3}$ **c.** $m - 5$ **d.** $m + 8$
Written **1.** Abe, 21; Mindy, 18 **3.** 2 **5.** 25 **7.** 11 **9.** 1 **11.** 30 **13.** not enough information **15.** 19

5-6 Page 161

Exploratory **1.** 10 **3.** 8 **5.** 45 **7.** 36 **9.** 40
Written **1.** 10 **3.** 28 **5.** \$3.00 **7. a.** D **b.** \$6 **c.** 6% **9. a.** \$28 **b.** \$172 **11. a.** \$48 **b.** 25% **13.** 39 **15.** $54\frac{6}{11}\%$ **17.** 20% **19.** \$27 **21.** \$303.60 **23.** \$78.81 **25.** \$41.18 **27.** \$88.50 **29.** \$17.45 **31.** 4% **33.** 100

5-7 Page 166

Exploratory **1.** $60x + 75(16 - x) = 1095$ **3.** $6x + 36 = 4.20(12 + x)$ ***Written*** **1.** 1 L **3.** 250 mL **5.** 12 **7.** 5 pounds **9.** 110 at 95¢ and 154 at \$1.25 **11.** 6 nickels, 12 quarters, 9 dimes

CHAPTER REVIEW Page 170

1. b^9 **3.** a^4b^3 **5.** $-12a^3b^4$ **7.** $49a^2$ **9.** $\frac{1}{9}b^4$ **11.** $576x^5y^2$ **13.** y^4 **15.** $3b^3$ **17.** $\frac{a}{6b^2c^2}$ **19.** 2.4×10^5 **21.** 6×10^{11} **23.** 6×10^{-7} **25.** Jim, 27; Joe, 37 **27.** 18 **29.** \$22.11 **31.** 11 dimes, 21 nickels

CHAPTER 6 POLYNOMIALS

6-1 Page 176

Exploratory **1.** trinomial **3.** monomial **5.** not polynomial **7.** not polynomial **9.** binomial

11. trinomial **13.** monomial **15.** not polynomial **17.** 0 **19.** 2 **21.** 4 **23.** 7 **25.** 5 **27.** 5 **29.** 0 **31.** 4 **33.** 4 **35.** 29 **37.** 7 **39.** 6
Written **1.** 2 **3.** 3 **5.** 5 **7.** 4 **9.** 4 **11.** 7 **13.** 14 **15.** 7 **17.** 2 **19.** $3 + 2x^2 + x^4$ **21.** $a^3 + 5ax + 2x^2$ **23.** $11b^2x + 17bx^2 - x^3$ **25.** $x^2 + 4x - 7$ **27.** $-3x^3 + 5x^2 + 2x + 7$ **29.** $b^3x^2 + \frac{2}{3}bx + 5b$ **31.** $x + a$ **33.** $4x^3y - x^2y^3 + 3xy^4 + y^4$ **35.** $4x^4 + \frac{1}{3}s^2x^3 - \frac{2}{5}s^4x^2 + \frac{1}{4}x$ **37.** $2.4tx^5 + 5.1x^3 - 0.3tx^2 + 1.7t^3x$

6-2 Page 180

Exploratory **1.** -29 **3.** $-3x - 2y$ **5.** $-7a + 6b$ **7.** $8m - 7n$ **9.** $-x^2 - 3x - 7$ **11.** $-2y^2 + 7y - 12$ **13.** $10r^2 - 4rs + 6s$ **15.** $x^3 + x + 1$ **17.** $3ab^2 - 5a^2b + b^3$ **19.** $10n + 9p$ **21.** $7a + b$ **23.** $8x + 3y$ **25.** $12n - 2r$ **27.** $-6s$
Written **1.** $7x + 14y$ **3.** $7m + 5n$ **5.** $7x - 4y$ **7.** $3a - 11m$ **9.** $13m + 3n$ **11.** $13x - 2y$ **13.** $3n^2 + 13n + 11$ **15.** $-2 - 6a$ **17.** $x^2 + \frac{4}{3}x + \frac{1}{4}$ **19.** $7ax^2 + 3a^2x - 5ax + 2x$ **21.** $-3x^2y + 6xy^2 - 2y^2$ **23.** $-2mn^2 + 3mn - n - 3n^3$ **25.** $9a - 3b - 4c + 16d$ **27.** $7ax^2 - 5a^2x - 7a^3 + 4$ **29.** $x^2y^2 - 5xy - 10$ **31.** $6m^2n^2 + 8mn - 28$ **33.** $6m^2n^2 + 10mn - 23$ **35.** $-3y^2$ **37.** $2x + 6y$

6-3 Page 183

Exploratory **1.** $-60a^3$ **3.** $10x - 6$ **5.** $35a^2 + 56a$ **7.** $15a^2b - 9ab$ **9.** $-24a^4 - 56a^3$ **11.** $9x^2y^2 + 6x^2y$ ***Written*** **1.** $15a + 35$ **3.** $-24x - 15$ **5.** $4x^2 + 3x$ **7.** $15b^2 + 24b$ **9.** $-10x^2 - 22x$ **11.** $2.2a^2 + 7.7a$ **13.** $21a^3 - 14a^2$ **15.** $15s^3t + 6s^2t^2$ **17.** $35x^3y - 7xy^3$ **19.** $10a^4 - 14a^3 + 4a$ **21.** $35x^4y - 21x^3y^2 + 7x^2y^2$ **23.** $40y^4 + 35y^3 - 15y^2$ **25.** $-28x^3 + 16x^2 - 12x$ **27.** $15x^4y - 35x^3y^2 + 5x^2y^3$ **29.** $36m^4n + 4m^3n - 20m^2n^2$ **31.** $-32x^2y^2 - 56x^2y + 112xy^3$ **33.** $-3x^3 - \frac{1}{3}x^2 + \frac{5}{3}x$ **35.** $-16m^3n + 6m^2n^2 - 2mn^3$ **37.** $-\frac{1}{4}ab^4 + \frac{1}{3}ab^3 - \frac{3}{4}ab^2$ **39.** $7a^4 + 21a^3 - a + 25$ **41.** $-3m^3 + 41m^2 - 14m - 16$ **43.** $21a^3 - 6a^2 - 46a + 28$ **45.** $50.6t^2 + 21t - 102$ **47.** $6m^3 + 21m^2 - \frac{33}{2}m$

6-4 Page 187

Exploratory **1.** $8x$ **3.** $11a$ **5.** $22x$ **7.** $13x$ **9.** $-b$ **11.** $14m$ ***Written*** **1.** $a^2 + 10a + 21$ **3.** $m^2 - 16m + 55$ **5.** $x^2 + 7x - 44$ **7.** $2x^2 + 17x + 8$ **9.** $8a^2 + 2a - 3$ **11.** $40a^2 + 47a + 12$ **13.** $10a^2 + 11ab - 6b^2$ **15.** $14y^2 - 23y + 3$ **17.** $20r^2 - 13rs - 21s^2$ **19.** $20x^2 - 2x - \frac{2}{9}$ **21.** $12x^2 + \frac{3}{2}x - \frac{3}{8}$ **23.** $\frac{3}{16}y^2 + \frac{1}{12}xy - \frac{2}{9}x^2$ **25.** $6a^2 + 3.7a + 0.56$ **27.** $0.63x^2 + 3.9xy + 6y^2$ **29.** $0.24a^2 - 0.63a - 1.32$ **31.** $m^3 + 14m^2 + 47m - 14$ **33.** $5a^3 - 13a^2 + 49a + 22$ **35.** $12s^3 + 47s^2 + 4s - 45$ **37.** $30x^3 - 37x^2y + 55xy^2 - 18y^3$ **39.** $0.15n^3 + 3.07n^2 - 24.35n + 12$ **41.** $0.63t^3 - 0.40t^2 + 1.41t + 0.34$ **43.** $\frac{4}{9}x^3 - \frac{5}{6}x^2y + \frac{13}{24}xy^2 - \frac{1}{8}y^3$ **45.** $\frac{1}{12}a^3 + \frac{5}{12}a^2 + \frac{1}{12}a - \frac{5}{6}$ **47.** $a^4 - a^3 - 8a^2 - 29a - 35$ **49.** $20x^4 - 9x^3 + 73x^2 - 39x + 99$ **51.** $-24a^3 - 34a^2 + 19a + 15$ **53.** $5x^6 - 25x^5 + 13x^4 + 10x^3 - 5x^2 - 5x + 3$ **55.** $\frac{1}{4}a^6 + 2a^3 - 9a^2 + 4$ **57.** $0.6x^5 + 9x^4 - 6.06x^3 - 1.5x^2 - 8.4x + 6$

6-5 Page 191

Exploratory **1.** $a^2 + 4ab + 4b^2$ **3.** $4x^2 + 4xy + y^2$ **5.** $9m^4 + 12m^2n + 4n^2$ **7.** $a^4 + 2a^2b + b^2$ **9.** $9a^2 - 48ab + 64b^2$ **11.** $25a^2 - 30ab + 9b^2$ **13.** $9x^2 - 6xy + y^2$ **15.** $4x^4 - 12x^2y^3 + 9y^6$
Written **1.** $16x^2 + 8xy + y^2$ **3.** $4a^2 - 4ab + b^2$ **5.** $36m^2 + 24mn + 4n^2$ **7.** $16x^2 - 72xy + 81y^2$ **9.** $25a^2 - 120ab + 144b^2$ **11.** $25x^2 + 60xy + 36y^2$ **13.** $\frac{1}{4}a^2 + ab + b^2$ **15.** $\frac{16}{9}x^2 - \frac{16}{3}xy + 4y^2$ **17.** $0.49x^2 - 1.54x + 1.21$ **19.** $1.21x^2 + 2.2xy + y^2$ **21.** $a^4 - 6a^2b^3 + 9b^6$ **23.** $9x^2 - 25$ **25.** $10.24x^4 - 8.96x^2y + 1.96y^2$ **27.** $\frac{1}{9}a^2 - \frac{9}{4}b^2$ **29.** $49a^4 - 14a^2b + b^2$ **31.** $9a^4 + 6a^2b + b^2$ **33.** $36a^6 - 12a^5 + a^4$ **35.** $\frac{1}{16}a^4 - \frac{1}{6}a^2b^3 + \frac{1}{9}b^6$ **37.** $1.69x^4 + 5.46x^2y^3 + 4.41y^6$ **39.** $x^{6n} - 2x^{3n}y^{2n} + y^{4n}$ **41.** $x^{4n} - y^{2n}$ **43.** $x^{6n} + 2x^{3n}y^{5n} + y^{10n}$

6-6 Page 194

Exploratory **1.** $7x - 11$ **3.** $8y - 42$ **5.** $-36y + 34$ **7.** $10q + 48$ **9.** $2x - 79$ ***Written*** **1.** $-\frac{1}{5}$ **3.** $-\frac{13}{9}$ **5.** $\frac{15}{7}$ **7.** $-\frac{7}{2}$ **9.** 8 **11.** $\frac{21}{2}$ **13.** -3 **15.** -83 **17.** 2 **19.** $\frac{58}{33}$ **21.** $\frac{23}{4}$ **23.** $\frac{1}{3}$ **25.** 2 **27.** 0 **29.** $-\frac{3}{2}$ **31.** 17

6-7 Page 196

Exploratory **1. a.** $n + 5$ **b.** $2t + 5$ **c.** $2n + 8$ **d.** $t + 4$ ***Written*** **1.** 18 ft × 40 ft **3.** 11 in. × 11 in. **5.** 8 cm, 16 cm, 13 cm **7.** 60 yd × 80 yd

6-8 Page 199

Exploratory **1.** \$480 **3.** $3\frac{1}{2}$ yr **5.** 12% **7.** \$2400

9. $15\frac{1}{2}\%$ ***Written*** **1.** \$6400 at 8% and \$3600 at 12% **3.** \$3800 **5.** \$3000 **7.** $14\frac{1}{2}\%$

6-9 Page 202

Exploratory **1. a.** 160 km **b.** 480 km **c.** 80 h km **3. a.** 40 mph **b.** $\frac{240}{t}$ mph **5. a.** 8 hr **b.** 4 hr ***Written*** **1.** $3\frac{1}{2}$ **3.** 44 mph **5.** $2\frac{1}{2}$ **7.** 11:30 A.M. **9.** 240 km **11.** 450 mph and 530 mph **13.** 320 mph

CHAPTER REVIEW Page 206

1. $3x^4 + x^2 - x - 5$ **3.** $-3x^3 + x^2 - 5x + 5$ **5.** $x^2 - 8x + 7$ **7.** $7x^3y + 28x^2y^2 - 56xy^3$ **9.** $x^3 + 56x^2 - 30x + 3$ **11.** $4x^2 + 13x - 12$ **13.** $\frac{3}{8}a^3 - \frac{1}{4}a^2 + 4$ **15.** $25x^2 - 9y^2$ **17.** $64x^2 - 80x + 25$ **19.** 4 **21.** 18 in. **23.** \$5240 **25.** 2 hours later

CHAPTER 7 FACTORING

7-1 Page 212

Exploratory 1. yes **3.** no **5.** yes **7.** composite; $3 \cdot 13$ **9.** composite; $7 \cdot 13$ **11.** 4 **13.** 9 **15.** 11 **17.** 10 **19.** 36 ***Written*** **1.** $3 \cdot 7$ **3.** $2^2 \cdot 3 \cdot 5$ **5.** $3 \cdot 13$ **7.** $2 \cdot 17$ **9.** $2^4 \cdot 7$ **11.** $2^4 \cdot 19$ **13.** $2^2 \cdot 3 \cdot 5^2$ **15.** $2^2 \cdot 5 \cdot 7 \cdot 11$ **17.** $-1 \cdot 2 \cdot 13$ **19.** $-1 \cdot 2 \cdot 2 \cdot 5 \cdot 5 \cdot 5$ **21.** $2 \cdot 7 \cdot 7 \cdot a \cdot a \cdot b$ **23.** $2 \cdot 2 \cdot 7 \cdot 7 \cdot b \cdot b$ **25.** $-1 \cdot 2 \cdot 3 \cdot 17 \cdot x \cdot x \cdot x \cdot y$ **27.** 5 **29.** 29 **31.** 126 **33.** 19 **35.** $17a$ **37.** $2x$ **39.** 5 **41.** 6 **43.** 2 **45.** $4a$ **47.** $4a$ **49.** mn **51.** 1 **53.** $2ab$ **55.** $4a$

7-2 Page 215

Exploratory **1.** 3 **3.** y **5.** 1 **7.** $3y$ **9.** $4x$ **11.** $7a$ **13.** xy **15.** 1 ***Written*** **1.** $4(4x + y)$ **3.** $12x(y + x)$ **5.** $11x(1 + 4xy)$ **7.** $5ab^2(5a + 6b)$ **9.** $2xy(7 - 9y)$ **11.** $12ab(3ab - 1)$ **13.** $2xy(7y + 1)$ **15.** $x(x^4y - 1)$ **17.** $a(17 - 41a^2b)$ **19.** $3(x^2y + 3y^2 + 2)$ **21.** $2ab(a^2b - 8ab^2 + 4)$ **23.** $x(24xy^2 + 12y + 1)$ **25.** $4x(3a + 5b + 8c)$ **27.** $x^3(a + 5b + 9c)$ **29.** $3x(2x - 3y + 8xy^2)$ **31.** $6x(4ab + 2ax + x^2)$ **33.** $\frac{1}{2}x\left(x - \frac{1}{2}a\right)$ **35.** $\frac{1}{5}y\left(4x^2 + 3y\right)$

7-3 Page 218

Exploratory **1.** yes **3.** yes **5.** no **7.** no **9.** no **11.** yes ***Written*** **1.** $(a - 3)(a + 3)$ **3.** $(2x - 3y)(2x + 3y)$ **5.** $(a - 2b)(a + 2b)$ **7.** $(4a - 3b)(4a + 3b)$ **9.** $(4a - 5)(4a + 5)$ **11.** $6(x^2 + 2)$ **13.** $2(z - 7)(z + 7)$ **15.** $2(2x - 3)(2x + 3)$ **17.** $(5y - 7z^2)(5y + 7z^2)$ **19.** $17(1 - 2a)(1 + 2a)$ **21.** $(5ax - 1)(5ax + 1)$ **23.** $12(a - 1)(a + 1)$ **25.** $9(2x - 3y^2)(2x + 3y^2)$ **27.** $9(3 - x)(3 + x)$ **29.** $(2y - 3)(2y + 3)$ **31.** $(x - y)(x + y)(x^2 + y^2)$ **33.** $(x - 1)(x + 1)(x^2 + 1)$ **35.** $\frac{1}{2}(3x - 7y)(3x + 7y)$ **37.** $(x - 1)(x + 1)(x^2 + 1)(x^4 + 1)$ **39.** $(x - y - y)(x - y + y)$ or $(x - 2y)x$ **41.** $[a - (b - c)][a + (b - c)]$ or $(a - b + c)(a + b - c)$ **43.** $(x - 2)(x + 2)(x^2 + 4)$ **45.** $2(7x^3 - 8y^4)(7x^3 + 8y^4)$

7-4 Page 222

Exploratory **1.** yes, $(a + 2)^2$ **3.** no **5.** no **7.** no **9.** yes, $(2x - 1)^2$ **11.** yes, $(3b - 1)^2$ ***Written*** **1.** $(a + 6)^2$ **3.** $(x + 8)^2$ **5.** $(2a + 1)^2$ **7.** $(1 - 5a)^2$ **9.** not factorable **11.** $(5x + 2)^2$ **13.** $(11y + 1)^2$ **15.** $(5b - 3)^2$ **17.** not factorable **19.** $(m + 8n)^2$ **21.** $(3x + 4y)^2$ **23.** $9(4p + 1)^2$ **25.** $(8x - 5)^2$ **27.** $3(x + 6)^2$ **29.** $6(x^2 + 3x + 4)$ **31.** $(2x + z^2)^2$ **33.** $\left(\frac{1}{2}a + 3\right)^2$ **35.** $(m^2 + 6n^2)^2$ **37.** $\left(3a + \frac{4}{5}\right)^2$ **39.** 42 or -42 **41.** 4 **43.** y^2 **45.** $4y^2$

7-5 Page 226

Exploratory **1.** 1, 14 **3.** $-2, -6$ **5.** $9, -4$ **7.** $3, -10$ **9.** 3, 15 **11.** $3, -20$ **13.** 10 **15.** 7 **17.** $+8$ **19.** $+7m$ ***Written*** **1.** $(y + 3)(y + 9)$ **3.** $(c + 3)(c - 1)$ **5.** $(y - 3)(y - 5)$ **7.** $(m - 5)(m + 4)$ **9.** prime **11.** $(z - 13)(z + 3)$ **13.** $(r - 12)(r + 2)$ **15.** $(y + 10)(y - 2)$ **17.** $(p + 2)(p + 8)$ **19.** $(b - 4)(b - 7)$ **21.** $(15 + m)(1 + m)$ **23.** $(y + 4)(y + 7)$ **25.** $(a - 20)(a + 2)$ **27.** $(x - 6)(x + 3)$ **29.** prime **31.** $(m - 11)(m + 5)$ **33.** $(11 - j)(6 - j)$ **35.** $(21 - a)(2 - a)$ **37.** $(c - 4d)(c + 2d)$ **39.** $(a + 3b)(a - b)$ **41.** $(a + b - 6)(a + b + 1)$ **43.** $7, -7, 11, -11$ **45.** $12, -12$ **47.** $8, -8$ **49.** $4, -4$ **51.** $10, -10, 22, -22$ **53.** $13, 5, -5, -13$

7-6 Page 229

Exploratory **1.** $(k - m)(r + s)$ **3.** $(4b + y)(x - y)$ **5.** $(7rp - 4q)(r - 3p)$ **7.** $(8m + 1)(x + y)$ **9.** $(5mp - 3bc)(2m + 3p)$ **11.** $(4m - 3p)(y - 5)$ **13.** $(7x + 3m - 4p)(a + b)$ **15.** $(a + b - c)(8r - 3y)$ ***Written*** **1.** $x + y$ **3.** $a + 3b$ **5.** $a - 4c$ **7.** $4m - 3p$ or $3p - 4m$ **9.** $6x + 7y$ **11.** $3a + 2b^2$ **13.** $(2x + b)(a + 3c)$ **15.** $(y + 4)(a + p)$ **17.** $(m + x)(2y + 7)$ **19.** $(a - c)(y - b)$ **21.** $(a + 1)(a - 2b)$ **23.** $(4x + 3y)(a + b)$ **25.** $(m^2 + p^2)(3 - 5p)$ **27.** $(a - 3b^2)(5a - 4b)$ **29.** $3(2a - c)(a - b)$ **31.** $a(x + ax - 1 - 2a)$ **33.** $(7m + 2n)(p + q)(p - q)$ **35.** $(x + 1)(x - 1)(x + 2)$ **37.** $(a^2 + b^2)(a - b)$ **39.** $(7m - 5y)(a + b)(a - b)$ **41.** $(m - 2)(a + 2)(a + 3)$ **43.** $(2x + 3y)(a - 4)(a - 3)$

7-7 Page 234

Exploratory **1.** 2, 9 **3.** 3, 8 **5.** $-2, -6$ **7.** $-5, -4$ **9.** $-4, 10$ **11.** $-15, 2$ **13.** $-4, 15$ ***Written*** **1.** $(3y + 5)(y + 1)$ **3.** $(3x + 2)(x + 2)$ **5.** $(4m - 3)(2m - 1)$ **7.** $(2h - 3)(h + 1)$ **9.** $(3k - 2)(k + 3)$ **11.** $(3p - 2)(2p + 1)$ **13.** $(2a + 7)(a - 2)$ **15.** $(3t - 2)(2t + 3)$ **17.** $(2y - 3)(y - 1)$ **19.** $(2q + 3)(q - 6)$ **21.** $(3m + 2)(2m + 5)$ **23.** $(3x - 11)(2x + 1)$ **25.** prime **27.** $(4p - 3)(2p - 3)$ **29.** $(5p - 2)(3p + 4)$ **31.** $(2y - 3)(3y - 5)$ **33.** $(2s + 5)(3s - 4)$ **35.** $(9x + 5)(2x + 5)$ **37.** $(9c - 2)(2c + 5)$ **39.** $(2r - 3)(9r + 4)$ **41.** $(4m - n)(2m - 3n)$ **43.** $(4x - 5y)(4x + y)$ **45.** $(4p - q)(5p + 4q)$ **47.** $(3k + 5m)(3k + 5m)$ **49.** $(5s + 8t)(4s - 3t)$ **51.** $(4x - 3q)(2x - 9q)$ **53.** $(2c + 5d)(7c + 3d)$

7-8 Page 237

Exploratory **1.** $3(x^2 + 5)$ **3.** $a(5x + 6y)$ **5.** $6a(2x^2 + 3y^2)$ **7.** $5x^2(1 - 2y)$ **9.** $3ab(a + 2 + 3b)$ **11.** $(a + 3b)(a - 3b)$ ***Written*** **1.** $2(a + 6)(a - 6)$ **3.** $m(m + 3)^2$ **5.** $4a(a + 3)(a - 3)$ **7.** $(m^2 + p)(m^2 - p)$ **9.** $(2k + 1)(k + 1)$ **11.** $6(y + 2x)(y - 2x)$ **13.** $3(y + 8)(y - 1)$ **15.** $2(5y + 1)(2y + 3)$ **17.** $2(b^2 + 3b + 1)$ **19.** $(m + 4n)^2$ **21.** $3(2p - 7)(p + 5)$ **23.** $a(3x + 7)(x + 3)$ **25.** $3a(3a - 2)(a + 8)$ **27.** $3(x - 5y)(x + 2y)$ **29.** $m(mn + 7)(mn - 7)$ **31.** $3(x + 11y)(x - 3y)$ **33.** $4(x^2 + 9)(x + 3)(x - 3)$ **35.** $(3y + 1)(3y - 1)(y^2 + 1)$ **37.** $6r(5r - 2s)(2r - s)$ **39.** $(y^2 + z^2)(x + 1)(x - 1)$ **41.** $0.7(y + 3)(y - 3)$ **43.** $\frac{1}{3}(b + 3)^2$ **45.** $\frac{1}{4}(r + 4)(r + 2)$ **47.** $0.7(y + 2)(y + 3)$

CHAPTER REVIEW Page 240

1. $2 \cdot 3 \cdot 7$ **3.** $2 \cdot 2 \cdot 2 \cdot 2 \cdot 5$ or $2^4 \cdot 5$ **5.** $11 \cdot 11 \cdot a \cdot a \cdot a \cdot a \cdot b \cdot b \cdot b \cdot b$ **7.** 5 **9.** $2ab$ **11.** $6(x^2y + 2xy + 1)$ **13.** $2a(12b + 9c + 16a)$ **15.** $\frac{1}{4}(3x + y)$ **17.** $(5 + 3y)(5 - 3y)$ **19.** $2(y + 8)(y - 8)$ **21.** $(x^2 + 1)(x + 1)(x - 1)$ **23.** $(4x - 1)^2$ **25.** $2(4x - 5)^2$ **27.** $\left(y - \frac{3}{4}\right)^2$ **29.** $(b - 3)(b - 5)$ **31.** $(b - 1)(b + 6)$ **33.** $(a - 7)(a + 4)$ **35.** $(s + 13)(s - 3)$ **37.** $(m - 7n)(m + 6n)$ **39.** $(r - 13s)(r + 5s)$ **41.** $(m - 1)(4m - 3p)$ **43.** $(m + 7)(b + r)$ **45.** $(4k - p)(4k - 7p)$ **47.** $(4m + 3)(m + 2)$ **49.** $(2r + 5)(r - 4)$ **51.** $(2m - 1)(2m - 3)$ **53.** $(3x - 4)(x + 5)$ **55.** $3(x + 2)(x - 2)$ **57.** $a(3y + 4)(4y + 5)$ **59.** $(x + 7)(x - 7)(m + b)$

CHAPTER 8 APPLICATION OF FACTORING

8-1 Page 246

Exploratory **1.** $x = 0$ or $x + 3 = 0$ **3.** $3r = 0$ or $r - 4 = 0$ **5.** $3t = 0$ or $4t - 32 = 0$ **7.** $x - 6 = 0$ or $x + 4 = 0$ **9.** $a + 3 = 0$ or $3a - 12 = 0$ **11.** $2y + 8 = 0$ or $3y + 24 = 0$ **13.** $x - 3 = 0$ **15.** $3x + 2 = 0$ or $x - 7 = 0$ **17.** $4x - 7 = 0$ or $3x + 5 = 0$ **19.** $2x + 3 = 0$ or $x + 7 = 0$ ***Written*** **1.** $\{0, -3\}$ **3.** $\{0, 4\}$ **5.** $\{0, 8\}$ **7.** $\{6, -4\}$ **9.** $\{-3, 4\}$ **11.** $\{-4, -8\}$ **13.** $\{3\}$ **15.** $\left\{-\frac{2}{3}, 7\right\}$ **17.** $\left\{\frac{7}{4}, -\frac{5}{3}\right\}$ **19.** $\left\{-\frac{3}{2}, -7\right\}$ **21.** Sue is 16 years old. **23.** The number is -2.

8-2 Page 249

Exploratory **1.** GCF **3.** Perfect Square Trinomial **5.** Trinomial with two binomial factors **7.** GCF **9.** Difference of Squares **11.** Trinomial with two binomial factors ***Written*** **1.** $x^2(x - 5)$ **3.** $(3x - 2y)^2$ **5.** $(7z - 3)(5z + 4)$ **7.** $2(2x - 9)^2$ **9.** $(1 + 7k)(1 - 7k)$ **11.** $(x - 2)(x - 3)$ **13.** $2x(3x + 5y)$ **15.** $(2a + 3b)(2a - 3b)$ **17.** $(x + 6)^2$ **19.** $(4y - 3)(3y + 7)$ **21.** $(x + 5)(x + 3)(x - 3)$ **23.** $m^2(m + n)(m - n)$ **25.** $5a^2(2c + 3)^2$ **27.** $3x^2y(5x - 2y^2 + z)$ **29.** $8x(x + 2y^2)(x - 2y^2)$ **31.** $(x + 2)(x^2 + 8)$ **33.** $12(x + 2y)(x - 3y)$ **35.** $[(x + y) + (a - b)][(x + y) - (a - b)]$ **37.** $2(3x - 2y)(x + y)$ **39.** $\left(\frac{m}{2} + \frac{3}{5}\right)\left(\frac{m}{2} - \frac{3}{5}\right)$ **41.** $(2x - 7y)(3a + 5b)$ **43.** prime **45.** $[(x + 3) + y][(x + 3) - y]$

8-3 Page 252

Exploratory **1.** $n(n - 3) = 0$ **3.** $8c(c + 4) = 0$ **5.** $3x\left(x - \frac{1}{4}\right) = 0$ **7.** $7y(y - 2) = 0$ **9.** $13y(2y - 1) = 0$ **11.** $(y + 4)(y - 4) = 0$ **13.** $(y - 8)(y - 8) = 0$ **15.** $(y - 5)(y - 5) = 0$ ***Written*** **1.** $\{0, 6\}$ **3.** $\{0, -2\}$ **5.** $\{9, -9\}$ **7.** $\{0, -9\}$ **9.** $\{0, -8\}$ **11.** $\{6, -6\}$ **13.** $\{0, -5\}$ **15.** $\left\{\frac{2}{3}, -\frac{2}{3}\right\}$ **17.** $\{3, -3\}$ **19.** $\{2, -2\}$ **21.** $\{-5\}$ **23.** $\left\{-\frac{2}{9}\right\}$ **25.** $\left\{\frac{4}{9}, -\frac{4}{9}\right\}$ **27.** $\left\{0, \frac{1}{6}\right\}$ **29.** 2 meters **31.** 5 **33.** 10 or -10 **35.** 25 or -25

8-4 Page 255

Exploratory **1.** $\{-7, 3\}$ **3.** $\{8, 2\}$ **5.** $\left\{-\frac{3}{2}, -1\right\}$ **7.** $\left\{\frac{4}{3}, \frac{1}{2}\right\}$ **9.** $\{5, -3\}$ **11.** $\left\{0, -\frac{7}{2}, -\frac{4}{3}\right\}$ ***Written*** **1.** $\{-4, -9\}$ **3.** $\{-7, 3\}$ **5.** $\{-9, 7\}$

7. $\{-7, 7\}$ **9.** $\{0, 12\}$ **11.** $\{8, -3\}$ **13.** $\left\{\frac{3}{2}, -5\right\}$ **15.** $\left\{-\frac{1}{3}, -\frac{5}{2}\right\}$ **17.** $\left\{\frac{3}{2}, -8\right\}$ **19.** $\{0, 9, -9\}$ **21.** $\{0, 2, 4\}$ **23.** $\left\{0, \frac{1}{5}, -7\right\}$ **25.** $\left\{0, -\frac{5}{2}, 3\right\}$ **27.** $\{-4, 12\}$ **29.** $\left\{\frac{10}{3}, 30\right\}$ **31.** $\{-4, -5\}$ **33.** $\left\{\frac{2}{3}, -4\right\}$ **35.** $\{-2, -1, 2\}$ **37.** $\left\{\frac{1}{3}, -\frac{1}{3}, 2\right\}$ **39.** $\{2, -2\}$ **41.** $\{(x, y) \mid x = 3 \text{ or } y = -4\}$ **43.** $\left\{(r, s) \mid r = -2 \text{ or } s = \frac{4}{3}\right\}$

8-5 Page 257

Exploratory **1.** Let x = smaller integer; $x(x + 1) = 110$ **3.** Let x = smaller integer; $x(x + 2) = 168$ **5.** Let x = one integer; $x(15 - x) = 44$ **7.** Let x = smaller integer; $x^2 + (x + 1)^2 = 181$ **9.** Let x = smaller integer; $(x + 1)^2 - x^2 = 17$ or $x^2 - (x + 1)^2 = 17$ ***Written*** **1.** 10, 11; -10, -11 **3.** 12, 14; -12, -14 **5.** 4, 11 **7.** 9, 10; -9, -10 **9.** 8, 9; -8, -9 **11.** 10, 12; -10, -12 **13.** 13, 15 **15.** 13, 14 **17.** 3, 8 **19.** 8, 11; -8, -11 **21.** 9, 11 **23.** 4, 9 **25.** 5, 6; -8, -7 **27.** 6, 8; -12, -10 **29.** 13, 15, 17 **31.** Colleen is 48; Jim is 49

8-6 Page 261

Exploratory **1.** Let x = width; $x(x + 3) = 108$ **3.** Let x = length; $x(x - 9) = 90$ **5.** Let x = side of smaller square; $(x + 7)^2 = x^2 + 231$ **7.** Let x = side of smaller square; $(2x)^2 = x^2 + 363$ ***Written*** **1.** length, 12 ft; width, 9 ft **3.** length, 15 ft; width, 6 ft **5.** 13 in., 20 in. **7.** 11 in. **9.** 5m × 8m **11.** 6″ × 9″ **13.** $2\frac{1}{2}$ ft **15.** 176 ft/s **17.** 156 ft **19.** 12 s **21.** 20 s, 125 s **23.** $\frac{1}{2}$s, 102 s

8-7 Page 264

Exploratory **1.** 24; $\frac{1}{3}$ **3.** $6y$; $\frac{7}{3x}$ **5.** $2a$; $\frac{19a}{21b}$ **7.** $3x^2y^2$; $\frac{-y^3}{6x^3}$ **9.** x; $\frac{y + 1}{y - 2}$ **11.** $m + 5$; $\frac{1}{2}$ ***Written*** **1.** $\frac{1}{y + 4}$ **3.** $\frac{1}{a - b}$ **5.** $\frac{c - 2}{c + 2}$ **7.** a **9.** $\frac{1}{x^2 - 4}$ **11.** $\frac{x^2 + 3}{x + 3}$ **13.** $\frac{-2}{1 - 2y}$ **15.** $\frac{-4a}{a^2 + 6ab + 9b}$ **17.** $\frac{1}{x + y}$ **19.** $\frac{1}{x + 4}$ **21.** $\frac{m - 4}{m + 4}$ **23.** $\frac{y - 3}{y + 3}$ **25.** $x + 5$ **27.** $\frac{6x}{x + 4}$ **29.** $\frac{2k + 5}{2k - 5}$ **31.** $\frac{1}{2 - 5x}$ **33.** $\frac{5}{3}$ **35.** $\frac{m - 4}{m - 3}$ **37.** $\frac{p - 2}{p + 6}$ **39.** $\frac{2x - 3}{3x - 2}$ **41.** $y + 3$ **43.** $\frac{2x + 8}{x - 11}$ **45.** $\frac{1}{x}$

CHAPTER REVIEW Page 266

1. $\{0, -11\}$ **3.** $\{0, 5\}$ **5.** $\left\{\frac{2}{3}, -\frac{7}{4}\right\}$ **7.** $\{6\}$ **9.** $m(2m - 3)(3m + 5)$ **11.** $(8m - 3n)(3a + 5b)$ **13.** $(7m + 5n)^2$ **15.** $8(a^2 + 4b^2)$ **17.** $\{0, -7\}$ **19.** $\{0, 3\}$ **21.** $\left\{-\frac{2}{5}\right\}$ **23.** $\{-5, -8\}$ **25.** $\left\{\frac{3}{2}, -8\right\}$ **27.** $\left\{\frac{4}{3}, \frac{1}{2}\right\}$ **29.** 11, 13; -11, -13 **31.** 8″ × 10″ **33.** $\frac{x}{4y^2z}$ **35.** $\frac{x + 3}{x - 3}$

CHAPTER 9 FUNCTIONS AND GRAPHS

9-1 Page 273

Exploratory **1.** (1, 4) **3.** $(-3, 3)$ **5.** $(-1, -2)$ **7.** (4, 0) **9.** (1, 1) **11.** $(-1, -1)$ **13.** $(-2, 2)$ **15.** $(-3, -4)$ **17.** (2, 2) **19.** 3, -1 **21.** I **23.** II **25.** none **27.** IV **29.** IV **31.** II **33.** II **35.** III **37.** IV **39.** none **41.** I **43.** III

9-2 Page 275

Exploratory **1.** $D = \{0, 1, 2\}$, $R = \{2, -2, 4\}$ **3.** $D = \{-3, -2, -1, 0\}$, $R = \{1, 0, 2\}$ **5.** $D = \{7, -2, 4, 5, -9\}$, $R = \{5, -3, 0, -7, 2\}$ **7.** $D = \{-3, -2, 3, 0\}$, $R = \{0, -5, 6, 7, -17\}$ **9.** $D = \{3.1, 4.7, 2.4, -9\}$, $R = \{-1, 3.9, -3.6, 12.12\}$ **11.** $D = \left\{\frac{1}{2}, 1\frac{1}{2}, -3, -5\frac{1}{4}\right\}$, $R = \left\{\frac{1}{4}, -\frac{2}{3}, \frac{2}{5}, -7\frac{2}{7}\right\}$ **13.** $\{(1, 5), (2, 7), (3, 9), (4, 11)\}$, $D = \{1, 2, 3, 4\}$, $R = \{5, 7, 9, 11\}$ **15.** $\{(-4, 1), (-2, 3), (0, 1) (2, 3) (4, 1)\}$, $D = \{-4, -2, 0, 2, 4\}$ $R = \{1, 3\}$ ***Written*** **1.** $\{(1, 5), (2, 3), (3, -2), (4, 9)\}$ **3.** $\{(1, 7), (-2, 7), (3, 7)\}$ **5.** $\{(4, 2), (-3, 2), (8, 2), (8, 9), (7, 5)\}$ **7.** $\{(1, 3), (2, 4), (3, 5), (4, 6), (5, 7)\}$ **9.** $\{(8, 1), (7, 3), (6, 5), (2, -2)\}$ **11.** $\{(1, 3), (2, 5), (1, -7), (2, 9)\}$ **13.** $\{(-2, 2), (-1, 1), (0, 1), (1, 1), (1, -1), (2, -1), (3, 1)\}$; $D = \{-2, -1, 0, 1, 2, 3\}$, $R = \{2, 1, -1\}$ **15.** $\{(-3, 0), (-2, 2), (-1, 3), (0, 1), (1, -1), (1, -2), (1, -3), (3, -2)\}$; $D = \{-3, -2, -1, 0, 1, 3\}$, $R = \{-3, -2, -1, 0, 1, 2, 3\}$ **17.** $\{(-3, 3), (-1, 2), (1, 1), (1, 3), (2, 0), (2, -1), (3, -1)\}$, $D = \{-3, -1, 1, 2, 3\}$, $R = \{-1, 0, 1, 2, 3\}$

9-3 Page 281

Exploratory **1.** $\{(-4, -12), (-2, -6), (0, 0), (1, 3), (2, 6), (3, 9)\}$ **3.** $\left\{(-4, -1), \left(-2, \frac{1}{3}\right), \left(0, \frac{5}{3}\right), \left(1, \frac{7}{3}\right), \left(3, \frac{11}{3}\right)\right\}$ ***Written*** **1.** $y = 5 - x$ **3.** $b = 3 + 5a$ **5.** $y = 3 - 4x$ **7.** $b = \frac{7 - 4a}{3}$ **9.** $y = \frac{6x - 2}{3}$ **11.** $q = \frac{4p - 7}{-2}$ **13.** a, c **15.** a, d **17.** a, c **19.** $\{(-2, -3), (-1, -1), (0, 1), (2, 5), (5, 11)\}$

21. $\{(-2, 9), (-1, 8), (0, 7), (2, 5), (5, 2)\}$
23. $\{(-2, 11), (-1, 9), (0, 7), (2, 3), (5, -3)\}$
25. $\left\{\left(-2, \frac{17}{3}\right), (-1, 5), \left(0, \frac{13}{3}\right), (2, 3), (5, 1)\right\}$
27. $\left\{\left(-2, -\frac{11}{2}\right), (-1, -4), \left(0, -\frac{5}{2}\right), \left(2, \frac{1}{2}\right), (5, 5)\right\}$
29. $\left(-2, \frac{9}{2}\right), \left(-1, \frac{13}{4}\right), (0, 2), \left(2, -\frac{1}{2}\right), \left(5, -\frac{17}{4}\right)\}$
31. $\left\{(-2, 5), \left(-1, \frac{31}{6}\right), \left(0, \frac{16}{3}\right), \left(2, \frac{17}{3}\right), \left(5, \frac{37}{6}\right)\right\}$
33. $\left\{\left(-2, \frac{23}{5}\right), \left(-1, \frac{17}{5}\right), \left(0, \frac{11}{5}\right), \left(2, -\frac{1}{5}\right), \left(5, -\frac{19}{5}\right)\right\}$
35. $\left\{\left(-2, \frac{2}{5}\right), (-1, 1), \left(0, \frac{8}{5}\right), \left(2, \frac{14}{5}\right), \left(5, \frac{23}{5}\right)\right\}$

9-4 Page 285

Exploratory **1.** yes **3.** no **5.** yes **7.** no **9.** no **11.** no **13.** no **15.** yes **17.** yes ***Written*** **1.** no **3.** no **5.** yes **7.** no **9.** yes **11.** no

9-5 Page 289

Exploratory **1.** no **3.** yes **5.** yes **7.** yes **9.** no **11.** no **13.** no **15.** yes **17.** 3 **19.** -13 **21.** 0 ***Written*** **1.** yes **3.** no **5.** no **7.** yes **9.** yes **11.** yes **13.** no **15.** yes **17.** yes **19.** yes **21.** 1 **23.** 20 **25.** $-3\frac{1}{2}$ **27.** $\frac{3}{4}$ **29.** $6a - 5$ **31.** 30 **33.** 11.5 **35.** $3a + 4$

9-6 Page 293

Exploratory **1.** a **3.** a, b, c **5.** a, c, **7.** b

9-7 Page 295

Exploratory **1.** 11, 13, 15 **3.** 0, -3, -6 **5.** 55, 62, 69 **7.** 28, 39, 52 **9.** 25, 36, 49 **11.** 34, 55, 89 ***Written*** **1.** 13, 15, 17 **3.** -13, -16, **5.** 16, 21, 31 **7.** $\frac{1}{9}, \frac{1}{27}, \frac{1}{81}$ **9.** 1, 0, 1 **11.** -5, -7, -9 **13.** $y = 4x$; 20, 24 **15.** $n = 2m + 1$; 1, 3, 5, 7 **17.** $b = 4a + 3$; 19, 27, 35 **19.** $y = -x + 15$; 11, 9, 8 **21.** $b = -5a + 3$; -2, -7, -17, -32

CHAPTER REVIEW Page 299

1. IV **3.** I **9.** $D = \{3, 2, 5\}$, $R = \{5, 6, 7\}$ **11.** $D = \{-3, 4\}$, $R = \{5, 6\}$ **13.** $\{(-1, 4), (3, 4), (4, 6), (0, -4)\}$; $D = \{-1, 3, 4, 0\}$, $R = \{4, 6, -4\}$ **15.** $\{(3, 6), (3, -3), (4, 8), (5, 8), (5, 11), (5, 3)\}$; $D = \{3, 4, 5\}$, $R = \{6, -3, 8, 11, 3\}$ **17.** $\{(-2, -1), (-1, -1), (0, 1), (1, 2), (3, 2)\}$; $D = \{-2, -1, 0, 1, 3\}$, $R = \{-1, 1, 2\}$ **19.** $y = 7 - 3x$ **21.** c **23.** $\{(-4, -11), (-2, -3), (0, 5), (2, 13), (4, 21)\}$ **25.** $\left\{\left(-4, \frac{21}{2}\right), \left(-2, \frac{15}{2}\right), \left(0, \frac{9}{2}\right), \left(2, \frac{3}{2}\right), \left(4, -\frac{3}{2}\right)\right\}$ **27.** no **29.** yes **35.** no **37.** yes **39.** 3 **41.** 1 **43.** 3 **49.** $y = 3x + 5$

CHAPTER 10 LINES AND SLOPES

10-1 Page 307

Exploratory **1.** 3 **3.** -1 **5.** $\frac{2}{1} = 2$ **7.** $\frac{-1}{1} = -1$ **9.** $\frac{-4}{3} = -\frac{4}{3}$ **11.** 1 **13.** $\frac{7}{3}$ **15.** $\frac{2}{3}$ ***Written*** **1.** 3 **3.** $\frac{4}{3}$ **5.** undefined **7.** $\frac{1}{3}$ **13.** 2 **15.** $-\frac{3}{2}$ **17.** $\frac{2}{5}$ **19.** $\frac{3}{2}$ **21.** $\frac{1}{2}$ **23.** 0 **25.** $\frac{1}{2}$ **27.** $-\frac{2}{3}$ **29.** 9 **31.** 2 **33.** -23 **35.** 6

10-2 Page 311

Exploratory **1.** 3, (5, 2) **3.** -2, $(-1, -5)$ **5.** $-\frac{3}{2}$, $(-5, -6)$ **7.** 2, $\left(3, -\frac{3}{2}\right)$ **9.** $-\frac{3}{5}$, $\left(-6, -\frac{3}{8}\right)$ **11.** 0, $(0, -2)$ **13.** $2x - y = -6$ **15.** $2x - 3y = -1$ **17.** $5x - 3y = -37$ ***Written*** **1.** $2x + 3y = 22$ **3.** $2x - 3y = 15$ **5.** $2x - y = 11$ **7.** $3x - 2y = -20$ **9.** $y = 7$ **11.** $x = 1$ **13.** $x + y = 9$ **15.** $5x + y = 31$ **17.** $x + 4y = -4$ **19.** $x + 2y = -4$ **21.** $4x + 7y = -18$ **23.** $x = 5$ **25.** $4x + 10y = 13$ **27.** $2x + 3y = 6$ **29.** $2y = 1$ **31.** $x = -2$ **33.** $x = -2$

10-3 Page 314

Exploratory **1.** 5, 3, **3.** 3, -7 **5.** $\frac{1}{3}$, 0 **7.** $\frac{3}{5}$, $-\frac{1}{4}$ **9.** $\frac{3}{5}, \frac{1}{4}$ **11.** $\frac{1}{4}, \frac{3}{4}$ **13.** 6, 5 **15.** $\frac{3}{2}, \frac{7}{2}$ **17.** $\frac{1}{2}, \frac{7}{4}$ **19.** $x = c$ (c is any number) ***Written*** **1.** $y = 3x + 1$ **3.** $y = -3x + 5$ **5.** $y = 4x - 2$ **7.** $y = \frac{1}{2}x + 5$ **9.** $y = -\frac{5}{4}x + 3$ **11.** 3, 5 **13.** 3, 6 **15.** 14, -4 **17.** $-\frac{11}{2}$, $-\frac{11}{5}$ **19.** $\frac{1}{2}, \frac{3}{2}$ **21.** -1, $-\frac{8}{3}$ **23.** $y = -\frac{3}{4}x + 3$; $m = -\frac{3}{4}$, $b = 3$ **25.** $y = 3x - 9$; $m = 3$, $b = -9$ **27.** $y = -\frac{7}{4}x + 2$; $m = -\frac{7}{4}$, $b = 2$ **29.** $y = \frac{7}{3}x - \frac{10}{3}$; $m = \frac{7}{3}$, $b = -\frac{10}{3}$ **31.** $y = 12x - 24$; $m = 12$, $b = -24$ **33.** $y = -4x + 12$; $m = -4$; $b = 12$ **35.** $y = -\frac{5}{2}x + 4$; $m = -\frac{5}{2}$, $b = 4$ **37.** $y = \frac{1}{2}x + \frac{3}{2}$; $m = \frac{1}{2}$, $b = \frac{3}{2}$ **39.** $y = \frac{3}{5}x - \frac{18}{5}$; $m = \frac{3}{5}$, $b = -\frac{18}{5}$

10-4 Page 318

Exploratory **1.** 2, 8 **3.** 2, $\frac{3}{2}$ **5.** 4, -8 **7.** $\frac{10}{7}$, 5 **9.** $\frac{2}{5}$, -4 **11.** $\frac{1}{3}$, -1 **13.** $\frac{5}{2}, \frac{5}{3}$ **15.** 0, 0 **17.** (1, 3), (2, 5) **19.** $(1, -3)$, $(2, -10)$ **21.** (7, 3), (9, 4) **23.** $(-2, 14)$, $(-1, 22)$

10-5 Page 321

Exploratory **1.** -5 **3.** 9 **5.** -5 **7.** $-\frac{1}{3}$ **9.** $\frac{1}{3}$ **11.** $2, -2, y = 2x - 2$ **13.** $-4, 4, y = -4x + 4$ **15.** $-\frac{2}{3}, 2, y = -\frac{2}{3}x + 2$ ***Written*** **1.** $y = 3x - 17$ **3.** $y = \frac{2}{3}x + \frac{2}{3}$ **5.** $y = -5x + 29$ **7.** $y = \frac{3}{4}x - \frac{5}{2}$ **9.** $y = \frac{1}{4}x + 8$ **11.** $y = \frac{1}{2}x - \frac{1}{2}$ **13.** $y = x - 3$ **15.** $y = -x + 1$ **17.** $y = -\frac{4}{5}x - \frac{1}{5}$ **19.** $y = \frac{1}{6}x + \frac{37}{6}$ **21.** $y = \frac{1}{5}x + 6$

10-6 Page 326

Exploratory **1.** $5, -\frac{1}{5}$ **3.** $2, -\frac{1}{2}$ **5.** $\frac{2}{3}, -\frac{3}{2}$ **7.** undefined, 0 **9.** $\frac{5}{3}, -\frac{3}{5}$ **11.** $4, -\frac{1}{4}$ **13.** $\frac{2}{3}, -\frac{3}{2}$ **15.** $-1, 1$ ***Written*** **1.** $y = -\frac{3}{5}x - 1$ **3.** $y = \frac{3}{4}x$ **5.** $y = \frac{5}{2}x - 4$ **7.** $y = \frac{3}{4}x + \frac{1}{2}$ **9.** $y = x - 9$ **11.** $y = \frac{8}{3}x + 4$ **13.** $y = \frac{1}{3}x + 2$ **15.** $y = -\frac{3}{2}x - \frac{9}{2}$ **17.** $y = \frac{9}{5}x$

10-7 Page 328

Exploratory **1.** 6 **3.** 14 **5.** 5 **7.** $1\frac{1}{2}$ **9.** $2\frac{1}{2}$ **11.** (5, 3) **13.** (9, -8) **15.** (1, 6) **17.** $\left(-1, -1\frac{1}{2}\right)$ **19.** $\left(5\frac{1}{2}, -5\frac{1}{2}\right)$ **21.** $(4x, 2y)$ ***Written*** **1.** (10, 3) **3.** (10, 3) **5.** (8, 2) **7.** (6, -2) **9.** (1, -3) **11.** $\left(10, -1\frac{1}{2}\right)$ **13.** $\left(8\frac{1}{2}, 2\frac{1}{2}\right)$ **15.** (3, 3) **17.** $\left(\frac{1}{2}, 5\frac{1}{2}\right)$, **19.** $\left(\frac{3}{4}, \frac{1}{2}\right)$, **21.** $\left(4\frac{5}{12}, 1\frac{7}{8}\right)$ **23.** $B(19, 9)$ **25.** $P(-1, 6)$ **27.** $A(13, 11)$ **29.** $B(6, -7)$ **31.** $P(-2, -3)$ **33.** $B(9, -18)$ **35.** $(-2, -7)$ **37.** $\left(\frac{1}{2}, 2\right)$ **39.** $\left(4\frac{1}{2}, -\frac{1}{2}\right)$ **41.** (4, 1) **43.** $\left(-1\frac{1}{2}, 1\right)$

CHAPTER REVIEW Page 332

1. 0 **3.** undefined **5.** $y - 1 = -\frac{4}{11}(x - 8)$; $4x + 11y = 43$ **7.** $y - 5 = \frac{5}{2}x$; $5x - 2y = -10$ **9.** $\frac{3}{2}, -\frac{7}{2}$ **11.** $-8, 4$ **13.** (5, 0)(0, 2) **19.** $y = 4x - 26$ **21.** $y = -\frac{5}{2}x + 7$ **23.** $y = 4x - 9$ **25.** $y = -\frac{7}{2}x - 14$ **27.** (6, 1)

CHAPTER 11 SYSTEMS OF OPEN SENTENCES

11-1 Page 338

Exploratory **1.** (6, 3) **3.** (1, 3) **5.** (0, 0) **7.** (6, 3) **9.** (3, -3) **11.** a, c, d **13.** a, d ***Written*** **1.** (3, 1) **3.** (5, 3) **5.** (2, 2) **7.** (2, 3) **9.** (0, 0) **11.** (-1, 2) **13.** (-1, -5) **15.** (-3, 1) **17.** (3, 4) **19.** (-2, -3) **21.** (0.75, 0.5) **23.** (4, -7)

11-2 Page 341

Exploratory **1.** $m = -1, b = 4; m = -\frac{2}{3}, b = 3$; one **3.** $m = -1, b = 6; m = -1, b = 1$; none **5.** $m = -\frac{1}{2}, b = \frac{5}{2}; m = -\frac{1}{2}, b = \frac{5}{2}$; infinitely many **7.** $m = -3, b = 0; m = \frac{1}{6}, b = -\frac{19}{3}$; one **9.** $m = \frac{3}{7}, b = \frac{6}{7}; m = -\frac{1}{2}, b = \frac{11}{2}$; one **11.** $m = \frac{3}{8}, b = -\frac{1}{2}; m = \frac{3}{8}, b = -\frac{21}{8}$; none ***Written*** **1.** (2, 0) **3.** none **5.** (2, 3) **7.** none **9.** infinitely many **11.** (7, -3) **13.** none **15.** infinitely many **17.** (0, -4)

11-3 Page 343

Exploratory **1.** $x = y + 1, y = x - 1$ **3.** $x = -\frac{3}{2}y + 3, y = -\frac{2}{3}x + 2$ **5.** $x = 6y - 14, y = \frac{1}{6}x + \frac{7}{3}$ **7.** $x = -y - 5, y = -x - 5$ **9.** $x = \frac{3}{4}y + 15, y = \frac{4}{3}x - 20$ **11.** $x = -\frac{0.8}{0.75}y - \frac{6}{0.75}, y = -\frac{0.75}{0.8}x - \frac{6}{0.8}$ ***Written*** **1.** $\left(\frac{8}{5}, \frac{16}{5}\right)$ **3.** $\left(3, \frac{3}{2}\right)$ **5.** infinitely many **7.** none **9.** (0, 0) **11.** (-9, -7) **13.** (4, 2) **15.** (4, 3) **17.** (1, 1)

11-4 Page 346

Exploratory **1.** yes **3.** no **5.** yes **7.** no **9.** yes ***Written*** **1.** (8, -1) **3.** $\left(\frac{11}{2}, -\frac{1}{2}\right)$ **5.** (3, 4) **7.** (1, 2) **9.** (2, 1.5) **11.** $\left(3, \frac{19}{3}\right)$ **13.** $\frac{14}{9}, -\frac{16}{3}$ **15.** (0, 25) **17.** 12, 36 **19.** 38, 63 **21.** 55 m, 45 m **23.** 14, 23 **25.** 18 men, 25 women

11-5 Page 349

Exploratory **1.** To eliminate x, multiply the first equation by 3 and subtract. To eliminate y, multiply the second equation by 2 and subtract. **3.** To eliminate x, multiply the first equation by 4 and subtract. To eliminate y, multiply the second equation by 4 and add. **5.** To eliminate x, multiply the second equation by 4 and subtract. To eliminate y, multiply the first equation by 2 and add. **7.** To eliminate x, multiply the second equation by 8 and subtract. To eliminate y, multiply the second equation by 3 and subtract. **9.** To eliminate x, multiply the second equation by 2 and subtract. To eliminate y, simply subtract. **11.** To eliminate x, multiply the second equation by 1.2 and subtract. To eliminate y, multiply the first equation by 2 and add. ***Written*** **1.** $\left(\frac{11}{2}, -\frac{1}{2}\right)$ **3.** $\left(\frac{5}{2}, 5\right)$ **5.** (1, 1) **7.** $\left(-1, \frac{9}{2}\right)$ **9.** $\left(\frac{20}{9}, \frac{5}{3}\right)$ **11.** (5, 1) **13.** (8, −2) **15.** (−9, −2) **17.** (6, 2) **19.** (3, 3) **21.** (10, 25) **23.** (4, 16) **25.** (0, 0) **27.** (0, −2) **29.** 14, 34 **31.** 6, 2 **33.** 28 m, 112 m **35.** Layla is 30; Diana is 10. **37.** 13 cm, 7 cm **39.** 19, 5 **41.** 16, 36

11-6 Page 353

Exploratory **1.** yes **3.** yes **5.** B **7.** C ***Written*** **21.** $x \leq 2, y \geq 2$ **23.** $y \geq -x, y < -x + 3$ **25.** $y < -\frac{1}{2}x + 2, y \geq 0$

11-7 Page 357

Written **1.** 24 **3.** 82 **5.** 16 **7.** 48 **9.** 51 **11.** 35 **13.** 94 **15.** 72 **17.** 603 **19.** 370 **21.** 39 **23.** 40, 51, 62, 73, 84, 95 **25.** $\frac{21}{12}, \frac{42}{24}, \frac{63}{36}, \frac{84}{48}$

11-8 Page 360

Written **1.** The rate of the boat is 4 mph. The rate of the current is 1 mph. **3.** 7 lb., 3 lb. **5.** 20 at 20¢, 10 at 35¢ **7.** The rate of the wind is 50 mph. The rate of the plane is 550 mph. **9.** $1000 at 10%, $3000 at 12% **11.** 280 miles **13.** 42 mph and 48 mph **15.** 4 mph **17.** 1 mph

CHAPTER REVIEW Page 365

1. (1, 1) **3.** (6, 5) **5.** none **7.** one **9.** (3, −5) **11.** (4, −2) **13.** $\left(\frac{1}{2}, \frac{1}{2}\right)$ **15.** (2, 0) **17.** (2, −1) **19.** (5, 1) **21.** (−9, −7) **23.** (2, −1) **25.** (−4, 6) **29.** 16 cm, 9 cm **31.** 35 **33.** 5 mph

CHAPTER 12 RADICALS

12-1 Page 373

Exploratory **1.** 100 **3.** 49 **5.** 009 **7.** $\frac{1}{4}$ **9.** $\frac{49}{64}$ **11.** 11 **13.** −9 **15.** $\frac{2}{3}$ **17.** $\pm\frac{7}{11}$ **19.** 0.04 ***Written*** **1.** 7 **3.** 4 **5.** $\frac{5}{6}$ **7.** $\frac{3}{7}$ **9.** 0.03 **11.** 0.06 **13.** 6 **15.** −10 **17.** ±5 **19.** 0.6 **21.** 23 **23.** 26 **25.** −21 **27.** ±32 **29.** 27 **31.** 42 **33.** $-\frac{17}{10}$ **35.** 3 **37.** $\frac{16}{19}$ **39.** $\pm\frac{31}{27}$ **41.** $\frac{8}{9}$ **43.** $\frac{3}{4}$

12-2 Page 376

Exploratory **1.** Z, Q **3.** N, W, Z, Q **5.** Q **7.** N, W, Z, Q **9.** I **11.** Q **13.** 1.414 **15.** 4.472 **17.** 256 **19.** 5.568 **21.** 9.434 **23.** 4,356 ***Written*** **7.** I **9.** Q **11.** Q **13.** I **15.** Q, 82 **17.** I, $61 < \sqrt{3{,}800} < 62$ **19.** I, $29 < \sqrt{888} < 30$ **21.** Q, 95 **23.** I, $84 < \sqrt{7{,}166} < 85$ **25.** Q, 38

12-3 Page 378

Exploratory **1.** 16 **3.** 49 **5.** 64 **7.** 49 **9.** 9 **11.** 81 **13.** 4 **15.** 6 **17.** 5 ***Written*** **1.** 8 **3.** 6 **5.** 10 **7.** 12 **9.** 13 **11.** 12 **13.** 9.219 **15.** 7.765 **17.** 12.206557 **19.** 13.892444 **21.** 11.462986 **23.** 13.251415 **25.** 0.781 **27.** 0.06797

12-4 Page 380

Exploratory **1.** $2\sqrt{2}$ **3.** $2\sqrt{5}$ **5.** $2\sqrt{6}$ **7.** $4\sqrt{3}$ **9.** $|m|$ **11.** $x^2\sqrt{x}$ **13.** $2a\sqrt{2a}$ **15.** $ab\sqrt{ab}$ **17.** 6 **19.** $5\sqrt{2}$ **21.** 11 **23.** $3 + \sqrt{6}$ **25.** $\sqrt{35} - 7$ ***Written*** **1.** $3\sqrt{3}$ **3.** $3\sqrt{5}$ **5.** $6\sqrt{2}$ **7.** $3\sqrt{10}$ **9.** $8\sqrt{2}$ **11.** $10\sqrt{5}$ **13.** $12\sqrt{5}$ **15.** 72 **17.** $4|x|\sqrt{2}$ **19.** $2b^2\sqrt{10}$ **21.** $6|ab|$ **23.** $2|a|\sqrt{30ab}$ **25.** $2|x|y^2\sqrt{15}$ **27.** $2|m|n^3\sqrt{5n}$ **29.** $8m^2|n^3|\sqrt{5}$ **31.** $|x|\sqrt{21y}$ **33.** $4\sqrt{3}$ **35.** 10 **37.** $14\sqrt{15}$ **39.** 150 **41.** $60\sqrt{3}$ **43.** $3 + 3\sqrt{2}$ **45.** $\sqrt{42} - 3\sqrt{2}$ **47.** $5\sqrt{2} - \sqrt{10}$ **49.** $\sqrt{15} + 9$ **51.** $12 + 4\sqrt{21}$ **53.** 49 cm^2

12-5 Page 384

Exploratory **1.** 2 **3.** 8 **5.** $\sqrt{3} - 4$; −13 **7.** $6 - \sqrt{8}$; 28 **9.** $\sqrt{2} - \sqrt{5}$; −3 **11.** $2\sqrt{5} + \sqrt{6}$; 14 **13.** $\frac{\sqrt{5}}{\sqrt{5}}$ **15.** $\frac{\sqrt{6}}{\sqrt{6}}$ **17.** $\frac{\sqrt{5}}{\sqrt{5}}$ **19.** $\frac{3 - \sqrt{7}}{3 - \sqrt{7}}$ ***Written*** **1.** $\sqrt{7}$ **3.** $\frac{\sqrt{70}}{7}$ **5.** $\frac{\sqrt{3}}{3}$ **7.** $\frac{\sqrt{21}}{7}$ **9.** $\frac{\sqrt{35}}{10}$ **11.** $\frac{\sqrt{15}}{3}$ **13.** $\frac{\sqrt{3a}}{3}$ **15.** $\frac{|a|\sqrt{5}}{5}$ **17.** $\frac{3\sqrt{3}}{|b|}$ **19.** $\frac{n^2\sqrt{5mn}}{2|m^3|}$ **21.** $\frac{7 + \sqrt{3}}{46}$ **23.** $\frac{55 - 11\sqrt{2}}{23}$ **25.** $6\sqrt{3} - 6\sqrt{2}$ **27.** $\frac{20a + 10a\sqrt{a}}{4 - a}$ **29.** $\frac{-6\sqrt{5} - 2\sqrt{30}}{3}$ **31.** $\frac{9\sqrt{2} + 9}{2}$ **33.** $\frac{5\sqrt{21} - 3\sqrt{35}}{15}$ **35.** $\frac{12\sqrt{10} + 3\sqrt{35}}{25}$

12-6 Page 386

Exploratory **1.** $5\sqrt{3}, 3\sqrt{3}$ **3.** $3\sqrt{12}, 5\sqrt{12}$ **5.** $-3\sqrt{3}, 12\sqrt{3}$ **7.** $2\sqrt{10}, -6\sqrt{10}, 7\sqrt{10}$ **9.** $2\sqrt{5}, 3\sqrt{5}, -5\sqrt{5}$ **11.** $-3\sqrt{3}$ **13.** $11\sqrt{6}$ **15.** $11\sqrt{y}$ **17.** $3\sqrt{15} - 2\sqrt{5}$ **19.** $26\sqrt{13}$ **21.** $40\sqrt{19}$ **23.** $-5\sqrt{3a}$ ***Written*** **1.** $11\sqrt{11}$ **3.** $-4\sqrt{13}$ **5.** $17\sqrt{7}$ **7.** $9\sqrt{3}$ **9.** $\sqrt{7}$ **11.** $-10\sqrt{5}$ **13.** $\sqrt{2} + 13\sqrt{3}$ **15.** $4\sqrt{6} + 5\sqrt{7} - 6\sqrt{2}$ **17.** $\sqrt{3} + \sqrt{5}$ **19.** $4\sqrt{3}$ **21.** $-\sqrt{7}$ **23.** $35\sqrt{5}$ **25.** $-18\sqrt{7}$ **27.** $8\sqrt{2} + 6\sqrt{3}$ **29.** $48\sqrt{2} + \sqrt{5}$ **31.** $20\sqrt{3}$ **33.** $\frac{8\sqrt{7}}{7}$ **35.** $\frac{2\sqrt{3}}{3}$ **37.** 0 **39.** $\frac{85\sqrt{7}}{14}$

12-7 Page 390

Exploratory **1.** 9 **3.** none **5.** 4 **7.** none ***Written*** **1.** 25 **3.** 3 **5.** none **7.** 25 **9.** none **11.** $\frac{9}{2}$ **13.** 2 **15.** 3 **17.** 144 **19.** 3 **21.** 80 **23.** 45 **25.** $\pm\sqrt{22}$ **27.** 11 **29.** 7 **31.** 6 **33.** 169

12-8 Page 393

Exploratory **1.** 5 **3.** 13 **5.** 8 **7.** 8 **9.** True **11.** False ***Written*** **1.** 13 **3.** $3\sqrt{3}$ **5.** 9 **7.** 16 **9.** 21 **11.** 4 **13.** no **15.** yes **17.** $\sqrt{89}\ m$ **19.** $4\sqrt{6}$ m

12-9 Page 395

Exploratory **1.** $x_1 = 3, x_2 = 6, y_1 = 4, y_2 = 8$ **3.** $x_1 = -4, x_2 = 4, y_1 = 2, y_2 = 17$ **5.** $x_1 = -3, x_2 = 2, y_1 = 5, y_2 = 7$ **7.** $x_1 = -8, x_2 = -3, y_1 = -4, y_2 = 8$ **9.** $x_1 = 3, x_2 = -2, y_1 = 7, y_2 = -5$ ***Written*** **1.** 17 **3.** $\sqrt{29}$ **5.** 13 **7.** $3\sqrt{5}$ **9.** $\sqrt{185}$ **11.** $\frac{5}{2}$ **13.** $\frac{\sqrt{85}}{3}$ **15.** $\frac{13}{10}$ **17.** 13, −9 **19.** 17, −13 **21.** 10, 0 **23.** −7, −3 **25.** −9, 5

CHAPTER REVIEW Page 399

1. 11 **3.** $\pm\frac{2}{9}$ **7.** Q **9.** $15 < \sqrt{250} < 16$ **11.** 4.359 **13.** 7.855 **15.** $6\sqrt{3}$ **17.** 54 **19.** $\frac{2\sqrt{15}}{|x|}$ **21.** $4\sqrt{3}$ **23.** $3 + 3\sqrt{2}$ **25.** $\sqrt{7}$ **27.** $14 + 7\sqrt{3}$ **29.** $\frac{30\sqrt{10} - 35\sqrt{6}}{33}$ **31.** $5\sqrt{13}$ **33.** $-7\sqrt{7}$ **35.** $\frac{9\sqrt{2}}{4}$ **37.** $x = 12$ **39.** $x = \frac{26}{7}$ **41.** $x = 12$ **43.** $c = 2\sqrt{34}$ **45.** $a = 6\sqrt{3}$ **47.** $\sqrt{130}$

CHAPTER 13 QUADRATICS

13-1 Page 408

Exploratory **1.** upward **3.** downward **5.** downward **7.** upward **9.** downward **11.** $x = 2$ **13.** $x = -3$; **15.** $x = -2$; **17.** $x = -\frac{7}{10}$; ***Written*** **1.** $x = \frac{5}{2}; \left(\frac{5}{2}, \frac{49}{4}\right)$ **3.** $x = 2$; (2, 9) **5.** $x = \frac{3}{2}; \left(\frac{3}{2}, \frac{137}{4}\right)$ **7.** $x = -1$; (−1, −20) **9.** $x = 0$; (0, −9) **11.** $x = -1$; (−1, 1) **13.** $x = 1$; (1, 22) **15.** $x = 0$; (0, 4) **17.** $x = 1$; (1, 42) **19.** $x = 2$; (2, −9) **21.** $x = 3$; (3, 14) **23.** $x = 0$; (0, −3) **25.** $x = 0$; (0, 3) **27.** $x = \frac{1}{2}; \left(\frac{1}{2}, -\frac{49}{4}\right)$ **29.** $x = -3$; (−3, 0)

13-2 Page 411

Exploratory **1.** 1, −1 **3.** no real roots **5.** 0, 2 ***Written*** **1.** −3, 4 **3.** 6, −1 **5.** −2, 2 **7.** 3, 7 **9.** no real roots **11.** between 0 and 1, between 3 and 4 **13.** 4 **15.** between −1 and −2, between 0 and 1 **17.** no real roots **19.** between 0 and −1, between 2 and 3 **21.** no real roots **23.** between 0 and 1, between 2 and 3

13-3 Page 415

Exploratory **1.** no **3.** yes **5.** no **7.** no **9.** yes **11.** 4 **13.** 9 **15.** 49 **17.** $\frac{25}{4}$ **19.** $\frac{49}{4}$ **21.** $\frac{169}{4}$ ***Written*** **1.** −1, −3 **3.** 7, −3 **5.** 4, 2 **7.** $2 \pm \sqrt{6}$ **9.** $4 \pm 2\sqrt{5}$ **11.** $4 \pm 2\sqrt{7}$ **13.** 1, −15 **15.** −2, −3 **17.** 12, −1 **19.** $3 \pm \sqrt{2}$ **21.** $2 \pm \sqrt{7}$ **23.** $\frac{7}{2}$, −3 **25.** 3, $\frac{1}{2}$ **27.** $-\frac{5}{3}$, 2 **29.** 1, $-\frac{1}{3}$

13-4 Page 418

Exploratory **1.** $a = 1, b = 7, c = 6$ **3.** $a = 1, b = 4, c = 3$ **5.** $a = 4, b = 8, c = 3$ **7.** $a = 1, b = 0, c = -25$ **9.** $a = 2, b = 8, c = 0$ **11.** $a = 3, b = 5, c = 2$ **13.** 49 **15.** 81 **17.** 36 **19.** 49 **21.** 2,500 **23.** 256 ***Written*** **1.** −6 **3.** −1, −3 **5.** $\frac{5}{2}$, −3 **7.** 3, $-\frac{5}{2}$ **9.** $-\frac{1}{2}$, −3 **11.** 7, −7 **13.** 6, 0 **15.** $\frac{1}{3}$, −4 **17.** −1, −9 **19.** 9, 4 **21.** 7, −5 **23.** 5, 0 **25.** $\frac{3}{2}$, −1 **27.** $-\frac{2}{3}$, −7 **29.** $\frac{2}{5}, \frac{1}{5}$ **31.** $-3 \pm 2\sqrt{3}$ **33.** 1.5, −0.4 **35.** $\frac{2 \pm \sqrt{7}}{2}$ **37.** $\frac{-5 \pm \sqrt{2}}{2}$ **39.** $\frac{4 \pm \sqrt{29}}{2}$ **41.** $-\frac{2}{3}, \frac{3}{7}$

13-5 Page 422

Exploratory **1.** 25; 2 **3.** 49; 2 **5.** 124; 2 **7.** 0; 1 **9.** 25; 2 **11.** 25; 2 **13.** 1.44; 2 **15.** 48; 2 **17.** 73; 2 **19.** 0; 1 **21.** 0; 1 ***Written*** **1.** $\frac{-5 \pm \sqrt{13}}{2}$ **3.** 2, −6 **5.** −1, −6 **7.** no real roots **9.** no real roots **11.** 1, $\frac{1}{2}$ **13.** 7 **15.** no real roots **17.** 8, −4 **19.** $\frac{2 \pm \sqrt{7}}{3}$ **21.** 1, $\frac{1}{3}$ **23.** $\frac{5 \pm \sqrt{193}}{12}$ **25.** $\frac{-1 \pm \sqrt{41}}{4}$ **27.** 1, $-\frac{1}{3}$ **29.** $\frac{-7 \pm \sqrt{33}}{4}$ **31.** $\frac{-7 \pm \sqrt{17}}{16}$ **33.** $\frac{-5 \pm \sqrt{133}}{6}$ **35.** $\frac{3 \pm \sqrt{14}}{5}$ **37.** 1, $\frac{2}{3}$ **39.** $\frac{7 \pm \sqrt{69}}{10}$

13-6 Page 425

Exploratory **1.** factoring **3.** formula or factoring **5.** formula **7.** factoring **9.** formula **11.** formula **13.** 18 **15.** 4 **17.** 10 ***Written*** **1.** 4, 5 **3.** $\frac{-2 \pm \sqrt{2}}{2}$ **5.** −2, −1 **7.** $\frac{1}{2}$, 2 **9.** $-\frac{1}{4}$, 2 **11.** $-\frac{1}{3}$, 2 **13.** $-2 \pm \sqrt{3}$ **15.** no real roots **17.** $\frac{-9 \pm \sqrt{21}}{10}$ **19.** $\frac{-2 \pm \sqrt{14}}{2}$ **21.** $1 \pm \sqrt{3}$ **23.** $\frac{3}{2}, \frac{5}{2}$ **25.** $\frac{1}{4}, \frac{3}{5}$

27. $-\frac{2}{5}, \frac{3}{2}$ **29.** $-\frac{1}{5}, \frac{3}{10}$

13-7 Page 427

Written **1.** 7 m, 8 m **3.** 10 yd, 13 yd **5.** 4 m, 14 m **7.** 10 in., 20 in. **9.** 5 in., 9 in. **11.** 10 m, 16 m **13.** 6, 8 **15.** −15, −9 or 9, 15 **17.** 12 cm, 36 cm **19.** 3 cm, 6 cm **21.** 1.6 m, 2.4 m **23.** 5.92 m, 9.47 m **25.** 1 m

13-8 Page 430

Exploratory **1.** 5; 6 **3.** −12; −28 **5.** −5; −4 **7.** $-1; -\frac{35}{4}$ **9.** $\frac{5}{6}; -\frac{21}{6}$ **11.** yes **13.** no **15.** no **17.** no **19.** no **21.** $x^2 - 7x + 10 = 0$ **23.** $x^2 - 11x + 24 = 0$ **25.** $x^2 + 5x - 6 = 0$ **27.** $4x^2 + 4x - 3 = 0$ **29.** $x^2 - 3.6x + 3.08 = 0$ **31.** $25x^2 + 5x - 6 = 0$ ***Written*** **1.** −15; 54 **3.** 1; −6 **5.** $-\frac{31}{6}; \frac{35}{6}$ **7.** −6; −1 **9.** $\frac{33}{7}; -\frac{10}{7}$ **11.** 3; 8 **13.** 1.2; 0.27 **15.** $\frac{1}{3}; -\frac{1}{12}$ **17.** $-4\sqrt{3}$; 9 **19.** $x^2 - 6x + 5 = 0$ **21.** $x^2 - x - 30 = 0$ **23.** $x^2 + 9x - 10 = 0$ **25.** $x^2 + 4x - 45 = 0$ **27.** $x^2 - 11x - 80 = 0$ **29.** $3x^2 + 2x - 21 = 0$ **31.** $6x^2 + 5x - 6 = 0$ **33.** $5x^2 - 29x - 6 = 0$ **35.** $x^2 - 2x\sqrt{3} + 3 = 0$ **37.** $x^2 - 10x + 23 = 0$ **39.** $x^2 - (q + r)x + qr = 0$

CHAPTER REVIEW Page 434

1. $x = \frac{3}{2}; \left(\frac{3}{2}, -\frac{25}{4}\right)$ **3.** $x = -\frac{9}{4}; \left(-\frac{9}{4}, -\frac{9}{8}\right)$ **5.** −3 **7.** 36 **9.** $\frac{25}{4}$ **11.** $-3 \pm \sqrt{5}$ **13.** $-\frac{3}{2}, -\frac{5}{2}$ **15.** $-5, \frac{3}{2}$ **17.** $\frac{2 \pm \sqrt{5}}{3}$ **19.** $\frac{9 \pm \sqrt{21}}{10}$ **21.** no real roots **23.** 2 **25.** formula; $-5 \pm 4\sqrt{2}$ **27.** formula; 1, $\frac{11}{9}$ **29.** formula; no real roots **31.** factoring; $\frac{3}{10}$, 2 **33.** 6, 15 **35.** $\frac{3}{2}, \frac{11}{4}$ **37.** $2x^2 + 5x - 12 = 0$

CHAPTER 14 RATIONAL EXPRESSIONS

14-1 Page 439

Exploratory **1.** $\frac{1}{y-4}; y \neq -4, 4$ **3.** $r; r \neq -3$ **5.** $\frac{x}{x^2-4}; x \neq 0, 2, -2$ **7.** $\frac{a+4}{a-2}; a \neq 2, 4$ **9.** $(t-2)(t+3); t \neq -2, 3$ **11.** $\frac{-1}{w+4}; w \neq \frac{2}{3}, -4$ ***Written*** **1.** $\frac{1}{y+3}; y \neq -3, 3$ **3.** $\frac{a-5}{a-2}; a \neq -5, 2$ **5.** $r^2; r \neq 1$ **7.** $\frac{n^2-2}{n-1}; n \neq 1$ **9.** $\frac{m^2}{2m-1}; m \neq 0, \frac{1}{2}$ **11.** $\frac{1}{x+3}; x \neq -3$ **13.** $\frac{g+2}{g-2}; g \neq 1, 2$ **15.** $\frac{m-6}{m-1}; m \neq -6, 1$ **17.** $\frac{2}{y+5}; y \neq -5, 2$ **19.** $\frac{k-1}{k+1}; k \neq -1$ **21.** $-\frac{a+3}{a+2}; a \neq -2, 3$ **23.** $-1; x \neq 3$ **25.** $\frac{4y-1}{8y-1}; y \neq -2, \frac{1}{8}$ **27.** $\frac{3}{4}; m \neq -2, -1$ **29.** $\frac{t+3}{t-4}; t \neq \frac{7}{2}, 4$ **31.** $\frac{1}{(b-4)(b+2)}; b \neq \pm 4, \pm 2$ **33.** $\frac{x+4}{2x-1}; x \neq \frac{1}{2}$ **35.** $\frac{x^2+1}{(x-2)(x+2)}, x \neq \pm 2$ **37.** $\frac{2s+7}{s-6}; s \neq \frac{2}{3}, 6$ **39.** $\frac{c-5}{c(c+6)}; c \neq -6, -4, 0$ **41.** $(a+2)(a-1); a \neq -1, 2$ **43.** $\frac{2x+3}{2(x+2)}; x \neq -2, 0, \frac{1}{2}$

14-2 Page 443

Exploratory **1.** $\frac{5}{24}$ **3.** $-\frac{35}{48}$ **5.** $\frac{35}{72}$ **7.** $\frac{a^2}{15}$ **9.** $\frac{1}{3}$ **11.** $-\frac{1}{6}$ **13.** $-\frac{1}{6}$ **15.** $\frac{11}{28}$

Written **1.** $\frac{2}{15}$ **3.** $\frac{2}{15}$ **5.** $\frac{1}{16}$ **7.** $\frac{64}{729}$ **9.** $\frac{b}{d}$ **11.** a **13.** $\frac{2m^2}{c^2}$ **15.** $\frac{4m^4}{15n^2}$ **17.** 2 **19.** $3a$ **21.** $\frac{5a+10b}{3a+4b}$ **23.** $\frac{x-y}{2}$ **25.** $2m + 4$ **27.** $\frac{r^3 + r^2s}{s^2}$ **29.** $\frac{x^2 + 8x + 16}{9}$ **31.** $\frac{x+y}{x+1}$ **33.** $\frac{m^2 + 16}{m^2 + 8m + 16}$ **35.** $\frac{3}{a^2 - 3a}$ **37.** $\frac{a^2 + 2ab + b^2}{-a}$ **39.** $\frac{x}{x^2 + 8x + 16}$ **41.** $\frac{1}{x^2 + 9x + 20}$ **43.** $\frac{b+7}{b+1}$ **45.** $b + 7$ **47.** $\frac{y^2}{y^2 - y - 2}$ **49.** $\frac{2y^2 + 11y - 21}{12y^3 + 19y^2 + 5y}$

14-3 Page 446

Exploratory **1.** $\frac{4}{3}$ **3.** $-\frac{3}{8}$ **5.** $\frac{2}{m}$ **7.** $\frac{2p}{5}$ **9.** $\frac{1}{6}$ **11.** $\frac{1}{a^2}$ **13.** $\frac{1}{2bc}$ **15.** $\frac{x-y}{x+y}$ **17.** $\frac{5}{3a^2b^2}$ **19.** $\frac{3}{8} \cdot \frac{4}{1}$ **21.** $\frac{1}{3} \cdot \left(-\frac{1}{6}\right)$ **23.** $\frac{2x}{4-2a} \cdot \frac{b-2}{a^2}$ ***Written*** **1.** $\frac{5}{4}$ **3.** $-\frac{5}{2}$ **5.** $\frac{a^4}{b^4}$ **7.** $\frac{xy^2}{a^2}$ **9.** a **11.** $\frac{3m}{m^2 - m - 2}$ **13.** $\frac{b+3}{4b}$ **15.** y **17.** $-\frac{p}{y+2}$ **19.** $-\frac{x}{7}$ **21.** $\frac{m^2-1}{2}$ **23.** $\frac{2m + 2n}{3m^2 - 3mn}$ **25.** $\frac{y-4}{y+8}$ **27.** $\frac{3t+12}{2w-6}$ **29.** $\frac{x+3}{x}$

31. $\frac{x^2 + x - 12}{x^2 - 3x - 18}$ 33. -1

14-4 Page 449

Exploratory **1.** a **3.** $4m^2$ **5.** x^2
Written **1.** $x + 4$ **3.** $a + 5$ **5.** $c + 3$ R9

7. $r - 5$ **9.** $2x + 3$ **11.** $x^2 + 2x - 3$

13. $t^2 + 4t - 1$ **15.** $3c - 2 + \frac{4}{9c - 2}$ **17.** $3n^2 - 2n + 3 + \frac{3}{2n + 3}$ **19.** $3s^2 + 2s - 3 - \frac{1}{s + 2}$

21. $5t^2 - 3t - 2$ **23.** $8x^2 - 9$

14-5 Page 453

Exploratory **1.** $\frac{7}{8}$ **3.** $\frac{7}{a}$ **5.** $\frac{b + 2}{x}$ **7.** $\frac{2k}{t}$ **9.** $\frac{1}{11}$
11. $\frac{a - b}{5}$ **13.** $-\frac{2}{x}$ **15.** $\frac{k}{m}$ ***Written*** **1.** y **3.** $\frac{10}{x}$
5. $\frac{y}{b}$ **7.** $\frac{9a}{7}$ **9.** $y - 3$ **11.** $-\frac{1}{6}$ **13.** $\frac{x + 1}{x - 1}$

15. $\frac{2}{y - 2}$ **17.** 1 **19.** 0 **21.** 0 **23.** $\frac{r^2 + s^2}{r - s}$

25. $m + n$ **27.** 4 **29.** 4 **31.** $a + b$ **33.** $r + 3$

35. $\frac{x + 1}{x - 1}$ **37.** 0 **39.** $\frac{1}{x + 1}$ **41.** $\frac{3}{m + 1}$

43. $-\frac{1}{t + 1}$ **45.** $\frac{b + 2}{b - 2}$

14-6 Page 456

Exploratory **1.** 24 **3.** $55a$ **5.** ab **7.** ab^3

9. $120a^2b^3$ **11.** $120a^4n^2$ **13.** $(a + 5)(a - 3)$

15. $3(x - 2)$ ***Written*** **1.** $\frac{7}{8}$ **3.** $\frac{35}{36}$ **5.** $-\frac{3}{8}$ **7.** $\frac{52}{77}$

9. $\frac{32 + 11y}{44}$ **11.** $\frac{5x - 28}{35}$

13. $\frac{2}{a}$ **15.** $\frac{14a - 3}{6a^2}$ **17.** $\frac{5bx + a}{7x^2}$ **19.** $\frac{3z - 14wz}{7w^2}$

21. $\frac{2s + t^2 + 3t}{st}$ **23.** $\frac{ab + 6a + 6b}{b(a + b)}$ **25.** $\frac{2a + 3}{a + 3}$

27. $\frac{3x + 2}{x - 5}$ **29.** $\frac{2}{b - 8}$ **31.** $\frac{m^2 - 2mn + n^2 - 1}{m^2 - n^2}$

33. $\frac{a^3 - a^2b + a^2 + ab}{(a + b)(a - b)^2}$ **35.** $\frac{2x + 1}{(x + 1)^2}$ **37.** $\frac{2(x^3 + 1)}{x^2 - 1}$ or $\frac{2(x^2 - x + 1)}{x - 1}$ **39.** $\frac{k^2 - 2k - 2}{(2k + 1)(k + 2)}$

41. $\frac{-7y - 39}{y^2 - 9}$ **43.** $\frac{4ab - 4b + a^3 - a^2}{16ab^2}$ or $\frac{(4b + a^2)(a - 1)}{16ab^2}$ **45.** $\frac{6 - 6a + 6b}{a^2 - 2ab + b^2}$ or $\frac{6(1 - a + b)}{(a - b)^2}$

47. $\frac{1}{m - 2}$

49. $\frac{-m^3 + m^2 - 3m^2n - 3mn^2 - 2nm + n^2 - n^3}{(m + n)^2(m - n)}$

51. $\frac{9m + 6}{(m + 2)^2(m + 1)}$ **53.** $\frac{x^2 + 12x + 2}{(x - 1)^2(x + 4)}$

55. $\frac{4a^2 + 7a + 2}{(6a + 1)(a + 3)(a - 3)}$

14-7 Page 461

Exploratory **1.** $\frac{4x + 2}{x}$ **3.** $\frac{xy + x}{y}$ **5.** $\frac{2m^2 + m + 4}{m}$
7. $\frac{b^3 - 2b^2 - 2}{b - 2}$ ***Written*** **1.** $\frac{14}{19}$ **3.** $\frac{y^3(x + 4)}{x^2(y - 2)}$

5. $\frac{a - b}{x - y}$ **7.** $\frac{x + y}{x - y}$ **9.** $\frac{x + 1}{x - 2}$ **11.** $\frac{1}{y + 4}$ **13.** $\frac{a + 2}{a + 3}$

15. $\frac{(x + 3)(x - 1)}{(x - 2)(x + 4)}$ **17.** $\frac{8x^2 - 27y^2}{(x - 2y)(x + 2y)}$ **19.** $\frac{x + 1}{x + 5}$

21. 1 **23.** $\frac{x + 3}{x + 1}$

CHAPTER REVIEW Page 464

1. $\frac{y}{4y^2z}$ **3.** $\frac{x + 3}{x - 3}$ **5.** $\frac{7a^2}{9b}$ **7.** $\frac{1}{x - 3}$ **9.** $\frac{5by}{4ax}$

11. $\frac{7ab(x + 9)}{3(x - 5)}$ **13.** $x^2 + 4x - 2$ **15.** $m^2 - 9m - 3 - \frac{40}{m - 4}$ **17.** $\frac{7 + a}{x^2}$ **19.** $\frac{2 - x}{x - y}$ **21.** $\frac{x^2 - 7x - 15}{(x + 3)(x - 2)}$

23. $\frac{3x^2 - 2x + 35}{(x + 5)(x - 2)(x - 3)}$ **25.** $\frac{x(x - 3)}{(x + 5)^2}$

27. $\frac{(x - 5)(x + 13)}{(x + 2)(x + 6)}$

CHAPTER 15 APPLICATIONS OF RATIONAL EXPRESSIONS

15-1 Page 470

Exploratory **1.** 6 **3.** 8 **5.** $r^2 - 1$ **7.** $4x(x - 1)$ or $4x^2 - 4x$ **9.** $(k + 5)(k + 3)$ or $k^2 + 8k + 15$
11. $(x + 1)(x - 1)$ or $x^2 - 1$ ***Written*** **1.** -3

3. $\frac{5}{4}$ **5.** $\frac{5}{4}$ **7.** $\frac{41}{10}$ **9.** $\frac{1}{4}$ **11.** $-\frac{1}{4}$ **13.** $-\frac{3}{2}$ **15.** $\frac{1}{2}$

17. 6, -1 **19.** -5, 3 **21.** 20, 10 **23.** -3 **25.** 1

27. 5, 10 **29.** 7 **31.** $\frac{27}{7}$ **33.** $-\frac{1}{2}$, 5 **35.** -1

15-2 Page 473

Exploratory **1. a.** $\frac{1}{8}$ **b.** $\frac{3}{8}$ **c.** $\frac{x}{8}$ **3. a.** $\frac{1}{8}$ **b.** $\frac{x}{8}$ **c.** $\frac{1}{10}$
d. $\frac{x}{10}$ **e.** $\frac{9}{40}$ **f.** $\frac{9x}{40}$ **5. a.** $r = 120 + s$; $t = \frac{420}{120 + s}$; $d = 420$; against the wind, $r = 120 - s$; $t = \frac{300}{120 - s}$; $d = 300$ **b.** $\frac{420}{120 + s} = \frac{300}{120 - s}$ ***Written*** **1.** $2\frac{2}{5}$ hours
3. $3\frac{1}{13}$ days **5.** 3 hours **7.** 30 minutes **9.** 12 mph
11. 30 mph

15-3 Page 477

Exploratory **1.** Multiply by 2, divide by n, then subtract t. **3.** Multiply by abf, subtract af, then divide by $b - f$ ***Written*** **1.** $t = \frac{v}{a}$

3. $v = \frac{2s - at^2}{2t}$ 5. $M = \frac{Fd^2}{Gm}$ 7. $V = \pm\sqrt{\frac{fgR}{W}}$

9. $P = \frac{A}{1 + rt}$ 11. $P = \frac{36{,}500}{IR + 365}$ 13. $y = \frac{r}{2a + 0.5}$

15. $R = \frac{H}{0.24I^2t}$ 17. $R_1 = \frac{R_TR_2}{R_2 - R_T}$ 19. $n = \frac{IR}{E - Ir}$

21. $m = \frac{y - b}{x}$ 23. $y_2 - mx_2 - mx_1 + y_1$

25. $R = P - DQ$ 27. $n = 2b$ 29. $n = \frac{acd - ab}{c}$

31. $n = \frac{ac + bc}{a + b + 1}$ 33. $n = \frac{b^2 - 2c}{c - b}$ 35. $n = \pm\sqrt{ab}$

37. $n = \frac{rk - k^2}{2}$

15-4 Page 480

Exploratory **1.** yes, 3 **3.** no **5.** yes, 7 **7.** no

Written **1.** 28 **3.** −5 **5.** 6 **7.** $26\frac{1}{4}$ **9.** 5.2

11. $-4, y = -4x$ **13.** $\frac{1}{2}, y = \frac{1}{2}x$ **15.** $\frac{8}{3}, y = \frac{8}{3}x$

17. $\frac{17}{22}, y = \frac{17}{22}x$ **19.** \$13.33 **21.** \$467.50

22. $6\frac{142}{143}$ or about 7 gallons **25.** $3\frac{1}{3}$ ft^3

15-5 Page 482

Exploratory **1.** yes, 6 **3.** yes, 50 **5.** yes, −13 **7.** yes, 1 **9.** yes, 40 **11.** no ***Written*** **1.** 48

3. $6\frac{2}{3}$ **5.** $-2\frac{2}{7}$ **7.** 4.48 **9.** 1.44 **11.** 192; $xy = 192$

13. $\frac{5}{3}$; $xy = \frac{5}{3}$ **15.** −16; $xy = -16$ **17.** 15.68;

$xy = 15.68$ **19.** 2.16; $xy = 2.16$ **21.** $3\frac{1}{3}$ hours

23. 30 m^3 **25.** $21\frac{1}{3}$ m^3 **27.** 4 in. **29.** 640 cycles per second

15-6 Page 485

Exploratory **1.** Emilio **3.** They must both sit the same distance away. **5.** Betty ***Written*** **1.** 8 ft

3. 8.02 ft **5.** 12 ft **7.** 3060 pounds

15-7 Page 490

Written **1.** $3\frac{3}{7}$ ohms **3.** 4 ohms **5.** 8 ohms, 4 ohms

7. 18 ohms **9.** $1\frac{1}{3}$ ohms **11.** $7\frac{1}{17}$ ohms

13. $2\frac{11}{12}$ ohms **15.** $7\frac{2}{3}$ ohms

15-8 Page 493

Exploratory **1.** joint **3.** combined **5.** quadratic direct **7.** joint ***Written*** **1.** 2 **3.** −10 **5.** 8 **7.** ±0.09 **9.** 4 **11.** 39.2 **13.** 12.5

CHAPTER REVIEW Page 496

1. $-\frac{14}{3}$ **3.** $\frac{25}{7}$ **5.** $-\frac{23}{2}$ **7.** $21\frac{9}{11}$ hours

9. $a = \frac{bf}{b - f}$ **11.** 12 **13.** $74\frac{2}{3}$ gallons **15.** 2.3

17. On the other end of the meter stick, 50 cm from the fulcrum. **19. a.** 10 ohms **b.** $2\frac{2}{5}$ ohms **21.** 3

23. 0.2

CHAPTER 16 TRIGONOMETRY

16-1 Page 501

Exploratory **1.** 5° **3.** 77° **5.** 35° **7.** $(50 - 2x)°$ **9.** 50° **11.** 53° **13.** 90° **15.** $(175 - 3x)°$
Written **1.** 48°; 138° **3.** none; 55° **5.** 0°; 90° **7.** 69°; 159° **9.** none; 81° **11.** $(90 - 3y)°$; $(180 - 3y)°$ **13.** $(128 - x)°$; $(218 - x)°$ **15.** $x°$; $(90 + x)°$ **17.** 122° **19.** 85° **21.** 1° **23.** $(160 - 2x)°$ **25.** 37°; 106° **27.** 26°; 64° **29.** 25°; 35°; 120° **31.** 50° **33.** 30°

16-2 Page 504

Exploratory **1.** 4 m **3.** 6.5 mm **5.** $2\frac{1}{4}$ in. **7.** 8.18 m **9.** 14 m **11.** 8.70 mm **13.** $12\frac{3}{4}$ in. **15.** 7.72 cm **17.** 4 ft; 8 ft **19.** 8 m; 16 m **21.** 7 yd; 14 yd
Written **1.** $4\sqrt{3}$ cm **3.** 10 units **5.** 3 m; $3\sqrt{3}$ m

7. $1\frac{3}{4}$ in; $\frac{7}{4}\sqrt{3}$ in. **9.** 13 m; $6.5\sqrt{3}$ m **11.** 4 m, 2 m

16-3 Page 507

Exploratory **1.** $\angle A$ and $\angle D$; $\angle B$ and $\angle E$; $\angle C$ and $\angle F$; AB and DE; BC and EF; AC and DF
3. $\angle Q$ and $\angle Q$; $\angle S$ and $\angle P$; $\angle T$ and $\angle R$; QS and QP; QT and QR; ST and PR ***Written*** **1.** $b = \frac{25}{7}$, $c = \frac{30}{7}$ **3.** $a \approx 2.8, c \approx 4.2$ **5.** $d = 10.2, e = 9$

7. $c = \frac{7}{2}, d = \frac{17}{8}$ **9.** $\frac{125}{9}$ ft **11.** 11 cm

16-4 Page 510

Exploratory **1.** $\frac{8}{17}$ **3.** $\frac{8}{15}$ **5.** $\frac{15}{17}$ **7.** $\frac{9}{41}$ **9.** $\frac{9}{40}$

11. $\frac{40}{41}$ **13.** $\frac{4}{5}$ **15.** $\frac{4}{3}$ **17.** $\frac{3}{5}$ **19.** yes; yes **21.** The sine of an angle is equal to the cosine of its complement. **Written** **1.** $\sin A = 0.690$. $\cos A = 0.724$, $\tan A = 0.952$; $\sin B = 0.724$, $\cos B = 0.690$, $\tan B = 1.050$ **3.** $\sin R = 0.324$, $\cos R = 0.946$, $\tan R = 0.343$; $\sin T = 0.946$, $\cos T = 0.324$, $\tan T = 2.917$ **5.** $\sin J = 0.899$, $\cos J = 0.438$, $\tan J = 2.051$, $\sin K = 0.438$, $\cos K = 0.899$, $\tan K = 0.488$ **7.** $\sin J = 0.800$, $\cos J = 0.600$, $\tan J = 1.333$, $\sin L = 0.600$, $\cos L = 0.800$, $\tan L = 0.750$ **9.** $\sin E = 0.220$, $\cos E = 0.976$, $\tan E = 0.225$, $\sin F = 0.976$, $\cos F = 0.220$, $\tan F = 4.444$

16-5 Page 513

Exploratory **1.** 90°; 0° **3.** 45° **5.** No maximum or minimum value. ***Written*** **1.** 0.9063 **3.** 0.9998 **5.** 0.2419 **7.** 0.1219 **9.** 0.9272 **11.** 0.4540 **13.** 0.7431 **15.** 0.7071 **17.** 0.5000 **19.** 0.8660 **21.** 26° **23.** 53° **25.** 79° **27.** 67° **29.** 16° **31.** 40° **33.** 89° **35.** 3° **37.** 30° **39.** 0 **41.** 2 **43.** −0.5000 **45.** 0.3660 **47.** 0

16-6 Page 515

Exploratory **1.** cos 46; sin 46° **3.** sin 42°; tan 42° **5.** $\sin A = \frac{4}{9}$; $\cos B = \frac{4}{9}$ **7.** $\cos A = \frac{8}{13}$; $\sin B = \frac{8}{13}$
Written **1.** $A = 44°$, $AC = 10.8$ ft, $BC = 10.4$ ft **3.** $A = 48°$, $AB = 9.0$ in, $BC = 6.7$ in. **5.** $A = 26°$, $B = 64°$, $AC = 8.1$ cm **7.** $A = 52°$, $B = 38°$, $BC = 10.2$ mm **9.** $B = 69°$, $AB = 13.9$ in., $BC = 5$ in. **11.** $B = 20°$, $AB = 9.6$ cm, $AC = 3.3$ cm **13.** $B = 68°$, $AB = 12.9$ ft, $BC = 4.8$ ft **15.** $B = 50°$, $AC = 12.3$ m, $BC = 10.3$ m **17.** $B = 59°$, $AB = 11.6$ m, $AC = 10m$ **19.** $A = 48°$, $AC = 6.7$ in., $BC = 7.4$ in. **21.** $A = 38°$, $B = 52°$, $BC = 8.7$ ft **23.** $A = 50°$, $AB = 9.3$ cm, $AC = 7.2$ cm **25.** $B = 45°$, $AC = 1$ ft, $BC = 1$ ft

16-7 Page 518

Exploratory **1.** $\angle CAB$; $\angle DCA$ **3.** $\angle JKL$; $\angle MJK$
Written **1.** 4° **3.** 65 ft **5.** 300 ft **7.** 65 ft **9.** 1716 cm **11.** 1394 ft

CHAPTER REVIEW Page 523

1. 24° **3.** $(90 - y)°$ **5.** 32° **7.** 92° **9.** $(180 - x - y)°$ **11.** 2.125 cm, 3.681 cm **13.** 7 in., 6.062 in. **15.** 9 m, 4.5 m **17.** $d = \frac{45}{8}$, $e = \frac{27}{4}$ **19.** $\frac{28}{53}$ **21.** $\frac{45}{53}$ **23.** $\frac{28}{45}$ **25.** 0.528 **27.** 1.607 **29.** 0.6691 **31.** 0.9945 **33.** 2.2460 **35.** 28° **37.** 3.7 m **39.** 7.9 m **41.** 173.2 m

Index

D

E

F

G

H

I

J

K

L

M

N

O

S

T

U

V

W

X

Y

Z